Voltage-Gated Calcium Channels

Gerald Werner Zamponi
Norbert Weiss
Editors

Voltage-Gated Calcium Channels

 Springer

Editors
Gerald Werner Zamponi
Department of Clinical Neurosciences
University of Calgary
Calgary, AB, Canada

Norbert Weiss
Department of Pathophysiology, Third
Faculty of Medicine, Charles University
Prague, Czech Republic

ISBN 978-3-031-08883-4   ISBN 978-3-031-08881-0   (eBook)
https://doi.org/10.1007/978-3-031-08881-0

This Springer imprint is published by the registered company Springer Nature Switzerland AG
The registered company address is: Gewerbestrasse 11, 6330 Cham, Switzerland

Preface

In 2004, we were asked whether we would be willing to organize a book on inactivation mechanisms in calcium channels. This seemed overly focused, and we made a counterproposal to have a comprehensive book on calcium channels. It took almost 18 months to collect the more than 20 chapters, but the end result was a comprehensive book that covered the state of knowledge on calcium channels at the time.

Much has happened globally since 2004. The world is ruled by social media, and Facebook, YouTube, Twitter, and Instagram are now part and parcel of our science communications, but they have also served as platforms for science denial and disinformation. The publishing world has been transformed through rapid online publications, but it has been inundated with a plethora of predatory journals that pollute the purity of scientific output. Through these changes, the calcium channel field has endured, but there have been a number of changes.

Prior to the early 2000s, large scientific conferences, such as the Society for Neuroscience annual conference or the Biophysical Society meetings, had multiple oral sessions dedicated to calcium channels, and this is now no longer the case. As a way of bringing the calcium channel field together, we organized five international calcium channel conferences, always in a different location across the globe. This was subsequently complemented by the European calcium channel meeting organized by colleagues at the University of Innsbruck. These types of activities have fostered scientific exchange and helped build a sense of camaraderie, and many of the regulars at these conferences are participants in this book. It is interesting to note that a dozen of the authors of the current book also contributed to the earlier book, in some cases as trainees who are now firmly established leaders in the field, and there are many contributors who constitute a newer cadre of calcium channel experts. Sadly, we lost one of the true greats of calcium channel physiology due to the unfortunate passing of Dr. David Yue in 2014, but his legacy lives on through the outstanding work by his former trainees such as Drs. Dick and Ben Johny.

There have been numerous advances in the calcium channel field over the past two decades. Due to rapid genetic sequencing, the field of calcium channelopathies has expanded dramatically, and has given rise to parent-led organizations such as the CACNA1A foundation. The availability of new technical approaches such as CRISPR has helped provide deeper insights into the physiological and pathophysiological roles of calcium channels. Perhaps, most impressively, the availability of crystal and cryo-EM structural information as highlighted by Dr. Catterall has dramatically advanced the field and will have a major impact on drug discovery for the treatment of calcium channel–related disorders.

The 2022 edition of *Voltage-Gated Calcium Channels* builds on the large body of information contained in the earlier book, but greatly extends this content to incorporate these new and exciting developments. The first section covers the most important structural and molecular aspects of voltage-gated calcium channels and their associated ancillary subunits, but also the importance of splice variation and RNA editing which contribute to extend their diversity and physiological roles. The second section deals with the different aspects of their regulation, ranging from their genetic control to their trafficking and regulation at the plasma membrane, and also their association with other channels and receptors. The third section is dedicated to their implication in the development of human disorders, a field that has expanded dramatically over the last years and continues to be one of the fastest growing research areas. Finally, the fourth section naturally covers the different aspects of their pharmacology, ranging from molecular pharmacology to a variety of small molecules and peptides that not only represent invaluable tools for the physiological study of calcium channels but also represent potential pharmacological tools for therapeutic intervention.

We would like to warmly thank all of the authors without whom this second edition would not have been possible. We believe that to have all of this information in a single volume provides a fantastic resource for both those actively involved in the field and for those wishing to find out more about particular aspects of calcium channels.

Calgary, AB, Canada   Gerald Werner Zamponi
Prague, Czech Republic   Norbert Weiss

Contents

Part III (Patho)physiology of VGCCs

A Lived History of Early Calcium Channel Discoveries Over the Past Half-Century

Emilio Carbone

Abstract

This chapter of the book is directed to PhD students, post-docs and young researchers who are attracted by the unique properties of voltage-gated calcium channels. This chapter aims to provide an overview of the most important discoveries that helped elucidate the structure and function of calcium channels in excitable cells. While systematically reviewing the numerous works in the field, I chose to write a personal story derived from my own experience on Ca^{2+} channels, as it developed in the lab and through the discussions with many colleagues working on ion channels. This occurred in a period in which Ca^{2+} channels reached maximal attention among scientists and brought the many astonishing achievements described in this book. Given the broad interdisciplinarity of Ca^{2+} channel discoveries, the present history may probably appear incomplete, but certainly, the other chapters of this book will cover all possible gaps on the matter.

Keywords

Ca-spikes · Ca^{2+} currents · L-type · N-type · P/Q-type · R-type · T-type channels · Calcium channels structure and function · GPCR-mediated inhibition of Cav2 channels

Ca²⁺ as Central Ion for Muscle Contraction

The relevance of calcium ions in myocardial function is about 140 years old. It coincides with Sydney Ringer's observations on the ionic constituents of blood on heart contraction published in the newly founded *Journal of Physiology* (Ringer, 1883). By changing the ionic composition of bath solutions used to keep frog ventricles contracting, Ringer discovered that Ca^{2+} is the key extracellular cation required to maintain the regular heart beating (see Fye, 1984 for curious aspects of Ringer's experiments). Ringer's observation was soon extended to smooth muscle contraction (Stiles, 1901), and 60 years later, T. Kamada (Japan) and L.V. Heilbrunn (USA) uncovered the role of Ca^{2+} on skeletal muscle contraction (Kamada & Kinoshita, 1943; Heilbrunn & Wiercinski, 1947). Following this, calcium was progressively identified as a determinant constituent of extracellular solutions in all excitable cells and recognized as the most

E. Carbone (✉)
Department of Drug Science, Laboratory of Cell Physiology and Molecular Neuroscience, N.I.S. Centre, Torino, Italy
e-mail: emilio.carbone@unito.it

widespread and versatile signalling element involved in many cellular processes (Carafoli et al., 2001). Intracellular Ca^{2+} is now considered the most ubiquitous second messenger capable of modulating numerous cell functions (Berridge et al., 2003). Initiation of intracellular Ca^{2+} signalling events obviously requires Ca^{2+} movement across the plasma membrane, and this, in the majority of cells, involves the presence of Ca^{2+} permeable channels.

The "Ca-Spikes" of Crustacean Skeletal Muscles

Following the four revolutionary papers by Alan Hodgkin and Andrew Huxley on the ionic basis of action potential (AP) generation in the squid giant axon (Hodgkin & Huxley, 1952a, b, c, d), it soon became evident that Na^+ currents alone were not sufficient to account for the generation of the AP upstroke in several excitable cells. Thanks to the strong driving force for Ca^{2+} at the threshold potential where APs originate, other ions, like Ca^{2+}, could potentially carry enough inward current to sustain the APs. Paul Fatt, Bernard Katz and Bernard Ginsborg were the first to recognize the importance of transmembrane Ca^{2+} fluxes to sustain APs in crustacean muscle fibres (Fatt & Katz, 1953) and to propose that a Ca^{2+} current was responsible for the APs in the absence of external Na^+ (Fatt & Ginsborg, 1958). Similar findings on Ca-dependent APs were reported in the following years in other crustacean muscle fibres using variable Na^+/Ca^{2+} concentration ratios in the external solution to test for the amplitude and shape of APs (for a review see Reuter, 1973). However, the most direct information regarding the ability of Ca^{2+} ions to serve as charge carriers during excitation in skeletal muscle came from the studies of Susumu Hagiwara (USA) on barnacle muscle fibres. Hagiwara's group showed that in the absence of external Ca^{2+} the APs stopped, regardless of the external $[Na^+]$ concentration. The overshoot of the spike increased with increasing extracellular $[Ca^{2+}]$ (Fig. 1a), and all-or-none spikes could be recorded when the intracellular

$[Ca^{2+}]$ was reduced by injecting various Ca^{2+}-binding agents (Hagiwara & Naka, 1964). Using TTX and local anaesthetics, Hagiwara and colleagues could clearly separate "Na-spikes" from "Ca-spikes" and identify manganese (Mn^{2+}) as an effective blocker of Ca^{2+} influx (Hagiwara & Nakajima, 1966).

"Ca-Spikes" in Heart and Neurons

The existence of a Na^+ and Ca^{2+} inward current system that regulates AP waveforms and their amplitudes became soon evident also in cardiac tissues and mollusc neurons. In the frog heart, for instance, it became apparent that the rising phase of the AP was composed of two phases: one fast, Na^+ driven and one slower, Ca^{2+} driven (Wright & Ogata, 1961; Niedergerke & Orkand, 1966). The same was observed in mammalians heart cells (Stanley & Reiter, 1965; Paes de Carvalho et al., 1966), confirming the existence of Ca-spikes in heart muscle fibres.

In parallel to skeletal muscle and heart cells, Ca-spikes were described also in several nerve preparations. Initial observations were made on frog spinal ganglion cells (Koketsu et al., 1959; Nishi et al., 1965) and pulmonate mollusc neurons (Oomura et al., 1961). In the first case, frog ganglion cells were shown to produce prolonged APs in Na^+-free solutions when quaternary ammonium ions were present. Ca-spikes were abolished when Ca^{2+} was withdrawn from the Na^+-free solution and were preserved when Ba^{2+} replaced Ca^{2+}. The AP overshoot increased in Na^+-free solutions in the presence of Ba^{2+}, suggesting that Ca^{2+} (or Ba^{2+}) carries charges across the membrane of spinal ganglion cells. In the second case, the giant nerve cells of pulmonate mollusc were shown to produce all-or-none APs in Na^+-free solutions. APs were abolished by removing Ca^{2+} from the solution. Following this, "Ca-spikes" became popular and were reported in many other nerve cell types (Gerasimov, 1965; Geduldig & Junge, 1968; Koketsu & Nishi, 1969).

It is also worth mentioning that Ca^{2+} fluxes were recorded not only at the soma but also at the axonal level. TTX-insensitive

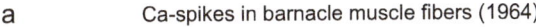

a   Ca-spikes in barnacle muscle fibers (1964)

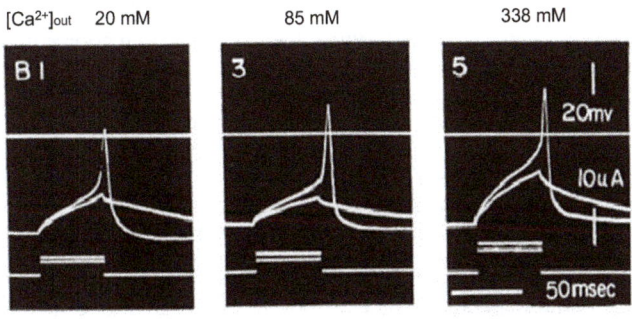

b   Ca²⁺ currents in cardiac Purkinje fibers (1967)

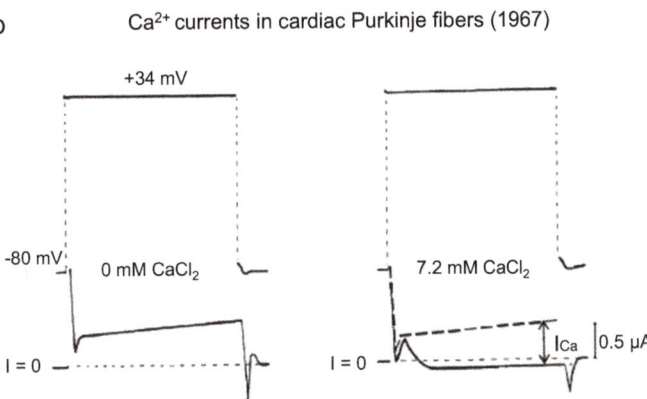

Fig. 1 "Ca-spikes" and Ca²⁺ currents in muscle fibres (1964–1967). (**a**) Effects of external Ca²⁺ on Ca²⁺-AP overshoot on barnacle muscle fibres. The muscle fibre was injected with K⁺-containing solutions. $[Ca^{2+}]_{out}$ is indicated on top of each panel. The top straight line indicates the 0 mV potential. (Redrawn with permission from Hagiwara & Naka, 1964). (**b**) Bottom, Ca²⁺ current recorded from a sheep Purkinje fibre bathed in Na⁺-free solution containing 0 mM (left) or 7.2 mM (right) $CaCl_2$, at +34 mV from Vh = −80 mV. The net inward Ca²⁺ current on the right panel (ICa) was calculated by subtracting the current trace of the left panel (dashed curve). (Redrawn with permission from Reuter, 1967)

Ca²⁺ fluxes in the squid giant axon were too small to generate Ca-spikes. Nevertheless, they could be detected through the light produced by the Ca²⁺-sensitive bioluminescent protein *aequorin* injected in the axon (Baker et al., 1971; Hallett & Carbone, 1972). Ca²⁺ entry during depolarizing voltage pulses was divided into an early TTX-sensitive component, plus a late TTX-insensitive component. Mn^{2+}, Co^{2+}, La^{3+} and organic Ca²⁺ antagonists (D-600 and iproveratril) effectively blocked this component (Baker et al., 1973). At variance with the squid axon, TTX-insensitive Ca-spikes were successfully recorded in *Aplysia* axons in Na⁺-free TTX-containing solutions (Horn, 1978). Under these conditions, Ca²⁺ currents were the only currents responsible for propagating APs. The overshoot and maximum rate of rise of Ca-spikes increased with increasing external Ca²⁺. Co^{2+} or Cd^{2+} blocked these currents and Sr^{2+} or Ba^{2+} could substitute for Ca²⁺ to sustain the AP. In conclusion, axons and cell bodies were able to generate and propagate Ca-spikes (see Hagiwara & Byerly, 1981 for a review).

How to Look at Ca²⁺ Currents Through Voltage-Clamp Recordings

Together with recording and characterizing Ca-spikes in a multitude of excitable cells, the increased availability and popularity of the "voltage-clamp" technique helped resolving the key properties of voltage-gated Ca²⁺ currents in excitable cells. As shown by Hodgkin and Huxley (1952a, b, c, d) for the Na⁺ and K⁺ conductances of squid axon, the "voltage-clamp" technique allows measuring the voltage-dependent kinetics of activation, inactivation and closing of active ion conductances.

Voltage-clamp recordings have great advantages with respect to current-clamp recordings but have strong limitations concerning the cell shape. Cells need to be either spherical (cell bodies without dendrites) or cylindrical (axons with no branches) and, most importantly, should have large dimensions (>150 μm diam) to allow the positioning of "two electrodes", one for voltage command and one for current recordings. The squid giant axon is ideal for voltage-clamp recordings. It has a quite homogeneous cylindrical shape of large diameter (300–800 μm), so that by using low-resistance platinum axial current electrodes and thin axial voltage glass microelectrodes, it is possible to record Na⁺ and K⁺ currents with high-time resolution (Armstrong, 1966, 1969). These conditions were quite limiting for recording Ca²⁺ currents from muscle and neuronal preparations that exhibited Ca-spikes. Therefore, the first critical issue to solve was to find proper cell preparations and voltage-clamp approaches to measure Ca²⁺ currents at fixed potentials.

Ca²⁺ Currents in the Heart and Mollusc Neurons: The Problem of Blocking K⁺ Currents

In 1967, Harald Reuter (Switzerland) succeeded to measure Ca²⁺ currents in cardiac Purkinje fibres bathed in Na⁺-free solution (Reuter, 1967), using the double sucrose voltage-clamp technique developed by Josef Dudel and Wolfang Trautwein in Homburg (Germany) (Dudel et al., 1966). Reuter's experiments gave clear indications that cardiac Ca²⁺ currents were robust and had a slower time course with respect to Na⁺ currents. Their identification was not trivial. Ca²⁺ currents in 7.2 mM Ca²⁺ were separated from the dominant outward currents recorded in 0 mM Ca²⁺ by subtraction (Fig. 1b). Interestingly, the isolated Ca²⁺ currents activated with sigmoidal voltage dependence between −60 and +10 mV and zeroed at ~ +150 mV, near the Ca²⁺ equilibrium potential. Since then, the "slow Ca²⁺ currents" were resolved and described in many other heart cell preparations (see Reuter, 1973; Trautwein, 1973 for a review).

The voltage-clamp technique was also successfully applied to measure the Ca²⁺ inward currents responsible for the Ca-spikes of mollusc neurons (Geduldig & Gruener, 1970; Krishtal & Magura, 1970; Kostyuk et al., 1974), including autorhythmic Helix neurons (Eckert & Lux, 1975). In parallel to this, Ca²⁺ currents were characterized in many different cells, establishing the importance of Ca²⁺ channels as molecules coupling cell excitation with intracellular Ca²⁺ signalling (see Hagiwara & Byerly, 1981 for a review).

Most of these works reported the existence of Ca²⁺ currents with rather similar time course and voltage-dependent activation. The currents reached a peak within few milliseconds at about +10 mV and then inactivated with variable time courses. However, the current decay after the peak (inactivation) depended strongly on the block of outward K⁺ currents, which were prominent in these neurons and incompletely blocked due to the difficulty to replace intracellular K⁺ with commonly used K⁺ channel blockers (Cs⁺, TEA⁺, TMA⁺). In spite of this, Paul Brehm and Roger Eckert (USA) could prove that Ca²⁺ current inactivation was effectively regulated by Ca²⁺ fluxes in *Paramecium* (Brehm & Eckert, 1978) and that this feedback signal was fundamental to autoregulate Ca²⁺ entry during cell stimulation to preserve low cytoplasmic Ca²⁺ levels. Eckert and his colleagues invented the "double-pulse" protocol that we still use routinely nowadays to define the degree of

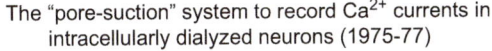

The "pore-suction" system to record Ca²⁺ currents in intracellularly dialyzed neurons (1975-77)

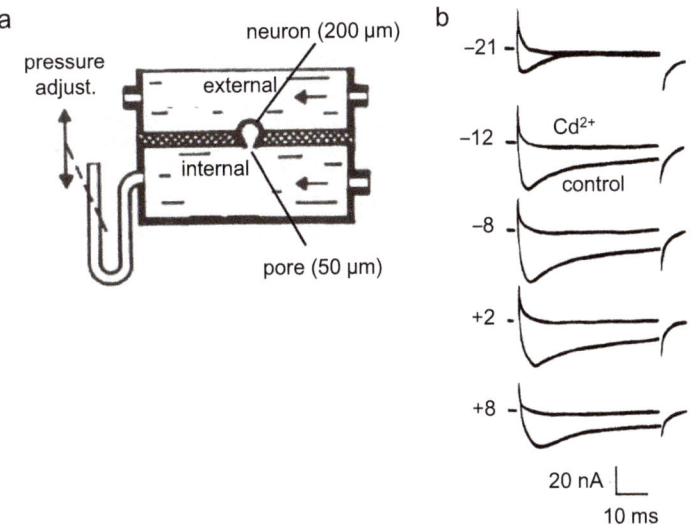

Fig. 2 The "pore-suction technique" to record Ca²⁺ currents in intracellularly dialysed mollusc neurons (1975–1977). (**a**) Experimental set-up for the intracellular dialysis of isolated neurons. The neuron is positioned in the conical hole of the partition. A normal negative pressure in the internal compartment stabilizes the neuron in the hole, while a strong transient negative suction breaks the membrane and creates a pore on the internal side of the neuron. Following this starts the intracellular dialyses and ion current recordings. (**b**) Ca²⁺ currents in a *Helix* neuron dialysed with Tris-phosphate and kept in Na-free solution (Ca²⁺ 10 mM) before (control) and after adding 2 mM Cd²⁺. (Redrawn with permission from Kostyuk et al., 1977)

Ca²⁺-dependent inactivation (CDI) of the different Ca²⁺ channels expressed in various excitable cells (see Eckert & Chad, 1984 for a review).

The problem of K⁺ current contamination was solved by Platon Kostyuk and Oleg Krishtal in Kiev (Ukraine) by developing an intracellular perfusion method on mollusc (Kostyuk et al., 1975) and mammalian neurons (Krishtal & Pidoplichko, 1980). Kostyuk and Krishtal invented an ingenious system in which a neuron was held firmly into a conical pore within a partition separating the external from the internal solution (Fig. 2a). A transiently applied suction from the intracellular compartment caused the rupture of the cell membrane and complete replacement of the cytoplasm with a solution containing Ca²⁺ chelators and Cs⁺ to block K⁺ channels (see Kostyuk et al., 1981). Complete block of voltage-gated and Ca²⁺-activated K⁺ currents in the presence of TTX allowed for the first time to record Ca²⁺ currents in isolation, displaying the real voltage-dependent characteristics of neuronal Ca²⁺ channels uncontaminated by K⁺ outward currents (Fig. 2b). Modifications of this method were soon proposed using large glass pipettes to hold cells of 150–250 μm diameter. In one case, a negative pressure was created on marine eggs (Takahashi & Yoshii, 1978) while, in another case, mollusc neurons were punched with a metallic wire (Lee et al., 1978) to break the membrane at one side and get access to the intracellular compartment.

A Convergent View on the Existence of "a" Ca²⁺ Channel in Excitable Cells

The works on mollusc neurons and the many others following in that period converged on a commonly accepted view that physiologically and pharmacologically isolated Ca²⁺ currents reflected the time course of Ca²⁺ fluxes through specific cell membrane pathways generically indicated as "Ca²⁺ channels" (see Reuter, 1979;

Kostyuk, 1980; Hagiwara & Byerly, 1981). Ca^{2+} current recordings made it possible to look closer at the role of these channels in the regulation of heart beat, muscle contraction, neuronal excitability and cell exocytosis, and also revealed modulatory effects of hormone and neurotransmitters on these channels (Tsien et al., 1972; Dunlap & Fischbach, 1978).

At this time, there was a general agreement among all groups that nearly all excitable cells expressed only "one type" of Ca^{2+} permeable channel with well-defined characteristics. The channel activated in a voltage-dependent manner at potentials more positive than Na^+ channels and had slower activation kinetics. Inactivation was also slower but, differently from Na^+ channels, was voltage and Ca^{2+} dependent. All these properties fit nicely with the idea that the Ca^{2+} channel was responsible for the broadening of APs in nearly all excitable cells. Broad APs are fundamentals to allow sustained Ca^{2+} entry during heart contraction, neurotransmitter secretion and neuronal signalling with respect to the narrow spikes of the squid axon that serve mainly to conduct excitability along the axonal cable. It was also well proved that Ca^{2+} channels were rather selective for divalent vs. monovalent cations and that Ba^{2+} and Sr^{2+} were more permeable than Ca^{2+} through the open pore (Hagiwara & Byerly, 1983). For more details on the permeability properties of Ca^{2+} channels, please see the review by (Sather and McCleskey, 2003).

The "Patch-Clamp" Technique and the Explosive Interest on Ca^{2+} Channels

All the above discoveries on isolated Ca^{2+} currents were rapidly overwhelmed by the advent of the "patch-clamp" technique that Erwin Neher and Bert Sakmann (Germany) developed to record single ACh receptor channels from "electrically isolated" membrane patches of skeletal muscles (Neher & Sakmann, 1976). Within the next few years, the technique was further refined by introducing a "negative-suction system" to form high-resistance seals (giga-seals) between

the glass pipette and the cell membrane that enabled higher current resolution, physical isolation of membrane patches and direct recording of whole-cell and single-channel currents from cells of a small diameter (<20 μm) (Hamill et al., 1981).

The patch-clamp technique boosted in an impressive way the interest on single Ca^{2+} channels and whole-cell Ca^{2+} currents in a variety of vertebrate and invertebrate cells. Snail neurons (Lux & Nagy, 1981), chick DRG neurons and PC12 cells (Brown et al., 1982), bovine chromaffin cells (Fenwick et al., 1982), rat and guinea pig ventricular myocytes (Reuter et al., 1982; Cavalié et al., 1983) and rat pituitary cells (Hagiwara & Ohmori, 1983) were the first preparations used to record single channel and macroscopic Ca^{2+} currents. Thanks to the "patch-clamp" technique, the interest moved quickly from the properties of Ca^{2+} channels in large-size mollusc and marine animal cells to small-size vertebrate cells. I also moved in March 1983 from the lab of Franco Conti and Enzo Wanke at the Institute of Cybernetics and Biophysics in Camogli (Italy), where I was working on Na^+ and K^+ channels of squid axons (Wanke et al., 1980; Carbone et al., 1982), to Hans Dieter Lux' lab of Neurophysiology at the Max Planck Institute for Psychiatry, in Martinsried (Germany).

At that time (1983–1985), Lux's and other groups were interested in the many unsolved problems related to the kinetics, permeability and modulatory properties of Ca^{2+} channels, as well as to the existence of new Ca^{2+} channel types that could be identified biophysically. The accepted paradigm was that probably only one Ca^{2+} channel existed, although the first evidence for a dual Ca^{2+} channel population (types I and II) was obtained already in 1975 using a two-electrode voltage-clamp method in starfish eggs (Hagiwara et al., 1975). Hagiwara's group could show that starfish eggs displayed two channel types: one activating at relatively negative potentials (−50 mV; type I) and one at more positive voltages (−7 mV, type II). Type I inactivated relatively fast with respect to type II, which closely resembled the "slow Ca^{2+} current" of cardiac cells and mollusc neurons. Type I channel caught immediately the attention since it activated at low voltages, inactivated rapidly and inactivated steadily following conditioning pre-pulses to −35 mV, similar to Na⁺ channels. Surprisingly the time course and amplitude of type I current was found sensitive to the external Na⁺ concentration. External Na⁺ slowed-down markedly the inactivation of type I channel, possibly due to an unexplained effect of Na⁺ on channel gating and permeation, unusual for a vertebrate Ca^{2+} channel.

The Discovery of the "Low-Voltage Activated" T-Type Channel

Evidence for two types of voltage-gated Ca^{2+} channels became soon evident while patching cultured chick sensory neurons under ionic conditions in which only pharmacologically isolated Ca^{2+} currents were recorded, using TTX and high $[Ca^{2+}]$ in the bath and high $[Cs^+]$ and Ca^{2+} chelators in the pipette (Carbone & Lux, 1984b). In nearly every neuron, a "low-voltage activated" (LVA) component emerged transiently at very negative potentials (~ −50 mV) with fast activation and complete inactivation, while a "high-voltage-activated" (HVA) component activated at potentials positive to −20 mV (Fig. 3a). The HVA current was fast activating and slowly inactivating. This current closely resembled the Ca^{2+} current described in other neurons and was therefore less of a focus in these experiments. LVA currents increased with Ca^{2+} (Fig. 3b) and occasionally could be recorded in isolation from HVA currents (Fig. 3c) (Carbone & Lux, 1984a).

More convincing evidence for the presence of a "low-voltage-activating and fully inactivating" Ca^{2+} channel came in parallel while recordings single Ca^{2+} channels in membrane-excised patches. Using the outside-out configuration that allows better control of transmembrane voltage and replacement of internal K⁺ ions with Cs⁺, Lux and I uncovered the fast bursting activity of single (or multiple) LVA Ca^{2+} channels (Carbone & Lux, 1984a). The channel activity was strictly holding potential (V_h) dependent. With V_h near resting potentials (−60 mV), the channel was

silent during depolarizations to −40 mV but was very active after holding the patch to −100 mV (Fig. 4a, left). Repeated openings started soon after the patch depolarization and terminated at increasingly shorter times with increasing voltages. By summing up a sufficient number of single-channel traces at a fix potential, it was evident that the channel fully inactivated during pulses of 200 ms at −20 mV (Fig. 4a, right), mimicking the time course of macroscopic LVA currents.

Lux and I were lucky that the LVA channel resisted to the outside-out patch dialysis so that we could record the activity of the channel for the time required to accumulate a sufficient number of traces for the analysis. We were on the contrary not careful enough in selecting the permeant ion to better separate LVA from HVA channels. Ba^{2+} would have been certainly a better choice to separate biophysically the two channels since Ba^{2+} is more permeable than Ca^{2+} through HVA but not through LVA channels (Fedulova et al., 1985) and would have made the separation during single channel recording easier. We preferred to use 20 mM Ca^{2+} simply to increase the size of the unitary events without excessively altering the external divalent cations concentration. In conclusion, we erroneously estimated the conductance of the LVA fully inactivating channel at twice as much of the real value (Carbone & Lux, 1987a), possibly due to the dual overlapping of more than one active channel in the outside-out patches. However, the permeability to Ca^{2+}, the low voltage-dependent activation, the fast and full inactivation, the sensitivity to the holding potential and the persistent activity in excised patches were the exact fingerprints of the expected single LVA channel, as we know today.

One year later, Dick Tsien's group (USA) found that in cell-attached patches, 110 mM Ba^{2+} and BayK8644 could better separate LVA and HVA channels in chick sensory neurons (Nowycky et al., 1985) and guinea pig ventricular cells (Nilius et al., 1985). Tsien's group brought also evidence for a third type of voltage-gated Ca^{2+} channel in neurons. The three channels were termed: T-type for transient (LVA), N-type for "neither T nor L" (HVA) and L-type for

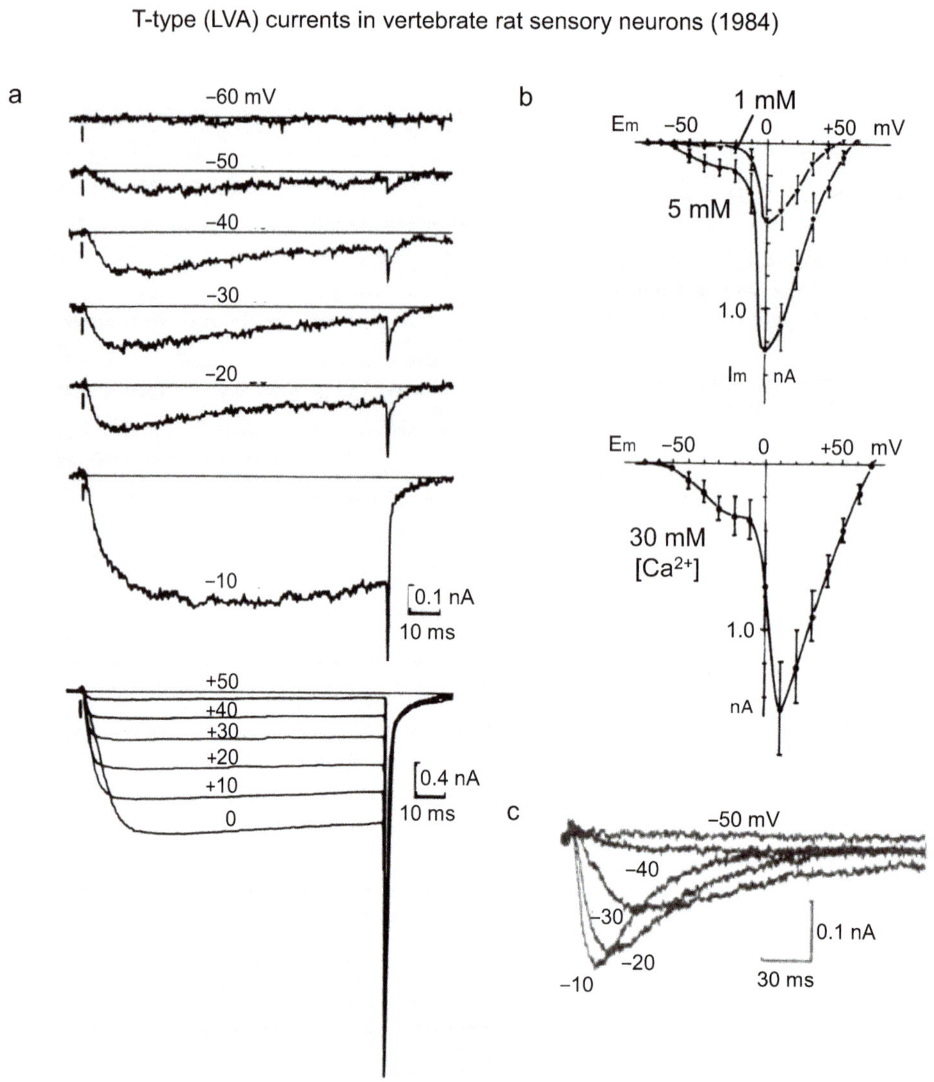

T-type (LVA) currents in vertebrate rat sensory neurons (1984)

Fig. 3 Whole-cell LVA (T-type) Ca²⁺ currents in chick and rat DRG neurons (1984). (**a**) LVA and HVA whole-cell clamp Ca²⁺ currents recorded from a chick sensory neuron bathed in 5 mM Ca²⁺ plus 3 µM TTX at the potential indicated (Vh −80 mV; 130 mM CsCl in the pipette). (**b**) I-V relationships in 1, 5 and 30 mM Ca²⁺ in chick DRGs. (Redrawn with permission from Carbone & Lux, 1984b). (**c**) LVA (T-type) Ca²⁺ currents recorded in isolation from a rat DRG neuron at the potential indicated (Vh −90 mV). (Redrawn with permission from Carbone & Lux, 1984a)

"long-lasting" (HVA). This terminology became soon popular and often used to indicate the existence of these three channels in a variety of cell preparations. Even nowadays, after the Ca²⁺ channel nomenclature has been drastically modified to account for the newly uncovered Ca²⁺ channel types (Ertel et al., 2000), the T-, N- and L-type terminology is still commonly used.

The Unique Properties of T-Type (LVA) Channels

Concurrently, it became evident that besides activating transiently at very negative membrane potential the neuronal T-type (LVA) channel possessed several other unique properties. Two of them concern the very slow deactivation rate

Single T-type (LVA) channels in outside-out patches (1984)

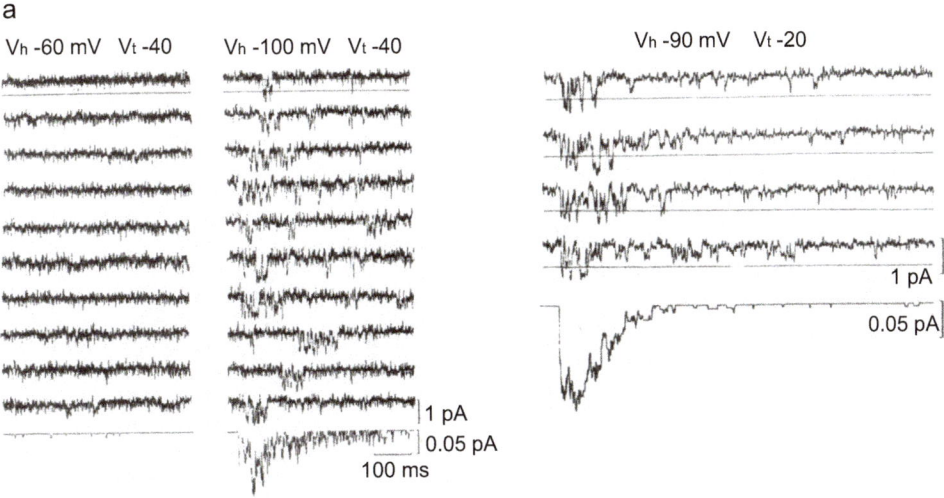

T-type (LVA) currents in outside-out macro-patches (1987)

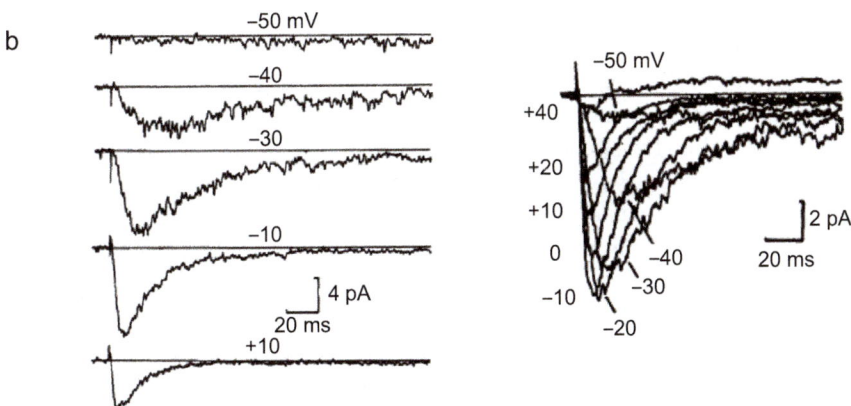

Fig. 4 Single LVA (T-type) channels in outside-out patches and macro-patches of sensory neurons. (**a**) Left, unitary Ca²⁺ currents recorded from an outside-patch of chick DRG neuron bathed in 20 mM CaCl$_2$, 120 mM Choline-Cl and 3 µM TTX (100 mM CsCl and 16 mM TEACl in the pipette) during sequential depolarizations to −40 mV (Vt). Average currents are shown at the bottom. T-type channel activity is visible only when the holding potential (Vh) is lowered to −100 mV (right panel). Right, transient activity of T-type channels recorded during sequential depolarizations to −20 mV. The averaged current (*bottom trace*) is fully inactivated after 200 ms. (Redrawn with permission from Carbone & Lux, 1984a). (**b**) Left, T-type Ca²⁺ currents recorded in an outside-out macro-patch from a rat DRG neuron in 20 mM Ca²⁺ at the potentials indicated. Right, superimposed average traces recorded as in the left panel. Each trace is the average of five records (Vh −80 mV). Apparently, the outside-out macro-patch of rat DRG neurons contains only active T-type channels. (Redrawn with permission from Carbone & Lux, 1987a)

(Carbone & Lux, 1984b; Armstrong & Matteson, 1985) and the effective recruitment of full size low-threshold Ca²⁺ currents following short hyperpolarization (Carbone & Lux, 1987b; Crunelli et al., 1989). These features could nicely account for the existence of Ca²⁺-mediated rebound spikes observed in the inferior olive and thalamic neurons (Llinas & Yarom, 1981; Llinas & Jahnsen, 1982), named "low-threshold Ca^{2+} spikes" (for a review see Dreyfus et al., 2010). It is curious how Lux and I became aware of Rodolfo Llinas' papers on the existence of a "low-threshold" Ca^{2+} channel conductance in olivary and thalamic neurons (1981–1982) while writing our manuscripts. We were discussing nearly every day with the lab people about a possible function for the new neuronal LVA channel. Unexpectedly, one day Arthur Konnerth (Germany), at that time post-doc in Lux's lab, came to my desk and very excitedly showed me the two papers by Llinas's group. Unforgettable was the excitation we had with Lux and Arthur while reading the two papers. We were among the first quoting these two excellent works.

Another important feature of the T-type channel regards its high resistance to run-down (loss of activity) following patch excision or dialysis of the cytosol. T-type channels are more resistant to run-down than any other HVA channel (L and N). This allowed, for instance, to easily record T-type channels in isolation by simply using macroscopic outside-out patches (Fig. 4b, left) or just waiting enough (>15 min) while recording Ca^{2+} currents in whole-cell conditions in the absence of intracellular phosphorylating agents (Fedulova et al., 1985; Carbone & Lux, 1987b). HVA currents progressively disappeared, while T-type currents persisted in isolation. T-type currents could be nicely visualized by averaging five records at each potential and superimposing the averaged traces to appreciate the strict voltage dependence of T-type channel activation and inactivation (Carbone & Lux, 1987a) (Fig. 4b, right).

In addition, T-type channels are especially sensitive to block by Ni^{2+} ions while they are resistant to block by Cd^{2+} ions. HVA channels have opposing sensitivities (Fox et al., 1987; Narahashi et al., 1987). T-type channels are equally permeable to Ca^{2+}, Ba^{2+} and Sr^{2+}, while HVA channels are more permeable to Ba^{2+} and Sr^{2+} than Ca^{2+} ions (Bean, 1985; Fedulova et al., 1985). Interestingly, T-type channels carry also large Na^+ currents in low $[Ca^{2+}]_o$ without losing

its unique activation/inactivation characteristics (Fukushima & Hagiwara, 1985; Carbone & Lux, 1987a).

The Explosive Interest on T-Type Channels

The discovery of the T-type (LVA) channel at the soma of sensory neurons and cardiac myocytes of vertebrates, as well as the easiness to perform biophysical tests to isolate them from the rest of HVA channels, had great impact on the future studies of T-type Ca^{2+} channel function. It revealed a new component of the family of voltage-gated Ca^{2+} channels with a clearly different role on cell functioning with respect to HVA channels. It also stimulated the interest to uncover additional novel Ca^{2+} channel types in the non-somatic regions of neurons and in other excitable tissues. The consequence of this was an explosive rise of reports from 1985 on the existence of T-type channels either in isolation or together with L, N or other not yet identified HVA channels.

The existence of T-type channels was reported in nearly all excitable tissues: cardiac cells (Bean, 1985; Nilius et al., 1985; Mitra & Morad, 1986; Hagiwara et al., 1988), skeletal and smooth muscle (Bean et al., 1986; Loirand et al., 1986; Beam & Knudson, 1988), GH3 pituitary cell line (Armstrong & Matteson, 1985; Cohen & McCarthy, 1987), mouse neuroblastoma (Narahashi et al., 1987), various types of peripheral and central neurons (Bossu et al., 1985, 1989; Fedulova et al., 1985; Yaari et al., 1987; Coulter et al., 1989; Crunelli et al., 1989), pituitary cells (Cota, 1986; Marchetti et al., 1987), adrenal gland (Cohen et al., 1988; Bossu et al., 1991; Mlinar et al., 1993) and pancreatic β-cells (Ashcroft et al., 1990; Sala & Matteson, 1990). The existence of T-type channels was also soon extended to non-excitable cells, i.e. spermatogenic cells (Arnoult et al., 1998), cultured astrocytes (Barres et al., 1988) and oligodendrocytes (Blankenfeld Gv et al., 1992). For a more complete list of tissues expressing T-type channels, please see Perez-Reyes (2003).

It became also very clear that T-type channels could carry significant "window Ca^{2+} currents" near resting potential to regulate Ca^{2+} homeostasis near resting potential (Coulter et al., 1989; Williams et al., 1997b) and drive Ca^{2+}-dependent basal activities, such as cell cycle, proliferation, differentiation, transcription and apoptosis. It became soon evident that an increase of T-type channel expression was often associated with cancer development and/or progression (Panner & Wurster, 2006; Monteith et al., 2007) and reported in numerous types of tumour cells (see Dziegielewska et al., 2014; Zhang et al., 2014 and chapter "Voltage-Gated Calcium Channels as Key Regulators of Cancer Progression" by Brackenbury in this book). The high expression of LVA T-type channels in sensory neurons have also uncovered the key role of these channels in nociceptive signalling, specifically in the development and maintenance of neuropathic pains (Jagodic et al., 2007) (see also chapter "Voltage-Gated Calcium Channels in the Afferent Pain Pathway" by Gerald Zamponi in this book).

The modern view of the LVA T-type channel is that of a channel involved in many key physiological processes (see Nilius & Carbone, 2014 for a recent review). Obviously, the physiological roles of T-type channels are primary linked to their unique low-threshold activation, which produces Ca^{2+} influx in non-excitable cells at rest and triggers all-or-none APs by activating sodium and HVA calcium channels in excitable cells. These signals regulate key processes, like muscle contraction, AP conduction, epileptic discharges, neurotransmission, hormone secretion, gamete interaction, gene expression and associated pathologies (see the chapters "T-Type Calcium Channels in Epilepsy" by Terry Snutch and "Functional Role and Plasticity of Voltage Gated Calcium Channels in the Control of Heart Automaticity" by Matteo Mangoni in this book).

The Ca^{2+} Channel Family Growths

The N-Type Channel

The N-type channel, originally identified in sensory neurons (Nowycky et al., 1985), was not found in the heart (Nilius et al., 1985) and in skeletal muscles (Beam & Knudson, 1988). It was described as a rapidly inactivating and DHP resistant with a single-channel conductance intermediate between L- and T-type channels. Since neurotransmitter release was insensitive to DHPs but sensitive to the peptide toxin ω-conotoxin (ω-CTx) GVIA from *Conus geographus* (Olivera et al., 1991), N-type channels were soon proposed as neuron specific and specialized for the release of neurotransmitters (Perney et al., 1986; Hirning et al., 1988). In these experiments, however, it was unclear whether ω-CTx GVIA blocked only the N-type channel or both N- and L-type channels as suggested by (McCleskey et al., 1987). This uncertainty of the blocking selectivity of ω-CTx GVIA derived from the simplified definition of the N-type channel as a "rapidly inactivating" HVA channel (Fox et al., 1987).

As outlined above, many groups disagreed on whether N- and L-type channels could be uniquely separated from one another based on their inactivation kinetics. The N-type channel was postulated to be responsible for the "fast inactivating" HVA current, while the L-type channel was responsible for the "steady-state plateau" current. There was indeed clear evidence that this was not the case (Swandulla et al., 1991; Jones & Elmslie, 1992). In bullfrog sympathetic neurons, the N-type channel (90% N, 10% L) carries most of the Ca^{2+} current. However, these currents inactivate slowly and more than 50% of the current persists even after 1-s step depolarization (Jones & Marks, 1989). Obviously, these currents carry an inactivating and a non-inactivating component that derive almost exclusively from the N-type channel. The same was evident in PC12 cells (Plummer et al., 1989) and sensory neurons (Aosaki & Kasai, 1989). Peter Hess and collaborators elegantly solved the issue by proposing that N-type channels were "slowly" (not "fast") inactivating channels and ω-CTx GVIA was the selective blocker of these channels (Plummer et al., 1989). After that, the general agreement was that N- and L-type currents were best identified using ω-CTx GVIA to block N-type channels and DHP antagonists or DHP agonists to either block or potentiate L-type channels. This new pharmacological approach allowed identifying the two HVA channels in combination or isolation in a variety of brain and peripheral neurons and other excitable cells (for a review see Catterall, 2011; Jurkovicova-Tarabova & Lacinova, 2019).

The P/Q-Type Channel

An additional HVA Ca^{2+} channel subtype with very slow inactivation and insensitive to both DHP and ω-CTx GVIA was soon identified in the Purkinje cells of cerebellum (Llinás et al., 1989). The new channel was named P-type (for Purkinje) and was found sensitive to the new peptide toxin of the venom from the American funnel web spider *Agelenopsis aperta*, ω-agatoxin (ω-Aga) IVA (Mintz et al., 1992). Another ω-Aga IVA-sensitive current component, which showed more rapid inactivation and had lower affinity for the toxin, was subsequently identified in cerebellar granule cells. The original proposal was that this new channel was a different HVA channel, named Q-type (Randall & Tsien, 1995). However, P- and Q-type pharmacological and biophysical properties were soon found to be recapitulated from different splicing of the same molecular entity (Bourinet et al., 1999). For simplicity, the channel was subsequently indicated as "P/Q-type" within the currently accepted terminology.

The R-Type Channel

After introduction of ω-Aga IVA as a P-type channel inhibitor, it was found that some neurons, including cortical, spinal cord and hippocampal CA1 neurons, had a "residual" Ca^{2+} current that was resistant to DHPs, ω-CTx GVIA and ω-Aga IVA (Mintz et al., 1992). This component of current could also be recognized using a new Conus peptide, ω-CTx MVIIC (Hillyard et al., 1992), which was found to block potently N- and P/Q-type channels (see McDonough et al., 1996) and was convenient to use in combination with the other Ca^{2+} channel blockers to uncover the "residual" component of HVA currents in brain neurons. The "residual" current persisting to all the available organic Ca^{2+} channel blockers was familiar to nearly all Ca^{2+} channelists at that time but it was Tsien's group (Zhang et al., 1993) that ultimately associated it with a newly-recognized cloned channel that received the name "R-type". This component of calcium current was soon separated into two R-channel subtypes (Forti et al., 1994). R-type channels were also found to be selectively blocked by the peptide SNX-482 derived from the tarantula *Hysterocrates gigas* (Newcomb et al., 1998). It became also clear that while L-type and T-type channels were expressed in a wide variety of cell, the N-, P-, Q- and R-type channels were most prominent in neurons (see Catterall, 2011).

From Ionic Currents to the Molecular Structure of Ca²⁺ Channels

In parallel with the biophysical identification of HVA channel types, several groups started purifying the Ca^{2+} channel components of skeletal muscle. Specifically, the DHP receptor (L-type) that is highly expressed in the T-tubules of skeletal muscles. The groups of Bill Catterall (USA), Franz Hofmann (Germany), Michel Lazdunski (France), Kevin Campbell (USA) and Hartmut Glossmann (Austria) were able to purify the DHP receptor complex that was composed by five protein subunits: $\alpha 1$, $\alpha 2$, β, δ and γ (Curtis & Catterall, 1984; Flockerzi et al., 1986; Hosey et al., 1987; Leung et al., 1987; Striessnig et al., 1987). The $\alpha 1$ subunit (190 kDa) contained the 1,4-DHP binding site and was identified as the pore-forming protein in association with a disulphide-linked $\alpha 2\delta$ dimer (170 kDa), an intra-

a The predicted topology of α1, α2-δ, β and γ subunits of Cav channels
(1987-1998)

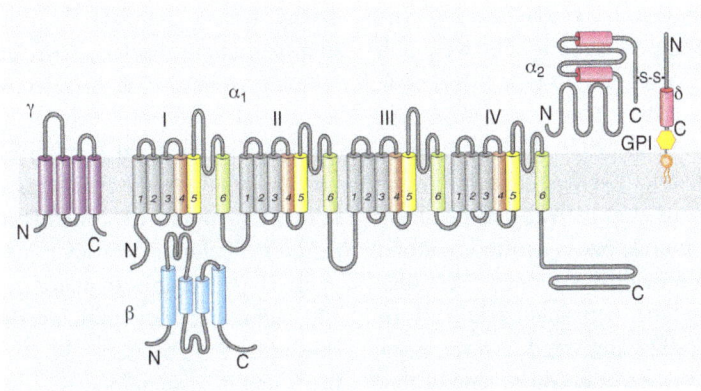

b New and past nomenclature of voltage-gated Ca²⁺ channels
(1985-2000)

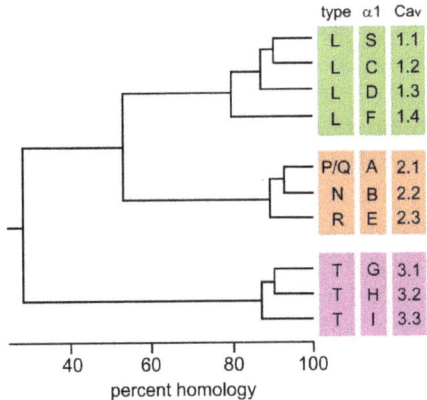

Fig. 5 Structure of Ca²⁺ channel subunits and classification of α1 subunits (1985–2000). (**a**) Subunit structure of Ca²⁺ channels. Predicted α helices are represented as cylinders. The lengths of lines correlate approximately to the lengths of the drawn polypeptide segments. The orange drawing on the δ subunit indicates the glycophosphatidylinositol (GPI) anchor to the membrane. (Redrawn after Catterall, 2011). (**b**) The three classifications used to indicate the ten isoforms of voltage-gated Ca²⁺ channels. The first (*type*) is based on biophysical properties and pharmacological sensitivity, while the last two (*α1* and *Cav*) are based on sequence homology. (Redrawn after Ertel et al., 2000)

cellular phosphorylated β subunit (55 kDa) and a transmembrane γ subunit (33 kDa) (Takahashi et al., 1987). Following this, Shosaku Numa's group in Japan was able to clone and sequence the cDNA of the skeletal muscle L-type channel α1 subunit (Tanabe et al., 1987). Through impressive team work, the group showed that by analogy with the previously cloned α1 subunit of the sodium channel (Noda et al., 1984), the amino acid sequence of the skeletal muscle α1 subunit (termed α1S) was organized into four homologous repeated domains (I–IV) with intracellular linkers and N- and C-terminals. Each domain contains six transmembrane segments (S1–S6) and a membrane-associated loop between transmembrane segments S5 and S6 (Fig. 5a).

This laid the foundation to clone and sequence other Ca²⁺ channel isoforms. The next was the cardiac L-type calcium channel, named α1C, that was cloned by homology with α1S (Mikami et al., 1989). Examining rat brain, Terry Snutch's group (Canada) next reported that multiple α1 subunits were expressed in the nervous system and encoded by distinct genes (Snutch et al., 1990). Designated α1A, α1B, α1C and α1D (the "Snutch" nomenclature – see Fig. 5b) were subsequently shown to encode P/Q-type (Mori et al., 1991; Starr et al., 1991; Bourinet et al., 1999), N-type (Dubel et al., 1992; Williams et al., 1992b) and two distinct L-type channels (Snutch et al., 1991; Williams et al., 1992a), respectively. The α1C subunit corresponding to the neuronal isoform of the previously cloned "cardiac" L-type. Curiously, the neuronal α1D isoform (Williams et al., 1992a) activated at more negative potentials and inactivated more slowly than the α1C subunit (Koschak et al., 2001; Xu & Lipscombe, 2001). A third neuronal L-type channel was later identified in the retina and named α1F (Strom et al., 1998). The channel had properties that were quite distinct from the α1D and α1C L-type isoforms (Koschak et al., 2003).

Snutch's group cloned also the α1E channel (Soong et al., 1993), which was classified as a LVA channel type, but soon became clear that it did not possess all the expected biophysical properties but rather those of the distinct R-type channel (Zhang et al., 1993). The last Ca^{2+} channels to be cloned were the T-type channels. Following an in silico approach Ed Perez-Reyes's group was able to clone the α1G, α1H and α1I subunits of rat brain (Perez-Reyes et al., 1998; Lee et al., 1999) and human heart (Cribbs et al., 1998). The cloning of T-type channels occurred after more or less 14 years from the 1984 functional discovery of the LVA T-type channel. It also ended the misconception of some colleagues that the T-type Ca^{2+} channel was a possible artefact of some sort. Following the cloning and initial study of all the calcium channel α1 subunits identified in the mammalian genome, a rationalized nomenclature was adopted in 2000, grouping the α1 subunits into Cav1 (L-type), Cav2 (N, P/Q, R-type) and Cav3 (T-type) (Ertel et al., 2000) (Fig. 5b). Since then, the distinctive pharmacological, biophysical and modulatory properties of alternative splicing of the α1 subunits have also been recognized.

Recent advances in the structural studies of calcium channels using X-ray crystallography and cryo-electron microscopy (cryo-EM) have brought new unprecedented insights into the molecular basis of their function and pharmacology. The first crystal structure for a Ca^{2+}-selective voltage-gated channel was obtained using the homotetrameric bacterial Nav channel (NavAb), which was mutated at the selectivity filter to form a Ca^{2+}-selective pore (CavAb) (Payandeh et al., 2011). The bacterial CavAb channel provided new insights into the mechanism of Ca^{2+} permeation (Tang et al., 2014) and binding locations of different calcium channel antagonists (Tang et al., 2016). Thanks to major progresses in cryo-EM technology, it is now available also a high-resolution 3D structure (3.6 Å) of the pore and the subunit arrangement of the rabbit skeletal muscle Cav1.1 channel (Wu et al., 2016). Following the same approach, the human Cav3.1 (T-type) and Cav2.2 (N-type) channel structures are also accessible for function and pharmacological studies (Zhao et al., 2019; Gao et al., 2021). This opens unexpected avenues on the function of Cav channel subunits. For more details on Cav channels structure, please refer to chapter "Pharmacology of Voltage-Gated Calcium Channels at Atomic Resolution" by Bill Catterall in this book.

The cAMP-Mediated Enhancement of Cardiac L-Type Channels as First Example of Ca^{2+} Channel Modulation

As for other voltage-gated ion channels, HVA Ca^{2+} channels are effectively up- or down-regulated by a variety of neurotransmitter-mediated second messenger pathways. Of the many modulatory pathways acting on Ca^{2+} channels, the first one discovered was the cAMP-dependent protein phosphorylation pathway that mediates the up-regulatory action of adrenaline (A) and noradrenaline (NA) on cardiac L-type channels. Both catecholamines were shown to cause an elevation and increased duration of the cardiac AP as a consequence of a marked potentiation of the Ca^{2+} current amplitudes (Reuter, 1967; Vassort et al., 1969). The action was mimicked either by injecting cAMP into myocardial fibres (Tsien et al., 1972), applying mono- or dibutyryl-cAMP analogues, phosphodiesterase

Fig. 6 The β_1-AR-agonist isoprenaline increases the size of cardiac L-type currents by increasing the channel open probability (1983–1984). (**a**) Isoprenaline (1 µM) increases the open probability (P) of single L-type channels of cultured rat heart cells, while the current amplitude of the single event remained the same (Ba^{2+} as charge carrier). (Redrawn with permission from Reuter, 1983). (**b**) Isoprenaline (0.5 µM) increases about fivefold the size of L-type currents recorded at +10 mV from frog ventricular heart cells (Vh −60 mV). (Redrawn with permission from Bean et al., 1984)

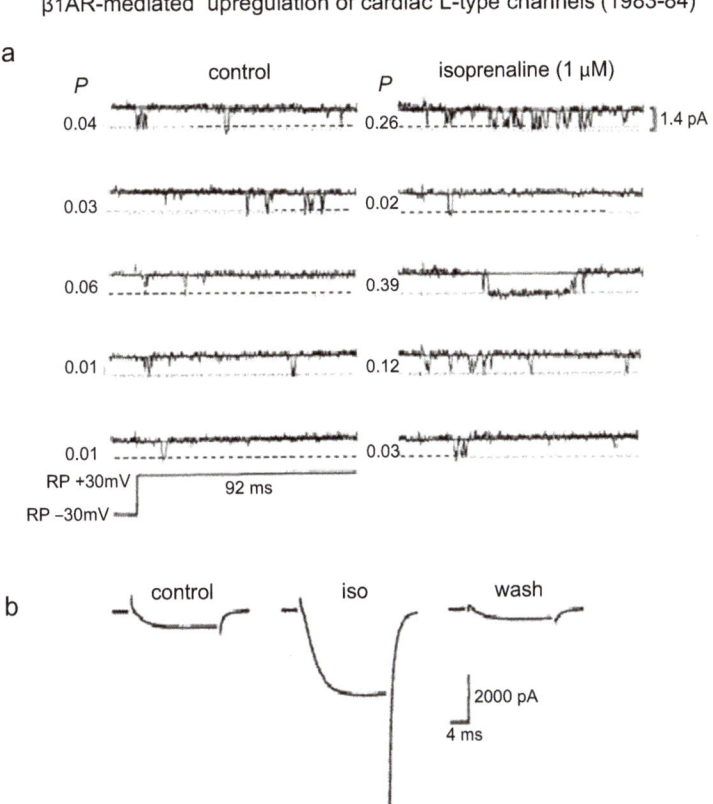

β1AR-mediated upregulation of cardiac L-type channels (1983-84)

inhibitors, or by injecting the catalytic subunit of cAMP-dependent protein kinase (PKA) (see Catterall, 2000 for a review).

Harald Reuter was the first to demonstrate that single-cardiac L-type channel activity is potentiated by cAMP analogues and β1-adrenoreceptor (β1-AR) agonists (Cachelin et al., 1983) (Fig. 6a). Addition of isoprenaline or 8-bromo-cAMP to the bath up-regulated the cardiac L-type channel activity by increasing the open channel probability (P) while preserving the size of single-channel conductance (Reuter, 1983). The action included also a three-fold increase in the number of active L-type channels, contributing to the impressive increase of the control current at +10 mV induced by isoprenaline (Fig. 6b) (Bean et al., 1984). These early findings identified the main role that the β1-AR-mediated up-regulation of cardiac L-type channels plays in the control of the "fight-or-flight" response of body function.

The experiments on cAMP-mediated enhancement of cardiac DHP-sensitive channels were soon extended to skeletal and smooth muscles, endocrine cells and neurons, where the distinctive up-regulatory effects of the cAMP/PKA phosphorylation pathway on L-type channels were observed at different degrees (see Armstrong et al., 1991 for a review). In the following years a multitude of second messenger pathways were discovered with a more or less directed up- or down-regulatory action on L-type, as well as on N-, P/Q- and R-type channels in excitable tissues. This made more complex and intriguing to understand how extracellular or intracellular second messengers modulate Ca^{2+} channels. Remarkably, recent work has shown that the cAMP-mediated enhancement of the cardiac calcium channel does not involve phosphorylation of the channel itself, as was long assumed, but rather a small interacting Gem ATPase that constitutively inhibits channel activation (Liu et al., 2020). Two chapters in this book cover the recent achievements on transmitter modulation of calcium channels (see chapters "Modulation of VGCCs by G-protein Coupled Receptors and Their Second Messengers" by Herlitze et al. and "Calmodulin Regulation of Voltage-Gated Calcium Channels" by Dick and Ben Johny).

Early Observations on the GPCR-Mediated Inhibition of Neuronal Ca2+ Channels

Nearly in parallel with the discovery of the β1-AR-mediated potentiation of cardiac L-type channels, Kathleen Dunlap and Gerald Fischbach (USA) reported a new completely different modulatory action of NA on neuronal HVA channels. They showed that NA reduces the duration of APs in chick dorsal root ganglion cells (Dunlap & Fischbach, 1978) as a consequence of a reduction of HVA Ca^{2+} currents (Dunlap & Fischbach, 1981) (Fig. 7a, left). The same occurred with serotonin and GABA. The inhibitory action of these neurotransmitters was proposed as a possible mechanism for presynaptic inhibition and attracted immediately the attention of many researchers interested on the function of presynaptic Ca^{2+} channels. Subsequent findings confirmed and extended these observations to rat sympathetic neurons (Galvan & Adams, 1982), suggesting a marked selectivity of the NA-mediated action for neuronal HVA channels but no requirements of

cAMP, cGMP or Ca^{2+} as second messengers (Forscher et al., 1986). In the following years (1986–1987) several groups were able to show that NA and other neurotransmitters inhibited the HVA Ca^{2+} currents, but it was not clear whether the action was on N- and L-type channels as the currents of the two channels were not yet clearly biophysically separated (see above).

Carla Marchetti, Dieter Lux and I were first to observe that dopamine (DA) and NA inhibited the HVA channels of sensory and sympathetic avian neurons in a "voltage-dependent" manner (Marchetti et al., 1986). DA and NA were found to produce a dramatic slowdown of HVA channel activation at low membrane potentials that was accelerated and partly relieved at higher membrane depolarizations (Fig. 7a, middle). Similar findings were reported by other groups using leu-enkephalin and somatostatin (Luini et al., 1986; Tsunoo et al., 1986; Ikeda et al., 1987).

We suggested that the slowdown of HVA channel activation was a consequence of the slow opening of HVA channels that were inhibited at rest and slowly unblocked while opening during mild depolarizations. Increasing depolarization accelerated the unblock rate, so that quickly repeated depolarizations were predicted to remove the block and recover the fast time course of HVA channel activation. This interpretation was in contrast with the more popular interpretation (at that time) that the inhibitory action of neurotransmitters was due to a "block" of a fast inactivating N-type current while preserving a

Fig. 7 (continued) (Redrawn with permission from Dunlap & Fischbach, 1978, 1981). Center, dopamine (DA, 10 μM) slows down the activation of HVA Ca^{2+} currents in 5 mM Ca^{2+}in chick DRG neurons. Activation of HVA currents accelerates with DA by increasing the step depolarization from 0 to +10 mV (Vh −70 mV). Calibration: 0.7 nA, 10 ms. (Redrawn with permission from Marchetti et al., 1986). Right, slow activation of HVA Ca^{2+} currents induced by GTP-γ-S (500 μM) at +10 mV (Vh −80 mV) in rat sensory neurons. Calibrations: top (50 pA, 10 ms), bottom (0.5 nA, 25 ms). (Redrawn with permission from Dolphin & Scott, 1987). (**b**) Left, the tails of Ca^{2+} currents at −50 mV after step depolarization to +10 mV are strongly reduced by NA (30 μM), while are fully preserved after step depolarization to +110 mV in bullfrog DRG neurons. *Inset:* the "willing-reluctant" model proposed by Bean (1989) with inclusion of the voltage-independent RO↔O transition introduced by Elmslie et al. (1990) to account for the pre-pulse induced "facilitation". (Redrawn with permission from Bean, 1989). Right, the "pre-pulse protocol" used to prove that reluctant N-type channels bound to G protein subunits recover their fast and complete activation by anticipating the test depolarization to 0 mV with a strong pre-pulse to +70 mV in bullfrog sympathetic neurons. (Redrawn with permission from Elmslie et al., 1990). (**c**) Left, prolonged delay of first openings during sequential recordings of single-N-type channel traces at +20 mV in control (left) and in the presence of NA (20 μM) + DPDPE (1 μM) (right) (Vh = −90 mV). Right, facilitation of single-N-type channel activity by a strong pre-pulse in the presence of NA + DPDPE. Same conditions as in the left panel. In most traces the delayed openings induced by the two agonists (left) were accelerated by the pre-pulse to +120 mV (right). (Redrawn with permission from Carabelli et al., 1996)

a   The GPCR-mediated inhibition of N-type channel (1978-1987)

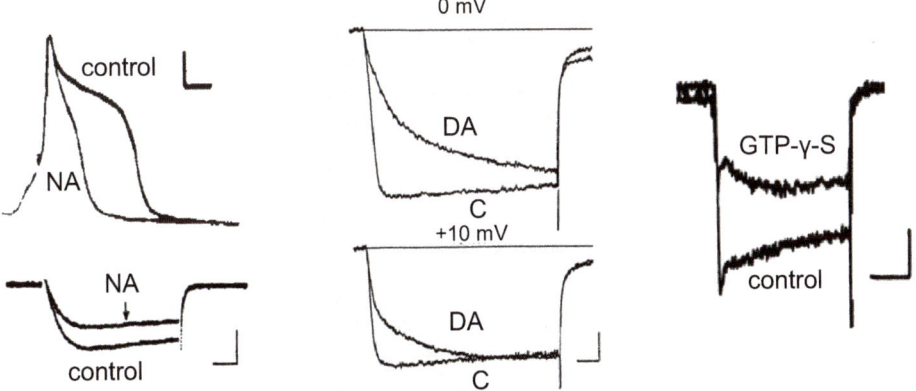

b   The "willing-reluctant" model and the "pre-pulse" protocol (1989-1990)

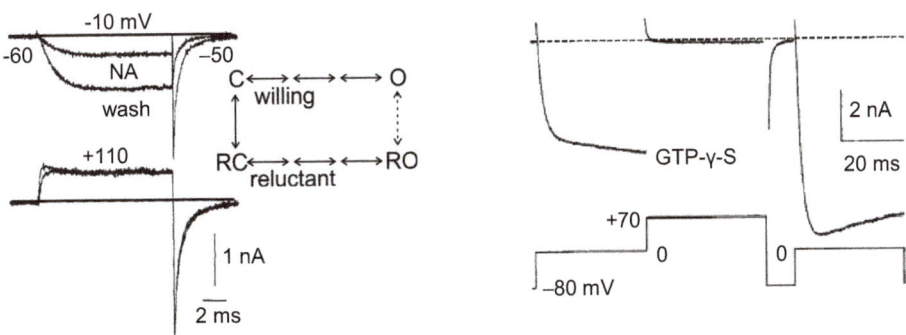

c   The delayed single channel openings induced by activated GPCRs (1996)

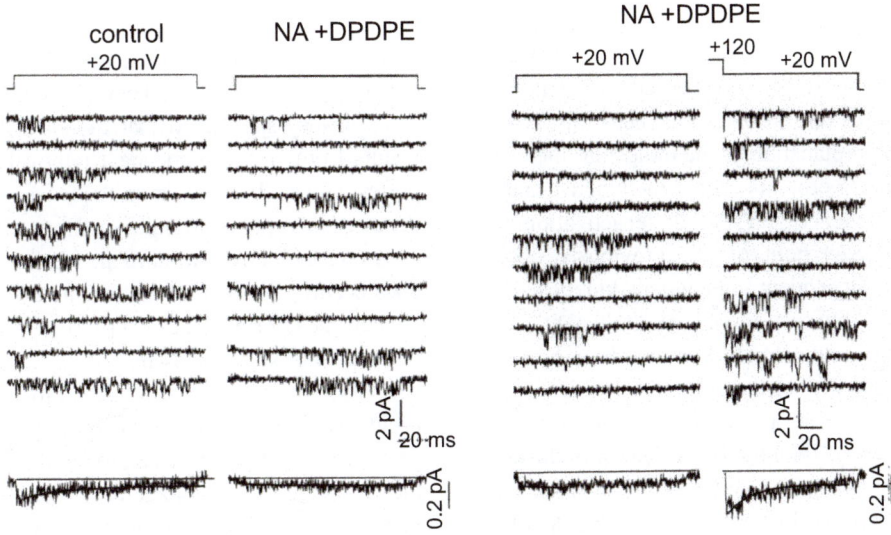

Fig. 7 The GPCR-mediated inhibition of N-type channels (1978–1996). (**a**) Left, NA (100 µM) shortens the duration of APs and reduces the Ca^{2+} currents in chick DRG neurons (10 mM Ca^{2+} and TTX in the bath, TEA in the pipette). Step depolarization to +60 mV from Vh −50 mV. Calibrations: top (20 mV, 2 ms), bottom (10 nA, 5 ms).

slowly activating L-type component (Dolphin & Scott, 1987; Gross & Macdonald, 1987; Wanke et al., 1987; Lipscombe et al., 1989).

While this controversy continued, evidence increased rapidly concerning the involvement of intracellular GTP-binding proteins through the activation of G-protein-coupled receptors (GPCRs) by neurotransmitters. Annette Dolphin's (UK) and Kathleen Dunlap's (USA) groups were among the first to prove that the non-hydrolysable GTP analogue, GTP-γ-S, mimics the inhibitory action of neurotransmitters (Dolphin & Scott, 1987) (Fig. 7a, right), while the non-hydrolysable GDP analogue, GDP-β-S, prevents the inhibition (Holz et al., 1986; Wanke et al., 1987). Ca²⁺ channel modulation by neurotransmitter and GTP-γ-S was prevented by cell incubation with pertussis toxin (PTX) (Holz et al., 1986; Lewis et al., 1986), while intracellular application of purified G protein subunits restored the GPCR-mediated response to PTX-treated cells (Toselli et al., 1989).

Towards a Full Understanding of the GPCR-Induced Delayed Activation of HVA Channels

In the years 1986–1989, many groups worked on the inhibitory action of G proteins on HVA channels and confirmed that the mechanism mediated by PTX-sensitive G proteins was unique in regulating the inhibitory action of most neurotransmitters in nearly all neurons tested (see Hille, 1994; Dolphin, 2003 for reviews). There was, however, no clear explanation of the molecular mechanism by which G proteins altered the kinetics and voltage dependence of HVA channel activation. Bruce Bean (USA) built on the previously observed "voltage-dependent" change in kinetics to propose an intuitive model of "reluctant" versus "willing" modes of channel activation based on a "positive shift" of the voltage dependence of tail currents in the presence of neurotransmitters (inset in Fig. 7b, left) (Bean, 1989). In the normal "willing" gating mode (C ↔ O) channels can be readily opened by small or moderate depolarizations while in the "reluctant" gating mode induced or stabilized by transmitters (RC ↔ RO), chan-

nels were postulated to require much larger depolarizations to activate. Using NA and bullfrog DRG neurons, Bean showed that the tail currents of N-type currents at −50 mV were strongly reduced following mild depolarizations (−10 mV) but fully recovered after strong depolarization to +110 mV (Fig. 7b, left). Assuming a voltage-independent equilibration between the two modes (RC ↔ C), the model could mimic the slow activation of HVA channels and supported the idea that neurotransmitters produce a "resting inhibition" that can be "slowly reversed" during depolarization rather than a "permanent block" of transiently inactivating channels.

However, the fully convincing evidence that the "delayed activation" of HVA channels derives from an "unblock" of G protein-inhibited channels at rest came from the work of Keith Elmslie and Steve Jones (USA). Using bullfrog sympathetic neurons, Keith and Steve showed that if moderate depolarizations to 0 mV were preceded by a strong pre-pulse depolarization to +70 mV, the HVA current inhibited by the luteinizing hormone-releasing hormone (LHRH) could fully recover its fast activation (Fig. 7b, right) (Elmslie et al., 1990). To account for this, it was necessary to include the voltage-independent transition RO ↔ O in the "willing-reluctant" model (dashed arrows in the inset of Fig. 7b, left).

As mentioned in the introduction of their Neuron paper, Keith and Steve nicely commented on the idea Hans Dieter Lux, Carla Marchetti and I had in our 1986 paper. They wrote, "Marchetti et al. (1986) speculated that the slow activation results from voltage-dependent removal of transmitter block. If this were the case, a strong depolarizing pulse should remove inhibition, returning the current to normal. We find that this does occur ….". Two issues became evident from the new data of Elmslie et al. (1990). First, as sympathetic neurons express primarily N-type channels (95% of the total), it was unequivocal that G protein subunits acted effectively on N-type channels. Second, the "pre-pulse protocol" used to test the existence of the voltage-dependent effects of neurotransmitter-induced inhibition was so ingenious and easy to apply that became immediately the routine test used by all groups to distinguish "voltage-dependent" from "voltage-independent" GPCR-mediated inhibitions. Indeed, a voltage-independent mechanism was shown to coexist in sympathetic (Bernheim et al., 1991) and sensory neurons (Luebke & Dunlap, 1994) but, at variance with the voltage-dependent one, it was mediated by soluble second messenger pathways (see Tedford & Zamponi, 2006 for review).

A voltage-dependent inhibition by neurotransmitters was soon reported also for the P-type channels of Purkinje neurons (Mintz & Bean, 1993), suggesting that the voltage-dependent relief of the inhibition (also termed "facilitation") is physiologically relevant at presynaptic terminals where N- and P/Q-type channels are highly expressed (see for review Tedford & Zamponi, 2006). Indeed, presynaptic facilitation was shown to occur at various degrees when delivering high-frequency trains of AP-like waveforms, giving a functional role to the G protein-mediated inhibition of presynaptic Ca^{2+} channels (Womack & McCleskey, 1995; Brody et al., 1997; Williams et al., 1997a), which likely underlies the widespread phenomenon of presynaptic inhibition of synaptic transmission by G-protein-coupled transmitter receptors.

Looking Deeper to the Structure and Function of Cav2 Channels Modulation by G proteins

Having clarified how G proteins inhibit N- and P/Q-type channels, the next issue was to identify which G protein subunit (α or βγ) was more effective on binding to the α1 subunit of Cav2 channels. Steve Ikeda (USA) and Bill Catterall with Bertil Hille (USA) were able to demonstrate independently that only the Gβγ dimer was involved in this modulation (Ikeda, 1996) and that the Gβγ subunit was effective on both N- and P/Q-type channels (Herlitze et al., 1996). Following this, the subsequent issue was to identify the α1 subunit regions to which the Gβγ dimer binds during the closed state of the channel. Terry Snutch's group showed both that a single Gβγ dimer interacts with the N-type channel to affect modulation (Zamponi & Snutch,

1998) and that Gβγ physically binds to the α1 subunit I-II linker (Zamponi et al., 1997). Annette Dolphin's (UK) and Lutz Birnbaumer (USA) groups further postulated that the N-terminal (Page et al., 1998) and the C-terminal (Qin et al., 1997) also represent potential intracellular regions of the α1 subunit of the channel where the Gβγ dimer could bind. Curiously enough, the intracellular I-II linker contains the highly conserved motif AID (α-interaction domain), to which the β subunit isoforms bind to regulate the α1 subunit (Pragnell et al., 1994). Thus, it was soon evident that the presence or absence of β subunits was an important determinant to up- or down-regulate the Gβγ-mediated inhibition of Cav2 channels (N, P/Q, R) (Bourinet et al., 1996; see Tedford & Zamponi, 2006; Zamponi & Currie, 2013; Dolphin, 2018 for reviews).

The modern view of G proteins inhibition is now enriched by further structural details. They include a better identification of the regions of the AID motif interacting with the β and Gβγ subunits, and the crosstalk with other cell signalling pathways, synaptic proteins and γ subunits. See chapter "Modulation of VGCCs by G-protein Coupled Receptors and Their Second Messengers" by S. Herlitze et al. in this book for a review on these issues.

In parallel to this, several groups were trying to identify the origin of the slowdown of Cav2 channel activation at the single-channel level. The prediction was that after GPCR activation, the inhibited N-type channels would have delayed their first opening (first latency) by a certain amount and strong depolarizing pre-pulses would have prevented the delay, recruiting the fast channel opening of control. Valentina Carabelli and I in Torino (Italy) were the first to show that slowdown of N-type channel activation by NA, and the δ-opioid agonist DPDPE in IMR32 cells was primarily due to a marked delay of the first single-channel opening with a consequent increase of the first latency of openings at moderate depolarization (Carabelli et al., 1996) (Fig. 7c, left). The delay had no effect on single-channel conductance, had minor effects on the mean open time and was effectively removed (facilitated) by large depolarizations (Fig. 7c, right). Several months later, similar findings were reported also by David Yue's group (USA) using HEK 293 cells transfected with recombinant M2 muscarinic receptors and N-type channels, confirming that an augmented latency of first opening was the cause of the delayed first openings of reconstituted N-type channel (Patil et al., 1996). Yue's and Elmslie's groups went further and showed that the N-type channel of sympathetic neurons (Lee & Elmslie, 2000) or transfected HEK 293 cells (Colecraft et al., 2001) exhibit "reluctant openings" that appeared as rare single brief openings during the low open probability interval that preceded the "willing openings". These later appeared in the form of repeated wide openings with increased open probability. Reluctant openings, however, were absent in reconstituted P/Q--type channels (Colecraft et al., 2001). Hence, after about 15 years, these studies settled ultimately the controversy concerning the molecular mechanisms by which the Gβγ dimer delays the activation of Cav2 channels by directly binding to the α1 subunit. GPCRs can also recruit several other distinct mechanisms, including phosphorylation, lipid signalling pathways and channel trafficking, that result in voltage-independent inhibition (for a review see Zamponi & Currie, 2013, chapters "Modulation of VGCCs by G-protein Coupled Receptors and Their Second Messengers" by Herlitze et al. and "Trafficking of Neuronal Calcium Channels" by Norbert Weiss in this book).

Take-Home Message

The first consideration, after having reviewed only a small part of the immense literature on voltage-gated Ca^{2+} channels (>50,000 papers; PubMed), is on the impressive work done by many groups over the past 50 years to take us to the present knowledge of the structure and function of Ca^{2+} channels. Thanks to the outstanding achievements of hundreds of scientists working in many countries, learning from and inspiring each other by papers and conference presentations, Cav channels are now widely recognized as

key regulatory molecules of many vital functions of the brain, sensory and motor neurons, heart, skeletal and smooth muscles and neuroendocrine and glial cells, as well as many vital functions of non-excitable cells (see Pitt et al., 2021). Ca^{2+} channels are now recognized as the targets of an increasing number of human pathologies, which derive either directly from genetic mutations of the channel subunits (α1, α2δ, β) (Ca^{2+} channelopathies) or indirectly from the altered second messenger pathways that up- or down-regulate Ca^{2+} channel function. All these new studies are in progress and described in details in this book.

A second consideration is on how the present status of Ca^{2+} channels has been reached thanks to the availability of new technical advances that occurred periodically and to the outstanding work of the many groups who exploited them to achieve milestone discoveries. As in any field of science, new advances created disputes and controversies that required time to be settled. Regardless of being right or wrong, it was crucial to participate to the debates that systematically led to new advances.

A third consideration is more personal and concerns how important it was for me to be at the right time and in the right place. Joining Lux's group at the MPI of Neurophysiology in Munich offered me the unique opportunity to deal with key unsolved issues on Ca^{2+} channel function that were completely open at that time and to discuss them with the leading personalities in the field at that moment. It was a great pleasure to have met and discussed with most of them, including those who are no longer with us: Roger Eckert, Susumu Hagiwara, Hans Dieter Lux, Platon Kostyuk, Wolfang Trautwein and David Yue.

Finally, I hope that this brief history towards the understanding of Ca^{2+} channel function will help young and new investigators realize and appreciate the pioneering work done by the many colleagues who worked with strength and passion on Ca^{2+} channel structure and function.

Acknowledgements I wish to thank Bruce Bean, Terry Snutch, Keith Elmslie and David Vandael for reading, correcting and commenting enthusiastically on the manuscript. I wish also to thank Valentina Carabelli, Andrea

Marcantoni and Giulia Tomagra for continuous support and discussions, and the editors Gerald Zamponi and Norbert Weiss for suggestions and encouragements.

References

Aosaki, T., & Kasai, H. (1989). Characterization of two kinds of high-voltage-activated Ca-channel currents in chick sensory neurons. Differential sensitivity to dihydropyridines and omega-conotoxin GVIA. *Pflügers Archiv/European Journal of Physiology, 414*, 150–156.

Armstrong, C. M. (1966). Time course of TEA(+)-induced anomalous rectification in squid giant axons. *The Journal of General Physiology, 50*, 491–503.

Armstrong, C. M. (1969). Inactivation of the potassium conductance and related phenomena caused by quaternary ammonium ion injection in squid axons. *The Journal of General Physiology, 54*, 553–575.

Armstrong, C. M., & Matteson, D. R. (1985). Two distinct populations of calcium channels in a clonal line of pituitary cells. *Science (New York, N.Y.), 227*, 65–67.

Armstrong, D. L., Rossier, M. F., Shcherbatko, A. D., & White, R. E. (1991). Enzymatic gating of voltage-activated calcium channels. *Annals of the New York Academy of Sciences, 635*, 26–34.

Arnoult, C., Villaz, M., & Florman, H. M. (1998). Pharmacological properties of the T-type Ca2+ current of mouse spermatogenic cells. *Molecular Pharmacology, 53*, 1104–1111.

Ashcroft, F. M., Kelly, R. P., & Smith, P. A. (1990). Two types of Ca channel in rat pancreatic beta-cells. *Pflügers Archiv/European Journal of Physiology, 415*, 504–506.

Baker, P. F., Hodgkin, A. L., & Ridgway, E. B. (1971). Depolarization and calcium entry in squid giant axons. *The Journal of Physiology, 218*, 709–755.

Baker, P. F., Meves, H., & Ridgway, E. B. (1973). Calcium entry in response to maintained depolarization of squid axons. *The Journal of Physiology, 231*, 527–548.

Barres, B. A., Chun, L. L., & Corey, D. P. (1988). Ion channel expression by white matter glia: I. Type 2 astrocytes and oligodendrocytes. *Glia, 1*, 10–30.

Beam, K. G., & Knudson, C. M. (1988). Calcium currents in embryonic and neonatal mammalian skeletal muscle. *The Journal of General Physiology, 91*, 781–798.

Bean, B. P. (1985). Two kinds of calcium channels in canine atrial cells. Differences in kinetics, selectivity, and pharmacology. *The Journal of General Physiology, 86*, 1–30.

Bean, B. P. (1989). Neurotransmitter inhibition of neuronal calcium currents by changes in channel voltage dependence. *Nature, 340*, 153–156.

Bean, B. P., Nowycky, M. C., & Tsien, R. W. (1984). Beta-adrenergic modulation of calcium channels in frog ventricular heart cells. *Nature, 307*, 371–375.

Bean, B. P., Sturek, M., Puga, A., & Hermsmeyer, K. (1986). Calcium channels in muscle cells isolated from rat mesenteric arteries: Modulation by dihydropyridine drugs. *Circulation Research, 59*, 229–235.

Bernheim, L., Beech, D. J., & Hille, B. (1991). A diffusible second messenger mediates one of the pathways coupling receptors to calcium channels in rat sympathetic neurons. *Neuron, 6*, 859–867.

Berridge, M. J., Bootman, M. D., & Roderick, H. L. (2003). Calcium signalling: Dynamics, homeostasis and remodelling. *Nature Reviews. Molecular Cell Biology, 4*, 517–529.

Blankenfeld Gv, G. V., Verkhratsky, A. N., & Kettenmann, H. (1992). Ca2+ channel expression in the oligodendrocyte lineage. *The European Journal of Neuroscience, 4*, 1035–1048.

Bossu, J. L., Feltz, A., & Thomann, J. M. (1985). Depolarization elicits two distinct calcium currents in vertebrate sensory neurones. *Pflügers Archiv/European Journal of Physiology, 403*, 360–368.

Bossu, J. L., Dupont, J. L., & Feltz, A. (1989). Calcium currents in rat cerebellar Purkinje cells maintained in culture. *Neuroscience, 30*, 605–617.

Bossu, J. L., De Waard, M., & Feltz, A. (1991). Two types of calcium channels are expressed in adult bovine chromaffin cells. *The Journal of Physiology, 437*, 621–634.

Bourinet, E., Soong, T. W., Stea, A., & Snutch, T. P. (1996). Determinants of the G protein-dependent opioid modulation of neuronal calcium channels. *Proceedings of the National Academy of Sciences of the United States of America, 93*, 1486–1491.

Bourinet, E., Soong, T. W., Sutton, K., Slaymaker, S., Mathews, E., Monteil, A., Zamponi, G. W., Nargeot, J., & Snutch, T. P. (1999). Splicing of alpha 1A subunit gene generates phenotypic variants of P- and Q-type calcium channels. *Nature Neuroscience, 2*, 407–415.

Brehm, P., & Eckert, R. (1978). Calcium entry leads to inactivation of calcium channel in Paramecium. *Science (New York, N.Y.), 202*, 1203–1206.

Brody, D. L., Patil, P. G., Mulle, J. G., Snutch, T. P., & Yue, D. T. (1997). Bursts of action potential waveforms relieve G-protein inhibition of recombinant P/Q-type Ca2+ channels in HEK 293 cells. *The Journal of Physiology, 499*(Pt 3), 637–644.

Brown, A. M., Camerer, H., Kunze, D. L., & Lux, H. D. (1982). Similarity of unitary Ca2+ currents in three different species. *Nature, 299*, 156–158.

Cachelin, A. B., de Peyer, J. E., Kokubun, S., & Reuter, H. (1983). Ca2+ channel modulation by 8-bromocyclic AMP in cultured heart cells. *Nature, 304*, 462–464.

Carabelli, V., Lovallo, M., Magnelli, V., Zucker, H., & Carbone, E. (1996). Voltage-dependent modulation of single N-type Ca2+ channel kinetics by receptor agonists in IMR32 cells. *Biophysical Journal, 70*, 2144–2154.

Carafoli, E., Santella, L., Branca, D., & Brini, M. (2001). Generation, control, and processing of cellular calcium signals. *Critical Reviews in Biochemistry and Molecular Biology, 36*, 107–260.

Carbone, E., & Lux, H. D. (1984a). A low voltage-activated, fully inactivating Ca channel in vertebrate sensory neurones. *Nature, 310*, 501–502.

Carbone, E., & Lux, H. D. (1984b). A low voltage-activated calcium conductance in embryonic chick sensory neurons. *Biophysical Journal, 46*, 413–418.

Carbone, E., & Lux, H. D. (1987a). Single low-voltage-activated calcium channels in chick and rat sensory neurons. *Journal of Physiology (London), 386*, 571–601.

Carbone, E., & Lux, H. D. (1987b). Kinetics and selectivity of a low-voltage-activated calcium current in chick and rat sensory neurons. *Journal of Physiology (London), 386*, 547–570.

Carbone, E., Wanke, E., Prestipino, G., Possani, L. D., & Maelicke, A. (1982). Selective blockage of voltage-dependent K+ channels by a novel scorpion toxin. *Nature, 296*, 90–91.

Catterall, W. A. (2000). Structure and regulation of voltage-gated Ca2+ channels. *Annual Review of Cell and Developmental Biology, 16*, 521–555.

Catterall, W. A. (2011). Voltage-gated calcium channels. *Cold Spring Harbor Perspectives in Biology, 3*, 23.

Cavalié, A., Ochi, R., Pelzer, D., & Trautwein, W. (1983). Elementary currents through Ca2+ channels in guinea pig myocytes. *Pflügers Archiv/European Journal of Physiology, 398*, 284–297.

Cohen, C. J., & McCarthy, R. T. (1987). Nimodipine block of calcium channels in rat anterior pituitary cells. *The Journal of Physiology, 387*, 195–225.

Cohen, C. J., McCarthy, R. T., Barrett, P. Q., & Rasmussen, H. (1988). Ca channels in adrenal glomerulosa cells: K+ and angiotensin II increase T-type Ca channel current. *Proceedings of the National Academy of Sciences of the United States of America, 85*, 2412–2416.

Colecraft, H. M., Brody, D. L., & Yue, D. T. (2001). G-protein inhibition of N- and P/Q-type calcium channels: Distinctive elementary mechanisms and their functional impact. *The Journal of Neuroscience, 21*, 1137–1147.

Cota, G. (1986). Calcium channel currents in pars intermedia cells of the rat pituitary gland. Kinetic properties and washout during intracellular dialysis. *The Journal of General Physiology, 88*, 83–105.

Coulter, D. A., Huguenard, J. R., & Prince, D. A. (1989). Calcium currents in rat thalamocortical relay neurones: Kinetic properties of the transient, low-threshold current. *The Journal of Physiology, 414*, 587–604.

Cribbs, L. L., Lee, J. H., Yang, J., Satin, J., Zhang, Y., Daud, A., Barclay, J., Williamson, M. P., Fox, M., Rees, M., & Perez-Reyes, E. (1998). Cloning and characterization of alpha1H from human heart, a member of the T-type Ca2+ channel gene family. *Circulation Research, 83*, 103–109.

Crunelli, V., Lightowler, S., & Pollard, C. E. (1989). A T-type Ca2+ current underlies low-threshold Ca2+ potentials in cells of the cat and rat lateral geniculate nucleus. *The Journal of Physiology, 413*, 543–561.

Curtis, B. M., & Catterall, W. A. (1984). Purification of the calcium antagonist receptor of the voltage-sensitive calcium channel from skeletal muscle transverse tubules. *Biochemistry, 23*, 2113–2118.

Dolphin, A. C. (2003). G protein modulation of voltage-gated calcium channels. *Pharmacological Reviews, 55*, 607–627.

Dolphin, A. C. (2018). Voltage-gated calcium channels: Their discovery, function and importance as drug targets. *Brain and Neuroscience Advances, 2*, 2398212818794805.

Dolphin, A. C., & Scott, R. H. (1987). Calcium channel currents and their inhibition by (-)-baclofen in rat sensory neurones: Modulation by guanine nucleotides. *The Journal of Physiology, 386*, 1–17.

Dreyfus, F. M., Tscherter, A., Errington, A. C., Renger, J. J., Shin, H. S., Uebele, V. N., Crunelli, V., Lambert, R. C., & Leresche, N. (2010). Selective T-type calcium channel block in thalamic neurons reveals channel redundancy and physiological impact of I(T)window. *The Journal of Neuroscience, 30*, 99–109.

Dubel, S. J., Starr, T. V., Hell, J., Ahlijanian, M. K., Enyeart, J. J., Catterall, W. A., & Snutch, T. P. (1992). Molecular cloning of the alpha-1 subunit of an omega-conotoxin-sensitive calcium channel. *Proceedings of the National Academy of Sciences of the United States of America, 89*, 5058–5062.

Dudel, J., Peper, K., Rüdel, R., & Trautwein, W. (1966). Excitatory membrane current in heart muscle (Purkinje fibers). *Pflügers Archiv für die Gesamte Physiologie des Menschen und der Tiere, 292*, 255–273.

Dunlap, K., & Fischbach, G. D. (1978). Neurotransmitters decrease the calcium component of sensory neurone action potentials. *Nature, 276*, 837–839.

Dunlap, K., & Fischbach, G. D. (1981). Neurotransmitters decrease the calcium conductance activated by depolarization of embryonic chick sensory neurones. *The Journal of Physiology, 317*, 519–535.

Dziegielewska, B., Gray, L. S., & Dziegielewski, J. (2014). T-type calcium channels blockers as new tools in cancer therapies. *Pflügers Archiv/European Journal of Physiology, 466*, 801–810.

Eckert, R., & Chad, J. E. (1984). Inactivation of Ca channels. *Progress in Biophysics and Molecular Biology, 44*, 215–267.

Eckert, R., & Lux, H. D. (1975). A non-inactivating inward current recorded during small depolarizing voltage steps in snail pacemaker neurons. *Brain Research, 83*, 486–489.

Elmslie, K. S., Zhou, W., & Jones, S. W. (1990). LHRH and GTP-gamma-S modify calcium current activation in bullfrog sympathetic neurons. *Neuron, 5*, 75–80.

Ertel, E. A., Campbell, K. P., Harpold, M. M., Hofmann, F., Mori, Y., Perez-Reyes, E., Schwartz, A., Snutch, T. P., Tanabe, T., Birnbaumer, L., Tsien, R. W., & Catterall, W. A. (2000). Nomenclature of voltage-gated calcium channels. *Neuron, 25*, 533–535.

Fatt, P., & Ginsborg, B. L. (1958). The ionic requirements for the production of action potentials in crustacean muscle fibres. *The Journal of Physiology, 142*, 516–543.

Fatt, P., & Katz, B. (1953). The electrical properties of crustacean muscle fibres. *The Journal of Physiology, 120*, 171–204.

Fedulova, S. A., Kostyuk, P. G., & Veselovsky, N. S. (1985). Two types of calcium channels in the somatic membrane of new-born rat dorsal root ganglion neurones. *The Journal of Physiology, 359*, 431–446.

Fenwick, E. M., Marty, A., & Neher, E. (1982). Sodium and calcium channels in bovine chromaffin cells. *The Journal of Physiology, 331*, 599–635.

Flockerzi, V., Oeken, H. J., Hofmann, F., Pelzer, D., Cavalié, A., & Trautwein, W. (1986). Purified dihydropyridine-binding site from skeletal muscle t-tubules is a functional calcium channel. *Nature, 323*, 66–68.

Forscher, P., Oxford, G. S., & Schulz, D. (1986). Noradrenaline modulates calcium channels in avian dorsal root ganglion cells through tight receptor-channel coupling. *The Journal of Physiology, 379*, 131–144.

Forti, L., Tottene, A., Moretti, A., & Pietrobon, D. (1994). Three novel types of voltage-dependent calcium channels in rat cerebellar neurons. *The Journal of Neuroscience, 14*, 5243–5256.

Fox, A. P., Nowycky, M. C., & Tsien, R. W. (1987). Kinetic and pharmacological properties distinguishing three types of calcium currents in chick sensory neurones. *The Journal of Physiology, 394*, 149–172.

Fukushima, Y., & Hagiwara, S. (1985). Currents carried by monovalent cations through calcium channels in mouse neoplastic B lymphocytes. *The Journal of Physiology, 358*, 255–284.

Fye, W. B. (1984). Sydney Ringer, calcium, and cardiac function. *Circulation, 69*, 849–853.

Galvan, M., & Adams, P. R. (1982). Control of calcium current in rat sympathetic neurons by norepinephrine. *Brain Research, 244*, 135–144.

Gao, S., Yao, X., & Yan, N. (2021). Structure of human Ca(v)2.2 channel blocked by the painkiller ziconotide. *Nature, 596*, 143–147.

Geduldig, D., & Gruener, R. (1970). Voltage clamp of the Aplysia giant neurone: Early sodium and calcium currents. *The Journal of Physiology, 211*, 217–244.

Geduldig, D., & Junge, D. (1968). Sodium and calcium components of action potentials in the Aplysia giant neurone. *The Journal of Physiology, 199*, 347–365.

Gerasimov, V. D. (1965). Effect of ion composition of medium on excitation processes in giant neurons of snail. *Federation Proceedings. Translation Supplement, 24*, 371–374.

Gross, R. A., & Macdonald, R. L. (1987). Dynorphin A selectively reduces a large transient (N-type) calcium current of mouse dorsal root ganglion neurons in cell culture. *Proceedings of the National Academy of Sciences of the United States of America, 84*, 5469–5473.

Hagiwara, S., & Byerly, L. (1981). Calcium channel. *Annual Review of Neuroscience, 4*, 69–125.

Hagiwara, S., & Byerly, L. (1983). The calcium channel. *Trends in Neurosciences, 6*, 189–193.

Hagiwara, S., & Naka, K. I. (1964). The initiation of spike potential in barnacle muscle fibers under low intracellular Ca++. *The Journal of General Physiology, 48*, 141–162.

Hagiwara, S., & Nakajima, S. (1966). Differences in Na and Ca spikes as examined by application of tetrodotoxin, procaine, and manganese ions. *The Journal of General Physiology, 49*, 793–806.

Hagiwara, S., & Ohmori, H. (1983). Studies of single calcium channel currents in rat clonal pituitary cells. *The Journal of Physiology, 336*, 649–661.

Hagiwara, S., Ozawa, S., & Sand, O. (1975). Voltage clamp analysis of two inward current mechanisms in the egg cell membrane of a starfish. *The Journal of General Physiology, 65*, 617–644.

Hagiwara, N., Irisawa, H., & Kameyama, M. (1988). Contribution of two types of calcium currents to the pacemaker potentials of rabbit sino-atrial node cells. *The Journal of Physiology, 395*, 233–253.

Hallett, M., & Carbone, E. (1972). Studies of calcium influx into squid giant axons with aequorin. *Journal of Cellular Physiology, 80*, 219–226.

Hamill, O. P., Marty, A., Neher, E., Sakmann, B., & Sigworth, F. J. (1981). Improved patch-clamp techniques for high-resolution current recording from cells and cell-free membrane patches. *Pflügers Archiv/European Journal of Physiology, 391*, 85–100.

Heilbrunn, L. V., & Wiercinski, F. J. (1947). The action of various cations on muscle protoplasm. *Journal of Cellular and Comparative Physiology, 29*, 15–32.

Herlitze, S., Garcia, D. E., Mackie, K., Hille, B., Scheuer, T., & Catterall, W. A. (1996). Modulation of Ca2+ channels by G-protein beta gamma subunits. *Nature, 380*, 258–262.

Hille, B. (1994). Modulation of ion-channel function by G-protein-coupled receptors. *Trends in Neurosciences, 17*, 531–536.

Hillyard, D. R., Monje, V. D., Mintz, I. M., Bean, B. P., Nadasdi, L., Ramachandran, J., Miljanich, G., Azimi-Zonooz, A., McIntosh, J. M., Cruz, L. J., et al. (1992). A new Conus peptide ligand for mammalian presynaptic Ca2+ channels. *Neuron, 9*, 69–77.

Hirning, L. D., Fox, A. P., McCleskey, E. W., Olivera, B. M., Thayer, S. A., Miller, R. J., & Tsien, R. W. (1988). Dominant role of N-type Ca2+ channels in evoked release of norepinephrine from sympathetic neurons. *Science (New York, N.Y.), 239*, 57–61.

Hodgkin, A. L., & Huxley, A. F. (1952a). A quantitative description of membrane current and its application to conduction and excitation in nerve. *The Journal of Physiology, 117*, 500–544.

Hodgkin, A. L., & Huxley, A. F. (1952b). The dual effect of membrane potential on sodium conductance in the giant axon of Loligo. *The Journal of Physiology, 116*, 497–506.

Hodgkin, A. L., & Huxley, A. F. (1952c). The components of membrane conductance in the giant axon of Loligo. *The Journal of Physiology, 116*, 473–496.

Hodgkin, A. L., & Huxley, A. F. (1952d). Currents carried by sodium and potassium ions through the membrane of the giant axon of Loligo. *The Journal of Physiology, 116*, 449–472.

Holz, G. G. T., Rane, S. G., & Dunlap, K. (1986). GTP-binding proteins mediate transmitter inhibition of voltage-dependent calcium channels. *Nature, 319*, 670–672.

Horn, R. (1978). Propagating calcium spikes in an axon of Aplysia. *The Journal of Physiology, 281*, 513–534. https://doi.org/10.1113/jphysiol.1978.sp012437 PMID: 702405.

Hosey, M. M., Barhanin, J., Schmid, A., Vandaele, S., Ptasienski, J., O'Callahan, C., Cooper, C., & Lazdunski, M. (1987). Photoaffinity labelling and phosphorylation of a 165 kilodalton peptide associated with dihydropyridine and phenylalkylamine-sensitive calcium channels. *Biochemical and Biophysical Research Communications, 147*, 1137–1145.

Ikeda, S. R. (1996). Voltage-dependent modulation of N-type calcium channels by G-protein beta gamma subunits. *Nature, 380*, 255–258.

Ikeda, S. R., Schofield, G. G., & Weight, F. F. (1987). Somatostatin blocks a calcium current in acutely isolated adult rat superior cervical ganglion neurons. *Neuroscience Letters, 81*, 123–128.

Jagodic, M. M., Pathirathna, S., Nelson, M. T., Mancuso, S., Joksovic, P. M., Rosenberg, E. R., Bayliss, D. A., Jevtovic-Todorovic, V., & Todorovic, S. M. (2007). Cell-specific alterations of T-type calcium current in painful diabetic neuropathy enhance excitability of sensory neurons. *The Journal of Neuroscience, 27*, 3305–3316.

Jones, S. W., & Elmslie, K. S. (1992). Separation and modulation of calcium currents in bullfrog sympathetic neurons. *Canadian Journal of Physiology and Pharmacology, 70*(Suppl), S56–S63.

Jones, S. W., & Marks, T. N. (1989). Calcium currents in bullfrog sympathetic neurons. II. Inactivation. *The Journal of General Physiology, 94*, 169–182.

Jurkovicova-Tarabova, B., & Lacinova, L. (2019). Structure, function and regulation of Ca(V) 2.2 N-type calcium channels. *General Physiology and Biophysics, 38*, 101–110.

Kamada, T., & Kinoshita, H. (1943). Disturbances initiated from naked surface of muscle protoplasm. *Japanese Journal of Zoology, 10*, 469–493.

Koketsu, K., & Nishi, S. (1969). Calcium and action potentials of bullfrog sympathetic ganglion cells. *The Journal of General Physiology, 53*, 608–623.

Koketsu, K., Cerf, J. A., & Nishi, S. (1959). Effect of quaternary ammonium ions on electrical activity of spinal ganglion cells in frogs. *Journal of Neurophysiology, 22*, 177–194.

Koschak, A., Reimer, D., Huber, I., Grabner, M., Glossmann, H., Engel, J., & Striessnig, J. (2001). alpha 1D (Cav1.3) subunits can form L-type Ca2+ channels activating at negative voltages. *Journal of Biological Chemistry, 276*, 22100–22106.

Koschak, A., Reimer, D., Walter, D., Hoda, J. C., Heinzle, T., Grabner, M., & Striessnig, J. (2003). Cav1.4alpha1 subunits can form slowly inactivating dihydropyridine-sensitive L-type Ca2+ channels lacking Ca2+-dependent inactivation. *The Journal of Neuroscience, 23*, 6041–6049.

Kostyuk, P. G. (1980). Calcium ionic channels in electrically excitable membrane. *Neuroscience, 5*, 945–959.

Kostyuk, P. G., Krishtal, O. A., & Doroshenko, P. A. (1974). Calcium currents in snail neurones. I. Identification of calcium current. *Pflügers Archiv/ European Journal of Physiology, 348*, 83–93.

Kostyuk, P. G., Krishtal, O. A., & Pidoplichko, V. I. (1975). Effect of internal fluoride and phosphate on membrane currents during intracellular dialysis of nerve cells. *Nature, 257*, 691–693.

Kostyuk, P. G., Krishtal, O. A., & Shakhovalov, Y. A. (1977). Separation of sodium and calcium currents in the somatic membrane of mollusc neurones. *The Journal of Physiology, 270*, 545–568.

Kostyuk, P. G., Krishtal, O. A., & Pidoplichko, V. I. (1981). Intracellular perfusion. *Journal of Neuroscience Methods, 4*, 201–210.

Krishtal, O. A., & Magura, I. S. (1970). Calcium ions as inward current carriers in mollusc neurones. *Comparative Biochemistry and Physiology, 35*, 857–866.

Krishtal, O. A., & Pidoplichko, V. I. (1980). A receptor for protons in the nerve cell membrane. *Neuroscience, 5*, 2325–2327.

Lee, H. K., & Elmslie, K. S. (2000). Reluctant gating of single N-type calcium channels during neurotransmitter-induced inhibition in bullfrog sympathetic neurons. *The Journal of Neuroscience, 20*, 3115–3128.

Lee, K. S., Akaike, N., & Brown, A. M. (1978). Properties of internally perfused, voltage-clamped, isolated nerve cell bodies. *The Journal of General Physiology, 71*, 489–507.

Lee, J. H., Daud, A. N., Cribbs, L. L., Lacerda, A. E., Pereverzev, A., Klöckner, U., Schneider, T., & Perez-Reyes, E. (1999). Cloning and expression of a novel member of the low voltage-activated T-type calcium channel family. *The Journal of Neuroscience, 19*, 1912–1921.

Leung, A. T., Imagawa, T., & Campbell, K. P. (1987). Structural characterization of the 1,4-dihydropyridine receptor of the voltage-dependent Ca2+ channel from rabbit skeletal muscle. Evidence for two distinct high molecular weight subunits. *The Journal of Biological Chemistry, 262*, 7943–7946.

Lewis, D. L., Weight, F. F., & Luini, A. (1986). A guanine nucleotide-binding protein mediates the inhibition of voltage-dependent calcium current by somatostatin in a pituitary cell line. *Proceedings of the National Academy of Sciences of the United States of America, 83*, 9035–9039.

Lipscombe, D., Kongsamut, S., & Tsien, R. W. (1989). Alpha-adrenergic inhibition of sympathetic neurotransmitter release mediated by modulation of N-type calcium-channel gating. *Nature, 340*, 639–642.

Liu, G., Papa, A., Katchman, A. N., Zakharov, S. I., Roybal, D., Hennessey, J. A., Kushner, J., Yang, L., Chen, B. X., Kushnir, A., Dangas, K., Gygi, S. P., Pitt, G. S., Colecraft, H. M., Ben-Johny, M., Kalocsay, M., & Marx, S. O. (2020). Mechanism of adrenergic Ca(V)1.2 stimulation revealed by proximity proteomics. *Nature, 577*, 695–700.

Llinas, R., & Jahnsen, H. (1982). Electrophysiology of mammalian thalamic neurones in vitro. *Nature, 297*, 406–408.

Llinas, R., & Yarom, Y. (1981). Properties and distribution of ionic conductances generating electroresponsiveness of mammalian inferior olivary neurones in vitro. *The Journal of Physiology, 315*, 569–584.

Llinás, R., Sugimori, M., Lin, J. W., & Cherksey, B. (1989). Blocking and isolation of a calcium channel from neurons in mammals and cephalopods utilizing a toxin fraction (FTX) from funnel-web spider poison. *Proceedings of the National Academy of Sciences of the United States of America, 86*, 1689–1693.

Loirand, G., Pacaud, P., Mironneau, C., & Mironneau, J. (1986). Evidence for two distinct calcium channels in rat vascular smooth muscle cells in short-term primary culture. *Pflügers Archiv/European Journal of Physiology, 407*, 566–568.

Luebke, J. I., & Dunlap, K. (1994). Sensory neuron N-type calcium currents are inhibited by both voltage-dependent and -independent mechanisms. *Pflügers Archiv/European Journal of Physiology, 428*, 499–507.

Luini, A., Lewis, D., Guild, S., Schofield, G., & Weight, F. (1986). Somatostatin, an inhibitor of ACTH secretion, decreases cytosolic free calcium and voltage-dependent calcium current in a pituitary cell line. *The Journal of Neuroscience, 6*, 3128–3132.

Lux, H. D., & Nagy, K. (1981). Single channel Ca2+ currents in Helix pomatia neurons. *Pflügers Archiv/European Journal of Physiology, 391*, 252–254.

Marchetti, C., Carbone, E., & Lux, H. D. (1986). Effects of dopamine and noradrenaline on Ca channels of cultured sensory and sympathetic neurons of chick. *Pflügers Archiv/European Journal of Physiology, 406*, 104–111.

Marchetti, C., Childs, G. V., & Brown, A. M. (1987). Membrane currents of identified isolated rat corticotropes and gonadotropes. *The American Journal of Physiology, 252*, E340–E346.

McCleskey, E. W., Fox, A. P., Feldman, D. H., Cruz, L. J., Olivera, B. M., Tsien, R. W., & Yoshikami, D. (1987). Omega-conotoxin: Direct and persistent blockade of specific types of calcium channels in neurons but not muscle. *Proceedings of the National Academy of Sciences of the United States of America, 84*, 4327–4331.

McDonough, S. I., Swartz, K. J., Mintz, I. M., Boland, L. M., & Bean, B. P. (1996). Inhibition of calcium channels in rat central and peripheral neurons by omega-conotoxin MVIIC. *The Journal of Neuroscience, 16*, 2612–2623.

Mikami, A., Imoto, K., Tanabe, T., Niidome, T., Mori, Y., Takeshima, H., Narumiya, S., & Numa, S. (1989). Primarystructure and functional expression of the cardiac dihydropyridine-sensitive calcium channel. *Nature, 340*, 230-233. https://doi.org/10.1038/340230a0 PMID: 2474130.

Mintz, I. M., & Bean, B. P. (1993). GABAB receptor inhibition of P-type Ca2+ channels in central neurons. *Neuron, 10*, 889–898.

Mintz, I. M., Venema, V. J., Swiderek, K. M., Lee, T. D., Bean, B. P., & Adams, M. E. (1992). P-type calcium channels blocked by the spider toxin omega-Aga-IVA. *Nature, 355*, 827–829.

Mitra, R., & Morad, M. (1986). Two types of calcium channels in Guinea pig ventricular myocytes. *Proceedings of the National Academy of Sciences of the United States of America, 83*, 5340–5344.

Mlinar, B., Biagi, B. A., & Enyeart, J. J. (1993). Voltage-gated transient currents in bovine adrenal fasciculata cells. I. T-type Ca2+ current. *The Journal of General Physiology, 102*, 217–237.

Monteith, G. R., McAndrew, D., Faddy, H. M., & Roberts-Thomson, S. J. (2007). Calcium and cancer: Targeting Ca2+ transport. *Nature Reviews. Cancer, 7*, 519–530.

Mori, Y., Friedrich, T., Kim, M. S., Mikami, A., Nakai, J., Ruth, P., Bosse, E., Hofmann, F., Flockerzi, V., Furuichi, T., et al. (1991). Primary structure and functional expression from complementary DNA of a brain calcium channel. *Nature, 350*, 398–402.

Narahashi, T., Tsunoo, A., & Yoshii, M. (1987). Characterization of two types of calcium channels in mouse neuroblastoma cells. *The Journal of Physiology, 383*, 231–249.

Neher, E., & Sakmann, B. (1976). Single-channel currents recorded from membrane of denervated frog muscle fibres. *Nature, 260*, 799–802.

Newcomb, R., Szoke, B., Palma, A., Wang, G., Chen, X., Hopkins, W., Cong, R., Miller, J., Urge, L., Tarczy-Hornoch, K., Loo, J. A., Dooley, D. J., Nadasdi, L., Tsien, R. W., Lemos, J., & Miljanich, G. (1998). Selective peptide antagonist of the class E calcium channel from the venom of the tarantula Hysterocrates gigas. *Biochemistry, 37*, 15353–15362.

Niedergerke, R., & Orkand, R. K. (1966). The dual effect of calcium on the action potential of the frog's heart. *The Journal of Physiology, 184*, 291–311.

Nilius, B., & Carbone, E. (2014). Amazing T-type calcium channels: Updating functional properties in health and disease. *Pflügers Archiv/European Journal of Physiology, 466*, 623–626.

Nilius, B., Hess, P., Lansman, J. B., & Tsien, R. W. (1985). A novel type of cardiac calcium-channel in ventricular cells. *Nature, 316*, 443–446.

Nishi, S., Soeda, H., & Koketsu, K. (1965). Effect of alkali-earth cations on frog spinal ganglion cell. *Journal of Neurophysiology, 28*, 457–472.

Noda, M., Shimizu, S., Tanabe, T., Takai, T., Kayano, T., Ikeda, T., Takahashi, H., Nakayama, H., Kanaoka, Y., Minamino, N., et al. (1984). Primary structure of Electrophorus electricus sodium channel deduced from cDNA sequence. *Nature, 312*, 121–127.

Nowycky, M. C., Fox, A. P., & Tsien, R. W. (1985). Three types of neuronal calcium channel with different calcium agonist sensitivity. *Nature, 316*, 440–443.

Olivera, B. M., Imperial, J. S., Cruz, L. J., Bindokas, V. P., Venema, V. J., & Adams, M. E. (1991). Calcium channel-targeted polypeptide toxins. *Annals of the New York Academy of Sciences, 635*, 114–122.

Oomura, Y., Ozaki, S., & Maeno, T. (1961). Electrical activity of a giant nerve cell under abnormal conditions. *Nature, 191*, 1265–1267.

Paes de Carvalho, A. P., Hoffman, B. F., & Langan, W. B. (1966). Two components of the cardiac action potential. *Nature, 211*, 938–940.

Page, K. M., Cantí, C., Stephens, G. J., Berrow, N. S., & Dolphin, A. C. (1998). Identification of the amino terminus of neuronal Ca2+ channel alpha1 subunits alpha1B and alpha1E as an essential determinant of G-protein modulation. *The Journal of Neuroscience, 18*, 4815–4824.

Panner, A., & Wurster, R. D. (2006). T-type calcium channels and tumor proliferation. *Cell Calcium, 40*, 253–259.

Patil, P. G., de Leon, M., Reed, R. R., Dubel, S., Snutch, T. P., & Yue, D. T. (1996). Elementary events underlying voltage-dependent G-protein inhibition of N-type calcium channels. *Biophysical Journal, 71*, 2509–2521.

Payandeh, J., Scheuer, T., Zheng, N., & Catterall, W. A. (2011). The crystal structure of a voltage-gated sodium channel. *Nature, 475*, 353–358.

Perez-Reyes, E. (2003). Molecular physiology of low-voltage-activated T-type calcium channels. *Physiological Reviews, 83*, 117–161.

Perez-Reyes, E., Cribbs, L. L., Daud, A., Lacerda, A. E., Barclay, J., Williamson, M. P., Fox, M., Rees, M., & Lee, J. H. (1998). Molecular characterization of a neuronal low-voltage-activated T-type calcium channel. *Nature, 391*, 896–900.

Perney, T. M., Hirning, L. D., Leeman, S. E., & Miller, R. J. (1986). Multiple calcium channels mediate neurotransmitter release from peripheral neurons. *Proceedings of the National Academy of Sciences of the United States of America, 83*, 6656–6659.

Pitt, G. S., Matsui, M., & Cao, C. (2021). Voltage-gated calcium channels in nonexcitable tissues. *Annual Review of Physiology, 83*, 183–203.

Plummer, M. R., Logothetis, D. E., & Hess, P. (1989). Elementary properties and pharmacological sensitivities of calcium channels in mammalian peripheral neurons. *Neuron, 2*, 1453–1463.

Pragnell, M., De Waard, M., Mori, Y., Tanabe, T., Snutch, T. P., & Campbell, K. P. (1994). Calcium channel beta-subunit binds to a conserved motif in the I-II cytoplasmic linker of the alpha 1-subunit. *Nature, 368*, 67–70.

Qin, N., Platano, D., Olcese, R., Stefani, E., & Birnbaumer, L. (1997). Direct interaction of gbetagamma with a C-terminal gbetagamma-binding domain of the Ca2+ channel alpha1 subunit is responsible for channel inhi-

bition by G protein-coupled receptors. *Proceedings of the National Academy of Sciences of the United States of America, 94*, 8866–8871.

Randall, A., & Tsien, R. W. (1995). Pharmacological dissection of multiple types of Ca2+ channel currents in rat cerebellar granule neurons. The Journal of neuroscience : the officialjournal of the Society for Neuroscience, 15, 2995–3012. https://doi.org/10.1523/jneurosci.15-04-02995 MID: 7722641.

Reuter, H. (1967). The dependence of slow inward current in Purkinje fibres on the extracellular calcium-concentration. *The Journal of Physiology, 192*, 479–492.

Reuter, H. (1973). Divalent cations as charge carriers in excitable membranes. *Progress in Biophysics and Molecular Biology, 26*, 1–43.

Reuter, H. (1979). Properties of two inward membrane currents in the heart. *Annual Review of Physiology, 41*, 413–424.

Reuter, H. (1983). Calcium channel modulation by neurotransmitters, enzymes and drugs. *Nature, 301*, 569–574.

Reuter, H., Stevens, C. F., Tsien, R. W., & Yellen, G. (1982). Properties of single calcium channels in cardiac cell culture. *Nature, 297*, 501–504.

Ringer, S. (1883). A further contribution regarding the influence of the different constituents of the blood on the contraction of the heart. *The Journal of Physiology, 4*(29–42), 23.

Sala, S., & Matteson, D. R. (1990). Single-channel recordings of two types of calcium channels in rat pancreatic beta-cells. *Biophysical Journal, 58*, 567–571.

Sather, W. A., & McCleskey, E. W. (2003). Permeation and selectivity in calcium channels. *Annual Review of Physiology, 65*, 133–159. https://doi.org/10.1146/annurev.physiol.65.092101.142345 PMID: 12471162.

Snutch, T. P., Leonard, J. P., Gilbert, M. M., Lester, H. A., & Davidson, N. (1990). Rat brain expresses a heterogeneous family of calcium channels. *Proceedings of the National Academy of Sciences of the United States of America, 87*, 3391–3395.

Snutch, T. P., Tomlinson, W. J., Leonard, J. P., & Gilbert, M. M. (1991). Distinct calcium channels are generated by alternative splicing and are differentially expressed in the mammalian CNS. *Neuron, 7*, 45–57.

Soong, T. W., Stea, A., Hodson, C. D., Dubel, S. J., Vincent, S. R., & Snutch, T. P. (1993). Structure and functional expression of a member of the low voltage-activated calcium channel family. *Science (New York, N.Y.), 260*, 1133–1136.

Stanley, E. J., & Reiter, M. (1965). The antagonistic effects of sodium and calcium on the action potential of guinea pig papillary muscle. *Naunyn-Schmiedebergs Archiv für Experimentelle Pathologie und Pharmakologie, 252*, 159–172.

Starr, T. V., Prystay, W., & Snutch, T. P. (1991). Primary structure of a calcium channel that is highly expressed in the rat cerebellum. *Proceedings of the National Academy of Sciences of the United States of America, 88*, 5621–5625.

Stiles, P. G. (1901). On the rhythmic activity of the œsophagus and the influence upon it of various media. *American Journal of Physiology, 5*, 338–357.

Striessnig, J., Knaus, H. G., Grabner, M., Moosburger, K., Seitz, W., Lietz, H., & Glossmann, H. (1987). Photoaffinity labelling of the phenylalkylamine receptor of the skeletal muscle transverse-tubule calcium channel. *FEBS Letters, 212*, 247–253.

Strom, T. M., Nyakatura, G., Apfelstedt-Sylla, E., Hellebrand, H., Lorenz, B., Weber, B. H., Wutz, K., Gutwillinger, N., Rüther, K., Drescher, B., Sauer, C., Zrenner, E., Meitinger, T., Rosenthal, A., & Meindl, A. (1998). An L-type calcium-channel gene mutated in incomplete X-linked congenital stationary night blindness. *Nature Genetics, 19*, 260–263.

Swandulla, D., Carbone, E., & Lux, H. D. (1991). Do calcium-channel classifications account for neuronal calcium-channel diversity. *Trends in Neurosciences, 14*, 46–51.

Takahashi, K., & Yoshii, M. (1978). Effects of internal free calcium upon the sodium and calcium channels in the tunicate egg analysed by the internal perfusion technique. *The Journal of Physiology, 279*, 519–549.

Takahashi, M., Seagar, M. J., Jones, J. F., Reber, B. F., & Catterall, W. A. (1987). Subunit structure of dihydropyridine-sensitive calcium channels from skeletal muscle. *Proceedings of the National Academy of Sciences of the United States of America, 84*, 5478–5482.

Tanabe, T., Takeshima, H., Mikami, A., Flockerzi, V., Takahashi, H., Kangawa, K., Kojima, M., Matsuo, H., Hirose, T., & Numa, S. (1987). Primary structure of the receptor for calcium channel blockers from skeletal muscle. *Nature, 328*, 313–318.

Tang, L., Gamal El-Din, T. M., Payandeh, J., Martinez, G. Q., Heard, T. M., Scheuer, T., Zheng, N., & Catterall, W. A. (2014). Structural basis for Ca2+ selectivity of a voltage-gated calcium channel. *Nature, 505*, 56–61.

Tang, L., Gamal El-Din, T. M., Swanson, T. M., Pryde, D. C., Scheuer, T., Zheng, N., & Catterall, W. A. (2016). Structural basis for inhibition of a voltage-gated Ca(2+) channel by Ca(2+) antagonist drugs. *Nature, 537*, 117–121.

Tedford, H. W., & Zamponi, G. W. (2006). Direct G protein modulation of Cav2 calcium channels. *Pharmacological Reviews, 58*, 837–862.

Toselli, M., Lang, J., Costa, T., & Lux, H. D. (1989). Direct modulation of voltage-dependent calcium channels by muscarinic activation of a pertussis toxin-sensitive G-protein in hippocampal neurons. *Pflügers Archiv/European Journal of Physiology, 415*, 255–261.

Trautwein, W. (1973). Membrane currents in cardiac muscle fibers. *Physiological Reviews, 53*, 793–835.

Tsien, R. W., Giles, W., & Greengard, P. (1972). Cyclic AMP mediates the effects of adrenaline on cardiac purkinje fibres. *Nature: New Biology, 240*, 181–183.

Tsunoo, A., Yoshii, M., & Narahashi, T. (1986). Block of calcium channels by enkephalin and somatostatin in neuroblastoma-glioma hybrid NG108-15 cells.

Proceedings of the National Academy of Sciences of the United States of America, 83, 9832–9836.

Vassort, G., Rougier, O., Garnier, D., Sauviat, M. P., Coraboeuf, E., & Gargouïl, Y. M. (1969). Effects of adrenaline on membrane inward currents during the cardiac action potential. *Pflügers Archiv/European Journal of Physiology, 309*, 70–81.

Wanke, E., Carbone, E., & Testa, P. L. (1980). The sodium channel and intracellular H+ blockage in squid axons. *Nature, 287*, 62–63.

Wanke, E., Ferroni, A., Malgaroli, A., Ambrosini, A., Pozzan, T., & Meldolesi, J. (1987). Activation of a muscarinic receptor selectively inhibits a rapidly inactivated Ca2+ current in rat sympathetic neurons. *Proceedings of the National Academy of Sciences of the United States of America, 84*, 4313–4317.

Williams, M. E., Feldman, D. H., McCue, A. F., Brenner, R., Velicelebi, G., Ellis, S. B., & Harpold, M. M. (1992a). Structure and functional expression of alpha 1, alpha 2, and beta subunits of a novel human neuronal calcium channel subtype. *Neuron, 8*, 71–84.

Williams, M. E., Brust, P. F., Feldman, D. H., Patthi, S., Simerson, S., Maroufi, A., McCue, A. F., Veliçelebi, G., Ellis, S. B., & Harpold, M. M. (1992b). Structure and functional expression of an omega-conotoxin-sensitive human N-type calcium channel. *Science (New York, N.Y.), 257*, 389–395.

Williams, S., Serafin, M., Mühlethaler, M., & Bernheim, L. (1997a). Facilitation of N-type calcium current is dependent on the frequency of action potential-like depolarizations in dissociated cholinergic basal forebrain neurons of the guinea pig. *The Journal of Neuroscience, 17*, 1625–1632.

Williams, S. R., Tóth, T. I., Turner, J. P., Hughes, S. W., & Crunelli, V. (1997b). The 'window' component of the low threshold Ca2+ current produces input signal amplification and bistability in cat and rat thalamocortical neurones. *The Journal of Physiology, 505*(Pt 3), 689–705.

Womack, M. D., & McCleskey, E. W. (1995). Interaction of opioids and membrane potential to modulate Ca2+ channels in rat dorsal root ganglion neurons. *Journal of Neurophysiology, 73*, 1793–1798.

Wright, E. B., & Ogata, M. (1961). Action potential of amphibian single auricular muscle fiber: A dual response. *The American Journal of Physiology, 201*, 1101–1108.

Wu, J., Yan, Z., Li, Z., Qian, X., Lu, S., Dong, M., Zhou, Q., & Yan, N. (2016). Structure of the voltage-gated calcium channel Ca(v)1.1 at 3.6 Å resolution. *Nature, 537*, 191–196.

Xu, W. F., & Lipscombe, D. (2001). Neuronal Ca(v)1.3 alpha(1) L-type channels activate at relatively hyperpolarized membrane potentials and are incompletely inhibited by dihydropyridines. *Journal of Neuroscience, 21*, 5944–5951.

Yaari, Y., Hamon, B., & Lux, H. D. (1987). Development of two types of calcium channels in cultured mammalian hippocampal neurons. *Science (New York, N.Y.), 235*, 680–682.

Zamponi, G. W., & Currie, K. P. (2013). Regulation of Ca(V)2 calcium channels by G protein coupled receptors. *Biochimica et Biophysica Acta, 1828*, 1629–1643.

Zamponi, G. W., & Snutch, T. P. (1998). Decay of prepulse facilitation of N type calcium channels during G protein inhibition is consistent with binding of a single Gbeta subunit. *Proceedings of the National Academy of Sciences of the United States of America, 95*, 4035–4039.

Zamponi, G. W., Bourinet, E., Nelson, D., Nargeot, J., & Snutch, T. P. (1997). Crosstalk between G proteins and protein kinase C mediated by the calcium channel alpha1 subunit. *Nature, 385*, 442–446.

Zhang, J. F., Randall, A. D., Ellinor, P. T., Horne, W. A., Sather, W. A., Tanabe, T., Schwarz, T. L., & Tsien, R. W. (1993). Distinctive pharmacology and kinetics of cloned neuronal Ca2+ channels and their possible counterparts in mammalian CNS neurons. *Neuropharmacology, 32*, 1075–1088.

Zhang, Y., Wang, H., Qian, Z., Feng, B., Zhao, X., Jiang, X., & Tao, J. (2014). Low-voltage-activated T-type Ca2+ channel inhibitors as new tools in the treatment of glioblastoma: The role of endostatin. *Pflügers Archiv/ European Journal of Physiology, 466*, 811–818.

Zhao, Y., Huang, G., Wu, Q., Wu, K., Li, R., Lei, J., Pan, X., & Yan, N. (2019). Cryo-EM structures of apo and antagonist-bound human Ca(v)3.1. *Nature, 576*, 492–497.

Part I

Structural and Molecular Aspects of VGCCs

Subunit Architecture and Atomic Structure of Voltage-Gated Ca²⁺ Channels

William A. Catterall

Abstract

Voltage-gated calcium channels mediate calcium entry into cells in response to membrane depolarization. The high-voltage-activated Ca^{2+} channels that have been characterized biochemically are complexes of a pore-forming $\alpha 1$ subunit of ~200–250 kD, a transmembrane, disulfide-linked complex of $\alpha 2$ and δ subunits, an intracellular β subunit, and in some cases a transmembrane γ subunit. The $\alpha 1$ subunits form the transmembrane pore. The $\alpha 2$ and δ subunits are glycoproteins encoded by the same gene and produced by posttranslational proteolytic processing followed by attachment of a lipid anchor. The γ subunits are transmembrane glycoproteins, whereas the β subunits are hydrophilic proteins located on the cytosolic face of the channel. In this chapter, I have reviewed early work that led to identification, purification, and characterization of the protein subunits of calcium channels and presented more recent studies that have given clear insights into the structural basis for calcium channel function from X-ray crystallography and cryoelectron microscopy.

Keywords

Calcium channel · Subunit architecture · Structure · X-ray crystallography · Cryogenic electron microscopy

Introduction

Calcium channels have been studied intensively by electrophysiological methods since the initial recordings of calcium currents by Reuter (1967). In the early 1980s, stimulated in part by success at purification and reconstitution of sodium channels (Catterall, 1984), a new focus of calcium channel research developed aimed at identification of the calcium channel proteins and analysis of their biochemical and structural properties. In this chapter, I briefly review the biochemical and molecular biological studies that led to identification of the calcium channel proteins and determination of their subunit composition, and then I present the modern views of their structures, subunit architectures, and functional elements from recent high-resolution structural studies.

W. A. Catterall (✉)
Department of Pharmacology, University of Washington, Seattle, WA, USA
e-mail: wcatt@uw.edu

Purification and Biochemical Characterization of Skeletal Muscle Calcium Channels

The calcium channels in the transverse tubule membranes of skeletal muscle served as the primary biochemical preparation for studies of calcium channels because of their high abundance. These channels serve two critical physiological roles. Like other calcium channels, they mediate calcium entry in response to depolarization. The primary voltage-gated calcium currents in skeletal muscle are L-type (Sanchez & Stefani, 1978; Almers et al., 1984; McCleskey & Almers, 1985), characterized by slow voltage-dependent inactivation, large single-channel conductance (10–25 pS), high voltage of activation, and specific inhibition by dihydropyridine calcium channel antagonists. These currents activate very slowly, and the calcium entering vertebrate skeletal muscle through voltage-gated calcium channels is not directly required for muscle contraction. It may serve to replenish cellular calcium during periods of rapid activity and to increase intracellular calcium in response to tetanic stimulation, leading to increased contractile force. The primary physiological role for the skeletal muscle calcium channel is to serve as a voltage sensor in excitation–contraction coupling (Tanabe et al., 1988; Catterall, 1991a). Voltage-gated calcium channels in the transverse tubule membranes interact physically with the calcium release channels located in the sarcoplasmic reticulum membrane. Voltage-driven conformational changes in the voltage-gated calcium channels then activate calcium release from the sarcoplasmic reticulum by protein–protein interactions and initiate contraction (Rios & Brum, 1987; Adams & Beam, 1990; Adams et al., 1990; Catterall, 1991a).

Purification of calcium channels from skeletal muscle began with isolation of the transverse tubule membranes, which are highly enriched in calcium channel protein, followed by specific labeling of the channel protein by high-affinity binding of dihydropyridine calcium channel antagonist drugs (Borsotto et al., 1984; Curtis & Catterall, 1984). Ca^{2+} channels were solubilized in the mild detergent digitonin to retain native subunit associations and purified by a combination of ion exchange chromatography, affinity chromatography on wheat germ agglutinin–Sepharose, and sedimentation through sucrose gradients (Curtis & Catterall, 1984). A heterogeneous α subunit band and associated β subunits of 50 kD and γ subunits of 33 kDa were identified as components of the calcium channel complex in the initial purification studies (Curtis & Catterall, 1984). Subsequent experiments demonstrated that the heterogeneous α subunit band contained not only the principal $\alpha1$ subunits with an apparent molecular mass of 175 kD but also a disulfide-linked dimer of $\alpha2$ and δ subunits with apparent molecular masses of 143 kD and 27 kD, respectively, as illustrated in the SDS-PAGE results in Fig. 1a (Hosey et al., 1987; Sieber et al., 1987; Striessnig et al., 1987; Takahashi et al., 1987; Leung et al., 1988). Reconstitution of the purified calcium channel preparation in phospholipid vesicles yielded calcium influx after activation with the dihydropyridine agonist Bay8644 (Fig. 1b), and incorporation into planar bilayers revealed single calcium channel currents (Curtis & Catterall, 1986; Flockerzi et al., 1986). The specific association of these proteins as a multisubunit complex (Fig. 2a) was supported by the co-purification of each subunit with the dihydropyridine binding activity and calcium conductance activity of the calcium channel (Curtis & Catterall, 1984, 1986; Flockerzi et al., 1986; Takahashi et al., 1987), by co-immunoprecipitation of all these proteins by antibodies directed against the $\alpha1$ subunits (Leung et al., 1987; Morton & Froehner, 1987; Takahashi et al., 1987), and by co-immunoprecipitation of the calcium channel complex with antibodies against each auxiliary subunit (Leung et al., 1988; Sharp & Campbell, 1989; Ahlijanian et al., 1990). Estimates of stoichiometry indicated that each mol of calcium channel complex contains approximately 1 mol of each of the 5 subunits (Fig. 2a).

α1 Subunit The $\alpha1$ subunit of skeletal muscle calcium channels was cloned by library screening based on amino acid sequence (Fig. 2b; Tanabe et al., 1987). The $\alpha1$ subunit resembled

a   **b**

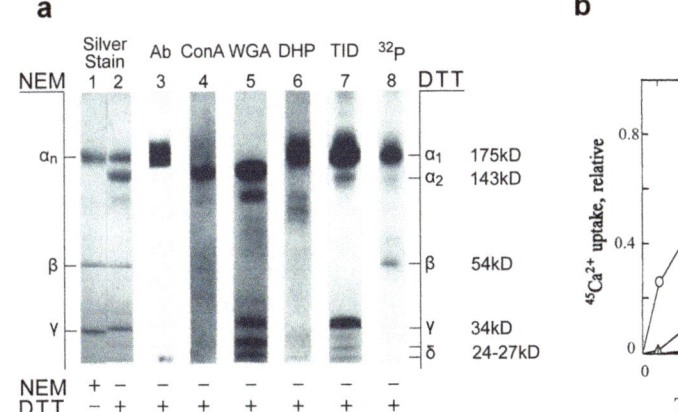

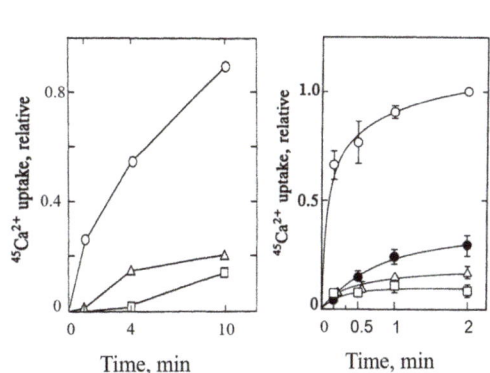

Fig. 1 Biochemical properties of skeletal muscle calcium channels. (**a**) Summary of the biochemical properties of purified skeletal muscle calcium channels. Lanes 1 and 2, silver stain of polypeptides; lane 3, staining with an antibody against the α1 subunit; lane 4, staining with concanavalin A, a lectin binding high-mannose N-linked carbohydrate chains; lane 5, staining with wheat germ agglutinin, a lectin staining N-linked complex carbohydrate chains; lane 6, photoaffinity labeling with azidopine, a photoreactive dihydropyridine; lane 7, photoaffinity labeling with TID, a hydrophobic probe of the transmembrane regions of proteins; lane 8, phosphorylation by cAMP-dependent protein kinase. (Adapted from Takahashi et al., 1987). (**b**) Functional reconstitution of skeletal muscle calcium channels. Purified skeletal muscle calcium channels were reconstituted into phospholipid vesicles at approximately one calcium channel per vesicle. Influx of ⁴⁵Ca²⁺ was measured by rapid centrifugation through a Sephadex-G50 column and scintillation counting. (Adapted from Curtis & Catterall, 1986; Nunoki et al., 1989)

a   **b**

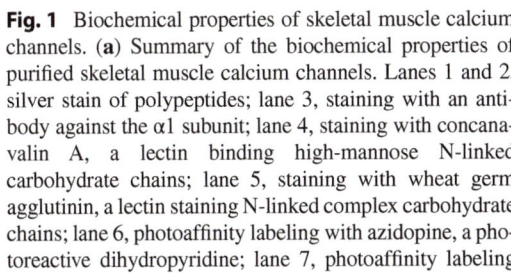

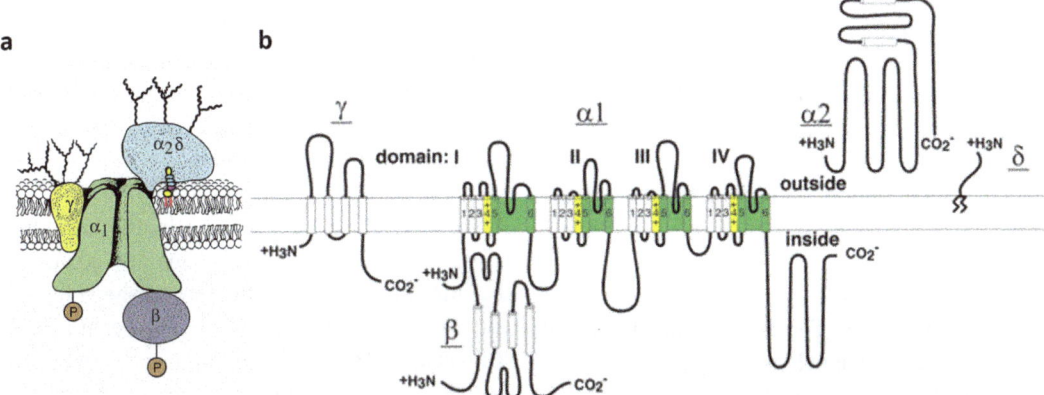

Fig. 2 The subunit structure of calcium channels purified from skeletal muscle. (**a**) A biochemical model of the skeletal muscle calcium channel taken from the original description of the subunit structure of skeletal muscle Ca²⁺ channels (Takahashi et al., 1987). P, sites of phosphorylation by cAMP-dependent protein kinase and protein kinase C. Ψ, sites of N-linked glycosylation. (**b**) Transmembrane folding models of the Ca$_V$1.1 subunits. Predicted alpha helices are depicted as cylinders. The lengths of lines correspond approximately to the lengths of the polypeptide segments represented. The δ subunit is depicted in proteolytically processed form with a glycophosphatidylinositol lipid anchor. (Adapted from Catterall, 1991b)

the α subunits of voltage-gated sodium channels with 24 transmembrane segments organized in four homologous domains of six transmembrane segments each (Catterall, 1991b). Later work showed that the S1–S4 segments form the voltage-sensing module, whereas the S5 and S6 segments and the P loop form the pore module (Fig. 2; reviewed in Catterall 2000).

The cDNA predicted a protein of 1873 amino acids with a molecular mass of 212 kD, considerably larger than the estimate of 175 kD for the $\alpha 1$ subunits of purified calcium channels. Analysis of the $\alpha 1$ subunits in both purified preparations of calcium channels and calcium channels in transverse tubule membranes using sequence-directed antibodies showed that most $\alpha 1$ subunits (>90%) were truncated in their carboxyl terminal domain between residues 1685 and 1699, resulting in a 190 kDa form that ran anomalously in SDS gels at 175 kDa (De Jongh et al., 1989, 1990, 1994; Lai et al., 1990). Only a small fraction (<10%) of skeletal muscle calcium channels contained the full-length $\alpha 1$ subunit encoded by the cDNA. Both size forms were detected in rat skeletal muscle cells in culture suggesting that both may be present in vivo (Lai et al., 1990). Since no mRNA encoding the more abundant, truncated form has been identified, the truncated form is thought to be produced by specific proteolytic processing in vivo. A similar cleavage product can be produced in vitro by calpain treatment, suggesting calpain as a candidate for in vivo processing of the $\alpha 1$ subunits (De Jongh et al., 1994). Antibody mapping of the C-terminal region of $\alpha 1_{190}$ placed the C-terminus between residues 1685 and 1699 (Röhrkasten et al., 1988; De Jongh et al., 1991; Rotman et al., 1992), and more precise mapping by mass spectrometry identified Ala1664 as the point of proteolytic truncation in vivo (Hulme et al., 2005).

α2 and δ Subunits The $\alpha 2$ subunit of skeletal muscle calcium channels is a large glycoprotein with an apparent molecular mass of 143 kD before deglycosylation and 105 kD after deglycosylation (Fig. 1a, lanes 1 and 2) (Takahashi et al., 1987; Vandaele et al., 1987; Burgess & Norman, 1988). It contains both high mannose and complex carbohydrate chains, as revealed by labeling with the lectins concanavalin A and wheat germ agglutinin, respectively (Fig. 1a, lanes 4 and 5). Cloning and sequencing cDNAs encoding the $\alpha 2$ subunit defined a protein of 1106 amino acids with a molecular mass of 125 kD, multiple potential transmembrane segments, and multiple consensus sites for N-linked glycosylation (Ellis et al., 1988) (Fig. 2b). The δ subunit appears on SDS gels as a doublet of 24 and 27 kD proteins, which are both hydrophobic and glycosylated (Takahashi et al., 1987; Vandaele et al., 1987). Determination of the amino acid sequences of peptides derived from the δ subunit showed that it was encoded by the same mRNA as the $\alpha 2$ subunit (De Jongh et al., 1990; Jay et al., 1991). The mature $\alpha 2$ subunit is truncated at alanine 934 of the $\alpha 2\delta$ precursor protein; residues 935–1106 constitute the disulfide-linked δ subunit. This sequence comprises a protein of 16 kD and contains a single predicted transmembrane segment and three consensus sequences for N-linked glycosylation. The doublet on SDS gels represents two differently glycosylated forms of the δ subunit. More recent work has shown that the δ subunit is further processed by proteolytic cleavage and attachment of a glycosylphosphatidylinositol lipid anchor that tethers the $\alpha 2\delta$ disulfide-linked dimer in the outer leaflet of the transverse tubule membrane (Davies et al., 2010). See chapter "Regulation of Calcium Channels and Synaptic Function by Auxiliary $\alpha_2\delta$ Subunits" by Dr. Annette Dolphin and Gerald Obermair for a comprehensive review of the cell biology and pharmacology of the $\alpha 2\delta$ subunits.

β Subunit The β subunits are hydrophilic proteins that are not glycosylated and are located on the intracellular side of the membrane (Figs. 1a and 2b; Takahashi et al., 1987; Leung et al., 1988; Ruth et al., 1989). cDNA cloning and sequencing revealed a skeletal muscle $\beta 1$ subunit protein of 524 amino acids with a predicted molecular mass of 58 kD (Ruth et al., 1989). Three additional β subunits are expressed in other tissues and modulate calcium channel function differentially (Perez-Reyes & Schneider, 1995). In agreement with biochemical data, the primary structure does not include any potential transmembrane segments but contains multiple consensus sites for phosphorylation by different protein kinases (Nastainczyk et al., 1987; Jahn et al., 1988).

γ Subunit The γ subunit of skeletal muscle calcium channels is a hydrophobic glycoprotein with an apparent molecular mass of 30 kD without deglycosylation and 20 kD following deglycosylation (Sharp et al., 1987; Takahashi et al., 1987). Cloning and sequencing cDNAs encoding γ subunits revealed a protein of 222 amino acid residues with a molecular mass of 25 kD (Bosse et al., 1990; Jay et al., 1990). The deduced primary structure contained four predicted hydrophobic transmembrane segments and multiple sites for N-linked glycosylation. A family of nine related γ subunits are expressed in brain and other tissues. They interact primarily with glutamate receptors as TARPs (Díaz-Alonso & Nicoll, 2021), although γ2 (also known as *stargazin*) may interact with neuronal calcium channels.

Membrane Association of Subunits The transmembrane organization of the calcium channel complex was initially investigated by labeling potential transmembrane segments with a hydrophobic photoaffinity probe (Takahashi et al., 1987). By this criterion, the α1, α2δ complex, and γ subunits were identified as probable integral membrane proteins, with the labeling of the α2 subunits the weakest despite its large size (Fig. 1a, lane 7; Takahashi et al., 1987). Subsequent hydropathy analysis of the primary structures revealed 24 transmembrane segments in the α1 subunit, four in the γ subunit, and one in the δ subunit (Fig. 2b) (Tanabe et al., 1987; Ellis et al., 1988; Ruth et al., 1989; Gurnett & Campbell, 1996; Gurnett et al., 1997). Analysis of the association of the α2 subunit with the membrane by expression and biochemical extraction procedures showed that it does not have a true transmembrane segment (Gurnett et al., 1997). Instead, it is associated with the membrane through its disulfide linkage to the δ subunit (Fig. 1b). Long after this initial analysis of the architecture of the calcium channel complex, the δ subunit was found to be further proteolytically processed and modified by addition of a glycophosphatidylinositol lipid anchor (Davies et al., 2010). See chapter "Regulation of Calcium Channels and Synaptic Function by Auxiliary α₂δ Subunits" in this volume by Drs. Annette Dolphin and Gerald Obermair for a complete account of the role of the α2δ subunit in the cell biology and regulation of calcium channels.

Structures of Na$_V$Ab and Ca$_V$Ab Channels

The first insights into the structure of voltage-gated calcium channels came from studies of homotetrameric bacterial sodium channels that are equally similar in amino acid sequence to mammalian sodium and calcium channels and are likely to be their common ancestor (Payandeh et al., 2011). DNA sequence analysis of the members of the ion channel superfamily suggests that sodium and calcium channels are derived from a common precursor, like the bacterial sodium channel that is separated from the ancestral proteins yielding other ion channel families (Yu & Catterall, 2004). The structure of the bacterial sodium channel Na$_V$Ab revealed a central pore formed by the S5 and S6 segments and the connecting P loop, surrounded by four symmetrically located voltage sensors formed by the S1–S4 segments (Fig. 3a, b; Payandeh et al., 2011). The voltage sensors are composed of a bundle of four transmembrane segments, S1–S4, which form outer and inner aqueous vestibules separated by a narrow hydrophobic constriction site (HCS; Fig. 3c). The S4 segments contain the gating charges (R1–R4) arrayed across the membrane in four repeated motifs of an Arg flanked by two hydrophobic residues (Fig. 3c, R1–R4; Payandeh et al., 2011). The voltage sensors were activated in the initial Na$_V$Ab structure, with three of the four S4 gating charges located on the extracellular side of the hydrophobic constriction site (Fig. 3c, HCS; Payandeh et al., 2011). The gating charges in the S4 segment are drawn ~12 Å inward in the resting state of the voltage sensor, following the sliding-helix mechanism of voltage-dependent gating (Catterall et al., 2017; Wisedchaisri et al., 2019).

The pore module of Na$_V$Ab forms an extracellular hydrophilic vestibule, a narrow ion selectivity filter, a large, water-filled central

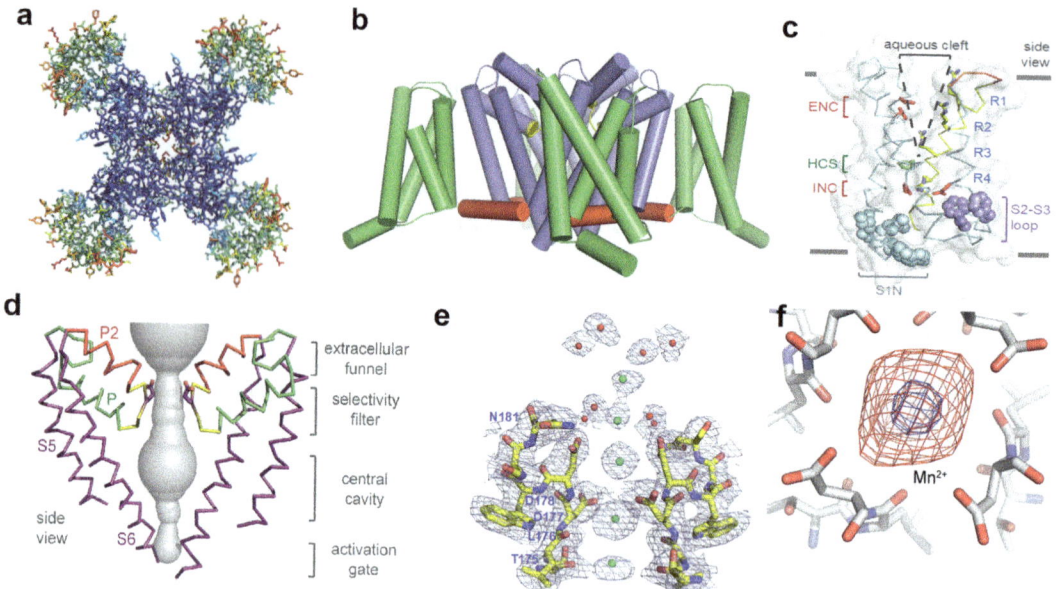

Fig. 3 Structures of ancestral Na$_V$/Ca$_V$ channels. (**a**) Structure of the bacterial sodium channel Na$_V$Ab viewed from the extracellular side. Dark blue indicates the pore module; green and red tones indicate the voltage sensor. (**b**) Transmembrane organization of Na$_V$Ab. Voltage-sensing module, green; pore, blue; S4–S5 linker, red. Note that the S4–S5 linker connects the voltage-sensing module in one subunit to the pore module in the subunit that is located to its counterclockwise side in a domain-swapped pattern. (**c**) Voltage sensor of Na$_V$Ab. ENC, extracellular negative cluster; HCS, hydrophobic constriction site; INC, intracellular negative cluster; S1N, N-terminal alpha helix preceding transmembrane segment S1. (**d**) Pore domain of Na$_V$Ab viewed from the membrane perspective. Only the pore-forming S5, P, P2, and S6 segments of two subunits are shown. Gray, water-accessible space in the pore. (**e**) Structure of the ion selectivity filter of Ca$_V$Ab with a line of Ca^{2+} ions (green) and immobilized water molecules (red). The narrow point in the selectivity filter is the high field-strength site formed by Asp177. (**f**) A Mn^{2+} ion (blue) in the ion selectivity filter with surrounding waters of hydration (red) bound at Site 3 in the selectivity filter, which is formed by the backbone carbonyls of Thr175. Electron density is illustrated in mesh. (Adapted from Payandeh et al., 2011; Tang et al., 2014)

cavity, and a closed activation gate at the intracellular ends of the S6 segments (Fig. 3d; Payandeh et al., 2011). The extracellular edge of the ion selectivity filter has a nearly square array of negatively charged Glu side chains, the High Field strength (HFS) site, which is directly involved in ion permeation. Conducted ions then pass through the central and inner ion binding sites, which are formed by backbone carbonyls (Fig. 3d; Payandeh et al., 2011). Consistent with the designation of the bacterial sodium channels as the ancestors of both sodium and calcium channels, mutations of three amino acids in the outer vestibule to negatively charged Asp residues to yield the construct Ca$_V$Ab create a ladder of negative charges across the membrane that bind Ca^{2+} and confer calcium selectivity similar to mammalian calcium channels (P$_{Ca}$:P$_{Na}$ ~ 400; Fig. 3e; Tang et al., 2014). Ca^{2+} is bound in a hydrated form with four equatorial waters of hydration visible in high-resolution structures (Fig. 3f; Tang et al., 2014). This structure of Ca$_V$Ab corresponds closely to the expectations of the classical knock-off models for calcium selectivity and permeation, in which Ca^{2+} ions approaching from the external solution "knock-off" Ca^{2+} ions resident in the pore and thereby create high conductance plus high selectivity (Almers & McCleskey, 1984; Almers et al., 1984; Hess & Tsien, 1984).

Structure of the Skeletal Muscle Ca$_V$1.1 Calcium Channel

The high abundance of Ca$_V$1.1 channels in skeletal muscle transverse tubules allowed purification of this channel in sufficient amounts for structural analysis by cryo-EM using essentially the same biochemical techniques developed in early purification studies to isolate native calcium channel complexes (Fig. 4; Wu et al., 2015, 2016). The overall subunit architecture fits closely with the model from the original biochemical studies (compare Figs. 2a and 4a). All five subunits are present in one-to-one stoichiometry, and their transmembrane arrangement fits closely with expectations from previous biochemical studies (Fig. 2a). The pore-forming

module has a conformation that is very similar to Na$_V$Ab and Ca$_V$Ab, with a high field-strength site formed by four Glu side chains and negatively charged residues strategically placed in the outer vestibule in positions analogous to those that confer calcium selectivity in Ca$_V$Ab (Fig. 4c; Wu et al., 2016). Ca²⁺ is seen bound to two of these sites, as in Ca$_V$Ab (Fig. 4c; Wu et al., 2016). All four voltage-sensing domains (VSDs) resemble Ca$_V$Ab in conformation (Fig. 4d; Wu et al., 2016). Four of the six gating charges in each S4 segment are located on the extracellular side of the HCS, consistent with an inactivated state of the channel that has activated voltage sensors but a closed pore (Wu et al., 2016).

The extracellular binding site for the α$_2$δ-1 subunit is revealed in detail (Fig. 4b). Intimate

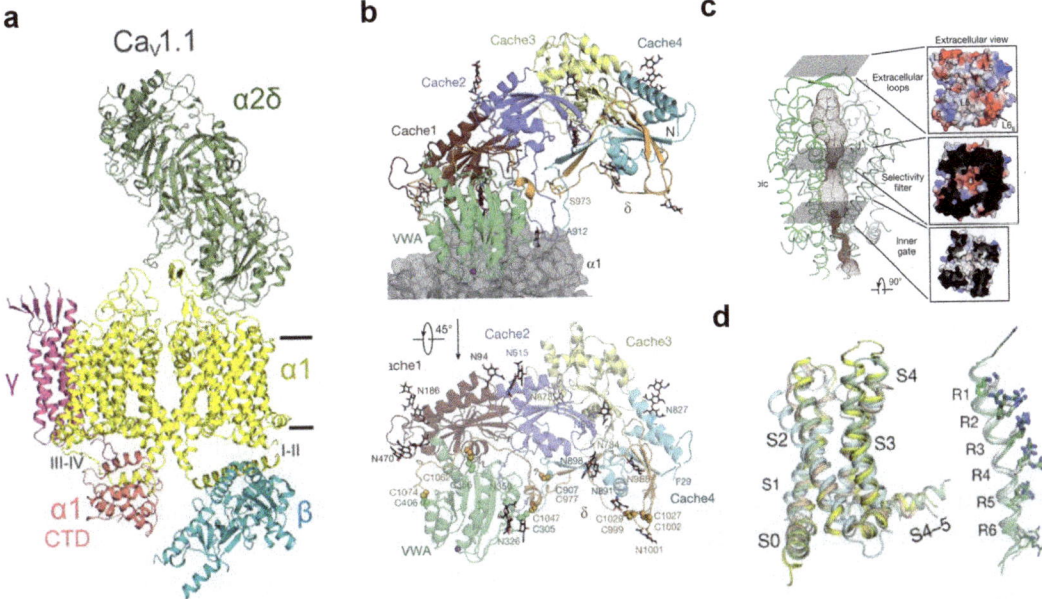

Fig. 4 Structure of the Ca$_V$1.1 channel. (**a**) Structure of the Ca$_V$1.1 complex modeled and refined at an overall resolution of 3.6 Å. The structure is color coded for distinct subunits. The four homologous repeats (repeats I–IV) of the α1 subunit are colored yellow. (**b**) The α$_2$δ-1 subunit comprises one Von Willebrand Factor Type A (VWA) domain (green) and four tandem Cache domains (brown, dark blue, yellow, medium blue). The extracellular region of the α1 subunit is shown in semi-transparent surface view. The δ-subunit is colored orange. The structure suggests glycophosphatidylinositol modification of the cleavage-exposed C-terminus (Cys1074) of the δ subunit because the 5σ EM map that extends beyond Cys1074 may correspond to the ethanolamine of the glycophosphatidylinositol. The glycosylation sites (black sticks) and

disulfide bonds (spheres) identified in the cryo-EM structure of the α$_2$δ-1 subunit. (**c**) The water-filled ion-conducting pore, calculated by HOLE, is illustrated by brown dots in the left panel. Cut-open extracellular views of the electrostatic potentials calculated in PyMol are shown for the indicated layers. (**d**) Structural comparison of the voltage-sensing domains (VSDs) between Ca$_V$1.1 VSDII, Na$_V$Rh, Na$_V$Ab, and the Na$_V$1.7 VSD4-Na$_V$Ab chimera. The VSDs of Na$_V$Rh, Na$_V$Ab, and the chimera (PDB accession numbers 4DXW, 3RW0, and 5EK0, respectively) are colored pale cyan, wheat, and yellow, respectively. The four S4 segments exhibit similar but nonidentical conformations. When the four VSDs are superimposed, the S4–S5 linkers deviate from each other. (Adapted from Wu et al., 2016) interactions between α_2 and δ result in one of the four Cache motifs being supplied in part by the δ subunit. The $\alpha_2\delta$ precursor protein is also proteolytically cleaved as shown in prior biochemical studies. The C-terminus of the δ subunit is further covalently modified by addition of a glycophosphatidylinositol anchor, as its poorly resolved cryo-EM density resembles lipid more than protein (Wu et al., 2016). The transmembrane segments of the γ subunit interact most closely with the S3 and S4 segments in the voltage sensor of Domain IV of the $\alpha1$ subunit, placing those segments in an optimal position for involvement in modulation of voltage-dependent gating by the γ subunit (Wu et al., 2016). The intracellular β subunit interacts with the Alpha Interaction Domain (AID) motif, which is part of a rigid helix that extends from the S6 segment in Domain I and binds the $Ca_V\beta$ subunit. The AID motif is positioned between the voltage sensor of Domain II and the β subunit. This close structural interaction places the β subunit in position to modulate voltage-dependent conformational changes through the voltage sensor and the L_{I-II} helix (Wu et al., 2016). Overall, the structure of $Ca_V1.1$ provided key insights into its protein-folding pattern, subunit interactions, pore function, and potential points of regulation of channel gating, and this high-resolution structure opened the door to structural studies of members of the other calcium channel families by cryo-EM.

Molecular Properties of the Ca_V2 Family of Calcium Channels

Multiple types of calcium channels, which differ in physiological and pharmacological properties, are expressed in neurons. Three types of high-voltage-activated calcium currents were distinguished in addition to L-type in neurons (Tsien et al., 1988; Llinas et al., 1992; Zhang et al., 1993). N-type, P/Q-type, and R-type calcium currents all have intermediate single-channel conductances (about 15 pS) and decay with varying rates of voltage-dependent inactivation depending on their subunit composition and on other factors (see below). They are best distinguished by their pharmacological properties: N-type calcium currents are specifically inhibited by ω-conotoxin GVIA from a cone snail *Conus geographus*, whereas P/Q-type currents are most sensitive to ω-agatoxin IVA from a funnel-web spider and to ω-conotoxin MVIIC from a different cone snail. These specific peptide toxins have provided unique experimental tools for analysis of the protein subunits of neuronal calcium channels.

$Ca_v2.2$ Channels The ω-conotoxin-sensitive $Ca_V2.2$/N-type calcium channels purified from rat brain contained an $\alpha1$ subunit, a 140 kD $\alpha2\delta$-like subunit, and β subunits of 60 kDa to 70 kD, as identified by antibodies against the skeletal muscle forms of these subunits (Ahlijanian et al., 1991; McEnery et al., 1991; Sakamoto & Campbell, 1991; Witcher et al., 1993; Leveque et al., 1994). Both Ca_V1 and $Ca_V2.2$ channels from the brain appear to lack a γ subunit, but a protein of approximately 100 kD is specifically associated with Ca_V1 and Ca_V2 channels from the brain and may be an additional, brain-specific associated protein (Ahlijanian et al., 1991; McEnery et al., 1991; Witcher et al., 1993; Leveque et al., 1994). The $\alpha1$ subunit of purified N-type calcium channels was identified as $Ca_V2.2$ by homology cDNA cloning, toxin labeling, co-immunoprecipitation, and functional expression (Dubel et al., 1992; Williams et al., 1992). As for the Ca_V1 family channels, analysis of the $Ca_V2.2$ $\alpha1$ subunit using sequence-specific antibodies revealed two size forms (240 kDa and 210 kDa) that differ in their C-termini and in their phosphorylation by specific protein kinases (Westenbroek et al., 1992; Hell et al., 1993). It is not known whether these $Ca_V2.2$ $\alpha1$ subunits of different sizes are derived from proteolytic truncation, alternatively spliced mRNAs, or both.

$Ca_v2.1$ Channels Agatoxin-sensitive $Ca_V2.1$ P/Q-type Ca^{2+} channels purified from the brain are also composed of $\alpha1$, $\alpha2\delta$, and β subunits (Martin-Moutot et al., 1995; Liu et al., 1996). In addition, a novel γ subunit, which is the target of the *stargazer* mutation in mice, may serve as a calcium channel subunit (Letts et al., 1998). This γ-subunit-like protein can modulate the voltage dependence of expressed Ca²⁺ channels containing Ca$_V$2.1 subunits, so they may be associated with these Ca²⁺ channels in vivo (Letts et al., 1998). If this γ subunit is indeed associated with neuronal Ca²⁺ channels, their subunit composition would be identical to that of skeletal muscle Ca²⁺ channels as defined in biochemical experiments (Takahashi et al., 1987). The cDNA encoding the Ca$_V$2.1 α1 subunit was isolated by homology cloning and identified by functional expression and labeling with neurotoxins and sequence-specific antibodies (Mori et al., 1991; Starr et al., 1991; Westenbroek et al., 1995). Analysis of the α1 subunit peptides of Ca$_V$2.1 channels present in the brain revealed multiple size forms, including C-terminal truncations and internal deletions, and these isoforms are differentially phosphorylated by protein kinases (Sakurai et al., 1995, 1996). These results continued the theme that calcium channels have multiple forms due to varying C-terminal domains that they are differentially phosphorylated by protein kinases.

Structure of Ca$_V$2.2 Calcium Channels and Implications for Regulation

Calcium channels in the Ca$_V$2 family exhibit approximately 50% amino acid sequence similarity to the Ca$_V$1 family in their conserved transmembrane core (Ertel et al., 2000), suggesting a close structural and functional relationship between these calcium channel families. As expected from this close similarity in amino acid sequence, the high-resolution structure of the human Ca$_V$2.2 channel has the same transmembrane architecture as Ca$_V$1.1, with a root mean square deviation (RMSD) between their structures of ~1 Å (compare Figs. 4a and 5a; Gao et al., 2021). Calcium channel β3 and α2δ-1 subunits were clearly observed (Gao et al., 2021). However, no γ subunit was detected in the structure, and the membrane-associated C-terminal end of the δ subunit was not resolved clearly, perhaps because of its posttranslational proteolysis and attachment of a glycophosphatidylinositol

lipid anchor that is mobile in the structure (Davies et al., 2010).

Uniquely among the wild-type sodium and calcium channels studied to date, the voltage sensor in Domain II of Ca$_V$2.2 was in a "down," putative resting state, position in the cryo-EM structure, with its S4 segment and gating charges drawn inward ~12 Å in comparison to the activated states of the S4 segments in the other three voltage sensors (Fig. 5b, c; Gao et al., 2021). This position of the S4 segment resembles the resting state of the ancestral sodium channel Na$_V$Ab. Remarkably, a single molecule of phosphatidylinositol 4,5-bisphosphate (PIP₂) forms a network of extensive polar interactions between the AID helix in the L$_{I-II}$ linker, which binds to the β3-subunit, and the S4–S5 linker and S6 segment in Domains II (Fig. 5d; Gao et al., 2021). The 5'-phosphate group of PIP₂ is coordinated by the R4 and K5 gating charges on the S4 segment in Domain II, suggesting that the AID helix may serve as a lever that couples the motion of the gating charges in the S4 segment of Domain II to the S6 segments in Domains I and II. Ca$_V$2.2 channels are inhibited by activation of muscarinic acetylcholine receptors and other G protein-coupled receptors (GPCRs), which release G protein βγ subunits that bind to a site in the L$_{I-II}$ intracellular linker and inhibit Ca$_V$2.2 channels in a voltage-dependent manner (Herlitze et al., 1996; Ikeda, 1996; Herlitze et al., 1997; Page et al., 1997; Zamponi et al., 1997; Dolphin, 1998; Zamponi & Snutch, 1998a, b). In parallel, GPCRs acting through Gq activate phospholipase C and hydrolyze PIP₂, which mediates voltage-independent regulation of Ca$_V$2.2 (Vivas et al., 2013). These molecular interactions observed at high resolution by cryo-EM may underlie the biphasic voltage-dependent and voltage-independent inhibition of Ca$_V$2.2 channels by muscarinic acetylcholine receptors acting through this bound PIP₂ complex. Voltage-independent regulation may be mediated by PIP₂ hydrolysis, whereas voltage-dependent inhibition may be relieved by strong depolarization that breaks the S4 segment loose from this complex and allows its outward movement that is required for channel activation.

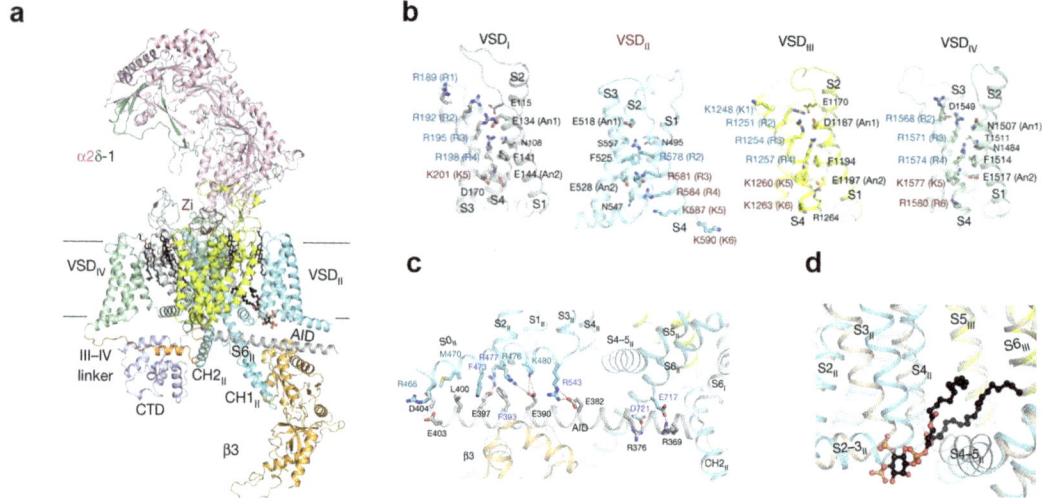

Fig. 5 Structure of the Ca$_V$2.2 channel. (**a**) Overall structure of the Ca$_V$2.2–ziconotide complex at an average resolution of 3.0 Å. CTD, C-terminal domain; Zi, ziconotide. The resolved lipid, cholesterol, and cholesterol hemisuccinate molecules are shown as black sticks. (**b**) Structure of the four VSDs. In each VSD, the gating charge residues on S4 and the surrounding residues that may facilitate gating charge transfer are shown as sticks. An1, An2, conserved acidic or polar residues on S2. The gating charge residues above and below the occluding Phe on S2 in the HCS are labeled dark cyan and brown, respectively. (**c**)

The AID helix is an organizing center for segments within and near VSD$_{II}$. The straight AID helix, in addition to mediating channel modulation by the β subunits, may serve as a lever that couples the motion of VSD$_{II}$ to S6$_I$ and S6$_{II}$. (**d**) The bound PIP$_2$ favors a down conformation of VSD$_{II}$ similar to the resting state. Left, VSD$_{II}$ and the ensuing S4–S5 $_{linker}$ in the up conformation as in Ca$_V$1.1 (wheat) would clash with PIP$_2$ in its current binding pose. Right, a cytosolic view of PIP$_2$ coordination by polar residues in Ca$_V$2.2. (Adapted from Gao et al., 2021)

Molecular Properties and Structure of Ca$_V$3 Calcium Channels

The Ca$_V$3 family of calcium channels is only 25% identical in amino acid sequence to the Ca$_V$1 and Ca$_V$2 families in their conserved transmembrane domains (Ertel et al., 2000; Yu & Catterall, 2004). They do not have tightly associated auxiliary subunits similar to the Ca$_V$β or α2δ subunits. The high-resolution structure of the Ca$_V$3.1 channels was resolved at 3.3 Å as a single protein without auxiliary subunits or other interacting partners, using a splice variant with a deletion in the L$_{I-II}$ linker designated Ca$_V$3.1-Δ8b, which increases cell surface expression (Fig. 6a; Zhao et al., 2019). Despite its major molecular differences, the transmembrane core of the Ca$_V$3.1 channel has the same alpha-helical folding pattern as Ca$_V$1 and Ca$_V$2 families of calcium channels, as well as the ancestral sodium channel Na$_V$Ab. The four voltage sensors are in activated

conformations, but the intracellular activation gate is closed, consistent with an inactivated state structure. Overall, the Ca$_V$3.1 structure can be overlaid with the transmembrane structures of the Ca$_V$1 and Ca$_V$2 channels with RMSD of ~2 Å (Zhao et al., 2019), remarkably close similarity in structure considering the substantial divergence in amino acid sequence.

Extended extracellular segments between the pore-forming segment S5 and the pore helix in Domains I and III contain a short helix and a pair of anti-parallel β-strands (Fig. 6a, b; Zhao et al., 2019). These extracellular loops are stabilized by multiple disulfide bonds, which are conserved in all Ca$_V$ channels (Zhao et al., 2019). A striking difference between the Ca$_V$3.1 channel and the Ca$_V$1 and Ca$_V$2 families of calcium channels is the lack of auxiliary subunits. The structure of Ca$_V$3.1 reveals molecular differences that may prevent interaction with α$_2$δ subunits (Fig. 6b; Zhao et al., 2019). The rigid structure of the

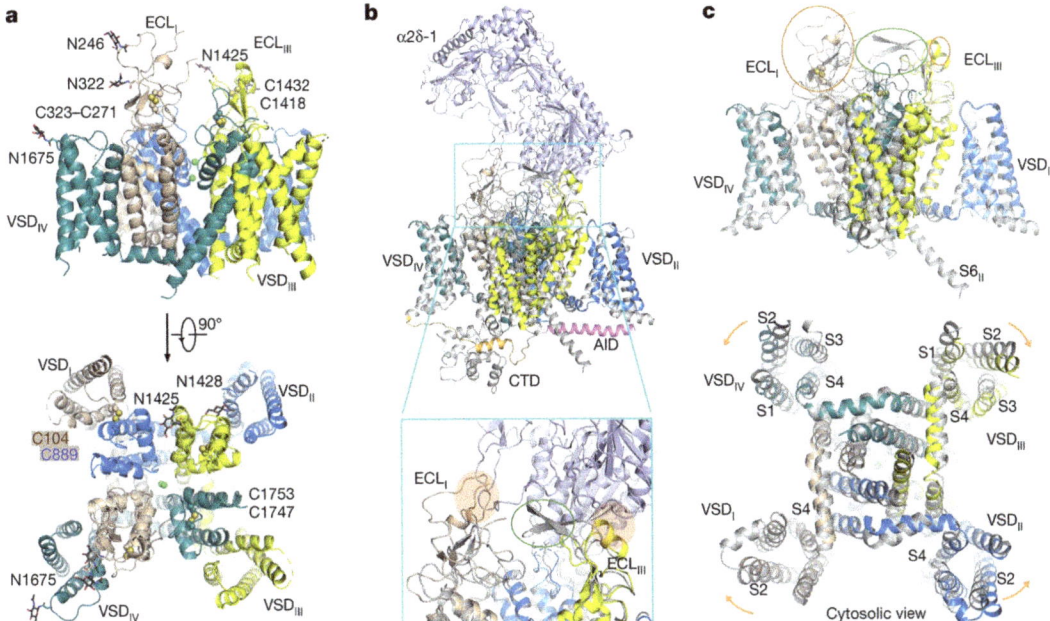

Fig. 6 Structure of the Ca$_v$3.1 channel. (**a**) Overall structure of human Ca$_v$3.1-Δ8b. Domains I–IV are colored by domain. Disulfide bonds and sugar moieties are shown as spheres and black sticks, respectively. The disulfide bond between Cys104 and Cys889 is unique to T-type VGCCs. Two potential Ca²⁺ ions in the selectivity filter are shown as green spheres. ECL, extracellular loop. (**b**) Structural basis for the incompatibility of α2δ association with the Ca$_v$3 channels. Structures of Ca$_v$3.1 and nifedipine-bound Ca$_v$1.1 (PDB code 6JP5) are superimposed. For visual clarity, the β1 and γ subunits in the Ca$_v$1.1 complex are omitted. The α2δ-1 and α1 subunits of Ca$_v$1.1 are colored light purple and gray, respectively. The segments that correspond to the AID motif (pink), III–IV linker (orange), and CTD in Ca$_v$1.1-α1 are not resolved in the Ca$_v$3.1 structure. The inset shows the structural basis for the incompatibility between Ca$_v$3.1 and α2δ-1. The semi-transparent ovals indicate the potential clashes between Ca$_v$3.1 and α2δ-1, and the green circle indicates the α2δ-1-interacting segments that are not resolved in Ca$_v$3.1. (**c**) Conformational differences between Ca$_v$3.1 and Ca$_v$1.1. Side and cytosolic views of the superimposed structures are shown. The orange and green circles correspond to those in the inset of **b**. The orange arrows indicate the slight rotations of the indicated VSDs from the positions in Ca$_v$1.1 to Ca$_v$3.1. (Adapted from Zhao et al., 2019)

extracellular loops of Ca$_v$3.1, connected by disulfide bonds, would clash with one of the Cache domains of bound α$_2$δ, which is an important site of intersubunit interactions in Ca$_v$2.2 channels (Fig. 6c, d). The intracellular β subunits of Ca$_v$1 and Ca$_v$2 channels bind to the AID in the L$_{I-II}$ intracellular linker. Unfortunately, the structure of the intracellular linkers and the N- and C-terminal domains of Ca$_v$3.1 were not resolved, preventing comparison with the portions of these segments whose structures are known in Ca$_v$1 and Ca$_v$2 channels. Thus, the structures responsible for regulation of Ca$_v$3 channels by extracellular interactions, intracellular subunits, other intracellular interacting proteins, and second messenger signaling pathways remain to be determined in future experiments.

Pharmacology of Calcium Channels

Voltage-gated calcium channels are important molecular targets for drugs and neurotoxins. The receptor sites for multiple classes of small molecule inhibitors and polypeptide toxins have now been resolved and imaged at the atomic level by X-ray crystallography and cryo-EM. The functional effects and the structural basis for the actions of these drugs and neurotoxins are considered in Chap. 25.

Conclusion

Voltage-gated calcium channels are complex proteins containing five distinct protein subunits: $\alpha 1$, $\alpha 2$, β, γ, and δ. These subunits specifically associate with each other, and their biosynthesis involves extensive glycosylation as well as proteolytic processing, disulfide linkage of subunits, and addition of a lipid anchor. Each $\alpha 1$ subunit that has been studied is present in multiple isoforms that differ in their C-terminal domains and in phosphorylation by specific kinases. Calcium channels provide intracellular calcium to initiate local signaling events, and they are directly associated with both effector proteins that initiate calcium-dependent processes and with regulatory proteins that control their activity. Calcium channel signaling complexes are a crucial element of local regulation of cellular events in response to hormones, neurotransmitters, and electrical signaling.

References

Adams, B. A., & Beam, K. G. (1990). Muscular dysgenesis in mice: A model system for studying excitation-contraction coupling. *The FASEB Journal, 4*, 2809–2816.

Adams, B. A., Tanabe, T., Mikami, A., Numa, S., & Beam, K. G. (1990). Intramembrane charge movement restored in dysgenic skeletal muscle by injection of dihydropyridine receptor cDNAs. *Nature, 346*, 569–572.

Ahlijanian, M. K., Striessnig, J., & Catterall, W. A. (1991). Phosphorylation of an α1-like subunit of an w-conotoxin-sensitive brain calcium channel by cAMP-dependent protein kinase and protein kinase C. *Journal of Biological Chemistry, 266*, 20192–20197.

Ahlijanian, M. K., Westenbroek, R. E., & Catterall, W. A. (1990). Subunit structure and localization of dihydropyridine-sensitive calcium channels in mammalian brain, spinal cord, and retina. *Neuron, 4*, 819–832.

Almers, W., & McCleskey, E. W. (1984). The nonselective conductance due to calcium channels in frog muscle: Calcium-selectivity in a single file pore. *The Journal of Physiology, 353*, 585–608.

Almers, W., McCleskey, E. W., & Palade, P. T. (1984). A nonselective cation conductance in frog muscle membrane blocked by micromolar external Ca++. *The Journal of Physiology, 353*, 565–583.

Borsotto, M., Barhanin, J., Norman, R. I., & Lazdunski, M. (1984). Purification of the dihydropyridine receptor of the voltage-dependent Ca2+ channel from skeletal muscle transverse tubules using (+) [³H] PN200-110. *Biochemical and Biophysical Research Communications, 122*, 1357–1365.

Bosse, E., Regulla, S., Biel, M., Ruth, P., Meyer, H. E., Flockerzi, V., & Hofmann, F. (1990). The cDNA and deduced amino acid sequence of the gamma subunit of the L-type calcium channel from rabbit skeletal muscle. *FEBS Letters, 267*, 153–156.

Burgess, A. J., & Norman, R. I. (1988). The large glycoprotein subunit of the skeletal muscle voltage-sensitive calcium channel. *European Journal of Biochemistry, 178*, 527–533.

Catterall, W. A. (1984). The molecular basis of neuronal excitability. *Science 223*(4637):653—661. https://doi.org/10.1126/science.6320365

Catterall, W. A. (1991a). Excitation-contraction coupling in vertebrate skeletal muscle: A tale of two calcium channels. *Cell, 64*, 871–874.

Catterall, W. A. (1991b). Structure and function of voltage-gated sodium and calcium channels. *Current Opinion in Neurobiology, 1*(1), 5–13. https://doi.org/10.1016/0959-4388(91)90004-q

Catterall, W. A. (2000). Structure and regulation of voltage-gated calcium channels. *Annual Review of Cell and Developmental Biology, 16*, 521–555.

Catterall, W. A., Wisedchaisri, G., & Zheng, N. (2017). The chemical basis for electrical signaling. *Nature Chemical Biology, 13*(5), 455–463. https://doi.org/10.1038/nchembio.2353

Curtis, B. M., & Catterall, W. A. (1984). Purification of the calcium antagonist receptor of the voltage-sensitive calcium channel from skeletal muscle transverse tubules. *The Biochemist, 23*, 2113–2118.

Curtis, B. M., & Catterall, W. A. (1986). Reconstitution of the voltage-sensitive calcium channel purified from skeletal muscle transverse tubules. *Biochemistry, 25*, 3077–3083.

Davies, A., Kadurin, I., Alvarez-Laviada, A., Douglas, L., Nieto-Rostro, M., Bauer, C. S., Pratt, W. S., & Dolphin, A. C. (2010). The α2δ subunits of voltage-gated calcium channels form GPI-anchored proteins, a posttranslational modification essential for function. *PNAS, 107*(4), 1255–1690. https://doi.org/10.1073/pnas.0908735107

De Jongh, K. S., Colvin, A. A., Wang, K. K. W., & Catterall, W. A. (1994). Differential proteolysis of the full-length form of the L-type calcium channel α1 subunit by calpain. *Journal of Neurochemistry, 63*, 1558–1564.

De Jongh, K. S., Merrick, D. K., & Catterall, W. A. (1989). Subunits of purified calcium channels: A 212-kDa form of α1 and partial amino acid sequence of a phosphorylation site of an independent β subunit. *Proceedings of the National Academy of Sciences of the United States of America, 86*, 8585–8589.

De Jongh, K. S., Warner, C., & Catterall, W. A. (1990). Subunits of purified calcium channels. α2 and δ are encoded by the same gene. *The Journal of Biological Chemistry, 265*, 14738–14741.

De Jongh, K. S., Warner, C., Colvin, A. A., & Catterall, W. A. (1991). Characterization of the two size forms of the α1 subunit of skeletal muscle L-type calcium channels. *Proceedings of the National Academy of Sciences of the United States of America, 88*(23), 10778–10782.

Dolphin, A. C. (1998). Mechanisms of modulation of voltage-dependent calcium channels by G proteins. *Journal of Physiology (London), 506*, 3–11.

Dubel, S. J., Starr, T. V., Hell, J., Ahlijanian, M. K., Enyeart, J. J., Catterall, W. A., & Snutch, T. P. (1992). Molecular cloning of the alpha-1 subunit of an omega-conotoxin-sensitive calcium channel. *Proceedings of the National Academy of Sciences of the United States of America, 89*(11), 5058–5062.

Díaz-Alonso, J., & Nicoll, R. A. (2021). AMPA receptor trafficking and LTP: Carboxy-termini, amino-termini and TARPs. *Neuropharmacology, 197*, 108710. https://doi.org/10.1016/j.neuropharm.2021.108710

Ellis, S. B., Williams, M. E., Ways, N. R., Brenner, R., Sharp, A. H., Leung, A. T., Campbell, K. P., McKenna, E., Koch, W. J., Hui, A., Schwartz, A., & Harpold, M. M. (1988). Sequence and expression of mRNAs encoding the α1 and α2 subunits of a DHP-sensitive calcium channel. *Science, 241*, 1661–1664.

Ertel, E. A., Campbell, K. P., Harpold, M. M., Hofmann, F., Mori, Y., Perez-Reyes, E., Schwartz, A., Snutch, T. P., Tanabe, T., Birnbaumer, L., Tsien, R. W., & Catterall, W. A. (2000). Nomenclature of voltage-gated calcium channels, *Neuron 25*(3), 533–535.

Flockerzi, V., Oeken, H. J., Hofmann, F., Pelzer, D., Cavalie, A., & Trautwein, W. (1986). Purified dihydropyridine-binding site from skeletal muscle t-tubules is a functional calcium channel. *Nature, 323*, 66–68.

Gao, S., Yao, X., & Yan, N. (2021). Structure of human Cav2.2 channel blocked by the painkiller ziconotide. *Nature, 596*(7870), 143–147. https://doi.org/10.1038/s41586-021-03699-6

Gurnett, C. A., & Campbell, K. P. (1996). Transmembrane auxiliary subunits of voltage-dependent ion channels. *The Journal of Biological Chemistry, 271*, 27975–27978.

Gurnett, C. A., Felix, R., & Campbell, K. P. (1997). Extracellular interaction of the voltage-dependent Ca²⁺ channel α2δ and α1 subunits. *Journal of Biological Chemistry, 272*, 18508–18512.

Hell, J. W., Appleyard, S. M., Yokoyama, C. T., Warner, C., & Catterall, W. A. (1993). Differential phosphorylation of two size forms of the N-type calcium channel α1 subunit which have different COOH-termini. *The Journal of Biological Chemistry, 269*, 7390–7396.

Herlitze, S., Garcia, D. E., Mackie, K., Hille, B., Scheuer, T., & Catterall, W. A. (1996). Modulation of Ca²⁺ channels by G protein beta/gamma subunits. *Nature, 380*, 258–262.

Herlitze, S., Hockerman, G. H., Scheuer, T., & Catterall, W. A. (1997). Molecular determinants of inactivation and G protein modulation in the intracelular loop connecting domains I and II of the calcium channel

α1A subunit. *Proceedings of the National Academy of Sciences of the United States of America, 94*, 1512–1516.

Hess, P., & Tsien, R. W. (1984). Mechanism of ion permeation through calcium channels. *Nature, 309*, 453–456.

Hosey, M. M., Barhanin, J., Schmid, A., Vandaele, S., Ptasienski, J., O'Callahan, C., Cooper, C., & Lazdunski, M. (1987). Photoaffinity labelling and phosphorylation of a 165 kilodalton peptide associated with dihydropyridine and phenylalkylamine-sensitive calcium channels. *Biochemical and Biophysical Research Communications, 147*, 1137–1145.

Hulme, J. T., Konoki, K., Lin, T. W., Gritsenko, M. A., Camp, D. G., 2nd, Bigelow, D. J., & Catterall, W. A. (2005). Sites of proteolytic processing and noncovalent association of the distal C-terminal domain of Ca_V1.1 channels in skeletal muscle. *Proceedings of the National Academy of Sciences of the United States of America, 102*(14), 5274–5279.

Ikeda, S. R. (1996). Voltage-dependent modulation of N-type calcium channels by G-protein bg subunits. *Nature, 380*, 255–258.

Jahn, H., Nastainczyk, W., Rohrkasten, A., Schneider, T., & Hofmann, F. (1988). Site-specific phosphorylation of the purified receptor for calcium-channel blockers by cAMP- and cGMP-dependent protein kinases, protein kinase C, calmodulin-dependent protein kinase II and casein kinase II. *European Journal of Biochemistry, 178*(2), 535–542.

Jay, S. D., Ellis, S. B., McCue, A. F., Williams, M. E., Vedvick, T. S., Harpold, M. M., & Campbell, K. P. (1990). Primary structure of the gamma subunit of the DHP-sensitive calcium channel from skeletal muscle. *Science, 248*, 490–492.

Jay, S. D., Sharp, A. H., Kahl, S. D., Vedvick, T. S., Harpold, M. M., & Campbell, K. P. (1991). Structural characterization of the dihydropyridine-sensitive calcium channel alpha-2 subunit and the associated delta peptides. *The Journal of Biological Chemistry, 266*, 3287–3293.

Lai, Y., Seagar, M. J., Takahashi, M., & Catterall, W. A. (1990). Cyclic AMP-dependent phosphorylation of two size forms of α1 subunits of L-type calcium channels in rat skeletal muscle cells. *The Journal of Biological Chemistry, 265*, 20839–20848.

Letts, V. A., Felix, R., Biddlecome, G. H., Arikkath, J., Mahaffey, C. L., Valenzuela, A., Bartlett, I. F. S., Mori, Y., Campbell, K. P., & Frankel, W. N. (1998). The mouse stargazer gene encodes a neuronal Ca²⁺−channel γ subunit. *Nature Genetics, 19*, 340–347.

Leung, A. T., Imagawa, T., Block, B., Franzini-Armstrong, C., & Campbell, K. P. (1988). Biochemical and ultrastructural characterization of the 1,4-dihydropyridine receptor from rabbit skeletal muscle evidence for a 52,000 Da subunit. *The Journal of Biological Chemistry, 263*, 994–1001.

Leung, A. T., Imagawa, T., & Campbell, K. P. (1987). Structural characterization of the 1,4-dihydropyridine receptor of the voltage-dependent Ca²⁺ channel from rabbit skeletal muscle. Evidence for two distinct high molecular weight subunits. *The Journal of Biological Chemistry, 262*, 7943–7946.

Leveque, C., El Far, O., Martin-Moutot, N., Sato, K., Kato, R., Takahashi, M., & Seagar, M. J. (1994). Purification of the N-type calcium channel associated with syntaxin and synaptotagmin: A complex implicated in synaptic vesicle exocytosis. *The Journal of Biological Chemistry, 269*, 6306–6312.

Liu, H., De Waard, M., Scott, V. E. S., Gurnett, C. A., Lennon, V. A., & Campbell, K. P. (1996). Identification of three subunits of the high affinity ω-conotoxin MVIIC-sensitive Ca²⁺ channel. *Journal of Biological Chemistry, 271*, 13804–13810.

Llinas, R., Sugimori, M., Hillman, D. E., & Cherksey, B. (1992). Distribution and functional significance of the P-type, voltage-dependent Ca²⁺ channels in the mammalian central nervous system. *Trends in Neurosciences, 15*, 351–355.

Martin-Moutot, N., Leveque, C., Sato, K., Kato, R., Takahashi, M., & Seagar, M. (1995). Properties of omega conotoxin MVIIC receptors associated with α 1A calcium channel subunits in rat brain. *FEBS Letters, 366*, 21–25.

McCleskey, E. W., & Almers, W. (1985). The Ca channel in skeletal muscle is a large pore. *Proceedings of the National Academy of Sciences of the United States of America, 82*, 7149–7153.

McEnery, M. W., Snowman, A. M., Sharp, A. H., Adams, M. E., & Snyder, S. H. (1991). Purified γ-conotoxin GVIA receptor of rat brain resembles a dihydropyridine-sensitive L-type calcium channel. *Proceedings of the National Academy of Sciences of the United States of America, 88*, 11095–11099.

Mori, Y., Friedrich, T., Kim, M. S., Mikami, A., Nakai, J., Ruth, P., Bosse, E., Hofmann, F., Flockerzi, V., Furuichi, T., Mikoshiba, K., Imoto, K., Tanabe, T., & Numa, S. (1991). Primary structure and functional expression from complementary DNA of a brain calcium channel. *Nature, 350*, 398–402.

Morton, M. E., & Froehner, S. C. (1987). Monoclonal antibody identifies a 200-kDa subunit of the dihydropyridine-sensitive calcium channel. *Journal of Biological Chemistry, 262*, 11904–11907.

Nastainczyk, W., Rohrkasten, A., Sieber, M., Rudolph, C., Schachtele, C., Marme, D., & Hofmann, F. (1987). Phosphorylation of the purified receptor for calcium channel blockers by cAMP kinase and protein kinase C. *European Journal of Biochemistry, 169*(1), 137–142.

Nunoki, K., Florio, V., & Catterall, W. A. (1989). Activation of purified calcium channels by stoichiometric protein phosphorylation. *Proceedings. National Academy of Sciences. United States of America, 86*, 6816–6820.

Page, K. M., Stephens, G. J., Berrow, N. S., & Dolphin, A. C. (1997). The intracellular loop between domains I and II of the B-type calcium channel confers aspects of G-protein sensitivity to the E-type calcium channel. *The Journal of Neuroscience, 17*, 1330–1338.

Payandeh, J., Scheuer, T., Zheng, N., & Catterall, W. A. (2011). The crystal structure of a voltage-gated sodium channel. *Nature, 475*(7356), 353–358. https://doi.org/10.1038/nature10238

Perez-Reyes, E., & Schneider, T. (1995). Molecular biology of calcium channels. *Kidney International, 48*, 1111–1124.

Reuter, H. (1967). The dependence of slow inward current in Purkinje fibres on the extracellular calcium-concentration. *Journal of Physiology (London), 192*, 479–492.

Rios, E., & Brum, G. (1987). Involvement of dihydropyridine receptors in excitation-contraction coupling in skeletal muscle. *Nature (London), 325*, 717–720.

Rotman, E. I., De Jongh, K. S., Florio, V., Lai, Y., & Catterall, W. A. (1992). Specific phosphorylation of a COOH-terminal site on the full-length form of the α1 subunit of the skeletal muscle calcium channel by cAMP-dependent protein kinase. *The Journal of Biological Chemistry, 267*, 16100–16105.

Ruth, P., Röhrkasten, A., Biel, M., Bosse, E., Regulla, S., Meyer, H. E., Flockerzi, V., & Hofmann, F. (1989). Primary structure of the β subunit of the DHP-sensitive calcium channel from skeletal muscle. *Science, 245*, 1115–1118.

Röhrkasten, A., Meyer, H. E., Nastainczyk, W., Sieber, M., & Hofmann, F. (1988). cAMP-dependent protein kinase rapidly phosphorylates serine- 687 of the skeletal muscle receptor for calcium channel blockers. *Journal of Biological Chemistry, 263*, 15325–15329.

Sakamoto, J., & Campbell, K. P. (1991). A monoclonal antibody to the β subunit of the skeletal muscle dihydropyridine receptor immunoprecipitates the brain ω-conotoxin GVIA receptor. *Journal of Biological Chemistry, 266*, 18914–18919.

Sakurai, T., Hell, J. W., Woppmann, A., Miljanich, G. P., & Catterall, W. A. (1995). Immunochemical identification and differential phosphorylation of alternatively spliced forms of the alpha-1A subunit of brain calcium channels. *Journal of Biological Chemistry, 270*, 21234–21242.

Sakurai, T., Westenbroek, R. E., Rettig, J., Hell, J., & Catterall, W. A. (1996). Biochemical properties and subcellular distribution of the BI and rbA isoforms of alpha-1A subunits of brain calcium channels. *The Journal of Cell Biology, 134*, 511–528.

Sanchez, J. A., & Stefani, E. (1978). Inward calcium current in twitch muscle fibers of the frog. *The Journal of Physiology, 283*, 197–209.

Sharp, A. H., & Campbell, K. P. (1989). Characterization of the 1,4-dihydropyridine receptor using subunit-specific polyclonal antibodies. Evidence for a 32,000-Da subunit. *Journal of Biological Chemistry, 264*, 2816–2825.

Sharp, A. H., Imagawa, T., Leung, A. T., & Campbell, K. P. (1987). Identification and characterization of the dihydropyridine-binding subunit of the skeletal muscle dihydropyridine receptor. *Journal of Biological Chemistry, 262*, 12309–12315.

Sieber, M., Nastainczyk, W., Zubor, V., Wernet, W., & Hofmann, F. (1987). The 165-kDa peptide of the purified skeletal muscle dihydropyridine receptor contains the known regulatory sites of the calcium channel. *European Journal of Biochemistry, 167*(1), 117–122.

Starr, T. V. B., Prystay, W., & Snutch, T. P. (1991). Primary structure of a calcium channel that is highly expressed in the rat cerebellum. *Proceedings of the National Academy of Sciences of the United States of America, 88*, 5621–5625.

Striessnig, J., Knaus, H. G., Grabner, M., Moosburger, K., Seitz, W., Lietz, H., & Glossmann, H. (1987). Photoaffinity labelling of the phenylalkylamine receptor of the skeletal muscle transverse-tubule calcium channel. *FEBS Letters, 212*, 247–253.

Takahashi, M., Seagar, M. J., Jones, J. F., Reber, B. F., & Catterall, W. A. (1987). Subunit structure of dihydropyridine-sensitive calcium channels from skeletal muscle. *Proceedings. National Academy of Sciences. United States of America, 84*, 5478–5482.

Tanabe, T., Beam, K. G., Powell, J. A., & Numa, S. (1988). Restoration of excitation-contraction coupling and slow calcium current in dysgenic muscle by dihydropyridine receptor complementary DNA. *Nature, 336*, 134–139.

Tanabe, T., Takeshima, H., Mikami, A., Flockerzi, V., Takahashi, H., Kangawa, K., Kojima, M., Matsuo, H., Hirose, T., & Numa, S. (1987). Primary structure of the receptor for calcium channel blockers from skeletal muscle. *Nature, 328*, 313–318.

Tang, L., Gamal El-Din, T. M., Payandeh, J., Martinez, G. Q., Heard, T. M., Scheuer, T., Zheng, N., & Catterall, W. A. (2014). Structural basis for Ca²⁺ selectivity of a voltage-gated calcium channel. *Nature, 505*(7481), 56–61. https://doi.org/10.1038/nature12775

Tsien, R. W., Lipscombe, D., Madison, D. V., Bley, K. R., & Fox, A. P. (1988). Multiple types of neuronal calcium channels and their selective modulation. *Trends in Neurosciences, 11*, 431–438.

Vandaele, S., Fosset, M., Galizzi, J. P., & Lazdunski, M. (1987). Monoclonal antibodies that coimmunoprecipitate the 1,4-dihydropyridine and phenylalkylamine receptors and reveal the Ca²⁺ channel structure. *Biochemistry, 26*, 5–9.

Vivas, O., Castro, H., Arenas, I., Elías-Viñas, D., & García, D. E. (2013). PIP₂ hydrolysis is responsible for voltage independent inhibition of Ca$_V$2.2 channels in sympathetic neurons. *Biochemical and Biophysical Research Communications, 432*(2), 275–280. https://doi.org/10.1016/j.bbrc.2013.01.117

Westenbroek, R. E., Hell, J. W., Warner, C., Dubel, S. J., Snutch, T. P., & Catterall, W. A. (1992). Biochemical properties and subcellular distribution of an N-type calcium channel α1 subunit. *Neuron, 9*, 1099–1115.

Westenbroek, R. E., Sakurai, T., Elliott, E. M., Hell, J. W., Starr, T. V. B., Snutch, T. P., & Catterall, W. A. (1995). Immunochemical identification and subcellular distribution of the α1A subunits of brain calcium channels. *The Journal of Neuroscience, 15*, 6403–6418.

Williams, M. E., Feldman, D. H., McCue, A. F., Brenner, R., Velicelebi, G., Ellis, S. B., & Harpold, M. M. (1992). Structure and functional expression of α1, α2, and β subunits of a novel human neuronal calcium channel subtype. *Neuron, 8*, 71–84.

Wisedchaisri, G., Tonggu, L., McCord, E., Gamal El-Din, T. M., Wang, L., Zheng, N., & Catterall, W. A. (2019). Resting-state structure and gating mechanism of a voltage-gated sodium channel. *Cell, 178*(4), 993–1003.e12. https://doi.org/10.1016/j.cell.2019.06.031

Witcher, D. R., De Waard, M., Sakamoto, J., Franzini-Armstrong, C., Pragnell, M., Kahl, S. D., & Campbell, K. P. (1993). Subunit identification and reconstitution of the N-type Ca²⁺ channel complex purified from brain. *Science, 261*, 486–489.

Wu, J., Yan, Z., Li, Z., Qian, X., Lu, S., Dong, M., Zhou, Q., & Yan, N. (2016). Structure of the voltage-gated calcium channel Ca$_V$1.1 at 3.6 a resolution. *Nature, 537*(7619), 191–196. https://doi.org/10.1038/nature19321

Wu, J., Yan, Z., Li, Z., Yan, C., Lu, S., Dong, M., & Yan, N. (2015). Structure of the voltage-gated calcium channel Ca$_V$1.1 complex. *Science, 350*(6267), aad2395. https://doi.org/10.1126/science.aad2395

Yu, F. H., & Catterall, W. A. (2004). The VGL-chanome: A protein superfamily specialized for electrical signaling and ionic homeostasis. *Science's STKE, 2004*(253), re15.

Zamponi, G. W., Bourinet, E., Nelson, D., Nargeot, J., & Snutch, T. P. (1997). Crosstalk between G proteins and protein kinase C mediated by the calcium channel α1 subunit. *Nature, 385*, 442–446.

Zamponi, G. W., & Snutch, T. P. (1998a). Decay of prepulse facilitation of N type calcium channels during G protein inhibition is consistent with binding of a single Gβγ subunit. *Proceedings of the National Academy of Sciences of the United States of America, 95*, 4035–4039.

Zamponi, G. W., & Snutch, T. P. (1998b). Modulation of voltage-dependent calcium channels by G proteins. *Current Opinion in Neurobiology, 8*, 351–356.

Zhang, J. F., Randall, A. D., Ellinor, P. T., Horne, W. A., Sather, W. A., Tanabe, T., Schwarz, T. L., & Tsien, R. W. (1993). Distinctive pharmacology and kinetics of cloned neuronal Ca²⁺ channels and their possible counterparts in mammalian CNS neurons. *Neuropharmacology, 32*, 1075–1088.

Zhao, Y., Huang, G., Wu, Q., Wu, K., Li, R., Lei, J., Pan, X., & Yan, N. (2019). Cryo-EM structures of apo and antagonist-bound human Ca$_V$3.1. *Nature, 576*(7787), 492–497. https://doi.org/10.1038/s41586-019-1801-3

Splicing and Editing to Fine-Tune Activity of High Voltage-Activated Calcium Channels

Hua Huang, Zhenyu Hu, Sean Qing Zhang Yeow, and Tuck Wah Soong

Abstract

The mRNA transcripts of voltage-gated calcium channels (VGCCs) are subjected to extensive alternative splicing giving rise to a plethora of channel variants with potentially altered biophysical properties to affect physiological functions. Notably, the discovery of *A*-to-*I* RNA editing within the IQ domain of Cav1.3 channel further expands on the post-transcriptional modification processes of VGCCs. Here, we highlight the functional influence alternative splicing and RNA editing has on fine-tuning the activity of High Voltage-Activated (HVA) calcium channels. We will discuss the limitations of employing heterologous expression systems and feature the extensive use of transgenic mouse models in the field, and suggest certain novel approaches that can be applied to further address the physiological consequences of these post-transcriptional modifications of HVA calcium channels.

Keywords

Calcium channel · Alternative splicing · RNA editing · Post-transcriptional modification · Calcium channelopathy

H. Huang · T. W. Soong (✉)
Department of Physiology, Yong Loo Lin School of Medicine, National University of Singapore, Singapore, Singapore

NUS Medicine Electrophysiology Core, Singapore, Singapore

Healthy Longevity Translational Research Programme, Singapore, Singapore

Cardiovascular Disease Translational Research Programme, Singapore, Singapore
e-mail: phsstw@nus.edu.sg

Z. Hu
Department of Physiology, Yong Loo Lin School of Medicine, National University of Singapore, Singapore, Singapore

Cardiovascular Disease Translational Research Programme, Singapore, Singapore

S. Q. Z. Yeow
Department of Physiology, Yong Loo Lin School of Medicine, National University of Singapore, Singapore, Singapore

Generation of mRNA Diversity by *A*-to-*I* RNA Editing and Alternative Splicing

The limited mammalian genome is diversified through dynamic and complex networks of post-transcriptional and post-translational modifications, giving rise to various mRNA transcripts and proteins with subtle to gross changes in structure to alter functions that are essential for adaptation and survival. The most prevalent form of RNA editing in mammals is mediated by *Adenosine Deaminases Acting on RNA* (ADAR) that converts adenosine to inosine (*A*-to-*I*). Of note, *A*-to-*I* RNA editing that occurs within the coding sequence could potentially result in a pinpoint change of the coding of an amino acid in the final protein product. Alternative splicing, on the other hand, is a post-transcriptional process by which the exons of primary RNA transcripts are subjected to combinatorial reassembly resulting in the transcripts and final protein products with various sequences and diverse functions. It is a choreographed process that is catalyzed by a multistep assembly of the spliceosome machinery upon recognition of specific intronic or exonic cis-element within the pre-mRNA. Several mechanisms for alternative splicing exist and they include the utilization of the following: (i) cassette exon – an alternate exon could either be included or excluded; (ii) mutually exclusive exons – one or more adjacent exons are spliced such that only one exon is retained at a time; (iii) different 5′ or 3′ alternative splice acceptor or donor sites allowing for either the lengthening or shortening of a particular exon; (iv) intron retention where an intron is included in the mature mRNA; and (v) alternative promoters or polyadenylation sites.

Splice Variations in High Voltage-Activated Calcium Channels

Here, we will highlight splice variations that have been characterized and demonstrated to alter the electrophysiological or pharmacological properties of the channel or that may affect interaction with cytoplasmic proteins. We will feature splice loci by which alterations in or deregulation of splicing levels may affect physiology or are implicated in disease progression or pathogenesis.

Ca$_V$1.1 Splice Variant in Myotonic Dystrophy

Unlike other VGCCs that have multiple splice variants with diversified functions, Ca$_V$1.1 calcium channel has only one reported alternatively

spliced site, and splice variants with either the inclusion or exclusion of exon 29 have been detected in rodent and human skeletal muscles (Tuluc et al., 2009). Exon 29 encodes 19 amino acids of the extracellular linker of domain IV transmembrane segments 3 and 4 (Bannister & Beam, 2013). Exon 29 deficiency in Ca$_V$1.1 channel was reported to induce a hyperpolarizing shift of 30 mV on voltage-dependent activation and an increase in open probability (P_o) as reflected by the increased slope in the plot of tail current amplitudes at the reversal potential against gating currents. As a result, the exclusion of exon 29 increases Ca$_V$1.1 current density by eightfold and triggers stronger skeletal muscle contraction in transfected myotubes of Ca$_V$1.1-deficient skeletal muscle cell line GLT (Tuluc et al., 2009). This finding was then further corroborated by two classical studies on Ca$_V$1.1 splicing: (1) In isolated flexor digitorum brevis (FDB) fibers from mice injected with an antisense oligonucleotide (ASO) that induced nearly complete exon 29 skipping, the voltage for half-maximal channel activation ($V_{0.5}$) had a dramatic 20 mV shift to hyperpolarizing direction and the peak current density was increased by about twofold (Tang et al., 2012). (2) In the FDB fibers isolated from exon 29-null mice, $V_{0.5}$ displayed an even larger shift of about 38.5 mV and the Ca^{2+} influx was also significantly increased during excitation–contraction coupling (Sultana et al., 2016).

Alternative splicing of exon 29 was found to be developmentally regulated. The transcripts of Ca$_V$1.1$_{\Delta e29}$ channels were reported to represent about 80% of total Ca$_V$1.1 expressed in myotubes, but was significantly shifted to exon 29 inclusion in late fetal and early postnatal development with 100% exon 29 inclusion at postnatal day 30 in mouse hind limb muscles (Tuluc et al., 2009; Flucher & Tuluc, 2011). These data suggest that larger Ca^{2+} currents from embryonic Ca$_V$1.1$_{\Delta e29}$ splice variant may be required for normal development of embryos and neonates. In contrast, patients with myotonic dystrophy type 1 exhibited the re-introduction of embryonic Ca$_V$1.1$_{\Delta e29}$ variant, with exon 29 inclusion level reduced to 30% in skeletal muscles from more than 93% in healthy individuals (Tang et al., 2012). Mechanistically, this switch may be due to altered expression levels of two splicing factors: down-regulation of muscle blind-like 1 (MBNL1) that promotes exon 29 inclusion (Andre et al., 2019) and the up-regulation of CUG-binding protein 1 (CUGBP1) that increases exon 29 skipping (Tang et al., 2012). More importantly, exon 29 skipping induced by an ASO significantly aggravated muscle weakness in mice (Tang et al., 2012), indicating that altered $Ca_V1.1$ function in human myotonic dystrophy may contribute to the exacerbation of myopathy.

Recently another new $Ca_V1.1$ splice variant with exon 29 exclusion and substitution of exons 1 and 2 with five new N-terminal exons was identified in T cells (Matza et al., 2016). This T-cell $Ca_V1.1$ variant was reported to be constitutively open at resting potential in transfected HEK 293 cells (Matza et al., 2016), which may contribute to the death of $Ca_V1.1$-overexpressing cells. As $Ca_V1.1$ channels are essential for T-cell receptors (TCR)-induced Ca^{2+} entry, thus further studies are needed to investigate the electrophysiological properties of this T-cell splice variant and the expression level in TCR-stimulated T cells.

$Ca_V1.2$ Splice Variants in Health and Disease

Exon 9* in Cardiovascular Diseases

The alternatively spliced exon 9* consists of 75 nucleotides downstream of exon 9 and the amino acid sequences encoded by these two exons contribute in part to the intracellular I-II loop of $Ca_V1.2$ channel (Fig. 1) (Liao et al., 2004). Recent studies have shown that the level of exon 9* inclusion is regulated by Rbfox2 and is increased in smooth muscle of hypertensive rat arteries (Zhou et al., 2017) and in cardiac muscles of patients with end-stage heart failure (Papa et al., 2021). The $Ca_V1.2_{e9*}$ channels display altered electrophysiological properties (Liao et al., 2004; Papa et al., 2021) and specific responses to a calcium channel blocker (Zhang

et al., 2010), diltiazem, and inclusion of exon 9* restricts the interaction of Galectin-1 with $Ca_V1.2$ channels (Hu et al., 2018; Wang et al., 2011).

Compared to $Ca_V1.2_{\Delta e9*}$ channels, whole-cell patch-clamp recordings demonstrated that human $Ca_V1.2_{e9*}$ channels displayed a hyperpolarized shift in voltage-dependent activation by 9 mV and $I–V$ relationships by 11 mV in transfected HEK 293 cells (Liao et al., 2004). Moreover, the cardiomyocytes isolated from rabbit $Ca_V1.2_{e9*}$-expressing transgenic mice showed a significant increase in conductance density with more robust open probability (P_o) of calcium channel, as compared to wild-type cardiomyocytes (Papa et al., 2021). These results suggest that exon 9* inclusion may contribute to larger Ca^{2+} influx, which may explain that exon 9* inclusion level was increased to 11% in cardiac muscles of spontaneously hypertensive rats (SHRs) from 2% in Wistar Kyoto (WKY) rats (Tang et al., 1783), and was also up-regulated to 50.4% in mesenteric arteries of SHR from 40.1% in WKY arteries (Zhou et al., 2017). However, it is noteworthy that there is no direct correlation of mRNA–protein expressions of $Ca_V1.2$ channels in arteries of SHRs (Pratt et al., 2002), and the function of increased exon 9* in hypertensive arteries remains to be investigated. Although we do not know about the protein level of $Ca_V1.2_{e9*}$channels in arteries under hypertensive conditions, a study has reported that an ASO targeting $Ca_V1.2_{e9*}$ channels reduced the maximal contractility by 75% in rabbit cerebral arteries (Nystoriak et al., 2009), suggesting that $Ca_V1.2_{e9*}$ channels play a dominant role in artery.

In addition to disease models, exon 9* was also developmentally regulated in mouse cortex by Rbfox1 and Rbfox2 proteins (Tang et al., 2009). The inclusion level dropped to 4% at embryonic day 18 from 25% at day 12 in mouse cortex, which was mediated by the up-regulated Rbfox1/2 protein as Rbfox1/2 knockdown significantly increased exon 9* inclusion level, while their overexpression suppressed exon 9* inclusion. This study indicates that Rbfox1 and Rbfox2 act as the splicing factors to inhibit exon 9* inclusion. Moreover, Rbfox2 was also reported

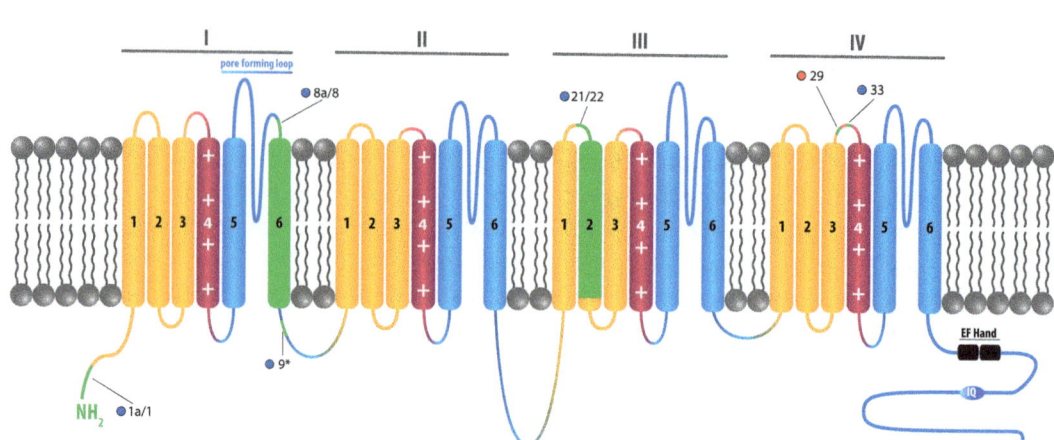

Fig. 1 Schematic diagram showing the positions of alternatively spliced exons in $Ca_V1.1$ and $Ca_V1.2$ channels discussed in this chapter. The only alternatively spliced exon 29 in $Ca_V1.1$ channel is marked as a red dot. Exon 29 exclusion induces a hyperpolarizing shift of 30 mV on voltage-dependent activation and an increase in open probability, and also contributes to the pathogenesis of myotonic dystrophy. The 5 functionally characterized and disease-associated mutually exclusive or alternatively spliced exons (exon 1a/1, 8a/8, 9*, 21/22, 33) of $Ca_V1.2$ channel are marked by blue dots

to regulate exon 9* inclusion in hypertensive arteries (Zhou et al., 2017). However, the up-regulated exon 9* inclusion was correlated to increased Rbfox2 protein in mesenteric arteries of SHR, which may be due to the increased expression of a dominant-negative Rbfox2 isoform lacking half of RNA recognition motif. As the inclusion level of exon 9* was increased in hypertensive arteries and failing hearts, this suggests that the reduced exon 9* level detected in certain rodent and human disease models may recapitulate the embryonal level. Potentially, $Cav1.2_{e9*}$ channels may play an essential role in the adaptation or pathogenesis of other diseases related to organs expressing high levels of $Ca_V1.2$.

Exon 9* is also able to mediate splice variant-selective modulation of $Ca_V1.2$ channel by its binding protein partner or a calcium channel blocker. Galectin-1, a member of β-galactoside-binding protein family (Camby et al., 2006), was reported to negatively modulate $Ca_V1.2$ channel function by interacting with the exon 9 fragment within the $Ca_V1.2$ I-II loop (Wang et al., 2011). However, inclusion of exon 9* completely abolishes the $Ca_V1.2$–Galectin-1 interaction and thereby the inhibitory effects of Galectin-1 on $Ca_V1.2$ channels (Wang et al., 2011). These results suggest that the inhibitory effects of Galectin-1 are specific to $Ca_V1.2_{\Delta e9*}$ channels. The positively charged amino acids in exon 9* could prevent Gal-1 binding to exon 9, or they could interact with the ER export signal that is composed of a few negatively charged residues, thereby attenuating the trafficking of the channels to the surface, mimicking the masking of the ER signal by Galectin-1. Additionally, diltiazem, a non-dihydropyridine calcium channel blocker, also displayed different inhibitory effects on $Ca_V1.2_{\Delta e9*}$ and $Ca_V1.2_{e9*}$ channels (Zhang et al., 2010). The IC_{50} of diltiazem for $Ca_V1.2_{\Delta e9*}$ was increased twofold as compared to $Ca_V1.2_{e9*}$ channels, suggesting that exon 9* inclusion may contribute to the $Ca_V1.2$ splice isoform-specific effects of diltiazem.

Mutually Exclusive Exons 1a/1 and Exons 8a/8

The expression of N-terminal exons 1a/1 and exons 8a/8 encoding domain I S6 segment of the $Ca_V1.2$ channel displays strong tissue selectivity (Bartels et al., 2018). Exons 1 and 8 are predominantly expressed in smooth muscle, while exons 1a and 8a are the primary combination in cardiac muscle, representing about 80% of cardiac $Ca_V1.2$ mRNA (Hu et al., 2017; Splawski et al., 2005) and exons 1 and 8a are expressed as the major brain isoform (Bartels et al., 2018). A study on transfected HEK 293 cells reported that $Ca_V1.2_{e1}$ channels possessed larger unitary gating current and Ca^{2+} current density as recorded by cell-attached single-channel recording and whole-cell patch-clamp recording, respectively, and higher expression level at cell surface, although exon 1 increased the Ca^{2+}-dependent inactivation (CDI) by more than 70% triggered by local Ca^{2+} sensing by the N-lobe of calmodulin (CaM) (Bartels et al., 2018).

The most well-studied disease associated with mutually exclusive exons 8a/8 is Timothy syndrome (TS), a multisystem disorder with arrhythmia and autism caused by G406R and/or G402S mutations (Splawski et al., 2004, 2005). In the two seminal TS articles, exon 8a (GRCh38/hg38, chr12: 2504436-2504539), upstream of exon 8 (GRCh38/hg38, chr12: 2504842-2504945) in human genomic sequence, was labeled as exon 8, which was labeled differently from the rest of calcium channel community (Hu et al., 2017; Wang et al., 2006; Tang et al., 2011). To avoid confusion, their nomenclature for exon 8 and 8a is still used in this chapter when these papers are cited.

One de novo missense mutation G406R in exon 8a was reported to induce classical Timothy syndrome (TS1) by specific loss of voltage-dependent inactivation (VDI) and thereby sustained depolarization (Splawski et al., 2004). On the other hand, patients with mutations of G402S and G406R within exon 8, associated with atypical Timothy syndrome (TS2), have extreme long QT syndrome and severe mental retardation, and they do not have the cardinal feature of simple syndactyly as in TS1 patients (Splawski et al., 2005). However, a recent study showed that, in addition to VDI, G402S mutation also led to loss of CDI because of a decrease in F_{CDI} (a function of Ca^{2+}), while G406R mutation mainly resulted in a reduction of CDI_{max} (a function of channel gating) (Dick et al., 2016). Intriguingly, further studies in iPSC-derived TS human cardiomyocytes showed that only ventricular-like cardiomyocytes displayed prolonged action potential durations (APDs), whereas both ventricular- and atrial-like cardiomyocytes from patients with long QT syndrome type 1 (LQTS1) had long APD (Yazawa et al., 2011). Moreover, the arrhythmia and delayed depolarizations were spontaneously recorded in beating iPSC-TS cardiomyocytes without any stimulation (Yazawa et al., 2011), while LQTS1 cardiomyocytes needed isoproterenol stimulation. More importantly, roscovitine, a cyclin-dependent kinase inhibitor that promotes VDI in transfected HEK 293 cells, was applied to prevent the Long QT phenotype by shortening the APD in iPSC-derived TS human cardiomyocytes (Yazawa et al., 2011), which provides the possibility of rescuing the cardiac phenotypes of TS patients.

One point to note is that, in addition to the classical G402S and G406R mutations, there have been another 12 missense mutations identified to be related to Timothy Syndrome with highly variable phenotypes (Bauer et al., 2021). As there are hundreds of $Ca_V1.2$ splice isoforms due to various combinations of multiple alternatively spliced exons, some TS mutations may not fully reproduce the G406R phenotypes due to tissue-selective expression of certain $Ca_V1.2$ isoforms with more or less abundance.

Moreover, the $Ca_V1.2_{e8}$ splice variant exhibited higher sensitivity to dihydropyridine (DHP) calcium channel blockers (Lei et al., 2020), which may be the reason why the smooth muscle splice variant $Ca_V1.2b$-expressing exon 8 is more sensitive to inhibition by DHPs than the heart variant $Ca_V1.2a$ containing exon 8a (Liao et al., 2005, 2007).

Alternative splicing of exons 8a/8 has been shown to be regulated by the splicing factor polypyrimidine tract-binding protein 1 (PTBP1).

PTBP1 strongly repressed exon 8a inclusion and shift the spliced exon to exon 8 by direct interaction with the conserved sequence elements upstream of exon 8a in mouse cortex (Tang et al., 2011). Exon 8a was largely inhibited in mouse embryonic brains, whereas it was gradually up-regulated during neuronal development with PTBP1 depletion (Tang et al., 2011). Moreover, PTBP1 protein was highly expressed in embryonic hearts and then dramatically reduced during cardiac development in both rats and mice (Zhang et al., 2009; Martí-Gómez et al., 2020), which may be associated with the dominant $Ca_V1.2_{e8a}$ channels in heart (Liao et al., 2005). Interestingly, PTBP1 protein expression was highly induced in failing hearts from mice subjected to transverse aortic constriction (TAC) or myocardial infarction (Martí-Gómez et al., 2020), suggesting there may be a pattern shift of exon 8a and 8 inclusion in failing and normal hearts. In contrast, PTBP1 protein was significantly down-regulated in mesenteric arteries of SHR as compared to normotensive arteries (Lei et al., 2020), which may also result in a shift from exon 8 to 8a utilization in smooth muscle $Ca_V1.2$ channels under hypertension. These results suggest that the fetal splicing program of exons 8a/8 in both smooth muscles and cardiac muscles may be reactivated under cardiovascular disease conditions, which remains to be further validated in human models and other diseases in $Ca_V1.2$-abundant organs, such as the brain.

Exons 21/22

The mutually exclusive exons 21 or 22 encode a short fragment of the $Ca_V1.2$ domain III segment S2. A nearly complete switch of exon 21 to exon 22 was identified in carotid and femoral arteries of human patients with atherosclerosis (Tiwari et al., 2006), a cardiovascular disorder featured by inflammation-mediated endothelial perturbation in medium- and large-size arteries (Libby et al., 2002). Compared to $Ca_V1.2_{e21}$ channels, the $Ca_V1.2_{e22}$ channels showed hyperpolarized shifts of~15 mV in the *I–V* relationship and ~15 mV in

the activation potential as recorded in *Xenopus* oocytes (Tiwari et al., 2006). This study suggests that human atherosclerotic smooth muscle cells may have larger Ca^{2+} influx, which may require Ca^{2+} imaging technique to further validate. However, the $Ca_V1.2$ channels used in this study do contain exon 9* which exists in about half of $Ca_V1.2$ channels expressed in smooth muscles. As such, a more comprehensive understanding of the role of alternative splicing of $Ca_V1.2$ channels in atherosclerosis remains to be investigated. In addition, the splicing factor responsible for this shift of exon 21 to exon 22 utilization has not been identified.

Additionally, a non-functional and developmentally regulated $Ca_V1.2$ splice variant containing both exon 21 and exon 22 in rodent and human hearts, $Ca_V1.2_{e21+22}$, was identified (Hu et al., 2016). Exon 21 + 22 inclusion level was reduced to 5.5% in adult heart from 14.3% in rat neonatal heart, which was then up-regulated 12.5-fold in adult mouse models of heart failure induced by TAC. This reactivation of the fetal splicing pattern is similar to that of exon 9* and exons 8a/8 in heart or smooth muscles. More importantly, co-expression of $Ca_V1.2_{e21+22}$ channels, which had a stronger interaction with β subunits, was able to promote the proteasomal degradation of wild-type $Ca_V1.2$ channels by competing for β-subunits.

Exon 33 and Exon 33L

Exon 33 is composed of 33 nucleotides encoding a portion of the extracellular loop linking domain IV segments 3 and 4. Loss of exon 33 led to hyperpolarized shifts for steady-state inactivation and activation potentials of $Ca_V1.2$ channel in transfected HEK 293 cells (Liao et al., 2009). Moreover, exon 33 inclusion level was reduced in the scar region of ischemic rat hearts subjected to myocardial infarction (Liao et al., 2009). More importantly, in exon 33-deficient mice cardiac contractility and output were significantly increased and ventricular tachyarrhythmia was also recorded due to larger Ca^{2+} influx in cardiomyocytes (Li et al., 2017a). Additionally, exon 33 inclusion was remarkably up-regulated by about 20% in human failing hearts from patients with ischemic or dilated cardiomyopathy (Li et al., 2017a). Although the molecular mechanisms in mouse failing hearts have been delineated, the exact role that exon 33 plays in the pathogenesis of human heart failure remains unclear. Deleting exon 33 in human patient-specific iPSC-derived cardiomyocytes may be useful for modelling of exon 33-associated heart failure.

Rbfox1/2 were reported to act as the splicing factor to enhance exon 33 inclusion by binding to the downstream UGCAUG elements (Tang et al., 2009). In comparison with the non-failing heart, *Rbfox2* mRNA level displayed no significant differences in both dilated and ischemic human hearts, while *Rbfox1* mRNA levels in dilated cardiomyopathy were significantly reduced, and this was positively correlated with *CACNA1C* mRNA levels and negatively correlated with the exon 33 inclusion level (Wang et al., 2018). This finding was inconsistent with the Rbfox1/2-mediated up-regulation of exon 33 inclusion, and a possible explanation could be compensatory or adaptive responses to heart failure. Similarly, Rbfox2 was found to be up-regulated in hypertensive arteries, which was also negatively correlated with reduced exon 33 inclusion levels (Zhou et al., 2017). This inconsistency in hypertensive arteries may be attributed to the reported increase in expression of a dominant-negative Rbfox2 isoform (Zhou et al., 2017).

Recently, a 66-nucleotide extension downstream of exon 33 was identified, which was named exon 33L. Exon 33L led to a frame-shift, and the production of a non-functional $Ca_V1.2$ truncated short form that arose from premature termination (Liao et al., 2015). Interestingly, exon 33 L inclusion was reduced by more than twofold in left ventricles of adult rat hearts as compared to neonatal hearts. More importantly, exon 33L induced dominant-negative suppression of wild-type $Ca_V1.2$ channel function by increasing the proteasomal degradation of $Ca_V1.2$ channels, but not T-type $Ca_V3.2$ channels, indicating that the inhibitory effects of exon 33L on

$Ca_V1.2$ channels may be β subunit dependent. By contrast, the human exon 33L only represents a single-nucleotide insertion, generating a functional full-length $Ca_V1.2$ channel with a much lower Ca^{2+}-conducting ability, which suggests that alternative splicing in exon 33L is species specific. Similar to exons 21/22, the splicing factor for exon 33L is still unknown.

As mentioned above, out of 50 exons in *CACNA1C* gene, in this chapter, 5 major alternative splicing loci (1a/1, 8a/8, 9*, 21/22 and 33) have been introduced, which have been reported to generate about 20 essential $Ca_V1.2$ splice variants with distinctive combinations in aorta or heart of rats. The combinatorial splicing patterns were determined by applying the transcript-scanning method on a mini library of full-length Cav1.2 cDNAs (Tang et al., 1783, 2007). However, a recent study on *CACNA1C* transcript profile of human brain identified 38 novel exons and 241 novel transcripts using long-range PCR and nanopore sequencing, which further substantiates the complexity of $Ca_V1.2$ alternative splicing (Clark et al., 2020). More importantly, the alternative splicing-mediated diversification of $Ca_V1.2$ function also displays a few unique features in tissue selectivity, DHP sensitivity, and reactivation under certain disease conditions. For example, the sensitivity of smooth muscle-specific $Ca_V1.2_{-1/8/9*/\Delta33}$ isoform to nifedipine was increased by 2.4-fold compared to the predominant smooth muscle $Ca_V1.2_{-1/8/9*/33}$ channel and by 9.3-fold compared to the cardiac isoform $Ca_V1.2_{-1a/8a/\Delta9*/33}$ (Liao et al., 2007), which is probably due to the large hyperpolarized shift of in $Ca_V1.2_{-1/8/9*/\Delta33}$ channel of VDI (Liao et al., 2007; Bean, 1984). This may help understand why smooth muscle $Ca_V1.2$ channels are particularly prone to nifedipine block, in addition to the more depolarized resting potential in smooth muscles and exon 8a/8 splicing-induced changes in VDI (Lei et al., 2020; Liao et al., 2007; Li et al., 2017a; Welling et al., 1997). Additionally, the effects of splice isoforms on the regulation of $Ca_V1.2$ channels by post-translational modifications and binding proteins remain to be comprehensively investigated (Loh et al., 2020).

Ca$_V$1.3 in Health and Disease

Among the four L-type channels, Ca$_V$1.2 and Ca$_V$1.3 are ubiquitously expressed in the central nervous system (CNS). The lack of selective blockers of Ca$_V$1.3 channels has hampered the understanding of the physiological roles of the channel. Nonetheless, extensive studies have suggested that as compared to Ca$_V$1.2, Ca$_V$1.3 channels play a significant role in gating low-threshold-activating Ca^{2+} currents that underlie neuronal pacemaking (Chan et al., 2007; Pennartz et al., 2002), excitation–transcription coupling (Wheeler et al., 2008; Zhang et al., 2005, 2006) normal synaptic function (Day et al., 2006; Sinnegger-Brauns et al., 2004), cardiac rhythm (Platzer et al., 2000), and hormone secretion (Marcantoni et al., 2007). It is widely expressed in the CNS, cochlea, sinoatrial node (SAN) of the heart, and in neuroendocrine tissues, including the beta cells of the pancreas and chromaffin cells of the adrenal gland.

The Pathophysiological Roles of Ca$_V$1.3

The Ca$_V$1.3 knockout mouse model discloses a plethora of phenotypic deficits as the Ca$_V$1.3 channels conduct significant inward current at the operating range of the hair cells of the cochlea and the pacemaking cells in SAN due to their low activation threshold (Koschak et al., 2001; Xu & Lipscombe, 2001). Correspondingly, deletion of Ca$_V$1.3 resulted in congenital deafness due to almost complete absence of calcium current (I_{Ca}) in the inner hair cells and degeneration of both outer and inner hair cells (Platzer et al., 2000). In addition, genetic knockout of Ca$_V$1.3 channels impairs the normal development of the auditory brain stem center. As the phenotype appears even before the onset of hearing (Hirtz et al., 2011; Satheesh et al., 2012), it is therefore suggestive that expression of Ca$_V$1.3 channels is essential for the development of both peripheral sensory cells and neurons. Furthermore, Ca$_V$1.3$^{-/-}$ mice exhibit bradycardia as a result of SAN dysfunction (Platzer et al., 2000). Moreover, in mouse

chromaffin cells, the pacemaking Ca$_V$1.3 current drives downstream SK channels activation and allows for sustained action potential firing even with prolonged stressful stimuli (Vandael et al., 2012). Within the CNS, it was shown that Ca$_V$1.3 deletion impaired the consolidation of conditioned fear (McKinney & Murphy, 2006) due to compromised long-term potentiation of the amygdala (McKinney et al., 2009). Notably, by employment of DHP-insensitive Ca$_V$1.2 mice, it was uncovered that acute selective pharmacological activation of Ca$_V$1.3 on the other hand results in depressive-like behavior (Sinnegger-Brauns et al., 2004).

In line with the findings in Ca$_V$1.3$^{-/-}$ mice, a loss-of-function mutation of human Ca$_V$1.3 was reported in two consanguineous Pakistani families (Baig et al., 2011). The mutation resulted in production of non-conducting Ca$_V$1.3 channels and expectedly subjects homozygous for such mutations suffered from sinoatrial node dysfunction and deafness (SANDD) syndrome (Baig et al., 2011), consistent with the phenotypes of Ca$_V$1.3$^{-/-}$ mice. While the neurological phenotypes of SANDD patients were not known, gain of function mutations of Ca$_V$1.3 channels in humans are associated with autism spectrum disorder (ASD) (Pinggera et al., 2015), aldosterone-producing adrenal adenomas (APAs) (Azizan et al., 2013), and primary aldosteronism, seizures, and neurological abnormalities (PASNA) (Scholl et al., 2013).

The Unique Biophysical and Pharmacological Properties of Ca$_V$1.3 Channels and Modulation

The property of the Ca$_V$1.3 channel is defined by its gating mechanisms. While the low activation threshold appears to be an intrinsic property of Ca$_V$1.3 channels, which is only starting to be understood, a variety of feedback mechanisms that inactivate the channel in response to either voltage-induced conformational change (VDI) or elevation of intracellular [Ca^{2+}] (CDI) have been well characterized. The process of VDI is initiated by the voltage-dependent conformational rearrangement of voltage-sensing domain comprising S1-to-S4 segments (Swartz, 2008) leading to subsequent opening of the S6 gate (Liu et al., 1997; Xie et al., 2005) and finally the occlusion of the gate by the I-II loop in a "hinge lid" mechanism. Interestingly, a "shield" that repels the closure of the channel gate by the I-II loop "lid" appears to be a unique feature of the $Ca_V1.3$ channel (Tadross et al., 2008) allowing the channel to remain open despite prolonged activation.

In comparison, CDI (calcium-dependent inactivation) is a negative feedback mechanism arising from the influx of Ca^{2+} ions. Calcium ions activate the bi-lobe calcium sensor, CaM, that is pre-associated with the preIQ-IQ domain of the C-terminus of the channel, subsequently triggering a series of conformational changes which lead eventually to channel inactivation (Dick et al., 2008; Erickson et al., 2003; Mori et al., 2004; Peterson et al., 1999; Pitt et al., 2001; Zuhlke et al., 1999). With a combination of technologies, such as patch-clamp electrophysiology and live-cell FRET (live-cell Foerster resonance energy transfer), conformational intermediates of CDI have gradually been elucidated. It is now clear that rather than acting as an effector site of CDI, the preIQ-IQ domain functions as a pre-association domain of apo-calmodulin. Activated CaM upon Ca^{2+} binding switches its binding partners, with its N-lobe binding to an N-terminal spatial Ca^{2+} transforming element (NSCaTE) module on the channel amino terminus and the C-lobe binding to the EF-hand domain on the carboxyl terminus (Ben Johny et al., 2013). Beyond regulating channel inactivation or calmodulation (Ben-Johny & Yue, 2014), pre-association of apo-calmodulin primary to the IQ domain enhances single-channel open probability of $Ca_V1.3$ in response to voltage stimuli, thus explaining the low threshold of activation of the short-form $Ca_V1.3$. Overexpression of CaM in dopaminergic neurons led to the broadening of action potential (Adams et al., 2014). Lastly, strong affinity of CaM to the IQ domain was found to be associated with attenuated sensitivity of the channels towards dihydropyridine (Huang et al., 2013).

Fitting with the diverse functional roles of the channel, the gating of $Ca_V1.3$ channel is often differentially modulated in a tissue-specific manner. The native $Ca_V1.3$ currents in pancreatic β-cells and SAN cells display substantial inactivation (Mangoni et al., 2003; Plant, 1988) matching the profile of $Ca_V1.3$ channels characterized in heterologous systems (Xu & Lipscombe, 2001; Song et al., 2003). In contrast, I_{Ca} recorded from hair cells in cochlea shows little inactivation (Platzer et al., 2000; Song et al., 2003) suitably allowing for persistent cellular activity even in the presence of the prolonged sound stimulus (Shen et al., 2006; Yang et al., 2006). Several mechanisms have been proposed to explain the tissue-specific specialization of $Ca_V1.3$ channels. Taking the cochlea as an example, selective co-localizations of $Ca_V1.3$ channels with various proteins, such as syntaxin, CaBP (calcium-binding protein), and Rab3-interacting molecule (RIM), were observed and co-expression of such proteins with $Ca_V1.3$ heterologous system were shown to slow channel inactivation (Song et al., 2003; Yang et al., 2006; Gebhart et al., 2010). Furthermore, it was shown that PDZ domain-containing protein, erbin, enhanced the $Ca_V1.3$ channel function via direct interaction with the C-terminus of the long-form channel (Calin-Jageman et al., 2007). Tissue-selective manipulation of these protein binding partners in the native system would provide more conclusive evidence of such regulation.

Regulation of α_{1D} Transcripts by Alternative Splicing and A-to-I RNA Editing

The $Ca_V1.3$ channels are subject to extensive alternative splicing, and a total of 16 exons have been reported to be alternatively spliced and some of them show tissue- and even species-specific distribution (Fig. 2). Despite the rich assortment of channel isoforms with possibly different functional characteristics, the functional impact of alternative splicing of the α_{1D} transcript is only partially understood.

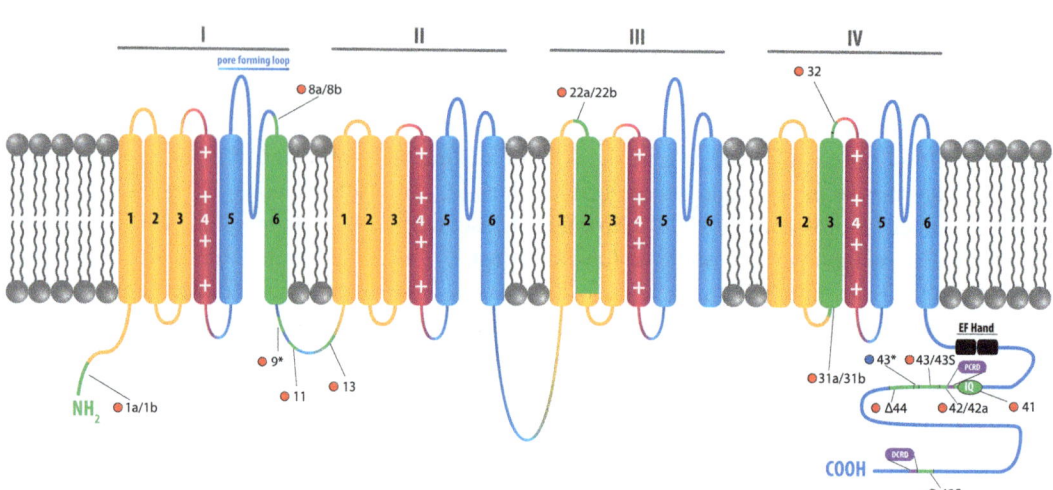

Fig. 2 Alternative splice loci of the $Ca_V1.3$ and $Ca_V1.4$. Segments of the channel colored in green denote the relative location of alternative splice sites in the $\alpha1$ subunit.

Red circles are used to indicate $Ca_V1.3$ splice loci; blue circles are used to indicate $Ca_V1.4$ splice loci

(a) *Alternative   Splicing   Pattern   Within N-terminal Domain*

Alternative splicing of the amino terminus (N-terminus) was known to affect the current density of $Ca_V1.3$ channels (Klugbauer et al., 2002; Xu et al., 2003). Inclusion of either exon 1a (Hui et al., 1991; Seino et al., 1992; Williams et al., 1992a) or 1b (Klugbauer et al., 2002) has been reported in mouse. Exon 1b appears to be mouse specific, while in rat and human, exon 1a is constitutively expressed. Although both splice variants support functional currents with similar gating properties in heterologous expression system, exon 1a confers much larger current density as compared to exon 1b (Klugbauer et al., 2002; Xu et al., 2003).

(b) *Alternative   Splicing   Pattern   Within   the Repeat Domains*

The IS6, IIIS2, and IVS3 segments of $Ca_V1.3$ are encoded by three pairs of mutually exclusive exons, including exons 8a/8b, 22a/22b, and 31a/31b, respectively. Interestingly, $Ca_V1.2$ channels display the same splicing patterns in the abovementioned regions, and relatively high

sequence conservation was observed between $Ca_V1.3$ and $Ca_V1.2$ channels in these three pairs of mutually exclusive exons (Tang et al., 2004). It is noteworthy that the exon nomenclature employed by many labs are different and this often causes confusion in comparing results from various papers. For instance, exon 8a occurs upstream of exon 8b and was named as such in papers from Joerg Striessnig's group (Koschak et al., 2001; Baig et al., 2011; Pinggera et al., 2015; Pinggera & Striessnig, 2016), while they were named as 8a and 8, respectively, in papers from Tuck Wah Soong's group (Liao et al., 2005; Huang et al., 2013; Tang et al., 2004). For the sake of consistency, they are termed 8a and 8b in the current discussion.

Alternative splicing in IS6, IIIS2 and IVS3 segments of $Ca_V1.2$ were known to alter the sensitivity of the channels towards DHP inhibition with exons 8, 22, and 31, conferring higher drug sensitivity (Liao et al., 2005). In contrast, mutually exclusive splice variants 8a/8b and 31a/31b do not dramatically alter the pharmacological properties of $Ca_V1.3$ channels (data no shown) when expressed in HEK293FT cells. In addition, exon 22b of $Ca_V1.3$ appeared to be expressed specifically in the rat organ of Corti with unknown functional role, while exon 22a is constitutively expressed in other tissues (Ramakrishnan et al., 2002). The alternate exon 32 encodes part of the extracellular loop between IVS3 and IVS4. Inclusion or exclusion of exon 32 in $Ca_V1.3$ channels has no effect on the gating properties of the channel and neither was sensitivity towards nitrendipine significantly changed (Xu & Lipscombe, 2001).

Notably, a number of disease-associated mutations were found to occur either in exon 8a or 8b. For example, the insertional mutation that results in loss of function of human $Ca_V1.3$ channel is located in exon 8b[61]. The ASD-linked mutation G407R only occurs in exon 8a[62], while G403D was identified in either exons 8a or 8b in APAs but was present only in exon 8b in PASNA (Azizan et al., 2013; Scholl et al., 2013). While dominant in heart tissue, exon 8b is included in almost 60% of rat brain (unpublished data). Therefore, understanding the tissue-specific distribution of exon 8a and 8b in different brain tissues could have profound implications for prognosis and possible target treatment of any neurophysiological disorders for patient suffering from mutations within this pair of mutually exclusive exons.

Given that $Ca_V1.3$ splice variants containing either exon 8a or 8b do not display distinct biophysical and pharmacological properties, it is therefore conceivable that understanding how exons 8a and 8b are regulated could contribute towards designing novel therapeutic strategy of treating the abovementioned diseases arising from mutations in either exon 8a or exon 8b. Interestingly, binding of splicing factor polypyrimidine tract-binding protein (PTB) in the intronic regions upstream of exon 8a was found to cause skipping of exon 8a of $Ca_V1.2$ (Tang et al., 2011). Of note, siRNA-mediated knockout of PTB leads to inclusion of exon 8a and developmental down-regulation of PTB correlated with up-regulation of exon 8a. Discovery of a similar mechanism that regulates the alternative splicing of 8a/8b in $Ca_V1.3$ would be highly relevant. Masking the binding of the splicing factors by using ASO could be employed to manipulate the level of alternative splicing of exon 8a/8b

relative to each other and therefore restore the normal channel function in the case of gain of function mutation or rescue of loss-of-function mutation in SANDD.

(c) *Alternative Splicing Pattern Within the I-II Loop*

The I-II loop of $Ca_V1.3$ contains three splice variations, including alternate exon 9*, 11, and 13. Exon 9* (Ramakrishnan et al., 2002) and 13 (Ihara et al., 1995) were identified in the rat organ of Corti and pancreas, respectively, with uncharacterized functional impact. On the other hand, exon 11 is more ubiquitously expressed in brain and pancreas and deletion of exon 11 was found not to affect the channel gating of $Ca_V1.3$ (Xu & Lipscombe, 2001). Inclusion of exon 9* introduces 26 amino acids into the I-II loop of the Cav1.3 channels. Sequence of exon 9* in chicken $Ca_V1.3$ contains a consensus sequence of serine surrounded by four basic amino acid residues and is therefore a potential substrate for protein kinase (Ramakrishnan et al., 2002). In contrast, no such consensus site was found in exon 9* of rat or human $Ca_V1.3$ (Ramakrishnan et al., 2002).

(d) *Alternative Splicing and A-to-I RNA Editing Within the C-terminus*

The carboxyl-terminus (C-terminus) of $Ca_V1.3$ represents another hotspot of alternative splicing that has been more extensively characterized. Of note, exon 41 encodes the IQ domain and truncation of exon 41 due to the alternative use of splice acceptor site in exon 41 results in complete removal of the IQ domain and early termination of the C-terminus (Shen et al., 2006). Although functional currents could still be recorded when this splice variant ($Cav1.3_{\Delta IQ}$) was expressed in HEK293 cells, deletion of IQ domain resulted in complete elimination of CDI (Shen et al., 2006). Selective localization of $Cav1.3_{\Delta IQ}$ channels in cochlear outer hair cell (Shen et al., 2006) was hypothesized to partially underlie the previous observation of slowly inactivating native $Ca_V1.3$ currents recorded in hair cells. Again, this result highlights the tissue-specific role of such splice isoforms in supporting normal function, in this case, of the cochlea. Moreover, exon 41 could also behave as a cassette exon. $Ca_V1.3$ transcript without the entire exon 41 has been reported in both rat and human brain (Bock et al., 2011; Tan et al., 2011). Deletion of exon 41 results in complete elimination of the IQ domain, leading to frame-shifting and early truncation of the C-terminus. Expectedly, $Ca_V1.3_{\Delta41}$ shows much lower current density and much slowing of CDI (Tan et al., 2011) given the important role of IQ domain as a pre-association site for apo-calmodulin. Notably, the expression level of $Ca_V1.3_{\Delta41}$ was very low as detected by RT-PCR assay (Bock et al., 2011; Tan et al., 2011).

Interestingly, a more recent study identified three closely spaced A-to-I RNA editing sites in the mRNA sequence which codes for tetrapeptide "IQDY" in the IQ domain (Huang et al., 2012). The editing is mediated by ADAR2, an isoform of the family of enzymes known as ADARs. Codon changes from ATA to ATG, CAG to CGG, and TAC to TGC result in amino acid changes from I to M, Q to R, and Y to C, respectively, giving rise to a total of 8 possible combinations of editing in the IQ domain. Peptide variants containing various edited amino acids in IQDY could also be detected by mass spectrometry. Biophysically, IQ→MR change arising from RNA editing resulted in the edited $Cav1.3_{MR}$ channels exhibiting slowed CDI, owing to the decreased binding affinity of apo-calmodulin. Expectedly, overexpression of wild-type CaM in substantia nigra pars compacta (SNc) neurons could effectively compensate for this decreased in binding affinity by mass effect to up-regulate CDI from a modest CDI recorded in native SNc neurons (Bazzazi et al., 2013). Physiologically, editing in the IQ domain is shown to regulate normal rhythmic firing activity of neurons in the suprachiasmatic nucleus, a hypothalamic region well known for its role as the master control of biological clock in mammalian system. Most importantly, RNA editing in the IQ domain is restricted to the CNS and is evolutionarily conserved across species from mouse and rat to human (Huang et al., 2012). More recently, the

molecular mechanism of A-to-I editing of the exon 41 has been elucidated whereby exon 41 was found to form a RNA duplex structure with an upstream intronic sequence termed Editing site Complementary Sequence (ECS). Notably, editing can be repressed by a splicing factor, Serine and Arginine-Rich Splicing Factor 9 (SRSF9). Down-regulation of SRSF9 in neurons appeared to correlate with tissue-selective editing of $Ca_V1.3$ in various CNS tissues (Huang et al., 2018).

Further downstream, alternative use of either exon 42 or 42a gives rise to long-form (LF) or short-form (SF) $Ca_V1.3$ channels, respectively (Singh et al., 2008). The stop codon in exon 42a results in expression of only 6 amino acids immediately after exon 41 and therefore the early termination of C-terminus. Although both variants are ubiquitously expressed in the brain, the LF $Ca_V1.3_{e42}$ channels display distinctive properties, such as more depolarized shift in window current, higher expression, lower current density, and significantly diminished CDI (Singh et al., 2008). The attenuated CDI in the long form was later explained by the presence of the distal C-terminal autoinhibitory domain (DCRD) at the distal carboxyl terminal which interacts intramolecularly to the proximal C-terminal autoinhibitory domain (PCRD) and leads to displacement of apo-calmodulin from binding to the IQ domain (Singh et al., 2008; Liu et al., 2010). The weakened binding between apo-calmodulin and $Ca_V1.3_{e42}$ channel therefore results in much slower channel inactivation, while the absence of the PCRD domain in the $SFCa_V1.3_{e42a}$ variant due to the early termination of C-terminus leads to fast CDI and much larger current density (Adams et al., 2014).

Moreover, half truncation of exon 43 due to the alternative use of splice site within exon 43 results in a frame-shift and early termination of the C-terminus (Seino et al., 1992; Bock et al., 2011; Tan et al., 2011; Williams et al., 1992b). The $Ca_V1.3_{e43s}$ variant represents the most dominant C-terminal splice variant and could be detected in diverse brain regions in human (Bock et al., 2011). As expected, the exclusion of PCRD domain in such a splice isoform would support rapid CDI similar to that of short-form $Ca_V1.3_{e42a}$ channels. However, it is worth mentioning that splice variants with premature stop codons are often considered faulty and subjected to degradation via mRNA surveillance mechanisms, such as nonsense-mediated mRNA decay (NMD). Briefly, the exon junctions of the mature mRNA are decorated with a protein complex known as Exon Junction Complex (EJC) after alternative splicing. During the first-round translation, ribosomes scan the mRNA and remove the EJC and the process continues towards the last exon. However, the presence of premature stop codon arrests the ribosome and if the unremoved EJC is located 50 nt downstream of the ribosome, the cell recognizes this transcript as aberrant and removes it via NMD (Jopling, 2014). By coupling alternative splicing to NMD, it is possible that a cell could functionally down-regulate expression of that gene under desired conditions by mediating alternative splicing. However, it remains to be tested if manipulation of exon 43 splicing could affect the expression level of $Ca_V1.3$ and in turn affect neuronal activity in a native system.

Lastly, deletion of exon 44 and use of a splice acceptor site within exon 48 result in shortening of C-terminus but do not result in early truncation of the channel. Interestingly, both $Ca_V1.3_{\Delta e44}$ and $Ca_V1.3_{e48s}$ variants display only slightly slower CDI as compared the channel with intact C-terminus (Tan et al., 2011).

Apart from regulation of biophysical and pharmacological properties, truncations within the C-terminus in variants, such as $Ca_V1.3_{\Delta e41}$, $Ca_V1.3_{e42a}$, and $Ca_V1.3_{e43s}$, have additional functional implications. Firstly, early truncation of the C-terminus effectively excludes two consensus sites for PKA activity. The two sites, identified using mass spectrometry, include serine 1743 and serine 1816 located on exon 43 (Ramadan et al., 2009). Phosphorylation of $Ca_V1.3$ channels by PKA is known to substantially increase $Ca_V1.3$ current which potentially contributes to the sympathetic control of heart rate (Qu et al., 2005). The C-terminal alternative splicing of the α_{1D} transcripts, particularly in SAN, could therefore regulate the responsiveness of heart rate to the regulation by activation of β-adrenergic receptors via cAMP-dependent PKA. Secondly shortening of $Ca_V1.3$ C-terminus omits C-terminal Src homology 3 (SH3) domain-binding motifs and postsynaptic density-95/discs large/zona occludens-1 (PDZ) binding motif which is crucial for interaction with the scaffold protein Shank (Zhang et al., 2005). Such interaction results in postsynaptic clustering of long-form $Ca_V1.3$ channels and was later found to be important for processes, such as $Ca_V1.3$-dependent phosphorylated cAMP response element-binding protein (pCREB) signaling (Zhang et al., 2005) and G-protein modulation of $Ca_V1.3$ channels by D2 dopaminergic and M1 muscarinic receptors (Ohlson et al., 2007). In addition, the PDZ binding motif of the $Ca_V1.3$ channel is also known to interact with PDZ domain-containing protein, erbin. The association of erbin with long-form $Ca_V1.3$ results in voltage-dependent facilitation of the current (Calin-Jageman et al., 2007).

$Ca_V1.4$ in Health and Disease

Among the four L-type channels, expression of $Ca_V1.4$ encoded by the gene *CACNA1F* is restricted to the photoreceptor of the eye whereby its current supports the release of glutamate from photoreceptor synaptic terminals in darkness, a process that is critical for vision. Antibodies to $Ca_V1.4$ label dominantly mammalian rod terminals, suggesting that $Ca_V1.4$ is the principal subtype in rods. Mutations in the channel were found to be associated with X-linked incomplete congenital stationary night blindness (xlCSNB) (Strom et al., 1998).

Structure–Function Relationship Learnt from Human Mutations and Alternative Splicing Patterns

In terms of biophysical properties, $Ca_V1.4$ channels display low ionic conductance and slow channel inactivation (Singh et al., 2006). Truncation of the entire C-terminus downstream of IQ domain in the xICSNB-linked K1591X mutation leads to enhanced CDI and hyperpolarizing shift in activation. The enhanced channel inactivation was deemed ill-suited for the role of $Ca_V1.4$ current to support tonic release of neurotransmitter in the photoreceptor and was thought of as the basis of pathogenesis in such xICSNB patients (Singh et al., 2006).

Patch-clamp electrophysiological characterization of K1591X was critical in the identification of C-terminal regulatory domain (CTM) that regulated many biophysical properties, such as channel open probability, channel inactivation kinetics, and voltage gating of the channels (Singh et al., 2006). A similar finding was published by Wahl-Schott et al., in the same year and the similar sequence was named as ICDI in their paper (Wahl-Schott et al., 2006). Subsequently, the molecular mechanism was further dissected to reveal that the interaction of DCRD with PCRD leads to the displacement of apocalmodulin from the IQ domain (Singh et al., 2006, 2008; Liu et al., 2010). Curiously, such a mechanism appeared to be highly conserved among the three L-type channels inclusive of $Ca_V1.2$, $Ca_V1.3$, and $Ca_V1.4$ owing to high sequence conservation among the three channel subtypes.

The transcript of $Ca_V1.4$ was reported to undergo extensive alternative splicing (Haeseleer et al., 2016; Tan et al., 2012) (Fig. 2). Notably, inclusion of novel exon 43*in between exon 43 and exon 44 leads to the truncation of C-terminal after PCRD domain as a result of a stop codon within the exon 43*. Remarkably, $Ca_V1.4_{e43*}$ occurs at frequency of 13.6% among all the $Ca_V1.4$ transcripts expressed in human retina. Expectedly, the $Ca_V1.4_{e43*}$ variant displays hyperpolarizing shift inactivation, enhanced CDI, and increased current density, hallmarks of lack of CTM. Of interest, co-expression of the ICDI peptide that spans the DCRD domain suppressed CDI and caused a depolarizing shift in the activation potential (Tan et al., 2012). Additional splice variants of $Ca_V1.4$ have been reported. For example, the exclusion of exon 47 leads to in-frame deletion of 66 amino acid sequence in the C-terminus which overlaps with the proximal portion of the DCRD domain. Remarkably the $Ca_V1.4_{\Delta e47}$ variant displays hyperpolarizing shift in the voltage dependence of activation, enhanced CDI, and increased current density, suggestive that the integrity of the DCRD domain is essential for its function.

The Ca_V2 Channel Family

$Ca_V2.1$, $Ca_V2.2$, and $Ca_V2.3$ are VGCCs, encoded by *CACNA1A*, *CACNA1B*, and *CACNA1E*, respectively, belonging to the Ca_V2 channel family. They are highly expressed in the presynaptic terminals of the central nervous system where they play a major role in neurotransmitter release (Gasparini et al., 2001; Mintz et al., 1995; Wu & Saggau, 1994). When an action potential invades the presynaptic terminal, Ca_V2 VGCCs open in response to membrane depolarization, allowing for Ca^{2+} influx which eventually results in the release of neurotransmitters across the synaptic cleft.

Pathophysiological Roles of Ca_V2 Channel Family

One can gain a deeper understanding of the function of each Ca_V2 channel isoform by examining Ca_V2 channel knockout mouse models. For example, $Ca_V2.1$-null mice display ataxia, and they exhibit absence seizures and die within 4 weeks after birth (Jun et al., 1999). Although $Ca_V2.1$ is the major Ca_V2 channel isoform at the neuromuscular junction, $Ca_V2.1$-null mice were not paralyzed presumably due to compensation by $Ca_V2.2$ and $Ca_V2.3$ (Jun et al., 1999; Urbano et al., 2003). Purkinje cell-specific postnatal conditional knockout of $Ca_V2.1$ (PC-$Ca_V2.1$ KO) recapitulated the ataxic phenotype observed in $Ca_V2.1$-null mice, but PC-$Ca_V2.1$ KO mice survived past 4 weeks (Todorov et al., 2012), which is not entirely unexpected given the cerebellum's role in motor control and $Ca_V2.1$ being the major

Ca_V2 channel isoform expressed in PCs (Stea et al., 1994).

In contrast to $Ca_V2.1$-null mice, $Ca_V2.2$-null mice have a normal life span albeit being accompanied by impairments in blood pressure control (Ino et al., 2001), are less anxious, and have decreased hyperalgesia and allodynia in an inflammatory context (Saegusa et al., 2001). The resultant phenotypes due to $Ca_V2.2$ deficiency could be attributed to $Ca_V2.2$ being highly expressed in the glomerular primary afferent terminals of the spinal dorsal horn, an important component of the pain pathway, where $Ca_V2.2$ is vital for neurotransmitter release (Nieto-Rostro et al., 2018).

$Ca_V2.3$-null mice are also viable and have decreased responses to inflammatory pain (Saegusa et al., 2000), and have deficits in second-phase insulin release (Jing et al., 2005). Chemical induction of seizures in $Ca_V2.3$-null mice revealed a decreased susceptibility to hippocampal seizures (Weiergraber et al., 2007). Ablation of $Ca_V2.3$ in mice also resulted in altered hippocampal theta oscillations (Muller et al., 2012).

Alternative Splicing in Ca_V2 Channel Family

Much like the Ca_V1 channel family, alternative splicing occurs extensively in the Ca_V2 channel family, generating various Ca_V2 variants that differ in channel electrophysiological and/or pharmacological properties as well as localization and expression (Fig. 3).

Of the different Ca_V2 channel isoforms, $Ca_V2.1$ has the most reported splice loci, seven in humans and 10 in mice (Bourinet et al., 1999; Soong et al., 2002; Allen et al., 2010). A total of eight splice loci were identified in rodent $Ca_V2.2$ (Lipscombe et al., 2002), and three splice loci were identified in rodent $Ca_V2.3$ (Fig. 3) (Schneider et al., 2020). As the functional consequences of each splice locus are not fully understood, only splice loci that have known functional

consequences will be discussed in the following sections.

(a) Alternative Splicing Within II-III Loop

Exon 18a of $Ca_V2.2$ is a cassette exon nested between exons 18 and 19 which encodes 21 amino acids within the cytoplasmic II-III loop (Fig. 3). Alternative splicing of exon 18a results in the inclusion or exclusion of 18a in $Ca_V2.2$ mRNA (Pan & Lipscombe, 2000). Expression levels of e18a-containing $Ca_V2.2$ mRNAs vary in different parts of the CNS and is more highly expressed in adult rat tissue as compared to newborn rats (Pan & Lipscombe, 2000; Gray et al., 2007). When $Ca_V2.2_{e18a}$ channels were co-expressed with β_1 and β_4 subunits, they were observed to be protected from entering into closed-state inactivation (Pan & Lipscombe, 2000; Thaler et al., 2004).

A 57-nucleotide exon was discovered to be nested within a 6221 bp intron of human $Ca_V2.3$ genomic DNA, which was also previously identified from previous studies involving $Ca_V2.3$ mRNA (Gray et al., 2007; Pereverzev et al., 1998; Williams et al., 1994). Although similar in size and position to e18a of $Ca_V2.2$, e18a of $Ca_V2.3$ encodes a distinct nucleotide sequence, with only 25% homology to $Ca_V2.2_{e18a}$[132]. The $Ca_V2.3_{e18a}$ channels were found to be more sensitive to Ca^{2+}-dependent modulation, where accumulation of $[Ca^{2+}]_i$ resulted in increased current density, slowed inactivation, and increased recovery from inactivation (Leroy et al., 2003; Pereverzev et al., 2002). $Ca_V2.3$ channels that include e18a were shown to be more stimulated by phorbol esters (Klockner et al., 2004).

The $Ca_V2.2_{e18a}$ mRNAs were found to be more highly expressed in the adult brain, spinal cord, and peripheral ganglia (Pan & Lipscombe, 2000; Gray et al., 2007), while $Ca_V2.3_{e18a}$ mRNAs were more highly expressed in fetal brains (Gray et al., 2007). What this reciprocal expression pattern means physiologically remains to be discovered as the effects of $Ca_V2.2_{e18}$ are dependent on $Ca_V\beta$ subunit expression. To better understand this,

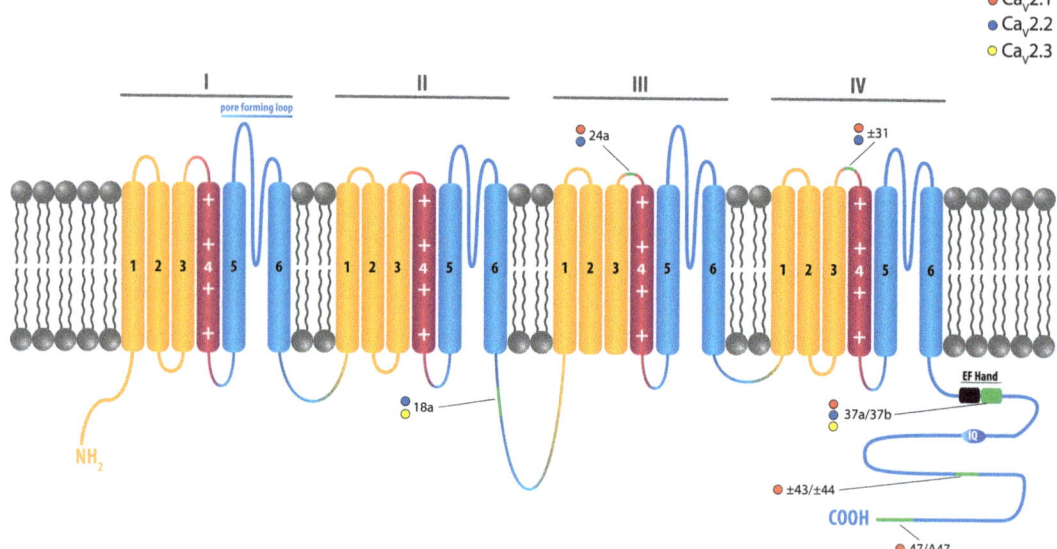

Fig. 3 Alternative splice loci of the Ca$_V$2 family. Segments of the channel colored in green denote the relative location of alternative splice sites in the α1 subunit. Red circles are used to indicate Ca$_V$2.1 splice loci, blue circles are used to indicate Ca$_V$2.2 splice loci, and yellow circles are used to indicate Ca$_V$2.3 splice loci

future studies will need to move away from bulk tissue expression and adopt a more targeted cell-specific expression study, such as single-cell RNA-seq. Previous studies involving the use of genome-wide analysis identified a Rbfox2 binding site upstream of e18a in *Cacna1b* (Weyn-Vanhentenryck et al., 2014; Gehman et al., 2012). This was confirmed by Lipscombe and colleagues where they showed that Rbfox2 represses e18a inclusion in Ca$_V$2.2 mRNAs (Allen et al., 2017).

(b) Alternative Splicing Within Extracellular Linkers of III and IV

E24a was first identified in *Cacna1b* cDNA obtained from rat sympathetic ganglia where it was found to encode four amino acids (SFMG) in the IIIS3-IIIS4 extracellular linker (Lin et al., 1997). Subsequent efforts to identify the functional consequence of e24a inclusion seem to suggest that e24a inclusion has minimal influence on Ca$_V$2.2 channel properties (Allen et al., 2010). In comparison to Ca$_V$2.2, Ca$_V$2.1$_{e24a}$ was discovered much later by Lipscombe and colleagues where it was also found to encode four amino acids (SSTR) in the IIIS3-IIIS4 extracel-

lular linker (Fig. 3). Given its relatively recent identification, the effect of e24a inclusion on Ca$_V$2.1 channel properties remains to be determined (Allen et al., 2010).

Exon 31 in both Ca$_V$2.1 and Ca$_V$2.2 (e31a for Ca$_V$2.2) encode two amino acids, NP in Ca$_V$2.1 [126] and ET in Ca$_V$2.2 (Lin et al., 1999), in the IVS3-IVS4 extracellular linker. Inclusion of NP in Ca$_V$2.1 results in decreased affinity for ω-Aga-IVA and the slowing of channel activation and deactivation kinetics (Bourinet et al., 1999; Soong et al., 2002). While both human and rat Ca$_V$2.1 splice variants have NP inclusion, the mechanism by which NP is included or excluded is different. In the human variant, NP is encoded by a mini-exon of six nucleotides (AATCCG) in the middle of GT/AG acceptor–donor sites within intron 31. On the other hand, the presence of an alternative 5′ splice donor site at the 3′ end of exon 31 allows for the inclusion of NP in the rat variant (Bourinet et al., 1999). Interestingly, the inclusion of ET in Ca$_V$2.2 in the rat variant is similar to how NP is included in the human Ca$_V$2.1 isoform. In rat *Cacna1b*, a six-base cassette exon (GAAACG) flanked by GT/AG acceptor–donor sites is located within intron 31 (Lin et al., 1999). Insertion of ET in IVS3-IVS4 of $Ca_V2.2$ causes the channel to activate at a slower rate (Lin et al., 1999).

The $Ca_V2.1$ mRNA transcripts containing e31 have been shown to be expressed in many regions of the brain and are the predominant variant in the cerebellum (Bourinet et al., 1999; Soong et al., 2002). On the other hand, $Ca_V2.2$ mRNA transcripts containing e31a were shown to be expressed at very low levels in the brain, but more highly expressed in the spinal cord (Lin et al., 1999).

Darnell, Lipscombe, and colleagues have identified Nova-2 to be the splice factor that regulates alternative splicing of e24 and e31 in *Cacna1a* and *Cacna1b*. In the context of e24, Nova-2 binds to the YCAY motif in introns downstream of e24 Ca_V2 pre-mRNA to enhance its inclusion, while binding of Nova-2 to the YCAY motif in introns upstream of e31 Ca_V2 pre-mRNA promotes the exclusion of e31 (Allen et al., 2010).

(c) *Alternative Splicing in C-terminal Region*

Exon 37

Exons 36 and 37 code for an EF-hand-like motif in the C-terminal region of Ca_V2 channels (Fig. 3). Alternative splicing of exon 37 is conserved in $Ca_V2.1$, $Ca_V2.2$ and $Ca_V2.3$ generating mutually exclusive splice variants, EFa and EFb (Gray et al., 2007).

Electrophysiological characterization of $Ca_V2.1$ EFa and EFb splice variants in a heterologous overexpression system showed that while both splice variants had similar channel kinetics, only $Ca_V2.1$ EFa splice variants exhibited Ca^{2+}-dependent facilitation (CDF). Subsequent single-channel recording experiments demonstrated CDF of EFa splice variants at a single-channel level where it was shown to be due to an increase in channel open probability and channel open duration, while EFb splice variants were shown to be locked in a normal gating mode (Chaudhuri et al., 2007). While the critical amino acids required for CDF have not been identified, chimeric channel experiments comprising $Ca_V2.1$ EF-hand region and the pre-IQ-IQ domain of

$Ca_V1.3$ and $Ca_V2.2$ demonstrated the necessity of both the EF-hand region and the pre-IQ-IQ domain of $Ca_V2.1$ for CDF to occur (Thomas et al., 2018; Mori et al., 2008).

While the physiological significance for EFa/EFb splice variants remains unknown, a reciprocal expression pattern of EFa/EFb splice variants in the developing rodent cerebellum has been observed. In the developing cerebellum, from the embryonic stage to P5, EFb was the dominantly expressed in the cerebellum. A developmental switch to an EFa dominant expression pattern was subsequently observed from P7 onwards (Chang et al., 2007; Chaudhuri et al., 2004), while recent RNA-seq data demonstrated that this switch occurred between P1 and P10 (Farini et al., 2020). Interestingly, this switch overlaps with the maturation timeline of the cerebellum, suggesting that the switch in splice variant expression could be vital for proper cerebellar circuitry maturation (Kano et al., 2018).

The $Ca_V2.1$ EFa/EFb splice variants have been found to play a role in homeostatic regulation of presynaptic plasticity in hippocampal neurons. Upon application of toxins or drugs to reduce hippocampal network activity, hippocampal neurons were found to selectively up-regulate $Ca_V2.1$ EFa splice variants, the isoform with higher synaptic efficacy, allowing for the regulation of presynaptic homeostatic plasticity (Thalhammer et al., 2017).

Similar to $Ca_V2.1$, alternative splicing of e37 of $Ca_V2.2$ mRNA results in the generation of mutually exclusive splice variants, e37a and e37b. Splicing of $Ca_V2.2_{e37}$ was first identified in various regions of the adult rat nervous system and it was later shown that $Ca_V2.2_{e37a}$ mRNA was only selectively expressed in the dorsal root ganglion, and preferentially within a subset of capsaicin-responsive neurons, while $Ca_V2.2_{e37b}$ mRNA was widely expressed throughout the CNS (Bell et al., 2004). Subsequent electrophysiological characterization of the $Ca_V2.2_{e37a}$ and $Ca_V2.2_{e37b}$ splice variants showed that the $Ca_V2.2_{e37a}$ variant had a higher membrane expression level, opened at slightly more hyperpolarized membrane potentials, had a longer open time, and is slower to deactivate and inactivate, resulting in a larger macroscopic Ca^{2+} current when compared to cells that only expressed $Ca_V2.2_{e37b}$ variant (Bell et al., 2004; Castiglioni et al., 2006). Differences in AP-1 binding motifs, [DE]xxxL[LI] and Yxxφ, between $Ca_V2.2_{e37a}$ and $Ca_V2.2_{e37b}$ splice variants, were demonstrated to underlie the difference in membrane trafficking of the e37 splice variants, with $Ca_V2.2_{e37a}$ being the more efficiently trafficked splice variant (Macabuag & Dolphin, 2015). Given that there is a difference in channel property among $Ca_V2.2$ e37a/e37b, and that $Ca_V2.2_{e37a}$ is differentially expressed in neurons involved in the pain pathway, functional consequences relating to pain processing have been demonstrated.

Exon 43/44

Alternative splicing of exons 43 and 44 of $Ca_V2.1$ results in a combination of splice variants where either exon can be present or absent, giving rise to four possible variants (±43/±44) (Fig. 3). Electrophysiological characterization of these splice variants revealed changes in CDI and Ca^{2+} current amplitude. The $43^-/44^-$ splice variant was shown to have the greatest degree of CDI and also produced the largest Ca^{2+} current amplitude (Soong et al., 2002). As CDI is affected by global increases in Ca^{2+}, it was not surprising that $43^-/44^-$ splice variant had the greatest CDI due to it having the largest Ca^{2+} current amplitude.

Exon 47

Insertion of GGCAG, a pentanucleotide, at the beginning of $Ca_V2.1$ exon 47 causes an in-frame translation of exon 47, producing a long variant of the C-terminus that contains a polyglutamine (polyQ) tract (Fig. 3). Absence of GGCAG in splice variants results in a frameshift, generating a stop codon near the beginning of exon 47, resulting in a shorter isoform of the channel (Δ47) (Soong et al., 2002; Zhuchenko et al., 1997). While $Ca_V2.1$ variants containing the polyQ tract do not have any changes in channel properties, the long form of $Ca_V2.1$ with the polyQ tract has been linked to spinocerebellar ataxia type 6 (SCA6), with longer repeats of the polyQ tract found in patients with an earlier onset of the disease (Ishikawa et al., 1997). Knock-in

animal models of SCA6 mutants also demonstrated no observable differences in $Ca_V2.1$ expressed in Purkinje cells (Aikawa et al., 2017; Saegusa et al., 2007; Watase et al., 2008).

In healthy control brains, it was demonstrated via in situ hybridization for the long variant of $Ca_V2.1$ mRNA containing the CAG repeat that $Ca_V2.1_{e47}$ was most highly expressed in Purkinje cells as compared to other brain regions (Ishikawa et al., 1999). This would partially explain why selective degeneration of the cerebellum is observed in SCA6 patients and not other brain regions.

It was first discovered that the C-terminus of α1A subunit (α1ACT) contained the polyQ tract and could exist as a stable fragment of 75kDA in cultured cells and Purkinje cells (Kubodera et al., 2003; Kordasiewicz et al., 2006). The α1ACT fragment was identified to possess nuclear localization signals, allowing it to be enriched in Purkinje cell nuclei (Kordasiewicz et al., 2006). α1ACT fragments containing the expanded polyQ tract were found to be toxic when localized to the nuclei of cultured cells and neurons (Kubodera et al., 2003; Kordasiewicz et al., 2006; Ishiguro et al., 2010).

The discovery of an internal ribosomal entry site (IRES) in the *CACNA1A* mRNA was later found to be responsible for the generation of α1ACT from full-length α1A mRNA (Du et al., 2013). α1ACT was also shown to be a transcription factor with the start of its amino acid sequence within the IQ-like domain of $Ca_V2.1$ (Du et al., 2013). Normal WT α1ACT is non-pathological and is in fact required for proper development and survival of cerebellar Purkinje cells (Du et al., 2019). Only when α1ACT contains exon 47 with a pathological expansion of the polyQ tract that the fragment becomes deleterious (Kordasiewicz et al., 2006).

Novel Approaches to Uncover the Pathophysiological Consequences of Alternative Splicing or Gene Mutations of Ion Channels: Moving Away from Heterologous Systems

Ca^{2+} ions are second messengers that are involved in many signaling pathways in the cell. Hence, tight coupling of Ca_V2 channels to neurotransmitter release machinery has been observed at the presynaptic terminal (Eggermann et al., 2011). Alternative splicing generates many different variants of Ca_V2 channels that affect channel kinetics and possibly neurotransmitter release. It will thus be beneficial to know the subcellular localization of each splice variant and determine how they affect neurotransmitter release. However, given that certain splice variants, such as $Ca_V2.1$ EFa/EFb, can be highly similar in sequence, generating antibodies that are able to distinguish the two isoforms can be extremely challenging. Nanobodies generated by commercial companies have been shown to be specific for the detection of a single amino acid change between two proteins and could possibly be one solution to detecting highly similar splice variants. Newer techniques, such as single-cell labeling of endogenous proteins by clustered regularly interspaced short palindromic repeats (CRISPR)-Cas9-mediated homology-directed repair (SLENDR), could be used to label a protein of interest with an epitope tag (Mikuni et al., 2016). This could be further combined with knockout of a specific splice variant and by comparing the difference between the wild-type and splice variant knockout, one could deduce the localization of the deleted splice variant.

In addition, many of the ion channel mutations and alternative splice variants discussed thus far have been studied in heterologous overexpression systems, such as in HEK293 cells, and expressing only specific cDNA clones of the channel. Although valuable knowledge regarding the structure–function relationship could be gleaned from such analysis, many aspects of RNA regulation were often missed or ignored. Taking $Ca_V1.3_{e43s}$ and $Ca_V1.4_{e43*}$ variants as examples, both transcripts contain the premature stop codon which could subject the mRNA transcripts to be degraded by NMD as discussed previously. If indeed true, such variants would not be expressed as functional proteins in native cells, such as neurons or photoreceptors, respectively, but instead serve to tune the expression level of the WT channels under certain physiological conditions. This also applies to nonsense

mutation K1519X of $Ca_V1.4$. While the enhanced CDI of the mutant was proposed to explain the functional deficit of the channel in photoreceptors, it awaits to be tested if aberrant transcripts containing such mutation are simply degraded resulting in the deficiency of $Ca_V1.4$ channel protein in the presynaptic terminal. Such information would be highly relevant to design targeted treatments for xICSNB; rather than focusing on slowing down the inactivation kinetics of the channel, overexpression of the WT channel would potentially help ameliorate the symptoms.

To overcome such shortcomings, many approaches could be taken involving transgenic mouse models whereby cassette exons could be deleted (Andrade et al., 2010; Li et al., 2017b) or mutations could be knocked in (Wu et al., 2011). Alternatively, human-induced Pluripotent Stem Cells (hiPSCs) could be engineered by CRISPR-Cas9-mediated genome editing and subsequently differentiated to either cardiomyocytes or neurons to study their physiological roles. It is noted that while such approaches provide a clean system for the functional role of alternative splicing to be studied, alternative splicing and RNA editing are often developmentally and tissue-selectively regulated events. Complete abolishment of such RNA regulation may result in unwanted side effects and exaggerate the interpretation the physiological consequences of such RNA processing. To further address such issues, RNA-binding ASOs could be employed to either disrupt formation of RNA duplex structure to reversibly abolish A-to-I RNA editing mediated by ADAR2 or mask the binding of splicing factor to its cis-element within pre-mRNA sequence so as to cause splice switching. Such ASO could be modified at the backbone, replacing the phosphodiester bond with phosphorothioate (PS) linkages to be nuclease resistance and at the base with 2'-O-methoxyethyl (2'-MOE) 2'-O-methyl (2'OMe) to prevent the activation of RNase H (Roberts et al., 2020). Overcoming the challenges of effective and cell-specific delivery of oligonucleotides may eventually provide a better approach to facilitate the analysis of many of these RNA processing mechanisms, such as A-to-I RNA and alternative splicing discussed here.

Summary

Alternative splicing and RNA editing are exquisite post-transcriptional mechanisms to expand calcium channel structures to potentially optimize function in development and for tissue-selective purposes. In this regard, careful consideration of combinatorial assortment of the various splice sites and the cellular context of the expression of the splice variations in health and disease may contribute towards better understanding of the genotype–phenotype relationship, especially with regard to mutation-causing disorders. As alternative splicing and RNA editing could be dynamically regulated, the search for the cues and upstream signals that govern these processes will be of scientific and clinical interests. With the advent of sophisticated technologies to visualize cellular localization of transcripts, to interrogate protein–protein interactions and to analyze massive single-cell transcriptomics and proteomics, the possibility to identify splice variant-selective interactors in specific cellular milieu in health and disease, together with the knowledge of altered channel biophysical properties, will probably help answer the question on why a cell expresses so many splice variants.

References

Adams, P. J., Ben-Johny, M., Dick, I. E., Inoue, T., & Yue, D. T. (2014). Apocalmodulin itself promotes ion channel opening and Ca(2+) regulation. *Cell, 159*, 608–622.

Aikawa, T., et al. (2017). Alternative splicing in the C-terminal tail of Cav2.1 is essential for preventing a neurological disease in mice. *Human Molecular Genetics, 26*, 3094–3104. https://doi.org/10.1093/hmg/ddx193

Allen, S. E., Darnell, R. B., & Lipscombe, D. (2010). The neuronal splicing factor Nova controls alternative splicing in N-type and P-type CaV2 calcium channels. *Channels (Austin, Tex.), 4*, 483–489, 12868 [pii]. https://doi.org/10.4161/chan.4.6.12868

Allen, S. E., et al. (2017). Cell-specific RNA binding protein Rbfox2 regulates CaV2.2 mRNA exon composition and CaV2.2 current size. *eNeuro, 4*, ENEURO.0332-16.2017. https://doi.org/10.1523/ENEURO.0332-16.2017

Andrade, A., Denome, S., Jiang, Y. Q., Marangoudakis, S., & Lipscombe, D. (2010). Opioid inhibition of N-type Ca2+ channels and spinal analgesia couple to alternative splicing. *Nature Neuroscience, 13*, 1249–1256.

Andre, L. M., van Cruchten, R. T. P., Willemse, M., & Wansink, D. G. (2019). (CTG)n repeat-mediated dysregulation of MBNL1 and MBNL2 expression during myogenesis in DM1 occurs already at the myoblast stage. *PLoS One, 14*, e0217317. https://doi.org/10.1371/journal.pone.0217317

Azizan, E. A., et al. (2013). Somatic mutations in ATP1A1 and CACNA1D underlie a common subtype of adrenal hypertension. *Nature Genetics, 45*, 1055–1060.

Baig, S. M., et al. (2011). Loss of Ca(v)1.3 (CACNA1D) function in a human channelopathy with bradycardia and congenital deafness. *Nature Neuroscience, 14*, 77–84. https://doi.org/10.1038/nn.2694

Bannister, R. A., & Beam, K. G. (2013). Ca(V)1.1: The atypical prototypical voltage-gated Ca(2)(+) channel. *Biochimica et Biophysica Acta, 1828*, 1587–1597. https://doi.org/10.1016/j.bbamem.2012.09.007

Bartels, P., et al. (2018). Alternative splicing at N terminus and domain I modulates CaV1.2 inactivation and surface expression. *Biophysical Journal, 115*, 163. https://doi.org/10.1016/j.bpj.2018.06.001

Bauer, R., Timothy, K. W., & Golden, A. (2021). Update on the molecular genetics of Timothy Syndrome. *Frontiers in Pediatrics, 9*, 668546. https://doi.org/10.3389/fped.2021.668546

Bazzazi, H., Ben Johny, M., Adams, P. J., Soong, T. W., & Yue, D. T. (2013). Continuously tunable Ca(2+) regulation of RNA-edited CaV1.3 channels. *Cell Reports, 5*, 367–377.

Bean, B. P. (1984). Nitrendipine block of cardiac calcium channels: High-affinity binding to the inactivated state. *Proceedings of the National Academy of Sciences of the United States of America, 81*, 6388–6392. https://doi.org/10.1073/pnas.81.20.6388

Bell, T. J., Thaler, C., Castiglioni, A. J., Helton, T. D., & Lipscombe, D. (2004). Cell-specific alternative splicing increases calcium channel current density in the pain pathway. *Neuron, 41*, 127–138.

Ben Johny, M., Yang, P. S., Bazzazi, H., & Yue, D. T. (2013). Dynamic switching of calmodulin interactions underlies Ca2+ regulation of CaV1.3 channels. *Nature Communications, 4*, 1717.

Ben-Johny, M., & Yue, D. T. (2014). Calmodulin regulation (calmodulation) of voltage-gated calcium channels. *The Journal of General Physiology, 143*, 679–692.

Bock, G., et al. (2011). Functional properties of a newly identified C-terminal splice variant of Cav1.3 L-type Ca2+ channels. *The Journal of Biological Chemistry, 286*, 42736–42748.

Bourinet, E., et al. (1999). Splicing of alpha 1A subunit gene generates phenotypic variants of P- and Q-type calcium channels. *Nature Neuroscience, 2*, 407–415. https://doi.org/10.1038/8070

Calin-Jageman, I., Yu, K., Hall, R. A., Mei, L., & Lee, A. (2007). Erbin enhances voltage-dependent facilitation of Ca(v)1.3 Ca2+ channels through relief of an auto-inhibitory domain in the Ca(v)1.3 alpha1 subunit. *The Journal of Neuroscience, 27*, 1374–1385.

Camby, I., Le Mercier, M., Lefranc, F., & Kiss, R. (2006). Galectin-1: A small protein with major functions. *Glycobiology, 16*, 137R–157R. https://doi.org/10.1093/glycob/cwl025

Castiglioni, A. J., Raingo, J., & Lipscombe, D. (2006). Alternative splicing in the C-terminus of CaV2.2 controls expression and gating of N-type calcium channels. *The Journal of Physiology, 576*, 119–134.

Chan, C. S., et al. (2007). 'Rejuvenation' protects neurons in mouse models of Parkinson's disease. *Nature, 447*, 1081–1086, nature05865 [pii]. https://doi.org/10.1038/nature05865

Chang, S. Y., et al. (2007). Age and gender-dependent alternative splicing of P/Q-type calcium channel EF-hand. *Neuroscience, 145*, 1026–1036, S0306-4522(06)01767-2 [pii]. https://doi.org/10.1016/j.neuroscience.2006.12.054

Chaudhuri, D., et al. (2004). Alternative splicing as a molecular switch for Ca2+/calmodulin-dependent facilitation of P/Q-type Ca2+ channels. *The Journal of Neuroscience, 24*, 6334–6342, 24/28/6334 [pii]. https://doi.org/10.1523/JNEUROSCI.1712-04.2004

Chaudhuri, D., Issa, J. B., & Yue, D. T. (2007). Elementary mechanisms producing facilitation of Cav2.1 (P/Q-type) channels. *The Journal of General Physiology, 129*, 385–401. https://doi.org/10.1085/jgp.200709749

Clark, M. B., et al. (2020). Long-read sequencing reveals the complex splicing profile of the psychiatric risk gene CACNA1C in human brain. *Molecular Psychiatry, 25*, 37–47. https://doi.org/10.1038/s41380-019-0583-1

Day, M., et al. (2006). Selective elimination of glutamatergic synapses on striatopallidal neurons in Parkinson disease models. *Nature Neuroscience, 9*, 251–259, nn1632 [pii]. https://doi.org/10.1038/nn1632

Dick, I. E., et al. (2008). A modular switch for spatial Ca2+ selectivity in the calmodulin regulation of CaV channels. *Nature, 451*, 830–834, nature06529 [pii]. https://doi.org/10.1038/nature06529

Dick, I. E., Joshi-Mukherjee, R., Yang, W., & Yue, D. T. (2016). Arrhythmogenesis in Timothy Syndrome is associated with defects in Ca(2+)-dependent inactivation. *Nature Communications, 7*, 10370. https://doi.org/10.1038/ncomms10370

Du, X., et al. (2013). Second cistron in CACNA1A gene encodes a transcription factor mediating cerebellar development and SCA6. *Cell, 154*, 118–133. https://doi.org/10.1016/j.cell.2013.05.059

Du, X., et al. (2019). alpha1ACT is essential for survival and early cerebellar programming in a critical neonatal window. *Neuron, 102*, 770–785 e777. https://doi.org/10.1016/j.neuron.2019.02.036

Eggermann, E., Bucurenciu, I., Goswami, S. P., & Jonas, P. (2011). Nanodomain coupling between Ca(2)(+)

channels and sensors of exocytosis at fast mammalian synapses. *Nature Reviews. Neuroscience, 13*, 7–21. https://doi.org/10.1038/nrn3125

Erickson, M. G., Liang, H., Mori, M. X., & Yue, D. T. (2003). FRET two-hybrid mapping reveals function and location of L-type Ca2+ channel CaM preassociation. *Neuron, 39*, 97–107, S0896627303003957 [pii]. https://doi.org/10.1016/s0896-6273(03)00395-7

Farini, D., et al. (2020). A dynamic splicing program ensures proper synaptic connections in the developing cerebellum. *Cell Reports, 31*, 107703. https://doi.org/10.1016/j.celrep.2020.107703

Flucher, B. E., & Tuluc, P. (2011). A new L-type calcium channel isoform required for normal patterning of the developing neuromuscular junction. *Channels (Austin, Tex.), 5*, 518–524. https://doi.org/10.4161/chan.5.6.17951

Gasparini, S., Kasyanov, A. M., Pietrobon, D., Voronin, L. L., & Cherubini, E. (2001). Presynaptic R-type calcium channels contribute to fast excitatory synaptic transmission in the rat hippocampus. *The Journal of Neuroscience, 21*, 8715–8721.

Gebhart, M., et al. (2010). Modulation of Cav1.3 Ca2+ channel gating by Rab3 interacting molecule. *Molecular and Cellular Neurosciences, 44*, 246–259. https://doi.org/10.1016/j.mcn.2010.03.011

Gehman, L. T., et al. (2012). The splicing regulator Rbfox2 is required for both cerebellar development and mature motor function. *Genes & Development, 26*, 445–460. https://doi.org/10.1101/gad.182477.111

Gray, A. C., Raingo, J., & Lipscombe, D. (2007). Neuronal calcium channels: Splicing for optimal performance. *Cell Calcium, 42*, 409–417.

Haeseleer, F., Williams, B., & Lee, A. (2016). Characterization of C-terminal splice variants of Cav1.4 Ca2+ channels in human retina. *The Journal of Biological Chemistry, 291*, 15663–15673.

Hirtz, J. J., et al. (2011). Cav1.3 calcium channels are required for normal development of the auditory brainstem. *The Journal of Neuroscience, 31*, 8280–8294.

Hu, Z., et al. (2016). Aberrant splicing promotes proteasomal degradation of L-type CaV1.2 calcium channels by competitive binding for CaVbeta subunits in cardiac hypertrophy. *Scientific Reports, 6*, 35247, srep35247 [pii]. https://doi.org/10.1038/srep35247

Hu, Z., Liang, M. C., & Soong, T. W. (2017). Alternative splicing of L-type CaV1.2 calcium channels: Implications in cardiovascular diseases. *Genes (Basel), 8*, 344. https://doi.org/10.3390/genes8120344

Hu, Z., et al. (2018). Regulation of blood pressure by targeting CaV1.2-Galectin-1 protein interaction. *Circulation, 138*, 1431–1445. https://doi.org/10.1161/CIRCULATIONAHA.117.031231

Huang, H., et al. (2012). RNA editing of the IQ domain in Ca(v)1.3 channels modulates their Ca(2)(+)-dependent inactivation. *Neuron, 73*, 304–316.

Huang, H., Yu, D., & Soong, T. W. (2013). C-terminal alternative splicing of CaV1.3 channels distinctively modulates their dihydropyridine sensitivity. *Molecular Pharmacology, 84*, 643–653.

Huang, H., et al. (2018). Tissue-selective restriction of RNA editing of CaV1.3 by splicing factor SRSF9. *Nucleic Acids Research, 46*, 7323–7338.

Hui, A., et al. (1991). Molecular cloning of multiple subtypes of a novel rat brain isoform of the alpha 1 subunit of the voltage-dependent calcium channel. *Neuron, 7*, 35–44.

Ihara, Y., et al. (1995). Molecular diversity and functional characterization of voltage-dependent calcium channels (CACN4) expressed in pancreatic beta-cells. *Molecular Endocrinology, 9*, 121–130.

Ino, M., et al. (2001). Functional disorders of the sympathetic nervous system in mice lacking the alpha 1B subunit (Cav 2.2) of N-type calcium channels. *Proceedings of the National Academy of Sciences of the United States of America, 98*, 5323–5328. https://doi.org/10.1073/pnas.081089398

Ishiguro, T., et al. (2010). The carboxy-terminal fragment of alpha(1A) calcium channel preferentially aggregates in the cytoplasm of human spinocerebellar ataxia type 6 Purkinje cells. *Acta Neuropathologica, 119*, 447–464. https://doi.org/10.1007/s00401-009-0630-0

Ishikawa, K., et al. (1997). Japanese families with autosomal dominant pure cerebellar ataxia map to chromosome 19p13.1-p13.2 and are strongly associated with mild CAG expansions in the spinocerebellar ataxia type 6 gene in chromosome 19p13.1. *American Journal of Human Genetics, 61*, 336–346. https://doi.org/10.1086/514867

Ishikawa, K., et al. (1999). Abundant expression and cytoplasmic aggregations of [alpha]1A voltage-dependent calcium channel protein associated with neurodegeneration in spinocerebellar ataxia type 6. *Human Molecular Genetics, 8*, 1185–1193. https://doi.org/10.1093/hmg/8.7.1185

Jing, X., et al. (2005). CaV2.3 calcium channels control second-phase insulin release. *The Journal of Clinical Investigation, 115*, 146–154. https://doi.org/10.1172/JCI22518

Jopling, C. L. (2014). Stop that nonsense! *eLife, 3*, e04300.

Jun, K., et al. (1999). Ablation of P/Q-type Ca(2+) channel currents, altered synaptic transmission, and progressive ataxia in mice lacking the alpha(1A)-subunit. *Proceedings of the National Academy of Sciences of the United States of America, 96*, 15245–15250.

Kano, M., Watanabe, T., Uesaka, N., & Watanabe, M. (2018). Multiple phases of climbing fiber synapse elimination in the developing cerebellum. *Cerebellum, 17*, 722–734. https://doi.org/10.1007/s12311-018-0964-z

Klockner, U., et al. (2004). The cytosolic II-III loop of Cav2.3 provides an essential determinant for the phorbol ester-mediated stimulation of E-type Ca2+ channel activity. *The European Journal of Neuroscience, 19*, 2659–2668. https://doi.org/10.1111/j.0953-816X.2004.03375.x

Klugbauer, N., Welling, A., Specht, V., Seisenberger, C., & Hofmann, F. (2002). L-type Ca2+ channels of the embryonic mouse heart. *European Journal of Pharmacology, 447*, 279–284.

Kordasiewicz, H. B., Thompson, R. M., Clark, H. B., & Gomez, C. M. (2006). C-termini of P/Q-type Ca2+ channel alpha1A subunits translocate to nuclei and promote polyglutamine-mediated toxicity. *Human Molecular Genetics, 15*, 1587–1599. https://doi.org/10.1093/hmg/ddl080

Koschak, A., et al. (2001). alpha 1D (Cav1.3) subunits can form l-type Ca2+ channels activating at negative voltages. *The Journal of Biological Chemistry, 276*, 22100–22106. https://doi.org/10.1074/jbc.M101469200

Kubodera, T., et al. (2003). Proteolytic cleavage and cellular toxicity of the human alpha1A calcium channel in spinocerebellar ataxia type 6. *Neuroscience Letters, 341*, 74–78. https://doi.org/10.1016/s0304-3940(03)00156-3

Lei, J., et al. (2020). Aberrant exon 8/8a splicing by downregulated PTBP (polypyrimidine tract-binding protein) 1 increases CaV1.2 dihydropyridine resistance to attenuate vasodilation. *Arteriosclerosis, Thrombosis, and Vascular Biology, 40*, 2440–2453. https://doi.org/10.1161/ATVBAHA.120.315010

Leroy, J., et al. (2003). Ca2+-sensitive regulation of E-type Ca2+ channel activity depends on an arginine-rich region in the cytosolic II-III loop. *The European Journal of Neuroscience, 18*, 841–855. https://doi.org/10.1046/j.1460-9568.2003.02819.x

Li, G., et al. (2017a). Exclusion of alternative exon 33 of CaV1.2 calcium channels in heart is proarrhythmogenic. *Proceedings of the National Academy of Sciences of the United States of America, 114*, E4288–E4295. https://doi.org/10.1073/pnas.1617205114

Li, G., et al. (2017b). Exclusion of alternative exon 33 of Ca(V)1.2 calcium channels in heart is proarrhythmogenic. *Proceedings of the National Academy of Sciences of the United States of America, 114*, E4288–E4295.

Liao, P., et al. (2004). Smooth muscle-selective alternatively spliced exon generates functional variation in Cav1.2 calcium channels. *The Journal of Biological Chemistry, 279*, 50329–50335. https://doi.org/10.1074/jbc.M409436200

Liao, P., Yong, T. F., Liang, M. C., Yue, D. T., & Soong, T. W. (2005). Splicing for alternative structures of Cav1.2 Ca2+ channels in cardiac and smooth muscles. *Cardiovascular Research, 68*, 197–203. https://doi.org/10.1016/j.cardiores.2005.06.024

Liao, P., et al. (2007). A smooth muscle Cav1.2 calcium channel splice variant underlies hyperpolarized window current and enhanced state-dependent inhibition by nifedipine. *The Journal of Biological Chemistry, 282*, 35133–35142.

Liao, P., et al. (2009). Molecular alteration of Ca(v)1.2 calcium channel in chronic myocardial infarction. *Pflügers Archiv, 458*, 701–711. https://doi.org/10.1007/s00424-009-0652-4

Liao, P., et al. (2015). Alternative splicing generates a novel truncated Cav1.2 channel in neonatal rat heart. *The Journal of Biological Chemistry, 290*, 9262–

9272, M114.594911 [pii]. https://doi.org/10.1074/jbc. M114.594911

Libby, P., Ridker, P. M., & Maseri, A. (2002). Inflammation and atherosclerosis. *Circulation, 105*, 1135–1143.

Lin, Z., Haus, S., Edgerton, J., & Lipscombe, D. (1997). Identification of functionally distinct isoforms of the N-type Ca2+ channel in rat sympathetic ganglia and brain. *Neuron, 18*, 153–166.

Lin, Z., et al. (1999). Alternative splicing of a short cassette exon in alpha1B generates functionally distinct N-type calcium channels in central and peripheral neurons. *The Journal of Neuroscience, 19*, 5322–5331.

Lipscombe, D., Pan, J. Q., & Gray, A. C. (2002). Functional diversity in neuronal voltage-gated calcium channels by alternative splicing of Ca(v)alpha1. *Molecular Neurobiology, 26*, 21–44.

Liu, Y., Holmgren, M., Jurman, M. E., & Yellen, G. (1997). Gated access to the pore of a voltage-dependent K+ channel. *Neuron, 19*, 175–184.

Liu, X., Yang, P. S., Yang, W., & Yue, D. T. (2010). Enzyme-inhibitor-like tuning of Ca(2+) channel connectivity with calmodulin. *Nature, 463*, 968–972. https://doi.org/10.1038/nature08766

Loh, K. W. Z., Liang, M. C., Soong, T. W., & Hu, Z. (2020). Regulation of cardiovascular calcium channel activity by post-translational modifications or interacting proteins. *Pflügers Archiv, 472*, 653–667. https://doi.org/10.1007/s00424-020-02398-x

Macabuag, N., & Dolphin, A. C. (2015). Alternative splicing in Ca(V)2.2 regulates neuronal trafficking via adaptor protein complex-1 adaptor protein motifs. *The Journal of Neuroscience, 35*, 14636–14652. https://doi.org/10.1523/JNEUROSCI.3034-15.2015

Mangoni, M. E., et al. (2003). Functional role of L-type Cav1.3 Ca2+ channels in cardiac pacemaker activity. *Proceedings of the National Academy of Sciences of the United States of America, 100*, 5543–5548. https://doi.org/10.1073/pnas.0935295100

Marcantoni, A., et al. (2007). L-type calcium channels in adrenal chromaffin cells: Role in pace-making and secretion. *Cell Calcium, 42*, 397–408. https://doi.org/10.1016/j.ceca.2007.04.015

Martí-Gómez, C., et al. (2020). PTBP1 promotes cardiac hypertrophy and diastolic dysfunction by modulating alternative splicing. *Biorxiv.* https://doi.org/10.1101/2020.06.30.171983

Matza, D., et al. (2016). T cell receptor mediated calcium entry requires alternatively spliced Cav1.1 channels. *PLoS One, 11*, e0147379. https://doi.org/10.1371/journal.pone.0147379

McKinney, B. C., & Murphy, G. G. (2006). The L-Type voltage-gated calcium channel Cav1.3 mediates consolidation, but not extinction, of contextually conditioned fear in mice. *Learning & Memory, 13*, 584–589. https://doi.org/10.1101/lm.279006

McKinney, B. C., Sze, W., Lee, B., & Murphy, G. G. (2009). Impaired long-term potentiation and enhanced neuronal excitability in the amygdala of Ca(V)1.3 knockout mice. *Neurobiology of Learning and Memory, 92*, 519–528. https://doi.org/10.1016/j.nlm.2009.06.012

Mikuni, T., Nishiyama, J., Sun, Y., Kamasawa, N., & Yasuda, R. (2016). High-throughput, high-resolution mapping of protein localization in mammalian brain by in vivo genome editing. *Cell, 165*, 1803–1817. https://doi.org/10.1016/j.cell.2016.04.044

Mintz, I. M., Sabatini, B. L., & Regehr, W. G. (1995). Calcium control of transmitter release at a cerebellar synapse. *Neuron, 15*, 675–688. https://doi.org/10.1016/0896-6273(95)90155-8

Mori, M. X., Erickson, M. G., & Yue, D. T. (2004). Functional stoichiometry and local enrichment of calmodulin interacting with Ca2+ channels. *Science, 304*, 432–435. https://doi.org/10.1126/science.1093490

Mori, M. X., Vander Kooi, C. W., Leahy, D. J., & Yue, D. T. (2008). Crystal structure of the CaV2 IQ domain in complex with Ca2+/calmodulin: High-resolution mechanistic implications for channel regulation by Ca2+. *Structure, 16*, 607–620. https://doi.org/10.1016/j.str.2008.01.011

Muller, R., et al. (2012). Atropine-sensitive hippocampal theta oscillations are mediated by Cav2.3 R-type Ca(2)(+) channels. *Neuroscience, 205*, 125–139. https://doi.org/10.1016/j.neuroscience.2011.12.032

Nieto-Rostro, M., Ramgoolam, K., Pratt, W. S., Kulik, A., & Dolphin, A. C. (2018). Ablation of alpha2delta-1 inhibits cell-surface trafficking of endogenous N-type calcium channels in the pain pathway in vivo. *Proceedings of the National Academy of Sciences of the United States of America, 115*, E12043–E12052. https://doi.org/10.1073/pnas.1811212115

Nystoriak, M. A., Murakami, K., Penar, P. L., & Wellman, G. C. (2009). Ca(v)1.2 splice variant with exon 9* is critical for regulation of cerebral artery diameter. *American Journal of Physiology. Heart and Circulatory Physiology, 297*, H1820–H1828. https://doi.org/10.1152/ajpheart.00326.2009

Ohlson, J., Pedersen, J. S., Haussler, D., & Ohman, M. (2007). Editing modifies the GABA(A) receptor subunit alpha3. *RNA, 13*, 698–703. https://doi.org/10.1261/rna.349107

Pan, J. Q., & Lipscombe, D. (2000). Alternative splicing in the cytoplasmic II-III loop of the N-type Ca channel alpha 1B subunit: Functional differences are beta subunit-specific. *The Journal of Neuroscience, 20*, 4769–4775.

Papa, A., et al. (2021). Adrenergic CaV1.2 activation via Rad phosphorylation converges at alpha1C I-II loop. *Circulation Research, 128*, 76–88. https://doi.org/10.1161/CIRCRESAHA.120.317839

Pennartz, C. M., de Jeu, M. T., Bos, N. P., Schaap, J., & Geurtsen, A. M. (2002). Diurnal modulation of pacemaker potentials and calcium current in the mammalian circadian clock. *Nature, 416*, 286–290, nature728 [pii]. https://doi.org/10.1038/nature728

Pereverzev, A., et al. (1998). Structural diversity of the voltage-dependent Ca2+ channel alpha1E-subunit. *The European Journal of Neuroscience, 10*, 916–925.

Pereverzev, A., et al. (2002). Alternate splicing in the cytosolic II-III loop and the carboxy terminus of human E-type voltage-gated Ca(2+) channels: Electrophysiological characterization of isoforms. *Molecular and Cellular Neurosciences, 21*, 352–365.

Peterson, B. Z., DeMaria, C. D., Adelman, J. P., & Yue, D. T. (1999). Calmodulin is the Ca2+ sensor for Ca2+-dependent inactivation of L-type calcium channels. *Neuron, 22*, 549–558.

Pinggera, A., & Striessnig, J. (2016). Ca(v) 1.3 (CACNA1D) L-type Ca(2+) channel dysfunction in CNS disorders. *The Journal of Physiology, 594*, 5839–5849.

Pinggera, A., et al. (2015). CACNA1D de novo mutations in autism spectrum disorders activate Cav1.3 L-type calcium channels. *Biological Psychiatry, 77*, 816–822.

Pitt, G. S., et al. (2001). Molecular basis of calmodulin tethering and Ca2+-dependent inactivation of L-type Ca2+ channels. *The Journal of Biological Chemistry, 276*, 30794–30802. https://doi.org/10.1074/jbc. M104959200

Plant, T. D. (1988). Properties and calcium-dependent inactivation of calcium currents in cultured mouse pancreatic B-cells. *The Journal of Physiology, 404*, 731–747.

Platzer, J., et al. (2000). Congenital deafness and sinoatrial node dysfunction in mice lacking class D L-type Ca2+ channels. *Cell, 102*, 89–97, S0092-8674(00)00013-1 [pii]. https://doi.org/10.1016/s0092-8674(00)00013-1

Pratt, P. F., Bonnet, S., Ludwig, L. M., Bonnet, P., & Rusch, N. J. (2002). Upregulation of L-type Ca2+ channels in mesenteric and skeletal arteries of SHR. *Hypertension, 40*, 214–219. https://doi. org/10.1161/01.hyp.0000025877.23309.36

Qu, Y., Baroudi, G., Yue, Y., El-Sherif, N., & Boutjdir, M. (2005). Localization and modulation of {alpha}1D (Cav1.3) L-type Ca channel by protein kinase A. *American Journal of Physiology. Heart and Circulatory Physiology, 288*, H2123–H2130. https:// doi.org/10.1152/ajpheart.01023.2004

Ramadan, O., et al. (2009). Phosphorylation of the consensus sites of protein kinase A on alpha1D L-type calcium channel. *The Journal of Biological Chemistry, 284*, 5042–5049. https://doi.org/10.1074/ jbc.M809132200

Ramakrishnan, N. A., et al. (2002). Voltage-gated Ca2+ channel Ca(V)1.3 subunit expressed in the hair cell epithelium of the sacculus of the trout Oncorhynchus mykiss: Cloning and comparison across vertebrate classes. *Brain Research. Molecular Brain Research, 109*, 69–83.

Roberts, T. C., Langer, R., & Wood, M. J. A. (2020). Advances in oligonucleotide drug delivery. *Nature Reviews. Drug Discovery, 19*, 673–694.

Saegusa, H., et al. (2000). Altered pain responses in mice lacking alpha 1E subunit of the voltage-dependent Ca2+ channel. *Proceedings of the National Academy of Sciences of the United States of America, 97*, 6132–6137.

Saegusa, H., et al. (2001). Suppression of inflammatory and neuropathic pain symptoms in mice lacking the N-type Ca2+ channel. *The EMBO Journal, 20*, 2349–2356.

Saegusa, H., et al. (2007). Properties of human Cav2.1 channel with a spinocerebellar ataxia type 6 mutation expressed in Purkinje cells. *Molecular and Cellular Neurosciences, 34*, 261–270. https://doi.org/10.1016/j. mcn.2006.11.006

Satheesh, S. V., et al. (2012). Retrocochlear function of the peripheral deafness gene Cacna1d. *Human Molecular Genetics, 21*, 3896–3909.

Schneider, T., Neumaier, F., Hescheler, J., & Alpdogan, S. (2020). Cav2.3 R-type calcium channels: From its discovery to pathogenic de novo CACNA1E variants: A historical perspective. *Pflügers Archiv, 472*, 811–816. https://doi.org/10.1007/s00424-020-02395-0

Scholl, U. I., et al. (2013). Somatic and germline CACNA1D calcium channel mutations in aldosterone-producing adenomas and primary aldosteronism. *Nature Genetics, 45*, 1050–1054.

Seino, S., et al. (1992). Cloning of the alpha 1 subunit of a voltage-dependent calcium channel expressed in pancreatic beta cells. *Proceedings of the National Academy of Sciences of the United States of America, 89*, 584–588.

Shen, Y., et al. (2006). Alternative splicing of the Ca(v)1.3 channel IQ domain, a molecular switch for Ca2+-dependent inactivation within auditory hair cells. *The Journal of Neuroscience, 26*, 10690–10699, 26/42/10690 [pii]. https://doi.org/10.1523/ JNEUROSCI.2093-06.2006

Singh, A., et al. (2006). C-terminal modulator controls Ca2+-dependent gating of Ca(v)1.4 L-type Ca2+ channels. *Nature Neuroscience, 9*, 1108–1116.

Singh, A., et al. (2008). Modulation of voltage- and Ca2+-dependent gating of CaV1.3 L-type calcium channels by alternative splicing of a C-terminal regulatory domain. *The Journal of Biological Chemistry, 283*, 20733–20744. https://doi.org/10.1074/jbc. M802254200

Sinnegger-Brauns, M. J., et al. (2004). Isoform-specific regulation of mood behavior and pancreatic beta cell and cardiovascular function by L-type Ca 2+ channels. *The Journal of Clinical Investigation, 113*, 1430–1439. https://doi.org/10.1172/JCI20208

Song, H., Nie, L., Rodriguez-Contreras, A., Sheng, Z. H., & Yamoah, E. N. (2003). Functional interaction of auxiliary subunits and synaptic proteins with Ca(v)1.3 may impart hair cell Ca2+ current properties. *Journal of Neurophysiology, 89*, 1143–1149. https://doi. org/10.1152/jn.00482.2002

Soong, T. W., et al. (2002). Systematic identification of splice variants in human P/Q-type channel alpha1(2.1) subunits: Implications for current density and Ca2+-dependent inactivation. *Journal of Neuroscience, 22*, 10142–10152, 22/23/10142 [pii]. https://doi. org/10.1523/JNEUROSCI.22-23-10142.2002

Splawski, I., et al. (2004). Ca(V)1.2 calcium channel dysfunction causes a multisystem disorder including arrhythmia and autism. *Cell, 119*, 19–31,

S0092867404008426 [pii]. https://doi.org/10.1016/j.cell.2004.09.011

Splawski, I., et al. (2005). Severe arrhythmia disorder caused by cardiac L-type calcium channel mutations. *Proceedings of the National Academy of Sciences of the United States of America, 102*, 8089–8096; discussion 8086–8088, 0502506102 [pii]. https://doi.org/10.1073/pnas.0502506102

Stea, A., et al. (1994). Localization and functional properties of a rat brain alpha 1A calcium channel reflect similarities to neuronal Q- and P-type channels. *Proceedings of the National Academy of Sciences of the United States of America, 91*, 10576–10580.

Strom, T. M., et al. (1998). An L-type calcium-channel gene mutated in incomplete X-linked congenital stationary night blindness. *Nature Genetics, 19*, 260–263.

Sultana, N., et al. (2016). Restricting calcium currents is required for correct fiber type specification in skeletal muscle. *Development, 143*, 1547–1559. https://doi.org/10.1242/dev.129676

Swartz, K. J. (2008). Sensing voltage across lipid membranes. *Nature, 456*, 891–897. https://doi.org/10.1038/nature07620

Tadross, M. R., Dick, I. E., & Yue, D. T. (2008). Mechanism of local and global Ca2+ sensing by calmodulin in complex with a Ca2+ channel. *Cell, 133*, 1228–1240. https://doi.org/10.1016/j.cell.2008.05.025

Tan, B. Z., et al. (2011). Functional characterization of alternative splicing in the C terminus of L-type CaV1.3 channels. *The Journal of Biological Chemistry, 286*, 42725–42735.

Tan, G. M., Yu, D., Wang, J., & Soong, T. W. (2012). Alternative splicing at C terminus of Ca(V)1.4 calcium channel modulates calcium-dependent inactivation, activation potential, and current density. *The Journal of Biological Chemistry, 287*, 832–847.

Tang, Z. Z., et al. (1783). Differential splicing patterns of L-type calcium channel Cav1.2 subunit in hearts of spontaneously hypertensive rats and Wistar Kyoto rats. *Biochimica et Biophysica Acta, 118-130*, 2008. https://doi.org/10.1016/j.bbamcr.2007.11.003

Tang, Z. Z., et al. (2004). Transcript scanning reveals novel and extensive splice variations in human l-type voltage-gated calcium channel, Cav1.2 alpha1 subunit. *The Journal of Biological Chemistry, 279*, 44335–44343, M407023200 [pii]. https://doi.org/10.1074/jbc.M407023200

Tang, Z. Z., Hong, X., Wang, J., & Soong, T. W. (2007). Signature combinatorial splicing profiles of rat cardiac- and smooth-muscle Cav1.2 channels with distinct biophysical properties. *Cell Calcium, 41*, 417–428. https://doi.org/10.1016/j.ceca.2006.08.002

Tang, Z. Z., Zheng, S., Nikolic, J., & Black, D. L. (2009). Developmental control of CaV1.2 L-type calcium channel splicing by Fox proteins. *Molecular and Cellular Biology, 29*, 4757–4765, MCB.00608-09 [pii]. https://doi.org/10.1128/MCB.00608-09

Tang, Z. Z., et al. (2011). Regulation of the mutually exclusive exons 8a and 8 in the CaV1.2 calcium channel transcript by polypyrimidine tract-binding protein.

The Journal of Biological Chemistry, 286, 10007–10016, M110.208116 [pii]. https://doi.org/10.1074/jbc.M110.208116

Tang, Z. Z., et al. (2012). Muscle weakness in myotonic dystrophy associated with misregulated splicing and altered gating of Ca(V)1.1 calcium channel. *Human Molecular Genetics, 21*, 1312–1324. https://doi.org/10.1093/hmg/ddr568

Thaler, C., Gray, A. C., & Lipscombe, D. (2004). Cumulative inactivation of N-type CaV2.2 calcium channels modified by alternative splicing. *Proceedings of the National Academy of Sciences of the United States of America, 101*, 5675–5679.

Thalhammer, A., et al. (2017). Alternative splicing of P/Q-type Ca(2+) channels shapes presynaptic plasticity. *Cell Reports, 20*, 333–343. https://doi.org/10.1016/j.celrep.2017.06.055

Thomas, J. R., Hagen, J., Soh, D., & Lee, A. (2018). Molecular moieties masking Ca(2+)-dependent facilitation of voltage-gated Cav2.2 Ca(2+) channels. *The Journal of General Physiology, 150*, 83–94. https://doi.org/10.1085/jgp.201711841

Tiwari, S., Zhang, Y., Heller, J., Abernethy, D. R., & Soldatov, N. M. (2006). Atherosclerosis-related molecular alteration of the human CaV1.2 calcium channel alpha1C subunit. *Proceedings of the National Academy of Sciences of the United States of America, 103*, 17024–17029. https://doi.org/10.1073/pnas.0606539103

Todorov, B., et al. (2012). Purkinje cell-specific ablation of Cav2.1 channels is sufficient to cause cerebellar ataxia in mice. *Cerebellum, 11*, 246–258. https://doi.org/10.1007/s12311-011-0302-1

Tuluc, P., et al. (2009). A CaV1.1 Ca2+ channel splice variant with high conductance and voltage-sensitivity alters EC coupling in developing skeletal muscle. *Biophysical Journal, 96*, 35–44. https://doi.org/10.1016/j.bpj.2008.09.027

Urbano, F. J., et al. (2003). Altered properties of quantal neurotransmitter release at endplates of mice lacking P/Q-type Ca2+ channels. *Proceedings of the National Academy of Sciences of the United States of America, 100*, 3491–3496. https://doi.org/10.1073/pnas.0437991100

Vandael, D. H., Zuccotti, A., Striessnig, J., & Carbone, E. (2012). Ca(V)1.3-driven SK channel activation regulates pacemaking and spike frequency adaptation in mouse chromaffin cells. *The Journal of Neuroscience, 32*, 16345–16359.

Wahl-Schott, C., et al. (2006). Switching off calcium-dependent inactivation in L-type calcium channels by an autoinhibitory domain. *Proceedings of the National Academy of Sciences of the United States of America, 103*, 15657–15662.

Wang, D., Papp, A. C., Binkley, P. F., Johnson, J. A., & Sadee, W. (2006). Highly variable mRNA expression and splicing of L-type voltage-dependent calcium channel alpha subunit 1C in human heart tissues. *Pharmacogenetics and Genomics, 16*, 735–745. https://doi.org/10.1097/01.fpc.0000230119.34205.8a

Wang, J., et al. (2011). Splice variant specific modulation of CaV1.2 calcium channel by galectin-1 regulates arterial constriction. *Circulation Research, 109*, 1250–1258. https://doi.org/10.1161/CIRCRESAHA.111.248849

Wang, J., et al. (2018). Characterization of CaV1.2 exon 33 heterozygous knockout mice and negative correlation between Rbfox1 and CaV1.2 exon 33 expressions in human heart failure. *Channels (Austin, Tex.), 12*, 51–57. https://doi.org/10.1080/19336950.2017.1381805

Watase, K., et al. (2008). Spinocerebellar ataxia type 6 knockin mice develop a progressive neuronal dysfunction with age-dependent accumulation of mutant CaV2.1 channels. *Proceedings of the National Academy of Sciences of the United States of America, 105*, 11987–11992. https://doi.org/10.1073/pnas.0804350105

Weiergraber, M., Henry, M., Radhakrishnan, K., Hescheler, J., & Schneider, T. (2007). Hippocampal seizure resistance and reduced neuronal excitotoxicity in mice lacking the Cav2.3 E/R-type voltage-gated calcium channel. *Journal of Neurophysiology, 97*, 3660–3669. https://doi.org/10.1152/jn.01193.2006

Welling, A., et al. (1997). Alternatively spliced IS6 segments of the alpha 1C gene determine the tissue-specific dihydropyridine sensitivity of cardiac and vascular smooth muscle L-type Ca2+ channels. *Circulation Research, 81*, 526–532.

Weyn-Vanhentenryck, S. M., et al. (2014). HITS-CLIP and integrative modeling define the Rbfox splicing-regulatory network linked to brain development and autism. *Cell Reports, 6*, 1139–1152. https://doi.org/10.1016/j.celrep.2014.02.005

Wheeler, D. G., Barrett, C. F., Groth, R. D., Safa, P., & Tsien, R. W. (2008). CaMKII locally encodes L-type channel activity to signal to nuclear CREB in excitation-transcription coupling. *The Journal of Cell Biology, 183*, 849–863. https://doi.org/10.1083/jcb.200805048

Williams, M. E., et al. (1992a). Structure and functional expression of alpha 1, alpha 2, and beta subunits of a novel human neuronal calcium channel subtype. *Neuron, 8*, 71–84.

Williams, M. E., et al. (1992b). Structure and functional expression of an omega-conotoxin-sensitive human N-type calcium channel. *Science, 257*, 389–395.

Williams, M. E., et al. (1994). Structure and functional characterization of neuronal alpha 1E calcium channel subtypes. *The Journal of Biological Chemistry, 269*, 22347–22357.

Wu, L. G., & Saggau, P. (1994). Pharmacological identification of two types of presynaptic voltage-dependent calcium channels at CA3-CA1 synapses of the hippocampus. *The Journal of Neuroscience, 14*, 5613–5622.

Wu, F., et al. (2011). A sodium channel knockin mutant (NaV1.4-R669H) mouse model of hypokalemic periodic paralysis. *The Journal of Clinical Investigation, 121*, 4082–4094.

Xie, C., Zhen, X. G., & Yang, J. (2005). Localization of the activation gate of a voltage-gated Ca2+ channel. *The Journal of General Physiology, 126*, 205–212. https://doi.org/10.1085/jgp.200509293

Xu, W., & Lipscombe, D. (2001). Neuronal Ca(V)1.3alpha(1) L-type channels activate at relatively hyperpolarized membrane potentials and are incompletely inhibited by dihydropyridines. *The Journal of Neuroscience, 21*, 5944–5951.

Xu, M., Welling, A., Paparisto, S., Hofmann, F., & Klugbauer, N. (2003). Enhanced expression of L-type Cav1.3 calcium channels in murine embryonic hearts from Cav1.2-deficient mice. *The Journal of Biological Chemistry, 278*, 40837–40841. https://doi.org/10.1074/jbc.M307598200

Yang, P. S., et al. (2006). Switching of Ca2+-dependent inactivation of Ca(v)1.3 channels by calcium binding proteins of auditory hair cells. *The Journal of Neuroscience, 26*, 10677–10689. https://doi.org/10.1523/jneurosci.3236-06.2006

Yazawa, M., et al. (2011). Using induced pluripotent stem cells to investigate cardiac phenotypes in Timothy syndrome. *Nature, 471*, 230–234. https://doi.org/10.1038/nature09855

Zhang, H., et al. (2005). Association of CaV1.3 L-type calcium channels with Shank. *The Journal of Neuroscience, 25*, 1037–1049. https://doi.org/10.1523/jneurosci.4554-04.2005

Zhang, H., et al. (2006). Ca1.2 and CaV1.3 neuronal L-type calcium channels: Differential targeting and signaling to pCREB. *The European Journal of Neuroscience, 23*, 2297–2310, EJN4734 [pii]. https://doi.org/10.1111/j.1460-9568.2006.04734.x

Zhang, J., Bahi, N., Llovera, M., Comella, J. X., & Sanchis, D. (2009). Polypyrimidine tract binding proteins (PTB) regulate the expression of apoptotic genes and susceptibility to caspase-dependent apoptosis in differentiating cardiomyocytes. *Cell Death and Differentiation, 16*, 1460–1468. https://doi.org/10.1038/cdd.2009.87

Zhang, H. Y., Liao, P., Wang, J. J., Yu, D. J., & Soong, T. W. (2010). Alternative splicing modulates diltiazem sensitivity of cardiac and vascular smooth muscle Ca(v)1.2 calcium channels. *British Journal of Pharmacology, 160*, 1631–1640. https://doi.org/10.1111/j.1476-5381.2010.00798.x

Zhou, Y., et al. (2017). Aberrant splicing induced by dysregulated Rbfox2 produces enhanced function of CaV1.2 calcium channel and vascular myogenic tone in hypertension. *Hypertension, 70*, 1183–1192. https://doi.org/10.1161/HYPERTENSIONAHA.117.09301

Zhuchenko, O., et al. (1997). Autosomal dominant cerebellar ataxia (SCA6) associated with small polyglutamine expansions in the alpha 1A-voltage-dependent calcium channel. *Nature Genetics, 15*, 62–69. https://doi.org/10.1038/ng0197-62

Zuhlke, R. D., Pitt, G. S., Deisseroth, K., Tsien, R. W., & Reuter, H. (1999). Calmodulin supports both inactivation and facilitation of L-type calcium channels. *Nature, 399*, 159–162. https://doi.org/10.1038/20200

Voltage-Gated Calcium Channel Auxiliary β Subunits

Sergej Borowik and Henry M. Colecraft

Abstract

High-voltage-activated Ca^{2+} channels (HVACCs) convert information encoded in action potentials into Ca^{2+} fluxes that control critical biological processes, such as muscle contraction, neurotransmitter or hormone release, and regulation of gene expression. HVACCs are hetero-multimeric proteins comprised minimally of a pore-forming α_1 subunit assembled with auxiliary cytosolic β and extracellular $\alpha_2\delta$ subunits. There are four distinct β subunit isoforms with multiple splice variants that are differentially expressed in different tissues and which exhibit some overlapping as well as unique physiological functions. The different $Ca_V\beta$s share a conserved central *src* homology 3 (SH3) and catalytically inactive guanylate kinase (GK) domain as well as variable N- and C-termini, and middle HOOK region. The conserved SH3-GK module is shared in common with the membrane-associated guanylate kinase (MAGUK) family of scaffold proteins that organize intracellular signaling pathways. $Ca_V\beta$s are important for surface trafficking of pore-forming α_1 subunits, and also regulate distinct aspects of channel gating. Some intracellular proteins, such as RGK proteins and Rab3-interacting molecule (RIM), regulate HVACCs via interacting with $Ca_V\beta$ subunits. $Ca_V\beta$ dysregulation is associated with human diseases, and they have been targeted with small molecules as well as engineered proteins to develop HVACC inhibitors.

Keywords

Voltage-gated calcium channel · Beta subunit · Ion channel regulation · RGK protein

S. Borowik
Department of Physiology and Cellular Biophysics, College of Physicians and Surgeons, Columbia University Medical Center, New York, NY, USA

H. M. Colecraft (✉)
Department of Physiology and Cellular Biophysics, College of Physicians and Surgeons, Columbia University Medical Center, New York, NY, USA

Department of Molecular Pharmacology and Therapeutics, College of Physicians and Surgeons, Columbia University Medical Center, New York, NY, USA
e-mail: hc2405@cumc.columbia.edu

Physiological Roles of High-Voltage-Activated Calcium Channels

High-voltage-activated Ca^{2+} channels (HVACCs) are a family of ion channel proteins that primarily function in the plasma membrane as passageways for Ca^{2+} influx into excitable cells to regulate their excitability and to convert action potentials into physiological responses (Fig. 1). There are

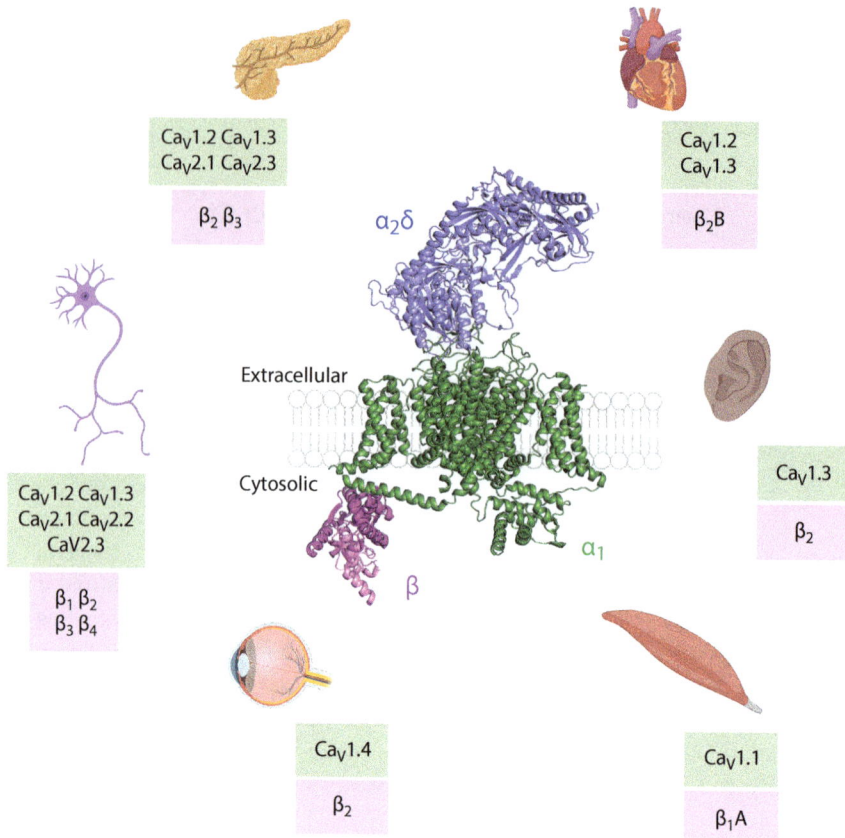

Fig. 1 Cryogenic electron microscopy structure of the mammalian voltage-gated calcium channel $Ca_V1.1$ complex (PDB accession code 5GJV) and distinctive distribution of HVACCs and $Ca_V\beta$ subunits in a variety of different tissues. (From BioRender)

seven distinct HVACCs ($Ca_V1.1$–$Ca_V1.4$; $Ca_V2.1$–$Ca_V2.3$), named on the basis of the pore-forming α_1 subunits. In neurons, Ca^{2+} influx through different classes of Ca_V1/Ca_V2 mediates neurotransmitter release at synapses, couples excitation to changes in gene expression, and regulates excitability by controlling Ca^{2+}-activated K^+ channels (Marrion & Tavalin, 1998; Dolphin & Lee, 2020). In skeletal muscle, sarcolemmal depolarization produces a change in $Ca_V1.1$ conformation that is mechanically transmitted to open intracellular type 1 ryanodine receptor (RyR1) channels and releases Ca^{2+} from intracellular stores to cause muscle contraction (Bannister & Beam, 2013). In the heart, sarcolemma depolarization opens $Ca_V1.2$ channels to let in a small amount of Ca^{2+} that triggers a larger Ca^{2+} release from sarcoplasmic reticulum stores through RyR2 intracellular Ca^{2+} release channels. This Ca^{2+}-induced Ca^{2+} release underlies cardiac excitation–contraction coupling (Bers, 2002). In endocrine cells, Ca^{2+} influx through distinct Ca_V1/Ca_V2 channels controls the release of hormones, including insulin in pancreatic beta cells and catecholamines from adrenal medulla chromaffin cells (Yang & Berggren, 2006; Marcantoni et al., 2010). $Ca_V1.3$ mediates synaptic transmission in inner hair cells in the ear and is necessary for hearing (Platzer et al., 2000), while $Ca_V1.4$ is present at ribbon synapses of photoreceptor cells where it enables neurotransmission that is important for normal visual function (Mansergh et al., 2005).

Subunits of Voltage-Gated Calcium Channels

Biochemical purification of the dihydropyridine (DPH) receptor from rabbit skeletal muscle identified a complex of three non-covalently linked subunits of 160, 53, and 32 kDa that were referred to as α, β, and γ, respectively (Curtis & Catterall, 1984). Further biochemical purification and characterization experiments of skeletal muscle dihydropyridine receptors delineated the presence of distinct α_1, $\alpha_2\delta$, β, and γ subunits (Leung et al., 1987; Takahashi et al., 1987). The α_1 subunit is an integral membrane protein that forms the channel pore and contains the voltage sensor, selectivity filter, and dihydropyridine binding site (Fig. 1). Topologically, it has four homologous domains (DI–DIV) each with six transmembrane segments (S1–S6). S1–S4 from DI–DIV form distinct voltage sensor domains, while S5–S6 segments form the channel pore and also contain the selectivity filter. DI–DIV are joined together by intracellular loops of varying lengths (I-II, II-III, and III-IV loops), and also preceded and followed by cytosolic N- and C-termini, respectively. There are seven distinct HVACC α_1 subunits that share a similar topology [$Ca_V1.1$ (α_{1S}); $Ca_V1.2$ (α_{1C}); $Ca_V1.3$ (α_{1D}); $Ca_V1.4$ (α_{1F}); $Ca_V2.1$ (α_{1A}); $Ca_V2.2$ (α_{1B}); and $Ca_V2.3$ (α_{1E})] all with multiple splice variants, distinctive tissue distributions, and functions (Fig. 1) (Zamponi et al., 2015). For additional information on the structure of α_1 subunits, please refer to chapter "Subunit Architecture and Atomic Structure of Voltage Gated Ca^{2+} Channels" by Dr. Catterall.

The $\alpha_2\delta$ subunit is a glycoprotein synthesized as a precursor protein that is post-translationally cleaved in the endoplasmic reticulum (ER) to generate separate α_2 and δ peptides. There are four distinct $\alpha_2\delta$ subunits ($\alpha_2\delta$-1–$\alpha_2\delta$-4) with characteristic distributions and functions. The α_2 protein is attached by disulfide bonds to the δ peptide which is tethered to the extracellular side of the plasma membrane by a glycosylphosphatidylinositol (GPI) anchor. $\alpha_2\delta$ subunits bind to extracellular loops of pore-forming α_1 subunits (Fig. 1) (Dolphin, 2013; Wu et al., 2016). For additional information on $\alpha_2\delta$ subunits, please

refer to chapter "Regulation of Calcium Channels and Synaptic Function by Auxiliary $\alpha_2\delta$ Subunits" by Drs. Dolphin and Obermair.

There are 8 distinct γ subunits (γ_1–γ_8) of which only γ_1 and γ_6 are believed to associate with HVGCCs (Chen et al., 2007). Structurally, the γ subunits have four transmembrane spanning regions and cytosolic N- and C-termini.

There are four $Ca_V\beta$ subunit isoforms (β1–β4) encoded by different genes – *CACNB1*, *CACNB2*, *CACNB3*, and *CACNB4* – and each with multiple splice variants (Buraei & Yang, 2010). Primary sequence alignments indicated $Ca_V\beta$ subunits displayed two conserved domains and three variable regions consisting of the N- and C-termini and a central HOOK region (Fig. 2a). Crystal structures elucidated that the $Ca_V\beta$-conserved regions comprise *src* homology 3 (SH3) and catalytically inactive guanylate-kinase (GK)-like domains, with the variable regions being unstructured (Chen et al., 2004; Opatowsky et al., 2004;

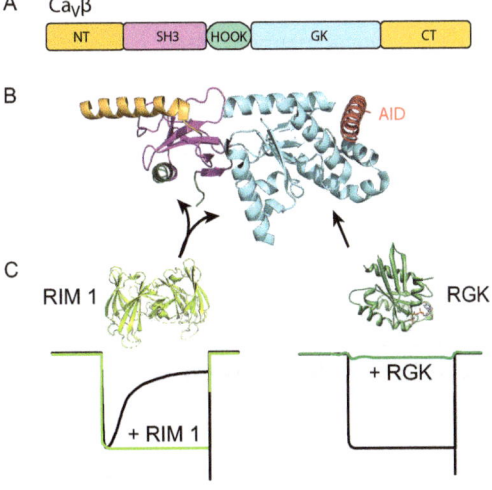

Fig. 2 (**a**) Topological domain structure of the $Ca_V\beta$ subunit. (**b**) Crystal structure of the $β_3$ core in complex with the AID (PDB accession code 1VYT). The following regions are highlighted: N-terminus in yellow, SH3 Domain in purple, HOOK domain in green, GK Domain in blue, and the AID region in red. Additionally, crystal structures of the $Ca_V\beta$ binding RIM1α C2B domain (PDB accession code 2Q3X) and the crystal structure of the mouse Rem2 GTPase (PDB accession code 3Q85) are shown. (**c**) Idealized current traces showing decreased inactivation with RIM1 bound to $Ca_V\beta$ (left) and deep current inhibition with RGK bound to $Ca_V\beta$ (right)

Van Petegem et al., 2004). The tandem SH3-GK module of $Ca_V\beta$s is shared by the superfamily of membrane-associated guanylate kinase (MAGUK) family of scaffold proteins that organize intracellular signaling pathways and include the postsynaptic density proteins PSD-95, PSD-93, SAP102, and SAP97 (Anderson, 1996; Craven, 1998).

$Ca_V\beta$s bind with nM affinity to a conserved 18-residue region, termed the α_1 interaction domain (AID), located in α_1-subunit intracellular I-II loops, utilizing an α_1-binding pocket (ABP) in the $Ca_V\beta$ GK domain (Pragnell et al., 1994; Chen et al., 2004; Opatowsky et al., 2004; Van Petegem et al., 2004). Consistent with their site of interaction with HVACC α_1 subunits, $Ca_V\beta$s are cytosolic proteins, except for two β_2 splice variants, β_{2a} and β_{2e}, which are autonomously membrane associated when expressed in cells (Chien et al., 1998; Takahashi et al., 2003). There are five β_2 variants (β_{2a}–β_{2e}) which are identical except for differences in their N-termini due to the use of alternate start sites (Takahashi et al., 2003). Membrane association of β_{2a} is enabled by post-translational palmitoylation of a cysteine residue in the N-terminus (Chien et al., 1996, 1998). By contrast, the N-terminus of β_{2e} forms an amphipathic helix that engages with the inner membrane using electrostatic interactions that is further stabilized by insertion of a tryptophan side chain into the lipid (Miranda-Laferte et al., 2014).

Physiological Roles of $Ca_V\beta$ Subunits

Data from knockout animals have been useful in discerning some of the physiological functions of HVACCs (Fig. 1). β_1-null mice die at birth due to asphyxiation as a consequence of β_{1a} being the sole $Ca_V\beta$ expressed in skeletal muscle (Gregg et al., 1996). β_1-null myotubes display a decrease in $Ca_V1.1$ expression and loss of excitation–contraction (EC) coupling. Similarly, loss of β_{1a} expression in zebrafish results in paralysis and death days after hatching due to loss of $Ca_V1.1$ assembly into tetrads and abolished EC coupling

(Schredelseker et al., 2005). Interestingly, replacement of β_{1a} with β_{2a} in the zebrafish led to full restoration of $Ca_V1.1$ charge movement; however, tetrad formation and EC coupling were still impaired (Schredelseker et al., 2009). Beyond skeletal muscle, β_1 splice variants are expressed in other tissues, including brain, spleen, and T cells, but their specific functional roles here are unknown (Buraei & Yang, 2010).

Knockout of $Ca_V\beta_2$ results in embryonic lethality in mice due to lack of proper heart development and cardiac contractions (Weissgerber et al., 2006). Transgenic expression of β_2 in β_2-knockout mice led to the rescue of viable animals, which, nevertheless, were deaf (Neef et al., 2009) and displayed visual deficits similar to patients with congenital stationary night blindness (CSNB2) (Ball et al., 2002).

$Ca_V\beta_3$-knockout mice are viable but displayed: a dampened response to pain initiated by chemical inflammation, while responses to thermal and mechanical stimuli were either moderately reduced or unaffected (Murakami et al., 2002); reduced L-type Ca^{2+} current in vascular smooth muscle and increased susceptibility to elevated blood pressure in response to a high salt diet (Murakami et al., 2003); enhanced Ca^{2+} oscillations and glucose-stimulated insulin release in pancreatic beta cells (Berggren et al., 2004); alterations in learning and memory tasks despite no changes in hippocampal neuron Ca^{2+} currents (Jeon et al., 2008); and decreased anxiety and increased aggression (Murakami et al., 2007).

A naturally occurring $Ca_V\beta_4$ knockout due to an insertion that causes exon skipping and a premature stop codon leads to the *lethargic* mouse phenotype characterized by ataxia and seizures (Burgess et al., 1997).

Functional Effects of $Ca_V\beta$ Subunits in HVACC Complexes

Heterologous expression studies have had a vital part in dissecting the functional roles and impact of $Ca_V\beta$ subunits on HVACC complexes. A key advantage of such studies is that the expression of HVACC subunits can be explicitly controlled and specified such that the functional contribution of individual subunits can be studied or deduced. Nevertheless, it is important to keep in mind that the intracellular milieu and nano-environment can have profound effects on the behavior of HVACC subunits; therefore, extrapolation of observations in heterologous studies to physiological interpretations in native cells is not always clear-cut and should be done with caution. In various heterologous cells, co-expressing $Ca_V\beta$s with pore-forming α_1 subunits results in a marked increase in whole-cell HVACC current (Singer et al., 1991; Jones et al., 1998; Takahashi et al., 2004; Leroy et al., 2005). In most cases, this is mediated at least in part by an increase in the surface density of channels. In addition, $Ca_V\beta$s modulate distinct aspects of channel gating, including regulating the voltage dependence of channel activation and steady-state inactivation and imparting distinctive inactivation signatures (Buraei & Yang, 2010).

$Ca_V\beta$ Regulation of HVACC Membrane Trafficking

Reconstitution experiments in mammalian heterologous expression systems generally find that $Ca_V\beta$s boost the trafficking of pore-forming α_1 subunits to the plasma membrane. Methods that have been employed to measure HVACC α_1-subunit surface density include detection of surface epitope-tagged channels by fluorescence or bioluminescence, gating current measurements, and surface biotinylation (Jones et al., 1998; Altier et al., 2002; Takahashi et al., 2004; Bourdin et al., 2010; Cassidy et al., 2014). In these studies, cells expressing α_1-subunit alone typically have a low surface density that is enhanced up to 12-fold by co-expressed $Ca_V\beta$ (Jones et al., 1998; Takahashi et al., 2004; Fang & Colecraft, 2011; Cassidy et al., 2014). By contrast, in *Xenopus* oocytes expression of $Ca_V1.2$ α_{1C} subunit alone yielded robust gating currents that were not further augmented by β_{2a} co-expression (Neely et al., 1993). This result may be explained by the expression of an endogenous $Ca_V\beta$ subunit in *Xenopus* oocytes at a concentration that is suffi-

cient to catalyze surface trafficking of α_1 subunits expressed following injection of cRNA (Tareilus et al., 1997; Canti et al., 2001).

The mechanism by which $Ca_V\beta$ subunits enhance surface density of HVACCs has been studied by several groups but there are still critical gaps in understanding this phenomenon. It was first suggested that binding of $Ca_V\beta$ to the AID masks an endoplasmic reticulum (ER) retention signal in the $Ca_V2.1$ I-II loop to promote forward trafficking of the channel (Bichet et al., 2000). However, no retention signal has been identified in the I-II loop of HVACC α_1 subunits, and deleting the AID does not promote α_1 trafficking to the surface (Maltez et al., 2005). Moreover, this model cannot account for observations that truncations in α_{1C} C-terminus that do not affect $Ca_V\beta$ binding, and point mutations in the C-terminus that disrupt apo-calmodulin but not $Ca_V\beta$ interaction, prevent $Ca_V1.2$ trafficking to the cell surface (Gao et al., 2000; Wang et al., 2007; Bourdin et al., 2010).

The Dolphin and Zamponi groups independently found that $Ca_V\beta$ binding to distinct α_1 subunits ($Ca_V2.2$ and $Ca_V1.2$, respectively) stabilized the protein by preventing ubiquitination and subsequent proteasomal degradation of the channel (Altier et al., 2011; Waithe et al., 2011). The two studies diverged on the impact of inhibiting the proteasome with small molecules on channel surface density. Waithe et al. found that inhibiting the proteasome with MG132 or lactacystin resulted in accumulation of ubiquitinated $Ca_V2.2$ which did not make it to the cell surface in tsA-201 cells in the absence of $Ca_V\beta$ (Waithe et al., 2011). By contrast, MG132 treatment was sufficient to promote surface trafficking of $Ca_V1.2$ in tsA-201 cells in the absence of $Ca_V\beta$, despite the α_{1C} protein being more ubiquitinated (Altier et al., 2011). It was not determined whether the $Ca_V1.2$ channels rescued by MG132 treatment gave rise to functional currents.

We undertook chimeric channel analyses to probe which intracellular regions of $Ca_V1.2$ were responsible for intracellular retention of the channel in the absence of $Ca_V\beta$, and were necessary for $Ca_V\beta$-dependent trafficking to the cell surface (Fang & Colecraft, 2011). We systematically swapped intracellular loops and termini of α_{1C} into the low-voltage-activated Ca^{2+} channel, $Ca_V3.1$ (α_{1G}), which shares the overall membrane topology of HVACC α_1 subunits, but does not require $Ca_V\beta$ for trafficking to the cell surface. We found that swapping in α_{1C} I-II loop into α_{1G} markedly increased ionic and gating currents, and this effect was abolished by mutating acidic residues just downstream of the AID to alanines. We thus concluded that the acidic region adjacent to the AID constituted an ER export signal. Interestingly, this putative ER export signal was found to be necessary for galectin interaction with $Ca_V1.2$ I-II loop (Wang et al., 2011). By contrast to the I-II loop, all other α_{1C} intracellular loops and termini caused a retention of the chimeric channel, albeit to different extents. When swapped in combination, the retention capabilities of these other α_{1C} intracellular regions were additive and opposed the export capability of the α_{1C} I-II loop. Finally, $Ca_V\beta$-dependent enhancement of current density was an emergent property that required at least four intracellular domains of α_{1C}, with the I-II loop and C-terminus being necessary. From these results, we proposed a model wherein in the absence of $Ca_V\beta$ the retention signals in $Ca_V1.2$ prevail to retain the channel inside the cell. When $Ca_V\beta$ binds to the AID, a conformational rearrangement of the intracellular domains occurs, with critical contributions from the I-II loop and C-terminus, and alters the balance of power between ER retention and export signals to favor forward trafficking of the channel to the cell surface. While consistent with the data obtained there are aspects of this model that remain to be explicitly verified. First, it is unknown what regions constitute the intracellular retention signals in $Ca_V1.2$ (and presumably other HVACC α_1 subunits). One possibility is that the retention is mediated by ubiquitination of specific lysine residues, in line with the observations that $Ca_V2.2$ and $Ca_V1.2$ undergo ubiquitination and proteasomal degradation in the absence of $Ca_V\beta$ (Altier et al., 2011; Waithe et al., 2011). Second, whether and how $Ca_V\beta$ binding to the AID induces conformational rearrangements of HVACC α_1-subunit intracellular loops and termini is yet to be directly observed. Unfortunately,

the bulk of these regions remains unresolved in available cryogenic electron microscopy (cryoEM) structures of HVACCs, presumably due to intrinsic mobility of these regions (Wu et al., 2016; Dong et al., 2021; Gao et al., 2021).

Heterologous reconstitution of HVACCs in mammalian cells established the paradigm that association with a $Ca_V\beta$ is obligatory for trafficking and functional expression of these channels. However, results from adult ventricular myocytes present a significant breach in this dogma. The first indication of this came from an experiment in which the dominant $Ca_V\beta$ in mouse heart, $Ca_V\beta_2$, was inducibly excised in adult heart cardiomyocytes using tamoxifen. Surprisingly, while cardiomyocyte $Ca_V\beta_2$ expression was decreased by >96% in the adult heart, the mice survived and displayed only a moderate <29% reduction in whole-cell L-type ($Ca_V1.2$) calcium current (Meissner et al., 2011). The primary conclusion of this result was corroborated by Marx and colleagues who used a transgenic mouse strategy to express a mutated $Ca_V1.2$ (α_{1C}) with mutations in the AID that prevented interaction with $Ca_V\beta$, and also with mutations that strongly reduce dihydropyridine sensitivity (Yang et al., 2019). Inhibition of endogenous $Ca_V1.2$ with nisoldipine revealed that the AID mutant dihydropyridine-resistant channels trafficked to the cell surface, gave rise to whole-cell currents, and effectively mediated cardiac excitation–contraction in adult heart ventricular cardiomyocytes (Yang et al., 2019). Notably, deleting cardiac β_2 has more severe consequences on the heart during embryonic development, resulting in compromised morphology and function of the heart and embryonic lethality (Weissgerber et al., 2006). Moreover, the embryonic cardiomyocytes from β_2-knockout mice displayed a substantial 75% decrease in $Ca_V1.2$ current compared to wild-type mice (Weissgerber et al., 2006). Thus, optimum functional expression of cardiac $Ca_V1.2$ in the mouse heart is highly dependent on $Ca_V\beta_2$ expression during development but dispensable in the adult heart. The mechanism for the switch in $Ca_V\beta$ dependence of cardiac $Ca_V1.2$ functional expression between embryonic and adult cardiomyocyte stages is unknown.

Possibilities include expression in the adult mouse heart of a protein that can replicate the $Ca_V1.2$ trafficking function of $Ca_V\beta_2$, or a change in the intracellular milieu that prevents the ER retention of α_{1C} that is not bound by $Ca_V\beta$. It is unknown whether similar $Ca_V\beta$-independent functional expression of HVACCs exists in other excitable cells at different developmental stages.

$Ca_V\beta$ Regulation of HVACC Activation Gating

Heterologous expression studies have consistently shown that $Ca_V\beta$ co-expression shifts the voltage dependence of channel activation in the hyperpolarizing direction for various HVACC types (Buraei & Yang, 2010; Neely & Hidalgo, 2014). In mammalian cells, the leftward shift is typically in the −5 to −20 mV range (Jones et al., 1998; Fang & Colecraft, 2011); in *Xenopus* oocytes, a wide variation in the magnitude of the hyperpolarizing shift has been reported for distinct α_1/β subunit combinations, ranging from ~ −8 mV for $\alpha_{1E} \pm \beta_{2a}$, ~ −15 mV for $\alpha_{1B}/\alpha_2\delta \pm \beta_3$, to a striking ~ −70 mV for $\alpha_{1C} \pm \beta_{2a}$ (De Waard & Campbell, 1995; Olcese et al., 1996; Canti et al., 2000, 2001). The impact of $Ca_V\beta$ on the voltage dependence of channel activation has been corroborated in knockout mice: rightward shifts in the voltage dependence of activation have been observed in $Ca_V1.1$ currents from β_1-null myotubes (Gregg et al., 1996); HVACC currents in DRG neurons from β_3-null mice (Murakami et al., 2002); and $Ca_V1.2$ currents recorded from embryonic cardiomyocytes in β_2-knockout mice (Weissgerber et al., 2006). What are the determinants of this β-mediated hyperpolarizing shift in voltage-dependent activation? A critical requirement is a rigid α-helical linker extending from α_1-subunit IS6 to the AID, the site for high-affinity $Ca_V\beta$ binding. The propensity of the IS6-AID linker to form an α-helix was first discerned in circular dichroism experiments (Arias et al., 2005), and has been confirmed by HVACC cryoEM structures (Wu et al., 2016; Gao et al., 2021). Replacing 6 amino acids with glycines disrupts the helix (Arias et al., 2005). In $Ca_V2.1$

and $Ca_V1.2$ channels with the IS6-AID helix interrupted by polyglycines, co-expressed $Ca_V\beta$s did not yield the hyperpolarizing shift in voltage-dependent activation, suggesting the necessity of this structural feature in mediating β effects on HVACC activation gating (Zhang et al., 2008; Findeisen & Minor, 2009; Buraei & Yang, 2010).

$Ca_V\beta$ Regulation of HVACC Open Probability

Binding of $Ca_V\beta$s to HVACC α_1 subunits typically results in an increase in channel open probability that has been demonstrated in two distinct ways. First, in whole-cell studies, the ratio of the maximal channel conductance (G_{max}) to maximal ON gating charge (Q_{max}) can provide an index of the relative open probability of an HVACC when compared under different conditions. These analyses when applied to $Ca_V2.3$ channels indicated an increase in the G_{max}/Q_{max} ratio when α_{1E} was co-expressed with $Ca_V\beta$ subunits, consistent with the auxiliary subunit producing an increase in channel P_o (Jones et al., 1998). Similarly, in $Ca_V1.2$ channels reconstituted with β_{2a} in HEK293 cells, scatter plots of G_{max} versus Q_{max} display a linear relationship with a slope that is higher than for channels reconstituted with α_{1C} alone, again consistent with a β-mediated increase in channel P_o (Takahashi et al., 2004).

Single-channel recordings offer the most direct way to assess the impact of β subunits on HVACC open probability. However, this type of experiment is complicated by several factors. First, the requirement for $Ca_V\beta$ for robust membrane trafficking of HVACCs in heterologous cells makes it technically challenging to record α_1-subunit alone channels. Second, most published single-channel recordings of HVACCs typically report P_o at one or two test pulse voltages that are well short of the voltage expected to elicit the maximum open probability ($P_{o,max}$). This is exacerbated by the fact that HVACC single-channel recordings are typically undertaken with high concentrations of charge carrier (e.g., 90 mM Ba^{2+}) that produce significant surface charge screening effects (Colecraft et al.,

2001). Thus, the test potentials at which the single-channel P_o of HVACCS expressed ±β expression are recorded is typically at the foot of the activation curve (Colecraft et al., 2001), and also do not take into account the leftward shifts in voltage-dependent activation produced by $Ca_V\beta s$. With these caveats in mind, there have been some single-channel studies in heterologous cells that suggest $Ca_V\beta$ increases P_o when co-expressed with HVACC α_1 subunits (Wakamori et al., 1999; Hullin et al., 2003; Herzig et al., 2007).

Single-channel recordings using ramp protocols of AID mutant dihydropyridine-resistant α_{1C} expressed in transgenic mouse hearts explicitly confirmed that β-bound $Ca_V1.2$ channels have a higher $P_{o,max}$ compared to α_{1C} alone channels (Papa et al., 2021). Several studies have also demonstrated that different $Ca_V\beta$ subunit isoforms or splice variants have distinctive effects on P_o of various HVACC subtypes (Colecraft et al., 2002; Hullin et al., 2003; Luvisetto et al., 2004; Herzig et al., 2007).

$Ca_V\beta$ Regulation of HVACC Inactivation Gating

HVACCs undergo various forms of inactivation that serve to limit Ca^{2+} entry into cells (Tadross et al., 2010; Zamponi et al., 2015). Inactivation of HVACCs is a crucial physiological phenomenon disruption of which leads to disease (Splawski et al., 2004, 2005). $Ca_V\beta$ subunits profoundly regulate different aspects of inactivation gating in HVACCs, including the voltage dependence of steady-state inactivation and the kinetics of voltage-dependent inactivation (VDI) during a test pulse. In heterologous expression studies, the impact of $Ca_V\beta s$ on the voltage dependence of steady-state inactivation of HVACCs depends on the identity of both the α_1 and $Ca_V\beta$ isoform. In Ca_V2 channels, β_1, β_3, and β_4 generally produce a hyperpolarizing shift in the voltage dependence of steady-state inactivation ranging from −5 to −30 mV when compared to channels that are not associated with a β subunit [reviewed in (Buraei & Yang, 2010)]. For β_2 subunits, the direction of the shift depends on the particular splice variant.

β_{2a} and β_{2e}, which are autonomously membrane-associated, produce depolarizing shifts in steady-state inactivation, whereas β_{2b}–β_{2d} behave similarly to β_1, β_3, and β_4 in yielding hyperpolarizing shifts. By contrast, the voltage dependence of $Ca_V1.2$ steady-state inactivation is relatively insensitive to differential regulation by distinct $Ca_V\beta$ subunits (Takahashi et al., 2003).

The rate of voltage-dependent inactivation of whole-cell currents elicited by a test pulse is also strongly influenced by the associated β subunit. In general, in reconstituted channel experiments, all β subunits speed rate of inactivation compared to α_1 alone channels except for β_{2a} and β_{2e} which strongly slow inactivation (Restituito et al., 2000; Takahashi et al., 2003; Buraei & Yang, 2010; Miranda-Laferte et al., 2014). Over-expression of distinct β subunits in adult rat ventricular cardiomyocytes imparts distinct rates of inactivation on endogenous $Ca_V1.2$ channels, with β_{2a} and β_4 producing a slowing of inactivation (Colecraft et al., 2002). Single-channel recordings revealed that β_{2a}-bound $Ca_V1.2$ channels in ventricular myocytes do not display microscopic inactivation and have virtually no null sweeps, in contrast to channels bound to β_{1b}, β_{2b}, or β_3 subunits (Colecraft et al., 2002). The impact of β_{2a} to slow the rate of HVACC inactivation is prevented by mutating two cysteines to serines (β_{2a}[C3,4S]) in the N-terminus that serve as sites for palmitoylation of the protein (Restituito et al., 2000). The slowed inactivation is retained when β_{2a}[C3,4S] is tethered to the membrane by adding a transmembrane segment to the N-terminus. Thus, membrane targeting, rather than palmitoylation per se, is responsible for the ability of β_{2a} to slow HVACC inactivation rates (Restituito et al., 2000). Consistent with this, the β_{2e} splice variant which also associates with the membrane also confers slow inactivation kinetics to HVACCs (Takahashi et al., 2003; Miranda-Laferte et al., 2014; Kim et al., 2016). The impact of β_{2a} to slow the inactivation of $Ca_V1.2$ is lessened by disrupting the α_1-subunit IS6 to AID helix with glycine residues, further supporting the importance of this structural feature to $Ca_V\beta$ regulation of distinct aspects of HVACC gating (Findeisen & Minor, 2009).

Structure–Function of $Ca_V\beta$ Regulation of HVACCs

The structure of $Ca_V\beta$ subunits suggests a structure–function paradigm in which the conserved SH3-GK module mediates functions common to all $Ca_V\beta$s (e.g., trafficking and regulation of activation gating), whereas the non-conserved N- and C-termini and middle HOOK region may contribute to isoform-specific functions (e.g., voltage dependence of steady-state inactivation and kinetics of inactivation) (Fig. 2a). There are some discrepancies in the literature regarding which parts of $Ca_V\beta$s are necessary and sufficient to fully reconstitute the trafficking and regulation of activation functions common to all $Ca_V\beta$s. Using reconstitution experiments in HEK293 cells, we found that introducing point mutations in either the SH3 (L93P in $Ca_V\beta_{2a}$) or GK (P234R) domains severely blunted (in the case of L93P) or prevented (in the case of P234R) $Ca_V\beta_{2a}$-mediated increase in whole-cell $Ca_V1.2$ current (Takahashi et al., 2004). Both mutations compromised the ability of $Ca_V\beta_{2a}$ to mediate channel trafficking to the cell surface, as indicated by gating charge measurements (Q_{max}). Co-expression of L93P and P234R resulted in full recovery of both whole-cell current and Q_{max}, indicating complementation. Consistent with this interpretation, split-$Ca_V\beta_{2a}$ containing either the SH3 or GK domain alone were ineffective in rescuing $Ca_V1.2$ trafficking and current when they were separately co-expressed with the channel, but fully recapitulated the whole-cell current when co-transfected together (Takahashi et al., 2004). The separated SH3 and GK domains interact when co-expressed, and this is dependent on a β5-strand (PYDVV) that follows the HOOK but is part of the SH3 domain (Chen et al., 2004; Opatowsky et al., 2004; Van Petegem et al., 2004; Takahashi et al., 2005). Deleting the β5-strand prevented the split SH3 and GK domains from interacting and ablated their capacity to traffic $Ca_V1.2$ and reconstitute current when co-expressed together (Takahashi et al., 2005). McGee et al. similarly found mutations in SH3 and GK that

disrupt the intramolecular interaction and showed complementation when co-expressed (McGee et al., 2001). In contrast to our results they showed that $\beta_{2a}[L93P]$ when expressed in *Xenopus* oocytes gave rise to current with similar amplitude to those reconstituted with WT β_{2a}. This discrepancy is likely explained by the presence of endogenous $Ca_V\beta$ in *Xenopus* oocytes at a concentration sufficient to traffic expressed HVACC α_1 subunits to the surface (Neely et al., 1993; Tareilus et al., 1997; Canti et al., 2001). In HEK293 cells the $Ca_V1.2$ channels reconstituted with $\beta_{2a}[L93P]$ that did make it to the cell surface displayed a voltage dependence of activation similar to channels reconstituted with WT β_{2a} (Takahashi et al., 2004) in agreement with observations in *Xenopus* oocytes (McGee et al., 2001).

The Dolphin group found that constructs containing β_{2a} or β_{1b} GK domains but lacking SH3 were able to reconstitute $Ca_V2.2$ currents in tsA201 cells, albeit at a significantly decreased current density compared to WT $Ca_V\beta$ subunits (Dresviannikov et al., 2009). A complication here is the inclusion of $\alpha_2\delta$-1 subunit, which is able to markedly augment surface expression of $Ca_V2.2$ (Cassidy et al., 2014), making it difficult to interpret which part of the current rescue is due to the co-expressed truncated $Ca_V\beta$ subunits (Dresviannikov et al., 2009). Similarly, He et al. found that β_3 GK domain reconstituted $Ca_V1.2$ currents similar to WT β_3 in HEK293T cells in the presence of $\alpha_2\delta$ expression (He et al., 2007). Overall, there remain ambiguities about the sufficiency of the $Ca_V\beta$ GK domain to fully reconstitute the trafficking functions of $Ca_V\beta$ subunits in heterologous expression systems. Factors contributing to the uncertainties include the presence of endogenous β subunits in *Xenopus* oocytes, the inclusion of $\alpha_2\delta$ subunits in some studies, and indications that the GK domain may have decreased stability when expressed in cells (Maltez et al., 2005; Gonzalez-Gutierrez et al., 2008).

There are hints that the SH3 domain may interact with HVACC α_1 subunits, but the precise putative binding sites remain undefined (Maltez et al., 2005; Takahashi et al., 2005).

$Ca_V\beta$-Interacting Proteins That Regulate HVACCs

Fitting with their distant relation to the MAGUK superfamily of scaffold proteins, several distinct proteins have been shown to associate with auxiliary $Ca_V\beta$ subunits and through this interaction be able to influence the activity or function of distinct HVACCs (Fig. 2b, c).

Regulation of HVACCs by RGK Proteins

RGK (Rad/Rem/Rem2/Gem/Kir) proteins are small G-proteins that belong to the Ras superfamily of monomeric GTPases. Gem was identified as a $Ca_V\beta$-binding protein in a yeast two-hybrid assay designed to fish out proteins in MIN6 cells that associate with $Ca_V\beta_3$ (Beguin et al., 2001). When co-expressed with reconstituted $Ca_V1.2$ and $Ca_V1.3$ channels in *Xenopus* oocytes, Gem virtually eliminated currents (Beguin et al., 2001). Subsequent experiments demonstrated that all four members of the RGK family potently and constitutively inhibit all HVACCS (Fig. 2c) (Finlin et al., 2003; Chen et al., 2005; Yang et al., 2007, 2010, 2012; Bannister et al., 2008; Wang et al., 2010; Xu et al., 2010). Surprisingly, the mechanisms by which RGK proteins inhibit HVACCs are varied and dependent on the identity of the GTPase, the channel subtype, and the cellular background (Yang & Colecraft, 2013). In HEK293 cells, Rem inhibited reconstituted $Ca_V1.2$ channels using three distinguishable mechanisms: (1) a decrease in the number of channels (N) at the cell surface due to enhanced dynamin-dependent endocytosis; (2) inhibition of the P_o of channels at the cell surface at a step beyond the movement of voltage sensors; and (3) inhibition of channel P_o by restraining voltage sensor movement (Yang et al., 2010). By contrast, over-expressing Rem2 in tsA201 or MIN6 cells inhibited $Ca_V2.2$ and $Ca_V1.2$ channels, respectively, without reducing channel surface density (Chen et al., 2005; Finlin et al., 2005; Flynn et al., 2008). Similarly, Rem

over-expression in cardiomyocytes strongly inhibited endogenous $Ca_V1.2$ channels; however, application of BAYK8644 yielded robust L-type calcium currents, indicating that the bulk of channels remained at the cell surface (Xu et al., 2010).

All four RGK proteins utilize a β-binding-dependent mechanism of channel inhibition. This is revealed by introducing point mutations in $Ca_V\beta$ GK domain (D243A, D319A, and D321A in β_{2a}) that selectively disrupt the interaction with RGKs without compromising the ability of $Ca_V\beta$ to regulate channel trafficking and gating (Beguin et al., 2007; Puckerin et al., 2018). However, Rem and Rad partially inhibited $Ca_V1.2$ channels (as well as $Ca_V2.2$ for Rad) reconstituted with β_{2a}[D243/319/321A], indicating an additional β-binding-independent mode of inhibition (Yang et al., 2012; Puckerin et al., 2018). In the case of Rem and $Ca_V1.2$, we found a direct interaction that involved Rem C-terminus binding to a distal region of the α_{1C} N-terminus (Yang et al., 2012; Puckerin et al., 2016). By contrast, Gem and Rem2 exclusively utilize a β-binding-dependent mechanism to inhibit Ca_V1/Ca_V2 channels (Yang et al., 2012; Puckerin et al., 2018).

Structurally, RGK proteins contain a central guanine nucleotide-binding domain (G-domain) that is conserved among the family of Ras-like G-proteins (Fig. 2b, c). In addition, by comparison to Ras, RGK proteins have N- and C-termini extensions, as well as non-conservative substitutions in the G-domain of residues important for GTP binding and hydrolysis (Colicelli, 2004; Yang & Colecraft, 2013). While no structures of $Ca_V\beta$ bound to an RGK protein have been obtained, extensive mutagenesis of both proteins and docking suggest RGK G-domains bind $Ca_V\beta$ GK domains at a surface distinct from the ABP which interacts with the AID (Fig. 2b, c) (Beguin et al., 2007). Mutations in Rem (R200A/L227A) and Rad (R208A/L235A) that disrupt the interaction with $Ca_V\beta$ yield proteins that are selective partial inhibitors of $Ca_V1.2$ and $Ca_V2.2$, owing to β-binding-independent inhibition (Puckerin et al., 2018). By contrast, similar mutations in Gem and Rem2 completely eliminate their capacity to inhibit HVACCs.

The C-terminus extensions of RGK proteins contain a distal conserved region separated from the respective G-domains by a variable linker sequence (12–22 residues). RGK distal C-termini contain basic and hydrophobic residues that attach these proteins to the inner leaflet of the plasma membrane. Deleting the distal C-terminus residues generates truncated RGK proteins that are cytosolic and no longer capable of inhibiting HVACCs (Finlin et al., 2000; Chen et al., 2005; Yang et al., 2007, 2010, 2013; Flynn et al., 2008; Puckerin et al., 2016). Redirecting C-terminus truncated RGK proteins back to the membrane using other membrane targeting motifs recovers the ability of these proteins to block HVACCs (Chen et al., 2005; Yang et al., 2007, 2010, 2013). Thus, membrane targeting and $Ca_V\beta$ binding is a dual requirement for β-binding-dependent inhibition of HVACCs by RGK proteins.

What is the physiological role and significance of RGK protein inhibition of HVACCs? This has been most clearly demonstrated for Rad inhibition of $Ca_V1.2$ channels in heart which we discuss next.

Roles of $Ca_V\beta$ and Rad in Sympathetic Regulation of Cardiac $Ca_V1.2$

Sympathetic up-regulation of cardiac $Ca_V1.2$ current via PKA-mediated phosphorylation is critical for the increased contractility of the heart during exercise or the fight-or-flight response (Reuter & Scholz, 1977; Cachelin et al., 1983; Bean et al., 1984; Tsien et al., 1986; Hartzell et al., 1991; Kamp & Hell, 2000). The mechanism of this physiologically crucial phenomenon was pursued for decades under the assumption it involved direct phosphorylation of $Ca_V1.2$ pore-forming α_{1C} and/or auxiliary β_{2b} subunits (Gao et al., 1997; Bunemann et al., 1999; Kamp & Hell, 2000; Ganesan et al., 2006; Lemke et al., 2008; Miriyala et al., 2008; Fuller et al., 2010; Brandmayr et al., 2012; Weiss et al., 2013; Fu et al., 2014; Hofmann et al., 2014). However, Marx and colleagues showed that transgenic

mice expressing DHP-resistant α_{1C} and β_{2B} with all potential consensus PKA consensus Ser and Thr residues mutated to Ala expressed channels that responded fully to PKA activation, explicitly ruling out phosphorylation sites on the channel subunits as being important for this modulation in heart (Liu et al., 2020). Nevertheless, transgenic mice expressing a DHP-resistant α_{1C}-AID mutant that could not bind $Ca_V\beta$ was still able to traffic to the cell surface in cardiomyocytes but was incapable of being up-regulated by PKA (Yang et al., 2019). Thus, in adult ventricular myocytes β binding to α_{1C} is not necessary for channel trafficking, but is obligatory for PKA modulation of $Ca_V1.2$. Using an ascorbate peroxidase (APEX2) proximity labeling method in transgenic mouse hearts combined with proteomics, they discovered that under basal conditions, Rad is enriched near $Ca_V1.2$ but is depleted from the channel neighborhood upon exposure to a β-adrenergic agonist (Liu et al., 2020). Moreover, co-expression of Rad with $Ca_V1.2$ in HEK293 cells enabled heterologous reconstitution of PKA modulation of L-type calcium current (Liu et al., 2020). This result has been replicated in *Xenopus* oocytes (Katz et al., 2021). The reconstituted modulation was prevented by mutation of Ser residues in Rad C-terminus (S272 and S300), and Forster resonance energy transfer (FRET) experiments indicated phosphorylation of Rad weakened its binding affinity for $Ca_V\beta$ (Liu et al., 2020). Altogether, these studies indicate that under basal conditions in cardiomyocytes a sub-population of $Ca_V1.2$ channels are held in a low P_o mode. Upon β-adrenergic stimulation, activated PKA phosphorylates Rad, including on the C-terminus, leading to disengagement from the membrane and a weakened affinity for β_2, resulting in relief of inhibition and enhancement of the whole-cell current (Liu et al., 2020; Papa et al., 2021). Consistent with the role of Rad as a constitutive inhibitor of cardiac $Ca_V1.2$ channels, Rad-knockout mice display a twofold increase in basal L-type current density in the heart, which is not further augmented by β-adrenergic activation (Ahern et al., 2019).

RIM-Binding Protein Regulation of HVACCs

A yeast two-hybrid screen of a mouse brain complementary DNA library using β_{4b} as bait identified the C-terminus of Rab3-interacting molecule 1 (RIM1) as an interacting partner (Fig. 2b) (Kiyonaka et al., 2007). The binding required an intact SH3-GK module as truncations or point mutations in either the SH3 or GK domain abolished the interaction with RIM1 (Kiyonaka et al., 2007). RIM1 belongs to a family of scaffold proteins that engage in protein–protein interactions in presynaptic active zones and are involved in Ca^{2+}-dependent neurotransmitter release in neurons (Sudhof, 2004; Schoch et al., 2006). Beyond neurons, RIM proteins are also present at the active zones in ribbon synapses of photoreceptors, inner hair cells in the cochlear, and other secretory cells (tom Dieck et al., 2005; Kiyonaka et al., 2007; Gebhart et al., 2010; Gandini & Felix, 2012). Structurally, RIM proteins are modular, containing a Zinc finger-like domain, a PDZ domain, a proline-rich region, and two C2 domains (C2A and C2B) (Gandini & Felix, 2012; Sudhof, 2012). RIM proteins interact with $Ca_V\beta s$ via the C2B domain at the C-terminus (Kiyonaka et al., 2007).

In heterologous expression studies RIM1 markedly reduced the rate of voltage-dependent inactivation and a substantial depolarizing shift in the voltage dependence of steady-state inactivation of $Ca_V2.1$–$Ca_V2.3$ and $Ca_V1.2$ channels reconstituted with β_{2a}, β_3, or β_4 subunits (Kiyonaka et al., 2007). In PC12 cells knockdown of endogenous RIM1 or expression of a dominant negative $Ca_V\beta$ construct that selectively disrupts the RIM1/β interaction results in a speeding up of the inactivation kinetics and hyperpolarizing shift in the voltage dependence of steady-state inactivation of endogenous HVACCs. Further, integrity of the RIM1-β interaction was found to enhance neurotransmitter release in PC12 cells (Kiyonaka et al., 2007). RIM2 was found co-expressed with $Ca_V1.3$ in the presynaptic compartment of cochlear inner hair cells; and in tsA201 cells co-expression of RIM2 with $Ca_V1.3$ also markedly suppressed the rate of inactivation and produced a depolarizing shift in the voltage dependence of steady-state inactivation (Gebhart et al., 2010). In addition to modulating channel gating, RIM binding to $Ca_V\beta$ aids in tethering HVACCs close to secretory vesicles (Kiyonaka et al., 2007).

Association of $Ca_V\beta$ Subunits with Disease

Given their central role in controlling physiological responses across many organs and tissues, it is not surprising that mutations in HVACC subunits are associated with a variety of diseases. Principally, this has been most prominently shown for pore-forming α_1 subunits: mutations in *CACNA1S* give rise to skeletal muscle disorders, including hypokalemic periodic paralysis type 1 and malignant hypothermia susceptibility type 5 (Flucher, 2020); mutations in *CACNA1C* are associated with cardiac rhythm abnormalities and also Timothy syndrome, a multisystemic disease characterized by a prolonged QT interval in the electrocardiogram, autism spectrum disorder (ASD), syndactyly, and facial dysmorphism (Splawski et al., 2004; Bauer et al., 2021); mutations in *CACNA1D* cause congenital deafness, bradycardia, and neurodevelopmental and endocrine disorders (Striessnig et al., 2010); *CACNA1F* mutations cause X-linked congenital stationary night blindness type 2 (CSNB2) and X-linked cone-rod dystrophy type 3; and mutations in *CACNA1A* lead to a variety of neurological diseases, such as epileptic encephalopathies, familial hemiplegic migraine type 1, episodic ataxia type 2, and spinocerebellar ataxia type 6.

Given the importance of auxiliary β subunits to the trafficking and gating modulation of HVACCs, it would be expected that mutations in them would also be linked with human diseases. Breitenkamp et al. identified rare missense mutations in *CANCB2* (G167S, S197F, and F240L) in patients with ASD but not in the control population. Recombinant $Ca_V1.2$ in HEK293 cells with either β_2[G167S] or β_2[S197F] displayed a significantly decelerated inactivation rate, reminiscent of gating changes in *CACNA1C* that cause Timothy syndrome (Breitenkamp et al., 2014).

Whole genome sequencing of 85 quartet families in which two siblings have ASD identified a *CACNB2* mutation (V2D) as an ASD-related gene (Yuen et al., 2015). Several mutations in $Ca_V\beta_{2b}$ have been linked to cardiac rhythm disorders, including Brugada and short QT syndromes (Burashnikov et al., 2010; Zhang et al., 2018). Mutations have been found distributed throughout the protein, though in most cases their impact on $Ca_V1.2$ channels has not been elucidated (Zhang et al., 2018). A loss of function mutation in $Ca_V\beta_{2b}$ (T11I) has been linked to Brugada Syndrome. Although peak I_{Ca} density, steady-state inactivation, and recovery from inactivation remained unchanged in WT compared to T11I, both fast and slow decays of I_{Ca} were significantly faster in mutant channels between 0 and + 20 mV. Action potential clamp experiments revealed a near 50% reduction in total charge in channels reconstituted with mutant β_{2b} compared to WT, while surface trafficking was not significantly affected. The reduction in $Ca_V1.2$ conductance most likely shortens phase 2 of the ventricular action potential, causing premature ventricular repolarization or J-waves on the electrocardiogram. Homozygous loss of function of β_4 in mouse generates the lethargic mouse, characterized by ataxia and seizures (Burgess et al., 1997). Mutations in *CACNB4* have been found in individuals with juvenile myoclonic epilepsy (R438X) and idiopathic generalized epilepsy/ataxia (C104F) (Escayg et al., 2000). Overall, while it is reasonable to expect that mutations in $Ca_V\beta$ subunits are likely to give rise to disease, the small numbers of patients with individual mutations make it difficult to establish causality and disease penetrance.

Pharmacological Targeting of $Ca_V\beta$ Subunits

HVACCs are important therapeutic targets for a variety of cardiovascular and neurological disorders, including, but not limited to, hypertension, cardiac arrhythmias, pain, and epilepsy (Kochegarov, 2003; Triggle, 2007; Zamponi et al., 2015). More specifically, blocking the activity of specific HVACCs is an important prevailing or aspirational approach to treat many diseases. Given the necessary role of binding to $Ca_V\beta$ for the functional expression of HVACCs in most cases, disrupting this interaction has long been pursued as an approach to develop Ca_V channel inhibitors. Over-expression of the AID peptide has been employed as a method to competitively disrupt HVACC α_1–β subunit interactions (Hohaus et al., 2000). Findeisen et al. utilized a stapled peptide approach to stabilize the AID α helix. They found that both the modified and unmodified AID peptides were able to inhibit $Ca_V1.2$ channels reconstituted with β_3, but not β_{2a}, in *Xenopus* oocytes (Findeisen et al., 2017). Treatment of ex vivo guinea pig heart, or rat heart in vivo with AID fused to TAT peptide, was reported to decrease ischemia–reperfusion injury and result in improved cardiac contractility (Viola et al., 2014). Over-expressing a YFP-tagged AID peptide in adult guinea pig ventricular cardiomyocytes did not substantively reduce basal L-type current, but strongly inhibited β-adrenergic stimulation of $Ca_V1.2$ (Yang et al., 2019). This is consistent with the notion that in adult cardiomyocytes $Ca_V\beta$ binding is not required for trafficking $Ca_V1.2$ to the sarcolemma, but is essential for PKA regulation of the channel (Weissgerber et al., 2006; Yang et al., 2019). Leveraging structure-based computational screening, small molecules designed to disrupt the HVACC α_1–β subunit interaction have been developed and shown to inhibit $Ca_V2.2$ channel trafficking and display analgesia in rodent models of pain (Chen et al., 2018; Khanna et al., 2019).

We have exploited understanding of the mechanisms of action of RGK proteins to develop nanobody-based genetically encoded HVACC inhibitors. Nanobodies against $Ca_V\beta$ subunits were obtained by llama immunization with purified β_1 and β_3, followed by generating a V_{HHS} phage library using mRNA extracted from lymphocytes, and identifying binders using three rounds of phage display and panning with purified β_1 as bait. We identified a nanobody, nb.F3, that indiscriminately binds all four β subunit isoforms with nM affinity but is functionally inert when co-expressed with recombinant HVACCs reconstituted in heterologous cells (Morgenstern et al., 2019). Fusing the catalytic HECT domain of the E3 ubiquitin ligase Nedd4L to nb.F3 generated a construct, termed Ca_V-aβlator, that potently inhibited both reconstituted and endogenous HVACCs in HEK293 cells, cardiac myocytes, and dorsal root ganglion neurons (Morgenstern et al., 2019). In some applications, genetically encoded HVACC inhibitors may have advantages over small molecules because the cells and tissues in which they are expressed may be precisely controlled, thereby limiting confounding off-target effects (Xu & Colecraft, 2009; Colecraft, 2020). A general consideration and challenge for all approaches that seek to modulate HVACCs by targeting β subunits is selectivity of action, given that the α_1–β interaction is a signature feature of this channel family.

Acknowledgments This work was supported by NIH grants R01 HL121253 and R01 HL122421 (to H.M.C.) and an American Heart Association predoctoral fellowship award (S.B.).

References

Ahern, B. M., Levitan, B. M., Veeranki, S., Shah, M., Ali, N., Sebastian, A., Su, W., Gong, M. C., Li, J., Stelzer, J. E., Andres, D. A., & Satin, J. (2019). Myocardial-restricted ablation of the GTPase RAD results in a pro-adaptive heart response in mice. *The Journal of Biological Chemistry, 294*, 10913–10927.

Altier, C., Dubel, S. J., Barrere, C., Jarvis, S. E., Stotz, S. C., Spaetgens, R. L., Scott, J. D., Cornet, V., De Waard, M., Zamponi, G. W., Nargeot, J., & Bourinet, E. (2002). Trafficking of L-type calcium channels mediated by the postsynaptic scaffolding protein AKAP79. *The Journal of Biological Chemistry, 277*, 33598–33603.

Altier, C., Garcia-Caballero, A., Simms, B., You, H., Chen, L., Walcher, J., Tedford, H. W., Hermosilla, T., & Zamponi, G. W. (2011). The Cavbeta subunit prevents RFP2-mediated ubiquitination and proteasomal degradation of L-type channels. *Nature Neuroscience, 14*, 173–180.

Anderson, J. M. (1996). Cell signalling: MAGUK magic. *Current Biology, 6*, 382–384.

Arias, J. M., Murbartian, J., Vitko, I., Lee, J. H., & Perez-Reyes, E. (2005). Transfer of beta subunit regulation from high to low voltage-gated Ca2+ channels. *FEBS Letters, 579*, 3907–3912.

Ball, S. L., Powers, P. A., Shin, H. S., Morgans, C. W., Peachey, N. S., & Gregg, R. G. (2002). Role of the beta(2) subunit of voltage-dependent calcium channels in the retinal outer plexiform layer. *Investigative Ophthalmology & Visual Science, 43*, 1595–1603.

Bannister, R. A., & Beam, K. G. (2013). Ca(V)1.1: The atypical prototypical voltage-gated Ca(2)(+) channel. *Biochimica et Biophysica Acta, 1828*, 1587–1597.

Bannister, R. A., Colecraft, H. M., & Beam, K. G. (2008). Rem inhibits skeletal muscle EC coupling by reducing the number of functional L-type Ca2+ channels. *Biophysical Journal, 94*, 2631–2638.

Bauer, R., Timothy, K. W., & Golden, A. (2021). Update on the molecular genetics of Timothy syndrome. *Frontiers in Pediatrics, 9*, 668546.

Bean, B. P., Nowycky, M. C., & Tsien, R. W. (1984). Beta-adrenergic modulation of calcium channels in frog ventricular heart cells. *Nature, 307*, 371–375.

Beguin, P., Nagashima, K., Gonoi, T., Shibasaki, T., Takahashi, K., Kashima, Y., Ozaki, N., Geering, K., Iwanaga, T., & Seino, S. (2001). Regulation of Ca2+ channel expression at the cell surface by the small G-protein kir/Gem. *Nature, 411*, 701–706.

Beguin, P., Ng, Y. J., Krause, C., Mahalakshmi, R. N., Ng, M. Y., & Hunziker, W. (2007). RGK small GTP-binding proteins interact with the nucleotide kinase domain of Ca2+-channel beta-subunits via an uncommon effector binding domain. *The Journal of Biological Chemistry, 282*, 11509–11520.

Berggren, P. O., Yang, S. N., Murakami, M., Efanov, A. M., Uhles, S., Kohler, M., Moede, T., Fernstrom, A., Appelskog, I. B., Aspinwall, C. A., Zaitsev, S. V., Larsson, O., de Vargas, L. M., Fecher-Trost, C., Weissgerber, P., Ludwig, A., Leibiger, B., Juntti-Berggren, L., Barker, C. J., … Flockerzi, V. (2004). Removal of Ca2+ channel beta3 subunit enhances Ca2+ oscillation frequency and insulin exocytosis. *Cell, 119*, 273–284.

Bers, D. M. (2002). Cardiac excitation-contraction coupling. *Nature, 415*, 198–205.

Bichet, D., Cornet, V., Geib, S., Carlier, E., Volsen, S., Hoshi, T., Mori, Y., & De Waard, M. (2000). The I-II loop of the Ca2+ channel alpha1 subunit contains an endoplasmic reticulum retention signal antagonized by the beta subunit. *Neuron, 25*, 177–190.

Bourdin, B., Marger, F., Wall-Lacelle, S., Schneider, T., Klein, H., Sauve, R., & Parent, L. (2010). Molecular determinants of the CaVbeta-induced plasma membrane targeting of the CaV1.2 channel. *The Journal of Biological Chemistry, 285*, 22853–22863.

Brandmayr, J., Poomvanicha, M., Domes, K., Ding, J., Blaich, A., Wegener, J. W., Moosmang, S., & Hofmann, F. (2012). Deletion of the C-terminal phosphorylation sites in the cardiac beta-subunit does not affect the basic beta-adrenergic response of the heart and the Ca(v)1.2 channel. *The Journal of Biological Chemistry, 287*, 22584–22592.

Breitenkamp, A. F., Matthes, J., Nass, R. D., Sinzig, J., Lehmkuhl, G., Nurnberg, P., & Herzig, S. (2014). Rare mutations of CACNB2 found in autism spectrum disease-affected families alter calcium channel function. *PLoS One, 9*, e95579.

Bunemann, M., Gerhardstein, B. L., Gao, T., & Hosey, M. M. (1999). Functional regulation of L-type calcium channels via protein kinase A-mediated phosphorylation of the beta(2) subunit. *The Journal of Biological Chemistry, 274*, 33851–33854.

Buraei, Z., & Yang, J. (2010). The {beta} subunit of voltage-gated Ca2+ channels. *Physiological Reviews, 90*, 1461–1506.

Burashnikov, E., Pfeiffer, R., Barajas-Martinez, H., Delpon, E., Hu, D., Desai, M., Borggrefe, M., Haissaguerre, M., Kanter, R., Pollevick, G. D., Guerchicoff, A., Laino, R., Marieb, M., Nademanee, K., Nam, G. B., Robles, R., Schimpf, R., Stapleton, D. D., Viskin, S., ... Antzelevitch, C. (2010). Mutations in the cardiac L-type calcium channel associated with inherited J-wave syndromes and sudden cardiac death. *Heart Rhythm, 7*, 1872–1882.

Burgess, D. L., Jones, J. M., Meisler, M. H., & Noebels, J. L. (1997). Mutation of the Ca2+ channel beta subunit gene Cchb4 is associated with ataxia and seizures in the lethargic (lh) mouse. *Cell, 88*, 385–392.

Cachelin, A. B., de Peyer, J. E., Kokubun, S., & Reuter, H. (1983). Ca2+ channel modulation by 8-bromocyclic AMP in cultured heart cells. *Nature, 304*, 462–464.

Canti, C., Bogdanov, Y., & Dolphin, A. C. (2000). Interaction between G proteins and accessory subunits in the regulation of 1B calcium channels in Xenopus oocytes. *The Journal of Physiology, 527*(Pt 3), 419–432.

Canti, C., Davies, A., Berrow, N. S., Butcher, A. J., Page, K. M., & Dolphin, A. C. (2001). Evidence for two concentration-dependent processes for beta-subunit effects on alpha1B calcium channels. *Biophysical Journal, 81*, 1439–1451.

Cassidy, J. S., Ferron, L., Kadurin, I., Pratt, W. S., & Dolphin, A. C. (2014). Functional exofacially tagged N-type calcium channels elucidate the interaction with auxiliary alpha2delta-1 subunits. *Proceedings of the National Academy of Sciences of the United States of America, 111*, 8979–8984.

Chen, Y. H., Li, M. H., Zhang, Y., He, L. L., Yamada, Y., Fitzmaurice, A., Shen, Y., Zhang, H., Tong, L., & Yang, J. (2004). Structural basis of the alpha1-beta subunit interaction of voltage-gated Ca2+ channels. *Nature, 429*, 675–680.

Chen, H., Puhl, H. L., 3rd, Niu, S. L., Mitchell, D. C., & Ikeda, S. R. (2005). Expression of Rem2, an RGK family small GTPase, reduces N-type calcium current without affecting channel surface density. *The Journal of Neuroscience, 25*, 9762–9772.

Chen, R. S., Deng, T. C., Garcia, T., Sellers, Z. M., & Best, P. M. (2007). Calcium channel gamma subunits: A functionally diverse protein family. *Cell Biochemistry and Biophysics, 47*, 178–186.

Chen, X., Liu, D., Zhou, D., Si, Y., Xu, D., Stamatkin, C. W., Ghozayel, M. K., Ripsch, M. S., Obukhov, A. G., White, F. A., & Meroueh, S. O. (2018). Small-molecule CaValpha1CaVbeta antagonist suppresses neuronal voltage-gated calcium-channel trafficking. *Proceedings of the National Academy of Sciences of the United States of America, 115*, E10566–E10575.

Chien, A. J., Carr, K. M., Shirokov, R. E., Rios, E., & Hosey, M. M. (1996). Identification of palmitoylation sites within the L-type calcium channel beta2a subunit and effects on channel function. *The Journal of Biological Chemistry, 271*, 26465–26468.

Chien, A. J., Gao, T., Perez-Reyes, E., & Hosey, M. M. (1998). Membrane targeting of L-type calcium channels. Role of palmitoylation in the subcellular localization of the beta2a subunit. *The Journal of Biological Chemistry, 273*, 23590–23597.

Colecraft, H. M. (2020). Designer genetically encoded voltage-dependent calcium channel inhibitors inspired by RGK GTPases. *The Journal of Physiology, 598*, 1683–1693.

Colecraft, H. M., Brody, D. L., & Yue, D. T. (2001). G-protein inhibition of N- and P/Q-type calcium channels: Distinctive elementary mechanisms and their functional impact. *The Journal of Neuroscience, 21*, 1137–1147.

Colecraft, H. M., Alseikhan, B., Takahashi, S. X., Chaudhuri, D., Mittman, S., Yegnasubramanian, V., Alvania, R. S., Johns, D. C., Marban, E., & Yue, D. T. (2002). Novel functional properties of Ca(2+) channel beta subunits revealed by their expression in adult rat heart cells. *The Journal of Physiology, 541*, 435–452.

Colicelli, J. (2004). Human RAS superfamily proteins and related GTPases. *Science's STKE, 2004*, RE13.

Craven, S. E., & Bredt, D. S. (1998). PDZ proteins organize synaptic signaling pathways. *Cell, 93*, 495–498.

Curtis, B. M., & Catterall, W. A. (1984). Purification of the calcium antagonist receptor of the voltage-sensitive calcium channel from skeletal muscle transverse tubules. *Biochemistry, 23*, 2113–2118.

De Waard, M., & Campbell, K. P. (1995). Subunit regulation of the neuronal alpha 1A Ca2+ channel expressed in Xenopus oocytes. *The Journal of Physiology, 485*, 619–634.

Dolphin, A. C. (2013). The alpha2delta subunits of voltage-gated calcium channels. *Biochimica et Biophysica Acta, 1828*, 1541–1549.

Dolphin, A. C., & Lee, A. (2020). Presynaptic calcium channels: Specialized control of synaptic neurotransmitter release. *Nature Reviews. Neuroscience, 21*, 213–229.

Dong, Y., Gao, Y., Xu, S., Wang, Y., Yu, Z., Li, Y., Li, B., Yuan, T., Yang, B., Zhang, X. C., Jiang, D., Huang, Z., & Zhao, Y. (2021). Closed-state inactivation and pore-blocker modulation mechanisms of human CaV2.2. *Cell Reports, 37*, 109931.

Dresviannikov, A. V., Page, K. M., Leroy, J., Pratt, W. S., & Dolphin, A. C. (2009). Determinants of the voltage dependence of G protein modulation within calcium channel beta subunits. *Pflügers Archiv, 457*, 743–756.

Escayg, A., De Waard, M., Lee, D. D., Bichet, D., Wolf, P., Mayer, T., Johnston, J., Baloh, R., Sander, T., & Meisler, M. H. (2000). Coding and noncoding variation of the human calcium-channel beta4-subunit gene

CACNB4 in patients with idiopathic generalized epilepsy and episodic ataxia. *American Journal of Human Genetics, 66*, 1531–1539.

Fang, K., & Colecraft, H. M. (2011). Mechanism of auxiliary beta-subunit-mediated membrane targeting of L-type (Ca(V)1.2) channels. *The Journal of Physiology, 589*, 4437–4455.

Findeisen, F., & Minor, D. L., Jr. (2009). Disruption of the IS6-AID linker affects voltage-gated calcium channel inactivation and facilitation. *The Journal of General Physiology, 133*, 327–343.

Findeisen, F., Campiglio, M., Jo, H., Abderemane-Ali, F., Rumpf, C. H., Pope, L., Rossen, N. D., Flucher, B. E., DeGrado, W. F., & Minor, D. L., Jr. (2017). Stapled voltage-gated calcium channel (CaV) alpha-interaction domain (AID) peptides act as selective protein-protein interaction inhibitors of CaV function. *ACS Chemical Neuroscience, 8*, 1313–1326.

Finlin, B. S., Shao, H., Kadono-Okuda, K., Guo, N., & Andres, D. A. (2000). Rem2, a new member of the Rem/Rad/Gem/Kir family of Ras-related GTPases. *The Biochemical Journal, 347*(Pt 1), 223–231.

Finlin, B. S., Crump, S. M., Satin, J., & Andres, D. A. (2003). Regulation of voltage-gated calcium channel activity by the Rem and Rad GTPases. *Proceedings of the National Academy of Sciences of the United States of America, 100*, 14469–14474.

Finlin, B. S., Mosley, A. L., Crump, S. M., Correll, R. N., Ozcan, S., Satin, J., & Andres, D. A. (2005). Regulation of L-type Ca2+ channel activity and insulin secretion by the Rem2 GTPase. *The Journal of Biological Chemistry, 280*, 41864–41871.

Flucher, B. E. (2020). Skeletal muscle CaV1.1 channelopathies. *Pflügers Archiv, 472*, 739–754.

Flynn, R., Chen, L., Hameed, S., Spafford, J. D., & Zamponi, G. W. (2008). Molecular determinants of Rem2 regulation of N-type calcium channels. *Biochemical and Biophysical Research Communications, 368*, 827–831.

Fu, Y., Westenbroek, R. E., Scheuer, T., & Catterall, W. A. (2014). Basal and beta-adrenergic regulation of the cardiac calcium channel CaV1.2 requires phosphorylation of serine 1700. *Proceedings of the National Academy of Sciences of the United States of America, 111*, 16598–16603.

Fuller, M. D., Emrick, M. A., Sadilek, M., Scheuer, T., & Catterall, W. A. (2010). Molecular mechanism of calcium channel regulation in the fight-or-flight response. *Science Signaling, 3*, ra70.

Gandini, M. A., & Felix, R. (2012). Functional interactions between voltage-gated Ca(2+) channels and Rab3-interacting molecules (RIMs): New insights into stimulus-secretion coupling. *Biochimica et Biophysica Acta, 1818*, 551–558.

Ganesan, A. N., Maack, C., Johns, D. C., Sidor, A., & O'Rourke, B. (2006). Beta-adrenergic stimulation of L-type Ca2+ channels in cardiac myocytes requires the distal carboxyl terminus of alpha1C but not serine 1928. *Circulation Research, 98*, e11–e18.

Gao, T., Yatani, A., Dell'Acqua, M. L., Sako, H., Green, S. A., Dascal, N., Scott, J. D., & Hosey, M. M. (1997). cAMP-dependent regulation of cardiac L-type Ca2+ channels requires membrane targeting of PKA and phosphorylation of channel subunits. *Neuron, 19*, 185–196.

Gao, T., Bunemann, M., Gerhardstein, B. L., Ma, H., & Hosey, M. M. (2000). Role of the C terminus of the alpha 1C (CaV1.2) subunit in membrane targeting of cardiac L-type calcium channels. *The Journal of Biological Chemistry, 275*, 25436–25444.

Gao, S., Yao, X., & Yan, N. (2021). Structure of human Cav2.2 channel blocked by the painkiller ziconotide. *Nature, 596*, 143–147.

Gebhart, M., Juhasz-Vedres, G., Zuccotti, A., Brandt, N., Engel, J., Trockenbacher, A., Kaur, G., Obermair, G. J., Knipper, M., Koschak, A., & Striessnig, J. (2010). Modulation of Cav1.3 Ca2+ channel gating by Rab3 interacting molecule. *Molecular and Cellular Neurosciences, 44*, 246–259.

Gonzalez-Gutierrez, G., Miranda-Laferte, E., Nothmann, D., Schmidt, S., Neely, A., & Hidalgo, P. (2008). The guanylate kinase domain of the beta-subunit of voltage-gated calcium channels suffices to modulate gating. *Proceedings of the National Academy of Sciences of the United States of America, 105*, 14198–14203.

Gregg, R. G., Messing, A., Strube, C., Beurg, M., Moss, R., Behan, M., Sukhareva, M., Haynes, S., Powell, J. A., Coronado, R., & Powers, P. A. (1996). Absence of the beta subunit (cchb1) of the skeletal muscle dihydropyridine receptor alters expression of the alpha 1 subunit and eliminates excitation-contraction coupling. *Proceedings of the National Academy of Sciences of the United States of America, 93*, 13961–13966.

Hartzell, H. C., Mery, P. F., Fischmeister, R., & Szabo, G. (1991). Sympathetic regulation of cardiac calcium current is due exclusively to cAMP-dependent phosphorylation. *Nature, 351*, 573–576.

He, L. L., Zhang, Y., Chen, Y. H., Yamada, Y., & Yang, J. (2007). Functional modularity of the beta-subunit of voltage-gated Ca2+ channels. *Biophysical Journal, 93*, 834–845.

Herzig, S., Khan, I. F., Grundemann, D., Matthes, J., Ludwig, A., Michels, G., Hoppe, U. C., Chaudhuri, D., Schwartz, A., Yue, D. T., & Hullin, R. (2007). Mechanism of Ca(v)1.2 channel modulation by the amino terminus of cardiac beta2-subunits. *The FASEB Journal, 21*, 1527–1538.

Hofmann, F., Flockerzi, V., Kahl, S., & Wegener, J. W. (2014). L-type CaV1.2 calcium channels: From in vitro findings to in vivo function. *Physiological Reviews, 94*, 303–326.

Hohaus, A., Poteser, M., Romanin, C., Klugbauer, N., Hofmann, F., Morano, I., Haase, H., & Groschner, K. (2000). Modulation of the smooth-muscle L-type Ca2+ channel alpha1 subunit (alpha1C-b) by the beta2a subunit: A peptide which inhibits binding of beta to the I-II linker of alpha1 induces functional uncoupling. *The Biochemical Journal, 348*(Pt 3), 657–665.

Hullin, R., Khan, I. F., Wirtz, S., Mohacsi, P., Varadi, G., Schwartz, A., & Herzig, S. (2003). Cardiac L-type calcium channel beta-subunits expressed in human heart have differential effects on single channel characteristics. *The Journal of Biological Chemistry, 278*, 21623–21630.

Jeon, D., Song, I., Guido, W., Kim, K., Kim, E., Oh, U., & Shin, H. S. (2008). Ablation of Ca2+ channel beta3 subunit leads to enhanced N-methyl-D-aspartate receptor-dependent long term potentiation and improved long term memory. *The Journal of Biological Chemistry, 283*, 12093–12101.

Jones, L. P., Wei, S. K., & Yue, D. T. (1998). Mechanism of auxiliary subunit modulation of neuronal alpha1E calcium channels. *The Journal of General Physiology, 112*, 125–143.

Kamp, T. J., & Hell, J. W. (2000). Regulation of cardiac L-type calcium channels by protein kinase A and protein kinase C. *Circulation Research, 87*, 1095–1102.

Katz, M., Subramaniam, S., Chomsky-Hecht, O., Tsemakhovich, V., Flockerzi, V., Klussmann, E., Hirsch, J. A., Weiss, S., & Dascal, N. (2021). Reconstitution of beta-adrenergic regulation of CaV1.2: Rad-dependent and Rad-independent protein kinase A mechanisms. *Proceedings of the National Academy of Sciences of the United States of America, 118*(21), e2100021118.

Khanna, R., Yu, J., Yang, X., Moutal, A., Chefdeville, A., Gokhale, V., Shuja, Z., Chew, L. A., Bellampalli, S. S., Luo, S., Francois-Moutal, L., Serafini, M. J., Ha, T., Perez-Miller, S., Park, K. D., Patwardhan, A. M., Streicher, J. M., Colecraft, H. M., & Khanna, M. (2019). Targeting the CaValpha-CaVbeta interaction yields an antagonist of the N-type CaV2.2 channel with broad antinociceptive efficacy. *Pain, 160*(7), 1644–1661.

Kim, D. I., Kweon, H. J., Park, Y., Jang, D. J., & Suh, B. C. (2016). Ca2+ controls gating of voltage-gated calcium channels by releasing the beta2e subunit from the plasma membrane. *Science Signaling, 9*, ra67.

Kiyonaka, S., Wakamori, M., Miki, T., Uriu, Y., Nonaka, M., Bito, H., Beedle, A. M., Mori, E., Hara, Y., De Waard, M., Kanagawa, M., Itakura, M., Takahashi, M., Campbell, K. P., & Mori, Y. (2007). RIM1 confers sustained activity and neurotransmitter vesicle anchoring to presynaptic Ca2+ channels. *Nature Neuroscience, 10*, 691–701.

Kochegarov, A. A. (2003). Pharmacological modulators of voltage-gated calcium channels and their therapeutical application. *Cell Calcium, 33*, 145–162.

Lemke, T., Welling, A., Christel, C. J., Blaich, A., Bernhard, D., Lenhardt, P., Hofmann, F., & Moosmang, S. (2008). Unchanged beta-adrenergic stimulation of cardiac L-type calcium channels in Ca v 1.2 phosphorylation site S1928A mutant mice. *The Journal of Biological Chemistry, 283*, 34738–34744.

Leroy, J., Richards, M. W., Butcher, A. J., Nieto-Rostro, M., Pratt, W. S., Davies, A., & Dolphin, A. C. (2005). Interaction via a key tryptophan in the I-II linker of N-type calcium channels is required for beta1 but not

for palmitoylated beta2, implicating an additional binding site in the regulation of channel voltage-dependent properties. *The Journal of Neuroscience, 25*, 6984–6996.

Leung, A. T., Imagawa, T., & Campbell, K. P. (1987). Structural characterization of the 1,4-dihydropyridine receptor of the voltage-dependent Ca2+ channel from rabbit skeletal muscle. Evidence for two distinct high molecular weight subunits. *The Journal of Biological Chemistry, 262*, 7943–7946.

Liu, G., Papa, A., Katchman, A. N., Zakharov, S. I., Roybal, D., Hennessey, J. A., Kushner, J., Yang, L., Chen, B. X., Kushnir, A., Dangas, K., Gygi, S. P., Pitt, G. S., Colecraft, H. M., Ben-Johny, M., Kalocsay, M., & Marx, S. O. (2020). Mechanism of adrenergic CaV1.2 stimulation revealed by proximity proteomics. *Nature, 577*, 695–700.

Luvisetto, S., Fellin, T., Spagnolo, M., Hivert, B., Brust, P. F., Harpold, M. M., Stauderman, K. A., Williams, M. E., & Pietrobon, D. (2004). Modal gating of human CaV2.1 (P/Q-type) calcium channels: I. The slow and the fast gating modes and their modulation by beta subunits. *The Journal of General Physiology, 124*, 445–461.

Maltez, J. M., Nunziato, D. A., Kim, J., & Pitt, G. S. (2005). Essential Ca(V)beta modulatory properties are AID-independent. *Nature Structural & Molecular Biology, 12*, 372–377.

Mansergh, F., Orton, N. C., Vessey, J. P., Lalonde, M. R., Stell, W. K., Tremblay, F., Barnes, S., Rancourt, D. E., & Bech-Hansen, N. T. (2005). Mutation of the calcium channel gene Cacna1f disrupts calcium signaling, synaptic transmission and cellular organization in mouse retina. *Human Molecular Genetics, 14*, 3035–3046.

Marcantoni, A., Vandael, D. H., Mahapatra, S., Carabelli, V., Sinnegger-Brauns, M. J., Striessnig, J., & Carbone, E. (2010). Loss of Cav1.3 channels reveals the critical role of L-type and BK channel coupling in pacemaking mouse adrenal chromaffin cells. *The Journal of Neuroscience, 30*, 491–504.

Marrion, N. V., & Tavalin, S. J. (1998). Selective activation of Ca2+-activated K+ channels by co-localized Ca2+ channels in hippocampal neurons. *Nature, 395*, 900–905.

McGee, A. W., Dakoji, S. R., Olsen, O., Bredt, D. S., Lim, W. A., & Prehoda, K. E. (2001). Structure of the SH3-guanylate kinase module from PSD-95 suggests a mechanism for regulated assembly of MAGUK scaffolding proteins. *Molecular Cell, 8*, 1291–1301.

Meissner, M., Weissgerber, P., Londono, J. E., Prenen, J., Link, S., Ruppenthal, S., Molkentin, J. D., Lipp, P., Nilius, B., Freichel, M., & Flockerzi, V. (2011). Moderate calcium channel dysfunction in adult mice with inducible cardiomyocyte-specific excision of the cacnb2 gene. *The Journal of Biological Chemistry, 286*, 15875–15882.

Miranda-Laferte, E., Ewers, D., Guzman, R. E., Jordan, N., Schmidt, S., & Hidalgo, P. (2014). The N-terminal domain tethers the voltage-gated calcium channel beta2e-subunit to the plasma membrane via electrostatic and hydrophobic interactions. *The Journal of Biological Chemistry, 289*, 10387–10398.

Miriyala, J., Nguyen, T., Yue, D. T., & Colecraft, H. M. (2008). Role of CaVbeta subunits, and lack of functional reserve, in protein kinase A modulation of cardiac CaV1.2 channels. *Circulation Research, 102*, e54–e64.

Morgenstern, T. J., Park, J., Fan, Q. R., & Colecraft, H. M. (2019). A potent voltage-gated calcium channel inhibitor engineered from a nanobody targeted to auxiliary CaVbeta subunits. *eLife, 8*, e49253.

Murakami, M., Fleischmann, B., De Felipe, C., Freichel, M., Trost, C., Ludwig, A., Wissenbach, U., Schwegler, H., Hofmann, F., Hescheler, J., Flockerzi, V., & Cavalie, A. (2002). Pain perception in mice lacking the beta3 subunit of voltage-activated calcium channels. *The Journal of Biological Chemistry, 277*, 40342–40351.

Murakami, M., Yamamura, H., Suzuki, T., Kang, M. G., Ohya, S., Murakami, A., Miyoshi, I., Sasano, H., Muraki, K., Hano, T., Kasai, N., Nakayama, S., Campbell, K. P., Flockerzi, V., Imaizumi, Y., Yanagisawa, T., & Iijima, T. (2003). Modified cardiovascular L-type channels in mice lacking the voltage-dependent Ca2+ channel beta3 subunit. *The Journal of Biological Chemistry, 278*, 43261–43267.

Murakami, M., Nakagawasai, O., Yanai, K., Nunoki, K., Tan-No, K., Tadano, T., & Iijima, T. (2007). Modified behavioral characteristics following ablation of the voltage-dependent calcium channel beta3 subunit. *Brain Research, 1160*, 102–112.

Neef, J., Gehrt, A., Bulankina, A. V., Meyer, A. C., Riedel, D., Gregg, R. G., Strenzke, N., & Moser, T. (2009). The Ca2+ channel subunit beta2 regulates Ca2+ channel abundance and function in inner hair cells and is required for hearing. *The Journal of Neuroscience, 29*, 10730–10740.

Neely, A., & Hidalgo, P. (2014). Structure-function of proteins interacting with the alpha1 pore-forming subunit of high-voltage-activated calcium channels. *Frontiers in Physiology, 5*, 209.

Neely, A., Wei, X., Olcese, R., Birnbaumer, L., & Stefani, E. (1993). Potentiation by the beta subunit of the ratio of the ionic current to the charge movement in the cardiac calcium channel. *Science, 262*, 575–578.

Olcese, R., Neely, A., Qin, N., Wei, X., Birnbaumer, L., & Stefani, E. (1996). Coupling between charge movement and pore opening in vertebrate neuronal alpha 1E calcium channels. *The Journal of Physiology, 497*, 675–686.

Opatowsky, Y., Chen, C. C., Campbell, K. P., & Hirsch, J. A. (2004). Structural analysis of the voltage-dependent calcium channel beta subunit functional core and its complex with the alpha 1 interaction domain. *Neuron, 42*, 387–399.

Papa, A., Kushner, J., Hennessey, J. A., Katchman, A. N., Zakharov, S. I., Chen, B. X., Yang, L., Lu, R., Leong, S., Diaz, J., Liu, G., Roybal, D., Liao, X., Del Rivero Morfin, P. J., Colecraft, H. M., Pitt, G. S., Clarke, O., Topkara, V., Ben-Johny, M., & Marx, S. O. (2021).

Adrenergic CaV1.2 activation via Rad phosphorylation converges at alpha1C I-II loop. *Circulation Research, 128*, 76–88.

Platzer, J., Engel, J., Schrott-Fischer, A., Stephan, K., Bova, S., Chen, H., Zheng, H., & Striessnig, J. (2000). Congenital deafness and sinoatrial node dysfunction in mice lacking class D L-type Ca2+ channels. *Cell, 102*, 89–97.

Pragnell, M., De Waard, M., Mori, Y., Tanabe, T., Snutch, T. P., & Campbell, K. P. (1994). Calcium channel beta-subunit binds to a conserved motif in the I-II cytoplasmic linker of the alpha 1-subunit. *Nature, 368*, 67–70.

Puckerin, A. A., Chang, D. D., Subramanyam, P., & Colecraft, H. M. (2016). Similar molecular determinants on Rem mediate two distinct modes of inhibition of CaV1.2 channels. *Channels (Austin, Tex.), 10*, 379–394.

Puckerin, A. A., Chang, D. D., Shuja, Z., Choudhury, P., Scholz, J., & Colecraft, H. M. (2018). Engineering selectivity into RGK GTPase inhibition of voltage-dependent calcium channels. *Proceedings of the National Academy of Sciences of the United States of America, 115*, 12051–12056.

Restituito, S., Cens, T., Barrere, C., Geib, S., Galas, S., De Waard, M., & Charnet, P. (2000). The [beta]2a subunit is a molecular groom for the Ca2+ channel inactivation gate. *The Journal of Neuroscience, 20*, 9046–9052.

Reuter, H., & Scholz, H. (1977). The regulation of the calcium conductance of cardiac muscle by adrenaline. *The Journal of Physiology, 264*, 49–62.

Schoch, S., Mittelstaedt, T., Kaeser, P. S., Padgett, D., Feldmann, N., Chevaleyre, V., Castillo, P. E., Hammer, R. E., Han, W., Schmitz, F., Lin, W., & Sudhof, T. C. (2006). Redundant functions of RIM1alpha and RIM2alpha in Ca(2+)-triggered neurotransmitter release. *The EMBO Journal, 25*, 5852–5863.

Schredelseker, J., Di Biase, V., Obermair, G. J., Felder, E. T., Flucher, B. E., Franzini-Armstrong, C., & Grabner, M. (2005). The beta 1a subunit is essential for the assembly of dihydropyridine-receptor arrays in skeletal muscle. *Proceedings of the National Academy of Sciences of the United States of America, 102*, 17219–17224.

Schredelseker, J., Dayal, A., Schwerte, T., Franzini-Armstrong, C., & Grabner, M. (2009). Proper restoration of excitation-contraction coupling in the dihydropyridine receptor beta1-null zebrafish relaxed is an exclusive function of the beta1a subunit. *The Journal of Biological Chemistry, 284*, 1242–1251.

Singer, D., Biel, M., Lotan, I., Flockerzi, V., Hofmann, F., & Dascal, N. (1991). The roles of the subunits in the function of the calcium channel. *Science, 253*, 1553–1557.

Splawski, I., Timothy, K. W., Sharpe, L. M., Decher, N., Kumar, P., Bloise, R., Napolitano, C., Schwartz, P. J., Joseph, R. M., Condouris, K., Tager-Flusberg, H., Priori, S. G., Sanguinetti, M. C., & Keating, M. T. (2004). Ca(V)1.2 calcium channel dysfunction causes a multisystem disorder including arrhythmia and autism. *Cell, 119*, 19–31.

Splawski, I., Timothy, K. W., Decher, N., Kumar, P., Sachse, F. B., Beggs, A. H., Sanguinetti, M. C., & Keating, M. T. (2005). Severe arrhythmia disorder caused by cardiac L-type calcium channel mutations. *Proceedings of the National Academy of Sciences of the United States of America, 102*, 8089–8096. discussion 8086-8088.

Striessnig, J., Bolz, H. J., & Koschak, A. (2010). Channelopathies in Cav1.1, Cav1.3, and Cav1.4 voltage-gated L-type Ca2+ channels. *Pflügers Archiv, 460*, 361–374.

Sudhof, T. C. (2004). The synaptic vesicle cycle. *Annual Review of Neuroscience, 27*, 509–547.

Sudhof, T. C. (2012). The presynaptic active zone. *Neuron, 75*, 11–25.

Tadross, M. R., Ben Johny, M., & Yue, D. T. (2010). Molecular endpoints of Ca2+/calmodulin- and voltage-dependent inactivation of Ca(v)1.3 channels. *The Journal of General Physiology, 135*, 197–215.

Takahashi, M., Seagar, M. J., Jones, J. F., Reber, B. F., & Catterall, W. A. (1987). Subunit structure of dihydropyridine-sensitive calcium channels from skeletal muscle. *Proceedings of the National Academy of Sciences of the United States of America, 84*, 5478–5482.

Takahashi, S. X., Mittman, S., & Colecraft, H. M. (2003). Distinctive modulatory effects of five human auxiliary beta 2 subunit splice variants on L-type calcium channel gating. *Biophysical Journal, 84*, 3007–3021.

Takahashi, S. X., Miriyala, J., & Colecraft, H. M. (2004). Membrane-associated guanylate kinase-like properties of beta-subunits required for modulation of voltage-dependent Ca2+ channels. *Proceedings of the National Academy of Sciences of the United States of America, 101*, 7193–7198.

Takahashi, S. X., Miriyala, J., Tay, L. H., Yue, D. T., & Colecraft, H. M. (2005). A CaVbeta SH3/guanylate kinase domain interaction regulates multiple properties of voltage-gated Ca2+ channels. *The Journal of General Physiology, 126*, 365–377.

Tareilus, E., Roux, M., Qin, N., Olcese, R., Zhou, J., Stefani, E., & Birnbaumer, L. (1997). A Xenopus oocyte beta subunit: Evidence for a role in the assembly/expression of voltage-gated calcium channels that is separate from its role as a regulatory subunit. *Proceedings of the National Academy of Sciences of the United States of America, 94*, 1703–1708.

tom Dieck, S., Altrock, W. D., Kessels, M. M., Qualmann, B., Regus, H., Brauner, D., Fejtova, A., Bracko, O., Gundelfinger, E. D., & Brandstatter, J. H. (2005). Molecular dissection of the photoreceptor ribbon synapse: Physical interaction of Bassoon and RIBEYE is essential for the assembly of the ribbon complex. *The Journal of Cell Biology, 168*, 825–836.

Triggle, D. J. (2007). Calcium channel antagonists: Clinical uses – Past, present and future. *Biochemical Pharmacology, 74*, 1–9.

Tsien, R. W., Bean, B. P., Hess, P., Lansman, J. B., Nilius, B., & Nowycky, M. C. (1986). Mechanisms of calcium channel modulation by beta-adrenergic agents and dihydropyridine calcium agonists. *Journal of Molecular and Cellular Cardiology, 18*, 691–710.

Van Petegem, F., Clark, K. A., Chatelain, F. C., & Minor, D. L., Jr. (2004). Structure of a complex between a voltage-gated calcium channel beta-subunit and an alpha-subunit domain. *Nature, 429*, 671–675.

Viola, H. M., Jordan, M. C., Roos, K. P., & Hool, L. C. (2014). Decreased myocardial injury and improved contractility after administration of a peptide derived against the alpha-interacting domain of the L-type calcium channel. *Journal of the American Heart Association, 3*, e000961.

Waithe, D., Ferron, L., Page, K. M., Chaggar, K., & Dolphin, A. C. (2011). Beta-subunits promote the expression of Ca(V)2.2 channels by reducing their proteasomal degradation. *The Journal of Biological Chemistry, 286*, 9598–9611.

Wakamori, M., Mikala, G., & Mori, Y. (1999). Auxiliary subunits operate as a molecular switch in determining gating behaviour of the unitary N-type Ca2+ channel current in Xenopus oocytes. *The Journal of Physiology, 517*(Pt 3), 659–672.

Wang, H. G., George, M. S., Kim, J., Wang, C., & Pitt, G. S. (2007). Ca2+/calmodulin regulates trafficking of Ca(V)1.2 Ca2+ channels in cultured hippocampal neurons. *The Journal of Neuroscience, 27*, 9086–9093.

Wang, G., Zhu, X., Xie, W., Han, P., Li, K., Sun, Z., Wang, Y., Chen, C., Song, R., Cao, C., Zhang, J., Wu, C., Liu, J., & Cheng, H. (2010). Rad as a novel regulator of excitation-contraction coupling and beta-adrenergic signaling in heart. *Circulation Research, 106*, 317–327.

Wang, J., Thio, S. S., Yang, S. S., Yu, D., Yu, C. Y., Wong, Y. P., Liao, P., Li, S., & Soong, T. W. (2011). Splice variant specific modulation of CaV1.2 calcium channel by galectin-1 regulates arterial constriction. *Circulation Research, 109*, 1250–1258.

Weiss, S., Oz, S., Benmocha, A., & Dascal, N. (2013). Regulation of cardiac L-type Ca(2)(+) channel CaV1.2 via the beta-adrenergic-cAMP-protein kinase A pathway: Old dogmas, advances, and new uncertainties. *Circulation Research, 113*, 617–631.

Weissgerber, P., Held, B., Bloch, W., Kaestner, L., Chien, K. R., Fleischmann, B. K., Lipp, P., Flockerzi, V., & Freichel, M. (2006). Reduced cardiac L-type Ca2+ current in Ca(V)beta2−/− embryos impairs cardiac development and contraction with secondary defects in vascular maturation. *Circulation Research, 99*, 749–757.

Wu, J., Yan, Z., Li, Z., Qian, X., Lu, S., Dong, M., Zhou, Q., & Yan, N. (2016). Structure of the voltage-gated calcium channel Ca(v)1.1 at 3.6 A resolution. *Nature, 537*, 191–196.

Xu, X., & Colecraft, H. M. (2009). Engineering proteins for custom inhibition of Ca(V) channels. *Physiology (Bethesda), 24*, 210–218.

Xu, X., Marx, S. O., & Colecraft, H. M. (2010). Molecular mechanisms, and selective pharmacological rescue, of Rem-inhibited CaV1.2 channels in heart. *Circulation Research, 107*, 620–630.

Yang, S. N., & Berggren, P. O. (2006). The role of voltage-gated calcium channels in pancreatic beta-cell physiology and pathophysiology. *Endocrine Reviews, 27*, 621–676.

Yang, T., & Colecraft, H. M. (2013). Regulation of voltage-dependent calcium channels by RGK proteins. *Biochimica et Biophysica Acta, 1828*, 1644–1654.

Yang, T., Suhail, Y., Dalton, S., Kernan, T., & Colecraft, H. M. (2007). Genetically encoded molecules for inducibly inactivating CaV channels. *Nature Chemical Biology, 3*, 795–804.

Yang, T., Xu, X., Kernan, T., Wu, V., & Colecraft, H. M. (2010). Rem, a member of the RGK GTPases, inhibits recombinant CaV1.2 channels using multiple mechanisms that require distinct conformations of the GTPase. *The Journal of Physiology, 588*, 1665–1681.

Yang, T., Puckerin, A., & Colecraft, H. M. (2012). Distinct RGK GTPases differentially use alpha(1)- and auxiliary beta-binding-dependent mechanisms to inhibit Ca(V)1.2/Ca(V)2.2 channels. *PLoS One, 7*, e37079.

Yang, T., He, L. L., Chen, M., Fang, K., & Colecraft, H. M. (2013). Bio-inspired voltage-dependent calcium channel blockers. *Nature Communications, 4*, 2540.

Yang, L., Katchman, A., Kushner, J., Kushnir, A., Zakharov, S. I., Chen, B. X., Shuja, Z., Subramanyam, P., Liu, G., Papa, A., Roybal, D., Pitt, G. S., Colecraft, H. M., & Marx, S. O. (2019). Cardiac CaV1.2 channels require beta subunits for beta-adrenergic-mediated modulation but not trafficking. *The Journal of Clinical Investigation, 129*, 647–658.

Yuen, R. K., Thiruvahindrapuram, B., Merico, D., Walker, S., Tammimies, K., Hoang, N., Chrysler, C., Nalpathamkalam, T., Pellecchia, G., Liu, Y., Gazzellone, M. J., D'Abate, L., Deneault, E., Howe, J. L., Liu, R. S., Thompson, A., Zarrei, M., Uddin, M., Marshall, C. R., … Scherer, S. W. (2015). Whole-genome sequencing of quartet families with autism spectrum disorder. *Nature Medicine, 21*, 185–191.

Zamponi, G. W., Striessnig, J., Koschak, A., & Dolphin, A. C. (2015). The physiology, pathology, and pharmacology of voltage-gated calcium channels and their future therapeutic potential. *Pharmacological Reviews, 67*, 821–870.

Zhang, Y., Chen, Y. H., Bangaru, S. D., He, L., Abele, K., Tanabe, S., Kozasa, T., & Yang, J. (2008). Origin of the voltage dependence of G-protein regulation of P/Q-type Ca2+ channels. *The Journal of Neuroscience, 28*, 14176–14188.

Zhang, Q., Chen, J., Qin, Y., Wang, J., & Zhou, L. (2018). Mutations in voltage-gated L-type calcium channel: Implications in cardiac arrhythmia. *Channels (Austin, Tex.), 12*, 201–218.

Regulation of Calcium Channels and Synaptic Function by Auxiliary $\alpha_2\delta$ Subunits

Annette C. Dolphin and Gerald J. Obermair

Abstract

Voltage-gated calcium channels of the Ca_V1 (L-type) and Ca_V2 (N-type, P/Q-type, R-type) classes associate with auxiliary β and $\alpha_2\delta$ subunits that are both important for the function of these channels. While the β subunits are cytoplasmic, $\alpha_2\delta$ subunits are entirely extracellular and provide a critical link between channel and extracellular signaling functions. Here we describe what is known about the structure of $\alpha_2\delta$ subunits, and their importance both for channel trafficking and for their physiological function. We also describe distinct roles of $\alpha_2\delta$ proteins in synapse development and synaptic transmission, potentially beyond their classical role as auxiliary calcium channel subunits. Dysregulation and mutations of specific $\alpha_2\delta$ subunits are associated with sev-

eral diseases, and $\alpha_2\delta$-1 represents an important therapeutic target for the anti-epileptic and anti-allodynic gabapentinoid drugs. A detailed understanding of specific and potentially redundant $\alpha_2\delta$ functions may pave the way for developing novel treatment options, particularly for neuropsychiatric disorders.

Keywords

Calcium channel auxiliary subunit · Calcium currents · Trafficking · Gabapentin · Excitable cells · Synaptogenesis · Synaptic function · Brain disorders · Drug target · Pain

A. C. Dolphin (✉)
Department of Neuroscience Physiology and Pharmacology, University College London, London, UK
e-mail: a.dolphin@ucl.ac.uk

G. J. Obermair (✉)
Division Physiology, Department Pharmacology, Physiology and Microbiology, Karl Landsteiner University of Health Sciences, Krems, Austria
e-mail: Gerald.Obermair@kl.ac.at

Abbreviations

AMPA	α-amino-3-hydroxy-5-methyl-4-isoxazolepropionic acid
Ca_V	voltage-dependent calcium
DRG	dorsal root ganglion
eGFP	enhanced green fluorescent protein
ER	endoplasmic reticulum
GABA	Gamma-aminobutyric acid
GPI	glycosylphosphatidylinositol

gSTED microscopy	gated stimulated emission depletion microscopy
LRP1	low-density lipoprotein receptor-related protein 1
mCherry	monomeric Cherry fluorescent protein
MIDAS	metal ion-dependent adhesion site
NMDA	N-methyl-D-aspartate
PSD95	postsynaptic density protein 95
TSP	thrombospondin
VWA	von Willebrand factor-A

Introduction

Voltage-gated calcium (Ca_V) channels mediate and regulate a variety of functions ranging from muscle contraction, hormone secretion, and synaptic transmission to gene regulation. To provide precise timing and control of the calcium ions entering through Ca_V channels, they operate, with some exceptions, in heteromultimeric complexes with auxiliary and cytoplasmic β and extracellular $\alpha_2\delta$ subunits. At least in skeletal muscle, a transmembrane γ subunit is also part of the channel structure (Wu et al., 2016). The traditional concept of Ca_V channel function envisions the pore-forming subunit as the major determinant of the basic biophysical, pharmacological, and physiological properties (see chapters "Subunit Architecture and Atomic Structure of Voltage-Gated Ca2+ Channels" by William A. Catterall, "Voltage-Gated Calcium Channel Auxiliary β Subunits" by Sergej Borowik and Henry M. Colecraft, "Pharmacology of Voltage-Gated Calcium Channels at Atomic Resolution" by William A. Catterall and "Pharmacology and Structure-Function of Venom Peptide Inhibitors of N-Type (Cav2.2) Calcium Channels" by Md. Mahadhi Hasan et al.). Auxiliary subunits are thought to fine-tune Ca_V channel functions by regulating or modulating the membrane expression, localization, trafficking, and biophysical properties of the channel complex (reviewed in Arikkath & Campbell, 2003; Buraei & Yang, 2010; Dolphin, 2009, 2016; Obermair et al., 2008).

Among the auxiliary subunits, the $\alpha_2\delta$ subunit is unique in that the entire protein is extracellular and highly glycosylated, and hence, it interacts with the extracellularly exposed surface of the α_1 subunit. Importantly, this distinctive position also theoretically enables $\alpha_2\delta$ subunits to link the Ca_V complex with other potential extracellular signaling or scaffolding proteins. Indeed, research over recent years has identified a number of extracellular interaction partners as well as physiological functions, some of which may be independent of the classical interaction with the Ca_V complex. A variety of disorders linked to defective $\alpha_2\delta$ subunit functions provide another line of evidence of key roles for these versatile proteins, and the widely prescribed anti-epileptic and anti-allodynic drugs gabapentin and pregabalin act by binding to specific $\alpha_2\delta$ isoforms. In this chapter, we provide a concise overview of the present state of knowledge about $\alpha_2\delta$ subunits, including $\alpha_2\delta$ structure and tissue distribution, the role of specific structural domains for protein function, and Ca_V channel-dependent and -independent functions. We particularly summarize the mechanisms of action of $\alpha_2\delta$ subunits and discuss recent developments concerning their pharmacology and disease association. Investigating and understanding the detailed pathophysiological mechanisms involving Ca_V complexes and $\alpha_2\delta$ subunits may open the path to the identification of new and specific treatment paradigms.

Discovery of $\alpha_2\delta$ Subunits

The Ca_V channel complex was purified, and its properties first investigated from skeletal and then cardiac muscle, by virtue of the ability of the channel to bind ^{3}H-dihydropyridines (DHPs), which were known to block L-type calcium channels (Cooper et al., 1987; Curtis & Catterall, 1984; De Jongh et al., 1989; Leung et al., 1987; Takahashi et al., 1987; Tanabe et al., 1987). The subunit to which the DHP calcium channel blocker bound was identified to be the channel itself, and termed the α_1 subunit (α_1S, with the "S" referring to skeletal muscle). This co-purified with a similar molecular weight but glycosylated protein, later named $\alpha_2\delta$ (Sharp et al., 1987), and smaller molecular weight β and γ subunits (De Jongh et al., 1989, 1990; Takahashi et al., 1987; Tanabe et al., 1987).

The $\alpha_2\delta$ subunit was so called as it was found to consist of the two proteins α_2 and δ that remain associated, except in reducing conditions when they are clearly separated (De Jongh et al., 1990). These two disulfide-bonded proteins were both highly glycosylated, suggesting they were largely extracellular proteins, and they were also membrane associated, and initially identified as transmembrane proteins, with δ containing a hydrophobic, potentially transmembrane, α-helix (Brickley et al., 1995; Gurnett et al., 1996).

Identification of $\alpha_2\delta$ Subunits

Cloning

The $\alpha_2\delta$-1 subunit was the first $\alpha_2\delta$ whose gene was identified and cloned (*Cacna2d1*). This was achieved by first obtaining some peptide sequences from the α_2 and δ proteins purified from skeletal muscle. Once obtained, the $\alpha_2\delta$ sequence then showed that both α_2 and δ subunits were encoded by the same gene, which is therefore translated as a pre-protein, and then enzymatically cleaved into α_2 and δ (De Jongh et al., 1990; Ellis et al., 1988). Two additional genes encoding $\alpha_2\delta$-2 and $\alpha_2\delta$-3 were identified by homology to $\alpha_2\delta$-1 (Ellis et al., 1988), and $\alpha_2\delta$-3 was then cloned (Klugbauer et al., 1999). $\alpha_2\delta$-2 was also cloned by virtue of the fact that it is mutated in a spontaneously arising mouse model of absence epilepsy and ataxia called *Ducky* (Barclay et al., 2001; Barclay & Rees, 2000), and $\alpha_2\delta$-4 was cloned by homology to other $\alpha_2\delta$ sequences (Qin et al., 2002).

Splice Variants

The *Cacna2d1* gene for $\alpha_2\delta$-1 was originally identified to contain three alternatively spliced regions (A, B, and C) within the α_2 moiety (Angelotti & Hofmann, 1996). The rat $\alpha_2\delta$-1 genomic sequence contains 39 exons, and further analysis of this sequence found that A is encoded by exon 18a, B is formed as a result of utilizing an alternative 3′ splice acceptor site for exon 19, while C represents the inclusion of exon 23 (Lana et al., 2014) (Fig. 1). In skeletal muscle, $\alpha_2\delta$-1 (+A + B ΔC) was the only splice variant detected, whereas in heart, multiple different splice variants were found (Angelotti & Hofmann, 1996; Lana et al., 2014). By contrast, in neuronal tissue, including dorsal root ganglion (DRG) neurons, the predominant splice variant is ΔA + B + C (Angelotti & Hofmann, 1996; Lana et al., 2014) (Fig. 1). The mouse *Cacna2d2* gene also contains 39 exons (Barclay & Rees, 2000), and splice region C has been found to be of functional relevance, as it regulates the trans-synaptic recruitment of postsynaptic $GABA_A$ receptors and axonal wiring (Geisler et al., 2019).

Tissue Distribution of $\alpha_2\delta$ Subunits

$\alpha_2\delta$-1

The distribution of $\alpha_2\delta$-1 is widespread, being the main $\alpha_2\delta$ in skeletal, smooth, and cardiac muscle, and it is also present in brain (Angelotti & Hofmann, 1996). It is strongly expressed in DRG neurons (Newton et al., 2001) and is the major $\alpha_2\delta$ isoform in cortical brain regions (Schlick et al., 2010).

$\alpha_2\delta$-2

This species was originally found in multiple tissues, including skeletal and cardiac muscle, pancreas, and brain (Klugbauer et al., 1999, 2003). Within the brain, it is found to be strongly expressed in cerebellum, particularly in Purkinje neurons (Barclay et al., 2001) and, at a lower expression level, also in the cortex (Schlick et al., 2010).

$\alpha_2\delta$-3

This isoform is found mainly in brain (Klugbauer et al., 2003), such as the cortex and hippocampus, and it is the major isoform expressed in the striatum (Geisler et al., 2019).

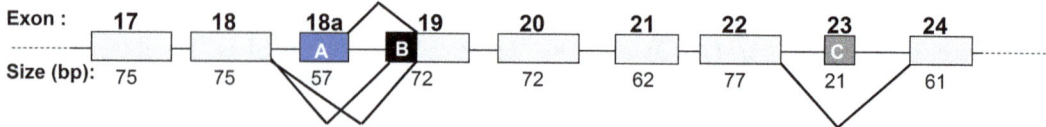

Fig. 1 Splice variants in α₂δ-1. Three alternatively spliced regions (**A**), (**B**), and (**C**) were originally described by Angelotti and Hofmann (1996), and the splice variants were further defined by Lana et al. (2014)

α₂δ-4

The α₂δ-4 isoform is strongly expressed in the retina, particularly at photoreceptor ribbon synapses (Knoflach et al., 2013; Lee et al., 2015; Wycisk et al., 2006a); its expression in the brain, however, seems negligible (Schlick et al., 2010).

Structure of α₂δ Subunits

Biochemical and Bioinformatic Studies

Following cloning, the primary α₂δ sequence showed that the N-terminus of α₂δ-1 had a predicted signal sequence; therefore signifying it was extracellular (Fig. 2a). The C-terminus of α₂δ-1 had a hydrophobic domain, suggesting it is transmembrane, with a very short potentially intracellular sequence. Furthermore, the existence of multiple (16–18) glycosylation sites (Fig. 2a, b), and antibody mapping indicated the protein is mainly extracellular (Brickley et al., 1995; Gurnett et al., 1996). Experimentally, both reducing conditions and deglycosylation are important tools to distinguish between α₂δ and α₂ in immunoblots, particularly in expression studies (Fig. 2b), and the low molecular weight δ can also be identified separately (Davies et al., 2010; De Jongh et al., 1990; Jay et al., 1991). Although almost all α₂δ in native tissues is proteolytically cleaved, uncleaved α₂δ-1 was identified in DRG neuron cell bodies (Kadurin et al., 2016). Biochemical and other studies also showed disulfide bonding between α₂ and δ (Calderon-Rivera et al., 2012; De Jongh et al., 1990).

GPI Anchoring

It was found that α₂δ subunits were strongly expressed in lipid rafts (Davies et al., 2006), and subsequently, it was identified that the α₂δ subunits were not transmembrane, but rather glycosylphosphatidylinositol (GPI) anchored, using both bioinformatic and biochemical techniques (Davies et al., 2010). This GPI anchoring has been confirmed from the structure of α₂δ-1 in the Ca$_V$1.1 and Ca$_V$2.2 channel complexes (Gao et al., 2021; Wu et al., 2016). GPI anchoring occurs immediately following translation in the endoplasmic reticulum (ER), when the preformed GPI anchor replaces the C-terminal hydrophobic sequence (Hooper, 2001) (Fig. 2a). GPI-anchored proteins have different properties from transmembrane proteins, for example, they are highly mobile and concentrated in cholesterol-rich membrane fractions (termed lipid rafts) (Hooper, 2001; Kadurin et al., 2012), and are rapidly endocytosed and recycled to the plasma membrane (Mayor & Riezman, 2004; Tran-Van-Minh & Dolphin, 2010).

VWA Domain

The presence of a von Willebrand factor-A (VWA) domain in the α₂δ subunits was identified bioinformatically (Whittaker & Hynes, 2002). The VWA domains were then found to be essential for the function of α₂δ to enhance Ca$_V$1 and Ca$_V$2 calcium currents (Canti et al., 2005) and for synaptic calcium channel localization (Schöpf et al., 2021). The VWA domain was predicted to be involved in interacting with the α₁ subunit, via its metal ion-dependent adhesion site (MIDAS)

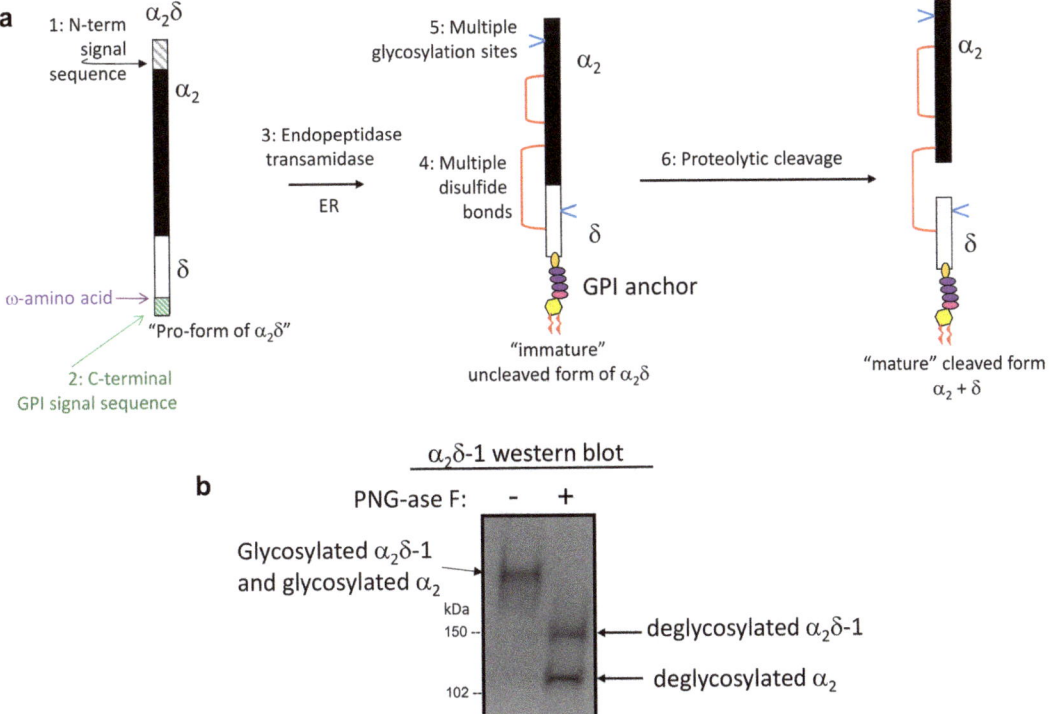

Fig. 2 Post-translational processing of $\alpha_2\delta$ subunits. (**a**) Diagram showing $\alpha_2\delta$ pre-protein (left) and subsequent post-translational processing steps (center), including removal of N-terminal signal sequence during translation (1), GPI anchor attachment within the ER (2, 3), formation of multiple disulfide bonds (4), glycosylation (5), and proteolytic cleavage of $\alpha_2\delta$ into α_2 and δ (6, right). (**b**) Western blot of material from tsA-201 cells transfected with $\alpha_2\delta$-1, showing that $\alpha_2\delta$-1 and cleaved α_2 cannot be clearly differentiated (left lane) unless deglycosylated with PNGase F (right lane). Primary antibody $\alpha_2\delta$-1 monoclonal recognizing an epitope within α_2. (Data courtesy of Dr. Ivan Kadurin)

motif (Canti et al., 2005). This was shown to be the case from the structure of the $Ca_V1.1$ and $Ca_V2.2$ channel complexes, where the MIDAS motif in $\alpha_2\delta$-1 interacts with an aspartate in the first extracellular loop of domain I of the α_1 subunit (Gao et al., 2021; Wu et al., 2016) (Fig. 3a–c). In confirmation of this, mutation of this interaction site in $Ca_V1.2$ and $Ca_V2.2$ also prevented the effect of $\alpha_2\delta$-1 (Bourdin et al., 2017; Dahimene et al., 2018).

Cache Domains

Two Cache domains were identified bioinformatically in $\alpha_2\delta$ subunits (Anantharaman & Aravind, 2000); these Cache domains are very similar in structure to those identified in bacterial chemore-

ceptors and chemotransducers, where they are often involved in nutrient sensing and chemotaxis (Gumerov et al., 2022). The cryo-EM structure has identified four Cache domains in $\alpha_2\delta$-1 (Fig. 3a, b).

Proteolytic Maturation of $\alpha_2\delta$

The $\alpha_2\delta$ pre-protein encodes α_2 at its N-terminus, followed by the shorter δ sequence. Following translation in the ER, the uncleaved $\alpha_2\delta$ begins to be cleaved in the Golgi apparatus (Kadurin et al., 2017), but the two "subunits" remain associated by pre-formed disulfide bonds, created as the protein folds in the ER. Proteolytic cleavage of $\alpha_2\delta$ appears to represent an activation step for

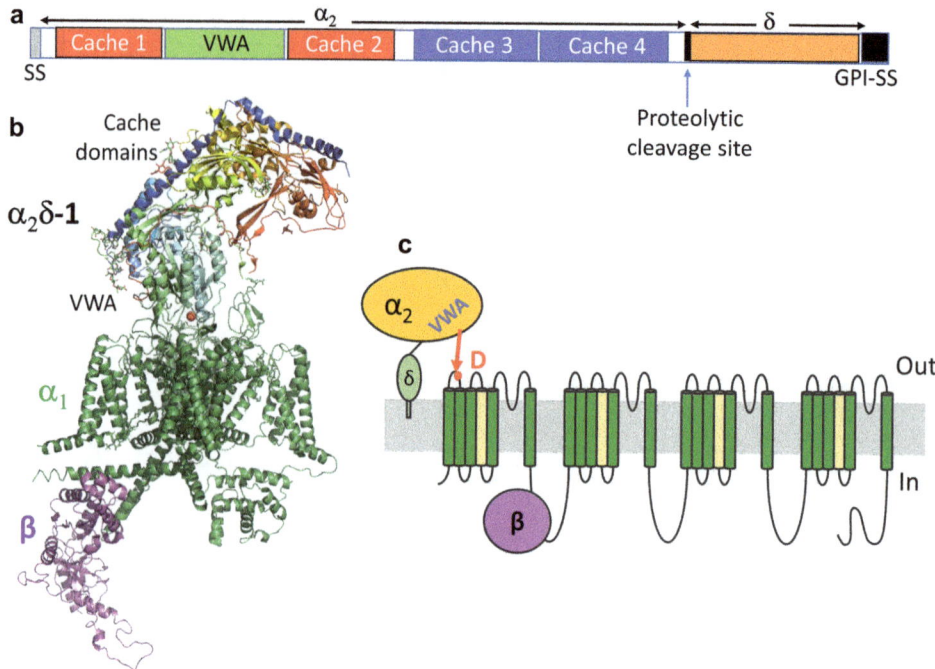

Fig. 3 Domains in $\alpha_2\delta$-1 and interaction with the α_1 subunit. (**a**) Domain structure for $\alpha_2\delta$ showing four Cache domains, with the VWA domain inserted between Cache 1 and Cache 2. (**b**) Cryo-EM structure of the $Ca_V2.2$ complex (Gao et al., 2021), showing the α_1 subunit (green), the β_3 subunit (magenta), and the $\alpha_2\delta$-1 subunit (rainbow). The MIDAS interaction site of the $\alpha_2\delta$-1 VWA domain with the α_1 subunit contains a co-ordinating divalent cation (red sphere). Image prepared from pdb 7miy using Pymol. (**c**) Diagram showing interaction site of the MIDAS motif in the VWA domain of α_2 with an aspartate residue (D) in the first extracellular loop of the α_1 subunit

Ca_V channel function (Ferron et al., 2018; Kadurin et al., 2016) (Fig. 2a).

Cachd1 Protein

Cachd1 was initially identified bioinformatically (although misnamed) (Whittaker & Hynes, 2002) and later confirmed as an $\alpha_2\delta$-like protein (Cottrell et al., 2018; Dahimene et al., 2018) containing an imperfect VWA domain and Cache domains. However, the MIDAS motif in the VWA domain of Cachd1 is highly disrupted and non-functional in terms of enhancement of $Ca_V2.2$ function (Dahimene et al., 2018). Cachd1 also produced differential functional effects on particular calcium channels (see section "Cachd1 function") (Cottrell et al., 2018; Dahimene et al., 2018).

Molecular Structure of $\alpha_2\delta$-1

The predicted domains and post-translational modifications in $\alpha_2\delta$ proteins are all supported by the structure of $\alpha_2\delta$-1 in the skeletal muscle calcium channel ($Ca_V1.1$) complex, which was obtained by cryo-electron microscopy (EM) (Wu et al., 2016). In summary of the salient points, the VWA domain was found to interact with an aspartate on the first extracellular loop on the α_1S calcium channel, which coordinated divalent cation binding with the MIDAS motif in $\alpha_2\delta$-1. No transmembrane domain was detected for $\alpha_2\delta$, in agreement with the evidence that it is GPI anchored (Davies et al., 2010). Four Cache domains were identified in $\alpha_2\delta$, with the VWA domain inserted between the first two (Wu et al., 2016). The structure of $\alpha_2\delta$-1 was found to be very similar in the $Ca_V2.2$ complex obtained following expression of the subunits in HEK293T cells, and subsequent purification of the complex (Gao et al., 2021).

Functions of $\alpha_2\delta$ Subunits as Calcium Channel Subunits

Effects of Cloned $\alpha_2\delta$ Subunits on Calcium Currents

In most studies $\alpha_2\delta$ subunits have been found to increase Ca_V1- and Ca_V2-mediated currents, although this function is partly dependent on the α_1 isoform, as knockdown of $\alpha_2\delta$-1 in skeletal muscle cells did not reduce $Ca_V1.1$-mediated currents (Meyer et al., 2019; Obermair et al., 2005), whereas the effect on Ca_V2 channel current density was more pronounced (Barclay et al., 2001; Canti et al., 2005; Hendrich et al., 2008; Obermair et al., 2008). Co-expression studies showed that $\alpha_2\delta$-1 increased Ca_V calcium currents that were mediated, for example, by $Ca_V2.1/\beta4$ (Gurnett et al., 1996, 1997), $Ca_V1.2/\beta2a$, and $Ca_V2.3/\beta3$ (Klugbauer et al., 1999; Yamaguchi et al., 2000). Moreover, for $Ca_V1.2$, $Ca_V2.2$, and $Ca_V2.3$, the different auxiliary β subunits were found to regulate Ca_V current properties synergistically with $\alpha_2\delta$-1 in a β subunit-specific manner (Yasuda et al., 2004). Similarly, $\alpha_2\delta$-2 increased $Ca_V2.1/\beta4$ calcium currents (Barclay et al., 2001; Brodbeck et al., 2002) as well as $Ca_V1.2/\beta1b$ and $Ca_V2.2/\beta1b$ currents (Canti et al., 2005). The $\alpha_2\delta$-3 subunit also increased $Ca_V1.2/\beta2a$ and $Ca_V2.3/\beta3$ (Klugbauer et al., 1999) and $\alpha_2\delta$-4, the least well-studied $\alpha_2\delta$ subunit, increased $Ca_V1.2/\beta3$-mediated calcium influx (Qin et al., 2002). The results of these studies always need to be prefaced by the finding that some heterologous expression systems may contain endogenous calcium currents (Berjukow et al., 1996), as well as auxiliary β (Canti et al., 2001) and $\alpha_2\delta$ subunits (Kadurin et al., 2012).

Effects of $\alpha_2\delta$ Subunits on Biophysical Properties of Calcium Currents

No change in single-channel conductance has been observed that can be attributed to $\alpha_2\delta$ subunits (Brodbeck et al., 2002; Wakamori et al., 1999), indicating the main effects are not on permeation. However, $\alpha_2\delta$-1 was found to reduce the percentage of null sweeps in single-channel recordings of $Ca_V2.2$ (Wakamori et al., 1999) and shRNA knockdown of $\alpha_2\delta$-1 slightly reduced the open probability of $Ca_V1.2$ (Tuluc et al., 2007). The $\alpha_2\delta$ subunits also affect the kinetics of activation and inactivation (Canti et al., 2003; Obermair et al., 2005, 2008; Tuluc et al., 2007; Wakamori et al., 1999), and voltage sensor movement for $Ca_V1.2$ (Savalli et al., 2016) and result in a hyperpolarization of both current activation (Felix et al., 1997; Savalli et al., 2016) and inactivation (Canti et al., 2003).

Effects of $\alpha_2\delta$ Subunits on Calcium Channel Trafficking

$\alpha_2\delta$ subunits have been found to increase the plasma membrane expression of $Ca_V2.2$ (Cassidy et al., 2014). This is the case for both exon 37a and 37b-containing C-terminal splice variants of $Ca_V2.2$ (Macabuag & Dolphin, 2015), which show differential gating (Castiglioni et al., 2006), cell surface expression (Macabuag & Dolphin, 2015), and expression in the pain pathway (Bell et al., 2004). Furthermore, presynaptic clustering of $Ca_V2.1$ and $Ca_V2.2$ was strongly reduced in $\alpha_2\delta$ triple knockout/knockdown neurons (Schöpf et al., 2021).

However, for $Ca_V2.2$, it has been shown that the increase in cell surface expression is generally not to the same extent as the increase in calcium current density, which may be up to 12-fold (Hoppa et al., 2012). The increase in cell surface expression has also been found to be greater for $\alpha_2\delta$-1/2 than for $\alpha_2\delta$-3, and is likely to stem from an increase in net forward trafficking of $Ca_V2.2$,

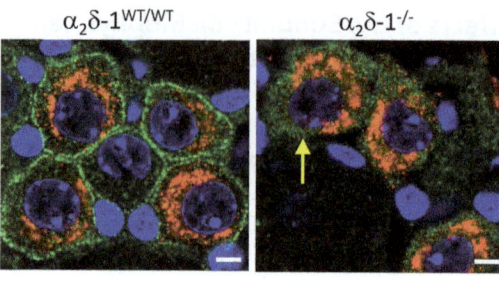

$\alpha_2\delta\text{-}1^{WT/WT}$ $\alpha_2\delta\text{-}1^{-/-}$

Ca$_V$2.2-HA CGRP DAPI

Fig. 4 Effect of $\alpha_2\delta$-1 knockout on Ca$_V$2.2-HA distribution in mouse DRG neurons. DRG sections, showing immunostaining for HA (green) in Ca$_V$2.2_HA$^{KI/KI}$ DRGs, co-stained with CGRP (red). Nuclei are stained with 4′,6-diamidino-2-phenylindole (DAPI, blue). Left: $\alpha_2\delta$-1$^{+/+}$ DRGs, showing Ca$_V$2.2 presence at the cell surface; right: $\alpha_2\delta$-1$^{-/-}$ DRGs, showing no Ca$_V$2.2 at the cell surface (e.g., yellow arrow). Scale bar 5 μm. (Images taken from Nieto-Rostro et al., 2018 under CC-BY 4.0 license)

which is much more pronounced for $\alpha_2\delta$-1/2 than for $\alpha_2\delta$-3 (Meyer & Dolphin, 2021), with no effect on endocytosis (Cassidy et al., 2014). Ca$_V$ channels with mutations in their selectivity filter, such that they do not conduct Ca^{2+}, show defective trafficking (Meyer et al., 2019). However, the trafficking of these permeation-defective channels is still enhanced by $\alpha_2\delta$-1 (Meyer et al., 2019), indicating that $\alpha_2\delta$ subunits do not represent a checkpoint such that only functional Ca$_V$ channels respond to them with increased trafficking.

The development of a knock-in mouse containing an epitope tag in endogenous Ca$_V$2.2 has allowed the effect of knockout of $\alpha_2\delta$-1 to be investigated on the distribution of endogenous Ca$_V$2.2 (Nieto-Rostro et al., 2018). This study showed that Ca$_V$2.2 is strongly expressed on the cell surface of particular types of nociceptive DRG neurons and in their presynaptic terminals in the dorsal horn (Fig. 4). Knockout of $\alpha_2\delta$-1 dramatically reduced the cell surface expression of Ca$_V$2.2 and presynaptic distribution (Fig. 4) (Nieto-Rostro et al., 2018), confirming the importance of $\alpha_2\delta$-1 proteins in Ca$_V$2.2 trafficking within the pain pathway.

In contrast, the Ca$_V$1 channels appear to show a smaller trafficking response to $\alpha_2\delta$ subunits than the Ca$_V$2 channels, which may reflect their greater membrane stability in cardiac and skeletal mus-

cle. For example, knockdown of $\alpha_2\delta$-1 in muscle cells did not affect targeting and membrane expression of Ca$_V$1.1 and Ca$_V$1.2 and only modestly decreased Ca$_V$1.2 current density, while it strongly affected activation and inactivation kinetics (Obermair et al., 2005; Tuluc et al., 2007).

Cachd1 Function

The roles of the Cache domains in $\alpha_2\delta$ subunits remain unclear. Relevant to this, the $\alpha_2\delta$-like Cachd1 protein was found to produce an increase in Ca$_V$2.2 but not Ca$_V$2.1 currents (Dahimene et al., 2018), and quite contrary to $\alpha_2\delta$ subunits, Cachd1 was also found to produce an increase in Ca$_V$3 T-type channel currents (Cottrell et al., 2018). The basis for this differential selectivity remains unclear and might relate to specific splice variants, or interactions with specific extracellular elements of the different Ca$_V$ channels. Although Cachd1 was found to increase Ca$_V$2.2 currents, this was to a much smaller extent than observed with $\alpha_2\delta$ subunits (Dahimene et al., 2018), and it also increased Ca$_V$2.2 cell surface expression in parallel (Dahimene et al., 2018). However, unlike $\alpha_2\delta$-1, the interaction with Cachd1 did not require its imperfect VWA domain interacting with the first extracellular loop of the Ca$_V$2.2 α_1 subunit (Dahimene et al., 2018). Therefore, for Ca$_V$2.2, it is likely to involve other interactions, for example, between the Cache domains of Cachd1 and extracellular domains of the α_1 subunit.

Synaptic Functions of $\alpha_2\delta$ Proteins Beyond Their Role as Calcium Channel Subunits

Importance of $\alpha_2\delta$ Proteins in Neuronal and Synaptic Functions

Several studies implicate an involvement of $\alpha_2\delta$ proteins in neuronal and synaptic functions, which partly go beyond their role as Ca$_V$ channel subunits (Chen et al., 2018; Eroglu et al., 2009;

Geisler et al., 2019; Kurshan et al., 2009; Schöpf et al., 2021). Most importantly, $\alpha_2\delta$ proteins have emerged as critical regulators of synapse formation and differentiation in central nervous system neurons (Eroglu et al., 2009; Schöpf et al., 2021) and expression of specific $\alpha_2\delta$ isoforms during distinct developmental phases may shape the structural and functional neuronal network connectivity (Bikbaev et al., 2020). While postsynaptically located $\alpha_2\delta$-1 subunits have been found to be relevant for basic neuronal differentiation and for modulating postsynaptic signaling (Eroglu et al., 2009), for example, by serving as a receptor for the astrocyte-secreted and synaptogenic thrombospondins (TSPs), presynaptic $\alpha_2\delta$ proteins serve dual purposes: on one hand, they regulate the abundance of presynaptic calcium channels and hence can directly regulate the efficacy of synaptic transmission (Ferron et al., 2018; Hoppa et al., 2012; Schöpf et al., 2021). On the other hand, they regulate synapse formation, differentiation, and the trans-synaptic recruitment of postsynaptic receptors (Fell et al., 2016; Geisler et al., 2019; Schöpf et al., 2021).

Postsynaptic Functions of $\alpha_2\delta$-1

In retinal ganglion cells the interaction of postsynaptic $\alpha_2\delta$-1 and astrocyte-secreted thrombospondin-1 (TSP-1) mediates excitatory synapse formation in a mechanism, which is independent of the presence and function of α_1 subunits (Eroglu et al., 2009) but involves the recruitment of NMDA receptors to the postsynaptic membrane (Risher et al., 2018). A role in synapse formation is further supported by impaired excitatory synaptogenesis and spine morphology in conditional (Risher et al., 2018) and constitutive (Bikbaev et al., 2020) $\alpha_2\delta$-1 knockout mice. TSP-4 was also identified as an $\alpha_2\delta$-1 ligand (Eroglu et al., 2009), although the interaction was subsequently found to be weak (El-Awaad et al., 2019; Lana et al., 2016), and less evident than for another known TSP ligand, low-density lipoprotein receptor-related protein 1 (LRP1) (Lana

et al., 2016). However, and in contrast to TSP-1, this interaction likely occurs with presynaptic $\alpha_2\delta$ subunits (Yu et al., 2018).

Presynaptic and Trans-synaptic Roles of $\alpha_2\delta$ Proteins

We recently showed that in cultured hippocampal neurons presynaptic triple knockout/knockdown of all brain $\alpha_2\delta$ isoforms severely compromised synapse formation and pre- and postsynaptic differentiation (Fig. 5) (Schöpf et al., 2021). This striking phenotype could be rescued by the expression of each individual $\alpha_2\delta$ isoform, suggesting a surprisingly redundant presynaptic and trans-synaptic role in this critical neuronal function. Trans-synaptic signaling via $\alpha_2\delta$ proteins is supported by other recent observations. Expression of an $\alpha_2\delta$-2 splice variant lacking exon 23 (splice site C, see above) in hippocampal neurons is sufficient to trigger the aberrant wiring of presynaptic excitatory (glutamatergic) axons to inhibitory (GABAergic) postsynaptic sites (Fig. 6). Furthermore, this $\alpha_2\delta$-2 splice variant, when expressed in presynaptic nerve terminals, regulates postsynaptic $GABA_A$ receptor ($GABA_AR$) abundance, both in aberrantly wired synapses as well as normally wired inhibitory synapses (Geisler et al., 2019). Also, in inner hair cell synapses of the cochlea, $\alpha_2\delta$-2 is necessary for the proper spatial alignment of presynaptic L-type $Ca_V1.3$ calcium channels and postsynaptic AMPA receptors (Fell et al., 2016). The $\alpha_2\delta$-2 isoform is also involved in modulating axonal regeneration after injury and establishing neuronal circuits (Tedeschi et al., 2016). Synaptic functions of $\alpha_2\delta$-3 so far have mainly been addressed in invertebrate model systems, where it was involved in regulating the size and morphology of motoneuron terminals (Caylor et al., 2013; Kurshan et al., 2009). A role for $\alpha_2\delta$-3 in regulating presynaptic differentiation has also been confirmed in vertebrates, by studying the consequences of $\alpha_2\delta$-3 knockout on auditory nerve fibers (Pirone et al., 2014) and the expression of $\alpha_2\delta$-3 in GABAergic neurons (Geisler

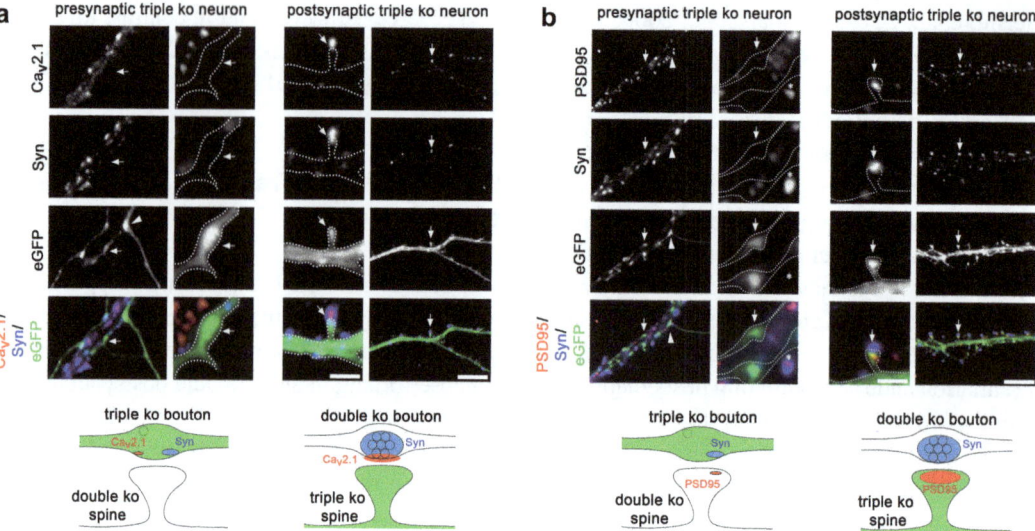

Fig. 5 Presynaptic $\alpha_2\delta$ subunits mediate excitatory synapse formation and trans-synaptic differentiation. Immunofluorescence labeling of eGFP-labeled axons with presynaptic boutons (**a** and **b**, left panels, arrows) and postsynaptic dendrites with dendritic spines (**a** and **b**, right panels, arrows) from presynaptic or postsynaptic $\alpha_2\delta$ triple knockout/knockdown hippocampal neurons. The sketches summarize the observed labelling patterns. (Images taken from Figure 4 of Schöpf et al., 2021 under a CC BY 4.0 license). Scale bars, 2 μm (selection) and 8 μm (overview). (**a**) Presynaptic triple knockout/knockdown of $\alpha_2\delta$-1, $\alpha_2\delta$-2, and $\alpha_2\delta$-3 results in failed calcium channel and synapsin clustering (left panel, arrows and

sketch). In contrast, dendritic spines opposite presynaptic boutons containing Cav2.1 and synapsin clusters develop normal when all $\alpha_2\delta$ subunits have been knocked out/down in the postsynaptic neuron (right panel, arrows and sketch). (**b**) Presynaptic triple knockout/knockdown of $\alpha_2\delta$-1, $\alpha_2\delta$-2, and $\alpha_2\delta$-3 also affects postsynaptic differentiation as seen by missing PSD95 clustering (left panel, arrows and sketch) opposite defective presynaptic terminals. In contrast, postsynaptic $\alpha_2\delta$ subunit triple knockout/knockdown does not affect PSD95 clustering opposite presynaptic terminals containing Cav2.1 and synapsin clusters (right panel, arrows and sketch)

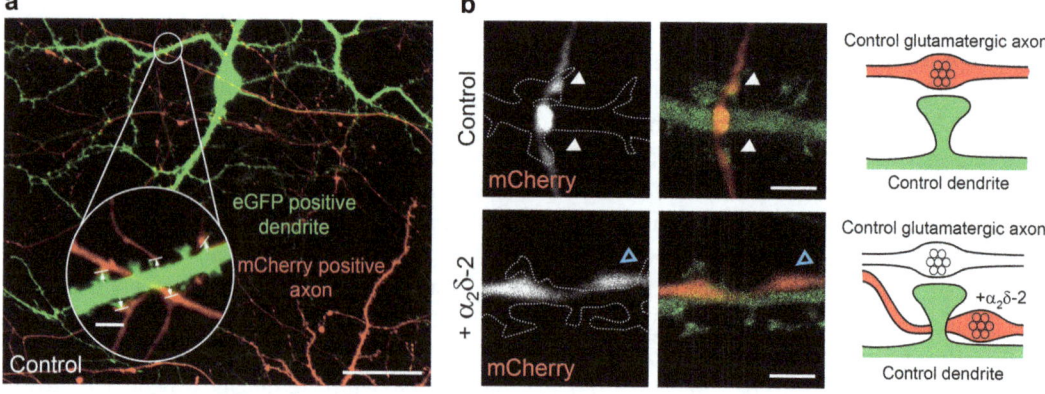

Fig. 6 Presynaptic $\alpha_2\delta$-2 induces aberrant wiring of excitatory synapses. (**a**) Experimental setup to analyze the position of synapses on postsynaptic dendrites of excitatory cultured hippocampal neurons: Presynaptic control or $\alpha_2\delta$-2 expressing neurons were labeled with mCherry (red) and postsynaptic neurons with eGFP (green). The magnified inset shows the contact of presynaptic axonal boutons (mCherry) with postsynaptic dendritic spines (eGFP), as expected for excitatory spine synapses. Scale bars, 50 μm (overview), 3 μm (inset). (**b**) Super-resolution

gSTED microscopy confirms the preferential location of excitatory synapses on dendritic spines of excitatory glutamatergic neurons (white arrowheads and sketch). Strikingly, presynaptic axons expressing $\alpha_2\delta$-2_ΔE23 aberrantly wire to postsynaptic sites along dendritic shafts (blue arrowheads and sketch), a position typically observed for GABAergic synapses. Scale bars, 2 μm. (Images taken from Figure 11 of Geisler et al., 2019 under a CC BY 4.0 license) et al., 2019). In the context of $\alpha_2\delta$ functions independent of the calcium channel complex, it is noteworthy that a quantitative study of the neuronal Ca$_V$2 channel proteome suggested a considerably weaker association of $\alpha_2\delta$ with α_1 subunits when compared to β subunits, although this was dependent on the detergent used for membrane solubilization (Müller et al., 2010). Furthermore, α_1 and $\alpha_2\delta$ subunit interactions in the neuronal membrane are likely dynamic and association may only be transient, as suggested by single-molecule live-cell imaging (Brockhaus et al., 2018; Schneider et al., 2015; Voigt et al., 2016). In retinal photoreceptor cells, where the L-type channel Ca$_V$1.4 serves as the main presynaptic channel of ribbon synapses, $\alpha_2\delta$-4 regulates functional membrane expression of α_1 subunits and synaptic transmission of rods (Wang et al., 2017) and cones (Kerov et al., 2018; Schlegel et al., 2019). In line with a trans-synaptic role of $\alpha_2\delta$ proteins, knockout $\alpha_2\delta$-4 affects not only presynaptic structure but also postsynaptic receptor clustering (Wang et al., 2017). Finally, the well-characterized mechanism of $\alpha_2\delta$-1 in neuropathic pain and the development of hyperalgesia are also mediated by presynaptic mechanisms (Bauer et al., 2009; Nieto-Rostro et al., 2018; Yu et al., 2018).

Diseases Associated with α₂δ Subunits

It is evident from the above that $\alpha_2\delta$ proteins are abundantly expressed in various organs and particularly in the brain, and that they are involved in critical calcium channel-dependent as well as-independent functions. Hence, it is not surprising that calcium channel dysfunctions are associated with a large variety of disorders. Over recent years an increasing number of disorders in humans have been linked to the genes encoding $\alpha_2\delta$ subunits (reviewed in Ablinger et al., 2020), however, so far detailed insights into pathophysiological mechanisms are only beginning to emerge.

Ducky Mice and Human Mutations of the α₂δ-2 Gene

A number of recessive mutations in *Cacna2d2* underlie the cerebellar ataxia and epilepsy phenotypes seen in spontaneously arising mouse mutants *ducky*, *ducky²ʲ*, and *entla*, as well as engineered *Cacna2d2* knockout mice (Barclay et al., 2001; Brill et al., 2004; Brodbeck et al., 2002; Donato et al., 2006; Geisler et al., 2021; Ivanov et al., 2004). Furthermore, mutations in human *CACNA2D2* (located on chromosome 3p21.31) are associated with recessive epileptic encephalopathy and mental retardation (Edvardson et al., 2013; Pippucci et al., 2013; Punetha et al., 2019) and may be also be linked to schizophrenia (Rodríguez-López et al., 2018).

Mutations in the α₂δ-1 Gene Associated with Cardiac Phenotypes

CACNA2D1 (located on chromosome 7q21.11) mutations have been reported to be associated with cardiac dysfunction, including short QT syndrome (Templin et al., 2011) and Brugada syndrome (Burashnikov et al., 2010), although the effects of these mutations measured experimentally are rather inconsistent (Bourdin et al., 2015). In agreement, homozygous *Cacna2d1* knockout causes a mild cardiac phenotype in mice (Fuller-Bicer et al., 2009) and additionally increases the susceptibility for diabetes (Mastrolia et al., 2017). However, several cases have been identified of epileptic encephalopathy involving copy number variants that usually contain multiple genes including *CACNA2D1* (Mefford et al., 2011; Vergult et al., 2015). A recent study also found $\alpha_2\delta$-1 was a target for autoimmune encephalitis (Lee et al., 2021), and another study also showed the presence of $\alpha_2\delta$-1 auto-antibodies in cases of amyotrophic lateral sclerosis associated with type 2 diabetes (Shi et al., 2019).

Neuropathic Injury and the Role of $\alpha_2\delta$-1

Peripheral sensory nerve injury in a variety of rodent models, including spinal nerve injury and administration of chemotherapeutic drugs, results in an increase of $\alpha_2\delta$-1 mRNA in damaged DRG neurons (Bauer et al., 2009; Lana et al., 2014; Newton et al., 2001; Wang et al., 2002; Xiao et al., 2007). This gives rise to a corresponding increase of $\alpha_2\delta$-1 protein within the injured DRG cell bodies, and also in their primary afferent terminals (Bauer et al., 2009; Luo et al., 2001). Furthermore, there was a differential upregulation of a splice variant of $\alpha_2\delta$-1 that shows a lower affinity for ^{3}H-gabapentin (Lana et al., 2014). In agreement with these results, *Cacna2d1* knockout mice showed reduced sensitivity to mechanical stimulation and delayed onset of neuropathic mechanical hypersensitivity following peripheral nerve injury (Patel et al., 2013). Furthermore, $\alpha_2\delta$-1 overexpressing mice have been generated, and they show increased baseline response to painful stimuli (Li et al., 2006).

Disorders Associated with $\alpha_2\delta$-3

From knockout models of *Cacna2d3*, a clear role for $\alpha_2\delta$-3 in hearing has been identified (Pirone et al., 2014), but no related human mutations in *CACNA2D3* (located on chromosome 3p21.1) have yet been reported. However, independent studies qualify *CACNA2D3* as risk gene for autism spectrum disorders (see below). Furthermore, a genome-wide Drosophila RNAi screen for heat nociception identified *Cacna2d3* as a pain-related gene, which was supported by the phenotype of $\alpha_2\delta$-3 knockout mice and human single-nucleotide polymorphisms (SNPs) associated with pain conditions, although potential human disease mechanisms are still unknown (Neely et al., 2010).

Disorders Associated with $\alpha_2\delta$-4

Mutations in both mouse and human *CACNA2D4* (located on chromosome 12p13.33 in humans) cause dysfunction of photoreceptors, resulting in certain forms of night blindness (Wycisk et al., 2006a, b).

Psychiatric Disorders

An interesting common feature of $\alpha_2\delta$ subunits is that over recent years SNPs or genomic variations in all *CACNA2D* ($\alpha_2\delta$ subunit) genes have been linked to a spectrum of psychiatric disorders (Ablinger et al., 2020; Consortium, 2013). Indeed, SNPs in *CACNA2D2* and *CACNA2D4* exhibited statistically significant associations with disease, across multiple psychiatric conditions. More recently, an excess of several rare disruptive mutations in *CACNA2D1, CACNA2D2, CACNA2D3, and CACNA2D4* were observed in cases of schizophrenia (Purcell et al., 2014), as reviewed recently (Ablinger et al., 2020; Heyes et al., 2015). Furthermore, several mutations in *CACNA2D3*, including a splice site mutation (Iossifov et al., 2012), are potentially linked to autism spectrum disorders (De Rubeis et al., 2014; Guo et al., 2018). Similarly in *CACNA2D1* a de novo mutation has been linked to autism (Iossifov et al., 2014), and a recent exome sequencing study suggests *CACNA2D1* as candidate risk gene for neurodevelopmental disorders (Valentino et al., 2021). Furthermore, a partial *CACNA2D4* deletion has been associated with rare cases of late onset bipolar disorder (Van Den Bossche et al., 2012).

Pharmacology Involving $\alpha_2\delta$ Subunits

Therapeutic Uses of Gabapentinoids

Gabapentin was first identified as an anti-epileptic drug (Crawford et al., 1987), and then as having efficacy in painful neuropathies (Wiffen et al.,

2005). There are currently three therapeutically available gabapentinoid $\alpha_2\delta$ ligands which are used in chronic neuropathic pain conditions: gabapentin, pregabalin (Field et al., 1999; Li et al., 2011), and mirogabalin (Domon et al., 2018).

Gabapentinoid Drug Binding to $\alpha_2\delta$ Subunits and Potential Mechanisms of Action

The therapeutic target for gabapentin was identified as $\alpha_2\delta$-1 by purifying a brain protein that binds to ^{3}H-gabapentin (Brown & Gee, 1998; Gee et al., 1996). However, gabapentin has very little effect acutely on calcium currents (Hendrich et al., 2008; Kang et al., 2002; Martin et al., 2002), but chronic incubation with gabapentin for 24–48 h reduces calcium currents (Biggs et al., 2014; Hendrich et al., 2008), synaptic transmission (Hendrich et al., 2012; Lempel et al., 2017), and cell surface expression of $\alpha_2\delta$-1, $\alpha_2\delta$-2, and Ca_V2.2 (Cassidy et al., 2014; Tran-Van-Minh & Dolphin, 2010).

The binding site for gabapentin was found to include an RRR motif, which is just upstream of the VWA domain (Brown & Gee, 1998; Wang et al., 1999). This is present in $\alpha_2\delta$-1/2 but not the other $\alpha_2\delta$ subunits which do not bind gabapentin (Gong et al., 2001; Marais et al., 2001). If the third R is mutated to A in $\alpha_2\delta$-1 or $\alpha_2\delta$-2, this reduces the affinity of gabapentinoid binding and $\alpha_2\delta$ function (Field et al., 2006; Hendrich et al., 2008; Tran-Van-Minh & Dolphin, 2010; Wang et al., 1999). It also reduces the response to gabapentinoids in chronic neuropathic pain models (Field et al., 2006) and in experimental epilepsy and anxiety models (Lotarski et al., 2011, 2014; Taylor et al., 2007). The skeletal muscle splice variant of $\alpha_2\delta$-1 (+A + BΔC) bound to ^{3}H-gabapentin with high affinity (Lana et al., 2014), in agreement with data from rat skeletal muscle $\alpha_2\delta$-1 (Gee et al., 1996). Thus, the lack of effect of gabapentin on skeletal muscle function is not a result of its inability to bind to the skeletal muscle $\alpha_2\delta$-1 isoform and could relate instead to the high stability of the calcium channel complex in skeletal muscle.

The First Double Cache Domain in $\alpha_2\delta$-1 Has Structural Homology to a Universal Amino Acid Binding Domain

A ubiquitous extracellular double Cache domain has been identified in bacterial chemoreceptors and chemotransducers that contains a simple amino acid recognition motif. In bacteria and archaea this has been found to exclusively bind a variety of amino acids, whose identity and binding affinity depends on the exact residues in the binding pocket (Gumerov et al., 2022). In eukaryotes the same motif is found only in $\alpha_2\delta$ proteins and Cachd1, and in $\alpha_2\delta$-1 the first double Cache domain (consisting of Cache 1 and Cache 2, Fig. 3a) contains the binding site for gabapentin (Gumerov et al., 2022). It has previously been shown that $\alpha_2\delta$-1 also binds leucine and isoleucine, and these compete with gabapentin (Brown et al., 1998). Whether endogenous amino acids have a physiological role in binding to $\alpha_2\delta$ proteins, for example, as positive or negative modulators of trafficking, is yet unknown.

Effect of $\alpha_2\delta$ Subunits on Ziconotide Binding

The presence of $\alpha_2\delta$ proteins was found to reduce the on-rate and equilibrium inhibition of the Ca_V2.2 channel blocking ω-conotoxins (Mould et al., 2004). This study included ω-conotoxin MVIIA or ziconotide, which is licensed for use in neuropathic pain conditions. A recent structural study has identified the mechanism for $\alpha_2\delta$ interfering with the drug binding site, as ziconotide alters the orientation of $\alpha_2\delta$-1 with respect to the channel (Gao et al., 2021).

Other Interactions of $\alpha_2\delta$ Proteins

The $\alpha_2\delta$s, being completely extracellular, and with structural domain similarities to other extracellular matrix and cell adhesion proteins (Whittaker & Hynes, 2002), are likely to have interactions with other proteins. For example, they may be associated with proteins involved in ion channel clustering. Indeed they have also variously been shown to either directly interact with or influence the function of, several other proteins, including the trafficking and endocytosis protein LRP1 (Kadurin et al., 2017), the extracellular matrix proteins, α-neurexins (Brockhaus et al., 2018; Tong et al., 2017), and other ion channels, such as BK channels (Zhang et al., 2018). Interactions of $\alpha_2\delta$-1 with NMDA receptors (Chen et al., 2018) and certain AMPA receptors (Li et al., 2021) have also been identified. The interaction of $\alpha_2\delta$-1 with these glutamate receptors was shown to involve the extreme C-terminus of $\alpha_2\delta$-1 which would normally be cleaved off during the process of GPI anchoring in the ER (Davies et al., 2010; Guizzunti & Zurzolo, 2014); thus, any interaction must either occur in the ER and prevent the formation of the GPI-anchored $\alpha_2\delta$-1, or the interaction occurs with the cleaved C-terminal GPI signal peptide, which has been shown to be rapidly degraded for other GPI-anchored proteins (Guizzunti & Zurzolo, 2014). This topic was recently reviewed (Dolphin, 2018). Generally, such potential interactions may also serve to co-locate multiple ion channels with neurotransmitter receptors and other signaling proteins within synaptic structures.

Summary and Outlook

Over the last 20 years, our appreciation of the Ca_V $\alpha_2\delta$ subunits has changed considerably: from a purely auxiliary channel subunit, which modulates some biophysical channel properties, to a ubiquitous, albeit enigmatic, signaling protein, which serves as an important drug target and regulates synaptic function. This remarkable evolution is based on several scientific developments: first, the study of calcium channels and hence $\alpha_2\delta$ subunits was conferred progressively from heterologous expression systems into native, differentiated cells and tissues. Second, knockout and mutant mouse models became available, which

highlighted the functional importance of $\alpha_2\delta$ proteins in distinct tissues and physiological functions. Third and particularly owing to the increasingly employed OMICs technologies, our insights into disease associations are expanding at an accelerated speed. Because $\alpha_2\delta$ proteins have been linked to a variety of diseases, ranging from hormone secretion to neuropsychiatric disorders (and many new disease associations may be identified in future years), pharmacological targeting of $\alpha_2\delta$ proteins bears a tremendous therapeutic potential. Indeed, gabapentin is used as an anti-epileptic drug, and all gabapentinoids are widely prescribed for treating neuropathic pain. However, before the full theoretical therapeutic potential can be exploited, a number of questions need to be addressed:

1. Several studies have identified and proposed $\alpha_2\delta$ isoform and even splice variant-specific functions and disease associations. However, the physiological importance of the functional redundancy between different $\alpha_2\delta$ isoforms is not yet understood. In this context it is noteworthy that the synaptic phenotype in a presynaptic $\alpha_2\delta$ triple knockout/knockdown model can be rescued by the expression of each brain $\alpha_2\delta$ isoform (Schöpf et al., 2021).

2. As discussed above, recent studies propose calcium channel-independent functions of $\alpha_2\delta$ proteins. However, any experimental condition or pathophysiological mechanism affecting the expression and function of $\alpha_2\delta$ subunits will also affect calcium channels; hence, the definitive distinction between calcium channel-dependent and -independent functions represents an experimental challenge. Yet, a thorough understanding is a prerequisite for novel therapeutic concepts, for example, for targeting trans-synaptic functions without altering/inhibiting calcium channels.

3. On one hand, OMICs technologies continue to identify novel disease associations for all $\alpha_2\delta$ isoforms (reviewed in Ablinger et al., 2020). On the other hand, detailed insights into the pathophysiological mechanisms are

limited to a few examples, such as the key role of α₂δ-1 upregulation in neuropathic pain (reviewed in Dolphin, 2016). The experimental challenge for future years thus lies in elucidating the mechanisms linking the α₂δ proteins to disease, both in terms of calcium channel-dependent and -independent functions.

4. Our current understanding about the α₂δ isoform and splice variant-specific expression, particularly during development and disease, is limited to tissues and brain regions. Investigating cell-type-specific expression patterns is another prerequisite for ultimately understanding specific functions, such as the establishment of network connectivity (Bikbaev et al., 2020). For example, in brain expression of the retinal α₂δ-4 isoform is extremely low, basically negligible (Schlick et al., 2010). Nevertheless, α₂δ-4 expression is strongly upregulated during development and disease (Schlick et al., 2010; van Loo et al., 2019). In theory this may be related to an important role of α₂δ-4 in a single and rare neuron type. Hence, establishing conditional α₂δ knockout mouse models may provide a first step into studying the cell-type-specific expression and function.

5. As discussed above, α₂δ proteins are emerging as critical regulators of synapse formation, trans-synaptic signaling, and axonal/synaptic wiring. Future novel therapeutic paradigms affecting the expression and function of α₂δ thus may have the potential to modulate synaptic wiring during and after development (Bikbaev et al., 2020; Geisler et al., 2019; Veroniki et al., 2017). Hence, the ethical basis for potentially interfering with neurodevelopment requires a discussion between scientists and affected individuals, for example, people with autism (Sanderson, 2021).

6. Finally, considering the newly discovered synaptic roles of α₂δ, the consequences on human behavior (e.g., learning) of the extremely widespread used gabapentinoid drugs need to be more thoroughly investigated. Nevertheless, it should be noted that a meta-analysis of adverse effects of antiepileptic drugs did not detect negative effects of gabapentin on cognitive development of children exposed during pregnancy or breast-feeding (Veroniki et al., 2017). Direct effects on synaptic transmission, synaptic wiring, and postsynaptic receptor clustering may ultimately qualify gabapentinoid drugs to be classified as a psychiatric medication. Because these drugs are predominantly prescribed for chronic pain conditions, they also bear a considerable abuse potential, which so far has not been sufficiently investigated (Goins et al., 2021).

Taken together, understanding the multiple functions of α₂δ proteins, which are ubiquitously expressed in excitable cells, has come a long way. Future research efforts will be aimed at elucidating pathophysiological mechanisms of associated disorders and may lead to novel therapeutic paradigms.

References

Ablinger, C., Geisler, S. M., Stanika, R. I., Klein, C. T., & Obermair, G. J. (2020). Neuronal alpha2delta proteins and brain disorders. *Pflügers Archiv, 472*, 845–863.

Anantharaman, V., & Aravind, L. (2000). Cache-a signalling domain common to animal Ca channel subunits and a class of prokaryotic chemotaxis receptors. *Trends in Biochemical Sciences, 25*, 535–537.

Angelotti, T., & Hofmann, F. (1996). Tissue-specific expression of splice variants of the mouse voltage-gated calcium channel α2/δ subunit. *FEBS Letters, 397*, 331–337.

Arikkath, J., & Campbell, K. P. (2003). Auxiliary subunits: Essential components of the voltage-gated calcium channel complex. *Current Opinion in Neurobiology, 13*, 298–307.

Barclay, J., & Rees, M. (2000). Genomic organization of the mouse and human α2δ2 voltage-dependent calcium channel subunit genes. *Mammalian Genome, 11*, 1142–1144.

Barclay, J., Balaguero, N., Mione, M., Ackerman, S. L., Letts, V. A., Brodbeck, J., Canti, C., Meir, A., Page, K. M., Kusumi, K., et al. (2001). Ducky mouse phenotype of epilepsy and ataxia is associated with mutations in the *Cacna2d2* gene and decreased calcium channel current in cerebellar Purkinje cells. *Journal of Neuroscience, 21*, 6095–6104.

Bauer, C. S., Nieto-Rostro, M., Rahman, W., Tran-Van-Minh, A., Ferron, L., Douglas, L., Kadurin, I., Sri Ranjan, Y., Fernandez-Alacid, L., Millar, N. S., et al. (2009). The increased trafficking of the calcium channel subunit α2δ-1 to presynaptic terminals in neuropathic pain is inhibited by the α2δ ligand pregabalin. *Journal of Neuroscience, 29,* 4076–4088.

Bell, T. J., Thaler, C., Castiglioni, A. J., Helton, T. D., & Lipscombe, D. (2004). Cell-specific alternative splicing increases calcium channel current density in the pain pathway. *Neuron, 41,* 127–138.

Berjukow, S., Doring, S., Froschmayr, M., Grabner, M., Glossmann, H., & Hering, S. (1996). Endogenous calcium channels in human embryonic kidney (HEK293) cells. *British Journal of Pharmacology, 118,* 748–754.

Biggs, J. E., Boakye, P. A., Ganesan, N., Stemkowski, P. L., Lantero, A., Ballanyi, K., & Smith, P. A. (2014). Analysis of the long-term actions of gabapentin and pregabalin in dorsal root ganglia and substantia gelatinosa. *Journal of Neurophysiology, 112,* 2398–2412.

Bikbaev, A., Ciuraszkiewicz-Wojciech, A., Heck, J., Klatt, O., Freund, R., Mitlohner, J., Enrile Lacalle, S., Sun, M., Repetto, D., Frischknecht, R., et al. (2020). Auxiliary alpha2delta1 and alpha2delta3 subunits of calcium channels drive excitatory and inhibitory neuronal network development. *The Journal of Neuroscience, 40,* 4824–4841.

Bourdin, B., Shakeri, B., Tetreault, M. P., Sauve, R., Lesage, S., & Parent, L. (2015). Functional characterization of CaValpha2delta mutations associated with sudden cardiac death. *The Journal of Biological Chemistry, 290,* 2854–2869.

Bourdin, B., Briot, J., Tetreault, M. P., Sauve, R., & Parent, L. (2017). Negatively charged residues in the first extracellular loop of the L-type CaV1.2 channel anchor the interaction with the CaValpha2delta1 auxiliary subunit. *The Journal of Biological Chemistry, 292,* 17236–17249.

Brickley, K., Campbell, V., Berrow, N., Leach, R., Norman, R. I., Wray, D., Dolphin, A. C., & Baldwin, S. (1995). Use of site-directed antibodies to probe the topography of the α_2 subunit of voltage-gated Ca^{2+} channels. *FEBS Letters, 364,* 129–133.

Brill, J., Klocke, R., Paul, D., Boison, D., Gouder, N., Klugbauer, N., Hofmann, F., Becker, C. M., & Becker, K. (2004). entla, a novel epileptic and ataxic Cacna2d2 mutant of the mouse. *The Journal of Biological Chemistry, 279,* 7322–7330.

Brockhaus, J., Schreitmuller, M., Repetto, D., Klatt, O., Reissner, C., Elmslie, K., Heine, M., & Missler, M. (2018). alpha-Neurexins together with alpha2delta-1 auxiliary subunits regulate Ca(2+) influx through Cav2.1 channels. *The Journal of Neuroscience, 38,* 8277–8294.

Brodbeck, J., Davies, A., Courtney, J.-M., Meir, A., Balaguero, N., Canti, C., Moss, F. J., Page, K. M., Pratt, W. S., Hunt, S. P., et al. (2002). The ducky mutation in *Cacna2d2* results in altered Purkinje cell morphology and is associated with the expression of a truncated *a2d*-2 protein with abnormal function. *Journal of Biological Chemistry, 277,* 7684–7693.

Brown, J. P., & Gee, N. S. (1998). Cloning and deletion mutagenesis of the α2δ calcium channel subunit from porcine cerebral cortex. *Journal of Biological Chemistry, 273,* 25458–25465.

Brown, J. P., Dissanayake, V. U., Briggs, A. R., Milic, M. R., & Gee, N. S. (1998). Isolation of the [3H] gabapentin-binding protein/alpha 2 delta Ca2+ channel subunit from porcine brain: Development of a radioligand binding assay for alpha 2 delta subunits using [3H]leucine. *Analytical Biochemistry, 255,* 236–243.

Buraei, Z., & Yang, J. (2010). The beta subunit of voltage-gated Ca2+ channels. *Physiological Reviews, 90,* 1461–1506.

Burashnikov, E., Pfeiffer, R., Barajas-Martinez, H., Delpon, E., Hu, D., Desai, M., Borggrefe, M., Haissaguerre, M., Kanter, R., Pollevick, G. D., et al. (2010). Mutations in the cardiac L-type calcium channel associated with inherited J-wave syndromes and sudden cardiac death. *Heart Rhythm, 7,* 1872–1882.

Calderon-Rivera, A., Andrade, A., Hernandez-Hernandez, O., Gonzalez-Ramirez, R., Sandoval, A., Rivera, M., Gomora, J. C., & Felix, R. (2012). Identification of a disulfide bridge essential for structure and function of the voltage-gated Ca(2+) channel alpha(2)delta-1 auxiliary subunit. *Cell Calcium, 51,* 22–30.

Canti, C., Davies, A., Berrow, N. S., Butcher, A. J., Page, K. M., & Dolphin, A. C. (2001). Evidence for two concentration-dependent processes for β subunit effects on α1B calcium channels. *Biophysical Journal, 81,* 1439–1451.

Canti, C., Davies, A., & Dolphin, A. C. (2003). Calcium channel alpha2delta subunits: Structure, function and target site for drugs. *Current Neuropharmacology, 1,* 209–217.

Canti, C., Nieto-Rostro, M., Foucault, I., Heblich, F., Wratten, J., Richards, M. W., Hendrich, J., Douglas, L., Page, K. M., Davies, A., et al. (2005). The metal-ion-dependent adhesion site in the Von Willebrand factor-A domain of alpha2delta subunits is key to trafficking voltage-gated Ca2+ channels. *Proceedings of the National Academy of Sciences of the United States of America, 102,* 11230–11235.

Cassidy, J. S., Ferron, L., Kadurin, I., Pratt, W. S., & Dolphin, A. C. (2014). Functional exofacially tagged N-type calcium channels elucidate the interaction with auxiliary alpha2delta-1 subunits. *Proceedings of the National Academy of Sciences of the United States of America, 111,* 8979–8984.

Castiglioni, A. J., Raingo, J., & Lipscombe, D. (2006). Alternative splicing in the C-terminus of CaV2.2 controls expression and gating of N-type calcium channels. *The Journal of Physiology, 576,* 119–134.

Caylor, R. C., Jin, Y., & Ackley, B. D. (2013). The Caenorhabditis elegans voltage-gated calcium channel subunits UNC-2 and UNC-36 and the calcium-dependent kinase UNC-43/CaMKII regulate neuromuscular junction morphology. *Neural Development, 8,* 10.

Chen, J., Li, L., Chen, S. R., Chen, H., Xie, J. D., Sirrieh, R. E., MacLean, D. M., Zhang, Y., Zhou, M. H., Jayaraman, V., et al. (2018). The alpha2delta-1-NMDA receptor complex is critically involved in neuropathic pain development and gabapentin therapeutic actions. *Cell Reports, 22,* 2307–2321.

Consortium, C.-d.g.o.P.G. (2013). Identification of risk loci with shared effects on five major psychiatric disorders: A genome-wide analysis. *Lancet, 381,* 1371–1379.

Cooper, C. L., Vandaele, S., Barhanin, J., Fosset, M., Lazdunski, M., & Hosey, M. M. (1987). Purification and characterization of the dihydropyridine-sensitive voltage-dependent calcium channel from cardiac tissue. *The Journal of Biological Chemistry, 262,* 509–512.

Cottrell, G. S., Soubrane, C. H., Hounshell, J. A., Lin, H., Owenson, V., Rigby, M., Cox, P. J., Barker, B. S., Ottolini, M., Ince, S., et al. (2018). CACHD1 is an alpha2delta-like protein that modulates CaV3 voltage-gated calcium channel activity. *The Journal of Neuroscience, 38*(43), 9186–9201.

Crawford, P., Ghadiali, E., Lane, R., Blumhardt, L., & Chadwick, D. (1987). Gabapentin as an antiepileptic drug in man. *Journal of Neurology, Neurosurgery, and Psychiatry, 50,* 682–686.

Curtis, B. M., & Catterall, W. A. (1984). Purification of the calcium antagonist receptor of the voltage-sensitive calcium channel from skeletal muscle transverse tubules. *Biochemistry, 23,* 2113–2118.

Dahimene, S., Page, K. M., Kadurin, I., Ferron, L., Ho, D. Y., Powell, G. T., Pratt, W. S., Wilson, S. W., & Dolphin, A. C. (2018). The α₂δ-like protein Cachd1 increases N-type calcium currents and cell surface expression and competes with α₂δ-1. *Cell Reports, 25,* 1610–1621.

Davies, A., Douglas, L., Hendrich, J., Wratten, J., Tran-Van-Minh, A., Foucault, I., Koch, D., Pratt, W. S., Saibil, H., & Dolphin, A. C. (2006). The calcium channel α₂δ-2 subunit partitions with CaV2.1 in lipid rafts in cerebellum: Implications for localization and function. *The Journal of Neuroscience, 26,* 8748–8757.

Davies, A., Kadurin, I., Alvarez-Laviada, A., Douglas, L., Nieto-Rostro, M., Bauer, C. S., Pratt, W. S., & Dolphin, A. C. (2010). The α₂δ subunits of voltage-gated calcium channels form GPI-anchored proteins, a post-translational modification essential for function. *Proceedings of the National Academy of Sciences of the United States of America, 107,* 1654–1659.

De Jongh, K. S., Merrick, D. K., & Catterall, W. A. (1989). Subunits of purified calcium channels: A 212-kDa form of α₁ and partial amino acid sequence of a phosphorylation site of an independent β subunit. *Proceedings of the National Academy of Sciences of the United States of America, 86,* 8585–8589.

De Jongh, K. S., Warner, C., & Catterall, W. A. (1990). Subunits of purified calcium channels. α2 and δ are encoded by the same gene. *Journal of Biological Chemistry, 265,* 14738–14741.

De Rubeis, S., He, X., Goldberg, A. P., Poultney, C. S., Samocha, K., Cicek, A. E., Kou, Y., Liu, L., Fromer, M., Walker, S., et al. (2014). Synaptic, transcriptional and chromatin genes disrupted in autism. *Nature, 515,* 209–215.

Dolphin, A. C. (2009). Calcium channel diversity: Multiple roles of calcium channel subunits. *Current Opinion in Neurobiology, 19,* 237–244.

Dolphin, A. C. (2016). Voltage-gated calcium channels and their auxiliary subunits: Physiology and pathophysiology and pharmacology. *The Journal of Physiology, 594,* 5369–5390.

Dolphin, A. C. (2018). Voltage-gated calcium channel alpha 2delta subunits: An assessment of proposed novel roles. *F1000Res, 7,* 1830.

Domon, Y., Arakawa, N., Inoue, T., Matsuda, F., Takahashi, M., Yamamura, N., Kai, K., & Kitano, Y. (2018). Binding characteristics and analgesic effects of mirogabalin, a novel ligand for the alpha2delta subunit of voltage-gated calcium channels. *The Journal of Pharmacology and Experimental Therapeutics, 365,* 573–582.

Donato, R., Page, K. M., Koch, D., Nieto-Rostro, M., Foucault, I., Davies, A., Wilkinson, T., Rees, M., Edwards, F. A., & Dolphin, A. C. (2006). The ducky2J mutation in Cacna2d2 results in reduced spontaneous Purkinje cell activity and altered gene expression. *The Journal of Neuroscience, 26,* 12576–12586.

Edvardson, S., Oz, S., Abulhijaa, F. A., Taher, F. B., Shaag, A., Zenvirt, S., Dascal, N., & Elpeleg, O. (2013). Early infantile epileptic encephalopathy associated with a high voltage gated calcium channelopathy. *Journal of Medical Genetics, 50,* 118–123.

El-Awaad, E., Pryymachuk, G., Fried, C., Matthes, J., Isensee, J., Hucho, T., Neiss, W. F., Paulsson, M., Herzig, S., Zaucke, F., et al. (2019). Direct, gabapentin-insensitive interaction of a soluble form of the calcium channel subunit alpha2delta-1 with thrombospondin-4. *Scientific Reports, 9,* 16272.

Ellis, S. B., Williams, M. E., Ways, N. R., Brenner, R., Sharp, A. H., Leung, A. T., Campbell, K. P., McKenna, E., Koch, W. J., Hui, A., et al. (1988). Sequence and expression of mRNAs encoding the α₁ and α₂ subunits of a DHP-sensitive calcium channel. *Science, 241,* 1661–1664.

Eroglu, C., Allen, N. J., Susman, M. W., O'Rourke, N. A., Park, C. Y., Ozkan, E., Chakraborty, C., Mulinyawe, S. B., Annis, D. S., Huberman, A. D., et al. (2009). Gabapentin receptor alpha2delta-1 is a neuronal thrombospondin receptor responsible for excitatory CNS synaptogenesis. *Cell, 139,* 380–392.

Felix, R., Gurnett, C. A., De Waard, M., & Campbell, K. P. (1997). Dissection of functional domains of the voltage-dependent Ca2+ channel alpha2delta subunit. *Journal of Neuroscience, 17,* 6884–6891.

Fell, B., Eckrich, S., Blum, K., Eckrich, T., Hecker, D., Obermair, G. J., Munkner, S., Flockerzi, V., Schick, B., & Engel, J. (2016). alpha2delta2 controls the function and trans-synaptic coupling of Cav1.3 channels in mouse inner hair cells and is essential for normal hearing. *The Journal of Neuroscience, 36,* 11024–11036.

Ferron, L., Kadurin, I., & Dolphin, A. C. (2018). Proteolytic maturation of alpha2delta controls the probability of synaptic vesicular release. *eLife, 7,* e37507.

Field, M. J., McCleary, S., Hughes, J., & Singh, L. (1999). Gabapentin and pregabalin, but not morphine and amitriptyline, block both static and dynamic components of mechanical allodynia induced by streptozocin in the rat. *Pain, 80*, 391–398.

Field, M. J., Cox, P. J., Stott, E., Melrose, H., Offord, J., Su, T. Z., Bramwell, S., Corradini, L., England, S., Winks, J., et al. (2006). Identification of the α2δ-1 subunit of voltage-dependent calcium channels as a novel molecular target for pain mediating the analgesic actions of pregabalin. *Proceedings of the National Academy of Sciences of the United States of America, 103*, 17537–17542.

Fuller-Bicer, G. A., Varadi, G., Koch, S. E., Ishii, M., Bodi, I., Kadeer, N., Muth, J. N., Mikala, G., Petrashevskaya, N. N., Jordan, M. A., et al. (2009). Targeted disruption of the voltage-dependent Ca2+ channel {alpha}2/{delta}-1 subunit. *American Journal of Physiology. Heart and Circulatory Physiology, 297*, H117–H124.

Gao, S., Yao, X., & Yan, N. (2021). Structure of human Cav2.2 channel blocked by the painkiller ziconotide. *Nature, 596*(7870), 143–147.

Gee, N. S., Brown, J. P., Dissanayake, V. U. K., Offord, J., Thurlow, R., & Woodruff, G. N. (1996). The novel anticonvulsant drug, gabapentin (Neurontin), binds to the α2δ subunit of a calcium channel. *Journal of Biological Chemistry, 271*, 5768–5776.

Geisler, S., Schopf, C. L., Stanika, R., Kalb, M., Campiglio, M., Repetto, D., Traxler, L., Missler, M., & Obermair, G. J. (2019). Presynaptic alpha2delta-2 calcium channel subunits regulate postsynaptic GABAA receptor abundance and axonal wiring. *The Journal of Neuroscience, 39*, 2581–2605.

Geisler, S. M., Benedetti, A., Schopf, C. L., Schwarzer, C., Stefanova, N., Schwartz, A., & Obermair, G. J. (2021). Phenotypic characterization and brain structure analysis of calcium channel subunit alpha2delta-2 mutant (ducky) and alpha2delta double knockout mice. *Frontiers in Synaptic Neuroscience, 13*, 634412.

Goins, A., Patel, K., & Alles, S. R. A. (2021). The gabapentinoid drugs and their abuse potential. *Pharmacology & Therapeutics, 227*, 107926.

Gong, H. C., Hang, J., Kohler, W., Li, L., & Su, T. Z. (2001). Tissue-specific expression and gabapentin-binding properties of calcium channel alpha2delta subunit subtypes. *Journal of Membrane Biology, 184*, 35–43.

Guizzunti, G., & Zurzolo, C. (2014). The fate of PrP GPI-anchor signal peptide is modulated by P238S pathogenic mutation. *Traffic, 15*, 78–93.

Gumerov, V. M., Andrianova, E. P., Matilla, M. A., Page, K. M., Dolphin, A. C., Krell, T., & Zhulin, I. B. (2022). Amino acid sensor conserved from bacteria to humans. *PNAS (USA). 119, e2110415119.*

Guo, H., Wang, T., Wu, H., Long, M., Coe, B. P., Li, H., Xun, G., Ou, J., Chen, B., Duan, G., et al. (2018). Inherited and multiple de novo mutations in autism/developmental delay risk genes suggest a multifactorial model. *Molecular Autism, 9*, 64.

Gurnett, C. A., De Waard, M., & Campbell, K. P. (1996). Dual function of the voltage-dependent Ca2+ channel α2δ subunit in current stimulation and subunit interaction. *Neuron, 16*, 431–440.

Gurnett, C. A., Felix, R., & Campbell, K. P. (1997). Extracellular interaction of the voltage-dependent Ca2+ channel α2δ and α1 subunits. *Journal of Biological Chemistry, 272*, 18508–18512.

Hendrich, J., Tran-Van-Minh, A., Heblich, F., Nieto-Rostro, M., Watschinger, K., Striessnig, J., Wratten, J., Davies, A., & Dolphin, A. C. (2008). Pharmacological disruption of calcium channel trafficking by the α2δ ligand gabapentin. *Proceedings of the National Academy of Sciences of the United States of America, 105*, 3628–3633.

Hendrich, J., Bauer, C. S., & Dolphin, A. C. (2012). Chronic pregabalin inhibits synaptic transmission between rat dorsal root ganglion and dorsal horn neurons in culture. *Channels (Austin, Tex.), 6*, 124–132.

Heyes, S., Pratt, W. S., Rees, E., Dahimene, S., Ferron, L., Owen, M. J., & Dolphin, A. C. (2015). Genetic disruption of voltage-gated calcium channels in psychiatric and neurological disorders. *Progress in Neurobiology, 134*, 36–54.

Hooper, N. M. (2001). Determination of glycosylphosphatidylinositol membrane protein anchorage. *Proteomics, 1*, 748–755.

Hoppa, M. B., Lana, B., Margas, W., Dolphin, A. C., & Ryan, T. A. (2012). alpha2delta expression sets presynaptic calcium channel abundance and release probability. *Nature, 486*, 122–125.

Iossifov, I., Ronemus, M., Levy, D., Wang, Z., Hakker, I., Rosenbaum, J., Yamrom, B., Lee, Y. H., Narzisi, G., Leotta, A., et al. (2012). De novo gene disruptions in children on the autistic spectrum. *Neuron, 74*, 285–299.

Iossifov, I., O'Roak, B. J., Sanders, S. J., Ronemus, M., Krumm, N., Levy, D., Stessman, H. A., Witherspoon, K. T., Vives, L., Patterson, K. E., et al. (2014). The contribution of de novo coding mutations to autism spectrum disorder. *Nature, 515*, 216–221.

Ivanov, S. V., Ward, J. M., Tessarollo, L., McAreavey, D., Sachdev, V., Fananapazir, L., Banks, M. K., Morris, N., Djurickovic, D., Devor-Henneman, D. E., et al. (2004). Cerebellar ataxia, seizures, premature death, and cardiac abnormalities in mice with targeted disruption of the Cacna2d2 gene. *The American Journal of Pathology, 165*, 1007–1018.

Jay, S. D., Sharp, A. H., Kahl, S. D., Vedvick, T. S., Harpold, M. M., & Campbell, K. P. (1991). Structural characterization of the dihydropyridine-sensitive calcium channel α2-subunit and the associated δ peptides. *Journal of Biological Chemistry, 266*, 3287–3293.

Kadurin, I., Alvarez-Laviada, A., Ng, S. F., Walker-Gray, R., D'Arco, M., Fadel, M. G., Pratt, W. S., & Dolphin, A. C. (2012). Calcium currents are enhanced by alpha2delta-1 lacking its membrane anchor. *The Journal of Biological Chemistry, 1287*, 33554–33566.

Kadurin, I., Ferron, L., Rothwell, S. W., Meyer, J. O., Douglas, L. R., Bauer, C. S., Lana, B., Margas, W.,

Alexopoulos, O., Nieto-Rostro, M., et al. (2016). Proteolytic maturation of α2δ represents a checkpoint for activation and neuronal trafficking of latent calcium channels. *eLife, 5*, e21143.

Kadurin, I., Rothwell, S. W., Lana, B., Nieto-Rostro, M., & Dolphin, A. C. (2017). LRP1 influences trafficking of N-type calcium channels via interaction with the auxiliary alpha2delta-1 subunit. *Scientific Reports, 7*, 43802.

Kang, M. G., Felix, R., & Campbell, K. P. (2002). Long-term regulation of voltage-gated Ca(2+) channels by gabapentin. *FEBS Letters, 528*, 177–182.

Kerov, V., Laird, J. G., Joiner, M. L., Knecht, S., Soh, D., Hagen, J., Gardner, S. H., Gutierrez, W., Yoshimatsu, T., Bhattarai, S., et al. (2018). alpha2delta-4 is required for the molecular and structural organization of rod and cone photoreceptor synapses. *The Journal of Neuroscience, 38*, 6145–6160.

Klugbauer, N., Lacinova, L., Marais, E., Hobom, M., & Hofmann, F. (1999). Molecular diversity of the calcium channel α2-δ subunit. *Journal of Neuroscience, 19*, 684–691.

Klugbauer, N., Marais, E., & Hofmann, F. (2003). Calcium channel alpha2delta subunits: Differential expression, function, and drug binding. *Journal of Bioenergetics and Biomembranes, 35*, 639–647.

Knoflach, D., Kerov, V., Sartori, S. B., Obermair, G. J., Schmuckermair, C., Liu, X., Sothilingam, V., Garcia Garrido, M., Baker, S. A., Glosmann, M., et al. (2013). Cav1.4 IT mouse as model for vision impairment in human congenital stationary night blindness type 2. *Channels (Austin, Tex.), 7*, 503–513.

Kurshan, P. T., Oztan, A., & Schwarz, T. L. (2009). Presynaptic alpha(2)delta-3 is required for synaptic morphogenesis independent of its Ca(2+)-channel functions. *Nature Neuroscience, 12*, 1415–1423.

Lana, B., Schlick, B., Martin, S., Pratt, W. S., Page, K. M., Goncalves, L., Rahman, W., Dickenson, A. H., Bauer, C. S., & Dolphin, A. C. (2014). Differential up-regulation in DRG neurons of an alphadelta-1 splice variant with a lower affinity for gabapentin after peripheral sensory nerve injury. *Pain, 155*, 522–533.

Lana, B., Page, K. M., Kadurin, I., Ho, S., Nieto-Rostro, M., & Dolphin, A. C. (2016). Thrombospondin-4 reduces binding affinity of [3H]-gabapentin to calcium-channel α2δ-1-subunit but does not interact with α2δ-1 on the cell-surface when co-expressed. *Scientific Reports, 6*, 24531.

Lee, A., Wang, S., Williams, B., Hagen, J., Scheetz, T. E., & Haeseleer, F. (2015). Characterization of Cav1.4 complexes (alpha11.4, beta2, and alpha2delta4) in HEK293T cells and in the retina. *The Journal of Biological Chemistry, 290*, 1505–1521.

Lee, S. T., Lee, B. J., Bae, J. Y., Kim, Y. S., Han, D. H., Shin, H. S., Kim, S., Park, D. K., Seo, S. W., Chu, K., et al. (2021). CaV alpha2delta autoimmune encephalitis: A novel antibody and its characteristics. *Annals of Neurology, 89*, 740–752.

Lempel, A. A., Coll, L., Schinder, A. F., Uchitel, O. D., & Piriz, J. (2017). Chronic pregabalin treatment decreases excitability of dentate gyrus and accelerates maturation of adult-born granule cells. *Journal of Neurochemistry, 140*, 257–267.

Leung, A. T., Imagawa, T., & Campbell, K. P. (1987). Structural characterization of the 1,4-dihydropyridine receptor of the voltage-dependent Ca2+ channel from rabbit skeletal muscle. Evidence for two distinct high molecular weight subunits. *The Journal of Biological Chemistry, 262*, 7943–7946.

Li, C. Y., Zhang, X. L., Matthews, E. A., Li, K. W., Kurwa, A., Boroujerdi, A., Gross, J., Gold, M. S., Dickenson, A. H., Feng, G., et al. (2006). Calcium channel alpha(2)delta(1) subunit mediates spinal hyperexcitability in pain modulation. *Pain, 125*, 20–34.

Li, Z., Taylor, C. P., Weber, M., Piechan, J., Prior, F., Bian, F., Cui, M., Hoffman, D., & Donevan, S. (2011). Pregabalin is a potent and selective ligand for alpha(2)delta-1 and alpha(2)delta-2 calcium channel subunits. *European Journal of Pharmacology, 667*, 80–90.

Li, L., Chen, S. R., Zhou, M. H., Wang, L., Li, D. P., Chen, H., Lee, G., Jayaraman, V., & Pan, H. L. (2021). alpha2delta-1 switches the phenotype of synaptic AMPA receptors by physically disrupting heteromeric subunit assembly. *Cell Reports, 36*, 109396.

Lotarski, S. M., Donevan, S., El Kattan, A., Osgood, S., Poe, J., Taylor, C. P., & Offord, J. (2011). Anxiolytic-like activity of pregabalin in the Vogel conflict test in alpha2delta-1 (R217A) and alpha2delta-2 (R279A) mouse mutants. *The Journal of Pharmacology and Experimental Therapeutics, 338*, 615–621.

Lotarski, S., Hain, H., Peterson, J., Galvin, S., Strenkowski, B., Donevan, S., & Offord, J. (2014). Anticonvulsant activity of pregabalin in the maximal electroshock-induced seizure assay in alphadelta (R217A) and alphadelta (R279A) mouse mutants. *Epilepsy Research, 108*, 833–842.

Luo, Z. D., Chaplan, S. R., Higuera, E. S., Sorkin, L. S., Stauderman, K. A., Williams, M. E., & Yaksh, T. L. (2001). Upregulation of dorsal root ganglion α₂δ calcium channel subunit and its correlation with allodynia in spinal nerve-injured rats. *Journal of Neuroscience, 21*, 1868–1875.

Macabuag, N., & Dolphin, A. C. (2015). Alternative splicing in CaV2.2 regulates neuronal trafficking via adaptor protein complex-1 adaptor protein binding motifs. *Journal of Neuroscience, 35*, 14636–14652.

Marais, E., Klugbauer, N., & Hofmann, F. (2001). Calcium channel alpha(2)delta subunits – Structure and gabapentin binding. *Molecular Pharmacology, 59*, 1243–1248.

Martin, D. J., McClelland, D., Herd, M. B., Sutton, K. G., Hall, M. D., Lee, K., Pinnock, R. D., & Scott, R. H. (2002). Gabapentin-mediated inhibition of voltage-activated Ca2+ channel currents in cultured sensory neurones is dependent on culture conditions and channel subunit expression. *Neuropharmacology, 42*, 353–366.

Mastrolia, V., Flucher, S. M., Obermair, G. J., Drach, M., Hofer, H., Renstrom, E., Schwartz, A., Striessnig, J., Flucher, B. E., & Tuluc, P. (2017). Loss of alpha2delta-1 calcium channel subunit function increases the susceptibility for diabetes. *Diabetes, 66*, 897–907.

Mayor, S., & Riezman, H. (2004). Sorting GPI-anchored proteins. *Nature Reviews. Molecular Cell Biology, 5*, 110–120.

Mefford, H. C., Yendle, S. C., Hsu, C., Cook, J., Geraghty, E., McMahon, J. M., Eeg-Olofsson, O., Sadleir, L. G., Gill, D., Ben-Zeev, B., et al. (2011). Rare copy number variants are an important cause of epileptic encephalopathies. *Annals of Neurology, 70*, 974–985.

Meyer, J. O., & Dolphin, A. C. (2021). Rab11-dependent recycling of calcium channels is mediated by auxiliary subunit alpha2delta-1 but not alpha2delta-3. *Scientific Reports, 11*, 10256.

Meyer, J. O., Dahimene, S., Page, K. M., Ferron, L., Kadurin, I., Ellaway, J. I. J., Zhao, P., Patel, T., Rothwell, S. W., Lin, P., et al. (2019). Disruption of the key Ca^{2+} binding site in the selectivity filter of neuronal voltage-gated calcium channels inhibits channel trafficking. *Cell Reports, 29*, 22–33.

Mould, J., Yasuda, T., Schroeder, C. I., Beedle, A. M., Doering, C. J., Zamponi, G. W., Adams, D. J., & Lewis, R. J. (2004). The alpha2delta auxiliary subunit reduces affinity of omega-conotoxins for recombinant N-type (Cav2.2) calcium channels. *The Journal of Biological Chemistry, 279*, 34705–34714.

Müller, C. S., Haupt, A., Bildl, W., Schindler, J., Knaus, H. G., Meissner, M., Rammner, B., Striessnig, J., Flockerzi, V., Fakler, B., et al. (2010). Quantitative proteomics of the Cav2 channel nano-environments in the mammalian brain. *Proceedings of the National Academy of Sciences, 107*, 14950–14957.

Neely, G. G., Hess, A., Costigan, M., Keene, A. C., Goulas, S., Langeslag, M., Griffin, R. S., Belfer, I., Dai, F., Smith, S. B., et al. (2010). A genome-wide Drosophila screen for heat nociception identifies alpha2delta3 as an evolutionarily conserved pain gene. *Cell, 143*, 628–638.

Newton, R. A., Bingham, S., Case, P. C., Sanger, G. J., & Lawson, S. N. (2001). Dorsal root ganglion neurons show increased expression of the calcium channel alpha2delta-1 subunit following partial sciatic nerve injury. *Brain Research. Molecular Brain Research, 95*, 1–8.

Nieto-Rostro, M., Ramgoolam, K., Pratt, W. S., Kulik, A., & Dolphin, A. C. (2018). Ablation of alpha2delta-1 inhibits cell-surface trafficking of endogenous N-type calcium channels in the pain pathway in vivo. *Proceedings of the National Academy of Sciences of the United States of America, 115*, E12043–E12052.

Obermair, G. J., Kugler, G., Baumgartner, S., Tuluc, P., Grabner, M., & Flucher, B. E. (2005). The Ca2+ channel alpha2delta-1 subunit determines Ca2+ current kinetics in skeletal muscle but not targeting of alpha1S or excitation-contraction coupling. *The Journal of Biological Chemistry, 280*, 2229–2237.

Obermair, G. J., Tuluc, P., & Flucher, B. E. (2008). Auxiliary Ca(2+) channel subunits: Lessons learned from muscle. *Current Opinion in Pharmacology, 8*, 311–318.

Patel, R., Bauer, C. S., Nieto-Rostro, M., Margas, W., Ferron, L., Chaggar, K., Crews, K., Ramirez, J. D., Bennett, D. L., Schwartz, A., et al. (2013). alpha2delta-1 gene deletion affects somatosensory neuron function and delays mechanical hypersensitivity in response to peripheral nerve damage. *The Journal of Neuroscience, 33*, 16412–16426.

Pippucci, T., Parmeggiani, A., Palombo, F., Maresca, A., Angius, A., Crisponi, L., Cucca, F., Liguori, R., Valentino, M. L., Seri, M., et al. (2013). A novel null homozygous mutation confirms CACNA2D2 as a gene mutated in epileptic encephalopathy. *PLoS One, 8*, e82154.

Pirone, A., Kurt, S., Zuccotti, A., Ruttiger, L., Pilz, P., Brown, D. H., Franz, C., Schweizer, M., Rust, M. B., Rubsamen, R., et al. (2014). alpha2delta3 is essential for normal structure and function of auditory nerve synapses and is a novel candidate for auditory processing disorders. *The Journal of Neuroscience, 34*, 434–445.

Punetha, J., Karaca, E., Gezdirici, A., Lamont, R. E., Pehlivan, D., Marafi, D., Appendino, J. P., Hunter, J. V., Akdemir, Z. C., Fatih, J. M., et al. (2019). Biallelic CACNA2D2 variants in epileptic encephalopathy and cerebellar atrophy. *Annals of Clinical Translational Neurology, 6*, 1395–1406.

Purcell, S. M., Moran, J. L., Fromer, M., Ruderfer, D., Solovieff, N., Roussos, P., O'Dushlaine, C., Chambert, K., Bergen, S. E., Kahler, A., et al. (2014). A polygenic burden of rare disruptive mutations in schizophrenia. *Nature, 506*, 185–190.

Qin, N., Yagel, S., Momplaisir, M. L., Codd, E. E., & D'Andrea, M. R. (2002). Molecular cloning and characterization of the human voltage-gated calcium channel $\alpha_2\delta$-4 subunit. *Molecular Pharmacology, 62*, 485–496.

Risher, W. C., Kim, N., Koh, S., Choi, J. E., Mitev, P., Spence, E. F., Pilaz, L. J., Wang, D., Feng, G., Silver, D. L., et al. (2018). Thrombospondin receptor alpha2delta-1 promotes synaptogenesis and spinogenesis via postsynaptic Rac1. *The Journal of Cell Biology, 217*(10), 3747–3765.

Rodríguez-López, J., Sobrino, B., Amigo, J., Carrera, N., Brenlla, J., Agra, S., Paz, E., Carracedo, Á., Páramo, M., Arrojo, M., et al. (2018). Identification of putative second genetic hits in schizophrenia carriers of high-risk copy number variants and resequencing in additional samples. *European Archives of Psychiatry and Clinical Neuroscience, 268*, 585–592.

Sanderson, K. (2021). High-profile autism genetics project paused amid backlash. *Nature, 598*, 17–18.

Savalli, N., Pantazis, A., Sigg, D., Weiss, J. N., Neely, A., & Olcese, R. (2016). The alpha2delta-1 subunit remodels CaV1.2 voltage sensors and allows Ca2+ influx at physiological membrane potentials. *The Journal of General Physiology, 148*, 147–159.

Schlegel, D. K., Glasauer, S. M. K., Mateos, J. M., Barmettler, G., Ziegler, U., & Neuhauss, S. C. F. (2019). A new zebrafish model for CACNA2D4-dysfunction. *Investigative Ophthalmology & Visual Science, 60,* 5124–5135.

Schlick, B., Flucher, B. E., & Obermair, G. J. (2010). Voltage-activated calcium channel expression profiles in mouse brain and cultured hippocampal neurons. *Neuroscience, 167,* 786–798.

Schneider, R., Hosy, E., Kohl, J., Klueva, J., Choquet, D., Thomas, U., Voigt, A., & Heine, M. (2015). Mobility of calcium channels in the presynaptic membrane. *Neuron, 86,* 672–679.

Schöpf, C. L., Ablinger, C., Geisler, S. M., Stanika, R. I., Campiglio, M., Kaufmann, W. A., Nimmervoll, B., Schlick, B., Brockhaus, J., Missler, M., et al. (2021). Presynaptic alpha2delta subunits are key organizers of glutamatergic synapses. *Proceedings of the National Academy of Sciences of the United States of America, 118,* e1920827118.

Sharp, A. H., Imagawa, T., Leung, A. T., & Campbell, K. P. (1987). Identification and characterization of the dihydropyridine-binding subunit of the skeletal muscle dihydropyridine receptor. *The Journal of Biological Chemistry, 262,* 12309–12315.

Shi, Y., Park, K. S., Kim, S. H., Yu, J., Zhao, K., Yu, L., Oh, K. W., Lee, K., Kim, J., Chaggar, K., et al. (2019). IgGs from patients with amyotrophic lateral sclerosis and diabetes target CaValpha2delta1 subunits impairing islet cell function and survival. *Proceedings of the National Academy of Sciences of the United States of America, 116,* 26816–26822.

Takahashi, M., Seager, M. J., Jones, J. F., Reber, B. F. X., & Catterall, W. A. (1987). Subunit structure of dihydropyridine-sensitive calcium channels from skeletal muscle. *Proceedings of the National Academy of Sciences of the United States of America, 84,* 5478–5482.

Tanabe, T., Takeshima, H., Mikami, A., Flockerzi, V., Takahashi, H., Kangawa, K., Kojima, M., Matsuo, H., Hirose, T., & Numa, S. (1987). Primary structure of the receptor for calcium channel blockers from skeletal muscle. *Nature, 328,* 313–318.

Taylor, C. P., Angelotti, T., & Fauman, E. (2007). Pharmacology and mechanism of action of pregabalin: The calcium channel alpha2-delta (alpha2-delta) subunit as a target for antiepileptic drug discovery. *Epilepsy Research, 73,* 137–150.

Tedeschi, A., Dupraz, S., Laskowski, C. J., Xue, J., Ulas, T., Beyer, M., Schultze, J. L., & Bradke, F. (2016). The calcium channel subunit Alpha2delta2 suppresses axon regeneration in the adult CNS. *Neuron, 92,* 419–434.

Templin, C., Ghadri, J. R., Rougier, J. S., Baumer, A., Kaplan, V., Albesa, M., Sticht, H., Rauch, A., Puleo, C., Hu, D., et al. (2011). Identification of a novel loss-of-function calcium channel gene mutation in short QT syndrome (SQTS6). *European Heart Journal, 32,* 1077–1088.

Tong, X.-J., López-Soto, E. J., Li, L., Liu, H., Nedelcu, D., Lipscombe, D., Hu, Z., & Kaplan, J. M. (2017). Retrograde synaptic inhibition is mediated by α-neurexin binding to the α2δ subunits of N-type calcium channels. *Neuron, 95,* 1–15.

Tran-Van-Minh, A., & Dolphin, A. C. (2010). The alpha2delta ligand gabapentin inhibits the Rab11-dependent recycling of the calcium channel subunit alpha2delta-2. *The Journal of Neuroscience, 30,* 12856–12867.

Tuluc, P., Kern, G., Obermair, G. J., & Flucher, B. E. (2007). Computer modeling of siRNA knockdown effects indicates an essential role of the Ca2+ channel alpha2delta-1 subunit in cardiac excitation-contraction coupling. *Proceedings of the National Academy of Sciences of the United States of America, 104,* 11091–11096.

Valentino, F., Bruno, L. P., Doddato, G., Giliberti, A., Tita, R., Resciniti, S., Fallerini, C., Bruttini, M., Lo Rizzo, C., Mencarelli, M. A., et al. (2021). Exome sequencing in 200 intellectual disability/autistic patients: New candidates and atypical presentations. *Brain Sciences, 11,* 936.

Van Den Bossche, M. J., Strazisar, M., De Bruyne, S., Bervoets, C., Lenaerts, A. S., De Zutter, S., Nordin, A., Norrback, K. F., Goossens, D., De Rijk, P., et al. (2012). Identification of a CACNA2D4 deletion in late onset bipolar disorder patients and implications for the involvement of voltage-dependent calcium channels in psychiatric disorders. *American Journal of Medical Genetics. Part B, Neuropsychiatric Genetics, 159B,* 465–475.

van Loo, K. M. J., Rummel, C. K., Pitsch, J., Muller, J. A., Bikbaev, A. F., Martinez-Chavez, E., Blaess, S., Dietrich, D., Heine, M., Becker, A. J., et al. (2019). Calcium channel subunit alpha2delta4 is regulated by early growth response 1 and facilitates epileptogenesis. *The Journal of Neuroscience, 39,* 3175–3187.

Vergult, S., Dheedene, A., Meurs, A., Faes, F., Isidor, B., Janssens, S., Gautier, A., Le Caignec, C., & Menten, B. (2015). Genomic aberrations of the CACNA2D1 gene in three patients with epilepsy and intellectual disability. *European Journal of Human Genetics: EJHG, 23,* 628–632.

Veroniki, A. A., Rios, P., Cogo, E., Straus, S. E., Finkelstein, Y., Kealey, R., Reynen, E., Soobiah, C., Thavorn, K., Hutton, B., et al. (2017). Comparative safety of antiepileptic drugs for neurological development in children exposed during pregnancy and breast feeding: A systematic review and network meta-analysis. *BMJ Open, 7,* e017248.

Voigt, A., Freund, R., Heck, J., Missler, M., Obermair, G. J., Thomas, U., & Heine, M. (2016). Dynamic association of calcium channel subunits at the cellular membrane. *Neurophotonics, 3,* 041809.

Wakamori, M., Mikala, G., & Mori, Y. (1999). Auxiliary subunits operate as a molecular switch in determining gating behaviour of the unitary N-type Ca2+ channel current in Xenopus oocytes. *Journal of Physiology (London), 517,* 659–672.

Wang, M. H., Offord, J., Oxender, D. L., & Su, T. Z. (1999). Structural requirement of the calcium-channel subunit $\alpha_2\delta$ for gabapentin binding. *Biochemical Journal, 342*, 313–320.

Wang, H., Sun, H., Della, P. K., Benz, R. J., Xu, J., Gerhold, D. L., Holder, D. J., & Koblan, K. S. (2002). Chronic neuropathic pain is accompanied by global changes in gene expression and shares pathobiology with neurodegenerative diseases. *Neuroscience, 114*, 529–546.

Wang, Y., Fehlhaber, K. E., Sarria, I., Cao, Y., Ingram, N. T., Guerrero-Given, D., Throesch, B., Baldwin, K., Kamasawa, N., Ohtsuka, T., et al. (2017). The auxiliary calcium channel subunit alpha2delta4 is required for axonal elaboration, synaptic transmission, and wiring of rod photoreceptors. *Neuron, 93*, 1359–1374 e1356.

Whittaker, C. A., & Hynes, R. O. (2002). Distribution and evolution of von Willebrand/integrin A domains: Widely dispersed domains with roles in cell adhesion and elsewhere. *Molecular Biology of the Cell, 13*, 3369–3387.

Wiffen, P. J., McQuay, H. J., Edwards, J. E., & Moore, R. A. (2005). Gabapentin for acute and chronic pain. *Cochrane Database of Systematic Reviews, 2005*, CD005452.

Wu, J., Yan, Z., Li, Z., Qian, X., Lu, S., Dong, M., Zhou, Q., & Yan, N. (2016). Structure of the voltage-gated calcium channel Cav1.1 at 3.6 A resolution. *Nature, 537*, 191–196.

Wycisk, K. A., Budde, B., Feil, S., Skosyrski, S., Buzzi, F., Neidhardt, J., Glaus, E., Nurnberg, P., Ruether, K., & Berger, W. (2006a). Structural and functional abnormalities of retinal ribbon synapses due to Cacna2d4 mutation. *Investigative Ophthalmology & Visual Science, 47*, 3523–3530.

Wycisk, K. A., Zeitz, C., Feil, S., Wittmer, M., Forster, U., Neidhardt, J., Wissinger, B., Zrenner, E., Wilke, R., Kohl, S., et al. (2006b). Mutation in the auxiliary calcium-channel subunit CACNA2D4 causes autosomal recessive cone dystrophy. *American Journal of Human Genetics, 79*, 973–977.

Xiao, W., Boroujerdi, A., Bennett, G. J., & Luo, Z. D. (2007). Chemotherapy-evoked painful peripheral neuropathy: Analgesic effects of gabapentin and effects on expression of the alpha-2-delta type-1 calcium channel subunit. *Neuroscience, 144*, 714–720.

Yamaguchi, H., Okuda, M., Mikala, G., Fukasawa, K., & Varadi, G. (2000). Cloning of the β_{2a} subunit of the voltage-dependent calcium channel from human heart: Cooperative effect of α_2/δ and β_{2a} on the membrane expression of the α_{1c} subunit. *Biochemical and Biophysical Research Communications, 267*, 156–163.

Yasuda, T., Chen, L., Barr, W., Mcrory, J. E., Lewis, R. J., Adams, D. J., & Zamponi, G. W. (2004). Auxiliary subunit regulation of high-voltage activated calcium channels expressed in mammalian cells. *The European Journal of Neuroscience, 20*, 1–13.

Yu, Y. P., Gong, N., Kweon, T. D., Vo, B., & Luo, Z. D. (2018). Gabapentin prevents synaptogenesis between sensory and spinal cord neurons induced by thrombospondin-4 acting on pre-synaptic Cav alpha2 delta1 subunits and involving T-type Ca(2+) channels. *British Journal of Pharmacology, 175*, 2348–2361.

Zhang, F. X., Gadotti, V. M., Souza, I. A., Chen, L., & Zamponi, G. W. (2018). BK potassium channels suppress Cavalpha2 delta subunit function to reduce inflammatory and neuropathic pain. *Cell Reports, 22*, 1956–1964.

Voltage-Gated Calcium Channels in Invertebrates

Adriano Senatore and J. David Spafford

Abstract

The α_1, β, and $\alpha_2\delta$ subunits of voltage-gated calcium channels emerged before animals, and are present in all animal genomes, including those that lack nervous systems and muscle. Considerable variability is evident in the gene copy numbers of these subunits, attributable to independent gene duplications and losses. Over more than seven decades of research, invertebrates have provided important contributions toward our understanding of calcium channel function, through studies of endogenous Ca^{2+} currents recorded from various tissues, *in vitro* studies of cloned channel subunits, and *in vivo* studies in species amenable to genetic manipulation. A picture emerges that calcium channel types exhibit conservation across all animals with respect to their ion-conducting properties, alternative splicing, and functions in homologous cell types. Unique adaptations in properties and cellular functions often appear because of lineage-specific gene duplications. Much less is known about the functional and physiological properties of voltage-gated calcium channels in the earliest-diverging animals, which provide exciting opportunities for exploring the origins of electrical and synaptic signaling in the nervous system.

Keywords

Calcium channel · Invertebrate · Evolution · $\alpha_1 \cdot \beta \cdot \alpha_2\delta$ · CACHD1 · CACHD0

Abbreviations

AID	Alpha interaction domain
CACHD0	CACHE domain-containing protein 0
CACHD1	CACHE domain-containing protein 1
CACHE	calcium channel and chemotaxis
CDF	Ca^{2+}-dependent facilitation
CDI	Ca^{2+}-dependent inactivation
CICR	Ca^{2+}-induced Ca^{2+} release
DHP	Dihydropyridine
ER	Endoplasmic reticulum
GK	guanylate kinase
GPCR	G protein-coupled receptor
HVA	High voltage activated
LVA	Low voltage activated

A. Senatore (✉)
Department of Cell and Systems Biology, University of Toronto, Toronto, ON, Canada

Department of Biology, University of Toronto Mississauga, Mississauga, ON, Canada
e-mail: adriano.senatore@utoronto.ca

J. D. Spafford
Department of Biology, University of Waterloo, Waterloo, ON, Canada

MIDAS	metal ion-dependent adhesion site
NATE	NSCaTE associated transduction element
NSCaTE	N-terminal spatial Ca^{2+} transforming element
PAS domain	Per-Arnt-Sim domain
PDZ	PSD95, Dlg1, and zo-1
PZQ	praziquantel
SH3	Src Homology 3
synprint	synaptic protein interaction
TRPN	transient receptor potential type N
vWA	von Willebrand factor A

Introduction: Invertebrates and Calcium Channel Discovery

Early electrophysiological studies on invertebrates played an important role toward the discovery of voltage-gated calcium channels, and in establishing their physiological importance. For example, work in crustacean muscle (phylum Arthropoda, Fig. 1a) provided the first evidence that Ca^{2+} influx occurs upon membrane depolarization, and that this process is required for contraction (Fatt & Katz, 1953; Fatt & Ginsborg, 1958). In the squid giant synapse (phylum Mollusca), it was shown that Ca^{2+} influx through voltage-gated calcium channels serves to translate the electrical signal of the action potential, propagated by the flux of Na^+ and K^+ across the cell membrane, into the regulated release of neurotransmitter (Katz & Miledi, 1967; Llinás et al., 1976; Augustine et al., 1985). Invertebrates also provided first insights into the existence of multiple calcium channel types, with the identification of distinct low- and high-voltage-activated Ca^{2+} currents (LVA and HVA) in starfish eggs (phylum Echinodermata) (Hagiwara et al., 1975), and a year later, in our own phylum Chordata in eggs of the invertebrate tunicate (Okamoto et al., 1976).

Crucial steps forward involved the identification of pharmacological compounds that could selectively block specific calcium chan-

nel currents recorded from neuronal, cardiac, and muscle cells, providing convincing evidence for the existence of five classes of calcium channels in mammals: T-type, L-type, N-type, P/Q-type, and R-type (Tsien & Barrett, 2005; Dolphin, 2006). Subsequently, biochemical and genetic identification of calcium channels (Takahashi et al., 1987; Tanabe et al., 1987) opened the door for cloning and *in vitro* expression for electrophysiology (Mori et al., 1991), allowing researchers to link pharmacologically defined calcium currents with specific calcium channel genes. This era of molecular discovery culminated with the genome sequences of the nematode worm *C. elegans* in 1998 (Consortium, 1998), the vinegar fly *Drosophila melanogaster* in 2000 (Adams et al., 2000), and of humans in 2003 (Venter et al., 2001; Lander et al., 2001), permitting the complete identification and phylogenetic classification of calcium channels. Thus, L-type channels were phylogenetically classified as the Ca_V1 family, N-, P/Q-, and R-type channels as the Ca_V2 family, and T-type channels as the Ca_V3 family (Ertel et al., 2000). Importantly, along with the biochemical and molecular identification of calcium channel α_1 subunits came the discovery of the ancillary β and $\alpha_2\delta$ subunits, which interact with and regulate HVA Ca_V1 and Ca_V2 channels (Fig. 1b).

Recently, the ever-expanding availability of genome sequences from a diversity of organisms has revealed that numerous genes important for complex animal traits, including calcium channels and their ancillary subunits, are present in the genomes of the simplest animals, and even single-celled eukaryotes, such as choanoflagellates (Fig. 1a). This has fuelled questions about the evolutionary origins of animal genomic and phenotypic complexity, and perhaps most intriguing, evolution of the nervous system and synaptic transmission. In their ability to translate electrical signals at the cell membrane into cytoplasmic Ca^{2+} signals, voltage-gated calcium channels would have played an integral role in the evolution of excitable cell types and neural communication.

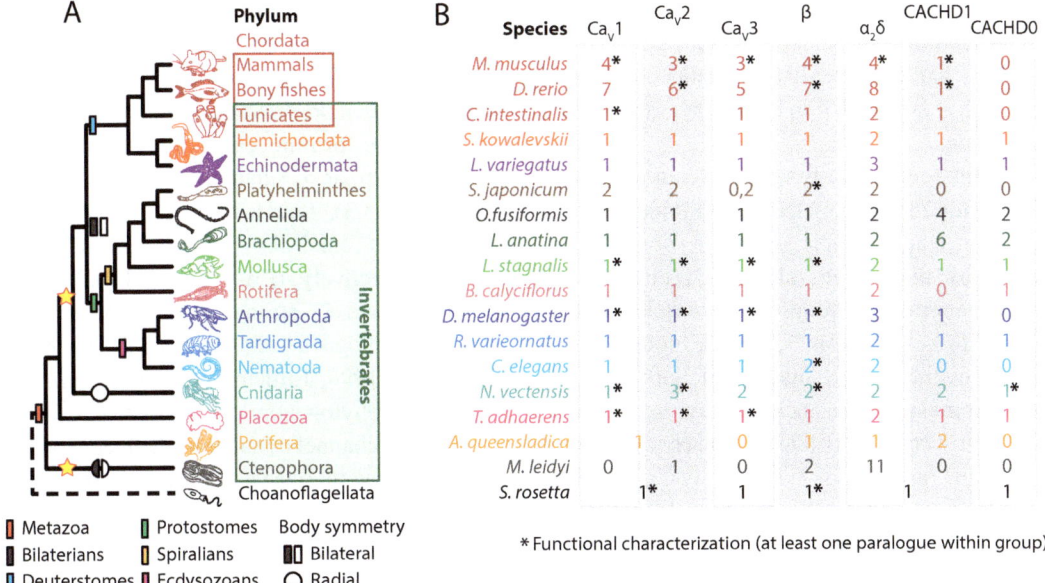

Fig. 1 (**a**) Cladogram depicting the phylogenetic relationships of some of the major animal phyla, rooted against the unranked clade of choanoflagellates (dotted line), which are unicellular organisms that are the most closely related to animals. Three subgroups of the phylum Chordata are shown to separate representative vertebrates (mammals, bony fishes) from invertebrates (tunicates). Invertebrates comprise all other animal phyla. The depicted relationships are derived from a collection of phylogenomic studies (Kapli & Telford, 2020; Pisani et al., 2015; Simion et al., 2017; Whelan et al., 2017; Marlétaz et al., 2019; Delsuc et al., 2006). The relationship between Porifera and Ctenophora is shown as unresolved, as it is still uncertain which of these two phyla was the first to diverge from other animals. The colored boxes denote some of the major groupings within the Metazoa, and the stars denote animals with nervous systems. Animals can also be subdivided according to body symmetry, where the protostomes and deuterostomes have bilateral body symmetry, cnidarians have radial symmetry, ctenophores have a combination of bilateral and radial symmetry (i.e., biradial), and placozoans and poriferans lack body symmetry. (**b**) Table of gene counts of voltage-gated calcium channel subunits found in the genomes of the different phyla depicted in panel A. Names of representative species discussed throughout the text are given, but the reported gene counts correspond to observations made for multiple species within each phylum. The $\alpha_2\delta$ subunits are divided into three groups, based on the phylogenetic tree presented in Fig. 4: canonical $\alpha_2\delta$, CACHD1, and $\alpha_2\delta$ like. Numbers present between columns indicate sister relationships on phylogenetic trees. For example, the choanoflagellate $Ca_V1/2$ channel forms a sister relationship with metazoan Ca_V1 and Ca_V2 channels, and one of the two $\alpha_2\delta$ subunits forms a sister relationship with canonical $\alpha_2\delta$ and CACHD1 (Fig. 4). Asterisks denote subunits from each phylum that have been cloned and functionally characterized *in vitro*, as described throughout the text (for invertebrates only)

So, what does continued research on invertebrate calcium channels have to offer? From an applied perspective, calcium channels provide viable pharmacological targets for invertebrate pest control (King, 2007), or management of parasites and pathogens (Salvador-Recatalà & Greenberg, 2012). From a basic science perspective, identification of deeply conserved structural and functional features of calcium channels highlights core innovations that have persisted in diverse animal lineages because of their importance in cellular physiology. A phylogenetic understanding of this homology, or alternatively, divergence or homoplasy, is important as it can inform the validity of genetic inferences across model organisms. Also, invertebrates often have single-copy genes for important neuronal proteins, including Ca_V1 to Ca_V3 channels which are duplicated in vertebrates. Thus, in addition to advantages conferred through diminished genetic redundancy, invertebrates provide a reference for understanding how gene duplication of calcium channels and their ancillary subunits contributed to their evolution in humans and other vertebrates.

In this review, we first provide an update on the phylogenetic properties and molecular evolution of Ca_V channels and their subunits, followed by discussions about Ca_V channels in three general groupings of invertebrate animals: (1) deuterostomes, which include chordates, echinoderms, and hemichordates, (2) protostomes, which include key invertebrate research subjects, such as molluscs (e.g., squid and snail), arthropods (vinegar fly), and nematode worms, and (3) the most early-diverging animal lineages of cnidarians and ctenophores, that have the most divergent nervous systems, and sponges and placozoans, that lack nervous systems and synapses (Fig. 1a). Although calcium currents have been recorded extensively from an array of invertebrate species, including early-diverging animals, the specific ion channels involved remain largely undefined (for a review see (Senatore et al., 2016)). Here, we will focus our discussion mainly on studies where specific ion channel subunits were directly implicated, for example, through *in vitro* expression and functional characterization, or through direct genetic manipulation *in vivo*.

Phylogenetic Properties of the Ca_V Channel α_1 Subunit

Genome sequencing revealed that humans possess ten pore-forming calcium channel genes, also referred to as α_1 subunits (Fig. 1b): 3 T-type ($Ca_V3.1$–$Ca_V3.3$), four L-type ($Ca_V1.1$–$Ca_V1.4$), and N-, P/Q-, and R-type channels ($Ca_V2.1$–$Ca_V2.3$, respectively) (Ertel et al., 2000). Instead, *Drosophila* and *C. elegans* were found to have single homologues of Ca_V1–Ca_V3 channels (Bargmann, 1998; Littleton & Ganetzky, 2000). More broad phylogenomic analysis has revealed that calcium channel α_1 subunits evolved from an ancient eukaryotic lineage of four-domain cation channels, which includes voltage-gated sodium (Na_V1, Na_V2) channels and the Na^+ leak channel NALCN (Pozdnyakov et al., 2018). A close kinship between Ca_V1–Ca_V3 and Na_V channel genes is also evident in their genomic features, sharing five intron splice sites, including a rare U12-type (AT-AC) found in only ~4% of mammalian genes (Fux et al., 2018). All of these channels share a common transmembrane topology of four repeat Domains (I–IV), each bearing six transmembrane alpha helices or segments (Fig. 2a). The first four helices/segments of each Domain (S1–S4) make up the voltage sensor modules, with S4 helices bearing positively charged lysine and arginine residues that are crucial for voltage sensing (Bezanilla, 2000; Catterall, 2010). The four extracellular pore-loops of each Domain, along

Fig. 2 (continued) are indicated on the X axes. Regions of the alignments that correspond to Domains I to IV and conserved cytoplasmic structures are depicted with colored bars above each plot. The insets on the right of each plot are consensus sequence logos derived by aligning the last 7 amino acids (i.e., C-termini) of the same Ca_V channel orthologues used in each plot, generated with the program Seq2Logo 2.0 (Thomsen & Nielsen, 2012). Large letters indicate frequently observed amino acids, and lower bits (y axis) indicate sequence variability at each C-terminal position. Putatively conserved PDZ ligand signatures for Ca_V1 channels ($VTTL_{COOH}$), Ca_V2 channels ($DDWC_{COOH}$), and Ca_V3 channels ($DDYV_{COOH}$) are evident, most striking for Ca_V2 channels with an invariable tryptophan in the second last position. (**c**) Illustration of C-terminal interactions of Ca_V2 channels with the presynaptic scaffolding proteins RIM, RIM-BP, Mint, and CASK, defined in vertebrates and invertebrates. RIM and Mint both have PDZ domains that can bind the conserved $DDWC_{COOH}$ signature of Ca_V2 channels, while RIM-BP and CASK use SH3 domains to bind internal proline-rich regions of the C-terminus. CASK and Mint interact via a calmodulin kinase (CaMK) domain on CASK binding and N-terminal region of Mint (Wu et al., 2020). (**d**) Illustration of C-terminal interactions of $Ca_V1.3$ channels with the postsynaptic scaffolding proteins Shank and Erbin, which have PDZ domains that recognize the C-terminal $ITTV_{COOH}$ signature. The interaction between Ca_V2 and Shank has also been described in *C. elegans*. (**e**) Illustration of C-terminal interactions of $Ca_V1.2$ channels with the postsynaptic scaffolding proteins CIPP and NIL-16, which have PDZ domains that recognize the C-terminal $VSSL_{COOH}$ signature. For panels C to E, symbols for domains that mediate protein–protein interaction are depicted in a legend below panel E

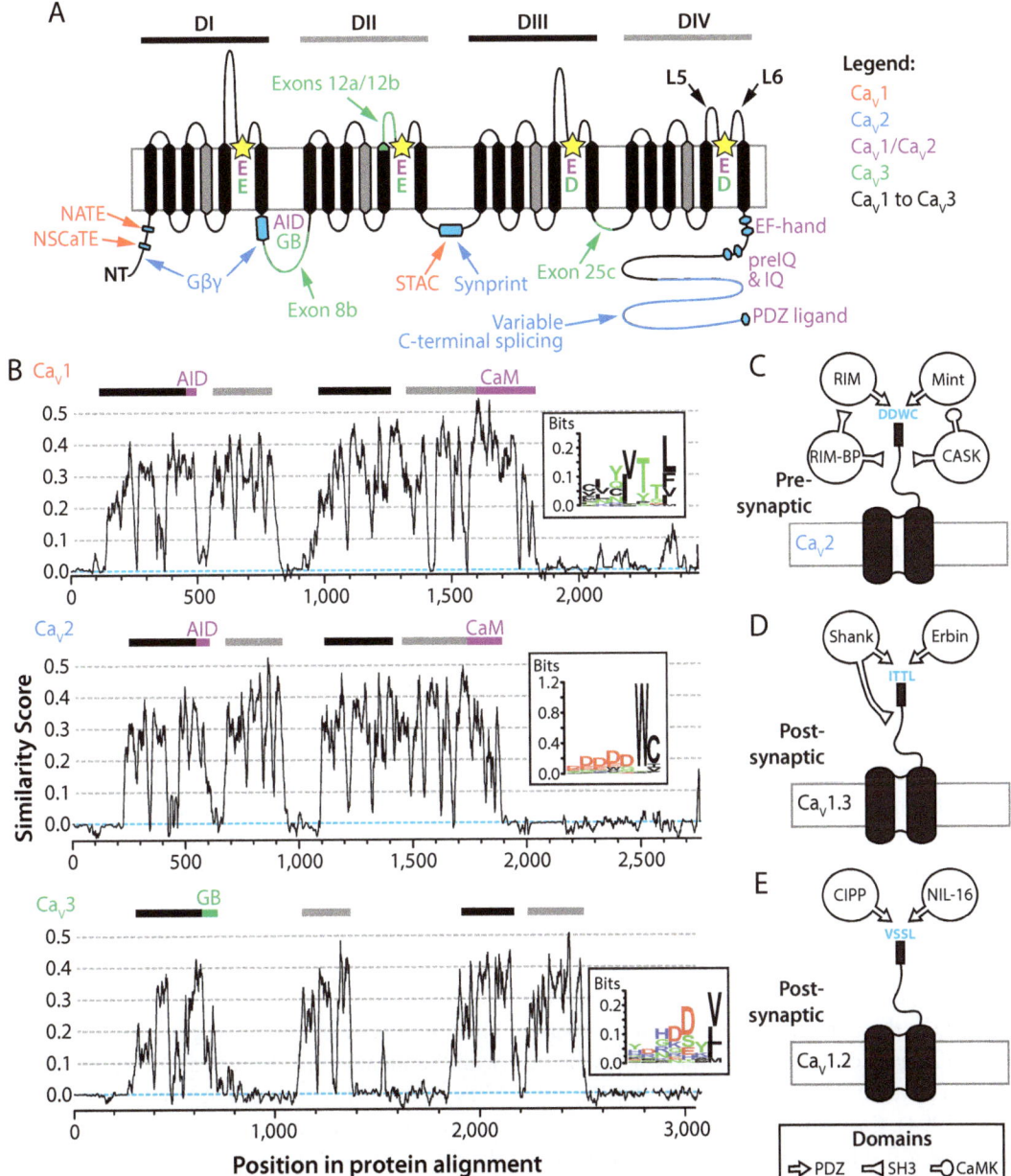

Fig. 2 (a) Illustration of the membrane topology of the Ca$_V$ channel α$_1$ subunit and several key areas described throughout the text. The extracellular structures are located on the top and cytoplasmic structures on the bottom. Voltage sensor S4 helices within each of the four repeat domains (DI to DIV) are colored in gray, and all other transmembrane segments are in black. The stars denote the locations of the four selectivity filter residues located in the extracellular pore-loops of the four repeat domains. Features described in the text that are unique to Ca$_V$1 channels are shown in red, Ca$_V$2 channels in blue, both Ca$_V$1 and Ca$_V$2 in purple, Ca$_V$3 channels in green, and to all channel types in black. (B) Plots of phylum-level protein sequence similarity among Ca$_V$1, Ca$_V$2, and Ca$_V$3 channel orthologues, generated with the program EMBOSS Plotcon (Rice et al., 2000). The protein alignment was generated using single representative sequences of Ca$_V$1–Ca$_V$3 orthologues from each phylum depicted in Fig. 1a, using the program Muscle (Edgar, 2004), and trimmed with trimAl (Capella-Gutiérrez et al., 2009) to remove regions where only one sequence was represented. The Y axes represent the Plotcon EBLOSUM62 score for a sliding window of 11 amino acids along the length of each alignment, and the positions within the alignments

with flanking S5 and S6 segments, assemble in the center of the channel protein to form the ion-conducting pathway (Wu et al., 2016). For calcium channels, selectivity for Ca^{2+} over other cations is attributed to a motif of four key amino acids from each of the four pore-loops that form a high-affinity Ca^{2+} binding site called the ion selectivity filter (Sather & Mccleskey, 2003). For Ca_V1 and Ca_V2 channels, the selectivity filter motif comprises four glutamate/E residues (i.e., EEEE), while for Ca_V3 channels, two glutamate and two aspartate/D residues (EEDD) (for additional review, please see Chapter XX by Dr. Tsien). All four-domain cation channels possess a highly specific asymmetrical patterning of extracellular loops/turrets lining the pore regions. This includes a rising extracellular turret from the S5 transmembrane helix (named L5 for loop 5) before the selectivity filter, which is the longest in length in Domain I, followed by Domain III, and a longest and most variable L6 extracellular turret rising after the pore before the S6 transmembrane helix in Domain IV (Fig. 2a) (Stephens et al., 2015).

Ca_V1 and Ca_V2 channels are distinct from Ca_V3 channels in having conserved C-terminal EF-hand, pre-IQ and IQ domains where the calcium sensor protein calmodulin binds and regulates channel gating (Ben-Johny & Yue, 2014). Also different are respective structures in the cytoplasmic linker between Domains I and II, where Ca_V1 and Ca_V2 channels possess an alpha interaction domain (AID), required for association with the ancillary cytoplasmic $Ca_V\beta$ subunit in both invertebrates and vertebrates (Pragnell et al., 1994). Ca_V3 channels do not require ancillary β or $\alpha_2\delta$ subunits for their function (Perez-Reyes, 2003). Instead, Ca_V3 channels possess a "gating brake" in position of the AID structure (Fig. 2a) (Vitko et al., 2007), which in both invertebrates and vertebrates also serves as a primary calmodulin-binding motif, in lieu of an IQ motif present in the C-terminus of Ca_V1 and Ca_V2 channels (Chemin et al., 2017). Notably, calmodulin also associates with $Ca_V3.1$ channel at their C-terminus in a Ca^{2+}-dependent manner, although the precise location within the channel

has not been identified (Asmara et al., 2017). Beyond their structural differences, Ca_V1/Ca_V2 channels and Ca_V3 channels also differ in their voltage sensitivity, where Ca_V3 channels begin activating in response to only slight membrane depolarization at low voltages (hence low-voltage-activated or LVA), while Ca_V1 and Ca_V2 channels require strong depolarization (hence high-voltage-activated or HVA). Interestingly, both the Ca_V1/Ca_V2 channel selectivity filter motif of EEEE and the C-terminal calmodulin-binding regions were present in the pre-metazoan ancestor of all four-domain channels, while the AID structure evolved strictly in Ca_V3 channels of animals and choanoflagellates (Pozdnyakov et al., 2018).

Phylogenetic studies have suggested that while Ca_V3 channels predate animals, Ca_V1 and Ca_V2 channels evolved strictly in animals via gene duplication of an ancestral channel type called $Ca_V1/2$ (Moran & Zakon, 2014; Piekut et al., 2020). Accordingly, the choanoflagellate *Salpingoeca rosetta* has both a Ca_V3 and a $Ca_V1/2$ channel (Fig. 1b), the latter forming a sister relationship with Ca_V1 and Ca_V2 channels on phylogenetic trees. Both ctenophores (e.g., *Mnemiopsis leidyi*) and sponges (*Amphimedon queenslandica*) lack Ca_V1 and Ca_V3 channels, the latter likely lost given its presence in choanoflagellates. The precise origins of Ca_V1 and Ca_V2 channels remain unclear, but it is evident that these are animal specific, and their close kinship is evident in their genomic structure with the identical placement of 31 intron splice sites shared across vertebrate and invertebrate homologues (Fux et al., 2018). Evident is that the most early-diverging animals to possess Ca_V1 to Ca_V3 channels together are the placozoans (e.g., *Trichoplax adhaerens*), which like the majority of invertebrates, possess single homologues of each channel type (Senatore et al., 2012; Moran & Zakon, 2014). Interestingly, similar to vertebrates, cnidarians (e.g., the sea anemone *Nematostella vectensis*; Fig. 1b) extensively duplicated Ca_V channel genes to give rise to three Ca_V2 and two Ca_V3 channels, while retaining a single Ca_V1 homologue (Piekut et al., 2020; Moran & Zakon, 2014). Gene duplications are also apparent in flatworms (Phylum Platyhelminthes), with the human blood fluke *Schistosoma japonicum* possessing two gene copies each of Ca_V1 and Ca_V2 channels (Salvador-Recatalà & Greenberg, 2012), while appearing to have lost Ca_V3 channel homologues, retained as two genes in non-parasitic free-living species, such as *Schmidtea mediterranea* (Fig. 1b). Perhaps the most extreme example of Ca_V channel duplication is in the ray-finned fish (Actinopterygii), which underwent an additional round of genome duplication relative to other vertebrates (Meyer & Schartl, 1999). *Danio rerio* (zebrafish), for example, possesses an impressive set of eighteen Ca_V channel α_1 subunit genes: seven L-type ($Ca_V1.1a$, $Ca_V1.1b$, $Ca_V1.2$, $Ca_V1.3a$, $Ca_V1.3b$, $Ca_V1.4a$, and $Ca_V1.4b$), six P/Q-, N-, and R-type ($Ca_V2.1a$, $Ca_V2.1b$, $Ca_V2.2a$, $Ca_V2.2b$, $Ca_V2.3a$, and $Ca_V2.3b$), and five T-type channels ($Ca_V3.1$, $Ca_V3.2a$, $Ca_V3.2b$, $Ca_V3.3a$, $Ca_V3.3b$) (Fig. 1b) (Haverinen et al., 2018).

With respect to primary sequence, it is notable that at the phylum level, Ca_V1 to Ca_V3 channels are highly divergent along intracellular regions outside of the gating brake structure that is unique to Ca_V3 channels, and the AID, EF-hand, pre-IQ, and IQ domains that are shared between Ca_V1 and Ca_V2 channels; Fig. 2a, b). These divergent cytoplasmic regions are generally devoid of secondary structure, but are important loci for calcium channel modulation, for example, through interactions with modulatory proteins or as sites for post-translational modification, such as phosphorylation (Kamp & Hell, 2000; Catterall & Few, 2008; Huc et al., 2009). Altogether, it is unclear if this extreme sequence divergence reflects divergence in modulation, especially given recent indications that disordered proteins can harbor obscured functional homology that is not apparent in sequence alignments, but rather is manifested through similarities in amino acid composition, length, and net charge (Zarin et al., 2019).

Phylogenetic Properties of the Ca_V Channel β Subunit

The calcium channel β subunit plays a crucial role in the membrane expression and gating properties of Ca_V1 and Ca_V2 channel α_1 subunits, and mediates important modulatory processes, such as G protein inhibition and phosphorylation (Buraei & Yang, 2010; Catterall, 2000; Dolphin, 2016). β subunits possess conserved guanylate kinase (GK) and Src Homology 3 (SH3) domains (Hanlon et al., 1999; Birnbaumer et al., 1998), the former mediating the interaction with the α_1 subunit AID structure (Opatowsky et al., 2004; Van Petegem et al., 2004; Chen et al., 2004) (Fig. 2a) (for more details, please see chapter "Voltage-Gated Calcium Channel Auxiliary β Subunits" by Dr. Colecraft). Previously, limited gene sequence data led to the suggestion that the β subunit evolved in the common ancestor of animals and choanoflagellates, concurrent with the emergence of $Ca_V1/2$ channels, but was lost in ctenophores and poriferans (Moran & Zakon, 2014). Nevertheless, a search through updated genome sequence data for the sponge *A. queenslandica* and the ctenophore *M. leidyi* revealed the presence of one and two homologues, respectively (Figs. 1b and 3a). Like the choanoflagellate *S. rosetta*, most invertebrates possess a single β gene. Platyhelminths, nematodes, cnidarians, and ctenophores each possess two (Fig. 1b), which given their distributed phylogenetic relationships relative to each other (Fig. 3a), likely emerged via independent duplication in each corresponding lineage. As with the significant expansion of α_1 subunits in vertebrates, mammals and fish possess four and seven β genes, respectively (e.g., mouse β1–4, and zebrafish β1, β2a, β2b, β3a, β3b, β4a, and β4b) (Zhou et al., 2008). Instead, early-diverging invertebrate chordates, like the tunicate *Ciona intestinalis,* only possess one β subunit gene.

Alignment of single orthologues from each phylum corroborates previous reports of high sequence divergence within the β subunit N- and C-termini, as well as the flexible HOOK domain

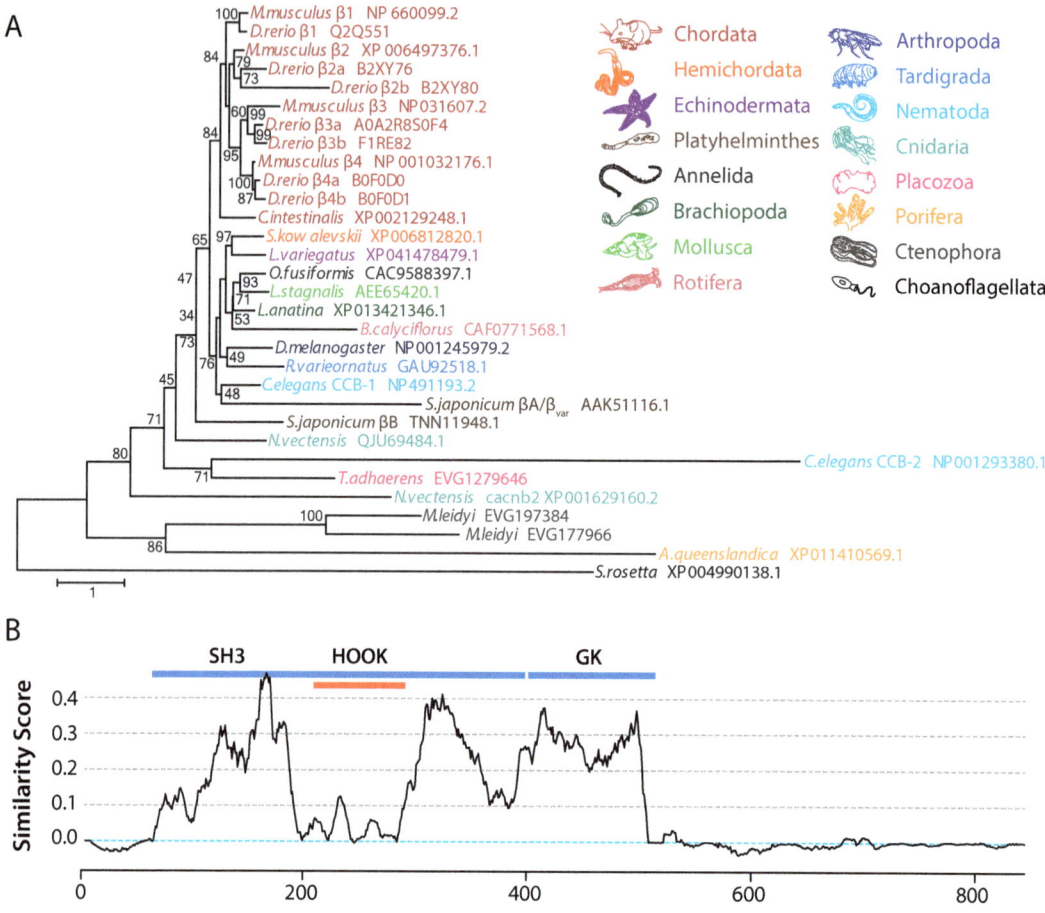

Fig. 3 (**a**) Maximum likelihood phylogenetic tree of Ca$_V$ channel β subunits, inferred from a protein alignment and rooted against the divergent homologue from the choanoflagellate *Salpingoeca rosetta*. Names of select subunits described in the text are provided, as are NCBI accession numbers, except sequences with EVG identifiers that were extracted from an unpublished transcriptome study that is available as a preprint (Wong et al., 2019). Sequences from species within each phylum depicted in Fig. 1a were obtained by searching through corresponding genome sequences using the local alignment program NCBI BLAST (Johnson et al., 2008), and these were manually curated and corroborated by confirming the presence of orthologous sequences in other species from the same phylum. Identified protein sequences were aligned with the program Muscle (Edgar, 2004), trimmed with trimAl (Capella-Gutiérrez et al., 2009) using a gap threshold of 0.8, and the phylogenetic tree inferred with the program IQ-Tree (Nguyen et al., 2015) using a best fit model of LG+F+R5. Node support values from 1,000 ultrafast bootstrap replicates are given. The color-coded legend on the right indicates phylum, and the species and sequence identifiers on the phylogenetic tree are colored accordingly. (**b**) Plots of phylum-level protein sequence similarity score of β subunits generated with EMBOSS Plotcon (Rice et al., 2000) (window of 11 amino acids). The protein alignment was generated using single representative sequences of β subunit orthologues from each phylum depicted in Fig. 1a, using the program Muscle (Edgar, 2004), and trimmed with trimAl (Capella-Gutiérrez et al., 2009) to remove regions where only one sequence was represented. Regions in the alignment corresponding to the SH3 domain, GK domain, and HOOK region are depicted with colored bars above the plot

located between the SH3 and GK domains (Fig. 3b) (Hanlon et al., 1999; Dawson et al., 2014). Given the role of the N-terminus and HOOK region in modulating channel gating (Richards et al., 2007; He et al., 2007), variability at these sites might reflect lineage-specific adaptations that served to alter Ca_V channel function for specific needs within the different animals (Dawson et al., 2014; Weir et al., 2020).

Phylogenetic Properties of the Ca_V Channel $\alpha_2\delta$ Subunit

When first biochemically isolated, the auxiliary $\alpha_2\delta$ subunit was co-purified as two separate proteins, α_2 and δ, named according to their decreasing molecular weight relative to the α_1 subunit (Tanabe et al., 1987). Subsequently, it was determined that these are derived from a single gene encoding a polypeptide that is subject to post-translational cleavage (Jay et al., 1991; De Jongh et al., 1990). Nevertheless, the α_2 and δ proteins remain tightly associated via extensive disulfide linkages (Wu et al., 2015). Like the β subunit, $\alpha_2\delta$ plays a crucial role in the membrane trafficking and gating of Ca_V1 and Ca_V2 channels (Dolphin, 2016), but is located on the outside of the cell where it forms numerous contacts with extracellular loops in Domains I and III of the α_1 subunit (Wu et al., 2015). These contacts are mostly mediated by conserved *v*on *W*illebrand factor *A* (vWA) and *ca*lcium channel and *che*motaxis (CACHE) domains of the $\alpha_2\delta$ protein (Wu et al., 2015). Notably, the vWA domain contains a divalent cation-binding site made up of five polar/charged amino acids, called the *m*etal *i*on-*d*ependent *a*dhesion *s*ite (MIDAS), that likely binds Mg^{2+} or Ca^{2+} in the ER lumen to regulate trafficking of the α_1 subunit to the cell membrane (Canti et al., 2005). CACHE domains on the other hand originate in bacteria where they bind small molecules and are involved in sensory signal transduction (Upadhyay et al., 2016), but whether they retain this capacity in the $\alpha_2\delta$ protein is unknown (for more details, please see chapter "Regulation of Calcium Channels and Synaptic Function by Auxiliary $\alpha_2\delta$ Subunits" by Drs. Dolphin and Obermair).

Although Ca_V3 channels generally operate independently of the β and $\alpha_2\delta$ subunits *in vivo*, both subunits enhance Ca_V3 channels currents *in vitro*, likely through an indirect mechanism (Dubel et al., 2004; Gao et al., 2000; Dolphin et al., 1999; Arteaga-Tlecuitl et al., 2018). Recently, however, the $\alpha_2\delta$-related protein CACHD1 was found to be endogenously co-expressed with Ca_V3 channels in mammalian neurons and to significantly enhance Ca_V3 channel current in cell lines and neurons through an increase in both channel open probability and membrane trafficking (Cottrell et al., 2018). Furthermore, the protein was found to form a complex with the $Ca_V3.1$ channel at the cell membrane, and in rodents but not humans, to mildly enhance $Ca_V2.2$ channel current and membrane expression *in vitro* (Dahimene et al., 2018; Soubrane et al., 2012). More recently, knockout of CACHD1 in mice was shown to impair hearing and balance, perhaps through dysregulation of calcium channels (Tian et al., 2021). Phylogenetically, the relationship between $\alpha_2\delta$ and CACHD1 has not been comprehensively examined, although the latter possesses a very similar domain architecture composed of a signal peptide for co-translational insertion into the endoplasmic reticulum, a vWA domain with a variant MIDAS motif, CACHE domains, and a C-terminal transmembrane helix (Stephens & Cottrell, 2019). However, unlike $\alpha_2\delta$, CACHD1 does not appear to be post-translationally cleaved into two separate proteins, and lacks C-terminal glycosylphosphatidylinositol (GPI) anchors (Davies et al., 2010), instead harboring a large cytoplasmic domain downstream of the C-terminal transmembrane helix (Cottrell et al., 2018).

Previously, the $\alpha_2\delta$ subunit was reported to be absent in the genomes of sponges, ctenophores, and non-metazoans, suggesting this subunit evolved in the common ancestor of placozoans, cnidarians, and bilaterians (Moran & Zakon, 2014) (Fig. 1a). However, analysis of updated genome datasets indicates more ancient origins with homologues present for both sponges and ctenophores, and two $\alpha_2\delta$-like genes present in the genome of the choanoflagellate *S. rosetta*

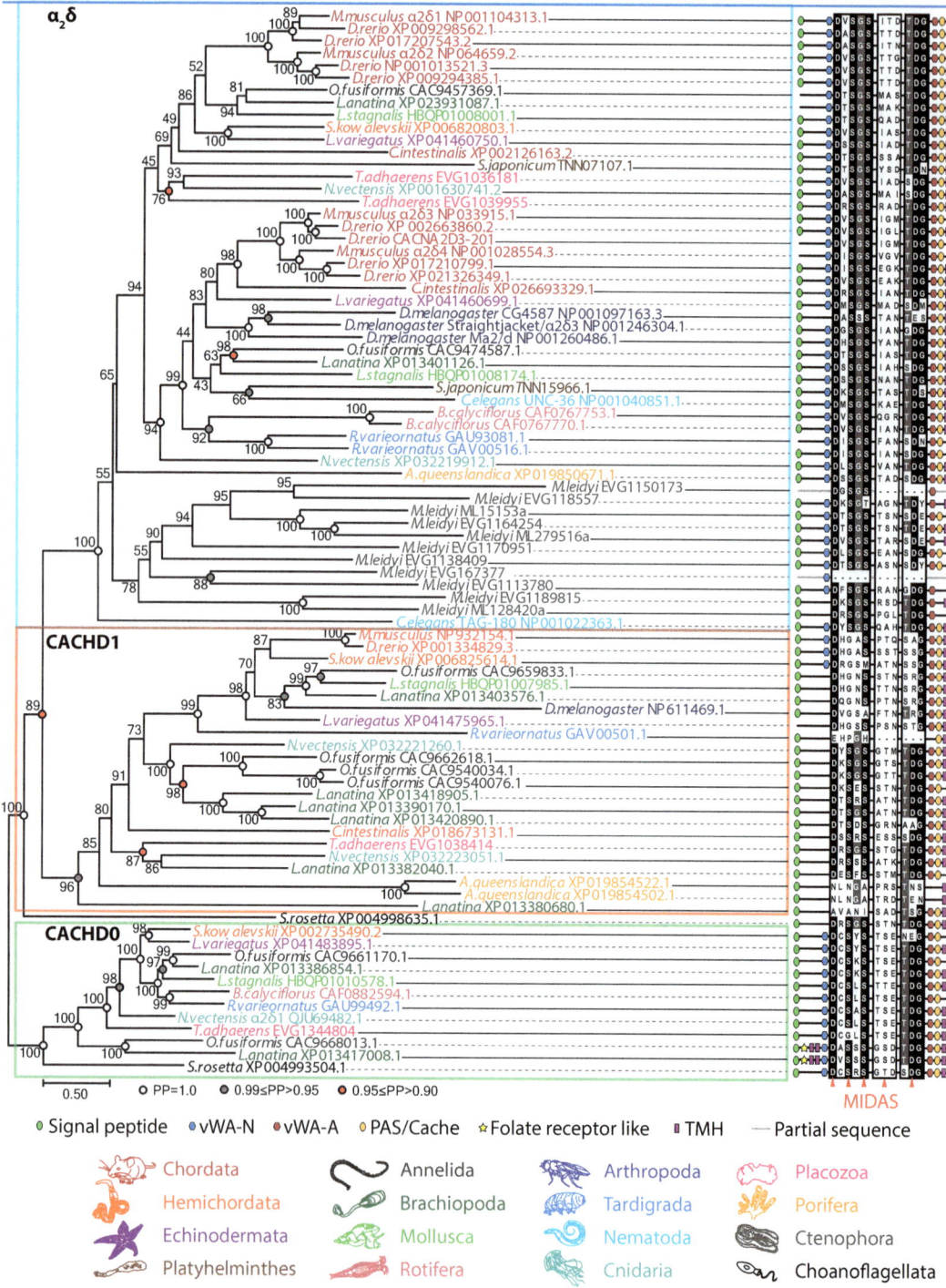

Fig. 4 Maximum likelihood phylogenetic tree of Ca_V channel $\alpha_2\delta$ and related subunits, inferred from a protein alignment and rooted against a clade containing a divergent homologue from the choanoflagellate *Salpingoeca rosetta*. Names of select subunits described in the text are provided, as are NCBI accession numbers, except sequences with EVG identifiers that were extracted from

an unpublished transcriptome study that is available as a preprint (Wong et al., 2019), and those with ML identifiers extracted from the *Mnemiopsis* Genome Portal (Moreland et al., 2014). Sequences from species within each phylum depicted in Fig. 1a were obtained by searching through corresponding genome sequences using the local alignment program NCBI BLAST (Johnson et al., 2008),

(Figs. 1b and 4). Based on phylogenetic analysis of the proteins (Fig. 4), animals possess three distinct classes of $\alpha_2\delta$ genes: the canonical $\alpha_2\delta$ subunit, CACHD1, and a third class not found in chordates but present among various bilaterian and early-diverging animals, which we name here CACHD0 for *cache domain*-containing protein *0*. All animals examined possess $\alpha_2\delta$, while CACHD1 and CACHD0 appear to have been lost in some lineages. For example, CACHD1 is absent in the genomes of rotifers, nematodes, and platyhelminths, the latter concurrent with the noted loss of Ca_V3 channels (Figs. 1b and 4). In contrast to α_1 and β subunits, $\alpha_2\delta$ is generally present as two copies in most invertebrate genomes (Fig. 1b), expanding in vinegar flies to three copies (i.e., Ma2/d, straightjacket, and CG4587), and in vertebrates to four copies in mammals (mouse $\alpha_2\delta1$–4) and eight copies in fish (zebrafish $\alpha_2\delta1a$–4a and $\alpha_2\delta1b$–4b) (Schlegel et al., 2019). The most striking expansion of $\alpha_2\delta$ genes is in the ctenophore *Mnemiopsis leidyi*, which has eleven copies despite having only a single Ca_V channel gene.

Sequence homology and hidden Markov analysis reveal broad conservation of N-terminal signal peptides, vWA domains (including an N-terminal domain often found upstream of vWA domains), CACHE/PAS domains, and predicted C-terminal transmembrane helices among all three classes of $\alpha_2\delta$ proteins (Fig. 4). Notable is that the absence of a predicted C-terminal transmembrane helix as observed for the mouse $\alpha_2\delta3$ and $\alpha_2\delta4$ homologues is a rare feature and that

the N-terminal domain associated with vWA is absent in all protostome and early-diverging CACHD1 homologues. Furthermore, the five signature residues of the MIDAS motif are broadly conserved, even in numerous CACHD1 and CACHD0 homologues, with a subset of sequences bearing "variant" motifs lacking at least one of the five charged or polar residues. The two sponge CACHD1 homologues are perhaps the most structurally divergent, lacking most of the MIDAS residues, and predictions for vWA and CACHE domains. Lastly, the CACHD0 proteins from the annelid *Owenia fusiformis* and the brachiopod *Lingula anatina* possess atypical tandem alpha helices predicted at their N-termini, preceded by a unique folate receptor-like domain.

Ca_V Channels in Deuterostome Invertebrates

Since Hagiwara's Discovery of LVA and HVA Ca²⁺ Currents in Starfish Eggs

Over 100 years ago, experiments on sexual reproduction of the echinoderm sand dollar, a burrowing sea urchin (Just, 1919), led to the general prediction that during fertilization eggs use two strategies to prevent entry of more than one sperm (polyspermy), which can have catastrophic consequences for embryonic development: a slow "obstructive" component, mediated by the egg extracellular matrix and cell membrane (Bianchi

---

Fig. 4 (continued) and these were manually curated and corroborated by confirming the presence of orthologous sequences in other species from the same phylum. Identified protein sequences were aligned with the program Muscle (Edgar, 2004), trimmed with trimAl (Capella-Gutiérrez et al., 2009) with a gap threshold of 0.9, and the phylogenetic tree inferred with the program IQ-Tree (Nguyen et al., 2015) using a best fit model of WAG+F+R7. Node support values from 1,000 ultrafast bootstrap replicates are given. To provide more information on the nodes, Bayesian inference was done on the trimmed alignment with the program MrBayes (Huelsenbeck & Ronquist, 2001), using the best fit model WAG+F+G4. Colored circles represent posterior probability (PP) support values for nodes that are common to the Bayesian and maximum likelihood trees. The color-coded legend on the bottom indicates phylum, and the species and sequence identifiers on the phylogenetic tree are colored accordingly. To the right of the phylogenetic tree are illustrations of protein domains predicted with the program InterProScan (Jones et al., 2014) (legend at bottom), flanking an alignment of the three protein regions that contain the 5 amino acids of the MIDAS motif (indicated by red chevrons below). Evident from the tree is that there are three strongly supported clades: canonical $\alpha_2\delta$, CACHD1, and what we refer to as CACHD0

& Wright, 2016), and a fast "electrical" component, involving fertilization-induced depolarization of the cell membrane that acts as an electrical barrier for subsequent sperm entry (Wozniak & Carlson, 2020). Fast polyspermy block is found broadly in oviparous (egg laying) organisms that use a multitude of sperm to fertilize one or a small number of eggs, including echinoderms (Jaffe, 1976), amphibians (Charbonneau et al., 1983), tunicates (Goudeau et al., 1994), marine worms (Kline et al., 1985), and algae (Brawley, 1991). Extensive work has revealed that in sea urchins, this membrane depolarization is mediated by voltage-gated calcium channels, most likely corresponding to the classic LVA and HVA calcium currents recorded by Susumu Hagiwara in the starfish species *Mediaster aequalis* (Hagiwara et al., 1975). Specifically, contact of a sperm with the egg leads to a slight membrane depolarization from the typical resting voltage of -70 mV, toward the activation range of LVA channels (between -60 and -50 mV). This triggers a fast and pronounced membrane depolarization mediated by Ca^{2+} influx (Chambers & De Armendi, 1979), most likely through the combined activation of the LVA and HVA calcium channels. Evidence that this process is indeed dependent on voltage-gated calcium channels comes from electrophysiological studies showing that depletion of external Ca^{2+} disrupts sperm-induced depolarization, that evoked depolarization to the activation range of LVA channels (-59 mV) can trigger the Ca^{2+} response in the absence of fertilization (Chambers & De Armendi, 1979), and that the response will not occur if the existing resting voltage is more positive than -40 mV, attributed to calcium channel inactivation (Nuccitelli & Grey, 1984). Of note, the molecular identity of the observed LVA and HVA channels in echinoderm eggs has not been determined, nor have any echinoderm calcium channel subunits been cloned and functionally expressed *in vitro*, or subject to genetic disruption to assess phenotypic contributions. However, mRNAs of both Ca_V1 and Ca_V2 channels have been detected in sperm of the sea urchin *Strongylocentrotus purpuratus*, where they are suggested to play a role in the acrosome reaction (Granados-Gonzalez et al., 2005).

The discovery of unique LVA Ca^{2+} currents by Hagiwara in starfish eggs (Hagiwara et al., 1975), and then in tunicate eggs (Okamoto et al., 1976), eventually led to the molecular cloning and *in vitro* expression of T-type/Ca_V3 channels, a quarter-century later (Perez-Reyes et al., 1998) (for a history on key calcium channel discoveries, please see chapter "A Lived History of Early Calcium Channel Discoveries Over the Past Half-Century" by Dr. Carbone). Like echinoderms, tunicates are oviparous (although some are viviparous) and exhibit fast polyspermy block (Goudeau et al., 1994). In both echinoderms and tunicates, the LVA and HVA components of the calcium current became evident when external recordings solutions were depleted of Na^+ ions, and both studies noted that the LVA channel was permeable to Na^+ and Ca^{2+}, consistent with our current understanding that LVA Ca_V3 channels are markedly permeable to Na^+, unlike Ca_V1 and Ca_V2 channels that are highly Ca^{2+} selective (Shcheglovitov et al., 2007; Hess et al., 1986). The two studies differed, however, in their interpretation of the possible underlying ion channels. Whereas Hagiwara and colleagues attributed the LVA current to a novel class of calcium channel dubbed "type I," distinct from previously recorded "type II" HVA channels, Okamoto and colleagues attributed it to Ca^{2+} influx through a Na_V channel, citing previous suggestions made for the squid axon (Meves & Vogel, 1973; Baker et al., 1971). Interestingly, all of these animals (molluscs, echinoderms, and tunicates) possess in their genomes two types of voltage-gated sodium channels: Na_V1 channels, with ion selectivity filter motifs of DEKA that make them highly selective for Na^+, and Na_V2 channels, that have Ca_V-like filter motifs of DEEA, reported often to be highly Ca^{2+} permeable and in some cases highly Ca^{2+} selective *in vitro* (Fig. 2a) (Liebeskind et al., 2011; Barzilai et al., 2012; Gosselin-Badaroudine et al., 2016). Thus, it is reasonable to suggest that the observed LVA currents in these various electrophysiological preparations could be attributed to Na_V2 channels rather than Ca_V3 channels, as they can also operate at low voltages and conduct mixed Na^+/Ca^{2+} currents.

Tunicates and the Evolution of $Ca_V1.1$ Channels and Ryanodine Receptor Tetrad Coupling

Early-diverging chordates, like tunicates (Fig. 1a) and cephalochordates (lancelets), resemble most invertebrates in possessing just single homologues of each of the three Ca_V channels α_1 subunit types. Likely, the duplication of calcium channel genes in vertebrates (Fig. 1b), and the resulting genetic/functional redundancy, would have relaxed selective constraints permitting diversification of channel function (Ohno, 2013). Here, interesting questions emerge about the distribution of core, ancestral functions, and the mechanisms for the emergence of novel functions among the duplicated paralogues.

Conserved among chordates, molluscs, arthropods, and nematodes, Ca_V2 channels are the predominant drivers for neurotransmitter release at the presynaptic terminal, representing an ancestral function for this channel type (Catterall, 2011; Spafford et al., 2003b; Saheki & Bargmann, 2009; Kawasaki et al., 2004; Smith et al., 1996; Mathews et al., 2003). In vertebrates, the duplicated $Ca_V2.1$ (P/Q-type) and $Ca_V2.2$ (N-type) channels retain this predominant presynaptic function (Kaeser et al., 2011; Wen et al., 2013), with varying contributions in different synapse types and developmental stages (Dolphin & Lee, 2020). Instead, $Ca_V2.3$ (R-type) channels evolved to play only minor roles in neurotransmitter release (Wormuth et al., 2016), taking up a diversity of other specialized functions, including regulating synaptic plasticity at mossy fiber synapses (Dietrich et al., 2003), neural development (Nishiyama et al., 2011), and regulating neural excitability (Weiergräber et al., 2006). A similar core function is apparent for Ca_V1 channels, being the predominant drivers for muscle contraction in chordates (Catterall, 2011), molluscs (Senatore et al., 2014; Kits & Mansvelder, 1996), nematodes (Jospin et al., 2002; Lainé et al., 2011), arthropods (Hara et al., 2015; Eberl et al., 1998), and platyhelminths (Chan et al., 2017). In vertebrates, $Ca_V1.2$ channels drive contraction in smooth and cardiac muscle, $Ca_V1.1$ channels drive contraction in striated skeletal muscle,

while $Ca_V1.3$ and $Ca_V1.4$ channels are less involved in contraction but play specialized roles in other cell types, for example, in auditory transduction in hair cells ($Ca_V1.3$) and visual transduction in the retina ($Ca_V1.4$) (Catterall, 2011; Joiner & Lee, 2015).

In being the only non-vertebrate deuterostome calcium channel to be cloned and functionally expressed *in vitro* (Okagaki et al., 2001) (Fig. 1b), the Ca_V1 channel from the tunicate *Halocynthia roretzi* provides an interesting perspective on calcium channel neofunctionalization after gene duplication, in this case with respect to muscle contraction. *In vitro* expression of the cloned tunicate Ca_V1 channel produces characteristic high-voltage-activated currents that are subject to Ca^{2+}/calmodulin feedback regulation, in line with vertebrate Ca_V1 channels. Furthermore, in situ hybridization experiments reveal Ca_V1 channel mRNA expression first occurs in muscle early during embryonic development, suggesting a conserved role in muscle contraction, followed by expression in developing neural tissues (Okagaki et al., 2001).

Interestingly, the evolutionary origin of both smooth/cardiac and striated (skeletal) muscle predates the duplication of Ca_V1 channels in vertebrates, as both muscle cell types were present in the common ancestor of protostome and deuterostome animals (i.e., bilaterians; Fig. 1a). These two cell types have distinct and conserved transcriptional/developmental programs, as well as mechanisms for contraction (i.e., troponin dependent for striated, calponin dependent for smooth/cardiac) (Brunet et al., 2016). Notably though, striations are not indicative of homology, where, for example, they evolved independently in smooth muscle cells of arthropod and vertebrate heart, and were lost in striated muscle of platyhelminths (Brunet et al., 2016).

In most bilaterian muscle, Ca_V1 channels drive contraction through the process of Ca^{2+}-induced Ca^{2+} release (CICR), which activates ryanodine receptors in the sarcoplasmic reticulum to trigger the release of intracellular Ca^{2+}, that floods the contractile filaments causing them to contract (Roderick et al., 2003; Dirksen & Beam, 1999). Importantly, concurrent with the duplication of calcium channel genes in vertebrates was the duplication of ryanodine receptors, with a single ancestral homologue duplicating to three copies in mammals (RyR1–3) and five in fish (RyR1a, RyR1b, RyR2a, RyR2b, and RyR3) (Alzayady et al., 2015; Darbandi & Franck, 2009). This permitted the diversification in Ca_V1-RyR functional coupling, with $Ca_V1.1$ coupled to RyR1 in striated/skeletal muscle (Takeshima et al., 1989), and $Ca_V1.2$ coupled with RyR2 in cardiac muscle (Nakai et al., 1990). Furthermore, in vertebrate striated/skeletal muscle, $Ca_V1.1$ channels evolved the unique capacity to physically interact with ryanodine receptors in the sarcoplasmic reticulum at transverse tubules, forming a characteristic tetrad arrangement of four calcium channels coupled to a single RyR1 receptor (Block et al., 1988). This particular adaptation circumvented canonical excitation–contraction coupling involving CICR, permitting $Ca_V1.1$ channels to directly activate ryanodine receptors through conformational changes in their voltage sensors, in a process that is independent of Ca^{2+} influx (i.e., mechanical coupling) (Dayal et al., 2017). The precise mechanism for this mechanical coupling is not fully understood, but the cytoplasmic proteins STAC and calmodulin appear to be important (Niu et al., 2018; Yuen et al., 2017), the former interacting directly with the channel II–III linker (Fig. 2a) (Nakai et al., 1998). Of note, $Ca_V1.2$ channels also interact with STAC, but with a 10-fold lower affinity than $Ca_V1.3$ (Yuen et al., 2017), as well as RyR2 receptors via the C-terminal calmodulin-binding regions (Mouton et al., 2001). However, $Ca_V1.2$ channels do not form tetrads at the cell membrane, and based on electron microscopy, do not physically interact with RyR2 *in vivo* (Franzini-Armstrong, 2004).

Interestingly, within the chordate phylum there appears to be a continuum with respect to the mode of striated muscle contraction, with a switch from CICR to direct Ca_V1-RyR coupling occurring concurrently with the gene duplication of Ca_V1 channels in vertebrates. That is, early-diverging chordates, such as tunicates and cephalochordates utilize non-tetrad coupling and CICR, utilizing the single Ca_V1 channel ortho-

logue. Vertebrates on the other hand utilize direct tetrad coupling, with an extreme condition evident in fish where the duplicated $Ca_V1.1a$ and $Ca_V1.1b$ channels lost the ability to conduct Ca^{2+} due to mutations in the pore (Schredelseker et al., 2010; Di Biase & Franzini-Armstrong, 2005). Thus, the small amplitude Ca^{2+} current observed for mammalian $Ca_V1.1$ channels may represent a vestigial function, although it has been suggested to contribute toward the replenishment of intracellular Ca^{2+} stores during elevated levels of excitability (Bannister & Beam, 2013).

Ca_V Channels in Protostome Invertebrates

The Atypical Nature of Calcium Channels in Parasitic Flatworms

Parasitic flatworms of the phylum Platyhelminthes, genus *Schistosoma*, are causative agents in an array of tropical diseases collectively referred to as schistosomiasis, with at least 230 million people affected worldwide (Colley et al., 2014). Calcium channels have been identified as potential targets for pharmacological intervention, especially in muscle, with the animals showing adverse, Ca^{2+}-dependent responses to dihydropyridines (DHPs) (Silva-Moraes et al., 2013), the broadly used anthelmintic drug praziquantel (PZQ) (Jeziorski & Greenberg, 2006), and the Ca_V1 channel agonist FPL-64176 (Mccusker & Chan, 2019). Here, we briefly discuss some of the research that has been done on platyhelminth calcium channels.

Parasitic platyhelminths distinguish themselves from most invertebrates by having lost Ca_V3 channels and CACHD1, and duplicating Ca_V1, Ca_V2, β, and $\alpha_2\delta$ subunits to produce two gene copies of each (Figs. 1b, 3a, and 4) (Salvador-Recatalà & Greenberg, 2012). Efforts to express and electrophysiologically record platyhelminth Ca_V channels have been unsuccessful (Salvador-Recatalà & Greenberg, 2012; Kohn, Lea, et al., 2001a), precluding direct electrophysiological and pharmacological studies. Furthermore, studies of endogenous voltage-gated calcium currents, especially in muscle, are hampered by their small amplitudes and large background potassium currents, and a rapid decay (i.e., rundown) of calcium currents during electrophysiological recording (Salvador-Recatalà & Greenberg, 2012).

Nevertheless, both β subunits from the blood fluke species *S. mansoni*, dubbed βA and βB, were functionally expressed *in vitro*, providing some unexpected insights into potential mechanisms for PZQ action, and the significance of the β subunit for lineage-specific adaptation in calcium channel function. Specifically, the βB subunit not only increases current amplitude of the human $Ca_V2.3$ channel when co-expressed in *Xenopus* oocytes, consistent with other β subunits, but also causes a pronounced hyperpolarizing shift in the channel's steady-state inactivation (Kohn et al., 2003). This subunit also possesses an atypical cluster of acidic residues at its N-terminus, which are deterministic for rundown of $Ca_V2.3$ currents recorded in mammalian cells (Salvador-Recatalà et al., 2008), as well as acceleration of Ca^{2+}/calmodulin-dependent inactivation (Salvador-Recatalà & Greenberg, 2010). Instead, the βA subunit causes only a moderate effect on steady-state inactivation of $Ca_V2.3$ (in *Xenopus* oocytes), but a dramatic reduction in current amplitude relative to control conditions without a co-expressed β subunit. Since β subunits are endogenously expressed in oocytes (Tareilus et al., 1997), this observation might indicate that the platyhelminth βA subunit has an atypical property of suppressing calcium channel current, causing a dominant-negative effect *in vitro* by competing with the endogenous *Xenopus* β subunit. Altogether, the two schistosome β subunits bear structural variations in their N-termini and regions close to the HOOK domain that impose suppressive effects on calcium channel current, suggested to accord with the attenuated (and difficult to record) calcium currents observed *in vivo*, and perhaps, resulting from the parasitic life strategy of these animals (Salvador-Recatalà & Greenberg, 2012). Interestingly, βA also conferred PZQ sensitivity to the human $Ca_V2.3$ channel, since application of the drug in its presence led to a doubling of current amplitude (Kohn,

Anderson, et al., 2001b), perhaps through the release of a background inhibitory effect. This phenomenon was subsequently attributed to a cluster of unique amino acids at the junction between the SH3 and GK domains of βA, just a few amino acids downstream of the HOOK domain (Kohn et al., 2003).

Functional Studies of Calcium Channels of the Pond Snail Lymnaea Stagnalis

Animals from the phylum Mollusca have a long history of contributing to neuroscience research, including the important work done by Hodgkin, Huxley, and others on electrical conduction in the squid giant axon (Hodgkin & Huxley, 1952), Katz, Miledi, and others on synaptic transmission in the squid giant synapse (Katz & Miledi, 1967), and Eric Kandel and others on learning and memory in the sea hare *Aplysia californica* (Kandel, 2001). The freshwater pond snail *Lymnaea stagnalis* belongs to this legacy, with extensive research done on rhythmic neural circuits that control breathing (Syed et al., 1990) and feeding (Straub et al., 2002), and neuromodulatory circuits that control heartbeat (Buckett et al., 1990) and egg laying (Geraerts, 1976).

The most detailed structure–function profiling of voltage-gated calcium channels outside of vertebrates has been done for *Lymnaea*, with all three of its Ca_V channel α_1 subunits, as well as its single β subunit (Fig. 1b), having been cloned and expressed *in vitro* for electrophysiological and pharmacological characterization (Spafford et al., 2006, Senatore & Spafford, 2010, Spafford et al., 2003a, Dawson et al., 2014). Notably, the tissue expression of calcium channels in *Lymnaea* resemble general patterns seen in mammals (Ertel et al., 2000), with Ca_V2 channels expressed in the central nervous system and secretory glands, while Ca_V1, Ca_V3, and $Ca_V\beta$ are expressed more broadly, also found in the heart and buccal muscle (Dawson et al., 2014; Senatore et al., 2014).

The first *Lymnaea* α_1 subunit to be functionally characterized was the Ca_V2 channel, which produces characteristic high-voltage-activated Ca^{2+} currents *in vitro* when co-expressed with rat β and α$_2$δ subunits (Spafford et al., 2003a). This channel resembles mammalian Ca$_V$2.2 in its voltage sensitivity for activation and inactivation, minimal Ca^{2+}/calmodulin-dependent inactivation, and sensitivity to cadmium (Cd^{2+}) block, but differs in its kinetics and lack of sensitivity to the potent Ca$_V$2.2 channel peptide blocker ω-conotoxin GVIA (Huang et al., 2010). Also different is its regulation by G proteins, a mechanism that links neuromodulatory G protein-coupled receptor (GPCR) signaling with Ca$_V$2 channel function, permitting precise control of neurotransmitter release and hormone secretion (Tedford & Zamponi, 2006; Herlitze et al., 1996). Specifically, in mammals cytoplasmic G protein βγ heterodimers become activated by a ligand-bound GPCR, to form a physical association with Ca$_V$2.1 and Ca$_V$2.2 channels. This leads to inhibition of the channels, which can be alleviated by a train of fast action potentials or prolonged depolarization (hence the term voltage-dependent inhibition). This latter feature permits neurons to temporarily override G protein inhibition during periods of heightened excitability, contributing to a form of short-term plasticity (Park & Dunlap, 1998). Interestingly, the *Lymnaea* Ca$_V$2 channel was found to be inhibited by G proteins, but in a voltage-independent manner (Spafford et al., 2003a; Huang et al., 2010), resembling observations made for the mammalian Ca$_V$2.1 channel in the absence of a co-expressed Ca$_V$ β subunit (Zhang et al., 2008). Furthermore, replacing the N-terminus and I–II linker of the *Lymnaea* channel with sequences from rat Ca$_V$2.2, regions where Gβγ proteins bind, did not confer G protein voltage sensitivity (Fig. 2a) (Tedford & Zamponi, 2006; Agler et al., 2005). Possibly, the lack of voltage sensitivity for the *Lymnaea* Ca$_V$2 channel could have resulted from differences in β and/or Gβγ proteins, since only mammalian proteins were used in these *in vitro* experiments. This notion is supported by recent studies revealing that the neuromodulatory ligands dopamine and retinoic acid elicit voltage-dependent G protein inhibition of Ca$_V$ channel currents in isolated

Lymnaea neurons, affecting presynaptic neurotransmitter release and thus consistent with direct modulation of the Ca$_V$2 channel (de Hoog et al., 2019; Dong et al., 2018) (although see (Dunn & Sossin, 2013, Dunn et al., 2018)).

Separate work has established that like in mammals, the *Lymnaea* Ca$_V$2 channel is the main driver for neurotransmitter release, while Ca$_V$1 plays a minor role in synaptic transmission and instead contributes to neural excitability (Spafford et al., 2003b, 2006). Notably, the *Lymnaea* Ca$_V$2 channel lacks a *synaptic protein interaction* (synprint) site in its domain II–III linker (Fig. 2a), which in mammalian Ca$_V$2.1 and Ca$_V$2.2 channels mediates direct interaction with the exocytotic proteins syntaxin-1 and SNAP-25 (Sheng et al., 1994; Rettig et al., 1996). This interaction contributes to the proximal and functional coupling of the calcium channels with primed presynaptic vesicles, but its absence in invertebrates suggests a secondary adaptation that is unique to vertebrates. Instead, a key determinant for presynaptic function of the *Lymnaea* Ca$_V$2 channel was identified in the extreme C-terminus (Spafford et al., 2003b), composed of a deeply conserved acidic amino motif of DDWC$_{-COOH}$, present even in cnidarian Ca$_V$2 channel orthologues (Piekut et al., 2020) (Fig. 2b, c). As we discuss in more detail later, subsequent work has established the significance and deep phylogenetic conservation of this motif, which mediates interaction with the presynaptic scaffolding protein *Rab3-interacting molecule* or RIM (Südhof, 2012).

The next *Lymnaea* calcium channel to be functionally characterized was the Ca$_V$1 channel (Spafford et al., 2006), producing high-voltage-activated Ca^{2+} currents very similar to rat Ca$_V$1.2, with overlapping voltage sensitivities and kinetics (Senatore et al., 2011), and similar pronounced Ca^{2+}/calmodulin-dependent inactivation (Taiakina et al., 2013; Peterson et al., 1999; Zühlke et al., 1999). The latter feature is important in that it permits negative feedback regulation of Ca^{2+} influx (DeMaria et al., 2001). For mammalian channels, extensive research has established that the Ca^{2+} sensor protein calmodulin pre-associates with Ca$_V$1 and Ca$_V$2 channels at their C-terminal

EF-hand, pre-IQ, and IQ domains (Fig. 2a). Channel opening and Ca^{2+} influx activate calmodulin, which in turn causes structural changes in the channel most often resulting in faster/more pronounced inactivation (i.e., Ca^{2+}-dependent inactivation or CDI) (Ben-Johny & Yue, 2014). $Ca_V2.1$ also exhibits a second form of calmodulin regulation: Ca^{2+}-dependent facilitation (CDF) (DeMaria et al., 2001).

The calmodulin protein consists of two lobes, an N-lobe and a C-lobe, with different Ca^{2+}-binding affinities and hence sensitivities to cytoplasmic Ca^{2+} buffering. For mammalian $Ca_V1.2$, $Ca_V1.3$, and $Ca_V1.4$ channels, the C-lobe of calmodulin confers pronounced CDI that is buffer resistant, while all Ca_V2 channels lack C-lobe mediated CDI (Ben-Johny & Yue, 2014). Instead, for mammalian $Ca_V2.1$-$Ca_V2.3$ channels, as well as $Ca_V1.4$, the N-lobe of calmodulin confers buffer-sensitive CDI (Ben-Johny & Yue, 2014). Clearly then, a key experimental distinction with respect to calmodulin regulation of Ca_V1 and Ca_V2 channels is reflected by sensitivity to buffering, where Ca_V1 channel CDI mediated by the C-lobe is buffer resistant, while Ca_V2 channel CDI mediated by the N-lobe is buffer sensitive. Notably, this distinction is also apparent for the *Lymnaea* calcium channel homologues, where under high buffering conditions, the Ca_V2 channel shows little to no CDI when the permeating cation is switched from Ba^{2+} to Ca^{2+}, while the Ca_V1 channel shows pronounced CDI that is resistant to strong buffering (Taiakina et al., 2013; Spafford et al., 2003a). Thus, work on the *Lymnaea* calcium channels suggests that the bifurcated N-lobe and C-lobe regulation of Ca_V1 and Ca_V2 channel types by calmodulin is an ancestral feature that was present in the common ancestor of all bilaterians (Fig. 1) (for more information on calcium regulation, please refer to chapter "Calmodulin Regulation of Voltage-Gated Calcium Channels" by Drs. Dick and Ben-Johny).

Importantly, the N-lobe of calmodulin also modulates $Ca_V1.2$ and $Ca_V1.3$ channels through interactions with two N-terminal motifs named NSCaTE (*N*-terminal *s*patial *Ca*$^{2+}$ *t*ransforming

element) and NATE (*N*SCaTE *a*ssociated *t*rans-duction *e*lement; Fig. 2a). These interactions cause N-lobe mediated CDI to become resistant to strong buffering (Dick et al., 2008; Simms et al., 2014). Mutation of key amino acids within the NSCaTE and NATE motifs causes N-lobe CDI of $Ca_V1.2$ and $Ca_V1.3$ channel to become buffer sensitive, thus resembling $Ca_V2.1$-$Ca_V2.3$ and $Ca_V1.4$ channels (Dick et al., 2008). Interestingly, the NSCaTE motif is conserved in Ca_V1 channels from protostome invertebrates (Taiakina et al., 2013), while the downstream NATE motif appears to have more ancient origins present in Ca_V1 channels from cnidarians, and Ca_V2 channels from placozoans (Gauberg et al., 2022). For *Lymnaea* Ca_V1, alternatively spliced N-terminal transcripts with and without the NSCaTE motif (but with an invariable NATE motif) exhibit similar resistance to strong cytoplasmic Ca^{2+} buffering (Taiakina et al., 2013), which contrasts observations of mammalian $Ca_V1.2$ and $Ca_V1.3$ channels where mutation of NSCaTE renders N-lobe modulation buffer sensitive. Hence, there may be differences in N-lobe regulation between vertebrates and invertebrates, or rather different reliance on the NATE motif, which operates somewhat redundantly with NSCaTE (Simms et al., 2014).

The final *Lymnaea* α_1 subunit to be functionally expressed was the Ca_V3 channel, producing characteristic LVA Ca^{2+} currents with similar fast kinetics as mammalian $Ca_V3.1$ and $Ca_V3.2$ channels (Senatore & Spafford, 2010). Also conserved with mammals are developmental changes in mRNA expression and alternative splicing. For example, like rat $Ca_V3.1$, the *Lymnaea* Ca_V3 channel bears an optional exon within the I–II linker, just downstream of the gating brake that encodes a large insert (i.e., exon 8b; Fig. 2a). *In vitro*, for both rat $Ca_V3.1$ and *Lymnaea* Ca_V3, exon 8b imposes a decrease in total protein expression levels and membrane localization, resulting in smaller Ca^{2+} currents (Shcheglovitov et al., 2008; Senatore & Spafford, 2012). In the III–IV linker, mammalian $Ca_V3.1$ and $Ca_V3.2$ and the *Lymnaea* Ca_V3 channels possess a conserved optional exon (exon 25c; Fig. 2a), encoding a small insert that imposes similar hyperpolarizing shifts in channel activation and inactivation, faster activation and inactivation kinetics, and slower deactivation kinetics (Senatore & Spafford, 2012; Emerick et al., 2006; Zhong et al., 2006; Chemin et al., 2001; Ohkubo et al., 2005). Furthermore, the expression patterns of the Ca_V3 channel and its splice variants in the snail match those in mammals. Specifically, like mammalian Ca_V3 channels, total mRNA levels of *Lymnaea* Ca_V3 decline sharply from embryo, to juvenile, to adult, especially in the heart (Ono & Iijima, 2010; Senatore & Spafford, 2012). Rodent and snail Ca_V3 channel variants with and without exon 8b are roughly equally expressed in the brain (Shcheglovitov et al., 2008), and during development of the heart, $Ca_V3.1$, $Ca_V3.2$, and *Lymnaea* Ca_V3 variants bearing exon 25c become significantly enriched (Latour et al., 2004; Emerick et al., 2006; Monteil et al., 2000; Zhong et al., 2006; David et al., 2010). In general, phylogenetic conservation of alternative splicing appears to be common for Ca_V1–Ca_V3 channels (Tyson & Snutch, 2013). For example, the *Lymnaea* Ca_V2 channel and mammalian $Ca_V2.1$ and $Ca_V2.2$ channels have common C-terminal splice variants that lack key elements for interactions with synaptic scaffolding proteins, such as RIM and Mint (Hirano et al., 2017; Spafford et al., 2003b; Lu & Dunlap, 1999), resulting in diminished roles in driving neurotransmitter release (Spafford et al., 2003b; Lübbert et al., 2017; Maximov & Bezprozvanny, 2002). Furthermore, the *Lymnaea* $Ca_V\beta$ subunit exhibits alternative splicing at its N-terminus and HOOK regions, with similar alternate/optional exons found in mammalian homologues, the latter imposing effects on the inactivation kinetics of the *Lymnaea* Ca_V2 and Ca_V3 channels (Dawson et al., 2014).

Interestingly, work on the *Lymnaea* Ca_V3 channel also revealed that adaptations in gene structure and alternative splicing can give rise to novel, highly divergent function (Senatore et al., 2014). Here, the gene was found to contain a duplicated exon homologous to exon 12 of the human $Ca_V3.1$ channel gene, permitting alternate structures in the extracellular pore-loop of Domain II, just five amino acids upstream of the selectivity filter glutamate (Fig. 2a). Genomic analysis revealed that these mutually exclusive exons are absent in chordates and deuterostomes, but ubiquitous among protostome invertebrates (Fig. 1a), with one exon (exon 12a) being shorter by roughly ten amino acids and having fewer cysteine residues than the other (exon 12b). Functionally, exon 12a variants were found to be highly permeable to Na^+ ions, so much so that removal of external Na^+ from the recording solution caused a >15-fold reduction in current amplitude (leaving the residual Ca^{2+} current), compared to only ~2.5-fold for the 12b variant. Thus, whereas the exon 12b variant of *Lymnaea* Ca_V3 conducts currents that are moderately selective for Ca^{2+}, in line with mammalian homologues (Shcheglovitov et al., 2007), the 12a variant conducts mixed currents with a major component being Na^+. This observation is particularly interesting when inferred for nematodes, such as *C. elegans*, which lost Na_V channels (Bargmann, 1998), and might instead utilize Ca_V3 channel exon 12a splice variants to generate fast electrical impulses driven by Na^+ influx. In *Lymnaea*, both splice variants are expressed in the brain, while the heart exclusively expresses the Na^+ permeable 12a variant, without Na_V channels, suggesting that the exon 12a channel variant drives Na^+-dependent depolarization in cardiomyocytes for excitation and pacemaking (Senatore et al., 2014). Subsequently, exploration into the mechanism for increased Na^+ permeation revealed a key functional interaction between the 12a exon region and the domain IV pore-loop of the channel (DIV-L6; Fig. 2a), where replacement of these two elements from *Lymnaea* Ca_V3 into human $Ca_V3.2$ rendered the mammalian channel highly Na^+ permeable (Guan et al., 2020).

Genetic Studies of Ecdysozoan β and α₂δ Subunits Highlight Gene Duplication as a Mechanism for Functional Diversification and the Emergence of Pleiotropic Functions

Genetic, developmental, and molecular studies of the vinegar fly *Drosophila melanogaster* (phylum Arthropoda) and the worm *Caenorhabditis elegans* (Nematoda), both molting animals in the clade ecdysozoa (Fig. 1a), have given rise to nine Nobel prizes and led to the discovery of numerous genes that are critical for neuronal and synaptic signaling. Indeed, one or more review papers could be devoted solely to calcium channel research done in these two species. However, for this review, we focus our attention on genetic studies exploring the duplication and phenotypic functions of the β and $\alpha_2\delta$ ancillary subunits, on the Ca_V3 channel from *Drosophila*, being the only α_1 subunit from *C. elegans* and *D. melanogaster* to be functionally characterized *in vitro*, and genetic studies that highlight the distinct roles and evolution of Ca_V1 and Ca_V2 channel function across the synapse.

Considerable genetic research has been done on the β and $\alpha_2\delta$ subunits of *D. melanogaster* and *C. elegans*. With respect to the β subunit, *C. elegans* is somewhat unique among invertebrates in possessing two paralogues (CCB-1 and CCB-2), that likely arose through gene duplication (Fig. 1b). One of these, CCB-1, appears to play a predominant role in regulating Ca_V1 and Ca_V2 channel function *in vivo*, where its loss of function is homozygous lethal, and heterozygote animals are smaller and sluggish in their movement, and Ca_V1 channel Ca^{2+} currents recorded from body wall muscles are severely disrupted (Lainé et al., 2011). Instead, the CCB-2 gene has a divergent protein sequence, especially in the GK domain, forming a clade with the homologue from *Trichoplax adhaerens* attributable to long branch attraction that brings highly divergent sequences together on phylogenetic trees (Fig. 3a) (Bergsten, 2005). Given the absence of a predicted guanylate kinase domain, CCB-2 might interact weakly, or perhaps not at all, with the AID region of Ca_V1/Ca_V2 channels, and this is also suggested by the lack of obvious phenotype when the gene is knocked out, and by its restricted expression in just a few neurons (Lainé et al., 2011). The single $Ca_V\beta$ subunit in *D. melanogaster* has also been knocked out, leading to reduced expression of Ca_V1 and Ca_V2 channels in dendrites, which interestingly disrupts the pruning process that is critical for establishing and maturing neural circuits (Kanamori et al., 2013).

Normally during development, sensory neuron dendrites targeted for pruning exhibit increased excitability leading to the generation of Ca^{2+} spikes driven by Ca_V1 and Ca_V2 channels. These in turn activate the Ca^{2+} sensitive protease calpain, triggering a cascade of downstream processes that include caspase enzymes and the ubiquitin-proteasome complex, ultimately causing dendritic pruning (Kuo et al., 2005; Williams et al., 2006). In this particular study, the significance of the β was made evident by its loss of function, which caused a more significant disruption to dendritic pruning than genetic lesion of Ca_V1 or Ca_V2 channels separately, instead resembling the disruption of both channels together (Kanamori et al., 2013).

Not surprisingly, the $\alpha_2\delta$ subunits in both *D. melanogaster* and *C. elegans* are also essential for the sub-cellular localization, membrane expression, and function of calcium channels *in vivo*. In *C. elegans*, similar to the β subunit, the $\alpha_2\delta$ subunits duplicated such that one plays a central role in regulating calcium channels *in vivo*, while the other does not. Specifically, the $\alpha_2\delta$ paralogue TAG-180 is highly divergent in its protein sequence, forming a sister clade with other $\alpha_2\delta$ homologues, and lacking key amino acids within the MIDAS motif (Fig. 4). Like CCB-2, loss of function of TAG-180 produces no obvious phenotype, or effects on Ca_V1 channel Ca^{2+} currents recorded from body wall muscles (Lainé et al., 2011). An interesting question, in light of the parallel divergence of CCB-2 and TAG-180, is whether these two subunits couple *in vivo* to exert atypical regulation of Ca_V1 and Ca_V2 channels, or alternatively, evolved non-canonical (pleiotropic) functions that are independent of calcium channels (Dolphin, 2018). Instead, the *C. elegans* $\alpha_2\delta$ homologue UNC-36 regulates the biophysical properties of the Ca_V1 channel in muscle (Lainé et al., 2011), and the localization and function of Ca_V2 channels at presynaptic terminals (Mathews et al., 2003; Saheki & Bargmann, 2009; Tong et al., 2017; Caylor et al., 2013). UNC-36 also plays an important role in the dynamic regulation of synapse morphology and function (Caylor et al., 2013), including retrograde signaling at the neuromuscular junction that fine-tunes the presynaptic neurotransmitter release via direct interaction with neurexin on the postsynaptic membrane (Tong et al., 2017). Notably, this particular homeostatic process was subsequently observed for the mammalian $Ca_V2.2$ channel, mediated specifically by the $\alpha_2\delta3$ subunit, at least *in vitro* (Tong et al., 2017).

In *D. melanogaster*, the $\alpha_2\delta$ subunits expanded to give rise to three paralogues named Ma2/d, straightjacket, and CG4587 (Figs. 1b and 4). The mRNA tissue expression of these genes is different, with straightjacket and CG4587 being highly expressed in the nervous system and eyes, and Ma2/d having a broader expression, including in muscle-bearing tissues, such as the heart (cardiac muscle), digestive tract (smooth muscle), and thorax (striated muscle) (Leader et al., 2018). Accordingly, the muscle-specific Ma2/d plays a key role in striated muscle function, where it regulates Ca_V1 channels at the cell membrane. Interestingly, this gene plays an additional and perhaps pleiotropic function, where it localizes to the sarcoplasmic reticulum and mediates molecular interactions that bridge T-tubules with the sarcoplasmic reticulum, resulting in the proper positioning of nuclei close to T-tubules (Reuveny et al., 2018). Indeed, such localization of nuclei within muscle is an important arrangement, which when perturbed in humans leads to pathologies, such as muscular dystrophy, skeletal muscle wasting and weakness, and cardiac disease (Jungbluth & Gautel, 2014). Whether this function for $\alpha_2\delta$ is conserved in other animals or is unique to *Drosophila* is unknown.

Interestingly, despite the overlapping expression of straightjacket and CG4587 in the nervous system, these two paralogues evolved to differentially regulate Ca_V1 and Ca_V2 channels. That is, straightjacket performs the canonical function of promoting Ca_V1 and Ca_V2 channel membrane expression, and targeting Ca_V2 channels to axonal and presynaptic compartments (Dickman et al., 2008; Heinrich & Ryglewski, 2020; Ly et al., 2008), while CG4587 counters axonal/presynaptic localization of Ca_V2 channels, instead promoting their enrichment in dendrites (Heinrich & Ryglewski, 2020).

Altogether, work on the β and $\alpha_2\delta$ subunits in *D. melanogaster* and *C. elegans* corroborates *in vivo* functions observed in mammals. In addition, this work suggests that gene duplication of ancillary subunits permitted lineage-specific adaptation in calcium channel regulation. In *C. elegans*, the duplicated paralogues CCA-B1 and UNC-36 retained the ancestral function of regulating Ca_V1 and Ca_V2 channels in neurons and muscle, while CCB-2 and TAG-180 diversified toward an unknown function. In *D. melanogaster*, triplication of the $\alpha_2\delta$ subunits permitted the Ma2/d paralogue to evolve a specific role in regulating Ca_V1 channels in muscle, where it also serves a noncanonical function in bridging sub-cellular membrane compartments (Reuveny et al., 2018). Straightjacket retained the ancestral function of promoting Ca_V1 and Ca_V2 channel membrane expression and localization in neurons, while CG4587 evolved the distinct capacity of countering the axonal/presynaptic localization of Ca_V2 channels for enriched expression in dendrites (Heinrich & Ryglewski, 2020).

Genetic and Functional Studies of Ecdysozoan Ca_V1–Ca_V3 Channels Reveal Deep Conservation of C-Terminal Protein Interactions and Sub-cellular Localization

Besides *Lymnaea*, the only other bilaterian invertebrate species to have its complete set of voltage-gated calcium channels functionally characterized (i.e., Ca_V1–Ca_V3), as well as a $Ca_V\beta$ subunit, is the arthropod honeybee *Apis mellifera* (Cens et al., 2013, 2015). Thus, we will briefly summarize this work as it provides the most proximal phylogenetic perspective on calcium channel homologues from the fellow ecdysozoans *D. melanogaster* and *C. elegans*. Together, the two studies on *Apis* calcium channels revealed similar features as reported for *Lymnaea*: (1) conserved LVA and HVA Ca^{2+} currents for the Ca_V3 and Ca_V1/Ca_V2 channels, respectively (expressed in *Xenopus* oocytes), (2) conserved gene structures and alternative splicing, including short C-terminal variants of the Ca_V2 channel lacking motifs for presynaptic targeting, and Ca_V3 channel variants with alternate domain II pore-loops associated with altered Na^+/Ca^{2+} selectivity (i.e., exons 12a and 12b; Fig. 2a), (3) downregulation of Ca_V3 mRNA expression during development, and (4) alternative spicing of the $Ca_V\beta$ subunit, producing short and long N-terminal variants that confer unique sensitivity to kinase modulation and altered calcium channel inactivation (Cens et al., 2013, Cens et al., 2015). Altogether, the shared features reported for *Lymnaea* and *Apis* Ca_V channels likely reflect a set of conserved properties among protostome invertebrates, and perhaps more broadly among all bilaterians and beyond.

Like the *Apis* calcium channels, *Drosophila* Ca_V3 was expressed in oocytes, producing characteristic voltage-gated Ca^{2+} currents that slightly differed from rat $Ca_V3.1$, more resembling the snail Ca_V3 channel with slower activation and inactivation kinetics, faster deactivation kinetics, and larger macroscopic currents for Ba^{2+} than Ca^{2+} (Jeong et al., 2015; Senatore & Spafford, 2010). One unexpected feature of the *Drosophila* channel was its strong sensitivity to block by the cation Ni^{2+}, a unique feature of mammalian $Ca_V3.2$ attributed to a histidine residue in domain I S3–S4 extracellular loop that is absent from mammalian $Ca_V3.1$, $Ca_V3.3$, and *Lymnaea* Ca_V3 channels (Kang et al., 2006; Senatore & Spafford, 2010). Accordingly, the *Drosophila* homologue bears a histidine residue just two amino acids upstream of the one found in $Ca_V3.2$, which likely evolved independently and perhaps similarly forms a high-affinity binding site for Ni^{2+} that confers low micromolar sensitivity to block.

This same study utilized transgenic flies with the endogenous Ca_V3 channel tagged at its N-terminus with green fluorescent protein to reveal localization of the channel in dendrite-rich regions of the *D. melanogaster* brain (Jeong et al., 2015), thus resembling mammalian Ca_V3 channels that are enriched in soma and dendrites in various central neuron types (McKay et al., 2006). Lastly, homozygous knockout of the fly Ca_V3 channel produced only a mild phenotype of prolonged sleep, in contrast to genetic lesions to

Ca_V1 and Ca_V2 channels that are embryonic lethal (Smith et al., 1996; Eberl et al., 1998). Thus, although important *in vivo*, Ca_V3 channels in mammals and flies appear similarly dispensable for viability (Lory et al., 2020), unlike the HVA channels whose disruption produces much more severe phenotypes.

Because of the extensive genetic tools available for *D. melanogaster* and *C. elegans*, these two animals have been at the forefront of scientific discovery pertaining to the synapse, with many essential genes identified through genetic screens of individuals with compromised behaviors, such as locomotion (e.g., uncoordinated or Unc mutants in *C. elegans*). For example, the exocytotic Ca^{2+} sensor synaptotagmin was discovered in *D. melanogaster* (Littleton et al., 1994), and the SNARE protein regulators Unc13 and Unc18, critical for priming presynaptic vesicles at the active zone, were discovered in *C. elegans* (Hosono et al., 1992; Maruyama & Brenner, 1991). This work extends to the precise mechanisms by which Ca_V2 calcium channels are positioned within nanometers of primed presynaptic vesicles, where combined with studies in mice, a picture is emerging of an ancestral arrangement of core proteins that critically link Ca_V2 channels with the exocytotic machinery.

As noted earlier, Ca_V2 channels from chordates to cnidarians possess conserved C-terminal motifs with a consensus sequence of DDWC$_{COOH}$, with the tryptophan/W in the second last position being the most invariable (Fig. 2b, c) (Piekut et al., 2020). The first notion that this motif mediates important interactions was with Mint (Fig. 2c), a presynaptic modular adaptor protein whose PDZ domain selectively binds to the mammalian $Ca_V2.1$ and $Ca_V2.2$ channel C-termini *in vitro* (DDWC$_{COOH}$ and DHWC$_{COOH}$, respectively), but not $Ca_V2.3$ channels that bear a DDKC$_{COOH}$ motif, lacking the invariable tryptophan (Maximov et al., 1999). Conserved in mammals, *D. melanogaster*, and *C. elegans*, Mint also interacts with the synaptic scaffolding protein CASK (Borg et al., 1998; Butz et al., 1998; Mukherjee et al., 2014), which in turn binds an internal proline-rich motif in the $Ca_V2.2$ channel C-terminus via an SH3 domain (a motif also found in $Ca_V2.1$) (Maximov et al., 1999) (Fig. 2c). Subsequently, this interaction was observed *in vitro* for the $Ca_V2.2$ channel in chick (Wong et al., 2014) and the *Lymnaea* Ca_V2 channel (Spafford et al., 2003b).

Notably, although the *in vitro* interaction between Ca_V2 channels and Mint is biochemically robust, indicative of high-affinity binding, its physiological significance is unknown. For example, $Ca_V2.2$ and Mint were found to not colocalize at presynaptic loci in chick calyx synapses (Khanna et al., 2006). Furthermore, in mice, knockout of the two neuronal Mint paralogues (Mint-1 and Mint-2) perturbed presynaptic function, but not through dysregulation of $Ca_V2.1$ and $Ca_V2.2$ channel localization and presynaptic function (Ho et al., 2006; Kaeser et al., 2011). In the snail, presynaptic injection of a $DDWC_{-COOH}$ peptide disrupted synaptic transmission mediated by Ca_V2 channels, but RNA interference of Mint had no effect (Spafford et al., 2003b). Lastly, loss of function of the Mint homologue in *C. elegans*, Lin-10, had no effect on the presynaptic localization of Ca_V2 channels *in vivo* (Saheki & Bargmann, 2009). Altogether, Mint does not appear to play a significant role in localizing Ca_V2 channels at the synapse active zone in vertebrates and invertebrates. Instead, work on the two Mint paralogues in *Drosophila* suggests that this gene plays a more broad and general role in the polarized trafficking of proteins away from dendrites and the cell soma toward axons and presynaptic compartments (Gross et al., 2013), and in rodents, a role in activity-dependent regulation of membrane protein expression, for example, of presenilin and amyloid precursor protein (Sullivan et al., 2014).

As noted, the presynaptic scaffolding protein RIM also binds to the Ca_V2 C-terminus at its $DDWC_{-COOH}$ motif, also with a PDZ domain (Fig. 2b, c). This interaction, in contrast to Mint, appears much more critical for the localization of Ca_V2 channels at neurotransmitter release sites. RIM also interacts with the vesicular protein Rab3 and thus provides a molecular bridge between Ca_V2 channels and presynaptic vesicles (Südhof, 2012; Wang et al., 1997). The first evidence that RIM directly interacts with Ca_V2

channels came from a yeast II hybrid screen of a rat brain cDNA library, subsequently narrowed to a specific interaction between the RIM PDZ domain and the respective $Ca_V2.1$ and $Ca_V2.2$ channel C-terminal $DDWC_{-COOH}$ and $DHWC_{-COOH}$ motifs (Kaeser et al., 2011). In mice, conditional knockout of RIM1 and RIM2 significantly diminished presynaptic Ca^{2+} influx and localization of $Ca_V2.1$ channels (Han et al., 2011; Kaeser et al., 2011). In these experiments, the specific requirement for the PDZ domain of RIM was established by the observation that efficient genetic rescue of RIM loss of function required the PDZ domain to be present (Kaeser et al., 2011).

Subsequently, similar findings were made in *D. melanogaster* and *C. elegans*, where genetic lesion of their single respective RIM orthologues reduced Ca_V2 channel clustering at neurotransmitter release sites and disrupted synaptic transmission (Graf et al., 2012; Oh et al., 2021; Kushibiki et al., 2019). Importantly, however, a direct *in vitro* interaction between a RIM PDZ domain and a Ca_V2 channel C-terminus has not yet been verified for any invertebrate species. This is noteworthy because although the *D. melanogaster* Ca_V2 channel C-terminus bears a canonical C-terminal ligand motif of $EDWC_{-COOH}$, the *C. elegans* orthologue has a divergent sequence of $WAIV_{-COOH}$ (Piekut et al., 2020). This motif is inconsistent with the expected binding specificity of the RIM PDZ domain, for which all known C-terminal ligand sequences have a tryptophan/W in the second last position, which in addition to Ca_V2 channels includes the presynaptic protein CAST/ELKS with a C-terminal $GIWA_{-COOH}$ motif (Ohtsuka et al., 2002; Wang et al., 2002). Notably, the *C. elegans* CAST orthologue bears a canonical RIM PDZ binding signature of $GIWA_{-COOH}$, suggesting that the RIM PDZ domain recognizes canonical ligands, so the divergence of the Ca_V2 channel C-terminal sequence in this species is altogether unclear.

In addition to RIM, work in *D. melanogaster* revealed that another protein, RIM-binding protein or RIM-BP, also binds Ca_V2 channels and is important for their presynaptic localization (Liu et al., 2011). This interaction was subsequently observed for the RIM-BP1 and RIM-BP2 para-logues in rodents (Kaeser et al., 2011; Acuna et al., 2015; Grauel et al., 2016) and the single orthologue in *C. elegans* (Kushibiki et al., 2019), identifying a conserved tripartite complex between RIM, RIM-BP, and Ca_V2 channels (Fig. 2c). In essence, a core function of RIM-BP is to bolster the interaction between RIM and Ca_V2 channels, using SH3 domains to bind separate proline motifs in the Ca_V2 channel C-terminus, and a region downstream of the PDZ domain in the RIM protein (Fig. 2c) (Wang et al., 2000; Hibino et al., 2002; Kaeser et al., 2011). Accordingly, the most severe disruption in pre-synaptic localization of Ca_V2 channels emerges when both RIM and RIM-BP are knocked out (Kushibiki et al., 2019; Kaeser et al., 2011).

Like Ca_V2 channels, Ca_V1 channels also appear to have deeply conserved C-terminal sequences that mediate interactions with PDZ domains of synaptic scaffolding proteins, but instead, for localization at the postsynaptic membrane (Piekut et al., 2020). Specifically, verte-brate $Ca_V1.3$ channels bear conserved C-terminal signatures of ITTL$_{-COOH}$ which are bound by the PDZ domains of the postsynaptic scaffolding proteins Shank and Erbin (Zhang et al., 2005; Calin-Jageman et al., 2007). Shank also binds an internal proline-rich motif in the $Ca_V1.3$ channel C-terminus using an SH3 domain, akin to the binding of Ca_V2 channels by RIMP-BP and CASK (Fig. 2b, d). $Ca_V1.2$ also bears a conserved hydrophobic C-terminal motif of VSSL$_{-COOH}$, that is not bound by Shank (Zhang et al., 2005) but associates with the PDZ domains of *n*euronal *i*nter*l*eukin-*16* (NIL-16) (Kurschner & Yuzaki, 1999) and *c*hannel-*i*nteracting *P*DZ domain pro-tein (CIPP) (Kurschner et al., 1998) (Fig. 2e). For both $Ca_V1.3$ and $Ca_V1.2$ channels, C-terminal interactions with Shank and NIL-16, respec-tively, are necessary for their contributions to postsynaptic excitation–transcription coupling (Weick et al., 2003; Kurschner & Yuzaki, 1999; Zhang et al., 2005; Zhang et al., 2006), an impor-tant dendritic function that links synaptic activity and membrane excitation with changes in gene

expression associated with synaptic plasticity, learning, and memory (Deisseroth et al., 1996). Interestingly, despite losing the SH3 domain, the PDZ domain of the *C. elegans* Shank homologue binds to its cognate Ca_V1 channel (egl-19), via a C-terminal ligand sequence identical to that of $Ca_V1.3$ (i.e., ITTL$_{-COOH}$) (Pym et al., 2017). Like $Ca_V1.3$–Shank in mammals (Zhang et al., 2005), disruption of the Ca_V1–Shank interaction in *C. elegans* diminishes the postsynaptic membrane expression of the channel, and CREB-mediated changes in gene expression (Pym et al., 2017). In *D. melanogaster*, a postsynaptic interaction between Ca_V1 and Shank has also been reported, strictly at the genetic level (Hogg, 2018), but notably the C-terminal sequence of the channel more resembles that of $Ca_V1.2$ with a sequence of TYSS$_{-COOH}$ (Piekut et al., 2020). Nevertheless, the fly Shank homologue retains an SH3 domain (Wu et al., 2017), which could mediate an inter-nal interaction as occurs in mammals. Lastly, the postsynaptic scaffolding protein Densin also binds and regulates $Ca_V1.3$ channels via the ITTL$_{-COOH}$ sequence (Jenkins et al., 2010), and through a separate mechanism interacts with the N-terminus of $Ca_V1.2$ to promote its expression at the postsynaptic membrane (Wang et al., 2017).

Altogether, the various studies in rodents, *Drosophila* and *C. elegans,* suggest that early during evolution, distinct C-terminal sequences emerged to differentially couple Ca_V1 and Ca_V2 channels with protein complexes across the synapse, mediated by PDZ domains of scaf-folding proteins with different ligand-binding properties. This is certainly apparent for Ca_V2 channels, where conserved interactions with Mint, RIM, and RIM-BP have been reported widely in vertebrates and invertebrates. However, this is also apparent for Ca_V1 chan-nels, and at least with respect to Shank. Ca_V3 channels also bear C-terminal sequences that are consistent with PDZ ligands (Piekut et al., 2020; Mulatz, 2013) (Fig. 2b), so they too might exhibit conserved interactions with PDZ-bearing proteins.

Ca$_V$ Channels in Early-Diverging Animals

The phyla Cnidaria, Placozoa, Porifera, and Ctenophora diverged from bilaterians between 615 and 1,200 million years ago (Wang et al., 1999; Peterson et al., 2004). Several species from each of these groups have had their genome sequenced (Srivastava et al., 2008; Eitel et al., 2018; Moroz et al., 2014; Ryan et al., 2013; Putnam et al., 2007; Chapman et al., 2010; Leclère et al., 2019; Chang et al., 2015; Srivastava et al., 2010; Kenny et al., 2020), providing insights into the evolution of genes associated with cellular excitability and neural signaling, including voltage-gated calcium channels (Fig. 1b). Also, voltage-gated Ca^{2+} currents have also been electrophysiologically recorded from various tissues and cell types in these animals, although the molecular identity of underlying channels is mostly unknown. Since the topic of voltage-gated calcium channels in early-diverging animals has been previously reviewed (Senatore et al., 2016), here, we aim to complement this work by focusing on some of the more recent findings.

Cnidaria: Our Most Distant Neural Ancestors?

Cnidarians are radially symmetrical animals (Fig. 1a) that are categorized into two main groups: anthozoans, such as sea anemones and corals, which as adults form polyps and have sessile life strategies, and medusozoans, such as jellyfish and hydra, which also form sessile polyps but can develop into free-swimming medusa. Both have nervous systems and smooth muscle; however, anthozoans generally lack fast-twitching striated muscle, consistent with their sessile lifestyles (Seipel & Schmid, 2005; Mackie, 1990). Instead, medusozoans have extensive striated musculature that evolved independently from that of bilaterians (Steinmetz et al., 2012). The general arrangement of the cnidarian nervous system is as a diffuse nerve net largely devoid of neural condensation, contrast-

ing the neural structures of bilaterian brains and ganglia. Nevertheless, the nervous systems of cnidarians and bilaterians share many common features, from genomics to development, and it is generally accepted that they are of a single phyletic origin (Arendt et al., 2016; Moroz, 2021).

A considerable number of studies have reported voltage-gated Ca^{2+} currents in cnidarian tissues (e.g., Anderson & Schwab, 1983; Meech & Mackie, 1993; Przysiezniak & Spencer, 1992), and both synaptic transmission and muscle contraction require Ca^{2+} flux across the cell membrane (for a recent review see Senatore et al., 2016). However, outside of a few examples described below, little is known about the molecular identity of the calcium channel subunits that are involved. In total, cnidarians possess at least 11 pore-forming subunit genes that could generate voltage-gated Ca^{2+} currents: 6 Ca$_V$ channels (Fig. 1b), and 4 Na$_V$2 channels with Ca^{2+}-selective selectivity filters of DEEA or DEET (Barzilai et al., 2012). Interestingly, a fifth cnidarian Na$_V$2 channel, named Na$_V$2.5, independently evolved high Na$^+$ selectivity through alterations in its pore region, including a key glutamate to lysine switch in the domain II selectivity filter producing a DKEA motif, akin to the highly Na$^+$-selective DEKA motif of the Na$_V$1 channels (Barzilai et al., 2012).

Indeed, the very first invertebrate voltage-gated calcium channel to be cloned and functionally expressed *in vitro* was the single Ca$_V$1 channel from the jellyfish *Cyanea capillata* (Jeziorski et al., 1998), as well as its single β subunit (Jeziorski et al., 1999). Unlike Ca$_V$2 and Ca$_V$3 channels which duplicated, most cnidarians possess just a single Ca$_V$1 channel, whose physiological function remains unknown (Fig. 1b). A suggested function comes from *in situ* hybridization experiments in the sea anemone *Nematostella vectensis*, revealing mRNA expression in the endoderm and ectoderm of developing tentacles, regions that are enriched in neurons and muscle (Moran & Zakon, 2014). Expressed in *Xenopus* oocytes, the *Cyanea* Ca$_V$1 channel produces characteristic HVA currents that inactivate faster with Ca^{2+} vs. Ba^{2+} as the charge carrier, indicative of Ca^{2+}/calmodulin-dependent inactivation. The channel is also largely insensitive to dihydropyridines (DHPs), which are selective blockers of mammalian Ca_V1 channels. Furthermore, the DHP enantiomer and L-type channel agonist $S(-)$-BayK 8644 has the opposite effect in blocking the channel with low affinity, rather than potentiating currents. These observations are altogether consistent with other invertebrate tissues where DHPs generally do not selectively block presumed Ca_V1 channel currents. Subsequently, detailed molecular analysis of the cloned *Lymnaea stagnalis* Ca_V1 channel suggested that insensitivity to DHPs can be attributed to divergence at key amino acids in the channel structure where DHPs bind mammalian orthologues (Senatore et al., 2011). Indeed, converting three divergent residues of *Lymnaea* Ca_V1 to match those of mammalian $Ca_V1.2$ rendered it highly DHP sensitive. However, the *Cyanea* Ca_V1 channel completely resembles $Ca_V1.2$ in these same amino acid positions, and yet is much less sensitive to DHPs than the *Lymnaea* orthologue (Jeziorski et al., 1999), even the wild-type channel bearing three amino acids that are divergent from $Ca_V1.2$. Perhaps this inconsistency is not surprising, where additional channel structures likely contribute to DHP binding and the pharmacological responses therein. Indeed, it is altogether apparent that compounds that are identified as selective for mammalian calcium channels often do not exert the same affinity on their invertebrate orthologues, and even on channels from more closely related species. For example, the zebrafish $Ca_V2.1$ and $Ca_V2.2$ channels are both highly sensitive to the peptide blocker ω-conotoxin GVIA, which in mammals is strictly selective for $Ca_V2.2$, and are insensitive to ω-Agatoxin IVA, which in mammals is selective for $Ca_V2.1$ (Wen et al., 2013) (for additional information on peptide toxin inhibitors please see chapter "Pharmacology and Structure-Function of Venom Peptide Inhibitors of N-Type (Cav2.2) Calcium Channels" by Drs. Lewis and Adams).

Currently, in cnidarians, we do not know whether the generalized postsynaptic functions Ca_V1 channels for excitation–transcription and excitation–contraction coupling are conserved, nor the specialized roles for Ca_V2 channels in driving presynaptic excitation–secretion coupling. In jellyfish striated muscle, which evolved independently from that of bilaterians (Steinmetz et al., 2012), only LVA Ca^{2+} currents have been observed (Lin & Spencer, 2001), suggesting a role for Ca_V3 or perhaps Na_V2 channels. To the best of our knowledge, the activation properties of calcium channels that drive contraction in smooth muscle, thought to be homologous to that of bilaterians (Steinmetz et al., 2012), have not been examined. It is notable that cnidarian Ca_V1 and Ca_V2 channels bear conserved C-terminal sequences that as noted above mediate distinct interactions with respective post- and presynaptic scaffolding proteins in bilaterians (Piekut et al., 2020). That is, all three of the triplicated cnidarian Ca_V2 channel paralogues (i.e., Ca_V2a, Ca_V2b, Ca_V2c; Fig. 1b) (Moran & Zakon, 2014) bear C-termini with invariable tryptophan residues in the second last position, for interactions with the PDZ domains of RIM and Mint (i.e., Ca_V2a: $DDWC_{-COOH}$, Ca_V2b: $DDWC_{-COOH}$, and Ca_V2c: $ETWC_{-COOH}$). Cnidarian Ca_V1 channels also have conserved C-terminal sequences that are enriched in hydrophobic residues (i.e., $ITDL_{-COOH}$ in most species, $TSYL_{-COOH}$ in *Cyanea*), resembling those of mammalian $Ca_V1.2$ ($ISSL_{-COOH}$) and $Ca_V1.3$ ($ITTL_{-COOH}$) for PDZ domain-mediated interactions with Shank, Erbin, CIPP, and NIL-16 (Fig. 2d, e). Indeed, the conservation of these C-terminal signatures is even more striking when considering their location after extended stretches of C-terminal sequences that are highly disordered and divergent among Ca_V channel orthologues (Fig. 2b). Clearly, deciphering protein–protein interactions of Ca_V channels in early-diverging animals, including cnidarians, will be useful for understanding calcium channel evolution.

Interestingly, recent evidence suggests that the duplication of calcium channel subunit genes in cnidarians led to neofunctionalization within the phylum-specific cell type called the cnidocyte (Weir et al., 2020). Cnidocytes, also referred to as stinging cells, are positioned along the tentacles of polyps and medusae, where they discharge venom-laced barbs for the purpose of predation and defense. There are two types of cnidocytes, spirocytes and nematocytes, the former unique to anthozoans and the latter common to all cnidarians (Babonis & Martindale, 2014). Both are thought to have evolved from neurons or neuroendocrine cells and to utilize similar machinery for barb discharge as that used for regulated vesicular exocytosis (Babonis & Martindale, 2014). In nematocytes, a single-use exocytotic organelle, called a nematocyst, encloses a barb covered with pore-forming toxins and peptide neurotoxins (Jouiaei et al., 2015). Being single-use in nature, nematocytes prevent erroneous discharge by integrating both chemical and mechanical sensory signals, such that contact alone is not sufficient to trigger barb release.

The precise mechanism by which cnidocytes integrate mechano- and chemosensory signals was recently determined and found to involve the Ca_V2a channel isotype (Weir et al., 2020). Previously, mRNA *in situ* hybridization revealed that Ca_V2a is expressed exclusively in cnidocytes, compared to Ca_V2b and Ca_V2c, which exhibit broader expression in areas enriched in neurons and muscle (Moran & Zakon, 2014). Whole-cell recordings of nematocytes isolated from *Nematostella vectensis* revealed voltage-gated Ca^{2+} currents with dramatically left-shifted steady-state inactivation properties, such that the channel population was completely inactivated at the typical resting voltage of -65 mV (Weir et al., 2020). This is in contrast to Ca^{2+} currents recorded from isolated neurons, which had more typical steady-state inactivation properties. A tentacle-specific transcriptome confirmed enriched mRNA expression of Ca_V2a, along with one of the two duplicated β subunit homologues (cacnb2), and interestingly, the *Nematostella* $\alpha_2\delta$-like CACHD0 homologue (Weir et al., 2020) (Figs. 1b, 3a, and 4). *In vitro* functional expression of the cloned Ca_V2a, cacnb2, and CACHD0 subunits revealed an atypical feature of the β subunit, where it dramatically accelerated the inactivation kinetics of the channel Ca^{2+} currents and imposed a strong hyperpolarizing shift in its steady-state inactivation resembling the atypical

currents recorded in cnidocytes. This channel complex recovers quickly from inactivation, such that transient hyperpolarization of the cell membrane can quickly reanimate the channel population to an activatable state. Thus, the mechanism for sensory integration was identified: in the absence of chemical signals, direct mechanical stimulation of cnidocytes activates a depolarizing cation current through a mechanically gated transient receptor potential type N (TRPN) channel. However, in the absence of chemosensory signals, and due to the inactivation properties imposed by the cacnb2 subunit, the Ca_V2a channel population is inactivated and unable to respond to membrane depolarization. However, in the presence of appropriate chemosensory signals, sensory neurons that synapse with cnidocytes secrete the transmitter acetylcholine, causing the activation of a Ca^{2+}-activated K^+ current that hyperpolarizes the cnidocyte cell membrane. Now, mechanical transduction through the TRPN channel is able to depolarize the membrane and activate Ca_V2a channels, whose Ca^{2+} influx triggers nematocyst exocytosis and barb discharge (Weir et al., 2020). Notably, because the *Nematostella* cacnb2 subunit imposes a similar hyperpolarizing shift in the steady-state inactivation of the mouse $Ca_V2.1$ channel, the evolution of this unique cellular feature is attributable to the duplicated β subunit, rather than structural/functional changes in the Ca_V2a channel (Fig. 1b). Nonetheless, as noted, voltage-gated Ca^{2+} currents recorded from isolated *Nematostella* neurons (Weir et al., 2020), as well as other cnidarian tissues (Senatore et al., 2016), more resemble those of other animals with respect to their steady-state voltage properties. Perhaps, the second cnidarian β subunit (Fig. 3a), in conjunction with the Ca_V2b and Ca_V2c α_1 subunits, underlies these more canonical currents observed in cnidarian neurons. However, this second β subunit bears a significant deletion at its N-terminus, such that the entire SH3 domain is absent. Lastly, to the best of our knowledge, the described study is the first to functionally characterize a CACHD0 homologue (Figs. 1b and 4). This is notable because this particular subunit caused a roughly 7 mV depolarizing shift in the activation curve of $Ca_V2.1$ (Weir et al., 2020), indicative of a functional (and perhaps physical) interaction, despite mammals and chordates lacking CACHD0 orthologues.

Evolutionary Insights from Functional Studies of the Placozoan Ca_V1 to Ca_V3 Channels

Placozoans are amoeboid animals that lack body symmetry and have the simplest body plan composed of only six cell types arranged in a flat, pancake-like arrangement (Smith et al., 2014). These animals lack genes for gap junctions, and combined with detailed ultrastructural studies (Smith et al., 2014), it is apparent that they do not possess chemical or electrical synapses and hence a nervous system. Still, cells are clearly able to sense environmental signals and communicate with each other, evident in the various locomotive behaviors that would require cellular coordination, such as feeding (Smith et al., 2015), chemotaxis (Smith et al., 2019; Romanova, Heyland, et al., 2020a), phototaxis (Heyland et al., 2014), and gravitaxis (Mayorova et al., 2018).

Despite their cellular and morphological simplicity, placozoan genomes contain a large repertoire of genes associated with complex animal traits (Srivastava et al., 2008; Eitel et al., 2018), including body patterning and development, and genes associated with membrane excitation, synaptic transmission, and Ca^{2+} signaling. For example, placozoans uniquely expanded Na_V2 channel genes giving rise to five copies in the species *Trichoplax adhaerens*, and eight in the species *Hoilungia hongkongensis* (Romanova, Smirnov, et al., 2020b), contrasting other invertebrates that typically possess only one Na_V1 and one Na_V2 channel gene. Placozoans also possess an expanded repertoire of voltage-gated potassium (K_V) channels which emerged via gene duplication from a Shaker-like channel in the common ancestor of placozoans, cnidarians, and bilaterians (i.e., Shab, Shal, and Shaw) (Li et al., 2015). As noted, placozoans are also the most

early-diverging animals to possess genes for Ca_V1–Ca_V3 channels, as well as one β subunit, two $α_2δ$ subunits, and one each of CACHD1 and CACHD0 (Figs. 1b, 3a, and 4). Remarkably, it was only recently that direct evidence for electrical signaling was provided for placozoans, with the first observation of Na^+-dependent action potentials from immobilized animals and dissociated cells that could be elicited by stimulation with an extracellular recording electrode (Romanova, Smirnov, et al., 2020b). Combined with the extensive gene content, it is evident that placozoans utilize fast electrical and Ca^{2+} signaling for their physiology and behavior, although many questions remain about the nature and function of this signaling in the different cell types, and furthermore, of the homology of placozoans cell types to those of other animals (Sebé-Pedrós et al., 2018).

In being the most early-diverging animals to possess Ca_V1 to Ca_V3 channels (Fig. 1a, b), placozoans represent an important group for inferring primordial functions of Ca_V channels and gaining perspectives on their distinct evolutionary histories. All three Ca_V channels from *Trichoplax adhaerens* have been cloned and functionally expressed *in vitro* (Gauberg et al., 2020; Gauberg et al., 2022; Smith et al., 2017), making it only the third invertebrate species to have this done after *Lymnaea stagnalis* and *Apis mellifera*. A combination of immunolocalization and fluorescence *in situ* hybridization indicates that the *Trichoplax* Ca_V1 to Ca_V3 channels are co-expressed in a cell type called type II gland cells (Gauberg et al., 2020, Gauberg et al., 2022, Smith et al., 2017), which resemble neuroendocrine cells in their expression of gene homologues for regulated exocytosis, including syntaxin, SNAP-25, and complexin (Smith et al., 2014; Smith et al., 2017; Gauberg et al., 2020). Type II gland cells secrete both mucus and the neuropeptide *Ta*ELP (*Trichoplax adhaerens* endomorphin-*l*ike *p*eptide). Secreted mucus is thought to facilitate the animal's adhesion to the substrate, as well as locomotion mediated by asynchronously beating cilia on the ventral epithelium (Mayorova et al., 2019). Instead, the peptide *Ta*ELP, named so because of an internal YPFF amino acid sequence also found in vertebrate endomorphin, causes ventral cilia to stop beating and hence pauses ciliary locomotion (Senatore et al., 2017). Placozoans express an array of additional peptides that regulate ciliary beating to pause or accelerate locomotion (Senatore et al., 2017; Varoqueaux et al., 2018), within cell types that are transcriptionally similar to type II gland cells (Sebé-Pedrós et al., 2018). Thus, it may be that gland cells resemble neuroendocrine cells of other animals in their combined utilization of Ca_V1–Ca_V3 channels for regulated secretion of signaling molecules (Mahapatra et al., 2012). Interestingly, both peptides and glycine cause placozoans to contract, most likely mediated by dorsal epithelial cells that exhibit ultrafast contractions requiring transient rises in cytoplasmic Ca^{2+}, thus resembling muscle cells (Armon et al., 2018). Interestingly, immunolocalization revealed that the *Trichoplax* Ca_V1 channel is uniquely expressed in these dorsal cells, where perhaps, as in bilaterians, the channel serves the specialized function of excitation–contraction coupling (Gauberg et al., 2022).

The first *Trichoplax* Ca_V channel to be functionally characterized *in vitro* was the Ca_V3, which despite its significant divergence in protein sequence possesses hallmark structural features, including an EEDD selectivity filter motif and a gating brake (Fig. 2a). At the genomic level, the *Trichoplax* Ca_V3 channel gene resembles that of other animals, but lacks a few introns and the alternatively spliced exons 8b, 12a/12b, 25c, and 26 (Fig. 2a). Accordingly, and like the *Trichoplax* Ca_V1 and Ca_V2 channels, not a single alternatively spliced transcript was identified during cloning of the Ca_V3 channel protein-coding sequence from a whole-animal cDNA library. Thus, Ca_V channels in placozoans are not subject to extensive functional diversification via alternative splicing, a common feature of bilaterian Ca_V channels, which is consistent with bioinformatic observations that alternative splicing is much less prevalent in early-diverging animals, especially in placozoans (Grau-Bové et al., 2018).

In vitro, the *Trichoplax* Ca_V3 channel conducts LVA Ca^{2+} currents that resemble those of mammalian Ca_V3 channels, but is perhaps most similar to *Lymnaea* Ca_V3 (Smith et al., 2017). Unexpectedly, efficient heterologous expression of the placozoan channel was found to require co-expression with the HVA channel ancillary $α_2δ$ and $β$ subunits (i.e., rat $α_2δ1$ and $β1b$), which increased expression of the GFP-tagged channel by 75-fold. Also interesting, the domain II pore region of the *Trichoplax* channel, where *Lymnaea* and other protostome Ca_V3 channels bear alternatively spliced exons 12a and 12b (Fig. 2a), resembles exon 12a in its shorter length and reduced cysteine content. As noted earlier, *Lymnaea* Ca_V3 channels harboring exon 12a are highly Na^+ permeable, while exon 12b variants have Ca^{2+} selectivity more similar to mammalian channels (Senatore et al., 2014). Nevertheless, *Trichoplax* Ca_V3 is much less permeable to Na^+ than *Lymnaea* Ca_V3, attributable to a potent block of Na^+ currents by extracellular Ca^{2+} ions that bind the channel pore (Sather & Mccleskey, 2003). Compared to the *Trichoplax* Ca_V3 and mammalian $Ca_V3.1$ channels, the two *Lymnaea* exon 12 variants exhibit significantly reduced Ca^{2+} block, especially those with exon 12a, accounting for the high Na^+ permeability (Senatore et al., 2014). Altogether, Ca^{2+} block appears to be a locus for evolutionary change in Ca^{2+} vs. Na^+ permeation, perhaps broadly applicable to other channel types with Ca_V-like selectivity filter motifs. This includes Na_V2 channels (Barzilai et al., 2012) and most pre-metazoan four-domain channels (Pozdnyakov et al., 2018).

Like Ca_V3, the *Trichoplax* Ca_V1 and Ca_V2 channels also resemble their bilaterian counterparts in bearing hallmark structures, namely, EEEE selectivity filter motifs, an AID structure in the domain I–II cytoplasmic linker, and C-terminal EF-hand, pre-IQ, and IQ domains (Fig. 1a). However, unlike the channels from cnidarians and bilaterians, both placozoan channels lack C-terminal PDZ ligand sequences associated with synaptic localization (Fig. 2) (Piekut et al., 2020). As expected from their shared structural features, the two *Trichoplax* channels conduct HVA Ca^{2+} currents *in vitro* (Gauberg et al., 2020;

Gauberg et al., 2022), but exhibit some notable differences from their bilaterian orthologues. For example, the Ca_V1 channel is insensitive to block by DHPs, and the Ca_V2 channel to the peptide blockers ω-conotoxin GVIA and ω-agatoxin IVA. They also differ in their voltage sensitivities, for example, activation and inactivation of the *Trichoplax* Ca_V2 channel is considerably left-shifted compared to human $Ca_V2.1$, especially activation with a half-maximal voltage that is roughly 13 mV more negative (Gauberg et al., 2020). Gating of this *Trichoplax* channel is also slower compared to human $Ca_V2.1$, both in terms of recovery from inactivation and kinetics for current activation and deactivation. Similar left-shifted voltage sensitivity is apparent for the *Trichoplax* Ca_V1 channel compared to rat $Ca_V1.2$, especially steady-state inactivation that is hyperpolarized by roughly 27 mV. However, unlike Ca_V2 channels, recovery from inactivation and current kinetics are much more similar between *Trichoplax* Ca_V1 and rat $Ca_V1.2$. Altogether, outside of the clear distinction of Ca_V3 vs. Ca_V1/Ca_V2 in terms of low and high activation voltages, respectively, the three *Trichoplax* channels exhibit considerable differences in their ion-conducting properties from their bilaterian counterparts.

When compared to each other, the three *Trichoplax* Ca_V channels span a voltage range for activation between roughly −70 mV (Ca_V3), to −50 mV (Ca_V1), to −25 mV (Ca_V2) (Gauberg et al., 2020, Gauberg et al., 2022, Smith et al., 2017). The three channels also provide avenues for constitutive Ca^{2+} influx through their window currents, through a range of resting membrane voltages between −70 and −20 mV, especially prominent for the Ca_V1 channel where roughly 3% of the total population remains open at a resting voltage of −40 mV (Gauberg et al., 2022).

Indeed, the placozoan Ca_V1 and Ca_V2 channels orthologues are the most divergent set relative to those of other animals and are hence useful for exploring the early evolutionary history of Ca_V1 and Ca_V2 channels. Since placozoans would have diverged from other animals shortly after the emergence of Ca_V1 and Ca_V2 channels in animals (Moran & Zakon, 2014), identification of conserved features with their bilaterian orthologues would reflect early adaptations that occurred shortly after their emergence. Perhaps, the most distinguishing functional features of bilaterians HVA channels pertain to their unique regulation by calmodulin and G protein βγ subunits. That is, only Ca_V1 channels exhibit CDI that is resistant to strong cytoplasmic Ca^{2+} buffering, indicative of C-lobe modulation (Ben-Johny & Yue, 2014), while Ca_V2 channels are uniquely inhibited by neuromodulatory Gβγ proteins in a voltage-dependent manner (Tedford & Zamponi, 2006).

In accordance with the shared calmodulin-binding sites in the *Trichoplax* Ca_V1 and Ca_V2 channel C-termini (Fig. 1a), both exhibit CDI mediated by calmodulin, evidenced by the loss of modulation when co-expressed with a mutant calmodulin lacking the ability to bind Ca^{2+} (Gauberg et al., 2022). Like its bilaterian orthologues, CDI of the *Trichoplax* Ca_V1 channel is buffer resistant (i.e., persists in 10 mM BAPTA), indicative of C-lobe modulation, while that of *Trichoplax* Ca_V2 is not (Gauberg et al., 2022). Thus, calmodulin regulation of Ca_V1 and Ca_V2 channels is an ancestral feature, which is perhaps not surprising given the prevalence of calmodulin-binding sites at the C-termini of eukaryotic four-domain channels (Pozdnyakov et al., 2018), including Na_V channels that are also subject to CDI (Ben-Johny et al., 2015). However, the unique C-lobe regulation of Ca_V1 channels, which permits strong CDI sensitive to local Ca^{2+} signals (i.e., those emanating from the channel pore itself), emerged shortly after the emergence of these two channel types, before placozoans diverged from other animals, likely contributing to their adaptation for distinct roles in Ca^{2+} signaling in different cellular contexts.

Instead, unlike bilaterian Ca_V2 channels, the *Trichoplax* Ca_V2 channel was found to lack voltage-dependent G protein inhibition, even when co-expressed with its conspecific Gβ and $Gγ_{1-3}$ proteins. Nevertheless, the *Trichoplax* G proteins could impose voltage-dependent inhibition on the human $Ca_V2.1$ channel (Gauberg et al., 2020). Thus, direct, voltage-dependent G protein modulation of Ca_V2 channels appears to have evolved after placozoans diverged from other animals, perhaps concurrent with the emergence of synapses and nervous systems in cnidarians and bilaterians.

A Calcium Channel Regulates Choanocyte Ciliary Beating and the Feeding Current in Sponges

Poriferans (sponges; Fig. 1a) are unique in that all species have motile larval stage, and an exclusively sessile adult life stage (Dunn et al., 2015). Sponges feed by collecting food particles from the environment using a feeding current, which is generated by beating cilia of choanocyte cells that line the internal epithelium. Interestingly, choanocyte cells resemble single-celled choanoflagellates, both bearing a single flagellum surrounded by a membranous skirt supported by actin (James-Clark, 1868) (Fig. 1a) and both taking up food particles such as bacteria through endocytosis (Leys & Eerkes-Medrano, 2006). Like placozoans, poriferans lack body symmetry, nervous systems, chemical synapses, and genes for gap junctions (Slivko-Koltchik et al., 2019), but the cells of hard (glass) sponges can form extensive electrical linkages through fusions of the cell membrane, such that all cells in the body exist as a single interconnected syncytium (Leys et al., 2006). Despite their limited cell types (Sebé-Pedrós et al., 2018) and morphological simplicity, sponges can elicit behaviors in response to environmental signals. For example, soft sponges trigger whole-body contractions and hard (glass) sponges propagate Ca^{2+} action potentials that cause pausing of choanocyte ciliary beating and the feeding current, both triggered by chemosensory and mechanosensory signals indicating the presence of excess particulates in the environment (e.g., disruption of the seafloor), and both serving to prevent the uptake of particles that would obstruct internal chambers (Nickel, 2010; Leys, 2015).

Interestingly, poriferans secondarily lost voltage-gated Na^+ and K^+ channels (Moran et al., 2015), indicative of a reduced reliance on fast electrical signaling, and perhaps consistent with their sessile life strategy. Indeed, the only cation channel identified to date that bears voltage sensors, and is thus sensitive to membrane depolarization, is the single $Ca_V1/2$ channel, along with single β and $α_2δ$ subunits, and two CACHD1 homologues (Figs. 1b, 3a, and 4). Structurally, the poriferan $Ca_V1/2$ channel possesses the hallmark features of HVA Ca_V1 and Ca_V2 channels, including an EEEE selectivity filter motif, an AID, and C-terminal EF-hand, pre-IQ, and IQ domains (Fig. 2a). Altogether, inferred from its structural similarity and phylogenetic proximity to Ca_V1 and Ca_V2 channels, this channel likely conducts high-voltage-activated currents that are Ca^{2+} selective, and regulated by calmodulin through CDI.

Likely, this single calcium channel underlies the noted Ca^{2+}-dependent action potentials that have been observed in glass sponges which cause the arrest of choanocyte ciliary beating. These action potentials are strictly dependent on Ca^{2+} for depolarization, on K^+ for repolarization (likely through a K^+ leak channel given the absence of K_V channels), are sensitive to block by the cations Co^{2+} and Mn^{2+}, and are also blocked by the DHP nimodipine (24 μM) (Leys et al., 1999). Soft-bodied sponges on the other hand do not form syncytia, nor have they been reported to generate Ca^{2+}-dependent action potentials. As noted, soft sponges respond to excess particulates in the environment by contracting, which both prevent objects from entering the body, and expels them (Elliott & Leys, 2007). In this case, however, a role for the $Ca_V1/2$ channel is unlikely, given that membrane excitation/depolarization is not required for contraction in sponges (Prosser, 1967) and is instead likely controlled by slow intracellular signaling pathways (Leys, 2015).

Ctenophores

Comb bearers, or ctenophores in Greek, are gelatinous animals with a combination of bilateral and radial body symmetry (i.e., biradial; Fig. 1a), named according to their unique beating rows of amalgamated cilia (combs) used for swimming (Tamm, 2014; Pang & Martindale, 2008). Morphologically, ctenophores resemble cnidarians (jellyfish) in many respects, including soft translucent tissues, a hollow body cavity (coelom) with a single opening, diffuse nervous systems arranged in a nerve net, tentacles, and the shared presence of various cell types, including neurons, smooth muscle, and striated muscle (Mackie et al., 1988; Tamm & Tamm, 1989; Hernandez-Nicaise, 1973; Steinmetz et al., 2012). Thus, for over 100 years, these two animal types were grouped together into the clade Coelenterata (Hertwig, 1880; Leuckart, 1848), for hollow intestine or coelom.

However, recent phylogenomic studies have provided strong evidence that ctenophores diverged before cnidarians, perhaps being the first phylum to diverge from all other animals. Accordingly, the ctenophore genome is dramatically different from that of cnidarians and bilaterians, for example, lacking most genes associated with development of mesoderm and muscle (Ryan et al., 2013). Ctenophores also lack numerous genes involved in neurotransmitter biosynthesis, and corresponding neurotransmitter receptors (Moroz et al., 2014). However, they uniquely expanded select receptor types, such as ionotropic glutamate receptors (i.e., homologous to NMDA, AMPA, and Kainate receptors) (Ramos-Vicente et al., 2018), and degenerin/epithelial sodium channels (gated by protons, mechanical stimuli, or neuropeptides) (Gründer & Assmann, 2015). Accordingly, both glutamate and neuropeptides have been implicated in synaptic signaling in ctenophores (Moroz et al., 2014; Sachkova et al., 2021). Ctenophore synapses also have an atypical ultrastructure, with features suggesting that presynaptic vesicles emerge directly from the smooth endoplasmic reticulum (Hernandez-Nicaise, 1973; Sachkova et al., 2021), thus bypassing the Golgi and axonal vesicular transport system. Because of these various striking differences, ctenophores have been proposed to have independently evolved synapses and the nervous system (Moroz, 2021; Moroz et al., 2014), although more certainly, they have the most divergent of all nervous systems.

In terms of genes for electrical signaling, ctenophores possess two Na_V2 channels, and Shaker K_V channels but lack the Shab, Shal, and Shaw types (Li et al., 2015). Ctenophores exhibit action potentials that are carried by mixed Na^+/Ca^{2+} currents, and these have been observed in various cell types, including neurons and muscle (reviewed in (Senatore et al., 2016)). As noted above, ctenophores lack Ca_V1 and Ca_V3 channels and possess only a single Ca_V2 channel homologue (Fig. 1b). They also lack CACHD1 and CACHD0 homologues, which were likely lost given the presence of gene homologues in the choanoflagellate *S. rosetta* (Figs. 2b and 4). However, ctenophores uniquely expanded their repertoire of ancillary subunits, having two $Ca_V\beta$ and an unprecedented eleven $\alpha_2\delta$ subunit genes (Figs. 2b, 3a, and 4). This significant expansion might reflect a unique strategy to generate functional diversity in lieu of single Ca_V2 channel, or alternatively, adaptation of these genes for novel functions that do not involve the calcium channel.

Importantly, ctenophores are the only animals with synapses for which the role of presynaptic Ca^{2+} influx for neurotransmitter release has not been experimentally established. However, single-cell transcriptome analysis has identified a neuron-like cell type that expresses gene homologues consistent with Ca^{2+}-dependent excitation–secretion coupling as observed at the synapse active zone (Sebé-Pedrós et al., 2018; Burkhardt & Jékely, 2021). This includes the Ca_V2 channel, the $\alpha_2\delta$ subunit ML15153a (Fig. 4), K_V channels, elements of the core SNARE complex, the Ca^{2+}-sensitive exocytotic proteins complexin and CADPS/CAPS, various neuropeptide precursor genes, and interestingly the presynaptic proteins Unc18, Unc13, RIM, and RIM-BP (Fig. 2c). The authors of this preprint suggested that these transcriptional features reflect homology with bilaterian neurons, which would argue against the independent evolution of nervous systems and synapses in ctenophores. However, there is one caveat worth considering. A recent phylogenetic analysis revealed that animals actually possess two RIM homologues, classified as type I and II RIM, the former corresponding to the canonical RIM defined as a key regulator of Ca_V2 channels at the presynaptic terminal (Fig. 2c) and the latter a previously undescribed clade of RIM orthologues (Piekut et al., 2020). All animals, except for ctenophores, possess a type I RIM homologue. Ctenophores are therefore the only animals with synapses that lack a canonical RIM, having only type II. Furthermore, the PDZ domain of type II RIM differs from type I at key amino acid positions associated with ligand specificity, so it is unlikely that type II RIM binds $DDWC_{-COOH}$ sequences. Lastly, ctenophore Ca_V2 channels lack a $DDWC_{-COOH}$ motif (Fig. 2b, c), instead harboring C-terminal sequences of $SGDL_{COOH}$ (*Mnemiopsis leidyi*) or $TDNL_{COOH}$ (*Beroe ovata*). Altogether, even if Ca_V2 channels and type II RIM interact in ctenophores, it would not necessarily indicate homology given that the latter gene is a phylogenetically distinct paralogue from type I RIM.

Conclusions

Studies of invertebrate calcium channels reveal that some features are broadly conserved, while others represent unique adaptations within specific animal phyla. A theme emerges that gene duplication of α_1, β, and $\alpha_2\delta$ and related subunits occurred frequently and independently, facilitating the emergence of new or modified roles. Although we have come to understand much about calcium channels in vertebrates and bilaterian invertebrates, we know less about the most early-diverging animals, and further still about homologues of single-celled choanoflagellates which diverged prior to the emergence of animals. Phylogenetically, the most underexamined calcium channel subunits are $\alpha_2\delta$, and especially the related CACHD1 and CACHD0 subunits.

References

Acuna, C., Liu, X., Gonzalez, A., & Südhof, T. C. (2015). Rim-Bps mediate tight coupling of action potentials to Ca^{2+}-triggered neurotransmitter release. *Neuron, 87,* 1234–1247.

Adams, M. D., Celniker, S. E., Holt, R. A., Evans, C. A., Gocayne, J. D., Amanatides, P. G., Scherer, S. E., Li, P. W., Hoskins, R. A., & Galle, R. F. (2000). The genome sequence of drosophila melanogaster. *Science, 287,* 2185–2195.

Agler, H. L., Evans, J., Tay, L. H., Anderson, M. J., Colecraft, H. M., & Yue, D. T. (2005). G protein-gated inhibitory module of N-type (Cav2. 2) Ca^{2+} channels. *Neuron, 46,* 891–904.

Alzayady, K. J., Sebé-Pedrós, A., Chandrasekhar, R., Wang, L., Ruiz-Trillo, I., & Yule, D. I. (2015). Tracing the evolutionary history of inositol, 1, 4, 5-Trisphosphate receptor: Insights from analyses of capsaspora Owczarzaki Ca^{2+} release channel orthologs. *Molecular Biology and Evolution, 32,* 2236–2253.

Anderson, P. A., & Schwab, W. E. (1983). Action potential in neurons of motor nerve net of cyanea (Coelenterata). *Journal of Neurophysiology, 50,* 671–683.

Arendt, D., Tosches, M. A., & Marlow, H. (2016). From nerve net to nerve ring, nerve cord and brain—Evolution of the nervous system. *Nature Reviews Neuroscience, 17,* 61–72.

Armon, S., Bull, M. S., Aranda-Diaz, A., & Prakash, M. (2018). Ultrafast epithelial contractions provide insights into contraction speed limits and tissue integrity. *Proceedings of the National Academy of Sciences, 115,* E10333–E10341.

Arteaga-Tlecuitl, R., Sanchez-Sandoval, A. L., Ramirez-Cordero, B. E., Rosendo-Pineda, M. J., Vaca, L., & Gomora, J. C. (2018). Increase of Cav3 channel activity induced by Hva B1b-subunit is not mediated by a physical interaction. *BMC Research Notes, 11,* 1–9.

Asmara, H., Micu, I., Rizwan, A. P., Sahu, G., Simms, B. A., Zhang, F.-X., Engbers, J. D., Stys, P. K., Zamponi, G. W., & Turner, R. W. (2017). A T-type channel-calmodulin complex triggers Acamkii activation. *Molecular Brain, 10,* 1–15.

Augustine, G., Charlton, M., & Smith, S. (1985). Calcium entry into voltage-clamped presynaptic terminals of squid. *The Journal of Physiology, 367,* 143–162.

Babonis, L. S., & Martindale, M. Q. (2014). Old cell, new trick? Cnidocytes as a model for the evolution of novelty. *American Zoologist, 54,* 714–722.

Baker, P., Hodgkin, A., & Ridgway, E. (1971). Depolarization and calcium entry in squid giant axons. *The Journal of Physiology, 218,* 709–755.

Bannister, R. A., & Beam, K. G. (2013). Cav1. 1: The atypical prototypical voltage-gated Ca^{2+} channel. *Biochimica et Biophysica Acta (BBA) - Biomembranes, 1828,* 1587–1597.

Bargmann, C. I. (1998). Neurobiology of the caenorhabditis elegans genome. *Science, 282,* 2028–2033.

Barzilai, M. G., Reitzel, A. M., Kraus, J. E., Gordon, D., Technau, U., Gurevitz, M., & Moran, Y. (2012). Convergent evolution of sodium ion selectivity in metazoan neuronal signaling. *Cell Reports, 2*, 242–248.

Ben-Johny, M., & Yue, D. T. (2014). Calmodulin regulation (Calmodulation) of voltage-gated calcium channels. *Journal of General Physiology, 143*, 679–692.

Ben-Johny, M. E., Dick, I., Sang, L., Limpitikul, W., Wei Kang, P., Niu, J., Banerjee, R., Yang, W., Babich, J., & Issa, J. (2015). Towards a unified theory of calmodulin regulation (calmodulation) of voltage-gated calcium and sodium channels. *Current Molecular Pharmacology, 8*, 188–205.

Bergsten, J. (2005). A review of long-branch attraction. *Cladistics, 21*, 163–193.

Bezanilla, F. (2000). The voltage sensor in voltage-dependent ion channels. *Physiological Reviews, 80*, 555–592.

Bianchi, E., & Wright, G. J. (2016). Sperm meets egg: The genetics of mammalian fertilization. *Annual Review of Genetics, 50*, 93–111.

Birnbaumer, L., Qin, N., Olcese, R., Tareilus, E., Platano, D., Costantin, J., & Stefani, E. (1998). Structures and functions of calcium channel B subunits. *Journal of Bioenergetics and Biomembranes, 30*, 357–375.

Block, B. A., Imagawa, T., Campbell, K. P., & Franzini-Armstrong, C. (1988). Structural evidence for direct interaction between the molecular components of the transverse tubule/sarcoplasmic reticulum junction in skeletal muscle. *The Journal of Cell Biology, 107*, 2587–2600.

Borg, J.-P., Straight, S. W., Kaech, S. M., De Taddéo-Borg, M., Kroon, D. E., Karnak, D., Turner, R. S., Kim, S. K., & Margolis, B. (1998). Identification of an evolutionarily conserved heterotrimeric protein complex involved in protein targeting. *Journal of Biological Chemistry, 273*, 31633–31636.

Brawley, S. H. (1991). The fast block against polyspermy in fucoid algae is an electrical block. *Developmental Biology, 144*, 94–106.

Brunet, T., Fischer, A. H., Steinmetz, P. R., Lauri, A., Bertucci, P., & Arendt, D. (2016). The evolutionary origin of bilaterian smooth and striated myocytes. *eLife, 5*, e19607.

Buckett, K., Peters, M., Dockray, G., Van Minnen, J., & Benjamin, P. (1990). Regulation of heartbeat in lymnaea by motoneurons containing Fmrfamide-like peptides. *Journal of Neurophysiology, 63*, 1426–1435.

Buraei, Z., & Yang, J. (2010). The B subunit of voltage-gated Ca^{2+} channels. *Physiological Reviews, 90*, 1461–1506.

Burkhardt, P., & Jékely, G. (2021). Evolution of synapses and neurotransmitter systems: The divide-and-conquer model for early neural cell-type evolution.

Butz, S., Okamoto, M., & Südhof, T. C. (1998). A tripartite protein complex with the potential to couple synaptic vesicle exocytosis to cell adhesion in brain. *Cell, 94*, 773–782.

Calin-Jageman, I., Yu, K., Hall, R. A., Mei, L., & Lee, A. (2007). Erbin enhances voltage-dependent facilitation of Cav1. 3 Ca^{2+} channels through relief of an autoinhibitory domain in the Cav1. 3 A1 subunit. *Journal of Neuroscience, 27*, 1374–1385.

Canti, C., Nieto-Rostro, M., Foucault, I., Heblich, F., Wratten, J., Richards, M., Hendrich, J., Douglas, L., Page, K., & Davies, A. (2005). The metal-ion-dependent adhesion site in the Von Willebrand factor-a domain of A2δ subunits is key to trafficking voltage-gated Ca^{2+} channels. *Proceedings of the National Academy of Sciences, 102*, 11230–11235.

Capella-Gutiérrez, S., Silla-Martínez, J. M., & Gabaldón, T. (2009). Trimal: A tool for automated alignment trimming in large-scale phylogenetic analyses. *Bioinformatics, 25*, 1972–1973.

Catterall, W. A. (2000). Structure and regulation of voltage-gated Ca^{2+} channels. *Annual Review of Cell and Developmental Biology, 16*, 521–555.

Catterall, W. A. (2010). Ion channel voltage sensors: Structure, function, and pathophysiology. *Neuron, 67*, 915–928.

Catterall, W. A. (2011). Voltage-gated calcium channels. *Cold Spring Harbor Perspectives in Biology, 3*, a003947.

Catterall, W. A., & Few, A. P. (2008). Calcium channel regulation and presynaptic plasticity. *Neuron, 59*, 882–901.

Caylor, R. C., Jin, Y., & Ackley, B. D. (2013). The caenorhabditis elegans voltage-gated calcium channel subunits Unc-2 and Unc-36 and the calcium-dependent kinase Unc-43/Camkii regulate neuromuscular junction morphology. *Neural Development, 8*, 1–13.

Cens, T., Rousset, M., Collet, C., Raymond, V., Démares, F., Quintavalle, A., Bellis, M., Le Conte, Y., Chahine, M., & Charnet, P. (2013). Characterization of the first honeybee Ca 2+ channel subunit reveals two novel species-and splicing-specific modes of regulation of channel inactivation. *Pflügers Archiv - European Journal of Physiology, 465*, 985–996.

Cens, T., Rousset, M., Collet, C., Charreton, M., Garnery, L., Le Conte, Y., Chahine, M., Sandoz, J.-C., & Charnet, P. (2015). Molecular characterization and functional expression of the apis mellifera voltage-dependent Ca^{2+} channels. *Insect Biochemistry and Molecular Biology, 58*, 12–27.

Chambers, E., & De Armendi, J. (1979). Membrane potential, action potential and activation potential of eggs of the sea urchin, lytechinus variegatus. *Experimental Cell Research, 122*, 203–218.

Chan, J. D., Zhang, D., Liu, X., Zarowiecki, M., Berriman, M., & Marchant, J. S. (2017). Utilizing the planarian voltage-gated ion channel transcriptome to resolve a role for a Ca^{2+} channel in neuromuscular function and regeneration. *Biochimica et Biophysica Acta (BBA)-Molecular Cell Research, 1864*, 1036–1045.

Chang, E. S., Neuhof, M., Rubinstein, N. D., Diamant, A., Philippe, H., Huchon, D., & Cartwright, P. (2015). Genomic insights into the evolutionary origin of myxozoa within cnidaria. *Proceedings of the National Academy of Sciences, 112*, 14912–14917.

Chapman, J. A., Kirkness, E. F., Simakov, O., Hampson, S. E., Mitros, T., Weinmaier, T., Rattei, T., Balasubramanian, P. G., Borman, J., & Busam, D. (2010). The dynamic genome of hydra. *Nature, 464*, 592–596.

Charbonneau, M., Moreau, M., Picheral, B., Vilain, J., & Guerrier, P. (1983). Fertilization of amphibian eggs: A comparison of electrical responses between anurans and urodeles. *Developmental Biology, 98*, 304–318.

Chemin, J., Taiakina, V., Monteil, A., Piazza, M., Guan, W., Stephens, R. F., ... & Spafford, J. D. (2017). Calmodulin regulates Cav3 T-type channels at their gating brake. *Journal of Biological Chemistry, 292*(49), 20010–20031.

Chemin, J., Monteil, A., Bourinet, E., Nargeot, J., & Lory, P. (2001). Alternatively spliced A1g (Cav3. 1) intracellular loops promote specific T-type Ca^{2+} channel gating properties. *Biophysical Journal, 80*, 1238–1250.

Chen, Y.-H., Li, M.-H., Zhang, Y., He, L.-L., Yamada, Y., Fitzmaurice, A., Shen, Y., Zhang, H., Tong, L., & Yang, J. (2004). Structural basis of the A 1–B subunit interaction of voltage-gated Ca 2+ channels. *Nature, 429*, 675–680.

Colley, D. G., Bustinduy, A. L., Secor, W. E., & King, C. H. (2014). Human schistosomiasis. *The Lancet, 383*, 2253–2264.

Consortium, C. E. S. (1998). Genome sequence of the nematode *C. Elegans*: A platform for investigating biology. *Science, 282*, 2012–2018.

Cottrell, G. S., Soubrane, C. H., Hounshell, J. A., Lin, H., Owenson, V., Rigby, M., Cox, P. J., Barker, B. S., Ottolini, M., & Ince, S. (2018). Cachd1 is an A2δ-like protein that modulates Cav3 voltage-gated calcium channel activity. *Journal of Neuroscience, 38*, 9186–9201.

Dahimene, S., Page, K. M., Kadurin, I., Ferron, L., Ho, D. Y., Powell, G. T., Pratt, W. S., Wilson, S. W., & Dolphin, A. C. (2018). The A2δ-like protein Cachd1 increases N-type calcium currents and cell surface expression and competes with A2δ-1. *Cell Reports, 25*, 1610–1621. e5.

Darbandi, S., & Franck, J. P. (2009). A comparative study of ryanodine receptor (Ryr) gene expression levels in a Basal Ray-Finned Fish, Bichir (Polypterus Ornatipinnis) and the Derived Euteleost Zebrafish (Danio Rerio). *Comparative Biochemistry and Physiology Part B: Biochemistry and Molecular Biology, 154*, 443–448.

David, L. S., Garcia, E., Cain, S. M., Thau, E., Tyson, J. R., & Snutch, T. P. (2010). Splice-variant changes of the Cav3. 2 T-type calcium channel mediate voltage-dependent facilitation and associate with cardiac hypertrophy and development. *Channels, 4*, 375–389.

Davies, A., Kadurin, I., Alvarez-Laviada, A., Douglas, L., Nieto-Rostro, M., Bauer, C. S., Pratt, W. S., & Dolphin, A. C. (2010). The A2δ subunits of voltage-gated calcium channels form Gpi-anchored proteins, a posttranslational modification essential for function.

Proceedings of the National Academy of Sciences, 107, 1654–1659.

Dawson, T. F., Boone, A. N., Senatore, A., Piticaru, J., Thiyagalingam, S., Jackson, D., Davison, A., & Spafford, J. D. (2014). Gene splicing of an invertebrate beta subunit (Lcavβ) in the N-terminal and Hook domains and its regulation of Lcav1 and Lcav2 calcium channels. *PLoS One, 9*, e92941.

Dayal, A., Schrötter, K., Pan, Y., Föhr, K., Melzer, W., & Grabner, M. (2017). The Ca 2+ influx through the mammalian skeletal muscle dihydropyridine receptor is irrelevant for muscle performance. *Nature Communications, 8*, 1–14.

De Hoog, E., Lukewich, M. K., & Spencer, G. E. (2019). Retinoid receptor-based signaling plays a role in voltage-dependent inhibition of invertebrate voltage-gated Ca^{2+} channels. *Journal of Biological Chemistry, 294*, 10076–10093.

De Jongh, K. S., Warner, C., & Catterall, W. A. (1990). Subunits of purified calcium channels. Alpha 2 and delta are encoded by the same gene. *Journal of Biological Chemistry, 265*, 14738–14741.

Deisseroth, K., Bito, H., & Tsien, R. W. (1996). Signaling from synapse to nucleus: Postsynaptic creb phosphorylation during multiple forms of hippocampal synaptic plasticity. *Neuron, 16*, 89–101.

Delsuc, F., Brinkmann, H., Chourrout, D., & Philippe, H. (2006). Tunicates and not cephalochordates are the closest living relatives of vertebrates. *Nature, 439*, 965–968.

Demaria, C. D., Soong, T. W., Alseikhan, B. A., Alvania, R. S., & Yue, D. T. (2001). Calmodulin bifurcates the local Ca 2+ signal that modulates P/Q-type Ca 2+ channels. *Nature, 411*, 484–489.

Di Biase, V., & Franzini-Armstrong, C. (2005). Evolution of skeletal type E–C coupling: A novel means of controlling calcium delivery. *The Journal of Cell Biology, 171*, 695–704.

Dick, I. E., Tadross, M. R., Liang, H., Tay, L. H., Yang, W., & Yue, D. T. (2008). A modular switch for spatial Ca 2+ selectivity in the calmodulin regulation of Ca V channels. *Nature, 451*, 830–834.

Dickman, D. K., Kurshan, P. T., & Schwarz, T. L. (2008). Mutations in a drosophila A2δ voltage-gated calcium channel subunit reveal a crucial synaptic function. *Journal of Neuroscience, 28*, 31–38.

Dietrich, D., Kirschstein, T., Kukley, M., Pereverzev, A., Von Der Brelie, C., Schneider, T., & Beck, H. (2003). Functional specialization of presynaptic Cav2. 3 Ca^{2+} channels. *Neuron, 39*, 483–496.

Dirksen, R. T., & Beam, K. G. (1999). Role of calcium permeation in dihydropyridine receptor function: Insights into channel gating and excitation–contraction coupling. *The Journal of General Physiology, 114*, 393–404.

Dolphin, A. C. (2006). A short history of voltage-gated calcium channels. *British Journal of Pharmacology, 147*, S56–S62.

Dolphin, A. C. (2016). Voltage-gated calcium channels and their auxiliary subunits: Physiology and pathophysiology and pharmacology. *The Journal of Physiology, 594*, 5369–5390.

Dolphin, A. C. (2018). Voltage-gated calcium channel A 2δ subunits: An assessment of proposed novel roles *F1000Research, 7*.

Dolphin, A. C., & Lee, A. (2020). Presynaptic calcium channels: Specialized control of synaptic neurotransmitter release. *Nature Reviews Neuroscience, 21*, 213–229.

Dolphin, A. C., Wyatt, C. N., Richards, J., Beattie, R., Craig, P., Lee, J., Cribbs, L., Volsen, S. & Perez-Reyes, E. 1999. The effect of the calcium channel A2-Δ accessory subunit on expression of the low voltage-activated calcium channel A1g.

Dong, N., Lee, D. W., Sun, H.-S., & Feng, Z.-P. (2018). Dopamine-mediated calcium channel regulation in synaptic suppression in L. Stagnalis interneurons. *Channels, 12*, 153–173.

Dubel, S. J., Altier, C., Chaumont, S., Lory, P., Bourinet, E., & Nargeot, J. (2004). Plasma membrane expression of T-type calcium channel A1 subunits is modulated by high voltage-activated auxiliary subunits. *Journal of Biological Chemistry, 279*, 29263–29269.

Dunn, T. W., & Sossin, W. S. (2013). Inhibition of the aplysia sensory neuron calcium current with dopamine and serotonin. *Journal of Neurophysiology, 110*, 2071–2081.

Dunn, C. W., Leys, S. P., & Haddock, S. H. (2015). The hidden biology of sponges and ctenophores. *Trends in Ecology & Evolution, 30*, 282–291.

Dunn, T. W., Fan, X., Ase, A. R., Séguéla, P., & Sossin, W. S. (2018). The Ca V 2α1 Ef-hand F helix tyrosine, a highly conserved locus for Gpcr inhibition of Ca V 2 channels. *Scientific Reports, 8*, 1–16.

Eberl, D. F., Ren, D., Feng, G., Lorenz, L. J., Van Vactor, D., & Hall, L. M. (1998). Genetic and developmental characterization of Dmca1d, a calcium channel A1 subunit gene in drosophila melanogaster. *Genetics, 148*, 1159–1169.

Edgar, R. C. (2004). Muscle: Multiple sequence alignment with high accuracy and high throughput. *Nucleic Acids Research, 32*, 1792–1797.

Eitel, M., Francis, W. R., Varoqueaux, F., Daraspe, J., Osigus, H.-J., Krebs, S., Vargas, S., Blum, H., Williams, G. A., & Schierwater, B. (2018). Comparative genomics and the nature of placozoan species. *PLoS Biology, 16*, e2005359.

Elliott, G. R., & Leys, S. P. (2007). Coordinated contractions effectively expel water from the aquiferous system of a freshwater sponge. *Journal of Experimental Biology, 210*, 3736–3748.

Emerick, M. C., Stein, R., Kunze, R., Mcnulty, M. M., Regan, M. R., Hanck, D. A., & Agnew, W. S. (2006). Profiling the array of Cav3. 1 variants from the human T-type calcium channel gene Cacna1g: Alternative structures, developmental expression, and biophysical variations. *Proteins: Structure, Function, and Bioinformatics, 64*, 320–342.

Ertel, E. A., Campbell, K. P., Harpold, M. M., Hofmann, F., Mori, Y., Perez-Reyes, E., Schwartz, A., Snutch, T. P.,

Tanabe, T., & Birnbaumer, L. (2000). Nomenclature of voltage-gated calcium channels. *Neuron, 25*, 533–535.

Fatt, P., & Ginsborg, B. L. (1958). The ionic requirements for the production of action potentials in crustacean muscle fibres. *The Journal of Physiology, 142*, 516–543.

Fatt, P., & Katz, B. (1953). The electrical properties of crustacean muscle fibres. *The Journal of Physiology, 120*, 171–204.

Franzini-Armstrong, C. (2004). Functional implications of Ryr-Dhpr relationships in skeletal and cardiac muscles. *Biological Research, 37*, 507–512.

Fux, J. E., Mehta, A., Moffat, J., & Spafford, J. D. (2018). Eukaryotic voltage-gated sodium channels: On their origins, asymmetries, losses, diversification and adaptations. *Frontiers in Physiology, 9*, 1406.

Gao, B., Sekido, Y., Maximov, A., Saad, M., Forgacs, E., Latif, F., Wei, M. H., Lerman, M., Lee, J.-H., & Perez-Reyes, E. (2000). Functional properties of a new voltage-dependent calcium channel A2δ auxiliary subunit gene (Cacna2d2). *Journal of Biological Chemistry, 275*, 12237–12242.

Gauberg, J., Abdallah, S., Elkhatib, W., Harracksingh, A. N., Piekut, T., Stanley, E. F., & Senatore, A. (2020). Conserved biophysical features of the Cav2 presynaptic Ca^{2+} channel homologue from the early-diverging animal trichoplax adhaerens. *Journal of Biological Chemistry, 295*, 18553–18578.

Gauberg, J., Elkhatib, W., Smith, C. L., Singh, A., & Senatore, A. (2022). Divergent Ca^{2+}/calmodulin feedback regulation of Ca_v1 and Ca_v2 voltage-gated calcium channels evolved in the common ancestor of placozoa and bilateria. *Journal of Biological Chemistry, 298*.

Geraerts, W. (1976). The control of ovulation in the hermaphroditic freshwater snail lymnaea stagnalis by the neurohormone of the caudodorsal cells. *General and Comparative Endocrinology, 28*, 350–357.

Gosselin-Badaroudine, P., Moreau, A., Simard, L., Cens, T., Rousset, M., Collet, C., Charnet, P., & Chahine, M. (2016). Biophysical characterization of the honeybee Dsc1 orthologue reveals a novel voltage-dependent Ca^{2+} channel subfamily: Cav4. *Journal of General Physiology, 148*, 133–145.

Goudeau, H., Depresle, Y., Rosa, A., & Goudeau, M. (1994). Evidence by a voltage clamp study of an electrically mediated block to polyspermy in the egg of the ascidian phallusia mammillata. *Developmental Biology, 166*, 489–501.

Graf, E. R., Valakh, V., Wright, C. M., Wu, C., Liu, Z., Zhang, Y. Q., & Diantonio, A. (2012). Rim promotes calcium channel accumulation at active zones of the drosophila neuromuscular junction. *Journal of Neuroscience, 32*, 16586–16596.

Granados-Gonzalez, G., Mendoza-Lujambio, I., Rodriguez, E., Galindo, B., Beltrán, C., & Darszon, A. (2005). Identification of voltage-dependent Ca^{2+} channels in sea urchin sperm. *FEBS Letters, 579*, 6667–6672.

Grau-Bové, X., Ruiz-Trillo, I., & Irimia, M. (2018). Origin of exon skipping-rich transcriptomes in animals driven by evolution of gene architecture. *Genome Biology, 19*, 1–21.

Grauel, M. K., Maglione, M., Reddy-Alla, S., Willmes, C. G., Brockmann, M. M., Trimbuch, T., Rosenmund, T., Pangalos, M., Vardar, G., & Stumpf, A. (2016). Rim-binding protein 2 regulates release probability by fine-tuning calcium channel localization at murine hippocampal synapses. *Proceedings of the National Academy of Sciences, 113*, 11615–11620.

Gross, G. G., Lone, G. M., Leung, L. K., Hartenstein, V., & Guo, M. (2013). X11/mint genes control polarized localization of axonal membrane proteins in vivo. *Journal of Neuroscience, 33*, 8575–8586.

Gründer, S., & Assmann, M. (2015). Peptide-gated ion channels and the simple nervous system of hydra. *Journal of Experimental Biology, 218*, 551–561.

Guan, W., Stephens, R. F., Mourad, O., Mehta, A., Fux, J., & Spafford, J. D. (2020). Unique cysteine-enriched, D2l5 and D4l6 extracellular loops in Ca V 3 T-type channels alter the passage and block of monovalent and divalent ions. *Scientific Reports, 10*, 1–24.

Hagiwara, S., Ozawa, S., & Sand, O. (1975). Voltage clamp analysis of two inward current mechanisms in the egg cell membrane of a starfish. *The Journal of General Physiology, 65*, 617–644.

Han, Y., Kaeser, P. S., Südhof, T. C., & Schneggenburger, R. (2011). Rim determines Ca^{2+} channel density and vesicle docking at the presynaptic active zone. *Neuron, 69*, 304–316.

Hanlon, M., Berrow, N., Dolphin, A., & Wallace, B. (1999). Modelling of a voltage-dependent Ca^{2+} channel B subunit as a basis for understanding its functional properties. *FEBS Letters, 445*, 366–370.

Hara, Y., Koganezawa, M., & Yamamoto, D. (2015). The Dmca1d channel mediates Ca^{2+} inward currents in drosophila embryonic muscles. *Journal of Neurogenetics, 29*, 117–123.

Haverinen, J., Hassinen, M., Dash, S. N., & Vornanen, M. (2018). Expression of calcium channel transcripts in the zebrafish heart: Dominance of T-type channels. *Journal of Experimental Biology, 221*, jeb179226.

He, L.-L., Zhang, Y., Chen, Y.-H., Yamada, Y., & Yang, J. (2007). Functional modularity of the B-subunit of voltage-gated Ca^{2+} channels. *Biophysical Journal, 93*, 834–845.

Heinrich, L., & Ryglewski, S. (2020). Different functions of two putative drosophila A 2 Δ subunits in the same identified motoneurons. *Scientific Reports, 10*, 1–17.

Herlitze, S., Garcia, D., Mackie, K., Hille, B., Scheuer, T., & Catterall, W. A. (1996). Modulation of Ca2 channels by G-protein ß subunits. *Nature, 380*, 258–262.

Hernandez-Nicaise, M.-L. (1973). The nervous system of ctenophores Iii. Ultrastructure of synapses. *Journal of Neurocytology, 2*, 249–263.

Hertwig, R. (1880). *Ueber Den Bau Der Ctenophoren*. Fischer.

Hess, P., Lansman, J. B., & Tsien, R. W. (1986). Calcium channel selectivity for divalent and monovalent cat-

ions. Voltage and Concentration dependence of single channel current in ventricular heart cells. *The Journal of General Physiology, 88*, 293–319.

Heyland, A., Croll, R., Goodall, S., Kranyak, J., & Wyeth, R. (2014). Trichoplax adhaerens, an enigmatic basal metazoan with potential. In *Developmental biology of the sea urchin and other marine invertebrates*. Springer.

Hibino, H., Pironkova, R., Onwumere, O., Vologodskaia, M., Hudspeth, A., & Lesage, F. (2002). Rim binding proteins (Rbps) couple Rab3-interacting molecules (Rims) to voltage-gated Ca^{2+} channels. *Neuron, 34*, 411–423.

Hirano, M., Takada, Y., Wong, C. F., Yamaguchi, K., Kotani, H., Kurokawa, T., Mori, M. X., Snutch, T. P., Ronjat, M., & De Waard, M. (2017). C-terminal splice variants of P/Q-type Ca^{2+} channel Cav2. 1 A1 subunits are differentially regulated by Rab3-interacting molecule proteins. *Journal of Biological Chemistry, 292*, 9365–9381.

Ho, A., Morishita, W., Atasoy, D., Liu, X., Tabuchi, K., Hammer, R. E., Malenka, R. C., & Südhof, T. C. (2006). Genetic analysis of mint/X11 proteins: Essential presynaptic functions of a neuronal adaptor protein family. *Journal of Neuroscience, 26*, 13089–13101.

Hodgkin, A. L., & Huxley, A. F. (1952). A quantitative description of membrane current and its application to conduction and excitation in nerve. *The Journal of Physiology, 117*, 500–544.

Hogg, A. (2018). *Genetic screen to characterize shank interactors at the drosophila neuromuscular junction*. University of Toronto (Canada).

Hosono, R., Hekimi, S., Kamiya, Y., Sassa, T., Murakami, S., Nishiwaki, K., Miwa, J., Taketo, A., & Kodaira, K. I. (1992). The Unc-18 gene encodes a novel protein affecting the kinetics of acetylcholine metabolism in the nematode caenorhabditis elegans. *Journal of Neurochemistry, 58*, 1517–1525.

Huang, X., Senatore, A., Dawson, T. F., Quan, Q., & Spafford, J. D. (2010). G-proteins modulate invertebrate synaptic calcium channel (Lcav2) differently from the classical voltage-dependent regulation of mammalian Cav2. 1 and Cav2. 2 channels. *Journal of Experimental Biology, 213*, 2094–2103.

Huc, S., Monteil, A., Bidaud, I., Barbara, G., Chemin, J., & Lory, P. (2009). Regulation of T-type calcium channels: Signalling pathways and functional implications. *Biochimica et Biophysica Acta (BBA)-Molecular Cell Research, 1793*, 947–952.

Huelsenbeck, J. P., & Ronquist, F. (2001). Mrbayes: Bayesian inference of phylogenetic trees. *Bioinformatics, 17*, 754–755.

Jaffe, L. A. (1976). Fast block to polyspermy in sea urchin eggs is electrically mediated. *Nature, 261*, 68–71.

James-Clark, H. (1868). Xxvii.—on the Spongiæ Ciliatæ as infusoria flagellata; or observations on the structure, animality, and relationship of leucosolenia botryoides, bowerbank. *Annals and Magazine of Natural History, 1*, 188–215.

Jay, S., Sharp, A., Kahl, S., Vedvick, T., Harpold, M., & Campbell, K. (1991). Structural characterization of the dihydropyridine-sensitive calcium channel alpha 2-subunit and the associated delta peptides. *Journal of Biological Chemistry, 266*, 3287–3293.

Jenkins, M. A., Christel, C. J., Jiao, Y., Abiria, S., Kim, K. Y., Usachev, Y. M., Obermair, G. J., Colbran, R. J., & Lee, A. (2010). Ca^{2+}-dependent facilitation of Cav1. 3 Ca^{2+} channels by densin and Ca^{2+}/calmodulin-dependent protein kinase Ii. *Journal of Neuroscience, 30*, 5125–5135.

Jeong, K., Lee, S., Seo, H., Oh, Y., Jang, D., Choe, J., Kim, D., Lee, J.-H., & Jones, W. D. (2015). Ca-Alt, a Fly T-type Ca 2+ channel, negatively modulates sleep. *Scientific Reports, 5*, 1–13.

Jeziorski, M. C., & Greenberg, R. M. (2006). Voltage-gated calcium channel subunits from platyhelminths: Potential role in praziquantel action. *International Journal for Parasitology, 36*, 625–632.

Jeziorski, M. C., Greenberg, R. M., Clark, K. S., & Anderson, P. A. (1998). Cloning and functional expression of a voltage-gated calcium channel A1 subunit from jellyfish. *Journal of Biological Chemistry, 273*, 22792–22799.

Jeziorski, M. C., Greenberg, R. M., & Anderson, P. (1999). Cloning and expression of a jellyfish calcium channel beta subunit reveal functional conservation of the Alpha1-beta interaction. *Receptors & Channels, 6*, 375–386.

Johnson, M., Zaretskaya, I., Raytselis, Y., Merezhuk, Y., Mcginnis, S., & Madden, T. L. (2008). Ncbi blast: A better web interface. *Nucleic Acids Research, 36*, W5–W9.

Joiner, M.-L. A., & Lee, A. (2015). Voltage-gated Cav1 channels in disorders of vision and hearing. *Current Molecular Pharmacology, 8*, 143–148.

Jones, P., Binns, D., Chang, H.-Y., Fraser, M., Li, W., Mcanulla, C., Mcwilliam, H., Maslen, J., Mitchell, A., & Nuka, G. (2014). Interproscan 5: Genome-scale protein function classification. *Bioinformatics, 30*, 1236–1240.

Jospin, M., Jacquemond, V., Mariol, M.-C., Ségalat, L., & Allard, B. (2002). The L-type voltage-dependent Ca^{2+} channel Egl-19 controls body wall muscle function in caenorhabditis elegans. *The Journal of Cell Biology, 159*, 337–348.

Jouiaei, M., Yanagihara, A. A., Madio, B., Nevalainen, T. J., Alewood, P. F., & Fry, B. G. (2015). Ancient venom systems: A review on Cnidaria toxins. *Toxins, 7*, 2251–2271.

Jungbluth, H., & Gautel, M. (2014). Pathogenic mechanisms in centronuclear myopathies. *Frontiers in Aging Neuroscience, 6*, 339.

Just, E. (1919). The fertilization reaction in echinarachnius parma: I. Cortical response of the egg to insemination. *The Biological Bulletin, 36*, 1–10.

Kaeser, P. S., Deng, L., Wang, Y., Dulubova, I., Liu, X., Rizo, J., & Südhof, T. C. (2011). Rim proteins tether Ca^{2+} channels to presynaptic active zones via a direct Pdz-domain interaction. *Cell, 144*, 282–295.

Kamp, T. J., & Hell, J. W. (2000). Regulation of cardiac L-type calcium channels by protein kinase A and protein kinase C. *Circulation Research, 87*, 1095–1102.

Kanamori, T., Kanai, M. I., Dairyo, Y., Yasunaga, K.-I., Morikawa, R. K., & Emoto, K. (2013). Compartmentalized calcium transients trigger dendrite pruning in drosophila sensory neurons. *Science, 340*, 1475–1478.

Kandel, E. R. (2001). The molecular biology of memory storage: A dialogue between genes and synapses. *Science, 294*, 1030–1038.

Kang, H.-W., Park, J.-Y., Jeong, S.-W., Kim, J.-A., Moon, H.-J., Perez-Reyes, E., & Lee, J.-H. (2006). A molecular determinant of nickel inhibition in Cav3. 2 T-type calcium channels. *Journal of Biological Chemistry, 281*, 4823–4830.

Kapli, P., & Telford, M. J. (2020). Topology-dependent asymmetry in systematic errors affects phylogenetic placement of ctenophora and xenacoelomorpha. *Science Advances, 6*, eabc5162.

Katz, B., & Miledi, R. (1967). A study of synaptic transmission in the absence of nerve impulses. *The Journal of Physiology, 192*, 407–436.

Kawasaki, F., Zou, B., Xu, X., & Ordway, R. W. (2004). Active zone localization of presynaptic calcium channels encoded by the cacophony locus of drosophila. *Journal of Neuroscience, 24*, 282–285.

Kenny, N. J., Francis, W. R., Rivera-Vicéns, R. E., Juravel, K., De Mendoza, A., Díez-Vives, C., Lister, R., Bezares-Calderón, L. A., Grombacher, L., & Roller, M. (2020). Tracing animal genomic evolution with the chromosomal-level assembly of the freshwater sponge ephydatia muelleri. *Nature Communications, 11*, 1–11.

Khanna, R., Sun, L., Li, Q., Guo, L., & Stanley, E. (2006). Long splice variant N type calcium channels are clustered at presynaptic transmitter release sites without modular adaptor proteins. *Neuroscience, 138*, 1115–1125.

King, G. F. (2007). Modulation of insect Cav channels by peptidic spider toxins. *Toxicon, 49*, 513–530.

Kits, K. S., & Mansvelder, H. D. (1996). Voltage gated calcium channels in molluscs: Classification, Ca 2+ dependent inactivation, modulation and functional roles. *Invertebrate Neuroscience, 2*, 9–34.

Kline, D., Jaffe, L. A., & Tucker, R. P. (1985). Fertilization potential and polyspermy prevention in the egg of the nemertean, cerebratulus lacteus. *Journal of Experimental Zoology, 236*, 45–52.

Kohn, A., Lea, J., Roberts-Misterly, J., Anderson, P., & Greenberg, R. (2001a). Structure of three high voltage-activated calcium channel A1 subunits from schistosoma mansoni. *Parasitology, 123*, 489–497.

Kohn, A. B., Anderson, P. A., Roberts-Misterly, J. M., & Greenberg, R. M. (2001b). Schistosome calcium channel B subunits: Unusual modulatory effects and potential role in the action of the antischistosomal drug praziquantel. *Journal of Biological Chemistry, 276*, 36873–36876.

Kohn, A., Roberts-Misterly, J., Anderson, P., Khan, N., & Greenberg, R. (2003). Specific sites in the beta interaction domain of a schistosome Ca²⁺ channel B subunit are key to its role in sensitivity to the anti-schistosomal drug praziquantel. *Parasitology, 127*, 349–356.

Kuo, C. T., Jan, L. Y., & Jan, Y. N. (2005). Dendrite-specific remodeling of drosophila sensory neurons requires matrix metalloproteases, ubiquitin-proteasome, and ecdysone signaling. *Proceedings of the National Academy of Sciences, 102*, 15230–15235.

Kurschner, C., & Yuzaki, M. (1999). Neuronal interleukin-16 (Nil-16): A dual function Pdz domain protein. *Journal of Neuroscience, 19*, 7770–7780.

Kurschner, C., Mermelstein, P. G., Holden, W. T., & Surmeier, D. J. (1998). Cipp, a novel multivalent Pdz domain protein, selectively interacts with Kir4. 0 family members, Nmda receptor subunits, neurexins, and neuroligins. *Molecular and Cellular Neuroscience, 11*, 161–172.

Kushibiki, Y., Suzuki, T., Jin, Y., & Taru, H. (2019). Rimb-1/Rim-binding protein and Unc-10/Rim redundantly regulate presynaptic localization of the voltage-gated calcium channel in caenorhabditis elegans. *Journal of Neuroscience, 39*, 8617–8631.

Lainé, V., Frøkjær-Jensen, C., Couchoux, H., & Jospin, M. (2011). The A1 subunit Egl-19, the A2/Δ subunit Unc-36, and the B subunit Ccb-1 underlie voltage-dependent calcium currents in caenorhabditis elegans striated muscle. *Journal of Biological Chemistry, 286*, 36180–36187.

Lander, E. S., Linton, L. M., Birren, B., Nusbaum, C., Zody, M. C., Baldwin, J., Devon, K., Dewar, K., Doyle, M., & Fitzhugh, W. (2001). Initial sequencing and analysis of the human genome.

Latour, I., Louw, D. F., Beedle, A. M., Hamid, J., Sutherland, G. R., & Zamponi, G. W. (2004). Expression of T-type calcium channel splice variants in human glioma. *Glia, 48*, 112–119.

Leader, D. P., Krause, S. A., Pandit, A., Davies, S. A., & Dow, J. A. T. (2018). Flyatlas 2: A new version of the drosophila melanogaster expression atlas with Rna-Seq, Mirna-Seq and sex-specific data. *Nucleic Acids Research, 46*, D809–D815.

Leclère, L., Horin, C., Chevalier, S., Lapébie, P., Dru, P., Péron, S., Jager, M., Condamine, T., Pottin, K., & Romano, S. (2019). The genome of the jellyfish clytia hemisphaerica and the evolution of the cnidarian lifecycle. *Nature Ecology & Evolution, 3*, 801–810.

Leuckart, R. (1848). *Ueber Die Morphologie Und Verwandtschaftsverhältnisse Der Wirbellosen Thiere: Ein Beitrag Zur Charakteristik Und Classification Der Thierischen Formen.* F. Vieweg und Sohn.

Leys, S. P. (2015). Elements of a 'nervous system'in sponges. *Journal of Experimental Biology, 218*, 581–591.

Leys, S. P., & Eerkes-Medrano, D. I. (2006). Feeding in a calcareous sponge: Particle uptake by pseudopodia. *The Biological Bulletin, 211*, 157–171.

Leys, S. P., Mackie, G. O., & Meech, R. W. (1999). Impulse conduction in a sponge. *Journal of Experimental Biology, 202*, 1139–1150.

Leys, S., Cheung, E., & Boury-Esnault, N. (2006). Embryogenesis in the glass sponge oopsacas minuta: Formation of syncytia by fusion of blastomeres. *Integrative and Comparative Biology, 46*, 104–117.

Li, X., Liu, H., Luo, J. C., Rhodes, S. A., Trigg, L. M., Van Rossum, D. B., Anishkin, A., Diatta, F. H., Sassic, J. K., & Simmons, D. K. (2015). Major diversification of voltage-gated K+ channels occurred in ancestral parahoxozoans. *Proceedings of the National Academy of Sciences, 112*, E1010–E1019.

Liebeskind, B. J., Hillis, D. M., & Zakon, H. H. (2011). Evolution of sodium channels predates the origin of nervous systems in animals. *Proceedings of the National Academy of Sciences, 108*, 9154–9159.

Lin, Y.-C. J., & Spencer, A. N. (2001). Calcium currents from jellyfish striated muscle cells: Preservation of phenotype, characterisation of currents and channel localisation. *Journal of Experimental Biology, 204*, 3717–3726.

Littleton, J. T., & Ganetzky, B. (2000). Ion channels and synaptic organization: Analysis of the drosophila genome. *Neuron, 26*, 35–43.

Littleton, J. T., Stern, M., Perin, M., & Bellen, H. J. (1994). Calcium dependence of neurotransmitter release and rate of spontaneous vesicle fusions are altered in drosophila synaptotagmin mutants. *Proceedings of the National Academy of Sciences, 91*, 10888–10892.

Liu, K. S., Siebert, M., Mertel, S., Knoche, E., Wegener, S., Wichmann, C., Matkovic, T., Muhammad, K., Depner, H., & Mettke, C. (2011). Rim-binding protein, a central part of the active zone, is essential for neurotransmitter release. *Science, 334*, 1565–1569.

Llinás, R., Steinberg, I. Z., & Walton, K. (1976). Presynaptic calcium currents and their relation to synaptic transmission: Voltage clamp study in squid giant synapse and theoretical model for the calcium gate. *Proceedings of the National Academy of Sciences, 73*, 2918–2922.

Lory, P., Nicole, S., & Monteil, A. (2020). Neuronal Cav3 channelopathies: Recent progress and perspectives. *Pflügers Archiv - European Journal of Physiology*, 1–14.

Lu, Q., & Dunlap, K. (1999). Cloning and functional expression of novel N-type Ca²⁺ channel variants. *Journal of Biological Chemistry, 274*, 34566–34575.

Lübbert, M., Goral, R. O., Satterfield, R., Putzke, T., Van Den Maagdenberg, A. M., Kamasawa, N., & Young, S. M., Jr. (2017). A novel region in the Cav2. 1 A1 subunit C-terminus regulates fast synaptic vesicle fusion and vesicle docking at the mammalian presynaptic active zone. *eLife, 6*, e28412.

Ly, C. V., Yao, C.-K., Verstreken, P., Ohyama, T., & Bellen, H. J. (2008). Straightjacket is required for the synaptic stabilization of cacophony, a voltage-gated calcium channel A1 subunit. *The Journal of Cell Biology, 181*, 157–170.

Mackie, G. (1990). The elementary nervous system revisited. *American Zoologist, 30,* 907–920.

Mackie, G., Mills, C., & Singla, C. (1988). Structure and function of the prehensile tentilla of euplokamis (Ctenophora, Cydippida). *Zoomorphology, 107,* 319–337.

Mahapatra, S., Calorio, C., Vandael, D. H. F., Marcantoni, A., Carabelli, V., & Carbone, E. (2012). Calcium channel types contributing to chromaffin cell excitability, exocytosis and endocytosis. *Cell Calcium, 51,* 321–330.

Marlétaz, F., Peijnenburg, K. T., Goto, T., Satoh, N., & Rokhsar, D. S. (2019). A new spiralian phylogeny places the enigmatic arrow worms among gnathiferans. *Current Biology, 29,* 312–318. e3.

Maruyama, I. N., & Brenner, S. (1991). A phorbol ester/diacylglycerol-binding protein encoded by the Unc-13 gene of caenorhabditis elegans. *Proceedings of the National Academy of Sciences, 88,* 5729–5733.

Mathews, E. A., García, E., Santi, C. M., Mullen, G. P., Thacker, C., Moerman, D. G., & Snutch, T. P. (2003). Critical residues of the caenorhabditis elegans Unc-2 voltage-gated calcium channel that affect behavioral and physiological properties. *Journal of Neuroscience, 23,* 6537–6545.

Maximov, A., & Bezprozvanny, I. (2002). Synaptic targeting of N-type calcium channels in hippocampal neurons. *Journal of Neuroscience, 22,* 6939–6952.

Maximov, A., Südhof, T. C., & Bezprozvanny, I. (1999). Association of neuronal calcium channels with modular adaptor proteins. *Journal of Biological Chemistry, 274,* 24453–24456.

Mayorova, T. D., Smith, C. L., Hammar, K., Winters, C. A., Pivovarova, N. B., Aronova, M. A., Leapman, R. D., & Reese, T. S. (2018). Cells containing aragonite crystals mediate responses to gravity in trichoplax adhaerens (Placozoa), an animal lacking neurons and synapses. *PLoS One, 13,* e0190905.

Mayorova, T. D., Hammar, K., Winters, C. A., Reese, T. S., & Smith, C. L. (2019). The ventral epithelium of trichoplax adhaerens deploys in distinct patterns cells that secrete digestive enzymes, mucus or diverse neuropeptides. *Biology Open, 8,* bio045674.

Mccusker, P., & Chan, J. D. (2019). Anti-schistosomal action of the calcium channel agonist Fpl-64176. *International Journal for Parasitology: Drugs and Drug Resistance, 11,* 30–38.

Mckay, B. E., Mcrory, J. E., Molineux, M. L., Hamid, J., Snutch, T. P., Zamponi, G. W., & Turner, R. W. (2006). Cav3 T-type calcium channel isoforms differentially distribute to somatic and dendritic compartments in rat central neurons. *European Journal of Neuroscience, 24,* 2581–2594.

Meech, R., & Mackie, G. (1993). Ionic currents in giant motor axons of the jellyfish, aglantha digitale. *Journal of Neurophysiology, 69,* 884–893.

Meves, H., & Vogel, W. (1973). Calcium inward currents in internally perfused giant axons. *The Journal of Physiology, 235,* 225–265.

Meyer, A., & Schartl, M. (1999). Gene and genome duplications in vertebrates: The one-to-four (-to-eight in fish) rule and the evolution of novel gene functions. *Current Opinion in Cell Biology, 11,* 699–704.

Monteil, A., Chemin, J., Bourinet, E., Mennessier, G., Lory, P., & Nargeot, J. (2000). Molecular and functional properties of the human A1g subunit that forms T-type calcium channels. *Journal of Biological Chemistry, 275,* 6090–6100.

Moran, Y., & Zakon, H. H. (2014). The evolution of the four subunits of voltage-gated calcium channels: Ancient roots, increasing complexity, and multiple losses. *Genome Biology and Evolution, 6,* 2210–2217.

Moran, Y., Barzilai, M. G., Liebeskind, B. J., & Zakon, H. H. (2015). Evolution of voltage-gated ion channels at the emergence of metazoa. *Journal of Experimental Biology, 218,* 515–525.

Moreland, R. T., Nguyen, A.-D., Ryan, J. F., Schnitzler, C. E., Koch, B. J., Siewert, K., Wolfsberg, T. G., & Baxevanis, A. D. (2014). A customized web portal for the genome of the ctenophore mnemiopsis leidyi. *BMC Genomics, 15,* 1–13.

Mori, Y., Friedrich, T., Kim, M.-S., Mikami, A., Nakai, J., Ruth, P., Bosse, E., Hofmann, F., Flockerzi, V., & Furuichi, T. (1991). Primary structure and functional expression from complementary DNA of a brain calcium channel. *Nature, 350,* 398–402.

Moroz, L. L. (2021). Multiple origins of neurons from secretory cells. *Frontiers in Cell and Development Biology, 9.*

Moroz, L. L., Kocot, K. M., Citarella, M. R., Dosung, S., Norekian, T. P., Povolotskaya, I. S., Grigorenko, A. P., Dailey, C., Berezikov, E., & Buckley, K. M. (2014). The ctenophore genome and the evolutionary origins of neural systems. *Nature, 510,* 109–114.

Mouton, J., Ronjat, M., Jona, I., Villaz, M., Feltz, A., & Maulet, Y. (2001). Skeletal and cardiac ryanodine receptors bind to the Ca^{2+}-sensor region of dihydropyridine receptor A1c subunit. *FEBS Letters, 505,* 441–444.

Mukherjee, K., Slawson, J. B., Christmann, B. L., & Griffith, L. C. (2014). Neuron-specific protein interactions of drosophila Cask-B are revealed by mass spectrometry. *Frontiers in Molecular Neuroscience, 7,* 58.

Mulatz, K. J. (2013). *A Pdz-3 mediated physical and functional interaction between the Cav3. 2 T-type calcium channel and neuronal nitric oxide synthase.* University of British Columbia.

Nakai, J., Imagawa, T., Hakamata, Y., Shigekawa, M., Takeshima, H., & Numa, S. (1990). Primary structure and functional expression from Cdna of the cardiac ryanodine receptor/calcium release channel. *FEBS Letters, 271,* 169–177.

Nakai, J., Tanabe, T., Konno, T., Adams, B., & Beam, K. G. (1998). Localization in the Ii-Iii loop of the dihydropyridine receptor of a sequence critical for excitation-contraction coupling. *Journal of Biological Chemistry, 273,* 24983–24986.

Nguyen, L.-T., Schmidt, H. A., Von Haeseler, A., & Minh, B. Q. (2015). Iq-tree: A fast and effective stochastic algorithm for estimating maximum-likelihood phylogenies. *Molecular Biology and Evolution, 32*, 268–274.

Nickel, M. (2010). Evolutionary emergence of synaptic nervous systems: What can we learn from the non-synaptic, nerveless porifera? *Invertebrate Biology, 129*, 1–16.

Nishiyama, M., Togashi, K., Von Schimmelmann, M. J., Lim, C.-S., Maeda, S.-I., Yamashita, N., Goshima, Y., Ishii, S., & Hong, K. (2011). Semaphorin 3a induces Ca V 2.3 channel-dependent conversion of axons to dendrites. *Nature Cell Biology, 13*, 676–685.

Niu, J., Yang, W., Yue, D. T., Inoue, T., & Ben-Johny, M. (2018). Duplex signaling by Cam and Stac3 enhances Cav1. 1 Function and provides insights into congenital myopathy. *Journal of General Physiology, 150*, 1145–1161.

Nuccitelli, R., & Grey, R. D. (1984). Controversy over the fast, partial, temporary block to polyspermy in sea urchins: A reevaluation. *Developmental Biology, 103*, 1–17.

Oh, K. H., Krout, M. D., Richmond, J. E., & Kim, H. (2021). Unc-2 Cav2 channel localization at pre-synaptic active zones depends on Unc-10/Rim and Syd-2/Liprin-A in caenorhabditis elegans. *Journal of Neuroscience, 41*, 4782–4794.

Ohkubo, T., Inoue, Y., Kawarabayashi, T., & Kitamura, K. (2005). Identification and electrophysiological characteristics of isoforms of T-type calcium channel Cav3. 2 expressed in pregnant human uterus. *Cellular Physiology and Biochemistry, 16*, 245–254.

Ohno, S. (2013). *Evolution by gene duplication*. Springer Science & Business Media.

Ohtsuka, T., Takao-Rikitsu, E., Inoue, E., Inoue, M., Takeuchi, M., Matsubara, K., Deguchi-Tawarada, M., Satoh, K., Morimoto, K., & Nakanishi, H. (2002). Cast a novel protein of the cytomatrix at the active zone of synapses that forms a ternary complex with Rim1 and Munc13-1. *The Journal of Cell Biology, 158*, 577–590.

Okagaki, R., Izumi, H., Okada, T., Nagahora, H., Nakajo, K., & Okamura, Y. (2001). The maternal transcript for truncated voltage-dependent Ca^{2+} channels in the ascidian embryo: A potential suppressive role in Ca^{2+} channel expression. *Developmental Biology, 230*, 258–277.

Okamoto, H., Takahashi, K., & Yoshii, M. (1976). Two components of the calcium current in the egg cell membrane of the tunicate. *The Journal of Physiology, 255*, 527–561.

Ono, K., & Iijima, T. (2010). Cardiac T-type Ca^{2+} channels in the heart. *Journal of Molecular and Cellular Cardiology, 48*, 65–70.

Opatowsky, Y., Chen, C.-C., Campbell, K. P., & Hirsch, J. A. (2004). Structural analysis of the voltage-dependent calcium channel B subunit functional core and its complex with the A1 interaction domain. *Neuron, 42*, 387–399.

Pang, K., & Martindale, M. Q. (2008). Ctenophores. *Current Biology, 18*, R1119–R1120.

Park, D., & Dunlap, K. (1998). Dynamic regulation of calcium influx by G-proteins, action potential waveform, and neuronal firing frequency. *Journal of Neuroscience, 18*, 6757–6766.

Perez-Reyes, E. (2003). Molecular physiology of low-voltage-activated T-type calcium channels. *Physiological Reviews, 83*, 117–161.

Perez-Reyes, E., Cribbs, L. L., Daud, A., Lacerda, A. E., Barclay, J., Williamson, M. P., Fox, M., Rees, M., & Lee, J.-H. (1998). Molecular characterization of a neuronal low-voltage-activated T-type calcium channel. *Nature, 391*, 896–900.

Peterson, B. Z., Demaria, C. D., & Yue, D. T. (1999). Calmodulin is the Ca^{2+} sensor for Ca^{2+}-dependent inactivation of L-type calcium channels. *Neuron, 22*, 549–558.

Peterson, K. J., Lyons, J. B., Nowak, K. S., Takacs, C. M., Wargo, M. J., & Mcpeek, M. A. (2004). Estimating metazoan divergence times with a molecular clock. *Proceedings of the National Academy of Sciences, 101*, 6536–6541.

Piekut, T., Wong, Y. Y., Walker, S. E., Smith, C. L., Gauberg, J., Harracksingh, A. N., Lowden, C., Novogradac, B. B., Cheng, H.-Y. M., Spencer, G. E., & Senatore, A. (2020). Early metazoan origin and multiple losses of a novel clade of rim presynaptic calcium channel scaffolding protein homologs. *Genome Biology and Evolution, 12*, 1217–1239.

Pisani, D., Pett, W., Dohrmann, M., Feuda, R., Rota-Stabelli, O., Philippe, H., Lartillot, N., & Wörheide, G. (2015). Genomic data do not support comb jellies as the sister group to all other animals. *Proceedings of the National Academy of Sciences, 112*, 15402–15407.

Pozdnyakov, I., Matantseva, O., & Skarlato, S. (2018). Diversity and evolution of four-domain voltage-gated cation channels of eukaryotes and their ancestral functional determinants. *Scientific Reports, 8*, 1–10.

Pragnell, M., De Waard, M., Mori, Y., Tanabe, T., Snutch, T. P., & Campbell, K. P. (1994). Calcium channel B-subunit binds to a conserved motif in the I–Ii cytoplasmic linker of the A1-subunit. *Nature, 368*, 67–70.

Prosser, C. L. (1967). Ionic analyses and effects of ions on contractions of sponge tissues. *Zeitschrift für Vergleichende Physiologie, 54*, 109–120.

Przysiezniak, J., & Spencer, A. N. (1992). Voltage-activated calcium currents in identified neurons from a hydrozoan jellyfish, polyorchis penicillatus. *Journal of Neuroscience, 12*, 2065–2078.

Putnam, N. H., Srivastava, M., Hellsten, U., Dirks, B., Chapman, J., Salamov, A., Terry, A., Shapiro, H., Lindquist, E., & Kapitonov, V. V. (2007). Sea anemone genome reveals ancestral eumetazoan gene repertoire and genomic organization. *Science, 317*, 86–94.

Pym, E., Sasidharan, N., Thompson-Peer, K. L., Simon, D. J., Anselmo, A., Sadreyev, R., Hall, Q., Nurrish, S., & Kaplan, J. M. (2017). Shank is a dose-dependent regulator of Cav1 calcium current and creb target expression. *eLife, 6*, e18931.

Ramos-Vicente, D., Ji, J., Gratacòs-Batlle, E., Gou, G., Reig-Viader, R., Luís, J., Burguera, D., Navas-Perez,

E., García-Fernández, J., & Fuentes-Prior, P. (2018). Metazoan evolution of glutamate receptors reveals unreported phylogenetic groups and divergent lineage-specific events. *eLife, 7,* e35774.

Rettig, J., Sheng, Z.-H., Kim, D. K., Hodson, C. D., Snutch, T. P., & Catterall, W. A. (1996). Isoform-specific interaction of the Alpha1 subunits of brain Ca^{2+} channels with the presynaptic proteins syntaxin and Snap-25. *Proceedings of the National Academy of Sciences, 93,* 7363–7368.

Reuveny, A., Shnayder, M., Lorber, D., Wang, S., & Volk, T. (2018). Ma2/D promotes myonuclear positioning and association with the sarcoplasmic reticulum. *Development, 145,* dev159558.

Rice, P., Longden, I., & Bleasby, A. (2000). Emboss: The European molecular biology open software suite. *Trends in Genetics, 16,* 276–277.

Richards, M. W., Leroy, J., Pratt, W. S., & Dolphin, A. C. (2007). The Hook-Domain between the Sh3-and the Gk-domains of Cavβ subunits contains key determinants controlling calcium channel inactivation. *Channels, 1,* 92–101.

Roderick, H. L., Berridge, M. J., & Bootman, M. D. (2003). Calcium-induced calcium release. *Current Biology, 13,* R425–R425.

Romanova, D. Y., Heyland, A., Sohn, D., Kohn, A. B., Fasshauer, D., Varoqueaux, F., & Moroz, L. L. (2020a). Glycine as a signaling molecule and chemoattractant in trichoplax (placozoa): Insights into the early evolution of neurotransmitters. *NeuroReport, 31,* 490–497.

Romanova, D. Y., Smirnov, I. V., Nikitin, M. A., Kohn, A. B., Borman, A. I., Malyshev, A. Y., Balaban, P. M., & Moroz, L. L. (2020b). Sodium action potentials in placozoa: Insights into behavioral integration and evolution of nerveless animals. *Biochemical and Biophysical Research Communications, 532,* 120–126.

Ryan, J. F., Pang, K., Schnitzler, C. E., Nguyen, A.-D., Moreland, R. T., Simmons, D. K., Koch, B. J., Francis, W. R., Havlak, P., & Program, N. C. S. (2013). The genome of the ctenophore mnemiopsis leidyi and its implications for cell type evolution. *Science, 342,* 1242592.

Sachkova, M. Y., Nordmann, E.-L., Soto-Àngel, J. J., Meeda, Y., Górski, B., Naumann, B., Dondorp, D., Chatzigeorgiou, M., Kittelmann, M., & Burkhardt, P. (2021). Neuropeptide repertoire and 3d anatomy of the ctenophore nervous system. *Current Biology*

Saheki, Y., & Bargmann, C. I. (2009). Presynaptic Cav2 calcium channel traffic requires Calf-1 and the A 2 Δ subunit Unc-36. *Nature Neuroscience, 12,* 1257–1265.

Salvador-Recatalà, V., & Greenberg, R. M. (2010). The N terminus of a schistosome B subunit regulates inactivation and current density of a Cav2 channel. *Journal of Biological Chemistry, 285,* 35878–35888.

Salvador-Recatalà, V., & Greenberg, R. M. (2012). Calcium channels of schistosomes: Unresolved questions and unexpected answers. *Wiley Interdisciplinary Reviews: Membrane Transport and Signaling, 1,* 85–93.

Salvador-Recatalà, V., Schneider, T., & Greenberg, R. M. (2008). Atypical properties of a conventional calcium channel B subunit from the platyhelminth schistosoma mansoni. *BMC Physiology, 8,* 1–11.

Sather, W. A., & Mccleskey, E. W. (2003). Permeation and selectivity in calcium channels. *Annual Review of Physiology, 65,* 133–159.

Schlegel, D. K., Glasauer, S. M., Mateos, J. M., Barmettler, G., Ziegler, U., & Neuhauss, S. C. (2019). A new zebrafish model for Cacna2d4-dysfunction. *Investigative Ophthalmology & Visual Science, 60,* 5124–5135.

Schredelseker, J., Shrivastav, M., Dayal, A., & Grabner, M. (2010). Non–Ca^{2+}-conducting Ca^{2+} channels in fish skeletal muscle excitation-contraction coupling. *Proceedings of the National Academy of Sciences, 107,* 5658–5663.

Sebé-Pedrós, A., Chomsky, E., Pang, K., Lara-Astiaso, D., Gaiti, F., Mukamel, Z., Amit, I., Hejnol, A., Degnan, B. M., & Tanay, A. (2018). Early metazoan cell type diversity and the evolution of multicellular gene regulation. *Nature Ecology & Evolution, 2,* 1176–1188.

Seipel, K., & Schmid, V. (2005). Evolution of striated muscle: Jellyfish and the origin of triploblasty. *Developmental Biology, 282,* 14–26.

Senatore, A., & Spafford, J. D. (2010). Transient and big are key features of an invertebrate T-type channel (Lcav3) from the central nervous system of lymnaea stagnalis. *Journal of Biological Chemistry, 285,* 7447–7458.

Senatore, A., & Spafford, J. D. (2012). Gene transcription and splicing of T-type channels are evolutionarily-conserved strategies for regulating channel expression and gating. *PLoS One, 7,* e37409.

Senatore, A., Boone, A., Lam, S., Dawson, T. F., Zhorov, B., & Spafford, J. (2011). Mapping of dihydropyridine binding residues in a less sensitive invertebrate L-type calcium channel (Lcav1). *Channels, 5,* 173–187.

Senatore, A., Zhorov, B. S., & Spafford, J. D. (2012). Cav3 T-type calcium channels. *Wiley Interdisciplinary Reviews: Membrane Transport and Signaling, 1,* 467–491.

Senatore, A., Guan, W., Boone, A. N., & Spafford, J. D. (2014). T-type channels become highly permeable to sodium ions using an alternative extracellular turret region (S5-P) Outside the selectivity filter*♦. *Journal of Biological Chemistry, 289,* 11952–11969.

Senatore, A., Raiss, H., & Le, P. (2016). Physiology and evolution of voltage-gated calcium channels in early diverging animal phyla: Cnidaria, placozoa, porifera and ctenophora. *Frontiers in Physiology, 7,* 481.

Senatore, A., Reese, T. S., & Smith, C. L. (2017). Neuropeptidergic integration of behavior in trichoplax adhaerens, an animal without synapses. *Journal of Experimental Biology, 220,* 3381–3390.

Shcheglovitov, A., Kostyuk, P., & Shuba, Y. (2007). Selectivity signatures of three isoforms of recombinant T-type Ca^{2+} channels. *Biochimica et Biophysica Acta (BBA) - Biomembranes, 1768,* 1406–1419.

Shcheglovitov, A., Vitko, I., Bidaud, I., Baumgart, J. P., Navarro-Gonzalez, M. F., Grayson, T. H., Lory, P., Hill, C. E., & Perez-Reyes, E. (2008). Alternative splicing within the I–Ii loop controls surface expression of T-type Cav3. 1 calcium channels. *FEBS Letters, 582,* 3765–3770.

Sheng, Z.-H., Rettig, J., Takahashi, M., & Catterall, W. A. (1994). Identification of a syntaxin-binding site on N-type calcium channels. *Neuron, 13,* 1303–1313.

Silva-Moraes, V., Couto, F. F. B., Vasconcelos, M. M., Araújo, N., Coelho, P. M. Z., Katz, N., & Grenfell, R. F. Q. (2013). Antischistosomal activity of a calcium channel antagonist on schistosomula and adult schistosoma mansoni worms. *Memórias do Instituto Oswaldo Cruz, 108,* 600–604.

Simion, P., Philippe, H., Baurain, D., Jager, M., Richter, D. J., Di Franco, A., Roure, B., Satoh, N., Quéinnec, É., & Ereskovsky, A. (2017). A large and consistent phylogenomic dataset supports sponges as the sister group to all other animals. *Current Biology, 27,* 958–967.

Simms, B. A., Souza, I. A., & Zamponi, G. W. (2014). A novel calmodulin site in the Cav1. 2 N-terminus regulates calcium-dependent inactivation. *Pflügers Archiv - European Journal of Physiology, 466,* 1793–1803.

Slivko-Koltchik, G. A., Kuznetsov, V. P., & Panchin, Y. V. (2019). Are there gap junctions without connexins or pannexins? *BMC Evolutionary Biology, 19,* 5–12.

Smith, L. A., Wang, X., Peixoto, A. A., Neumann, E. K., Hall, L. M., & Hall, J. C. (1996). A drosophila calcium channel A1 subunit gene maps to a genetic locus associated with behavioral and visual defects. *Journal of Neuroscience, 16,* 7868–7879.

Smith, C. L., Varoqueaux, F., Kittelmann, M., Azzam, R. N., Cooper, B., Winters, C. A., Eitel, M., Fasshauer, D., & Reese, T. S. (2014). Novel cell types, neurosecretory cells, and body plan of the early-diverging metazoan trichoplax adhaerens. *Current Biology, 24,* 1565–1572.

Smith, C. L., Pivovarova, N., & Reese, T. S. (2015). Coordinated feeding behavior in trichoplax, an animal without synapses. *PLoS One, 10,* e0136098.

Smith, C. L., Abdallah, S., Wong, Y. Y., Le, P., Harracksingh, A. N., Artinian, L., Tamvacakis, A. N., Rehder, V., Reese, T. S., & Senatore, A. (2017). Evolutionary insights into T-type Ca^{2+} channel structure, function, and ion selectivity from the trichoplax adhaerens homologue. *Journal of General Physiology, 149,* 483–510.

Smith, C. L., Reese, T. S., Govezensky, T., & Barrio, R. A. (2019). Coherent directed movement toward food modeled in trichoplax, a ciliated animal lacking a nervous system. *Proceedings of the National Academy of Sciences, 116,* 8901–8908.

Soubrane, C., Stevens, E., & Stephens, G. (2012). Expression and functional studies of the novel Cns protein Cachd1. *Physiology.*

Spafford, J. D., Chen, L., Feng, Z.-P., Smit, A. B., & Zamponi, G. W. (2003a). Expression and modulation of an invertebrate presynaptic calcium channel A1

subunit homolog. *Journal of Biological Chemistry, 278,* 21178–21187.

Spafford, J. D., Munno, D. W., Van Nierop, P., Feng, Z.-P., Jarvis, S. E., Gallin, W. J., Smit, A. B., Zamponi, G. W., & Syed, N. I. (2003b). Calcium channel structural determinants of synaptic transmission between identified invertebrate neurons. *Journal of Biological Chemistry, 278,* 4258–4267.

Spafford, J. D., Dunn, T., Smit, A. B., Syed, N. I., & Zamponi, G. W. (2006). In vitro characterization of L-type calcium channels and their contribution to firing behavior in invertebrate respiratory neurons. *Journal of Neurophysiology, 95,* 42–52.

Srivastava, M., Begovic, E., Chapman, J., Putnam, N. H., Hellsten, U., Kawashima, T., Kuo, A., Mitros, T., Salamov, A., & Carpenter, M. L. (2008). The trichoplax genome and the nature of placozoans. *Nature, 454,* 955–960.

Srivastava, M., Simakov, O., Chapman, J., Fahey, B., Gauthier, M. E., Mitros, T., Richards, G. S., Conaco, C., Dacre, M., & Hellsten, U. (2010). The amphimedon Queenslandica genome and the evolution of animal complexity. *Nature, 466,* 720–726.

Steinmetz, P. R., Kraus, J. E., Larroux, C., Hammel, J. U., Amon-Hassenzahl, A., Houliston, E., Wörheide, G., Nickel, M., Degnan, B. M., & Technau, U. (2012). Independent evolution of striated muscles in cnidarians and bilaterians. *Nature, 487,* 231–234.

Stephens, R. F., Guan, W., Zhorov, B. S., & Spafford, J. D. (2015). Selectivity filters and cysteine-rich extracellular loops in voltage-gated sodium, calcium, and NALCN channels. *Frontiers in Physiology, 6,* 153.

Stephens, G. J., & Cottrell, G. S. (2019). Cachd1: A new activity-modifying protein for voltage-gated calcium channels. *Channels, 13,* 120–123.

Straub, V. A., Staras, K., Kemenes, G. R., & Benjamin, P. R. (2002). Endogenous and network properties of lymnaea feeding central pattern generator interneurons. *Journal of Neurophysiology, 88,* 1569–1583.

Südhof, T. C. (2012). The presynaptic active zone. *Neuron, 75,* 11–25.

Sullivan, S. E., Dillon, G. M., Sullivan, J. M., & Ho, A. (2014). Mint proteins are required for synaptic activity-dependent amyloid precursor protein (App) trafficking and amyloid B generation. *Journal of Biological Chemistry, 289,* 15374–15383.

Syed, N., Bulloch, A., & Lukowiak, K. (1990). In vitro reconstruction of the respiratory central pattern generator of the mollusk lymnaea. *Science, 250,* 282–285.

Taiakina, V., Boone, A. N., Fux, J., Senatore, A., Weber-Adrian, D., Guillemette, J. G., & Spafford, J. D. (2013). The calmodulin-binding, short linear motif, nscate is conserved in L-type channel ancestors of vertebrate Cav1. 2 and Cav1. 3 channels. *PLoS One, 8,* e61765.

Takahashi, M., Seagar, M. J., Jones, J. F., Reber, B., & Catterall, W. A. (1987). Subunit structure of dihydropyridine-sensitive calcium channels from skeletal muscle. *Proceedings of the National Academy of Sciences, 84,* 5478–5482.

Takeshima, H., Nishimura, S., Matsumoto, T., Ishida, H., Kangawa, K., Minamino, N., Matsuo, H., Ueda, M., Hanaoka, M., & Hirose, T. (1989). Primary structure and expression from complementary DNA of skeletal muscle ryanodine receptor. *Nature, 339,* 439–445.

Tamm, S. L. (2014). Cilia and the life of ctenophores. *Invertebrate Biology, 133,* 1–46.

Tamm, S., & Tamm, S. L. (1989). Extracellular ciliary axonemes associated with the surface of smooth muscle cells of ctenophores. *Journal of Cell Science, 94,* 713–724.

Tanabe, T., Takeshima, H., Mikami, A., Flockerzi, V., Takahashi, H., Kangawa, K., Kojima, M., Matsuo, H., Hirose, T., & Numa, S. (1987). Primary structure of the receptor for calcium channel blockers from skeletal muscle. *Nature, 328,* 313–318.

Tareilus, E., Roux, M., Qin, N., Olcese, R., Zhou, J., Stefani, E., & Birnbaumer, L. (1997). A xenopus oocyte B subunit: Evidence for a role in the assembly/expression of voltage-gated calcium channels that is separate from its role as a regulatory subunit. *Proceedings of the National Academy of Sciences, 94,* 1703–1708.

Tedford, H. W., & Zamponi, G. W. (2006). Direct G protein modulation of Cav2 calcium channels. *Pharmacological Reviews, 58,* 837–862.

Thomsen, M. C. F., & Nielsen, M. (2012). Seq2logo: A method for construction and visualization of amino acid binding motifs and sequence profiles including sequence weighting, pseudo counts and two-sided representation of amino acid enrichment and depletion. *Nucleic Acids Research, 40,* W281–W287.

Tian, C., Johnson, K. R., Lett, J. M., Voss, R., Salt, A. N., Hartsock, J. J., Steyger, P. S., & Ohlemiller, K. K. (2021). Cachd1-deficient mice exhibit hearing and balance deficits associated with a disruption of calcium homeostasis in the inner ear. *Hearing Research, 409,* 108327.

Tong, X.-J., López-Soto, E. J., Li, L., Liu, H., Nedelcu, D., Lipscombe, D., Hu, Z., & Kaplan, J. M. (2017). Retrograde synaptic inhibition is mediated by A-neurexin binding to the A2δ subunits of N-type calcium channels. *Neuron, 95,* 326–340. e5.

Tsien, R. W., & Barrett, C. F. (2005). A brief history of calcium channel discovery. In *Voltage-gated calcium channels.* Springer.

Tyson, J. R., & Snutch, T. P. (2013). Molecular nature of voltage-gated calcium channels: Structure and species comparison. *Wiley Interdisciplinary Reviews: Membrane Transport and Signaling, 2,* 181–206.

Upadhyay, A. A., Fleetwood, A. D., Adebali, O., Finn, R. D., & Zhulin, I. B. (2016). Cache domains that are homologous to, but different from pas domains comprise the largest superfamily of extracellular sensors in prokaryotes. *PLoS Computational Biology, 12,* e1004862.

Van Petegem, F., Clark, K. A., Chatelain, F. C., & Minor, D. L. (2004). Structure of a complex between a voltage-gated calcium channel B-subunit and an A-subunit domain. *Nature, 429,* 671–675.

Varoqueaux, F., Williams, E. A., Grandemange, S., Truscello, L., Kamm, K., Schierwater, B., Jékely, G., & Fasshauer, D. (2018). High cell diversity and complex peptidergic signaling underlie placozoan behavior. *Current Biology, 28,* 3495–3501. e2.

Venter, J. C., Adams, M. D., Myers, E. W., Li, P. W., Mural, R. J., Sutton, G. G., Smith, H. O., Yandell, M., Evans, C. A., & Holt, R. A. (2001). The sequence of the human genome. *Science, 291,* 1304–1351.

Vitko, I., Bidaud, I., Arias, J. M., Mezghrani, A., Lory, P., & Perez-Reyes, E. (2007). The I–Ii loop controls plasma membrane expression and gating of Cav3. 2 T-type Ca^{2+} channels: A paradigm for childhood absence epilepsy mutations. *Journal of Neuroscience, 27,* 322–330.

Wang, Y., Okamoto, M., Schmitz, F., Hofmann, K., & Südhof, T. C. (1997). Rim is a putative Rab3 effector in regulating synaptic-vesicle fusion. *Nature, 388,* 593–598.

Wang, D. Y.-C., Kumar, S., & Hedges, S. B. (1999). Divergence time estimates for the early history of animal phyla and the origin of plants, animals and fungi. *Proceedings of the Royal Society of London. Series B: Biological Sciences, 266,* 163–171.

Wang, Y., Sugita, S., & SüDhof, T. C. (2000). The Rim/Nim family of neuronal C2 domain proteins: Interactions with Rab3 and a new class of Src homology 3 domain proteins. *Journal of Biological Chemistry, 275,* 20033–20044.

Wang, Y., Liu, X., Biederer, T., & Südhof, T. C. (2002). A family of rim-binding proteins regulated by alternative splicing: Implications for the genesis of synaptic active zones. *Proceedings of the National Academy of Sciences, 99,* 14464–14469.

Wang, S., Stanika, R. I., Wang, X., Hagen, J., Kennedy, M. B., Obermair, G. J., Colbran, R. J., & Lee, A. (2017). Densin-180 controls the trafficking and signaling of L-type voltage-gated Cav1. 2 Ca^{2+} channels at excitatory synapses. *Journal of Neuroscience, 37,* 4679–4691.

Weick, J. P., Groth, R. D., Isaksen, A. L., & Mermelstein, P. G. (2003). Interactions with Pdz proteins are required for L-type calcium channels to activate camp response element-binding protein-dependent gene expression. *Journal of Neuroscience, 23,* 3446–3456.

Weiergräber, M., Henry, M., Krieger, A., Kamp, M., Radhakrishnan, K., Hescheler, J., & Schneider, T. (2006). Altered seizure susceptibility in mice lacking the Cav2. 3 E-type Ca^{2+} channel. *Epilepsia, 47,* 839–850.

Weir, K., Dupre, C., Van Giesen, L., Lee, A. S., & Bellono, N. W. (2020). A molecular filter for the cnidarian stinging response. *eLife, 9,* e57578.

Wen, H., Linhoff, M. W., Hubbard, J. M., Nelson, N. R., Stensland, D., Dallman, J., Mandel, G., & Brehm, P. (2013). Zebrafish calls for reinterpretation for the roles of P/Q calcium channels in neuromuscular transmission. *Journal of Neuroscience, 33,* 7384–7392.

Whelan, N. V., Kocot, K. M., Moroz, T. P., Mukherjee, K., Williams, P., Paulay, G., Moroz, L. L., & Halanych,

K. M. (2017). Ctenophore relationships and their placement as the sister group to all other animals. *Nature Ecology & Evolution, 1*, 1737–1746.

Williams, D. W., Kondo, S., Krzyzanowska, A., Hiromi, Y., & Truman, J. W. (2006). Local caspase activity directs engulfment of dendrites during pruning. *Nature Neuroscience, 9*, 1234–1236.

Wong, F. K., Nath, A. R., Chen, R. H., Gardezi, S. R., Li, Q., & Stanley, E. F. (2014). Synaptic vesicle tethering and the Cav2. 2 distal C-terminal. *Frontiers in Cellular Neuroscience, 8*, 71.

Wong, Y. Y., Le, P., Elkhatib, W., Piekut, T., & Senatore, A. (2019). Transcriptome profiling of trichoplax adhaerens highlights its digestive epithelium and a rich set of genes for fast electrogenic and slow neuromodulatory cellular signaling.

Wormuth, C., Lundt, A., Henseler, C., Müller, R., Broich, K., Papazoglou, A., & Weiergräber, M. (2016). Cav2. 3 R-type voltage-gated Ca²⁺ channels-functional implications in convulsive and non-convulsive seizure activity. *The Open Neurology Journal, 10*, 99.

Wozniak, K. L., & Carlson, A. E. (2020). Ion Channels and signaling pathways used in the fast polyspermy block. *Molecular Reproduction and Development, 87*, 350–357.

Wu, J., Yan, Z., Li, Z., Yan, C., Lu, S., Dong, M., & Yan, N. (2015). Structure of the voltage-gated calcium channel Cav1.1 complex. *Science, 350*.

Wu, J., Yan, Z., Li, Z., Qian, X., Lu, S., Dong, M., Zhou, Q., & Yan, N. (2016). Structure of the voltage-gated calcium channel Ca V 1.1 at 3.6 Å resolution. *Nature, 537*, 191–196.

Wu, S., Gan, G., Zhang, Z., Sun, J., Wang, Q., Gao, Z., Li, M., Jin, S., Huang, J., & Thomas, U. (2017). A presynaptic function of shank protein in drosophila. *Journal of Neuroscience, 37*, 11592–11604.

Wu, X., Cai, Q., Chen, Y., Zhu, S., Mi, J., Wang, J., & Zhang, M. (2020). Structural basis for the high-affinity interaction between Cask and Mint1. *Structure, 28*, 664–673. e3.

Yuen, S. M. W. K., Campiglio, M., Tung, C.-C., Flucher, B. E., & Van Petegem, F. (2017). Structural insights into binding of stac proteins to voltage-gated calcium channels. *Proceedings of the National Academy of Sciences, 114*, E9520–E9528.

Zarin, T., Strome, B., Ba, A. N. N., Alberti, S., Forman-Kay, J. D., & Moses, A. M. (2019). Proteome-wide signatures of function in highly diverged intrinsically disordered regions. *eLife, 8*, e46883.

Zhang, H., Maximov, A., Fu, Y., Xu, F., Tang, T.-S., Tkatch, T., Surmeier, D. J., & Bezprozvanny, I. (2005). Association of Cav1. 3 L-type calcium channels with shank. *Journal of Neuroscience, 25*, 1037–1049.

Zhang, H., Fu, Y., Altier, C., Platzer, J., Surmeier, D. J., & Bezprozvanny, I. (2006). Cav1. 2 and Cav1. 3 neuronal L-type calcium channels: Differential targeting and signaling to Pcreb. *European Journal of Neuroscience, 23*, 2297–2310.

Zhang, Y., Chen, Y.-H., Bangaru, S. D., He, L., Abele, K., Tanabe, S., Kozasa, T., & Yang, J. (2008). Origin of the voltage dependence of G-protein regulation of P/Q-type Ca²⁺ channels. *Journal of Neuroscience, 28*, 14176–14188.

Zhong, X., Liu, J. R., Kyle, J. W., Hanck, D. A., & Agnew, W. S. (2006). A profile of alternative RNA splicing and transcript variation of Cacna1h, a human T-channel gene candidate for idiopathic generalized epilepsies. *Human Molecular Genetics, 15*, 1497–1512.

Zhou, W., Horstick, E. J., Hirata, H., & Kuwada, J. Y. (2008). Identification and expression of voltage-gated calcium channel B subunits in zebrafish. *Developmental Dynamics: An Official Publication of the American Association of the Anatomists, 237*, 3842–3852.

Zühlke, R. D., Pitt, G. S., Deisseroth, K., Tsien, R. W., & Reuter, H. (1999). Calmodulin supports both inactivation and facilitation of L-type calcium channels. *Nature, 399*, 159–162.

Part II

Regulation of VGCCs

Modulation of VGCCs by G-Protein Coupled Receptors and Their Second Messengers

Melanie D. Mark, Jan Claudius Schwitalla, and Stefan Herlitze

Abstract

Voltage-gated calcium channels (VGCCs) are essential for transforming electrical signals such as action potentials into biochemical and eventually physiological responses through the control of intracellular calcium. They mediate the depolarization of cells and increase the intracellular Ca^{2+} levels to regulate a variety of physiological events including neurotransmission, secretion, enzyme activity, cellular differentiation, gene expression, smooth muscle contraction, and pacemaker activity. Deficits in VGCCs function can lead to epilepsy, migraine, ataxia, autism, cardiac arrythmias, and pain. Therefore, careful modulation of VGCCs is vital. In this chapter, we will discuss the modulation of VGCCs by different G protein-coupled receptors and their downstream effectors such as G proteins, lipids, kinases, and synaptic associated proteins focusing primarily on the central nervous system and heart. Moreover, we will describe the underlying mechanisms that G proteins and their downstream second messengers use to control VGCCs.

M. D. Mark (✉) · J. C. Schwitalla
Behavioral Neuroscience, Ruhr-University Bochum, Bochum, Germany

S. Herlitze
Department of Zoology and Neurobiology, Ruhr-University Bochum, Bochum, Germany

Keywords

Ca_v1 · Ca_v2 · Ca_v3 · L-type calcium channels · P/Q-type calcium channels · N-type calcium channels · R-type calcium channels · T-type calcium channels · G protein · GPCR · Arachidonic acid · Kinases · PIP_2 · Calmodulin · SNARES · G protein-coupled second messengers · Lipids

Abbreviations

5-HT	Serotonin
AA	Arachidonic acid
AC	Adenylyl cyclase
AGS	Activator of G protein signaling
AID	α interaction domain
AKAP	A-kinase anchoring protein
AMPA	α-Amino-3-hydroxy-5-methyl-4-isoxazolepropionic acid
AR	adrenoceptor
BDNF	Brain-derived neurotrophic factor
Ca^{2+}	Calcium
CaBP	Ca^{2+} binding protein
CaM	Calmodulin
CaMKII	Ca^{2+}/calmodulin-dependent protein kinase II
cAMP	Cyclic adenosine monophosphate
CaS	Ca^{2+} sensor
CB	Cannabinoid receptor

G. W. Zamponi, N. Weiss (eds.), *Voltage-Gated Calcium Channels*,
https://doi.org/10.1007/978-3-031-08881-0_7

cGMP	Cyclic guanosine monophosphate	PMA	Phorbol-myristate 13-acetate
CNS	Central nervous system	PP	protein phosphatase
CRF	Corticotropin-releasing factor	PSD-95	Postsynaptic density protein-95
CTX	Cholera toxin	PTK	Protein tyrosine kinase
D1R	Dopamine D1 receptor 2	PTX	Pertussis toxin
D2R	Dopamine D2 receptor	RGS	Regulators of G protein signaling
DAG	Diacylglycerol	ROCK	Rho-associated protein kinase
DRG	Dorsal root ganglion	SNAP-25	Synaptosomal-associated protein
ER	Endoplasmatic reticulum	SNAREs	Soluble N-ethylmaleimide-sensitive
FGF	Fibroblast growth factor		factor attachment protein receptors
FGFR1	Fibroblast growth factor type 1 receptor	SST	Somatostatin receptor
		Thr	Threonine
FRET	Fluorescence resonance energy transfer	VGCCs	Voltage-gated calcium channels
		VILIP-2	Visinin-like protein-2
GABA	Gamma-aminobutyric acid		
GDP	Guanosine diphosphate		
Ghrelin	Growth hormone release inducing		
GHS-R1a	Growth hormone secretagogue receptor 1a		
GID	G protein interaction domain		
GPCRs	G protein-coupled receptors		
GRK	G protein coupled receptor kinase		
GTP	Guanosine triphosphate		
HEK	Human embryonic kidney		
I_{Ca}	Ca^{2+} current		
IGF-1	Insulin-like growth factor 1		
IP$_3$	Inositol triphosphate		
Kir	Inward rectifying potassium channel		
LPA	Lysophosphatidic acid		
LTD	Long term depression		
LTP	Long-term potentiation		
M1R	Muscarinic receptor 1		
MAPK	Mitogen-activated protein kinase		
mGluR	Metabotropic glutamate receptor		
NA	noradrenaline		
Na	sodium		
NCS-1	Neural Ca^{2+} sensor-1		
NO	Nitric oxide		
OAG	1,2-Oleoylacetyl-glycerol		
PDE	Phosphodiesterase		
PI$_3$K	Phosphatidylinositol 3-kinase		
PIP$_2$	Phosphatidylinositol 4,5-bisphosphate		
PKA	Protein kinase A		
PKC	Protein kinase C		
PKG	Protein kinase G		
PKI	Protein kinase A inhibitor		
PLC	Phospholipase C		

Introduction

Voltage-gated calcium channels (VGCCs) have three major family members: Ca$_v$1, Ca$_v$2, and Ca$_v$3. High-voltage-activated Ca$_v$1 channels also known as L-type channels for their long, lasting activation come in four flavors distinguished according to their pore-forming α$_1$ subunit, for example, Ca$_v$1.1 expressed in skeletal muscle, Ca$_v$1.2 expressed in cardiac, smooth muscle, and neurons, Ca$_v$1.3 expressed in the sinoatrial node, cochlear hairs cells and neurons, and Ca$_v$1.4 expressed in the retina. Similarly, Ca$_v$2 are high-voltage-activated Ca^{2+} channels consisting of three family members, Ca$_v$2.1 (P/Q-type), Ca$_v$2.2 (N-type), and Ca$_v$2.3 (R-type) and are important for neurotransmitter release and synaptic integration in neurons. Lastly, the low-voltage-activated Ca$_v$3 or T-type channels for transient channel opening are expressed in a variety of cells in the central nervous system (CNS), peripheral nervous system, sperm, heart, smooth muscle, bone, and endocrine system. Each channel comprises a Ca$_v$α subunit consisting of four homologous domains (I-IV) each of which has six transmembrane regions (S1-S6) connected by cytoplasmic loops (Fig. 3) and associated auxiliary subunits β and α$_2$δ. Auxiliary subunits support the trafficking and stabilization of Ca$_v$α subunits at the plasma membrane in addition to the controlling its gating kinetics (Catterall, 2000). These intracellular, cytoplasmic loops serve as binding regions for the auxiliary subunit β or G proteins and their second messengers for further regulation of VGCCs. Seven transmembrane GPCRs regulate Ca_v1-3 activity either directly through G protein binding or indirectly through the recruitment of downstream second messenger cascades. GPCRs bind ligands to transform extracellular signals into a cascade of intracellular responses via coupling to heterotrimeric G proteins (for review: Masuho et al., 2021; Tennakoon et al., 2021). G proteins are composed of an α, β, and γ subunit, each of which is encoded by 16, 5, and 12 genes, respectively (Oldham & Hamm, 2008). The G_α subunit has four major family members, G_s, $G_{i/o}$, $G_{q/11}$, and G_{12}, and G_t found in the retina. Ligand binding to the extracellular binding domain on a GPCR catalyzes the exchange of GDP to GTP on G_α initiating a conformational change and dissociation of G_α from $G_{\beta\gamma}$. G_α bound to GTP is free to activate second messengers including the cAMP, DAG, IP_3, and Ca^{2+}, as well as $G_{\beta\gamma}$ to regulate Ca_v channels. Termination and reassociation of heterotrimeric G proteins occur when GTP is hydrolyzed to GDP by the intrinsic GTPase activity of G_α or regulators of G protein signaling (RGS) proteins. Termination can also occur in the presence of agonists when GPCRs uncouple to their G proteins due to receptor desensitization or endocytic removal of GPCRs by G protein-coupled receptor kinase (GRK) (Proft & Weiss, 2015). In some incidences, internalization of Ca_v channels from the plasma membrane due to their direct interaction with GPCRs occurs during desensitization and internalization (Altier et al., 2006; Evans et al., 2010; Kitano et al., 2003). Since the internalization of GPCR-Ca_v complexes does not reduce I_{Ca}, the functional significance of these mechanisms still remains to be determined (Murali et al., 2012). Dysfunction or misregulation of VGCCs can lead to cardiac problems, chronic/acute pain, epilepsy, ataxia, migraines, neuropsychiatric diseases, retinal disorders, or cognitive deficits. In this chapter, we will focus on the G protein regulation involved in the regulation of Ca_v1 (section "Modulation of Ca_v1 channels by GPCRs and second messengers"), Ca_v2 (section "Modulation of Ca_v2 channels by GPCRs and second messengers"), and

Ca_v3 (section "Modulation of Ca_v3 channels by GPCRs and second messengers") by G proteins and their downstream effectors.

Modulation of Ca_v1 Channels by GPCRs and Second Messengers

β-Adrenergic Modulation of L-Type Channels: Activation of the G_s Signaling Pathway

Activation of the sympathetic nervous system during fight-or-flight response leads to the release of noradrenaline (NA) from postganglionic neurons onto the target tissue causing for example the increase in contractile force of skeletal muscles, acceleration of the heart rate, or constriction of blood vessels. Noradrenaline is also a neurotransmitter in the brain released by specialized neurons in the locus coeruleus in the pons. Here, NA is involved in regulating attention, arousal, anxiety, pain, mood, brain metabolism, and learning and memory (Bear et al., 2015). Noradrenaline mediates its effect via three groups of adrenoceptors (AR). α_1-AR couple to $G_{q/11}$, α_2-AR to $G_{i/o}$ and β-AR to G_s/G_i pathways leading to different modulatory actions within the cell. L-type Ca^{2+} channels in skeletal muscle, heart, and brain are in particular modulated by β_1 and β_2-ARs, coupling to the G_s signaling pathway (β_1-AR) and G_s/G_i pathways (β_2-AR) in highly specific subcellular signaling domains (Harvey & Hell, 2013; Xiao et al., 2004).

The skeletal muscle L-type channel Ca_v1.1 mediates the excitation-contraction coupling via direct interaction with the ryanodine-sensitive calcium release channel (RyR1) located in the sarcoplasmic reticulum (Catterall, 2000). Ca_v1.1 channels are slowly activating leading to a sustained Ca^{2+} influx. This Ca^{2+} influx does not directly participate in the excitation-contraction coupling but contributes to the increase in contractile force during high-frequency stimulation of skeletal muscle fibers, which for example can be increased by fight-or-flight responses involving the sympathetic nervous system. The increase in contractile force involves the enhanced trans-

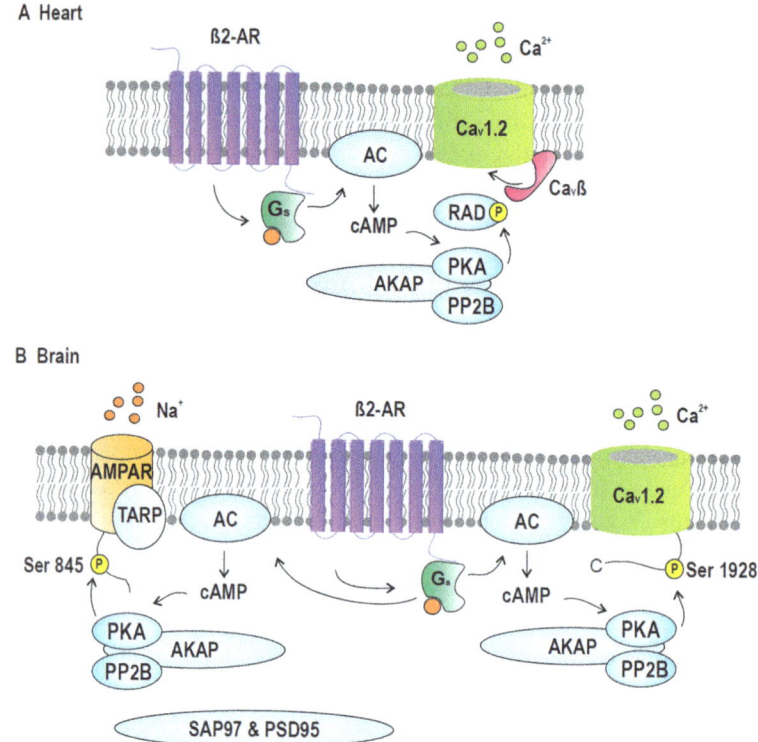

Fig. 1 Differences in β-adrenergic modulation of $Ca_v1.2$ in heart and brain. Activation of β2-AR in the heart and brain leads to the activation of the G_s protein and adenylyl cyclase (AC) leading to the production of cyclic adenosine monophosphate (cAMP). Cyclic AMP activates protein kinase A (PKA). PKA and the protein phosphatases 2B (PP2B) are anchored to the channel via A-kinase anchoring proteins (AKAP). (**a**) In cardiomyocytes, PKA phosphorylates the monomeric, small Ras-like G protein RAD leading to a decrease in the affinity to the $Ca_v\beta_2$ subunit and relieves constitutive inhibition of the channel by increasing the channel open probability and therefore calcium (Ca^{2+}) influx and an increase in heart rate. (**b**) In postsynaptic spines in neurons PKA phosphorylates serine (Ser)[1928] of the pore-forming $Ca_v1.2\alpha$ subunit and Ser845 of the colocalized α-amino-3-hydroxy-5-methyl-4-isoxazolepropionic acid receptor (AMPAR) leading to an increase in open probability of both channels. The increase in AMPAR mediated sodium (Na^+) and L-type channels mediated Ca^{2+} influx has been suggested to underly the potentiation of synaptic strength in prolonged theta tetanus long-term potentiation (PTT-LTP). *PSD95* postsynaptic density protein 95, *Ras* rat sarcoma virus, *SAP97* synapse associated protein 97, *TARP* transmembrane AMPAR regulatory protein

sarcolemmal Ca^{2+} influx through phosphorylated $Ca_v1.1$ (Cairns & Borrani, 2015). Upregulation of $Ca_v1.1$ current is mediated via β-adrenergic stimulation of cyclic adenosine monophosphate (cAMP)-dependent protein kinase A (PKA). Activation of PKA leads to the phosphorylation of a specific serine residue (S1575) located within the distal and proximal C-terminal domains of the pore-forming $Ca_v\alpha_1$ subunit. This phosphorylation site is also conserved in cardiac $Ca_v1.2$ channels. It has therefore been suggested that phosphorylation of this site is directly involved in $Ca_v1.1$ regulation by the sympathetic nervous system (Emrick et al., 2010). Phosphorylation by PKA and interaction with the cytoskeleton is required for frequency- and voltage-dependent potentiation of $Ca_v1.1$ activity (Murphy & Scott, 1998; Gray et al., 1998; Johnson et al., 2005). PKA is anchored to the $Ca_v1.1$ by an A-kinase anchoring protein (AKAP-15), which directly binds to $Ca_v1.1$ via a leucine zipper motif (Hulme et al., 2002) (Fig. 1).

$Ca_v1.2$ and $Ca_v1.3$ are expressed in the heart, brain, and endocrine system, where they regulate heart rate, gene expression, neuronal excitability, and plasticity such as long-term potentiation (LTP) and long-term depression (LTD) and hormone secretion (Zamponi et al., 2015). Cardiac and neuronal L-type channel activity is augmented during the flight-or-fight response, excitement and exercise, mediated by NA and specifically β-ARs (Harvey & Hell, 2013; Xiao et al., 2004). Modulation by β-ARs leads to an activation of PKA and to an increase in channel activity in native cardiac cells and to a lesser extent in heterologous expression systems. G protein-coupled receptors (GPCRs) mediated Ca_v1 modulation is organized in specific signaling complexes containing all proteins of the signaling cascade for up and down-regulation of Ca_v1 activity. Therefore, $Ca_v1.2$ signaling complexes contain β2-ARs and $Ca_v1.2$ (including $Ca_v1.2$ and $Ca_v\beta_{2b}$), adenylyl cyclase (AC5/6), the guanine-nucleotide binding protein (G_s protein), A-kinase anchoring protein (AKAP5/7), phosphodiesterases (PDE) 4B and PDE4D, protein phosphatases (PP) 2A and PP2B and PKA (Fig. 1) (Harvey & Hell, 2013; Sanderson & Dell'Acqua, 2011). The C-terminal tail of most of the cardiac $Ca_v1.2$ protein is post-translationally proteolytically cleaved and noncovalently attached via two putative α-helices with the remaining C-terminus. The interaction between the two C-terminal fragments inhibits channel activity. The signaling complex also contains AKAP for PKA anchoring and PP2A and PP2B/calcineurin for dephoshorylation. Therefore, phosphorylation and dephosphorylation can spatially be regulated (Zamponi et al., 2015). Surprisingly, PKA-mediated phosphorylation does not involve the channel ($Ca_v1.2$ or $Ca_v\beta_2$) subunits (Roybal et al., 2020). This has been shown in knock-in mice ($Ca_v1.2$ [Ser1928Ala]) and transgenic mice with alanine substitution of all conserved serine (Ser) and threonine (Thr) residues suggested to be phosphorylated by PKA within the intracellular domains of $Ca_v1.2$. These mutations do not show changes in basal L-type current nor regulation by PKA or β-ARs in cardiomyocytes (Katchman et al., 2017; Lemke et al., 2008; Roybal et al., 2020). Using a quantitative, proteomic approach it could be shown that

the augmentation of $Ca_v1.2$ channels is mediated by phosphorylation of a Ser residue on the monomeric G protein Rad, a member of the Rad and Gem/Kir (inward rectifying potassium channel), Ras-related GTP-binding protein family of GTP binding proteins acting as a Ca^{2+} channel inhibitor. Phosphorylation of Rad by PKA during β-adrenergic stimulation decreases its affinity to the $Ca_v\beta$ subunit and relieves constitutive inhibition by increasing channel open probability (Liu et al., 2020). β-adrenergic stimulation requires the binding of β subunits to the linker region in loop I-II of $Ca_v1.2$ (Papa et al., 2021; Roybal et al., 2020). The Rad-dependent modulation seems to be a general mechanism for PKA/$Ca_v\beta$ subunit-mediated activation of other VGCCs such as $Ca_v1.3$ and $Ca_v2.2$ (Liu et al., 2020). A Rad-independent modulation of $Ca_v1.2$ activity has also been identified and accounts for about 20% of the modulation in heterologous expression systems. The Rad-independent modulation of $Ca_v1.2$ does not depend on the $Ca_v\beta$ subunit and involves the cleaved C-terminus inhibitory domain and the initial segment (20 amino acids) of the N-terminus, which acts as an inhibitory module on $Ca_v1.2$ channels (Katz et al., 2021).

In the brain, NA release is associated with arousal and learning in emotional situations, which is correlated with specific changes in synaptic plasticity. NA activates also β1- and β2-AR coupling to the G_s/PKA signaling pathway, which can modulate L-type Ca^{2+} currents (I_{Ca}) contributing to changes in synaptic plasticity. For example, in neurons β2-AR activation induces a certain form of prolonged theta tetanus LTP involving $Ca_v1.2$ channels (Patriarchi et al., 2016). Like in the heart, β2-AR and $Ca_v1.2$ are assembled into a signaling complex containing AC, G_s protein, AKAPs and PKA, phosphatases, but are also colocalized with PSD-95 (postsynaptic density protein-95) and auxiliary transmembrane AMPA receptor regulatory protein subunits to α-amino-3-hydroxy-5-methyl-4-isoxazolepropionic acid (AMPA) receptors (Patriarchi et al., 2018; Sanderson & Dell'Acqua, 2011). Within the signaling complex, β2-AR binds to amino acids 1923–1942 of the pore-forming $Ca_v\alpha_1$ subunit, which is required for receptor-mediated channel regulation (Patriarchi et al., 2016). Activation of the β2-AR-Ca_v1.2 signaling complex activates PKA, which phosphorylates Ser^{1928} of the pore-forming $Ca_v\alpha_1$ subunit. The functional PKA-mediated phosphorylation is specific for the regulation of Ca_v1.2 in neurons and vascular smooth muscle cells but not in the heart (see above and Fig. 1) (Patriarchi et al., 2018). The PKA-mediated phosphorylation leads to the displacement/disruption of the interaction between β2-AR and the channel subunit causing the signaling complex to be refractory for a certain period of time (Patriarchi et al., 2016). The displacement mechanism leading to the desensitization of β2-AR-L-type channel regulation has been suggested to be a negative feedback loop for cellular overexcitation (Patriarchi et al., 2016).

G_q-Coupled GPCR Modulation of L-Type Channels

Activation of $G_{q/11}$ coupled GPCRs leads to the activation of phospholipase C (PLC), the breakdown of phosphatidylinositol 4,5-bisphosphate (PIP_2) into inositol triphosphate (IP_3) and diacylglycerol (DAG), IP_3 mediated Ca^{2+} release from internal stores and Ca^{2+}/DAG dependent activation of protein kinase C (PKC) (Wettschureck & Offermanns, 2005). In most studies, activation of the G_q pathway leads to the inhibition of L-type Ca^{2+} channels and involves different molecules of the $G_{q/11}$ signaling cascade.

Medium spiny neurons of the striatum express Ca_v1.3 L-type channels. The long splice variant of the channels containing a cytoplasmic Src homology 3 and PDZ binding domain are inhibited via muscarinic M1 receptor (M1R) and dopamine D2 receptor (D2R) resulting in the reduction of excitability (Hernádez-López et al., 2000; Howe & Surmeier, 1995; Olson et al., 2005; Stanika et al., 2015). D2R mediated inhibition involves $G_{\beta\gamma}$-mediated activation of PLC, the release of Ca^{2+} from internal stores and the Ca^{2+}-dependent phosphatase calcineurin (Hernádez-López et al., 2000). D2R- and M1R-mediated inhibition is also dependent on PDZ binding and

the interaction between Shank and Homer suggesting the functional formation of D2R/M1R/mGluR (metabotropic glutamate receptor)-Ca_v1.3 signaling complexes at corticostriatal synapses (Olson et al., 2005; Stanika et al., 2015). Additional PDZ protein interaction partner such as densin and erbin also interact with the C-terminus of Ca_v1.3 and modulate the Ca^{2+}-dependent facilitation of the channel (Fig. 2) (Calin-Jageman et al., 2007; Jenkins et al., 2010). In superior cervical ganglion cells G_q/PLC coupled M1R also inhibit L-type channels. Inhibition of L-type channels in superior cervical ganglion cells (most likely Ca_v1.3) is mediated by arachidonic acid (AA) produced by the cytosolic phospholipase A2 from phospholipids. The amount of M1R and AA mediated Ca_v1.3 inhibition in HEK293 cells is dependent on the $Ca_v\beta$ subunit, with pronounced inhibition in the presence of β_{1b}, β_3, and β_4 and moderate inhibition with β_{2a} (Mandy & Rittenhouse, 2009; Roberts-Crowley & Rittenhouse, 2018).

M1R activation also leads to the inhibition of Ca_v1 (Ca_v1.2 and Ca_v1.3) via PLC-mediated breakdown of PIP_2 (Hille et al., 2015; Suh et al., 2010). The M1R and also angiotensin II inhibition is dependent on $Ca_v\beta$ subunits, where inhibition is smaller in the presence of β_{2a} subunits. This effect is dependent on membrane anchoring of the β_{2a} subunit via its palmitoylation sites and AA (Hermosilla et al., 2011; Suh et al., 2012). It has been suggested that PIP_2 stabilizes an active channel conformation by tethering channel domains to the plasma membrane and that AA can occupy the fatty acid-binding site of PIP_2 and therefore interfering with channel stabilization. Thus the breakdown of PIP_2 and competition for binding with AA subunits may destabilize the active states of the channel (Fig. 2) (Zamponi et al., 2015). However, a polybasic plasma binding motif (a putative phosphoinositide binding site) has been identified in the loop I-II of Ca_v1.2. This motif forms a positively charged α-helix facing the negatively charged lipid phosphates of the inner site of the plasma membrane. This binding is disrupted by PLC, which causes phosphoinositide breakdown. Surprisingly, elimina-

A Brain

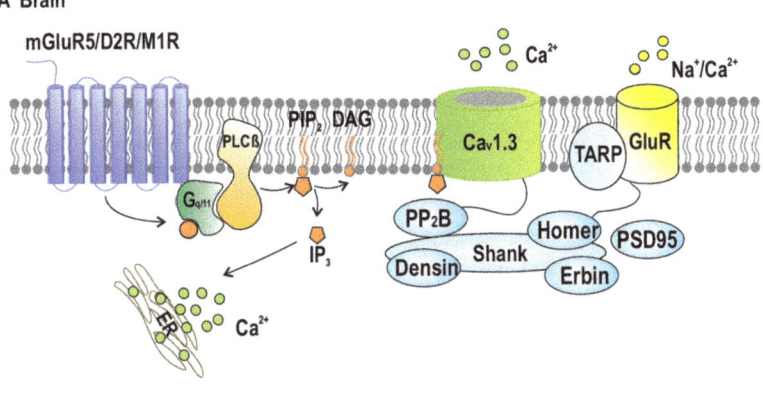

B Mechanism of G$_{q/11}$ mediated Inhibition

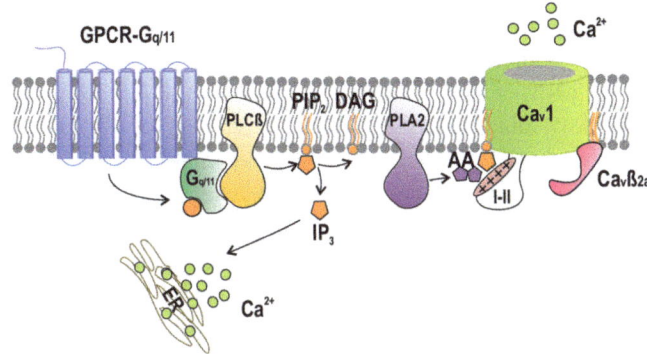

Fig. 2 G$_{q/11}$ mediated inhibition of Ca$_v$1.2 and Ca$_v$1.3 channels in heart, brain, and heterologous expression systems. (**a**) G$_{q/11}$ mediated inhibition of L-type channels in dendritic spines. The long splice variant of Ca$_v$1.3 channels contains a cytoplasmic sarcoma (Src) homology 3 and PDZ binding domain. G$_{q/11}$ mediated inhibition by dopamine D2 receptor (D2R), muscarinic M1 receptor (M1R) or metabotropic glutamate receptor 5 (mGluR5) involves G$_{\beta\gamma}$-mediated activation of phosphoplipase C (PLC) β, the release of calcium (Ca^{2+}) from internal endoplasmic reticulum (ER) stores and the Ca^{2+}-dependent phosphatase calcineurin 2B (PP2B). D2R and M1R mediated inhibition is dependent on PDZ binding and the interaction between Shank and Homer. Additional PDZ protein interaction partners such as densin and erbin also interact with the C-terminus of Ca$_v$1.3. The Ca$_v$1.3 signalosome allows the precise spatio-temporal control of Ca^{2+} influx next to ionotropic glutamate receptors (GluRs) into dendritic spines through G$_{q/11}$ coupled G protein coupled receptors (GPCRs). (Adapted from Olson et al., 2005; Stanika et al., 2015) (**b**) Molecular mechanisms involved in G$_{q/11}$-mediated L-type channel inhibition. M1R-mediated inhibition of L-type channels involves the breakdown of phosphatidylinositol 4,5-bisphosphate (PIP$_2$) and arachidonic acid (AA) produced by the cytosolic phospholipase A2 (PLA2) from phospholipids. This inhibition is dependent on the Ca$_v$β subunits, where inhibition is smaller in the presence of membrane-anchored Ca$_v$β$_{2a}$ subunits. It has been suggested that PIP$_2$ stabilizes an active channel conformation by tethering channel domains to the plasma membrane while AA occupies the fatty acid-binding site of phosphatidylinositol 4,5-phosphate (PIP$_2$) and therefore interfering with channel stabilization. According to this model breakdown of PIP$_2$ and competition for binding with AA subunits may destabilize the active states of the channel. Another mechanism for G$_{q/11}$-mediated modulation of L-type channels involves a polybasic plasma binding motif within loop I-II linker of L-type channels, consisting of four positively charged arginines. This motif forms a positively charged α-helix facing the negatively charged lipid phosphates of for example PIP$_2$ at the inner side of the plasma membrane. This binding is disrupted by PLC together with the PLC activator Ca^{2+} and is supposed to increase channel activity. (Kaur et al., 2015). *DAG* diacylglycerol, *IP$_3$* inositol triphosphate, *PSD95* postsynaptic density protein 95, *TARP* transmembrane AMPAR regulatory protein

tion of the plasma membrane binding leads to facilitated opening of the channels and weaker inhibition by PLC (Fig. 2) (Kaur et al., 2015).

Neuronal $Ca_v1.2$ channels synergistically work together with mGluRs to induce LTD in hippocampal neurons (Bolshakov & Siegelbaum, 1994). Metabotropic glutamate receptor-dependent LTD involves AMPA receptor mediated-depolarization induced Ca^{2+} influx through L-type channels and facilitation of the $mGluR5-G_{q/11}/PLC$ signaling cascade (Fig. 2) (Kim et al., 2015). In addition, it has been shown that activation of mGluR5, Ca^{2+} entry through postsynaptic $Ca_v1.2$ channels, the release of Ca^{2+} from intracellular stores and postsynaptic NO (nitric oxide) is involved in the induction of presynaptic LTP (Zamponi et al., 2015).

PKC has also been shown to modulate L-type channels in the heart, vascular smooth muscles, and brain (Weiss & Dascal, 2015). In general, activation of the $G_{q/11}$-PKC pathway via α1AR, endothelin, or angiotensin receptors in the heart and smooth muscle leads to the increased force of contraction and Ca^{2+} influx. For example, activation of the $G_{q/11}$ pathways by angiotensin II, phorbol esters or DAG in cardiomyocytes or smooth muscle cells enhance L-type currents involving the activation of PKC. However, PK-mediated inhibition or enhancement followed by inhibition of I_{Ca} has also been described, which may reflect cell-type specific localization, pathway, or isoform-specific differences (Harvey & Hell, 2013; Weiss & Dascal, 2015).

$G_{i/o}$-Coupled GPCR Modulation of L-Type Channels

The primary output of the sympathetic nervous system is the chromaffin cells of the adrenal medulla, which express $Ca_v1.2$ and $Ca_v1.3$ channels (Mahapatra et al., 2012; Polo-Parada et al., 2006). Under stress conditions, L-type channels are in particular recruited for triggering Ca^{2+}-dependent catecholamine release (Polo-Parada et al., 2006), which is under the voltage-independent control of auto- and heteroreceptors involving adrenergic, opioid, and purinergic

GPCRs (Mahapatra et al., 2012). Activation of the PTX (pertussis toxin) sensitive $G_{i/o}$ pathways causes the fast inhibition of L-type channels, while slow potentiation of L-type currents involves PKA-mediated phosphorylation (Cesetti et al., 2003; Mahapatra et al., 2012; Polo-Parada et al., 2006; Vandael et al., 2013). L-type channels in chromaffin cells are also tonically inhibited by NO via cGMP-dependent protein kinases (PKG) (Carabelli et al., 2002; Vandael et al., 2013). NO-cGMP-PKG signaling pathway also inhibits L-type channels in insulin-secreting cell lines. PKG phosphorylation sites in $Ca_v1.3$ have been described to be located in the intracellular loop II-III of the $Ca_v\alpha_1$ subunit (Sandoval et al., 2017).

GPCR-Mediated Modulation of $Ca_v1.4$ in the Retina

$Ca_v1.4$ L-type channels are primarily expressed in rod and cone photoreceptors in the retina (Heidelberger et al., 2005; Masland, 2012), but $Ca_v1.4$ splice variants are also found in B- and T-lymphocytes and probably other immune cells (note, over 20 different splice variants have been identified with different biophysical properties) (Waldner et al., 2018). In photoreceptors, $Ca_v1.4$ is responsible for triggering presynaptic glutamate release at the ribbon synapse at least in mammals (Heidelberger et al., 2005; Masland, 2012). Synaptic ribbons are specialized structures for tethering and releasing a large number of synaptic vesicles during sustained Ca^{2+} influx (Waldner et al., 2018). In non-mammalian vertebrates such as salamander and chicken photoreceptors, $Ca_v1.3$ seems to be the predominant L-type channel isoform (Waldner et al., 2018). Since various studies on L-type channel modulation in photoreceptor cells have been performed in the salamander, we will discuss L-type channel modulation in general and specifically identify when $Ca_v1.3$ was studied as the GPCR target. In vivo, the pore-forming $Ca_v\alpha_{1F}$ subunit is associated with a specific splice variant of $Ca_v\beta_2$ (β_{2X13}) and $\alpha_2\delta_4$ subunits, which contribute to the unique biophysical properties of $Ca_v1.4$ in rods and cones, i.e. minimal Ca^{2+} and voltage-dependent inactivation (Thoreson, 2021; Waldner et al., 2018). $Ca_v1.4$ is also directly associated with the Ca^{2+} binding protein (CaBP4)-4 at the C-terminus. Binding leads to a shift in the voltage-dependence of activation to more hyperpolarized potentials in heterologous expression systems, which is similar to what is seen in native cells (Haeseleer et al., 2004). The shift in the voltage-dependent activation guarantees the constant Ca^{2+} influx necessary for sustained glutamate release during the dark at relatively negative membrane potentials of about -40 mV.

Calcium currents in photoreceptors are modulated by various transmitters such as dopamine, adenosine, cannabinoids, somatostatin, insulin, NO as well as retinoids and polyunsaturated fats (Heidelberger et al., 2005). Dopamine in the retina is released by a specific set of amacrine cells. Dopamine release, which activates dopamine D1 receptors (D1Rs) and D2Rs, shows a circadian fluctuation and is higher during the day than at night (Witkovsky, 2004). Dopamine mediates its effects on calcium currents (I_{Ca}) in photoreceptors via dopamine D2-like receptors (D2 or dopamine D4 receptor). Activation of dopamine D2-like receptors in rods increases I_{Ca} but inhibits I_{Ca} in the largest cone population in the salamander retina. The enhancement of I_{Ca} in rods leads to the suppression of glutamate release from rods. It has been suggested that a negative feedback loop between I_{Ca} and Ca^{2+}-activated Cl^- conductance is responsible for this effect. D2-like receptors are coupled to the PTX sensitive $G_{i/o}$ pathway and inhibit AC, the production of cAMP and therefore reducing the activity of PKA. PKA phosphorylation of $Ca_v1.4$ modulates the activity of the channel. Phosphorylation occurs within the inhibitor of the Ca^{2+}-dependent inactivation motif which promotes the binding of calmodulin (CaM) to the channel, which increases the channel open probability and voltage-dependent inactivation (Sang et al., 2016). According to this mechanism, regulation of the activity of PKA by dopaminergic receptors or more generally by $G_s/G_{i/o}$ coupled GPCRs within photoreceptors will upregulate L-type channel activity when PKA is activated and downregulate channel activity

when PKA is inhibited. Therefore, the integration of $G_{i/o}$ and G_s pathways on PKA activity in photoreceptors will determine Ca^{2+} influx and glutamate release in rods and cones.

Several other GPCRs such as the cannabinoid receptor (CB) 1 have also been described in rod and cone terminals. Like D2-like receptors, CB1 couples to the PTX sensitive $G_{i/o}$ protein pathway, and activation enhances I_{Ca} in rods but inhibits I_{Ca} in cones of the salamander retina. In the zebrafish retina, CB1 leads to inhibition of I_{Ca} at high and enhancement of I_{Ca} at low agonist concentrations. The enhancement of I_{Ca} is blocked by PKA inhibitor and cholera toxin (CTX) but not by PTX, suggesting the involvement of G_s proteins at low agonist concentrations. The $G_{i/o}$ coupled somatostatin receptor (SST) 2a is also expressed in rod and cone terminals, but has opposite effects on rod and cone I_{Ca} in comparison to D2 and CB1 receptors. Somatostatin receptor 2a activation leads to inhibition of I_{Ca} in rods and increase in cones. Moreover, insulin inhibits L-type Ca^{2+} channels in rods. Insulin activates a class II receptor tyrosine kinase and modulates intracellular signals via various adaptor proteins. Lastly, adenosine receptors are also localized at photoreceptor cells. Application of adenosine and/or activation of A2A-like adenosine receptors, which is a G_s-coupled GPCR stimulating AC and PKA, leads to the inhibition of I_{Ca} at the photoreceptor synapse in rods and cones of salamander retinae (Heidelberger et al., 2005; Stella et al., 2007).

Nitric oxide synthase is expressed in the photoreceptor inner segments. Nitric oxide seems to inhibit I_{Ca} and cone output but increases I_{Ca} and rod output. It has been suggested that the reciprocal action of NO on rods versus cones is involved in shifting to rod-dominated vision during low light conditions. Photoreceptor (rod and cone opsin) activation leads to the activation of the G protein transducin and phospholipase A2, which also causes the production of the polyunsaturated fatty acids such as AA and docosahexaenoic acid. In addition, light also leads to the production of all-trans-retinal (retinoid) within the photoreceptors. Retinoids and polyunsaturated fatty acids have the potential to inhibit I_{Ca} and therefore rod and cone output, which may contribute to light/activity-dependent adaptation processes.

In summary, modulation of $Ca_v1.4$ channels by GPCRs has not been very well characterized. It is surprising that GPCR effects on I_{Ca} are opposite in rods versus cones and that the classical G protein coupling and signals described for the GPCRs does not explain the modulation of $Ca_v1.4$.

Modulation of Ca_v2 Channels by GPCRs and Second Messengers

Mechanisms $G_{i/o}$ Mediated Inhibition: Fast, Direct, Voltage-Dependent

Ca_v2 channels play an important role in neurotransmitter release and short-term plasticity at synapses via their second messenger Ca^{2+} in the CNS. Spatiotemporal regulation of Ca^{2+} release is in part controlled via the inhibition of Ca_v2 channels by GPCRs coupling to the $G_{i/o}$ and G_q pathways. In 1978 Dunlap and Fischbach found the first evidence that the neurotransmitter NA decreased action potential (AP) duration and I_{Ca} amplitude in sensory neurons (Dunlap & Fischbach, 1978). Since this initial finding more than 20 neurotransmitters including NA, somatostatin, acetylcholine, and gamma-aminobutyric acid (GABA) acting through GPCRs have been demonstrated to predominantly inhibit $Ca_v2.1$ and $Ca_v2.2$ channels and to a lesser extent $Ca_v2.3$ (Currie, 2010; Proft & Weiss, 2015; Zamponi & Currie, 2013). Ligand binding to GPCRs dissociates $G_{\beta\gamma}$ from G_α to allow direct binding of $G_{\beta\gamma}$ to $Ca_v2\alpha_1$ subunit, leading to conformational changes which alter four basic features of its whole cell voltage-dependence: [1]peak amplitudes of I_{Ca} are decreased, and this decrease is reduced at depolarizing membrane potentials; [2]activation and inactivation kinetics are slowed; [3]voltage-dependence of activation is shifted to more depolarizing potentials; [4]conditioning prepulse to non-physiological depolarized potentials relieve most of the inhibition and normalize channel kinetics (also known as prepulse facilitation), and physiological depolarizations or high-frequency

trains of AP recover the channel kinetics to a lesser extent (Currie, 2010; Zamponi & Currie, 2013). The mechanism of current inhibition is proposed to be attributed to one $G_{\beta\gamma}$ dimer binding directly to one $Ca_v2\alpha_1$ subunit and recovery to the dissociation of the $Ca_v2\alpha_1$ subunit to the $G_{\beta\gamma}$ dimer, presumably due to a conformational shift in $Ca_v2\alpha_1$ subunit which is unfavorable to $G_{\beta\gamma}$ binding in the "willing" state (Herlitze et al., 1996; Ikeda, 1996; Zamponi & Snutch, 1998). To describe the mechanism of G protein inhibition of Ca_v2 channels, Bruce Bean (1989) proposed that Ca_v2 channels are either in a "willing" or "reluctant" gating state, where the "reluctant" state is stabilized by the activated G protein and can be relieved from G protein inhibition by depolarizing membrane potentials to return to the "willing" state (Fig. 4, Jones & Elmslie, 1997). $G_{\beta\gamma}$ binding to the Ca_v2 channel appears to slow or disrupt the voltage sensor movement and thereby causing a conformation shift in the channel to the more "reluctant" state (Hernández-Ochoa et al., 2007a, b; Jones & Elmslie, 1997; Rebolledo-Antúnez et al., 2009). Since the voltage-dependent kinetics of relief and inhibition correlates with binding and unbinding kinetics of $G_{\beta\gamma}$, increasing $G_{\beta\gamma}$ levels in the cell does not change the rate of relief during a prepulse but does accelerate the rate of reinhibition after the prepulse. Moreover, inhibition does not involve a diffusible second messenger (membrane delimited) and is PTX-sensitive indicating a direct interaction between Ca_v2 channel and $G_{\beta\gamma}$ via a $G_{i/o}$ mediated pathway (Bernheim et al., 1991; Elmslie, 2003; Elmslie & Jones, 1994; Forscher et al., 1986; Herlitze et al., 1996; Hille, 1994; Rosenthal et al., 1990; Zamponi & Currie, 2013; Zamponi & Snutch, 1998).

Ca_v2 Channel Interaction Sites with $G_{\beta\gamma}$ Subunits

Three regions in the $Ca_v2\alpha_1$ subunit, the intracellular loop I-II, N-terminus and C-terminus have been identified to interact with the $G_{\beta\gamma}$ heterodimer (Fig. 3). The $Ca_v2\alpha_1$ subunit intracellular loop I-II contains two G protein interaction sites (Pragnell et al., 1994; De Waard et al., 1997). The first site partially overlaps with the main interac-

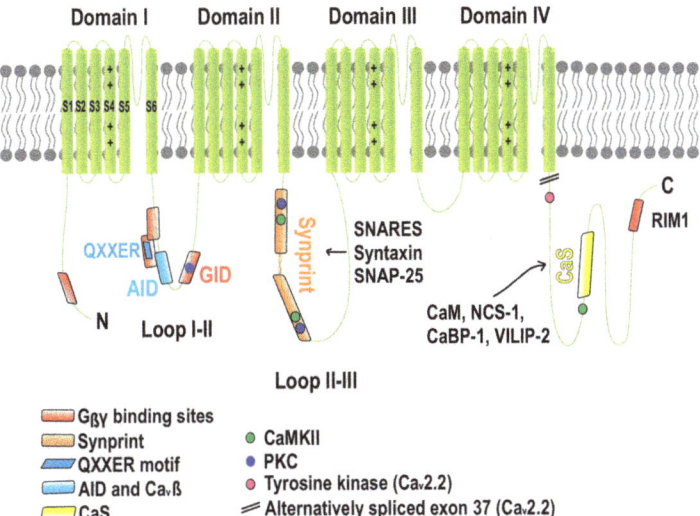

Fig. 3 Schematic topology of pore-forming $Ca_v2\alpha_1$ subunit. The $Ca_v2\alpha_1$ subunit is composed of four homologous domains (I-IV). Each domain consists of 6 α helical, transmembrane segments (S1-S6), where S4 is the charged (+), voltage-sensing domain and S5-S6 form the pore. The N- and C-termini and the cytoplasmic loops are important for interactions with regulatory proteins such as their auxiliary $Ca_v\beta$ subunit that binds at the AID (α interaction domain) within loop I-II. The AID overlaps with $G_{\beta\gamma}$ binding domain (depicted as red boxes) containing a QXXER consensus motif within loop I-II. Additional $G_{\beta\gamma}$ binding sites are located in the $Ca_v2\alpha_1$ loop I-II (GID, G protein interaction domain), N- and C-termini. Synaptic-associated proteins mediate Ca_v2 channel activity via the synprint region located in loop II-III, and calcium-binding proteins through the calcium sensor (CaS) protein regulatory site in the C-terminus. These cytoplasmic regions are also targeted by downstream kinases such as protein kinase C (PKC), Ca/calmodulin-dependent protein kinase II (CaMKII), and tyrosine kinases. Alternative splicing of the C-terminus (exon 37) of $Ca_v2.2$ channel is responsible for G protein-coupled receptor (GPCR) modulation. *CaBP-1* Ca^{2+} binding protein-1, *CaM* calmodulin, *GID* G protein interaction domain, *NCS-1* neural Ca^{2+} sensor-1, *RIM* rab3-interacting molecule 1, *VILIP-2* visinin-like protein-2

tion region for $Ca_v\beta$ subunits also known as α interaction domain (AID) and contains the same QXXER consensus motif where $G_{\beta\gamma}$ binds AC and PLC (Campbell et al., 1995; Chen et al., 1995; Witcher et al., 1995; Pragnell et al., 1991, 1994). A QQIER motif was identified in all three Ca_v2 family members. When a $Ca_v\beta$ subunit is bound to Ca_v2 channel, the first $G_{\beta\gamma}$ interaction site is only partially accessible to $G_{\beta\gamma}$. Peptides in the recording pipette to 2 separate binding sites within intracellular loop I-II in the human embryonic kidney (HEK293) cells blocked $G_{\beta\gamma}$-mediated inhibition of $Ca_v2.2/\alpha_2\delta/\beta_{1b}$ suggesting that the peptides interacted $G_{\beta\gamma}$ to abolish its effects on the channel (Zamponi et al., 1997). In addition, $G_{\beta1\gamma2}$ directly bound to a glutathione S-transferase-AID fusion protein with K_d of 63 nM compared to 5 nM for $Ca_v\beta_{1b}$ (De Waard et al., 1995, 1997). Site-directed mutagenesis of the QXXER motif within the glutathione

S-transferase-AID fusion protein prevented $G_{\beta1\gamma2}$ binding and GTPγS induced inhibition of $Ca_v2.1/\alpha_2\delta/\beta_4$ in Xenopus oocytes but fosters inhibition $Ca_v2.1/\alpha_2\delta/\beta_{1b}$ in tsA-201 cells (Herlitze et al., 1997; De Waard et al., 1997). Surprisingly, exchange of G protein-insensitive Ca_v1 QQLEE motif to QXXER did not make Ca_v1 channel more sensitive to G-protein inhibition, indicating that the QXXER motif is not the sole binding site for driving G protein inhibition of Ca_v2 channels (Herlitze et al., 1997). A second site within the $Ca_v2\alpha_1$ subunit intracellular loop I-II, located downstream of the AID and designated as G protein interaction domain (GID) was identified (Fig. 3). In vitro binding studies with GID fusion proteins demonstrated direct binding of $G_{\beta\gamma}$ to GID with a K_d of 24 nM (De Waard et al., 1997). GID peptides and PKC phosphorylation of a threonine within the GID domain prevented $G_{\beta\gamma}$ binding to $Ca_v2\alpha_1$ subunit and G protein-mediated inhibition of Ca_v2 channels, suggesting crosstalk between direct $G_{i/o}$ and indirect $G_{q/11}$ pathways (Currie, 2010; Herlitze et al., 2001; Zamponi & Currie, 2013; Zamponi et al., 1997). Moreover, site-directed mutagenesis of arginine[376] and valine[416] to alanine in the $Ca_v2.2$ subunit loop I-II reduced voltage-dependent inhibition by $G_{\beta\gamma}$. In contrast mutating arginine[376] to phenylalanine enhanced the voltage-dependent inhibition by $G_{\beta\gamma}$ (Tedford et al., 2010). These studies indicate that the intracellular loop I-II serves as a binding site for $G_{\beta\gamma}$ in addition to serving as the modulatory role in the voltage-dependent inhibition via $Ca_v\beta$ subunits which will be discussed in the next section.

The second $G_{\beta\gamma}$ interaction domain on Ca_v2 N-terminus is described as a glycine-rich stretch followed by a highly conserved 11 amino acid domain (YKQSIAQRART). It forms an α helix which is important for direct binding and functional modulation of $G_{\beta\gamma}$ to Ca_v2 channels (Page et al., 2010). Peptides corresponding to this highly conserved 11 amino acid region in the Ca_v2 N-terminus blocked NA-induced G protein-mediated inhibition in superior cervical ganglion neurons (Bucci et al., 2011; Canti et al., 1999). Adjacent to this conserved 11 amino acid $G_{\beta\gamma}$ binding region, $Ca_v2.2$ channels can directly bind intracellular loop I-II at their N-terminus to mediate voltage-dependent inhibition (Agler et al., 2005). Replacing the $Ca_v2.2$ N-terminus into the $Ca_v1.2$ backbone was sufficient to produce moderate voltage-dependent inhibition (Canti et al., 1999; Page et al., 1998). In support of the Ca_v2 N-terminus contributing to $G_{\beta\gamma}$ binding, $Ca_v2.2$ channels have 2 splice variants, one with a truncated N-terminus which lacks G protein inhibition and the other containing the full-length N-terminus which demonstrates voltage-dependent inhibition by $G_{\beta\gamma}$ (Page et al., 1998; Fig. 3). Therefore, the N-terminus of Ca_v2 channels may contribute to $G_{\beta\gamma}$ binding in addition to serving as an "inhibitory module" through its interactions with intracellular loop I-II from $Ca_v2.2$ to mediate a shift from "willing" to "reluc-

tant" gating states (Currie, 2010; Zamponi & Currie, 2013).

The interaction between the Ca_v2 C-terminus and $G_{\beta\gamma}$ is not as well studied and seems to be less prominent for $Ca_v2.1$ and $Ca_v2.2$ channels. However, the C-terminus of $Ca_v2.3$ channels is crucial for G protein modulation. Exchange of the $Ca_v2.1$ C-terminus in the $Ca_v2.3$ backbone abolished M2 muscarinic receptor (M2R)-mediated inhibition of $Ca_v2.3$ chimera (Qin et al., 1997). Fluorescence resonance energy transfer (FRET) studies demonstrated an increase in association between the $Ca_v2.1$ C-terminus and $Ca_v\beta_{1b}$ subunit which was even more pronounced in the presence of $G_{\beta2}$, implicating its role in G protein inhibition (Hümmer et al., 2003). Although the $Ca_v2.2$ C-terminus does not seem to affect GTPγS mediated inhibition, it diminished somatostatin-mediated inhibition of $Ca_v2.2$ channels by potentially coupling channel-GPCR interactions via $G_{\beta\gamma}$ (Hamid et al., 1999; Meza & Adams, 1998; Proft & Weiss, 2015). For $Ca_v2.2$ channels, the C-terminus may also serve as an auxiliary role to increase the affinity of $G_{\beta\gamma}$ binding to $Ca_v2.2$ channels and a binding site for other second messenger molecules such as CaM, Ca^{2+}/calmodulin-dependent protein kinase II (CaMKII), PKC, tyrosine kinases and G_α subunits to facilitate cross-talk between different G proteins.

All three G protein interaction sites (intracellular loop I-II, N-terminus, and C-terminus) on Ca_v2 channels form a dynamic structure known as the G protein binding pocket to bring Ca_v2 channels in close proximity with GPCR and to mediate the association and dissociation of $G_{\beta\gamma}$ (De Waard et al., 2005). The $Ca_v2\alpha_1$ subunit N-terminus seems to be the critical domain for G protein inhibition (Canti et al., 1999; Page et al., 1998). While the $Ca_v2\alpha_1$ subunit intracellular loop I-II may represent a "hinged lid" to mediate voltage-dependent inactivation which is disrupted by the binding of $G_{\beta\gamma}$ and ultimately causes inhibition (Currie, 2010; Stotz & Zamponi, 2001; Zamponi & Currie, 2013). The C-terminus is critical for the G protein-mediated inhibition of

$Ca_v2.3$ channels, however, has a less important role for $Ca_v2.1$ and $Ca_v2.2$ channels.

$G_{\beta\gamma}$ and $Ca_v\beta$ Subunit Diversity and Interaction Sites with Ca_v2 Channels

An additional dimension of complexity determining G protein inhibition of Ca_v2 channels is the subunit diversity of G_β, G_γ, and $Ca_v\beta$. For example, $G_{\gamma2}$ modulates G protein inhibition the greatest compared to $G_{\gamma1}$, $G_{\gamma3}$ and $G_{\gamma13}$. Moreover the degree of inhibition, reinhibition, and recovery from inhibition is different depending on the G_β subunit and which Ca^{2+} channel, $Ca_v2.1$ versus $Ca_v2.2$ are present (Arnot et al., 2000). Site-directed mutagenesis of asparagine[35,36] on $G_{\beta1}$ supports PKC's ability to antagonize $Ca_v2.2$ channel inhibition (Doering et al., 2004; Hamid et al., 1999). Furthermore, $Ca_v2.2$ recovery from G protein inhibition is affected by which $Ca_v\beta$ subtype ($Ca_v\beta_3 > Ca_v\beta_4 > Ca_v\beta_{1b} > Ca_v\beta_{2a}$) is expressed with $Ca_v\beta_{2a}$ providing the strongest relief from G protein-mediated inhibition (Canti et al., 2000; Feng et al., 2001; Weiss et al., 2007). The $Ca_v\beta$ subunit binding to the AID in $Ca_v2\alpha_1$ intracellular loop I-II is required for the voltage-dependence of G protein inhibition. The absence of $Ca_v\beta$ subunit expression with $Ca_v2\alpha_1$ and $Ca_v\alpha_2\delta$ subunits or diminished affinity of $Ca_v\beta$ for $Ca_v2\alpha_1$ subunit due to mutagenesis in the $Ca_v2.2\alpha_{1(W391A)}$ or $Ca_v\beta$ subunit resulted in a loss of voltage-dependence. Although $G_{\beta\gamma}$ continued to inhibit the I_{Ca}, in the absence of $Ca_v\beta$, the pre-pulse facilitation was abolished with the exception of $Ca_v\beta_{2a}$ (Dresviannikov et al., 2009; Leroy et al., 2005). $Ca_v\beta_{2a}$ was able to restore voltage-dependent relief in the presence of $Ca_v2.2_{\alpha1(W391A)}$ due to two palmitoylation sites at its N-terminus which target the subunit to the plasma membrane in closer proximity to $Ca_v2.2_{\alpha1(W391A)}$ subunit, and thus allowing for partial voltage-dependent inhibition by $G_{\beta\gamma}$. Although past FRET studies were contradictory in line of new findings, it appears that binding of $Ca_v\beta$ and $G_{\beta\gamma}$ to $Ca_v2.1$ channels are not exclusive but synergistic to allow voltage-dependent inhibition by

G proteins (Hümmer et al., 2003). The $Ca_v\beta$ subunits contain two conserved domains, a src homology 3 and guanylate kinase-like domains which are separated by a variable HOOK region. The guanylate kinase-like domain is solely responsible and sufficient for binding the AID region on the Ca_v2 intracellular loop I-II and restoring voltage-dependent inhibition by $G_{\beta\gamma}$. In the presence of $Ca_v\beta$, the AID domain conforms from a random coil to an α helix that extends to the IS6 voltage sensor region on the channel. It has been proposed that this conformational change in AID transmits movement of the voltage sensor and activation gate to shift the intracellular loop I-II linker and ultimately altering the $G_{\beta\gamma}$ binding pocket at depolarizing potentials to the more "willing" state (Fig. 4). The HOOK region located between these conserved domains is important for modulating the inactivation kinetics of $Ca_v2.2$ channels. Deleting the HOOK domain results in tonic inhibition of $Ca_v2.2$ channels and increased inactivation due to a higher affinity for $G_{\beta\gamma}$ (Currie, 2010; Zamponi & Currie, 2013).

In general, the direct binding and modulation of Ca_v2 channels by $G_{\beta\gamma}$ in particular $Ca_v2.1$ and $Ca_v2.2$ are similar but there are some noteworthy differences. $Ca_v2.2$ channels demonstrate very brief duration, low probability of "reluctant openings" but not $Ca_v2.1$ channels and also a greater reduction in peak amplitude of I_{Ca} compared to $Ca_v2.1$ channels (Carabelli et al., 1996; Lee & Elmslie, 2000). Moreover, $Ca_v2.1$ channels are more sensitive to trains of action potential-like stimuli for reversal of inhibition than $Ca_v2.2$ channels (Currie & Fox, 2002). Their differences in binding affinities for the different $G_{\beta\gamma}$ subtype combinations influenced the intensity of G protein inhibition (Agler et al., 2003; Arnot et al., 2000). Site-directed mutagenesis studies in $G_{\beta1}$ caused distinct outcomes on inhibition between $Ca_v2.1$ and $Ca_v2.2$. $G_{\beta1}$ binding is disturbed by the PKC phosphorylation of threonine 422 located in loop I-II of $Ca_v2.2$ subunit (Barrett & Rittenhouse, 2000; Cooper et al., 2000; Swartz, 1993; Zamponi et al., 1997).

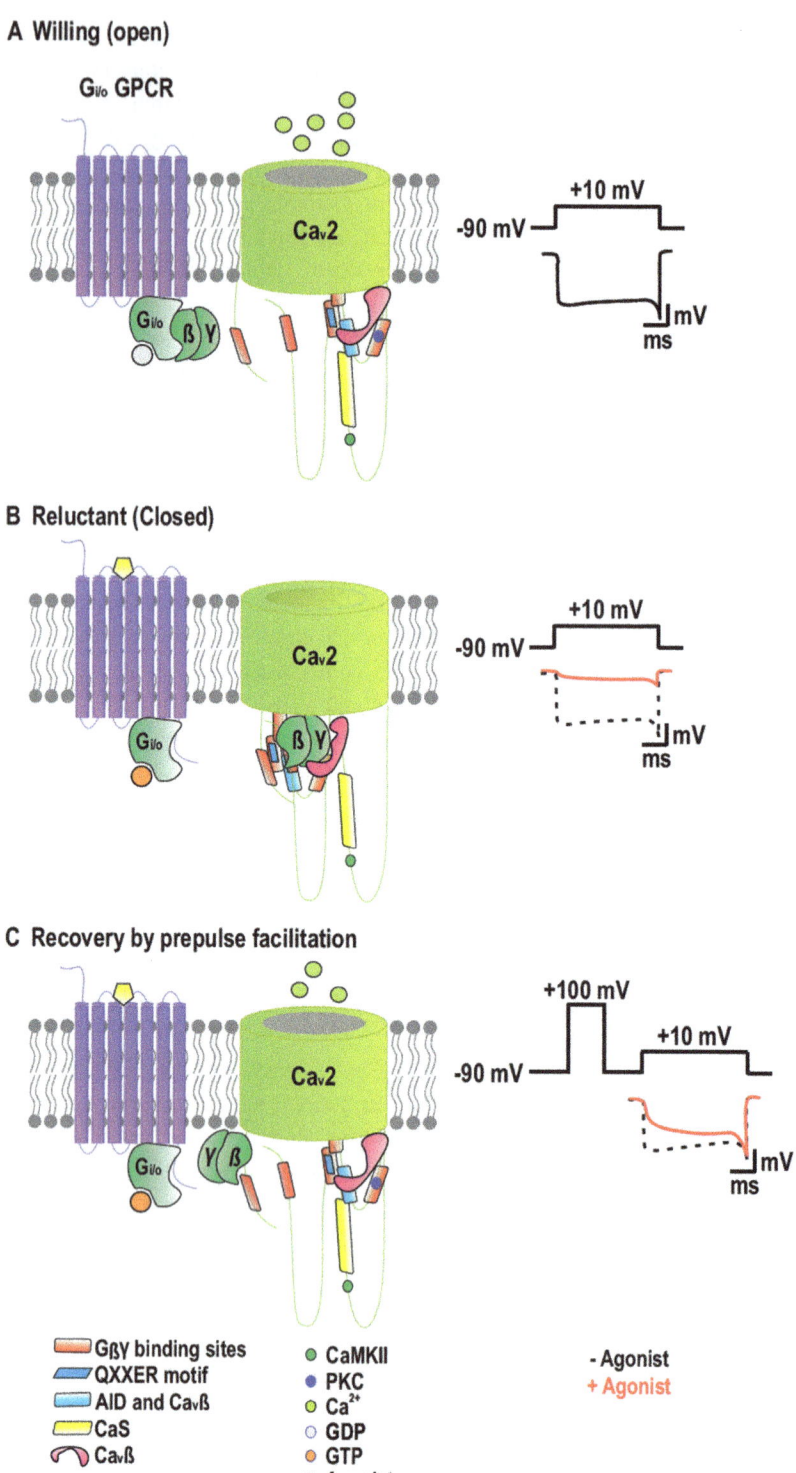

Fig. 4 A putative model representing the molecular interactions between Ca$_v$2 channels and G$_{\beta\gamma}$ during G$_{i/o}$-mediated inhibition. (**a**) Ca$_v$2 channels are active and in the "willing" state in the G$_{\beta\gamma}$ unbound state. Ca$_v\beta$ binds to the α interaction domain (AID) located in loop I-II and is important for the membrane targeting and gating

Modulation of G Protein-Mediated Inhibition by Synaptic Associated and Other Proteins

Voltage-gated calcium channels, in particular $Ca_v2.1$ and $Ca_v2.2$, associate with presynaptic vesicles through the vesicular release proteins also known as soluble N-ethylmaleimide-sensitive factor attachment protein receptors (SNAREs) to control Ca^{2+}- and voltage-dependent neurotransmitter release. $Ca_v2.1$ and $Ca_v2.2$ channels directly bind to the SNARE proteins syntaxin and synaptosomal-associated protein 25 kDa (SNAP-25) via a synaptic protein interaction site called synprint located within the $Ca_v2\alpha_1$ intracellular loop I-II (Rettig et al., 1996; Sheng et al., 1994). The synprint site contains two regions separated by a flexible linker that can independently bind syntaxin 1A and SNAP-25. Within each of these peptide regions in the synprint are PKC and CaMKII phosphorylation sites which decrease the binding of syntaxin 1A and SNAP-25, suggesting that PKC and CaMKII phosphorylation may serve as a biochemical switch for controlling SNARE-synprint interactions and ultimately fast neurotransmitter release (Mochida, 2018; Fig. 3). This Ca_v2-SNARE complex is critical for localizing Ca_v2 channels to the presynaptic vesicle release sites for efficient and fast neurotransmitter release (Catterall, 1999; Nanou & Catterall, 2018; Sheng et al., 1998). Disruption of this Ca_v2-SNARE complex alters the Ca^{2+}-dependent release of neurotransmitters. Binding of syntaxin 1A and SNAP-25 to Ca_v2 channels shifts their voltage-dependence of inactivation to more negative membrane potentials and decreases the probability of channels to open (Mochida, 2018; Proft & Weiss, 2015; Zamponi & Currie, 2013). In addition, syntaxin 1A increases $G_{\beta\gamma}$-mediated tonic inhibition of $Ca_v2.2$ channels by directly binding and colocalizing $G_{\beta\gamma}$ and $Ca_v2.2$ channels. Cysteine string proteins also interact with G proteins via the Ca_v2 synprint site to enhance tonic inhibition. On the other hand, SNAP-25 reverses the tonic inhibition of syntaxin 1A on $Ca_v2.2$ channels by reducing the magnitude of prepulse current needed for channel recovery. Similarly, rab3-interacting molecule 1 promotes the recovery from $G_{\beta\gamma}$-mediated inhibition by slowing down Ca_v2 channel inactivation (Mochida, 2018; Nanou & Catterall, 2018; Proft & Weiss, 2015; Zamponi & Currie, 2013). Another family of proteins that alter the inhibition of Ca_v2 is RGS proteins. RGS proteins change the inhibition of Ca_v2 by accelerating the intrinsic GTPase activity of G_α, and thereby limiting the available $G_{\beta\gamma}$ and reducing $G_{\beta\gamma}$-mediated inhibition (Mark et al., 2000; Currie, 2010; Mochida, 2018; Proft & Weiss, 2015). In contrast, the activator of G protein signaling (AGS) increases the G_α nucleotide exchange of GDP to GTP in a GPCR-independent manner, resulting in dissociation of $G_{\beta\gamma}$ and induced tonic inhibition (Currie, 2010; Fig. 5). The ultimate effects of SNAREs and their associated synaptic proteins have on Ca^{2+} channel-mediated synaptic transmission will be discussed in more detail in chapter "Pancreatic β cell CaV channels in health and disease" VGCCs and Synaptic Transmission.

Regulation of Ca_v2 Channels by Lipids

The initial characterization of G protein-mediated inhibition of $Ca_v2.2$ channels in sympathetic neurons revealed two mechanisms of inhibition: [1]a

---

Fig. 4 (continued) kinetics of the Ca_v2 channel. The binding of $Ca_v\beta$ transforms the AID domain into a helix that extends to the IS6 voltage sensor region on the channel. (**b**) Agonist stimulation of $G_{i/o}$-coupled G protein-coupled receptors (GPCRs) release $G_{\beta\gamma}$, allowing $G_{\beta\gamma}$ to bind to domains in Ca_v2 loop I-II, N- and C-termini (shown as red boxes), resulting in a conformational shift to the "reluctant" state and inhibition of the channel. Since there is overlap between the G protein interaction domain (GID) and AID domains in loop I-II, $Ca_v\beta$ binding to the Ca_v2 channel is weakened however sufficient to confer the voltage-dependence of G protein-mediated inhibition. (**c**) Strong depolarizing membrane potentials (prepulse facilitation) release the binding of $G_{\beta\gamma}$ to the Ca_v2 channel even in the presence of an agonist and results in conformational change which strengthens $Ca_v\beta$ binding to loop I-II AID and recovers Ca_v2 channel activity. Ca^{2+} calcium, *CaS* calcium sensor; *PKC* protein kinase C, *CaMKII* Ca^{2+}/calmodulin-dependent protein kinase II, *GDP* guanosine diphosphate, *GTP* guanosine triphosphate, *PKC* protein kinase C

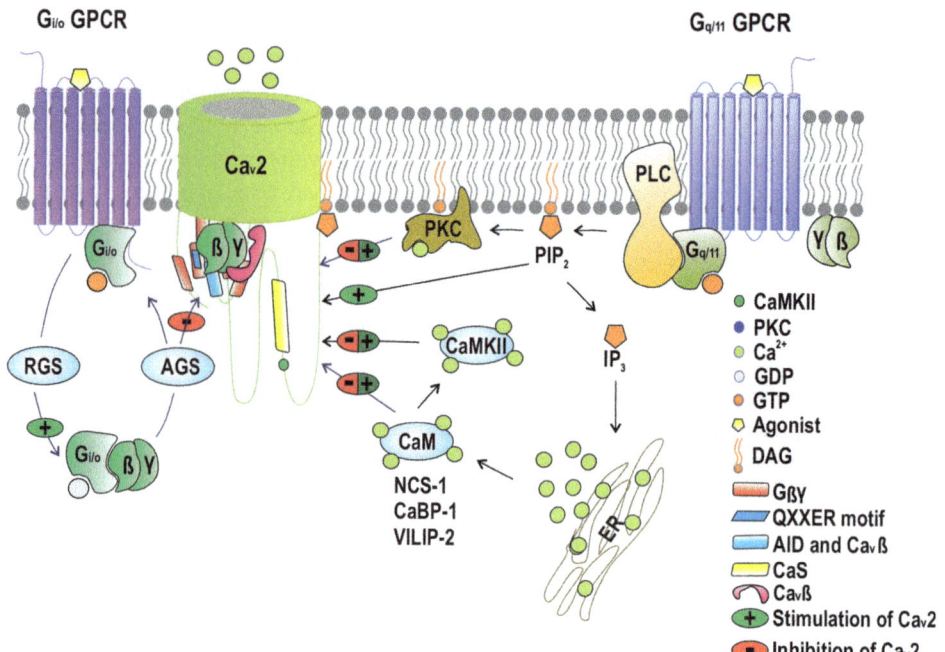

Fig. 5 A schematic overview depicting the differential modulation of Ca$_v$2 channels by G$_{i/o}$- and G$_{q/11}$-coupled pathways and their downstream messengers. Upon agonist stimulation of certain G$_{i/o}$-coupled G protein-coupled receptors (GPCRs), guanosine diphosphate (GDP) is exchanged for guanosine triphosphate (GTP) on G$_\alpha$, and G$_{\beta\gamma}$ dimer is released to directly inhibit Ca$_v$2 channels. Regulators of G protein signaling (RGS) protein indirectly reduce the G$_{\beta\gamma}$-mediated inhibition of Ca$_v$2 channels by accelerating the intrinsic GTPase activity of G$_\alpha$, and thereby limiting the available G$_{\beta\gamma}$. On the other hand, the activator of G protein signaling (AGS) increases the G$_\alpha$ nucleotide exchange of GDP to GTP independently of GPCRs, increasing G$_{\beta\gamma}$ levels to induce tonic inhibition of the channel. Additional downstream messengers from the G$_{q/11}$-coupled pathway indi- rectly inhibit Ca$_v$2 channels by hydrolyzing phosphatidylinositol 4,5-phosphate (PIP$_2$) into diacylglyc- erol (DAG) and inositol 1,4,5-triphosphate (IP$_3$) via PLC (phospholipase C). DAG together with calcium (Ca^{2+}) acti- vates protein kinase C (PKC) to phosphorylate the G protein interaction domain (GID) and synprint domains that differ- entially modulate Ca$_v$2 channels. IP$_3$ increases intracellular Ca stores from the endoplasmic reticulum (ER) to mediate Ca^{2+}-dependent effectors such as calmodulin (CaM), Ca^{2+}/ calmodulin-dependent protein kinase II (CaMKII), neural Ca^{2+} sensor-1 (NCS-1), Ca^{2+} binding protein-1 (CaBP-1) and visinin-like protein-2 (VILIP-2) which subsequently bind the calcium sensor (CaS) protein regulatory site at the Ca$_v$2 C-terminus to control Ca$_v$2 channel function. PIP$_2$ hampers G protein-mediated inhibition

fast, voltage-dependent inhibition mediated through G$_{i/o}$-coupled GPCRs and direct binding to G$_{\beta\gamma}$ and [2]a slow, voltage-independent inhibition (Zamponi & Currie, 2013; Fig. 5). This slow component of inhibition by G proteins was observed to be correlated to the hydrolysis kinet- ics of PIP$_2$, where the addition of PIP$_2$ into the cell prevented G protein-mediated inhibition (Gamper et al., 2004). PIP$_2$ was initially observed to slow down a time-dependent "rundown" and shift the voltage-dependence of Ca$_v$2.1 channels in excised membrane patches which was later confirmed in Ca$_v$2.2 channels in sympathetic neurons (Gamper et al., 2004; Wu et al., 2002). Depletion of PIP$_2$ by increasing its hydrolysis with a voltage-sensitive 5 phosphatase or block- ing the replenishment of PIP$_2$ with phosphatidylinositol-4-kinase antagonists reduced Ca$_v$2.1 and Ca$_v$2.2 currents and repre- sented the slow component of I$_{Ca}$ inhibition (Gamper et al., 2004; Okamura et al., 2009; Suh et al., 2010). Another cleavage product of PIP$_2$, AA, has also been implicated to bidirectionally modulate Ca$_v$2.2 channels depending on the membrane potential, however, the evidence is controversial (Zamponi & Currie, 2013).

Although the role of PIP_2 in Ca_v2 channel regulation is unknown, it may involve a dynamic between lower and higher affinity interaction domains which may crosslink hydrophobic and hydrophilic domains favoring protein conformations conducive to active channel states (Delmas et al., 2005).

Mechanisms of $G_{q/11}$-Mediated Inhibition: Slow, Indirect, Voltage-Independent

Activation of $G_{q/11}$-coupled receptors stimulates the downstream kinase PKC. Phosphorylation of the $Ca_v2\alpha_1$ subunit by PKC increases the current amplitude and accelerates the current inactivation. Moreover, PKC phosphorylation in the intracellular loop I-II of $Ca_v2\alpha_1$ subunit blocks G protein-mediated inhibition of the channel and thereby augmenting the intracellular Ca^{2+} levels in the cells. The $Ca_v2.2$ channels are also modulated by tyrosine kinases at its C-terminus splice variant 37a via a voltage-independent pathway (Fig. 3). This novel, tyrosine phosphorylation site in $Ca_v2.2$ exon 37a has been proposed to control voltage-independent inhibition in nociceptive neurons in the dorsal root ganglia to modulate pain transmission (Andrade et al., 2010; Lin et al., 1997; Pan & Lipscombe, 2000).

All Ca_v2 channels undergo Ca^{2+}-dependent modulation but $Ca_v2.1$ channel activity is more affected by local changes in intracellular Ca^{2+} concentrations (Nanou & Catterall, 2018). The C-terminus of $Ca_v2.1$ channel contains a Ca^{2+} sensor (CaS) protein regulatory site which binds calmodulin (CaM) during the upregulation of channel activity. Calmodulin is an EF hand protein containing 2 high and 2 low affinity Ca^{2+} binding sites in its N- and C-termini. Local increases in intracellular Ca^{2+} are sufficient to bind the high affinity, C-terminus EF hands of CaM leading to the facilitation of the channels. Prolonged or repetitive activation of $Ca_v2.1$ results in a continual inactivation of the channel as Ca^{2+} binds the lower affinity N-terminus EF hands of CaM at the CaS regulatory site (DeMaria et al., 2001; Lee et al., 1999, 2000). Mutagenesis

of the CaS regulatory site which abolishes CaM binding to the $Ca_v2.1$ channel also impedes the facilitation and inactivation of the channel (DeMaria et al., 2001; Lee et al., 2000; Zühlke et al., 2000). In addition to CaM other CaS proteins such as neural Ca^{2+} sensor-1 (NCS-1), Ca^{2+} binding protein (CaBP)-1 and visinin-like protein-2 (VILIP-2) compete for binding at the CaS regulatory site in $Ca_v2.1$ channels to modulate the balance between facilitation and inactivation (Figs. 3 and 5). They are structurally similar to CaM containing Ca^{2+} binding EF hands but are differentially expressed in the brain and have distinct effects on channel activity. For example, VILIP-2 and NCS-1 increase the facilitation and decrease the inactivation of the channel, whereas CaBP-1 decreases the facilitation and increases the rapid inactivation of $Ca_v2.1$ (Nanou & Catterall, 2018).

Conclusion for Ca_v2 Channels

GPCRs and their downstream messengers are one of the prominent regulatory mechanisms to keep intracellular Ca^{2+} levels in balance. The control of Ca_v2 channels by G proteins and their downstream effectors can be mediated by either a fast, direct, voltage-dependent $G_{i/o}$-coupled pathway or a slow, indirect, voltage-independent mechanism that includes lipids and G_q-coupled messengers. Pertussis toxin-sensitive $G_{i/o}$-coupled GPCRs directly inhibit Ca_v2 channels through the release of $G_{\beta\gamma}$, and then $G_{\beta\gamma}$ mediates its inhibitory effects on $Ca_v2\alpha_1$ subunit via its intracellular cytoplasmic loop I-II and N- and C-termini. Likewise, the stimulation of various G_q-coupled GPCRs activates various downstream effectors such as kinases, lipids, and Ca^{2+} binding proteins which bidirectionally modulate Ca_v2 channels via their $Ca_v2\alpha_1$ subunit intracellular cytoplasmic domains. Both $G_{i/o}$- and G_q-coupled GPCRs synergistically tightly control Ca_v2 channels and thereby intracellular Ca^{2+} levels in the brain. The biochemical and functional details on how these complex signaling pathways control Ca_v2 channels under different physiological behaviors still remain a puzzle. To add to this level of complexity new interaction partners have been identified and will bring new insights into the regulation of Ca_v2 channels during various behavior and disease states (Lacinova et al., 2020).

Modulation of Ca_v3 Channels by GPCRs and Second Messengers

Like all VGCCs, T-type channels are primarily regulated by changes in membrane potential. Nevertheless, alternative pathways of modulation exist that are often mediated by GPCRs and second messenger pathways. While it is relatively well understood how GPCRs interact with VGCC of the Ca_v1 and Ca_v2 families, there is less agreement about the modulation of members of the Ca_v3 family. Most studies demonstrated a G protein-dependent modulation by various Ser/Thr kinases or direct interaction with either the G_α or $G_{\beta\gamma}$ subunit of the heterotrimeric G protein. However, some endogenous neurotransmitters directly modulate T-currents independent of G protein interaction such as AA or anandamide (Chemin et al., 2001; Schmitt & Meves, 1995; Talavera et al., 2004; Zhang et al., 2000).

The mechanisms and pathways of neurotransmitter-mediated T-current modulation are often not well understood but several different sometimes synergistic pathways were described. For example, in rat dorsal root ganglion (DRG) neurons nickel-sensitive T-currents depicted a mixed response to the intracellular photorelease of an activator of G protein signaling, GTPγS. Low doses (6μM) resulted in an enhanced T-current while higher doses (up to 20μM) resulted in inhibition of T-currents (Scott et al., 1990). Interestingly, only the inhibitory response was sensitive to the inhibitor of $G_{i/o}$ signaling PTX, indicating that the increased T-type currents are dependent on another G protein. Similarly, the application of low doses of the $GABA_B$ ($G_{i/o}$ coupled receptor) agonist, baclofen, (2μM) mimicked the enhancing effect while high doses (100μM) mimicked the inhibitory effect (Scott et al., 1990). Further inhibitory and PTX-sensitive effects were demonstrated after the application of low doses of baclofen (2μM) in

ND7–23 cells (Kobrinsky et al., 1994), vasopressin via V1a receptors in rat glomerulosa cells (Grazzini et al., 1996) or dopamine (Lledo et al., 1990a, b, 1992). The dopamine-induced decrease in T-currents was found in several cell lines such as DRG neurons (Marchetti et al., 1986), pituitary intermediate cells (Nussinovitch & Kleinhaus, 1992), adrenal glomerulosa cells (Osipenko et al., 1994), rat pituitary melanotrophs (Keja et al., 1992; Williams et al., 1990), and lactotroph cells (Lledo et al., 1990a, b, 1992) and could be blocked by antagonists of the $G_{i/o}$-coupled D2Rs (Keja et al., 1992; Lledo et al., 1990a, b; Osipenko et al., 1994) suggesting the involvement of the $G_{i/o}$ signaling pathway.

Interestingly, PTX-insensitive inhibitory effects were also described. For example, inhibition of T-currents was insensitive to PTX in ND7-23 cells after bradykinin (Kobrinsky et al., 1994) or in the cultured nucleus basalis neurons after neurotensin or substance P (Margeta-Mitrovic et al., 1997) application. Whether the inhibitory effect was PTX-sensitive or not is not clear for activation of μ-opioid receptor (Schroeder et al., 1991; Schroeder & McCleskey, 1993), oxytocin receptors (Liu et al., 2005), or prostaglandin E1 via iloprost-sensitive prostanoid receptors (Meves, 2005). While most of the effects were described for nickel-sensitive currents also modulation of $Ca_v3.3$ was described. In HEK-293 cells, recombinantly expressed $Ca_v3.3$-currents (all isoforms were tested) were inhibited by activation of G_q-coupled M1Rs. Using a pharmacologic approach, it was found that $G_{\beta\gamma}$ and PKC were not involved but specifically, the $G\alpha q/11$ subunit was responsible for the effect. Generation of chimeric receptors consisting of parts of $Ca_v3.3$ and $Ca_v3.1$ (not sensitive to acetylcholine) suggested that two different regions of the $Ca_v3.3$ are responsible for the acetylcholine-induced inhibition of T-currents (Hildebrand et al., 2007).

Not only inhibition of T-currents was found after modulation of GPCRs but also enhancement has been described. Activation of angiotensin II receptors increases T-currents in atrial frog cells (Bonvallet & Rougier, 1989) in a G protein-dependent (Lu et al., 1996; McCarthy et al., 1993) and PTX sensitive pathway in adrenal glomerulosa cells (Lu et al., 1996). Noteworthy, in adrenal glomerulosa cells the nickel-sensitive $Ca_V3.2$ isoform is predominantly expressed (Schrier et al., 2001). Also, in rat nodose ganglion neurons, dynorphin (binds to κ-opioid receptors, $G_{i/o}$-coupled) reduced T-type currents in a PTX-sensitive manner (Gross et al., 1990a, b). These studies demonstrated a diverse interaction of T-type channels with different second-messenger pathways. A more detailed description of the pathways involved will be described in the following sections.

Direct Modulation by $G_{\beta\gamma}$ Subunits

As described for P/Q-type Ca^{2+} channels a direct modulation by $G_{\beta\gamma}$ subunits (Herlitze et al., 1996) has also been demonstrated for T-type channels. In adrenal glomerulosa cells activation of the D1Rs (G_s-coupled receptor) by dopamine or by a D1-dopamine receptor agonist (SKF-38393) reversibly decreases nickel-sensitive T-type currents while the D1R antagonist (SCH-23390) blocked the dopamine-induced decrease in T-currents (Drolet et al., 1997). This effect was abolished when an inhibitor of G protein activation, GDPβS, was included in the pipette solution demonstrating the involvement of G protein signaling. The inhibitor of $G_{i/o}$ signaling PTX, as well as the D2R antagonist (spiperone), had no effect which suggests that D2R ($G_{i/o}$ coupled receptor) signaling is not involved. Activation of G_s-coupled receptors such as D1Rs involves AC and PKA second-messenger pathways, but neither inhibition of the AC (2′,3′-dideoxyadenosine) nor activation of PKA (8-Br-cAMP) mimicked the effect of dopamine receptor activation. However, both prevented the dopamine-induced inhibition of the T-type current. This suggests that although cyclic adenosine monophosphate (cAMP) and PKA do not directly affect T-currents, they are necessary for dopamine-induced inhibition. In addition, another element of the signaling pathway is needed for channel inhibition. When $G_{\beta\gamma}$ was included in the pipette solution in combination with a PKA activator (8-Br-cAMP) the dopamine-induced inhibition

of T-currents was mimicked. Therefore, a combination of PKA activity and $G_{\beta\gamma}$ effects was suggested as necessary for the dopamine-induced inhibition (Drolet et al., 1997).

In further studies, the role of different $G_{\beta\gamma}$ subunit dimer combinations on dopamine-induced inhibition of T-type currents was investigated. Only dimers containing the $G_{\beta2}$ subunit (except for a dimer containing $G_{\beta2\gamma11}$) reduced the current density by up to 50% in HEK-293 cells stably expressing $Ca_V3.2$ but not $Ca_V3.1$ (Wolfe et al., 2003). This was further supported by the generation of two chimeric receptors where the intracellular II-III loop was exchanged between $Ca_V3.1$ and $Ca_V3.2$. Only the $Ca_V3.1$ chimeric receptors that contained the II-III loop of $Ca_V3.2$ demonstrated a reduced current density which suggests that the II-III loop of $Ca_V3.2$ is necessary for the inhibitory effect (Wolfe et al., 2003). Noteworthy, the inhibitory effect does not involve a decrease in the number of channels at the surface but relies on the reduced channel open probability in the membrane due to the binding of the $G_{\beta2\gamma}$ to the intracellular II-III loop in the $Ca_V3.2$ channels (DePuy et al., 2006). Similarly, the inhibitory effect (up to 30%) in overexpressed $Ca_V3.2$ and $Ca_V3.1$ chimeric receptors (containing the II-III loop of $Ca_V3.2$) was also found after the application of a D1R agonist (SKF-38393) in human adenocarcinoma cells (H295R) that naturally express high levels of $G_{\beta2}$ subunits and D1Rs (Wolfe et al., 2003). As described by Drolet et al. (1997) the inhibitory effect does not solely rely on $G_{\beta\gamma}$ but also on the co-activation of PKA. Indeed, T-type channels phosphorylation of Ser^{1107} in II-III loop via PKA is required for the $G_{\beta2\gamma2}$ mediated dopamine-induced inhibition of the T-type currents (Hu et al., 2009). A more detailed description of the inhibitory effect was proposed, that D1R activation increases cAMP levels which in turn activates PKA while D2R activation provides $G_{\beta2\gamma}$ dimers (Hu et al., 2009). A similar decrease in T-type currents requiring $G_{\beta\gamma}$ dimers and PKA phosphorylation was found after activation of muscarinic receptors (Zhang et al., 2012). In mouse DRG neurons, activation of muscarinic M4 receptors ($G_{i/o}$ coupled receptor) produced a T-type current inhibition, an effect that was mimicked by activation of AC (forskolin) and blocked by application of PKA inhibitors (H89 or PKI 6-22), the inhibitor of $G_{i/o}$ signaling (PTX) or blocking of $G_{\beta\gamma}$ signaling (QEHA or G-specific subunit antibody) suggesting that a combination of PKA and $G_{\beta\gamma}$ (originating from $G_{o\alpha}$) is required for the effect (Zhang et al., 2012).

Not only PKA and $G_{\beta\gamma}$ together can inhibit T-type currents but also phosphorylation by PKC together with $G_{\beta\gamma}$ are responsible for T-type current modulation. For example, in trigeminal ganglion neurons, melatonin receptor 2 activation inhibits $Ca_v3.2$ currents in a PTX-sensitive pathway which is associated with a negative shift in the steady-state inactivation (Zhang et al., 2018). The effect was abolished by blocking $G_{\beta\gamma}$ signaling (QEHA) and the specific PKCη inhibitor (PKCη-PSI). In contrast, in mouse DRG neurons activation of insulin-like growth factor 1 (IGF-1) receptor increased T-type currents (mostly nickel-sensitive), an effect that was sensitive to GDPβS and PTX but not cholera toxin (CTX) and could be mimicked by the specific IGF-1 receptor agonist (JB-1) (Zhang et al., 2014). The IGF-1-mediated increase in T-type currents was blocked by inhibitors of $G_{\beta\gamma}$ signaling (QEHA) and antagonists of PKCα but not PKCβ suggesting a role for $G_{\beta\gamma}$ (originating from $G_{i/o}$) activates PKCα in a Ca^{2+}-dependent manner (Zhang et al., 2014). Furthermore, in HEK-293 cells activation of the native corticotropin-releasing factor (CRF) receptor 1 by CRF and urocortin exclusively inhibits $Ca_v3.2$ currents (all isoforms were tested). The CRF receptor 1 activation resulted in a negative shift in the steady-state inactivation and was blocked by the specific CRF receptor 1 antagonist astressin. Surprisingly, the effect was independent of kinase activity such as PKC, tyrosine kinase, or CaMKII although activation of the G_s-coupled receptor (CTX sensitive, not PTX sensitive) was involved suggesting that another element of G protein signaling was necessary. Indeed, blocking of $G_{\beta\gamma}$ signaling (QEHA) abolished CRF-mediated inhibition of T-type currents providing evidence for the involvement of $G_{\beta\gamma}$ dimers (Tao et al., 2008).

Modulation by Kinases

Protein Kinase A

Activation of G_s-coupled receptors is classically associated with an AC-mediated increase in cAMP, an activator of PKA (see Fig. 6). For example, in white bass retinal horizontal cells, a D1R-induced inhibition of T-currents could be mimicked by a cAMP analog (8-CPT-cAMP) and blocked by a PKA inhibitor (Pfeiffer-Linn & Lasater, 1993, 1998). Whether this response is dependent on the $G_{\beta\gamma}$ dimer as described before is not known (Drolet et al., 1997; Wolfe et al., 2003). A similar inhibitory effect was found in *Cynops pyrrhogaster* newt (family member of salamanders) olfactory receptor and rat cerebral smooth muscle cells (Harraz & Welsh, 2013) after the application of adrenaline (Kawai et al., 1999) or isoproterenol (agonist for G_s-coupled β-adrenergic receptors) which could be mimicked by activation of AC (forskolin) or PKA (8-Br-cAMP or db-cAMP). The inhibitory effect was blocked by PKA inhibitors (PKI 14-22 and KT5720) suggesting the involvement of PKA in the β-adrenergic-induced inhibition of T-currents (Harraz & Welsh, 2013). Interestingly, inhibition of T-type currents involving PKA but independent of G_s signaling was also described. In NG108-15 cells db-cAMP (cAMP analog) inhibits nickel-sensitive T-type currents (up to 40%), the effect was prevented by preincubation with PTX which suggests the involvement of $G_{i/o}$ signaling since inhibition of G_s signaling by CTX had no effect (Kasai, 1992). Furthermore, a $G_{\beta\gamma}$ independent but PKA-dependent inhibition of nickel-sensitive T-type currents was found in mouse DRG neurons (Wang et al., 2011). The neuromedin U-induced inhibition of T-currents was blocked by PTX while blocking of $G_{\beta\gamma}$ mediated signaling (QEHA) had no effect. Furthermore, the application of PKA inhibitors (H89 or PKI 5-24) blocked the inhibitory effect suggesting a $G_{i/o}$ dependent PKA pathway (Wang et al., 2011).

In contrast, enhanced T-type currents due to activation of PKA have also been described after activation of β-adrenergic receptors (G_s-coupled receptor) by isoproterenol in isolated bullfrog

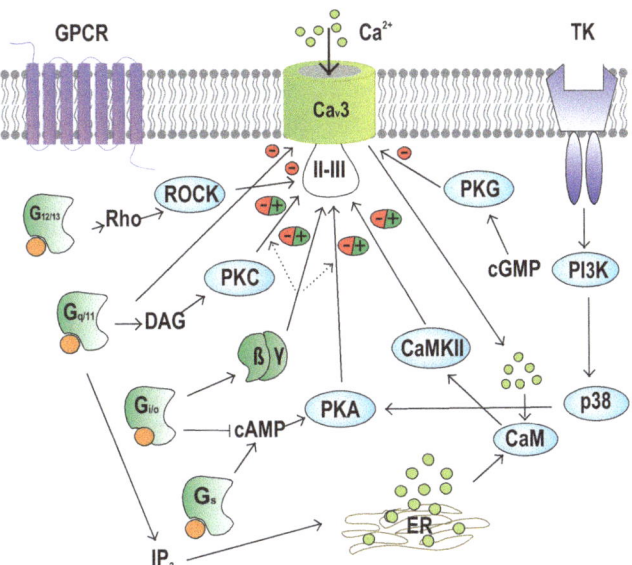

Fig. 6 Schematic representation of signaling pathways involved in T-type current modulation. Abbrevations: *GPCR* G protein-coupled receptor, *TK* Tyrosine kinase, *ROCK* Rho-associated protein kinase, *DAG* Diacylglycerine, *PKC* Protein kinase C, *cAMP* Cyclic adenosine monophosphate, *PKA* Protein kinase A, *CaMKII* Ca^{2+}/calmodulin-dependent protein kinase II, *cGMP* Cyclic guanosine monophosphate, *PKG* Protein kinase G, *CaM* Calmodulin, *PI3K* Phosphoinositide 3-kinases, *p38* p38 mitogen-activated protein kinase, *IP3* Inositol triphosphate, *ER* Endoplasmatic reticulum

atrial cardiomyocytes (Alvarez et al., 1996) or mouse sinoatrial nodal cells for Ca$_v$3.1 currents (Li et al., 2012). In both cell types, the effect was mimicked by cAMP and blocked by PKA inhibitors. Similarly, 5-HT (serotonin) 7 receptor (G$_s$-coupled) activation increases nickel-sensitive T-type currents via a PKA-dependent signaling pathway in rat adrenal glomerulosa cells (Lenglet et al., 2002) and for recombinant Ca$_v$3.2 expressed in *Xenopus* oocytes (Kim et al., 2006). The latter study also described a negative shift in the steady-state inactivation curve and demonstrated by the exchange of intracellular loops of Ca$_v$3.2 with Na$_v$1.4 (not sensitive to PKA phosphorylation) an important function for the II-III loop of the Ca$_v$3.2 for PKA-mediated increase in T-type currents (Kim et al., 2006). Whether G$_{\beta\gamma}$ is necessary for the 5-HT-induced T-type current increase is not known yet. Additionally, a similar G$_s$-coupled receptor-induced increase in T-type currents was found in human adrenocortical cells after the application of 5-HT, a cAMP analog (db-cAMP), and the agonist for 5-HT4 (zacopride) suggesting

that 5-HT4 is responsible for the effect (Louiset et al., 2017). Furthermore, application of prosta-glandin in NG108-15 cells (Sekiguchi et al., 2013), growth hormone-releasing hormone in ovine pituitary somatotropes (Chen et al., 2000) or activation of muscarinic M3 and M5 receptors (Gq coupled receptors) (Pemberton et al., 2000) resulted in a PKA dependent increase in T-type currents since the effect could be blocked by PKA inhibitors. Also, in mouse trigeminal ganglia cells activation of the orphan receptor GPR30 with a selective agonist (GPR-specific compound 1) increases T-type currents. The effect was blocked by pretreatment with GDPβS (inhibitor of G protein activation) and CTX (blocker of G$_s$ signaling) indicating the involvement of G$_s$ proteins. Indeed, the application of two PKA inhibitors (PKI 6-22 and H89) abolished the GPR30-mediated increase in T-type currents (Yue et al., 2014).

Other pathways including the activation of the phosphatidylinositol 3-kinase (PI3K)/Akt pathway activate PKA and thus increase in T-type currents were demonstrated. In mouse DRG neurons activation of the fibroblast growth factor (FGF) type 1 receptor (FGFR1) by its agonist FGF-23 increased T-type currents (mostly nickel sensitive) and induced a depolarizing shift in steady-state activation. FGFR1 activates the PI3K pathway since the FGF-23 induced increase was blocked by a PI3K inhibitor (LY294002) but not the inhibitor of AKT kinase. PKA inhibition (KT-5720) abolished the effect suggesting that PI3K activates PKA and leads to T-type channel increase (Si et al., 2018). Similarly, in trigeminal ganglion neurons activation of tropomyosin receptor kinase B by the brain-derived neurotrophic factor (BDNF) increased T-type currents through a PI3K-p38 mitogen-activated protein kinase (MAPK)-PKA signaling pathway (Wang et al., 2019). The BDNF-induced increase in T-type currents was blocked by an inhibitor of PI_3K but not its downstream target AKT kinase suggesting another pathway. Indeed, blocking of p38 MAPK another target of PI_3K as well as PKA resulted in a loss of BDNF-induced increase in T-type currents demonstrating activation of PKA by PI_3K (Wang et al., 2019).

Protein Kinase C

PKCs are typically activated by DAG and Ca^{2+} (some do not require additional Ca^{2+}) that originates from the activation of $G_{q/11}$ coupled receptors via PLC-mediated cleavage of PIP_2 into IP_3 and DAG and IP_3 induced Ca^{2+} release from the endoplasmic reticulum (see Fig. 6). For example, T-type currents were reduced after the application of PKC activators 1,2-oleoylacetyl-glycerol (OAG) in pituitary GH3 cells (Marchetti & Brown, 1988), or phorbol-myristate 13-acetate (PMA) in canine ventricular and Purkinje cells (Tseng & Boyden, 1991). Inhibition of PKC by H-8 abolished the reduction in T-type currents providing evidence for the involvement of PKC (Tseng & Boyden, 1991). Interestingly, this effect seems to be temperature-dependent since a PMA-mediated decrease in T-type currents was only found if the temperature was higher than 29 °C in rat DRG neurons (Schroeder et al., 1990). Although the activators of PKC, OAG and phorbol esters (phorbol-12,13-diacetate or 4-a-

phorbol-12,13-didecanoate) demonstrate a similar decrease in T-type currents in chick DRG neurons, the effect is independent of PKC activity since PKC inhibitors (staurosporine, sphingosine, or H-7[12]) had no effect (Hockberger et al., 1989). Also, the application of various neurotransmitters resulted in PKC-dependent reduction of T-type currents. In rat hippocampal neurons acetylcholine reduces nickel-sensitive T-type currents via a PKC-dependent pathway since OAG and PMA (PKC activator) mimic the acetylcholine-induced depression of T-type currents. The involvement of G protein signaling was demonstrated by blocking of $G_{i/o}$ signaling (PTX) as well as inhibitors for PKC (H7) and acetylcholine (atropine) which abolished the effect (Toselli & Lux, 1989). A similar PTX-sensitive and PKC-dependent pathway was described in mouse spermatogenic cells via growth hormone release inducing (ghrelin) which activates the selective growth hormone secretagogue receptor 1a (GHS-R1a) (Liu et al., 2011). Activation of these receptors by ghrelin reversibly inhibits nickel-sensitive T-type currents (Arnoult et al., 1998) (up to 32%) and was antagonized by the specific GHS-R1a receptor blocker (Liu et al., 2011). Ghrelin-induced inhibition was dependent on G protein signaling as demonstrated by GDPβS and more specifically involves $G_{i/o}$ signaling since PTX blocked the effect. Interestingly, the effect was also abolished by inhibitors of PLC (U73122) and PKC (chelerythrine) suggesting a PTX-sensitive PKC signaling pathway (Liu et al., 2011). Similarly, activation of the muscarinic M3 receptor in mouse DRG neurons inhibits nickel-sensitive-currents and involves a PTX-sensitive PKC signaling pathway since blocking of G protein signaling (GDPβS), $G_{i/o}$ signaling (PTX), and PKC inhibition (GF109203X and chelerythrine, U73211 but not Ro 31-8220) abolished the effect (Zhang et al., 2011). In NIH 3 T3 cells expressing M1R, acetylcholine application after blocking of PKC (calphostin C) increased T-type currents via PKA activation suggesting that PKC inhibits the T-type currents (Pemberton et al., 2000). Furthermore, in MN9D cells a native ligand for CRF receptor (Urocortin 1) and PMA reversibly inhibited nickel-sensitive currents (Kim et al., 2007). Specific CRF receptor 1 antagonists (astressin or antalarmin) and the PKC inhibitor, chelerythrine, blocked the effect while PMA application mimicked the CRF-induced inhibition of T-type currents (Kim et al., 2007). A similar inhibitory effect was demonstrated for recombinantly expressed $Ca_v3.2$ in HEK-293 cells after neurokinin-1 (Rangel et al., 2010) or ethanol (Shan et al., 2013) application. In both studies, PMA mimicked the effect while inhibition of PKC (Rangel et al., 2010; Shan et al., 2013) and PLCβ (Shan et al., 2013) blocked it. Furthermore, ethanol negatively shifted the steady-state inactivation (Shan et al., 2013). Interestingly, the neurokinin-induced inhibition was independent of $G_{βγ}$ signaling since buffering of $G_{βγ}$ (coexpression of human transducin) had no effect (Rangel et al., 2010). Angiotensin II in adrenal glomerulosa cells decreases nickel-sensitive T-current (Schrier et al., 2001) but positively shifts the activation curve (Rossier et al., 1995). The specific angiotensin II type 1 receptor blocker, losartan, and PKC inhibitor (chelerythrine) blocked the effect of angiotensin II (Rossier et al., 1995). In white bass retinal cells dopamine inhibits T-type currents by activation of the PKA and PKC pathways. This effect was mimicked by OAG, PMA and inhibited by PKC inhibitors, staurosporine, and 8-cpt (Pfeiffer-Linn & Lasater, 1998).

In contrast, increases in nickel-sensitive T-type currents via angiotensin II type 1 receptor in human and chick myocytes (Bkaily et al., 2005) or endothelin-1 in rat neonatal ventricular myocytes (Furukawa et al., 1992) which was mimicked by phorbol 12,13-dibutyrate (PKC activator) (Bkaily et al., 2005) were demonstrated. The endothelin1-mediated increase in T-type currents was antagonized by staurosporine and H-7 (PKC inhibitor) providing evidence for the involvement of PKC (Furukawa et al., 1992). Also, recombinantly expressed T-type channels in *Xenopus* oocytes demonstrated an increase in currents in all isoforms after PKC activation via PMA (Park et al., 2003, 2006). Furthermore, exchanging the intracellular loops

with $Ca_v2.1$ demonstrated that loop II-III is responsible for the PMA-induced increase in T-type currents in $Ca_v3.1$ (Park et al., 2006).

Calmodulin-Dependent Kinase II

In adrenal glomerulosa cells, elevated levels of intracellular Ca^{2+} enhance nickel-sensitive T-type currents by negatively shifting the activation curve. Specific inhibitors for CaMKII (KN-62) blocked the enhanced T-type current demonstrating that phosphorylation of CaMKII is necessary (Barrett et al., 2000; Fern et al., 1995; Lu et al., 1994). Indeed, a mutated Ca^{2+}-independent version of CaMKII (T286D-CaMKII) mimicked the effect of elevated intracellular Ca^{2+} levels and increased T-type currents (Barrett et al., 2000). Similarly, the expression of recombinant $Ca_v3.2$ in HEK-293 cells demonstrated an increase in T-type currents including the negative shift in the activation curve due to elevated intracellular Ca^{2+} levels while the $Ca_v3.1$ was unaffected (Wolfe et al., 2002). Generation of chimeric $Ca_v3.1$ constructs by exchanging loop II-III between $Ca_v3.1$ and $Ca_v3.2$ demonstrated the importance of loop II-III of $Ca_v3.2$ for the enhanced T-type currents. More specifically, Ser^{1198} in the loop II-III is preferably phosphorylated by CaMKII, and mutation of this site abolished the CaMKII-mediated increase in T-type currents (Welsby et al., 2003). Interestingly, intracellular Ca^{2+} increase after acetylcholine application in mouse oocytes was in part dependent on CaMKII-mediated increase in T-type currents. The T-type current increase was blocked by CaMKII blocker, KN-93, and was PKC independent (Kang et al., 2007). Not only enhancing effects were found but also suppression of T-type currents mediated by CaMKII signaling. In rat retinal ganglion cells cannabinoid receptor CB1 activation (Win55212-2 or arachidonyl-2′-chloroethylamide) suppressed T-type currents and negatively shifted the inactivation curve by activation of CaMKII (Qian et al., 2017). Furthermore, activation of CB2 (CB65 or HU308) similarly suppressed T-type currents via cAMP/PKA and CaMKII signaling (Qian et al., 2017).

Protein Kinase G

In newt olfactory receptor cells cyclic guanosine monophosphate (cGMP) increased the nickel-sensitive T-type currents (up to 40%) and induced a hyperpolarizing shift in the activation curve (Kawai & Miyachi, 2001). The specific cGMP-dependent protein kinase inhibitor, KT5823, blocked the cGMP-induced increase while CPT-cGMP (permeant cGMP analog) and the inhibitor of PDE, zaprinast, mimicked the cGMP-mediated increase (Kawai & Miyachi, 2001). In contrast, in acutely isolated rat retinal ganglion cells somatostatin inhibits T-type currents via the SST5 which was verified by a specific SST5 agonist (L-817,818). The effect of L-817,818 persisted in the presence of inhibitors for most signaling pathways including PLC/PKC and cAMP/PKA. However, block of the cGMP/PKG pathway by KT 5823 or inhibitor of NO production blocked the effect suggesting the involvement of NO/cGMP/PKG signaling (Li et al., 2019).

Rho-Kinase

The endogenous lysophosphatidic acid (LPA) is the native ligand for a family of GPCR (LPA receptors) which can bind to various signaling pathways including the activation of Rho via $G_{12/13}$ (Anliker & Chun, 2004). In ts-201 cells, LPA inhibits T-type currents (all isoforms) and shifts inactivation and activation curves to more depolarized potentials for $Ca_v3.2$ (Iftinca et al., 2007). The effect is dependent on the activation of a G protein since GDPβS (inhibitor of G protein signaling) or lack of GTP in the solution blocked the effect. Indeed, the LPA-induced inhibition was abolished by dominant-negative inhibitors of $G_{\alpha12}$ and $G_{\alpha13}$, antagonists for the GPCR LPA1/LPA3 (VOC322183) receptors, inactivation of RhoA (exoenzyme C3), and inhibitors for Rho-associated protein kinase (ROCK) (Y27362 and fasudil) which demonstrates the involvement of Rho kinase pathway. More specifically, mutagenesis of the II-III loop (Ser^{1031}-Thr^{1034} and Thr^{1041}-Ser^{1047} were mutated to alanine) revealed two distinct phosphorylation consensus sites

responsible for the LPA-induced reduction in $Ca_v3.1$-currents (Iftinca et al., 2007).

Tyrosine Kinase

In mouse, spermatogenic cells inhibitors for a protein tyrosine kinase (PTK; tyrphostin A47 or A25) increase nickel-sensitive T-type currents (Arnoult et al., 1998) by up to 50% while the antagonist for protein tyrosine phosphatase (phenylarsine oxide or sodium (Na) orthovanadate) blocked a voltage-dependent increase in T-type currents (Arnoult et al., 1997). In contrast, a G protein-independent (GDPβS had no effect) suppression of T-type currents was demonstrated in NG108–15 cells after the application of three different inhibitors for PTK (genistein, lavendustin A, and herbimycin A) (Morikawa et al., 1998). Surprisingly, the inhibitory effect of genistein was also independent of PTK activity since extracellular application similarly suppressed the $Ca_v3.1$ current in HEK-293 cells while another PTK inhibitor (tyrphostin AG213) had no effect. Indeed, intracellular application of the catalytically active PTK p60c-^{SRC} also had no effect (Kurejová & Lacinová, 2006) suggesting that genistein directly affects T-type channels. The same effect was demonstrated in mouse spermatogenic cells where extracellular but not intracellular genistein effectively suppressed nickel-sensitive T-type currents while another PTK inhibitor (lavendustin A) was ineffective (Tao et al., 2009). A non-related PTK inhibitor, imatinib-mesylate, inhibited $Ca_v3.3$ channels stably expressed in HEK-293 after extracellular but not intracellular application suggesting a PTK-independent mechanism (Cataldi et al., 2004).

GPCR-Mediated Change in T-Type Channel Expression

Most of the described mechanisms change the T-type current via direct or indirect modulation without affecting the expression level of T-type channels. However, activation of β-ARs in rat pinealocytes increases the expression of $Ca_v3.1$ leading to an increased current density following 24 h incubation with NA (Yu et al., 2015). More specifically the signal was mimicked via the β-AR agonist (isoproterenol) and blocked after incubation with β blockers (propranolol). Similar results were demonstrated by incubation with a cAMP activator (forskolin) or cAMP analogs (db-cAMP, parachloropenylthio-cAMP) while the NA-mediated T-type current density increase was blocked by simultaneous incubation with two PKA inhibitors (RP-8-CPT-cAMP or H89) (Yu et al., 2015).

Conclusion for T-Type Channels

The detailed mechanisms of how specific second-messenger pathways regulate the activity of T-type Ca^{2+} channels are still not well understood and the described studies demonstrate the diversity of responses to various neurotransmitters or agonists (Table 1). Most of the studies demonstrate results for nickel-sensitive ($Ca_v3.2$) currents, but also in $Ca_v3.1$ (Kurejová & Lacinová, 2006; Li et al., 2012; Yu et al., 2015), $Ca_v3.3$ (Hildebrand et al., 2007) or all isoforms (Park et al., 2006; Chemin et al., 2007; Iftinca et al.,

Table 1 Summary of pathways and effects involved in T-type calcium channel modulation

Pathway	Receptor/Neurotransmitter	Effects	References
PKA	β-adrenergic receptor; 5-HT7 and 5-HT4 receptors; Growth hormone-releasing hormone; Prostaglandin; Muscarinic M3 and M5 receptors; G protein-coupled receptor 30	Enhancement	Alvarez et al. (1996), Chen et al. (2000), Kim et al. (2006), Lenglet et al. (2002), Li et al. (2012), Pemberton et al. (2000), Sekiguchi et al. (2013) and Yue et al. (2014)
PKA	β-adrenergic receptor; Neuromedin U	Inhibition	Chen et al. (2000) and Harraz and Welsh (2013)
PI3K & PKA	Fibroblast growth factor 23; Brain-derived neurotrophic factor	Enhancement	Si et al. (2018) and Wang et al. (2019)
PKC	Angiotensin II type 1 receptor; Endothelin-1	Enhancement	Bkaily et al. (2005) and Furukawa et al. (1992)
PKC	Acetylcholine; Growth hormone secretagogue receptor 1a; Muscarinic M3 receptor; Corticotropin-releasing factor receptor 1; Neurokinin-1 receptor; Angiotensin II type 1 receptor	Inhibition	Kim et al. (2007), Liu et al. (2011), Rangel et al. (2010), Rossier et al. (1995), Toselli and Lux (1989) and Zhang et al. (2011)
PKG	cGMP	Enhancement	Kawai and Miyachi (2001)
PKG	Somatostatin receptor 5	Inhibition	Li et al. (2019)
CaMKII	Intracellular calcium; Acetylcholine	Enhancement	Barrett et al. (2000), Fern et al. (1995), Kang et al. (2007), Lu et al. (1994) and Welsby et al. (2003)
CaMKII	Cannabinoid receptor 1 and 2	Inhibition	Qian et al. (2017)
Rho kinase	Lysophosphatidic acid receptor G12/13	Inhibition	Iftinca et al. (2007)
Gq/11	Muscarinic M1 receptor	Inhibition	Hildebrand et al. (2007)
Gβ2γ & PKA Gβγ & PKA	Dopamine D1 and D2 receptor; Muscarinic M4 receptor	Inhibition	Hu et al. (2009) and Zhang et al. (2012)
Gβγ & PKCη	Melatonin receptor 2	Inhibition	Zhang et al. (2018)
Gβγ & PKCα	Insulin-like growth factor 1 receptor	Enhancement	Zhang et al. (2014)
Gβγ	Corticotropin-releasing factor receptor 1	Inhibition	Tao et al. (2008)

2007). Moreover, it has been shown that loop II-III is a conserved phosphorylation site in all isoforms and has a fundamental role for the phosphorylation of PKA (Ser[1107]), CaMKII (Ser[1198]), Rho-kinase and PKC (Welsby et al., 2003; Wolfe et al., 2003; Kim et al., 2006; Park et al., 2006; Iftinca et al., 2007).

Still, several differences even for the same neurotransmitter were observed. An explanation could be the use of different cell types, native or recombinant receptors, and experimental conditions. Indeed, in mammalian cells the current density for all isoforms of Ca_v3 was elevated following activation of PKC or PKA at physiological temperature (37 °C) but not at room temperature (Chemin et al., 2007). Furthermore, the direct phosphorylation of $Ca_v3.2$ by PKA is temperature-dependent and was only demonstrated at 37 °C and not at room temperature (Chemin et al., 2007).

Summary

In summary, the intracellular, cytoplasmic loops connecting the transmembrane segments of the $Ca_v\alpha1$ subunit are major targets for G protein subunits, kinases, phosphatases, lipids, and Ca^{2+} binding proteins. In general, the $G_{i/o}$- and also G_s-coupled $G_{\beta\gamma}$ heterodimers bind directly to VGCCs to mediate its timely inhibitory effects while G_q-coupled mediated regulatory actions are slower and involve secondary messengers and their targets such as PKC, CaMKII, CaM, and Ca^{2+} itself and its binding proteins. However, G_s-coupled second messengers targets such as PKA can also phosphorylate cytoplasmic sites on the $Ca_v\alpha1$ subunit to bidirectionally modulate VGCCs. To add to this complexity, the different subtypes have various splice variants which respond differently to the various $Ca_v\beta$ and $G_{\beta\gamma}$ subtypes. In addition, these effects are mediated by potentially 40 different neurotransmitters stimulating more than 1000 GPCRs in various areas of our CNS, peripheral nervous system, heart, smooth muscle, and endocrine system to properly respond to our environment during fight-or-flight response and cognitive or motor tasks. Coordination of these different G protein pathways to tightly control VGCCs is important to maintain equilibrium Ca^{2+} levels in the cell. Dysfunction or dysregulation of VGCCs in any of these regions can lead to deficits in our cognition, motor coordination, and proper response during threatening situations.

References

Agler, H. L., Evans, J., Colecraft, H. M., & Yue, D. T. (2003). Custom distinctions in the interaction of G-protein beta subunits with N-type (CaV2.2) versus P/Q-type (CaV2.1) calcium channels. *The Journal of General Physiology, 121,* 495–510.

Agler, H., Evans, J., Tay, L., Anderson, M., Colecraft, H., & Yue, D. (2005). G protein-gated inhibitory module of N-type (ca(v)2.2) ca2+ channels. *Neuron, 46*(6), 891–904.

Altier, C., Khosravani, H., Evans, R. M., Hameed, S., Peloquin, J. B., Vartian, B. A., et al. (2006). ORL1 receptor-mediated internalization of N-type calcium channels. *Nature Neuroscience, 9*(1), 31–40.

Alvarez, J. L., Rubio, L. S., & Vassort, G. (1996). Facilitation of T-type calcium current in bullfrog atrial cells: Voltage-dependent relief of a G protein inhibitory tone. *The Journal of Physiology, 491*(2), 321–334.

Andrade, A., Denome, S., Jiang, Y., Marangoudakis, S., & Lipscombe, D. (2010). Opioid inhibition of N-type Ca2+ channels and spinal analgesia couple to alternative splicing. *Nature Neuroscience, 13*(10), 1249–1256.

Anliker, B., & Chun, J. (2004). Lysophospholipid G protein-coupled receptors. *The Journal of Biological Chemistry, 279*(20), 20555–20558.

Arnot, M. I., Stotz, S. C., Jarvis, S. E., & Zamponi, G. W. (2000). Differential modulation of N-type 1B and P/Q-type 1A calcium channels by different G protein subunit isoforms. *The Journal of Physiology, 527,* 203–212.

Arnoult, C., Lemos, J. R., & Florman, H. M. (1997). Voltage-dependent modulation of T-type calcium channels by protein tyrosine phosphorylation. *The EMBO Journal, 16*(7), 1593–1599.

Arnoult, C., Villaz, M., & Florman, H. M. (1998). Pharmacological properties of the T-type Ca2+ current mouse spermatogenic cells. *Molecular Pharmacology, 53*(6), 1104–1111.

Barrett, C., & Rittenhouse, A. (2000). Modulation of N-type calcium channel activity by G-proteins and protein kinase C. *The Journal of General Physiology, 115*(3), 277–286.

Barrett, P. Q., Lu, H. K., Colbran, R., Czernik, A., & Pancrazio, J. J. (2000). Stimulation of unitary T-type Ca2+ channel currents by calmodulin-dependent protein kinase II. *American Journal of Physiology. Cell Physiology, 279*(6), 48–46.

Bean, B. P. (1989). Neurotransmitter inhibition of neuronal calcium currents by changes in channel voltage dependence. *Nature, 340*, 153–156.

Bear, M. F., Connors, B. W., & Paradiso, M. A. (2015). *Neuroscience: Exploring the brain: Fourth edition.* Jones & Bartlett Learning.

Bernheim, L., Beech, D., & Hille, B. (1991). A diffusible second messenger mediates one of the pathways coupling receptors to calcium channels in rat sympathetic neurons. *Neuron, 6*(6), 859–867.

Bkaily, G., Sculptoreanu, A., Wang, S., Nader, M., Hazzouri, K. M., Jacques, D., et al. (2005). Angiotensin II-induced increase of T-type Ca2+ current and decrease of L-type Ca2+ current in heart cells. *Peptides, 26*(8), 1410–1417.

Bolshakov, V., & Siegelbaum, S. (1994). Postsynaptic induction and presynaptic expression of hippocampal long-term depression. *Science, 264*(5162), 1148–1152.

Bonvallet, R., & Rougier, O. (1989). Existence of two calcium currents recorded at normal calcium concentrations in single frog atrial cells. *Cell Calcium, 10*(7), 499–508.

Bucci, G., Mochida, S., & Stephens, G. (2011). Inhibition of synaptic transmission and G protein modulation by synthetic CaV2.2 Ca^2+ channel peptides. *The Journal of Physiology, 589*(13), 3085–3101.

Carabelli, V., D'Ascenzo, M., Carbone, E., Grassi, C. (2002). Nitric oxide inhibits neuroendocrine Ca(V)1 L-channel gating via cGMP-dependent protein kinase in cell-attached patches of bovine chromaffin cells. *The Journal of Physiology, 541*(2), 351–366. https://doi.org/10.1113/jphysiol.2002.017749

Cairns, S., & Borrani, F. (2015). β-Adrenergic modulation of skeletal muscle contraction: Key role of excitation-contraction coupling. *The Journal of Physiology, 593*(21), 4713–4727.

Calin-Jageman, I., Yu, K., Hall, R., Mei, L., & Lee, A. (2007). Erbin enhances voltage-dependent facilitation of Ca(v)1.3 Ca2+ channels through relief of an autoinhibitory domain in the Ca(v)1.3 alpha1 subunit. *The Journal of Neuroscience, 27*(6), 1374–1385.

Campbell, V., Berrow, N. S., Fitzgerald, E. M., Brickley, K., & Dolphin, A. C. (1995). Inhibition of the interaction of G protein G(o) with calcium channels by the calcium channel beta-subunit in rat neurones. *The Journal of Physiology, 485*, 365–372.

Canti, C., Page, K. M., Stephens, G. J., & Dolphin, A. C. (1999). Identification of residues in the N terminus of alpha1B critical for inhibition of the voltage-dependent calcium channel by Gbeta gamma. *The Journal of Neuroscience, 19*, 6855–6864.

Canti, C., Bogdanov, Y., & Dolphin, A. C. (2000). Interaction between G proteins and accessory subunits in the regulation of 1B calcium channels in Xenopus oocytes. *The Journal of Physiology, 527*, 419–432.

Carabelli, V., Lovallo, M., Magnelli, V., Zucker, H., & Carbone, E. (1996). Voltage-dependent modulation of single N-Type Ca2+ channel kinetics by receptor

agonists in IMR32 cells. *Biophysical Journal, 70*(5), 2144–2154.

Cataldi, M., Gaudino, A., Lariccia, V., Russo, M., Amoroso, S., Di Renzo, G., et al. (2004). Imatinib-mesylate blocks recombinant T-type calcium channels expressed in human embryonic kidney-293 cells by a protein tyrosine kinase-independent mechanism. *The Journal of Pharmacology and Experimental Therapeutics, 309*(1), 208–215.

Catterall, W. A. (1999). Interactions of presynaptic Ca2+ channels and snare proteins in neurotransmitter release. *Annals of the New York Academy of Sciences, 868*, 144–159.

Catterall, W. A. (2000). Structure and regulation of voltage-gated Ca2+ channels. *Annual Review of Cell and Developmental Biology, 16*, 521–555.

Cesetti, T., Hernández-Guijo, J., Baldelli, P., Carabelli, V., & Carbone, E. (2003). Opposite action of beta1- and beta2-adrenergic receptors on Ca(V)1 L-channel current in rat adrenal chromaffin cells. *The Journal of Neuroscience, 23*(1), 73–83.

Chemin, J., Monteil, A., Perez-Reyes, E., Nargeot, J., & Lory, P. (2001). Direct inhibition of T-type calcium channels by the endogenous cannabinoid anandamide. *The EMBO Journal, 20*(24), 7033–7040.

Chemin, J., Mezghrani, A., Bidaud, I., Dupasquier, S., Marger, F., Barrère, C., et al. (2007). Temperature-dependent modulation of CaV3 T-type calcium channels by protein kinases C and A in mammalian cells. *The Journal of Biological Chemistry, 282*(45), 32710–32718.

Chen, J., DeVivo, M., Dingus, J., Harry, A., Li, J., Sui, J., et al. (1995). A region of adenylyl cyclase 2 critical for regulation by G protein beta gamma subunits. *Science, 268*, 1166–1169.

Chen, C., Xu, R., Clarke, I. J., Ruan, M., Loneragan, K., & Roh, S. G. (2000). Diverse intracellular signalling systems used by growth hormone-releasing hormone in regulating voltage-gated Ca2+ or K+ channels in pituitary somatotropes. *Immunology and Cell Biology, 78*(4), 356–368.

Cooper, C., Arnot, M., Feng, Z., Jarvis, S., Hamid, J., & Zamponi, G. (2000). Cross-talk between G-protein and protein kinase C modulation of N-type calcium channels is dependent on the G-protein beta subunit isoform. *The Journal of Biological Chemistry, 275*(52), 40777–40781.

Currie, K. P. M. (2010). G protein inhibition of CaV2 calcium channels. *Channels, 4*(6), 497–509.

Currie, K., & Fox, A. (2002). Differential facilitation of N- and P/Q-type calcium channels during trains of action potential-like waveforms. *The Journal of Physiology, 539*(2), 419–431.

De Waard, M., Witcher, D., Pragnell, M., Liu, H., & Campbell, K. (1995). Properties of the alpha 1-beta anchoring site in voltage-dependent Ca2+ channels. *The Journal of Biological Chemistry, 270*(20), 12056–12064.

De Waard, M., Liu, H., Walker, D., Scott, V. E., Gurnett, C. A., & Campbell, K. P. (1997). Direct binding of

G-protein betagamma complex to voltage-dependent calcium channels. *Nature, 385*, 446–450.

De Waard, M., Hering, J., Weiss, N., & Feltz, A. (2005). How do G proteins directly control neuronal Ca2+ channel function? *Trends in Pharmacological Sciences, 26*(8), 427–436.

Delmas, P., Coste, B., Gamper, N., & Shapiro, M. (2005). Phosphoinositide lipid second messengers: New paradigms for calcium channel modulation. *Neuron, 47*(2), 179–182.

DeMaria, C., Soong, T., Alseikhan, B., Alvania, R., & Yue, D. (2001). Calmodulin bifurcates the local Ca2+ signal that modulates P/Q-type Ca2+ channels. *Nature, 411*(6836), 484–489.

DePuy, S. D., Yao, J., Hu, C., McIntire, W., Bidaud, I., Lory, P., et al. (2006). The molecular basis for T-type Ca2+ channel inhibition by G protein beta2gamma2 subunits. *Proceedings of the National Academy of Sciences, 103*(39), 14590–14595.

Doering, C. J., Kisilevsky, A. E., Feng, Z. P., Arnot, M. I., Peloquin, J., Hamid, J., et al. (2004). A single G beta subunit locus controls cross-talk between protein kinase C and G protein regulation of N-type calcium channels. *The Journal of Biological Chemistry, 279*, 29709–29717.

Dresviannikov, A., Page, K., Leroy, J., Pratt, W., & Dolphin, A. (2009). Determinants of the voltage dependence of G protein modulation within calcium channel beta subunits. *Pflügers Archiv, 457*(4), 743–756.

Drolet, P., Bilodeau, L., Chorvatova, A., Laflamme, L., Gallo-Payet, N., & Payet, M. D. (1997). Inhibition of the T-type Ca2+ current by the dopamine D1 receptor in rat adrenal glomerulosa cells: Requirement of the combined action of the G betagamma protein subunit and cyclic adenosine 3′,5′-monophosphate. *Molecular Endocrinology, 11*(4), 503–514.

Dunlap, K., & Fischbach, G. (1978). Neurotransmitters decrease the calcium ocmponent of sensory neurone action potentials. *Nature, 276*(5690), 837–839.

Elmslie, K. (2003). Neurotransmitter modulation of neuronal calcium channels. *Journal of Bioenergetics and Biomembranes, 35*(6), 477–489.

Elmslie, K. S., & Jones, S. W. (1994). Concentration dependence of neurotransmitter effects on calcium current kinetics in frog sympathetic neurones. *The Journal of Physiology, 481*, 35–46.

Emrick, M., Sadilek, M., Konoki, K., & Catterall, W. (2010). Beta-adrenergic-regulated phosphorylation of the skeletal muscle Ca(V)1.1 channel in the fight-or-flight response. *Proceedings of the National Academy of Sciences of the United States of America, 107*(43), 18712–18717.

Evans, R. M., You, H., Hameed, S., Altier, C., Mezghrani, A., Bourinet, E., et al. (2010). Heterodimerization of ORL1 and opioid receptors and its consequences for N-type calcium channel regulation. *The Journal of Biological Chemistry, 285*(2), 1032–1040.

Feng, Z., Arnot, M., Doering, C., & Zamponi, G. (2001). Calcium channel beta subunits differentially regu-

late the inhibition of N-type channels by individual Gbeta isoforms. *The Journal of Biological Chemistry, 276*(48), 45051–45058.

Fern, R. J., Hahm, M. S., Lu, H. K., Liu, L. P., Gorelick, F. S., & Barrett, P. Q. (1995). Ca2+/calmodulin-dependent protein kinase II activation and regulation of adrenal glomerulosa Ca2+ signaling. *The American Journal of Physiology, 269*(6), 751–760.

Forscher, P., Oxford, G. S., & Schulz, D. (1986). Noradrenaline modulates calcium channels in avian dorsal root ganglion cells through tight receptor-channel coupling. *The Journal of Physiology, 379*, 131–144.

Furukawa, T., Ito, H., Nitta, J., Tsujino, M., Adachi, S., Hiroe, M., et al. (1992). Endothelin-1 enhances calcium entry through T-type calcium channels in cultured neonatal rat ventricular myocytes. *Circulation Research, 71*(5), 1242–1253.

Gamper, N., Reznikov, V., Yamada, Y., Yang, J., & Shapiro, M. (2004). Phosphatidylinositol [correction] 4,5-bisphosphate signals underlie receptor-specific Gq/11-mediated modulation of N-type Ca2+ channels. *The Journal of Neuroscience, 24*(48), 10980–10992.

Gray, P., Scott, J., & Catterall, W. (1998). Regulation of ion channels by cAMP-dependent protein kinase and A-kinase anchoring proteins. *Current Opinion in Neurobiology, 8*(3), 330–334.

Grazzini, E., Durroux, T., Payet, M. D., Bilodeau, L., Gallo-Payet, N., & Guillon, G. (1996). Membrane-delimited G protein-mediated coupling between V(1a) vasopressin receptor and dihydropyridine binding sites in rat glomerulosa cells. *Molecular Pharmacology, 50*(5), 1273–1283.

Gross, R. A., Moises, H. C., Uhler, M. D., & Macdonald, R. L. (1990a). Dynorphin a and cAMP-dependent protein kinase independently regulate neuronal calcium currents. *Proceedings of the National Academy of Sciences of the United States of America, 87*(18), 7025–7029.

Gross, R. A., Uhler, M. D., & Macdonald, R. L. (1990b). The cyclic AMP-dependent protein kinase catalytic subunit selectively enhances calcium currents in rat nodose neurones. *The Journal of Physiology, 429*(1), 483–496.

Haeseleer, F., Imanishi, Y., Maeda, T., Possin, D., Maeda, A., Lee, A., et al. (2004). Essential role of Ca2+-binding protein 4, a Cav1.4 channel regulator, in photoreceptor synaptic function. *Nature Neuroscience, 7*(10), 1079–1087.

Hamid, J., Nelson, D., Spaetgens, R., Dubel, S., Snutch, T., & Zamponi, G. (1999). Identification of an integration center for cross-talk between protein kinase C and G protein modulation of N-type calcium channels. *The Journal of Biological Chemistry, 274*(10), 6195–6202.

Harraz, O. F., & Welsh, D. G. (2013). Protein kinase A regulation of T-type Ca2+ channels in rat cerebral arterial smooth muscle. *Journal of Cell Science, 126*(13), 2944–2954.

Harvey, R., & Hell, J. (2013). CaV1.2 signaling complexes in the heart. *Journal of Molecular and Cellular Cardiology, 58*(1), 143–152.

Heidelberger, R., Thoreson, W. B., & Witkovsky, P. (2005). Synaptic transmission at retinal ribbon synapses. *Progress in Retinal and Eye Research, 24*(6), 682–720.

Herlitze, S., Garcia, D. E., Mackie, K., Hille, B., Scheuer, T., & Catterall, W. A. (1996). Modulation of Ca2+ channels by G-protein beta gamma subunits. *Nature, 380*(6571), 258–262. https://doi.org/10.1038/380258a0

Herlitze, S., Hockerman, G. H., Scheuer, T., & Catterall, W. A. (1997). Molecular determinants of inactivation and G protein modulation in the intracellular loop connecting domains I and II of the calcium channel alpha1A subunit. *Proceedings of the National Academy of Sciences of the United States of America, 94*(4), 1512–1516.

Herlitze, S., Zhong, H., Scheuer, T., & Catterall, W. A. (2001). Allosteric modulation of Ca2+ channels by G proteins, voltage-dependent facilitation, protein kinase C, and Ca(v)beta subunits. *Proceedings of the National Academy of Sciences of the United States of America, 98*(8), 4699–4704.

Hermosilla, T., Moreno, C., Itfinca, M., Altier, C., Armisén, R., Stutzin, A., et al. (2011). L-type calcium channel β subunit modulates angiotensin II responses in cardiomyocytes. *Channels (Austin, Tex.), 5*(3), 280–286.

Hernádez-López, S., Tkatch, T., Perez-Garci, E., Galarraga, E., Bargas, J., Hamm, H., et al. (2000). D2 dopamine receptors in striatal medium spiny neurons reduce L-type Ca2+ currents and excitability via a novel PLCβ1-IP3-Calcineurin-signaling cascade. *The Journal of Neuroscience, 20*(24), 8987–8995.

Hernández-Ochoa, E., García-Ferreiro, R., & García, D. (2007a). G protein activation inhibits gating charge movement in rat sympathetic neurons. *American Journal of Physiology. Cell Physiology, 292*(6), C2226–C2238.

Hernández-Ochoa, E., García-Ferreiro, R., & García, D. (2007b). G protein activation inhibits gating charge movement in rat sympathetic neurons. *American Journal of Physiology. Cell Physiology, 292*(6), 2226–2238.

Hildebrand, M. E., David, L. S., Hamid, J., Mulatz, K., Garcia, E., Zamponi, G. W., et al. (2007). Selective inhibition of Cav3.3 T-type calcium channels by Gαq/11-coupled muscarinic acetylcholine receptors. *The Journal of Biological Chemistry, 282*(29), 21043–21055.

Hille, B. (1994). Modulation of ion-channel function by G-protein-coupled receptors. *Trends in Neurosciences, 17*(12), 531–536.

Hille, B., Dickson, E., Kruse, M., Vivas, O., & Suh, B. (2015). Phosphoinositides regulate ion channels. *Biochimica et Biophysica Acta, 1851*(6), 844–856.

Hockberger, P., Toselli, M., Swandulla, D., & Lux, H. D. (1989). A diacylglycerol analogue reduces neuronal calcium currents independently of protein kinase C activation. *Nature, 338*(6213), 340–342.

Howe, A., & Surmeier, D. (1995). Muscarinic receptors modulate N-, P-, and L-type Ca2+ currents in rat striatal neurons through parallel pathways. *The Journal of Neuroscience, 15*(1), 458–469.

Hu, C., DePuy, S. D., Yao, J., McIntire, W. E., & Barrett, P. Q. (2009). Protein kinase A activity controls the regulation of T-type CaV3.2 channels by Gbetagamma dimers. *The Journal of Biological Chemistry, 284*(12), 7465–7473.

Hulme, J., Ahn, M., Hauschka, S., Scheuer, T., & Catterall, W. (2002). A novel leucine zipper targets AKAP15 and cyclic AMP-dependent protein kinase to the C terminus of the skeletal muscle Ca2+ channel and modulates its function. *The Journal of Biological Chemistry, 277*(6), 4079–4087.

Hümmer, A., Delzeith, O., Gomez, S. R., Moreno, R. L., Mark, M. D., & Herlitze, S. (2003). Competitive and synergistic interactions of G protein beta(2) and Ca(2+) channel beta(1b) subunits with Ca(v)2.1 channels, revealed by mammalian two-hybrid and fluorescence resonance energy transfer measurements. *The Journal of Biological Chemistry, 278*(49), 49386–49400.

Iftinca, M., Hamid, J., Chen, L., Varela, D., Tadayonnejad, R., Altier, C., et al. (2007). Regulation of T-type calcium channels by Rho-associated kinase. *Nature Neuroscience, 10*(7), 854–860.

Ikeda, S. R. (1996). Voltage-dependent modulation of N-type calcium channels by G-protein beta gamma subunits. *Nature, 380*(6571), 255–258.

Jenkins, M., Christel, C., Jiao, Y., Abiria, S., Kim, K., Usachev, Y., et al. (2010). Ca2+-dependent facilitation of Cav1.3 Ca2+ channels by densin and Ca2+/calmodulin-dependent protein kinase II. *The Journal of Neuroscience, 30*(15), 5125–5135.

Johnson, B., Scheuer, T., & Catterall, W. (2005). Convergent regulation of skeletal muscle Ca2+ channels by dystrophin, the actin cytoskeleton, and cAMP-dependent protein kinase. *Proceedings of the National Academy of Sciences of the United States of America, 102*(11), 4191–4196.

Jones, S. W., & Elmslie, K. S. (1997). Transmitter modulation of neuronal calcium channels. *The Journal of Membrane Biology, 155*, 1–10.

Kang, D., Hur, C. G., Park, J. Y., Han, J., & Hong, S. G. (2007). Acetylcholine increases Ca2+ influx by activation of CaMKII in mouse oocytes. *Biochemical and Biophysical Research Communications, 360*(2), 476–482.

Kasai, H. (1992). Voltage- and time-dependent inhibition of neuronal calcium channels by a GTP-binding protein in a mammalian cell line. *The Journal of Physiology, 448*(1), 189–209.

Katchman, A., Yang, L., Zakharov, S. I., Kushner, J., Abrams, J., Chen, B. X., et al. (2017). Proteolytic cleavage and PKA phosphorylation of α1C subunit are not required for adrenergic regulation of CaV1.2 in the heart. *Proceedings of the National Academy of Sciences of the United States of America, 114*(34), 9194–9199.

Katz, M., Subramaniam, S., Chomsky-Hecht, O., Tsemakhovich, V., Flockerzi, V., Klussmann, E., et al. (2021). Reconstitution of β-adrenergic regulation of Ca V 1.2: Rad-dependent and Rad-independent protein kinase A mechanisms. *Proceedings of the National Academy of Sciences of the United States of America, 118*(21), e2100021118.

Kaur, G., Pinggera, A., Ortner, N., Lieb, A., Sinnegger-Brauns, M., Yarov-Yarovoy, V., et al. (2015). A polybasic plasma membrane binding motif in the I-II linker stabilizes voltage-gated CaV1.2 calcium channel function. *The Journal of Biological Chemistry, 290*(34), 21086–21100.

Kawai, F., & Miyachi, E. I. (2001). Modulation by cGMP of the voltage-gated currents in newt olfactory receptor cells. *Neuroscience Research, 39*(3), 327–337.

Kawai, F., Kurahashi, T., & Kaneko, A. (1999). Adrenaline enhances odorant contrast by modulating signal encoding in olfactory receptor cells. *Nature Neuroscience, 2*(2), 133–138.

Keja, J. A., Stoof, J. C., & Kits, K. S. (1992). Dopamine D2 receptor stimulation differentially affects voltage-activated calcium channels in rat pituitary melanotropic cells. *The Journal of Physiology, 450*(1), 409–435.

Kim, J. A., Park, J. Y., Kang, H. W., Huh, S. U., Jeong, S. W., & Lee, J. H. (2006). Augmentation of Cav3.2 T-type calcium channel activity by cAMP-dependent protein kinase A. *The Journal of Pharmacology and Experimental Therapeutics, 318*(1), 230–237.

Kim, H. H., Lee, K. H., Lee, D., Han, Y. E., Lee, S. H., Sohn, J. W., & Ho, W. K. (2015). Costimulation of AMPA and metabotropic glutamate receptors underlies phospholipase C activation by glutamate in hippocampus. *The Journal of Neuroscience, 35*(16), 6401–6412. https://doi.org/10.1523/JNEUROSCI.4208-14.2015

Kim, Y., Park, M. K., Uhm, D. Y., & Chung, S. (2007). Modulation of T-type Ca2+ channels by corticotropin-releasing factor through protein kinase C pathway in MN9D dopaminergic cells. *Biochemical and Biophysical Research Communications, 358*(3), 796–801.

Kitano, J., Nishida, M., Itsukaichi, Y., Minami, I., Ogawa, M., Hirano, T., et al. (2003). Direct interaction and functional coupling between metabotropic glutamate receptor subtype 1 and voltage-sensitive Cav2.1 Ca2+ channel. *The Journal of Biological Chemistry, 278*(27), 25101–25108.

Kobrinsky, E. M., Pearson, H. A., & Dolphin, A. C. (1994). Low- and high-voltage-activated calcium channel currents and their modulation in the dorsal root ganglion cell line ND7-23. *Neuroscience, 58*(3), 539–552.

Kurejová, M., & Lacinová, L. (2006 Feb). Effect of protein tyrosine kinase inhibitors on the current through the CaV3.1 channel. *Archives of Biochemistry and Biophysics, 446*(1), 20–27.

Lacinova, L., Mallmann, R. T., Jurkovičová-Tarabová, B., & Klugbauer, N. (2020). Modulation of voltage-gated CaV2.2 Ca2+ channels by newly identified interaction partners. *Channels, 14*(1), 380–392.

Lee, H. K., & Elmslie, K. S. (2000). Reluctant gating of single N-type calcium channels during neurotransmitter-induced inhibition in bullfrog sympathetic neurons. *The Journal of Neuroscience, 20*, 3115–3128.

Lee, A., Wong, S. T., Gallagher, D., Li, B., Storm, D. R., Scheuer, T., et al. (1999). Ca2+/calmodulin binds to and modulates P/Q-type calcium channels. *Nature, 399*(6732), 155–159.

Lee, A., Scheuer, T., & Catterall, W. A. (2000). Ca2+/calmodulin-dependent facilitation and inactivation of P/Q-type Ca2+ channels. *The Journal of Neuroscience, 20*(18), 6830–6838.

Lemke, T., Welling, A., Christel, C. J., Blaich, A., Bernhard, D., Lenhardt, P., et al. (2008). Unchanged β-adrenergic stimulation of cardiac L-type calcium channels in Cav1.2 phosphorylation site S1928A mutant mice. *The Journal of Biological Chemistry, 283*(50), 34738–34744.

Lenglet, S., Louiset, E., Delarue, C., Vaudry, H., & Contesse, V. (2002). Activation of 5-HT(7) receptor in rat glomerulosa cells is associated with an increase in adenylyl cyclase activity and calcium influx through T-type calcium channels. *Endocrinology, 143*(5), 1748–1760.

Leroy, J., Richards, M., Butcher, A., Nieto-Rostro, M., Pratt, W., Davies, A., et al. (2005). Interaction via a key tryptophan in the I-II linker of N-type calcium channels is required for beta1 but not for palmitoylated beta2, implicating an additional binding site in the regulation of channel voltage-dependent properties. *The Journal of Neuroscience, 25*(30), 6984–6996.

Li, Y., Wang, F., Zhang, X., Qi, Z., Tang, M., Szeto, C., et al. (2012). ß-Adrenergic stimulation increases Cav3.1 activity in cardiac myocytes through protein kinase A. *PLoS One, 7*(7), e39965.

Li, Q., Zhang, Y., Wu, N., Yin, N., Sun, X. H., & Wang, Z. (2019). Activation of somatostatin receptor 5 suppresses T-type Ca2+ channels through NO/cGMP/PKG signaling pathway in rat retinal ganglion cells. *Neuroscience Letters, 708*, 134337.

Lin, Z., Haus, S., Edgerton, J., & Lipscombe, D. (1997). Identification of functionally distinct isoforms of the N-type Ca2+ channel in rat sympathetic ganglia and brain. *Neuron, 18*, 153–166.

Liu, B., Hill, S. J., & Khan, R. N. (2005). Oxytocin inhibits T-type calcium current of human decidual stromal cells. *The Journal of Clinical Endocrinology and Metabolism, 90*(7), 4191–4197.

Liu, K., Jiang, D., Zhang, T., Tao, J., Shen, L., & Sun, X. (2011). Activation of growth hormone secretagogue type 1a receptor inhibits T-type Ca2+ channel currents through pertussis toxin-sensitive novel protein kinase C pathway in mouse spermatogenic cells. *Cellular Physiology and Biochemistry, 27*(5), 613–624.

Liu, G., Papa, A., Katchman, A., Zakharov, S., Roybal, D., Hennessey, J., et al. (2020). Mechanism of adrenergic Ca V 1.2 stimulation revealed by proximity proteomics. *Nature, 577*(7792), 695–700.

Lledo, P. M., Israel, J. M., & Vincent, J. D. (1990a). A guanine nucleotide-binding protein mediates the inhibition of voltage-dependent calcium currents by dopamine in rat lactotrophs. *Brain Research, 528*(1), 143–147.

Lledo, P. M., Legendre, P., Israel, J. M., & Vincent, J. D. (1990b). Dopamine inhibits two characterized voltage-dependent calcium currents in identified rat lactotroph cells. *Endocrinology, 127*(3), 990–1001.

Lledo, P. M., Homburger, V., Bockaert, J., & Vincent, J. D. (1992). Differential G protein-mediated coupling of D2 dopamine receptors to K+ and Ca2+ currents in rat anterior pituitary cells. *Neuron, 8*(3), 455–463.

Louiset, E., Duparc, C., Lenglet, S., Gomez-Sanchez, C. E., & Lefebvre, H. (2017). Role of cAMP/PKA pathway and T-type calcium channels in the mechanism of action of serotonin in human adrenocortical cells. *Molecular and Cellular Endocrinology, 441*, 99–107.

Lu, H. K., Fern, R. J., Nee, J. J., & Barrett, P. Q. (1994). Ca2+-dependent activation of T-type Ca2+ channels by calmodulin- dependent protein kinase II. *The American Journal of Physiology, 267*(1), 183–189.

Lu, H. K., Fern, R. J., Luthin, D., Linden, J., Liu, L. P., Cohen, C. J., et al. (1996). Angiotensin II stimulates T-type Ca2+ channel currents via activation of a G protein, G(i). *The American Journal of Physiology, 271*(4), 1340–1349.

Mahapatra, S., Calorio, C., Vandael, D., Marcantoni, A., Carabelli, V., & Carbone, E. (2012). Calcium channel types contributing to chromaffin cell excitability, exocytosis and endocytosis. *Cell Calcium, 51*(3–4), 321–330.

Mandy, L. R. C., & Rittenhouse, A. R. (2009). Arachidonic acid inhibition of L-type calcium (Ca V 1.3b) channels varies with accessory Ca Vβ subunits. *The Journal of General Physiology, 133*(4), 387–403.

Marchetti, C., & Brown, A. M. (1988). Protein kinase activator 1-oleoyl-2-acetyl-sn-glycerol inhibits two types of calcium currents in GH3 cells. *The American Journal of Physiology, 254*(1), 206–210.

Marchetti, C., Carbone, E., & Lux, H. D. (1986). Effects of dopamine and noradrenaline on Ca channels of cultured sensory and sympathetic neurons of chick. *Pflügers Archiv, 406*(2), 104–111.

Margeta-Mitrovic, M., Grigg, J. J., Koyano, K., Nakajima, Y., & Nakajima, S. (1997). Neurotensin and substance P inhibit low- and high-voltage-activated Ca2+ channels in cultured newborn rat nucleus basalis neurons. *Journal of Neurophysiology, 78*(3), 1341–1352.

Mark, M. D., Wittemann, S., & Herlitze, S. (2000). G protein modulation of recombinant P/Q-type calcium channels by regulators of G protein signalling proteins. *The Journal of Physiology, 528*(1), 65–77.

Masland, R. (2012). The neuronal organization of the retina. *Neuron, 76*(2), 266–280.

Masuho, I., Skamangas, N., Muntean, B., & Martemyanov, K. (2021). Diversity of the Gβγ complexes defines spatial and temporal bias of GPCR signaling. *Cell Systems, 12*(4), 324–337.

McCarthy, R. T., Isales, C., & Rasmussen, H. (1993). T-type calcium channels in adrenal glomerulosa

cells: GTP-dependent modulation by angiotensin II. *Proceedings of the National Academy of Sciences of the United States of America, 90*(8), 3260–3264.

Meves, H. (2005). The effect of prostaglandin E1 on ion currents of NG108-15 cells. *Prostaglandins & Other Lipid Mediators, 76*(1–4), 117–132.

Meza, U., & Adams, B. (1998). G-Protein-dependent facilitation of neuronal alpha1A, alpha1B, and alpha1E Ca channels. *The Journal of Neuroscience, 18*(14), 5240–5252.

Mochida, S. (2018). Presynaptic calcium channels. *Neuroscience Research, 127*, 33–44.

Morikawa, H., Fukuda, K., Mima, H., Shoda, T., Kato, S., & Mori, K. (1998). Tyrosine kinase inhibitors suppress N-type and T-type Ca2+ channel currents in NG108-15 cells. *Pflügers Archiv, 436*(1), 127–132.

Murali, S., Napier, I., Rycroft, B., & Christie, M. (2012). Opioid-related (ORL1) receptors are enriched in a subpopulation of sensory neurons and prolonged activation produces no functional loss of surface N-type calcium channels. *The Journal of Physiology, 590*(7), 1655–1667.

Murphy, B., & Scott, J. (1998). Functional anchoring of the cAMP-dependent protein kinase. *Trends in Cardiovascular Medicine, 8*(2), 89–95.

Nanou, E., & Catterall, W. (2018). Calcium channels, synaptic plasticity, and neuropsychiatric disease. *Neuron, 98*(3), 466–481.

Nussinovitch, I., & Kleinhaus, A. L. (1992). Dopamine inhibits voltage-activated calcium channel currents in rat pars intermedia pituitary cells. *Brain Research, 574*(1–2), 49–55.

Okamura, Y., Murata, Y., & Iwasaki, H. (2009). Voltage-sensing phosphatase: Actions and potentials. *The Journal of Physiology, 587*(3), 513–520.

Oldham, W., & Hamm, H. (2008). Heterotrimeric G protein activation by G-protein-coupled receptors. *Nature Reviews. Molecular Cell Biology, 9*(1), 60–71.

Olson, P. A., Tkatch, T., Hernandez-Lopez, S., Ulrich, S., Ilijic, E., Mugnaini, E., et al. (2005). G-protein-coupled receptor modulation of striatal Cav1.3 L-type Ca2+ channels is dependent on a shank-binding domain. *The Journal of Neuroscience, 25*(5), 1050–1062.

Osipenko, O. N., Várnai, P., Mike, A., Spät, A., & Vizi, E. S. (1994). Dopamine blocks T-type calcium channels in cultured rat adrenal. *Endocrinology, 134*(1), 511–514.

Page, K., Cantí, C., Stephens, G., Berrow, N., & Dolphin, A. (1998). Identification of the amino terminus of neuronal Ca2+ channel alpha1 subunits alpha1B and alpha1E as an essential determinant of G-protein modulation. *The Journal of Neuroscience, 18*(13), 4815–4824.

Page, K., Heblich, F., Margas, W., Pratt, W., Nieto-Rostro, M., Chaggar, K., et al. (2010). N terminus is key to the dominant negative suppression of Ca(V)2 calcium channels: Implications for episodic ataxia type 2. *The Journal of Biological Chemistry, 285*(2), 835–844.

Pan, J. Q., & Lipscombe, D. (2000). Alternative splicing in the cytoplasmic II-III loop of the N-type Ca channel alpha 1B subunit: Functional differences are beta subunit-specific. *The Journal of Neuroscience, 20,* 4769–4775.

Papa, A., Kushner, J., Hennessey, J. A., Katchman, A. N., Zakharov, S. I., Chen, B. X., et al. (2021). Adrenergic CaV1.2 activation via Rad phosphorylation converges at α1CI-II loop. *Circulation Research, 128*(1), 76–88.

Park, J. Y., Jeong, S. W., Perez-Reyes, E., & Lee, J. H. (2003). Modulation of Cav3.2 T-type Ca2+ channels by protein kinase C. *FEBS Letters, 547*(1–3), 37–42.

Park, J. Y., Kang, H. W., Moon, H. J., Huh, S. U., Jeong, S. W., Soldatov, N. M., et al. (2006). Activation of protein kinase C augments T-type Ca2+ channel activity without changing channel surface density. *The Journal of Physiology, 577*(2), 513–523.

Patriarchi, T., Qian, H., Di Biase, V., Malik, Z., Chowdhury, D., Price, J., et al. (2016). Phosphorylation of Cav1.2 on S1928 uncouples the L-type Ca2+ channel from the β2 adrenergic receptor. *The EMBO Journal, 35*(12), 1330–1345.

Patriarchi, T., Buonarati, O., & Hell, J. (2018). Postsynaptic localization and regulation of AMPA receptors and Cav1.2 by β2 adrenergic receptor/PKA and Ca 2+/CaMKII signaling. *The EMBO Journal, 37*(20), e99771.

Pemberton, K., Hill-Eubanks, L., & Penelope, J. S. (2000). Modulation of low-threshold T-type calcium channels by the five muscarinic receptor subtypes in NIH 3T3 cells. *Pflügers Archiv, 440*(3), 452–461.

Pfeiffer-Linn, C., & Lasater, E. M. (1993). Dopamine modulates in a differential fashion T- and L-type calcium currents in bass retinal horizontal cells. *The Journal of General Physiology, 102*(2), 277–294.

Pfeiffer-Linn, C. L., & Lasater, E. M. (1998). Multiple second-messenger system modulation of voltage-activated calcium currents in teleost retinal horizontal cells. *Journal of Neurophysiology, 80*(1), 377–388.

Polo-Parada, L., Chan, S. A., & Smith, C. (2006). An activity-dependent increased role for L-type calcium channels in exocytosis is regulated by adrenergic signaling in chromaffin cells. *Neuroscience, 143*(2), 445–459.

Pragnell, M., Sakamoto, J., Jay, S. D., & Campbell, K. P. (1991). Cloning and tissue-specific expression of the brain calcium channel beta-subunit. *FEBS Letters, 291,* 253–258.

Pragnell, M., De Waard, M., Mori, Y., Tanabe, T., Snutch, T. P., & Campbell, K. P. (1994). Calcium channel beta-subunit binds to a conserved motif in the I-II cytoplasmic linker of the alpha 1-subunit. *Nature, 368,* 67–70.

Proft, J., & Weiss, N. (2015). G protein regulation of neuronal calcium channels: Back to the future. *Molecular Pharmacology, 87*(6), 890–906.

Qian, W.-J., Yin, N., Gao, F., Miao, Y., Li, Q., Li, F., et al. (2017). Cannabinoid CB1 and CB2 receptors differentially modulate L- and T-type Ca2+ channels in rat retinal ganglion cells. *Neuropharmacology, 124,* 143–156.

Qin, N., Platano, D., Olcese, R., Stefani, E., & Birnbaumer, L. (1997). Direct interaction of gbetagamma with a C-terminal Gbetagamma-binding domain of the Ca2+ channel alpha1 subunit is responsible for channel inhibition by G protein-coupled receptors. *Proceedings of the National Academy of Sciences of the United States of America, 94,* 8866–8871.

Rangel, A., Sánchez-Armass, S., & Meza, U. (2010). Protein kinase C-mediated inhibition of recombinant T-type CaV3.2 channels by neurokinin 1 receptors. *Molecular Pharmacology, 77*(2), 202–210.

Rebolledo-Antúnez, S., Farías, J., Arenas, I., & García, D. (2009). Gating charges per channel of Ca(V)2.2 channels are modified by G protein activation in rat sympathetic neurons. *Archives of Biochemistry and Biophysics, 486*(1), 51–57.

Rettig, J., Sheng, Z. H., Kim, D. K., Hodson, C. D., Snutch, T. P., & Catterall, W. A. (1996). Isoform-specific interaction of the alpha1A subunits of brain Ca2+ channels with the presynaptic proteins syntaxin and SNAP-25. *Proceedings of the National Academy of Sciences of the United States of America, 93,* 7363–7368.

Roberts-Crowley, M., & Rittenhouse, A. (2018). Modulation of Ca V 1.3b L-type calcium channels by M 1 muscarinic receptors varies with Ca V β subunit expression. *BMC Research Notes, 11*(1), 681.

Rosenthal, W., Hescheler, J., Eckert, R., Offermanns, S., Schmidt, A., Hinsch, K. D., et al. (1990). Pertussis toxin-sensitive G-proteins: Participation in the modulation of voltage-dependent Ca2+ channels by hormones and neurotransmitters. *Advances in Second Messenger and Phosphoprotein Research, 24,* 89–94.

Rossier, M. F., Aptel, H. B. C., Python, C. P., Burnay, M. M., Vallotton, M. B., & Capponi, A. M. (1995). Inhibition of low threshold calcium channels by angiotensin II in adrenal glomerulosa cells through activation of protein kinase C. *The Journal of Biological Chemistry, 270*(25), 15137–15142.

Roybal, D., Hennessey, J. A., & Marx, S. O. (2020). The quest to identify the mechanism underlying adrenergic regulation of cardiac Ca2+ channels. *Channels, 14*(1), 123.

Sanderson, J., & Dell'Acqua, M. (2011). AKAP signaling complexes in regulation of excitatory synaptic plasticity. *The Neuroscientist, 17*(3), 321–336.

Sandoval, A., Duran, P., Gandini, M., Andrade, A., Almanza, A., Kaja, S., et al. (2017). Regulation of L-type Ca V 1.3 channel activity and insulin secretion by the cGMP-PKG signaling pathway. *Cell Calcium, 66,* 1–9.

Sang, L., Dick, I., & Yue, D. (2016). Protein kinase A modulation of CaV1.4 calcium channels. *Nature Communications, 7,* 12239.

Schmitt, H., & Meves, H. (1995). Modulation of neuronal calcium channels by arachidonic acid and related substances. *The Journal of Membrane Biology, 145*(3), 233–244.

Schrier, A. D., Wang, H., Talley, E. M., Perez-Reyes, E., & Barrett, P. Q. (2001). α1H T-type Ca2+ channel is the predominant subtype expressed in bovine and rat zona glomerulosa. *The American Journal of Physiology, 280*(2), 265–272.

Schroeder, J. E., & McCleskey, E. W. (1993). Inhibition of Ca2+ currents by a μ-opioid in a defined subset of rat sensory neurons. *The Journal of Neuroscience, 13*(2), 867–873.

Schroeder, J. E., Fischbach, P. S., & McCleskey, E. W. (1990). T-type calcium channels: Heterogeneous expression in rat sensory neurons and selective modulation by phorbol esters. *The Journal of Neuroscience, 10*(3), 947–951.

Schroeder, J. E., Fischbach, P. S., Zheng, D., & McCleskey, E. W. (1991). Activation of μ opioid receptors inhibits transient high- and low-threshold Ca 2+ currents, but spares a sustained current. *Neuron, 6*(1), 13–20.

Scott, R. H., Wootton, J. F., & Dolphin, A. C. (1990). Modulation of neuronal T-type calcium channel currents by photoactivation of intracellular guanosine 5′-0(3-thio) triphosphate. *Neuroscience, 38*(2), 285–294.

Sekiguchi, F., Aoki, Y., Nakagawa, M., Kanaoka, D., Nishimoto, Y., Tsubota-Matsunami, M., et al. (2013). AKAP-dependent sensitization of Cav3.2 channels via the EP 4 receptor/cAMP pathway mediates PGE2-induced mechanical hyperalgesia. *British Journal of Pharmacology, 168*(3), 734–745.

Shan, H. Q., Hammarback, J. A., & Godwin, D. W. (2013). Ethanol inhibition of a T-type Ca2+ channel through activity of protein kinase C. *Alcoholism, Clinical and Experimental Research, 37*(8), 1333–1342.

Sheng, Z. H., Rettig, J., Takahashi, M., & Catterall, W. A. (1994). Identification of a syntaxin-binding site on N-type calcium channels. *Neuron, 13*, 1303–1313.

Sheng, Z. H., Westenbroek, R. E., & Catterall, W. A. (1998). Physical link and functional coupling of presynaptic calcium channels and the synaptic vesicle docking/fusion machinery. *Journal of Bioenergetics and Biomembranes, 30*(4), 335–345.

Si, W., Zhang, Y., Chen, K., Hu, D., Qian, Z., Gong, S., et al. (2018). Fibroblast growth factor type 1 receptor stimulation of T-type Ca 2+ channels in sensory neurons requires the phosphatidylinositol 3-kinase and protein kinase A pathways, independently of Akt. *Cellular Signalling, 45*, 93–101.

Stanika, R., Flucher, B., & Obermair, G. (2015). Regulation of postsynaptic stability by the L-type calcium channel CaV1.3 and its interaction with PDZ proteins. *Current Molecular Pharmacology, 8*(1), 95–101.

Stella, S. L., Hu, W. D., Vila, A., & Brecha, N. C. (2007). Adenosine inhibits voltage-dependent Ca2+ influx in cone photoreceptor terminals of the tiger salamander retina. *Journal of Neuroscience Research, 85*(5), 1126–1137.

Stotz, S., & Zamponi, G. (2001). Structural determinants of fast inactivation of high voltage-activated Ca(2+) channels. *Trends in Neurosciences, 24*(3), 176–182.

Suh, B., Leal, K., & Hille, B. (2010). Modulation of high-voltage activated Ca(2+) channels by membrane phosphatidylinositol 4,5-bisphosphate. *Neuron, 67*(2), 224–238.

Suh, B., Kim, D., Falkenburger, B., & Hille, B. (2012). Membrane-localized β-subunits alter the PIP2 regulation of high-voltage activated Ca2+ channels. *Proceedings of the National Academy of Sciences of the United States of America, 109*(8), 3161–3166.

Swartz, K. (1993). Modulation of Ca2+ channels by protein kinase C in rat central and peripheral neurons: Disruption of G protein-mediated inhibition. *Neuron, 11*(2), 305–320.

Talavera, K., Staes, M., Janssens, A., Droogmans, G., & Nilius, B. (2004). Mechanism of arachidonic acid modulation of the T-type Ca2+ channel α1G. *The Journal of General Physiology, 124*(3), 225–238.

Tao, J., Hildebrand, M. E., Liao, P., Mui, C. L., Tan, G., Li, S., et al. (2008). Activation of corticotropin-releasing factor receptor 1 selectively inhibits CaV3.2 T-type calcium channels. *Molecular Pharmacology, 73*(6), 1596–1609.

Tao, J., Zhang, Y., Li, S., Sun, W., & Soong, T. W. (2009). Tyrosine kinase-independent inhibition by genistein on spermatogenic T-type calcium channels attenuates mouse sperm motility and acrosome reaction. *Cell Calcium, 45*(2), 133–143.

Tedford, H. W., Kisilevsky, A. E., Vieira, L. B., Varela, D., Chen, L., & Zamponi, G. W. (2010). Scanning mutagenesis of the I-II loop of the Cav2.2 calcium channel identifies residues Arginine 376 and Valine 416 as molecular determinants of voltage dependent G protein inhibition. *Molecular Brain, 3*, 6.

Tennakoon, M., Senarath, K., Kankanamge, D., Ratnayake, K., Wijayaratna, D., Olupothage, K., et al. (2021). Subtype-dependent regulation of Gβγ signalling. *Cellular Signalling, 82*, 109947.

Thoreson, W. B. (2021). Transmission at rod and cone ribbon synapses in the retina. *Pflügers Archiv, 473*(9), 1469–1491.

Toselli, M., & Lux, H. D. (1989). Opposing effects of acetylcholine on the two classes of voltage-dependent calcium channels in hippocampal neurons. *EXS, 57*, 97–103.

Tseng, G. N., & Boyden, P. A. (1991). Different effects of intracellular Ca and protein kinase C on cardiac T and L Ca currents. *The American Journal of Physiology, 261*(2), 364–379.

Vandael, D. H. F., Mahapatra, S., Calorio, C., Marcantoni, A., & Carbone, E. (2013). Cav1.3 and Cav1.2 channels of adrenal chromaffin cells: Emerging views on cAMP/cGMP-mediated phosphorylation and role in pacemaking. *Biochimica et Biophysica Acta, 1828*(7), 1608–1618.

Waldner, D., Bech-Hansen, N., & Stell, W. (2018). Channeling vision: Ca V 1.4-A critical link in retinal signal transmission. *BioMed Research International, 2018*, 7272630.

Wang, F., Zhang, Y., Jiang, X., Zhang, Y., Zhang, L., Gong, S., et al. (2011). Neuromedin U inhibits T-type Ca2+ channel currents and decreases membrane excitability in small dorsal root ganglia neurons in mice. *Cell Calcium, 49*(1), 12–22.

Wang, H., Wei, Y., Pu, Y., Jiang, D., Jiang, X., Zhang, Y., et al. (2019). Brain-derived neurotrophic factor stimulation of T-type Ca2+ channels in sensory neurons contributes to increased peripheral pain sensitivity. *Science Signaling, 12*(600), eaaw2300.

Weiss, S., & Dascal, N. (2015). Molecular aspects of modulation of L-type calcium channels by protein kinase C. *Current Molecular Pharmacology, 8*(1), 43–53.

Weiss, N., Tadmouri, A., Mikati, M., Ronjat, M., & De Waard, M. (2007). Importance of voltage-dependent inactivation in N-type calcium channel regulation by G-proteins. *Pflügers Archiv, 454*(1), 115–129.

Welsby, P. J., Wang, H., Wolfe, J. T., Colbran, R. J., Johnson, M. L., & Barrett, P. Q. (2003). A mechanism for the direct regulation of T-type calcium channels by Ca 2+/calmodulin-dependent kinase II. *The Journal of Neuroscience, 23*(31), 10116–10121.

Wettschureck, N., & Offermanns, S. (2005). Mammalian G proteins and their cell type specific functions. *Physiological Reviews, 85*(4), 1159–1204.

Williams, P. J., MacVicar, B. A., & Pittman, Q. J. (1990). Synaptic modulation by dopamine of calcium currents in rat pars intermedia. *The Journal of Neuroscience, 10*(3), 757–763.

Witcher, D., De Waard, M., Liu, H., Pragnell, M., & Campbell, K. (1995). Association of native Ca2+ channel beta subunits with the alpha 1 subunit interaction domain. *The Journal of Biological Chemistry, 270*(30), 18088–18093.

Witkovsky, P. (2004). Dopamine and retinal function. *Documenta Ophthalmologica, 108*(1), 17–39.

Wolfe, J. T., Wang, H., Perez-Reyes, E., & Barrett, P. Q. (2002). Stimulation of recombinant CaV3.2, T-type, Ca2+ channel currents by CaMKIIγC. *The Journal of Physiology, 538*(2), 343–355.

Wolfe, J. T., Wang, H., Howard, J., Garrison, J. C., & Barrett, P. Q. (2003). T-type calcium channel regulation by specific G-protein betagamma subunits. *Nature, 424*(6945), 209–213.

Wu, X., Kushwaha, N., Albert, P., & Penington, N. (2002). A critical protein kinase C phosphorylation site on the 5-HT(1A) receptor controlling coupling to N-type calcium channels. *The Journal of Physiology, 538*(1), 41–51.

Xiao, R., Zhu, W., Zheng, M., Chakir, K., Bond, R., Lakatta, E., et al. (2004). Subtype-specific beta-adrenoceptor signaling pathways in the heart and their potential clinical implications. *Trends in Pharmacological Sciences, 25*(7), 358–365.

Yu, H., Seo, J. B., Jung, S. R., Koh, D. S., & Hille, B. (2015). Noradrenaline upregulates T-type calcium channels in rat pinealocytes. *The Journal of Physiology, 593*(4), 887–904.

Yue, J., Zhang, Y., Li, X., Gong, S., Tao, J., & Jiang, X. (2014). Activation of G-protein-coupled receptor 30 increases t-type calcium currents in trigeminal ganglion neurons via the cholera toxin-sensitive protein kinase a pathway. *Die Pharmazie, 69*(11), 804–808.

Zamponi, G., & Currie, K. (2013). Regulation of Ca(V)2 calcium channels by G protein coupled receptors. *Biochimica et Biophysica Acta, 1828*(7), 1629–1643.

Zamponi, G. W., & Snutch, T. P. (1998). Modulation of voltage-dependent calcium channels by G proteins. *Current Opinion in Neurobiology, 8*, 351–356.

Zamponi, G. W., Bourinet, E., Nelson, D., Nargeot, J., & Snutch, T. P. (1997). Crosstalk between G proteins and protein kinase C mediated by the calcium channel alpha1 subunit. *Nature, 385*, 442–446.

Zamponi, G., Striessnig, J., Koschak, A., & Dolphin, A. (2015). The physiology, pathology, and pharmacology of voltage-gated calcium channels and their future therapeutic potential. *Pharmacological Reviews, 67*(4), 821–870.

Zhang, Y., Cribbs, L. L., & Satin, J. (2000). Arachidonic acid modulation of α1H, a cloned human T-type calcium channel. *The American Journal of Physiology, 278*(1), 184–193.

Zhang, Y., Zhang, L., Wang, F., Zhang, Y., Wang, J., Qin, Z., et al. (2011). Activation of M3 muscarinic receptors inhibits T-type Ca2+ channel currents via pertussis toxin-sensitive novel protein kinase C pathway in small dorsal root ganglion neurons. *Cellular Signalling, 23*(6), 1057–1067.

Zhang, L., Zhang, Y., Jiang, D., Reid, P. F., Jang, X., Qin, Z., et al. (2012). Alpha-cobratoxin inhibits T-type calcium currents through muscarinic M4 receptor and G o-protein βγ subunits-dependent protein kinase A pathway in dorsal root ganglion neurons. *Neuropharmacology, 62*(2), 1062–1072.

Zhang, Y., Qin, W., Qian, Z., Liu, X., Wang, H., Gong, S., et al. (2014). Peripheral pain is enhanced by insulin-like growth factor 1 through a G protein-mediated stimulation of T-type calcium channels. *Science Signaling, 7*(346), ra94.

Zhang, Y., Ji, H., Wang, J., Sun, Y., Qian, Z., Jiang, X., et al. (2018). Melatonin-mediated inhibition of Cav3.2 T-type Ca2+ channels induces sensory neuronal hypoexcitability through the novel protein kinase C-eta isoform. *Journal of Pineal Research, 64*(4), 12476.

Zühlke, R., Pitt, G., Tsien, R., & Reuter, H. (2000). Ca2+-sensitive inactivation and facilitation of L-type Ca2+ channels both depend on specific amino acid residues in a consensus calmodulin-binding motif in the(alpha)1C subunit. *The Journal of Biological Chemistry, 275*(28), 21121–21129.

Trafficking of Neuronal Calcium Channels

Norbert Weiss and Rajesh Khanna

Abstract

Neuronal voltage-gated calcium channels (VGCCs) serve a wide range of complex yet critical physiological functions by converting electrical signals into intracellular calcium increase and subsequent activation of downstream signaling pathways. The magnitude to which VGCCs affect neuronal activities largely depends on their density in the plasma membrane. Ectopic expression at the cell surface can lead to severe neurological conditions. Hence, numerous efforts have been dedicated to identifying the molecular mechanisms underlying the regulation of VGCCs which has led to the discovery of several interacting proteins and posttranslational modifications that contribute to their trafficking to and from the plasma membrane. In this chapter, we synthesize the current state of knowledge of these underlying mechanisms that drive the expression of VGCCs at the plasma membrane and address their implications in pathophysiological circumstances, and their potential as therapeutic targets.

Keywords

Calcium channels · Voltage-gated calcium channels · Trafficking · Ancillary subunit · Posttranslational modification

Abbreviations

AID	alpha-interaction domain
CaM	calmodulin
CRMP2	collapsin response mediator protein 2
DRG	dorsal root ganglion
ER	endoplasmic reticulum
ERAD	endoplasmic reticulum-associated protein degradation
GK	guanylate kinase
HEK	human embryonic kidney
HVA	high-voltage-activated
KLHL1	Kelch-like 1
KO	knockout
LVA	low-voltage-activated
MIDAS	metal ion adhesion
NSF	N-ethylmaleimide-sensitive fusion protein
rack-1	receptor for activated C kinase 1

N. Weiss (✉)
Department of Pathophysiology,
Third Faculty of Medicine, Charles University,
Prague, Czech Republic

R. Khanna (✉)
Department of Molecular Pathobiology, College of Dentistry, New York University, New York, NY, USA

NYU Pain Research Center, New York University,
New York, NY, USA
e-mail: rk4272@nyu.edu

RRR   triple arginine
SCAMP   secretory carrier-associated mem-
           brane protein
SNAP   soluble NSF attachment protein
SNARE   SNAP receptor
UPS   ubiquitin-proteasome system
USP   ubiquitin-specific protease
VGCC   voltage-gated calcium channel
VWA   Von Willebrand factor A

Introduction

Within nerve cells, calcium ions (Ca^{2+}) are required for a variety of cellular activities and, ultimately, physiological functions (Berridge, 1998). Much has been learnt about the mechanisms and molecular machinery involved in Ca^{2+} mobilization (Berridge et al., 2003). Plasma membrane voltage-gated Ca^{2+} channels (VGCCs) that convert surface electrical signals into Ca^{2+} influx and subsequent intracellular Ca^{2+} elevations support a variety of neuronal functions, including synaptic integration, Ca^{2+}-evoked gene transcription, and neurotransmitter release (Catterall, 2011; Zamponi et al., 2015). VGCCs are macromolecular complexes composed of a pore-forming $Ca_v\alpha_1$ subunit and in some circumstances several ancillary subunits including $Ca_v\beta$, $Ca_v\alpha_2\delta$, and $Ca_v\gamma$ (Zamponi et al., 2015). Here, we refer the readers to chapter "Subunit Architecture and Atomic Structure of Voltage Gated Ca2+ Channels" by Catterall for more details on the diversity, structure, and molecular composition of VGCCs.

A multitude of regulatory mechanisms acts on VGCCs to ensure that the amplitude, duration, and subcellular localization of the Ca^{2+} signal is dynamically controlled. These include regulations by ancillary subunits (chapters "Voltage-Gated Calcium Channel Auxiliary β Subunits" by Colecraft, and "Regulation of Calcium Channels and Synaptic Function by Auxiliary α2δ Subunits" by Dolphin and Obermair), G-protein coupled receptors and second messengers (chapter "Modulation of VGCCs by G-protein Coupled Receptors and Their Second Messengers" by Mark et al.), calcium itself (chapter "Calmodulin

Regulation of Voltage-gated Calcium Channels" by Ben-Johny and Dick), formation of signaling complexes with other ion channels (chapter "Cav3 Calcium Channel Interactions with Potassium Channels" by Turner), and mRNA editing (chapter "Splicing and Editing to Fine-Tune Activity of High Voltage-Activated Calcium Channels" by Huang et al.). In this chapter, we focus on the molecular mechanisms controlling the expression of neuronal VGCCs in the plasma membrane. Although these mechanisms are by far the least understood, several recent studies have identified critical factors and signaling pathways that ensure an optimal density of channels in the plasma membrane. We look at how to channel ancillary subunits and other interacting proteins affect channel expression. The importance of post-translational modifications such as glycosylation and ubiquitination in the trafficking and stability of the channel at the cell surface is also discussed. Finally, we discuss the implication of these regulatory systems in the context of chronic disorders associated with VGCCs and assess their potential as therapeutic targets.

Regulation of VGCCs by Ancillary Subunits

$Ca_v\beta$-Subunit

The ancillary $Ca_v\beta$-subunit belongs to the membrane-associated guanylate kinase (GK) family of proteins. It contains SH3 and GK domains that are substantially conserved among $Ca_v\beta$ isoforms ($Ca_v\beta_{1-4}$) and are linked by a variable HOOK region (Maltez et al., 2005; McGee et al., 2004; Buraei & Yang, 2010). The $Ca_v\beta$-subunit directly binds with high affinity (in vitro Kd of about 5 nM) to a conserved motif (QQ-E–L–GY–WI–E) known as the alpha-interaction domain (AID) within the cytoplasmic linker between domains I and II of the $Ca_v\alpha_1$ subunit of high-voltage-activated (HVA) channels (Pragnell et al., 1994) . The initial evidence for a role of $Ca_v\beta$ in the expression of VGCCs came from co-expression studies in *Xenopus* oocytes and mammalian cells, which revealed a range of effects including an increase in Ca^{2+} current amplitude by orders of magnitude in the presence of $Ca_v\beta$ subunits (Singer et al., 1991; Hullin et al., 1992; Stea et al., 1993). Conversely, global knockdown of $Ca_v\beta$ subunits using a pan oligonucleotide antisense in dorsal root ganglion (DRG) neurons resulted in a considerable, albeit partial, drop in the Ca^{2+} current (Berrow et al., 1995). In contrast, in cardiomyocytes lacking $Ca_v\beta_2$, only a minor decrease in $Ca_v1.2$ currents was observed (Meissner et al., 2011). While other $Ca_v\beta$ isoforms did not appear to have compensated for the lack of $Ca_v\beta_2$, it is a possibility that alternative compensatory mechanisms may have taken place, therefore, masking the real effect of $Ca_v\beta_2$ deficiency on $Ca_v1.2$ currents. Despite the fact that $Ca_v\beta$-dependent potentiation of the Ca^{2+} current is multifactorial and partly results from an increase in the channel opening probability (Wakamori et al., 1999; Jones et al., 1998; Shistik et al., 1995; Luvisetto et al., 2004; Colecraft et al., 2002), a role in the trafficking of $Ca_v\alpha_1$ to the cell surface quickly emerged as one of the major underlying mechanisms. Hence, based on the finding that a CD8 fusion construct containing the I-II linker of $Ca_v2.1$ is retained in the endoplasmic reticulum (ER) unless co-expressed with a $Ca_v\beta$-subunit prompted the authors to propose that $Ca_v\beta$ could shield an ER retention signal within the I-II linker of $Ca_v\alpha_1$ thus enhancing the sorting of the channel to the plasma membrane (Bichet et al., 2000). However, several subsequent studies cast doubt on this hypothesis. First, no ER retention signal was found in the AID or adjacent areas, and removal of the AID domain in $Ca_v2.1$ failed to potentiate the expression of the channel at the cell surface (Maltez et al., 2005). Second, despite harboring the AID motif, not all fusion proteins containing the I-II linkers from diverse HVA $Ca_v\alpha_1$ are retained in the ER (Cornet et al., 2002; Altier et al., 2011). Third, replacing the I-II linker of $Ca_v3.1$ (which does not require the presence of $Ca_v\beta$ to translocate to the cell surface) with the I-II linker of different HVA $Ca_v\alpha_1$ including $Ca_v1.2$, $Ca_v2.1$, and $Ca_v2.2$ did not result in the ER retention of the chimeric $Ca_v3.1$channel but in contrast resulted in an increase of the T-type current (Arias et al.,

2005; Fang & Colecraft, 2011). This effect was further evaluated in an elegant study where $Ca_v3.1$ intracellular linkers were sequentially swapped with corresponding regions of $Ca_v1.2$ and surface expression of chimeric $Ca_v3.1$ was assessed in the presence of $Ca_v\beta$ (Fang & Colecraft, 2011). In contrast to initial observations, this study suggested that the I-II linker of $Ca_v1.2$ may rather contain an ER "export-like" signal. In an attempt to reconcile these contradictory findings, it was postulated that binding of $Ca_v\beta$ onto $Ca_v\alpha_1$ could restructure the molecular organization of intracellular linkers rendering the channel more "willing" to exit the ER. In addition, several more recent studies have shown that $Ca_v\beta$ plays an important role in regulating proteasome-dependent degradation of the channel. This aspect is expanded below in the section dealing with the regulation of VGCCs by the ubiquitin-proteasome system.

Overall, these data suggest that $Ca_v\beta$ represents an important factor for the expression of the channel at the cell surface, on the one hand by potentiating the sorting of $Ca_v\alpha_1$ from the ER, and on the other hand by protecting $Ca_v\alpha_1$ from proteasomal degradation. Further research is needed to determine whether these two mechanisms are independent or related. However, one possibility is that a pool of channels may sit dormant in the ER, protected from degradation, and readily available to translocate to the cell surface in response to appropriate stimuli.

$Ca_v\alpha_2\delta$-Subunit

The $Ca_v\alpha_2\delta$ subunit is an extracellular component of the VGCC complex that contributes to the surface expression and pharmacological regulation of HVA $Ca_v\alpha_1$ (Dolphin, 2013; Geisler et al., 2015; Mould et al., 2004). Each of the four $Ca_v\alpha_2\delta$ isoforms ($Ca_v\alpha_2\delta_{1-4}$) is transcribed and translated from a single gene and goes through a series of post-translational modifications (see also chapter "Regulation of Calcium Channels and Synaptic Function by Auxiliary $\alpha_2\delta$ Subunits" by Dolphin and Obermair). Proteolytic cleavage of $Ca_v\alpha_2\delta$ produces α_2 and δ peptides that remain linked together by disulfide bonds. Second, a glycosylphosphatidylinositol (GPI) moiety is attached to the C-terminal region of the δ peptide which allows for the anchoring of $Ca_v\alpha_2\delta$ to the plasma membrane after it has been secreted (Davies et al., 2010). Third, $Ca_v\alpha_2\delta$ undergoes extensive glycosylation which accounts for almost a third of its total molecular weight (the importance of $Ca_v\alpha_2\delta$ glycosylation in the expression of $Ca_v\alpha_1$ is discussed in the section related to glycosylation) (Marais et al., 2001). A Von Willebrand factor A (VWF-A or VWA) domain is found in all $Ca_v\alpha_2\delta$ subunits and represents a dinucleotide binding fold with a metal ion adhesion (MIDAS) motif that is involved in several divalent cation interactions (Whittaker & Hynes, 2002). Two bacterial chemosensory-like domains (Cache) are also found downstream of the VWA domain (Anantharaman & Aravind, 2000). $Ca_v\alpha_2\delta$ is found in skeletal (Takahashi et al., 1987), cardiac (Chang & Hosey, 1988), and neuronal channel complexes (Witcher et al., 1993; Liu et al., 1996), and all HVA channels are predicted to bind to $Ca_v\alpha_2\delta$. In co-expression studies as well as in native preparations, $Ca_v\alpha_2\delta$ boosts surface expression of several HVA $Ca_v\alpha_1$ isoforms including $Ca_v1.2$ (Shistik et al., 1995; Felix et al., 1997; Gao et al., 2000a; Yasuda et al., 2004), $Ca_v2.1$ (Yasuda et al., 2004; Barclay et al., 2001; Brodbeck et al., 2002; Cantí et al., 2005; Davies et al., 2006), and $Ca_v2.2$ (Gao et al., 2000a; Yasuda et al., 2004; Cantí et al., 2005; Hendrich et al., 2008), while this effect is less pronounced for $Ca_v2.3$ (Jones et al., 1998; Qin et al., 1998). In addition, $Ca_v\alpha_2\delta$ appears to slow down the internalization of the channels from the cell surface (Bernstein & Jones, 2007). The exact molecular mechanism by which $Ca_v\alpha_2\delta$ potentiates HVA $Ca_v\alpha_1$ surface expression is still unknown. The MIDAS motif in the VWA domain of $Ca_v\alpha_2\delta_1$ and $Ca_v\alpha_2\delta_2$ has been shown to be important in mediating this process (Cantí et al., 2005; Hoppa et al., 2012). Given that the VWA domain is engaged in protein–protein interactions, it is possible that $Ca_v\alpha_2\delta$ interacts with proteins involved in channel trafficking, or directly with $Ca_v\alpha_1$ to promote the trafficking of the channel complex. The cryo-EM structure of the skeletal muscle $Ca_v1.1$ channel complex [61] supports this notion by shedding information on the molecular determinants involved in $Ca_v\alpha_2\delta/Ca_v\alpha_1$ interaction. In this structure, $Ca_v\alpha_2\delta$ interacts with the extracellular loops of domains I to III of $Ca_v\alpha_1$ via VWA and Cache1 domains, which is consistent with prior biochemical investigations (Gurnett et al., 1997). Furthermore, there is pharmacological evidence for the role of $Ca_v\alpha_2\delta/Ca_v\alpha_1$ interaction in the trafficking of HVA channels. For instance, the antiepileptic/antiallodynic drugs gabapentin and pregabalin (Field et al., 2007; Taylor et al., 2007; Moore et al., 2014) target $Ca_v\alpha_2\delta$ (Gee et al., 1996; Gong et al., 2001; Brown & Gee, 1998; Wang et al., 1999) not only prevent the trafficking of $Ca_v2.1$ and $Ca_v2.2$ to the cell surface (Hendrich et al., 2008) but also counteract the pathological increased expression of $Ca_v2.2$ that occurs during chronic pain conditions (Bauer et al., 2009). Gabapentin also inhibits rab11-dependent recycling of $Ca_v\alpha_2\delta_2$ from post-Golgi compartments to the cell surface which may in turn diminish cell surface expression of the channel complex (Tran-Van-Minh & Dolphin, 2010). Furthermore, the ability of $Ca_v\alpha_2\delta$ to enhance surface expression of the channel is abolished by mutation of the third arginine in the RRR motif (involved in the binding of gabapentinoids) located upstream of the VWA domain, suggesting that $Ca_v\alpha_2\delta$-dependent expression of the channel relies on the coupling with $Ca_v\alpha_1$, and that gabapentinoid drugs exert their effect by disrupting $Ca_v\alpha_1/Ca_v\alpha_2\delta$ interaction (Hendrich et al., 2008; Field et al., 2006).

Overall, there is unequivocal evidence that $Ca_v\alpha_2\delta$ is required for the trafficking and surface expression of HVA channels. However, and despite the fact that a biochemical contact between $Ca_v\alpha_2\delta$ and the $Ca_v\alpha_1$ appears to be essential, the cellular mechanism by which the $Ca_v\alpha_1/Ca_v\alpha_2\delta$ protein complex traffics to the cell surface and whether ER retention signals are involved in this regulation are yet to be fully elucidated.

Ca$_v$γ-Subunit

The Ca$_v$γ-subunit is a transmembrane protein that was first identified as a component of the skeletal muscle channel complex (Takahashi et al., 1987), but it was later shown that nerve cells also express multiple Ca$_v$γ isoforms. To date, eight mammalian isoforms (Ca$_v$γ$_{1-8}$) have been identified, although compared to Ca$_v$β- and Ca$_v$α$_2$δ-subunits, the Ca$_v$γ-subunit has received little attention, owing to its reportedly minor effect on recombinant channels expressed in *Xenopus* oocytes and mammalian cells (Klugbauer et al., 2000; Rousset et al., 2001; Green et al., 2001; Kang et al., 2001; Moss et al., 2003; Black, 2003). However, there is evidence that Ca$_v$γ plays a role in modulating native channels. Indeed, in the stargazer mouse model of absence seizures where the expression of Ca$_v$γ$_2$ is severely reduced due to a transposon insertion in the corresponding gene (Letts, 2005; Letts et al., 1998, 2003; Sharp et al., 2001), the current density of both HVA and LVA channels in thalamocortical relay nuclei is enhanced (Zhang et al., 2002). In vitro investigations, on the other hand, tend to point to a direct modification of channel activity rather than a change in channel trafficking (Moss et al., 2002). While co-expression of Ca$_v$γ$_7$ with recombinant Ca$_v$2.2 reduced the Ca^{2+} conductance, this effect was not associated with a decreased channel expression in the plasma membrane. Similarly, the expression of Ca$_v$γ$_6$ in an atrial cell line reduced Ca$_v$3.1 currents without altering the expression of the channel at the cell surface (Hansen et al., 2004; Lin et al., 2008; Chen & Best, 2009). In the stargazer mouse model, however, altered trafficking of AMPA receptors from the Golgi apparatus to the plasma membrane was observed, demonstrating that Ca$_v$γ$_2$ is required for surface expression and postsynaptic targeting of the receptor (Chen et al., 2000; Tomita et al., 2004). In addition, Ca$_v$γ$_4$ which presents a high degree of homology with Ca$_v$γ$_2$ is significantly more effective at trafficking AMPA receptors (Körber et al., 2007). Hence, neuronal Ca$_v$γ subunits appear to be a major element of AMPA and kainate receptors rather than having a primary role in the regulation of VGCCs.

Regulation of VGCCs by Other Interacting Proteins

Calmodulin

Calmodulin (CaM) is a multifunctional intermediate Ca^{2+}-binding protein that is necessary for neuronal signaling (Xia & Storm, 2005). CaM associates with nearly all HVA Ca$_v$α$_1$ which has led to the notion that CaM should be considered as an intrinsic component of the Ca^{2+} channel complex per se, prompting some scientists to coin the term "calmodulation" to describe any CaM-dependent ion channel regulation (Xia & Storm, 2005). Over the last few years, the multidimensional regulation of VGCCs by CaM has been widely explored (Ben-Johny & Yue, 2014; Ben-Johny et al., 2015; Soong & Mori, 2016). Here, we concentrate solely on its involvement in the trafficking of the channel. Although several domains located in the C-terminal region of HVA Ca$_v$α$_1$ have been shown to be involved in the binding of Ca^{2+}/CaM, the IQ domain has emerged as a prerequisite for the binding of Ca^{2+} free Apo-CaM. However, the exact involvement of CaM in the trafficking of Ca$_v$α$_1$ remains contentious. Indeed, CaM has no influence on the surface expression of recombinant Ca$_v$1.2 in HEK-293 cells regardless of the presence or absence of ancillary Ca$_v$ subunits (Bourdin et al., 2010). However, the removal of the Pre-IQ3 region located upstream of the IQ domain and crucial for the binding of Ca^{2+}/CaM totally abolished the surface expression of Ca$_v$1.2 (Gao et al., 2000b). Moreover, expression of Ca$_v$1.2 in distal dendrites of hippocampal neurons is increased by Ca^{2+}/CaM but not by Apo-CaM, indicating that CaM plays a role in the subcellular targeting of the channel (Brunet et al., 2005). Moreover, mutations in the EF-hand domain of Ca$_v$1.2 channels have been shown to alter the Ca^{2+} conductance in a variety of experimental conditions (Dolmetsch et al., 2001; Peterson et al., 2000; Bernatchez et al., 1998). Although these alterations undoubtedly arose, at least partly, from the effect of CaM on the gating of the channel, it is also a possibility that CaM-dependent trafficking of the channel was impeded (Brunet et al., 2005,

2009). In addition to the CaM-binding domains in the C-terminal region, $Ca_v1.2$ contains two additional CaM binding loci within the N-terminal domain that are not present in $Ca_v2.x$ channels (Benmocha et al., 2009; Dick et al., 2008; Simms et al., 2014). However, there is no evidence that these loci play a role in the trafficking of the channel. In contrast, a CaM kinase II (CaMKII) binding domain is located in the N-terminal region of $Ca_v1.2$ and plays a critical role in the expression of the channel at the cell surface (Simms et al., 2015). It remains to be seen whether there is any regulatory cross-talk between Ca^{2+}/CaM and CaMKII binding. It is worth noting that many studies looking into the role of CaM and/or CaMKII interactions rely on point mutations in the key binding regions of $Ca_v\alpha_1$. In that regard, it is crucial to consider the likelihood that these mutations may have broader effects on channel folding and/or trafficking when evaluating functional data acquired with mutant constructs. Nonetheless, these findings suggest that Ca^{2+}/CaM may play a role in regulating the trafficking of VGCCs although the underlying mechanisms are yet to be established.

Collapsin Response Mediator Proteins

Collapsin response mediator protein 2, or CRMP2, is the protein product of the gene Dihydropyrimidinase-related protein 2 (*Dpysl2*) found on chromosome 8p21, which is also known by the alternative names TOAD64/ULIP-2/Unc-33-like phosphoprotein (Khanna et al., 2012). The homolog of CRMP2 in *C. elegans*, unc33, was first identified in a screen for mutations in genes relating to axonal outgrowth (Hedgecock et al., 1985). The functional effects of CRMP2 were first explored using *Xenopus laevis* oocytes, where inward currents in response to the extracellular application of what was then called collapsin were first recorded. These currents were blocked with intracellular administration of an anti-CRMP2 antibody, which pointed to an intracellular protein that mediated the downstream action of collapsin binding (Hedgecock et al., 1985; Strittmatter et al., 1995). From these stud-

ies, we learned that CRMP2 is a cytosolic phosphoprotein of approximately 63 kDa that mediates semaphorin 3a (collapsin) induced axonal growth cone collapse. This protein is expressed highly during development where it plays a pivotal role in the neurite fate determination and axonogenesis (Inagaki et al., 2001; Quach et al., 2004). As a microtubule-binding protein, it is intimately involved in the control of the cytoskeletal network and the transport of cargo along with this network (Gu & Ihara, 2000; Fukata et al., 2002). Specifically, CRMP2 binds tubulin heterodimers and promotes assembly at the microtubule plus end, which is thought to stabilize growing microtubule networks during neuronal polarization (Gu & Ihara, 2000). Overexpression of CRMP2 causes neurons to develop multiple axons-like neurites, which highlights the importance of CRMP2 for proper neuronal development and assembly of neuronal networks (Yoshimura et al., 2006). In addition to its binding of microtubules, CRMP2 has also been associated with the actin regulatory protein Rac1-associated protein 1 (Sra-1)/WASP family verprolin-homologous protein 1 (WAVE1), where it mediates the transport of this protein to the developing axon through an interaction with the light chain of kinesin −1 (Kawano et al., 2005). Loss of CRMP2 and kinesin-1 delocalized Sra-1 and WAVE1 away from the axonal growth cone, which suggests that CRMP2 transports the Sra-1/WAVE1 complex to the distal axonal compartment in a kinesin-1 dependent fashion. Furthermore, CRMP2 has been linked to the regulation of trafficking events via its association with dynein motors, where it functions to connect endocytic regulatory elements with target proteins (Rahajeng et al., 2010). These findings point toward the involvement of CRMP2 in cytoskeletal dynamics and therefore its potential involvement in channel internalization, which relies upon cytoskeletal anchoring proteins.

CRMP2 is a single member of a larger group of proteins, including CRMPs 1–5, that all share homology with the dihydropyrimidinase (DHP) enzyme but lack the DHP catalytic site. These homologous proteins are expressed throughout the nervous system with developmental regulation of their expression profile controlling neuronal patterning (Wang & Strittmatter, 1996). There is significant homology between the members of the CRMP protein family and there is speculation that these proteins can form heterotetrameric complexes (Wang & Strittmatter, 1997; Ponnusamy & Lohkamp, 2013). They have complementary roles in the formation of neuronal processes, with CRMP2 controlling the formation of axonal processes, while its complementary partner CRMP3 is found at higher levels in dendritic compartments (Quach et al., 2008). After birth, the levels of CRMP2 expression drop dramatically, however expression is still retained in neural tissues, such as in dorsal root ganglia and the hippocampus. This evidence points to CRMP2, and the CRMP family in general, as critical mediators of neuronal assembly, which could potentially be targeted to control the excitability of neuronal networks.

All CRMP proteins are developmentally regulated with high expression during embryogenesis that peaks during the first postnatal week of life followed by steep downregulation in adulthood (Minturn et al., 1995b). Despite the common pattern, each CRMP protein has unique profiles of expression that are distinct. For example, while the expression of other CRMP proteins becomes weaker in adulthood, the expression level of CRMP2 remains relatively high. An example of the distinct expression profiles can be found in the external granule layer of the cerebellum, where CRMP2 is expressed but CRMP5 is never observed (Wang & Strittmatter, 1996). However, these two proteins do co-express in the internal granule layers of the cerebellum and are present in the fasciculi of neuronal fibers (Ricard et al., 2001). CRMP2 is the most highly expressed CRMP in the adult brain, found most often in post-mitotic neurons, and is expressed in the olfactory system, cerebellum, hippocampus, and spinal dorsal horn (Veyrac et al., 2005). Early studies showed that despite downregulation in adulthood after most axonal pathfinding is complete, CRMP2 expression is enhanced following axotomy of the sciatic nerve, which has implications for CRMP2 in the development of pain following nerve injury (Minturn et al., 1995a). In the hippocampus, CRMP2 expression remains abundant after maturation where it has been dem-

onstrated to participate in the formation of synaptic connections (Nishimura et al., 2003). Since CRMP2 has an essential role in the regulation of the axonal growth cone, which is a highly specialized sensory apparatus for detecting external growth cues, it is reasonable to speculate that CRMP2 could serve a similar role in the specialized nociceptive sensory apparatus found in primary sensory neurons.

In addition to the established roles in regulating neuronal polarization during development, recent evidence has pointed to an important role for CRMP2 in regulating the trafficking of ion channels. Previous work has provided a post-translational modification code that directs CRMP2 toward different cellular functions, as outlined above. Since CRMP2 is primarily a synaptic protein, it is logical to conclude that once the nervous system has developed and the role of CRMP2 in determining neuronal polarity is no longer necessary, this protein could be reoriented toward the regulation of other synaptic events, such as neurotransmitter release and electrogenesis. At synapses, the release of neurotransmitters is driven by the opening of calcium channels that allow calcium ions into the cytoplasm, which triggers the fusion of vesicles and the release of neurotransmitters into the synaptic cleft (Kavalali, 2015). The voltage-gated calcium channel subtype $Ca_v2.2$ is a primary contributor to the flow of calcium ions across the membrane that triggers transmitter release in nociceptive synapses (Castiglioni et al., 2006). A connection between CRMP2 and the regulation of $Ca_v2.2$ channels was identified through phosphorylation of CRMP2 at Ser522 by the kinase Cdk5 (Brittain et al., 2012). Normally, this phosphorylation event is thought to inactivate CRMP2 but investigation of this regulatory event in primary sensory neurons revealed that it is instead a switch for directing the affinity of CRMP2 away from its function involving microtubules and toward a role in regulating protein interactions. The association between CRMP2 and $Ca_v2.2$ was first identified in a screen for $Ca_v2.2$ interacting partners, which was followed up by a confirmatory study that demonstrated CRMP2 and $Ca_v2.2$ associate to form a complex (Khanna et al., 2007; Brittain et al., 2009). The interaction was confirmed using a short 15 amino acid peptide derived from the $Ca_v2.2$ channel that was modified with a TAT domain (HIV-derived transactivator of transcription), which allows the peptide to become cell penetrant. When primary sensory neurons were treated with this peptide a significant reduction in $Ca_v2.2$ current density was observed in conjunction with reduced neurotransmitter release and impaired neurotransmission in vitro and in vivo (Brittain et al., 2011). Furthermore, $Ca_v2.2$ localization in the membrane was impaired and nociceptive responses to formalin treatment were significantly reduced. Altogether, these findings implicate CRMP2 in the surface trafficking of $Ca_v2.2$ and point toward the relevance of this finding as a potential therapeutic avenue for treating chronic pain.

Stac Adaptor Proteins

The functional role of Stac adaptor proteins is still not well characterized (Suzuki et al., 1996). Stacs are a family of SH3 and cysteine-rich domain-containing proteins. Stac1 and Stac2 are broadly distributed in the peripheral and central nervous systems, while Stac3 is mostly expressed in the skeletal muscle (Legha et al., 2010). Initial studies on the role of Stac3 in the assembly and function of the excitation–contraction coupling machinery ($Ca_v1.1/RyR1$), as well as in the development of skeletal muscles have shed light on their potential role in the regulation of VGCCs (Horstick et al., 2013; Nelson et al., 2013; Bower et al., 2012; Polster et al., 2016). Hence, Stac3 was found to play a role in the trafficking of recombinant $Ca_v1.1$ channels to the cell surface (Polster et al., 2015; Weiss, 2015). These findings led to studies investigating whether Stac proteins could also have a role in the expression of neuronal VGCCs, particularly T-type channels. In vitro studies revealed that Stac1 binds to the distal region of the N-terminus of $Ca_v3.2$ channels and potentiates T-type currents in tsA-201 cells (Rzhepetskyy et al., 2016). Although Stac1 had no effect on the gating properties of $Ca_v3.2$, it did increase surface expression of $Ca_v3.2$ implying that the increased T-type conductance was mostly

due to an increased channel expression at the plasma membrane. It remains to be seen whether Stac1 enhances $Ca_v3.2$ trafficking to the cell surface or stabilizes the channel in the plasma membrane (Iftinca & Altier, 2017).

Kelch-like 1

Kelch-like 1 (KLHL1) is a member of the actin-organizing protein family. It is expressed mostly in neural tissues and interacts with actin. When recombinant $Ca_v2.1$, $Ca_v3.1$, and $Ca_v3.2$ were co-expressed with KLHL1 in tsA-201 cells, the Ca^{2+} conductance was increased (Aromolaran et al., 2009, 2010). This effect was investigated further for $Ca_v3.1$ and $Ca_v3.2$ channels and was found to be attributable to an enhanced channel expression at the plasma membrane. In addition, KLHL1 can be immunoprecipitated with the channels, although the detailed molecular determinants of the interaction have yet to be identified. Furthermore, whereas pharmacological disruption of actin filaments did not impair T-type channel activity per se, it did reduce KLHL1-dependent channel potentiation (Aromolaran et al., 2009). Similar findings were made when endosomal recycling was disrupted, implying that KLHL1 promotes T-type channel surface expression by potentiating their re-insertion into the plasma membrane from recycling endosomes (Aromolaran et al., 2009). The observation that shRNA knock-down of KLHL1 in cultured hippocampal neurons caused a dramatic decrease of both HVA and LVA Ca^{2+} currents further supports the role of KLHL1 in the expression of VGCCs (Perissinotti et al., 2014b). Moreover, hippocampal neurons isolated from KLHL1 knock-out mice show decreased expression of $Ca_v2.1$ and $Ca_v3.2$ (Perissinotti et al., 2014a). Interestingly, this alteration appears to be counterbalanced by the upregulation of $Ca_v1.2$ and $Ca_v3.1$ indicating the occurrence of compensatory mechanisms. Nonetheless, and consistent with the presynaptic role of $Ca_v2.1$ and $Ca_v3.2$ channels, the synaptic activity of $KLHL1^{-/-}$ neurons remained altered (Perissinotti et al., 2014a).

Overall, these findings highlight the role of KLHL1 and actin-binding proteins in the expression of VGCCs at the cell surface. It remains to be seen whether this regulation involves any of the channel molecular factors involved in their trafficking.

Calnexin

Calnexin is a type I endoplasmic reticulum integral membrane protein and molecular chaperone involved in the folding, quality control, and sorting of newly-synthetized (glycol)proteins (Ellgaard & Helenius, 2003). Consistent with the observation that T-type Ca^{2+} channels undergo asparagine-linked glycosylation (see next section), it was shown that calnexin binds to and modulates the trafficking of $Ca_v3.2$ channels to the cell surface (Proft et al., 2017). This interaction relies on the binding of calnexin onto the III-IV linker of the channel to retain the channel in the ER, therefore, limiting its expression in the plasma membrane. Importantly, this effect is more pronounced for exon 25-containing $Ca_v3.2$ channels indicating that calnexin-dependent expression of $Ca_v3.2$ can be further modulated by alternative splicing of $Ca_v3.2$ III-IV linker. Along these lines, the GAERS mutation in the *Cacna1H* gene located in the III-IV linker of $Ca_v3.2$ and responsible for seizure-like behavior in the GAERS model of absence epilepsy disrupts calnexin-dependent regulation of channel leading to an increased expression of the channel at the plasma membrane (Proft et al., 2017).

Altogether, these data indicate that calnexin is an important determinant limiting the expression of $Ca_v3.2$ channels at the cell surface, a mechanism that may have important pathophysiological implications.

Secretory Carrier-Associated Membrane Proteins

Secretory carrier-associated membrane proteins (SCAMPs) are a family of integral membrane proteins found mostly in the trans-Golgi network and recycling endosome membranes, where they regulate vesicular trafficking and recycling (Castle & Castle, 2005). Of the five known mammalian SCAMPs, SCAMP2 is ubiquitously expressed and several reports have documented the role of SCAMP2 in the expression of ion channels and transporters including the serotonin transporter (Müller et al., 2006), the Na^+/H^+ exchanger (Diering et al., 2009), the NKCC2 cotransporter (Zaarour et al., 2011), and the dopamine transporter (Fjorback et al., 2011). Recently, we have shown that SCAMP2 also plays an important role in the trafficking of T-type Ca^{2+} channels at the plasma membrane. For instance, we reported that SCAMP2 interacts with $Ca_v3.2$ and co-expression of SCAMP2 in tsA-201 cells expressing recombinant $Ca_v3.1$, $Ca_v3.2$, or $Ca_v3.3$ channels resulted in an almost complete reduction in whole cell T-type current and intramembrane charge movement, indicating that SCAMP2 negatively regulates the trafficking of the channels to the cell surface (Cmarko et al., 2022).

Receptor for Activated C kinase 1

Receptor for activated C kinase 1 (Rack-1) is a member of the tryptophan-aspartate repeat (WD-repeat) family of proteins and represents an important adaptor protein to (1) shuttle binding partners to intracellular sites and (2) modulate the enzymatic activity of its binding partners (Adams et al., 2011). Recently, it was reported that Rack-1 forms a molecular complex with $Ca_v3.2$ channels and co-expression of Rack-1 in tsA-201 cells expressing recombinant $Ca_v3.2$ led to a reduction in the magnitude of the whole-cell Ca^{2+} conductance and expression of the channel at the cell surface (Gandini et al., 2021). Interestingly, the effect was abolished upon co-expression of PKCβII suggesting that Rack-1 regulates $Ca_v3.2$ channels in a PKC-dependent manner.

Regulation of VGCCs by Posttranslational Modifications

Ubiquitination

The ubiquitin-proteasome system (UPS) plays an essential role in the degradation of VGCCs in a ubiquitin-dependent manner, allowing the removal of channels expressed at the cell surface (Felix & Weiss, 2017). Ubiquitination is the process of attaching a ubiquitin moiety to a protein by a ubiquitin ligase, which determines the destiny of the target protein (Hershko & Ciechanover, 1998). Polyubiquitinated proteins are targeted to the proteasome system for proteolysis, whereas monoubiquitinated proteins are frequently internalized and either degraded in lysosomes or recycled back to the plasma membrane. Although ubiquitination is a common means of controlling protein destiny, it was only recently discovered that VGCCs are ubiquitinated. For example, the ubiquitin ligase RFP2 ubiquitinates $Ca_v1.2$ channels, and ubiquitinated channels interact with key proteins of the endoplasmic reticulum-associated protein degradation (ERAD) complex, such as derlin-1 and p97, which in turn triggers the targeting of the channel to the proteasome system (Altier et al., 2011). In cells expressing a dominant-negative RFP2 construct, the Ca^{2+} current density and surface expression of $Ca_v1.2$ channels is dramatically increased (Altier et al., 2011). Furthermore, the $Ca_v\beta$-subunit has been shown to prevent RFP2-mediated ubiquitination of $Ca_v1.2$, preventing the channel from being degraded by the proteasome system (Altier et al., 2011). Likewise, the proteasome inhibitor MG132 partially restores surface expression of the channel in the absence of $Ca_v\beta$. According to other reports, $Ca_v1.2$ channels can be ubiquitinated by the ubiquitin ligase Nedd4-1 (Rougier et al., 2011), which can be counteracted by the ubiquitin-specific protease (USP) 2-45 (Rougier et al., 2015). Despite the fact that USP2-45 may deubiquitinate the channel, Ca^{2+} currents were drastically reduced in tsA-201 cells expressing $Ca_v1.2$ with USP2-45, implying a more complex regulatory mechanism (Rougier et al., 2015). $Ca_v2.2$ channel ubiquitination has also been described, a process that the $Ca_v\beta$-subunit modu-

lates as part of its protective function (Waithe et al., 2011; Page et al., 2016). $Ca_v2.2$ has also been shown to interact with a macromolecular complex involving the microtubule-associated protein 1B (MAP 1B) light chain 1 subunit (LC1) and the ubiquitin conjugase UBE2L3 to modulate the expression of the channel at the cell surface (Gandini et al., 2014a, b). Furthermore, binding of the fragile X mental retardation protein (FMRP) to $Ca_v2.2$ enhances proteasome-dependent degradation of the channel, a mechanism that is deficient in fragile X syndrome (Macabuag & Dolphin, 2015). In addition, alternative splicing of the pre-mRNA influences ubiquitin-dependent degradation of the channel. For instance, exon 37b-containing $Ca_v2.2$ channels exhibit higher ubiquitination levels than exon 37a-containing channels and is associated with a lower expression of the channel (Marangoudakis et al., 2012). Because nociceptors exhibit cell-specific alternative splicing of exon 37a, reduced ubiquitination of $Ca_v2.2$ e37a could have significant implications for pain processing. The fact that the adaptor protein complex-1 (AP-1) potentiates the trafficking of $Ca_v2.2$ e37a channels from the trans-Golgi to the cell surface, but not of the e37b variant that lacks the AP-1 binding motif, may amplify this effect (Macabuag & Dolphin, 2015). Along these lines, ubiquitination of $Ca_v3.2$ channels has been shown to be an important mechanism for modulating pain signaling in primary afferent nociceptive neurons. For example, the ubiquitin ligase WWP1 ubiquitinates $Ca_v3.2$ channels within the III-IV linker, whereas the ubiquitin protease USP5 removes ubiquitin moieties, thus influencing the ubiquitination/deubiquitination levels of the channel and its expression at the cell surface (García-Caballero et al., 2014). Furthermore, in vivo suppression of USP5 caused analgesia in both inflammatory and neuropathic rodent pain models of mechanical hypersensitivity, confirming the importance of $Ca_v3.2$ channels in pain signaling (García-Caballero et al., 2014). Importantly, analgesia was induced in vivo by disrupting the $Ca_v3.2$/USP5 interaction with interfering peptides or small organic molecules, revealing the therapeutic potential of targeting ubiquitination signaling as a new analgesic avenue (Garcia-Caballero et al., 2016; Gadotti et al., 2015). Taken together, these findings strongly support the role of the UPS in the regulation of VGCC expression. While the pore-forming $Ca_v\alpha_1$ subunit has received the most attention, Ca_v ancillary subunits are likely to be ubiquitinated as well, providing an extra degree of control over the expression of the channel complex.

Glycosylation

Another post-translational modification is known to be important for the expression and regulation of ion channels in general, and VGCCs in particular is asparagine (N)-linked glycosylation (Lazniewska & Weiss, 2014, 2017). The orchestrated enzymatic conjugation of an oligosaccharide tree (glycan) to an asparagine (N) residue in a consensus motif N-X-S/T (X being any residue, though the nature of this residue strongly influences the probability of glycosylation) within the target protein is referred to as N-glycosylation. While the involvement of N-glycosylation in the folding and ER quality control of nascent proteins has received a lot of attention, the importance of N-glycosylation in the trafficking and stability of VGCCs at the plasma membrane has also been documented. Virtually all ten $Ca_v\alpha_1$-subunits are subject to N-glycosylation, and pharmacological or molecular disruption of canonical glycosylation sites in $Ca_v\alpha_1$ has revealed the relevance of the glycan tree in the trafficking of the channel. For example, disrupting $Ca_v1.2$ glycosylation in *Xenopus* oocytes resulted in a significant drop in the surface expression of the channel (Park et al., 2015). There is also biochemical evidence for glycosylation of the skeletal channel homolog $Ca_v1.1$, but no functional data has been published yet (Wang et al., 2004). T-type channels have seen the most complete characterization so far. Biochemical studies of $Ca_v3.1$ and $Ca_v3.3$ provided the first evidence for T-type channel glycosylation. While the apparent molecular weight of $Ca_v3.1$ and $Ca_v3.3$ varies with their regional brain expression and during neuronal development (Yunker et al., 2003), enzymatic deglycosylation of the channels with PNGase F was enough to eliminate these differences, implying that $Ca_v3.1$ and $Ca_v3.3$ are glycosylated, but that the degree of glycosylation varies with their regional and developmental expression patterns (Chen et al., 2007). The functional importance of N-glycosylation in T-type channel trafficking has only recently started to emerge. Glycosylation of $Ca_v3.2$ at certain loci is required for surface expression of the channel protein, as reported by us and others (Weiss et al., 2013; Orestes et al., 2013). For instance, disruption of the glycosylation loci at asparagine N192 and N1466 in the human $Ca_v3.2$ channel resulted in a significant decrease in the channel surface density without alteration of the total expression implying that N-glycosylation plays a selective role in the trafficking of the channel to the plasma membrane. Furthermore, studies of the kinetics of the channels at the plasma membrane revealed that increased internalization of glycosylation-deficient channels was mostly responsible for the decreased surface expression of $Ca_v3.2$ (Lazniewska et al., 2016). In addition, we found that increased $Ca_v3.2$ surface expression after the persistent elevation of external glucose levels is dependent on the glycosylation of the channel (Weiss et al., 2013; Lazniewska et al., 2016). This finding could have important implications in chronic diseases such as peripheral painful diabetic neuropathy, where increased T-type channel expression is thought to have a role in the development and maintenance of the disease (Jagodic et al., 2007; Messinger et al., 2009; Latham et al., 2009; Duzhyy et al., 2015). The fact that peripheral injection of neuraminidase alleviated neuropathic pain in an animal model of diabetes supports this theory (Orestes et al., 2013; Joksimovic et al., 2020). Although the mechanism by which glycosylation potentially enhances the expression of T-type channels in diabetic conditions remains to be analyzed, altered expression of glycan-processing enzymes could represent one of the potential mechanisms (Stringer et al., 2020). Overall, these findings suggest that, in addition to affecting channel gating properties (Ondacova et al., 2016), N-glycosylation of $Ca_v3.2$ contributes to the expression of the channel in the plasma membrane, which could have substantial pathophysiological implications. Glycosylation has also been reported for Ca_v1 and Ca_v2 channels as well as for Ca_v ancillary subunits (Takahashi et al., 1987). As previously stated, $Ca_v\alpha_2\delta$ is heavily glycosylated, and early results showed that N-glycosylation might help to strengthen the interaction with $Ca_v\alpha_1$ (Gurnett et al., 1996). Furthermore, it has been postulated that glycosylation of $Ca_v\alpha_2\delta$ is required for the production of $Ca_v1.3$ and $Ca_v2.2$ channels (Sandoval et al., 2004; Andrade et al., 2009). Sequential disruption of a number of glycosylation sites in the $Ca_v\alpha_2\delta_1$ followed by the analysis of its surface expression has revealed a number of glycosylation loci that are required for $Ca_v\alpha_2\delta_1$ expression at the cell surface (Tétreault et al., 2016). However, it is unclear whether these effects are due to the trafficking of $Ca_v\alpha_2\delta_1$ to the cell surface or to its stability in the plasma membrane. In addition, Ca^{2+} currents in HEK-293 cells expressing $Ca_v1.2$ were found to be decreased in the presence of glycosylation-deficient $Ca_v\alpha_2\delta_1$, confirming its chaperone role. However, it is not clear if this effect was due to the channel complex being transported to the cell surface or due to direct modulation of the channel activity. Indeed, a number of additional glycosylation sites, such as asparagine N136 and N184, which were previously thought to be required for the functional production of $Ca_v2.2$ (Sandoval et al., 2004), had minimal effects on the expression of $Ca_v\alpha_2\delta_1$ at the plasma membrane (Tétreault et al., 2016; Lazniewska & Weiss, 2016). These glycosylation sites are interestingly positioned inside the VWA-N region, suggesting that they may have a role in the functional interaction of $Ca_v\alpha_2\delta_1$ with the channel. The disruption of a glycosylation locus in the VWA domain (N348) affected the expression of $Ca_v1.2$, which supports this theory (Tétreault et al., 2016). Although a parallel analysis of Ca_v1 subunit surface expression in the presence of glycosylation-deficient $Ca_v\alpha_2\delta_1$ would be required to uncover the exact mechanisms by which glycosylation of $Ca_v\alpha_2\delta_1$ influences channel expression, these findings clearly show that N-glycosylation is an important factor contributing to surface expression of $Ca_v\alpha_2\delta$. It is worth mentioning that several glycosylation loci are found near the binding sites of gabapentinoid medicines, which could affect their pharmacological actions on these channels. Likewise, glycosylation of the $Ca_v\gamma$-subunit appears to contribute to its expression at the plasma membrane, and removal of the asparagine N48 in the first extracellular loop connecting the first and second transmembrane segments hindered cell surface trafficking (Price et al., 2005).

Subcellular Targeting of VGCCs

Although VGCC trafficking and dynamics at the plasma membrane are important, their particular targeting of subcellular loci is critical for supporting their neuronal functions. While VGCCs channels are found throughout cells in some cases (Obermair et al., 2004; Tippens et al., 2008), they typically have a more restricted subcellular expression pattern that permits to perform specialized roles. Ca_v1 channels, for example, are mostly found in proximal dendrites, where they play a role in dendritic Ca^{2+} signaling arising from back-propagating action potentials and synaptic plasticity, and on cell bodies, where they promote activity-dependent gene transcription (Barbado et al., 2009). $Ca_v2.2$ and $Ca_v2.1$ channels, on the other hand, are largely found at presynaptic terminals and promote voltage-dependent neurotransmitter release (Westenbroek et al., 1992, 1995, 1998; Hell et al., 1993). The specific molecular mechanisms underlying subcellular localization of VGCCs are still under study. The interaction of VGCCs with the $Ca_v\beta$-subunit appears to contribute to the channel complex's subcellular localization. $Ca_v2.1$ channel interaction with $Ca_v\beta_4$-subunit appears to contribute to the presynaptic targeting of the channel (Brice & Dolphin, 1999; Wittemann et al., 2000; Etemad et al., 2014). The precise mechanisms of $Ca_v\beta$-mediated channel targeting, on the other hand, are unknown. $Ca_v2.1$ and $Ca_v2.2$ channel interactions with a number of presynaptic proteins were revealed to be critical for channel targeting into nerve terminals. Some vesicular release machinery (SNARE) proteins, such as syntaxin 1A, SNAP-25, and synaptogamin, bind to the synprint area within the cytosolic II-III linker of $Ca_v\alpha$, allowing proper targeting of the channel (Jarvis & Zamponi, 2005; Szabo et al., 2006; Zamponi, 2003; Khanna et al., 2007). $Ca_v2.2$ splice mutants lacking the majority of the synprint region fail to cluster into presynaptic terminals and, instead, exhibit an axonal expression profile (Szabo et al., 2006; Mochida et al., 2003). $Ca_v3.2$ also contains a synprint-like domain within the C-terminus that interacts with syntaxin 1A and SNAP-25 (Weiss et al., 2012a) and may contribute to T-type channel targeting to presynaptic terminals (Jacus et al., 2012) where they may enable activity-dependent low-threshold exocytosis (Weiss et al., 2012a, b). Despite evidence that $Ca_v\alpha_1$/SNARE interactions play a role in channel targeting, transplanting the synprint region of $Ca_v2.2$ into $Ca_v1.2$ did not result in axonal expression or synaptic clustering of the chimera channel, implying that other targeting mechanisms are involved in the incorporation of Ca_v2 channels into nerve terminals (Szabo et al., 2006). The C-terminus of $Ca_v2.2$ contains PDZ and SH3 binding domains, which are necessary for its interaction with presynaptic adaptor proteins Mint-1 and CASK, and contribute to, but are not required, for synaptic targeting (Maximov & Bezprozvanny, 2002; Hu et al., 2005). The processes governing L-type channel somatodendritic targeting are less well understood, but they may involve interactions with structural proteins. $Ca_v1.3$ channels, for example, colocalize in postsynaptic locations in hippocampal neurons and associate with the postsynaptic adaptor protein Shank via a motif located within the channel C-terminus (Zhang et al., 2005). $Ca_v1.3$ channels can also be detected in sensory hair cells at presynaptic locations, where they interact with Ribeye (Sheets et al., 2011). $Ca_v1.4$ channels colocalize with the presynaptic scaffolding protein bassoon at the ribbon synapse in photoreceptors (Morgans, 2001). It was also reported that the $Ca_v\beta$-subunit has a role in the localization of L-type channels. For instance, point mutations in the AID region of $Ca_v1.2$ that disrupt the interaction with the $Ca_v\beta$-subunit abolished dendritic clustering of the channel in hippocampal neurons (Obermair et al., 2010). Interestingly, $Ca_v1.2$

channels are widely produced in the growth cone of growing nerve cells, despite the fact that they are localized to the soma and dendrites of adult neurons (Obermair et al., 2004). This change in the expression pattern of $Ca_v1.2$ could indicate the presence of a dynamic and reversible interaction between the channel and various $Ca_v\beta$-isoforms (Tanaka et al., 1995; McEnery et al., 1998). Along these lines, the molecular makeup of the $Ca_v2.2$ channel complex is flipped during postnatal development, from $Ca_v\beta_{1b}$ > $Ca_v\beta_3$ >> $Ca_v\beta_2$ at P2 to $Ca_v\beta_3$ > $Ca_v\beta_{1b}$ = $Ca_v\beta_4$ at P14 and the adult stage (Vance et al., 1998). In skeletal muscles, a dynamic interaction of $Ca_v\beta$-subunit with $Ca_v1.1$ also exists (Campiglio et al., 2013). Furthermore, it has been demonstrated that the assembly and disassembly of the Ca_v1/$Ca_v\beta_{2-1a}$ complex modulate the channel (Voigt et al., 2016). Within the I-II linker of $Ca_v1.2$ channels, a polybasic plasma membrane binding motif comprised of a cluster of four arginine residues was recently discovered. The positive charges face and interact with negatively charged phospholipids in the plasma membrane, forming a straight helix. This cluster was discovered to be critical for maintaining the channel at the cell surface, and neutralization of the arginine residues resulted in a decrease in the expression of the channel in the plasma membrane (Kaur et al., 2015). Given the plasma membrane phospholipid asymmetry and associated lipid-translocating enzymes, it is plausible that this polybasic motif in $Ca_v1.2$ contributes to the channel subcellular localization, similar to the clustering reported for cholinergic receptors (Scher & Bloch, 1993).

Concluding Remarks and Perspectives

The trafficking of VGCCs to the plasma membrane, and eventually into intracellular loci, is a crucial part of their dynamics. Nerve cells have access to large machinery that allows them to pick and create distinct signaling systems with the spatial and temporal features required to control certain functions. In this chapter, we have highlighted some of the components of this machinery that underpins the expression of VGCCs, but there are likely to be many more actors that have yet to be discovered. Whether or not some level of redundancy exists to protect critical neurological functions, it is clear that these systems are not perfect, and that altered signaling pathways frequently result in changes in VGCC expression and, as a result, contribute to multiple neurological disorders. However, new therapeutic avenues have opened up as a result of knowledge gained from studying the fundamental mechanisms that control the expression of these channels. One notable example is the treatment of neuropathic pain with gabapentinoid medications, and recent findings have shown intriguing evidence that selective targeting of calcium channel complex trafficking might be used as the basis for therapeutic intervention. This strategy differs from the traditional and frequently ineffective ways of focusing on channel activity per se, and it likely offers new opportunities for the development of future medications for a variety of neurological illnesses associated with calcium channel malfunction (Zamponi, 2016).

References

Adams, D. R., Ron, D., & Kiely, P. A. (2011). RACK1, A multifaceted scaffolding protein: Structure and function. *Cell Communication and Signaling: CCS, 9*, 22.

Altier, C., Garcia-Caballero, A., Simms, B., You, H., Chen, L., Walcher, J., Tedford, H. W., Hermosilla, T., & Zamponi, G. W. (2011). The Cavβ subunit prevents RFP2-mediated ubiquitination and proteasomal degradation of L-type channels. *Nature Neuroscience, 14*, 173–180.

Anantharaman, V., & Aravind, L. (2000). Cache – A signaling domain common to animal Ca(2+)-channel subunits and a class of prokaryotic chemotaxis receptors. *Trends in Biochemical Sciences, 25*, 535–537.

Andrade, A., Sandoval, A., González-Ramírez, R., Lipscombe, D., Campbell, K. P., & Felix, R. (2009). The alpha(2)delta subunit augments functional expression and modifies the pharmacology of Ca(V)1.3 L-type channels. *Cell Calcium, 46*, 282–292.

Arias, J. M., Murbartián, J., Vitko, I., Lee, J. H., & Perez-Reyes, E. (2005). Transfer of beta subunit regulation from high to low voltage-gated Ca2+ channels. *FEBS Letters, 579*, 3907–3912.

Aromolaran, K. A., Benzow, K. A., Cribbs, L. L., Koob, M. D., & Piedras-Rentería, E. S. (2009). Kelch-like 1 protein upregulates T-type currents by an actin-F dependent increase in α(1H) channels via the recycling endosome. *Channels (Austin, Tex.), 3*, 402–412.

Aromolaran, K. A., Benzow, K. A., Cribbs, L. L., Koob, M. D., & Piedras-Rentería, E. S. (2010). T-type current modulation by the actin-binding protein Kelch-like 1. *American Journal of Physiology. Cell Physiology, 298*, C1353–C1362.

Barbado, M., Fablet, K., Ronjat, M., & De Waard, M. (2009). Gene regulation by voltage-dependent calcium channels. *Biochimica et Biophysica Acta, 1793*, 1096–1104.

Barclay, J., Balaguero, N., Mione, M., Ackerman, S. L., Letts, V. A., Brodbeck, J., Canti, C., Meir, A., Page, K. M., Kusumi, K., Perez-Reyes, E., Lander, E. S., Frankel, W. N., Gardiner, R. M., Dolphin, A. C., & Rees, M. (2001). Ducky mouse phenotype of epilepsy and ataxia is associated with mutations in the Cacna2d2 gene and decreased calcium channel current in cerebellar Purkinje cells. *The Journal of Neuroscience, 21*, 6095–6104.

Bauer, C. S., Nieto-Rostro, M., Rahman, W., Tran-Van-Minh, A., Ferron, L., Douglas, L., Kadurin, I., Sri Ranjan, Y., Fernandez-Alacid, L., Millar, N. S., Dickenson, A. H., Lujan, R., & Dolphin, A. C. (2009). The increased trafficking of the calcium channel subunit alpha2delta-1 to presynaptic terminals in neuropathic pain is inhibited by the alpha2delta ligand pregabalin. *The Journal of Neuroscience, 29*, 4076–4088.

Ben-Johny, M., & Yue, D. T. (2014). Calmodulin regulation (calmodulation) of voltage-gated calcium channels. *The Journal of General Physiology, 143*, 679–692.

Ben-Johny, M., Dick, I. E., Sang, L., Limpitikul, W. B., Kang, P. W., Niu, J., Banerjee, R., Yang, W., Babich, J. S., Issa, J. B., Lee, S. R., Namkung, H., Li, J., Zhang, M., Yang, P. S., Bazzazi, H., Adams, P. J., Joshi-Mukherjee, R., Yue, D. N., & Yue, D. T. (2015). Towards a unified theory of calmodulin regulation (calmodulation) of voltage-gated calcium and sodium channels. *Current Molecular Pharmacology, 8*, 188–205.

Benmocha, A., Almagor, L., Oz, S., Hirsch, J. A., & Dascal, N. (2009). Characterization of the calmodulin-binding site in the N terminus of CaV1.2. *Channels (Austin, Tex.), 3*, 337–342.

Bernatchez, G., Talwar, D., & Parent, L. (1998). Mutations in the EF-hand motif impair the inactivation of barium currents of the cardiac alpha1C channel. *Biophysical Journal, 75*, 1727–1739.

Bernstein, G. M., & Jones, O. T. (2007). Kinetics of internalization and degradation of N-type voltage-gated calcium channels: Role of the alpha2/delta subunit. *Cell Calcium, 41*, 27–40.

Berridge, M. J. (1998). Neuronal calcium signaling. *Neuron, 21*, 13–26.

Berridge, M. J., Bootman, M. D., & Roderick, H. L. (2003). Calcium signalling: Dynamics, homeostasis and remodelling. *Nature Reviews. Molecular Cell Biology, 4*, 517–529.

Berrow, N. S., Campbell, V., Fitzgerald, E. M., Brickley, K., & Dolphin, A. C. (1995). Antisense depletion of beta-subunits modulates the biophysical and pharmacological properties of neuronal calcium channels. *The Journal of Physiology, 482*, 481–491.

Bichet, D., Cornet, V., Geib, S., Carlier, E., Volsen, S., Hoshi, T., Mori, Y., & De Waard, M. (2000). The I-II loop of the Ca2+ channel alpha1 subunit contains an endoplasmic reticulum retention signal antagonized by the beta subunit. *Neuron, 25*, 177–190.

Black, J. L. (2003). The voltage-gated calcium channel gamma subunits: A review of the literature. *Journal of Bioenergetics and Biomembranes, 35*, 649–660.

Bourdin, B., Marger, F., Wall-Lacelle, S., Schneider, T., Klein, H., Sauvé, R., & Parent, L. (2010). Molecular determinants of the CaVbeta-induced plasma membrane targeting of the CaV1.2 channel. *The Journal of Biological Chemistry, 285*, 22853–22863.

Bower, N. I., de la Serrana, D. G., Cole, N. J., Hollway, G. E., Lee, H. T., Assinder, S., & Johnston, I. A. (2012). Stac3 is required for myotube formation and myogenic differentiation in vertebrate skeletal muscle. *The Journal of Biological Chemistry, 287*, 43936–43949.

Brice, N. L., & Dolphin, A. C. (1999). Differential plasma membrane targeting of voltage-dependent calcium channel subunits expressed in a polarized epithelial cell line. *The Journal of Physiology, 515*, 685–694.

Brittain, J. M., Piekarz, A. D., Wang, Y., Kondo, T., Cummins, T. R., & Khanna, R. (2009). An atypical role for collapsin response mediator protein 2 (CRMP-2) in neurotransmitter release via interaction with presynaptic voltage-gated calcium channels. *The Journal of Biological Chemistry, 284*, 31375–31390.

Brittain, J. M., Duarte, D. B., Wilson, S. M., et al. (2011). Suppression of inflammatory and neuropathic pain by uncoupling CRMP-2 from the presynaptic Ca^{2+} channel complex. *Nature Medicine, 17*, 822–829.

Brittain, J. M., Wang, Y., Eruvwetere, O., & Khanna, R. (2012). Cdk5-mediated phosphorylation of CRMP-2 enhances its interaction with CaV2.2. *FEBS Letters, 586*, 3813–3818.

Brodbeck, J., Davies, A., Courtney, J. M., Meir, A., Balaguero, N., Canti, C., Moss, F. J., Page, K. M., Pratt, W. S., Hunt, S. P., Barclay, J., Rees, M., & Dolphin, A. C. (2002). The ducky mutation in Cacna2d2 results in altered Purkinje cell morphology and is associated with the expression of a truncated alpha 2 delta-2 protein with abnormal function. *The Journal of Biological Chemistry, 277*, 7684–7693.

Brown, J. P., & Gee, N. S. (1998). Cloning and deletion mutagenesis of the alpha2 delta calcium channel subunit from porcine cerebral cortex. Expression of a soluble form of the protein that retains [3H]gabapentin binding activity. *The Journal of Biological Chemistry, 273*, 25458–25465.

Brunet, S., Scheuer, T., Klevit, R., & Catterall, W. A. (2005). Modulation of CaV1.2 channels by Mg2+ acting at an EF-hand motif in the COOH-terminal domain. *The Journal of General Physiology, 126*, 311–323.

Brunet, S., Scheuer, T., & Catterall, W. A. (2009). Cooperative regulation of Ca(v)1.2 channels by intracellular Mg(2+), the proximal C-terminal EF-hand, and the distal C-terminal domain. *The Journal of General Physiology, 134*, 81–94.

Buraei, Z., & Yang, J. (2010). The ß subunit of voltage-gated Ca2+ channels. *Physiological Reviews, 90*, 1461–1506.

Campiglio, M., Di Biase, V., Tuluc, P., & Flucher, B. E. (2013). Stable incorporation versus dynamic exchange of β subunits in a native Ca2+ channel complex. *Journal of Cell Science, 126*, 2092–2101.

Cantí, C., Nieto-Rostro, M., Foucault, I., Heblich, F., Wratten, J., Richards, M. W., Hendrich, J., Douglas, L., Page, K. M., Davies, A., & Dolphin, A. C. (2005). The metal-ion-dependent adhesion site in the Von Willebrand factor-A domain of alpha2delta subunits is key to trafficking voltage-gated Ca2+ channels. *Proceedings of the National Academy of Sciences of the United States of America, 102*, 11230–11235.

Castiglioni, A. J., Raingo, J., & Lipscombe, D. (2006). Alternative splicing in the C-terminus of CaV2.2 controls expression and gating of N-type calcium channels. *The Journal of Physiology, 576*, 119–134.

Castle, A., & Castle, D. (2005). Ubiquitously expressed secretory carrier membrane proteins (SCAMPs) 1-4 mark different pathways and exhibit limited constitutive trafficking to and from the cell surface. *Journal of Cell Science, 118*, 3769–3780.

Catterall, W. A. (2011). Voltage-gated calcium channels. *Cold Spring Harbor Perspectives in Biology, 3*, a003947.

Chang, F. C., & Hosey, M. M. (1988). Dihydropyridine and phenylalkylamine receptors associated with cardiac and skeletal muscle calcium channels are structurally different. *The Journal of Biological Chemistry, 263*, 18929–18937.

Chen, R. S., & Best, P. M. (2009). A small peptide inhibitor of the low voltage-activated calcium channel Cav3.1. *Molecular Pharmacology, 75*, 1042–1051.

Chen, L., Chetkovich, D. M., Petralia, R. S., Sweeney, N. T., Kawasaki, Y., Wenthold, R. J., Bredt, D. S., & Nicoll, R. A. (2000). Stargazin regulates synaptic targeting of AMPA receptors by two distinct mechanisms. *Nature, 408*, 936–943.

Chen, Y., Sharp, A. H., Hata, K., Yunker, A. M., Polo-Parada, L., Landmesser, L. T., & McEnery, M. W. (2007). Site-directed antibodies to low-voltage-activated calcium channel CaV3.3 (alpha1I) subunit also target neural cell adhesion molecule-180. *Neuroscience, 145*, 981–996.

Cmarko, L., Stringer, R. N., Jurkovicova-Tarabova, B., Vacik, T., Lacinova, L., & Weiss, N. (2022). Secretory carrier-associated membrane protein 2 (SCAMP2) regulates cell surface expression of T-type calcium channels. *Molecular Brain, 15*, 1.

Colecraft, H. M., Alseikhan, B., Takahashi, S. X., Chaudhuri, D., Mittman, S., Yegnasubramanian, V., Alvania, R. S., Johns, D. C., Marbán, E., & Yue, D. T. (2002). Novel functional properties of Ca(2+) channel beta subunits revealed by their expression in adult rat heart cells. *The Journal of Physiology, 541*, 435–452.

Cornet, V., Bichet, D., Sandoz, G., Marty, I., Brocard, J., Bourinet, E., Mori, Y., Villaz, M., & De Waard, M. (2002). Multiple determinants in voltage-dependent P/Q calcium channels control their retention in the endoplasmic reticulum. *The European Journal of Neuroscience, 16*, 883–895.

Davies, A., Douglas, L., Hendrich, J., Wratten, J., Tran Van Minh, A., Foucault, I., Koch, D., Pratt, W. S., Saibil, H. R., & Dolphin, A. C. (2006). The calcium channel alpha2delta-2 subunit partitions with CaV2.1 into lipid rafts in cerebellum: Implications for localization and function. *The Journal of Neuroscience, 26*, 8748–8757.

Davies, A., Kadurin, I., Alvarez-Laviada, A., Douglas, L., Nieto-Rostro, M., Bauer, C. S., Pratt, W. S., & Dolphin, A. C. (2010). The alpha2delta subunits of voltage-gated calcium channels form GPI-anchored proteins, a posttranslational modification essential for function. *Proceedings of the National Academy of Sciences of the United States of America, 107*, 1654–1659.

Dick, I. E., Tadross, M. R., Liang, H., Tay, L. H., Yang, W., & Yue, D. T. (2008). A modular switch for spatial Ca2+ selectivity in the calmodulin regulation of CaV channels. *Nature, 451*, 830–834.

Diering, G. H., Church, J., & Numata, M. (2009). Secretory carrier membrane protein 2 regulates cell-surface targeting of brain-enriched Na+/H+ exchanger NHE5. *The Journal of Biological Chemistry, 284*, 13892–13903.

Dolmetsch, R. E., Pajvani, U., Fife, K., Spotts, J. M., & Greenberg, M. E. (2001). Signaling to the nucleus by an L-type calcium channel-calmodulin complex through the MAP kinase pathway. *Science, 294*, 333–339.

Dolphin, A. C. (2013). The α2δ subunits of voltage-gated calcium channels. *Biochimica et Biophysica Acta, 1828*, 1541–1549.

Duzhyy, D. E., Viatchenko-Karpinski, V. Y., Khomula, E. V., Voitenko, N. V., & Belan, P. V. (2015). Upregulation of T-type Ca2+ channels in long-term diabetes determines increased excitability of a specific type of capsaicin-insensitive DRG neurons. *Molecular Pain, 11*, 29.

Ellgaard, L., & Helenius, A. (2003). Quality control in the endoplasmic reticulum. *Nature Reviews. Molecular Cell Biology, 4*, 181–191.

Etemad, S., Obermair, G. J., Bindreither, D., Benedetti, A., Stanika, R., Di Biase, V., Burtscher, V., Koschak, A., Kofler, R., Geley, S., Wille, A., Lusser, A., Flockerzi, V., & Flucher, B. E. (2014). Differential neuronal targeting of a new and two known calcium channel β4 subunit splice variants correlates with their regulation

of gene expression. *The Journal of Neuroscience, 34*, 1446–1461.

Fang, K., & Colecraft, H. M. (2011). Mechanism of auxiliary β-subunit-mediated membrane targeting of L-type (Ca(V)1.2) channels. *The Journal of Physiology, 589*, 4437–4455.

Felix, R., & Weiss, N. (2017). Ubiquitination and proteasome-mediated degradation of voltage-gated Ca2+ channels and potential pathophysiological implications. *General Physiology and Biophysics, 36*, 1–5.

Felix, R., Gurnett, C. A., De Waard, M., & Campbell, K. P. (1997). Dissection of functional domains of the voltage-dependent Ca2+ channel alpha2delta subunit. *The Journal of Neuroscience, 17*, 6884–6891.

Field, M. J., Cox, P. J., Stott, E., Melrose, H., Offord, J., Su, T. Z., Bramwell, S., Corradini, L., England, S., Winks, J., Kinloch, R. A., Hendrich, J., Dolphin, A. C., Webb, T., & Williams, D. (2006). Identification of the alpha2-delta-1 subunit of voltage-dependent calcium channels as a molecular target for pain mediating the analgesic actions of pregabalin. *Proceedings of the National Academy of Sciences of the United States of America, 103*, 17537–17542.

Field, M. J., Li, Z., & Schwarz, J. B. (2007). Ca2+ channel alpha2-delta ligands for the treatment of neuropathic pain. *Journal of Medicinal Chemistry, 50*, 2569–2575.

Fjorback, A. W., Müller, H. K., Haase, J., Raarup, M. K., & Wiborg, O. (2011). Modulation of the dopamine transporter by interaction with Secretory Carrier Membrane Protein 2. *Biochemical and Biophysical Research Communications, 406*, 165–170.

Fukata, Y., Itoh, T. J., Kimura, T., Ménager, C., Nishimura, T., Shiromizu, T., Watanabe, H., Inagaki, N., Iwamatsu, A., Hotani, H., & Kaibuchi, K. (2002). CRMP-2 binds to tubulin heterodimers to promote microtubule assembly. *Nature Cell Biology, 4*, 583–591.

Gadotti, V. M., Caballero, A. G., Berger, N. D., Gladding, C. M., Chen, L., Pfeifer, T. A., & Zamponi, G. W. (2015). Small organic molecule disruptors of Cav3.2 - USP5 interactions reverse inflammatory and neuropathic pain. *Molecular Pain, 11*, 12.

Gandini, M. A., Henríquez, D. R., Grimaldo, L., Sandoval, A., Altier, C., Zamponi, G. W., Felix, R., & González-Billault, C. (2014a). CaV2.2 channel cell surface expression is regulated by the light chain 1 (LC1) of the microtubule-associated protein B (MAP 1B) via UBE2L3-mediated ubiquitination and degradation. *Pflügers Archiv, 466*, 2113–2126.

Gandini, M. A., Sandoval, A., Zamponi, G. W., & Felix, R. (2014b). The MAP 1B-LC1/UBE2L3 complex catalyzes degradation of cell surface CaV2.2 channels. *Channels (Austin, Tex.), 8*, 452–457.

Gandini, M. A., Souza, I. A., Khullar, A., Gambeta, E., & Zamponi, G. W. (2021). Regulation of Ca$_V$3.2 channels by the receptor for activated C kinase 1 (Rack-1). *Pflügers Archiv, 474*(4), 447–454.

Gao, B., Sekido, Y., Maximov, A., Saad, M., Forgacs, E., Latif, F., Wei, M. H., Lerman, M., Lee, J. H., Perez-Reyes, E., Bezprozvanny, I., & Minna, J. D. (2000a).

Functional properties of a new voltage-dependent calcium channel alpha(2)delta auxiliary subunit gene (CACNA2D2). *The Journal of Biological Chemistry, 275*, 12237–12242.

Gao, T., Bunemann, M., Gerhardstein, B. L., Ma, H., & Hosey, M. M. (2000b). Role of the C terminus of the alpha 1C (CaV1.2) subunit in membrane targeting of cardiac L-type calcium channels. *The Journal of Biological Chemistry, 275*, 25436–25444.

García-Caballero, A., Gadotti, V. M., Stemkowski, P., Weiss, N., Souza, I. A., Hodgkinson, V., Bladen, C., Chen, L., Hamid, J., Pizzoccaro, A., Deage, M., François, A., Bourinet, E., & Zamponi, G. W. (2014). The deubiquitinating enzyme USP5 modulates neuropathic and inflammatory pain by enhancing Cav3.2 channel activity. *Neuron, 83*, 1144–1158.

Garcia-Caballero, A., Gadotti, V. M., Chen, L., & Zamponi, G. W. (2016). A cell-permeant peptide corresponding to the cUBP domain of USP5 reverses inflammatory and neuropathic pain. *Molecular Pain, 12*, 1744806916642444.

Gee, N. S., Brown, J. P., Dissanayake, V. U., Offord, J., Thurlow, R., & Woodruff, G. N. (1996). The novel anticonvulsant drug, gabapentin (Neurontin), binds to the alpha2delta subunit of a calcium channel. *The Journal of Biological Chemistry, 271*, 5768–5776.

Geisler, S., Schöpf, C. L., & Obermair, G. J. (2015). Emerging evidence for specific neuronal functions of auxiliary calcium channel $\alpha_2\delta$ subunits. *General Physiology and Biophysics, 34*, 105–118.

Gong, H. C., Hang, J., Kohler, W., Li, L., & Su, T. Z. (2001). Tissue-specific expression and gabapentin-binding properties of calcium channel alpha2delta subunit subtypes. *The Journal of Membrane Biology, 184*, 35–43.

Green, P. J., Warre, R., Hayes, P. D., McNaughton, N. C., Medhurst, A. D., Pangalos, M., Duckworth, D. M., & Randall, A. D. (2001). Kinetic modification of the alpha(1I) subunit-mediated T-type Ca(2+) channel by a human neuronal Ca(2+) channel gamma subunit. *The Journal of Physiology, 533*, 467–478.

Gu, Y., & Ihara, Y. (2000). Evidence that collapsin response mediator protein-2 is involved in the dynamics of microtubules. *The Journal of Biological Chemistry, 275*, 17917–17920.

Gurnett, C. A., De Waard, M., & Campbell, K. P. (1996). Dual function of the voltage-dependent Ca2+ channel alpha 2 delta subunit in current stimulation and subunit interaction. *Neuron, 16*, 431–440.

Gurnett, C. A., Felix, R., & Campbell, K. P. (1997). Extracellular interaction of the voltage-dependent Ca2+ channel alpha2delta and alpha1 subunits. *The Journal of Biological Chemistry, 272*, 18508–18512.

Hansen, J. P., Chen, R. S., Larsen, J. K., Chu, P. J., Janes, D. M., Weis, K. E., & Best, P. M. (2004). Calcium channel gamma6 subunits are unique modulators of low voltage-activated (Cav3.1) calcium current. *Journal of Molecular and Cellular Cardiology, 37*, 1147–1158.

Hedgecock, E. M., Culotti, J. G., Thomson, J. N., & Perkins, L. A. (1985). Axonal guidance mutants of Caenorhabditis elegans identified by filling sensory neurons with fluorescein dyes. *Developmental Biology, 111*, 158–170.

Hell, J. W., Westenbroek, R. E., Warner, C., Ahlijanian, M. K., Prystay, W., Gilbert, M. M., Snutch, T. P., & Catterall, W. A. (1993). Identification and differential subcellular localization of the neuronal class C and class D L-type calcium channel alpha 1 subunits. *The Journal of Cell Biology, 123*, 949–962.

Hendrich, J., Van Minh, A. T., Heblich, F., Nieto-Rostro, M., Watschinger, K., Striessnig, J., Wratten, J., Davies, A., & Dolphin, A. C. (2008). Pharmacological disruption of calcium channel trafficking by the alpha2delta ligand gabapentin. *Proceedings of the National Academy of Sciences of the United States of America, 105*, 3628–3633.

Hershko, A., & Ciechanover, A. (1998). The ubiquitin system. *Annual Review of Biochemistry, 67*, 425–479.

Hoppa, M. B., Lana, B., Margas, W., Dolphin, A. C., & Ryan, T. A. (2012). α2δ expression sets presynaptic calcium channel abundance and release probability. *Nature, 486*, 122–125.

Horstick, E. J., Linsley, J. W., Dowling, J. J., Hauser, M. A., McDonald, K. K., Ashley-Koch, A., Saint-Amant, L., Satish, A., Cui, W. W., Zhou, W., Sprague, S. M., Stamm, D. S., Powell, C. M., Speer, M. C., Franzini-Armstrong, C., Hirata, H., & Kuwada, J. Y. (2013). Stac3 is a component of the excitation-contraction coupling machinery and mutated in Native American myopathy. *Nature Communications, 4*, 1952.

Hu, Q., Saegusa, H., Hayashi, Y., & Tanabe, T. (2005). The carboxy-terminal tail region of human Cav2.1 (P/Q-type) channel is not an essential determinant for its subcellular localization in cultured neurones. *Genes to Cells, 10*, 87–96.

Hullin, R., Singer-Lahat, D., Freichel, M., Biel, M., Dascal, N., Hofmann, F., & Flockerzi, V. (1992). Calcium channel beta subunit heterogeneity: Functional expression of cloned cDNA from heart, aorta and brain. *The EMBO Journal, 11*, 885–890.

Iftinca, M. C., & Altier, C. (2017). Stacking up Ca,3.2 channels. *Channels (Austin, Tex.), 11*, 1–2.

Inagaki, N., Chihara, K., Arimura, N., Ménager, C., Kawano, Y., Matsuo, N., Nishimura, T., Amano, M., & Kaibuchi, K. (2001). CRMP-2 induces axons in cultured hippocampal neurons. *Nature Neuroscience, 4*, 781–782.

Jacus, M. O., Uebele, V. N., Renger, J. J., & Todorovic, S. M. (2012). Presynaptic Cav3.2 channels regulate excitatory neurotransmission in nociceptive dorsal horn neurons. *The Journal of Neuroscience, 32*, 9374–9382.

Jagodic, M. M., Pathirathna, S., Nelson, M. T., Mancuso, S., Joksovic, P. M., Rosenberg, E. R., Bayliss, D. A., Jevtovic-Todorovic, V., & Todorovic, S. M. (2007). Cell-specific alterations of T-type calcium current in painful diabetic neuropathy enhance excitability of sensory neurons. *The Journal of Neuroscience, 27,* 3305–3316.

Jarvis, S. E., & Zamponi, G. W. (2005). Masters or slaves? Vesicle release machinery and the regulation of presynaptic calcium channels. *Cell Calcium, 37,* 483–488.

Joksimovic, S. L., Evans, J. G., McIntire, W. E., Orestes, P., Barrett, P. Q., Jevtovic-Todorovic, V., & Todorovic, S. M. (2020). Glycosylation of Ca$_V$3.2 channels contributes to the hyperalgesia in peripheral neuropathy of type 1 diabetes. *Frontiers in Cellular Neuroscience, 14,* 605312.

Jones, L. P., Wei, S. K., & Yue, D. T. (1998). Mechanism of auxiliary subunit modulation of neuronal alpha1E calcium channels. *The Journal of General Physiology, 112,* 125–143.

Kang, M. G., Chen, C. C., Felix, R., Letts, V. A., Frankel, W. N., Mori, Y., & Campbell, K. P. (2001). Biochemical and biophysical evidence for gamma 2 subunit association with neuronal voltage-activated Ca2+ channels. *The Journal of Biological Chemistry, 276,* 32917–32924.

Kaur, G., Pinggera, A., Ortner, N. J., Lieb, A., Sinnegger-Brauns, M. J., Yarov-Yarovoy, V., Obermair, G. J., Flucher, B. E., & Striessnig, J. (2015). A polybasic plasma membrane binding motif in the I-II linker stabilizes voltage-gated CaV1.2 calcium channel function. *The Journal of Biological Chemistry, 290,* 21086–21100.

Kavalali, E. T. (2015). The mechanisms and functions of spontaneous neurotransmitter release. *Nature Reviews. Neuroscience, 16,* 5–16.

Kawano, Y., Yoshimura, T., Tsuboi, D., Kawabata, S., Kaneko-Kawano, T., Shirataki, H., Takenawa, T., & Kaibuchi, K. (2005). CRMP-2 is involved in kinesin-1-dependent transport of the Sra-1/WAVE1 complex and axon formation. *Molecular and Cellular Biology, 25,* 9920–9935.

Khanna, R., Zougman, A., & Stanley, E. F. (2007). A proteomic screen for presynaptic terminal N-type calcium channel (CaV2.2) binding partners. *Journal of Biochemistry and Molecular Biology, 40,* 302–314.

Khanna, R., Wilson, S. M., Brittain, J. M., Weimer, J., Sultana, R., Butterfield, A., & Hensley, K. (2012). Opening Pandora's jar: A primer on the putative roles of CRMP2 in a panoply of neurodegenerative, sensory and motor neuron, and central disorders. *Future Neurology, 7,* 749–771.

Klugbauer, N., Dai, S., Specht, V., Lacinová, L., Marais, E., Bohn, G., & Hofmann, F. (2000). A family of gamma-like calcium channel subunits. *FEBS Letters, 470,* 189–197.

Körber, C., Werner, M., Kott, S., Ma, Z. L., & Hollmann, M. (2007). The transmembrane AMPA receptor regulatory protein gamma 4 is a more effective modulator of AMPA receptor function than stargazin (gamma 2). *The Journal of Neuroscience, 27,* 8442–8447.

Latham, J. R., Pathirathna, S., Jagodic, M. M., Choe, W. J., Levin, M. E., Nelson, M. T., Lee, W. Y., Krishnan, K., Covey, D. F., Todorovic, S. M., & Jevtovic-Todorovic, V. (2009). Selective T-type calcium channel blockade alleviates hyperalgesia in ob/ob mice. *Diabetes, 58,* 2656–2665.

Lazniewska, J., & Weiss, N. (2014). The "sweet" side of ion channels. *Reviews of Physiology, Biochemistry and Pharmacology, 167,* 67–114.

Lazniewska, J., & Weiss, N. (2016). Glycosylation of α2δ1 subunit: A sweet talk with Cav1.2 channels. *General Physiology and Biophysics, 35,* 239–242.

Lazniewska, J., & Weiss, N. (2017). Glycosylation of voltage-gated calcium channels in health and disease. *Biochimica et Biophysica Acta - Biomembranes, 1859,* 662–668.

Lazniewska, J., Rzhepetskyy, Y., Zhang, F. X., Zamponi, G. W., & Weiss, N. (2016). Cooperative roles of glucose and asparagine-linked glycosylation in T-type calcium channel expression. *Pflügers Archiv, 468,* 1837–1851.

Legha, W., Gaillard, S., Gascon, E., Malapert, P., Hocine, M., Alonso, S., & Moqrich, A. (2010). stac1 and stac2 genes define discrete and distinct subsets of dorsal root ganglia neurons. *Gene Expression Patterns, 10,* 368–375.

Letts, V. A. (2005). Stargazer – A mouse to seize. *Epilepsy Currents, 5,* 161–165.

Letts, V. A., Felix, R., Biddlecome, G. H., Arikkath, J., Mahaffey, C. L., Valenzuela, A., Bartlett, F. S., Mori, Y., Campbell, K. P., & Frankel, W. N. (1998). The mouse stargazer gene encodes a neuronal Ca2+–channel gamma subunit. *Nature Genetics, 19,* 340–347.

Letts, V. A., Kang, M. G., Mahaffey, C. L., Beyer, B., Tenbrink, H., Campbell, K. P., & Frankel, W. N. (2003). Phenotypic heterogeneity in the stargazin allelic series. *Mammalian Genome, 14,* 506–513.

Lin, Z., Witschas, K., Garcia, T., Chen, R. S., Hansen, J. P., Sellers, Z. M., Kuzmenkina, E., Herzig, S., & Best, P. M. (2008). A critical GxxxA motif in the gamma6 calcium channel subunit mediates its inhibitory effect on Cav3.1 calcium current. *The Journal of Physiology, 586,* 5349–5366.

Liu, H., De Waard, M., Scott, V. E., Gurnett, C. A., Lennon, V. A., & Campbell, K. P. (1996). Identification of three subunits of the high affinity omega-conotoxin MVIIC-sensitive Ca2+ channel. *The Journal of Biological Chemistry, 271,* 13804–13810.

Luvisetto, S., Fellin, T., Spagnolo, M., Hivert, B., Brust, P. F., Harpold, M. M., Stauderman, K. A., Williams, M. E., & Pietrobon, D. (2004). Modal gating of human CaV2.1 (P/Q-type) calcium channels: I. The slow and the fast gating modes and their modulation by beta subunits. *The Journal of General Physiology, 124,* 445–461.

Macabuag, N., & Dolphin, A. C. (2015). Alternative splicing in Ca(V)2.2 regulates neuronal trafficking via adaptor protein complex-1 adaptor protein motifs. *The Journal of Neuroscience, 35,* 14636–14652.

Maltez, J. M., Nunziato, D. A., Kim, J., & Pitt, G. S. (2005). Essential Ca(V)beta modulatory properties are AID-independent. *Nature Structural & Molecular Biology, 12,* 372–377.

Marais, E., Klugbauer, N., & Hofmann, F. (2001). Calcium channel alpha(2)delta subunits-structure and Gabapentin binding. *Molecular Pharmacology, 59,* 1243–1248.

Marangoudakis, S., Andrade, A., Helton, T. D., Denome, S., Castiglioni, A. J., & Lipscombe, D. (2012). Differential ubiquitination and proteasome regulation of Ca(V)2.2 N-type channel splice isoforms. *The Journal of Neuroscience, 32,* 10365–10369.

Maximov, A., & Bezprozvanny, I. (2002). Synaptic targeting of N-type calcium channels in hippocampal neurons. *The Journal of Neuroscience, 22,* 6939–6952.

McEnery, M. W., Vance, C. L., Begg, C. M., Lee, W. L., Choi, Y., & Dubel, S. J. (1998). Differential expression and association of calcium channel subunits in development and disease. *Journal of Bioenergetics and Biomembranes, 30,* 409–418.

McGee, A. W., Nunziato, D. A., Maltez, J. M., Prehoda, K. E., Pitt, G. S., & Bredt, D. S. (2004). Calcium channel function regulated by the SH3-GK module in beta subunits. *Neuron, 42,* 89–99.

Meissner, M., Weissgerber, P., Londoño, J. E., Prenen, J., Link, S., Ruppenthal, S., Molkentin, J. D., Lipp, P., Nilius, B., Freichel, M., & Flockerzi, V. (2011). Moderate calcium channel dysfunction in adult mice with inducible cardiomyocyte-specific excision of the cacnb2 gene. *The Journal of Biological Chemistry, 286,* 15875–15882.

Messinger, R. B., Naik, A. K., Jagodic, M. M., Nelson, M. T., Lee, W. Y., Choe, W. J., Orestes, P., Latham, J. R., Todorovic, S. M., & Jevtovic-Todorovic, V. (2009). In vivo silencing of the Ca(V)3.2 T-type calcium channels in sensory neurons alleviates hyperalgesia in rats with streptozocin-induced diabetic neuropathy. *Pain, 145,* 184–195.

Minturn, J. E., Fryer, H. J., Geschwind, D. H., & Hockfield, S. (1995a). TOAD-64, a gene expressed early in neuronal differentiation in the rat, is related to unc-33, a C. elegans gene involved in axon outgrowth. *The Journal of Neuroscience, 15,* 6757–6766.

Minturn, J. E., Geschwind, D. H., Fryer, H. J., & Hockfield, S. (1995b). Early postmitotic neurons transiently express TOAD-64, a neural specific protein. *The Journal of Comparative Neurology, 355,* 369–379.

Mochida, S., Westenbroek, R. E., Yokoyama, C. T., Zhong, H., Myers, S. J., Scheuer, T., Itoh, K., & Catterall, W. A. (2003). Requirement for the synaptic protein interaction site for reconstitution of synaptic transmission by P/Q-type calcium channels. *Proceedings of the National Academy of Sciences of the United States of America, 100,* 2819–2824.

Moore, R. A., Wiffen, P. J., Derry, S., Toelle, T., & Rice, A. S. (2014). Gabapentin for chronic neuropathic pain and fibromyalgia in adults. *Cochrane Database of Systematic Reviews, 2014*(4), CD007938.

Morgans, C. W. (2001). Localization of the alpha(1F) calcium channel subunit in the rat retina. *Investigative Ophthalmology & Visual Science, 42,* 2414–2418.

Moss, F. J., Viard, P., Davies, A., Bertaso, F., Page, K. M., Graham, A., Cantí, C., Plumpton, M., Plumpton, C., Clare, J. J., & Dolphin, A. C. (2002). The novel product of a five-exon stargazin-related gene abolishes Ca(V)2.2 calcium channel expression. *The EMBO Journal, 21,* 1514–1523.

Moss, F. J., Dolphin, A. C., & Clare, J. J. (2003). Human neuronal stargazin-like proteins, gamma2, gamma3 and gamma4; an investigation of their specific localization in human brain and their influence on CaV2.1 voltage-dependent calcium channels expressed in Xenopus oocytes. *BMC Neuroscience, 4,* 23.

Mould, J., Yasuda, T., Schroeder, C. I., Beedle, A. M., Doering, C. J., Zamponi, G. W., Adams, D. J., & Lewis, R. J. (2004). The alpha2delta auxiliary subunit reduces affinity of omega-conotoxins for recombinant N-type (Cav2.2) calcium channels. *The Journal of Biological Chemistry, 279,* 34705–34714.

Müller, H. K., Wiborg, O., & Haase, J. (2006). Subcellular redistribution of the serotonin transporter by secretory carrier membrane protein 2. *The Journal of Biological Chemistry, 281,* 28901–28909.

Nelson, B. R., Wu, F., Liu, Y., Anderson, D. M., McAnally, J., Lin, W., Cannon, S. C., Bassel-Duby, R., & Olson, E. N. (2013). Skeletal muscle-specific T-tubule protein STAC3 mediates voltage-induced Ca2+ release and contractility. *Proceedings of the National Academy of Sciences of the United States of America, 110,* 11881–11886.

Nishimura, T., Fukata, Y., Kato, K., Yamaguchi, T., Matsuura, Y., Kamiguchi, H., & Kaibuchi, K. (2003). CRMP-2 regulates polarized Numb-mediated endocytosis for axon growth. *Nature Cell Biology, 5,* 819–826.

Obermair, G. J., Szabo, Z., Bourinet, E., & Flucher, B. E. (2004). Differential targeting of the L-type Ca2+ channel alpha 1C (CaV1.2) to synaptic and extrasynaptic compartments in hippocampal neurons. *The European Journal of Neuroscience, 19,* 2109–2122.

Obermair, G. J., Schlick, B., Di Biase, V., Subramanyam, P., Gebhart, M., Baumgartner, S., & Flucher, B. E. (2010). Reciprocal interactions regulate targeting of calcium channel beta subunits and membrane expression of alpha1 subunits in cultured hippocampal neurons. *The Journal of Biological Chemistry, 285,* 5776–5791.

Ondacova, K., Karmazinova, M., Lazniewska, J., Weiss, N., & Lacinova, L. (2016). Modulation of Cav3.2 T-type calcium channel permeability by asparagine-linked glycosylation. *Channels (Austin, Tex.), 10,* 175–184.

Orestes, P., Osuru, H. P., McIntire, W. E., Jacus, M. O., Salajegheh, R., Jagodic, M. M., Choe, W., Lee, J., Lee, S. S., Rose, K. E., Poiro, N., Digruccio, M. R., Krishnan, K., Covey, D. F., Lee, J. H., Barrett, P. Q., Jevtovic-Todorovic, V., & Todorovic, S. M. (2013). Reversal of neuropathic pain in diabetes by targeting glycosylation of Ca(V)3.2 T-type calcium channels. *Diabetes, 62,* 3828–3838.

Page, K. M., Rothwell, S. W., & Dolphin, A. C. (2016). The CaVβ subunit protects the I-II loop of the voltage-gated calcium channel CaV2.2 from proteasomal degradation but not oligoubiquitination. *The Journal of Biological Chemistry, 291*, 20402–20416.

Park, H. J., Min, S. H., Won, Y. J., & Lee, J. H. (2015). Asn-linked glycosylation contributes to surface expression and voltage-dependent gating of Cav1.2 Ca^{2+} channel. *Journal of Microbiology and Biotechnology, 25*, 1371–1379.

Perissinotti, P. P., Ethington, E. A., Almazan, E., Martínez-Hernández, E., Kalil, J., Koob, M. D., & Piedras-Rentería, E. S. (2014a). Calcium current homeostasis and synaptic deficits in hippocampal neurons from Kelch-like 1 knockout mice. *Frontiers in Cellular Neuroscience, 8*, 444.

Perissinotti, P. P., Ethington, E. G., Cribbs, L., Koob, M. D., Martin, J., & Piedras-Rentería, E. S. (2014b). Down-regulation of endogenous KLHL1 decreases voltage-gated calcium current density. *Cell Calcium, 55*, 269–280.

Peterson, B. Z., Lee, J. S., Mulle, J. G., Wang, Y., de Leon, M., & Yue, D. T. (2000). Critical determinants of Ca(2+)-dependent inactivation within an EF-hand motif of L-type Ca(2+) channels. *Biophysical Journal, 78*, 1906–1920.

Polster, A., Perni, S., Bichraoui, H., & Beam, K. G. (2015). Stac adaptor proteins regulate trafficking and function of muscle and neuronal L-type Ca2+ channels. *Proceedings of the National Academy of Sciences of the United States of America, 112*, 602–606.

Polster, A., Nelson, B. R., Olson, E. N., & Beam, K. G. (2016). Stac3 has a direct role in skeletal muscle-type excitation-contraction coupling that is disrupted by a myopathy-causing mutation. *Proceedings of the National Academy of Sciences of the United States of America, 113*, 10986–10991.

Ponnusamy, R., & Lohkamp, B. (2013). Insights into the oligomerization of CRMPs: Crystal structure of human collapsin response mediator protein 5. *Journal of Neurochemistry, 125*, 855–868.

Pragnell, M., De Waard, M., Mori, Y., Tanabe, T., Snutch, T. P., & Campbell, K. P. (1994). Calcium channel beta-subunit binds to a conserved motif in the I-II cytoplasmic linker of the alpha 1-subunit. *Nature, 368*, 67–70.

Price, M. G., Davis, C. F., Deng, F., & Burgess, D. L. (2005). The alpha-amino-3-hydroxyl-5-methyl-4-isoxazolepropionate receptor trafficking regulator "stargazin" is related to the claudin family of proteins by its ability to mediate cell-cell adhesion. *The Journal of Biological Chemistry, 280*, 19711–19720.

Proft, J., Rzhepetskyy, Y., Lazniewska, J., Zhang, F. X., Cain, S. M., Snutch, T. P., Zamponi, G. W., & Weiss, N. (2017). The Cacna1h mutation in the GAERS model of absence epilepsy enhances T-type Ca^{2+} currents by altering calnexin-dependent trafficking of Ca,3.2 channels. *Scientific Reports, 7*, 11513.

Qin, N., Olcese, R., Stefani, E., & Birnbaumer, L. (1998). Modulation of human neuronal alpha 1E-type calcium channel by alpha 2 delta-subunit. *The American Journal of Physiology, 274*, C1324–C1331.

Quach, T. T., Duchemin, A. M., Rogemond, V., Aguera, M., Honnorat, J., Belin, M. F., & Kolattukudy, P. E. (2004). Involvement of collapsin response mediator proteins in the neurite extension induced by neurotrophins in dorsal root ganglion neurons. *Molecular and Cellular Neurosciences, 25*, 433–443.

Quach, T. T., Massicotte, G., Belin, M. F., Honnorat, J., Glasper, E. R., Devries, A. C., Jakeman, L. B., Baudry, M., Duchemin, A. M., & Kolattukudy, P. E. (2008). CRMP3 is required for hippocampal CA1 dendritic organization and plasticity. *The FASEB Journal, 22*, 401–409.

Rahajeng, J., Giridharan, S. S., Naslavsky, N., & Caplan, S. (2010). Collapsin response mediator protein-2 (Crmp2) regulates trafficking by linking endocytic regulatory proteins to dynein motors. *The Journal of Biological Chemistry, 285*, 31918–31922.

Ricard, D., Rogemond, V., Charrier, E., Aguera, M., Bagnard, D., Belin, M. F., Thomasset, N., & Honnorat, J. (2001). Isolation and expression pattern of human Unc-33-like phosphoprotein 6/collapsin response mediator protein 5 (Ulip6/CRMP5): Coexistence with Ulip2/CRMP2 in Sema3a- sensitive oligodendrocytes. *The Journal of Neuroscience, 21*, 7203–7214.

Rougier, J. S., Albesa, M., Abriel, H., & Viard, P. (2011). Neuronal precursor cell-expressed developmentally down-regulated 4-1 (NEDD4-1) controls the sorting of newly synthesized Ca(V)1.2 calcium channels. *The Journal of Biological Chemistry, 286*, 8829–8838.

Rougier, J. S., Albesa, M., Syam, N., Halet, G., Abriel, H., & Viard, P. (2015). Ubiquitin-specific protease USP2-45 acts as a molecular switch to promote α2δ-1-induced downregulation of Cav1.2 channels. *Pflügers Archiv, 467*, 1919–1929.

Rousset, M., Cens, T., Restituito, S., Barrere, C., Black, J. L., McEnery, M. W., & Charnet, P. (2001). Functional roles of gamma2, gamma3 and gamma4, three new Ca2+ channel subunits, in P/Q-type Ca2+ channel expressed in Xenopus oocytes. *The Journal of Physiology, 532*, 583–593.

Rzhepetskyy, Y., Lazniewska, J., Proft, J., Campiglio, M., Flucher, B. E., & Weiss, N. (2016). A Ca,3.2/Stac1 molecular complex controls T-type channel expression at the plasma membrane. *Channels (Austin, Tex.), 10*, 346–354.

Sandoval, A., Oviedo, N., Andrade, A., & Felix, R. (2004). Glycosylation of asparagines 136 and 184 is necessary for the alpha2delta subunit-mediated regulation of voltage-gated Ca2+ channels. *FEBS Letters, 576*, 21–26.

Scher, M. G., & Bloch, R. J. (1993). Phospholipid asymmetry in acetylcholine receptor clusters. *Experimental Cell Research, 208*, 485–491.

Sharp, A. H., Black, J. L., Dubel, S. J., Sundarraj, S., Shen, J. P., Yunker, A. M., Copeland, T. D., & McEnery, M. W. (2001). Biochemical and anatomical evidence for specialized voltage-dependent calcium channel gamma isoform expression in the epileptic and ataxic mouse, stargazer. *Neuroscience, 105*, 599–617.

Sheets, L., Trapani, J. G., Mo, W., Obholzer, N., & Nicolson, T. (2011). Ribeye is required for presynaptic Ca(V)1.3a channel localization and afferent innervation of sensory hair cells. *Development, 138*, 1309–1319.

Shistik, E., Ivanina, T., Puri, T., Hosey, M., & Dascal, N. (1995). Ca2+ current enhancement by alpha 2/delta and beta subunits in Xenopus oocytes: Contribution of changes in channel gating and alpha 1 protein level. *The Journal of Physiology, 489*, 55–62.

Simms, B. A., Souza, I. A., & Zamponi, G. W. (2014). A novel calmodulin site in the Cav1.2 N-terminus regulates calcium-dependent inactivation. *Pflügers Archiv, 466*, 1793–1803.

Simms, B. A., Souza, I. A., Rehak, R., & Zamponi, G. W. (2015). The Cav1.2 N terminus contains a CaM kinase site that modulates channel trafficking and function. *Pflügers Archiv, 467*, 677–686.

Singer, D., Biel, M., Lotan, I., Flockerzi, V., Hofmann, F., & Dascal, N. (1991). The roles of the subunits in the function of the calcium channel. *Science, 253*, 1553–1557.

Soong, T. W., & Mori, M. X. (2016). Post-transcriptional modifications and "Calmodulation" of voltage-gated calcium channel function: Reflections by two collaborators of David T Yue. *Channels (Austin, Tex.), 10*, 14–19.

Stea, A., Dubel, S. J., Pragnell, M., Leonard, J. P., Campbell, K. P., & Snutch, T. P. (1993). A beta-subunit normalizes the electrophysiological properties of a cloned N-type Ca2+ channel alpha 1-subunit. *Neuropharmacology, 32*, 1103–1116.

Stringer, R. N., Lazniewska, J., & Weiss, N. (2020). Transcriptomic analysis of glycan-processing genes in the dorsal root ganglia of diabetic mice and functional characterization on Ca$_v$3.2 channels. *Channels (Austin, Tex.), 14*, 132–140.

Strittmatter, S. M., Fankhauser, C., Huang, P. L., Mashimo, H., & Fishman, M. C. (1995). Neuronal pathfinding is abnormal in mice lacking the neuronal growth cone protein GAP-43. *Cell, 80*, 445–452.

Suzuki, H., Kawai, J., Taga, C., Yaoi, T., Hara, A., Hirose, K., Hayashizaki, Y., & Watanabe, S. (1996). Stac, a novel neuron-specific protein with cysteine-rich and SH3 domains. *Biochemical and Biophysical Research Communications, 229*, 902–909.

Szabo, Z., Obermair, G. J., Cooper, C. B., Zamponi, G. W., & Flucher, B. E. (2006). Role of the synprint site in presynaptic targeting of the calcium channel CaV2.2 in hippocampal neurons. *The European Journal of Neuroscience, 24*, 709–718.

Takahashi, M., Seagar, M. J., Jones, J. F., Reber, B. F., & Catterall, W. A. (1987). Subunit structure of dihydropyridine-sensitive calcium channels from skeletal muscle. *Proceedings of the National Academy of Sciences of the United States of America, 84*, 5478–5482.

Tanaka, O., Sakagami, H., & Kondo, H. (1995). Localization of mRNAs of voltage-dependent Ca(2+)-channels: Four subtypes of alpha 1- and beta-subunits in developing and mature rat brain. *Brain Research. Molecular Brain Research, 30*, 1–16.

Taylor, C. P., Angelotti, T., & Fauman, E. (2007). Pharmacology and mechanism of action of pregabalin: The calcium channel alpha2-delta (alpha2-delta) subunit as a target for antiepileptic drug discovery. *Epilepsy Research, 73*, 137–150.

Tétreault, M. P., Bourdin, B., Briot, J., Segura, E., Lesage, S., Fiset, C., & Parent, L. (2016). Identification of glycosylation sites essential for surface expression of the CaVα2δ1 subunit and modulation of the cardiac CaV1.2 channel activity. *The Journal of Biological Chemistry, 291*, 4826–4843.

Tippens, A. L., Pare, J. F., Langwieser, N., Moosmang, S., Milner, T. A., Smith, Y., & Lee, A. (2008). Ultrastructural evidence for pre- and postsynaptic localization of Cav1.2 L-type Ca2+ channels in the rat hippocampus. *The Journal of Comparative Neurology, 506*, 569–583.

Tomita, S., Fukata, M., Nicoll, R. A., & Bredt, D. S. (2004). Dynamic interaction of stargazin-like TARPs with cycling AMPA receptors at synapses. *Science, 303*, 1508–1511.

Tran-Van-Minh, A., & Dolphin, A. C. (2010). The alpha2delta ligand gabapentin inhibits the Rab11-dependent recycling of the calcium channel subunit alpha2delta-2. *The Journal of Neuroscience, 30*, 12856–12867.

Vance, C. L., Begg, C. M., Lee, W. L., Haase, H., Copeland, T. D., & McEnery, M. W. (1998). Differential expression and association of calcium channel alpha1B and beta subunits during rat brain ontogeny. *The Journal of Biological Chemistry, 273*, 14495–14502.

Veyrac, A., Giannetti, N., Charrier, E., Reymond-Marron, I., Aguera, M., Rogemond, V., Honnorat, J., & Jourdan, F. (2005). Expression of collapsin response mediator proteins 1, 2 and 5 is differentially regulated in newly generated and mature neurons of the adult olfactory system. *The European Journal of Neuroscience, 21*, 2635–2648.

Voigt, A., Freund, R., Heck, J., Missler, M., Obermair, G. J., Thomas, U., & Heine, M. (2016). Dynamic association of calcium channel subunits at the cellular membrane. *Neurophotonics, 3*, 041809.

Waithe, D., Ferron, L., Page, K. M., Chaggar, K., & Dolphin, A. C. (2011). Beta-subunits promote the expression of Ca(V)2.2 channels by reducing their proteasomal degradation. *The Journal of Biological Chemistry, 286*, 9598–9611.

Wakamori, M., Mikala, G., & Mori, Y. (1999). Auxiliary subunits operate as a molecular switch in determining gating behaviour of the unitary N-type Ca2+ channel current in Xenopus oocytes. *The Journal of Physiology, 517*, 659–672.

Wang, L. H., & Strittmatter, S. M. (1996). A family of rat CRMP genes is differentially expressed in the nervous system. *The Journal of Neuroscience, 16*, 6197–6207.

Wang, L. H., & Strittmatter, S. M. (1997). Brain CRMP forms heterotetramers similar to liver dihydropyrimidinase. *Journal of Neurochemistry, 69*, 2261–2269.

Wang, M., Offord, J., Oxender, D. L., & Su, T. Z. (1999). Structural requirement of the calcium-channel subunit alpha2delta for gabapentin binding. *The Biochemical Journal, 342*, 313–320.

Wang, M. C., Collins, R. F., Ford, R. C., Berrow, N. S., Dolphin, A. C., & Kitmitto, A. (2004). The three-dimensional structure of the cardiac L-type voltage-gated calcium channel: Comparison with the skeletal muscle form reveals a common architectural motif. *The Journal of Biological Chemistry, 279*, 7159–7168.

Weiss, N. (2015). Stac gets the skeletal L-type calcium channel unstuck. *General Physiology and Biophysics, 34*, 101–103.

Weiss, N., Hameed, S., Fernández-Fernández, J. M., Fablet, K., Karmazinova, M., Poillot, C., Proft, J., Chen, L., Bidaud, I., Monteil, A., Huc-Brandt, S., Lacinova, L., Lory, P., Zamponi, G. W., & De Waard, M. (2012a). A Ca(v)3.2/syntaxin-1A signaling complex controls T-type channel activity and low-threshold exocytosis. *The Journal of Biological Chemistry, 287*, 2810–2818.

Weiss, N., Zamponi, G. W., & De Waard, M. (2012b). How do T-type calcium channels control low-threshold exocytosis. *Communicative & Integrative Biology, 5*, 377–380.

Weiss, N., Black, S. A., Bladen, C., Chen, L., & Zamponi, G. W. (2013). Surface expression and function of Cav3.2 T-type calcium channels are controlled by asparagine-linked glycosylation. *Pflügers Archiv, 465*, 1159–1170.

Westenbroek, R. E., Hell, J. W., Warner, C., Dubel, S. J., Snutch, T. P., & Catterall, W. A. (1992). Biochemical properties and subcellular distribution of an N-type calcium channel alpha 1 subunit. *Neuron, 9*, 1099–1115.

Westenbroek, R. E., Sakurai, T., Elliott, E. M., Hell, J. W., Starr, T. V., Snutch, T. P., & Catterall, W. A. (1995). Immunochemical identification and subcellular distribution of the alpha 1A subunits of brain calcium channels. *The Journal of Neuroscience, 15*, 6403–6418.

Westenbroek, R. E., Hoskins, L., & Catterall, W. A. (1998). Localization of Ca2+ channel subtypes on rat spinal motor neurons, interneurons, and nerve terminals. *The Journal of Neuroscience, 18*, 6319–6330.

Whittaker, C. A., & Hynes, R. O. (2002). Distribution and evolution of von Willebrand/integrin A domains: Widely dispersed domains with roles in cell adhesion and elsewhere. *Molecular Biology of the Cell, 13*, 3369–3387.

Witcher, D. R., De Waard, M., Sakamoto, J., Franzini-Armstrong, C., Pragnell, M., Kahl, S. D., & Campbell, K. P. (1993). Subunit identification and reconstitution of the N-type Ca2+ channel complex purified from brain. *Science, 261*, 486–489.

Wittemann, S., Mark, M. D., Rettig, J., & Herlitze, S. (2000). Synaptic localization and presynaptic function of calcium channel beta 4-subunits in cultured hippocampal neurons. *The Journal of Biological Chemistry, 275*, 37807–37814.

Xia, Z., & Storm, D. R. (2005). The role of calmodulin as a signal integrator for synaptic plasticity. *Nature Reviews. Neuroscience, 6*, 267–276.

Yasuda, T., Chen, L., Barr, W., McRory, J. E., Lewis, R. J., Adams, D. J., & Zamponi, G. W. (2004). Auxiliary subunit regulation of high-voltage activated calcium channels expressed in mammalian cells. *The European Journal of Neuroscience, 20*, 1–13.

Yoshimura, T., Arimura, N., & Kaibuchi, K. (2006). Molecular mechanisms of axon specification and neuronal disorders. *Annals of the New York Academy of Sciences, 1086*, 116–125.

Yunker, A. M., Sharp, A. H., Sundarraj, S., Ranganathan, V., Copeland, T. D., & McEnery, M. W. (2003). Immunological characterization of T-type voltage-dependent calcium channel CaV3.1 (alpha 1G) and CaV3.3 (alpha 1I) isoforms reveal differences in their localization, expression, and neural development. *Neuroscience, 117*, 321–335.

Zaarour, N., Defontaine, N., Demaretz, S., Azroyan, A., Cheval, L., & Laghmani, K. (2011). Secretory carrier membrane protein 2 regulates exocytic insertion of NKCC2 into the cell membrane. *The Journal of Biological Chemistry, 286*, 9489–9502.

Zamponi, G. W. (2003). Regulation of presynaptic calcium channels by synaptic proteins. *Journal of Pharmacological Sciences, 92*, 79–83.

Zamponi, G. W. (2016). Targeting voltage-gated calcium channels in neurological and psychiatric diseases. *Nature Reviews. Drug Discovery, 15*, 19–34.

Zamponi, G. W., Striessnig, J., Koschak, A., & Dolphin, A. C. (2015). The physiology, pathology, and pharmacology of voltage-gated calcium channels and their future therapeutic potential. *Pharmacological Reviews, 67*, 821–870.

Zhang, Y., Mori, M., Burgess, D. L., & Noebels, J. L. (2002). Mutations in high-voltage-activated calcium channel genes stimulate low-voltage-activated currents in mouse thalamic relay neurons. *The Journal of Neuroscience, 22*, 6362–6371.

Zhang, H., Maximov, A., Fu, Y., Xu, F., Tang, T. S., Tkatch, T., Surmeier, D. J., & Bezprozvanny, I. (2005). Association of CaV1.3 L-type calcium channels with Shank. *The Journal of Neuroscience, 25*, 1037–1049.

Calmodulin Regulation of Voltage-Gated Calcium Channels

Manu Ben-Johny and Ivy E. Dick

Abstract

The ubiquitous Ca^{2+} sensor calmodulin (CaM) serves as a preeminent modulator of voltage-gated Ca^{2+} channels, exerting rapid and powerful feedback control of Ca^{2+} entry into cardiomyocytes and neurons. Over the past three decades, this modulation has emerged as a prototype for ion channel regulation with important physiological and pathophysiological implications. In this chapter, we summarize key findings pertaining to structural and biological mechanisms of CaM regulation, as well as elaborate on how Ca_V channel misregulation underlies a wide range of human diseases.

Keywords

Calmodulin · Calmodulation · Ca^{2+} regulation · calcium channels

M. Ben-Johny (✉)
Department of Physiology and Cellular Biophysics, Columbia University, New York, NY, USA
e-mail: mbj2124@cumc.columbia.edu

I. E. Dick (✉)
Department of Physiology, University of Maryland School of Medicine, Baltimore, MD, USA
e-mail: ied@som.umaryland.edu

Introduction

Voltage-gated calcium channels (VGCCs) are critical conduits for Ca^{2+} entry into excitable cells and play essential roles in numerous physiological functions including muscle contraction, neurotransmission, gene transcription, and cardiac rhythm. As such, Ca^{2+} entry through these channels is precisely controlled both spatially and temporally by feedback regulation. In the classical Hodgkin-Huxley paradigm, such feedback regulation for voltage-gated ion channels typically entails a voltage-dependent conformational change that obstructs ion influx following depolarization, a process known as inactivation (Hodgkin & Huxley, 1952). For Ca^{2+} currents, however, pioneering studies in *Paramecia* demonstrated that, in addition to the well-established voltage-dependent inactivation (VDI), the Ca^{2+} ions that entered through the VGCC could itself inactivate the channel and prevent further Ca^{2+} entry (Brehm & Eckert, 1978). Subsequently, this powerful mode of autoregulation was substantiated in Ca^{2+} channels in the mammalian heart (Kass & Sanguinetti, 1984; Lee et al., 1985; Mentrard et al., 1984) and elsewhere (Tillotson, 1979; Xu & Wu, 2005), with important roles for physiology (Alseikhan et al., 2002; Weyrer et al., 2019) and pathophysiology (Limpitikul et al., 2014; Mahajan et al., 2008; Morotti et al., 2012).

More broadly, this phenomenon of Ca^{2+}-dependent feedback regulation is now well established for nearly all Ca_V channels (Chemin et al., 2017; DeMaria et al., 2001b; Lee et al., 1999; Peterson et al., 1999a; Qin et al., 1999; Singh et al., 2006; Wahl-Schott et al., 2006; Williams et al., 2018; Xu & Lipscombe, 2001; Yang et al., 2006; Zuhlke et al., 1999) and other ion channel superfamilies in diverse physiological settings (Ben-Johny et al., 2015; Saimi & Kung, 2002).

At the molecular level, voltage-dependent feedback for voltage-gated ion channels is easily envisaged, as one or more voltage-sensing domains could be specialized to initiate block the pore domain, as has been demonstrated for voltage-gated sodium channels (Capes et al., 2013). By contrast, coupling cytosolic Ca^{2+} to the pore domain requires specialized cytosolic structural elements that harbor either an intrinsic or an extrinsic Ca^{2+} sensor. Indeed, some ion channels such as the big conductance K^+ channels or the ryanodine receptor are equipped with elements that bind Ca^{2+} ions and allosterically modify the channel pore (Tao & MacKinnon, 2019; Zalk et al., 2015). For VGCCs, however, this task is instead accomplished by recruiting an extrinsic Ca^{2+} sensor, calmodulin (CaM), as a *de facto* channel subunit (Lee et al., 1999; Peterson et al., 1999a; Zuhlke et al., 1999). CaM is a 17 kDa ubiquitous protein that is highly conserved in eukaryotes. Structurally, CaM is comprised of two globular domains joined together by a flexible linker (Babu et al., 1985; Kuboniwa et al., 1995). Each domain contains two EF hand motifs that bind Ca^{2+} cooperatively, such that CaM, in total, is capable of binding up to 4 Ca^{2+} ions (Babu et al., 1985). Despite its seemingly simplistic architecture, CaM is a versatile modulator that adopts a wide-range of binding conformations, which allows it to recognize and modulate an impressive repertoire of targets including ion channels, receptors, and cytosolic proteins (Tidow & Nissen, 2013). In this context, CaM regulation of VGCCs serves as a prototypic modulation, exemplifying a rich scheme of spatial Ca^{2+} decoding, complex structural mechanisms to control ion channel activity, and sophisticated

biological governance. Furthermore, recent studies point to altered CaM feedback to VGCC as a prominent pathological mechanism underlying life-threatening cardiac arrhythmias and neurological disorders.

In this chapter, we summarize key mechanisms underlying calmodulation of VGCCs and discuss their physiological and pathophysiological importance.

Ca²⁺/CaM-Dependent Regulation of VGCCs

Ca^{2+} regulation of VGCCs may manifest as either positive or negative feedback, where channel activity is either upregulated or downregulated, respectively, in response to cytosolic Ca^{2+} elevation. In fact, these contrasting modes of channel regulation can even be observed within a single channel type (Liang et al., 2003). This phenomenon is exemplified in neuronal $Ca_V2.1$ channels where Ca^{2+} entry into the cell initially results in a rapid Ca^{2+} dependent facilitation (CDF), which is seen as an initial increase in Ca^{2+} current amplitude during a repetitive train of action potentials (Fig. 1a) (Chaudhuri et al., 2007; DeMaria et al., 2001b; Lee et al., 1999). As the AP stimulus continues, the peak amplitude of the Ca^{2+} current declines, indicative of Ca^{2+} dependent inactivation (CDI). Importantly, both these processes are absent when Ba^{2+} is used as the charge carrier through the channel, thus demonstrating the exquisite Ca^{2+} dependence of both processes.

Experimentally, the magnitude of CDF and CDI can be quantified using step pulse protocols designed to isolate each component within HEK 293 cells (DeMaria et al., 2001b; Thomas & Lee, 2016). For CDF, a paired-pulse protocol can be used (Fig. 1b, **left**), where a 50 ms test pulse (black) gives rise to a Ca^{2+} current, which exhibits biphasic kinetics in reaching the maximal amplitude (Chaudhuri et al., 2007; DeMaria et al., 2001b; Thomas & Lee, 2016). The initial phase corresponds to the fast voltage-dependent activation of the channel; however, the latter phase corresponds to a slower Ca^{2+}-dependent upregulation of channel activity (DeMaria et al.,

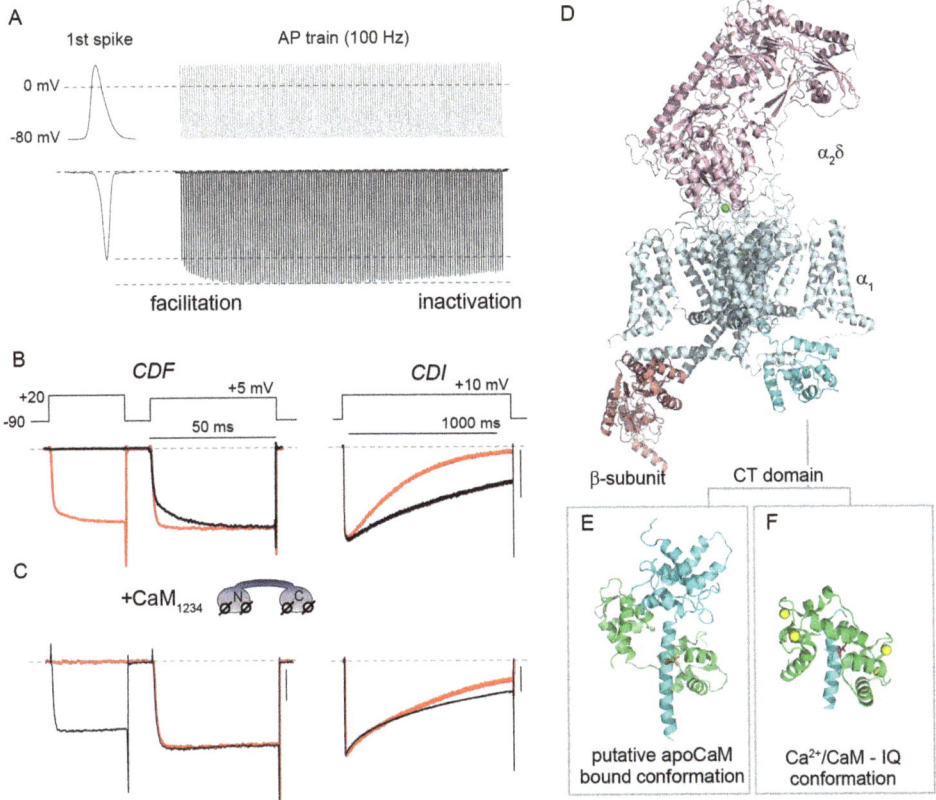

Fig. 1 CaM is the Ca^{2+} sensor for CDI and VDI. (**a**), **Top:** 100 Hz AP waveform stimulus applied to HEK 293 cells expressing $Ca_V2.1$ channels, left side shows a single AP on an expanded timescale for visualization of the AP morphology. **Bottom:** Ca^{2+} current recorded in response to the AP train demonstrates initial facilitation of the current, followed by a subsequent inactivation response. (Reproduced from Chaudhuri et al. (2004) with permission). (**b**), **Left:** CDF of $Ca_V2.1$ channels measured using a paired-pulse protocol in HEK293 cells. Without a prepulse (black), Ca^{2+} current slowly reaches peak amplitude within 50 ms. Application of a prepulse (red) results in rapid activation of the Ca^{2+} current to the facilitated level. Right: A prolonged depolarization induces a stronger decay in the Ca^{2+} current (red) as compared to the Ba^{2+} current, indicative of CDI. (Reproduced from Chaudhuri et al. (2004) with permission). (**c**), Overexpression of CaM_{1234} abolishes both CDF (left) and CDI (right) of $Ca_V2.1$ channels, demonstrating that CaM is the Ca^{2+} sensor for both processes. (Reproduced from DeMaria et al. (2001b)). (**d**), Top, Cryo-EM structure of $Ca_V1.1$ channels (PDB code 5GJW) shows transmembrane domains and a partial segment of the carboxy-terminus that is critical for CDI. (**e**), a homology model of $Ca_V1.3$ carboxy-terminus in complex with CaM based on structures of $Na_V1.4$/CaM complex. (**f**), the Crystal structure of $Ca_V1.2$ IQ domain in complex with Ca^{2+}/CaM (PDB code: 2BE6)

2001b). If this test pulse were preceded by a brief depolarizing step, the Ca^{2+} entry during the pre-pulse drives the channel into the facilitated state (red). As a consequence, the subsequent test pulse demonstrates the response of a channel which is Ca^{2+} facilitated, as can be seen by the rapid activation to the maximal current amplitude. CDF can then be quantified by the area between the non-facilitated (black) and facilitated (red) traces, i.e. the excess charge entry dur-

ing facilitation. CDI, on the other hand, is far slower and evolves over a time course of 100 ms. Inactivation is seen in $Ca_V2.1$ channels as the decay in Ca^{2+} current (red) in response to a depolarizing step (Fig. 1b, **right**). However, as voltage dependent inactivation (VDI) is also present in these cells, this inactivation profile represents the combined effect of both CDI and VDI. The specific contribution of VDI can be dissected by using Ba^{2+} as charge carrier (black), as CaM binds poorly to Ba^{2+} ions. CDI can then be quantified as the excesses inactivation present with Ca^{2+} currents as compared to Ba^{2+}.

Early studies of Ca^{2+} regulation of VGCCs considered multiple mechanisms for the initiation of CDI or CDF including the direct binding of Ca^{2+} to the channel (Eckert & Chad, 1984; Plant et al., 1983; Standen & Stanfield, 1982). In particular, a section of the channel carboxy-tail (termed the Ca^{2+} inactivating (CI) region) was noted to contain two putative EF hand like domains that are classically associated with Ca^{2+} binding (Babitch, 1990; Wingo et al., 2004). Although the channel EF hand domains lack key oxygen-bearing residues necessary for high affinity Ca^{2+}-coordination, chimeric analysis pointed to a vital role for this domain. Indeed, subsequent mutagenesis of putative Ca^{2+}-coordinating aspartate residues firmly ruled out direct Ca^{2+} binding to the channel EF domains as a mechanism for CDI (Peterson et al., 2000; Zhou et al., 1997), instead hinting at a possible transduction role for this domain as elaborated in subsequent sections. Further mutagenesis studies identified an isoleucine-glutamine (IQ) domain, also located on the carboxy-tail, as a critical locus involved in CDI (Qin et al., 1999; Zuhlke & Reuter, 1998). As the IQ motif is a well-known CaM binding site (Jurado et al., 1999), it is not surprising that CaM was soon identified as the Ca^{2+} sensor for calmodulation of VGCCs. The definitive role for CaM as the Ca^{2+} sensor for VGCC (Lee et al., 1999; Peterson et al., 1999a; Zuhlke et al., 1999) was established by using Ca^{2+}-insensitive CaM, CaM_{1234}, whose key aspartate residues in each of the four EF hand domains are substituted with alanine (Xia et al., 1998). Of note, since its initial application with SK channels (Xia et al., 1998), this strategy of using Ca^{2+}-insensitive CaM has been employed extensively to probe calmodulatory effects of ion channels. Figure 1c demonstrates the effect of CaM_{1234} overexpression with $Ca_V2.1$ channels—both CDF and CDI are sharply shunted (DeMaria et al., 2001b; Lee et al., 1999). Importantly, early studies of VGCC regulation had actually excluded a possible role for CaM on the basis that pharmacological inhibitors of $Ca^{2+}/$ CaM failed to appreciably alter inactivation

kinetics (Imredy & Yue, 1994; Zuhlke & Reuter, 1998). These findings along with the striking dominant negative effect of CaM_{1234} revealed a fundamental principle of calmodulation of ion channels—the notion that Ca^{2+}-free CaM (apoCaM) pre-associates with the channel and serves as a resident Ca^{2+} sensor (Erickson et al., 2001; Lee et al., 2003; Pitt et al., 2001). Classically, apoCaM is thought of as an inert Ca^{2+}-sensor that is mobilized to engage its target only following Ca^{2+} binding (Jurado et al., 1999). In this scheme, a pharmacological inhibitor has ample opportunity to bind $Ca^{2+}/$ CaM and preclude activation of the downstream target. For nearly all VGCCs, however, CaM is constitutively associated with the channel, listening for Ca^{2+} signals (Erickson et al., 2001; Lee et al., 2003; Liang et al., 2003; Pitt et al., 2001; Yang et al., 2006). Upon Ca^{2+} elevation, this resident CaM binds to Ca^{2+} and rebinds to the channel (Fig. 1d), resulting in a conformational change that supports CDI or CDF. In this scheme, as CaM is initially preassociated, the bound conformation prevents accessibility of the pharmacological CaM inhibitor, leaving Ca^{2+}-dependent regulation intact. By contrast, when CaM_{1234} is overexpressed, the mutant CaM can occupy the preassociation site (i.e. the IQ domain for VGCC) just as well, and prevent endogenous wild-type CaM (CaM_{WT}), from interacting with the channel. However, as mutant CaM_{1234} is incapable of binding Ca^{2+}, it fails to trigger feedback regulation. In this manner, pharmacological CaM inhibitors versus CaM_{1234} provides complementary tools to probe distinct calmodulatory mechanisms.

The exact structural mechanisms underlying this modulation has yet to be fully determined in the context of the full length VGCC. Cryo-EM structures of $Ca_V1.1$, a channel that binds poorly to apoCaM, have revealed only partial segments of the carboxy-tail domain including the dual vestigial EF hand segments (Fig. 1d) (Wu et al., 2015). Similarly, structures of $Ca_V2.2$ also revealed only partial segments of the carboxy-tail despite its ability to bind CaM with a higher affinity, potentially pointing to a high mobility of CaM binding segments (Dong et al., 2021; Gao et al., 2021). Biochemical studies, live-cell FRET assays, and structural analysis of channel peptides have been valuable in outlining key calmodulatory landmarks of VGCCs. First, biochemical analysis and FRET 2-hybrid assays show that a single apoCaM interacts with the carboxyterminal domain of Ca_V1 channels (Erickson et al., 2001, 2003; Evans et al., 2011; Pitt et al., 2001; Tang et al., 2003), and complementary alanine scanning mutagenesis suggests that the C-lobe engaging a canonical IQ domain, while the N-lobe weakly associates with upstream EF hand elements (Bazzazi et al., 2013; Ben Johny et al., 2013). A similar arrangement of apoCaM has been observed in $Na_V1.4$ channels (Yoder et al., 2019) which also undergoes a conserved CDI mechanism (Ben-Johny et al., 2014). Figure 1e depicts a homology model of $Ca_V1.3$ bound to apoCaM preassociation based on the Na_V channel structures (Banerjee et al., 2018). Second, multiple Ca^{2+}/CaM binding sites have been identified, including the IQ domain (Erickson et al., 2003; Halling et al., 2009; Pitt et al., 2001; Qin et al., 1999; Tang et al., 2003), upstream segments of the carboxy-terminus (Asmara et al., 2010; Ben Johny et al., 2013; Fallon et al., 2009; Kim et al., 2004, 2010), the NSCaTE domain (N-terminal Spatial Ca^{2+}-transforming element) (Benmocha et al., 2009; Dick et al., 2008; Ivanina et al., 2000) and nearby residues in the channel amino terminus (Simms et al., 2014), and other cytosolic loops such as the I-II linker and the III-IV linker (Zhou et al., 2005). X-ray crystallographic studies have shown that the two lobes of Ca^{2+}/CaM can engulf the IQ peptide (Mori et al., 2004; Van Petegem et al., 2005) (Fig. 1f); however, it is likely that CaM makes additional contacts with other cytosolic domains to evoke CDI (Ben Johny et al., 2013). At holo-channel level, FRET stoichiometric analysis suggests that only a single apoCaM associates with the $Ca_V1.2$ channel, although prolonged Ca^{2+}-elevation can recruit multiple CaM to the channel complex (Ben-Johny et al., 2016). Functional studies indicate that a single CaM, that is initially pre-associated to the channel carboxy-terminus mediates channel regulation (Chakouri et al., 2020; Mori et al., 2004).

Although Ca_V3 channels also undergo Ca^{2+}/CaM dependent regulation, this channel lacks the conserved carboxy-tail domain (Chemin et al., 2017). Instead, this feedback modulation depends on Ca^{2+}/CaM interaction with the gating brake encoded in the I-II linker (Asmara et al., 2017; Chemin et al., 2017). The consensus molecular mechanisms underlying CaM regulation of $Ca_V1/2$ are discussed further in subsequent sections.

Ca²⁺/CaM-Dependent Regulation Responds Differentially to Spatially Distinct Ca²⁺ Sources

The bilobal structure of CaM results in a remarkable functional bipartition, where each lobe of CaM is capable of imparting its own disparate effects (Kink et al., 1990). For VGCCs this means that the N- and C- lobes of CaM can each independently drive distinct forms of feedback regulation (DeMaria et al., 2001a; Yang et al., 2006; Zuhlke et al., 1999). For example, in the case of $Ca_V2.1$ overexpression of CaM_{12}, in which C-lobe Ca^{2+} binding has been disrupted, results in a selective loss of CDF but not CDI, indicating that CDF in this channel is controlled selectively by the C-lobe of CaM. Conversely, overexpression of CaM_{34}, in which the N-lobe EF-hands have been disrupted, results in a loss of CDI but not CDF, indicating that N-lobe CaM acts as the sensor for CDI in these channels. This functional bipartition has been demonstrated across multiple Ca_V1-2 channels, with N- and C- lobe CaM controlling distinct forms of CDI or CDF, often with varying kinetics (Liang et al., 2003).

The bilobal architecture of CaM also enables channels to respond differentially to Ca^{2+} signals from distinct Ca^{2+} sources (Dick et al., 2008; Tadross et al., 2008). Pre-association of CaM with the channel carboxy-tail means that it resides within nanometers of channel, where channel openings produce large Ca^{2+} oscillations. Within the nanodomain, the Ca^{2+} signal is the sum of two distinct components (Fig. 2a). The local Ca^{2+} signal is driven by Ca^{2+} entry through the channel itself and consists of large Ca^{2+} spikes

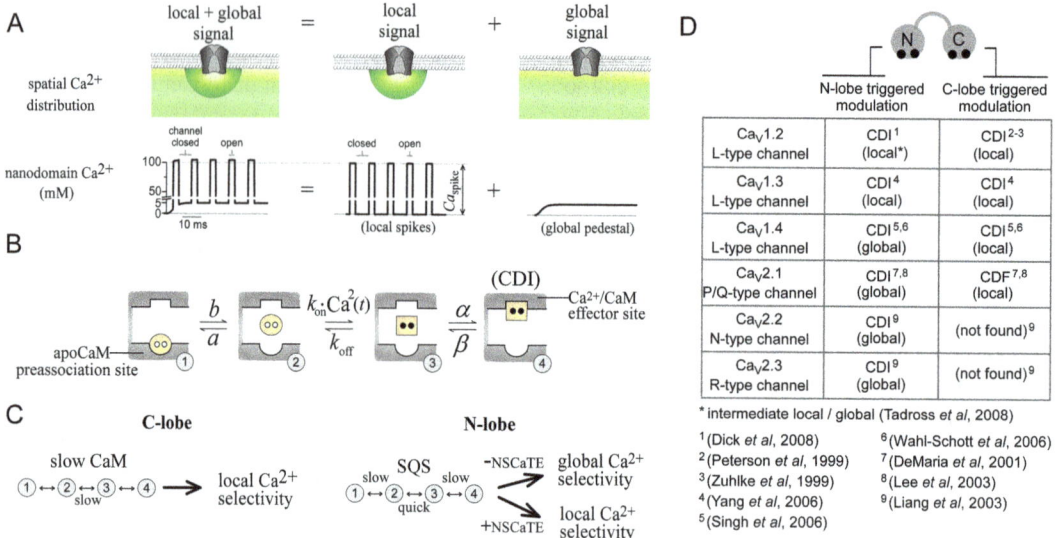

Fig. 2 Distinguishing between spatially distinct Ca²⁺ sources. (**a**), **Top:** Cartoon demonstrating the spatial distribution of Ca²⁺. Dark green hemisphere indicates the large local Ca²⁺ entering through the channel, while the diffuse green shading corresponds to the global Ca²⁺ signal. **Bottom:** The corresponding local and global Ca²⁺ signals that would be detected within the nanodomain of the channel where CaM resides. (**b**), Simplified model of a single lobe of CaM (yellow) binding to a channel (gray) in response to Ca²⁺ (black circles). (**c**), **Left:** The C-lobe of CaM features slow on/off kinetics of Ca²⁺ binding, resulting in an invariably local selectivity. **Right:** The N-lobe of CaM utilizes an 'SQS' mechanism, enabling the selectivity of N-lobe CaM to be tuned depending on the binding affinity of the channel for apoCaM vs. Ca²⁺/CaM. The inclusion of an additional Ca²⁺ CaM binding domain known as NSCaTE alters this ratio and acts as a switch between local and global selectivity. (**a–c** adapted from Tadross et al. (2008)). (**d**). The local/global selectivity of each lobe of CaM in the context of various VGCCs (Reproduced from Ben-Johny & Yue, 2014)

(~100 µM) resulting from channel openings (Neher, 1998; Sherman et al., 1990; Stern, 1992). In addition, a global Ca²⁺ signal results from the cumulative entry of Ca²⁺ through multiple distant Ca²⁺ sources within the cell. This global component comprises a far smaller (~5 µM) Ca²⁺ signal which is spatially distributed across the cytosol. Experimentally these Ca²⁺ sources can be isolated by adjusting the intracellular Ca²⁺ buffering, such that the global signal persists in physiological buffering conditions but is eliminated by high intracellular Ca²⁺ buffers (DeMaria et al., 2001a; Liang et al., 2003). Such experiments have revealed that for the Ca_V1-2 channel family, the C-lobe of CaM invariably acts as a local Ca²⁺ sensor, while the N-lobe may act as either a global sensor or local sensor depending on the channel context (Fig. 2d) (DeMaria et al., 2001b; Dick et al., 2008; Lee et al., 2003; Liang et al., 2003; Peterson et al., 1999b; Singh et al., 2006; Wahl-

Schott et al., 2006; Williams et al., 2018; Yang et al., 1993; Zuhlke et al., 1999).

This spatial selectivity of each lobe of CaM is imparted by the distinct kinetics and binding affinities of each channel/lobe interaction (Faas et al., 2011; Linse et al., 1991), enabling each lobe to respond to the distinct temporal features of each Ca²⁺ source (Tadross et al., 2008). The mechanism by which this occurs can be reduced to the scheme illustrated in Fig. 2b. State 1 depicts apoCaM (yellow) bound to the channel carboxy-tail (gray). The lobe of CaM transiently disengages from the channel but remains within the alcove of the channel (state 2) where it is available to bind Ca²⁺ (state 3). Upon Ca²⁺ binding, CaM binds the Ca²⁺/CaM effector site within each channel, inducing CDI (state 4). For the C-lobe of CaM, Ca²⁺ release from the EF hands is slow relative to the brief duration of channel closings. As a result, once CaM binds Ca²⁺ in response to the local Ca²⁺ spikes, this binding will be maintained during brief channel closures, promoting occupancy of state 3, which in turn drives entry into state 4. This 'slow CaM' mechanism (Fig. 2c) explains the invariable local selectivity of C-lobe CDI across Ca_V1-2 channels. For the N-lobe, however, the kinetics of Ca^{2+} binding and unbinding are much more rapid, allowing channels to quickly transition between states 2 and 3 on the timescale of channel openings. This rapid switching between states 2 and 3 positions the kinetics of entry and exit from states 1 and 4 as critical determinants of spatial selectivity. If transitions between states 1 and 2 and states 3 and 4 are slow, the resulting slow-quick-slow (SQS) mechanism produces a global selectivity. While in state 1, the slow exit from this state will inhibit a large but rapid Ca^{2+} spike from driving the CaM/channel complex to state 3. When a channel does transition to state 2 a brief spike that may drive entry into state 3, however rapid release of Ca^{2+} from CaM during the channel closing will quickly drive a return to state 2, preventing rapid Ca^{2+} signals from inducing state 4 occupancy. Conversely, the slow temporal signature of the global Ca^{2+} signal will outlast the relatively long occupancy of state 1. Once in state 2, the ~5 µM global signal is sufficiently larger than the Ca^{2+} binding affinity of N-lobe CaM, promoting entry into states 3 and 4. Interestingly, this SQS mechanism does not inevitably produce a global response and may be tuned to respond to distinct Ca^{2+} signals depending on the relative affinity of apoCaM *versus* Ca^{2+}/CaM for the channel. In other words, because transitions between states 2 and 3 are rapid, the bias of the scheme towards state 1 *vs.* state 4 becomes critical. Because Ca_V2 channels have a higher apoCaM affinity as compared to Ca^{2+}/CaM, they display a global spatial selectivity for N-lobe regulation (Fig. 2d). However, engineering a second Ca^{2+}/CaM binding site into the amino-terminus of these channels is sufficient to increase the Ca^{2+}/CaM affinity of the channel and enable a rapid Ca^{2+} signal to drive entry into state 4, resulting in CDI driven by local Ca^{2+} entry (Dick et al., 2008; Tadross et al., 2008). Thus, the spatial selectivity of N-lobe CaM regulation may be tuned depending on the channel context. Such is the case for LTCCs,

where Ca_V1.2 and Ca_V1.3 channels harbor an additional Ca^{2+}/CaM binding within their amino-terminus termed the N-terminal spatial Ca^{2+} transforming element (NSCaTE) (Dick et al., 2008; Ivanina et al., 2000). The presence of NSCaTE converts the spatial selectivity of N-lobe CaM to local for these two channels (Fig. 2d). In this way, channels may be tuned to respond to the specific needs of each cell type.

Molecular Mechanism of Calmodulation

Drawing upon insights from Ca_V1.3 channels, we here summarize the overall molecular mechanism of calmodulation, highlight key CaM/channel configurations, and elaborate on mechanisms that fine-tune these interactions.

In molecular terms, conventional VDI of ion channels (such as Na_V and shaker K^+) is thought to result from a cytosolic inactivation particle occluding the transmembrane ion conduction pathway, thereby rendering the channel to be constitutively nonconductive (Dong et al., 2021; Eaholtz et al., 1994; Zagotta et al., 1990). In sharp contrast, pioneering single-molecule studies demonstrated that CDI, in fact, reduced the frequency of channel openings without completely silencing the channels (Imredy & Yue, 1994). These findings point to an allosteric model of channel inhibition which favors channel closure yet permits intermittent channel openings. With this scheme, channels may be conceptualized to switch between distinct 'modes of gating,' with each mode endowed with a different intrinsic capacity for channel opening (Hess et al., 1984; Imredy & Yue, 1994). Interaction of apoCaM *versus* Ca^{2+}/CaM stabilizes distinct gating modes (Fig. 3a–c). Of note, this scheme of modal gating is as a general theme for understanding Ca^{2+} channel modulation by various mechanisms, including adrenergic regulation (Hess et al., 1984; Yue et al., 1990).

Accordingly, Fig. 3d–f illustrates simplified configurations relevant for the overall calmodulatory system. First, channels devoid of CaM exhibit a diminished open probability (P_O) cor-

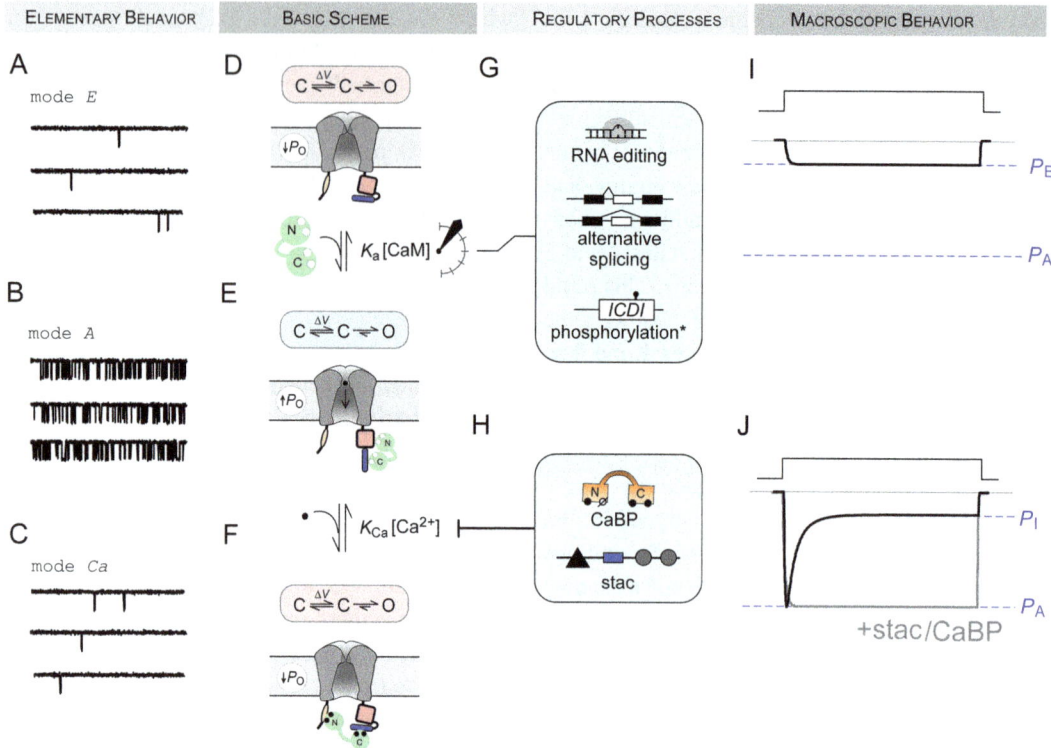

Fig. 3 Basic schematic of CDI mechanism. (**a–c**), Cartoon depicting the single-channel behavior for channels occupying each different gating mode. Mode E (panel A) and Mode Ca²⁺ (pane C) exhibit low open probabilities, while channels in mode A (panel B) display a high open probability. (**d–f**), Cartoon demonstrating the 3 basic modes of channel opening underlying CDI: mode E (top) is characterized by lack of CaM on the channel, mode A features apoCaM bound to the channel and mode Ca²⁺ represents channels bound to Ca²⁺/CaM. (**g–h**), Depiction of some of the known modulators of CDI and where they act within the calmodulation scheme. (**i–j**), Illustration of the macroscopic behavior of a channel in mode E (top) as compared to one transitioning between Mode A and Mode Ca²⁺ (black) or held in mode A due to stac/CaBP binding (gray)

responding to a pre-inhibited configuration (Adams et al., 2014; Bock et al., 2011; Kuzmenkina et al., 2019). This gating configuration is referred to as mode E. Simulated single-channel recordings show brief and rare openings for channels lacking CaM (Fig. 3a). The structural underpinnings of this configuration are not fully established. One possibility is that without CaM bound, the IQ domain may be unstable, potentially resulting in a collapsed conformation of this domain (Banerjee et al., 2018) (Fig. 3d). Second, apoCaM binding to the channel carboxytail promotes channel openings (Fig. 3c) by switching channels to mode A with increased P_O (Fig. 3e) (Adams et al., 2014; Kuzmenkina et al., 2019). NMR structures have shown an antiparallel of CaM C-lobe with an isolated IQ peptide

(Turner et al., 2020). Beyond this, the overall orientation of CaM with the carboxy-terminus or the holo-channel remains to be fully established. In the related voltage-gated Na channels and the NALCN Na leak channel, CaM associates with a conserved carboxy-terminus in a similar arrangement with the C-lobe binding the IQ domain and the N-lobe interacting with the vestigial EF-hand segments (Kschonsak et al., 2021; Yoder et al., 2019). Consistent with this scheme, alanine-scanning mutagenesis also identified residues in both the IQ domain and the upstream dual vestigial EF-hand segment that disrupts apoCaM binding (Ben Johny et al., 2013). Functionally, these mutations both reduce the P_O and diminish CDI (Adams et al., 2014; Bazzazi et al., 2013; Ben Johny et al., 2013). Relevant to this, recent work shows that α-Actinin also binds to the IQ domain (Turner et al., 2020). This interaction also upregulates basal P_O, suggesting that both CaM and α-Actinin may play a similar role (Turner et al., 2020). Third, following Ca^{2+}-binding, CaM switches its binding conformation triggering a rearrangement of the channel cytosolic domain (Benmocha Guggenheimer et al., 2016), resulting in the C-lobe binding to the IQ domain (Van Petegem et al., 2005) and potentially forming a tripartite complex with the upstream EF-hand segments (Banerjee et al., 2018; Ben Johny et al., 2013); the Ca^{2+}-bound N-lobe, on the other hand, associates with the NSCaTE binding site (Benmocha et al., 2009; Dick et al., 2008; Ivanina et al., 2000) (Fig. 3f). Structurally, these conformational changes switch channels to mode Ca where channel openings are sparse (Fig. 3c), yielding a sharply reduced P_O (Imredy & Yue, 1994). Of note, unlike the antiparallel arrangement of the apo-C-lobe, Ca^{2+}/CaM interacts with a parallel arrangement for Ca_V1 channels (Van Petegem et al., 2005). Ultimately, the Ca^{2+}/CaM induced conformational change is thought to be coupled to the selectivity filter, resulting in a collapse of the selectivity filter akin to C-type inactivation in K^+ channels (Abderemane-Ali et al., 2019). Importantly, how changes in the conformation of the cytosolic domain are coupled to the distant selectivity filter is largely unknown and may involve other channel domains such as the S6 gate (Tadross et al., 2010). For $Ca_V2.1$ that undergoes CDF, this overall scheme can be modified to include a 'facilitated' gating mode with an increased propensity for channel openings (Chaudhuri et al., 2007; Lee et al., 2015). Structurally, CaM can adopt distinct binding orientations to engage the IQ domain of Ca_V2 channels, and these differences may be important to understand structural mechanisms of CDF (Kim et al., 2008; Mori et al., 2008).

Tuning Feedback. Consistent with its importance in physiology, CaM regulation of Ca^{2+} channels is tuned by several cellular processes including post-transcriptional and post-translational modification, as well as protein-protein interactions. We here discuss key mechanisms that alter the magnitude of Ca^{2+}-feedback.

1. Post-transcriptional and post-translational regulation—Ca_V channels are subject to extensive alternative-splicing that alters key structural domains relevant for calmodulation. In particular, alternative splicing of the carboxy-termini of both $Ca_V1.3$ (Bock et al., 2011; Tan et al., 2011) and $Ca_V1.4$ (Singh et al., 2006; Tan et al., 2012; Wahl-Schott et al., 2006) channels results in the inclusion of an autoinhibitory domain (encoded by exon 47) within the distal carboxy-terminus, known as the inhibitor of CDI (ICDI), and also referred to as the C-terminal modulatory domain (CTM) (Liu et al., 2010; Singh et al., 2006; Wahl-Schott et al., 2006). Splicing inclusion of the ICDI domain is tightly regulated in a cell-type-specific manner (Scharinger et al., 2015). Biochemically, the ICDI domain has been shown to bind the IQ domain and the closely juxtaposed A-region (Liu et al., 2010; Singh et al., 2006). Importantly, this binding interface overlaps with that of the C-lobe of apoCaM, thus allowing the ICDI domain to competitively dislodge apoCaM from the channel carboxy-tail (Liu et al., 2010; Sang et al., 2021) (Fig. 3g, i). The intramolecular nature of this interaction along with the high local concentration of the ICDI domain results in a potent reduction in the apoCaM binding affinity (Liu et al., 2010; Sang et al., 2021). Functionally, the reduction in apoCaM binding affinity stabilizes channels in mode E, thus reducing the basal P_O and preventing CDI (Adams et al., 2014). A similar autoinhibitory interaction has been reported for $Ca_V1.2$ between the DCRD region in the distal carboxy-terminus and PCRD domain, albeit with a weaker affinity (Hulme et al., 2006; Sang et al., 2016). Of further relevance, the ICDI domain of $Ca_V1.4$ includes a consensus Protein Kinase A (PKA) phosphorylation site at residue Ser[1883] (Sang et al., 2016) (Fig. 3g). Phosphorylation of this residue weakens the association of the $Ca_V1.4$ ICDI with the IQ domain and restores apoCaM

preassociation (Sang et al., 2016). As a consequence, $Ca_V1.4$ channels switch to mode A resulting in an increase in P_O and enhanced CDI. Curiously, CDI of $Ca_V1.2$ in hippocampal neurons has also been shown to be tuned by PKA phosphorylation and calcineurin-dependent dephosphorylation anchored to the channel complex by AKAP79 (Dittmer et al., 2014; Oliveria et al., 2012). In line with this, phosphorylation of $Ca_V1.2$ carboxy-terminus has been suggested to upregulate CaM association (Lei et al., 2018) and helps slow the rundown of channel activity in excised patches (Xu et al., 2016). Importantly, it is worth noting that adrenergic upregulation of $Ca_V1.2$ that supports cardiac inotropy during fight-or-flight response, utilizes an entirely orthogonal mechanism involving PKA-phosphorylation of the G-protein Rad (Liu et al., 2020).

Alternative splicing of the Ca_V2 carboxy-terminus also tunes Ca^{2+}- regulation. In particular, for $Ca_V2.1$ channels, developmentally-regulated alternative splicing of exon 37 generates two distinct EF-hand domains (EF_a, EF_b) (Chaudhuri et al., 2005). The inclusion of EF_a supports strong CDF, while EF_b prevents CDF (Chaudhuri et al., 2004). Although the EF-hand domain of $Ca_V2.2$ is alternatively-spliced, this molecular change does not impact CDF (Thomas et al., 2018). Additionally, $Ca_V2.1$ channels also contain an alternatively-spliced distal carboxy terminus that alters CDF (Chaudhuri et al., 2004). However, this domain is structurally unrelated to Ca_V1 channels and is not known to compete with CaM. In all, post-transcriptional and phospho-regulation of calmodulatory domains in Ca_V channels can yield complex orchestration of Ca^{2+}-dependent feedback modulation.

Beyond alternative splicing, transcripts of $Ca_V1.3$ channels are also regulated by brain-selective adenosine to inosine RNA editing by ADAR2, which modifies key amino acids within the carboxy-terminal IQ domain (Huang et al., 2012). This precise editing of $Ca_V1.3$ mRNA requires the formation of an RNA duplex between exon 41 that encodes the IQ domain and a conserved intronic editing site complementary

sequence and is inhibited by serine/arginine-rich splicing factor 9, which confers tissue specificity (Huang et al., 2018). At the protein level, this change yields sequence variation in the IQ domain, including the substitution of the central isoleucine residue with methionine (Huang et al., 2012). Live cell FRET analysis suggests that RNA-edited $Ca_V1.3$ variants exhibit diminished apoCaM binding affinity, although Ca^{2+}/CaM binding affinity is largely unaffected (Bazzazi et al., 2013). This suggests that, in the overall calmodulatory scheme, RNA editing of $Ca_V1.3$ only alters the transition from configuration E to configuration A (Fig. 3g). Consistent with this, electrophysiological analysis demonstrates that RNA-edited $Ca_V1.3$ variants have reduced basal P_O and diminished CDI (Adams et al., 2014) (Fig. 3i). Moreover, CaM overexpression reverses both deficits in channel function in heterologous cells, further corroborating the notion that RNA-edited channels bear weakened CaM preassociation. Physiologically, RNA editing of $Ca_V1.3$ tunes action potential firing rates in the suprachiasmatic nucleus and may impact circadian rhythm (Huang et al., 2012).

2. Protein-protein interactions. The strength of Ca_V1 Ca^{2+}-feedback is also regulated by the interaction of other regulatory proteins including the family of CaM-like Ca^{2+}-binding proteins (CaBPs), and SH3 and cysteine-rich domain protein (Stac1-3).

CaBPs (1, 2, 4, and 5) have ~50% homology to CaM, and share a bilobal architecture with two EF-hand like domains constituting each lobe (Haeseleer et al., 2002). Unlike CaM, only certain EF-hands of CaBPs bind Ca^{2+} (Haeseleer et al., 2002). Furthermore, CaBPs show restricted and isoform-specific expression in the brain, retina, and inner ear (Haeseleer et al., 2002; Hardie & Lee, 2016). The physiological importance of CaBPs has been reviewed in the past (see review, (Hardie & Lee, 2016)). In brief, coexpression of CaBP1 and CaBP4 strongly inhibits CDI of Ca_V1 channels (Shaltiel et al., 2012; Yang et al., 2006; Zhou et al., 2005) (Fig. 3h, j). In terms of its molecular function, disabling Ca^{2+} binding to

CaBPs does not prevent CDI inhibition, suggesting Ca^{2+}-independent regulation of Ca_V1 channels (Findeisen & Minor, 2010). In this regard, multiple CaBP binding sites have been identified including an amino-terminal segment in close proximity to the NSCaTE segment, distal portions of the III-IV linker, and multiple segments in the carboxy-terminus (Yang et al., 2014; Zhou et al., 2005). Both the CaBP N-lobe and the inter-lobe linker are important for CDI inhibition. In comparison, the CaBP C-lobe is suggested to be a high affinity anchor that binds to the IQ domain to potentially compete with apoCaM (Findeisen et al., 2013; Oz et al., 2013). In-depth mechanistic analysis, however, points to a mixed allosteric scheme of CDI inhibition, where at low cytosolic concentrations, CaBPs can interact with binding sites outside of the apoCaM binding interface and potentially occlude accessibility of Ca^{2+}/CaM effector interfaces (Oz et al., 2013; Yang et al., 2014). At higher concentrations, it is possible that CaBPs may completely dislodge apoCaM (Oz et al., 2013; Yang et al., 2014). It is currently unknown whether such competitive inhibition alters baseline P_O.

Stac proteins are skeletal muscle and neuron-specific adaptor proteins that were recently identified as a Ca_V1 modulators (Flucher & Campiglio, 2019; Polster et al., 2015; Rufenach & Van Petegem, 2021). In skeletal muscle, stac3 is an obligatory component of the excitation-contraction coupling machinery that links the surface-membrane $Ca_V1.1$ with the ryanodine receptor 1 (RyR1) (Horstick et al., 2013; Linsley et al., 2017; Polster et al., 2016). In neurons, however, stac1-2 proteins associate with Ca_V1 channels to prevent CDI (Campiglio et al., 2018; Niu et al., 2018a; Polster et al., 2018a) (Fig. 3h, j). At the molecular level, stac proteins bind to both the Ca_V1 II-III linker (Polster et al., 2018b) and the proximal segments of carboxy-terminus (Campiglio et al., 2018; Niu et al., 2018a, b). The later interaction is thought to be critical for mediating CDI inhibition (Campiglio et al., 2018; Niu et al., 2018a, b), while the former is essential for EC coupling in the skeletal muscle (Polster et al., 2018b). Furthermore, in-depth mechanistic analysis suggests that stac proteins utilize an allosteric mechanism to decouple the pore domain from CaM-dependent conformational changes. This effect has been localized to an ~22 amino acid segment known as the U-domain (Niu et al., 2018a). The physiological importance of stac inhibition of CDI in neurons is yet to be fully determined (Polster et al., 2018a); however, studies in invertebrates suggest that stac may be essential for normal circadian rhythm (Hsu et al., 2018).

Ca^{2+} Regulation of VGCCs in Disease

CDI and CDF play critical roles in the normal function of excitable cells, including cardiac myocytes and neurons. In the heart, the inactivation of the $Ca_V1.2$ channel is critical to shaping the plateau phase of the cardiac action potential (AP) (Faber et al., 2007; Livshitz & Rudy, 2009; Morotti et al., 2012). When $Ca_V1.2$ CDI is disrupted, the enhanced opening of the channels during an AP drives the membrane potential in a depolarized direction, resulting in a significant prolongation of the AP. This has been clearly demonstrated in guinea pig ventricular myocytes, where overexpression of CaM_{1234} led to the generation of prolonged APs in the range of 2 s (Alseikhan et al., 2002). Thus, complete ablation of CDI in a cardiac AP would be incompatible with life, and it appeared that mutations that disrupted $Ca_V1.2$ CDI would likely be embryonic lethal. Nonetheless, in 2004, the first mutation in $Ca_V1.2$ was reported to underlie a severe multi-system disorder known as Timothy syndrome (TS). In addition to variable effects on channel activation and VDI (Splawski et al., 2004, 2005 ; Wemhoner et al., 2015; Yarotskyy et al., 2009), TS mutations often cause a significant deficit in CDI (Dick et al., 2008). As a result, TS mutations cause a profound prolongation of the cardiac AP, resulting in one of the most severe forms of a long-QT syndrome (LQTS type 8) described to-date. Interestingly, the first TS mutation described (G406R) occurred within the mutually exclusive exon 8a, which is contained within only 20% of $Ca_V1.2$ splice variants in the heart, resulting in only about 10% of channels harboring the deleterious mutation in heterozygous patients (Splawski et al., 2004). This low expression may help explain why the mutation is not embryonic lethal; however, since this first description numerous additional mutations have been identified in $Ca_V1.2$, often occurring within constitutive exons and resulting in a wide spectrum of effects on the channel (Barrett & Tsien, 2008; Boczek et al., 2015; Fukuyama et al., 2014; Landstrom et al., 2016; Splawski et al., 2004, 2005; Wemhoner et al., 2015; Yarotskyy et al., 2009). Beyond the heart, $Ca_V1.2$ is widely expressed in multiple tissues including the brain, smooth muscle, and the immune system. As a result, mutations in this channel produce a myriad of symptoms including autism spectrum disorder (ASD), developmental delay, immune deficits, syndactyly and facial dysmorphism (Landstrom et al., 2016; Ozawa et al., 2018; Splawski et al., 2004, 2005; Wemhoner et al., 2015). Given the overlapping effects of these mutations on multiple channel properties, it is difficult to ascribe CDI effects as directly causative in each case, nonetheless, it is clear that disruption of CDI results in a significant gain-of-function effect with profound implications in the heart, brain, and other tissues.

Beyond $Ca_V1.2$, disruption of Ca^{2+} dependent regulation of multiple VGCC subtypes has been implicated in disease. Mutations similar to the TS mutations have been identified in $Ca_V1.3$ channels, resulting in altered channel activation and inactivation, including a deficit in CDI (Limpitikul et al., 2016; Scholl et al., 2013; Wemhoner et al., 2015; Yarotskyy et al., 2009). Such mutations have been identified in aldosterone-producing adenomas, resulting in primary aldosteronism (Scholl et al., 2013; Tan et al., 2017). Moreover, when these types of mutations occur within the germline, they produce an array of phenotypes including ASD, intellectual disability, and hypotonia (Pinggera et al., 2015, 2017; Pinggera & Striessnig, 2016). In $Ca_V1.4$ channels, multiple truncation mutations result in a loss of the ICDI/CTM module, resulting in multiple effects on channel gating, including a dramatic increase in channel CDI. Patients harboring these mutations suffer from incomplete congenital stationary night blindness type 2 (CSNB2) (Burtscher et al.,

2014; Singh et al., 2006), implicating CDI in the pathogenesis of this disease. Beyond L-type channels, mutations that disrupt the CDF of $Ca_V2.1$ channels have been found to underly episodic ataxia type 2 (EA2) (Graves et al., 2008), or familial hemiplegic migraine type 1 (FHM-1) (Di Guilmi et al., 2014; Uchitel et al., 2014).

In addition to mutations occurring directly on the channel alpha subunit, mutations within CaM have the potential to significantly impact normal physiological function through a loss of Ca^{2+} regulation of VGCCs. Calmodulinopathies comprise a growing class of cardiac disorders in which patients exhibit a variety of symptoms including LQTS, catecholaminergic polymorphic ventricular tachycardia (CPVT), idiopathic ventricular fibrillation (IVF), sudden cardiac death, and neurological disruption (Crotti et al., 2013; Jensen et al., 2018; Nyegaard & Overgaard, 2019; Nyegaard et al., 2012; Crotti, 2019; Pipilas et al., 2016), with disrupted CDI of LTCCs primarily implicated in the LQTS phenotype (Limpitikul et al., 2014). Interestingly, these effects occur despite a small fraction of mutant CaM. Three separate genes (CALM 1-3) each encode for identical CaM proteins, and calmodulinopathy mutations can occur within any of these genes (Boczek et al., 2016). As a result, these heterozygous mutations are expressed in only one of 6 alleles. Nonetheless, calmodulinopathy mutations have been shown to exert a significant impact on LTCCs (Limpitikul et al., 2014). This may be explained by the pre-association of VGCCs with Ca^{2+} free CaM, which is absent in many CaM binding partners (Fig. 4a). To understand this, we first consider that calmodulinopathy mutations occur within the EF hands of CaM, resulting in a significant decrease in the affinity for Ca^{2+} binding to CaM (Crotti et al., 2013; Nyegaard et al., 2012). With both mutant and WT CaM present, this loss of Ca^{2+} binding would be unlikely to alter the function of effector proteins in which only the calcified form of CaM binds (Fig. 4a, cytosolic CaM). However, Ca_V1-2 VGCCs are known to pre-associate with apoCaM, and modulation of the channel occurs following Ca^{2+} binding to this resident CaM (Fig. 4a, pre-associated CaM) (Ben-Johny et al., 2015; Christel

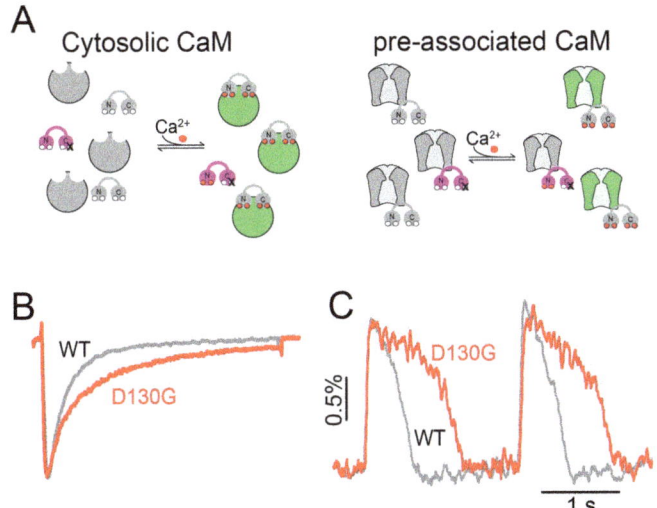

Fig. 4 Disruption of CDI in long-QT syndrome. (**a**), **Left:** Typical CaM binding scheme where cytosolic Ca^{2+}/CaM binds its effector site. Mutant CaM (pink) does not interfere with protein modulation (green) by WT Ca^{2+} CaM (gray). Red circles represent Ca^{2+}. **Right:** Pre-association of CaM to VGCC selectively allows modulation (green) in channels harboring CaM WT (gray) and not mutant CaM (pink), resulting in a portion of channels unable to undergo CDI. (**b**), Calcium currents recorded from iPSC-CMs harboring a CaM_{D130G} mutation display significantly less CDI (red) as compared with WT cardiomyocytes (gray). (**c**), The CaM_{D130G} iPSC-CMs exhibit significant prolongation of their action potentials (red) as compared to WT iPSC-CMs (gray), consistent with the long-QT syndrome.

& Lee, 2012; DeMaria et al., 2001b; Erickson et al., 2001; Peterson et al., 1999a; Zuhlke & Reuter, 1998). As a result, channels pre-associated with a mutant CaM that is unable to bind Ca^{2+} will fail to undergo Ca^{2+}/CaM dependent regulation (Erickson et al., 2001; Limpitikul et al., 2014, 2017; Mori et al., 2004). Thus, mutations that eliminate Ca^{2+} binding to CaM, but do not alter the binding between apoCaM and the channel, will exhibit a dominant-negative effect on the channel to which they are pre-associated (Fig. 4a) (Dick et al., 2016; Mori et al., 2004). In this way, VGCCs are particularly vulnerable to calmodulinopathy mutations even when expression levels of the mutant CaM are significantly less than that of WT (Dick et al., 2016; Limpitikul et al., 2014). This is demonstrated by the loss of CDI seen in LTCC currents recorded in iPSC-derived cardiomyocytes (iPSC-CMs) derived from a patient harboring the D130G calmodulinopathy mutation (Fig. 4b) (Limpitikul et al., 2017). This loss of CDI translates into a significant prolongation of the AP of the iPSC-derived cardiomyocytes

(Fig. 4c), consistent with the LQTS phenotype seen in these patients. Overall, deficits in $Ca_V1.2$ CDI are sufficient to explain the LQTS phenotype seen in many calmodulinopathy patients.

Overall, calmodulation of VGCCs is a critical process required for the proper function of numerous excitable tissues. As the number of identified mutations that alter VGCC calmodulation continues to grow, so too will the known physiological impacts. As such, CDI and CDF stand as critical elements of pathophysiology and may represent important therapeutic targets for a wide array of diseases.

References

Abderemane-Ali, F., Findeisen, F., Rossen, N. D., & Minor, D. L., Jr. (2019). A selectivity filter gate controls voltage-gated calcium channel calcium-dependent inactivation. *Neuron, 101*(1134-1149), e1133.

Adams, P. J., Ben-Johny, M., Dick, I. E., Inoue, T., & Yue, D. T. (2014). Apocalmodulin itself promotes ion channel opening and Ca(2+) regulation. *Cell, 159,* 608–622.

Alseikhan, B. A., DeMaria, C. D., Colecraft, H. M., & Yue, D. T. (2002). Engineered calmodulins reveal the unexpected eminence of Ca2+ channel inactivation in controlling heart excitation. *Proceedings of the National Academy of Sciences of the United States of America, 99,* 17185–17190.

Asmara, H., Minobe, E., Saud, Z. A., & Kameyama, M. (2010). Interactions of calmodulin with the multiple binding sites of Cav1.2 Ca2+ channels. *Journal of Pharmacological Sciences, 112,* 397–404.

Asmara, H., Micu, I., Rizwan, A. P., Sahu, G., Simms, B. A., Zhang, F. X., Engbers, J. D. T., Stys, P. K., Zamponi, G. W., & Turner, R. W. (2017). A T-type channel-calmodulin complex triggers alphaCaMKII activation. *Molecular Brain, 10,* 37.

Babitch, J. (1990). Channel hands. *Nature, 346,* 321–322.

Babu, Y. S., Sack, J. S., Greenhough, T. J., Bugg, C. E., Means, A. R., & Cook, W. J. (1985). Three-dimensional structure of calmodulin. *Nature, 315,* 37–40.

Banerjee, R., Yoder, J. B., Yue, D. T., Amzel, L. M., Tomaselli, G. F., Gabelli, S. B., & Ben-Johny, M. (2018). Bilobal architecture is a requirement for calmodulin signaling to CaV1.3 channels. *Proceedings of the National Academy of Sciences of the United States of America, 115,* E3026–E3035.

Barrett, C. F., & Tsien, R. W. (2008). The Timothy syndrome mutation differentially affects voltage- and calcium-dependent inactivation of CaV1.2 L-type calcium channels. *Proceedings of the National Academy of Sciences of the United States of America, 105,* 2157–2162.

Bazzazi, H., Ben Johny, M., Adams, P. J., Soong, T. W., & Yue, D. T. (2013). Continuously tunable Ca(2+) regulation of RNA-edited CaV1.3 channels. *Cell Reports, 5,* 367–377.

Ben Johny, M., Yang, P. S., Bazzazi, H., & Yue, D. T. (2013). Dynamic switching of calmodulin interactions underlies Ca2+ regulation of CaV1.3 channels. *Nature Communications, 4,* 1717.

Ben-Johny, M., & Yue, D. T. (2014). Calmodulin regulation (calmodulation) of voltage-gated calcium channels. *The Journal of General Physiology, 143,* 679–692.

Ben-Johny, M., Yang, P. S., Niu, J., Yang, W., Joshi-Mukherjee, R., & Yue, D. T. (2014). Conservation of Ca2+/calmodulin regulation across Na and Ca2+ channels. *Cell, 157,* 1657–1670.

Ben-Johny, M., Dick, I. E., Sang, L., Limpitikul, W. B., Kang, P. W., Niu, J., Banerjee, R., Yang, W., Babich, J. S., Issa, J. B., et al. (2015). Towards a unified theory of calmodulin regulation (calmodulation) of voltage-gated calcium and sodium channels. *Current Molecular Pharmacology, 8,* 188–205.

Ben-Johny, M., Yue, D. N., & Yue, D. T. (2016). Detecting stoichiometry of macromolecular complexes in live cells using FRET. *Nature Communications, 7,* 13709.

Benmocha Guggenheimer, A., Almagor, L., Tsemakhovich, V., Tripathy, D. R., Hirsch, J. A.,

& Dascal, N. (2016). Interactions between N and C termini of alpha1C subunit regulate inactivation of CaV1.2 L-type Ca(2+) channel. *Channels (Austin, Tex.), 10,* 55–68.

Benmocha, A., Almagor, L., Oz, S., Hirsch, J. A., & Dascal, N. (2009). Characterization of the calmodulin-binding site in the N terminus of CaV1.2. *Channels (Austin, Tex.), 3,* 337–342.

Bock, G., Gebhart, M., Scharinger, A., Jangsangthong, W., Busquet, P., Poggiani, C., Sartori, S., Mangoni, M. E., Sinnegger-Brauns, M. J., Herzig, S., et al. (2011). Functional properties of a newly identified C-terminal splice variant of Cav1.3 L-type Ca2+ channels. *The Journal of Biological Chemistry, 286,* 42736–42748.

Boczek, N. J., Miller, E. M., Ye, D., Nesterenko, V. V., Tester, D. J., Antzelevitch, C., Czosek, R. J., Ackerman, M. J., & Ware, S. M. (2015). Novel Timothy syndrome mutation leading to increase in CACNA1C window current. *Heart Rhythm: The Official Journal of the Heart Rhythm Society, 12,* 211–219.

Boczek, N. J., Gomez-Hurtado, N., Ye, D., Calvert, M. L., Tester, D. J., Kryshtal, D., Hwang, H. S., Johnson, C. N., Chazin, W. J., Loporcaro, C. G., et al. (2016). Spectrum and prevalence of CALM1-, CALM2-, and CALM3-encoded calmodulin variants in long QT syndrome and functional characterization of a novel long QT syndrome-associated calmodulin missense variant, E141G. *Circulation. Cardiovascular Genetics, 9,* 136–146.

Brehm, P., & Eckert, R. (1978). Calcium entry leads to inactivation of calcium channel in Paramecium. *Science, 202,* 1203–1206.

Burtscher, V., Schicker, K., Novikova, E., Pohn, B., Stockner, T., Kugler, C., Singh, A., Zeitz, C., Lancelot, M. E., Audo, I., et al. (2014). Spectrum of Cav1.4 dysfunction in congenital stationary night blindness type 2. *Biochimica et Biophysica Acta, 1838,* 2053–2065.

Campiglio, M., Coste de Bagneaux, P., Ortner, N. J., Tuluc, P., Van Petegem, F., & Flucher, B. E. (2018). STAC proteins associate to the IQ domain of CaV1.2 and inhibit calcium-dependent inactivation. *Proceedings of the National Academy of Sciences of the United States of America, 115,* 1376–1381.

Capes, D. L., Goldschen-Ohm, M. P., Arcisio-Miranda, M., Bezanilla, F., & Chanda, B. (2013). Domain IV voltage-sensor movement is both sufficient and rate limiting for fast inactivation in sodium channels. *The Journal of General Physiology, 142,* 101–112.

Chakouri, N., Diaz, J., Yang, P. S., & Ben-Johny, M. (2020). CaV channels reject signaling from a second CaM in eliciting Ca(2+)-dependent feedback regulation. *The Journal of Biological Chemistry, 295,* 14948–14962.

Chaudhuri, D., Chang, S. Y., DeMaria, C. D., Alvania, R. S., Soong, T. W., & Yue, D. T. (2004). Alternative splicing as a molecular switch for Ca2+/calmodulin-dependent facilitation of P/Q-type Ca2+ channels. *The Journal of Neuroscience, 24,* 6334–6342.

Chaudhuri, D., Alseikhan, B. A., Chang, S. Y., Soong, T. W., & Yue, D. T. (2005). Developmental activation of calmodulin-dependent facilitation of cerebellar

P-type Ca2+ current. *The Journal of Neuroscience, 25,* 8282–8294.

Chaudhuri, D., Issa, J. B., & Yue, D. T. (2007). Elementary mechanisms producing facilitation of Cav2.1 (P/Q-type) channels. *The Journal of General Physiology, 129,* 385–401.

Chemin, J., Taiakina, V., Monteil, A., Piazza, M., Guan, W., Stephens, R. F., Kitmitto, A., Pang, Z. P., Dolphin, A. C., Perez-Reyes, E., et al. (2017). Calmodulin regulates Cav3 T-type channels at their gating brake. *The Journal of Biological Chemistry, 292,* 20010–20031.

Christel, C., & Lee, A. (2012). Ca2+-dependent modulation of voltage-gated Ca2+ channels. *Biochimica et Biophysica Acta, 1820,* 1243–1252.

Crotti, L., Johnson, C. N., Graf, E., De Ferrari, G. M., Cuneo, B. F., Ovadia, M., Papagiannis, J., Feldkamp, M. D., Rathi, S. G., Kunic, J. D., et al. (2013). Calmodulin mutations associated with recurrent cardiac arrest in infants. *Circulation, 127,* 1009–1017.

Crotti, L., Spazzolini, C., Tester, D. J., Ghidoni, A, Baruteau, A. E,, Beckmann, B. M., Behr, E. R., Bennett, J. S., Bezzina, C. R., Bhuiyan, Z. A., Celiker, A., Cerrone, M., Dagradi, F., De Ferrari, G. M., Etheridge, S. P., Fatah, M., Garcia-Pavia, P., Al-Ghamdi, S., Hamilton, R. M., Al-Hassnan, Z. N., Horie, M., Jimenez-Jaimez, J., Kanter, R. J., Kaski, J. P., Kotta, M. C., Lahrouchi, N., Makita, N., Norrish, G., Odland, H. H., Ohno, S., Papagiannis, J., Parati, G., Sekarski, N., Tveten, K., Vatta, M., Webster, G., Wilde, A. A. M., Wojciak, J., George, A. L., Ackerman, M. J., Schwartz, P. J. (2019). Calmodulin mutations and life-threatening cardiac arrhythmias: insights from the International Calmodulinopathy Registry. *Eur Heart J, 40,* 2964–2975.

DeMaria, C. D., Soong, T. W., Alseikhan, B. A., Alvania, R. S., & Yue, D. T. (2001a). Calmodulin bifurcates the local Ca2+ signal that modulates P/Q-type Ca2+ channels. *Nature, 411,* 484–489.

DeMaria, C. D., Soong, T. W., Alseikhan, B. A., Alvania, R. S., & Yue, D. T. (2001b). Calmodulin bifurcates the local Ca2+ signal that modulates P/Q-type Ca2+ channels. *Nature, 411,* 484–489.

Di Guilmi, M. N., Wang, T., Inchauspe, C. G., Forsythe, I. D., Ferrari, M. D., van den Maagdenberg, A. M., Borst, J. G., & Uchitel, O. D. (2014). Synaptic gain-of-function effects of mutant Cav2.1 channels in a mouse model of familial hemiplegic migraine are due to increased basal [Ca2+]i. *The Journal of Neuroscience, 34,* 7047–7058.

Dick, I. E., Tadross, M. R., Liang, H., Tay, L. H., Yang, W., & Yue, D. T. (2008). A modular switch for spatial Ca2+ selectivity in the calmodulin regulation of CaV channels. *Nature, 451,* 830–834.

Dick, I. E., Joshi-Mukherjee, R., Yang, W., & Yue, D. T. (2016). Arrhythmogenesis in Timothy Syndrome is associated with defects in Ca(2+)-dependent inactivation. *Nature Communications, 7,* 10370.

Dittmer, P. J., Dell'Acqua, M. L., & Sather, W. A. (2014). Ca2+/calcineurin-dependent inactivation of neuronal L-type Ca2+ channels requires priming by

AKAP-anchored protein kinase A. *Cell Reports, 7,* 1410–1416.

Dong, Y., Gao, Y., Xu, S., Wang, Y., Yu, Z., Li, Y., Li, B., Yuan, T., Yang, B., Zhang, X. C., et al. (2021). Closed-state inactivation and pore-blocker modulation mechanisms of human CaV2.2. *Cell Reports, 37,* 109931.

Eaholtz, G., Scheuer, T., & Catterall, W. A. (1994). Restoration of inactivation and block of open sodium channels by an inactivation gate peptide. *Neuron, 12,* 1041–1048.

Eckert, R., & Chad, J. (1984). Inactivation of Ca channels. *Progress in Biophysics and Molecular Biology (London), 44,* 215–267.

Erickson, M. G., Alseikhan, B. A., Peterson, B. Z., & Yue, D. T. (2001). Preassociation of calmodulin with voltage-gated Ca(2+) channels revealed by FRET in single living cells. *Neuron, 31,* 973–985.

Erickson, M. G., Liang, H., Mori, M. X., & Yue, D. T. (2003). FRET two-hybrid mapping reveals function and location of L-type Ca2+ channel CaM preassociation. *Neuron, 39,* 97–107.

Evans, T. I., Hell, J. W., & Shea, M. A. (2011). Thermodynamic linkage between calmodulin domains binding calcium and contiguous sites in the C-terminal tail of Ca(V)1.2. *Biophysical Chemistry, 159,* 172–187.

Faas, G. C., Raghavachari, S., Lisman, J. E., & Mody, I. (2011). Calmodulin as a direct detector of Ca2+ signals. *Nature Neuroscience, 14,* 301–304.

Faber, G. M., Silva, J., Livshitz, L., & Rudy, Y. (2007). Kinetic properties of the cardiac L-type Ca2+ channel and its role in myocyte electrophysiology: A theoretical investigation. *Biophysical Journal, 92,* 1522–1543.

Fallon, J. L., Baker, M. R., Xiong, L., Loy, R. E., Yang, G., Dirksen, R. T., Hamilton, S. L., & Quiocho, F. A. (2009). Crystal structure of dimeric cardiac L-type calcium channel regulatory domains bridged by Ca2+* calmodulins. *Proceedings of the National Academy of Sciences of the United States of America, 106,* 5135–5140.

Findeisen, F., & Minor, D. L., Jr. (2010). Structural basis for the differential effects of CaBP1 and calmodulin on Ca(V)1.2 calcium-dependent inactivation. *Structure, 18,* 1617–1631.

Findeisen, F., Rumpf, C. H., & Minor, D. L., Jr. (2013). Apo states of calmodulin and CaBP1 control CaV1 voltage-gated calcium channel function through direct competition for the IQ domain. *Journal of Molecular Biology, 425,* 3217–3234.

Flucher, B. E., & Campiglio, M. (2019). STAC proteins: The missing link in skeletal muscle EC coupling and new regulators of calcium channel function. *Biochimica et Biophysica Acta-Molecular Cell Research, 1866,* 1101–1110.

Fukuyama, M., Wang, Q., Kato, K., Ohno, S., Ding, W. G., Toyoda, F., Itoh, H., Kimura, H., Makiyama, T., Ito, M., et al. (2014). Long QT syndrome type 8: Novel CACNA1C mutations causing QT prolongation and variant phenotypes. *Europace, 16,* 1828–1837.

Gao, S., Yao, X., & Yan, N. (2021). Structure of human Cav2.2 channel blocked by the painkiller ziconotide. *Nature, 596*, 143–147.

Graves, T. D., Imbrici, P., Kors, E. E., Terwindt, G. M., Eunson, L. H., Frants, R. R., Haan, J., Ferrari, M. D., Goadsby, P. J., Hanna, M. G., et al. (2008). Premature stop codons in a facilitating EF-hand splice variant of CaV2.1 cause episodic ataxia type 2. *Neurobiology of Disease, 32*, 10–15.

Haeseleer, F., Imanishi, Y., Sokal, I., Filipek, S., & Palczewski, K. (2002). Calcium-binding proteins: Intracellular sensors from the calmodulin superfamily. *Biochemical and Biophysical Research Communications, 290*, 615–623.

Halling, D. B., Georgiou, D. K., Black, D. J., Yang, G., Fallon, J. L., Quiocho, F. A., Pedersen, S. E., & Hamilton, S. L. (2009). Determinants in CaV1 channels that regulate the Ca2+ sensitivity of bound calmodulin. *The Journal of Biological Chemistry, 284*, 20041–20051.

Hardie, J., & Lee, A. (2016). Decalmodulation of Cav1 channels by CaBPs. *Channels (Austin, Tex.), 10*, 33–37.

Hess, P., Lansman, J. B., & Tsien, R. W. (1984). Different modes of Ca channel gating behaviour favoured by dihydropyridine Ca agonists and antagonists. *Nature, 311*, 538–544.

Hodgkin, A. L., & Huxley, A. F. (1952). The dual effect of membrane potential on sodium conductance in the giant axon of Loligo. *The Journal of Physiology, 116*, 497–506.

Horstick, E. J., Linsley, J. W., Dowling, J. J., Hauser, M. A., McDonald, K. K., Ashley-Koch, A., Saint-Amant, L., Satish, A., Cui, W. W., Zhou, W., et al. (2013). Stac3 is a component of the excitation-contraction coupling machinery and mutated in Native American myopathy. *Nature Communications, 4*, 1952.

Hsu, I. U., Linsley, J. W., Varineau, J. E., Shafer, O. T., & Kuwada, J. Y. (2018). Dstac is required for normal circadian activity rhythms in Drosophila. *Chronobiology International, 35*, 1016–1026.

Huang, H., Tan, B. Z., Shen, Y., Tao, J., Jiang, F., Sung, Y. Y., Ng, C. K., Raida, M., Kohr, G., Higuchi, M., et al. (2012). RNA editing of the IQ domain in Ca(v)1.3 channels modulates their Ca(2)(+)-dependent inactivation. *Neuron, 73*, 304–316.

Huang, H., Kapeli, K., Jin, W., Wong, Y. P., Arumugam, T. V., Koh, J. H., Srimasorn, S., Mallilankaraman, K., Chua, J. J. E., Yeo, G. W., et al. (2018). Tissue-selective restriction of RNA editing of CaV1.3 by splicing factor SRSF9. *Nucleic Acids Research, 46*, 7323–7338.

Hulme, J. T., Yarov-Yarovoy, V., Lin, T. W., Scheuer, T., & Catterall, W. A. (2006). Autoinhibitory control of the CaV1.2 channel by its proteolytically processed distal C-terminal domain. *The Journal of Physiology, 576*, 87–102.

Imredy, J. P., & Yue, D. T. (1994). Mechanism of Ca(2+)-sensitive inactivation of L-type Ca2+ channels. *Neuron, 12*, 1301–1318.

Ivanina, T., Blumenstein, Y., Shistik, E., Barzilai, R., & Dascal, N. (2000). Modulation of L-type Ca2+ channels by gbeta gamma and calmodulin via interactions with N and C termini of alpha 1C. *The Journal of Biological Chemistry, 275*, 39846–39854.

Jensen, H. H., Brohus, M., Nyegaard, M., & Overgaard, M. T. (2018). Human calmodulin mutations. *Frontiers in Molecular Neuroscience, 11*, 396.

Jurado, L. A., Chockalingam, P. S., & Jarrett, H. W. (1999). Apocalmodulin. *Physiological Reviews, 79*, 661–682.

Kass, R. S., & Sanguinetti, M. C. (1984). Inactivation of calcium channel current in the calf cardiac Purkinje fiber. Evidence for voltage- and calcium-mediated mechanisms. *The Journal of General Physiology, 84*, 705–726.

Kim, J., Ghosh, S., Nunziato, D. A., & Pitt, G. S. (2004). Identification of the components controlling inactivation of voltage-gated Ca2+ channels. *Neuron, 41*, 745–754.

Kim, E. Y., Rumpf, C. H., Fujiwara, Y., Cooley, E. S., Van Petegem, F., & Minor, D. L., Jr. (2008). Structures of CaV2 Ca2+/CaM-IQ domain complexes reveal binding modes that underlie calcium-dependent inactivation and facilitation. *Structure, 16*, 1455–1467.

Kim, E. Y., Rumpf, C. H., Van Petegem, F., Arant, R. J., Findeisen, F., Cooley, E. S., Isacoff, E. Y., & Minor, D. L., Jr. (2010). Multiple C-terminal tail Ca(2+)/CaMs regulate Ca(V)1.2 function but do not mediate channel dimerization. *The EMBO Journal, 29*, 3924–3938.

Kink, J. A., Maley, M. E., Preston, R. R., Ling, K. Y., Wallen-Friedman, M. A., Saimi, Y., & Kung, C. (1990). Mutations in paramecium calmodulin indicate functional differences between the C-terminal and N-terminal lobes in vivo. *Cell, 62*, 165–174.

Kschonsak, M., Chua, H. C., Weidling, C., Chakouri, N., Noland, C. L., Schott, K., Chang, T., Tam, C., Patel, N., Arthur, C. P., et al. (2021). Structural architecture of the human NALCN channelosome. *Nature, 603*, 180–186.

Kuboniwa, H., Tjandra, N., Grzesiek, S., Ren, H., Klee, C. B., & Bax, A. (1995). Solution structure of calcium-free calmodulin. *Nature Structural Biology, 2*, 768–776.

Kuzmenkina, E., Novikova, E., Jangsangthong, W., Matthes, J., & Herzig, S. (2019). Single-channel resolution of the interaction between C-terminal CaV1.3 isoforms and calmodulin. *Biophysical Journal, 116*, 836–846.

Landstrom, A. P., Boczek, N. J., Ye, D., Miyake, C. Y., De la Uz, C. M., Allen, H. D., Ackerman, M. J., & Kim, J. J. (2016). Novel long QT syndrome-associated missense mutation, L762F, in CACNA1C-encoded L-type calcium channel imparts a slower inactivation tau and increased sustained and window current. *International Journal of Cardiology, 220*, 290–298.

Lee, K. S., Marban, E., & Tsien, R. W. (1985). Inactivation of calcium channels in mammalian heart cells: Joint dependence on membrane potential and intracellular calcium. *The Journal of Physiology, 364*, 395–411.

Lee, A., Wong, S. T., Gallagher, D., Li, B., Storm, D. R., Scheuer, T., & Catterall, W. A. (1999). Ca2+/calmodulin binds to and modulates P/Q-type calcium channels. *Nature, 399*, 155–159.

Lee, A., Zhou, H., Scheuer, T., & Catterall, W. A. (2003). Molecular determinants of Ca(2+)/calmodulin-dependent regulation of Ca(v)2.1 channels. *Proceedings of the National Academy of Sciences of the United States of America, 100*, 16059–16064.

Lee, S. R., Adams, P. J., & Yue, D. T. (2015). Large Ca(2)(+)-dependent facilitation of Ca(V)2.1 channels revealed by Ca(2)(+) photo-uncaging. *The Journal of Physiology, 593*, 2753–2778.

Lei, M., Xu, J., Gao, Q., Minobe, E., Kameyama, M., & Hao, L. (2018). PKA phosphorylation of Cav1.2 channel modulates the interaction of calmodulin with the C terminal tail of the channel. *Journal of Pharmacological Sciences, 137*, 187–194.

Liang, H., DeMaria, C. D., Erickson, M. G., Mori, M. X., Alseikhan, B. A., & Yue, D. T. (2003). Unified mechanisms of Ca2+ regulation across the Ca2+ channel family. *Neuron, 39*, 951–960.

Limpitikul, W. B., Dick, I. E., Joshi-Mukherjee, R., Overgaard, M. T., George, A. L., Jr., & Yue, D. T. (2014). Calmodulin mutations associated with long QT syndrome prevent inactivation of cardiac L-type Ca(2+) currents and promote proarrhythmic behavior in ventricular myocytes. *Journal of Molecular and Cellular Cardiology, 74*, 115–124.

Limpitikul, W. B., Dick, I. E., Ben-Johny, M., & Yue, D. T. (2016). An autism-associated mutation in CaV1.3 channels has opposing effects on voltage- and Ca(2+)-dependent regulation. *Scientific Reports, 6*, 27235.

Limpitikul, W. B., Dick, I. E., Tester, D. J., Boczek, N. J., Limphong, P., Yang, W., Choi, M. H., Babich, J., DiSilvestre, D., Kanter, R. J., et al. (2017). A precision medicine approach to the rescue of function on malignant calmodulinopathic long-QT syndrome. *Circulation Research, 120*, 39–48.

Linse, S., Helmersson, A., & Forsen, S. (1991). Calcium binding to calmodulin and its globular domains. *The Journal of Biological Chemistry, 266*, 8050–8054.

Linsley, J. W., Hsu, I. U., Groom, L., Yarotskyy, V., Lavorato, M., Horstick, E. J., Linsley, D., Wang, W., Franzini-Armstrong, C., Dirksen, R. T., et al. (2017). Congenital myopathy results from misregulation of a muscle Ca2+ channel by mutant Stac3. *Proceedings of the National Academy of Sciences of the United States of America, 114*, E228–E236.

Liu, X., Yang, P. S., Yang, W., & Yue, D. T. (2010). Enzyme-inhibitor-like tuning of Ca(2+) channel connectivity with calmodulin. *Nature, 463*, 968–972.

Liu, G., Papa, A., Katchman, A. N., Zakharov, S. I., Roybal, D., Hennessey, J. A., Kushner, J., Yang, L., Chen, B. X., Kushnir, A., et al. (2020). Mechanism of adrenergic CaV1.2 stimulation revealed by proximity proteomics. *Nature, 577*, 695–700.

Livshitz, L., & Rudy, Y. (2009). Uniqueness and stability of action potential models during rest, pacing, and conduction using problem-solving environment. *Biophysical Journal, 97*, 1265–1276.

Mahajan, A., Sato, D., Shiferaw, Y., Baher, A., Xie, L. H., Peralta, R., Olcese, R., Garfinkel, A., Qu, Z., & Weiss, J. N. (2008). Modifying L-type calcium current kinetics: Consequences for cardiac excitation and arrhythmia dynamics. *Biophysical Journal, 94*, 411–423.

Mentrard, D., Vassort, G., & Fischmeister, R. (1984). Calcium-mediated inactivation of the calcium conductance in cesium-loaded frog heart cells. *The Journal of General Physiology, 83*, 105–131.

Mori, M. X., Erickson, M. G., & Yue, D. T. (2004). Functional stoichiometry and local enrichment of calmodulin interacting with Ca2+ channels. *Science, 304*, 432–435.

Mori, M. X., Vander Kooi, C. W., Leahy, D. J., & Yue, D. T. (2008). Crystal structure of the CaV2 IQ domain in complex with Ca2+/calmodulin: High-resolution mechanistic implications for channel regulation by Ca2+. *Structure, 16*, 607–620.

Morotti, S., Grandi, E., Summa, A., Ginsburg, K. S., & Bers, D. M. (2012). Theoretical study of L-type Ca(2+) current inactivation kinetics during action potential repolarization and early afterdepolarizations. *The Journal of Physiology, 590*, 4465–4481.

Neher, E. (1998). Vesicle pools and Ca^{2+} microdomains: New tools for understanding their roles in neurotransmitter release. *Neuron, 20*, 389–399.

Niu, J., Dick, I. E., Yang, W., Bamgboye, M. A., Yue, D. T., Tomaselli, G., Inoue, T., & Ben-Johny, M. (2018a). Allosteric regulators selectively prevent Ca(2+)-feedback of CaV and NaV channels. *eLife, 7*, e35222.

Niu, J., Yang, W., Yue, D. T., Inoue, T., & Ben-Johny, M. (2018b). Duplex signaling by CaM and Stac3 enhances CaV1.1 function and provides insights into congenital myopathy. *The Journal of General Physiology, 150*, 1145–1161.

Nyegaard, M., & Overgaard, M. T. (2019). The International Calmodulinopathy Registry: Recording the diverse phenotypic spectrum of un-CALM hearts. *European Heart Journal, 40*, 2976–2978.

Nyegaard, M., Overgaard, M. T., Sondergaard, M. T., Vranas, M., Behr, E. R., Hildebrandt, L. L., Lund, J., Hedley, P. L., Camm, A. J., Wettrell, G., et al. (2012). Mutations in calmodulin cause ventricular tachycardia and sudden cardiac death. *American Journal of Human Genetics, 91*, 703–712.

Oliveria, S. F., Dittmer, P. J., Youn, D. H., Dell'Acqua, M. L., & Sather, W. A. (2012). Localized calcineurin confers Ca2+-dependent inactivation on neuronal L-type Ca2+ channels. *The Journal of Neuroscience, 32*, 15328–15337.

Oz, S., Benmocha, A., Sasson, Y., Sachyani, D., Almagor, L., Lee, A., Hirsch, J. A., & Dascal, N. (2013). Competitive and non-competitive regulation of calcium-dependent inactivation in CaV1.2 L-type Ca2+ channels by calmodulin and Ca2+-binding protein 1. *The Journal of Biological Chemistry, 288,* 12680–12691.

Ozawa, J., Ohno, S., Saito, H., Saitoh, A., Matsuura, H., & Horie, M. (2018). A novel CACNA1C mutation identified in a patient with Timothy syndrome without syndactyly exerts both marked loss- and gain-of-function effects. *HeartRhythm Case Rep, 4,* 273–277.

Peterson, B. Z., DeMaria, C. D., Adelman, J. P., & Yue, D. T. (1999a). Calmodulin is the Ca^{2+} sensor for Ca2+-dependent inactivation of L- type calcium channels. *Neuron, 22,* 549–558.

Peterson, B. Z., DeMaria, C. D., Adelman, J. P., & Yue, D. T. (1999b). Calmodulin is the Ca2+ sensor for Ca2+ -dependent inactivation of L-type calcium channels. *Neuron, 22,* 549–558.

Peterson, B. Z., Lee, J. S., Mulle, J. G., Wang, Y., DeLeon, M., & Yue, D. T. (2000). Critical determinants of Ca^{2+}-dependent inactivation within an EF-hand motif of L-type Ca^{2+} channels. *Biophysical Journal, 78,* 1906–1920.

Pinggera, A., & Striessnig, J. (2016). Cav 1.3 (CACNA1D) L-type Ca2+ channel dysfunction in CNS disorders. *The Journal of Physiology, 594,* 5839–5849.

Pinggera, A., Lieb, A., Benedetti, B., Lampert, M., Monteleone, S., Liedl, K. R., Tuluc, P., & Striessnig, J. (2015). CACNA1D de novo mutations in autism spectrum disorders activate Cav1.3 L-type calcium channels. *Biological Psychiatry, 77,* 816–822.

Pinggera, A., Mackenroth, L., Rump, A., Schallner, J., Beleggia, F., Wollnik, B., & Striessnig, J. (2017). New gain-of-function mutation shows CACNA1D as recurrently mutated gene in autism spectrum disorders and epilepsy. *Human Molecular Genetics, 26,* 2923–2932.

Pipilas, D. C., Johnson, C. N., Webster, G., Schlaepfer, J., Fellmann, F., Sekarski, N., Wren, L. M., Ogorodnik, K. V., Chazin, D. M., Chazin, W. J., et al. (2016). Novel calmodulin mutations associated with congenital long QT syndrome affect calcium current in human cardiomyocytes. *Heart Rhythm: The Official Journal of the Heart Rhythm Society, 13,* 2012–2019.

Pitt, G. S., Zuhlke, R. D., Hudmon, A., Schulman, H., Reuter, H., & Tsien, R. W. (2001). Molecular basis of calmodulin tethering and Ca2+-dependent inactivation of L-type Ca2+ channels. *The Journal of Biological Chemistry, 276,* 30794–30802.

Plant, T., Standen, N., & Ward, T. (1983). The effects of injection of calcium ions and calcium chelators on calcium channel cnactivation in helix neurones. *Journal of Physiology, 334,* 189–212.

Polster, A., Perni, S., Bichraoui, H., & Beam, K. G. (2015). Stac adaptor proteins regulate trafficking and function of muscle and neuronal L-type Ca2+ channels. *Proceedings of the National Academy of Sciences of the United States of America, 112,* 602–606.

Polster, A., Nelson, B. R., Olson, E. N., & Beam, K. G. (2016). Stac3 has a direct role in skeletal muscle-type excitation-contraction coupling that is disrupted by a myopathy-causing mutation. *Proceedings of the National Academy of Sciences of the United States of America, 113,* 10986–10991.

Polster, A., Dittmer, P. J., Perni, S., Bichraoui, H., Sather, W. A., & Beam, K. G. (2018a). Stac proteins suppress Ca(2+)-dependent inactivation of neuronal l-type Ca(2+) channels. *The Journal of Neuroscience, 38,* 9215–9227.

Polster, A., Nelson, B. R., Papadopoulos, S., Olson, E. N., & Beam, K. G. (2018b). Stac proteins associate with the critical domain for excitation-contraction coupling in the II-III loop of CaV1.1. *The Journal of General Physiology, 150,* 613–624.

Qin, N., Olcese, R., Bransby, M., Lin, T., & Birnbaumer, L. (1999). Ca2+-induced inhibition of the cardiac Ca2+ channel depends on calmodulin. *Proceedings of the National Academy of Sciences of the United States of America, 96,* 2435–2438.

Rufenach, B., & Van Petegem, F. (2021). Structure and function of STAC proteins: Calcium channel modulators and critical components of muscle excitation-contraction coupling. *The Journal of Biological Chemistry, 297,* 100874.

Saimi, Y., & Kung, C. (2002). Calmodulin as an ion channel subunit. *Annual Review of Physiology, 64,* 289–311.

Sang, L., Dick, I. E., & Yue, D. T. (2016). Protein kinase A modulation of CaV1.4 calcium channels. *Nature Communications, 7,* 12239.

Sang, L., Vieira, D. C. O., Yue, D. T., Ben-Johny, M., & Dick, I. E. (2021). The molecular basis of the inhibition of CaV1 calcium-dependent inactivation by the distal carboxy tail. *The Journal of Biological Chemistry, 296,* 100502.

Scharinger, A., Eckrich, S., Vandael, D. H., Schonig, K., Koschak, A., Hecker, D., Kaur, G., Lee, A., Sah, A., Bartsch, D., et al. (2015). Cell-type-specific tuning of Cav1.3 Ca(2+)-channels by a C-terminal automodulatory domain. *Frontiers in Cellular Neuroscience, 9,* 309.

Scholl, U. I., Goh, G., Stolting, G., de Oliveira, R. C., Choi, M., Overton, J. D., Fonseca, A. L., Korah, R., Starker, L. F., Kunstman, J. W., et al. (2013). Somatic and germline CACNA1D calcium channel mutations in aldosterone-producing adenomas and primary aldosteronism. *Nature Genetics, 45,* 1050–1054.

Shaltiel, L., Paparizos, C., Fenske, S., Hassan, S., Gruner, C., Rotzer, K., Biel, M., & Wahl-Schott, C. A. (2012). Complex regulation of voltage-dependent activation and inactivation properties of retinal voltage-gated Cav1.4 L-type Ca2+ channels by Ca2+-binding protein 4 (CaBP4). *The Journal of Biological Chemistry, 287,* 36312–36321.

Sherman, A., Keizer, J., & Rinzel, J. (1990). Domain model for Ca2(+)-inactivation of Ca2+ channels at low channel density. *Biophysical Journal, 58,* 985–995.

Simms, B. A., Souza, I. A., & Zamponi, G. W. (2014). A novel calmodulin site in the Cav1.2 N-terminus regulates calcium-dependent inactivation. *Pflügers Archiv, 466,* 1793–1803.

Singh, A., Hamedinger, D., Hoda, J. C., Gebhart, M., Koschak, A., Romanin, C., & Striessnig, J. (2006). C-terminal modulator controls Ca2+-dependent gating of Ca(v)1.4 L-type Ca2+ channels. *Nature Neuroscience, 9*, 1108–1116.

Splawski, I., Timothy, K. W., Sharpe, L. M., Decher, N., Kumar, P., Bloise, R., Napolitano, C., Schwartz, P. J., Joseph, R. M., Condouris, K., et al. (2004). Ca(V)1.2 calcium channel dysfunction causes a multisystem disorder including arrhythmia and autism. *Cell, 119*, 19–31.

Splawski, I., Timothy, K. W., Decher, N., Kumar, P., Sachse, F. B., Beggs, A. H., Sanguinetti, M. C., & Keating, M. T. (2005). Severe arrhythmia disorder caused by cardiac L-type calcium channel mutations. *Proceedings of the National Academy of Sciences of the United States of America, 102*, 8089–8096; discussion 8086-8088.

Standen, N., & Stanfield, P. (1982). A binding-site model for calcium channel inactivation that depends on calcium entry. *Proceedings of the Royal Society of London. Series B, Biological Sciences, 217*, 101–110.

Stern, M. D. (1992). Buffering of calcium in the vicinity of a channel pore. *Cell Calcium, 13*, 183–192.

Tadross, M. R., Dick, I. E., & Yue, D. T. (2008). Mechanism of local and global Ca2+ sensing by calmodulin in complex with a Ca2+ channel. *Cell, 133*, 1228–1240.

Tadross, M. R., Ben Johny, M., & Yue, D. T. (2010). Molecular endpoints of Ca2+/calmodulin- and voltage-dependent inactivation of Ca(v)1.3 channels. *The Journal of General Physiology, 135*, 197–215.

Tan, B. Z., Jiang, F., Tan, M. Y., Yu, D., Huang, H., Shen, Y., & Soong, T. W. (2011). Functional characterization of alternative splicing in the C terminus of L-type CaV1.3 channels. *The Journal of Biological Chemistry, 286*, 42725–42735.

Tan, G. M., Yu, D., Wang, J., & Soong, T. W. (2012). Alternative splicing at C terminus of Ca(V)1.4 calcium channel modulates calcium-dependent inactivation, activation potential, and current density. *The Journal of Biological Chemistry, 287*, 832–847.

Tan, G. C., Negro, G., Pinggera, A., Tizen Laim, N. M. S., Mohamed Rose, I., Ceral, J., Ryska, A., Chin, L. K., Kamaruddin, N. A., Mohd Mokhtar, N., et al. (2017). Aldosterone-producing adenomas: Histopathology-genotype correlation and identification of a novel CACNA1D mutation. *Hypertension, 70*, 129–136.

Tang, W., Halling, D. B., Black, D. J., Pate, P., Zhang, J. Z., Pedersen, S., Altschuld, R. A., & Hamilton, S. L. (2003). Apocalmodulin and Ca2+ calmodulin-binding sites on the CaV1.2 channel. *Biophysical Journal, 85*, 1538–1547.

Tao, X., & MacKinnon, R. (2019). Molecular structures of the human Slo1 K(+) channel in complex with beta4. *eLife, 8*, e51409.

Thomas, J. R., & Lee, A. (2016). Measuring Ca2+-dependent modulation of voltage-gated Ca2+ channels in HEK-293T cells. *Cold Spring Harbor Protocols, 2016*, 762–767.

Thomas, J. R., Hagen, J., Soh, D., & Lee, A. (2018). Molecular moieties masking Ca(2+)-dependent facilitation of voltage-gated Cav2.2 Ca(2+) channels. *The Journal of General Physiology, 150*, 83–94.

Tidow, H., & Nissen, P. (2013). Structural diversity of calmodulin binding to its target sites. *The FEBS Journal, 280*, 5551–5565.

Tillotson, D. (1979). Inactivation of Ca conductance dependent on entry of Ca ions in molluscan neurons. *Proceedings of the National Academy of Sciences of the United States of America, 76*, 1497–1500.

Turner, M., Anderson, D. E., Bartels, P., Nieves-Cintron, M., Coleman, A. M., Henderson, P. B., Man, K. N. M., Tseng, P. Y., Yarov-Yarovoy, V., Bers, D. M., et al. (2020). alpha-Actinin-1 promotes activity of the L-type Ca(2+) channel Cav 1.2. *The EMBO Journal, 39*, e102622.

Uchitel, O. D., Gonzalez Inchauspe, C., & Di Guilmi, M. N. (2014). Calcium channels and synaptic transmission in familial hemiplegic migraine type 1 animal models. *Biophysical Reviews, 6*, 15–26.

Van Petegem, F., Chatelain, F. C., & Minor, D. L., Jr. (2005). Insights into voltage-gated calcium channel regulation from the structure of the CaV1.2 IQ domain-Ca2+/calmodulin complex. *Nature Structural & Molecular Biology, 12*, 1108–1115.

Wahl-Schott, C., Baumann, L., Cuny, H., Eckert, C., Griessmeier, K., & Biel, M. (2006). Switching off calcium-dependent inactivation in L-type calcium channels by an autoinhibitory domain. *Proceedings of the National Academy of Sciences of the United States of America, 103*, 15657–15662.

Wemhoner, K., Friedrich, C., Stallmeyer, B., Coffey, A. J., Grace, A., Zumhagen, S., Seebohm, G., Ortiz-Bonnin, B., Rinne, S., Sachse, F. B., et al. (2015). Gain-of-function mutations in the calcium channel CACNA1C (Cav1.2) cause non-syndromic long-QT but not Timothy syndrome. *Journal of Molecular and Cellular Cardiology, 80*, 186–195.

Weyrer, C., Turecek, J., Niday, Z., Liu, P. W., Nanou, E., Catterall, W. A., Bean, B. P., & Regehr, W. G. (2019). The role of CaV2.1 channel facilitation in synaptic facilitation. *Cell Reports, 26*, 2289–2297 e2283.

Williams, B., Haeseleer, F., & Lee, A. (2018). Splicing of an automodulatory domain in Cav1.4 Ca(2+) channels confers distinct regulation by calmodulin. *The Journal of General Physiology, 150*, 1676–1687.

Wingo, T. L., Shah, V. N., Anderson, M. E., Lybrand, T. P., Chazin, W. J., & Balser, J. R. (2004). An EF-hand in the sodium channel couples intracellular calcium to cardiac excitability. *Nature Structural & Molecular Biology, 11*, 219–225.

Wu, J., Yan, Z., Li, Z., Yan, C., Lu, S., Dong, M., & Yan, N. (2015). Structure of the voltage-gated calcium channel Cav1.1 complex. *Science, 350*, aad2395.

Xia, X. M., Fakler, B., Rivard, A., Wayman, G., Johnson-Pais, T., Keen, J. E., Ishii, T., Hirschberg, B., Bond, C. T., Lutsenko, S., et al. (1998). Mechanism of calcium gating in small-conductance calcium-activated potassium channels. *Nature, 395*, 503–507.

Xu, W., & Lipscombe, D. (2001). Neuronal Ca(V)1.3alpha(1) L-type channels activate at relatively hyperpolarized membrane potentials and are incompletely inhibited by dihydropyridines. *The Journal of Neuroscience, 21*, 5944–5951.

Xu, J., & Wu, L. G. (2005). The decrease in the presynaptic calcium current is a major cause of short-term depression at a calyx-type synapse. *Neuron, 46*, 633–645.

Xu, J., Yu, L., Minobe, E., Lu, L., Lei, M., & Kameyama, M. (2016). PKA and phosphatases attached to the Ca(V)1.2 channel regulate channel activity in cell-free patches. *American Journal of Physiology-Cell Physiology, 310*, C136–C141.

Yang, J., Ellinor, P. T., Sather, W. A., Zhang, J. F., & Tsien, R. W. (1993). Molecular determinants of Ca2+ selectivity and ion permeation in L-type Ca2+ channels. *Nature, 366*, 158–161.

Yang, P. S., Alseikhan, B. A., Hiel, H., Grant, L., Mori, M. X., Yang, W., Fuchs, P. A., & Yue, D. T. (2006). Switching of Ca2+-dependent inactivation of Ca(v)1.3 channels by calcium binding proteins of auditory hair cells. *The Journal of Neuroscience, 26*, 10677–10689.

Yang, P. S., Johny, M. B., & Yue, D. T. (2014). Allostery in Ca(2)(+) channel modulation by calcium-binding proteins. *Nature Chemical Biology, 10*, 231–238.

Yarotskyy, V., Gao, G., Peterson, B. Z., & Elmslie, K. S. (2009). The Timothy syndrome mutation of cardiac CaV1.2 (L-type) channels: Multiple altered gating mechanisms and pharmacological restoration of inactivation. *The Journal of Physiology, 587*, 551–565.

Yoder, J. B., Ben-Johny, M., Farinelli, F., Srinivasan, L., Shoemaker, S. R., Tomaselli, G. F., Gabelli, S. B., & Amzel, L. M. (2019). Ca(2+)-dependent regulation of sodium channels NaV1.4 and NaV1.5 is controlled by the post-IQ motif. *Nature Communications, 10*, 1514.

Yue, D. T., Herzig, S., & Marban, E. (1990). Beta-adrenergic stimulation of calcium channels occurs by potentiation of high-activity gating modes. *Proceedings of the National Academy of Sciences of the United States of America, 87*, 753–757.

Zagotta, W. N., Hoshi, T., & Aldrich, R. W. (1990). Restoration of inactivation in mutants of Shaker potassium channels by a peptide derived from ShB. *Science, 250*, 568–571.

Zalk, R., Clarke, O. B., des Georges, A., Grassucci, R. A., Reiken, S., Mancia, F., Hendrickson, W. A., Frank, J., & Marks, A. R. (2015). Structure of a mammalian ryanodine receptor. *Nature, 517*, 44–49.

Zhou, J., Olcese, R., Qin, N., Noceti, F., Birnbaumer, L., & Stefani, E. (1997). Feedback inhibition of Ca2+ channels by Ca2+ depends on a short sequence of the C terminus that does not include the Ca2+ – Binding function of a motif with similarity to Ca2+ -binding domains. *Proceedings of the National Academy of Sciences of the United States of America, 94*, 2301–2305.

Zhou, H., Yu, K., McCoy, K. L., & Lee, A. (2005). Molecular mechanism for divergent regulation of Cav1.2 Ca2+ channels by calmodulin and Ca2+-binding protein-1. *The Journal of Biological Chemistry, 280*, 29612–29619.

Zuhlke, R. D., & Reuter, H. (1998). Ca2+-sensitive inactivation of L-type Ca2+ channels depends on multiple cytoplasmic amino acid sequences of the α_{1c} subunit. *Proceedings of the National Academy of Sciences of the United States of America, 95*, 3287–3294.

Zuhlke, R. D., Pitt, G. S., Deisseroth, K., Tsien, R. W., & Reuter, H. (1999). Calmodulin supports both inactivation and facilitation of L-type calcium channels. *Nature, 399*, 159–162.

Cav3 Calcium Channel Interactions with Potassium Channels

Ray W. Turner

Abstract

The influence of potassium channels on cell activity is markedly enhanced by exhibiting the low voltage for activation, fast activation, and fast inactivation that characterize an A-type channel profile. Previous work had identified only a subset of members of the voltage-gated K_v1, K_v3, and K_v4 families in which alpha subunits are able to exhibit A-type biophysical properties. Recent work has shown that Cav3 channels closely associate with different potassium channels to impart their biophysical and kinetic properties and effectively create three new forms of A-type potassium channel. Ca_v3-K channel interactions identified to date include Ca_v3-K_v4, Ca_v3-IK, and Ca_v3-BK that enable novel aspects of synaptic processing and spike discharge in cerebellar granule, stellate, Purkinje, and medial vestibular neurons. This chapter will highlight data obtained in each of these cell classes of how Ca_v3 associations with different potassium channels extend the number of channels capable of exhibiting an A-type phenotype to control cell activity.

Keywords

$Ca_v3.x$ · T-type · K_v4 · KCa3.1 · KCa1.1 · A-type

Introduction

Potassium channels control an enormous array of functions in setting the intrinsic excitability of neurons and shaping the pattern and frequency of spike output that defines the neural code used for signal processing. The most common biophysical phenotype of these channels is that of a relatively high voltage-activated (HVA) and non-inactivating outward current first recognized as a "delayed rectifier" in axon membranes. A distinct set of potassium channel properties came to be recognized with reports of a fast activating and fast inactivating Drosophila Shaker channel and its mammalian homolog, $K_v1.4$. This transient "A-type" outward current further exhibited properties of low voltage-activation (LVA) that allowed it to function well within the resting membrane potential range and provide unique contributions to spike frequency and timing following membrane hyperpolarizations. The importance of expressing A-type potassium channels is evident in an almost ubiquitous expression of this form of outward current in central neurons. Additional classes of voltage-gated potassium channels in which the alpha subunit

R. W. Turner (✉)
Hotchkiss Brain Institute, Alberta Children's Hospital Research Institute, Department of Cell Biology and Anatomy, Cumming School of Medicine, University of Calgary, Calgary, AB, Canada
e-mail: rwturner@ucalgary.ca

alone exhibits A-type properties were eventually recognized, but it includes a relatively limited subset of members of the K_v1 ($K_v1.4$), K_v3 ($K_v3.3$, $K_v3.4$) and K_v4 ($K_v4.1$, 4.2, 4.3) families.

By comparison, the fast activation and inactivation of voltage-gated A-type channels were rarely attributed to calcium-gated potassium channels. The properties of calcium-gated potassium channels had been shown amenable to change by establishing a close association and functional coupling to voltage-gated calcium channels. One illustration of this was in the functional coupling of KCa1.1 (big conductance, BK) potassium channels to high voltage-activated (HVA) but slow inactivating calcium channels (Robitaille et al., 1993; Poolos & Johnston, 1999; Grunnet & Kaufmann, 2004; Berkefeld et al., 2006, 2010; Loane et al., 2007; Müller et al., 2007; Berkefeld & Fakler, 2008; Marcantoni et al., 2010; Vandael et al., 2010). Other work established functional coupling at even the nanometer level between HVA calcium channels and both SK (KCa2.x) and IK (KCa3.1) potassium channels (Marrion & Tavalin, 1998; Lima & Marrion, 2007; Sahu et al., 2017, 2019; Sahu & Turner, 2021). A similar functional coupling between the LVA class of Ca_v3 calcium channels and potassium channels also came to be recognized (Smith et al., 2002; Wolfart & Roeper, 2002; Cueni et al., 2008; Gittis et al., 2010).

One region that received extensive work on the role of Ca_v3 channels in controlling cell output is the cerebellum. The cerebellum is comprised of highly organized arrays of four main classes of interneurons (granule, Golgi, stellate, basket) and 3 classes of output cells in cortical Purkinje cells and the deep cerebellar nuclear and vestibular neurons that receive Purkinje cell input. The expression pattern of Cav3 channel isoforms in cerebellar neurons has been reported in numerous studies (Bossu et al., 1989; Mouginot et al., 1997; Talley et al., 1999; Isope & Murphy, 2005; McKay et al., 2006; Molineux et al., 2006; Hildebrand et al., 2009; Anderson et al., 2010a, 2013; Engbers et al., 2012; Isope et al., 2012; Heath et al., 2014; Aguado et al., 2016; Ly et al., 2016). Through this work, it became apparent

that all Ca_v3 channel isoforms are widely expressed across the seven cell types in the cerebellum. The extent of Ca_v3 channel expression was surprising as it did not hold a clear relationship to the ability of different cells to exhibit a rebound discharge thought characteristic of cells that express T-type channels (Molineux et al., 2006). A series of studies investigating how Ca_v3 and potassium channels might work in tandem to control spike output revealed a rich repertoire of Ca_v3 influence over spike discharge through coupling with potassium channels. This chapter will focus on Ca_v3 channel interactions that have been found with either voltage- or calcium-gated potassium channels in cerebellar stellate and granule cells, Purkinje cells, and medial vestibular cells that receive input from Purkinje cells.

Ca_v3 channel isoforms have now been recognized to form a close association at the molecular level with three different potassium channel classes (Fig. 1). The end result of the Ca_v3 channel link to these different channels is unique in effectively creating three new classes of LVA, fast activating and fast inactivating forms of potassium channel that take on specific roles in different cerebellar cell types.

Ca_v3-K_v4 Complex

The presence of immunolabel for Ca_v3 isoforms in cerebellar stellate and granule cells of lobule 9 despite a lack of rebound burst response following membrane hyperpolarizations encouraged further study of the role of Ca_v3 channels in these cells (Molineux et al., 2005; Heath et al., 2014; Rizwan et al., 2016). Recordings in stellate and granule cells confirmed the expression of both T-type calcium and K_v4-mediated A-type potassium current. Surprisingly, it was found that the $K_v4.2$ A-type current was sensitive to T-type channel blockers. Specifically, bath perfusion of mibefradil (0.5 μM) or low calcium medium (0.1 mM) selectively shifted the half inactivation voltage (Vh) of K_v4 current to more negative potentials ("left-shifted Vh") (Fig. 2a) (Molineux et al., 2005; Anderson et al., 2010a, 2013). No effect was detected on the half activation voltage

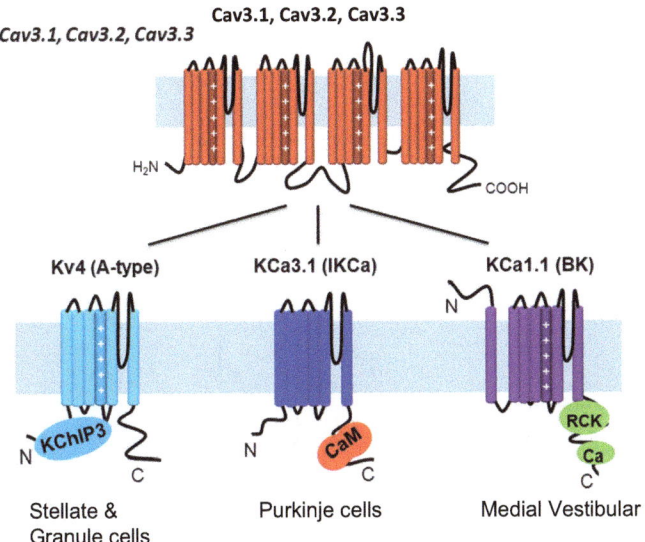

Fig. 1 T-type calcium channels couple at the molecular level to distinct potassium channels to invoke A-type current properties. Cartoon representation of Ca$_v$3.x channel isoforms and three potassium channels they link to in cerebellar neurons to modulate spike output: voltage-gated K$_v$4.x A-type channels in stellate and granule cell interneurons, calcium-gated channels KCa3.1 (IK) in Purkinje cells, and KCa1.1 (BK) channels in medial vestibular neurons. Abbreviations: *N* N-terminus, *C* C-terminus, *KChIP3* potassium channel interacting protein 3, *CaM* calmodulin, *RCK* the regulator of conductance of potassium. (Figure is modified from Turner and Zamponi (2014))

(Va) of K$_v$4 current (Fig. 2a), current density, kinetics, or rate of recovery from inactivation. The net effect of reducing Ca$_v$3 calcium conductance was to decrease K$_v$4 current amplitude over a range of holding membrane potentials of ~−50 to −90 mV (Fig. 2a). By focally ejecting low calcium medium (0.1 mM) the actions of Ca$_v$3 calcium influx on Kv4 Vh were further shown to be rapidly induced and fully reversible (Anderson et al., 2013). The application of a dynamic clamp of both Ca$_v$3 and K$_v$4 current in granule cells allowed for a reduction of K$_v$4 current amplitude over the appropriate voltage range in otherwise untreated cells (Heath et al., 2014). The importance of the Cav3-mediated control over K$_v$4 Vh was immediately evident in a substantial increase in spike firing rate gain (Fig. 2b, c).

These studies were the first to report a calcium-dependent change in K$_v$4 channel properties, a channel that had only been recognized as exhibiting voltage-dependent properties. This raised interest in the potential role of a class of calcium-sensing KChIP proteins that had been found to associate with K$_v$4 channels, but which had no identified role. Stellate cells were found to exhibit immunolabel for the KChIP3 isoform as well as a specific isoform of a second K$_v$4 accessory protein dipeptidyl peptidase 10c (DPP10c). Recordings of K$_v$4 current were used to assess the interaction between these subunits in stellate cells or when coexpressed in human embryonic kidney (HEK) cells. These and other tests revealed co-immunoprecipitation between Ca$_v$3.1 or Ca$_v$3.3 with K$_v$4 channel isoforms and KChIP3. Moreover, all of the effects of reducing Ca$_v$3 channel conductance on K$_v$4 Vh could be recapitulated by coexpressing these proteins in HEK cells together with DPP10c (Anderson et al., 2010a). Infusing through the electrode a Pan-KChIP or KChIP3 antibody, but not an antibody against KChIP1 reproduced the left-shift in K$_v$4 Vh by ~−10 mV. These recordings thus established an absolute dependence of the left-shift in K$_v$4 Vh on coexpression of Ca$_v$3.3 and KChIP3, with similar but slightly less pronounced effects by Ca$_v$3.1 (Anderson et al., 2010a, b). It was

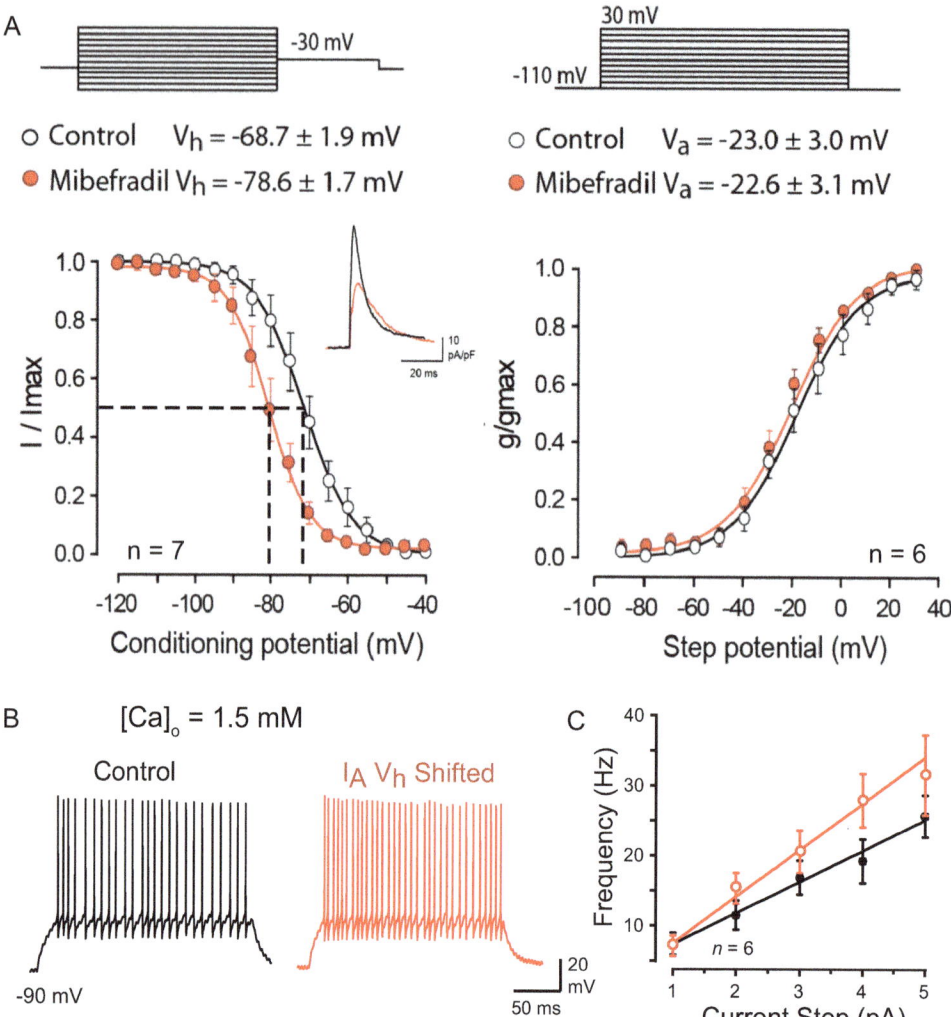

Fig. 2 A Ca_v3-$Kv4$ complex shifts Kv4 availability to reduce spike output. (**a**) Mean plots of the voltage of inactivation and activation of K_v4 A-type potassium current in whole-cell recordings from cerebellar stellate cells in 1.5 mM [Ca] in the bathing medium under resting conditions. *Inset* shows a representative A-type current recorded for a step from −100 to −30 mV. Values for Vh and Va are shown above plots for recordings in control conditions and after perfusion of 0.5 μM mibefradil. Blocking T-type current selectively left-shifts K_v4 Vh ∼ −10 mV ($p < 0.001$) with no effect on Va ($p > 0.05$) to reduce current amplitude (*inset*). (**b, c**) Whole-cell recordings of spike discharge (**b**) and frequency-current plots (**c**) from cerebellar lobule 9 granule cells that express the Ca_v3-K_v4 complex. Recordings were made in a dynamic clamp to reduce K_v4 current online according to measured biophysical properties of Ca_v3 and K_v4 current, with the external level of calcium maintained at 1.5 mM throughout. Dynamic clamp reduction of A-type current increases the rate of spike firing (**b**) and gain on a frequency-current plot (**c**). Average values are mean ± SEM. (Figures are modified from (**a**) Anderson et al. (2010a), and (**b, c**) Heath et al. (2014))

further shown that these results could not be reproduced if $K_v4.2$ was coexpressed with any of the HVA classes of calcium channels, revealing the specificity of K_v4 modulation by Ca_v3 channels.

Together these data established that the Ca_v3-K_v4 complex normally promotes a calcium-dependent rightward shift in K_v4 Vh and thus window current that is centered near resting potential and spike threshold (Fig. 3a, b). Since

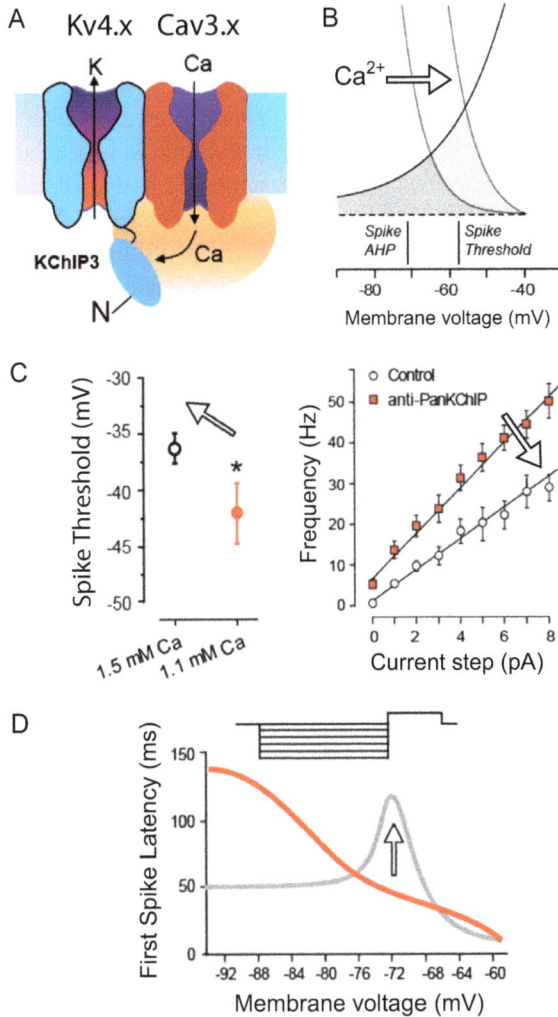

Fig. 3 Ca_v3 calcium influx shifts K_v4 Vh to reduce spike firing and delay first spike latency. (**a**) Cartoon depiction of the Ca_v3-KChIP3-K_v4 complex assembled to function with nanodomain interactions in cerebellar stellate and granule cells. (**b**) Schematic diagram depicting the role of Cav3 calcium influx in normally shifting the K_v4 Vh to the right to increase window current and availability of K_v4 for activation over a voltage range that straddles spike threshold and a fast AHP. *White filled arrow* in this and lower panels depicts the direction of action by the Cav3-K_v4 complex on cell activity. (**c**) Mean scatter plots of the change in spike threshold in stellate cells, and current-frequency plots of firing rate in granule cells upon interfering with Ca_v3-K_v4 complex function. White arrows depict the influence of a Cav3-K_v4 complex on cell function under normal conditions. (**d**) Fits of the mean values for FSL following membrane hyperpolarization in cerebellar stellate cells with (*grey* trace) and without (*red* trace) Ca_v3-K_v4 complex actions. The current step protocol used to activate spike responses to measure FSL following a hyperpolarization is shown above. *$p < 0.05$. (Figures are modified from (**a**) Turner and Zamponi (2014), (**b, c**) Anderson et al. (2010b) and Heath et al. (2014), and (**d**) Molineux et al. (2005))

this would change the extent of K_v4 control over the spike threshold, the effects of interfering with the Ca_v3-K_v4 complex on spike output were tested. In stellate cells, a decrease in the level of external calcium from 1.5 to 1.1 mM lowered the spike threshold to increase the firing rate (Fig. 3c). In cerebellar granule cells, infusing an anti-PanKChIP antibody through the electrode increased spike firing rate gain (Fig. 3c). These data confirmed that the normal role of the Ca_v3-

K$_v$4 complex is to reduce excitability and the rate of spike firing in both stellate and granule cells. But an additional influence in regulating spike output became apparent in cerebellar stellate cells. One role for A-type current is to modify the latency to first spike discharge following membrane hyperpolarization. First spike latency (FSL) typically takes on a graded, inverse relationship where an increase in the magnitude or duration of a preceding hyperpolarization produces more A-type current upon return to resting potential to delay the firing of the first spike. This is important as FSL represents a key element of encoding membrane hyperpolarizations and sensory input. In these cells, the Ca$_v$3-K$_v$4 complex was found to promote an unusual voltage-FSL relationship. Instead of an inverse relationship between membrane hyperpolarization and spike latency, stellate cells instead exhibit a non-monotonic relationship where a delay in FSL was restricted to ~10 mV window of membrane voltage around resting potential (−76 to −65 mV) (Fig. 3d). Reducing Ca$_v$3 calcium conductance by perfusing low calcium or calcium channel blockers removed the peak of FSL found near resting potentials, revealing a role for Ca$_v$3 calcium channels in producing this pattern of spike output (Fig. 3d) (Molineux et al., 2005).

Ca$_v$3.2-IK Complex

Cerebellar Purkinje cells express calcium-gated potassium channels that are activated by the large complex spike response generated by climbing fiber input. HVA calcium channels triggered by climbing fiber input control the duration of the complex spike and a subsequent AHP by activating calcium-gated potassium channels (Schmolesky et al., 2002; Fernandez et al., 2007; McKay et al., 2007; Davie et al., 2008; Ohtsuki et al., 2012; Engbers et al., 2013a; Ait Ouares et al., 2019). It was shown that Purkinje cells also express Cav3 channels and T-type current that can be activated synaptically and during spike discharge (Mouginot et al., 1997; Talley et al., 1999; Swensen & Bean, 2003; Engbers et al., 2012; Isope et al., 2012; Ly et al., 2016). Recently,

work on a parallel fiber EPSP-evoked slow afterhyperpolarization (sAHP) uncovered the additional expression of an intermediate conductance calcium-gated potassium channel (KCa3.1, IK) in Purkinje cells (Engbers et al., 2012). This was surprising in that IK channels were not believed to be expressed in central neurons. Importantly, IK channels are known to be voltage-insensitive and activated by calcium binding to calmodulin (Fig. 1) to evoke a non-inactivating potassium current. But the data of Engbers et al. (2012) revealed that Purkinje cells exhibit a new Ca$_v$3-IK channel complex (Fig. 4a) formed at the molecular level to effectively convert IK channels to calcium- and voltage-gated A-type channel profile to control several aspects of spike discharge.

Single-pulse stimulation of parallel fibers in the slice preparation established that EPSPs even 5 mV in amplitude are followed by a calcium-dependent sAHP of ~400 msec (Fig. 4b). Pharmacological tests revealed that this sAHP was insensitive to multiple toxins and blockers against the families of HVA calcium channels, including 200 nM AgTx against the P-type calcium channel that is heavily expressed in Purkinje cells. Instead, the Ca$_v$3 channel blockers Ni^{2+} (100 μM) or mibefradil (1.0 μM) slowed the rate of EPSP decay and blocked the subsequent sAHP (Fig. 4b). In attempting to identify the calcium-gated potassium channel responsible for the sAHP, it became clear that neither SK nor BK channels were involved in the slow AHP in its being insensitive to apamin, iberiotoxin, TEA, or paxilline. Instead, the parallel fiber EPSP rate of decay was reduced, and the slow AHP was blocked by charybdotoxin (ChTx, 100 nM) or the IK-selective blocker TRAM-34 (100 nM) (Fig. 4a, b) (Engbers et al., 2012).

To assess the interaction between Ca$_v$3 and IK channels they examined the currents evoked in Purkinje cells by a ramp command (500 msec) from a holding potential of −100 to −40 mV to scan from a subthreshold potential to near the peak voltage for T-type current. Ca$_v$3 current was also pharmacologically isolated using a suite of blockers against sodium, HVA calcium, and potassium channels, including SK, BK, and IK channels. To further identify the Ca$_v$3-dependent

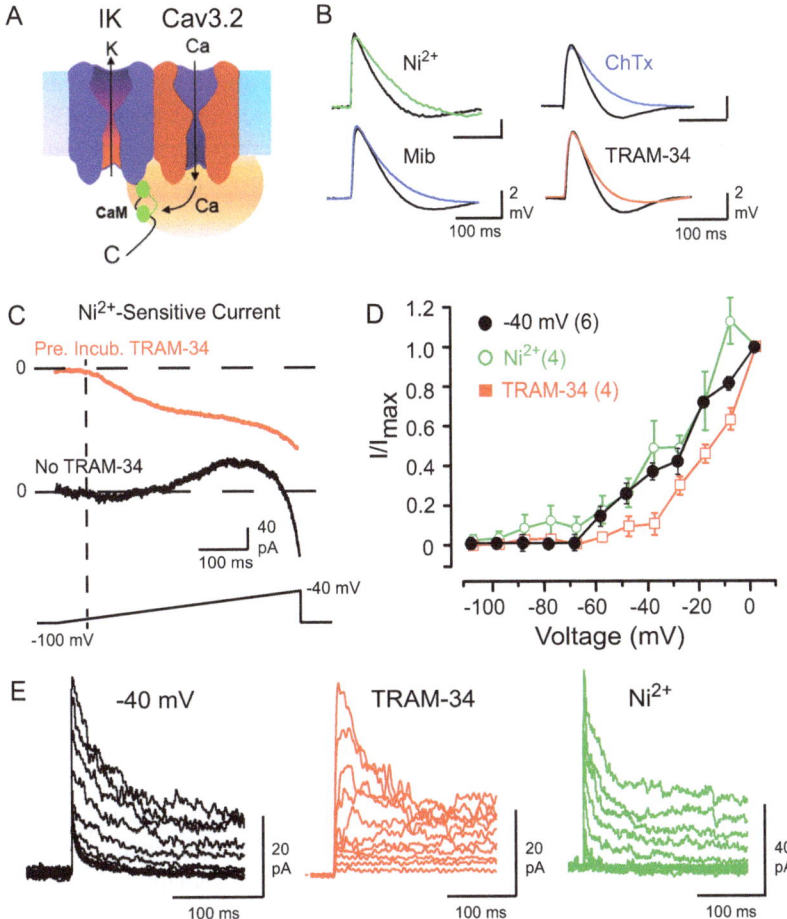

Fig. 4 A Ca$_v$3.2-IK channel complex in Purkinje cells imparts low voltage-dependence and an A-type channel profile to IK current. (**a**) Cartoon depiction of a nanodomain interaction between Ca$_v$3.2 and IK (KCa3.1) channels. (**b**) Representative subthreshold simulated parallel fiber EPSPs (simEPSPs) evoked by somatic current injection to test postsynaptic contributions to the generation of a sAHP. Superimposed recordings show a reduction in the decay phase of the simEPSP and sAHP by 100 μM Ni^{2+} or 1.0 μM mibefradil to block Cav3 calcium channels, or by 100 nM ChTx or 100 nM TRAM-34 to block IK channels. (**c**) Whole-cell currents evoked by a ramp command under conditions that pharmacologically isolate Ca$_v$3 current (*red* trace) or Ca$_v$3 and IK current (*black* trace) in the presence or absence of 100 nM TRAM-34 to block IK channels. Ca$_v$3-dependent inward or outward currents were

identified by subtracting traces following 100 μM Ni^{2+} application. (**d, e**) Outside-out patch recordings from separate Purkinje cell somata evoked by step commands from −110 to 0 mV under conditions that isolate Ca$_v$3 and IK current (Engbers et al., 2012). Shown in (**d**) are mean I-V plots for outward current calculated as the difference from those evoked after a preceding step to −40 mV to inactivate Ca$_v$3 channels, or those blocked by TRAM-34 (100 nM) or Ni^{2+} (100 μM). Shown in (**e**) are superimposed outward current traces for each of the conditions in (**d**), revealing a calcium-dependent, low voltage-activated, and fast activating/inactivating IK current. Sample values are shown in brackets in (**d**) and average values are mean ± SEM. (Figure is modified from (**a**) Turner and Zamponi (2014) and (**b–e**) from Engbers et al. (2012))

component of evoked currents they subtracted from control recordings the current blocked by perfusion of 100 nM Ni^{2+}. These tests uncovered a Ni^{2+}-sensitive inward Ca$_v$3 calcium current that

was triggered from a voltage of ~−90 mV that grew steadily through the voltage ramp to −40 mV (Fig. 4c). If the same test was repeated in the absence of TRAM-34 to allow activation of

IK channels, Ni^{2+} application instead uncovered a slowly activating net outward current that transitioned into a rapidly increasing inward calcium current as the voltage approached −40 mV (Fig. 4c). These data indicated that the Ca_v3 channel current is sufficient to activate IK channels even at potentials normally subthreshold for spike activation.

To further explore the voltage-dependence for IK activation outside-out patch recordings were obtained from separate Purkinje cells using a step command from −110 to 0 mV under conditions with Ca_v3 and IK currents pharmacologically isolated. Potassium currents were then calculated as the difference from those evoked from a preceding step to −40 mV to inactivate the Ca_v3 current, or after perfusing 100 μM Ni^{2+} to block Ca_v3 current or 100 nM TRAM-34 to block IK channels. Plotting potassium currents isolated in this manner revealed a set of I-V relations expected for an LVA outward current over the range of ~−60 to–0 mV (Fig. 4d). Moreover, superimposing the current traces isolated under each of these conditions revealed a low voltage-activated and fast activating/fast inactivating IK current (Fig. 4e). These results were remarkable in showing that a non-activating calcium-dependent IK channel can be transformed into a low voltage-dependent and rapidly activating/inactivating current by virtue of their association with Ca_v3 calcium channels.

All isoforms of Ca_v3 channels are reportedly expressed in Purkinje cells and with selective expression of one $Ca_v3.1$ isoform in dendritic spines (McKay et al., 2006; Molineux et al., 2006; Hildebrand et al., 2009; Isope et al., 2012; Ly et al., 2016). Parallel fiber stimulation has also been reported to activate Cav3 calcium current presumably in spines as a significant source of calcium influx given the relative lack of calcium-conducting ligand-gated receptors (Isope & Murphy, 2005; Hildebrand et al., 2009; Isope et al., 2012; Ly et al., 2016). Protein biochemical and immunocytochemical work identified at least the $Ca_v3.2$ channel isoform as colocalizing and undergoing coimmunoprecipitation with IK channels (Engbers et al., 2012). To assess the influence of the Ca_v3-IK complex on parallel fiber-evoked spike output they recorded under voltage clamp the EPSC in Purkinje cell somata in response to direct parallel fiber stimulation. This allowed for subsequent injection of the EPSC waveform through the electrode in other cells to evoke a simulated parallel fiber EPSP as a single response or as a stimulus train (Fig. 5a). In this way the simEPSP was restricted to the postsynaptic cell, allowing bath perfusion of drugs that block Cav3 channels without concern for interfering with presynaptic roles of Ca_v3 channels. Activation of a simEPSP adjusted to just subthreshold for spike discharge as a 25 Hz train in Purkinje cells revealed an initial temporal summation over the first few pulses followed by a gradual decay towards resting membrane potential over the course of 50 pulses (Fig. 5b, c). As a result, the peak EPSP remained below the spike threshold throughout a stimulus train (Fig. 5b, c). However, perfusing 100 μM Ni^{2+} or 100 nM TRAM-34 to interfere with the Ca_v3-IK complex uncovered an EPSP temporal summation that rapidly surpassed the spike threshold throughout the entire stimulus train (Fig. 5b, c). Recordings of this nature thus revealed a key role for the Ca_v3-IK complex in suppressing temporal summation of postsynaptic parallel fiber EPSPs (Engbers et al., 2012).

Cerebellar Purkinje cells project to postsynaptic cells in the deep cerebellar nuclei or the vestibular nuclei and discharge over a range of frequencies to signal the outcome of signal processing in the cerebellar cortex. It is thus important to ensure faithful propagation of spike discharge down Purkinje cell axons. This requires the coordinated activation of sodium and potassium channels to repolarize spikes to prevent their inactivation at high-frequency rates. The finding of a Ca_v3-IK complex in Purkinje cells led to a sophisticated set of tests of its potential role in maintaining spike propagation past nodes of Ranvier in Purkinje cell axons (Grundemann & Clark, 2015). Previous work had established that Purkinje cells faithfully transmit spike discharge down the axon at frequencies up to ~250 Hz, including branch points where nodes of Ranvier have been localized (Khaliq et al., 2003; Khaliq & Raman, 2005; Monsivais et al., 2005).

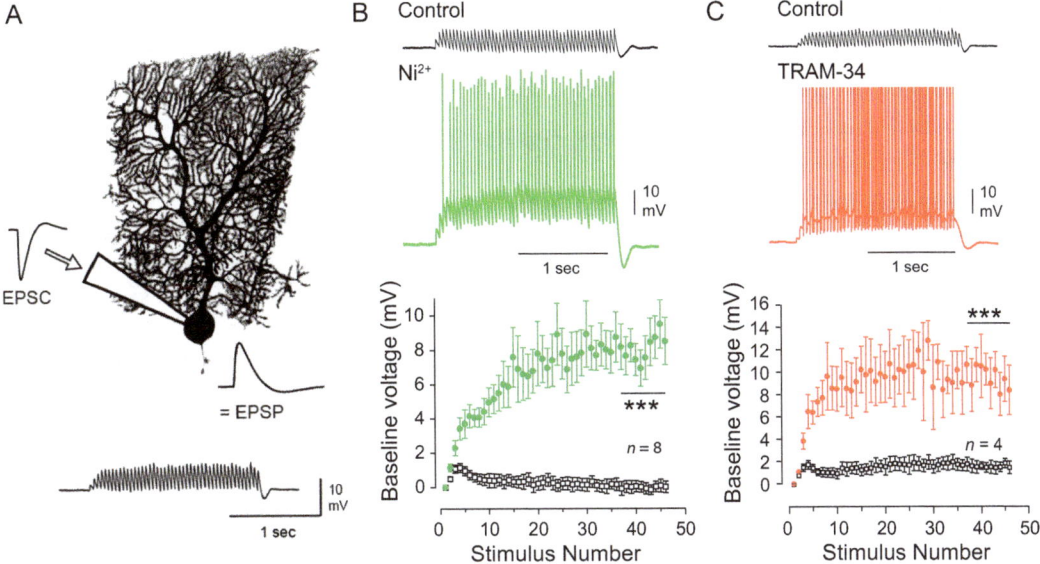

Fig. 5 The Ca$_v$3-IK complex suppresses temporal summation to reduce somatic spike firing in Purkinje cells. (**a**) Schematic diagram of a Purkinje cell and generation of a simulated parallel fiber EPSP by injecting the EPSC recorded in another cell under voltage clamp in response to direct parallel fiber stimulation. *Lower trace*, a compiled train of simEPSPs generated at 25 Hz and subthreshold for firing to inject into Purkinje cell somata. (**b, c**) Whole-cell recordings from two separate Purkinje cells of the response to a train of simEPSPs injected at the soma (Control) and after perfusion of 100 μM Ni^{2+} (**b**) or 100 nM TRAM-34 (**c**). The mean values of baseline voltage preceding each simEPSP are plotted below for control and drug-treated conditions. Spikes were truncated in (**c**). Average values are mean ± SEM. Statistical significance was tested for the last 10 pulses of stimulus trains in (**b, c**). ***$p < 0.001$. (Figure is modified from Engbers et al. (2012))

Using whole-cell recordings at Purkinje cell somata to dye fill the axon, a second on-cell recording could be obtained beyond an axonal branchpoint and node of Ranvier (Fig. 6a) (Grundemann & Clark, 2015). Successful transmission of spontaneous spike discharge at the soma could then be monitored in the axon beyond the node of Ranvier while disrupting the Cav3-IK complex. These tests showed that focal perfusion of 100 μM Ni^{2+} or 1 μM TRAM-34 near the branchpoint led to a rapid failure of spike propagation past the axon branchpoint without apparent effect on the baseline rate of somatic spike firing (Fig. 6b, c). Surprisingly, similar applications of apamin or iberiotoxin to block BK or SK channels, or even TEA at 10 mM did not induce failure of spike propagation despite the expected block by TEA of BK, K$_v$1, K$_v$3, and K$_v$7 channel subtypes.

Immunocytochemistry revealed IK immunolabel over extended lengths of Purkinje cell axons without apparent specific localization to nodes of Ranvier. To assess the distribution of Ca$_v$3 channel activation Grundemann and Clark (2015) used 2-photon imaging of calcium influx to detect increases in intracellular calcium during trains of evoked spikes, which proved to be centered within 5 μm of the nodes of Ranvier (Fig. 6d). The change in intracellular calcium was activity-dependent and cumulative in proportion to the rate of spike discharge at the Purkinje cell soma, with a decrease in nodal calcium signal upon somatic hyperpolarization (Fig. 6e). The calcium increase at nodes of Ranvier was further reduced by a low calcium medium or mibefradil but not by the P-type channel blocker AgTx. Further modeling studies suggested that the Ca$_v$3-IK interaction is sufficient to sustain spike conduction in Purkinje cell axons.

The work on Purkinje cells thus established the presence of a Ca$_v$3.2-IK channel complex where Ca$_v$3 channel activation imparts a low

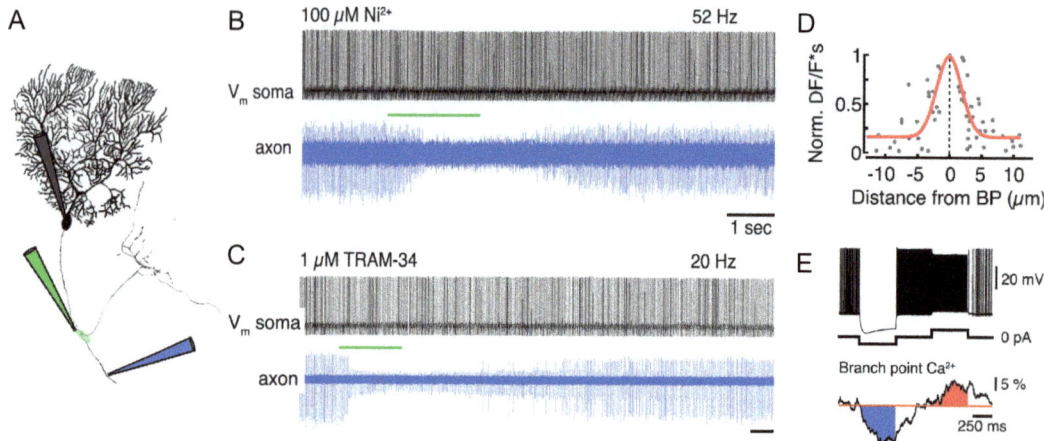

Fig. 6 The Ca_v3-IK complex ensures axonal spike propagation at Purkinje cell axon nodes of Ranvier. (**a**) Schematic diagram of a Purkinje cell with a simultaneous whole-cell recording at the soma (*grey*), a cell-attached axon recording site (*blue*), and drug ejection electrode positioned near a node of Ranvier and axonal branch site (*green*). (**b, c**) Simultaneous whole-cell somatic and cell-attached axon recording of a spontaneously firing Purkinje cell before and after ejection of Ni^{2+} (**b**) or TRAM-34 (**c**) (*green bar*), as per recording configuration shown in (**a**). Baseline Purkinje cell firing rate is indicated above somatic traces. (**d**) Normalized integrated DF/F (DF/F*s) of calcium increases against distance from an axonal branchpoint reveals calcium transients highly localized around the node of Ranvier. (**e**) Step changes in current injected at a somatic whole-cell recording site modulates spike firing frequency and the corresponding DF/F measured calcium transients at a distant node of Ranvier/axon branchpoint reveal activity-dependent changes in DF/F. (Figure is modified from Grundemann and Clark (2015))

voltage-activation of IK channel current, and with fast activating, fast inactivating kinetics. The end result is to allow Ca_v3 channels to be activated even by parallel fiber input to generate a sAHP that modulates temporal summation during repetitive inputs. Through specific localization of Ca_v3-IK influence at nodes of Ranvier, this complex further acts as a key factor in maintaining spike propagation down the axons of the principal output neuron of the cerebellar cortex.

$Ca_v3.2$-BK Complex

Another calcium-gated potassium channel with a large conductance of up to 200 pS (big conductance, BK, KCa1.1) is distinct from either Kv4 or IK channels in exhibiting seven transmembrane domains and structural features of a voltage-gated channel (Fig. 1). The N-terminus is positioned on an S0 subunit that faces the extracellular space, and an internal C-terminus contains an RCK domain for calcium sensing (Fig. 1). BK channels are responsive to both internal calcium

and membrane voltage in a synergistic manner, typically responding to relatively higher levels of internal calcium increases (1–10 μM) and activating near −20 mV as a non-inactivating current (Womack & Khodakhah, 2002; Latorre & Brauchi, 2006; Berkefeld et al., 2010; Lee & Cui, 2010). As a result, BK channels are expected to be activated by spike-associated depolarizations to contribute to spike repolarization and generation of a fast AHP, with specific interactions with $Ca_v2.1$ (P/Q-type) and $Ca_v2.2$ (N -type) calcium channels (Shao et al., 1999; Smith et al., 2002; Womack & Khodakhah, 2002; Sun et al., 2003; Goldberg & Wilson, 2005; Berkefeld et al., 2006; Berkefeld & Fakler, 2008). An association between an HVA calcium channel and BK channels also extends to the lower voltage-activated $Ca_v1.3$ channel isoform in dopaminergic neurons and chromaffin cells (Marcantoni et al., 2010; Vandael et al., 2010).

In the brainstem vestibular nuclei, BK channel activation has been associated with modifying the gain of the vestibular ocular reflex (VOR) (Smith et al., 2002). This is relevant to cerebellar function in that Purkinje cells of caudal cerebellar lobules project directly to cells in the medial vestibular nucleus (MVN). Interestingly, a potential association between the LVA $Ca_v3.2$ calcium and BK channels had been suggested in detecting coimmunoprecipitation between these proteins in brain lysates (Chen et al., 2003). A functional coupling in MVN neurons had further been implicated in a report that the BK-mediated AHP was sensitive to the block of Ca_v3 conductance by Ni^{2+} application (Smith et al., 2002). Recent work now establishes that $Ca_v3.2$ channels also form a close association with BK channels at the molecular level to invoke LVA activation and fast inactivation of BK outward current (Rehak et al., 2013).

The interactions between $Ca_v3.2$ and BK channels were investigated using tsA-201 cells and in MVN cells of rat tissue slices in vitro. Here it was found that expressing the BK channel alpha subunit alone in tsA-201 cells resulted in a small amplitude non-inactivating outward current in response to a step from −100 to +40 mV (Fig. 7a). To test the potential influence of $Ca_v3.2$ calcium influx, $Ca_v3.2$ and BK channels were coexpressed (Rehak et al., 2013). This process alone proved to be insufficient to promote any further activation of BK current using a step to +40 mV (250 msec). Yet, adding a second step command to +40 mV (250 msec interval) that was immediately preceded by a step to −30 mV to activate $Ca_v3.2$ calcium conductance evoked a BK current 2.7 fold greater in amplitude (Fig. 7a). Moreover, the BK current exhibited a rapid rate of decay ($\tau = 117$ msec) over the 250 msec pulse duration (Fig. 7a). Current density-voltage plots further showed a dramatic leftward shift in the activation voltage of BK current (Fig. 7b). By comparison, these effects were not found if BK channels were coexpressed with a $Ca_v3.2$ pore mutant that could not conduct calcium (Fig. 7a). Comparing the $Ca_v3.2$ current evoked by a series of preceding step commands and the P2/P1 ratio for BK current revealed a very close correspondence between the voltage-dependence and magnitude of BK current to that of Ca_v3 current over a range of −90 to +20 mV (Fig. 7c). Inspection of the BK or $Ca_v3.2$ current by calculating the dif-

ference currents after applying Ca_v3 or BK blockers showed that the close relationship between $Ca_v3.2$ and BK activation extended to kinetic properties. Specifically, when $Ca_v3.2$ and BK channels were coexpressed BK current adopted a fast activation, fast inactivation profile characteristic of the LVA T-type Ca_v3 current (Fig. 7d). A closer examination of the initial voltage for activation of either current was gained using a ramp command and pharmacological isolation of either current (Fig. 7e). With this, it was apparent that the initial activation of $Ca_v3.2$ current at ~−70 mV was very closely associated with the initial voltage for activation of BK current. This is important in revealing a substantial shift in BK activation voltage that will allow it to contribute to both LVA and HVA-related cellular events. By using current-clamp recording conditions in MVN cells in vitro, the application of mibefradil reduced both the rate of spike repolarization and the amplitude of the subsequent fast AHP (Fig. 7f), increasing the gain of firing frequency output by 1.4 times (Rehak et al., 2013).

Together these data indicate that the characteristics of BK current were closely linked to the activation voltage, time-course, and rate of inactivation of $Ca_v3.2$ calcium channels as the source of internal calcium concentration increases. The result was to effectively convert a BK channel current from a non-inactivating and relatively high voltage-activated current to one more characteristic of an LVA A-type potassium current.

Perspectives and Conclusions

The existence of an A-type potassium channel current phenotype has long been recognized for its unique biophysical properties of low voltage for activation and fast activation/inactivation that underlies its important role in regulating spike output. While the expression of an A-type current has been considered almost ubiquitous in central neurons, it was seemingly represented by only a small cohort of isoforms across 3 of the 4 most common voltage-gated potassium channel families ($K_v1.4$, $K_v3.4$, and $K_v4.1$–3). This chapter has highlighted how Ca_v3 channels can effectively

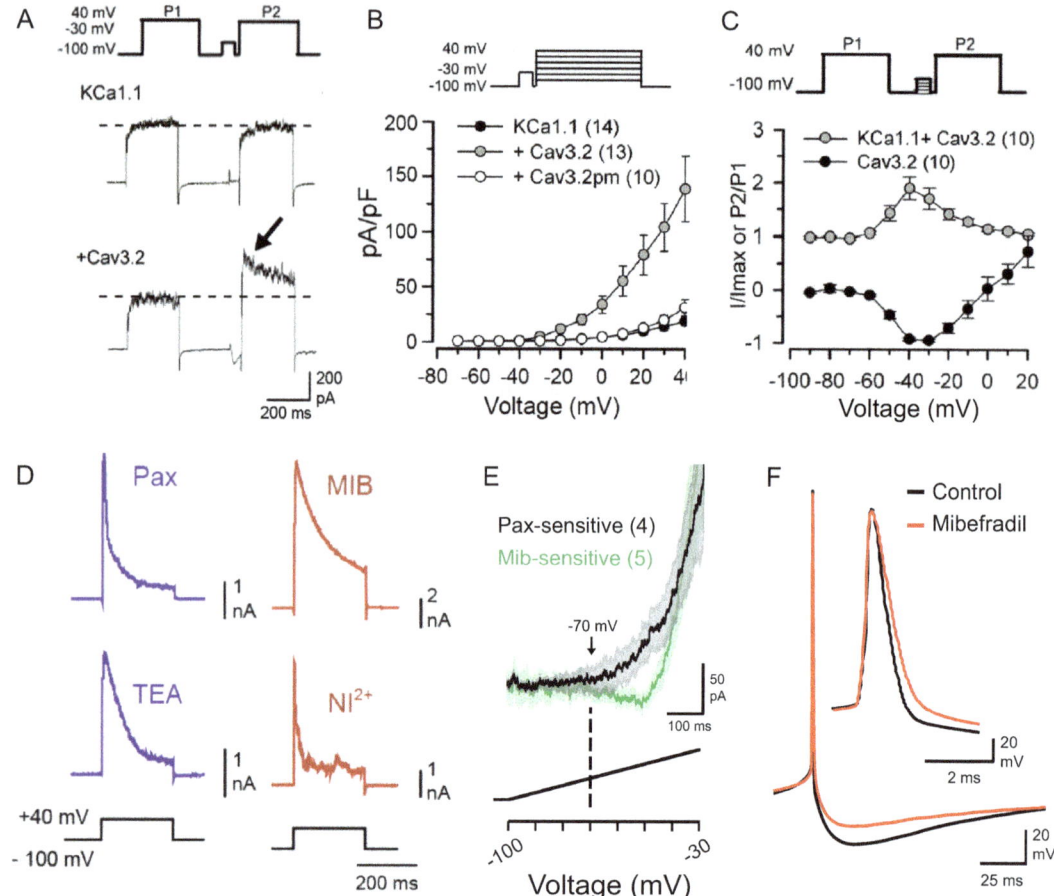

Fig. 7 A Ca$_v$3.2-BK channel complex in medial vestibu- lar neurons imparts low voltage-dependence and an A-type channel profile to BK current. (**a**) Whole-cell recordings of BK current in tsA-201 cells with or without coexpression of Ca$_v$3.2. BK channel activity was tested using two 250 msec steps to +40 mV (P1, P2) with a 50 msec prepulse to −30 mV immediately prior to P2 to activate Ca$_v$3 current. In the presence of Ca$_v$3.2 BK activa- tion is facilitated on P2 (*arrow*) by the prepulse used to maximally activate Ca$_v$3.2 current. (**b**) Average current- voltage plots of BK current on P2 indicate that a prepulse to −30 mV in the presence of Cav3 channels significantly left-shifts activation voltage and increases BK current. BK activation is unaltered when coexpressed with a non- conducting calcium channel mutant (Ca$_v$3.2 pm). (**c**) Plots of whole-cell current in tsA-201 cells for Ca$_v$3.2 expressed in isolation and the P2/P1 ratio of BK current as a function of pre-pulse command voltage stepped in 10 mV incre-

ments from −90 to +20 mV. Note the close correspon- dence in voltage dependence and magnitude of Ca$_v$3.2 and BK currents. (**d**) Whole-cell recordings from MVN cells in vitro under conditions that pharmacologically isolate Ca$_v$3 and BK channel activity. Shown are difference cur- rents in separate cells after selective block of BK current by 1 mM paxilline (Pax) or 1 mM TEA, and block of Ca$_v$3 channels by 1 μM mibefradil (MIB) or 300 μM Ni^{2+}. All recordings included 30 μM Cd^{2+} to prevent calcium con- ductance through HVA calcium channels. (**e**) Mean I-V plots of different currents in MVN cells recorded in response to a ramp command to identify paxilline- sensitive BK current or mibefradil-sensitive Ca$_v$3 current. (**f**) Current clamp recordings of spike discharge in MVN cells show a reduction of spike repolarization (*inset*) and a subsequent fast AHP by 1 μM mibefradil. Average values are mean ± SEM, shown as *shaded regions* on current traces in (**e**). (Figure is modified from Rehak et al. (2013))

create calcium-dependent A-type current in three additional classes of potassium channels expressed in different cerebellar cell types. By

forming a close association at the molecular level, one can detect an effective transfer of bio- physical properties from Ca$_v$3 channels to each of

K_v4, IK, and BK potassium channel isoforms. These different potassium channels then function effectively as calcium-dependent A-type potassium channels with the rich set of characteristics inherent to a fast activating, fast inactivating channel. Thus, the voltage-dependent class of K_v4 channels acquires a calcium dependence through Ca_v3 calcium interaction with KChIP3 to regulate cell excitability in stellate and granule cells. The IK channel in Purkinje cells instead transforms from a calcium-dependent but a voltage-independent channel to one that exhibits a very low voltage for activation with fast inactivation to modulate synaptic input and spike output. Finally, pairing $Ca_v3.2$ channels with the voltage-dependent but non-inactivating BK channel imparts a low voltage for activation and fast inactivation to control the gain of spike firing by modulating spike repolarization and a fast AHP in MVN neurons.

The current review focused on the effects of these channel interactions on spike output. But the existence of these Ca_v3-K channel pairings extends well beyond this to other aspects of signal processing. For instance, the influence of a Ca_v3-K channel coupling on cell activity can differ even within even a single cell according to regional localization. Thus, the Ca_v3-IK interaction in the soma of Purkinje cells produces a sAHP that reduces the temporal summation of parallel fiber input. In contrast, the same interaction centered on the nodes of Ranvier instead ensures faithful propagation of sodium spikes over Purkinje cell axons. The Ca_v3-K_v4-KChIP3 complex proves to be highly sensitive to shifts in external calcium concentrations during repetitive climbing fiber input that reduces external calcium. The reduction in external calcium (~0.4 mM) then lowers Ca_v3 calcium conductance to promote the left shift in K_v4 Vh to increase the rate of stellate cell firing (Anderson et al., 2013). In granule cells, a Ca_v3 calcium-induced shift in K_v4 Vh can be induced on a long-term basis through an ERK-dependent process to promote long-term potentiation of granule cell excitability to mossy fiber input (Rizwan et al., 2016). The influence of the Ca_v3-K_v4-KChIP3 complex on signal processing is further deter-

mined by the relative expression pattern of each of these subunits across cerebellar lobules. Thus, a high density of $Ca_v3.1$ and KChIP3 expression in lobule 9 tunes granule cells to respond to oscillatory-like signals related to vestibular input. But a lower expression of Ca_v3 channels in lobule 2 granule cells reduces the effectiveness of the Ca_v3-K_v4 complex to instead select for burst-like patterns of mossy fiber input (Heath et al., 2014). Since these subunits exhibit different expression patterns in various brain regions the role of Ca_v-K_v4 interactions could well be cell-specific or adjusted to produce different outcomes in the same class of cells.

K_v4, IK, and BK channels differ in terms of the calcium sensor and thus relative sensitivity to changes in calcium concentration. The N-terminus of K_v4 channels is known to complex with KChIP3, IK channels bind calmodulin at the C-terminus, while BK channels employ an RCK domain on the C-terminus (Fig. 1). Nanometer proximity of Ca_v3 channels with K_v4 and IK channels that incorporate the relatively sensitive KChIP3 and calmodulin as calcium sensors ensure rapid responses to changes in cell activity. Interestingly, the RCK domain of BK channels can require elevations of intracellular calcium in the order of 1–10 µM before modulating channel activity. The relatively low conductance and fast inactivation of Ca_v3 channels reduces the domain of calcium increase associated with channel opening. As a result, interesting interactions can occur in which virtually no effect is detected on BK channel properties with simple coexpression of $Ca_v3.2$ (Fig. 7a). Yet providing a prepulse to initiate Ca_v3 calcium influx immediately before activating BK channels reveals a distinct influence by imparting a low voltage for BK channel activation. This interaction thus provides the ability to participate in cell functions at membrane voltages not previously achieved even with the reciprocal interaction between HVA calcium influx and voltage inherent to BK channels. Modeling to investigate this process revealed that it reflects a microdomain interaction that requires the concerted actions of multiple Ca_v3 channels (Engbers et al., 2013b; Rehak et al., 2013). The functions enabled by an association between

Ca$_v$3.2 and BK channels are thus distinctly different from that found for coupling to HVA Ca$_v$1.2 and Ca$_v$1.3, Ca$_v$2.1 and Ca$_v$2.2 channels (Prakriya & Lingle, 1999; Grunnet & Kaufmann, 2004; Berkefeld et al., 2006; Loane et al., 2007; Berkefeld & Fakler, 2008).

The full slate of potassium channel subtypes that might be modulated by colocalizing Ca$_v$3 calcium and potassium channels remains to be determined. The current review focused only on cell types within the cerebellum and one of its primary output targets in the MVN. Ca$_v$3 channels have also been shown to at least functionally couple to SK calcium-dependent potassium channels in Purkinje cells as well as dopaminergic midbrain neurons (Wolfart et al., 2001; Wolfart & Roeper, 2002; Cueni et al., 2009; Ait Ouares et al., 2019). But the extent to which this reflects a direct interaction at the molecular level to impart Ca$_v$3 channel LVA properties and fast inactivation on SK channels is not fully known. The widespread expression of A-type potassium current in cells across the CNS speaks to the importance of this phenotype of potassium channel output. The ability for Ca$_v$3 calcium channels to effectively generate 3 new forms of A-type potassium channels that are both calcium- and voltage-dependent will substantially increase the range of functions that can be realized by A-type potassium currents.

Acknowledgments We gratefully acknowledge J. Forden, L. Chen, and M Kruskic for their technical assistance in these studies and numerous trainees in the Turner lab who contributed to this work. These studies were supported by operating grants from the Canadian Institutes for Health Research.

References

Aguado, C., García-Madrona, S., Gil-Minguez, M., & Luján, R. (2016). Ontogenic changes and differential localization of T-type Ca(2+) channel subunits Cav3.1 and Cav3.2 in mouse hippocampus and cerebellum. *Frontiers in Neuroanatomy, 10*(83), 1–16.

Ait Ouares, K., Filipis, L., Tzilivaki, A., Poirazi, P., & Canepari, M. (2019). Two distinct sets of Ca2+ and K+ channels are activated at different membrane potentials by the climbing fiber synaptic potential in Purkinje

neuron dendrites. *The Journal of Neuroscience, 39*, 1969–1981.

Anderson, D., Mehaffey, W. H., Iftinca, M., Rehak, R., Engbers, J. D., Hameed, S., Zamponi, G. W., & Turner, R. W. (2010a). Regulation of neuronal activity by Cav3-Kv4 channel signaling complexes. *Nature Neuroscience, 13*, 333–337.

Anderson, D., Rehak, R., Hameed, S., Mehaffey, W. H., Zamponi, G. W., & Turner, R. W. (2010b). Regulation of the KV4.2 complex by CaV3.1 calcium channels. *Channels, 4*, 163–167.

Anderson, D., Engbers, J. D., Heath, N. C., Bartoletti, T. M., Mehaffey, W. H., Zamponi, G. W., & Turner, R. W. (2013). The Cav3-Kv4 complex acts as a calcium sensor to maintain inhibitory charge transfer during extracellular calcium fluctuations. *The Journal of Neuroscience, 33*, 7811–7824.

Berkefeld, H., & Fakler, B. (2008). Repolarizing responses of BKCa-Cav complexes are distinctly shaped by their Cav subunits. *The Journal of Neuroscience, 28*, 8238–8245.

Berkefeld, H., Sailer, C. A., Bildl, W., Rohde, V., Thumfart, J.-O., Eble, S., Klugbauer, N., Reisinger, E., Bischofberger, J., Oliver, D., Knaus, H.-G., Schulte, U., & Fakler, B. (2006). BKCa-Cav channel complexes mediate rapid and localized Ca2+-activated K+ signaling. *Science, 314*, 615–620.

Berkefeld, H., Fakler, B., & Schulte, U. (2010). Ca2+-activated K+ channels: From protein complexes to function. *Physiological Reviews, 90*, 1437–1459.

Bossu, J. L., Fagni, L., & Feltz, A. (1989). Voltage-activated calcium channels in rat Purkinje cells maintained in culture. *Pflügers Archiv, 414*, 92–94.

Chen, C.-C., Lamping, K. G., Nuno, D. W., Barresi, R., Prouty, S. J., Lavoie, J. L., Cribbs, L. L., England, S. K., Sigmund, C. D., Weiss, R. M., Williamson, R. A., Hill, J. A., & Campbell, K. P. (2003). Abnormal coronary function in mice deficient in alpha1H T-type Ca2+ channels. *Science, 302*, 1416–1418.

Cueni, L., Canepari, M., Lujan, R., Emmenegger, Y., Watanabe, M., Bond, C. T., Franken, P., Adelman, J. P., & Luthi, A. (2008). T-type Ca2+ channels, SK2 channels and SERCAs gate sleep-related oscillations in thalamic dendrites. *Nature Neuroscience, 11*, 683–692.

Cueni, L., Canepari, M., Adelman, J. P., & Luthi, A. (2009). Ca(2+) signaling by T-type Ca(2+) channels in neurons. *Pflügers Archiv, 457*, 1161–1172.

Davie, J. T., Clark, B. A., & Hausser, M. (2008). The origin of the complex spike in cerebellar Purkinje cells. *The Journal of Neuroscience, 28*, 7599–7609.

Engbers, J. D., Anderson, D., Asmara, H., Rehak, R., Mehaffey, W. H., Hameed, S., McKay, B. E., Kruskic, M., Zamponi, G. W., & Turner, R. W. (2012). Intermediate conductance calcium-activated potassium channels modulate summation of parallel fiber input in cerebellar Purkinje cells. *Proceedings of the National Academy of Sciences of the United States of America, 109*, 2601–2606.

Engbers, J. D. T., Fernandez, F. R., & Turner, R. W. (2013a). Bistability in Purkinje neurons: Ups and downs in cerebellar research. *Neural Networks, 47,* 18–31.

Engbers, J. D., Zamponi, G. W., & Turner, R. W. (2013b). Modeling interactions between voltage-gated Ca (2+) channels and KCa1.1 channels. *Channels, 7,* 524–529.

Fernandez, F. R., Engbers, J. D., & Turner, R. W. (2007). Firing dynamics of cerebellar purkinje cells. *Journal of Neurophysiology, 98,* 278–294.

Gittis, A. H., Moghadam, S. H., & du Lac, S. (2010). Mechanisms of sustained high firing rates in two classes of vestibular nucleus neurons: Differential contributions of resurgent Na, Kv3, and BK currents. *Journal of Neurophysiology, 104,* 1625–1634.

Goldberg, J. A., & Wilson, C. J. (2005). Control of spontaneous firing patterns by the selective coupling of calcium currents to calcium-activated potassium currents in striatal cholinergic interneurons. *The Journal of Neuroscience, 25,* 10230–10238.

Grundemann, J., & Clark, B. A. (2015). Calcium-activated potassium channels at nodes of Ranvier secure axonal spike propagation. *Cell Reports, 12,* 1715–1722.

Grunnet, M., & Kaufmann, W. A. (2004). Coassembly of big conductance Ca2+-activated K+ channels and L-type voltage-gated Ca2+ channels in rat brain. *The Journal of Biological Chemistry, 279,* 36445–36453.

Heath, N. C., Rizwan, A. P., Engbers, J. D., Anderson, D., Zamponi, G. W., & Turner, R. W. (2014). The expression pattern of a Cav3-Kv4 complex differentially regulates spike output in cerebellar granule cells. *The Journal of Neuroscience, 34,* 8800–8812.

Hildebrand, M. E., Isope, P., Miyazaki, T., Nakaya, T., Garcia, E., Feltz, A., Schneider, T., Hescheler, J., Kano, M., Sakimura, K., Watanabe, M., Dieudonne, S., & Snutch, T. P. (2009). Functional coupling between mGluR1 and Cav3.1 T-type calcium channels contributes to parallel fiber-induced fast calcium signaling within Purkinje cell dendritic spines. *The Journal of Neuroscience, 29,* 9668–9682.

Isope, P., & Murphy, T. H. (2005). Low threshold calcium currents in rat cerebellar Purkinje cell dendritic spines are mediated by T-type calcium channels. *The Journal of Physiology, 562,* 257–269.

Isope, P., Hildebrand, M. E., & Snutch, T. P. (2012). Contributions of T-type voltage-gated calcium channels to postsynaptic calcium signaling within Purkinje neurons. *Cerebellum, 11,* 651–665.

Khaliq, Z. M., & Raman, I. M. (2005). Axonal propagation of simple and complex spikes in cerebellar Purkinje neurons. *The Journal of Neuroscience, 25,* 454–463.

Khaliq, Z. M., Gouwens, N. W., & Raman, I. M. (2003). The contribution of resurgent sodium current to high-frequency firing in Purkinje neurons: An experimental and modeling study. *The Journal of Neuroscience, 23,* 4899–4912.

Latorre, R., & Brauchi, S. (2006). Large conductance Ca2+-activated K+ (BK) channel: Activation by Ca2+ and voltage. *Biological Research, 39,* 385–401.

Lee, U. S., & Cui, J. (2010). BK channel activation: Structural and functional insights. *Trends in Neurosciences, 33,* 415–423.

Lima, P. A., & Marrion, N. V. (2007). Mechanisms underlying activation of the slow AHP in rat hippocampal neurons. *Brain Research, 1150,* 74–82.

Loane, D. J., Lima, P. A., & Marrion, N. V. (2007). Co-assembly of N-type Ca2+ and BK channels underlies functional coupling in rat brain. *Journal of Cell Science, 120,* 985–995.

Ly, R., Bouvier, G., Szapiro, G., Prosser, H. M., Randall, A. D., Kano, M., Sakimura, K., Isope, P., Barbour, B., & Feltz, A. (2016). Contribution of postsynaptic T-type calcium channels to parallel fibre-Purkinje cell synaptic responses. *The Journal of Physiology, 594,* 915–936.

Marcantoni, A., Vandael, D. H., Mahapatra, S., Carabelli, V., Sinnegger-Brauns, M. J., Striessnig, J., & Carbone, E. (2010). Loss of Cav1.3 channels reveals the critical role of L-type and BK channel coupling in pacemaking mouse adrenal chromaffin cells. *The Journal of Neuroscience, 30,* 491–504.

Marrion, N. V., & Tavalin, S. J. (1998). Selective activation of Ca2+-activated K+ channels by co-localized Ca2+ channels in hippocampal neurons. *Nature, 395,* 900–905.

McKay, B. E., McRory, J. E., Molineux, M. L., Hamid, J., Snutch, T. P., Zamponi, G. W., & Turner, R. W. (2006). Ca(V)3 T-type calcium channel isoforms differentially distribute to somatic and dendritic compartments in rat central neurons. *The European Journal of Neuroscience, 24,* 2581–2594.

McKay, B. E., Engbers, J. D. T., Mehaffey, W. H., Gordon, G. R. J., Molineux, M. L., Bains, J. S., & Turner, R. W. (2007). Climbing fiber discharge regulates cerebellar functions by controlling the intrinsic characteristics of purkinje cell output. *Journal of Neurophysiology, 97,* 2590–2604.

Molineux, M. L., Fernandez, F. R., Mehaffey, W. H., & Turner, R. W. (2005). A-type and T-type currents interact to produce a novel spike latency-voltage relationship in cerebellar stellate cells. *The Journal of Neuroscience, 25,* 10863–10873.

Molineux, M. L., McRory, J. E., McKay, B. E., Hamid, J., Mehaffey, W. H., Rehak, R., Snutch, T. P., Zamponi, G. W., & Turner, R. W. (2006). Specific T-type calcium channel isoforms are associated with distinct burst phenotypes in deep cerebellar nuclear neurons. *Proceedings of the National Academy of Sciences of the United States of America, 103,* 5555–5560.

Monsivais, P., Clark, B. A., Roth, A., & Hausser, M. (2005). Determinants of action potential propagation in cerebellar Purkinje cell axons. *The Journal of Neuroscience, 25,* 464–472.

Mouginot, D., Bossu, J. L., & Gahwiler, B. H. (1997). Low-threshold Ca2+ currents in dendritic recordings from Purkinje cells in rat cerebellar slice cultures. *The Journal of Neuroscience, 17,* 160–170.

Müller, A., Kukley, M., Uebachs, M., Beck, H., & Dietrich, D. (2007). Nanodomains of single Ca2+

channels contribute to action potential repolarization in cortical neurons. *The Journal of Neuroscience, 27,* 483–495.

Ohtsuki, G., Piochon, C., Adelman, J. P., & Hansel, C. (2012). SK2 channel modulation contributes to compartment-specific dendritic plasticity in cerebellar Purkinje cells. *Neuron, 75,* 108–120.

Poolos, N. P., & Johnston, D. (1999). Calcium-activated potassium conductances contribute to action potential repolarization at the soma but not the dendrites of hippocampal CA1 pyramidal neurons. *The Journal of Neuroscience, 19,* 5205–5212.

Prakriya, M., & Lingle, C. J. (1999). BK channel activation by brief depolarizations requires Ca2+ influx through L- and Q-type Ca2+ channels in rat chromaffin cells. *Journal of Neurophysiology, 81,* 2267–2278.

Rehak, R., Bartoletti, T. M., Engbers, J. D., Berecki, G., Turner, R. W., & Zamponi, G. W. (2013). Low voltage activation of KCa1.1 current by Cav3-KCa1.1 complexes. *PLoS One, 8,* e61844.

Rizwan, A. P., Zhan, X., Zamponi, G. W., & Turner, R. W. (2016). Long-term potentiation at the mossy fiber-granule cell relay invokes postsynaptic second-messenger regulation of Kv4 channels. *The Journal of Neuroscience, 36,* 11196–11207.

Robitaille, R., Garcia, M. L., Kaczorowski, G. J., & Charlton, M. P. (1993). Functional colocalization of calcium and calcium-gated potassium channels in control of transmitter release. *Neuron, 11,* 645–655.

Sahu, G., & Turner, R. W. (2021). The molecular basis for the calcium-dependent sAHP in CA1 hippocampal pyramidal cells. *Frontiers in Physiology, 1,* 1–25.

Sahu, G., Asmara, H., Zhang, F. X., Zamponi, G. W., & Turner, R. W. (2017). Activity-dependent facilitation of CaV1.3 calcium channels promotes KCa3.1 activation in hippocampal neurons. *The Journal of Neuroscience, 37,* 11255–11270.

Sahu, G., Wazen, R.-M., Colarusso, P., Chen, S. R. W., Zamponi, G. W., & Turner, R. W. (2019). Junctophilin proteins tether a Cav1-RyR2-KCa3.1 tripartite complex to regulate neuronal excitability. *Cell Reports, 28,* 2427–2442.e6.

Schmolesky, M. T., Weber, J. T., De Zeeuw, C. I., & Hansel, C. (2002). The making of a complex spike: Ionic composition and plasticity. *Annals of the New York Academy of Sciences, 978,* 359–390.

Shao, L. R., Halvorsrud, R., Borg-Graham, L., & Storm, J. F. (1999). The role of BK-type Ca2+-dependent K+ channels in spike broadening during repetitive firing in rat hippocampal pyramidal cells. *The Journal of Physiology, 521*(Pt 1), 135–146.

Smith, M. R., Nelson, A. B., & Du Lac, S. (2002). Regulation of firing response gain by calcium-dependent mechanisms in vestibular nucleus neurons. *Journal of Neurophysiology, 87,* 2031–2042.

Sun, X., Gu, X. Q., & Haddad, G. G. (2003). Calcium influx via L- and N-type calcium channels activates a transient large-conductance Ca2+-activated K+ current in mouse neocortical pyramidal neurons. *The Journal of Neuroscience, 23,* 3639–3648.

Swensen, A. M., & Bean, B. P. (2003). Ionic mechanisms of burst firing in dissociated Purkinje neurons. *The Journal of Neuroscience, 23,* 9650–9663.

Talley, E. M., Cribbs, L. L., Lee, J. H., Daud, A., Perez-Reyes, E., & Bayliss, D. A. (1999). Differential distribution of three members of a gene family encoding low voltage-activated (T-type) calcium channels. *The Journal of Neuroscience, 19,* 1895–1911.

Turner, R. W., & Zamponi, G. W. (2014). T-type channels buddy up. Invited review for Special issue, T-type (Cav3) Calcium channels in health and disease, *European Journal of Physiology, 466*(4), 661–675.

Vandael, D. H., Marcantoni, A., Mahapatra, S., Caro, A., Ruth, P., Zuccotti, A., Knipper, M., & Carbone, E. (2010). Ca(v)1.3 and BK channels for timing and regulating cell firing. *Molecular Neurobiology, 42,* 185–198.

Wolfart, J., & Roeper, J. (2002). Selective coupling of T-type calcium channels to SK potassium channels prevents intrinsic bursting in dopaminergic midbrain neurons. *The Journal of Neuroscience, 22,* 3404–3413.

Wolfart, J., Neuhoff, H., Franz, O., & Roeper, J. (2001). Differential expression of the small-conductance, calcium-activated potassium channel SK3 is critical for pacemaker control in dopaminergic midbrain neurons. *The Journal of Neuroscience, 21,* 3443–3456.

Womack, M. D., & Khodakhah, K. (2002). Characterization of large conductance Ca2+-activated K+ channels in cerebellar Purkinje neurons. *The European Journal of Neuroscience, 16,* 1214–1222.

Part III

(Patho)physiology of VGCCs

Voltage-Gated Ca²⁺ Channels. Lessons from Knockout and Knock-in Mice

Jörg Striessnig, Akito Nakao, and Yasuo Mori

Abstract

Functional diversification of voltage-gated Ca^{2+} channels is underlain by the existence of multiple α_1-subunit-encoding genes. To define Ca^{2+} channel types and to understand their physiological significance, sensitivity to pharmacological agents has been the most prevailing criterion. However, not all the types enjoyed the merit of pharmacology, as we didn't have selective high-affinity blockers for T-, Q-, and certain L-types, when they were first distinguished on the basis of biophysical properties and resistance to selective blockers of L-, N-, and P-types. Moreover, the number of α_1-subunit genes, which doubles that of well-established types, as well as channel modulation by auxiliary subunits, strongly suggested that further functional diversity should be considered to study the physiology of each Ca^{2+} channel type. To address these issues, genetically engineered knockout mice were instrumental to reveal that Ca^{2+} channels with distinct α_1-subunits not only play specific functions but also share common roles with other Ca^{2+} channels in controlling physiological processes. Knock-in mice with human mutations have allowed us to integratively understand how specific mutations cause neurological phenotypes by altering Ca^{2+} channel function and neuronal processes such as conduction and synaptic transmission. In this chapter, we will encyclopedically describe how existing Ca^{2+} channel α_1-subunit mutant mice have deepened our insights into the functional diversity of Ca^{2+} channels and their individual physiological and pathophysiological functions.

Keywords

α_1-subunit · Knockout mice · Knock-in mouse · Disease model · Channelopathy

J. Striessnig (✉)
Pharmacology and Toxicology, Institute of Pharmacy, Center of Molecular Biosciences, Center for Chemistry and Biomedicine, University of Innsbruck, Innsbruck, Austria
e-mail: joerg.striessnig@uibk.ac.at

A. Nakao · Y. Mori (✉)
Laboratory of Molecular Biology, Department of Synthetic Chemistry and Biological Chemistry, Graduate School of Engineering, Kyoto University, Kyoto, Japan
e-mail: mori@sbchem.kyoto-u.ac.jp

Introduction

Voltage-gated Ca^{2+} channels (VGCCs) are expressed in all electrically excitable cells and even in many other cells in which graded changes in membrane potential are coupled to intracellular Ca^{2+} signals. Due to their high relevance for vital physiological functions, David T. Yue, a pioneer of VGCC research, even regarded them as the "queen of ion channels" as follows: "Voltage-gated Ca²⁺ channels are arguably the queen of ion channels from the standpoint of neurobiological function—they not only sculpt membrane electrical waveforms as do other channels but also serve as gatekeepers of the ubiquitous second messenger Ca²⁺"(Yue, 2004). Indeed, VGCCs are key signaling molecules not only by functioning as Ca²⁺-selective pores opened by membrane depolarization, but the conformational rearrangements of their voltage sensors can also directly trigger Ca²⁺ release from intracellular stores, as prototypically shown for the initiation of skeletal muscle contraction (Tanabe et al., 1988).

As described in other chapters of this book, for proper physiological function, these channels must be targeted to specialized subcellular compartments and require sophisticated fine-tuning by alternative splicing, accessory subunits, calmodulin, cell-specific protein interactions, and posttranslational modifications (such as phosphorylation and glycosylation) (Zamponi et al., 2015). In addition, one channel cannot

serve the many VGCC functions, not even in the same cell. For example, in a neuron, different Ca²⁺ channels are required for fast neurotransmitter release from presynaptic active zones ("excitation-secretion coupling"), for Ca²⁺-dependent modulation of long-term changes in gene expression ("excitation–transcription coupling") and for shaping electrical firing patterns. To support this diversity, 10 distinct pore-forming α₁-subunit genes defining different subtypes of VGCCs have evolved, which—based on sequence homology (see Fig. 1) and pharmacological similarities (Zamponi et al., 2015)—can be grouped into three families: Ca$_V$1 (L-type), Ca$_V$2, and (low-voltage-activated) Ca$_V$3 (T-type) channels. For more details on the different families, including their encoding genes, the association with modulatory subunits or protein interaction partners, their pharmacological and functional properties, please refer to the Guide to Pharmacology (Alexander et al., 2021; www.guidetopharmacology.org) and other chapters in this book.

Given this biochemical complexity of VGCCs, only genetically modified mice could help to

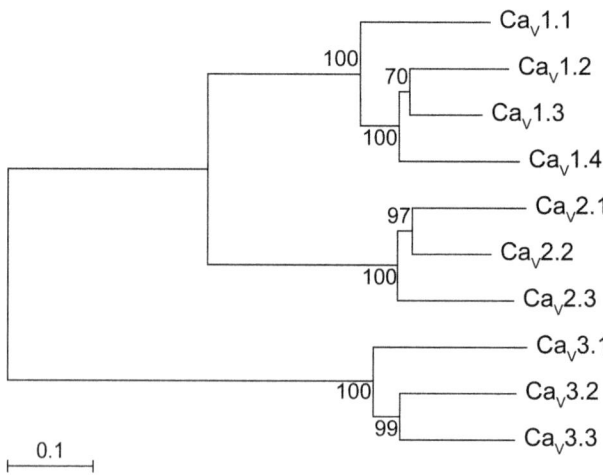

Fig. 1 Phylogenetic tree for mouse VGCC α₁-subunit proteins

The phylogram was calculated using the neighbor-joining method by MEGA11 (Tamura et al., 2021). Repeat I–IV excluding II–III linker predicted in the NCBI database of each α₁-subunit protein was used for the calculation. Bootstrap values from 1000 replications are indicated. The scale bar represents 0.1 substitutions per amino acid

site. Accession numbers of α₁-subunit proteins for the phylogenetic tree calculation are following: NP_001074492.1 for Ca$_V$1.1; NP_001153005.1 for Ca$_V$1.2; NP_001077085.1 for Ca$_V$1.3; NP_062528.2 for Ca$_V$1.4; NP_001238988.1 for Ca$_V$2.1; NP_001035993.1 for Ca$_V$2.2; NP_033912.2 for Ca$_V$2.3; NP_001106284.1 for Ca$_V$3.1; NP_001157163.1 for Ca$_V$3.2; NP_001037773.2 for Ca$_V$3.3

prove the physiological roles of individual VGCC subtypes predicted from pharmacological and biochemical studies, to discover unexpected novel functions and to explore their significance as potential novel drug targets. Here we summarize the major findings from mice with genetic modifications altering the expression or sequence of Ca^{2+} channel α_1-subunits, including the introduction of human pathogenic mutations for generating human disease models. We do not cover genetic modifications of accessory subunits that can associate with multiple VGCC subtypes and may even mediate channel-independent actions. We also provide a list of the original mouse strains employed, which are summarized in tables (Tables 1, 2, 3, and 4) and indicated as underlined references in the text.

Ca_V1

The Ca_V1 (L-type) Ca^{2+} channel family comprises four members. Their pore-forming α_1-subunits assemble with modulatory accessory subunits ($\alpha_2\delta$ and β plus γ_1 in $Ca_V1.1$), like other "high-voltage" activated Ca^{2+} channels (Zamponi et al., 2015) and carry high-affinity binding pockets for organic L-type Ca^{2+} channel blockers (Zhao et al., 2019; Catterall et al., 2020). Among those, dihydropyridines (DHPs) are the most potent and L-type channel selective drugs. Unfortunately, differences in the DHP sensitivity between the four different L-type Ca^{2+} channel isoforms are small and no other isoform-selective drugs are available to date (Sinnegger-Brauns et al., 2009; Ortner et al., 2017). Therefore, their individual contributions to cellular functions could not be dissected by pharmacological means and mouse animal models were crucial to study the role of L-type channel heterogeneity in health and disease. This was especially important for $Ca_V1.2$ and $Ca_V1.3$ channels, which are expressed together in most electrically excitable cells but can serve different cellular functions due to their different biophysical properties (Zamponi et al., 2015).

In contrast to $Ca_V1.2$ and $Ca_V1.3$, $Ca_V1.1$ and $Ca_V1.4$ exert highly specialized functions with restricted expression in skeletal muscle ($Ca_V1.1$) and the retina ($Ca_V1.4$). Their physiological importance outside these tissues (such as in the immune system, see below) is less clear.

$Ca_V1.1$

Muscle Contraction and Development: *Cacna1s⁻/⁻* (Dysgenic) Mice

In skeletal muscle transverse tubules, $Ca_V1.1$ channels serve as the voltage sensors for excitation–contraction (EC) coupling by triggering ryanodine receptor type 1 (RyR1)-mediated Ca^{2+} release from the SR (see chapter "The Skeletal Muscle Calcium Channel" by Flucher and Beam). In the triad junctions, this mechanical coupling requires the assembly of channel complexes as tetrads opposite of the RyR arrays in the SR membrane. Their especially high density in skeletal muscle made $Ca_V1.1$ α_1-subunits the first to be biochemically purified, cloned, and their structure and drug-binding domains resolved by cryo-electron microscopy at high resolution (Zhao et al., 2019). It was also the first Ca_V for which a α_1-subunit knockout (KO) model was available even before the advent of homologous recombination gene targeting techniques. These muscular dysgenic mice (Powell, 1990) lack EC coupling due to the absence of the $Ca_V1.1$ α_1-subunit caused by a truncating mutation (Chaudhari, 1992) but still express $\alpha_2\delta_1$ and β_{1a} accessory subunits. Cultured muscle cells and especially the corresponding immortalized myoblast cell line (GLT cells, Flucher et al., 1991) derived from these null mutants served as an ideal experimental platform for the discovery of the key role of $Ca_V1.1$ channels (often referred to with the pharmacologically incorrect term: skeletal muscle "dihydropyridine receptor") for muscle contraction and for studying the protein interaction machinery participating in the bidi-

Table 1 Ca_V1 mutant mice

Genotype	References	Studied biological processes
$Cacna1s^{-/-}$ (Dysgenic mice)	Powell (1990)	Skeletal muscle function
$Cacna1s^{N617D}$ KI	Dayal et al. (2017)	$Ca_V1.1$ conduction, skeletal muscle function
$Cacna1s^{E1014K}$ KI	Lee et al. (2015)	$Ca_V1.1$ conduction, skeletal muscle function
$Cacna1s^{\Delta Exon29}$	Sultana et al. (2016)	$Ca_V1.1$ α_1-subunit splicing, skeletal muscle function
$Cacna1s^{R528H}$ KI	Wu et al. (2012)	Hypokalemic (HypoPP) periodic paralysis
$Cacna1s^{N617D}$ KI/$RyR1^{-/-}$ double mutants	Kaplan et al. (2018)	Neuromuscular junction development
$Cacna1s^{\Delta Exon29}$/$RyR1^{-/-}$ double mutants	Kaplan et al. (2018)	Neuromuscular junction development
$Cacna1c^{-/-}$	Seisenberger et al. (2000)	Heart function
$Cacna1c^{flox/flox}$	Seisenberger et al. (2000), White et al. (2008) and Jeon et al. (2010)	Embryonic cortical neurons Observational fear Drug taking behaviors, see also conditional $Cacna1c$ Kos
Cerebral cortex- and hippocampus-specific KO of $Ca_V1.2$ ($Cacna1c^{flox/flox}$/Nex-Cre)	$Cacna1c^{flox/flox}$ from Seisenberger et al. (2000) and Moosmang et al. (2005)	Hippocampal memory, LTP
Full-brain KO of $Ca_V1.2$ ($Cacna1c^{flox/flox}$/Nestin-Cre)	$Cacna1c^{flox/flox}$ from Seisenberger et al. (2000) and Langwieser et al. (2010)	Brain, chronic stress Brain, fear Drug taking behaviors
Cardiomyocyte-specific $Ca_V1.2$ α_1 KO, ($Cacna1c^{flox/flox}$/αMHC-MerCreMer)	Rosati et al. (2011), $Cacna1c^{flox/flox}$ from White et al. (2008) and Goonasekera et al. (2012), $Cacna1c^{flox/flox}$ from Seisenberger et al. (2000)	Heart function
Smooth muscle-specific $Ca_V1.2$ α_1 KO ($Cacna1c^{flox/flox}$/SM22 Cre ERT2)	$Cacna1c^{flox/flox}$ from Seisenberger et al. (2000) and Moosmang et al. (2003); SMACKO mice	Arteries Bladder Intestinal smooth muscle
$Cacna1c^{+/lacZ\ reporter}$	B6.129P2-$Cacna1c^{tm1Dgen}$/J mouse strain (Deltagen; JAX stock #005783)	$Ca_V1.2$ α_1 expression during embryonic development Age-dependent cognitive decline

(continued)

Table 1 (continued)

Genotype	References	Studied biological processes
Cacna1c^T1066Y KI (Ca$_V$1.2DHP$^{-/-}$)	Sinnegger-Brauns et al. (2004)	Dissect Ca$_V$1.2 and Ca$_V$1.3 channel function Heart Pancreatic β-cells Neurons Neural progenitor cells Arterial smooth muscle
Cacna1c G1793-STOP (Ca$_V$1.2ΔDCT) *Cacna1c* D1874-STOP (Ca$_V$1.2^D1874stop)	Fu et al. (2011) and Domes et al. (2011)	Regulatory function of Ca$_V$1.2 α$_1$ C-terminus (protein kinase A regulation)
Cacna1c^S1700A/T1704A KI (Ca$_V$1.2^S1700A/T1704A) *Cacna1c* S1700 KI (Ca$_V$1.2^S1700A)	Fu et al. (2013), Poomvanicha et al. (2017), and Fu et al. (2014)	Heart (Ca$_V$1.2 protein kinase A regulation)
Cardiomyocyte-specific transgenic expression of DHP-insensitive rabbit Ca$_V$1.2 α$_1$ phosphorylation site or C-terminal truncation mutants: α-MHC doxycyclin-inducible-promoter vector/α-MHC promoter-driven reverse tetracycline-controlled transactivator (TET-ON)	Liu et al. (2020) and Katchman et al. (2017)	Heart (Ca$_V$1.2 protein kinase A regulation)
Cardiomyocyte-specific transgenic expression of DHP-insensitive rabbit Ca$_V$1.2 α$_1$ fused to ascorbate peroxidase (APEX): α-MHC doxycyclin-inducible-promoter vector/α-MHC promoter-driven reverse tetracycline-controlled transactivator (TET-ON)	Liu et al. (2020)	Heart (Ca$_V$1.2 protein kinase A regulation)
Cacna1c^I1624E KI (*Cacna1c*^flox/I1624E)	*Cacna1c*^flox/flox from Seisenberger et al. (2000) and Poomvanicha et al. (2011)	Heart (calmodulin regulation of Ca$_V$1.2)
Cacna1c^G406R KI	Bader et al. (2011) (TS2-NEO)	Timothy syndrome
Conditional transgenic expression of Ca$_V$1.2 α$_1$ G405R (or Ca$_V$1.2 α$_1$ WT as a control) with floxed STOP knocked into the Rosa26 locus/Cell specific Cre Chondrocyte- and osteoblast-specific (limb bud mesenchyme): Prx1-Cre Osteochondral progenitors: Col2a1-Cre Mature osteoblast-lineage cells: Col1a1-Cre	Ramachandran et al. (2013) and Cao et al. (2017, 2019)	Timothy syndrome Jaw and bone development
Cacna1c^Ser 1928 KI	Lemke et al. (2008)	Brain (PKA-regulation) Heart (PKA-regulation) Arterial smooth muscle (PKA regulation)
Cacna1c^ΔExon33	Li et al. (2017)	Heart
Conditional deletion of Ca$_V$1.2 in D1R-expressing cells (*Cacna1c*^flox/flox/D1R-Cre)	*Cacna1c*^flox/flox from Seisenberger et al. (2000)	Brain, drug taking behaviors
Conditional deletion of Ca$_V$1.2 in cochlear spiral ganglion neurons (*Cacna1c*^flox/flox/Pax2-Cre; *Cacna1c*^Pax2) Conditional deletion of Ca$_V$1.2 in the auditory brainstem nuclei (*Cacna1c*^flox/flox/Egr2-Cre; *Cacna1c*^Egr2)	Zuccotti et al. (2013)	Hearing pathway

(continued)

Table 1 (continued)

Genotype	References	Studied biological processes
Forebrain-specific *Cacna1c* KO in glutamateric neurons (*Cacna1c^flox/flox*/CaMKIIα-Cre)	White et al. (2008) *Cacna1c^flox/flox* from Seisenberger et al. (2000)	Brain function
Type-1 NPC-specific KO of Ca$_V$1.2 in the dentate gyrus with eGFP reporter (*Cacna1c^flox/flox*/Glast-CreERT2 and RCE: loxP)	Völkening et al. (2017)	Neural progenitor cells
Neuron-specific KO of Ca$_V$1.2 (*Cacna1c^flox/flox*/synapsin1-Cre	*Cacna1c^flox/flox* from White et al. (2008) and Temme et al. (2016)	Brain function Neural progenitor cells
Oligodendrocyte progenitor cell-specific *Cacna1c* KO (*Cacna1c^flox/flox*/NG2-creERT) (*Cacna1c^flox/flox*/PDGFRA-creERT) Astrocyte-specific *Cacna1c* KO (*Cacna1c^flox/flox*/GFAP-creERT)	*Cacna1c^flox/flox* from White et al. (2008) *Cacna1c^flox/flox* from Seisenberger et al. (2000) *Cacna1c^flox/flox* from White et al. (2008)	Glial function
T-lymphocyte-specific KO (*Cacna1c^flox/flox*/CD4-Cre)	*Cacna1c^flox/flox* from White et al. (2008)	Immune function
Heterozygous *Cacna1c^+/-* rats (Sprague-Dawley background)	generated using zinc finger technology by SAGE Labs (now Horizon Discovery Ltd, Cambridge, UK; TGR16930)	Brain function
Cacna1d^-/-	Platzer et al. (2000), Namkung et al. (2001), and Tatsuki et al. (2016)	Cochlear inner and outer hair cells Brain Heart Sinoatrial node Pancreatic β-cells Bone Circadian neural activity rhythm Retinal function
Cacna1d^eGFPflex (Cre/loxP-based FLEX-system (Flip excision) replaces *Cacna1d* expression by eGFP expression)	Satheesh et al. (2012) and Berger and Bartsch (2014)	eGFP marker for Ca$_V$1.3 α$_1$ expression
Conditional deletion of Ca$_V$1.3 in the auditory brainstem nuclei (*Cacna1d^eGFPflex*/Egr2-Cre) Conditional deletion of Ca$_V$1.3 in cochlear spiral ganglion neurons (*Cacna1c^flox/flox*/Pax2-Cre)	*Cacna1d^eGFPflex* from Satheesh et al. (2012)	Hearing pathway Spinal motoneurons/spasticity
Conditional deletion of Ca$_V$1.3 in excitatory spinal cord neurons (*Cacna1d^eGFPflex*/Vglut2-Cre) Conditional deletion of Ca$_V$1.3 in excitatory spinal cord neurons (*Cacna1d^eGFPflex*/VIAAT-Cre)	*Cacna1d^eGFPflex* from Satheesh et al. (2012)	
Cacna1d-HA (*DCRD disruption; Cacna1dDCRD^HA*)	Scharinger et al. (2015)	Brain (alternative splicing) Chromaffin cells Hearing
Transgenic brain overexpression of Ca$_V$1.3 α$_1$-subunit under control of CaMKIIα promoter	Krueger et al. (2017)	Aging brain
Cacna1f^-/-	Mansergh et al. (2005)	Retinal function Immune cells

Table 2 Ca$_\text{V}$2 mutant mice

Genotype	References	Studied biological processes
Cacna1a⁻/⁻	Jun et al. (1999)	Neuronal processes Exocytosis and neurotransmitter release Biochemical characterization Cerebellar function and maturation Cognition Presynapse
Cacna1a⁻/⁻	Fletcher et al. (2001)	Neuronal processes Pain
Cacna1a^{ΔE4(−/−)} (*Cacna1a^{flox/flox}*/EIIA Cre)	*Cacna1a^{flox/flox}* from Todorov et al. (2006)	Neuronal processes
Cacna1a^{purk(−/−)} (*Cacna1a^{Citrine}*/PCP/L7 Cre)	*Cacna1a^{Citrine}* from Mark et al. (2011)	Neuronal processes Episodic ataxia type 2
Genetic manipulation of Ca$_\text{V}$2.1 channels at the calyx (*Cacna1a^{flox/flox}*/Helper-dependent adenoviral vectors expressing Cre)	*Cacna1a^{flox/flox}* from Todorov et al. (2006)	Presynaptic protein interaction
Purkinje cell-selective Ca$_\text{V}$2.1 KO (Ca$_\text{v}$2.1^{flox/flox}/D2CreN)	Ca$_\text{V}$2.1^{flox/flox} from Hashimoto et al. (2011)	Cerebellar function and maturation
Cerebellar granule cell-specific Ca$_\text{V}$2.1 KO (Ca$_\text{v}$2.1^{flox/flox}/E3CreN)	Ca$_\text{V}$2.1^{flox/flox} from Hashimoto et al. (2011)	Cerebellar function and maturation
Cacna1a^{quirk(−/−)} (*Cacna1a^{Citrine}*/Gabra6 Cre)	*Cacna1a^{Citrine}* from Mark et al. (2011)	Cerebellar function and maturation Episodic ataxia type 2
Ca$_\text{V}$2.1 IM-AA KI	Nanou et al. (2016b)	Synaptic plasticity
Forebrain-specific Ca$_\text{V}$2.1 KO (*Cacna1a^{flox/flox}*/NEX Cre)	*Cacna1a^{flox/flox}* from Todorov et al. (2006)	Cognition
Hippocampal pyramidal neuron-selective Ca$_\text{V}$2.1 KO (*Cacna1a^{flox/flox}*/CamK2α Cre)	*Cacna1a^{flox/flox}* from Hashimoto et al. (2011)	Cognition
Parvalbumin and somatostatin interneuron-specific Ca$_\text{V}$2.1 KO (*Cacna1a^{flox/flox}*/Nkx2.1-BAC Cre)	*Cacna1a^{flox/flox}* from Todorov et al. (2006)	Epilepsy
Somatostatin interneuron-specific Ca$_\text{V}$2.1 KO (*Cacna1a^{flox/flox}*/Somatostatin Cre)	*Cacna1a^{flox/flox}* from Todorov et al. (2006)	Epilepsy
Cortical pyramidal cell-specific Ca$_\text{V}$2.1 KO (*Cacna1a^{flox/flox}*/Emx1 Cre)	*Cacna1a^{flox/flox}* from Todorov et al. (2006)	Epilepsy
Parvalbumin neuron-specific Ca$_\text{V}$2.1 KO (*Cacna1a^{flox/flox}*/Parvalbumin Cre)	*Cacna1a^{flox/flox}* from Todorov et al. (2006)	Epilepsy
Layer VI pyramidal neuron-specific Ca$_\text{V}$2.1 KO (*Cacna1a^{Citrine}*/neurotensin receptor 1 Cre)	*Cacna1a^{Citrine}* from Mark et al. (2011)	Epilepsy
Cacna1a^{R192Q} KI	van den Maagdenberg et al. (2004)	Familial hemiplegic migraine type 1
Cacna1a^{S218L} KI	Gao et al. (2012)	Familial hemiplegic migraine type 1
Cacna1a^{F1406C} KI	Rose et al. (2014)	Episodic ataxia type 2
Ca$_\text{V}$2.1 human Ca$_\text{V}$2.1 with 28 polyglutamine repeats KI	Saegusa et al. (2007)	Spinocerebellar ataxia type 6

(continued)

Table 2 (continued)

Genotype	References	Studied biological processes
Cacna1b⁻/⁻	Ino et al. (2001)	Sympathetic regulation Pain Neurotransmitter release Behavior and memory Cardiovascular systems Kidney function Presynapse
Cacna1b⁻/⁻	Kim et al. (2001a)	Pain Pharmacology Behavior and memory
Cacna1e⁻/⁻	Wilson et al. (2000)	R-type identification Synaptic plasticity Sleep
Cacna1e⁻/⁻	Sochivko et al. (2002) and Pereverzev et al. (2002)	R-type identification Pancreatic function and glucose metabolism Distribution in the brain Synaptic plasticity Seizure Epilepsy Parkinson's disease Cardiovascular systems Presynapse Retina Mendelian inheritance
Cacna1e⁻/⁻	Saegusa et al. (2000)	Pain Pancreatic function and glucose metabolism Distribution in the brain Synaptic plasticity Sperm Ischemic neuronal injury Narcotic effects Anesthesia
Cacna1e⁻/⁻	Lee et al. (2002)	Amygdala function Epilepsy

rectional coupling of $Ca_V1.1$ and RyR1 for proper EC coupling (Grabner et al., 1999). Dysgenic GLT cells are ideally suited to reconstitute L-type current, Ca^{2+} transients, EC coupling, and $Ca_V1.1$ triad clustering by expressing wild-type (WT) channels and to test the effect of specific mutations for these processes. In combination with molecular modeling, this revealed the molecular details responsible for the differential voltage dependence of ionic current and EC coupling of $Ca_V1.1$ (Tanabe et al., 1991; Tuluc et al., 2009).

$Ca_V1.1$ Ca^{2+} Conductance and Skeletal Muscle Function: *Cacna1s*[N617D], *Cacna1s*[E1014K] Knock-In (KI) Mice

$Ca_V1.1$ triggers muscle contraction not by Ca^{2+} entry through its slowly activating Ca^{2+} inward current but instead by fast conformational changes of its voltage sensors transmitted to the EC coupling machinery through the II–III cytoplasmic linker (Grabner et al., 1999). This raised the question about the physiological relevance of its Ca^{2+} conductance, in particular because

Table 3 Ca$_V$3 mutant mice

Genotype	References	Studied biological processes
Cacna1g$^{-/-}$	Kim et al. (2001b)	Epilepsy Cardiovascular systems Cerebellar function Behavior and hippocampal oscillation Sleep Pain Sperm Sympathetic activity Kidney function Lymphatic vessel
Cacna1g$^{-/-}$	Petrenko et al. (2007)	Anesthesia Cerebellar function
Cacna1g$^{-/-}$	Le Quang et al. (2011)	Cardiac function
Cacna1g$^{-/-}$	Wang et al. (2016)	Immune system
Thalamic neuron-selective Ca$_V$3.1 KO (*Cacna1g$^{flox/flox}$*/K$_V$3.2 Cre)	*Cacna1g$^{flox/flox}$* from Anderson et al. (2005)	Sleep
T-cell specific Ca$_V$3.1 KO (*Cacna1g$^{flox/flox}$*/Lck-Cre)	*Cacna1g$^{flox/flox}$* from Anderson et al. (2005)	Immune system
Overexpressing *Cacna1g*	Ernst et al. (2009)	Epilepsy
Cacna1g-overexpression (cardiomyocyte-specific)	Double transgenic: tetracycline transactivator (α-MHC promoter); Nakayama et al. (2009) and Li et al. (2012)	Cardiac function
Cacna1g^{R1723H} KI	Hashiguchi et al. (2019)	Spinocerebellar ataxia 42
Cacna1h$^{GFPtag/GFPtag}$	François et al. (2015)	Sensory systems
Small-diameter DRG neuron-specific *Cacna1h* O	*Cacna1h$^{GFP-tag/GFP-tag}$* from François et al. (2015)	Sensory systems
Cacna1h$^{EGFP/EGFP}$	Bernal Sierra et al. (2017)	Sensory systems
Cacna1h$^{Cre/Cre}$	Bernal Sierra et al. (2017)	Sensory systems
Cacna1h$^{-/-}$	Chen et al. (2003)	Sensory systems Pain Analgesia and anesthesia Itch Epilepsy Memory Neuronal excitability Expression distribution in the brain Neurogenesis Retina Carotid body and hypoxia sensing Gastrointestinal tract Fertilization Bone Sleep Sympathetic activity Anesthesia Kidney function Lymphatic vessel Sperm Mendelian inheritance Arterial vascular tone Cardiac function Hearing

(continued)

Table 3 (continued)

Genotype	References	Studied biological processes
Cacna1h$^{-/-}$	Seidel et al. (2021)	Hypertension
Cacna1h$^{-/-}$	Gibbons et al. (2009)	Gastrointestinal tract
Cacna1h$^{-/-}$	Le Quang et al. (2011)	Cardiac function
Cacna1h^{M1560V} KI	Seidel et al. (2021)	Primary aldosteronism Hypertension
Cacna1h^{H191Q} KI	Nelson et al. (2007)	Pain
Cacna1h$^{-/-}$/NFAT-Luciferase reporter double transgenic	*Cacna1h$^{-/-}$* mice from Chen et al. (2003)	Cardiac function
Cacna1i$^{-/-}$	Astori et al. (2011)	Sleep
Cacna1i$^{-/-}$	Lee et al. (2014)	Epilepsy
Cacna1i^{R1305H} KI	Ghoshal et al. (2020)	Sleep
Cacna1i$^{-/-}$	Ghoshal et al. (2020)	Sleep Anesthesia

Table 4 Multiple Ca$_V$s mutant mice

Genotype	References	Studied biological processes
Ca$_V$2 conditional triple-KO (*Cacna1a$^{flox/flox}$/Cacna1b$^{flox/flox}$/Cacna1e$^{flox/flox}$/* lentiviruses expressing *Cre*)	*Cacna1a$^{flox/flox}$* from Todorov et al. (2006) *Cacna1b$^{flox/flox}$* from Held et al. (2020) *Cacna1e$^{flox/flox}$* from Pereverzev et al. (2002)	Presynapse
Cacna1h$^{-/-}$/Cacna1i$^{-/-}$	*Cacna1i$^{-/-}$* from Astori et al. (2011) *Cacna1h$^{-/-}$* from Chen et al. (2003)	Sleep
Conventional KO for Ca$_V$1.1~1.4, Ca$_V$2.1~2.3, or Ca$_V$3.1~3.3	Tatsuki et al. (2016)	Sleep
Cacna1a$^{-/-}$/Cacna1g$^{-/-}$	*Cacna1a$^{-/-}$* from Jun et al. (1999) *Cacna1g$^{-/-}$* from Kim et al. (2001b)	Epilepsy
Tamoxifen-inducible conditional Ca$_V$2.1 KO (iKO$^{p/q}$) (*Cacna1a^{Citrine}/CAGGcre-ER^{TM}*)	*Cacna1a^{Citrine}* from Mark et al. (2011)	Epilepsy
iKO$^{p/q}$/*Cacna1g$^{-/-}$*	*Cacna1g$^{-/-}$* (*Sox2-Cre/Cacna1g$^{flox/flox}$* from Anderson et al., 2005)	Epilepsy
iKO$^{p/q}$/*Cacna1h$^{-/-}$*	*Cacna1h$^{-/-}$* from Chen et al. (2003)	Epilepsy
Cacna1e$^{-/-}$/Cacna1h$^{-/-}$	*Cacna1e$^{-/-}$* from Pereverzev et al. (2002) *Cacna1h$^{-/-}$* from Chen et al. (2003)	Retina

zebrafish and all other euteleost fishes express non-Ca^{2+}-conducting Ca$_V$1.1 channels in their skeletal muscles (Schredelseker et al., 2010; Dayal et al., 2017). Introduction of the selectivity-filter-adjacent point mutation N617D, which renders zebrafish Ca$_V$1.1 α_1-subunits non-conducting, into mice (*Cacna1s^{N617D}*) also abolished depolarization-induced Ca^{2+} influx without affecting SR Ca^{2+} release, SR Ca^{2+} store content, muscle fiber size, volume, histology, and contractility. Also, muscle function, coordination, strength, and fatigue susceptibility were not different from WT (Dayal et al., 2017). Moreover, no adaptive transcriptional or translational regu-

lation of Ca^{2+}-handling proteins was detected in *Cacna1s^{N617D}* mice, which could explain unaltered muscle performance. These data strongly suggest that the Ca$_V$1.1-mediated Ca^{2+} conductance is completely dispensable for mammalian skeletal muscle function and most likely an evolutionary remnant from the Ca^{2+} influx-dependent skeletal muscle EC coupling found in early chordates. This work also revealed an important difference between *Cacna1s^{N617D}* and another Ca^{2+}-permeation-deficient mutant *Cacna1s^{E1014K}* (Lee et al., 2015). While N617D enhances Ca^{2+}-binding affinity and thus blocks Ca^{2+} permeation by locking Ca^{2+} inside the selectivity filter,

E1014K reduces Ca^{2+} affinity for the selectivity filter and thus increases permeability for monovalent cations (Lee et al., 2015). In contrast to homozygous $Cacna1s^{N617D}$, $Cacna1s^{E1014K}$ mice display smaller muscle fiber size and reduced muscle endurance, which is likely due to changes in membrane excitability due to aberrant monovalent cation conductance causing K^+-outward current through the mutant channel (Beqollari et al., 2018).

Non-conducting $Cacna1s^{N617D}$ mice also provided novel insight into the role of Ca^{2+} signals for the proper formation of the neuromuscular junction (NMJ), a highly specialized synapse critical for efficient electric coupling between a very small region of the surface of muscle fibers and its innervating motor nerve terminal.

Neuromuscular Junction Development: $Cacna1s^{N617D}$, $RyR1^{-/-}/Cacna1s^{N617D}$, $Cacna1s^{\Delta E29}$, $Cacna1s^{\Delta E29}/RyR1^{-/-}$ Mutant Mice

An early step of this localized synapse formation is initiated by nicotinic acetylcholine receptor (AChR) clustering in the center of muscle fibers (AChR prepatterning) even before nerve arrival. This is driven by muscle-specific kinase (MuSK) signaling and the local transcription of AChR and MuSK in the central band of the muscle fibers (Kaplan et al., 2018 for references). In this muscle-intrinsic mechanism, Ca^{2+} signaling through $Ca_V1.1$ Ca^{2+} channels plays a key role as recently shown in a systematic study using different $Ca_V1.1$ mouse models. In WT mice, AChRs cluster within a central band perpendicular to the long axis of the muscle fibers, and the ingrowing motor nerve branches (e.g., of the phrenic nerve in this study) are restricted to this region. At E14.5, endplates were widened in the muscle of dysgenic ($Cacna1s^{-/-}$) mice with excessive nerve branching and a wider distribution of the phrenic nerve confirming earlier observations (Kaplan et al., 2018). Interestingly, at E14.5, early NMJ was completely normal in $Cacna1s^{N617D}$ mutant mice (Dayal et al., 2017) demonstrating that, in the presence of intact EC coupling and RyR1-

mediated Ca^{2+} release, $Ca_V1.1$-mediated inward Ca^{2+} current is physiologically not relevant (Kaplan et al., 2018). However, Ca^{2+} current through $Ca_V1.1$ became relevant when RyR1 was eliminated as a $Ca_V1.1$-triggered Ca^{2+} source in the muscle of $RyR1^{-/-}$ mice. While $RyR1^{-/-}$ muscle showed regular AChR patterning at E14.5 in the presence of WT $Ca_V1.1$ channels, elimination of this Ca^{2+} source in $RyR1^{-/-}/Cacna1s^{N617D}$ double KO mice caused severe defects in AChR patterning. AChR scattered along the entire length of the muscle fiber and nerve branch growth across the entire width of the muscle. In these mice, $Ca_V1.1$ α_1-N617D protein was still present in the triads suggesting that the absence of its Ca^{2+} signal rather than the absence of non-channel functions regulates AChR patterning. In summary, these data provide strong evidence that $Ca_V1.1$ Ca^{2+} conductance is not required for AChR patterning but that in this initial step of NMJ formation Cav1.1 can generate sufficient Ca^{2+} signals for normal AChR patterning and synaptic targeting of motor axons either by triggering Ca^{2+} release or, in the absence of SR Ca^{2+} release, by Ca^{2+} influx.

The observed defects in early NMJ development progressed into later fetal development. At E18.5, WT diaphragm muscle is innervated at a single NMJ in the center of each muscle fiber, aneural AChR clusters are absent, and nerve branches are restricted to the narrow central endplate band. In $Cacna1s^{-/-}$ and $RyR1^{-/-}/Cacna1s^{N617D}$ mice innervation patterns were severely distorted with up to eight NMJs innervating a muscle fiber and severely disrupted AChR patterning and nerve branching (Kaplan & Flucher, 2019). Interestingly, at this later developmental stage, $RyR1^{-/-}$ mice exhibited ectopic AChR clusters and innervation to some extent also outside the central band of AChR clusters. This suggested that the magnitude of $Ca_V1.1$-mediated Ca^{2+} current is sufficient for the normal regulation of AChR patterning up to E14.5 but, due to the known developmental decline of L-type current, not up to E18.5 (Kaplan & Flucher, 2019). To prove this hypothesis, another elegant mouse model came into play, $Ca_V1.1$ exon 29-deficient mice ($Cacna1s^{\Delta E29}$, Sultana et al., 2016; see below for details). Indeed, enhanced Ca²⁺ currents in homozygous *Cacna1s^{ΔE29}* mice (due to the expression of the embryonic splice variant of Cav1.1 α1-subunits, Cav1.1e, see below) could rescue normal NMJ development in *Ryr1⁻ᐟ⁻* mice (see above). In homozygous *Cacna1s^{ΔE29}/Ryr1⁻ᐟ⁻* mice, the *Ryr1⁻ᐟ⁻* phenotype was largely rescued, whereas *Cacna1s^{ΔE29/+}/Ryr1⁻ᐟ⁻* mice with one functional WT copy displayed an intermediate phenotype. Therefore, AChR patterning and innervation appear to be strictly dependent on the magnitude of muscle Ca²⁺ channel signaling (Kaplan & Flucher, 2019).

Notably, during aberrant NMJ development in these mutant mice, excessive outgrowth of motor axons is also observed beyond AChR clusters, i.e., independent of the distribution of their target structures. Experimental evidence using the same mouse models (Kaplan & Flucher, 2019) strongly points to an additional role of Ca_V1.1-mediated activity-dependent Ca²⁺ signaling in regulating motor axon differentiation and outgrowth beyond its above-described role for clustering and patterning of AChRs.

Developmentally Regulated Alternative Splicing in Muscle Function: *Cacna1s^{ΔE29}* Mutant Mice

Ca_V1.1 channels exist as two major, functionally distinct splice variants. The pore-forming α₁-subunit of the adult isoform (Ca_V1.1a) includes exon 29, which is absent in Ca_V1.1e (Ca_V1.1 ΔE29). This corresponds to an embryonic Cav1.1 variant, because it accounts for 50–80% of Ca_V1.1 transcripts at E17 but decreases to less than 5% in adult muscle (Flucher & Tuluc, 2011; Sultana et al., 2016). Exon 29 encodes 19 amino acids in the extracellular linker connecting transmembrane segments S3 and S4 in the fourth voltage-sensing domain. The shorter IVS3-S4 linker of Ca_V1.1e strongly increases the channel's Ca²⁺ conductance through a ~30-mV negative shift in its voltage sensitivity and a ~6-fold increase of current amplitude when expressed in GLT myotubes (Tuluc et al., 2009). The shorter

S3–S4 linker alters interactions of voltage sensor IVS4-positive charges with an aspartate D1196 residue likely facilitating and stabilizing the activated channel state (Tuluc et al., 2016). The switch from Ca_V1.1e to Ca_V1.1a is physiologically relevant as shown in mice in which exon 29 was deleted (*Cacna1s^{ΔE29}*; Sultana et al., 2016). Muscles of these mice express similar overall levels of Ca_V1.1 transcripts but all without exon 29, i.e. Ca_V1.1e only. Ca²⁺ currents and Ca²⁺ transients showed the expected strong negative shift (about −38 mV) of half-maximal activation with the expected overall increased Ca²⁺ influx during EC coupling and the appearance of aberrant spontaneous focal Ca²⁺ transients. Homozygous *Cacna1s^{ΔE29}* mice developed normally, displayed only mild motor deficits evident as decreased front paw muscle strength, and reduced voluntary distance running (Sultana et al., 2016). In comparison to WT, the contractile force was reduced, fatigue resistance increased, and tetanic fusion frequency was reduced in slow and fast muscles. Interestingly, the persistence of Ca_V1.1e after development substantially affected the relative abundance of fast- and slow-twitch muscle fibers. The distribution was shifted toward slower-twitch fiber types in both soleus (predominantly slow/oxidative) and extensor digitorum longus (predominantly fast/glycolytic) muscles (Sultana et al., 2016). In adult *Cacna1s^{ΔE29}* mice, the number of mitochondria was reduced and mitochondrial damage was evident. This suggested that increased Ca_V1.1e-mediated Ca²⁺ influx during EC coupling may contribute to the activation of the slow muscle fiber pathway (with some evidence of the involvement of calcineurin and CaMKII signaling (Sultana et al., 2016) and Ca²⁺-mediated mitochondrial dysfunction, thus emphasizing the important physiological relevance of this developmental switch in alternative splicing. Interestingly, myotonic dystrophy type 1 (DM1) in humans is caused by splicing defects of several important muscle protein patients. This includes aberrant expression of the Ca_V1.1e splice variant, which correlates with the degree of muscle weakness (Sultana et al., 2016). The mitochondrial damage in *Cacna1s^{ΔE29}* mice provides additional support for a pathogenic role of defective $Ca_V1.1$ splicing in DM1.

$Ca_V1.1$ in Human Disease: *Cacna1s*[R528H] KI Mice

In addition to the potential role of $Ca_V1.1$ alternative splicing in DM1 (see above), $Ca_V1.1$ has been associated with several human genetic disorders. These channelopathies include hypokalemic (HypoPP) and normokalemic (NormoPP) periodic paralysis, and malignant hyperthermia susceptibility. The molecular basis of these disorders has been reviewed recently (Flucher, 2020; Striessnig, 2021). Strikingly, almost all HypoPP mutations are missense mutations within the voltage sensors of structurally related pore-forming subunits of $Ca_V1.1$ (*CACNA1S* gene*)* or the skeletal muscle voltage-gated Na⁺-channel $Na_V1.4$ (*SCN4A*) (Striessnig, 2021 for review). This led to the discovery of the so-called pathogenic gating pore currents, anomalous inward currents through the mutated voltage-sensor. Replacement of one of the positive charges in the S4-helix or of stabilizing counter charges by a smaller or hydrophilic amino acid can allow water molecules to enter and create a pore large enough to conduct either measurable Na⁺ and K⁺ currents or proton currents (Striessnig, 2021). While such tiny gating pore "leak" currents could be demonstrated for $Ca_V1.1$ and $Na_V1.4$ in heterologous expression systems, KI mice expressing human mutations allowed a more detailed study not only of abnormal current in the native muscle fiber environment but also its consequences on muscle excitability, strength, and morphology. Mice heterozygous or homozygous for the most frequently occurring human HypoPP mutation R528H (*Cacna1s*[R528H]; Wu et al., 2012) displayed normal feeding behavior, weight gain, locomotor activity, and no spontaneous paralytic attacks. However, attacks could be triggered by low K⁺ challenge (such as observed in humans after strenuous exercise or high carbohydrate intake), the expected HypoPP phenotype. This challenge led to a paradoxic depolarization rather than hyperpolarization of the resting membrane rest-

ing potential associated with loss of muscle excitability and a steep decline in muscle force as the K⁺ concentration was lowered below 3 mM in heterozygous animals. In KI muscle fibers, a small anomalous inward current could be recorded at the resting potential which was absent in WT fibers. This current decreased in amplitude at more depolarized potentials which can be explained by the outward movement of S4 helices in activated voltage sensors leading to closure of the gating pore (for mechanistic explanations, see Striessnig, 2021). This inward leak current overwhelms outward K_{ir}-mediated K⁺ currents (which are diminished as external K+ is lowered), depolarizes the cell, and leads to Na⁺-channel inactivation and action-potential failure and nicely explains the paradoxical depolarization and HypoPP susceptibility (Wu et al., 2012).

$Ca_V1.2$ and $Ca_V1.3$

As outlined above, $Ca_V1.1$ and $Ca_V1.4$ are largely restricted to skeletal muscle and the retina, whereas $Ca_V1.2$ and $Ca_V1.3$ are present in all electrically excitable cells. In most of them, they are expressed together in the same cell. And although $Ca_V1.3$ channels are about an order of magnitude less sensitive to DHP Ca²⁺ channel blockers, this limited selectivity does not allow to reliably dissect the physiological roles of these channels *in vitro* or *in vivo*. However, despite their overlapping expression patterns, substantial differences in the biophysical properties of these channels also implied their different contributions to physiology and disease, a prediction that was ultimately confirmed with appropriate KO and KI mouse (and rat) models.

$Ca_V1.2$

Homozygous *Cacna1c* KO mice are lethal and embryos die before day 14.5 post coitum (p.c.), indicating that the $Ca_V1.2$ channel is required for heart development after this period (Seisenberger et al., 2000). Therefore, in addition to constitutive *Cacna1c* KO mice (Seisenberger et al.,

2000; Dao et al., 2010), conditional *Cacna1c* KO mouse strains were developed (Seisenberger et al., 2000) to allow the cell type and time-dependent inducible KO of $Ca_V1.2$ channels or its local inactivation using AAV injection-mediated *Cre* recombination. These mouse models demonstrated that $Ca_V1.2$ is the major L-type channel not only in cardiomyocytes but also in different types of smooth muscles and in the brain (Zamponi et al., 2015).

Cav1.2 in the Heart: Conventional and Conditional *Cacna1c* KO

In mice with a constitutive (germline) KO of *Cacna1c* (*Cacna1c*$^{-/-}$; Seisenberger et al., 2000), the amplitude of L-type current in beating isolated embryonic cardiomyocytes was reduced only by about 60% at day 12.5 p.c. indicating the prominent expression of another L-type current at this developmental stage. This current component is likely $Ca_V1.3$ mediated by an embryonic $Ca_V1.3$ splice variant upregulated in *Cacna1c*$^{-/-}$ cardiomyocytes (Xu et al., 2003). However, some uncertainty remains because this "$Ca_V1.3$-like" current was also present in $Ca_V1.3$-deficient mice (*Cacna1d*$^{-/-}$; Platzer et al., 2000; Seisenberger et al., 2000). It is therefore possible that a yet unknown embryonic splice variant is not silenced by the $Ca_V1.3$ α_1 gene-targeting construct in this *Cacna1d*$^{-/-}$ strain.

$Ca_V1.2$ is expressed in all regions of the adult heart. It is the major isoform in the atrial and ventricular working myocardium. As shown in different mouse models, it is the only L-type channel relevant for cardiac excitation–contraction coupling. Accordingly, it also mediates the negative inotropic effects of Ca^{2+} channel blockers (Sinnegger-Brauns et al., 2004). $Ca_V1.2$ inward current also contributes to myocardial action potential duration. In contrast, $Ca_V1.3$, but not $Ca_V1.2$, serves as a major channel for driving pacemaking in the sinoatrial node (SAN) and for supporting conduction in the AV node (Mesirca et al., 2015).

In 10-week old mice, constitutive heterozygous *Cacna1c* KO (Seisenberger et al., 2000) reduced heart $Ca_V1.2$ α_1 protein levels by 40% compared to WT, reduced whole cell L-type current (by 25%) and intracellular Ca^{2+} transients in freshly isolated adult ventricular myocytes, impaired contractility, and led to cardiac hypertrophy at 32 weeks of age (Goonasekera et al., 2012). Pathologic (e.g., transverse aortic constriction, TAC) or physiologic stress (forced swimming) reduced cardiac performance and increased cardiac hypertrophy. Since $Ca_V1.2$ is also expressed in arterial smooth muscle (see below) which could reduce peripheral arterial resistance and therefore confound the cardiac phenotype, cardiomyocyte-specific KO mice were generated. This was achieved by crossing floxed *Cacna1c* mice (independently generated by Seisenberger et al., 2000 and White et al., 2008) with transgenic mice carrying a transgene comprising an α-myosin heavy chain promoter driving a tamoxifen-inducible *Cre* recombinase, Mer-Cre-Mer (Rosati et al., 2011; Goonasekera et al., 2012).

Reduction of cardiomyocyte $Ca_V1.2$ α_1 protein expression to 19% of control by induction of KO in 7-week old homozygous *Cacna1c*^{merCremer} mice caused severe cardiac dysfunction and reduced the average survival time to 11 days (Rosati et al., 2011). Induction of heterozygous conditional KO reduced $Ca_V1.2$ α_1 protein expression by about 20% but this did not affect 2-month survival rates and cardiac function. However, cardiac-specific heterozygous KO worsened cardiac remodeling and decompensation upon pathologic stimulation (TAC, swimming exercise) (Goonasekera et al., 2012). Mice harboring different combinations of constitutive and floxed KO alleles allowed a graded reduction of $Ca_V1.2$ α_1-subunit protein expression by tamoxifen injection in 10-day-old mice (Goonasekera et al., 2012). Reduction of $Ca_V1.2$ α_1 protein expression to about 30% (*Cacna1c*$^{merCremer/-}$) or 40% (*Cacna1c*$^{merCremer/flox}$) of control (resulting in about 30% and 50% of current density, respectively) also reduced 80-day survival rates to about 40% and 70% with hearts showing dramatic left ventricular dilatation, cardiac hypertrophy, and heart failure. Strong biochemical evidence was provided that the reduction in L-type Ca^{2+} channel (LTCC) current and reduced electrically evoked intracellular Ca^{2+} transients induce neuroendocrine stress, with leaky sarcoplasmic reticulum Ca^{2+} release leading to enhanced calcineurin/NFAT (nuclear factor of activated T-cells) signaling that promotes hypertrophy and cardiac failure (Goonasekera et al., 2012).

Dissecting $Ca_V1.2$ and $Ca_V1.3$ Channel Function: Cav1.2DHP⁻/⁻ (*Cacna1c^{T1066Y}*) KI Mice

As mentioned above, L-type Ca^{2+} channel blockers inhibit $Ca_V1.2$ and $Ca_V1.3$ with similar affinity, although some DHPs are slightly selective for $Ca_V1.2$. In order to make DHPs selective for $Ca_V1.3$, a mouse model was constructed in which threonin-1066 in the DHP-binding domain of the $Ca_V1.2$ α_1-subunit was mutated to tyrosine ($Ca_V1.2$DHP⁻/⁻ mice). The tyrosine side chain prevents proper docking of DHPs, like isradipine, to the DHP-binding domain, decreasing affinity by more than two orders of magnitude (Sinnegger-Brauns et al., 2004). Importantly, this single mutation does not affect the expression and functional properties of Cav1.2 channels (Sinnegger-Brauns et al., 2004) thus preventing any compensatory changes in gene expression. In these homozygous animals, $Ca_V1.2$-mediated DHP effects are absent and, in the absence of $Ca_V1.1$ and $Ca_V1.4$ in most cells, must therefore be mediated solely by $Ca_V1.3$. The negative inotropic effect of DHP Ca^{2+} channel blockers is lost in isolated cardiac ventricles from homozygous $Ca_V1.2$DHP⁻/⁻ mice (Sinnegger-Brauns et al., 2004). Since $Ca_V1.3$ channels are still inhibited in these mice by DHPs, this demonstrates that $Ca_V1.3$ does not significantly contribute to myocardial contractility in mice, in accordance with the absence of $Ca_V1.3$ α_1 mRNA and protein expression in the ventricular myocardium.

$Ca_V1.2$ in Smooth Muscle: Conditional *Cacna1c* KO and $Ca_V1.2$DHP⁻/⁻ KI Mice

Cav1.2 is also the major L-type channel in smooth muscle cells. Smooth muscle-specific

conditional KO mice (SMACKO mice) were produced by crossing mice carrying one constitutive and one floxed KO allele (Seisenberger et al., 2000) with mice expressing tamoxifen-dependent *Cre* recombinase under the control of the smooth muscle-specific SM22 promoter (SM-Cre ER^{T2} Moosmang et al., 2003). In adult SMACKO mice L-type current in arterial smooth muscle cells was completely eliminated leading to reduced arterial blood pressure and attenuation of the response to vasoconstrictors (phenylephrine, angiotensin II) on blood pressure (Moosmang et al., 2003). Depolarization-induced vasoconstriction was completely absent in SMACKO arteries, like in isradipine-treated control muscle. Myogenic tone, the vasoconstriction-induced response to increased vascular pressure, was also almost absent in SMACKO arteries. This prominent role of $Ca_V1.2$ for regulating myogenic tone and the inhibition of myogenic tone by DHP Ca^{2+} channel blockers exclusively through $Ca_V1.2$ was also confirmed using $Ca_V1.2$DHP⁻/⁻ KI mice (Sinnegger-Brauns et al., 2004; Zhang et al., 2007).

SMACKO mice also show bladder dysfunction with reduced micturition, urine retention, bladder distension, and hypertrophy. This was due to an about 90% reduction of depolarization-induced contractility and about 70% reduction of carbachol-induced detrusor muscle contractility (Wegener et al., 2004). Likewise, elimination of $Ca_V1.2$ in intestinal smooth muscle resulted in reduced feces excretion and in small and large bowel distension, eventually leading to paralytic ileus. This was due to the absence of rhythmic contractile activity in about 90% of the smooth muscle preparations, strongly reduced electrically evoked contractions but with smaller effects on carbachol-induced contractions (Wegener et al., 2006). In summary, $Ca_V1.2$ channels underlie the vast majority of L-type Ca^{2+} currents in the working myocardium and in various smooth muscles. Accordingly, the clinically relevant pharmacological actions of Ca^{2+} channel blockers, vasodilation and cardiodepressant effects, are mediated by $Ca_V1.2$.

Endocrine Pancreas: Conditional *Cacna1c* KO and Ca$_V$1.2DHP$^{-/-}$ KI Mice

Data from two different mouse models support a prominent contribution of Ca$_V$1.2 channels for L-type current in first-phase insulin secretion in adult mice. Conditional KO of Ca$_V$1.2 in pancreatic β-cells (floxed *Cacna1c* mice (Seisenberger et al., 2000) expressing the *Cre* recombinase under the control of the rat insulin 2 promoter Schulla et al., 2003) completely eliminated DHP-, sensitive L-type currents, which comprise about 30% of total Ca^{2+} current in these cells. Fasting glucose levels were slightly increased and glucose tolerance was reduced due to reduced insulin secretion especially during the early phase after the glucose challenge of in situ perfused pancreatic glands (Schulla et al., 2003). The initial phase of Ca^{2+}-dependent exocytosis quantified by cell capacitance measurements was also reduced. Intracellular Ca^{2+} handling and electrical activity remained unaffected. Experiments with Ca$_V$1.2DHP$^{-/-}$ KI mice (Sinnegger-Brauns et al., 2004) confirmed and extended these findings. In these mice, Ca^{2+} currents, including L-type currents, were unaffected but the partial inhibition of Ca^{2+} current by isradipine or stimulation by BAYK8644 was completely eliminated, as expected for Ca$_V$1.2 DHP-insensitive channels (Sinnegger-Brauns et al., 2004). The absence of any response to these drugs also ruled out a substantial contribution of another L-type channel, in particular Ca$_V$1.3, to Ca^{2+} current, in accordance with results from *Cacna1d$^{-/-}$* mice (Platzer et al., 2000; Barg et al., 2001; see section "Ca$_V$1.3"). The contribution of the latter for endocrine pancreatic function is still less clear because there is evidence for a role in pancreatic development from an independent *Cacna1d$^{-/-}$* mouse (Namkung et al., 2001). In any case, *de novo* gain of function mutations in humans in both Ca$_V$1.2 (*CACNA1C*) and Ca$_V$1.3 (*CACNA1D*) is associated with congenital hyperinsulinemic hypoglycemia (Ortner et al., 2020; Striessnig, 2021) supporting an important role of both channels for pancreatic islet function also in humans.

Ca$_V$1.2 in Timothy Syndrome: *Cacna1c^{G406R}* KI (TS2-NEO) and *Cacna1c^{G406R}*-Expressing Transgenic Mice

The above physiological functions revealed in mouse models also allow us to understand disease symptoms arising from gain- and loss-of-function mutations in humans. Due to its key role in cardiac excitability, functional alterations of Ca$_V$1.2 result in changes of action potential duration and the QT interval. Loss-of-function changes appear to result in Brugada syndrome and short-QT syndrome (Striessnig, 2021 for review). In contrast, gain-of-function mutations cause a non-syndromic (LQT8) or syndromic (Timothy syndrome, TS) form of the long-QT syndrome, Timothy syndrome. This is a rare systemic disorder with typical diagnostic findings, in particular LQTS, often leading to fatal arrhythmias at a young age, and unilateral or bilateral cutaneous syndactyly of fingers (Bauer et al., 2021; Striessnig, 2021). In surviving individuals, other characteristic clinical features compatible with a multi-organ disease are facial anomalies, congenital heart defects, and developmental delay. Autism, seizures, and intellectual disability as well as hypoglycemia were also reported. Atypical forms of TS also exist (Bauer et al., 2021).

While effects of Ca$_V$1.2 gain-of-function mutations on the cardiac action potential and insulin secretion can be explained based on the role of Ca$_V$1.2 in these tissues, its contribution to other symptoms is less clear. *In vitro* and *in vivo* animal models harboring human TS mutations provided important answers to this question (Han et al., 2019 for review).

TS2-NEO Mice Heterozygous KI mice expressing the human G406R mutation (TS type-2, exon 8) only survived when an unexcised neomycin cassette was retained, likely moderating the expression of the mutant allele (TS2-NEO mice; Bader et al., 2011). Behavioral tests revealed impaired social interaction and vocalization, with restricted and repetitive/perseverative behavior, features similar to the spectrum of ASD behaviors in humans. Using an IntelliCage-based paradigm with WT and TS2-NEO mice group-housed together, mutant mice showed a high social competitive dominance status, without obvious deficits in acquiring spatial learning-based behavioral tasks and reversal learning (Horigane et al., 2020). Evidence for an altered synaptic excitatory/inhibitory balance in the neocortex of young adolescent TS2-NEO mice was also found as predicted from an excess of smaller-sized inhibitory presynaptic terminals and higher numbers of migrating inhibitory neurons from the medial ganglionic eminence during embryonic development (Horigane et al., 2020). Pleiotropic effects in the 5-HT system were also reported, including enhanced forebrain 5-HT axonal innervation of the dorsal striatum, reduced 5-HT turnover in the amygdala, and evidence for overactivity/disrupted feedback inhibition of dorsal raphe 5-HT neurons (Ehlinger & Commons, 2017).

The glial function also appears to be changed in TS2-NEO mice (for a role of Cav1.2 in glial function, see below). L-type channels modulate intracellular Ca^{2+} and the development of oligodendrocyte progenitor cells (OPCs), a process that was also affected by the TS mutations (Cheli et al., 2018). Cultured OPCs isolated from the cortex of TS mice not only showed larger DHP-sensitive Ca^{2+} influx but also biochemical and morphological characteristics of advanced maturation. In multiple regions of TS2-NEO brains, expression of myelin proteins was significantly higher as was the density of myelinating oligodendrocytes, the percentage of myelinated axons, and the thickness of myelin sheaths (Cheli et al., 2018).

Even more evidence for neurodevelopmental defects induced by TS mutations came from experiments with human iPSC-derived neurons, TS2-NEO brains, and/or expression of TS mutant Cav1.2 channels during embryonic development. These showed developmental abnormalities such as abnormal regulation of Cav1.2 exon utilization, changes in activity-dependent gene expression, and in genetic programs affecting neuronal subtype specifications and inducing activity-dependent dendrite retraction (Paşca et al., 2011; Krey et al., 2013; Panagiotakos et al., 2019). These data underscore the importance of Cav1.2 for brain function in mice and humans and emphasize the importance of mechanisms fine-tuning Cav1.2 channels.

***Cacna1c*TS Transgenic Mice** In Cav1.2TS transgenic mice, a floxed STOP *Cacna1c*G405R allele (or *Cacna1c*WT as a control) has been knocked into the Rosa26 locus (Paşca et al., 2011). This allowed the tissue-specific activation of the expression of these transgenes. Using these mice, a novel role of Cav1.2 channels for mandibular development was discovered (Ramachandran et al., 2013). Cav1.2 is expressed in the first and second pharyngeal arches within the subset of cells that form jaw primordia as shown using a *Cacna1c*+/lacZ reporter strain (*JAX stock #005783*) (Ramachandran et al., 2013). Mandible size was increased in Cav1.2TS expressing the mutant channel under the control of Prx1-dependent *Cre* recombinase in limb bud mesenchyme, in both chondrocytes and osteoblasts (Ramachandran et al., 2013). Genetic and pharmacological manipulations in zebrafish confirmed that Cav1.2 currents control jaw development and that *Cacna1c*TS channels induce hypertrophy and hyperplasia in developing chondrocytes (Ramachandran et al., 2013). These findings can explain the facial dysmorphisms, including macrognathia, observed in many TS patients. However, using additional Cre drivers (Col2a1-Cre, Col2a-Cre) for *Cacna1c*TS xpression in chondrogenic and/or osteogenic precursors (Cao et al., 2017, 2019) during bone development *in vivo* revealed the formation of thickened bone (including abnormal bone growth of the auditory ossicles and the otic capsule, Cao et al., 2019) due to a stimulation of osteogenesis and inhibition of osteoclast activity. Bone formation was also enhanced by activating *Cacna1c*TS expression postnatally. Interestingly, in postnatal bone, this could also prevent estrogen deficiency-induced bone loss in female mice. These studies revealed a novel anabolic mechanism for enhancing bone mass in mice which, if also associated with increased bone stability, could represent a novel therapeutic target also in humans.

Molecular Mechanisms Fine-Tuning Ca$_V$1.2 Activity: *Cacna1c* KI, Exon Deletion and Transgenic Mice

A number of mouse models were generated to confirm the relevance of molecular mechanisms controlling Ca$_V$1.2 channel activity as predicted from heterologous expression studies. In particular, the molecular mechanisms leading to the activation of cardiac Ca$_V$1.2 channels during sympathetic β-adrenergic receptor stimulation during the physiologically important fight-or-flight response remained unresolved despite enormous efforts over more than 25 years to answer this question. Recently, protein kinase A (PKA) phosphorylation sites within the C-terminus of Ca$_V$1.2 α$_1$ were identified as potential targets *in vitro*, waiting for confirmation of their physiological significance in *in vivo* animal models.

Germline Distal C-Terminus Truncation (*Cacna1c^{D1874stop}*, *Cacna1c^{ΔDCT}*) In Ca$_V$1.2, Ca$_V$1.3, and Ca$_V$1.4 Ca^{2+} channels, α-helical regions of the distal (distal C-terminal regulatory domain, DCRD) and proximal (PCRD) C-terminus undergo a non-covalent intramolecular interaction and thereby form a regulatory domain that reduces channel open probability and modulates the voltage- and Ca^{2+}-dependent gating properties of these channels (Zamponi et al., 2015). The distal C-terminus, in addition to DCRD, also harbors other regulatory domains including a terminal PDZ-binding domain and a leucine zipper-like domain able to bind A-protein kinase-anchoring protein (AKAPs, Gray et al., 1998). Moreover, the C-terminus can undergo truncation by proteolytic processing or alternative splicing *in vivo* (Zamponi et al., 2015). In the case of Ca$_V$1.2 α$_1$, this can generate two size forms (~240 and ~210 kDa) of α$_1$-subunits. In heterologous expression systems, the truncated form enhances activity of Ca$_V$1.2. It appears that the short C-terminal

fragment can still bind to the PCRD and thereby play an autoinhibitory role, which is relieved by PKA phosphorylation of residues within the PCRD (see below). Based on *in vitro* data, it was therefore expected that Ca$_V$1.2 activity would be increased, e.g., in heart and vascular smooth muscle, in mice expressing only the truncated variant of Ca$_V$1.2 and activation by PKA would be lost.

The role of the distal C-terminus in Ca$_V$1.2 α$_1$ has been studied independently in two mouse models expressing C-termini truncated close to the proposed endogenous cleavage site (glycine-1793, *Cacna1c^{ΔDCT}*, reference sequence UniProt Q01815-1; Fu et al., 2011) or 81 residues downstream (aspartate-1874; corresponds to aspartate-1904 in original publication *Cacna1c^{D1874stop}*; Domes et al., 2011). Homozygous offspring from both mouse models died in utero (*Cacna1c^{ΔDCT}*) or immediately after birth due to severe cardiac hypertrophy, contractile dysfunction, and the failure to upregulate heart rate after birth (Domes et al., 2011) leading to early heart failure. Regulation by PKA was also lost (Fu et al., 2011). In both cases, Ca$_V$1.2 α$_1$-subunit protein (but not mRNA) expression and L-type currents were severely reduced (to ~25% of WT) in cardiomyocytes from embryos or newborn pups. No reduction of Ca$_V$1.2 α$_1$ expression was found in vascular smooth muscle, (which predominantly expresses a "smooth muscle" splice variant of Ca$_V$1.2 α$_1$ with different N-terminus and IVS3-S4 linker; Domes et al., 2011). This could be explained in part by enhanced proteolytic degradation of the truncated cardiac but not the smooth muscle variant (*Cacna1c^{D1874stop}*; Domes et al., 2011). The development of heart failure in hearts with a reduction of L-type current to about 25% is in line with the observations in KO mice as discussed above. Major take-home messages from these KO models were that truncation of the C-terminus has splice variant and therefore tissue-specific effects and that the observation of enhanced activity of truncated Ca$_V$1.2 channels observed in heterologous systems cannot be translated into the situation *in vivo*.

Germline C-Terminal PKA Phosphorylation Site Mutations (*Cacna1c^{S1700A} Cacna1c^{S1700A/T1704A}*, KI) The discovery that phosphorylation of two sites in the C-terminal tail, Ser-1700 and Thr-1704, is required for normal basal activity and PKA stimulation by blocking the autoinhibitory intramolecular interaction of the distal with the proximal C-terminal tail (see above, Fuller et al., 2010) prompted the independent generation of $Ca_V1.2$ mutant mice with either both sites (*Cacna1c^{S1700A/T1704A}*; Fu et al., 2013; Poomvanicha et al., 2017) or only Ser-1700 (*Cacna1c^{S1700A}*; Fu et al., 2014; Yang et al., 2016a) mutated to alanine. In cardiomyocytes from homozygous $Ca_V1.2^{S1700A/T1704A}$ mice, basal L-type currents and $Ca_V1.2$ α_1 protein expression were markedly (30–40%) reduced. However, the percent stimulation of these smaller currents by isoproterenol was only slightly reduced or similar to WT channels in cardiomyocytes of mice > 4 weeks of age (Fu et al., 2013; Poomvanicha et al., 2017). Mutant channels appeared less responsive to low isoproterenol concentrations. The isoproterenol-induced increase in cell shortening of electrically stimulated myocytes was significantly reduced in *Cacna1c^{S1700A/T1704A}* myocytes. Mutant mice showed reduced exercise capacity, developed cardiac hypertrophy, and (in one of the models, Poomvanicha et al., 2017) died of heart failure prematurely. This phenotype may, at least in part, be due to the smaller Ca^{2+} currents as expected from KO mice (see above). Similar findings on Ca^{2+} current and adrenergic modulation were seen in *Cacna1c^{S1700A}* mice, although with a slightly less severe cardiac phenotype (Yang, Dai, et al., 2016a). A major open question remaining from these studies is if mutations of these sites reduced Ca^{2+} current by preventing basic phosphorylation of $Ca_V1.2$ or by reducing $Ca_V1.2$ α_1-subunit protein expression leaving less channels for (almost intact) stimulation by PKA. A pronounced reduction of cardiomyocyte $Ca_V1.2$ α_1-subunit protein expression was found in one (Poomvanicha et al., 2017) but not the other (Fu et al., 2013) of the two independently generated *Cacna1c^{S1700A/T1704A}* mouse models. Taken together these data showed that Ser-1700 and Thr-1704 are required for normal $Ca_V1.2$ activity in the heart but still permit PKA modulation of the channel.

Transgenic Cardiomyocyte-Specific Expression of $Ca_V1.2$ PKA-Phosphorylation Site Mutations The hypothesis that Ser-1700 and Thr-1704 are key determinants of $Ca_V1.2$ regulation by protein phosphorylation was also challenged by transgenic mouse models with cardiomyocyte-specific expression of $Ca_V1.2$ α_1-subunits in which 51 residues at 35 potential phospho-regulatory domains or 37 residues at 28 putative phospho-regulatory domains of β_{2b}-subunits (Liu et al., 2020) were replaced by alanine. Transgenic constructs were generated by fusing rabbit $Ca_V1.2$ α_1 cDNA to the modified murine α-MHC tetracycline-inducible-promoter vector with a reverse tetracycline-controlled transactivator also driven by the α-MHC promoter (TET-ON). $Ca_V1.2$ α_1 constructs were epitope-tagged and rendered DHP-insensitive (T1066Y and Q1070M mutations) allowing inhibition of endogenous but not exogenous channels with nisoldipine (Liu et al., 2020). These mutations did not affect typical modulation of L-type currents by isoproterenol or forskolin (an adenylyl cyclase activator). Using the same strategy, prevention of the C-terminal cleavage of $Ca_V1.2$ did not affect adrenergic stimulation (Katchman et al., 2017). This strongly suggested that neither direct phosphorylation of $Ca_V1.2$ α_1 nor of β_{2b} subunits is required.

Therefore, another transgenic mouse model was developed to identify components of the $Ca_V1.2$ macromolecular complex that could serve as PKA substrates. Transgenic mice with cardiomyocyte-specific expression of a DHP-resistant $Ca_V1.2$ α_1 ($Ca_V1.2$-APEX2 mice) or β_{2b} subunits (β_{2b}-APEX2 mice) (Liu et al., 2020) conjugated with ascorbate peroxidase (APEX2) to the amino termini enabled biotin labeling of proteins within around 20 nm of the Ca^{2+} channels. This proximity-based labeling approach eventually led to the identification of the small Ras-like GTP-binding protein Rad known to inhibit Ca^{2+} channels by binding to β-subunits. This allowed to discover a novel regulatory mechanism in which PKA-phosphorylation of Rad induces its dissociation from the channel complex causing current stimulation by relieving its tonic inhibition of $Ca_V1.2$ channels (Liu et al., 2020).

Germline Calmodulin-Interaction Site Mutations (*Cacna1c^{J1624E}*) Mutant mice also allowed to address the important question about the role of calmodulin binding to the proximal C-terminus ("IQ-motif"), which affects Ca^{2+}-dependent inactivation and facilitation of $Ca_V1.2$, two processes essential for the short-term fine-tuning of cardiac function. Disruption of CaM binding to the IQ motif by mutation of Ile-1624 (Uniprot Q01815) to Glu in the IQ motif abolished regulation of the channel by CDI and CDF when expressed in heterologous systems (Zühlke et al., 1999; Zamponi et al., 2015). *Cacna1c^{J1624E}* mice were generated but homozygous animals were not viable. Therefore, KI of the mutation into mouse cardiomyocytes was achieved in mice heterozygous for the *Cacna1c^{J1624E}* mutation also carrying a floxed WT allele, which could be inactivated by tamoxifen-induced cardiac-specific activation of the MerCreMer *Cre* recombinase (Poomvanicha et al., 2011; Blaich et al., 2012). These mice developed dilated cardiomyopathy with signs of myocyte apoptosis and cardiac fibrosis within 10 days after the first tamoxifen injection and died within 3 weeks. Survival rates were similar to *Cacna1c* KO mice. After 10 days, $Ca_V1.2$ α_1-subunit expression and Ca^{2+} current were reduced to about 50% in ventricular cardiomyocytes as compared to heterozygous KO controls. As predicted from heterologous expression, not only current density was decreased but the mutation also abolished CDI and facilitation (Poomvanicha et al., 2011; Blaich et al., 2012). Interestingly, it seemed to stabilize fast inactivation kinetics as seen in the presence of Ca^{2+} in WT. Reduced Ca^{2+} influx resulted in a smaller global Ca^{2+} transient, which was compensated by neuroendocrine activation resulting also in an enhanced of E–C coupling gain. It is likely that the consequences of altered gating (no CDI and facilitation) were confounded by the low expression of $Ca_V1.2$ protein (even enhanced by the KO of the WT allele), which by itself can explain the development of dilated cardiomyopathy and death in these mice.

Germline Serine-S1928 Phosphorylation-Site Mutation (*Cacna1c^{S1928A}*) Ser-1928 is located in the distal C-terminus of the full-length $Ca_V1.2$ α_1-subunit and is a major site phosphorylated by PKA in cardiomyocytes (Hulme et al., 2006; Lemke et al., 2008) and neurons (Hall et al., 2007). However, in contrast to predictions from heterologous systems (Gao et al., 1997), Ser-1928 turned out to be not relevant for adrenergic modulation in cardiomyocytes. This was demonstrated in mice in which Ser-1928 was mutated to an alanine (Lemke et al., 2008). They showed normal isoproterenol and forskolin stimulation of Ca^{2+} currents in cardiomyocytes, unchanged association of PKA subunits with $Ca_V1.2$ channels, and no change of basal heart rate or of the ISO-induced stimulation of heart rate and cardiac contractility. However, this mouse model allowed demonstrating the physiological relevance of Ser-1928 for fine-tuning $Ca_V1.2$ channel function in neurons and in vascular smooth muscle. Like in the heart $Ca_V1.2$ in neurons forms a signaling complex with the PKA-signaling machinery, including β_2-adrenergic receptor (β_2AR), adenylate cyclase, heterotrimeric G-protein Gs, an A-kinase anchoring protein (AKAP), and PKA (Qian et al., 2017). β_2AR stimulation activates $Ca_V1.2$ currents in neurons and thereby permits $Ca_V1.2$ to contribute to prolonged theta-tetanus - induced long-term potentiation (LTP) in hippocampal neurons. This regulation is lost in *Cacna1c^{S1928A}* but not in *Cacna1c^{S1700A/T1704A}* (Fu et al., 2013) neurons (Qian et al., 2017). Moreover, phosphorylation of Ser-1928 interrupts the protein–protein interaction of the $Ca_V1.2$ α_1 C-terminus with the β_2AR and this temporary displacement makes the channel transiently insensitive to adrenergic stimulation (Patriarchi et al., 2016).

This delicate role of Ser-1928 raises the important question about its physiological role in regulating more complex brain functions. A first answer for this question already came from studies investigating the role of $Ca_V1.2$ for drug-taking behaviors. $Ca_V1.2$ in dorsal dentate gyrus neurons, including a dopamine D1-receptor-expressing subpopulation, was found to be required for the extinction but not acquisition of cocaine conditioned-place preference (CPP) (Burgdorf et al., 2020). Interestingly, phosphorylation of Ser-1928 is not required for extinction but, unlike WT mice, *Cacna1c^{S1928A}* mutants failed to reinstate cocaine CPP, following a cocaine challenge after successful extinction.

In arterial smooth muscle cells of different vascular beds, elevated glucose levels (20 mM) increase Ca$_V$1.2-mediated Ca²⁺ currents, intracellular Ca²⁺ levels, and induce vasoconstriction, a finding considering the development of hypertension in diabetic patients. This glucose stimulation requires the association of the channel with PKA through a bound AKAP and it is inhibited by PKA inhibitors (Nystoriak et al., 2017). It is associated with increased phosphorylation of Ser-1928. In *Cacna1c^{S1928A}* mice (Lemke et al., 2008), this glucose-dependent enhancement of phosphorylation of Ser-1928 is absent as are the stimulation of L-type current and of intracellular Ca²⁺ levels. Myogenic tone in *Cacna1c^{S1928A}* arteries is normal but high glucose-induced contraction is absent. Interestingly, this glucose-induced PKA-dependent modulation was also observed in human arterial myocytes and myocytes of human diabetic subjects (Nystoriak et al., 2017) underscoring the significance of these findings for human disease.

Taken together, these data show that Ca$_V$1.2 PKA is essential for the regulation of many physiological processes but the molecular mechanisms coupling PKA activity to Ca$_V$1.2 channel function are diverse.

Germline Exon33 Deletion (*Cacna1c^{ΔE33}*)

In rat models of chronic myocardial infarction, changes in alternative splicing can affect the gating of Ca$_V$1.2 channels. In particular, Ca$_V$1.2 transcripts excluding alternative exon 33 was increased in the scar regions of the infarction area (Liao et al., 2009). This exon encodes part of the IVS3-S4 linker and its absence in Ca$_V$1.2 results in a more negative voltage-dependence of activation and inactivation. In agreement with these findings, in cardiomyocytes of mice with a homozygous deletion of this exon (*Cacna1c^{ΔE33}*; Li et al., 2017), voltage dependence of activation and inactivation was shifted by more than 10 mV to negative voltages, and Ca²⁺ current amplitude was increased at negative voltages resulting in larger inward current during action potentials. This resulted in increased contractility and a prolongation of the action potential. *In vivo* telemetric ECG recordings revealed tachycardia, prolonged QT interval, and increased susceptibility to premature ventricular contractions and ventricular tachycardia. Chronic administration of ISO reduced survival rates of homozygous *Cacna1c^{ΔE33}* mice to about 30% (about 80% in WT) indicating their higher susceptibility to adrenergic stimulation-induced heart pathology. Interestingly, exon 33-containing Ca$_V$1.2 α1-subunit transcripts were significantly upregulated in human myocardium from individuals with heart failure as compared to samples from non-failing hearts (Li et al., 2017). Although the pathogenetic significance of this finding remains speculative, it could represent a compensatory mechanism to reduce Ca$_V$1.2-induced susceptibility to cardiac arrhythmias.

Auditory Function: Conditional KI Mice

In addition to Ca$_V$1.3, Ca$_V$1.2 channels are also expressed in spiral ganglion neurons, and heterozygous deficiency of either channel leads to a reduced number of these neurons at 2–4 months of age (Lv et al., 2014). Both are also expressed in neurons of auditory nuclei. To investigate the role of Ca$_V$1.2 for auditory function and its effects on BDNF signaling, Cav1.2 was conditionally deleted under the Pax2 promoter (*Cacna1c^{flox/flox}/Pax2-Cre; Cacna1c^{Pax2}*) with deletion in the cochlear spiral ganglion neurons (but also in the dorsal cochlear nucleus, inferior colliculus and cerebellum) or under the Egr2 promotor (*Cacna1c^{flox/flox}/Egr2-Cre; Cacna1c^{Egr2}*) for deletion of Ca$_V$1.2 in the auditory brainstem nuclei (lateral superior olive and medial nucleus of the trapezoid body) (Zuccotti et al., 2013). Both KO mouse models showed normal hearing thresholds, an intact number of inner hair cell (IHC) release sites and normal outer hair cell (OHC) function. Detailed analysis of ABR waveforms revealed a slight deficit of wave I amplitudes at higher sound pressure levels, indicating impaired neural activity from high-response threshold afferent nerve fibers in *Cacna1c^{Pax2}* mice. This beneficial effect of $Ca_V1.2$ was in contrast to the finding that *Cacna1c^{Pax2}* mice had less pronounced noise-induced hearing loss than their WT controls and, unlike WT, no decrease in the number of ribbon synapses in IHCs 7 days after exposure to 1 h of broadband noise at 120 dB of sound pressure level. A reduced expression of BDNF was also found in the cochlea of *Cacna1c^{Pax2}* mice. Together, these data indicate a beneficial role of $Ca_V1.2$ in the cochlea (but not in the brainstem nuclei) for normal function of afferent nerve fibers and a likely BDNF-mediated harmful effect on hearing function during acoustic trauma (Zuccotti et al., 2013).

A closer examination of the brainstem of *Cacna1c^{Egr2}* mice demonstrated that $Ca_V1.2$, like $Ca_V1.3$ (see below) is necessary for neuronal survival of auditory brain stem nuclei (Ebbers et al., 2015). Despite essentially normal auditory brainstem responses in these mice, the loss of $Ca_V1.2$ caused a significant decrease in the volume and cell number in auditory nuclei, including the lateral superior olive (LSO) and medial nucleus of the trapezoid body (MNTB). Like in *Cacna1d* KOs (see below) cell loss occurred between postnatal days P0 and P4, suggesting that both channel types are required for neuronal survival and that they, despite their co-expression these cells, cannot substitute for one another. Unlike *Cacna1d* KO, $Ca_V1.2$ deficiency did not affect the firing behavior of LSO neurons but only narrowed the width of action potentials.

Brain Function of $Ca_V1.2$: Conditional KO and KI Mice

Early studies using radiolabeled and unlabeled Ca^{2+} channel blockers (such as DHPs) and activators (e.g., the DHP BAYK8644) demonstrated the presence of DHP-sensitive Ca^{2+} current components not only in their primary therapeutic target, the cardiovascular system, but also in essentially all electrically excitable cells, including the brain (Bellemann et al., 1982; Glossmann & Striessnig, 1988). The role of LTCCs for *in vivo* brain function was difficult to study for several reasons. First, no CNS side effects have been reported for brain permeable DHPs in humans. Second, *in vivo* effects of Ca^{2+} channel blockers are confounded by their prominent cardiodepressant and blood pressure lowering effects and the short half-life of these drugs in rodents (Waltereit et al., 2008). Third, $Ca_V1.2$ and $Ca_V1.3$ are both expressed throughout the brain (see below) but selective inhibitors reliably discriminating between them do not exist (Ortner et al., 2014). For these reasons, mouse animal models became instrumental in depicting the different roles of these channels. Due to their overlapping expression and function in the brain, neuronal $Ca_V1.2$ and $Ca_V1.3$ are discussed together in a separate section below.

L-type Channels in Glial Function, Myelination and Remyelination: Conditional *Cacna1c* KO and Conditional Transgenic $Ca_V1.2$ α_1 shRNA Expression

As described above, a gain-of-function $Ca_V1.2$ Timothy syndrome mutation appears to increase the percentage of myelinated axons and the thickness of myelin sheaths (Cheli et al., 2018). This, however, is not the only hint for a role of L-type channels in glial function. In Ca^{2+} imaging studies of astrocytes in culture or cortical slices and of cultured oligodendrocyte progenitor cells (OPCs), K^+ depolarization induces a rise in intracellular Ca^{2+} concentrations which is inhibited by the Ca^{2+} channel blockers nifedipine or verapamil (Cheli et al. 2016a, b; Zamora et al., 2020). In OPCs, $Ca_V1.2$ α_1-subunit siRNA knockdown (KD) also reduces this Ca^{2+} transient, decreases the proportion of myelin-expressing and enhances the proportion of immature oligodendrocytes, and impairs OPC proliferation (Cheli et al., 2015). Moreover, in astrocytes, $Ca_V1.2$ α_1 (but not $Ca_V1.3$ α_1) subunit expression is upregulated in reactive astrocytes after brain injury, hypomyelination, and ischemia (Westenbroek et al., 1998).

For further studies, several animal models were generated. Floxed *Cacna1c* mice (White et al., 2008) were crossed with NG2-creERT transgenic mice to conditionally delete $Ca_V1.2$ in OPCs (*Cacna1c^{NG2-creERT}*) and with GFAP-creERT mice for ablation in astrocytes (*Cacna1c^{GFAP-creERT}*). In addition, triple transgenic mice were generated by crossing *Cacna1c^{NG2-creERT}* mice with Cre-reporter mice leading to eGFP expression in *Cre* expressing cells. This provides a marker to monitor *Cacna1c^{NG2-creERT}* OPCs in the postnatal brain (Cheli et al., 2016a).

Tamoxifen-induced *Cacna1c* KO in OPCs during the first two postnatal weeks caused hypomyelination in the brain, a decreased number of mature oligodendrocytes, reduced expression of myelin proteins in several areas of the brain, and an important decrease in the percentage of myelinated axons. Depolarization-induced Ca^{2+} transients were also diminished in KO OPCs. A significant decline in the density of Olig2-positive cells and of myelinating oligodendrocytes was observed in the corpus callosum and cortex, due to the production of fewer myelinating oligodendrocytes in progenitors from KO mice than from controls. These changes were more pronounced in mice treated with tamoxifen at P4 and P10 but not at P30. Treatment at P10 resulted in reduced myelin protein expression when analyzed at P60, indicating that the hypomyelination persists throughout adulthood. Therefore, $Ca_V1.2$ channels appear to be important for oligodendrocyte maturation (Cheli et al., 2016a). Using these mouse lines, the same authors showed that $Ca_V1.2$ in oligodendrocytes also controls remyelination (Santiago González et al., 2017). In the cuprizone model of demyelination, $Ca_V1.2$ deletion in OPCs caused less efficient remyelination in the adult brain. During recovery from cuprizone-induced demyelination, $Ca_V1.2$-deficient OPCs matured more slowly and produced less myelin than control oligodendrocytes. The rate of OPC proliferation and maturation, the

number of myelinating oligodendrocytes, and the percentage of myelinated axons in the corpus callosum were reduced during the remyelination process (Santiago González et al., 2017). In NG2-positive OPCs, activation of $Ca_V1.2$ channels may occur by small synaptic inputs causing membrane depolarization from their relatively negative resting membrane potential (Cheli et al., 2016a for references).

Since OPCs proliferate and differentiate throughout life, Pitman et al. (Pitman et al., 2020) extended these studies by conditional deletion of $Ca_V1.2$ in adult mice by generating *Cacna1c^{PDGFRA-creERT}* mice to delete $Ca_V1.2$ α_1 in adult OPCs by administration of tamoxifen at P60. Deletion was verified by showing an about 60% reduction of L-type current by whole-cell patch-clamp recoding analysis in brain slices. They found that conditional deletion in adult OPCs significantly increased proliferation but did not affect the number of newly produced oligodendrocytes. Moreover, they found that $Ca_V1.2$ deletion was required for survival in a subpopulation of adult OPCs in the corpus callosum (but not in motor cortex). These data clearly show differences in the contribution of $Ca_V1.2$ channels for developing and adult OPCs.

However, the function of OPCs appears to be linked to an additional role of $Ca_V1.2$ channels in astrocytes. Ablation of $Ca_V1.2$ in astrocytes in *Cacna1c^{GFAP-creERT}* mice reduced depolarization-induced intracellular Ca^{2+} transients to about 28% in cultured astrocytes as compared to WT controls. Astrogliosis was reduced in the cuprizone-induced demyelination model (Zamora et al., 2020). The number of reactive astrocytes and of mitotic astrocytes declined in the cortex and corpus callosum of KO brains and also microglial activation markers were reduced. Interestingly, *Cacna1c* KO in astrocytes enhanced remyelination during 2- and 4-week recovery following cuprizone treatment and lead to thicker myelin sheaths in the corpus callosum apparently by affecting oligodendrocyte function. Enhanced remyelination in KO brains could be explained by a significant increase in the number of myelinating oligodendrocytes and in the density of proliferating OPCs, indicating an expansion in the OPC pool required for remyelination *Cacna1c* KO not only reduced astrocyte and microglia activation during demyelination but also favored myelin regeneration and for the development of new OPCs in the adult brain. A similar increase in remyelination was also observed by treating WT animals with the Ca^{2+} channel blocker nimodipine (s.c. administration every other day; Zamora et al., 2020) only during the last 3 weeks of cuprizone treatment. Again, the observed drug-induced reduction of astrocyte and microglia activation during demyelination seems to be beneficial for myelin regeneration and for the development of new OPCs in the adult brain.

These data nicely confirm and extend previously published evidence for nimodipine treatment attenuating symptoms of experimental autoimmune encephalomyelitis (a model for multiple sclerosis) and spinal cord degeneration in mice which was also associated with enhanced remyelination and increased numbers of oligodendrocytes (Schampel et al., 2017). Nimodipine treatment also promoted microglia apoptosis accompanied by decreased production of nitric oxide and reactive oxygen species. No evidence for $Ca_V1.2$ α_1 expression was obtained in this study, suggesting an effect independent of Ca^{2+} channels in microglia (Schampel et al., 2017).

However, the role of $Ca_V1.2$ channels in microglia is still unresolved because another study demonstrated that nifedipine treatment or KD of $Ca_V1.2$ in MG6 cells, a murine microglial cell line, promoted expression of inducible NO synthase and promoted neuroinflammatory microglia (Wang et al., 2019). To further test this *in vivo*, a conditional KD mouse was generated, where $Ca_V1.2$ expression was suppressed specifically in microglia/macrophages by tamoxifen-inducible RNA interference, whereby creERT2 was driven by a CD11b promoter, controlling the expression unit for shRNA targeting $Ca_V1.2$ (*Cacna1c^{KD}*). Microglia prepared from tamoxifen-treated *Cacna1c^{KD}* mice showed a ~40% decrease in $Ca_V1.2$ α_1 transcripts compared to that in the controls without affecting $Ca_V1.3$ α_1 expression. In a MPTP Parkinson disease model,

$Ca_V1.2$ KD increased the number of iNOS-positive, proinflammatory microglia, promoted degeneration of dopaminergic neurons, and worsened behavioral deficits (Wang et al., 2019).

While these data provide evidence for a neuroprotective/anti-neuroinflammatory role of $Ca_V1.2$ in microglia, $Ca_V1.2$ channels also control glial functions in astrocytes and oligodendrocytes. The favorable effect of pharmacological L-type channel inhibition on neurodegeneration and remyelination will therefore depend on the complex interplay of these glial cells in disease as seen from the protective effects of selective KO of $Ca_V1.2$ through positive effects on OPC development and perhaps indirect inhibition of inflammatory microglia. It is therefore possible that inhibition of astrocyte activation through $Ca_V1.2$ inhibition during demyelinating or neurodegenerative lesions (Zamora et al., 2020) outweighs the negative effects of $Ca_V1.2$ inhibition in OPCs and microglia cells.

$Ca_V1.3$

In heterologous expression systems, $Ca_V1.3$ LTCCs show a much more negative operation voltage range as compared to $Ca_V1.2$ (Koschak et al., 2001; Zamponi et al., 2015; Ortner et al., 2017). Like for $Ca_V1.2$, an intramolecular PCRD and DCRD interaction exists in the long splice variants of the α_1-subunit, which also forms a C-terminal modulatory domain. Its modulatory properties are slightly different as compared to $Ca_V1.2$. C-terminally short splice variants of $Ca_V1.3$ α_1- subunits lacking the DCRD domain are generated by alternative splicing (Bock et al., 2011; Tan et al., 2011). In these short variants the activation voltage range is shifted to even more negative potentials, and open probability and Ca^{2+}-dependent inactivation (due to higher affinity of the channel for calmodulin) are increased (Bock et al., 2011; Tan et al., 2011). Indeed, mutant mice provided valuable insight into $Ca_V1.3$ function and show that, largely due to their different gating properties, $Ca_V1.2$ and $Ca_V1.3$ cannot compensate for each other functions.

Cardiac Excitability and Inotropy: Constitutive KO and Ca$_V$1.2DHP$^{-/-}$ Mice

Constitutive (Platzer et al., 2000; Namkung et al., 2001) and conditional (Satheesh et al., 2012) *Cacna1d* KO mice revealed the key role of this channel for cellular functions requiring Ca²⁺ channel activity at subthreshold voltages. Unlike for Ca$_V$1.2, homozygous loss of Ca$_V$1.3 function is not lethal in mice (Platzer et al., 2000; Namkung et al., 2001), although prenatal mortality appears to vary in different KO strains (Berger & Bartsch, 2014; Tatsuki et al., 2016). It is therefore tempting to speculate that this could be due to the residual expression of some embryonic splice variants lacking a gene-disrupting neocassette (Tatsuki et al., 2016). *Cacna1d*$^{-/-}$ mice (Platzer et al., 2000) reach a normal life-span but are congenitally deaf and exhibit a pronounced SAN dysfunction at rest (Platzer et al., 2000; Namkung et al., 2001). Human individuals with a homozygous loss-of-function of exon 8-containing Ca$_V$1.3 α_1-subunits, which are the predominant variant in SAN and cochlear IHCs, show the same phenotype (Sinoatrial Node Dysfunction and Deafness, SANDD, OMIM #614896) confirming the translational value of the findings in mice (Baig et al., 2011).

In the SAN, Ca$_V$1.3 is one of the essential pacemaker channels promoting spontaneous diastolic depolarization which is enabled by its negative activation threshold (Platzer et al., 2000; Namkung et al., 2001). It is embedded in a complex pacemaking framework of different ion channels and their functional coupling to RyR2, the SR Ca²⁺ pump, and the Na⁺/Ca²⁺ exchanger (Torrente et al., 2020 for review). Mouse models also were essential to depict the structural basis for the highly DHP-sensitive sustained inward Na⁺ current (I$_{ST}$), which provides an additional mechanism for Ca$_V$1.3 to drive diastolic depolarization. Unexpectedly, Ca$_V$1.3 α_1-subunits also form the molecular substrate of I$_{ST}$ as independently confirmed in two mouse models. In SAN cells of *Cacna1d*$^{-/-}$ mice (Platzer et al., 2000), I$_{ST}$ was abolished. In contrast, its DHP sensitivity

was preserved in Ca$_V$1.2DHP$^{-/-}$ mice (Sinnegger-Brauns et al., 2004; Toyoda et al., 2017), in which DHP sensitivity (but not function) was eliminated in Ca$_V$1.2, ruling out a contribution of Ca$_V$1.2 for I$_{ST}$ (Toyoda et al., 2017). Unfortunately, these studies did not provide more insight about how Ca$_V$1.3 channels either become Na⁺-permeable or indirectly contribute to a Na⁺-conductance.

In the embryonic heart, Ca$_V$1.3 is the predominant L-type current but its expression decreases during development while Ca$_V$1.2 increases (see also section "Ca$_V$1.2", Xu et al., 2003; Takemura et al., 2005). In the adult heart, Ca$_V$1.3 is absent in *Cacna1d*$^{-/-}$ ventricular myocytes but expressed at a low level in the atria (Mancarella et al., 2008), where it contributes about 25% of the total inward Ca²⁺ current (Mancarella et al., 2008). In atrial myocytes of *Cacna1d*$^{-/-}$ mice, AP-triggered intracellular Ca²⁺ transients and Ca²⁺-induced SR-Ca²⁺ release are reduced and SR Ca²⁺ content is lower than in WT controls. *Cacna1d*$^{-/-}$ hearts (Platzer et al., 2000; Namkung et al., 2001) are more susceptible to atrial fibrillation induced by electrical stimulation (Zhang et al., 2005; Mancarella et al., 2008). Action potentials are broadened due to smaller SK-channel currents in the mutants (Namkung et al., 2001; Lu et al., 2007). These data clearly demonstrate an important role of Ca$_V$1.3 for normal atrial excitability.

Atrial Ca$_V$1.3 channels also appear to tightly control brain natriuretic peptide (BNP) production as evident form a >90% decrease of BNP mRNA abundance in atria of heterozygous Ca$_V$1.3$^{+/-}$ mice (Platzer et al., 2000) associated with an about 80-fold decrease in serum BNP concentrations (Srivastava et al., 2017). Interestingly, in failing human hearts Ca$_V$1.3 is re-expressed also in the ventricles, which may serve as a mechanism to sustain contractile function and compensate for reduced expression of Ca$_V$1.2 (Srivastava et al., 2020). This observation is also in accordance with Ca$_V$1.3 expression in fetal, but not adult, cardiomyocytes in *Cacna1c* KO mice (see above) and the fact that fetal gene expression is commonly observed in chronic heart failure.

Cochlear Inner Hair Cell Function and Hearing: Constitutive and Conditional KO

Deafness in mice and humans is explained by the predominant role of $Ca_V1.3$ for tonic glutamate release from IHCs. In contrast to fast neurotransmitter release at neuronal nerve terminals mediated by Ca_V2 channels (Catterall, 2011), L-type channels (predominantly $Ca_V1.3$ in IHCs and $Ca_V1.4$ in photoreceptors) closely associate with presynaptic vesicles at the synaptic ribbons in sensory cells (Brandt et al., 2005; Scharinger et al., 2015). With its negative operation range, $Ca_V1.3$ channels provide graded Ca^{2+} influx in response to the tonic sound-induced changes in membrane potential between about -60 and -35 mV in these cells and thereby trigger the release of glutamate to afferent fibers of the auditory nerve. $Ca_V1.3$-deficient mice are deaf, show a rapid postnatal loss of afferent nerve fibers, IHC and outer hair cells, and stay immature (Platzer et al., 2000; Namkung et al., 2001). This is evident from their failure to upregulate voltage- and Ca^{2+}-activated BK channels but with persistent expression of small conductance Ca^{2+}-activated SK2 channels (Eckrich et al., 2019). In addition, $Ca_V1.3$ is also required for normal neuronal development of nuclei in the auditory brainstem (Hirtz et al., 2011). In $Cacna1d^{-/-}$ mice (Platzer et al., 2000), the volume of all auditory brainstem centers was reduced with fewer neurons and striking malformations of the LSO (Hirtz et al., 2011).

This developmental deficit was also present upon conditional deletion of $Ca_V1.3$ only in the auditory brainstem, and not in IHCs. For this conditional KO of $Ca_V1.3$, $Cacna1d^{eGFPflex}$ mice were generated (Satheesh et al., 2012; Berger & Bartsch, 2014) in which the conditional allele is based on the Cre/loxP-based FLEX system (Flip excision, Schnütgen et al., 2003). By ablation of the $Ca_V1.3$ α_1 protein, expression of eGFP is induced under the endogenous $Cacna1d$ promoter. This allows the germline or conditional deletion of $Ca_V1.3$ α_1 by crossing animals with an appropriate Cre-deleter mouse line. eGFP fluorescence or immunoreactivity then allows identi-

fication of $Ca_V1.3$ expressing cells. Homozygous germline inactivation of the $Cacna1d$ gene did not yield viable offspring but heterozygous animals allowed the detection of eGFP-positive (expressed from one allele) $Ca_V1.3$ α_1-expressing cells in the brain and other tissues (Berger & Bartsch, 2014). Using this conditional model, cochlea-specific $Cacna1d$ KO mice induced by breeding $Cacna1d^{eGFPflex}$ mice with $Egr2$-Cre mice specifically deleted $Ca_V1.3$ α_1 in the auditory brainstem, not touching IHCs and leaving inner ear function intact (Satheesh et al., 2012). This model recapitulated the pathological morphological changes in the superior olivary complex observed in constitutive KO mice and therefore confirmed the key role of $Ca_V1.3$ channels in the auditory system beyond its role in cochlear hair cells. Conversely, cochlea-specific $Ca_V1.3$ α_1 KO was induced by crossing $Cacna1d^{eGFP-flex}$ mice with $Pax2$-Cre mice, inducing Cre activation in the mature organ of Corti and SGNs but not in the ventral cochlear nucleus and the superior olivary complex (Eckrich et al., 2019). Accordingly, GFP fluorescence was present in acutely dissected IHCs of 3-week old KO mice. At three weeks of age, the switch of both flex alleles was incomplete and resulted in IHCs with different levels of GFP fluorescence which correlated with the observed current densities. It was concluded that only those IHCs with strong fluorescence represented true KO cells with two switched flex alleles. Instead, weak fluorescence indicated cells with only one switched flex allele. Mice carrying one KO and one conditional allele ($Cacna1d^{-/eGFPflex-cre}$, not crossed with Cre mice) had profound hearing loss, BK channels were absent, and SK2 channels persisted in IHCs, as observed previously in $Cacna1d^{-/-}$ mice. Extensive characterization of this mouse model revealed further potential drawbacks. First, GFP expression can induce cellular toxicity in IHCs evident from the degeneration of IHCs of the basal cochlear turn in $Cacna1d^{eGFPflex}/Pax2$-Cre mice not observed in $Cacna1d$ $-/-$ at 4-5 weeks of age. GFP toxicity has also been described in other cells, including neurons (Eckrich et al., 2019 for reference). This has to be considered also when studying the phenotype of other cell types with this model. Second, the flex allele alone affected $Ca_V1.3$ α_1-subunit expression and current density in *Cre*-negative homozygous $Ca_V1.3$ $\alpha_1{}^{eGFPflex}$ mice.

Intestinal Ca²⁺ Absorption and Bone Development: Constitutive KO

$Ca_V1.3$ also appears to affect skeleton development in male mice. Body weight was lower, and both femoral bone mineral density and bone mineral content are significantly lower in adult male *Cacna1d⁻ᐟ⁻* (Platzer et al., 2000) than in WT mice (Li et al., 2010). This difference was not seen in females. Responses to osteogenic or bone anabolic stimuli (mechanical load, parathyroid hormone) were normal (Li et al., 2010). In P14 pups, femur growth plate thickness is larger in *Cacna1d⁻ᐟ⁻* mice (Platzer et al., 2000) than in WT, whereas no other differences were observed in trabecular or cortical bone (Beggs et al., 2019). These findings suggest a delayed bone mineralization in *Cacna1d⁻ᐟ⁻* mice. This is likely due to an important but unexpected role of $Ca_V1.3$ channels (together with TRPV6 channels) for net jejunal Ca²⁺ absorption in mice before weaning (Beggs et al., 2019). It is therefore possible that bone mineral content differences in adult mice result from reduced intestinal Ca²⁺ absorption early in life.

Endocrine Function: Constitutive KO and Ca_V1.2DHP⁻ᐟ⁻ KI

Together with $Ca_V1.2$, $Ca_V1.2$ is also expressed in endocrine cells. While in mouse pancreatic β-cells, $Ca_V1.2$ appears to be the predominant L-type channel underlying the L-type current component and early-phase insulin secretion (see above), the role of $Ca_V1.3$ is less clear. It does not appear to contribute much to L-type current and insulin vesicle release as shown in conditional *Cacna1c⁻ᐟ⁻* (see above) and *Cacna1d⁻ᐟ⁻* mice (Platzer et al., 2000; Barg et al., 2001). However, in another *Cacna1d⁻ᐟ⁻* mouse model, Ca²⁺ current amplitude was also unaffected and mice were

normoglycemic but glucose tolerance was reduced and insulin levels were decreased in the mutants (Namkung et al., 2001). This could be explained by a marked reduction in total islet mass to about 24–30% in the KO mice compared to WT. The reduction was attributed to an impairment of postnatal β-cell generation, a lower total islet number and a lower number of cells in individual islets (Namkung et al., 2001). Interestingly, this histological phenotype was more pronounced at P14 than at P30. It is unclear if reduced β-cell mass persists into adulthood because no data were provided for mice older than P30. Nevertheless, these data imply a special role of $Ca_V1.3$ during postnatal β-cell expansion. This is a highly relevant finding in the context of the pathology of diabetes mellitus and should be pursued in further studies using animal models and human islet (Tuluc et al., 2021).

Mouse models also allowed to answer the important question about the roles of $Ca_V1.2$ and $Ca_V1.3$ in chromaffin cells. In these cells, L-type channels trigger adrenaline and noradrenaline release and support spontaneous electrical activity. Mouse chromaffin cells in culture spontaneously fire action potentials at a frequency of about 1.5 Hz (Marcantoni et al., 2010). This firing involves L-type channels because the firing rate is reduced by nifedipine and increased by the selective L-type Ca²⁺ channel activator BayK-8644. These DHPs also alter the shape of the action potentials, resulting from a close coupling of LTCCs to BK channels (Marcantoni et al., 2010). In WT chromaffin cells, $Ca_V1.3$ and $Ca_V1.2$ represent 80% and 20% of the α_1-subunit transcripts (no evidence for substantial expression of $Ca_V1.1$ or $Ca_V1.4$) (Marcantoni et al., 2010). As shown in *Cacna1d⁻ᐟ⁻* mice (Platzer et al., 2000), $Ca_V1.3$ carries about 60% of total L-type current in these cells (i.e., about 30% of the total Ca²⁺ current, Marcantoni et al., 2010). Spontaneous firing rate is mainly controlled by $Ca_V1.3$ channels. In chromaffin cells from *Cacna1d⁻ᐟ⁻* mice, the number of normally firing cells is reduced from 80% in WT to 30%. The subthreshold voltage activation range of $Ca_V1.3$ channels allows them to carry most of an interspike DHP-sensitive inward pacemaker current.

This current is drastically reduced in *Cacna1d*$^{-/-}$ mice as well as the corresponding Ca^{2+}-activated BK current during spikes. These findings again demonstrate that the negative activation voltage range of $Ca_V1.3$ channels can sustain pacemaking activity.

$Ca_V1.3$ channels are also expressed in the aldosterone-producing zona granulosa of the adrenal cortex. While mouse models have so far not provided any insight into the physiological role for aldosterone secretion, human *de novo* gain-of-function missense have clearly demonstrated that excess activity of these channels causes congenital hyperaldosteronism (Ortner et al., 2020 for review). Interestingly, hyperinsulinemic hypoglycemia is also often observed in these patients (Ortner et al., 2020), emphasizing an important role of $Ca_V1.3$ for endocrine pancreatic function not only in mice (see above) but also in humans.

Pharmacological Separation of $Ca_V1.2$ from $Ca_V1.3$ Functions: Cav1.2DHP$^{-/-}$ KI Mice

Some of the findings obtained with this mouse model have already been described in the previous chapters but additional findings are summarized here. As already mentioned above, L-type Ca^{2+} channel blockers inhibit $Ca_V1.2$ and $Ca_V1.3$ with similar affinity, although some DHPs are slightly selective for $Ca_V1.2$. In $Ca_V1.2DHP^{-/-}$ mice, the side chain of the tyrosine residue replacing threonine-1066 in the DHP-binding pocket sterically prevents proper docking of DHPs like isradipine and thereby decreases their affinity by more than two orders of magnitude (Sinnegger-Brauns et al., 2004). In the homozygous animals, $Ca_V1.2$-mediated DHP effects are prevented therefore allowing the ablation of DHP effects mediated through this channel. In the absence of $Ca_V1.1$ and $Ca_V1.4$ in most cells, any remaining DHP effect must therefore be mediated by $Ca_V1.3$, and DHPs therefore mimic $Ca_V1.3$-selective drugs. In $Ca_V1.2DHP^{-/-}$ mice, the smooth muscle relaxant and cardiac negative inotropic actions of isradipine were completely absent, confirming the key role of $Ca_V1.2$ for these processes (see above). In contrast, the bradycardic effects of drugs were unaffected, in agreement with the importance of $Ca_V1.3$ for SAN (Sinnegger-Brauns et al., 2004). Likewise, isradipine-sensitive L-type current components and isradipine-induced inhibition of glucose-dependent insulin secretion observed in pancreatic β-cells of WT mice were eliminated in $Ca_V1.2DHP^{-/-}$ mice, arguing against a major contribution of $Ca_V1.3$ to β-cell currents and excitation–secretion coupling. These data nicely complement and extend the data from cell-specific *Cacna1c* KO mice.

$Ca_V1.2DHP^{-/-}$ mice were also an excellent tool to determine which of the two channel isoforms underlies the so-called anomalous L-type channels in neurons. Unlike for classical ("cardiac") LTCC gating, anomalous LTCC gating in single channel recordings is mainly characterized by low open probability during depolarization and typical long re-openings after repolarization following strong depolarization. Such anomalous L-type channels were described in cerebellar granule cells, hippocampal neurons, sensory neurons, and motoneurons (Koschak et al., 2007). These re-openings are prolonged by DHP channel activators. Anomalous LTCCs were observed in cerebellar neurons of *Cacna1d*$^{-/-}$ mice (Platzer et al., 2000) with a frequency not different from WT (Koschak et al., 2007). However, the modulation of re-openings by DHP activators was absent in neurons of $Ca_V1.2DHP^{-/-}$ mice. These data show that $Ca_V1.2$ $α_1$-subunits are the pore-forming subunits of anomalous LTCCs in mouse cerebellar granule cells. The molecular mechanism inducing these re-openings is unclear and has so far not been observed in heterologous expression systems.

In radioligand-binding studies, the density of high affinity (+)-[^{3}H]isradipine binding sites is reduced by about 90% in $Ca_V1.3DHP^{-/-}$ mouse brain membranes as compared to WT (Sinnegger-Brauns et al., 2009). These data show that a vast majority of LTCCs in the brain are $Ca_V1.2$, although brain region-dependent differences exist (Sinnegger-Brauns et al., 2009; Ortner et al., 2017). Nevertheless, $Ca_V1.3$ controls neuronal excitability and a number of brain functions as described in a separate section below.

Modulatory Role of the $Ca_V1.3$ α_1 CTM Domain: *Cacna1d* DCRD*ᴴᴬ* KI Mice

Like $Ca_V1.2$ (and $Ca_V1.4$, see separate section; Singh et al., 2006), the long C-terminal tail of $Ca_V1.3$ also harbors a C-terminal modulatory (CTM) domain formed by the non-covalent intramolecular association of a distal (DCRD) with a proximal (PCRD) α-helical domain. This CTM stabilizes slower Ca^{2+}-dependent inactivation (CDI) and shifts activation and inactivation voltage dependence to more positive voltages (Singh et al., 2008; Bock et al., 2011; Tan et al., 2011). This CTM is disrupted by alternative splicing-induced shortening of the C-terminus by removal of the DCRD. A number of short splice variants exist that are predicted to activate at even more negative potentials than the abundant long variants and also inactivate faster with Ca^{2+} as the charge carrier. To determine the functional role of this CTM *in vivo*, mice were created in which the DCRD–PCRD interaction was disrupted by the integration of a hemagglutinin (HA) tag in the DCRD (Scharinger et al., 2015) thereby replacing critical negative charges (Bock et al., 2011; Scharinger et al., 2015). As expected, the tagged channel displayed the gating kinetics of short splice variants when expressed in tsA-201 cells.

In the brain of homozygous *Cacna1d* DCRD*ᴴᴬ* mice, the HA-tagged $Ca_V1.3$ α_1-subunit was expressed as a full-length protein without altering α_1-subunit protein expression density. This work also showed that long and short splice variants of the $Ca_V1.3$ α_1-subunit protein are expressed at about equal ratios in mouse brain. In addition, Compelling indirect evidence was provided that the lower molecular mass ("short") α1-subunit protein species is generated by alternative splicing rather than by C-terminal proteolysis as observed in $Ca_V1.2$ α_1 (Bock et al., 2011; Scharinger et al., 2015).

So far, the functional consequence of CTM disruption was studied in mouse chromaffin cells and cochlear inner hair cells, two tissues in which $Ca_V1.3$ contributes substantially to L-type current (see above). $Ca_V1.3$ LTCC currents account for only about 30% of the total I_{Ca} in chromaffin cells (see above). However, interruption of CTM function caused a profound change in their electrical activity. Cells cultured from homozygous *Cacna1d* DCRD*ᴴᴬ* mice showed a more hyperpolarized resting potentials, reduced Ca^{2+}-influx during spontaneous pacemaking, less spontaneous pacemaking activity, and less spike-frequency adaptation (Scharinger et al., 2015). These changes could be explained by the altered gating changes due to the absence of CTM modulation in the long splice variant expressed in these cells. This included the predicted enhanced CDI with reduced signaling to SK channels (Scharinger et al., 2015). In IHCs, CDI is very small because calmodulin is competed off from the channel by other Ca^{2+}-binding proteins, which do not support CDI but are abundantly expressed in IHCs (Schrauwen et al., 2012). In contrast to predictions and to the findings in chromaffin cells, this weak CDI observed in the IHCs of WT mice was further reduced and was essentially absent in homozygous Cav1.3 DCRD*ᴴᴬ* IHCs. It is therefore possible that the absence of the CTM in the long splice variant not only facilitates CaM binding to the channel but also favors the binding of other CaBPs thereby inhibiting rather than strengthening CDI (Scharinger et al., 2015). These data clearly show that $Ca_V1.3$ fine-tuning by the CTM is strongly influenced by a cell's intracellular environment. Anti-HA staining showed the presence of the long splice variant in all IHC ribbon synapses. CTM ablation did not affect hearing function (Scharinger et al., 2015).

This mouse model was also used to address the question if the different voltage dependence of Ca^{2+} influx in the active zones of inner hair cells, which is important for proper sound decomposition, can be explained by the differences in voltage dependence induced by the PCRD–DCRD interaction. However, no evidence for a role of this C-terminal regulatory mechanism was found (Ohn et al., 2016).

Ca$_V$1.3 Channels in Spinal Cord Injury: Constitutive/Conditional *Cacna1d* KO Mice

In the absence of validated subtype-selective L-type channel blockers, animal models were also instrumental to reveal an important pathophysiological role of Ca$_V$1.3 in spinal cord injury. After spinal cord injury (SCI), most individuals develop involuntary muscle contractions, including spasms, as a complication (Marcantoni et al., 2020). In motoneurons, L-type Ca^{2+} currents generate persistent inward currents thus supporting plateau potentials from which neurons can sustain firing. Enhanced activity of plateau potentials after SCI correlates with spasticity (Marcantoni et al., 2020). Due to their more negative activation range, Ca$_V$1.3 channels have been implicated in this process (Li & Bennett, 2003) which has recently been confirmed in an elegant study (Marcantoni et al., 2020). Prolonged treatment of mice with the DHP Ca^{2+} channel blocker nimodipine starting early after SCI prevented the development of increased muscle tone and spontaneous spasms. In contrast to WT controls, lesioned *Cacna1d$^{-/-}$* mice (Platzer et al., 2000) exhibited a severity index not significantly different from unlesioned WT animals. This provided strong evidence that silencing of Ca$_V$1.3 channels is sufficient to improve spasticity after SCI and that nimodipine may act primarily through inhibition of Ca$_V$1.3. Since both excitatory (Vglut2-positive) and inhibitory (VIAAT-positive) interneurons are recruited during spasms, Ca$_V$1.3 was conditionally deleted in these neuronal subsets in *Cacna1d^{eGFPflex}* mice (Satheesh et al., 2012) by breeding to Vglut2-cre- and VIAAT-cre transgenic mice. Compared to lesioned WT controls, tonic muscle contraction and muscle spasm duration were reduced in both homozygous *Cacna1d$^{Vglut2-cre}$* and *Cacna1d$^{VIAAT-cre}$* mice, but

the reduction of symptom severity and duration in these mice was less pronounced than in the constitutive *Cacna1d* KO mice. This was interpreted as evidence for the role of multiple subpopulations of spinal neurons in the generation and maintenance of spasticity after SCI with a possible contribution of Ca$_V$1.3 (Marcantoni et al., 2020). A pharmacological study confirmed the involvement of LTCCs in SCI-induced hyperexcitability, long-lasting root reflexes, and plateau potentials (Jiang et al., 2021). However, the selectivity of the L-type channel blocker employed in this study is highly controversial (Ortner et al., 2014; Huang et al., 2014) allowing no conclusions about a specific role of Ca$_V$1.3 versus Ca$_V$1.2.

Other Functions of Ca$_V$1.3: Constitutive KO Mice

The role of Ca$_V$1.3 channels in retinal function is not entirely clear and reviewed in a separate section in this issue. The light peak of the electroretinogram, which is caused by depolarization of the basolateral plasma membrane of the retinal pigment epithelium, is reduced in *Cacna1d$^{-/-}$* mice (Platzer et al., 2000; Wu et al., 2007). Other functional and morphological changes in the retina of *Cacna1d$^{-/-}$* mice (Platzer et al., 2000) were reported in one (Shi et al., 2017) but not in other studies (Wu et al., 2007; Busquet et al., 2010).

Ca$_V$1.3 has also been implicated in the histamine-elicited phase delay of the circadian neural activity rhythm. Histamine receptor activation-induced shifts of circadian neural activity recorded from SCN slices are blocked by nimodipine and the KO of Ca$_V$1.3 channels (Platzer et al., 2000). Detailed analysis of Ca^{2+} signaling in these neurons showed that to reset the circadian clock, histamine increases Ca^{2+}i in SCN neurons by activating Ca$_V$1.3 channels through H1R, and secondarily by causing Ca^{2+}-induced Ca^{2+} release from RyR-mediated internal stores (Kim et al., 2015).

Ca$_V$1.2 and Ca$_V$1.3

Neuron and Brain Function of Ca$_V$1.2 and Ca$_V$1.3 Channels: Conditional *Cacna1c* KO Mice, *Cacna1d⁻/⁻*, and Ca$_V$1.2DHP⁻/⁻ Mice

Due to the overlapping expression of these channels in the brain, animal models depicting the function of neuronal Ca$_V$1.2 and Ca$_V$1.3 channels are discussed together in this chapter.

Ca$_V$1.2 and Ca$_V$1.3 are both widely expressed in the brain and in most cases even in the same neuron. Although these channels may also be located in nerve terminals (Sinnegger-Brauns et al., 2004; Tippens et al., 2008), there is no evidence for their clustering at active zones and therefore they play no major role for presynaptic control of fast neurotransmitter release. Instead, these channels are primarily located postsynaptically on dendrites, synaptic spines, and cell bodies where they contribute to synapse-to-nucleus signaling and synaptic plasticity. Ca$_V$1.2 appears to be especially privileged in activation of gene transcription through signaling molecules including CaMKII/CaMKIV, CREB and NFAT, which also involves the generation of LTCC-dependent spikes that can propagate from distal dendrites to the soma (Nanou & Catterall, 2018; Wild et al., 2019).

Ca$_V$1.2 and Ca$_V$1.3 mouse models have been instrumental to depict the subtle and distinct roles of brain Ca$_V$1.2 and Ca$_V$1.3 channels since pharmacological separation of their functions is impossible. No CNS-related side effects have been reported upon treatment with therapeutic doses of Ca²⁺ channel blockers for cardiovascular indications in humans. However, application of LTCC activators, like BAYK8644, in rodents induces a severe dystonic syndrome and neurobehavioral abnormalities, including self-biting behaviors (Sinnegger-Brauns et al., 2004). This is associated with intense and widespread neuronal activation as evident from Fos expression throughout the neuroaxis in 77 of 80 brain regions quantified (Hetzenauer et al., 2006). Administration of submicromolar concentrations of BayK8644 into the ventral striatum stimulate the efflux of dopamine, serotonin, glutamate, and noradrenaline. In Ca$_V$1.2DHP⁻/⁻ mice, these behavioral abnormalities are completely absent (Sinnegger-Brauns et al., 2004). In these mice, the BAYK8644-induced neurotransmitter overflow in the ventral striatum is also abolished or reduced (noradrenaline) and enhanced Fos expression is limited only to a small number of mainly limbic, hypothalamic, and brainstem areas, which are also associated with the integration of emotion-related behavior (Hetzenauer et al., 2006). Since in Ca$_V$1.2DHP⁻/⁻ mice BAYK8644 selectively activates only Ca$_V$1.3 channels, it is obvious that the dystonic syndrome, neurotransmitter overflow, and widespread neuronal activation are predominantly mediated by Ca$_V$1.2. More detailed behavioral studies showed that the more restricted activation of brain circuits by systemic BAYK8644 in these mice are associated with increased "depression-like" behaviors. This observation was in excellent agreement with a reduced passive coping ("antidepressant-like") behavior in *Cacna1d⁻/⁻* mice (Platzer et al., 2000; Sinnegger-Brauns et al., 2004; Busquet et al., 2010).

Mouse models also demonstrated that Ca$_V$1.2 and Ca$_V$1.3 participate in the regulation of neuronal activity. Due to their lower activation threshold, subthreshold Ca$_V$1.3-mediated currents can support pacemaking (as also shown in adrenal chromaffin cells, see above) and the stabilization of upstate potentials. In Ca$_V$1.3-deficient mice (Platzer et al., 2000), pacemaking robustness is decreased in spontaneously active SN DA neurons (Guzman et al., 2009) and NMDA-induced upstate potentials supporting the firing of striatal spiny neurons is absent (Olson et al., 2005). In neurons of the lateral nucleus of the basolateral amygdala (BLA), synaptic transmission is intact but LTP induced by high-frequency stimulation of the external capsule was significantly reduced in *Cacna1d* KO mice (Platzer et al., 2000). In these mice, BLA principal neurons were hyperexcitable upon prolonged somatic depolarization with enhanced firing rates and shorter interspike intervals (McKinney et al.,

2009). A reduction in the slow component of the post-burst afterhyperpolarization indicates a role of $Ca_V1.3$ for the normal expression or their functional coupling to Ca^{2+}-activated K^+-channels as also observed in chromaffin cells, atrial myocytes, IHCs, and the superior olivary complex (see above). Together these changes in the excitability of BLA neurons and $Ca_V1.3$ channels in the dorsal hippocampus (Kim et al., 2017) may account for the impaired ability of *Cacna1d* KO mice (Platzer et al., 2000) to consolidate conditioned fear observed in behavioral studies (McKinney et al., 2009; Kim et al., 2017).

$Ca_V1.2$ and $Ca_V1.3$ also have different roles in hippocampal development and function. With low (i.e., more physiological) intracellular Ca^{2+} buffering, *Cacna1d* KO does not alter the action potential and firing of hippocampal CA1 neurons but reduces the slow post-burst afterhyperpolarization, whereas forebrain-specific deletion of *Cacna1c* KO (CaMKIIα-Cre; White et al., 2008) has no effect (Gamelli et al., 2011). This is especially interesting because, as shown in conditional *Cacna1c* KO brains, $Ca_V1.3$ current appears to only contribute minimally to L-type current in these neurons (Moosmang et al., 2005, see below). $Ca_V1.3$ deficiency (Platzer et al., 2000) does not affect NMDA receptor-dependent and NMDA receptor-independent LTP within the CA1 region of the hippocampus (Clark et al., 2003). In contrast, conditional ablation (Nex-Cre; Moosmang et al., 2005) of $Ca_V1.2$ abolished Schaffer collateral/CA1 late-phase LTP and impaired the Ca^{2+}-dependent activation of the ERK/MAPK pathway and CRE-dependent gene transcription. Forebrain-specific *Cacna1c* KO (CaMKIIα-Cre; Lee et al., 2012) ablates theta-burst-induced NMDA-receptor-dependent LTP (Sridharan et al., 2020) and LTP that is induced by prolonged theta-tetanus and is regulated by channel-associated β2AR (White et al., 2008; Patriarchi et al., 2016, see above).

Ca^{2+} influx through LTCCs also controls neurogenesis and neuronal differentiation. Both $Ca_V1.2$ and $Ca_V1.3$ are expressed within the murine adult neurogenic niches. $Ca_V1.3$ expression was evident from eGFP expression in Nestin$^+$ NSCs and newborn mature NeuN$^+$ granu-

lar neurons (Marschallinger et al., 2015) in *Cacna1d$^{+/flex-cre}$* mice (EIIa-Cre; Satheesh et al., 2012; Berger & Bartsch, 2014). In 3-month-old *Cacna1d$^{-/-}$* mice (Platzer et al., 2000), dentate gyrus volume was decreased, the number of Nestin$^+$ neural stem cells (NSCs) was lower, survival of newly generated cells was reduced, and their differentiation into mature neurons impaired (Marschallinger et al., 2015; Kim et al., 2017). In addition, dendritic arborization and complexity were reduced in the KO mice compared to WT (Kim et al., 2017). These functions of $Ca_V1.3$ in developing and mature hippocampal neurons can also explain preserved normal working memory in *Cacna1d$^{-/-}$* mice (Platzer et al., 2000) but deficits in hippocampal-dependent memory tasks, including recent and remote memories of contextually conditioned fear (see also below), object location memory and memory in context-dependent passive avoidance tasks (McKinney & Murphy, 2006; McKinney et al., 2008; Marschallinger et al., 2015; Kim et al., 2017).

Several conditional mouse models using floxed $Ca_V1.2$ mice also demonstrated the requirement of $Ca_V1.2$ for normal neurogenesis. Specific deletion of *Cacna1c* in type-1 neural progenitor cells (NPCs) in the dentate gyrus was achieved by crossing *Cacna1c$^{flox/flox}$* mice with transgenic Glast-creERT2 and RCE:loxP mice (eGFP reporter allele with loxP-flanked STOP-cassette) allowing simultaneous eGFP expression as a reporter (Völkening et al., 2017). In these mice, $Ca_V1.2$ expression was detected at all stages of granule cell development. $Ca_V1.2$ deletion in NPCs decreased their proliferation and reduced the neuronal fate-choice decision of newly born cells, resulting in reduced net hippocampal neurogenesis and more astroglia (Völkening et al., 2017). Therefore, $Ca_V1.2$ appears to control postnatal hippocampal neurogenesis in a cell-autonomous fashion. Conditional KO of $Ca_V1.2$ in all neurons (synapsin1-Cre; Temme et al., 2016) or in forebrain glutamatergic neurons (White et al., 2008; Lee et al., 2016) also impaired neurogenesis by reducing the survival of young hippocampal neurons. However, these studies could not rule out a non-cell-autonomous effect from reduced BDNF protein levels also observed in these mice (Lee et al., 2016). Impaired hippocampal neurogenesis is implicated in the pathology of neurologic and psychiatric disorders. It is therefore tempting to speculate that this process also underlies to some extent the neuropsychiatric disease risk associated with human genetic variants in the $Ca_V1.2$ (*CACNA1C*) and $Ca_V1.3$ (*CACNA1D*) genes (Nanou & Catterall, 2018; Striessnig, 2021 for recent reviews).

However, the role of these channels is not only limited to hippocampal development but also extends to other brain areas. As outlined above, $Ca_V1.3$ is essential for the normal development of brainstem auditory nuclei (Hirtz et al., 2011; Satheesh et al., 2012) but other brain regions are also affected. In *Cacna1d*⁻/⁻ (Platzer et al., 2000) brains, the number of SN DA neurons is reduced (Ortner et al., 2017). Moreover, synaptic pruning and refinement are defective in physiologic as well as pathologic conditions. Dopamine depletion in rodent models of Parkinson's disease leads to a rapid and profound loss of spines and glutamatergic synapses on striatopallidal but not on neighboring striatonigral MSNs. This synaptic pruning also requires $Ca_V1.3$ because it is absent in *Cacna1d*⁻/⁻ mice (Platzer et al., 2000; Day et al., 2006). It has been proposed that the observed spine loss supports a disconnection of striatopallidal neurons from motor command structures that could contribute to symptoms in Parkinson's disease. During postnatal development at hearing onset in mice, the MNTB–LSO projection undergoes strengthening of synapses with concurrent elimination of individual synaptic connections to provide topographic refinement in this brainstem pathway. Both processes are severely impaired in *Cacna1d*⁻/⁻ mice (Platzer et al., 2000) and the number of boutons on single MNTB axons is reduced (Hirtz et al., 2012).

In the developing cortex, $Ca_V1.3$ and $Ca_V1.2$ are expressed and appear to contribute to spontaneous regenerative Ca²⁺ transients associated with downstream neurite growth effects (Kamijo et al., 2018). Transfection of *Cre* recombinase into embryonic cortical neurons of floxed *Cacna1c* mice (Seisenberger et al., 2000) by in utero electroporation reduced total neurite length

in neurons cultured from these mice (Kamijo et al., 2018) consistent with a role of these channels on immature neurite elongation. In accordance with results obtained from *Cacna1c*^G406R (TS2-NEO) mice (see above), introduction of this Timothy Syndrome mutation increased amplitude and spread and reduced duration of spontaneous regenerative Ca²⁺ transients and impaired radial migration of layer 2/3 excitatory neurons (Kamijo et al., 2018). Defects in hippocampal neurogenesis, impaired LTP, and altered brain development in brain *Cacna1c* KO mice can explain the deficits in hippocampal- and also extrahippocampal-dependent contextual and spatial learning and memory tasks as observed in reduced performance in the Morris Water Maze, context discrimination tasks, and a spatial learning labyrinth-based paradigm (Moosmang et al., 2005; White et al., 2008; Temme et al., 2016).

Taken together, the proper function of both L-type channels is essential for normal brain development. These findings in mice are of translational significance to better understand disease risk and perhaps even treatment of neurodevelopmental disorders resulting from gain- and loss-of-function variants in the genes of the pore-forming subunits of these channels (Kabir et al., 2017; Nanou & Catterall, 2018; Striessnig, 2021 for reviews).

In addition to impaired contextual and spatial learning and memory, other behaviors are altered upon conditional brain KO of $Ca_V1.2$ and constitutive KO of $Ca_V1.3$. As mentioned above, homozygous (but not heterozygous) *Cacna1d* KO mice (Platzer et al., 2000) show a passive coping/antidepressant-like phenotype in the forced-swim and tail-suspension tests, which is not due to their deaf phenotype (Busquet et al., 2010). An anxiolytic phenotype has also been noted in the elevated plus-maze paradigm but this may also be due to deafness (Busquet et al., 2010). In contrast to $Ca_V1.3$, constitutive heterozygous *Cacna1c* KO (Seisenberger et al., 2000) induces an increased anxiety-like behavior in female (Dao et al., 2010) and male mice (Bader et al., 2011; Lee et al., 2012). These mice also appear to exhibit an "antidepressive-like" phenotype in the FST and the tail suspension test (Dao et al., 2010;

Kabir et al., 2017). Stronger reduction of $Ca_V1.2$ expression with forebrain conditional KO led to a more robust anxiety-like phenotype (Lee et al., 2012). Focal deletion of $Ca_V1.2$ in the prefrontal cortex (PFC) by stereotactic injection of adeno-associated viral (AAV) vector-expressing *Cre* in the floxed mice also reproduced this phenotype implicating a role of $Ca_V1.2$ in the PFC (Lee et al., 2012). However, this phenotype could not be reproduced in mice with a pan-neuronal *Cacna1c* KO (synapsin1-Cre), which also eliminates $Ca_V1.2$ in inhibitory neurons and results in a significant shift of the inhibitory/excitatory balance in the amygdala (Temme & Murphy, 2017). Selection of the cell-specific promotor driving *Cre*-expression and/or genetic background of the mice may therefore also affect the outcome of these studies (Temme & Murphy, 2017).

Chronic unescapable stress increases depressive-like and anxiety-like behaviors and reduces working memory (Bavley et al., 2017). In WT mice, these behavioral alterations persisted 5–7 days post stress but were no longer present in heterozygous brain *Cacna1c+/−* mice (Nes-Cre; Langwieser et al., 2010) indicating dependence on the activity of $Ca_V1.2$ (Bavley et al., 2017). Interestingly, in WT controls, but not in the heterozygous KO mice, the stress-induced behavioral long-term changes were paralleled by a delayed increase of $Ca_V1.2$ α_1-subunit expression in the PFC (but not amygdala and hippocampus) and activation of the p25/Cdk5-glucocorticoid receptor pathway (Bavley et al., 2017).

Changes in the expression of brain $Ca_V1.2$ α_1-subunits in the CA1 region and dentate gyrus of the hippocampus have also been described in contextual fear conditioning and fear memory retrieval on the mRNA level in rats (Sykes et al., 2018), during cocaine withdrawal and cocaine conditioned place preference (see below) and during aging in mice. In comparison to young mice (4–5 months of age), aged mice (17–18 months) show a reduced exploratory behavior in a Y-maze but normal working memory (Zanos et al., 2015). This was also the case in $Ca_V1.2+/−$ littermates (constitutive KO, *JAX stock #005783*). However, old WT mice per-

formed worse than young ones in the novel object recognition paradigm, a short-term memory recognition test (Zanos et al., 2015). Interestingly, this memory was preserved in heterozygous KOs. Similarly, in a passive avoidance paradigm, which tests memory based on emotional/contextual learning, memory retention was impaired in old male WT mice but was preserved in the male heterozygous KO mice. In contrast, $Ca_V1.2$ haploinsufficiency was not protective in the passive avoidance learning in aged female mice. $Ca_V1.2$ α_1-subunit mRNA expression in hippocampal tissue samples was negatively correlated with the observed memory deficits and the aging-induced increase in hippocampal $Ca_V1.2$ expression observed in WT was absent in the KO mice. Although these changes in mRNA expression were not verified on the protein or functional level, these experiments support findings from some (but not all, Zanos et al., 2015 for references) previous studies associating enhanced L-type current densities in hippocampal neutrons with aging-induced cognitive decline (Moore & Murphy, 2020). These data do not rule out a role also for $Ca_V1.3$, a question currently addressed in a transgenic mouse line.

Cognitive Function in a Mouse Model for an "Aged" Brain: Transgenic Overexpression of $Ca_V1.3$ α_1 in the Brain

To test if the aging-induced upregulation of $Ca_V1.3$ is causing the cognitive deficits associated with aging, $Ca_V1.3$ was overexpressed in a transgenic mouse model. A long splice variant of $Ca_V1.3$ α_1 was tagged with an extracellular HA epitope and used for overexpression in mouse forebrain glutamatergic neurons under the control of the CaMKIIα promoter (Krueger et al., 2017). Despite an about 50% higher $Ca_V1.3$ α_1 immunoreactivity in the forebrain of transgenic mice, they exhibited no behavioral phenotype in a battery of behavioral tests compared to their WT littermates. Unfortunately, data obtained in this mouse model have to be interpreted with caution because the $Ca_V1.3$ α_1-subunit construct employed in these experiments contained a missense mutation disrupting the C-terminal modulatory domain (similar to the $Ca_V1.3$ DCRD^HA channel), resulting in a long $Ca_V1.3$ channel with the gating properties of short splice variants (Lieb et al., 2012). In addition to stabilizing gating kinetics, the CTM is key in providing functionally relevant protein interactions further fine-tuning its function (e.g., activity-induced facilitation; Sahu et al., 2017). Therefore, the CTM plays a key role also for $Ca_V1.3$ coupling to Ca^{2+}-activated K^+-channels and thus the amplitude of the slow afterhyperpolarizations likely involved in learning and memory (Moore & Murphy, 2020; see also above). The CTM properties of $Ca_V1.3$ channels in this mouse model do not correspond to any of the known physiological C-terminal splice variants limiting its validity.

L-Type Channels in Fear Responses: Constitutive/Conditional $Ca_V1.2$ and Constitutive *Cacna1d* KO

As mentioned above, $Ca_V1.3$ is required for the consolidation and remote memories of conditioned fear (McKinney et al., 2008; Kim et al., 2017) but *Cacna1d* KO does not affect fear memory acquisition and extinction (Busquet et al., 2008; McKinney et al., 2008).

The role of $Ca_V1.2$ has been more complex to assess, likely due to the different promotors driving *Cre*-expression to eliminate $Ca_V1.2$. In different conditional KO models, effects on fear memory were only found after pan-neuronal (synapsin1-Cre) $Ca_V1.2$ conditional KO. While acquisition and fear memory consolidation were normal, fear memory extinction was affected. No difference was observed within extinction training sessions but extinction memory consolidation was impaired when tested 24 h later (Temme & Murphy, 2017), similar to observation with intra-BLA infusions of LTCC blockers in rats (Davis & Bauer, 2012). This is in contrast to other reports. I.c.v injection of the DHP isradipine impaired

cue-associated fear memory acquisition in mice together with Hebbian LTP in afferent inputs to the lateral amygdala (Langwieser et al., 2010). However, like in the other brain mouse models (synapsin1-Cre, CaMKIIa-Cre; McKinney et al., 2008; Temme & Murphy, 2017), conditional brain *Cacna1c* KO (Nes-Cre) (Langwieser et al., 2010) showed no phenotype. It appears that, in contrast to acute pharmacological inhibition, KO induces a homeostatic change with compensatory upregulation of Ca^{2+}-permeable glutamate AMPA receptors in the amygdala making LTP $Ca_V1.2$-independent (Langwieser et al., 2010).

The systemically applied DHP nifedipine has also been reported to strongly inhibit the fear extinction (Busquet et al., 2008 for reference). However, this effect is absent when applied i.c.v. and is not observed in $Ca_V1.2DHP^{-/-}$ mice (Busquet et al., 2008). It must therefore be mediated by $Ca_V1.2$ channels outside the CNS likely due to a strong and protracted stress response induced by acute blood pressure lowering (Schafe, 2008; Waltereit et al., 2008). Taken together, although the importance of $Ca_V1.2$ and $Ca_V1.3$ function for normal fear responses has been demonstrated, neither pharmacological treatments nor conditional KO models have provided us with a complete picture of the roles of these two LTCC isoforms in these processes so far.

In addition to conditioned fear, fear can also be induced by other triggers, such as socially acquired fear (e.g., through verbal information) or from observation of a conspecific's distress (observational fear). Human brain imaging studies suggest also an involvement of the anterior cingulate cortex (ACC) in this process (Jeon et al., 2010). $Ca_V1.2$ controls electrical excitability in the ACC and thereby seems to be involved in the acquisition but not 24-h memory of observational fear. Ablation of $Ca_V1.2$ α_1-subunit expression by injection of purified cell-permeable *Cre*-recombinase into the ACC of floxed *Cacna1c* mice impaired the development of freezing in an observational fear-conditioning task observing another mouse subjected to repetitive foot shocks (Jeon et al., 2010).

It is also important to note, however, that effects of $Ca_V1.2$ deficiency (and likely also of $Ca_V1.3$) on brain function appear to be complex and thus the interpretation of results from behavioral studies on learning tasks may also depend on experimental design. Koppe et al. (2017) studied the consequences of brain *Cacna1c* KO (Nes-Cre; floxed *Cacna1c*, Seisenberger et al., 2000) mice in a two-choice operant-cue learning task employing touchscreen-trained mice rewarded for correct responses. This allowed to test for different behavioral learning strategies in combination with computational model testing of the behavioral data. In this paradigm, *Cacna1c* KO appears not to impair learning in general but to have a much more specific effect: it may simply change learning strategy. While the majority of matched controls adopted an experimentally intended "cue-association rule," KO mice still managed to increase reward feedback across trials but did so by adapting an "outcome-based strategy." A standard behavioral analysis would have come to the different conclusion that KO mice would have performed significantly worse than their controls to learn the cue-discrimination rule. These results indicate that *Cacna1c* KO mice may not suffer from a general learning deficit (Koppe et al., 2017). To date, $Ca_V1.3$-deficient mice have not been analyzed in this paradigm.

The role of $Ca_V1.2$ and $Ca_V1.3$ for emotional and social behaviors is of special interest, because genetic variants altering L-type channel activity in the brain may predispose to neuropsychiatric and neurological disease, as mentioned for mutations causing Timothy syndrome above. Likewise, *de novo CACNA1D* missense variants cause high risk for autism spectrum disorder (ASD) as well as for a severe neurodevelopmental disorder associated with endocrine symptoms (for review, see Ortner et al., 2020). Several intronic *CACNA1C* SNPs (especially SNP rs1006737) have reproducibly shown a strong association with neuropsychiatric disorders (including bipolar disorder, schizophrenia, and major depressive disorders and also structural as well as functional alterations in the brain of SNP carriers)

(Kabir et al., 2017 for review). The functional consequences of these SNPs have yet to be elucidated but are in accordance with the role of $Ca_V1.2$ for behavior in animal models. While conditional KO or KD studies provide us with cell- or organ-specific insight into the role of $Ca_V1.2$ and $Ca_V1.3$, replication of complex phenotypes observed in humans requires constitutive KO or the knockin of disease mutations.

$Ca_V1.2$ deficiency also affects electrophysiological sleep architecture in mice. Constitutive heterozygous *Cacna1c* KO mice (Seisenberger et al., 2000) display significantly lower EEG spectral power than WT mice across high-frequency ranges (β to γ) during wake and rapid eye movement (REM) sleep (Kumar et al., 2015). They spend slightly more time asleep in the dark period, but daily amount of sleep is unchanged. Recovery sleep after exposure to challenging stress (acute sleep deprivation or restraint stress) reduced REM sleep recovery responses compared to WT. This suggests the involvement of $Ca_V1.2$ in REM sleep. Interestingly, *CACNA1C* gene variants associated with human neuropsychiatric disorders are also associated with sleep disorders (Kantojärvi et al., 2017) again pointing at the translational relevance of these findings.

L-Type Channels in Drug Taking Behaviors: Conditional *Cacna1c* KO, Constitutive *Cacna1c* KO and $Ca_V1.2DHP^{-/-}$ Mice

Mouse models for the first time allowed dissecting the diverse functions of $Ca_V1.2$ and $Ca_V1.3$ in reward and addictive behaviors. *Cacna1c*$^{-/-}$ (Platzer et al., 2000) and $Ca_V1.2DHP^{-/-}$ mice (Sinnegger-Brauns et al., 2004) revealed that $Ca_V1.3$ through ERK2 signaling in the VTA plays a key role in the acquisition of behavioral sensitization and of cocaine conditioned place preference (CPP) which involves recruitment of GluA1 receptors through phosphorylation of Ser-831 in the nucleus accumbens (NAc) (Giordano et al.,

2006, 2010; Schierberl et al., 2011; Martínez-Rivera et al., 2017). Instead, $Ca_V1.2$ activity is not required for acquisition but plays a central role in the expression, extinction, and reinstatement of drug taking behaviors. In the nucleus accumbens (NAc), $Ca_V1.2$ α_1 is upregulated after 21 days of withdrawal. Brain *Cacna1c* KO (Nes-Cre; floxed *Cacna1c*; Seisenberger et al., 2000) and AAV-Cre KD of *Cacna1c* specifically in the NAc showed that $Ca_V1.2$ is required for the expression of the psychostimulant-induced sensitized response (Schierberl et al., 2011; Kabir et al., 2017). This drug-induced long-term synaptic and behavioral plasticity in the NAc involves signaling of $Ca_V1.2$ *via* CaMKII and ERK2 (Schierberl et al., 2011; Kabir et al., 2017). In the dorsal hippocampus, $Ca_V1.2$ is upregulated during the development of cocaine-conditioned place preference and its deletion by AAV-*Cre* injection in floxed *Cacna1c* mice (Seisenberger et al., 2000) inhibits the extinction of cocaine-induced CPP (Burgdorf et al., 2017). Conditional KO of $Ca_V1.2$ in D1R-expressing (D1R-Cre) but not glutamatergic neurons (CaMKII-Cre) resulted in a deficit in extinction of cocaine CPP with corresponding decreases in PSD, CaMKII, and Ser-831-phosphorylated GluA1 in the hippocampus. $Ca_V1.2$ in the dorsal hippocampus had no role for the acquisition of cocaine CPP. Given the important role of the hippocampus for learning (see above), this suggests that $Ca_V1.2$ may play a differential role in cocaine-associated versus other forms of hippocampal-dependent learning. Identification of $Ca_V1.2$ as a key signaling molecule for this process is in line with earlier reports about the role of D1R in extinction learning of different types of memory (Burgdorf et al., 2017). The attenuated cocaine CPP extinction in mice lacking $Ca_V1.2$ channels in D1R-expressing cells can be rescued through chemogenetic activation of D1R-expressing cells within the dorsal dentate gyrus, indicating that $Ca_V1.2$ channels in excitatory cells of this region are required for cocaine CPP extinction. As described above in *Cacna1c*^{S1928A} mice, Ser-1928 phosphorylation of $Ca_V1.2$ is not required for extinction but for cocaine-primed reinstatement (Burgdorf et al.,

2020). In cells of the prelimbic cortex that project to the nucleus accumbens core, $Ca_V1.2$ channels drive the cocaine- and stress-primed reinstatement of drug-associated memories. This was demonstrated by KD of *Cacna1c* in floxed mice (Seisenberger et al., 2000) within bilateral NAc core-projecting cells of the prelimbic cortex using stereotaxic injections of AAV expressing a flp-dependent *Cre*-recombinase (into the prelimbic cortex) combined with a retrograde AAV expressing flp recombinase (into the NAc core) leading to selective expression of *Cre* recombinase in cells projecting to the NAc core from neurons in the prelimbic cortex (Bavley et al., 2020). These findings are of potential clinical relevance indicating a role of brain $Ca_V1.2$ channels as a pharmacological target to help attenuate craving, a critical factor leading to risk for relapse (Bavley et al., 2020).

A number of pharmacological studies also indicate a role of LTCCs in the development of dependence from more commonly abused drugs, such as nicotine and alcohol. This was implicated from the transcriptional regulation of $Ca_V1.2$ and $Ca_V1.3$ by nicotine and alcohol in rodents and effects of treatment with LTCC blockers (with the caveat of indirect peripheral effects in case of systemic administration) on nicotine-induced behavioral changes (Bernardi et al., 2014; Uhrig et al., 2017 for references). Animal models point to a role of $Ca_V1.2$, but not of $Ca_V1.3$ for nicotine-induced place preference (Liu et al., 2017). Its development could be prevented in WT mice by nifedipine pretreatment. In *Cacna1d*^{-/-} mice (Platzer et al., 2000), nicotine-induced place preference could also be inhibited by nifedipine. However, this drug effect was absent in $Ca_V1.2DHP^{-/-}$ mice providing indirect evidence for an involvement of $Ca_V1.2$ in this response. Confirmatory data using *Cacna1c* KO models have not been provided. Therefore, the absence of a nifedipine effect in the $Ca_V1.2DHP^{-/-}$ mice could also be due to peripheral $Ca_V1.2$ channels (see above) and, for example, causes acute neuroendocrine activation interfering with the development of place preference.

Alcohol dependence induced in rats by exposure to intermittent cycles of alcohol vapor intoxication increases $Ca_V1.2$ α_1 mRNA and protein in the hippocampus and amygdala and enhances L-type current density in hippocampal CA1 neurons. $Ca_V1.3$ α_1 expression remains unchanged (Uhrig et al., 2017). I.c.v. injection of the Ca^{2+} channel blocker verapamil (expected to inhibit $Ca_V1.2$ and $Ca_V1.3$) attenuated the seeking response in alcohol-dependent rats in a cue-induced reinstatement test (with optical and olfactory cues predicting alcohol availability). Conditional *Cacna1c* KO out in forebrain neurons in adult mice (CaMKIIα-Cre-ERT2, floxed *Cacna1c*; Seisenberger et al., 2000) reduced self-administration in alcohol-dependent mice to the level of non-dependent KO mice and in floxed controls. Similar to verapamil cue-induced seeking was also attenuated (Uhrig et al., 2017).

Taken together, $Ca_V1.2$ appears to be involved in neuronal and behavioral plasticity associated with drug dependence of different drugs of abuse.

L-Type Channels in Brain Dopamine Neurons: Constitutive *Cacna1d* KO and $Ca_V1.2DHP^{-/-}$ Mice

Mouse models were also instrumental to reveal distinct roles of these LTCCs for dopamine signaling. Neuroprotective effects of DHP Ca^{2+} channel blockers reported in rodent models of Parkinson's disease in some studies (Liss & Striessnig, 2019 for review) suggested their involvement in the high vulnerability of substantia nigra pars compacta (SNc) dopamine neurons to cell death in Parkinson's disease (Surmeier et al., 2011; Ilijic et al., 2011). SN dopamine neurons are permanently active pacemaking and bursting neurons. This electrical behavior leads also to activity-dependent dendritic Ca^{2+} transients that generate a permanent Ca^{2+} load, which contributes to oxidative stress and, consequently, degeneration of these vulnerable neurons (see also separate chapter "Calcium Channels and Selective Neuronal Vulnerability in Parkinson's Disease" by Surmeier and Liss in this book). Ca^{2+} transients could be mediated by $Ca_V1.2$ and $Ca_V1.3$ because transcripts for both channels have been detected in these cells with $Ca_V1.2$ transcripts being more abundant (Ortner et al., 2017; Liss & Striessnig, 2019). The activation of dendritic Ca^{2+} transients at low-voltages hints toward a role of $Ca_V1.3$ which is supported by $Ca_V1.3$ α_1 siRNA KD in these neurons (Guzman et al., 2018). Unfortunately, mouse models played a minor role in confirming the predominant role of $Ca_V1.3$ versus $Ca_V1.2$ for Ca^{2+} transients and neuronal vulnerability. No neuroprotection was observed in *Cacna1d*$^{-/-}$ mice (Platzer et al., 2000) in a 6-OHDA model of Parkinson's disease (Ortner et al., 2017). It could be argued that constitutive *Cacna1d*$^{-/-}$ mice are unsuitable to address this question due to the compensatory upregulation of other Ca^{2+} channels, such as $Ca_V3.1$ (Poetschke et al., 2015). However, so far, no attempts have been made to directly assess the role of $Ca_V1.2$ by using also $Ca_V1.2$ α_1 shRNA KD in addition to $Ca_V1.3$ α_1 (Guzman et al., 2018). Alternatively, $Ca_V1.2DHP^{-/-}$ mice would have been highly suitable and available to ask if a DHP-sensitive component of Ca^{2+} current or dendritic Ca^{2+} transients is absent in SN DA neurons. Unlike in WT, isradipine inhibition of dendritic Ca^{2+} transients would be limited to $Ca_V1.3$-dependent components and spare those mediated by $Ca_V1.2$ in these mice. In addition, no data using conditional *Cacna1c* KO mice were published so far. This leaves the important question unanswered, if only $Ca_V1.3$ or both channels should be targeted for neuroprotection in Parkinson's disease.

However, germline *Cacna1d* KO has shown that $Ca_V1.3$ channels participate in the modulation of firing rates by dendritic dopamine release. Slice electrophysiology recordings in midbrain slices from *Cacna1d*$^{-/-}$ (Platzer et al., 2000) and $Ca_V1.2DHP^{-/-}$ mice revealed that $Ca_V1.3$ Ca^{2+} channels regulate SNc firing rates through dopamine D2 autoreceptor-induced firing inhibition by activation of G protein-coupled K^+-channels (GIRKs) such as GIRK2 (KCNJ6) (Dragicevic et al., 2014). $Ca_V1.3$-mediated Ca^{2+} signals support sensitized D2-autoreceptor responses by activating the Ca^{2+}-dependent interaction of the neural Ca^{2+} sensor NCS-1 with postsynaptic D2 receptors. NCS-1 inhibits receptor desensitization by preventing their beta-arrestin-mediated receptor internalization.

L-Type Channels in the Immune System: Conditional *Cacna1c* KO, Constitutive Ca$_V$1.3 and Ca$_V$1.4 KO

Normal T-cells express a number of pore-forming and auxiliary voltage-gated Ca²⁺ channel subunits. This includes β_3-, β_4-, and α_1-subunits of all L-type channels (Ca$_V$1.1, Ca$_V$1.2, Ca$_V$1.3, Ca$_V$1.4), and the T-type channels Ca$_V$3.1 and Ca$_V$3.2 (Wang et al., 2016; Fenninger & Jefferies, 2019 for references). Immunological phenotypes were described in mice deficient for accessory subunits, such as β_3- and β_4-subunits (Fenninger & Jefferies, 2019). Whether these result from their effects on the function and expression of Ca²⁺ channel α_1-subunits or non-channel associated functions is unclear.

Very small, slowly inactivating voltage-activated Ba²⁺ currents (requiring 100 mM Ba²⁺ as conducting ion in the extracellular solution as driving force) could be recorded in naïve CD4⁺- and CD8⁺ T-cells isolated from lymph nodes and spleen after T-cell receptor (TCR) activation, which were completely absent in Ca$_V$1.4 deficient mice (*Cacna1f⁻/⁻* mice; Mansergh et al., 2005) (Omilusik et al., 2011). Their DHP sensitivity was not tested. In contrast, in primary splenic CD4⁺ T-cells, Wang et al. (Wang et al., 2016) recorded only small fast inactivating, DHP-insensitive T-type currents (10 mM Ca²⁺ as charge carrier) with a large fraction being carried by Ca$_V$3.1 α_1. Other evidence for DHP-sensitive current components has been published but due to inconsistencies with these recordings, the paper had to be retracted (Badou et al., 2005).

Experiments with *Cacna1f⁻/⁻* mice show that Ca$_V$1.4 is required for TCR-induced store-operated Ca²⁺-entry (SOCE) into naive CD4⁺ and CD8⁺ T cells and the activation of the ERK and NFAT pathways (Omilusik et al., 2011). *Cacna1f⁻/⁻* mice also displayed a memory T-cell phenotype and upregulated activation markers, suggesting that Ca$_V$1.4 is necessary for naïve T-cell maintenance. Sustained elevated levels of activation markers on B-lymphocytes suggested a chronic state of activation. The dysfunctional immune phenotype of the *Cacna1f*-deficient mice displayed several hallmarks of immunological exhaustion and was associated with severe immune deficiency upon Listeria monocytogenes infection, with a reduced number of functional antigen-specific CD4⁺ and CD8⁺ T-cells (Omilusik et al., 2011; Fenninger et al., 2019).

Recent studies using KO models confirmed and extended earlier observations with L-type Ca²⁺ channel activators and blockers and antisense strategies providing also evidence for the role of Ca$_V$1.2 and Ca$_V$1.3 for TCR-driven Th2 functions and type 2-mediated airway inflammation (Fenninger & Jefferies, 2019; Giang et al., 2021 for references). Deletion of *Cacna1c* in T-lymphocytes was achieved by crossing *Cacna1c^{flox}* mice (White et al., 2008) with mice expressing *Cre*-recombinase under the CD4-promoter (*Cacna1c^{T−/−}*; Giang et al., 2021). Constitutive *Cacna1d⁻/⁻* mice (Platzer et al., 2000) were studied in parallel. KO of Ca$_V$1.2 or Ca$_V$1.3 did not affect T-cell subsets in the thymus and the periphery indicating that T-cell development was unaffected. However, the increase in the number of early Ca²⁺ responses after TCR activation, visualized as rapid Ca²⁺ changes close to the plasma membrane (spatiotemporal single elementary Ca²⁺ events), observed in WT was reduced in *Cacna1c^{T−/−}* or *Cacna1d⁻/⁻* Th2-cells and overall protein tyrosine phosphorylation was also significantly reduced after TCR stimulation (Giang et al., 2021). This decreased Ca²⁺ response and impaired proximal signaling could also explain an impairment of TCR-mediated IL-5 and IL-13 production in Ca$_V$1.2-or Ca$_V$1.3-deficient Th2-cells. Interestingly, deficiency in either one of these channels was sufficient to impair Ca²⁺ responses and to inhibit cardinal features of type 2 airway inflammation. Furthermore, allergic airway inflammation could only be triggered if Ca$_V$1.2 and Ca$_V$1.3 were co-expressed within the same CD4⁺ T cell. It therefore appears that both Ca²⁺ channels act synergistically for optimal Th2-cell function. These findings are of potential pathological significance because the expression of both channels by activated CD4+ T-cells from asthmatic children was also associated with increased Th2-cytokine transcription (Giang et al., 2021).

The exact mechanism leading to the activation of these voltage-gated channels is not completely clear. Like $Ca_V3.1$ channels, the more negatively activating $Ca_V1.3$ channels may carry some constant window current at the resting membrane potential of about -50 mV (Cahalan & Chandy, 2009) in T-cells. This aspect is further complicated by the fact that the Ca^{2+} sensor STIM1 inhibits $Ca_V1.2$ and most likely also $Ca_V1.3$ channels (Park et al., 2010b; Wang et al., 2010) in T-cells upon TCR stimulation. On the other hand, due to the small size of T lymphocytes and their low resting Ca^{2+} concentration of approximately 50 nM, corresponding to fewer than 10,000 free Ca^{2+} ions, already very tiny Ca^{2+} currents near or even below electrophysiological detection limits should be able to substantially increase cytosolic Ca^{2+} concentrations (Cahalan & Chandy, 2009). Another possibility is that not Ca^{2+} influx but like in skeletal muscle (see above and separate chapter "The Skeletal Muscle Calcium Channel" in this book by Beam and Flucher), electro-mechanical coupling of voltage-sensor movement to RyRs of intracellular stores triggers intracellular Ca^{2+} release. This would not even require channels with functional pores (as detected in immune cells) as long as critical voltage-sensor domains are present (Omilusik et al., 2013).

A role of T-type channels for T-cell function different from L-type channels has also been described and is discussed in the chapter on $Ca_V3.1$.

Rat Model of $Ca_V1.2$ Haploinsufficiency

The involvement of $Ca_V1.2$ L-type Ca^{2+} channel signaling in more complex brain functions has also been demonstrated in a rat model of $Ca_V1.2$ haploinsufficiency. Germline $Cacna1c^{+/-}$ KO (*SAGE Laboratories*) was obtained in rats by introducing an early stop in exon 6 using zinc finger technology (Braun et al., 2018; Moon et al., 2020). Compared to adults (3–5 months of age) WT rats' object recognition, reward sensitivity in a spatial learning task, spatial memory, and reversal-learning capabilities were intact. In heterozygous males, initial cognitive flexibility was impaired but they showed better long-term learning. Heterozygous females displayed initial hypoactivity at a slightly better performance. It therefore appears that $Ca_V1.2$ haploinsufficiency has a small but positive impact on (spatial) memory functions in adult rats. In young male rats (1–2 months of age), socio-affective deficits in social behavior and pro-social 50-kHz ultrasonic communication were observed (Kisko et al., 2018, 2020). Male heterozygous rats showed normal rough-and-tumble play behavior but emitted fewer 50-kHz ultrasonic vocalizations during social play suggesting that they derive less reward from playful social interactions than WT littermates. *Cacna1c*$^{+/-}$ rats also showed reduced social approach behavior elicited by playback of 50-kHz ultrasonic vocalizations. Together these findings suggest impaired ultrasonic communication in both sender and receiver rats. Interestingly, female rats showed opposite behavior: heterozygous KO increased the amount of time they spent playing to almost double that of their WT littermate controls and even exceeding playing time in WT males. In females, prosocial 50-kHz ultrasonic vocalizations (USV) did not differ. These data suggest that $Ca_V1.2$ plays a prominent role in regulating socio-affective communication in rats in a sex-dependent manner (Kisko et al., 2018, 2020). In adult female rats, partial $Ca_V1.2$ α_1-depletion lead to strongly reduced emission of 50-kHz USV and mild social deficits during direct reciprocal social interaction with most prominent reductions of 50-kHz USV during non-social behavior. This suggested a reduced positive affect in a social context in heterozygous rats that are not specifically linked to social behavior (Redecker et al., 2019). Sex-specific effects were also seen when behavioral inhibition was measured in response to playback of an alarm 22-kHz USV in recipient rats. The strength of behavioral inhibition during playback was very similar but more long lasting in females than in males (Wöhr et al., 2020). Taken together, these data in rats were interpreted as behavioral evidence in line with the role of *CACNA1C* as a risk gene for neuropsychiatric disease in humans (Kisko et al., 2020). This also fits to the more recent observation of aberrantly enhanced fear responses to inappropriate cues in these rats (Moon et al., 2020).

This interesting translational aspect is also supported by other behavioral changes suggesting parallels between humans and rats. Healthy human subjects bearing *CACNA1C* alleles associated with risk for psychiatric disorders show poorer reversal performance on a probabilistic reversal-learning task than non-risk allele carriers. A similar impairment was seen in the heterozygous *Cacna1c* KO rats in a touchscreen reversal-learning paradigm (Sykes et al., 2019). This was paralleled by a decreased BDNF levels in the PFC, both in humans (postmortem tissue) and in heterozygous KO mice (Sykes et al., 2019).

Unfortunately, the apparent correlation between lower $Ca_V1.2$ expression in rats and human *CACNA1C* risk alleles has to be interpreted with caution. First, risk SNPs in the *CACNA1C* gene are intronic and it is not yet clear how that impacts $Ca_V1.2$ α_1 protein expression levels and/or splicing. Moreover, the extent of depletion of $Ca_V1.2$ α_1-subunit protein in this *Cacna1c$^{+/-}$* mouse strain remains unclear because the quality of Western blots trying to quantify reduced expression used in these studies is poor and antibody specificity was not properly controlled. While a 20–40% reduction of $Ca_V1.2$ α_1-subunit protein was determined in one study (Sykes et al., 2019), a >50% reduction was measured by another group (Braun et al., 2018). In both cases, specific antibody staining of the $Ca_V1.2$ α_1-subunits has not been documented beyond reasonable doubt in control experiments in parallel experiments with mouse tissue from

conditional *Cacna1c* KO mice as for example shown by Buonarati et al. (2018). Moreover, a > 50% reduction of the $Ca_V1.2$ α_1 protein expression is not limited to the brain but may also affect blood pressure (lower expression of $Ca_V1.2$ in arterial smooth muscle) and is expected to also increase the risk for heart failure as shown in mice (see above). Cardiovascular function in heterozygous KO rats has not been tested. Therefore, indirect effects from neuroendocrine activation on behavioral studies cannot be excluded.

Ca_V1.4

Mouse models enabling insight into $Ca_V1.4$ function in the retina are discussed by Lee and Koschak in a separate chapter in this book. A $Ca_V1.4$ mouse model identifying a role of this channel in the rodent immune system is summarized in the chapters above.

Ca_V2

Ca_V2 proteins are VGCC α_1-subunits responsible for high-voltage-activated non-L-types localized mainly in the central and peripheral nervous systems. Being rich in pharmacological and genetic knowledge, members of the Ca_V2 family, $Ca_V2.1$, $Ca_V2.2$, and $Ca_V2.3$ that mediate P/Q-, N-, and R-types, respectively, were subjected to thorough investigation to understand their significance at different hierarchical levels. A variety of neuronal processes, including neurotransmitter release, synaptic plasticity, and pain sensation, as well as neurological disorders, in particular epilepsy and neurodegenerative diseases, have been well recognized as important biological outputs of Ca_V2-containing VGCCs. These processes and disorders have been extensively studied by generating KO and KI mouse models to more precisely and comprehensively resolve the normal physiology and pathophysiology of individual Ca_V2 subtypes.

Ca$_V$2.1

Early Reports of Ca$_V$2.1 Mutant Mice

Neuronal Processes and Deficits: Conventional Ca$_V$2.1 KO

The VGCC α_1-subunit Ca$_V$2.1 has been explored using engineered mutant mice mainly to study two important aspects: the functional diversity of VGCCs in their essential physiological roles in the nervous system, and the neurology of ataxia, dystonia, epilepsy, and migraine (Catterall et al., 2005; Rajakulendran et al., 2012). The Ca$_V$2.1 subunits were thought, from studies comparing recombinant Ca$_V$2.1 currents and native VGCC currents, to support both P- and Q-type Ca^{2+} channel currents, but direct evidence was lacking. In the first report on a conventional Ca$_V$2.1 null (*Cacna1a$^{-/-}$*) KO mouse line, which was obtained by deleting the exon that encodes the nucleotide sequence 401–493 from the initiation codon, *Cacna1a$^{-/-}$* mice underwent a progressive neurological deficit of ataxia and dystonia, dying at 3–4 weeks of age (Jun et al., 1999). P-type currents in Purkinje neurons and P- and Q-type currents in cerebellar granule cells were lost completely, while synaptic transmission persisted in the hippocampus despite the lack of P/Q-type channels; however, it showed enhanced reliance on N-type and R-type Ca^{2+} entry. Another conventional *Cacna1a$^{-/-}$* line, generated by replacing exons 14–17 encoding part of repeat II with a neomycin selection cassette, developed dystonia and late-onset cerebellar degeneration, indicating a requirement of P/Q-type VGCCs for survival in a subset of cerebellar neurons, while L-type and N-type currents were enhanced in *Cacna1a$^{-/-}$* granule cells (Fletcher et al., 2001). Reduced P/Q-type current density is unlikely sufficient to cause the pathophysiology of spontaneous mouse mutants with ataxia and seizures, because heterozygous *Cacna1a$^{+/-}$* mice are phenotypically normal despite a 50% reduction in P/Q-type current density. These earliest studies point to the critical roles of P/Q-type VGCCs, ablation of which is not or only incompletely compensated by other types in certain neurobiological processes.

Neuronal Processes and Deficits: Conditional Ca$_V$2.1 KO

A mouse line with two *loxP* sites flanking exon 4 of the Ca$_V$2.1 gene *Cacna1a* was generated (*Cacna1a$^{flox/flox}$*), and showed no altered synaptic function (Todorov et al., 2006). Conditional Ca$_V$2.1 KO mice *Cacna1a$^{\Delta E4(-/-)}$*, which were obtained by crossing the above-floxed mice with transgenic mice expressing *Cre* recombinase under the control of the adenovirus EIIA promoter, exhibited progressively severe ataxia and dystonia around postnatal day P10–12 and died at P20–22 if left unaided. The resultant phenotype was in general consistent with the conventional Ca$_V$2.1 KO mice. In generating another floxed *Cacna1a* mice *Cacna1a^{Citrine}*, the *Cacna1a* gene was subjected to an insertion of the floxed GFP derivative Citrine into the first exon, based on the knowledge that this insertion does not influence Ca$_V$2.1 function in the recombinant expression system (Mark et al., 2011). *Cacna1a^{Citrine}* mice were then crossed with the transgenic Purkinje cell-specific *Cre* line, in which *Cre* expression is controlled by the *PCP/L7* promoter (activated around P6 and fully established by 2–3 weeks after birth), to yield a conditional Ca$_V$2.1 KO line named *purky* (*Cacna1a$^{purk(-/-)}$*). Purkinje cells in *purky* mice exhibited impaired neurotransmission, as well as the same phenotypes reported for the conventional *Cacna1a$^{-/-}$* mice (Jun et al., 1999; Fletcher et al., 2001), suggesting that neurological deficits caused by inherited disorders of the P/Q-type channel arise from signaling defects beginning in late infancy.

Extended Knowledge of Ca$_V$2.1 Mutant Mice

Exocytosis and Neurotransmitter Release: Conventional Ca$_V$2.1 KO

Simultaneous recordings of VGCC currents and increased membrane capacitance (ΔCm) as indications of exocytosis were performed in adrenal chromaffin cells from WT and conventional *Cacna1a$^{-/-}$* mice (Jun et al., 1999), using the perforated-patch configuration of the patch-clamp technique, along with VGCC-type-selective blockers, in order to distinguish the contribution of each VGCC type (Aldea et al., 2002). When adrenal chromaffin cells from *Cacna1a⁻/⁻* mice were compared with those from WT mice, the L-type component of VGCC currents rose from 43% to 53%, while the P/Q-type disappeared, and the N-type and R-type components were similar. ΔCm associated with L-type increased from 40% to 60%, while ΔCm associated with P/Q-type was abolished, and that associated with N-type and R-type was unaltered. In general, contributions to exocytosis were proportional to the relative current density of each VGCC type. A compensatory increase of L-type-associated exocytosis failed to completely maintain the overall exocytotic response that was diminished by 35% in *Cacna1a⁻/⁻* mice.

Neuromuscular transmission was carefully compared at the endplates of WT and *Cacna1a⁻/⁻* mice (Urbano et al., 2003). Presynaptic terminal projection was morphologically intact, and miniature endplate potentials displayed intact amplitudes but reduced frequency in *Cacna1a⁻/⁻* mice. Recording of nerve-evoked endplate potential revealed decreased quantal content. When quantal transmitter release dependent on N-type and R-type was examined for susceptibility to intercellular Ca²⁺ buffering and its relationship to Ca²⁺-activated K⁺ channels, compared to that of N-type Ca²⁺ influx, R-type Ca²⁺ influx showed stronger coupling with Ca²⁺ sensors responsible for exocytic membrane fusion and Ca²⁺-activated K⁺ channels. Strikingly, paired-pulse facilitation was nearly abolished in *Cacna1a⁻/⁻* mice. These data suggested a rank of P/Q>R>N for active zone structures to select among multiple VGCC types present in the presynaptic area of neuromuscular junction.

Presynaptic Protein Interaction: Conditional Ca$_V$2.1 KO

Roles played by proposed interactions of various active zone proteins with the Ca$_V$2.1 C-terminus in depolarization-triggered neurotransmitter release and targeting of Ca$_V$2.1 proteins at the presynaptic active zone remain controversial (Lübbert et al., 2017). To address this issue,

Ca$_V$2.1 expressed at the calyx of Held was modified by delivering helper-dependent adenoviral vectors co-expressing *Cre* recombinase (to create a *Cacna1a* null background) with control or mutated Ca$_V$2.1 α₁-subunits, into the cochlear nucleus. Mutations designed to disrupt interactions of RIM1/2, MINT1, Rim-binding proteins, and CASK proteins, and a secondary Ca$_V$β₄ interaction site, as well as RIM-binding PXXP motifs, were all ineffective. Instead, a new Ca$_V$2.1 C-terminus region was identified that directly controls fast synaptic vesicle release and vesicle docking at the active zone independent of Ca$_V$2.1 abundance.

Biochemical Characterization of the Two-Domain 95kD Short Form: Conventional Ca$_V$2.1 KO

The conventional Ca$_V$2.1 null *Cacna1a⁻/⁻* mouse (Jun et al., 1999) was used to characterize the neuronal two-domain 95kD short form of the Ca$_V$2.1 protein (Arikkath et al., 2002). This 95kD form was also absent in neuronal preparations from *Cacna1a⁻/⁻* mouse, confirming that the 95kD form of the Ca$_V$2.1 protein originates from the *Cacna1a* gene.

Cerebellar Function and Maturation: Conventional/Conditional Ca$_V$2.1 KO

Each Purkinje cell in the adult cerebellum is innervated by a single climbing fiber on proximal dendrites, and by 10^5–10^6 parallel fibers on distal dendrites. Heterosynaptic competition between parallel fibers and climbing fibers, and homosynaptic competition among multiple climbing fibers, has been known to postnatally regulate this organized synaptic wiring. In conventional Ca$_V$2.1 KO (*Cacna1a⁻/⁻*) mice, parallel fibers expanded the innervation territory proximally to innervate the ectopic spines (Jun et al., 1999), while climbing fibers limited the innervation territory to the basal portion of proximal dendrites and somata (Miyazaki et al., 2004). Multiple climbing fibers and parallel fibers, which should normally be expelled from the compartment innervated by the main climbing fiber, persisted in the majority of *Cacna1a⁻/⁻* Purkinje cells. These results suggest that P/Q-type VGCCs control heterosynaptic and homosynaptic competition, facilitating the distal extension of climbing fiber monoinnervation along the dendritic tree of Purkinje cells.

To study the specific contribution of postsynaptic P/Q-type VGCC to developmental organization of climbing fiber synapses in Purkinje cells, Purkinje cell-selective $Ca_v2.1$ (PC-$Ca_v2.1$) KO mice were generated by crossing the *Cacna1a* gene-floxed $Ca_v2.1^{flox/flox}$ mice carrying two *loxP* sequences flanking exon 4 encoding the first voltage-sensor domain in the *Cacna1a* gene with a D2CreN line, whose *Cre* gene was expressed in Purkinje cells under the control of the GluD2 promoter (Hashimoto et al., 2011; Miyazaki et al., 2012). In the PC-$Ca_v2.1$ KO mice, Purkinje cells showed markedly reduced Ca^{2+} transients induced by spontaneous climbing fiber inputs. Multiple climbing fibers, which were equally strengthened in each Purkinje cell, underwent translocation to dendrites in Purkinje cells, and subsequent synapse elimination was severely impaired in PC-$Ca_v2.1$ KO mice; innervation territory of the climbing fiber was limited to the soma and basal dendrites, whereas that of parallel fibers was reciprocally expanded. Also, PC-$Ca_v2.1$ KO Purkinje cells displayed hyperspiny transformation in the proximal somatodendritic domain of Purkinje cells, as well as hyperinnervation by multiple climbing fibers and numerous parallel fibers. PC-$Ca_v2.1$ KO mice also displayed patterned degeneration of Purkinje cells. Importantly, in cerebellar granule cell-specific Cav2.1 KO mice obtained by crossing $Ca_v2.1^{flox/flox}$ mice with E3CreN mice (*GluN2C^{iCre/+}*), in which the *Cre* gene was expressed in cerebellar granule cells under the control of the GluN2C (GluRε3 or NR2C) promoter, none of the phenotypes observed in PC-$Ca_v2.1$ KO mice were observed. Thus, P/Q-type Ca^{2+} channels in postsynaptic Purkinje cells mediate both homosynaptic competition among multiple climbing fibers and heterosynaptic competition between parallel fibers and climbing fibers in the developing cerebellum.

To establish a specific KO of P/Q-type from cerebellar granule cells and precerebellar nuclei

neurons in the brainstem, *Cacna1a^{Citrine}* mice (Mark et al., 2011) were crossed with transgenic Gabra6-*Cre* mice that express *Cre* recombinase under the control of a $GABA_A$ receptor α_6 subunit (Gabra6) promoter (Maejima et al., 2013). The obtained mouse line, named *quirky* (*Cacna1a^{quirk(−/−)}*), has a selective KO of P/Q-type channels in rhombic lip-derived neurons, including the parallel fiber and mossy fiber pathways. *Cacna1a^{quirk(−/−)}* mice displayed ataxia, dyskinesia, and absence epilepsy in phenotypic analysis, and reduced parallel fiber-Purkinje cell synaptic transmission during low-frequency stimulation. These results suggest that impaired synaptic transmission confined to one of the main cerebellar excitatory pathways has important implications for the manifestation of P/Q-type channel-associated disease.

By cross-breeding the floxed mouse line targeted for exon 4 of *Cacna1a* (Todorov et al., 2006) with $\alpha6^{Cre}$ mice, which express *Cre* recombinase under the control of a Gabra6 promoter, a conditional KO line $\alpha6^{Cre}$-*Cacna1a* that lacks P/Q-type channels in most, but not all of the cerebellar granule cells were generated (Galliano et al., 2013). $\alpha6^{Cre}$-*Cacna1a* KO mice showed no defects at the level of cytoarchitecture or motor performance, but did show a compromised capability for adaptation and consolidation. Granule cell output *via* parallel fibers to Purkinje cells and stellate cells, as well as long-term plasticity at parallel fiber-Purkinje cell synapses, was suppressed. These results suggest that a minority of functionally intact granule cells remaining after $\alpha6^{Cre}$-*Cacna1a* KO is sufficient for the maintenance of basic motor performance

Synaptic Plasticity at the Neuromuscular Junction and in the Hippocampus: $Ca_V2.1$ KI

Facilitation and inactivation of Ca^{2+} currents through the regulation of P/Q-type VGCCs by calmodulin and related Ca^{2+} sensor proteins contribute to the facilitation and rapid depression of synaptic transmission (Nanou et al., 2016b). A mutation was introduced in exon 40 of *Cacna1a*, which converted amino acid residues IM to AA in the CaS protein-binding site of the $Ca_V2.1$ C-terminus, to engineer IM-AA mutant mice. In neuromuscular junction synapses that exclusively use $Ca_V2.1$ for Ca^{2+} entry to trigger synaptic transmission in IM-AA mice, short-term facilitation, in response to paired stimuli at short interstimulus intervals, as well as synaptic facilitation at high stimulus frequencies (50–100 Hz), were reduced, whereas facilitation at lower frequencies (10–30 Hz) was increased and synaptic depression was slowed. IM-AA mice exhibited reduced peak force in response to 50 Hz stimulation, increased muscle fatigue in the hindlimb tibialis anterior muscle, and impaired motor control, exercise capacity, and grip strength, suggesting that regulation of P/Q-type by Ca^{2+} sensor proteins is essential for normal synaptic plasticity at the neuromuscular junction *in vivo*.

In IM-AA mice, LTP at the Schaffer collateral (CA3)-CA1 synapse in the hippocampus was critically weakened, although this form of synaptic plasticity has been primarily attributed to a postsynaptic mechanism (Nanou et al., 2016a). LTP in response to θ-burst stimulation was much reduced, while LTP was intact in response to repetitive 100-Hz stimulation or to postsynaptic depolarization which prevents Mg^{2+} block of NMDA receptors. The IM-AA mice (Nanou et al., 2016b) exhibited pronounced deficits in spatial learning and memory in context-dependent fear conditioning and in the Barnes circular maze, suggesting that regulation of P/Q-type by Ca^{2+} sensor proteins is critical for LTP and spatial learning and memory in mice. Interestingly and paradoxi cally, under conditions of physiological extracellular Ca^{2+} concentration (1.5 mM) and near-physiological temperatures (33–36 °C), synaptic facilitation at CA3-CA1 synapses was not attenuated in IM-AA mice, and facilitation was more prominent at cerebellar parallel fiber-Purkinje cell and Purkinje cell-deep cerebellar nuclei synapses (Weyrer et al., 2019). Enhanced facilitation at these synapses is consistent with a reduction of action-potential-evoked P/Q-type currents in Purkinje cells. Thus, contribution of facilitation of Ca^{2+} entry through P/Q-type channels to synaptic facilitation is likely minimal under physiological conditions.

Cognition and Neuronal Electrical Oscillations: Conventional/Conditional $Ca_V2.1$ KO

$Cacna1a^{-/-}$ mice (Jun et al., 1999) were examined for different aspects of *in vivo* and *in vitro* thalamocortical functions (Llinás et al., 2007). Thalamic neurons in $Cacna1a^{-/-}$ mice failed to demonstrate γ-band subthreshold oscillations and cortical γ-band-dependent columnar activation involving cortical inhibitory interneuron activity. Electroencephalogram recordings showed persistent absence status and a dramatic reduction of γ-band activity in $Cacna1a^{-/-}$ mice, suggesting that P/Q VGCCs are essential for γ-band activity and associated cognitive function.

The forebrain-specific $Ca_V2.1$ KO mouse line, which was generated by crossing $Cacna1a^{flox/flox}$ (Todorov et al., 2006) with mice expressing *Cre* under the control of the NEX promoter, revealed impaired synaptic transmission at hippocampal glutamatergic synapses (Mallmann et al., 2013). Also, these KO mice demonstrated deficits in spatial learning and reference memory, reduced recognition memory, increased exploratory behavior, and a strongly attenuated circadian rhythmicity.

Hippocampal pyramidal neurons play an essential role in processing spatial information (Jung et al., 2016). In hippocampal pyramidal neuron-selective $Ca_V2.1$ KO mice, generated by crossing the $Cacna1a^{flox/flox}$ mice (Hashimoto et al., 2011) with CamK2α-*Cre* driver line, which expressed *Cre* recombinase from an αCamK2 promoter (in mainly the CA1 region starting at P21 and spread to the neocortex in 2 months), behavioral tasks, requiring spatial and contextual information processing, were impaired at >2 months of age (Jung et al., 2016). The ability to recollect spatial representation of places on re-visit was also altered in terms of the cue recognition. CA1 pyramidal neurons of the conditional $Cacna1a$ KO mice showed reduced burst frequency as well as abnormal temporal patterns of burst spiking. These results suggest that P/Q-type plays important roles in spatial recognition ability.

Epilepsy: Conditional Ca$_V$2.1 KO

Epilepsy is a heterogeneous condition with a broad range of presentations. For example, idiopathic generalized epilepsies present during childhood with different combinations of generalized seizures, which include absence and generalized tonic-clonic seizures, and myoclonus (Rossignol et al., 2013). To understand neuronal populations and mechanisms responsible for generalized spike-wave absence seizures, a targeted *Cacna1a* exon 4 deletion in defined structures and neuronal populations was achieved by crossing mice carrying a *Cacna1a*$^{flox/flox}$ conditional allele (Todorov et al., 2006) with different *Cre* driver lines: *Cacna1a* gene was ablated in cortical parvalbumin (PV) and somatostatin (SST) GABAergic interneurons using the *Nkx2.1-BAC*^{Cre} line, in SST GABAergic interneurons of cortex and reticular thalamic using the *SST*^{Cre} line, or in cortical pyramidal cells using the *Emx1*^{Cre} allele, which in conjunction with the *Nkx2.1-BAC*^{Cre} removed Ca$_V$2.1 in cortical excitatory neurons and PV/ SST interneurons, while sparing the thalamus (Rossignol et al., 2013). Selective *Cacna1a* KO from a subset of cortical interneurons, including PV and SST interneurons, resulted in severe generalized epilepsy. Ca$_V$2.1 KO impaired GABA release from PV but not from SST interneurons, in which N-type compensates for the P/Q-type deficiency. In these mutants, thalamocortical projection neurons failed to show enhanced bursting, suggesting that this feature is not essential for generalized spike-wave seizures. Ablation of Ca$_V$2.1 channels in both cortical pyramidal cells and interneurons reduced seizure severity by decreasing cortical excitability. The data suggest that conditional disruption of the *Cacna1a* gene in cortical PV interneurons causes generalized seizures by impairing their ability to constrain cortical pyramidal cell excitability. This was supported by further identification of the cellular and network consequences of *Cacna1a* deletion in PV interneurons, using PV neuron-selective *Cacna1a* KO mice (*PV*^{Cre};*Cacna1a*$^{flox/flox}$) (Jiang et al., 2018), generated by crossing *PV* ^{Cre} mice with *Cacna1a*$^{flox/flox}$ mice (Todorov et al., 2006).

PV^{Cre};*Cacna1a*$^{flox/flox}$ mice displayed reduced cortical perisomatic inhibition and frequent absences, but only rare motor seizures, attributable to a net increase in cortical inhibition, with a gain of dendritic inhibition through sprouting of SST interneuron axons. Interestingly, this beneficial compensatory remodeling of cortical GABAergic innervation was mTORC1-dependent.

To elucidate the role of P/Q-type within the thalamocortical spike-wave circuit and its ability to regulate lasting thalamic excitability changes, neurotensin receptor 1 (Ntsr1) *Cre*-driver mice (*Ntsr-1 Cre*) were used to delete the floxed *Cacna1a* (*Cacna1a*^{Citrine}) (Mark et al., 2011) in layer VI pyramidal neurons, which supply the sole descending cortical synaptic input to thalamocortical relay cells and reticular interneurons, and activate intrathalamic circuits (Bomben et al., 2016). The obtained *Cacna1a*$^{Ntsr(-/-)}$ line showed a robust spontaneous spike-wave absence seizure phenotype accompanied by behavioral arrest. Despite *Cacna1a* disruption, intrinsic excitability of the layer VI neurons was unaltered; T-type Ca^{2+} currents in the postsynaptic thalamic relay and reticular cells were dramatically elevated, resulting in rebound bursting and seizure generation. Thus, P/Q-type deficiency, in a single cell-type within the thalamocortical circuit, is sufficient to remodel synchronized firing behavior and to stabilize generalized epilepsy phenotype.

Mouse Models for Neurological Diseases: Ca$_V$2.1 KI

Familial hemiplegic migraine type 1 (FHM-1) is a Mendelian subtype of migraine with aura that is caused by missense mutations in the *Cacna1a* gene (van den Maagdenberg et al., 2004). Functional impacts of FHM-1 mutations were initially examined with recombinant mutated Ca$_V$2.1 proteins in heterologous systems or in *Cacna1a*$^{-/-}$ cerebellar granule cells. However, the results differed between the various systems. To address this controversy, human FHM-1 mutations were introduced at the corresponding position in the mouse ortholog of the *Cacna1a* gene by homologous recombination. The KI mouse model $Cacna1a^{R192Q}$, carrying the human FHM-1 mutation R192Q, showed multiple gain-of-function effects, including increased P/Q-type current density in cerebellar neurons, enhanced neurotransmission at the neuromuscular junction, and a reduced threshold and increased velocity of cortical spreading depression, which likely corresponds to the migraine aura (van den Maagdenberg et al., 2004). Also, cortical pyramidal neurons from the R192Q KI ($Cacna1a^{R192Q}$) mice revealed gain-of-function of excitatory neurotransmission, due to increases in action-potential-evoked Ca²⁺ influx and probability of glutamate release at pyramidal cell synapses (Tottene et al., 2009). Furthermore, basal and algogenic mediator bradykinin-stimulated release of calcitonin gene-related peptide was higher in primary trigeminal cultures from the $Cacna1a^{R192Q}$ mouse, where bradykinin significantly upregulated the number of satellite glial cells expressing functional UTP-sensitive P2Y receptors (Ceruti et al., 2011). Considering the contribution of the crosstalk between trigeminal ganglion neurons and satellite glial cells in neuronal sensitization and transduction of painful stimuli, the $Cacna1a^{R192Q}$ mutation may mediate migraine pathophysiology at least partly through activation of purinergic receptor mechanisms.

Most FHM-1 mutations, such as $CACNA1A^{R192Q}$, cause only mild forms of FHM, while $CACNA1A^{S218L}$ causes a particularly dramatic clinical syndrome, consisting of attacks of hemiplegic migraine, slowly progressive cerebellar ataxia, epileptic seizures, and severe cerebral edema, which can be triggered by a trivial head trauma (van den Maagdenberg et al., 2010). The S218L KI $Cacna1a^{S218L}$ mice mimicked the clinical features of the human S218L syndrome. The $Cacna1a^{S218L}$ mice also displayed a prominent sensitivity to cortical spreading depression with a reduced threshold, an increased propagation velocity, and multiple cortical spreading depression events after a single stimulus, in contrast to $Cacna1a^{R192Q}$ mice. In cerebellar Purkinje cells, the $Cacna1a^{S218L}$ mutation caused a negative shift of voltage dependence of P-type currents, low-

ered thresholds for somatic action potentials and dendritic Ca²⁺ spikes, and disrupted firing patterns, leading to hyperexcitability (Gao et al., 2012). Thus, in general, KI mouse studies support gain-of-function impacts of FHM-1 mutations on P-type activity and downstream neurotransmission and excitability *in vivo*.

Episodic ataxia type 2 (EA2) is an autosomal dominant disorder characterized by a relatively normal neurological baseline, and by attacks of ataxia that are typically precipitated by stress or exercise (Rose et al., 2014). EA2 mutations most often exert loss-of-function or reduction of $Ca_V2.1$ VGCC in recombinant expression systems; however, the pathogenic mechanisms are still debated. To address this issue *in vivo*, a mouse line was generated that carries the orthologous mouse $Cacna1a$ gene with the EA2-causing F1406C missense human mutation (Rose et al., 2014). Homozygous $Cacna1a^{F1406C/F1406C}$ mice, which exhibited a ~70% reduction in P-type current density in Purkinje cells, showed a normal motor phenotype, while hemizygous $Cacna1a^{F1406C/-}$ mice exhibited motor dysfunction measurable by Rotarod and pole test. Interestingly, when conditional $Ca_V2.1$ KO lines $Cacna1a^{purk(-/-)}$ (postnatal $Cacna1a$ removal from Purkinje cells) and $Cacna1a^{quirk(-/-)}$ (postnatal $Cacna1a$ removal from rhombic lip-derived neurons such as granule cells) were examined as EA2 mouse models, they displayed anxiolytic behavior to lit, open areas in the open field and light/dark place preference tests, but enhanced anxiousness in the novel suppressed feeding test (Bohne et al., 2021). These model mice were indecisive and anxious to explore new territories and objects, showed increased latencies in the novel object recognition test, and demonstrated deficiencies in specific social traits, suggesting that P/Q-type VGCCs contribute to cognitive functions in anxiety, memory, and social behavior.

Spinocerebellar ataxia type 6 (SCA6), caused by the expansion of a polyglutamine stretch encoded by CAG repeats in $Ca_V2.1$, is characterized by predominant degeneration of cerebellar Purkinje cells. A KI SCA6 mouse model which expresses human $Ca_V2.1$ with 28 polyglutamine (polyQ) repeats was generated, and its phenotypes were compared with that of 13 polyglutamine repeats (representing a normal control) (Saegusa et al., 2007). Purkinje cells from control or homozygous SCA6 KI mice revealed a non-inactivating current which was highly sensitive to the spider toxin ω-agatoxin IVA, indicating that the human $Ca_V2.1$ expressed in Purkinje cells exhibits typical P-type properties, in contrast to Q-type properties, of $Ca_V2.1$ constructs expressed recombinantly in heterologous systems. Functional properties of P-type channels in Purkinje cells, as well as behavioral scores, were indistinguishable between normal control and the 28 polyQ KI mice. A KI SCA6 mouse model with further expansion of polyQ repeats was established by "humanizing" exon 47 of the mouse *Cacna1a* gene, which originally lacks the CAG repeats, with CAG repeats of different lengths (14, 30, and 84) (Watase et al., 2008). The KI mice expressing a hyperexpanded 84 polyQ stretch developed progressive motor impairment and aggregation of $Ca_V2.1$ proteins. Electrophysiological properties of P-type channels in cerebellar Purkinje cells were indistinguishable among the three KI models, suggesting that expanded CAG repeat *per se* does not affect the intrinsic electrophysiological properties of the channels. It is interesting to note that the C-terminus fragment carrying polyQ of the P/Q-type channel has been shown to be sufficient to cause SCA6 pathogenesis (Mark et al., 2015).

Pain: Conventional $Ca_V2.1$ KO

Pain mechanisms have been the major target of VGCC pharmacology. The conventional *Cacna1a⁻/⁻* line (Fletcher et al., 2001) was used to assess the role of P/Q-type channels in nociception and pain transmission (Luvisetto et al., 2006). Tested at ages between P8 and P20, when the motor deficit was either absent or very mild, *Cacna1a⁻/⁻* mice showed reduced tail withdrawal latencies in the tail-flick test and reduced abdominal writhing in the acetic acid writhing test. Adult heterozygous *Cacna1a⁺/⁻* mice, which failed to show alterations in the tail-flick and the acetic acid writhing tests, surprisingly showed a reduced licking response during the second phase of formalin-induced inflammatory pain, and a reduced mechanical allodynia in the chronic constriction injury model of neuropathic pain. Thus, P/Q-type channels may play an antinociceptive role in sensitivity to non-injurious noxious thermal stimuli, and a pronociceptive role in inflammatory and neuropathic pain states.

$Ca_V2.2$

Early Reports of $Ca_V2.2$ Mutant Mice

Sympathetic Regulation: Conventional $Ca_V2.2$ KO

N-type VGCCs, mainly localized in the nervous system, have been considered to play critical roles in a variety of neuronal functions, including neurotransmitter release at sympathetic nerve terminals (Ino et al., 2001). As a direct approach to establishing the physiological significance of N-type VGCCs, a conventional $Ca_V2.2$ null (*Cacna1b⁻/⁻*) KO mouse was generated by disrupting the exon that encodes the central portion of the cytoplasmic repeat II–III linker in the *Cacna1b* gene. A complete and selective elimination of ω-conotoxin GVIA-sensitive N-type VGCC currents was observed, without the elicitation of significant changes in the activity of other VGCC types in neuronal preparations of *Cacna1b⁻/⁻* mice. In *Cacna1b⁻/⁻* mice, the baroreflex response mediated by the sympathetic nervous system, as well as the positive inotropic responses to electrical sympathetic neuronal stimulation, was dramatically suppressed, whereas parasympathetic nervous activity remained nearly intact. Interestingly, the *Cacna1b⁻/⁻* mice showed compensatory sustained elevation of heart rate and blood pressure. These results provide direct evidence that the N-type VGCC formed by $Ca_V2.2$ is indispensable for the function of the sympathetic nervous system in circulatory regulation.

Pain: Conventional $Ca_V2.2$ KO

Specific and precise roles played by N-type VGCCs in pain perception were not fully elucidated due to the coexistence of other types of VGCCs in sensory neurons. Nociceptive transmission was examined in the above $Cacna1b^{-/-}$ mouse line (Ino et al., 2001; Hatakeyama et al., 2001). In the $Cacna1b^{-/-}$ mice, N-type channel activities in dorsal root ganglion neurons and spinal synaptoneurosomes were eliminated without compensation by other VGCC types. The $Cacna1b^{-/-}$ mice showed a diminution in the phase 2 nociceptive responses more extensively than in the phase 1 nociceptive responses of the formalin test, and increases of thermal nociceptive thresholds in the hot plate test, but showed intact mechanical nociceptive thresholds in the tail pinch test. To establish roles of N-type VGCCs in pain perception, two additional $Cacna1b^{-/-}$ mouse lines were engineered. The $Cacna1b^{-/-}$ mouse line, generated by disrupting the exon that encodes the repeat I S3 region of the mouse $Cacna1b$ gene with the GFP-Neo cassette, showed reduced responses to mechanical stimuli in the von Frey test, and increased tail-flick latency in response to radiant heat (Kim et al., 2001a). In the formalin paw test, this $Cacna1b^{-/-}$ line also exhibited a significantly attenuated response in Phase 2, but a normal response in Phase 1. The response to visceral inflammatory pain caused by acetic acid was also reduced. Another $Cacna1b^{-/-}$ mouse line, generated by replacing exon 1 with the $nLacZ$-neo cassette, displayed suppressed responses to a painful stimulus that induces inflammation, and showed markedly reduced symptoms of neuropathic pain caused by nerve injury (Saegusa et al., 2001). Thus, N-type $Ca_V2.2$ channels play a critical role in pain perception.

Extended Knowledge of $Ca_V2.2$ Mutant Mice

Pharmacological Effects: Conventional $Ca_V2.2$ KO

N-type VGCCs are modulated by acute and chronic ethanol exposure *in vitro* at concentrations known to affect humans, but it was unclear whether this ethanol effect has physiological sig-

nificance in behavioral ethanol responses *in vivo*. In the $Cacna1b^{-/-}$ mice (Kim et al., 2001a), voluntary ethanol consumption was reduced, and place preference was developed only at a low dose of ethanol (Newton et al., 2004). The hypnotic effects of ethanol were also substantially diminished, whereas ethanol-induced ataxia was mildly increased. Thus, N-type VGCCs modulate acute responses to ethanol, playing a role in ethanol reward and preference.

Neurotransmitter Release: Conventional $Ca_V2.2$ KO

Visceral sensory information from the internal organs is conveyed *via* the vagus and glossopharyngeal primary afferent fibers to the second-order neurons in the nucleus of the solitary tract (NTS). The presynaptic VGCC types that play major roles in the glutamate release from the solitary tract (TS) axons to the second-order NTS neurons (Yamazaki et al., 2006) remain unidentified. In the $Cacna1b^{-/-}$ mice (Ino et al., 2001), the P/Q-type blocker ω-agatoxin IVA failed to affect the EPSC amplitude, while in WT mice, the R-type blockers SNX-482 (500 nM) and Ni²⁺ (100 μM) also failed to significantly reduce EPSC amplitude. These results indicate that the glutamate release at the TS–NTS synapse in mice occurs through activation of non-L, non-P/Q, non-R, non-T, and non-N VGCCs.

Presynaptic VGCCs were characterized at axon terminals of olfactory sensory neurons in the main olfactory bulb (MOB), and at axon terminals of vomeronasal sensory neurons in the accessory olfactory bulb (AOB) (Weiss et al., 2014). $Ca_V2.2$ was localized to the glomerular neuropil of MOB and AOB with presynaptic molecules. Pharmacological characterization provided evidence for a central role of the N-type in presynaptic transmitter release at these synapses: EPSCs in mitral/tufted (M/T) and superficial tufted cells of the MOB, as well as mitral cells of the AOB, were predominantly triggered by N-type at these synapses, whereas L-type, P/Q-type, and R-type channels made only minor contributions. In the $Cacna1b^{-/-}$ mice (Ino et al., 2001), EPSCs in olfactory nerve-evoked M/T cell showed intact amplitudes, but became blocker-resistant, thus indicating a major compensatory reorganization of Ca_Vs responsible for neurotransmitter release. The *Cacna1b*$^{-/-}$ mice revealed that Ca$_V$2.2 is critically required for paired-pulse depression of olfactory nerve-evoked EPSCs in M/T cells of the MOB, and for vomeronasal nerve-evoked EPSCs in AOB mitral cells. These results may suggest that N-type VGCCs contribute to olfactory perception, such as olfactory sensitivity and social responses to chemical stimuli.

Depolarization-evoked glutamate release from cerebrocortical synaptosomes was mediated by P/Q-type VGCCs in the *Cacna1b*$^{-/-}$ mice (Ino et al., 2001), in contrast to glutamate release that depends on P/Q- and N-types in WT mice (Martín et al., 2008). Moreover, metabotropic glutamate receptor 7 (mGluR7) failed to inhibit the evoked Ca^{2+} influx and glutamate release from *Cacna1b*$^{-/-}$ synaptosomes. The failure of mGluR7 to modulate evoked glutamate release was unlikely attributable to a lack of receptors, as nerve terminals from mice lacking N-type expressed mGluR7 proteins were intact (Ladera et al., 2009). Thus, glutamate release from cerebrocortical terminals is supported differently by N- and P/Q-type channels, but P/Q-type cannot fully compensate for the N-type loss in *Cacna1b*$^{-/-}$ mice.

Sympathetic Regulation: Conventional Ca$_V$2.2 KO

In the *Cacna1b*$^{-/-}$ mice (Ino et al., 2001), immunoprecipitation analysis revealed a compensatory enhancement of association between β_3 (a major accessory subunit of N-type Ca$_V$2.3) and the R-type Ca$_V$2.3 (Murakami et al., 2007). Elongated and fluctuating R–R intervals reminiscent of disturbed sympathetic tonus were observed in electrocardiogram recordings. R-type blocker SNX-482 further enhanced R–R intervals in *Cacna1b*$^{-/-}$ mice but not in WT control mice. Echocardiography and Langendorff heart analysis confirmed a major role for R-type in the *Cacna1b*$^{-/-}$ mice, suggesting that the residual sympathetic tonus in the *Cacna1b*$^{-/-}$ mice is mediated by R-type compensation.

Behavior, Memory, and Cognition: Conventional Ca$_V$2.2 KO

Ca$_V$2.2 proteins are localized in the piriform cortex, hippocampus, hypothalamus, locus coeruleus, dorsal raphe, thalamic nuclei, and granular layer of the cortex, some of which have been previously implicated in metabolic and vigilance state control (Beuckmann et al., 2003). *Cacna1b*$^{-/-}$ mice (Ino et al., 2001) were hyperactive, demonstrating 20% and 95% increases in activity under novel conditions and under habituated conditions during the dark phase, respectively. *Cacna1b*$^{-/-}$ mice also displayed changes in vigilance state during the light phase, including increases in consolidation of rapid-eye movement (REM) sleep and intervals between non-REM and wakefulness episodes. These results revealed a role played by N-type VGCCs in activity and vigilance state control.

In the Morris water maze and the social transmission of food preference tasks, the *Cacna1b*$^{-/-}$ mutant mice (Kim et al., 2001a) exhibited impaired learning and memory (Jeon et al., 2007). Among activity-dependent long-lasting synaptic changes, θ burst- or 200-Hz-stimulation-induced LTP was decreased in the *Cacna1b*$^{-/-}$ mice. Brain-derived neurotrophic factor-induced potentiation of field excitatory postsynaptic potentials and facilitation of the frequency of miniature EPSC were also reduced in the *Cacna1b*$^{-/-}$ mice, suggesting that N-type VGCCs are required for hippocampus-dependent learning and memory. The same *Cacna1b*$^{-/-}$ mouse line showed markedly enhanced aggressive behaviors to an intruder mouse in the resident-intruder test (Kim et al., 2009). In the dorsal raphe nucleus, which contains serotonin neurons and is involved in aggression in animals, electrophysiological analysis showed increased firing activity of serotonin neurons with reduced inhibitory neurotransmission in the *Cacna1b*$^{-/-}$ mice. Also, the *Cacna1b*$^{-/-}$ mice displayed an elevated level of arginine vasopressin, an aggression-related hormone, in the cerebrospinal fluid, and an increase of serotonin in the hypothalamus, suggesting that N-type VGCCs in the dorsal raphe nucleus play a key role in aggression control.

In locomotion on an activity wheel during the dark phase, the *Cacna1b⁻ᐟ⁻* mice (Ino et al., 2001) exhibited a significant increase compared to WT (Nakagawasai et al., 2010). Cognitive function examined using a passive avoidance task for long-term memory revealed that the latency time in the mutant mice was significantly decreased. The *Cacna1b⁻ᐟ⁻* mice showed prepulse inhibition deficits reminiscent of the sensorimotor gating deficits observed in a large majority of schizophrenic patients.

Thus, N-type VGCCs can play critical roles in modulation of a wide variety of mental and behavioral traits.

Endothelium-Dependent Relaxation and Autonomic Regulation of Cardiovascular Function: Conventional Ca$_V$2.2 KO

Impairment of endothelium-dependent relaxation of the thoracic aorta induced by Ang II treatment was significantly attenuated in the *Cacna1b⁻ᐟ⁻* mice (Ino et al., 2001), while the Ang-induced blood pressure increase was comparable between the mutant and WT mice (Nishida et al., 2013). Importantly, the thoracic aorta of the *Cacna1b⁻ᐟ⁻* mice showed a smaller area of oxidative stress compared to the WT mice. These results suggested that N-type VGCCs in the vascular endothelial cells contribute to the ROS production in the Ang II-treated hypertension model.

Dysregulation of autonomic nervous system activity is an important cause of ventricular arrhythmias and sudden death in patients with heart failure (Yamada et al., 2014). When transgenic mice expressing a cardiac-specific, dominant-negative form of neuron-restrictive silencer factor (dnNRSF-Tg) were used as a model of dilated cardiomyopathy leading to sudden arrhythmic death, crossing dnNRSF-Tg with *Cacna1b⁻ᐟ⁻* mice (Ino et al., 2001) interestingly improved the survival rate and malignant arrhythmias. These results suggest that reestablishing autonomic nervous system balance through N-type VGCC inhibition can be an effective approach to preventing sudden arrhythmic death associated with heart failure.

Diabetic Kidney Dysfunction: Conventional Ca$_V$2.2 KO

To clarify the significance of N-type VGCC in diabetic nephropathy, the *Cacna1b⁻ᐟ⁻* mice (Ino et al., 2001) were mated with the *db/db* (diabetic) mice (Ohno et al., 2016). Immunohistochemical localization revealed Ca$_V$2.2 proteins localized in glomeruli including podocytes and distal tubular cells. *db/db Cacna1b⁻ᐟ⁻* mice exhibited reduction in urinary albumin excretion, glomerular hyperfiltration, blood glucose levels, histological deterioration, systolic blood pressure, and urinary catecholamine compared to control *db/db Cacna1b⁺ᐟ⁺* mice. Consistently, diabetic mice treated with cilnidipine, an N- and L-type blocker, showed a reduction in albuminuria and an improvement of glomerular changes. In cultured podocytes, depolarization-dependent Ca²⁺ responses were decreased by ω-conotoxin GVIA. Furthermore, reduction of nephrin by transforming growth factor-β in podocytes was reversed by cilnidipine and ω-conotoxin GVIA or by a mitogen-activated protein kinase inhibitor. Thus, Ca$_V$2.2 inhibition exerts renoprotective effects, countering the progression of diabetic nephropathy.

Ca$_V$2.3

Early Reports of Ca$_V$2.3 Mutant Mice

R-Type Identification: Conventional and Conditional Ca$_V$2.3 KO

It was hypothesized that R-type VGCC currents result from the expression of Ca$_V$2.3 encoded by the *Cacna1e* gene, on the basis of recombinant expression studies, but direct evidence was missing. *Cacna1e⁻ᐟ⁻* mice were generated by replacing the exon encoding the S4–S6 regions of repeat II with a neomycin/URA3 selection cassette (Wilson et al., 2000). Application of ω-conotoxin GVIA, ω-agatoxin IVA, and nimodipine to cultured cerebellar granule neurons from WT mice suppressed the VGCC Ba⁺ current, but left a residual blocker-resistant current with an amplitude of ~30% of the original total current. A minor portion of this residual current was inhibited by the $Ca_V2.3$-selective toxin SNX-482. This SNX-482-sensitive portion of the granule cell residual current was absent in $Cacna1e^{-/-}$ mice. These results suggest that there exists a SNX-482-sensitive "residual" R-type current that results from the expression of $Ca_V2.3$.

To investigate the contribution of the $Ca_V2.3$ subunits to the "residual" current, another group generated a floxed $Cacna1e^{flox/+}$ line by inserting the loxP-flanked neomycin cassette at the site between exons 2 and 3, and then this line was crossed with the deleter line, which expresses Cre-recombinase constitutively under the control of a cytomegalovirus promoter, to yield $Cacna1e^{-/-}$ mice (Sochivko et al., 2002). In $Cacna1e^{-/-}$ mice, the residual current was considerably reduced in CA1 neurons (79%) and cortical neurons (87%), with less reduction occurring in dentate granule neurons (47%). SNX-482 partially inhibited the residual currents in hippocampal CA1 pyramidal cells, dentate granule cells, and cortical neurons. These results suggest that the so-called R-type sensitive to SNX-482 underlies a significant fraction of the "residual" current in different central neurons.

Pain: Conventional $Ca_V2.3$ KO

Physiological function of $Ca_V2.3$ was elusive compared with other Ca_Vs because of the limited availability of specific blockers. To clarify the physiological roles of the $Ca_V2.3$ channel, conventional $Ca_V2.3$ KO ($Cacna1e^{-/-}$) mice were generated by an in-frame fusion of the coding sequence of exon 1 of the $Cacna1e$ gene with the $nLacZ$-neo cassette, which allows labeling of $Ca_V2.3$-expressing cells via β-gal activity (Saegusa et al., 2000). $Cacna1e^{-/-}$ mice showed reduced spontaneous locomotor activities in an open field test. β-gal staining revealed that $Ca_V2.3$ is expressed in various regions involved in the control of pain transmission. Although $Cacna1e^{+/-}$ and $Cacna1e^{-/-}$ mice exhibited normal pain behaviors from acute mechanical, thermal, and chemical stimuli, they both showed reduced responses to somatic inflammatory pain. Furthermore, $Cacna1e^{-/-}$

mice presensitized with a visceral noxious conditioning stimulus showed increased responses to somatic inflammatory pain, in marked contrast with WT mice, in which the inhibitory effect of the descending antinociceptive pathway was predominant. These results suggest that $Ca_V2.3$ controls pain behavior by both spinal and supraspinal mechanisms.

Amygdala Function: Conventional $Ca_V2.3$ KO

Unlike other types of VGCCs, it was unclear how $Ca_V2.3$-containing R-type VGCCs contribute to a variety of neuronal processes. To understand the role of R-type currents in central amygdala neurons that exhibit abundant $Ca_V2.3$ expression, $Cacna1e^{-/-}$ mice were generated by replacing the 0.5-kb $XbaI$–$ClaI$ fragment within the exon encoding the translation initiation methionine and a part of the N-terminus cytosolic domain with a GFP-neo cassette (Lee et al., 2002). The majority of R-type current in central amygdala neurons was eliminated in $Cacna1e^{-/-}$ mice, whereas other Ca^{2+} channel types were unaffected. $Cacna1e^{-/-}$ mice exhibited anxiety-related behavior in an open field test. These results suggest that $Ca_V2.3$ has a role in amygdala physiology associated with anxiety.

Extended Knowledge of $Ca_V2.3$ Mutant Mice

Pain: Conventional $Ca_V2.3$ KO

Morphine is the drug of choice to treat intractable pain, although its prolonged administration often causes undesirable side effects, including analgesic tolerance. The $Cacna1e^{-/-}$ mice (Saegusa et al., 2000) were subjected to evaluation of the role of R-type VGCCs in morphine side effects (Yokoyama et al., 2004). In the $Cacna1e^{-/-}$ mice, analgesia was enhanced, compared with WT mice, upon systemic morphine administration or exposure to warm water swim-stress that induces endogenous opioid release. In addition, the mutant mice showed resistance to morphine tolerance. These results suggest that R-type VGCCs ameliorate morphine analgesia and suppress morphine tolerance.

Pancreatic Function and Glucose Metabolism: Conventional Ca$_V$2.3 KO

Cacna1e⁻/⁻ mice (Sochivko et al., 2002) were also subjected to *in vivo* validation of the functional role suggested for Ca$_V$2.3-containing R-type VGCC in endocrine processes (Pereverzev et al., 2002). Intraperitoneal injection of D-glucose showed that glucose tolerance was markedly reduced, and insulin release into plasma was impaired, in *Cacna1e⁻/⁻* mice. In stressed *Cacna1e⁻/⁻* mice, the rate of glucose release into the blood was only 29% of that observed for control WT, suggesting that Ca$_V$2.3 plays a role in insulin secretion and glucose homeostasis. Also, in the study by a different group, *in vivo* glucose tolerance, insulin tolerance, and stress-induced glucose release tests were performed on *Cacna1e⁻/⁻* mice (Saegusa et al., 2000; Matsuda et al., 2001). The *Cacna1e⁻/⁻* mice showed increased bodyweight, and in glucose tolerance and insulin tolerance tests, an increased blood glucose level. These results suggest that the role played by R-type Ca$_V$2.3 VGCC play in glucose homeostasis is through reduction of insulin sensitivity.

It has been suggested that the kinetics of insulin release from pancreatic islets is regulated by concerted activation of different VGCCs. *Cacna1e⁻/⁻* mice (Pereverzev et al., 2002) showed a reduction in glucose-evoked insulin secretion in islets, being consistent with the inhibition by SNX-482 (Jing et al., 2005). Dynamic insulin-release measurements demonstrated that genetic *Cacna1e⁻/⁻* or SNX-482 inhibits the second-phase secretion, sparing the first-phase secretion. This suppression of the second phase coincided with a reduction in oscillatory Ca²⁺ signaling and granule recruitment after completion of the initial exocytotic burst in single *Cacna1e⁻/⁻* β-cells. Thus, R-type VGCCs play a role in controlling the second-phase insulin release from pancreatic islets.

Zn²⁺ has been implicated in pancreatic islet cell crosstalk, while sudden cessation of Zn²⁺ supply during hypoglycemia is known to trigger glucagon secretion in rodents. In *Cacna1e⁻/⁻* mice (Pereverzev et al., 2002), fasting glucose and glucagon levels were increased compared to WT mice, while the Zn²⁺ chelator diethyldithio-carbamate produced a significant and correlated increase of blood glucose and serum glucagon concentrations in WT mice, but not in *Cacna1e⁻/⁻* mice (Drobinskaya et al., 2015). Severe glucose intolerance was observed in Zn²⁺-depleted *Cacna1e⁻/⁻* mice, but not in the control vehicle-treated *Cacna1e⁻/⁻* mice or Zn²⁺-depleted WT mice. These results suggest that R-type VGCCs are involved in the Zn²⁺-mediated suppression of glucagon secretion during hyperglycemia.

Distribution in the Brain: Conventional Ca$_V$2.3 KO

Little was known previously about the subcellular distribution of Ca$_V$2.3 proteins across various neuronal compartments. Using the antibodies whose R-type specificity was confirmed by *Cacna1e⁻/⁻* mice (Saegusa et al., 2000; Pereverzev et al., 2002), the regional and subcellular localization of Ca$_V$2.3 was characterized at both light and electron microscopic levels (Parajuli et al., 2012). Ca$_V$2.3 proteins were expressed ubiquitously in the brain, with higher levels in the hippocampus, interpeduncular nucleus, striatum, pallidum, cortex, amygdala, olfactory tubercle, accumbens, and dorsal cochlear nucleus. With respect to subcellular localization, Ca$_V$2.3 was predominantly presynaptic in the interpeduncular nucleus, but mainly postsynaptic in other brain regions. In CA1 neurons of the hippocampus, individual spine heads contained 0–16 Ca$_V$2.3 particles in the adult mouse and 0–19 Cav2.3 particles in the P20 rat. Thus, R-type VGCCs display unique presynaptic and postsynaptic localization in the brain.

Synaptic Plasticity in the Hippocampus and Cerebellum: Conventional Ca$_V$2.3 KO

To investigate the functional roles of the R-type Ca$_V$2.3 channel in hippocampal CA1 pyramidal neurons, *Cacna1e⁻/⁻* mice (Saegusa et al., 2000) were analyzed for hippocampal synaptic properties and hippocampus-dependent behaviors

(Kubota et al., 2001). The $Ca_V2.3$ mRNA was identified in the hippocampus by *in situ* hybridization. The basic excitatory synaptic transmission and LTP by θ-burst stimulation were intact in the CA1 region of *Cacna1e*$^{-/-}$ mice, which were able to establish and maintain fear memories. The *Cacna1e*$^{-/-}$ mice retained improvement in the performance of the Morris water maze test but showed impairment in the probe test. The results suggest that R-type plays a role in accurate spatial memory but not in fear memory.

Ca^{2+} influx into presynaptic terminals *via* VGCCs triggers fast neurotransmitter release as well as different forms of synaptic plasticity. Compared to the NMDA receptor-dependent postsynaptic forms of LTP, much less was known about NMDA-independent presynaptic forms of LTP. *Cacna1e*$^{-/-}$ mice (Pereverzev et al., 2002) revealed a contribution of presynaptic Ca^{2+} entry through $Ca_V2.3$ to the induction of mossy fiber LTP and post-tetanic potentiation by brief trains of presynaptic action potentials (Dietrich et al., 2003). Fast synaptic transmission, paired-pulse facilitation, or frequency facilitation were unaffected at the mossy fiber-CA3 synapse in *Cacna1e*$^{-/-}$ mice. R-type VGCCs thus may boost presynaptic Ca^{2+} accumulation, triggering presynaptic LTP and post-tetanic potentiation, without affecting the low release probability at mossy fiber synapses. An independent study (Breustedt et al., 2003) on *Cacna1e*$^{-/-}$ mice (Wilson et al., 2000) revealed an increase in threshold for inducing mossy fiber presynaptic LTP; $Ca_V2.3$ was not involved in the expression mechanism, but it contributed to the Ca^{2+} influx during the LTP induction phase. SNX482-sensitive Ca^{2+} rises indicated the presence and function of $Ca_V2.3$-containing VGCCs at mossy fiber terminals, supporting the role of $Ca_V2.3$ in the presynaptic form of LTP.

$Ca_V2.3$ is abundantly expressed in the cerebellum and was subjected to evaluation of roles in cerebellar functions responsible for motor coordination and learning (Osanai et al., 2006). *Cacna1e*$^{+/-}$ and *Cacna1e*$^{-/-}$ mice (Saegusa et al., 2000) showed delayed motor learning in Rotarod tests. Electrophysiological analysis revealed paradoxical results: a deficit in long-term depression (LTD) at the parallel fiber Purkinje cell synapse in *Cacna1e*$^{+/-}$ mice, but apparently normal LTD in *Cacna1e*$^{-/-}$ mice. The number of spikes evoked by current injection in Purkinje cells under the current-clamp mode was diminished in these $Ca_V2.3$ mutant mice in a gene dosage-dependent manner, suggesting a contribution of R-type to spike generation in cerebellar Purkinje cells.

Sleep: Conventional $Ca_V2.3$ KO

The role of R-type VGCCs localized in the reticular thalamic nucleus remained unclear. The role of $Ca_V2.3$ was analyzed in *Cacna1e*$^{-/-}$ mice (Wilson et al., 2000) in comparison with WT controls in both spontaneous and artificial urethane-induced sleep, using implantable video-electroencephalogram radiotelemetry. The *Cacna1e*$^{-/-}$ mice (Wilson et al., 2000) exhibited reduced wake duration and increased slow-wave sleep (Siwek et al., 2014). While mean sleep stage durations remained unchanged, the total number of slow-wave sleep epochs was increased in *Cacna1e*$^{-/-}$ mice. Sleep analysis further pointed to an alteration of sleep architecture in *Cacna1e*$^{-/-}$ mice due to aberration in sleep stage transitions. These results suggest that R-type VGCCs in the thalamocortical loop regulate rodent sleep architecture.

T-Type in Sperm: Conventional $Ca_V2.3$ KO

In mammalian male germline cells, low-voltage-activated T-type Ca^{2+} currents have been identified and their electrophysiological properties were studied. Whole-cell VGCC currents in acutely dissociated pachytene spermatocytes from *Cacna1e*$^{-/-}$ (Saegusa et al., 2000) and WT mice displayed a typical profile of T-type Ca^{2+} currents with indistinguishable kinetics (Sakata et al., 2001). Single-cell RT-PCR revealed the expression of *Cacna1g* in *Cacna1e*$^{-/-}$ and WT pachytene spermatocytes, from which low-voltage-activated Ca^{2+} currents were recorded. Thus, the results failed to support contribution of $Ca_V2.3$ to T-type in the pachytene spermatocyte.

To study the function of the $Ca_V2.3$-carrying VGCCs in spermatozoa, Ca^{2+} transients and sperm motility were analyzed (Sakata et al., 2002) using *Cacna1e*$^{-/-}$ mice (Saegusa et al.,

2000). The averaged rising rates of Ca^{2+} transients evoked by solubilized zona pellucida were significantly reduced in the head region of $Cacna1e^{-/-}$ sperm. A computer-assisted sperm motility assay revealed that straight-line velocity and linearity were greater in the $Cacna1e^{-/-}$ sperm. These results suggest that the R-type channel plays a role in sperm Ca^{2+} transients and the control of flagellar movement.

Seizure and Epilepsy: Conventional Ca$_V$2.3 KO

To reveal the role of R-type VGCCs in seizure susceptibility, convulsive seizure activity was studied in $Cacna1e^{-/-}$ mice (Pereverzev et al., 2002; Weiergräber et al., 2006). Pentylenetetrazol-induced seizure susceptibility was dramatically reduced in $Cacna1e^{-/-}$ mice, suggesting that Ca$_V$2.3 plays an important role in seizure initiation and propagation. Furthermore, $Cacna1e^{-/-}$ mice showed clear alteration in behavioral seizure architecture and dramatic resistance to limbic seizures induced by kainic acid, compared with WT controls (Weiergräber et al., 2007). Histochemical analysis within the hippocampus revealed that excitotoxic effects after kainic acid administration were absent in $Cacna1e^{-/-}$ mice, in contrast to WT mice, which exhibited clear and typical signs of excitotoxic cell death. Thus, R-type plays a role in both ictogenesis and seizure generalization, and is important in neuronal degeneration after excitotoxic events.

Lamotrigine is an often-used modern antiepileptic drug; however, its mechanism of action is not yet fully understood. In heterologous expression systems, lamotrigine inhibits Ca$_V$2.3-mediated R-type currents, which contribute to kainic acid-induced epilepsy *in vivo*. In $Cacna1e^{-/-}$ mice (Pereverzev et al., 2002), lamotrigine pretreatment increased seizure activity and increased the percentage of degenerated CA1 pyramidal neurons upon kainic acid-induced seizures (Dibué-Adjei et al., 2017). Antiepileptic drugs including lamotrigine, topiramate, and lacosamide reduced seizure activity in WT mice; however, only the non-Ca^{2+} channel modulating antiepileptic drug, lacosamide exerted an anti-

convulsive effect in $Cacna1e^{-/-}$ mice. These findings provide the first *in vivo* evidence for an essential role of Ca$_V$2.3 in lamotrigine pharmacology.

Epilepsy: Conventional Ca$_V$2.3 KO

Although Ca$_V$2.3 is expressed in key structures of the thalamocortical circuitry, its functional relevance in absence seizure remains unknown. $Cacna1e^{-/-}$ mice (Pereverzev et al., 2002) exhibited increased spike-wave discharge activity, in terms of total events and duration and altered frequency distribution, following administration of the GABA$_B$ receptor agonist γ-hydroxybutyrolactone, compared to WT mice (Weiergräber et al., 2008), suggesting that Ca$_V$2.3 is important in the absence of seizure-specific spike-wave discharge susceptibility.

In brain slices prepared from $Cacna1e^{-/-}$ mice (Lee et al., 2002), a hyperpolarizing current injection initiated a low-threshold burst of spikes in thalamic reticular nucleus neurons; subsequent oscillatory burst discharges were severely suppressed, with a significantly reduced slow afterhyperpolarization (Zaman et al., 2011). $Cacna1e^{-/-}$ mice also exhibited a marked decrease in the sensitivity to γ-hydroxybutyrolactone-induced absence epilepsy. These results suggest that R-type Ca$_V$2.3 is important in the oscillatory burst discharge in thalamic reticular nucleus neurons and in the expression of absence epilepsy.

Ischemic Neuronal Injury: Conventional Ca$_V$2.3 KO

When a focal ischemia model with a complete occlusion of the middle cerebral artery was used, the $Cacna1e^{-/-}$ mice (Saegusa et al., 2000) showed increased infarct size (Toriyama et al., 2002). In *in vitro* Ca^{2+} imaging studies using hippocampal slices, oxygen-glucose deprivation induced an elevated intracellular Ca^{2+} increase in $Cacna1e^{-/-}$ mice, while tetrodotoxin or bicuculline application abolished the difference between the mutant and WT mice. These results suggest a protective role of R-type in ischemic neuronal injury through a mechanism involving GABAergic actions.

Narcotic Effects: Conventional Ca$_V$2.3 KO

There is much evidence that VGCCs play a role in cocaine-induced behavioral responses. *Cacna1e*$^{-/-}$ mice (Saegusa et al., 2000) were analyzed to assess the contribution of Ca$_V$2.3 to the effects of cocaine (Han et al., 2002). Acute administration of cocaine, which enhanced locomotor activity in WT mice, failed to induce this response in *Cacna1e*$^{-/-}$ mice. Repeated exposure to cocaine-induced behavioral sensitization and conditioned place preference in both genotypes. Pretreatment with a D1-receptor antagonist, SCH23390, blocked the cocaine-induced place preference in WT mice, while it exerted no significant effects in *Cacna1e*$^{-/-}$ mice. These data indicate that R-type VGCCs contribute to the locomotor-stimulating effect of cocaine, are involved in a novel pathway that is related to cocaine rewarding, and insensitive to D1 receptor antagonists.

Anesthesia: Conventional Ca$_V$2.3 KO

To clarify the inhibitory action of general anesthesia on R-type, sensitivity to propofol and halothane was examined in *Cacna1e*$^{-/-}$ mice (Saegusa et al., 2000; Takei et al., 2003). Propofol-induced sleep time was reduced, and administration of 50% effective halothane concentration to induce the loss of the righting reflex was increased in *Cacna1e*$^{-/-}$ mice compared with WT. Propofol inhibited population spikes at Schaffer collateral-CA1 synapses by potentiating GABAergic inhibition, which was reversed in *Cacna1e*$^{-/-}$ mice. Halothane-induced suppression of population spikes was pronounced in *Cacna1e*$^{-/-}$ mice in the presence of bicuculline, indicating that the depression of excitatory neuronal activities by halothane was accelerated in the neuronal circuitry by Cav2.3 deficiency. Thus, Cav2.3 plays an important regulatory role in anesthetic effects of propofol and halothane.

Parkinson Disease: Conventional Ca$_V$2.3 KO

Degeneration of dopaminergic neurons in the SN causes the motor symptoms of Parkinson disease. Involvement of R-type VGCCs in this age-dependent and region-selective neurodegeneration remained unclear. Ca$_V$2.3 transcripts were more abundant than other VGCCs in mouse nigral neurons and were upregulated during aging (Benkert et al., 2019). *Cacna1e*$^{-/-}$ mice (Pereverzev et al., 2002) showed reduced activity-associated nigral somatic Ca^{2+} transients and Ca^{2+}-dependent after-hyperpolarizations. When the MPTP Parkinson's disease mouse model was applied to *Cacna1e*$^{-/-}$ mice, protection of SN dopaminergic neurons was observed. Transcription of NCS-1, a Ca^{2+}-binding protein implicated in neuroprotection, was upregulated in *Cacna1e*$^{-/-}$ mice. R-type VGCCs may thus act as regulators of nigral neuronal viability in Parkinson's disease.

Cardiac Arrhythmia and Altered Autonomic Control: Conventional Ca$_V$2.3 KO

To investigate the role of R-type VGCCs in heartbeats, the spontaneous activity of the prenatal hearts from *Cacna1e*$^{-/-}$ mice (Pereverzev et al., 2002) was characterized (Lu et al., 2004). Although the mean heart rate is nearly indistinguishable between both genotypes, the heart from *Cacna1e*$^{-/-}$ mice had arrhythmic beats. In isolated WT hearts, arrhythmia was induced by SNX-482, indicating that R-type stabilizes regular heartbeat in prenatal mice.

Cacna1e$^{-/-}$ mice (Pereverzev et al., 2002) exhibited cardiac arrhythmia and altered autonomic control within the cardiovascular system (Weiergräber et al., 2005). In telemetric electrocardiogram recordings, *Cacna1e*$^{-/-}$ mice displayed subsidiary escape rhythm, altered atrial activation patterns, atrioventricular conduction disturbances, and alteration in QRS morphology. After blockage of the autonomous nervous system by atropine/propranolol administration, intrinsic heart rate was no longer different between *Cacna1e*$^{-/-}$ and WT mice, suggesting that the increased heart rate in *Cacna1e*$^{-/-}$ mice is due to increased sympathetic tonus. These obtained data point to an important role played by R-type VGCC in normal impulse generation and conduction in the murine heart.

Ca$_V$3

Ca$_V$3 proteins are the only family that mediates low-voltage-activated T-type. Due to the lack of selective blockers, T-type has been the most elusive VGCC type for elucidation of its biological significance, despite widespread recognition of important roles in low-threshold action potentials, for instance, in cardiac automaticity. Use of appropriate KO and KI mice of respective Ca$_V$3 members, Ca$_V$3.1, Ca$_V$3.2, and Ca$_V$3.3, alone and in combination, has turned out to be extremely effective for addressing the deficit of convincing pharmacology during early T-type studies, providing direct evidence for the involvement of Ca$_V$3 subtypes in specific biological responses.

Ca$_V$3.1

Early Reports of Ca$_V$3.1 Mutant Mice

Epilepsy: Conventional Ca$_V$3.1 KO

Involvement of Cav3.1 in low-voltage-activated T-type VGCC currents, namely in the genesis of spike-wave discharges, the hallmark of absence seizures, was proposed. To provide direct *in vivo* evidence for this hypothesis, conventional Ca$_V$3.1 KO (*Cacna1g⁻/⁻*) mice were generated by deleting the exon encoding amino acid residues 82–118, which comprise the N terminus of the Ca$_V$3.1 protein (Kim et al., 2001b). In intracellular recording with a current clamp, the thalamocortical relay neurons of the *Cacna1g⁻/⁻* mice lacked the burst mode firing of action potentials, whereas they showed a normal pattern of tonic mode firing. Electroencephalography and recordings of field activity of thalamic nuclei, using depth electrodes in freely moving animals, revealed specific resistance of the *Cacna1g⁻/⁻* thalamus to the generation of spike-wave discharges, in response to GABA$_B$ receptor activation induced by agonists. These results suggest that Ca$_V$3.1 T-type VGCCs modulate the intrinsic firing pattern and play a critical role in the genesis of absence seizures *via* the thalamocortical pathway.

Anesthesia: Conventional Ca$_V$3.1 KO

It has been generally recognized that volatile anesthetics exert their clinical effects through direct inhibition of T-type VGCCs at concentrations comparable to those required to produce anesthesia. However, considering the important role of T-type VGCCs in regulating neuronal excitation, the possibility of their indirect and modulatory involvement in response to intravenous anesthesia was not excluded. To address this issue, a conventional *Cacna1g⁻/⁻* line was generated by deleting exon 5 encoding S5 region of repeat I (Petrenko et al., 2007). In the *Cacna1g⁻/⁻* mice, the 50% effective concentration for loss of righting reflex in the behavioral endpoint test, and the minimum alveolar concentration values using the tail-clamp method, indicated that the action of volatile anesthetics remained intact. However, the *Cacna1g⁻/⁻* mice required significantly more time to develop anesthesia-induced hypnosis after exposure to isoflurane, halothane, and sevoflurane, and after intraperitoneal administration of pentobarbital. These results suggest that Ca$_V$3.1 is required for the timely induction of anesthesia-induced hypnosis by volatile and some intravenous anesthetics, but not for the maintenance of anesthetic hypnosis and immobility.

Sleep: Conditional Ca$_V$3.1 KO

Inhibition of sensory signal transmission in the brain has been an important clue for understanding stable sleep. To establish the molecular and cellular basis for sensory suppression during sleep, and its potential role in preventing arousal, Ca$_V$3.1 T-type VGCCs were focally deleted in thalamic projection neurons (Anderson et al., 2005). Thalamic neuron-selective Ca$_V$3.1 KO mice were produced by crossing the mouse line carrying the K$_V$3.2 transgene with a *Cre* recombinase, which was inserted into the translational start site of K$_V$3.2, with another mouse line (*Cacna1g^{flox/flox}*) carrying *loxP* sequences in flanking introns of *Cacna1g* exons 9–12. The K$_V$3.2 promoter was chosen because its transcript is highly restricted (>90% of transcripts) to thalamic projection neurons. T-type activation caused prolonged inhibition (>9 s) of action-potential firing in thalamic projection neurons of WT, but not in the global $Ca_v3.1$ KO mice. In the thalamic neuron-selective $Ca_v3.1$ KO mice, arousals were more frequent and prolonged, causing the fragmentation of sleep. These findings support the hypothesis that thalamic T-type VGCCs are required to block transmission of arousal signals through the thalamus and to stabilize sleep.

Extended Knowledge of $Ca_v3.1$ Mutant Mice

In the adult heart Cav3.1 and Cav3.2 expression is restricted to the SAN, the atrioventricular node (AVN) and Purkinje fibers, indicating a role in cardiac automaticity and conduction. However, they are expressed in cardiomyocytes during developmental stages and re-expression in adult cardiac myocytes, and abnormal reexpression is observed in the diseased heart (Li et a., 2012 and Chiang et al., 2009 for references). Moreover, cardiac pathology can be initiated indirectly through the important role of these channels for vascular function as outlined below.

Cardiac Pacemaker Activity and Vascular Dilation: Conventional $Ca_v3.1$ KO

To establish the functional role of T-type VGCCs in the pacemaker activity of the SAN, $Cacna1g^{-/-}$ mice (Kim et al., 2001b) were examined (Mangoni et al., 2006). In isolated cells from the SAN and the atrioventricular node in the $Cacna1g^{-/-}$ mice, T-type currents were abolished, but L-type currents were unaffected. Telemetric electrocardiograms on unrestrained mice, as well as intracardiac recordings, revealed that $Cacna1g^{-/-}$ inactivation causes bradycardia and delays atrioventricular conduction, without affecting the excitability of the right atrium, where T-type currents were undetectable in both WT and $Cacna1g^{-/-}$ mice. Furthermore, in $Cacna1g^{-/-}$ mice, the intrinsic heart rate was

reduced, the SAN recovery time was prolonged, and the pacemaker activity of individual SAN cells was slowed through a reduction of the slope of the diastolic depolarization. Thus, $Ca_v3.1$ T-type VGCCs contribute to SAN pacemaker activity and atrioventricular conduction.

It has been hypothesized that T-type VGCCs promote formation of nitric oxide (NO) from the endothelium. $Cacna1g^{-/-}$ mice (Kim et al., 2001b) were used to examine the involvement of T-type in depolarization-dependent dilation of mesenteric arteries and blood pressure regulation (Svenningsen et al., 2014). NO-dependent vasodilatation following depolarization-mediated vasoconstriction was reduced significantly in mesenteric arteries from young (2–3 months of age) $Cacna1g^{-/-}$ mice compared to WT. Systemic infusion of a NO synthase-inhibitor elicited a significant increase in mean arterial blood pressure in WT, but not in the $Cacna1g^{-/-}$ mice. This was consistent with the reduced NO release and cGMP levels in depolarized $Cacna1g^{-/-}$ arteries, suggesting that T-type $Ca_v3.1$ is coupled with endothelial NO synthase to promote NO production by the endothelium. This finding was further supported by demonstrating $Ca_v3.1$ α_1 expression not only in vascular smooth muscle but also of its co-clustering of with eNOS in endothelial cells. It has to be noted, however, that (as also observed for $Ca_v3.2$, see below), this effect was age-dependent, as demonstrated with young and old (12–14 months) $Cacna1g^{-/-}$ mice (Thuesen et al., 2018). While WT mice developed the expected endothelial dysfunction with age, the impaired NO-dependent vasodilatation in young $Cacna1g^{-/-}$ mice normalized with aging to the level of young WT mice. This endothelial protection in the aged $Cacna1g^{-/-}$ mice was explained by enhanced NO levels measured in mesenteric arteries, as also observed in $Cacna1h$ KO arteries (see section "$Ca_v3.2$" below). While these data clearly demonstrate a role of $Ca_v3.1$ for endothelial function in addition to control of myogenic tone (see section "$Ca_v3.2$" below), the implications of these age-dependent changes for cardiovascular health are currently unclear.

Cerebellar Functions: Conventional Ca$_V$3.1 KO

To explore how T-type VGCCs functionally interact with postsynaptic receptors and contribute to synaptic signaling, ultrafast two-photon Ca²⁺ imaging was combined with electrophysiological recordings (Hildebrand et al., 2009). In cerebellar Purkinje cells, mGluR1 activation potentiated T-type currents *via* a G-protein- and tyrosine-phosphatase-dependent pathway. Colocalization of Cav3.1 proteins with mGluR1s was observed in Purkinje cell dendritic spines. In *Cacna1g*$^{-/-}$ Purkinje cells (Petrenko et al., 2007), T-type currents were abolished, in contrast to WT Purkinje cells, which display robust T-type currents. Parallel fiber stimulation induced fast subthreshold Ca²⁺ signaling in dendritic spines, and the synaptic Ca$_V$3.1-mediated Ca²⁺ transients were potentiated by mGluR1 selectively during bursts of excitatory parallel fiber inputs, revealing a fast Ca²⁺ signaling pathway in Purkinje cell dendritic spines.

The timing of motor coordination is regulated *via* the rhythmic motor pathway activation by pacemaker neurons or circuits, whose abnormal potentiation may lead to pathological tremors (Park et al., 2010c). After administration of harmaline, a reversible monoamine oxidase A inhibitor that induces serotonin syndrome and hypertensive crises, 4–10 Hz synchronous neuronal activities, which arise from the inferior olive, propagated to cerebellar motor circuits in WT mice, but these rhythmic activities were absent in the *Cacna1g*$^{-/-}$ mice (Kim et al., 2001b). The *Cacna1g*$^{-/-}$ inferior olive neurons lacked the subthreshold oscillation of membrane potentials and 4–10 Hz rhythmic burst discharges upon treatment with harmaline. A mathematical model suggested that harmaline could efficiently potentiate Ca$_V$3.1 VGCCs by shifting their activation voltage dependence toward a hyperpolarizing direction. Thus, potentiation of Ca$_V$3.1 T-type VGCCs in the inferior olive contributes to the onset of tremor in a pharmacological model of essential tremor.

Behavior and Hippocampal Oscillation: Conventional Ca$_V$3.1 KO

Exploratory drive is a fundamental emotion evoked by novelty stimulation, and plays roles in motivation, learning, and the well-being of organisms. Understanding of heterogeneity of exploratory behaviors is elusive, due to lack of solid experimental distinguishing criteria (Gangadharan et al., 2016). Localized Ca$_V$3.1 KD with short hairpin RNA in the medial septum selectively enhanced object exploration, while the *Cacna1g*$^{-/-}$ mice (Kim et al., 2001b) showed enhanced-object exploration and open-field exploration. In the medial septum-targeted Ca$_V$3.1 KD mice, only type 2 hippocampal θ rhythm was enhanced, whereas both type 1 and type 2 θ rhythm were enhanced in the *Cacna1g*$^{-/-}$ mice. Optogenetic activation of the septo-hippocampal GABAergic pathway in WT mice selectively enhanced object exploration behavior and type 2 θ rhythm. These findings provide important insights into defining object exploration, in distinction from open-field exploration.

Sleep: Conventional Ca$_V$3.1 KO

A role for T-type VGCCs in sleep regulation has been suggested, since the low-threshold spikes are predominantly found in the thalamus during non-REM sleep, but are absent during wakefulness and REM sleep (Lee et al., 2004). However, the extent of contribution of T-type to the heterogeneity of non-REM sleep remained uncertain. In the *Cacna1g*$^{-/-}$ mice (Kim et al., 2001b), thalamic δ waves (1–4 Hz) were abolished and sleep spindles (7–14 Hz) were impaired, whereas slow rhythms (<1 Hz) were relatively intact during urethane anesthesia and non-REM sleep, compared with the WT mice (Lee et al., 2004). The mutant mice also exhibited a higher incidence of brief awakening episodes of >16 s during non-REM sleep, whereas those of <16 sec were intact. These results suggest that T-type Ca$_V$3.1 VGCCs play critical roles in the genesis of thalamocortical oscillations, the modulation of sleep states, and the transition between non-REM sleep and wake states.

Sleep spindles are one type of rhythmic brain waves detected by electroencephalography during normal non-REM sleep and consist of characteristic waxing-and-waning field potentials grouped, as mentioned above, into 7–14 Hz oscillations that last for 1–3 s and recur every 5–10 s in the thalamus and the cortex (Lee et al., 2013). These oscillations are generated in the thalamus *via* synaptic interactions between GABAergic thalamic reticular nucleus (TRN) neurons and glutamatergic thalamocortical (TC) neurons and are then transmitted from TC neurons to cortical neurons, contributing to the network of TC oscillations. T-type VGCCs have been believed to mediate low-threshold burst firings of neurons in the two nuclei. Unexpectedly, in comparison with WT mice, which express $Ca_V3.1$ exclusively in TC regions (Talley et al., 1999), $Cacna1g^{-/-}$ mice (Kim et al., 2001b) showed similar typical waxing-and-waning sleep spindle waves at a similar occurrence and with similar amplitudes and episode durations during non-REM sleep (Lee et al., 2013); burst firing was however absent in the $Cacna1g^{-/-}$ TC neurons. The tonic spike frequency in TC neurons was significantly increased during spindle periods, compared with non-spindle periods, in both genotypes, whereas burst firing frequency was indistinguishable between spindle and non-spindle periods in the WT mice. Furthermore, spindle-like oscillations were generated within intrathalamic circuits composed of TRN and TC neurons in both the WT and $Cacna1g^{-/-}$ mice. These results raise questions about the role of low-threshold burst firings in TC neurons, suggesting the importance of tonic firing in spindle oscillations in the TC circuit.

Epilepsy: Conventional $Ca_V3.1$ KO and $Ca_V3.1$ Transgenic Overexpression

To understand the physiological significance of gating variability induced by diverse chemical ligands in T-type channels, the dynamic-clamp method was used to re-incorporate artificial T-type conductance (a real-time mimicry of ionic currents) in thalamic neurons of $Cacna1g^{-/-}$ mice (Kim et al., 2001b) devoid of endogenous T-type currents (Tscherter et al., 2011). The large depo-

larizing T-type current, with its specific kinetics, induced the high frequency reached during burst firing, in both thalamocortical neurons and the GABAergic neurons in the TRN. Minimal modifications in the properties of the T-type current, which were overlooked so far in classical biophysical studies, drastically affected the physiological output of the thalamic neurons and conditioned the dynamics of thalamocortical information integration.

To examine whether enhanced thalamocortical T-type currents suffice to induce absence of epilepsy, two mouse lines overexpressing the $Cacna1g$ gene under the control of native regulatory elements were generated by microinjection of the $Cacna1g$-containing bacterial artificial chromosome (BAC) into the pronuclei of fertilized mouse eggs (Ernst et al., 2009). Electroencephalographic analysis of both lines revealed bursts of spike-wave discharges accompanied by behavioral arrest. Interestingly, despite overexpression of $Ca_V3.1$ in the cerebellar cortex, neither line displayed the ataxic nor behavioral phenotypes associated with spike-wave epilepsy reported in all of the other monogenic mutant mouse models, suggesting that these comorbid phenotypes are genetically separable and depend on the inherited upstream gene defect. Thus, enhanced $Ca_V3.1$ VGCC activity is sufficient to induce spike-wave oscillations characteristic of pure absence epilepsy.

Pain: Conventional $Ca_V3.1$ KO

In response to intraperitoneal administration of acetic acid and $MgSO_4$, the WT mice showed typical visceral pain behaviors, such as writhing and abdominal stretching and constriction, while the $Cacna1g^{-/-}$ mouse (Kim et al., 2001b) showed stronger writhing responses to the chemicals (Kim et al., 2003). In response to visceral pain, the WT ventroposterolateral thalamic neurons evoked a surge of single spikes, which slowly decayed, while T type-dependent burst spikes gradually increased. In contrast, in $Cacna1g^{-/-}$ neurons, the single-spike response persisted without burst spikes. These results suggest inhibition of visceral nociception by T-type $Ca_V3.1$ VGCCs. This was unexpected, as it had been hypothesized that T-type is involved in the strong excitatory pathway that boosts pain signals.

GABAergic and output neurons in the periaqueductal gray (PAG), which convey analgesic signals generated by endogenous opioids, were classified by their electrical properties (Park et al., 2010a). GABAergic PAG neurons were mostly fast-spiking cells and were further divided into two distinct classes: with and without low-threshold spikes driven by T-type VGCCs. This was in contrast to the PAG output neurons that lacked low-threshold spikes and showed heterogeneous firing patterns. The $Cacna1g^{-/-}$ mice (Kim et al., 2001b) lacked low-threshold spikes in the PAG GABAergic neurons and showed impaired opioid-dependent analgesia. These results suggest that T-type $Ca_V3.1$ VGCCs are critical for the opioidergic analgesia system *via* the PAG.

Anesthesia: Conventional Ca$_V$3.1 KO

To study the precise mechanisms underlying the regulation of the activity of the central medial nucleus (CeM) in the arousal system for sleep and anesthesia initiation, T-type VGCCs were examined for their role in the excitability and rhythmic activity of CeM neurons during isoflurane-induced anesthesia (Timic Stamenic et al., 2019). Patch-clamp slice recordings revealed that T-type in CeM was inhibited by prototypical volatile anesthetic isoflurane and selective T-channel blocker 3,5-dichloro-N-[1-(2,2-dimethyl-tetrahydro-pyran-4-ylmethyl)-4-fluoro-piperidin-4-ylmethyl]-benzamide (TTA-P2). The effects of TTA-P2 and isoflurane, which attenuated tonic and burst firing modes and hyperpolarized CeM neurons in WT mice, were greatly diminished in the $Cacna1g^{-/-}$ mice (Petrenko et al., 2007). *In vivo* local field potential recordings from CeM indicated that the δ oscillation-promoting action of TTA-P2 and isoflurane was weakened in the $Cacna1g^{-/-}$ mice. The results demonstrated for the first time the importance of $Ca_V3.1$ T-type VGCCs in thalamocortical oscillations from the non-specific thalamic nuclei which underlie clinically relevant effects of isoflurane.

Mouse Model for Spinocerebellar Ataxia: Ca$_V$3.1 KI

Spinocerebellar ataxia 42 (SCA42) is a neurodegenerative disorder caused by an R1715H human missense mutation in *CACNA1G*. In studying a large Japanese family with SCA42, postmortem pathological examination of a SCA42 patient revealed severe cerebellar degeneration with prominent Purkinje cell loss, without ubiquitin accumulation. To determine whether this mutation causes ataxic symptoms and neurodegeneration, KI mice, harboring the R1723H mutation that corresponds to the mutation identified in the SCA42 family in *Cacna1g*, were generated (Hashiguchi et al., 2019). Both $Cacna1g^{R1723H/+}$ and $Cacna1g^{R1723H/R1723H}$ mice developed an ataxic phenotype from 11–20 weeks of age and showed Purkinje cell loss after 50 weeks, and degenerative Purkinje cells and atrophic thinning of the molecular layer were pronounced in $Cacna1g^{R1723H/R1723H}$ mice. Electrophysiological analysis of $Cacna1g^{R1723H/R1723H}$ Purkinje cells showed altered voltage dependence of $Ca_V3.1$ channel activation and reduced rebound action potentials after hyperpolarization, whereas basic properties of synaptic transmission onto Purkinje cells were intact. The membrane potential resonance of neurons in the inferior olivary nucleus was decreased in the $Cacna1g^{R1723H/R1723H}$ mice. Thus, the SCA42 point mutation in *CACNA1G* disrupts the firing of inferior olivary nucleus neurons, thereby causing abnormalities of climbing fiber signaling to Purkinje cells at an early stage preceding Purkinje cell degeneration.

Sperm: Conventional Ca$_V$3.1 KO

Successive activation of three different types of Ca^{2+} channels (T-type VGCCs, inositol 1,4,5-trisphosphate receptors, and TRPC2 channels) is required for mammalian acrosome reaction. Using the $Cacna1g^{-/-}$ mice (Kim et al., 2001b), it was revealed that the T-type current density in spermatogenic cells is unaffected by $Ca_V3.1$ deficiency (Stamboulian et al., 2004). Biophysical and pharmacological characterization of spermatogenic T-type currents from WT and $Cacna1g^{-/-}$ mice suggest that $Ca_V3.2$ contributes to T-type currents in $Cacna1g^{-/-}$ spermatogenic cells.

Immune System: Conventional and conditional Ca$_V$3.1 KO

In addition to L-type channels (see section on LTCCs), T-type channels also affect immune function. Wang et al. (2016) recorded small DHP-insensitive voltage-activated Ca^{2+} currents in primary splenic CD4$^+$ T-cells. These currents were strongly reduced in cells from constitutive *Cacna1g* KO mice (Wang et al., 2016). The remaining T-type-like current was likely due to upregulated Ca$_V$3.2 and Ca$_V$3.3 expression in the KO mice. Ca$_V$3.1-deficiency decreased intracellular free Ca^{2+} concentrations, but, unlike observed for Ca$_V$1.4 deficiency (see above), store-operated Ca^{2+} entry and Ca^{2+} signals stimulated by TCR activation were not changed in *Cacna1g$^{-/-}$* mice, as were T-cell development and maturation. However, KO mice were largely protected in the experimental autoimmune encephalomyelitis disease model with reduced CNS-infiltrating CD4$^+$ cells. This was due to the effects of Ca$_V$3.1 in T-cells because the immunological changes were also observed in mice in which Ca$_V$3.1 was conditionally partially knocked out in T-cells of heterozygous floxed *Cacna1g* mice (Anderson et al., 2005) under the control of Lck-Cre. The production of GM-CSF by CNS-infiltrating Th1 and Th17 cells was reduced and may account for the protective effect in EAE. *In vitro* polarized Th17 cells from *Cacna1g$^{-/-}$* naive CD4$^+$ T cells showed decreased IL-17A, IL-17F, and IL-21 production. Compared to WT, *Cacna1g$^{-/-}$* mice had less nuclear NFATc1 and NFATc2 in both resting and stimulated CD4$^+$ cells. By controlling intracellular Ca^{2+} concentration and nuclear translocation of NFAT, Ca$_V$3.1 therefore appears to play a critical role in inducing transcriptional activation of GM-CSF (Wang et al., 2016).

Ca$_V$3.2

Ca$_V$3.2 in Arterial Vascular Tone: Conventional Ca$_V$3.2 KO (*Cacna1h$^{-/-}$*)

T-type currents are absent in normal adult ventricular myocytes but are expressed during embryonic development. Like other fetal genes (see also Ca$_V$1.3 above), they can be re-expressed upon cardiac remodeling in failing hearts. An important role of Ca$_V$3.2 for normal cardiovascular function is evident from findings in homozygous KO mice. Unlike their WT controls, these animals develop diffuse regions of cardiac fibrosis which worsens between week 10 and 1 year of age. At 1 year, they show large areas of fibrosis, necrosis, and lymphocyte infiltration but without differences in heart rate or evidence for cardiac arrhythmias (Chen et al., 2003). In homozygous, but not heterozygous KO mice also, abnormalities in coronary arteries were found which, in contrast to the smooth and regular vessels in WT, appeared constricted with irregular spiral shapes. While they responded normally to vasoconstrictors, the acetylcholine-induced vasodilation of preconstricted WT arteries, which is mediated by endothelial NO production, was absent in the KO mice. In contrast, acetylcholine even led to further vasoconstriction. NO-donor-induced vasodilation was also reduced in the KO mice but they were relaxed normally by nifedipine (Chen et al., 2003).

Interestingly, Ca$_V$3.2 α_1-subunits were found to associate with BK$_{Ca}$ channels when co-expressed in HEK-293 cells and co-immunoprecipitated with BK$_{Ca}$ in solubilized brain microsomes from WT but not from the KO mice. This suggests that NO-dependent relaxation of arteries involves Ca$_V$3.2 activation of BK$_{Ca}$ channels which induces vasodilation through hyperpolarization (Chen et al., 2003). It should be noted, however, that the genetic background appears to strongly affect this vasodilatory mechanism because in mesenteric arteries, Ca$_V$3.2 KO impaired the NO-dependent vasodilatory recovery following depolarization in *Cacna1h$^{-/-}$* mice bred on a C57BL6J/129S3 background (also used for the experiments in coronary arteries above, Chen et al., 2003) but not after backcrossing into a C57Bl/6 background (Svenningsen & Hansen, 2016; Thuesen et al., 2018).

Coupling between Ca$_V$3.2 and BK$_{Ca}$ channels was further confirmed by demonstrating that in rat cerebral arterial smooth muscle cells plasmalemmal $Ca_V3.2$ channels and SR ryanodine receptors reside in close proximity to each other within microdomains in which the SR is localized underneath caveolae. Within these domains, the channels drive ryanodine receptor-mediated Ca^{2+} sparks enabling BK_{Ca} activation with hyperpolarization-induced attenuation of cerebral arterial constriction (Harraz et al., 2014). This coupling was also demonstrated in mesenteric arteries of young (2–4 months) mice (Chen et al., 2003). KO arteries had slightly smaller basal diameter and developed significantly higher myogenic tone when intravascular pressure was increased from 20 to ≥ 100 mmHg (Harraz et al., 2015; Mikkelsen et al., 2016). Like in rat cerebral arteries, block of $Ca_V3.2$ (with Ni^{2+}) increased pressure-induced vasoconstriction. This was due to inhibition of $Ca_V3.2$ because the Ni^{2+} effect was absent in arteries of $Ca_V3.2$ KO animals. Inhibition of BK_{Ca} channels by paxilline increased myogenic tone in young WT but not in the KO mice (Mikkelsen et al., 2016). At negative voltages where T-type channels are already active, Ni^{2+} reduced spontaneous transient outward BK_{Ca}-mediated K^+ currents (Harraz et al., 2015).

Using $Cacna1h^{-/-}$ mice an interesting age-dependent effect and a potentially cardiovascular protective role of $Ca_V3.2$ could be demonstrated, because the above-described inhibition of myogenic tone by coupling to BK_{Ca} channels is present in mesenteric arteries of young (2–4 months) but not of mature adult (8–14 months) mice (Mikkelsen et al., 2016). This could not be explained by a reduced expression of $Ca_V3.2\ \alpha_1$ at both the transcript and protein level ($Ca_V3.2\ \alpha_1$ is detected primarily in smooth muscle cells but at very low levels also in endothelial cells; Mikkelsen et al., 2016).

In young animals, $Ca_V3.2$ KO not only increased the myogenic tone but also significantly reduced the flow-mediated vasodilatory response, implicating also a role of $Ca_V3.2$ in flow-mediated vasodilatation in young mice. The flow-mediated vasodilatory response did not decrease with age in both genotypes (Mikkelsen et al., 2016). It has, therefore, been proposed that due to this dual role in myogenic and flow-induced tone regulation, $Ca_V3.2$ channels may exert a protective role against cardiovascular disease, which is lost with advancing age.

Another study also found an aging-dependent role of $Ca_V3.2$ by demonstrating reduced NO production and vasodilation in aged (12–14 months) mesenteric arteries and aortae following contraction by KCl-depolarization as compared to preparations from young (2–3 months of age) WT mice. In young $Cacna1h^{-/-}$ mice (Chen et al., 2003), NO-dependent vasodilation was slightly less pronounced than in young WT mice but, unlike in WT, it did not decrease with aging and was similar in young and aged animals (Thuesen et al., 2018). This was explained by the observed age-dependent increase of endothelial NO production during KCl depolarization in the KO arteries, whereas it decreased with age in WT preparations. Therefore, T-type channel blockers may prevent age-dependent endothelial dysfunction. From these studies in mice, one could speculate that $Ca_V3.2$ channels have beneficial effects in young animals but, once these vanish with aging, inhibition of T-type channels may be beneficial to preserve endothelial function.

$Ca_V3.2$ vs. $Ca_V3.1$ in Arterial Vascular Tone: Conventional *Cacna1g* and *Cacna1h* KO

Interestingly, $Ca_V3.1$ and $Ca_V3.2$ have different roles in the arterial vasculature. While KO mice revealed that $Ca_V3.2$ prevents development of excess vascular tone in different arteries (also including renal efferent glomerular arterioles, see section "$Ca_V3.1$"; Thuesen et al., 2014), genetic deletion of $Ca_V3.1$ (Kim et al., 2001b) abolished myogenic tone in mouse coronary and small mesenteric arteries but only at low pressures (40–80 mmHg) associated with relatively hyperpolarized membrane potentials at which $Ca_V3.1$ are active (Björling et al., 2013). This indicates that $Ca_V3.1$ controls myogenic tone at low pressures, whereas higher voltage-activated $Ca_V1.2$ L-type channels take over at more positive voltages induced at higher vascular pressures (see section "$Ca_V1.2$").

Ca$_V$3.2 in the Heart: Conventional Ca$_V$3.2 KO (*Cacna1h$^{-/-}$*) and *Cacna1h$^{-/-}$*/NFAT-Luc Double Transgenic Mice

Ca$_V$3.1 and Ca$_V$3.2 have been reported to be upregulated in various animal models following cardiac hypertrophy. To study the role of Ca$_V$3.2 for the pathogenesis of cardiac hypertrophy, WT *Cacna1h$^{-/-}$* (Chen et al., 2003), and for comparison *Cacna1g$^{-/-}$* mice (Kim et al., 2001b), were subjected to arterial blood pressure overload by constriction of the aorta (transverse aortic banding/constriction, TAC). After two weeks, the heart weight/body weight ratio as a measure of cardiac hypertrophy increased significantly by 39.9% and 22.9%, respectively, in WT and *Cacna1g$^{-/-}$* mice as compared to their sham-operated controls (Chiang et al., 2009). In contrast, this increase was smaller in *Cacna1h$^{-/-}$* mice and did not reach statistical significance with the n-number analyzed (n = 10–11 per group). Similar changes were also found for the increase in left ventricle mass (LVM) and the LVM/body weight ratio. *Cacna1h* KO also reduced the increase in LVM/body weight ratio resulting from blood pressure increase induced by continuous Ang II infusion for 2 weeks. Crossing the *Cacna1h$^{-/-}$* mice with NFAT–luciferase reporter transgenic mice showed that NFAT-luciferase reporter activity failed to increase after pressure overload in the KO mice.

Ca$_V$3.2 vs. Ca$_V$3.1 in the Heart: Conventional *Cacna1g* and *Cacna1h* KO, *Cacna1g*-Overexpressing Mice (Cardiomyocyte-Specific)

Like in arterial smooth muscle, Ca$_V$3.1 and Ca$_V$3.2 appear to have different functions also in the heart, and data from KO mouse models are therefore discussed together. Ca$_V$3.2 deficiency (Le Quang et al., 2011) does not affect myocardial post-infarct remodeling, whereas after KO of Ca$_V$3.1 α_1 (Le Quang et al., 2011) impaired cardiac function and enhanced arrhythmia vulnerability post-MI were observed. Differential roles

of these T-type channels is also evident from pressure-overload models. Nakayama et al. (Nakayama et al., 2009) investigated if overexpression of T-type current per se is sufficient to induce cardiac hypertrophy. Therefore, they established a double transgenic mouse line permitting overexpression of Ca$_V$3.1 α_1 specifically in cardiomyocytes (α-MHC promotor) under the control of a tetracycline transactivator (tTA; tet-off). Ca$_V$3.1 overexpression was selected because pressure overload-induced hypertrophy resulted in a much stronger upregulation of Ca$_V$3.1 α_1 transcripts than that of Ca$_V$3.2 α_1 (Nakayama et al., 2009). Unexpectedly, despite a huge increase of T-type current in adult ventricular cardiomyocytes, enhanced fractional shortening and SR-Ca^{2+} content, Ca$_V$3.1 α_1 overexpressing mice showed no cardiac pathology. These mice even showed significantly less signs of cardiac hypertrophy after 2 and 8 weeks of TAC-induced pressure overload and also isoproterenol- and exercise-induced cardiac hypertrophy. In contrast, adult induction of L-type Ca^{2+} channel β_{2a} expression to induce enhanced L-type current promoted development of hypertrophy under the same experimental conditions (Nakayama et al., 2009 for reference). Findings with overexpression of Ca$_V$3.1 α_1 were also different from *Cacna1g* KO mice (Kim et al., 2001b), which developed greater cardiac hypertrophy after 2 weeks of TAC (Nakayama et al., 2009). They also developed pulmonary edema, decreased ventricular performance, and a substantial fibrotic response compared with WT controls. It therefore appears that the upregulation of Ca$_V$3.1 α_1 expression can exert a cardioprotective response. This was further supported by the finding that enhanced hypertrophy and disease in the *Cacna1g$^{-/-}$* mice were rescued by transgenic cardiomyocyte-specific overexpression thus also confirming the myocyte-autonomous effects of Ca$_V$3.1. The protective effect appeared to be due to Ca$_V$3.1-mediated Ca^{2+}-dependent activation of NO synthase and subsequent augmentation of cGMP-dependent protein kinase type I. Using an independently developed identical cardiomyocyte-specific (tet-off) Ca$_V$3.1 α_1 transgenic mouse, direct PKA-dependent stimulation of the induced Ca$_V$3.1 T-type current could be demonstrated (Li et al., 2012).

Taken together, these data show a complex role of Ca$_V$3.2 and Ca$_V$3.1 T-type channels for cardiac pathology. In the absence of pressure load, Ca$_V$3.2, ablation of the channel causes cardiac fibrosis and coronary pathology due to endothelial dysfunction (see above). On the other hand, the absence of Ca$_V$3.2 ameliorates the development of cardiac hypertrophy induced by pressure overload. Instead, Ca$_V$3.1 KO does not lead to cardiac pathology (Nakayama et al., 2009) but its (re-) expression is cardioprotective during pressure overload.

Recently, an important role of Ca$_V$3.2 has also been found for aldosterone secretion in the adrenal because Ca$_V$3.2 (*CACNA1H*) gain of function mutations causes primary hyperaldosteronism in humans (see below). At present, it appears unlikely that this can confound the interpretation of the above data because *Cacna1h* KO does not seem to affect aldosterone secretion at least under normal conditions in mice (Seidel et al., 2021).

Ca$_V$3.2 and Hearing Function: Conventional *Cacna1h* KO

Age-related hearing loss (presbycusis) is characterized by an age-dependent decline of auditory function associated with loss of cells in the inner ear including cochlear hair cells and spiral ganglion neurons. Ca$_V$1.3 channels are the predominant Ca²⁺ channel isoforms expressed in both inner and outer hair cells (see section on Ca$_V$1.3 L-type channels) and, together with Ca$_V$1.2 and Ca$_V$2.1, in spiral ganglion neurons at postnatal day P30 (see also Li et al., 2020 for transcriptome analysis). At P30, expression of Ca$_V$3.1 and Ca$_V$3.2 is also detected in C57Bl/6 mice but at an order of magnitude lower level. Expression of *Cacna1h* is high during embryonic development (E15.5) but decreases postnatally (Li et al., 2020). However, it appears that Ca$_V$3.2 expression increases again after P30. Lei et al. (2011) investigated the role of Ca$_V$3.2 in C57Bl/6 mice that exhibit several characteristics of the aging human inner ear and can serve as an animal model to study presbycusis.

They found an age-dependent increase of Ca$_V$3.2 transcripts in spiral ganglion cells from age 2 to 8 months. Auditory brainstem response (ABR) measurements of hearing thresholds at different frequencies revealed the expected age-related increase in ABR thresholds accelerating around 8–9 months of age, particularly above 12 kHz. When they therefore quantified the threshold shifts occurring between 9 and 11 months of age they found that, in contrast to age-matched WT controls, hearing thresholds were preserved in heterozygous and homozygous *Cacna1h⁻/⁻* mice (Chen et al., 2003). Moreover, the total number of spiral ganglion neurons was also larger in the *Cacna1h⁻/⁻* mice. These protective effects were also seen upon treatment with T-type Ca²⁺ channel blockers (e.g., ethosuximide) suggesting that inhibition of Ca$_V$3.2 channel activity can delay age-related hearing loss in mice. This prophylactic and therapeutic effect of T-type channel blockers was also independently reported for age-and noise-induced hearing loss in other studies. To determine the role of Ca$_V$3.2 before the start of hearing loss in these mice, Lundt et al. (Lundt et al., 2019) performed a comprehensive study of the gender-specific auditory function in young adult (20 and 40 weeks of age) *Cacna1h⁺/⁻* and *Cacna1h⁻/⁻* mice (Chen et al., 2003). Audiometric click ABR thresholds were increased in homozygous but not in heterozygous KO mice from both genders compared to control littermates at 20 and 40 weeks of age. Similar results were observed for tone burst-derived ABR thresholds in both genders at the age of 20 weeks. A detailed analysis of ABR waveforms revealed an effect of Ca$_V$3.2 ablation at different levels of the hearing pathway indicating that Ca$_V$3.2 is a prerequisite for precise auditory information processing in young adult C57Bl/6 mice. Together with the data observed in *Cacna1h⁻/⁻* mice from 9 to 11 months of age (Lei et al., 2011), this indicates that Ca$_V$3.2 ablation increases hearing thresholds but the hearing function is less prone to further age-induced hearing loss. It is therefore tempting to speculate that while pharmacological inhibition of T-type channels may delay hearing loss in aged animals (and perhaps humans), it could have negative effects on hearing at younger ages (Lundt et al., 2019).

Sensory Systems: Conventional/ Conditional Ca$_V$3.2 KO and Ca$_V$3.2 KI

The expression patterns of Ca$_V$3.2 proteins within primary afferent neurons and their contribution to modality-specific signaling are currently unclear. To enable the genetic marking of Ca$_V$3.2 channels, as well as tissue-specific loss of function, a KI mouse line Ca$_V$3.2$^{GFP-Flox}$ (designated as *Cacna1h*$^{GFP-flox/GFP-flox}$ in this review) was developed by inserting a two *loxP* site-flanked ecliptic-GFP tag into exon 6 of the *Cacna1h* locus (François et al., 2015). Immunohistochemical analyses revealed that Ca$_V$3.2 is a selective marker for low-threshold mechanoreceptors (LTMRs) of C-fibers expressing tyrosine hydroxylase, vesicular transporter of glutamate 3, and chemokine-like protein TAFA4 in combination, for LTMRs of the Aδ-fibers expressing the BDNF receptor TrkB, and for neuronal populations not co-expressing any of the classical markers for peptidergic and non-peptidergic C-fibers. The presence of Ca$_V$3.2 along LTMR-fiber trajectories along the axons of afferent C-LTMR fibers and in the nodes of Ranvier of Aδ-LTMRs was consistent with the critical role played by T-type VGCCs in inducing strong excitability. A conditional KO line deficient of Ca$_V$3.2 specifically in small-diameter DRG neurons (C-fibers) was generated by crossing the KI line with mice expressing *Cre* under the Na$_V$1.8 promoter. This KO line showed a profound impairment of light-touch perception and noxious mechanical cold/chemical sensations, in addition to an amelioration of the allodynia symptoms of neuropathic pain, strikingly contradicting the canonical view that attributes allodynia entirely to Aδ-LTMR.

The KI line *Cacna1h*$^{GFP-flox/GFP-flox}$ mice also demonstrated that Ca$_V$3.2 is localized in ~60% of the lamina II (LII) neurons of the spinal cord, a site for integration of sensory processing, in both excitatory and inhibitory interneurons (Candelas et al., 2019). Non-selective T-type blockers slowed the inhibitory but not the excitatory transmission in LII neurons, modified the intrinsic properties of LII neurons, abolished low-threshold-activated currents, rebound depolarization, and blunted excitability. Ca$_V$3.2 ablation

after the intraspinal injection of a virus encoding *Cre*-recombinase-IRES-mCherry under the control of the Cav3.2 promoter caused effects in the dorsal horn of *Cacna1h*$^{GFP-flox/GFP-flox}$ mice (François et al., 2015): it blunted the likelihood of transient firing patterns as well as the likelihood and amplitude of rebound depolarization, eliminated action-potential pairing, and remodeled the kinetics of the action potentials. In contrast, the properties of Ca$_V$3.2-positive neurons were only marginally modified in Ca$_V$3.1 KO mice. Thus, in addition to their previously established roles in the superficial spinal cord and in primary afferent neurons, Ca$_V$3.2 is likely to play multiple roles in LII neuron excitability.

The expression pattern of Ca$_V$3.2 was precisely determined in the nervous system by generating two KI mouse strains that express EGFP or *Cre* under the control of the Ca$_V$3.2 gene promoter (Bernal Sierra et al., 2017). In these *Cacna1h*$^{EGFP/EGFP}$ and *Cacna1h*$^{Cre/Cre}$ strains, the DNA sequence encoding EGFP fused with membrane targeting signal GAP43 or the *Cre* gene was inserted after the start codon of the *Cacna1h* gene located at the beginning of the second exon. The insertion of either the EGFP or *Cre* cassettes should produce a null mutation of the Ca$_V$3.2 gene as these cassettes are transcribed into mRNAs ending with a stop codon and a poly-A signal. *Cacna1h*$^{Cre/Cre}$ mice were bred with *Tau*^{mGFP} or Rosa26^{LacZ} mice to induce the expression of membrane-targeted myristoylated GFP (*mGFP*) or β-gal in neurons through CRE-mediated recombination. The observation of fluorescence signals of EGFP or mGFP and β-gal staining revealed that, in the brain, Ca$_V$3.2 is predominantly expressed in the dentate gyrus of the hippocampus. In the peripheral nervous system, the activation of the promoter starts at E9.5 in neural crest cells that will give rise to DRG neurons, but not sympathetic neurons. At E18.5, EGFP-positive cells were mainly a heterogeneous population of TrkB- and TrkC-positive cells. The activity of the Ca$_V$3.2 promoter in sensory progenitors was marked at many mechanoreceptor and nociceptor endings, but not in slowly adapting mechanoreceptors with endings associated with Merkel cells. Using *Cacna1h*$^{Cre/Cre}$

together with AAV viruses containing a conditional fluorescent tdTomato reporter, $Ca_V3.2$ promoter activity was found to be restricted to two functionally distinct sensory neuron types in the adult ganglia. $Ca_V3.2$-positive neurons innervating the skin only formed lanceolate endings on hair follicles, while another population of $Ca_V3.2$-positive nociceptive sensory neurons was deep tissue nociceptors that did not innervate the skin. Thus, the genetic tracing of the $Ca_V3.2$ expression pattern provides a powerful tool for distinguishing subpopulations of neurons that can play distinct roles in the sensory system.

Immunohistochemistry, either with or without antigen retrieval pretreatment, was compared for the detection of Ca_V3 in the spinal dorsal horn and DRG to elucidate the discrepancies regarding the localization of $Ca_V3.2$ (Cheng et al., 2019). Different expression patterns of $Ca_V3.2$, but not $Ca_V3.1$ and $Ca_V3.3$, in the spinal dorsal horn were observed between immunohistochemical experiments with and without antigen retrieval pretreatment, which displayed localization in the cell with a neuron-like and an astrocyte-like appearance, respectively. The specificity of the $Ca_V3.2$ signal was confirmed using $Cacna1h^{-/-}$ mice (Chen et al., 2003). $Ca_V3.2$ was mainly co-stained with the neuronal marker NeuN after antigen retrieval pretreatment, but with the astrocyte marker glial fibrillary acidic protein without antigen retrieval pretreatment. Importantly, $Ca_V3.2$ immunosignals after antigen retrieval pretreatment were largely colocalized with $Ca_V3.2$ mRNA, suggesting that the $Ca_V3.2$-positive cells were neurons but not astrocytes.

The mechanism underlying the high sensitivity and speed of skin mechanoreceptors has been studied from the perspective of T-type VGCCs (Wang & Lewin, 2011). D-hair receptors, which innervate hair follicles, from $Cacna1h^{-/-}$ mice (Chen et al., 2003) fire approximately 50% fewer spikes in response to ramp-and-hold displacement stimuli compared to WT receptors, which was attributed to an increase in the mechanical threshold and a temporal delay in the onset of high-frequency firing to moving stimuli. Other cutaneous mechanoreceptors and Aδ- and C-fiber nociceptors in $Cacna1h^{-/-}$ mice failed to show alterations in their mechanosensitivity compared to WT mice. These results suggest that the T-type VGCC $Ca_V3.2$ is essential for the normal temporal coding of moving stimuli by D-hair receptors, enabling animals to react very rapidly to tactile stimuli that guide movement.

The molecular mechanism that regulates $Ca_V3.2$ expression in mechanoreceptors is unknown. The effects of the neurotrophic factor neurotrophin-4 (NT-4) on $Ca_V3.2$ expression were studied in D-hair neurons cultured *in vitro* (Hilaire et al., 2011). The interruption of the supply of NT-4 by peripheral nerve axotomy or NT-4 KO induced a decrease in the T-type current amplitude, which was restored by incubation with NT-4. NT-4 had no effect on T-type currents in D-hair neurons from $Cacna1h^{-/-}$ mice (Chen et al., 2003). The role of phosphoinositide 3-kinase in the potentiation of T-type currents was identified after the pharmacological characterization of signaling pathways activated by the high-affinity NT-4 receptor TrkB. These results indicate that NT-4 acts as a regulator of the mechanosensitive function of D-hair neurons *via* the post-transcriptional upregulation of $Ca_V3.2$.

Pain: Conventional/Conditional $Ca_V3.2$ KO and $Ca_V3.2$ KI

To establish specific roles played by respective T-type VGCC Ca_V3 α_1-subunit subtypes in the processing of noxious stimuli *in vivo*, $Cacna1h^{-/-}$ mice (Chen et al., 2003) were compared with WT littermates in terms of pain susceptibility using behavioral models of pain (Choi et al., 2007). $Cacna1h^{-/-}$ mice showed attenuated pain responses after acute mechanical, thermal, and chemical stimuli, as well as tonic noxious stimuli, such as intraperitoneal injections of irritant agents and intradermal injections of formalin. However, in spinal nerve ligation-induced neuropathic pain, the behavioral responses of $Cacna1h^{-/-}$ mice were similar to those of WT mice. These results suggest that the $Ca_V3.2$ subtype plays a critical pronociceptive role in mechanical, thermal, chemical, and inflammatory pain, but not in the development of neuropathic pain.

The hyperexcitability of nociceptive DRG neurons due to the regulation of voltage-gated ion channels has been generally assumed to contribute strongly to neuropathic pain. Upon partial sciatic nerve ligation at the mid-thigh level, WT and *Cacna1h*$^{-/-}$ mice (Chen et al., 2003) showed similar neuropathic pain behavior and increased T-type Ca^{2+} currents in capsaicin-responsive neurons, but not in capsaicin-unresponsive neurons (Jeub et al., 2019). This upregulation was due to an increase in the Ni^{2+}-resistant VGCC component, which is different from Cav3.2, suggesting that partial sciatic nerve ligation induces an upregulation of Ni^{2+}-insensitive VGCC currents mediated by Ca$_V$3.1 or Ca$_V$3.3 in the sciatic nerve. More detailed analyses were performed for partial sciatic nerve ligation at the high-thigh level in the superficial spinal dorsal horn (Feng et al., 2019). Partial sciatic nerve ligation attenuated mechanical allodynia, but not thermal hyperalgesia, in *Cacna1h*$^{-/-}$ mice (Chen et al., 2003). Whole-cell patch-clamp recordings showed that both the overall proportion of T-type current-positive neurons and the T-type current density in individual neurons were elevated upon partial sciatic nerve ligation in the spinal lamina II neurons of rats. These phenotypes were not observed in *Cacna1h*$^{-/-}$ mice, suggesting that the elevated Ca$_V$3.2 VGCC activity in the superficial spinal dorsal horn contributes to mechanical allodynia in the partial sciatic nerve ligation-induced neuropathic pain model.

After injury, peripheral pain-sensing nociceptors are sensitized and become hyperexcitable, which lowers neuronal activation thresholds and elevates the frequency of action potentials under many pain conditions. The mechanisms underlying the initiation and maintenance of sensitization are poorly understood. Interestingly, sensitization was induced by reducing agents in C-type nociceptors that express Ca$_V$3.2 (Nelson et al., 2007). The electrophysical characterization of recombinant Ca$_V$3.2 constructs, including the mutant with a single residue replacement from histidine (H) to glutamine (Q) (H191Q), revealed that T-type currents are facilitated through a mechanism in which reducing agents chelate Zn^{2+} ions off H191 in the repeat I S3–S4 extracellular loop and relieve Ca$_V$3.2 from tonic channel inhibition. Reducing agents and various Zn^{2+} chelators sensitized Ca$_V$3.2 current-containing C-type nociceptors from WT mice, but not those from *Cacna1h*$^{-/-}$ mice (Chen et al., 2003) *in vitro* and *in vivo*, suggesting a role for Ca$_V$3.2 in peripheral nociception and nociceptor sensitization. The physiological significance of this H191-dependent modulation of Ca$_V$3.2 was confirmed by generating a KI mouse *Cacna1h*$^{H191Q/H191Q}$ line carrying the exon 4 mutation that introduces the H191Q replacement in the Ca$_V$3.2 protein (Voisin et al., 2016). Electrophysiological measurements performed on the D-hair cells (a subpopulation of DRG neurons), which express large Ca$_V$3.2 currents, indicated that the sensitivity of the T-type current to Ni^{2+}, Zn^{2+}, and ascorbate was reduced in *Cacna1h*$^{H191Q/H191Q}$ D-hair cells compared with WT. In *Cacna1h*$^{H191Q/H191Q}$ D-hair cells, a low concentration of Zn^{2+} and Ni^{2+} failed to modulate the rheobase threshold current, afterdepolarization amplitude, threshold potential necessary to trigger low-threshold Ca^{2+} spikes, and amplitude of low-threshold Ca^{2+} spikes. These results provide an important physiological basis for H191-dependent metal/redox regulation of Cav3.2, which regulates neuronal excitability and sensitization.

Hydrogen sulfide (H$_2$S) is recognized as a gasotransmitter that induces acceleration of somatic and visceral pain signals, at least in part, by targeting the Ca$_V$3.2 T-type VGCC (Matsui et al., 2019). In WT mice, the intraplantar and intracolonic administration of the H$_2$S donor, sodium sulfide Na$_2$S, evoked mechanical allodynia and visceral nociceptive behavior, which were abolished in *Cacna1h*$^{-/-}$ mice (Chen et al., 2003). This study provides evidence for the role played by Ca$_V$3.2 in H$_2$S-dependent somatic and colonic pain.

With regard to the physiopathology of irritable bowel syndrome (IBS), it has been recognized that visceral hypersensitivity involves the sensitization of the colonic primary afferent fibers through an overexpression of ion channels. The Ca$_V$3.2 mRNA levels were elevated in human colonic biopsies from the IBS group as well as in the colon of low-dose DSS-treated WT mice compared to the control group (Scanzi et al., 2016). Immunofluorescence staining revealed $Ca_V3.2$ protein localization in the human colonic mucosa, particularly in nerve fibers. WT, but not $Cacna1h^{-/-}$ mice (Chen et al., 2003), developed visceral hypersensitivity, as evaluated by the visceromotor responses to colorectal distension without showing colonic inflammation after treatment with a low dose (0.5%) dextran sodium sulfate. Interestingly, the conditional KO line deficient of Cav3.2, specifically in small-diameter DRG neurons (François et al., 2015), as well as WT mice subjected to spinal (not systemic) administration of T-type blockers, displayed suppression of colonic hypersensitivity induced by the same dextran sodium sulfate protocol (Picard et al., 2019). These studies support the critical contribution of $Ca_V3.2$ T-type VGCC to abdominal pain in patients with IBS.

The roles of T-type VGCCs in chronic musculoskeletal pain, such as lower back pain, fibromyalgia, and myofascial pain syndrome, remain unclear. Strikingly, acid-induced chronic mechanical hyperalgesia developed in $Cacna1g^{-/-}$ (Kim et al., 2001b) and WT mice, but not in $Cacna1h^{-/-}$ mice (Chen et al., 2003, 2010). The T-type blocker ethosuximide injected immediately or at 5 or 10 min but not at 15 min or later after the second acid injection prevented chronic mechanical hyperalgesia, suggesting that T-type VGCCs are required for the initiation, but not maintenance, of acid-induced chronic muscle pain. In addition, ethosuximide prevented chronic mechanical hyperalgesia administered intraperitoneally or intracerebroventricularly, but not intramuscularly or intrathecally in WT mice. Furthermore, an acid-induced increase in phosphorylated ERK (pERK) staining was observed 10 min after the second acid injection in the anterior paraventricular thalamic nucleus of WT but not $Cacna1h^{-/-}$ mice, and acid-induced pERK staining in the central nucleus of the amygdala, piriform cortex, and paraventricular hypothalamic nucleus was similar between WT and $Cacna1h^{-/-}$ mice. The prevention of ERK activation by the inhibition of upstream mitogen-activated kinases abolished chronic mechanical hyperalgesia. Thus, the $Ca_V3.2$-dependent activation of ERK in the paraventricular thalamus is required for the development of acid-induced chronic mechanical hyperalgesia.

T-type currents in primary afferents are known to be enhanced under various pathological conditions. However, the protein homeostasis underlying this upregulation remains unclear. WWP1 is a plasma-membrane-associated ubiquitin ligase that ubiquitinates $Ca_V3.2$ proteins by modifying specific lysine residues in the repeat III–IV intracellular linker region (García-Caballero et al., 2014). Proteomic screening identified the deubiquitinating enzyme USP5 as a $Ca_V3.2$ III–IV linker interacting partner. The KD of USP5 using specific shRNA increased $Ca_V3.2$ ubiquitination, decreased $Ca_V3.2$ protein levels, and $Ca_V3.2$ whole-cell currents in catecholaminergic neuronal tumor CAD cells. The *in vivo* KD or uncoupling of USP5 from $Ca_V3.2$ *via* the intrathecal delivery of Tat peptides mediated analgesia in both inflammatory and neuropathic mouse models of mechanical hypersensitivity. In $Cacna1h^{-/-}$ mice (Chen et al., 2003), mechanical withdrawal thresholds in response to complete Freund's adjuvant were intact but resistant to the Tat-3.2-III–IV linker peptide, in contrast to the reduced thresholds observed in WT mice, ensuring the effect of the tested Tat peptides. These experiments revealed an ubiquitin-mediated protein homeostasis pathway that regulates T-type VGCC activity in nociceptive signaling. Painful diabetic neuropathy occurs in approximately 20% of diabetic patients, and the underlying pathomechanisms are not fully understood. The modulation of the $Ca_V3.2$ channel by de-glycosylation enzymes, as well as their behavioral and biochemical effects in streptozotocin-induced type 1 painful peripheral diabetic neuropathy, was studied (Joksimovic et al., 2020a). In whole-cell recordings, de-glycosylation decreased $Ca_V3.2$ current densities and shifted the voltage dependence of activation and inactivation toward depolarizing potentials. The intrathecal injection of neuraminidase (NEU) completely reversed thermal and mechanical hyperalgesia in diabetic WT rats and mice, while NEU failed to alter baseline thermal and mechanical sensitivity and to develop painful peripheral diabetic neuropathy in

Cacna1h$^{-/-}$ mice (Chen et al., 2003). Interestingly, the unbiased electrophysiological screening of mechanosensitive singlecfibers in isolated hairy hind paw skin revealed a relative loss of polymodal heat sensing in favor of cold sensing in a streptozotocin-induced diabetes model (Hoffmann et al., 2021). In healthy *Cacna1h*$^{-/-}$ mice (Chen et al., 2003), both heat and cold sensitivity among the C-fibers seemed underrepresented in favor of exclusive mechanosensitivity, particularly low threshold; this deficit became significant in diabetic *Cacna1h*$^{-/-}$ mice. Diabetes also markedly increased the incidence of spontaneous discharge activity among the C-fibers in WT mice, which was largely absent in *Cacna1h*$^{-/-}$ mice. Diabetic WT mice showed calcitonin gene-related peptide release, which was reduced as much as in healthy and diabetic *Cacna1h*$^{-/-}$ mice. In contrast, diabetes markedly increased calcitonin gene-related peptide release from isolated hairy skin of WT but not *Cacna1h*$^{-/-}$ mice, strongly suggesting that the *de novo* expression of Ca$_V$3.2 in peptidergic cutaneous nerve endings contributes to enhanced spontaneous activity. De-glycosylation by NEU suppressed spontaneous activity and calcitonin gene-related peptide release but included actions independent of Ca$_V$3.2. These results suggest that, in sensory neurons, glycosylation-induced functional alterations in Ca$_V$3.2 VGCCs directly enhance diabetic hyperalgesia.

Analgesia and Anesthesia: Conventional Ca$_V$3.2 KO

KO mice have served as a powerful tool to validate the selectivity of targeted analgesic compounds. A mixed ligand, NMP-181, was developed for cannabinoid receptors and T-type VGCC, both of which are considered potential targets for treating pain (Gadotti et al., 2013). NMP-181 inhibited peak Ca$_V$3.2 currents with IC$_{50}$ values in the low µM range and activated CB$_2$ cannabinoid receptors. NMP-181 showed an antinociceptive effect in WT but not in *Cacna1h*$^{-/-}$ mice (Chen et al., 2003) in the formalin test, and reversed the mechanical hyperalgesia in the

inflammatory pain model induced by complete Freund's adjuvant. Another mixed ligand, NMP-7, which inhibits T-type VGCC and exerts mixed agonist activity on CB$_1$ and CB$_2$ receptors *in vitro*, elicited the inhibition of mechanical hyperalgesia and neuropathic pain induced by sciatic nerve injury, without altering spontaneous locomotor activity (Berger et al., 2014). The antinociception produced by NMP-7 in the inflammatory pain model was completely abolished in *Cacna1h*$^{-/-}$ mice (Chen et al., 2003), confirming that Ca$_V$3.2 is the key target. The analgesic action of NMP-7 was significantly attenuated by pretreatment with the CB$_2$ antagonist AM630, but not with the CB$_1$ antagonist AM281, suggesting that CB$_2$ receptors are involved in the action of NMP-7 *in vivo*. These studies open an avenue for suppressing pain through novel mixed T-type/cannabinoid receptor ligands. Remarkably, based on these previous findings, a new series of small organic compounds with a carbazole scaffold have been developed, whose physiological and therapeutic actions are mediated through a block of Ca$_V$3.2 (Bladen et al., 2015b). In particular, compound 9 mediated analgesia in acute inflammatory pain models and in reducing tactile allodynia in the partial nerve ligation model. Compound 9 was ineffective in the Ca$_V$3.2 T-type-deficient *Cacna1h*$^{-/-}$ mice (Chen et al., 2003).

Among a novel class of DHPs with preferential selectivity for T-type over L-type channels, M4 equipotently inhibited Ca$_V$1.2 L-type and Ca$_V$3.2 T-type VGCCs, and demonstrated antinociception in the formalin test, as well as the reversal of mechanical hyperalgesia in the partial sciatic nerve injury model (Gadotti et al., 2015). Interestingly, M4 retained partial activity in *Cacna1h*$^{-/-}$ mice (Chen et al., 2003), which is consistent with the electrophysiological observation that M4 effectively blocks T-type Ca$_V$3.3 and N-type Ca$_V$2.2 VGCC currents. Thus, the broad-spectrum inhibition of multiple VGCC subtypes can produce potent analgesic effects. A second series of compounds with attributes similar to M4 was synthesized to determine if they had an enhanced affinity for T-type channels (Bladen et al., 2015a). Whole-cell patch-clamp recordings were performed to screen these DHP derivatives for high affinity and selectivity for $Ca_V3.2$ over $Ca_V1.2$. The two lead compounds, termed N10 and N12, suppressed the current amplitude of $Ca_V3.2$ VGCC without affecting activation and inactivation gating, and mediated analgesia in the acute inflammatory pain model in WT mice but not in $Cacna1h^{-/-}$ mice (Chen et al., 2003). The best compound, N12, also reversed the mechanical hyperalgesia induced by nerve injury. Thus, promising T-type-blocking DHPs have been developed for potential pain therapies.

Based on the previous reports on piperazine-based compounds, including flunarizine, which inhibits T-type VGCCs, piperazine compounds were synthesized and examined for analgesic effects (Pudukulatham et al., 2016). A compound with a diphenyl methyl-piperazine core pharmacophore (compound 10e), which blocked Ca_V3s but only weakly affected $Ca_V1.2$ and $Ca_V2.2$, mediated relief from formalin- and complete Freund's adjuvant-induced inflammatory pain, which was ablated in $Cacna1h^{-/-}$ mice (Chen et al., 2003), suggesting that compound 10e owes its analgesic effects to $Ca_V3.2$ blockade.

The 3,4-dihydroquinazoline derivative KYS-05090S, known for its inhibitory action on Cav3.1 T-type and Cav2.2 N-type, reduced $Ca_V3.2$ currents and mediated a hyperpolarizing shift in the inactivation of voltage dependence (M'Dahoma et al., 2016). KYS-05090S reduced acute pain induced by formalin, and this antinociceptive effect was abolished in $Cacna1h^{-/-}$ mice (Chen et al., 2003), revealing a $Ca_V3.2$-dependent mechanism. KYS-05090S also reduced neuropathic pain in a model of partial sciatic nerve injury. The 3,4-dihydroquinazoline skeleton can serve as an important scaffold for the exploration of a new class of analgesic compounds.

Neuroactive steroids have been recognized to induce sedation/hypnosis by potentiating $GABA_A$ receptor-mediated currents. Strikingly, when the effects of epipregnanolone, the endogenous 5β-reduced neuroactive steroid (3β,5β)-3-hydroxypregnan-20-one, were examined using dissociated rat DRG neurons (Ayoola et al., 2014), epipregnanolone reversibly blocked T-type currents and stabilized the channel in the

inactive state, but failed to affect GABA-induced currents. Injections of epipregnanolone directly into peripheral receptive fields attenuated paw withdrawal responses to nociceptive thermal and mechanical stimuli in WT mice but failed to exert these effects in $Cacna1h^{-/-}$ mice (Chen et al., 2003). More recently, the synthetic epipregnanolone analog, (3β,5β,17β)-3-hydroxyandrostane-17-carbonitrile (3β-OH), was examined for its potential analgesic effects relevant to surgical procedures (Joksimovic et al., 2020b). A combination of isoflurane with 3β-OH suppressed cortical electroencephalogram and potentiated thermal and mechanical anti-hyperalgesia post-surgery, when compared to isoflurane alone and isoflurane plus morphine. 3β-OH exerted prominent thermal and mechanical antinociception in healthy rats and reduced T-type-dependent excitability of primary sensory neurons. The time to loss of righting reflex used as a surrogate measure of hypnosis induced by anesthetics was delayed in $Cacna1h^{-/-}$ mice (Chen et al., 2003) in comparison with the WT mice, confirming the role of $Ca_V3.2$ in steroid-induced hypnosis.

Anandamide-related endogenous molecules, lipoamino acids, were examined to determine whether they exert analgesic action *via* Ca_V3 VGCCs (Barbara et al., 2009). Various lipoamino acids, including N-arachidonoyl glycine, reversibly inhibited $Ca_V3.1$, $Ca_V3.2$, and $Ca_V3.3$ currents with potent effects on $Ca_V3.2$, but only weakly affected $Ca_V1.2$ and $Ca_V2.2$, on voltage-dependent Na⁺ channels $Na_V1.7$ and $Na_V1.8$, as well as on anandamide-sensitive cation channel TRPV1 and K⁺ channel TASK1. N-arachidonoyl glycine and N-arachidonoyl 3-OH-γ-aminobutyric acid produced strong thermal analgesia when hind paw withdrawal latency to a noxious thermal stimulus (46 °C) was measured, the effects of which were abolished in $Cacna1h^{-/-}$ mice (Chen et al., 2003). These data suggest that lipoamino acids, as a family of endogenous T-type inhibitors, modify pain sensation.

Although TTA-A2 was recognized as a pan Ca_V3 blocker, it displayed a potency for $Ca_V3.2$, which was higher than $Ca_V3.1$ (Francois et al., 2013). Interestingly, in whole-cell recordings, the inhibition of $Ca_V3.2$ currents by TTA-A2 was potentiated by depolarizing the holding potentials. When thermal nociception was tested by measuring tail withdrawal latency to a noxious thermal stimulus (46°C) in WT mice, TTA-A2 elicited thermal analgesia, which was abolished in *Cacna1h*$^{-/-}$ mice (Chen et al., 2003). TTA-A2 also produced a dose-dependent reduction in hypersensitivity in an irritable bowel syndrome model. These results suggest that the high potency of TTA-A2 in the depolarized state strengthens its analgesic efficacy and selectivity toward pathological pain syndromes.

How volatile anesthetics modulate T-type activity and exert physiological effects *in vivo* remains to be clarified. In patch-clamp recordings, the inhibition of Ca$_V$3.1-and Ca$_V$3.2 T-type VGCCs by a prototypical volatile anesthetic, isoflurane, was more prominent at depolarized holding potentials (−65 compared to −100 mV) (Orestes et al., 2009). Isoflurane slowed recovery from inactivation and enhanced deactivation in Ca$_V$3.1 and Ca$_V$3.2 but elicited a depolarizing shift of activation voltage dependence and greater use-dependent block in Ca$_V$3.2. During a behavioral examination of anesthetic endpoints, *Cacna1h*$^{-/-}$ mice (Chen et al., 2003) showed decreased minimum alveolar concentration in comparison with WT littermates. Both *Cacna1h*$^{-/-}$ and WT mice showed similar loss of righting reflex as a measure of anesthetic-induced hypnosis, while the onset of anesthetic induction was delayed in *Cacna1h*$^{-/-}$ mice. These results suggest that the state-dependent inhibition of T-type VGCC Ca$_V$3 subtypes contributes to the important clinical effects of isoflurane in the central and peripheral nervous systems.

Itch: Conventional Ca$_V$3.2 KO

The roles played by Ca$_V$3.2 T-type VGCC in the transmission of itch remain poorly understood. In *Cacna1h*$^{-/-}$ mice (Chen et al., 2003), scratching responses were suppressed during both histamine- and chloroquine-induced acute itch compared with WT mice (Gadotti et al., 2020). In addition, the T-type channel blocker DX332 inhibited the scratching responses of WT mice

treated with histamine or chloroquine. These data highlight Ca$_V$3.2 as a pharmacological target for the development of novel anti-pruritus therapies.

Epilepsy: Conventional Ca$_V$3.2 KO

A single episode of status epilepticus can trigger epileptogenesis. To elucidate the structural and functional neuronal alterations underlying epileptogenesis, sustained seizures were induced in experimental animals *via* the administration of pilocarpine, a muscarinic cholinergic agonist. In the epilepticus model, a transient and selective upregulation of Ca$_V$3.2 mRNA and protein, as well as an increase in cellular T-type Ca^{2+} currents and a transitional increase in intrinsic burst firing, were observed (Becker et al., 2008), and these changes by status epilepticus were absent in *Cacna1h*$^{-/-}$ mice (Chen et al., 2003). Intriguingly, the development of neuropathological hallmarks of chronic epilepsy, such as subfield-specific neuronal loss in the hippocampal formation and mossy fiber sprouting, was also absent, and the appearance of spontaneous seizures was reduced in *Cacna1h*$^{-/-}$ mice. Thus, the transcriptional induction of Ca$_V$3.2, which transforms neurons from regular to burst firing mode, is a critical step in epileptogenesis and neuronal vulnerability.

Genetic studies have linked a large number of gain-of-function mutations in the *CACNA1H* gene to human childhood absence epilepsy. The acute viral expression of fluorescence-tagged Ca$_V$3.2 constructs, including human childhood absence epilepsy-linked Ca$_V$3.2 C456S mutant and Ca$_V$3.2 C-terminus (its dominant-negative inhibitory action on endogenous Ca$_V$3.2), was confirmed in *Cacna1h*$^{-/-}$ mice (Chen et al., 2003), both in brain slices and *in vivo*, which revealed that Ca$_V$3.2 VGCCs control NMDA receptor-mediated transmission and subsequent NMDA receptor-dependent plasticity of AMPA receptor-mediated transmission (Wang et al., 2015). Ca$_V$3.2 C456S mutant channels had a higher open probability, induced more Ca^{2+} influx, and enhanced glutamatergic transmission. Remarkably, the cortical expression of Ca$_V$3.2

C456S channels in rats induced 2–4-Hz spike and wave discharges and absence-like epilepsy, characteristic of childhood-absence epilepsy patients, which was suppressed by AMPA and NMDA receptor antagonists but not by T-type antagonists. These results suggest that $Ca_V3.2$ channels play an unexpected role in regulating NMDA receptor-mediated transmission and a novel epileptogenic mechanism for human childhood absence epilepsy.

Memory: Conventional $Ca_V3.2$ KO

The human *CACNA1H* gene mutation is associated with schizophrenia, autism spectrum disorder, and absence epilepsy, all of which are characterized by abnormal memory function. In addition, abundant levels of $Ca_V3.2$ expression have been described in the hippocampus, where $Ca_V3.2$ is associated with contextual, temporal, and spatial learning and memory. *Cacna1h*⁻ᐟ⁻ mice (Chen et al., 2003) displayed normal performance in the Morris water maze and auditory trace fear-conditioning tasks, but impaired performance in the context-cued trace fear conditioning and step-down and step-through passive avoidance tasks compared with *Cacna1h*⁺ᐟ⁻ and WT mice (Chen et al., 2012). LTP persisted for at least 180 min in the hippocampal slices of *Cacna1h*⁺ᐟ⁻ and WT mice, but only for 120 min in *Cacna1h*⁻ᐟ⁻ mice. These behaviors of *Cacna1h*⁻ᐟ⁻ mice were reproduced in WT mice infused with mibefradil into the hippocampal formation, suggesting that $Ca_V3.2$ is important for the retrieval of context-associated memory. More recently, it was shown by the same group that a coral-derived natural product, excavatolide-B, enhanced contextual memory retrieval in both WT and *Cacna1h*⁻ᐟ⁻ mice (Chen et al., 2003) *via* delayed rectifier K⁺ channel current suppression, which lowered the threshold for action potential initiation and enhanced the induction of LTP (Huang et al., 2018). Hence, excavatolide-B may emerge as a potential natural product that enhances memory to treat symptoms of memory retrieval deficits associated with various brain disorders.

Several classes of drugs currently used in therapy (antipsychotics, antidepressants, and antiepileptics) also show activity on T-type VGCCs, enhancing the recognition of T-type as a potential target for brain dysfunction. *Cacna1h*⁻ᐟ⁻ mice (Chen et al., 2003) showed elevated anxiety, whereas novelty-induced spontaneous locomotor activity was not impaired (Gangarossa et al., 2014). Hippocampus-dependent recognition memory was impaired in *Cacna1h*⁻ᐟ⁻ mice, while acute and sensitized hyperlocomotion induced by d-amphetamine and cocaine was reduced in *Cacna1h*⁻ᐟ⁻ mice. This was consistent with the prevention of locomotor sensitization observed in WT mice following the administration of the T-type blocker TTA-A2. Thus, $Ca_V3.2$ is likely to be required for affective and cognitive behaviors.

To determine the role of $Ca_V3.2$ in global gene expression after trace fear conditioning, a microarray transcriptome study was carried out on the hippocampi of *Cacna1h*⁻ᐟ⁻ mice (Chen et al., 2003) and their WT littermates, either untrained or trace fear conditioned (Chung et al., 2015). When the untrained *Cacna1h*⁻ᐟ⁻ mice and their WT littermates were compared, 3522 differentially expressed genes were found in the left hippocampus, but only four differentially expressed genes were found in the right hippocampus. After trace fear conditioning, the number of differentially expressed genes in the left hippocampus was markedly reduced to 6. These results suggest the asymmetric impacts of $Ca_V3.2$ deficiency and its prominent reversal by behavioral training on the hippocampal transcriptome. Significant efforts are still required to establish the relationship between trace fear conditioning and Ca^{2+}-mediated transcription *via* $Ca_V3.2$, which are complex subjects.

Neuronal Excitability: Conventional Ca$_V$3.2 KO

Entorhinal cortical layer III neurons have been suggested to play roles in processes such as spatial navigation and learning and memory, while hyperpolarization-activated cyclic nucleotide-gated (HCN) channel function has been associated with the modification of neural network rhythms and memory formation. Electron microscopy revealed that the HCN1 and Ca$_V$3.2 subunits coexist on entorhinal cortical presynaptic terminals (Huang et al., 2011). HCN channels inhibited glutamate synaptic release by suppressing the activity of low-threshold Ca$_V$3.2 Ca^{2+} channels. HCN channel inhibition failed to affect miniature EPSCs in slices from *Cacna1h$^{-/-}$* mice (Chen et al., 2003). These results represent a previously unknown mechanism by which HCN and Ca$_V$3.2 VGCC cooperatively regulate synaptic strength and thereby neural information processing and network excitability.

Despite the recognition of the essential role played by T-type VGCCs in generating thalamocortical oscillations, little is known about their regulatory molecular mechanisms. In reticular thalamic neurons and recombinant Ca$_V$3.2 currents, reducing agents, including endogenous sulfur-containing amino acid L-cysteine, selectively enhanced native T-type currents, but not native and recombinant Ca$_V$3.1 and Ca$_V$3.3 currents (Joksovic et al., 2006). T-type currents of reticular thalamic neurons from *Cacna1h$^{-/-}$* mice (Chen et al., 2003) were not modulated by the reducing agents. In contrast, oxidizing agents inhibited the native and recombinant T-type currents. Thus, Ca$_V$3.2 is the most likely molecular substrate for the redox regulation of thalamic T-type VGCCs.

β-arrestins, which were originally characterized by their ability to "arrest" GPCR signaling by uncoupling receptors from G proteins, have recently emerged as important signaling effectors for G-protein-coupled receptors. Based on the recordings from cartwheel cells in the auditory brainstem dorsal cochlear nucleus, D3 dopamine receptors were found to regulate axon initial segment excitability through β-arrestin-dependent signaling, modifying the voltage dependence of T-type currents, and suppressing high-frequency action potential generation (Yang et al., 2016b). Since the axon initial segment Ca^{2+} influx was unaffected by quinpirole, a D3 dopamine receptor agonist in *Cacna1h$^{-/-}$* mice (Chen et al., 2003), D3 dopamine receptor-induced regulation is mediated, at least in part, by Ca$_V$3.2. These results suggest that Ca$_V$3.2 is important for non-canonical D3 dopamine receptor signaling to regulate axon initial segment excitability.

Mature dentate granule cells, characterized by a low-input resistance and a hyperpolarized resting membrane potential, are prominently inhibited by GABAergic innervation, and their dendrites strongly attenuate synaptic potentials. Consequently, they occasionally fire action potentials. Recently, Ca$_V$3.2 channels at the axon initial segment were shown to be responsible for burst firing of mature rodent granule cells (Dumenieu et al., 2018). In *Cacna1h$^{-/-}$* mice (Chen et al., 2003), tonic spikes were induced along with impaired bursting and synaptic plasticity, and abnormal oscillatory activity in the dentate gyrus and CA3. These results suggest that Ca$_V$3.2 is a modulator of bursting and can be considered a molecular switch that enables effective information transfer from mature dentate granule cells to CA3 pyramidal neurons.

Expression Distribution in the Brain: Conventional Ca$_V$3.2 KO

The contribution of T-type current to neuronal output is often attributed to the differential distribution of Ca$_V$3 subtypes to somato-dendritic compartments. However, this has yet to be fully examined and confirmed. In histoblot analysis, Ca$_V$3.1 showed the highest expression level in the molecular layer of the cerebellum, whereas Ca$_V$3.2 was expressed in the hippocampus and the molecular layer of the cerebellum (Aguado et al., 2016). Ca$_V$3.1 and Ca$_V$3.2, both of which showed increased expression with age, were markedly different in terms of region- and developmental stage-specific expression.

Immunoelectron microscopy revealed that Ca$_V$3.1 was present in the somato-dendritic domains of hippocampal interneurons and Purkinje cells, whereas Ca$_V$3.2 was present in the somato-dendritic domains of CA1 pyramidal cells, hippocampal interneurons, and Purkinje cells. *Cacna1h⁻/⁻* mice (Chen et al., 2003) were used to confirm signal specificity in electron microscopic analysis. Thus, the non-uniform distribution of Ca$_V$3.1 and Ca$_V$3.2 subunits may account for the functional heterogeneity of the T-type over the plasma membrane of central neurons.

Neuroprotection: Conventional Ca$_V$3.2 KO

Cognitive and functional decline with age has been correlated with the regulation of intracellular Ca²⁺ handling, which can lead to neurodegeneration in the brain. The possible protective effects of blockers for T-type VGCCs, such as trimethadione, approved by the FDA as an antiepileptic drug, were tested (Wildburger et al., 2009). Cultured hippocampal neurons showed an increase in viability when treated with either trimethadione or the L-type blocker nimodipine. Ca$_V$3.2 is highly expressed in the hippocampus and certain cortical regions; however, trimethadione maintained neuroprotective effects in *Cacna1h⁻/⁻* mice, raising the possibility that neuroprotection elicited by these drugs occurs through VGCCs other than Ca$_V$3.2. In addition, nimodipine, but not trimethadione, exerted protective effects on neurons in long-term cultures. These results suggest that different mechanisms underlie the regulation of neuronal survival *via* Ca²⁺ signaling pathways.

Neurogenesis: Conventional Ca$_V$3.2 KO

The roles played by Ca$_V$3.2 have yet to be fully elucidated in the developing brain compared to those in physiological and pathological processes in adults. At the onset of neuronal differentiation, neural progenitor cells exhibit spontaneous Ca²⁺ activity, which is strongly correlated with the upregulation of Ca$_V$3.2 mRNA (Rebellato et al., 2019). In cells with robust spontaneous Ca²⁺ signaling, caspase-3 activity increased independently of apoptosis. The inhibition of Ca$_V$3.2 by drugs or viral *CACNA1H* KD resulted in the downregulation of caspase-3 activity, which was followed by suppressed neurogenesis. Cortical slices from *Cacna1h⁻/⁻* embryos (Chen et al., 2003) showed decreased spontaneous Ca²⁺ activity, a lower cleaved caspase-3 level, and microanatomical abnormalities in the subventricular/ventricular and cortical plate zones. These results suggest an association between Ca$_V$3.2 and caspase-3 signaling, which plays a role in neurogenesis in the developing brain.

Retina: Conventional Ca$_V$3.2 KO

Although retinal bipolar cells and ganglion cells are known to express T-type VGCCs, the precise expression patterns and functional roles of the individual T-type Ca$_V$ in the retina need to be addressed. Immunohistochemical analysis indicated that the Ca$_V$3.2 antibody labeled a subgroup of type-3 cone bipolar cells (CBCs), the PKAβII-immunopositive type-3 CBCs (Cui et al., 2012). The specificity of an anti-Ca$_V$3.2 antibody was confirmed in a recombinant expression system using HEK293 cells and *Cacna1h⁻/⁻* mice (Chen et al., 2003). The Ca$_V$3.2 immunosignals were observed throughout the cell, including the dendrites and axon terminals. Whole-cell recordings from the type-3 CBCs of *Cacna1h⁻/⁻* mice demonstrated that Ca$_V$3.2 contributes to T-type Ca²⁺ currents in a subpopulation of type-3 CBCs. These results suggest the functional role of the Ca$_V$3.2 T-type in retinal processing.

Early in development, the retina establishes direction-selective responses before the onset of vision and spontaneously generates bursts of action potentials that propagate across its extent. The precise spatial and temporal properties of these "retinal waves" have been implicated in the formation of retinal projections to the brain. To determine the role played by retinal waves in the development of direction-selective circuits within the retina, spontaneous activity was compared between WT and *Cacna1h⁻/⁻* mice (Chen et al., 2003). *Cacna1h⁻/⁻* mice displayed altered wave-associated bursts of action potentials: the frequency of the bursts was significantly decreased, and the firing rate within each burst was reduced (Hamby et al., 2015). *Cacna1h⁻/⁻* mice also exhibited reduced eye-specific segregation of retinogeniculate projections in the dorsal lateral geniculate nucleus. However, after eye-opening, the direction-selective responses of direction-selective ganglion cells (DSGCs) from *Cacna1h⁻/⁻* mice were indistinguishable from those of WT DSGCs. These results suggest that $Ca_V3.2$ plays a role in retinal waves, which is not critical for establishing direction selectivity in the retina.

Carotid Body and Hypoxia Sensing: Conventional $Ca_V3.2$ KO

A specialized sensory organ, called the carotid body, detects the levels of O_2 in the arterial blood. The carotid body has been reported to express high levels of $Ca_V3.2$ localized to glomus cells, the primary O_2-sensing cells in the carotid body (Makarenko et al., 2015). Mibefradil and TTA-A2, selective blockers of the T-type VGCC, markedly attenuated the elevation of hypoxia-induced increases in intracellular Ca^{2+} concentration, the secretion of catecholamines from glomus cells, and the sensory excitation of the rat carotid body. A similar degree of attenuation was observed in the carotid body and glomus cells from *Cacna1h⁻/⁻* mice (Chen et al., 2003). *Cacna1h⁻/⁻* mice displayed attenuation of responses to H_2S, a critical mediator of hypoxia-induced carotid body responses, including increases in intracellular Ca^{2+} concentration in glomus cells and the activity of carotid body sensory nerves. In WT mice, TTA-A2 markedly attenuated the responses of glomus cells and carotid body sensory nerves to hypoxia, and these effects were absent in mice deficient in cystathionine-γ-lyase responsible for H_2S production in the carotid body. These results suggest

that the $Ca_V3.2$ channels contribute to the H_2S-mediated carotid body response to hypoxia.

Chronic intermittent hypoxia (CIH) is a hallmark of sleep apnea. Carotid body glomus cells were harvested from adult rats or mice treated with CIH (Makarenko et al., 2016). CIH-treated glomus cells exhibited enhanced Ca^{2+} responses to hypoxia, which was abolished in the presence of TTA-A2 and in *Cacna1h⁻/⁻* (Chen et al., 2003) glomus cells. *Cacna1h⁻/⁻* mice showed disruption of CIH-induced hypersensitivity in the carotid body. Scavenging ROS, which is produced during CIH, prevents the enhancement of TTA-A2-sensitive Ca^{2+} responses to hypoxia. CIH augmented recombinant $Ca_V3.2$ Ca^{2+} currents and increased $Ca_V3.2$ protein in plasma membrane fractions, which were prevented by a ROS scavenger or Brefeldin-A, an inhibitor of protein trafficking, suggesting that CIH facilitates $Ca_V3.2$ protein trafficking to the plasma membrane in a ROS-dependent manner and increases Ca^{2+} influx in the carotid body glomus cells.

Gastrointestinal Tract: Conventional $Ca_V3.2$ KO

The α_1-subunit of T-type VGCC that contributes to the generation of pacemaker potentials was investigated in myocytes from the external muscular layers of gastrointestinal smooth muscles and in interstitial cells of Cajal (Gibbons et al., 2009). $Ca_V3.2$ expression was detected in the single myocytes and interstitial cells of Cajal. Whole-cell recordings demonstrated that DHP- and Cd^{2+}-resistant but mibefradil-sensitive currents are present in myocytes dissociated from the jejunum. Mibefradil reduced the frequency and initial increase rate of the electrical slow wave. *Cacna1h⁻/⁻* mice generated by disrupting exons 3–6 encoding amino acid residues 138–373 were embryonic lethal by E13.5 *in utero*. The electrical slow wave recorded from smooth muscle cells in the *Cacna1h⁻/⁻* jejunum appeared abnormal: the slope of the initial rising phase of the slow wave was lower and the frequency of the slow waves was less than that of the WT. These results suggest that T-type conductance through $Ca_V3.2$ contributes to the upstroke of the pacemaker potential. T-type Ca^{2+} conductance essential for slow waves in the gastrointestinal muscles was more precisely characterized in isolated interstitial cells of Cajal (Zheng et al., 2014). The interstitial cells of Cajal displayed a Ni^{2+}- or mibefradil-sensitive, low-voltage-activated current component and a DHP-sensitive high-voltage-activated current component. Increasing the temperature (20–30 °C) augmented the current amplitude and decreased the activation time with a temperature coefficient Q10 value of 3.0. Reducing temperature decreased the upstroke velocity of slow waves in the muscles of WT mice, but exerted a minimal effect in those of $Cacna1h^{-/-}$ mice (Chen et al., 2003). Hence, a $Ca_V3.2$ T-type current is likely to be required for the entrainment of pacemaker activity within the interstitial cells of Cajal to actively propagate slow waves through their networks.

Hypertension: Conventional Ca_V3.2 KO and M1560V KI

When increased production of aldosterone, a key regulator of blood pressure and electrolyte homeostasis, is uncoupled from its main physiological stimuli, primary aldosteronism ensues. A gain-of-function *CACNA1H* mutation that causes the amino acid substitution M1560V in human $Ca_V3.2$ underlies autosomal-dominant familial hyperaldosteronism type IV (FH-IV), and patients with FH typically present with early onset primary aldosteronism and hypertension. The CRISPR/Cas9 technique was employed to generate $Cacna1h^{M1560V/+}$ KI mice as a model of the M1560V mutation, along with $Cacna1h^{-/-}$ mice (Seidel et al., 2021). $Cacna1h^{M1560V/+}$ mice showed elevated aldosterone: renin ratios indicative of primary aldosteronism and increased adrenal aldosterone synthase *Cyp11b2* expression, which remained elevated on a high-salt diet (characteristic of primary aldosteronism). The systolic blood pressure of $Cacna1h^{M1560V/+}$ mice increased and remained elevated on a high-salt diet. In $Cacna1h^{-/-}$ mice, renin-1 expression was

enhanced, but adrenal *Cyp11b2* levels were normal, most likely through compensatory Ca^{2+} entry to maintain normal aldosterone production. $Cacna1h^{M1560V/+}$ adrenal slices showed increased intracellular Ca^{2+} concentrations in the zona glomerulosa compared to controls, but the frequency of Ca^{2+} was intact. It is therefore possible that a germline gain-of-function *CACNA1H* mutation is sufficient to cause mild primary aldosteronism.

Fertilization: Conventional Ca_V3.2 KO

Ca^{2+}-permeable channels that mediate the Ca^{2+} influx required for egg activation remain unknown. The robust voltage-activated Ca^{2+} current observed in eggs from WT mice was abolished in eggs from $Cacna1h^{-/-}$ mice (Chen et al., 2003), females of which displayed reduced litter sizes (Bernhardt et al., 2015). Ca^{2+} oscillation patterns in $Cacna1h^{-/-}$ eggs following *in vitro* fertilization revealed reductions in the first transient length and oscillation persistence. Ca^{2+} store content was also reduced in $Cacna1h^{-/-}$ eggs. These results suggest the roles played by $Ca_V3.2$ T-type VGCCs in supporting the increase in Ca^{2+} stores associated with meiotic maturation and the Ca^{2+} influx required for the activation of development.

$Ca_V3.2$ is not the sole Ca^{2+} influx pathway responsible for egg maturation and development following fertilization (Bernhardt et al., 2018). In eggs isolated from $Cacna1h^{-/-}$ mice (Chen et al., 2003) crossed with mice carrying an oocyte-specific KO of Mg^{2+}-sensitive, Ca^{2+}-permeable TRPM7 channel ($Cacna1h^{-/-}/Trpm7^{flox/flox}/Gdf9$-cre), Ca^{2+} oscillations were severely impaired and their resumption was substantially delayed, indicating that $Ca_V3.2$ and TRPM7 are responsible for the majority of Ca^{2+} influx immediately following fertilization. TRPM7 and $Ca_V3.2$ channels almost exhaustively account for Ca^{2+} influx upon Ca^{2+} store depletion. When bred with WT male mice, females with double TRPM7/$Ca_V3.2$ KO were subfertile, and their offspring showed increased variance in postnatal weight. In mouse oocytes arrested at the germinal vesicle stage, TRPV3 also participated in the filling of the Ca^{2+}

stores, although this was a relatively minor contribution compared to $Ca_V3.2$ (Ardestani et al., 2020). Double TRPV3/$Ca_V3.2$ KO females were subfertile; their oocytes and eggs showed reduced internal Ca^{2+} stores and, with regard to sperm entry or *Plcz* cRNA injection, fewer dKO eggs displayed Ca^{2+} responses with lower oscillation frequency compared to WT eggs (Mehregan et al., 2021). These defects were rescued by removing extracellular Mg^{2+}, suggesting that the residual Ca^{2+} influx could be mediated by TRPM7. Thus, three different Ca^{2+} influx pathways across the plasma membrane cooperatively dictate the periodicity and persistence of Ca^{2+} oscillations during mammalian fertilization.

Bone: Conventional $Ca_V3.2$ KO

VGCCs are key to osteoblast plasma membrane Ca^{2+} permeability under the control of calcitropic hormones. Subtype-specific antibodies were used to examine L-type $Ca_V1.2$ and T-type $Ca_V3.2$ subunit expression during mouse skeletal development (Shao et al., 2005). From E14.5 through skeletal maturity, the immunoreactivity of $Ca_V1.2$ and $Ca_V3.2$ was evident in regions of rapid long bone growth, including the perichondrium, periosteum, chondro-osseous junction, and trabecular bones. Femurs from *Cacna1h*$^{-/-}$ mice (Chen et al., 2003) showed no immunoreactivity. Both α_1-subunits were observed in osteoblasts and chondrocytes under high magnification, whereas only $Ca_V3.2$ was present in osteocytes. Thus, L-type $Ca_V1.2$ and T-type $Ca_V3.2$ VGCCs appear to be dynamically regulated in bones and cartilages during endochondral bone development.

Chondrogenesis is not completely understood in tracheal cartilage rings that protect and maintain the airway. *Cacna1h*$^{-/-}$ mice (Chen et al., 2003) showed congenital tracheal stenosis attributable to an incomplete formation of cartilaginous tracheal support, while in the murine chondrogenic cell line ATDC5, $Ca_V3.2$ overexpression enhanced chondrogenesis, which was blunted by blocking T-type VGCCs and calcineurin (Lin et al., 2014). Interestingly, the expression of the sex determination region of Y chromosome-related high-mobility group-Box gene 9 (Sox9), one of the earliest markers of committed chondrogenic cells, was reduced in *Cacna1h*$^{-/-}$ tracheas. Luciferase reporter assay, gel shift, and ChIP studies in ATDC5 cells, using inhibitors of T-type VGCC and calcineurin, unveiled a mechanism that underlies tracheal chondrogenesis: Ca^{2+} influx *via* $Ca_V3.2$ activates the calcineurin/NFAT signaling pathway, and the NFAT-binding site was identified within the mouse Sox9 promoter.

Knees exhibited significantly lower focal articular cartilage damage in *Cacna1h*$^{-/-}$ mice (Chen et al., 2003) compared with WT mice, when osteoarthritis was induced by repetitive mechanical insult (Srinivasan et al., 2015). T-type inhibition reduced the expression of both early and late mechanoresponsive genes in cultured osteoblasts, but not in chondrocytes. The expression of markers of hypertrophy induced in chondrocytes by the conditioned media obtained from sheared osteoblasts was nearly abolished by the pretreatment of osteoblasts with the T-type inhibitor. These results suggest that T-type VGCCs play a role in the induction of osteoarthritis *via* the release of osteoblast-derived factors that promote an early osteoarthritis phenotype in chondrocytes.

$Ca_V3.3$

Early Reports of $Ca_V3.3$ Mutant Mice

Sleep: Conventional $Ca_V3.3$ KO

In TRN, where $Ca_V3.3$ protein is abundantly expressed, intrinsic burst and rhythmic burst discharges are elicited by activation of T-type VGCC. To study the role played by $Ca_V3.3$ in the TRN bursts critical for generation and maintenance of thalamocortical oscillations, the $Ca_V3.3$ KO (*Cacna1i*$^{-/-}$) mouse line was engineered by replacing exons 11–21, which encode the repeat II S2-repeat III S4 region of the $Ca_V3.3$ protein, with a β-galactosidase-neomycin cassette (Astori et al., 2011). The *Cacna1i*$^{-/-}$ mice revealed the requirement for $Ca_V3.3$ in TRN function and for sleep spindles, a hallmark of natural sleep. The absence of $Ca_V3.3$ prevented oscillatory bursting in the low-frequency (4–10 Hz) range in TRN cells, but spared tonic discharge. In contrast, adjacent thalamocortical neurons retained low-threshold bursts. The reduced inhibition of thalamocortical neurons *via* TRN neurons markedly weakened synchronized thalamic network oscillations underlying sleep-spindle waves in the *Cacna1i⁻/⁻* mice. Sleeping *Cacna1i⁻/⁻* mice showed a selective reduction in the power density of the σ frequency band (10–12 Hz) at transitions from non-REM to REM sleep, with other electroencephalogram waves remaining unaltered. These data identify a central role of Ca$_V$3.3 T-type VGCCs in the rhythmogenic properties of the sleep-spindle generator.

Epilepsy: Conventional Ca$_V$3.3 KO

Thalamocortical oscillations lead to spike-and-wave discharges, the hallmarks of absence seizures. In *Cacna1i⁻/⁻* mice generated by the deletion of exons 3 and 4, which encode the N-terminus of the Ca$_V$3.3 protein, rhythmic burst discharges were completely abolished, whereas tonic firing was increased in the TRN (Lee et al., 2014). Susceptibility to drug-induced spike-wave discharges was increased both in *Cacna1i⁻/⁻* mice and in mice in which the Ca$_V$3.3 gene was silenced predominantly in the TRN. Furthermore, a double KO (*Cacna1h⁻/⁻/Cacna1i⁻/⁻*) mouse of both Ca$_V$3.3 and Ca$_V$3.2 (Chen et al., 2003) showed complete elimination of burst firing and rhythmic burst discharges in TRN neurons and displayed enhanced drug-induced rhythmic burst discharges and absence seizures. These results raise questions about the proposed role of burst firing in TRN neurons in the genesis of spike-wave discharges.

Extended Knowledge of Ca$_V$3.3 Mutant Mice

Sleep: Ca$_V$3.3 KO and KI

Genetic analyses of large patient cohorts have identified *Cacna1i* as a gene associated with the risk of schizophrenia. A *Cacna1i⁻/⁻* line and a mouse line with the mutation R1305H (RH), which is orthologous to the human R1346H

mutation found in the schizophrenia proband, were generated, using the CRISPR/Cas9-mediated genome editing approach coupled with pronuclear zygote injection (Ghoshal et al., 2020). A 10 base-pair genomic deletion that causes a premature stop codon was introduced into the *Cacna1i⁻/⁻* line, while the KI mutation was introduced through homologous recombination of a DNA template harboring the R1305H alteration in the *Cacna1i^{RH/RH}* line. T-type currents were nearly absent in the *Cacna1i⁻/⁻* mice, and were reduced in the *Cacna1i^{RH/RH}* mice. The *Cacna1i⁻/⁻* and the *Cacna1i^{RH/RH}* mice revealed altered cellular excitability in the TRN and had marked deficits in sleep spindle occurrence and morphology throughout non-REM sleep. This study provides insights into the deficits of sleep spindle and TRN function in schizophrenia.

Anesthesia: Conventional Ca$_V$3.3 KO

Functional roles of Ca$_V$3.3 VGCCs in anesthetic-induced hypnosis and underlying neuronal oscillations were studied using the *Cacna1i⁻/⁻* mice (Ghoshal et al., 2020; Feseha et al., 2020). The mutant mice were induced to hypnosis faster than WT mice, while the percent isoflurane at which hypnosis and immobility occurred were indistinguishable between the two genotypes (Feseha et al., 2020). The T-type blocker TTA-P2 facilitated isoflurane induction of hypnosis in the *Cacna1i⁻/⁻* mice more robustly than in WT mice. Isoflurane-induced hypnosis, following injections of TTA-P2, was accompanied by more prominent δ and θ oscillations in the electroencephalogram of *Cacna1i⁻/⁻* mice, reaching a burst-suppression pattern earlier than in WT mice. These results suggest that Ca$_V$3.3 has a specific role in anesthetic mechanisms.

Multiple Ca$_V$s

Presynapse: Conventional KO for Ca$_V$2.1, Ca$_V$2.2, or Ca$_V$2.3

Local Ca²⁺ signaling induced within nanometer spaces of VGCCs is crucial for a variety of physiological processes. To understand the molecular composition of nano-environments that molecularly and functionally interact with Ca_V2 channels in the rodent brain, a proteomic strategy combining multiepitope affinity purifications with high-resolution quantitative mass spectrometry was performed, using KO mice for Ca_V2s (Jun et al., 1999; Ino et al., 2001; Pereverzev et al., 2002) as negative controls (Müller et al., 2010). The analysis showed that Ca_V2-containing VGCCs are embedded in protein networks assembled from a pool of ~200 proteins with distinct abundance, stability of assembly, and preference for respective Ca_V2 subtypes. A majority of these proteins had not been previously known to be linked to Ca_Vs; about two-thirds are dedicated to the control of intracellular Ca^{2+} concentration, including G protein-coupled receptor-mediated signaling, and activity-dependent cytoskeleton remodeling or Ca^{2+}-dependent effector systems that comprise the priming and release machinery of synaptic vesicles. The identified protein networks can reflect the cellular processes associated with the Ca^{2+} influx activity or protein scaffolding mediated by Ca_V2 VGCCs.

Presynapse: Conditional Triple Ca_V2 KO

With respect to presynaptic VGCCs critical for Ca^{2+}-triggered exocytosis, it was debated whether Ca_V2 proteins control synapse assembly or active zone proteins play the primary role in forming the presynapse structure, scaffolding Ca_V2s to the presynaptic neurotransmitter release sites. To resolve this issue, mouse lines with *Cacna1a^{flox/flox}* (Todorov et al., 2006) and *Cacna1e^{flox/flox}* (Pereverzev et al., 2002), as well as a newly engineered *Cacna1b^{flox/flox}* line with loxP sites flanking exons 5 and 6, were intercrossed, and the "triple floxed" line was generated (Held et al., 2020). When all three Ca_V2s were ablated with Cre recombinase, evoked exocytosis was abolished in cultured hippocampal neurons, or at the calyx of Held, while synapse and active zone structure, vesicle docking, and transsynaptic nano-organization were unimpaired. Although triple Ca_V2 KO mice had impaired localization of β-subunits, $α_2δ_1$ localized normally. Thus, it can

be concluded that Ca_V2 proteins are not necessary for synapse structure, but are recruited and targeted to the active zone by specific interactions with active zone proteins.

Sleep: Conventional Double $Ca_V3.2/Ca_V3.3$ KO

T-type VGCCs sustain oscillatory discharges of thalamocortical and TRN neurons. $Ca_V3.3$ was reported to dominate rhythmic bursting reticular thalamic neurons, mediating a substantial fraction of spindle power in the electroencephalogram during non-REM sleep (Astori et al., 2011). However, the contribution of $Ca_V3.2$ to reticular thalamus-dependent spindle generation was yet to be determined. In the *Cacna1i^{-/-}* mice (Astori et al., 2011), reticular thalamic neurons showed intact discharge properties. In the *Cacna1h^{-/-}/Cacna1i^{-/-}* double KO mice obtained by crossing *Cacna1i^{-/-}* mice with *Cacna1h^{-/-}* mice (Chen et al., 2003), these neurons displayed abolished low-threshold Ca^{2+} currents and bursting, while thalamocortical neurons showed suppressed burst-mediated inhibitory responses (Pellegrini et al., 2016). The *Cacna1h^{-/-}/Cacna1i^{-/-}* mice demonstrated fragmented sleep, short non-REM sleep episodes, and pronounced microarousals. Suppression of the σ frequency band (10–15 Hz), which was accompanied by an increase in the δ band (1–4 Hz), was observed for the non-REM sleep electroencephalogram power spectrum in the *Cacna1h^{-/-}/Cacna1i^{-/-}* mice. The results suggest that $Ca_V3.2$ can boost intrathalamic synaptic transmission and can play a modulatory role adjusting the relative presence of non-REM sleep electroencephalogram rhythms.

Sleep: Conventional KO for $Ca_V1.1~1.4$, $Ca_V2.1~2.3$, or $Ca_V3.1~3.3$

To understand the ion conduction mechanisms underlying the regulation of sleep duration, a simple computational model was constructed that recapitulates the electrophysiological characteristics of slow-wave sleep and awake states (Tatsuki et al., 2016). Comprehensive bifurcation analysis predicted a role played by the Ca^{2+}-

dependent hyperpolarization pathway in slow-wave sleep, which regulates sleep duration. Based on this prediction, 21 lines of KO mice for genes related to Ca^{2+}-dependent hyperpolarization pathway, including the genes for all of the 10 Ca_V channel α_1-subunits, were generated by the triple-target CRISPR method. The gene disruption of Ca^{2+}-dependent K^+ channels (*Kcnn2* and *Kcnn3*), VGCCs (*Cacna1g* and *Cacna1h*), and Ca^{2+}/calmodulin-dependent kinases (*Camk2a* and *Camk2b*) decreased sleep duration, while that of plasma membrane Ca^{2+} ATPase (*Atp2b3*) increased sleep duration. Since KO mice of two main subunits of the NMDAR family exhibited a lethal phenotype, pharmacological intervention was employed with whole-brain imaging, which confirmed that impaired NMDA receptors reduce sleep duration and directly increase the excitability of cells. These results point to the importance of the Ca^{2+}-dependent hyperpolarization pathway in the regulation of sleep duration in mammals.

Epilepsy: Conventional Double $Ca_V2.1/Ca_V3.1$ KO

The contribution of T-type VGCCs to the genesis of absence seizures, characterized by typical spike-wave discharges and behavioral arrests, was explored in the $Ca_V2.1$-null *Cacna1*$^{-/-}$ mice (Jun et al., 1999) and in other spontaneous mutant mice. In double KO *Cacna1a*$^{-/-}$/*Cacna1g*$^{-/-}$ mice obtained by crossing *Cacna1a*$^{-/-}$ mice with the $Ca_V3.1$-null *Cacna1g*$^{-/-}$ mice (Kim et al., 2001b), T-type Ca^{2+} currents in thalamocortical relay neurons and spike-wave discharges were ablated (Song et al., 2004). Similar results were obtained for *Cacna1g*$^{-/-}$ mice harboring additional spontaneous mutations, such as *lethargic* (*Cacnb4*$^{lh/lh}$), *tottering* (*Cacna1a*$^{tg/tg}$), or stargazer (*Cacng2*$^{stg/stg}$), as models for absence seizures. Interestingly, *Cacna1a*$^{-/-}$/*Cacna1g*$^{+/-}$ mice with a partial T-type Ca^{2+} current (75% of WT current) in thalamocortical relay neurons showed spike-wave discharges quantitatively similar to those of *Cacna1a*$^{-/-}$ mice. The results provide *in vivo* evidence that baseline T-type Ca^{2+} currents, but not their augmentation in thalamocortical relay neurons, are necessary and sufficient to promote

absence seizures in various genetic mouse models.

Epilepsy: Conditional $Ca_V2.1$ KO with Conventional $Ca_V3.1$ or $Ca_V3.2$ KO

Inborn genetic errors of the *Cacna1a* gene encoding P/Q-type $Ca_V2.1$ proteins impair synaptic transmission and produce early and life-long neurological deficits, such as childhood absence epilepsy, ataxia, and dystonia. It is unclear whether these pathological phenotypes are due to the defective P/Q-type channel function during the critical period for thalamic network stabilization in the immature brain. To address this question, a conditional *Cacna1a* KO mouse line iKO$^{p/q}$ (*Cacna1a*$^{flox/flox}$;*CAG-Cre-ER* (iKO$^{p/q}$)) was obtained by crossing the *Cacna1a*-floxed *Cacna1a*^{Citrine} mice (Mark et al., 2011) with the tamoxifen-inducible *Cre* line CAGGcre-ER™; tamoxifen injection to iKO$^{p/q}$ mice at postnatal age 6–10 weeks allowed examination of the biological impact of adult loss of P/Q-type VGCCs on various physiological responses (Miao et al., 2020). These mice displayed dysfunction patterns reminiscent of the inborn loss-of-function phenotypes, thereby demonstrating that the neurological defects do not rely upon a developmental abnormality. Unlike the inborn *Cacna1a*$^{-/-}$ mice, the adult-onset pattern of excitability changes believed to be pathogenic within the thalamic network turned out to be non-canonical in iKO$^{p/q}$ mice: adult ablation of P/Q-type VGCCs failed to promote *Cacna1g*-mediated burst firing or T-type currents in the thalamocortical relay neurons, while burst firing in thalamocortical relay neurons itself remained essential, as the iKO$^{p/q}$ mice bred on a *Cacna1g*$^{-/-}$ background (this conditional $Ca_V3.1$ KO was obtained by breeding mice carrying the *Sox2-Cre* gene to the *Cacna1g*$^{flox/flox}$ mice (Anderson et al., 2005)) showed diminished seizure generation. Moreover, in TRN neurons, impairment of burst firing was accompanied by attenuated T-type currents. Interestingly, inborn deletion of *Cacna1h* (Chen et al., 2003), which is known as TRN-enriched, human childhood absence epilepsy-linked gene, reduced TRN burst firing, and promoted, rather than suppressed, seizures in iKO$^{p/q}$;*Cacna1h*$^{-/-}$ mice, in comparison with iKO$^{p/q}$ mice, suggesting an epileptogenic role for loss-of-function *Cacna1h* gene variants in human childhood absence epilepsy. Thus, P/Q-type appears to be critical for normal thalamo-cortical oscillations and motor control in the adult brain.

Epilepsy: Conventional Ca$_V$3.1 or Ca$_V$3.2 KO

Using the maximal electroshock seizure as a model of tonic-clonic generalized seizures, treatment with TTA-A2 significantly protected mice from tonic seizures (Sakkaki et al., 2016). Although no major change was observed in the local field potential pattern during the maximal electroshock seizure, the late post-ictal period displayed a significant increase in the δ frequency power in WT mice treated with TTA-A2. This was similar to *Cacna1g*$^{-/-}$ mice (Kim et al., 2001b) but not *Cacna1h*$^{-/-}$ mice (Chen et al., 2003). Analysis of extracellular signal-regulated kinase 1/2 phosphorylation and c-Fos expression revealed a rapid and elevated neuronal activation in the hippocampus following clonic seizures; this response was resistant to TTA-A2. Thus, it is likely that Ca$_V$3.1, but not Ca$_V$3.2, mediates the protective effect of TTA-A2 against tonic seizure protection.

Sympathetic Activity: Conventional KO for Ca$_V$3.1 or Ca$_V$3.2

In vitro brainstem-spinal cord-splanchnic sympathetic nerve preparations were used to investigate whether T-type VGCCs are involved in sympathetic nerve discharge (Chen et al., 2011). Applications of NNC 55-0396, a blocker of T-type Ca^{2+} channels, reduced sympathetic nerve discharge in a concentration-dependent manner in WT mice. Comparable discharges were observed in the WT and *Cacna1g*$^{-/-}$ mice (Kim et al., 2001b), while the amount of discharge in *Cacna1h*$^{-/-}$ mice was only ∼40% of that in WT or *Cacna1g*$^{-/-}$ mice (Chen et al., 2003). The maximal responses of the initial increase in glutamate-induced discharge were similar between WT and

Cacna1h$^{-/-}$ sympathetic neurons, but significantly higher in *Cacna1g*$^{-/-}$ sympathetic neurons. These results suggest different roles played by the spinal cord Ca$_V$3.1 and the brainstem Ca$_V$3.2, the former being functionally inhibitory for the genesis of sympathetic nerve discharge, while the latter is essential for the genesis of the presympathetic drive.

Retina: Conventional KO for Ca$_V$2.3 or Ca$_V$3.2, Conventional Double Ca$_V$2.3/Ca$_V$3.2 KO

Although the importance of Ni^{2+}-sensitive VGCCs has been recognized in retinal signaling, it is unclear whether R-type or T-type plays a role. Electroretinogram recordings of retinas isolated from *Cacna1e*$^{-/-}$ (Pereverzev et al., 2002), *Cacna1h*$^{-/-}$ (Chen et al., 2003), and *Cacna1e*$^{-/-}$/*Cacna1h*$^{-/-}$ mice revealed that Ca$_V$2.3 contributes rather to a later component in the b-wave, the most prominent component of the electroretinogram (Alnawaiseh et al., 2011). In the absence of Ca$_V$3.2, the gain of the Ni^{2+}-mediated increase in the b-wave amplitude was significantly increased, most likely due to a loss of the reciprocal inhibition of photoreceptors. These results are consistent with the idea that Ni^{2+}-sensitive Ca$_V$2.3 and Ca$_V$3.2 differentially contribute to specific features of the b-wave response in the electroretinogram.

Anesthesia: Conventional KO for Ca$_V$3.1, Ca$_V$3.2, or Ca$_V$3.3

The roles played by Ca$_V$3 subtypes were examined in hypnosis induced by the endogenous neurosteroid epipregnanolone. Consistent with increased thalamocortical oscillations in slower electroencephalography frequencies, epipregnanolone induced a hypnotic state and effectively facilitated the anesthetic effects of isoflurane and sevoflurane in WT mice (Coulter et al., 2021). Sensitivity to epipregnanolone-induced hypnosis was reduced in *Cacna1g*$^{-/-}$ mice (Petrenko et al., 2007), but not in *Cacna1h*$^{-/-}$ mice (Chen et al., 2003) and *Cacna1i*$^{-/-}$ mice (Ghoshal et al., 2020) or WT mice, whereas the onset of epipregnanolone-induced hypnosis was delayed in *Cacna1h*$^{-/-}$ mice, but not in *Cacna1g*$^{-/-}$ or

Cacna1i⁻ᐟ⁻ mice. The differences observed among the three Ca$_V$3 KO mice in epipregnanolone-induced hypnotic phenotypes can be attributed to their different functional characteristics and distribution patterns in the thalamocortical circuitry.

Kidney Function: Conventional KO for Ca$_V$3.1 or Ca$_V$3.2

T-type Ca$_V$3 proteins may differentially impact renal blood flow and glomerular filtration rate, given their differential expression in pre- and post-glomerular vessels. *Cacna1g⁻ᐟ⁻* (Lee et al., 2002) and *Cacna1h⁻ᐟ⁻* mice (Chen et al., 2003) were characterized in comparison with WT mice to address this issue (Thuesen et al., 2014). In *Cacna1h⁻ᐟ⁻* mice, *in vitro* perfused efferent arterioles showed enhanced constriction in response to depolarization and *in vivo* increased glomerular filtration rate (likely resulting from elevated filtration pressure), with normal renal plasma flow, heart rate, and blood pressure. In *Cacna1g⁻ᐟ⁻* mice, renal plasma flow was enhanced, although glomerular filtration rate and *in vitro* constriction of afferent and efferent arterioles in response to depolarization were indistinguishable from those of WT mice. Heart rate was attenuated in *Cacna1g⁻ᐟ⁻* mice (see role in sinoatrial node function above), with no difference in blood pressure. Thus, Ca$_V$3.2 supports the dilatation of efferent arterioles and affects glomerular filtration rate, whereas Ca$_V$3.1 somehow contributes to renal vascular resistance, perhaps due to an attenuated sympathetic efferent nerve traffic (Thuesen et al., 2014).

Lymphatic Vessel: Conventional KO for Ca$_V$3.1 or Ca$_V$3.2

It is necessary to clarify exactly which ionic mechanisms drive spontaneous contractions of collecting lymphatic vessels *via* an intrinsic electrical pacemaker in providing a propulsive force to return lymph centrally. In mouse lymphatic muscle cells, Ca$_V$3.1 and Ca$_V$3.2 were expressed and produced functional T-type VGCC currents. However, either *Cacna1h⁻ᐟ⁻* (Chen et al., 2003) or *Cacna1g⁻ᐟ⁻* mice (Kim et al., 2001b) showed intact frequency, amplitude, or fractional pump flow of lymphatic collectors *ex vivo* from two different regions of the mouse (To et al., 2020). WT and *Cacna1h⁻ᐟ⁻/Cacna1g⁻ᐟ⁻* double KO lymphatic vessels responded similarly to mibefradil and Ni²⁺, which reduced contraction amplitudes and increased frequencies. In contrast, in a mouse line with smooth-muscle-specific deletion of the L-type Ca$_V$1.2, generated by crossing *Cacna1c^flox/flox* mice (*Cacna1c^{tm3Hfm}/J*) with smooth-muscle-specific *Cre* driver SMMHC-creER^T2 mice, lymphatic spontaneous contractions were abolished. These results suggest that L-type, but not T-type, plays an important role in controlling the frequency and strength of spontaneous contractions in mouse lymphatic smooth muscle. In addition, the observed inhibition by mibefradil and Ni²⁺ is likely to have an off-target effect on L-type VGCCs.

Sperm: Conventional KO for Ca$_V$3.1 or Ca$_V$3.2

In spermatozoa, involvement of VGCCs has been found for different cellular functions, including acrosome reaction and sperm motility. Based initially on pharmacological studies, T-type, which mediates the low-voltage-activated inward Ca²⁺ current, was deemed essential for acrosome reaction in sperm. In spermatogenic cells isolated from *Cacna1h⁻ᐟ⁻* mice (Chen et al., 2003), the low-voltage-activated Ca²⁺ current was abolished, and depolarization-induced Ca²⁺ influx was reduced (Escoffier et al., 2007). Characterization of sperm isolated from *Cacna1g⁻ᐟ⁻* mice (Kim et al., 2001b) failed to show the expression or contribution to Ca²⁺ influx of Ca$_V$3.1 channels in the sperm. Despite the Ca$_V$3.2 T-type channel contribution in depolarization-induced Ca²⁺ influx, reproduction parameters remained intact in *Cacna1h⁻ᐟ⁻* mice. These results suggest that other types, such as high-voltage-activated types of VGCC, are activated during the depolarization-induced Ca²⁺ influx for acrosome reaction, likely compensating the deficiency of Ca$_V$3.2 in the *Cacna1h⁻ᐟ⁻* sperm.

Mendelian Inheritance: Conventional KO for Ca$_V$2.3 or Ca$_V$3.2

A significant deviation of genotype distribution from Mendelian inheritance after the mating of heterozygous pairs of *Cacna1h*$^{+/-}$ mice (Chen et al., 2003) was observed in weaned pups (Alpdogan et al., 2020). The mating of pairs of *Cacna1h*$^{-/-}$ females and *Cacna1h*$^{+/-}$ males and pairs of *Cacna1h*$^{+/-}$ females and *Cacna1h*$^{-/-}$ males confirmed the reduction of deficient homozygous *Cacna1h*$^{-/-}$ pups, leading to the hypothesis that prenatal lethality occurs when one or both *Cacna1h* alleles are missing. On the other hand, the mating of heterozygous pairs of *Cacna1e*$^{+/-}$ mice (Sochivko et al., 2002) revealed a deviation of genotype distribution from Mendelian inheritance only in weaned heterozygous male pups, suggesting that compensatory mechanisms operate in *Cacna1e*$^{-/-}$ male mice. These results highlight the importance of considering the function of VGCCs during prenatal development.

References

Aguado, C., García-Madrona, S., Gil-Minguez, M., & Luján, R. (2016). Ontogenic changes and differential localization of T-type Ca^{2+} channel subunits Ca$_V$3.1 and Ca$_V$3.2 in mouse hippocampus and cerebellum. *Frontiers in Neuroanatomy, 1083.* https://doi.org/10.3389/fnana.2016.00083

Aldea, M., Jun, K., Shin, H.-S., Andrés-Mateos, E., Solís-Garrido, L. M., Montiel, C., et al. (2002). A perforated patch-clamp study of calcium currents and exocytosis in chromaffin cells of wild-type and α_{1A} knockout mice. *Journal of Neurochemistry, 81*(5), 911–921. https://doi.org/10.1046/j.1471-4159.2002.00845.x

Alexander, S. P., Mathie, A., Peters, J. A., Veale, E. L., Striessnig, J., Kelly, E., et al. (2021). The concise guide to pharmacology 2021/22: Ion channels. *British Journal of Pharmacology, 178*(S1), S157–S245. https://doi.org/10.1111/bph.15539

Alnawaiseh, M., Albanna, W., Chen, C.-C., Campbell, K. P., Hescheler, J., Lüke, M., et al. (2011). Two separate Ni^{2+}-sensitive voltage-gated Ca^{2+} channels modulate transretinal signalling in the isolated murine retina. *Acta Ophthalmologica, 89*(7), e579–e590. https://doi.org/10.1111/j.1755-3768.2011.02167.x

Alpdogan, S., Clemens, R., Hescheler, J., Neumaier, F., & Schneider, T. (2020). Non-Mendelian inheritance during inbreeding of Ca$_V$3.2 and Ca$_V$2.3 deficient mice. *Scientific Reports, 10*(1), 15993. https://doi.org/10.1038/s41598-020-72912-9

Anderson, M. P., Mochizuki, T., Xie, J., Fischler, W., Manger, J. P., Talley, E. M., et al. (2005). Thalamic Ca$_V$3.1 T-type Ca^{2+} channel plays a crucial role in stabilizing sleep. *Proceedings of the National Academy of Sciences of the United States of America, 102*(5), 1743–1748. https://doi.org/10.1073/pnas.0409644102

Ardestani, G., Mehregan, A., Fleig, A., Horgen, F. D., Carvacho, I., & Fissore, R. A. (2020). Divalent cation influx and calcium homeostasis in germinal vesicle mouse oocytes. *Cell Calcium, 87102181.* https://doi.org/10.1016/j.ceca.2020.102181

Arikkath, J., Felix, R., Ahern, C., Chen, C.-C., Mori, Y., Song, I., et al. (2002). Molecular characterization of a two-domain form of the neuronal voltage-gated P/Q-type calcium channel $\alpha_1$2.1 subunit. *FEBS Letters, 532*(3), 300–308. https://doi.org/10.1016/S0014-5793(02)03693-1

Astori, S., Wimmer, R. D., Prosser, H. M., Corti, C., Corsi, M., Liaudet, N., et al. (2011). The Ca$_V$3.3 calcium channel is the major sleep spindle pacemaker in thalamus. *Proceedings of the National Academy of Sciences of the United States of America, 108*(33), 13823–13828. https://doi.org/10.1073/pnas.1105115108

Ayoola, C., Hwang, S. M., Hong, S. J., Rose, K. E., Boyd, C., Bozic, N., et al. (2014). Inhibition of Ca$_V$3.2 T-type calcium channels in peripheral sensory neurons contributes to analgesic properties of epipregnanolone. *Psychopharmacology, 231*(17), 3503–3515. https://doi.org/10.1007/s00213-014-3588-0

Bader, P. L., Faizi, M., Kim, L. H., Owen, S. F., Tadross, M. R., Alfa, R. W., et al. (2011). Mouse model of Timothy syndrome recapitulates triad of autistic traits. *Proceedings of the National Academy of Sciences of the United States of America, 108*(37), 15432–15437. https://doi.org/10.1073/pnas.1112667108

Badou, A., Basavappa, S., Desai, R., Peng, Y.-Q., Matza, D., Mehal, W. Z., et al. (2005). Requirement of voltage-gated calcium channel ß$_4$ subunit for T lymphocyte functions. *Science.* https://doi.org/10.1126/science.1100582

Baig, S. M., Koschak, A., Lieb, A., Gebhart, M., Dafinger, C., Nürnberg, G., et al. (2011). Loss of Ca$_V$1.3 (*CACNA1D*) function in a human channelopathy with bradycardia and congenital deafness. *Nature Neuroscience, 14*(1), 77–84. https://doi.org/10.1038/nn.2694

Barbara, G., Alloui, A., Nargeot, J., Lory, P., Eschalier, A., Bourinet, E., et al. (2009). T-type calcium channel inhibition underlies the analgesic effects of the endogenous lipoamino acids. *The Journal of Neuroscience, 29*(42), 13106–13114. https://doi.org/10.1523/JNEUROSCI.2919-09.2009

Barg, S., Ma, X., Eliasson, L., Galvanovskis, J., Göpel, S. O., Obermüller, S., et al. (2001). Fast exocytosis with few Ca^{2+} channels in insulin-secreting mouse pancreatic B cells. *Biophysical Journal, 81*(6), 3308–3323. https://doi.org/10.1016/S0006-3495(01)75964-4

Bauer, R., Timothy, K. W., & Golden, A. (2021). Update on the molecular genetics of Timothy syndrome. *Frontiers in Pediatrics, 9435*. https://doi.org/10.3389/fped.2021.668546

Bavley, C. C., Fischer, D. K., Rizzo, B. K., & Rajadhyaksha, A. M. (2017). Ca$_V$1.2 channels mediate persistent chronic stress-induced behavioral deficits that are associated with prefrontal cortex activation of the p25/Cdk5-glucocorticoid receptor pathway. *Neurobiology of Stress*, 727–737. https://doi.org/10.1016/j.ynstr.2017.02.004

Bavley, C. C., Fetcho, R. N., Burgdorf, C. E., Walsh, A. P., Fischer, D. K., Hall, B. S., et al. (2020). Cocaine- and stress-primed reinstatement of drug-associated memories elicit differential behavioral and frontostriatal circuit activity patterns via recruitment of L-type Ca²⁺ channels. *Molecular Psychiatry, 25*(10), 2373–2391. https://doi.org/10.1038/s41380-019-0513-2

Becker, A. J., Pitsch, J., Sochivko, D., Opitz, T., Staniek, M., Chen, C.-C., et al. (2008). Transcriptional upregulation of Ca$_V$3.2 mediates epileptogenesis in the pilocarpine model of epilepsy. *The Journal of Neuroscience, 28*(49), 13341–13353. https://doi.org/10.1523/JNEUROSCI.1421-08.2008

Beggs, M. R., Lee, J. J., Busch, K., Raza, A., Dimke, H., Weissgerber, P., et al. (2019). TRPV6 and Ca$_V$1.3 mediate distal small intestine calcium absorption before weaning. *Cellular and Molecular Gastroenterology and Hepatology, 8*(4), 625–642. https://doi.org/10.1016/j.jcmgh.2019.07.005

Bellemann, P., Ferry, D., Lübbecke, F., & Glossmann, H. (1982). [3H]-Nimodipine and [3H]-nitrendipine as tools to directly identify the sites of action of 1,4-dihydropyridine calcium antagonists in guinea-pig tissues. Tissue-specific effects of anions and ionic strength. *Arzneimittel-Forschung, 32*(4), 361–363.

Benkert, J., Hess, S., Roy, S., Beccano-Kelly, D., Wiederspohn, N., Duda, J., et al. (2019). Ca$_V$2.3 channels contribute to dopaminergic neuron loss in a model of Parkinson's disease. *Nature Communications, 10*(1), 5094. https://doi.org/10.1038/s41467-019-12834-x

Beqollari, D., Dockstader, K., & Bannister, R. A. (2018). A skeletal muscle L-type Ca²⁺ channel with a mutation in the selectivity filter (Ca$_V$1.1 E1014K) conducts K⁺. *The Journal of Biological Chemistry, 293*(9), 3126–3133. https://doi.org/10.1074/jbc.M117.812446

Berger, S. M., & Bartsch, D. (2014). The role of L-type voltage-gated calcium channels Ca$_V$1.2 and Ca$_V$1.3 in normal and pathological brain function. *Cell and Tissue Research, 357*(2), 463–476. https://doi.org/10.1007/s00441-014-1936-3

Berger, N. D., Gadotti, V. M., Petrov, R. R., Chapman, K., Diaz, P., & Zamponi, G. W. (2014). NMP-7 inhibits chronic inflammatory and neuropathic pain via block of Ca$_V$3.2 T-type calcium channels and activation of CB₂ receptors. *Molecular Pain*, 101744-8069-10–77. https://doi.org/10.1186/1744-8069-10-77

Bernal Sierra, Y. A., Haseleu, J., Kozlenkov, A., Bégay, V., & Lewin, G. R. (2017). Genetic tracing of Ca$_V$3.2 T-type calcium channel expression in the peripheral nervous system. *Frontiers in Molecular Neuroscience, 1070*. https://doi.org/10.3389/fnmol.2017.00070

Bernardi, R. E., Uhrig, S., Spanagel, R., & Hansson, A. C. (2014). Transcriptional regulation of L-type calcium channel subtypes Ca$_V$1.2 and Ca$_V$1.3 by nicotine and their potential role in nicotine sensitization. *Nicotine & Tobacco Research, 16*(6), 774–785. https://doi.org/10.1093/ntr/ntt274

Bernhardt, M. L., Zhang, Y., Erxleben, C. F., Padilla-Banks, E., McDonough, C. E., Miao, Y.-L., et al. (2015). Ca$_V$3.2 T-type channels mediate Ca²⁺ entry during oocyte maturation and following fertilization. *Journal of Cell Science, 128*(23), 4442–4452. https://doi.org/10.1242/jcs.180026

Bernhardt, M. L., Stein, P., Carvacho, I., Krapp, C., Ardestani, G., Mehregan, A., et al. (2018). TRPM7 and Ca$_V$3.2 channels mediate Ca²⁺ influx required for egg activation at fertilization. *Proceedings of the National Academy of Sciences of the United States of America, 115*(44), E10370–E10378. https://doi.org/10.1073/pnas.1810422115

Beuckmann, C. T., Sinton, C. M., Miyamoto, N., Ino, M., & Yanagisawa, M. (2003). N-type calcium channel α_{1B} subunit (Ca$_V$2.2) knock-out mice display hyperactivity and vigilance state differences. *The Journal of Neuroscience, 23*(17), 6793–6797. https://doi.org/10.1523/JNEUROSCI.23-17-06793.2003

Björling, K., Morita, H., Olsen, M. F., Prodan, A., Hansen, P. B., Lory, P., et al. (2013). Myogenic tone is impaired at low arterial pressure in mice deficient in the low-voltage-activated Ca$_V$3.1 T-type Ca²⁺ channel. *Acta Physiologica, 207*(4), 709–720. https://doi.org/10.1111/apha.12066

Bladen, C., Gadotti, V. M., Gündüz, M. G., Berger, N. D., Şimşek, R., Şafak, C., et al. (2015a). 1,4-Dihydropyridine derivatives with T-type calcium channel blocking activity attenuate inflammatory and neuropathic pain. *Pflügers Archiv – European Journal of Physiology, 467*, 1237–1247. https://doi.org/10.1007/s00424-014-1566-3

Bladen, C., McDaniel, S. W., Gadotti, V. M., Petrov, R. R., Berger, N. D., Diaz, P., et al. (2015b). Characterization of novel cannabinoid based T-type calcium channel blockers with analgesic effects. *ACS Chemical Neuroscience, 6*(2), 277–287. https://doi.org/10.1021/cn500206a

Blaich, A., Pahlavan, S., Tian, Q., Oberhofer, M., Poomvanicha, M., Lenhardt, P., et al. (2012). Mutation of the calmodulin binding motif IQ of the L-type Ca$_V$1.2 Ca²⁺ channel to EQ induces dilated cardiomyopathy and death. *The Journal of Biological Chemistry, 287*(27), 22616–22625. https://doi.org/10.1074/jbc.M112.357921

Bock, G., Gebhart, M., Scharinger, A., Jangsangthong, W., Busquet, P., Poggiani, C., et al. (2011). Functional properties of a newly identified C-terminal splice variant of Ca$_V$1.3 L-type Ca²⁺ channels. *The Journal of Biological Chemistry, 286*(49), 42736–42748. https://doi.org/10.1074/jbc.M111.269951

Bohne, P., Mourabit, D. B.-E., Josten, M., & Mark, M. D. (2021). Cognitive deficits in episodic ataxia type 2 mouse models. *Human Molecular Genetics*. https://doi.org/10.1093/hmg/ddab149

Bomben, V. C., Aiba, I., Qian, J., Mark, M. D., Herlitze, S., & Noebels, J. L. (2016). Isolated P/Q calcium channel deletion in layer VI corticothalamic neurons generates absence epilepsy. *The Journal of Neuroscience, 36*(2), 405–418. https://doi.org/10.1523/JNEUROSCI.2555-15.2016

Brandt, A., Khimich, D., & Moser, T. (2005). Few CaV1.3 channels regulate the exocytosis of a synaptic vesicle at the hair cell ribbon synapse. *The Journal of Neuroscience, 25*(50), 11577–11585. https://doi.org/10.1523/JNEUROSCI.3411-05.2005

Braun, M. D., Kisko, T. M., Vecchia, D. D., Andreatini, R., Schwarting, R. K. W., & Wöhr, M. (2018). Sex-specific effects of Cacna1c haploinsufficiency on object recognition, spatial memory, and reversal learning capabilities in rats. *Neurobiology of Learning and Memory*, 155543–155555. https://doi.org/10.1016/j.nlm.2018.05.012

Breustedt, J., Vogt, K. E., Miller, R. J., Nicoll, R. A., & Schmitz, D. (2003). α_{1E}-containing Ca^{2+} channels are involved in synaptic plasticity. *Proceedings of the National Academy of Sciences of the United States of America, 100*(21), 12450–12455. https://doi.org/10.1073/pnas.2035117100

Buonarati, O. R., Henderson, P. B., Murphy, G. G., Horne, M. C., & Hell, J. W. (2018). Proteolytic processing of the L-type Ca^{2+} channel $\alpha_1$1.2 subunit in neurons. *F1000Research, 61166*. https://doi.org/10.12688/f1000research.11808.2

Burgdorf, C. E., Schierberl, K. C., Lee, A. S., Fischer, D. K., Kempen, T. A. V., Mudragel, V., et al. (2017). Extinction of contextual cocaine memories requires Ca$_V$1.2 within D1R-expressing cells and recruits hippocampal Ca$_V$1.2-dependent signaling mechanisms. *The Journal of Neuroscience, 37*(49), 11894–11911. https://doi.org/10.1523/JNEUROSCI.2397-17.2017

Burgdorf, C. E., Bavley, C. C., Fischer, D. K., Walsh, A. P., Martinez-Rivera, A., Hackett, J. E., et al. (2020). Contribution of D1R-expressing neurons of the dorsal dentate gyrus and Ca$_V$1.2 channels in extinction of cocaine conditioned place preference. *Neuropsychopharmacology, 45*(9), 1506–1517. https://doi.org/10.1038/s41386-019-0597-z

Busquet, P., Hetzenauer, A., Sinnegger-Brauns, M. J., Striessnig, J., & Singewald, N. (2008). Role of L-type Ca^{2+} channel isoforms in the extinction of conditioned fear. *Learning & Memory, 15*(5), 378–386. https://doi.org/10.1101/lm.886208

Busquet, P., Khoi Nguyen, N., Schmid, E., Tanimoto, N., Seeliger, M. W., Ben-Yosef, T., et al. (2010). Ca$_V$1.3 L-type Ca^{2+} channels modulate depression-like behaviour in mice independent of deaf phenotype. *The International Journal of Neuropsychopharmacology, 13*(4), 499–513. https://doi.org/10.1017/S1461145709990368

Cahalan, M. D., & Chandy, K. G. (2009). The functional network of ion channels in T lymphocytes. *Immunological Reviews, 231*(1), 59–87. https://doi.org/10.1111/j.1600-065X.2009.00816.x

Candelas, M., Reynders, A., Arango-Lievano, M., Neumayer, C., Fruquière, A., Demes, E., et al. (2019). Ca$_V$3.2 T-type calcium channels shape electrical firing in mouse Lamina II neurons. *Scientific Reports, 9*(1), 3112. https://doi.org/10.1038/s41598-019-39703-3

Cao, C., Ren, Y., Barnett, A. S., Mirando, A. J., Rouse, D., Mun, S. H., et al. (2017). Increased Ca^{2+} signaling through Ca$_V$1.2 promotes bone formation and prevents estrogen deficiency-induced bone loss. *JCI Insight, 2*(22), 95512. https://doi.org/10.1172/jci.insight.95512

Cao, C., Oswald, A. B., Fabella, B. A., Ren, Y., Rodriguiz, R., Trainor, G., et al. (2019). The Ca$_V$1.2 L-type calcium channel regulates bone homeostasis in the middle and inner ear. *Bone*, 125160–125168. https://doi.org/10.1016/j.bone.2019.05.024

Catterall, W. A. (2011). Voltage-gated calcium channels. *Cold Spring Harbor Perspectives in Biology, 3*(8), a003947. https://doi.org/10.1101/cshperspect.a003947

Catterall, W. A., Perez-Reyes, E., Snutch, T. P., & Striessnig, J. (2005). International Union of Pharmacology. XLVIII. Nomenclature and structure-function relationships of voltage-gated calcium channels. *Pharmacological Reviews, 57*(4), 411–425. https://doi.org/10.1124/pr.57.4.5

Catterall, W. A., Lenaeus, M. J., & Gamal El-Din, T. M. (2020). Structure and pharmacology of voltage-gated sodium and calcium channels. *Annual Review of Pharmacology and Toxicology*, 60133–60154. https://doi.org/10.1146/annurev-pharmtox-010818-021757

Ceruti, S., Villa, G., Fumagalli, M., Colombo, L., Magni, G., Zanardelli, M., et al. (2011). Calcitonin gene-related peptide-mediated enhancement of purinergic neuron/Glia communication by the algogenic factor bradykinin in mouse trigeminal ganglia from wild-type and R192Q Ca$_V$2.1 knock-in mice: Implications for basic mechanisms of migraine pain. *The Journal of Neuroscience, 31*(10), 3638–3649. https://doi.org/10.1523/JNEUROSCI.6440-10.2011

Chaudhari, N. (1992). A single nucleotide deletion in the skeletal muscle-specific calcium channel transcript of muscular dysgenesis (mdg) mice. *The Journal of Biological Chemistry, 267*(36), 25636–25639. https://doi.org/10.1016/S0021-9258(18)35650-3

Cheli, V. T., Santiago González, D. A., Spreuer, V., & Paez, P. M. (2015). Voltage-gated Ca^{2+} entry promotes oligodendrocyte progenitor cell maturation and myelination in vitro. *Experimental Neurology*, 26569–26583. https://doi.org/10.1016/j.expneurol.2014.12.012

Cheli, V. T., Santiago González, D. A., Lama, T. N., Spreuer, V., Handley, V., Murphy, G. G., et al. (2016a). Conditional deletion of the L-type calcium channel Cav1.2 in oligodendrocyte progenitor cells affects postnatal myelination in mice. *The Journal of Neuroscience, 36*(42), 10853–10869. https://doi.org/10.1523/JNEUROSCI.1770-16.2016

Cheli, V. T., Santiago González, D. A., Smith, J., Spreuer, V., Murphy, G. G., & Paez, P. M. (2016b). L-type voltage-operated calcium channels contribute to astrocyte activation *in vitro. Glia, 64*(8), 1396–1415. https://doi.org/10.1002/glia.23013

Cheli, V. T., Santiago González, D. A., Zamora, N. N., Lama, T. N., Spreuer, V., Rasmusson, R. L., et al. (2018). Enhanced oligodendrocyte maturation and myelination in a mouse model of Timothy syndrome. *Glia, 66*(11), 2324–2339. https://doi.org/10.1002/glia.23468

Chen, C.-C., Lamping, K. G., Nuno, D. W., Barresi, R., Prouty, S. J., Lavoie, J. L., et al. (2003). Abnormal coronary function in mice deficient in α₁ₕ T-type Ca²⁺ channels. *Science, 302*(5649), 1416–1418. https://doi.org/10.1126/science.1089268

Chen, W.-K., Liu, I. Y., Chang, Y.-T., Chen, Y.-C., Chen, C.-C., Yen, C.-T., et al. (2010). Ca_V3.2 T-type Ca²⁺ channel-dependent activation of ERK in para-ventricular thalamus modulates acid-induced chronic muscle pain. *The Journal of Neuroscience, 30*(31), 10360–10368. https://doi.org/10.1523/JNEUROSCI.1041-10.2010

Chen, C.-C., Fan, Y.-P., Shin, H.-S., & Su, C.-K. (2011). Basal sympathetic activity generated in neonatal mouse brainstem–spinal cord preparation requires T-type calcium channel subunit α₁ₕ. *Experimental Physiology, 96*(5), 486–494. https://doi.org/10.1113/expphysiol.2010.056085

Chen, C.-C., Shen, J.-W., Chung, N.-C., Min, M.-Y., Cheng, S.-J., & Liu, I. Y. (2012). Retrieval of context-associated memory is dependent on the Ca_V3.2 T-type calcium channel. *PLoS ONE, 7*(1), e29384. https://doi.org/10.1371/journal.pone.0029384

Cheng, X. E., Ma, L. X., Feng, X. J., Zhu, M. Y., Zhang, D. Y., Xu, L. L., et al. (2019). Antigen retrieval pre-treatment causes a different expression pattern of Ca_V3.2 in rat and mouse spinal dorsal horn. *European Journal of Histochemistry, 63*(1), 2988. https://doi.org/10.4081/ejh.2019.2988

Chiang, C.-S., Huang, C.-H., Chieng, H., Chang, Y.-T., Chang, D., Chen, J.-J., et al. (2009). The Ca_V3.2 T-type Ca²⁺ channel is required for pressure overload–induced cardiac hypertrophy in mice. *Circulation Research, 104*(4), 522–530. https://doi.org/10.1161/CIRCRESAHA.108.184051

Choi, S., Na, H. S., Kim, J., Lee, J., Lee, S., Kim, D., et al. (2007). Attenuated pain responses in mice lacking Ca_V3.2 T-type channels. *Genes, Brain, and Behavior, 6*(5), 425–431. https://doi.org/10.1111/j.1601-183X.2006.00268.x

Chung, N.-C., Huang, Y.-H., Chang, C.-H., Liao, J. C., Yang, C.-H., Chen, C.-C., et al. (2015). Behavior training reverses asymmetry in hippocampal transcriptome of the Ca_V3.2 knockout mice. *PLoS ONE, 10*(3), e0118832. https://doi.org/10.1371/journal.pone.0118832

Clark, N. C., Nagano, N., Kuenzi, F. M., Jarolimek, W., Huber, I., Walter, D., et al. (2003). Neurological phenotype and synaptic function in mice lacking the Ca_V1.3 alpha subunit of neuronal L-type voltage-dependent Ca²⁺ channels. *Neuroscience, 120*(2), 435–442. https://doi.org/10.1016/s0306-4522(03)00329-4

Coulter, I., Timic Stamenic, T., Eggan, P., Fine, B. R., Corrigan, T., Covey, D. F., et al. (2021). Different roles of T-type calcium channel isoforms in hypnosis induced by an endogenous neurosteroid epipregnano-lone. *Neuropharmacology*, 197108739. https://doi.org/10.1016/j.neuropharm.2021.108739

Cui, J., Ivanova, E., Qi, L., & Pan, Z.-H. (2012). Expression of Ca_V3.2 T-type Ca²⁺ channels in a subpopulation of retinal type-3 cone bipolar cells. *Neuroscience*, 22463–22469. https://doi.org/10.1016/j.neuroscience.2012.08.017

Dao, D. T., Mahon, P. B., Cai, X., Kovacsics, C. E., Blackwell, R. A., Arad, M., et al. (2010). Mood Disorder susceptibility gene *CACNA1C* modifies mood-related behaviors in mice and interacts with sex to influence behavior in mice and diagnosis in humans. *Biological Psychiatry, 68*(9), 801–810. https://doi.org/10.1016/j.biopsych.2010.06.019

Davis, S. E., & Bauer, E. P. (2012). L-type voltage-gated calcium channels in the basolateral amyg-dala are necessary for fear extinction. *The Journal of Neuroscience, 32*(39), 13582–13586. https://doi.org/10.1523/JNEUROSCI.0809-12.2012

Day, M., Wang, Z., Ding, J., An, X., Ingham, C. A., Shering, A. F., et al. (2006). Selective elimination of glutamatergic synapses on striatopallidal neurons in Parkinson disease models. *Nature Neuroscience, 9*(2), 251–259. https://doi.org/10.1038/nn1632

Dayal, A., Schrötter, K., Pan, Y., Föhr, K., Melzer, W., & Grabner, M. (2017). The Ca²⁺ influx through the mammalian skeletal muscle dihydropyridine recep-tor is irrelevant for muscle performance. *Nature Communications, 8*(1), 475. https://doi.org/10.1038/s41467-017-00629-x

Dibué-Adjei, M., Kamp, M. A., Alpdogan, S., Tevoufouet, E. E., Neiss, W. F., Hescheler, J., et al. (2017). Ca_V2.3 (R-type) Calcium channels are critical for medi-ating anticonvulsive and neuroprotective proper-ties of lamotrigine in vivo. *Cellular Physiology and Biochemistry, 44*(3), 935–947. https://doi.org/10.1159/000485361

Dietrich, D., Kirschstein, T., Kukley, M., Pereverzev, A., von der Brelie, C., Schneider, T., et al. (2003). Functional specialization of presynaptic Ca_V2.3 Ca²⁺ channels. *Neuron, 39*(3), 483–496. https://doi.org/10.1016/S0896-6273(03)00430-6

Domes, K., Ding, J., Lemke, T., Blaich, A., Wegener, J. W., Brandmayr, J., et al. (2011). Truncation of murine Ca_V1.2 at Asp-1904 results in heart failure after birth. *The Journal of Biological Chemistry, 286*(39), 33863–33871. https://doi.org/10.1074/jbc.M111.252312

Dragicevic, E., Poetschke, C., Duda, J., Schlaudraff, F., Lammel, S., Schiemann, J., et al. (2014). Ca_V1.3 channels control D2-autoreceptor responses via NCS-1 in substantia nigra dopamine neurons. *Brain, 137*(8), 2287–2302. https://doi.org/10.1093/brain/awu131

Drobinskaya, I., Neumaier, F., Pereverzev, A., Hescheler, J., & Schneider, T. (2015). Diethyldithiocarbamate-mediated zinc ion chelation reveals role of Ca$_V$2.3 channels in glucagon secretion. *Biochimica et Biophysica Acta – Molecular Cell Research, 1853*(5), 953–964. https://doi.org/10.1016/j.bbamcr.2015.01.001

Dumenieu, M., Senkov, O., Mironov, A., Bourinet, E., Kreutz, M. R., Dityatev, A., et al. (2018). The low-threshold calcium channel Ca$_V$3.2 mediates burst firing of mature dentate granule cells. *Cerebral Cortex, 28*(7), 2594–2609. https://doi.org/10.1093/cercor/bhy084

Ebbers, L., Satheesh, S. V., Janz, K., Rüttiger, L., Blosa, M., Hofmann, F., et al. (2015). L-type calcium channel Ca$_V$1.2 Is required for maintenance of auditory brainstem nuclei. *The Journal of Biological Chemistry, 290*(39), 23692–23710. https://doi.org/10.1074/jbc.M115.672675

Eckrich, S., Hecker, D., Sorg, K., Blum, K., Fischer, K., Münkner, S., et al. (2019). Cochlea-specific deletion of Ca$_V$1.3 calcium channels arrests inner hair cell differentiation and unravels pitfalls of conditional mouse models. *Frontiers in Cellular Neuroscience*, 13225. https://doi.org/10.3389/fncel.2019.00225

Ehlinger, D. G., & Commons, K. G. (2017). Altered Ca$_V$1.2 function in the Timothy syndrome mouse model produces ascending serotonergic abnormalities. *The European Journal of Neuroscience, 46*(8), 2416–2425. https://doi.org/10.1111/ejn.13707

Ernst, W. L., Zhang, Y., Yoo, J. W., Ernst, S. J., & Noebels, J. L. (2009). Genetic enhancement of thalamocortical network activity by elevating α_{1G}-mediated low-voltage-activated calcium current induces pure absence epilepsy. *The Journal of Neuroscience, 29*(6), 1615–1625. https://doi.org/10.1523/JNEUROSCI.2081-08.2009

Escoffier, J., Boisseau, S., Serres, C., Chen, C.-C., Kim, D., Stamboulian, S., et al. (2007). Expression, localization and functions in acrosome reaction and sperm motility of Ca$_V$3.1 and Ca$_V$3.2 channels in sperm cells: An evaluation from Ca$_V$3.1 and Ca$_V$3.2 deficient mice. *Journal of Cellular Physiology, 212*(3), 753–763. https://doi.org/10.1002/jcp.21075

Feng, X.-J., Ma, L.-X., Jiao, C., Kuang, H.-X., Zeng, F., Zhou, X.-Y., et al. (2019). Nerve injury elevates functional Ca$_V$3.2 channels in superficial spinal dorsal horn. *Molecular Pain*, 151744806919836569. https://doi.org/10.1177/1744806919836569

Fenninger, F., & Jefferies, W. A. (2019). What's bred in the bone: Calcium channels in lymphocytes. *Journal of Immunology, 202*(4), 1021–1030. https://doi.org/10.4049/jimmunol.1800837

Fenninger, F., Han, J., Stanwood, S. R., Nohara, L. L., Arora, H., Choi, K. B., et al. (2019). Mutation of an L-type calcium channel gene leads to T lymphocyte dysfunction. *Frontiers in Immunology*, 102473. https://doi.org/10.3389/fimmu.2019.02473

Feseha, S., Timic Stamenic, T., Wallace, D., Tamag, C., Yang, L., Pan, J. Q., et al. (2020). Global genetic deletion of Ca$_V$3.3 channels facilitates anaesthetic induction and enhances isoflurane-sparing effects of T-type calcium channel blockers. *Scientific Reports, 10*(1), 21510. https://doi.org/10.1038/s41598-020-78488-8

Fletcher, C. F., Tottene, A., Lennon, V. A., Wilson, S. M., Dubel, S. J., Paylor, R., et al. (2001). Dystonia and cerebellar atrophy in Cacna1a null mice lacking P/Q calcium channel activity. *The FASEB Journal, 15*(7), 1288–1290. https://doi.org/10.1096/fj.00-0562fje

Flucher, B. E. (2020). Skeletal muscle Ca$_V$1.1 channelopathies. *Pflügers Archiv, 472*(7), 739–754. https://doi.org/10.1007/s00424-020-02368-3

Flucher, B. E., & Tuluc, P. (2011). A new L-type calcium channel isoform required for normal patterning of the developing neuromuscular junction. *Channels, 5*(6), 518–524. https://doi.org/10.4161/chan.5.6.17951

Flucher, B. E., Phillips, J. L., & Powell, J. A. (1991). Dihydropyridine receptor alpha subunits in normal and dysgenic muscle in vitro: Expression of alpha 1 is required for proper targeting and distribution of alpha 2. *The Journal of Cell Biology, 115*(5), 1345–1356. https://doi.org/10.1083/jcb.115.5.1345

Francois, A., Kerckhove, N., Meleine, M., Alloui, A., Barrere, C., Gelot, A., et al. (2013). State-dependent properties of a new T-type calcium channel blocker enhance Ca$_V$3.2 selectivity and support analgesic effects. *Pain, 154*(2), 283–293. https://doi.org/10.1016/j.pain.2012.10.023

François, A., Schüetter, N., Laffray, S., Sanguesa, J., Pizzoccaro, A., Dubel, S., et al. (2015). The low-threshold calcium channel Ca$_V$3.2 determines low-threshold mechanoreceptor function. *Cell Reports, 10*(3), 370–382. https://doi.org/10.1016/j.celrep.2014.12.042

Fu, Y., Westenbroek, R. E., Yu, F. H., Clark, J. P., Marshall, M. R., Scheuer, T., et al. (2011). Deletion of the distal C terminus of CaV1.2 channels leads to loss of β-adrenergic regulation and heart failure in vivo. *The Journal of Biological Chemistry, 286*(14), 12617–12626. https://doi.org/10.1074/jbc.M110.175307

Fu, Y., Westenbroek, R. E., Scheuer, T., & Catterall, W. A. (2013). Phosphorylation sites required for regulation of cardiac calcium channels in the fight-or-flight response. *Proceedings of the National Academy of Sciences of the United States of America, 110*(48), 19621–19626. https://doi.org/10.1073/pnas.1319421110

Fu, Y., Westenbroek, R. E., Scheuer, T., & Catterall, W. A. (2014). Basal and β-adrenergic regulation of the cardiac calcium channel Ca$_V$1.2 requires phosphorylation of serine 1700. *Proceedings of the National Academy of Sciences of the United States of America, 111*(46), 16598–16603. https://doi.org/10.1073/pnas.1419129111

Fuller, M. D., Emrick, M. A., Sadilek, M., Scheuer, T., & Catterall, W. A. (2010). Molecular mechanism of calcium channel regulation in the fight-or-flight response. *Science Signaling, 3*(141), ra70. https://doi.org/10.1126/scisignal.2001152

Gadotti, V. M., You, H., Petrov, R. R., Berger, N. D., Diaz, P., & Zamponi, G. W. (2013). Analgesic effect of a mixed T-type channel inhibitor/CB2 receptor agonist. *Molecular Pain*, 91744-8069-9–32. https://doi.org/10.1186/1744-8069-9-32

Gadotti, V. M., Bladen, C., Zhang, F. X., Chen, L., Gündüz, M. G., Şimşek, R., et al. (2015). Analgesic effect of a broad-spectrum dihydropyridine inhibitor of voltage-gated calcium channels. *Pflügers Archiv – European Journal of Physiology, 467*(12), 2485–2493. https://doi.org/10.1007/s00424-015-1725-1

Gadotti, V. M., Kreitinger, J. M., Wageling, N. B., Budke, D., Diaz, P., & Zamponi, G. W. (2020). Ca$_V$3.2 T-type calcium channels control acute itch in mice. *Molecular Brain, 13*(1), 119. https://doi.org/10.1186/s13041-020-00663-9

Galliano, E., Gao, Z., Schonewille, M., Todorov, B., Simons, E., Pop, A. S., et al. (2013). Silencing the majority of cerebellar granule cells uncovers their essential role in motor learning and consolidation. *Cell Reports, 3*(4), 1239–1251. https://doi.org/10.1016/j.celrep.2013.03.023

Gamelli, A. E., McKinney, B. C., White, J. A., & Murphy, G. G. (2011). Deletion of the L-type calcium channel Ca$_V$1.3 but not Ca$_V$1.2 results in a diminished sAHP in mouse CA1 pyramidal neurons. *Hippocampus, 21*(2), 133–141. https://doi.org/10.1002/hipo.20728

Gangadharan, G., Shin, J., Kim, S.-W., Kim, A., Paydar, A., Kim, D.-S., et al. (2016). Medial septal GABAergic projection neurons promote object exploration behavior and type 2 theta rhythm. *Proceedings of the National Academy of Sciences of the United States of America, 113*(23), 6550–6555. https://doi.org/10.1073/pnas.1605019113

Gangarossa, G., Laffray, S., Bourinet, E., & Valjent, E. (2014). T-type calcium channel Ca$_V$3.2 deficient mice show elevated anxiety, impaired memory and reduced sensitivity to psychostimulants. *Frontiers in Behavioral Neuroscience*, 892. https://doi.org/10.3389/fnbeh.2014.00092

Gao, T., Yatani, A., Dell'Acqua, M. L., Sako, H., Green, S. A., Dascal, N., et al. (1997). cAMP-dependent regulation of cardiac L-type Ca²⁺ channels requires membrane targeting of PKA and phosphorylation of channel subunits. *Neuron, 19*(1), 185–196. https://doi.org/10.1016/S0896-6273(00)80358-X

Gao, Z., Todorov, B., Barrett, C. F., van Dorp, S., Ferrari, M. D., van den Maagdenberg, A. M. J. M., et al. (2012). Cerebellar ataxia by enhanced Ca$_V$2.1 currents is alleviated by Ca²⁺-dependent K⁺-channel activators in *Cacna1a* [S218L]-mutant mice. *The Journal of Neuroscience, 32*(44), 15533–15546. https://doi.org/10.1523/JNEUROSCI.2454-12.2012

García-Caballero, A., Gadotti, V. M., Stemkowski, P., Weiss, N., Souza, I. A., Hodgkinson, V., et al. (2014). The deubiquitinating enzyme USP5 modulates neuropathic and inflammatory pain by enhancing Ca$_V$3.2

channel activity. *Neuron, 83*(5), 1144–1158. https://doi.org/10.1016/j.neuron.2014.07.036

Ghoshal, A., Uygun, D. S., Yang, L., McNally, J. M., Lopez-Huerta, V. G., Arias-Garcia, M. A., et al. (2020). Effects of a patient-derived de novo coding alteration of CACNA1I in mice connect a schizophrenia risk gene with sleep spindle deficits. *Translational Psychiatry, 10*(1), 1–12. https://doi.org/10.1038/s41398-020-0685-1

Giang, N., Mars, M., Moreau, M., Mejia, J. E., Bouchaud, G., Magnan, A., et al. (2021). Separation of the Ca$_V$1.2-Ca$_V$1.3 calcium channel duo prevents type 2 allergic airway inflammation. *Allergy*. https://doi.org/10.1111/all.14993

Gibbons, S. J., Strege, P. R., Lei, S., Roeder, J. L., Mazzone, A., Ou, Y., et al. (2009). The α_{1H} Ca²⁺ channel subunit is expressed in mouse jejunal interstitial cells of Cajal and myocytes. *Journal of Cellular and Molecular Medicine, 13*(11–12), 4422–4431. https://doi.org/10.1111/j.1582-4934.2008.00623.x

Giordano, T. P., Satpute, S. S., Striessnig, J., Kosofsky, B. E., & Rajadhyaksha, A. M. (2006). Up-regulation of dopamine D2L mRNA levels in the ventral tegmental area and dorsal striatum of amphetamine-sensitized C57BL/6 mice: role of Ca$_V$1.3 L-type Ca²⁺ channels. *Journal of Neurochemistry, 99*(4), 1197–1206. https://doi.org/10.1111/j.1471-4159.2006.04186.x

Giordano, T. P., Tropea, T. F., Satpute, S. S., Sinnegger-Brauns, M. J., Striessnig, J., Kosofsky, B. E., et al. (2010). Molecular switch from L-type Ca$_V$1.3 to Ca$_V$1.2 Ca²⁺ channel signaling underlies long-term psychostimulant-induced behavioral and molecular plasticity. *The Journal of Neuroscience, 30*(50), 17051–17062. https://doi.org/10.1523/JNEUROSCI.2255-10.2010

Glossmann, H., & Striessnig, J. (1988). Calcium channels. *Vitamins and Hormones, 44*155–44328. https://doi.org/10.1016/s0083-6729(08)60695-0

Goonasekera, S. A., Hammer, K., Auger-Messier, M., Bodi, I., Chen, X., Zhang, H., et al. (2012). Decreased cardiac L-type Ca²⁺ channel activity induces hypertrophy and heart failure in mice. *The Journal of Clinical Investigation, 122*(1), 280–290. https://doi.org/10.1172/JCI58227

Grabner, M., Dirksen, R. T., Suda, N., & Beam, K. G. (1999). The II-III loop of the skeletal muscle dihydropyridine receptor is responsible for the bi-directional coupling with the ryanodine receptor. *The Journal of Biological Chemistry, 274*(31), 21913–21919. https://doi.org/10.1074/jbc.274.31.21913

Gray, P. C., Scott, J. D., & Catterall, W. A. (1998). Regulation of ion channels by cAMP-dependent protein kinase and A-kinase anchoring proteins. *Current Opinion in Neurobiology, 8*(3), 330–334. https://doi.org/10.1016/s0959-4388(98)80057-3

Guzman, J. N., Sánchez-Padilla, J., Chan, C. S., & Surmeier, D. J. (2009). Robust pacemaking in substantia nigra dopaminergic neurons. *The Journal of Neuroscience, 29*(35), 11011–11019. https://doi.org/10.1523/JNEUROSCI.2519-09.2009

Guzman, J. N., Ilijic, E., Yang, B., Sanchez-Padilla, J., Wokosin, D., Galtieri, D., et al. (2018). Systemic israidipine treatment diminishes calcium-dependent mitochondrial oxidant stress. *The Journal of Clinical Investigation, 128*(6), 2266–2280. https://doi.org/10.1172/JCI95898

Hall, D. D., Davare, M. A., Shi, M., Allen, M. L., Weisenhaus, M., McKnight, G. S., et al. (2007). Critical role of cAMP-dependent protein kinase anchoring to the L-type calcium channel $Ca_V1.2$ via A-kinase anchor protein 150 in neurons. *Biochemistry, 46*(6), 1635–1646. https://doi.org/10.1021/bi062217x

Hamby, A. M., Rosa, J. M., Hsu, C.-H., & Feller, M. B. (2015). $Ca_V3.2$ KO mice have altered retinal waves but normal direction selectivity. *Visual Neuroscience, 32.* https://doi.org/10.1017/S0952523814000364

Han, W., Saegusa, H., Zong, S., & Tanabe, T. (2002). Altered cocaine effects in mice lacking $Ca_V2.3$ (α_{1E}) calcium channel. *Biochemical and Biophysical Research Communications, 299*(2), 299–304. https://doi.org/10.1016/S0006-291X(02)02632-3

Han, D., Xue, X., Yan, Y., & Li, G. (2019). Dysfunctional $Ca_V1.2$ channel in Timothy syndrome, from cell to bedside. *Experimental Biology and Medicine (Maywood, NJ), 244*(12), 960–971. https://doi.org/10.1177/1535370219863149

Harraz, O. F., Abd El-Rahman, R. R., Bigdely-Shamloo, K., Wilson, S. M., Brett, S. E., Romero, M., et al. (2014). $Ca_V3.2$ channels and the induction of negative feedback in cerebral arteries. *Circulation Research, 115*(7), 650–661. https://doi.org/10.1161/CIRCRESAHA.114.304056

Harraz, O. F., Brett, S. E., Zechariah, A., Romero, M., Puglisi, J. L., Wilson, S. M., et al. (2015). Genetic ablation of $Ca_V3.2$ channels enhances the arterial myogenic response by modulating the RyR-BKCa axis. *Arteriosclerosis, Thrombosis, and Vascular Biology, 35*(8), 1843–1851. https://doi.org/10.1161/ATVBAHA.115.305736

Hashiguchi, S., Doi, H., Kunii, M., Nakamura, Y., Shimuta, M., Suzuki, E., et al. (2019). Ataxic phenotype with altered $Ca_V3.1$ channel property in a mouse model for spinocerebellar ataxia 42. *Neurobiology of Disease, 130*104516. https://doi.org/10.1016/j.nbd.2019.104516

Hashimoto, K., Tsujita, M., Miyazaki, T., Kitamura, K., Yamazaki, M., Shin, H.-S., et al. (2011). Postsynaptic P/Q-type Ca^{2+} channel in Purkinje cell mediates synaptic competition and elimination in developing cerebellum. *Proceedings of the National Academy of Sciences of the United States of America, 108*(24), 9987–9992. https://doi.org/10.1073/pnas.1101488108

Hatakeyama, S., Wakamori, M., Ino, M., Miyamoto, N., Takahashi, E., Yoshinaga, T., et al. (2001). Differential nociceptive responses in mice lacking the α_{1B} subunit of N-type Ca^{2+} channels. *NeuroReport, 12*(11), 2423–2427.

Held, R. G., Liu, C., Ma, K., Ramsey, A. M., Tarr, T. B., De Nola, G., et al. (2020). Synapse and active zone assembly in the absence of presynaptic Ca^{2+} channels and Ca^{2+} entry. *Neuron, 107*(4), 667–683.e9. https://doi.org/10.1016/j.neuron.2020.05.032

Hetzenauer, A., Sinnegger-Brauns, M. J., Striessnig, J., & Singewald, N. (2006). Brain activation pattern induced by stimulation of L-type Ca^{2+}-channels: Contribution of $Ca_V1.3$ and $Ca_V1.2$ isoforms. *Neuroscience, 139*(3), 1005–1015. https://doi.org/10.1016/j.neuroscience.2006.01.059

Hilaire, C., Lucas, O., Valmier, J., & Scamps, F. (2011). Neurotrophin-4 modulates the mechanotransducer $Ca_V3.2$ T-type calcium current in mice down-hair neurons. *Biochemical Journal, 441*(1), 463–471. https://doi.org/10.1042/BJ20111147

Hildebrand, M. E., Isope, P., Miyazaki, T., Nakaya, T., Garcia, E., Feltz, A., et al. (2009). Functional coupling between mGluR1 and $Ca_V3.1$ T-type calcium channels contributes to parallel fiber-induced fast calcium signaling within purkinje cell dendritic spines. *The Journal of Neuroscience, 29*(31), 9668–9682. https://doi.org/10.1523/JNEUROSCI.0362-09.2009

Hirtz, J. J., Boesen, M., Braun, N., Deitmer, J. W., Kramer, F., Lohr, C., et al. (2011). $Ca_V1.3$ calcium channels are required for normal development of the auditory brainstem. *The Journal of Neuroscience, 31*(22), 8280–8294. https://doi.org/10.1523/JNEUROSCI.5098-10.2011

Hirtz, J. J., Braun, N., Griesemer, D., Hannes, C., Janz, K., Löhrke, S., et al. (2012). Synaptic refinement of an inhibitory topographic map in the auditory brainstem requires functional $Ca_V1.3$ calcium channels. *The Journal of Neuroscience, 32*(42), 14602–14616. https://doi.org/10.1523/JNEUROSCI.0765-12.2012

Hoffmann, T., Kistner, K., Joksimovic, S. L. J., Todorovic, S. M., Reeh, P. W., & Sauer, S. K. (2021). Painful diabetic neuropathy leads to functional $Ca_V3.2$ expression and spontaneous activity in skin nociceptors of mice. *Experimental Neurology, 346*113838. https://doi.org/10.1016/j.expneurol.2021.113838

Horigane, S., Ozawa, Y., Zhang, J., Todoroki, H., Miao, P., Haijima, A., et al. (2020). A mouse model of Timothy syndrome exhibits altered social competitive dominance and inhibitory neuron development. *FEBS Open Bio, 10*(8), 1436–1446. https://doi.org/10.1002/2211-5463.12924

Huang, Z., Lujan, R., Kadurin, I., Uebele, V. N., Renger, J. J., Dolphin, A. C., et al. (2011). Presynaptic HCN1 channels regulate $Ca_V3.2$ activity and neurotransmission at select cortical synapses. *Nature Neuroscience, 14*(4), 478–486. https://doi.org/10.1038/nn.2757

Huang, H., Ng, C. Y., Yu, D., Zhai, J., Lam, Y., & Soong, T. W. (2014). Modest $Ca_V1.3_{42}$-selective inhibition by compound 8 is β-subunit dependent. *Nature Communications, 5*(1), 4481. https://doi.org/10.1038/ncomms5481

Huang, I. Y., Hsu, Y.-L., Chen, C.-C., Chen, M.-F., Wen, Z.-H., Huang, H.-T., et al. (2018). Excavatolide-B enhances contextual memory retrieval via repressing the delayed rectifier potassium current in the hippocampus. *Marine Drugs, 16*(11), 405. https://doi.org/10.3390/md16110405

Hulme, J. T., Westenbroek, R. E., Scheuer, T., & Catterall, W. A. (2006). Phosphorylation of serine 1928 in the distal C-terminal domain of cardiac Ca$_V$1.2 channels during β1-adrenergic regulation. *Proceedings of the National Academy of Sciences of the United States of America, 103*(44), 16574–16579. https://doi.org/10.1073/pnas.0607294103

Ilijic, E., Guzman, J., & Surmeier, D. (2011). The L-type channel antagonist isradipine is neuroprotective in a mouse model of Parkinson's disease. *Neurobiology of Disease, 43*(2), 364–371. https://doi.org/10.1016/j.nbd.2011.04.007

Ino, M., Yoshinaga, T., Wakamori, M., Miyamoto, N., Takahashi, E., Sonoda, J., et al. (2001). Functional disorders of the sympathetic nervous system in mice lacking the α$_{1B}$ subunit (Ca$_V$2.2) of N-type calcium channels. *Proceedings of the National Academy of Sciences of the United States of America, 98*(9), 5323–5328. https://doi.org/10.1073/pnas.081089398

Jeon, D., Kim, C., Yang, Y.-M., Rhim, H., Yim, E., Oh, U., et al. (2007). Impaired long-term memory and long-term potentiation in N-type Ca²⁺ channel-deficient mice. *Genes, Brain, and Behavior, 6*(4), 375–388. https://doi.org/10.1111/j.1601-183X.2006.00267.x

Jeon, D., Kim, S., Chetana, M., Jo, D., Ruley, H. E., Lin, S.-Y., et al. (2010). Observational fear learning involves affective pain system and Ca$_V$1.2 Ca²⁺ channels in ACC. *Nature Neuroscience, 13*(4), 482–488. https://doi.org/10.1038/nn.2504

Jeub, M., Taha, O., Opitz, T., Racz, I., Pitsch, J., Becker, A., et al. (2019). Partial sciatic nerve ligation leads to an upregulation of Ni²⁺-resistant T-type Ca²⁺ currents in capsaicin-responsive nociceptive dorsal root ganglion neurons. *Journal of Pain Research, 12635–12647.* https://doi.org/10.2147/JPR.S138708

Jiang, X., Lupien-Meilleur, A., Tazerart, S., Lachance, M., Samarova, E., Araya, R., et al. (2018). Remodeled cortical inhibition prevents motor seizures in generalized epilepsy. *Annals of Neurology, 84*(3), 436–451. https://doi.org/10.1002/ana.25301

Jiang, M. C., Birch, D. V., Heckman, C. J., & Tysseling, V. M. (2021). The involvement of Ca$_V$1.3 channels in prolonged root reflexes and its potential as a therapeutic target in spinal cord injury. *Frontiers in Neural Circuits, 1522.* https://doi.org/10.3389/fncir.2021.642111

Jing, X., Li, D.-Q., Olofsson, C. S., Salehi, A., Surve, V. V., Caballero, J., et al. (2005). Ca$_V$2.3 calcium channels control second-phase insulin release. *The Journal of Clinical Investigation, 115*(1), 146–154. https://doi.org/10.1172/JCI22518

Joksimovic, S. L., Evans, J. G., McIntire, W. E., Orestes, P., Barrett, P. Q., Jevtovic-Todorovic, V., et al. (2020a). Glycosylation of Ca$_V$3.2 channels contributes to the hyperalgesia in peripheral neuropathy of type 1 diabetes. *Frontiers in Cellular Neuroscience, 14432.* https://doi.org/10.3389/fncel.2020.605312

Joksimovic, S. L., Joksimovic, S. M., Manzella, F. M., Asnake, B., Orestes, P., Raol, Y. H., et al. (2020b). Novel neuroactive steroid with hypnotic and T-type

calcium channel blocking properties exerts effective analgesia in a rodent model of post-surgical pain. *British Journal of Pharmacology, 177*(8), 1735–1753. https://doi.org/10.1111/bph.14930

Joksovic, P. M., Nelson, M. T., Jevtovic-Todorovic, V., Patel, M. K., Perez-Reyes, E., Campbell, K. P., et al. (2006). Ca$_V$3.2 is the major molecular substrate for redox regulation of T-type Ca²⁺ channels in the rat and mouse thalamus. *The Journal of Physiology, 574*(2), 415–430. https://doi.org/10.1113/jphysiol.2006.110395

Jun, K., Piedras-Rentería, E. S., Smith, S. M., Wheeler, D. B., Lee, S. B., Lee, T. G., et al. (1999). Ablation of P/Q-type Ca²⁺ channel currents, altered synaptic transmission, and progressive ataxia in mice lacking the α$_{1A}$-subunit. *Proceedings of the National Academy of Sciences of the United States of America, 96*(26), 15245–15250. https://doi.org/10.1073/pnas.96.26.15245

Jung, D., Hwang, Y. J., Ryu, H., Kano, M., Sakimura, K., & Cho, J. (2016). Conditional knockout of Ca$_V$2.1 disrupts the accuracy of spatial recognition of CA1 place cells and spatial/contextual recognition behavior. *Frontiers in Behavioral Neuroscience, 10.* https://doi.org/10.3389/fnbeh.2016.00214

Kabir, Z. D., Martínez-Rivera, A., & Rajadhyaksha, A. M. (2017). From gene to behavior: L-type calcium channel mechanisms underlying neuropsychiatric symptoms. *Neurotherapeutics, 14*(3), 588–613. https://doi.org/10.1007/s13311-017-0532-0

Kamijo, S., Ishii, Y., Horigane, S., Suzuki, K., Ohkura, M., Nakai, J., et al. (2018). A critical neurodevelopmental role for L-type voltage-gated calcium channels in neurite extension and radial migration. *The Journal of Neuroscience, 38*(24), 5551–5566. https://doi.org/10.1523/JNEUROSCI.2357-17.2018

Kantojärvi, K., Liuhanen, J., Saarenpää-Heikkilä, O., Satomaa, A.-L., Kylliäinen, A., Pölkki, P., et al. (2017). Variants in calcium voltage-gated channel subunit Alpha1 C-gene (*CACNA1C*) are associated with sleep latency in infants. *PLoS ONE, 12*(8), e0180652. https://doi.org/10.1371/journal.pone.0180652

Kaplan, M. M., & Flucher, B. E. (2019). Postsynaptic Ca$_V$1.1-driven calcium signaling coordinates presynaptic differentiation at the developing neuromuscular junction. *Scientific Reports, 9*(1), 18450. https://doi.org/10.1038/s41598-019-54900-w

Kaplan, M. M., Sultana, N., Benedetti, A., Obermair, G. J., Linde, N. F., Papadopoulos, S., et al. (2018). Calcium influx and release cooperatively regulate AChR patterning and motor axon outgrowth during neuromuscular junction formation. *Cell Reports, 23*(13), 3891–3904. https://doi.org/10.1016/j.celrep.2018.05.085

Katchman, A., Yang, L., Zakharov, S. I., Kushner, J., Abrams, J., Chen, B.-X., et al. (2017). Proteolytic cleavage and PKA phosphorylation of α$_{1C}$ subunit are not required for adrenergic regulation of Ca$_V$1.2 in the heart. *Proceedings of the National Academy of*

Sciences of the United States of America, 114(34), 9194–9199. https://doi.org/10.1073/pnas.1706054114

Kim, C., Jun, K., Lee, T., Kim, S.-S., McEnery, M. W., Chin, H., et al. (2001a). Altered nociceptive response in mice deficient in the α_{1B} subunit of the voltage-dependent calcium channel. *Molecular and Cellular Neurosciences, 18*(2), 235–245. https://doi.org/10.1006/mcne.2001.1013

Kim, D., Song, I., Keum, S., Lee, T., Jeong, M.-J., Kim, S.-S., et al. (2001b). Lack of the burst firing of thalamocortical relay neurons and resistance to absence seizures in mice lacking α_{1G} T-type Ca^{2+} channels. *Neuron, 31*(1), 35–45. https://doi.org/10.1016/S0896-6273(01)00343-9

Kim, D., Park, D., Choi, S., Lee, S., Sun, M., Kim, C., et al. (2003). Thalamic control of visceral nociception mediated by T-type Ca^{2+} channels. *Science, 302*(5642), 117–119. https://doi.org/10.1126/science.1088886

Kim, C., Jeon, D., Kim, Y.-H., Lee, C. J., Kim, H., & Shin, H.-S. (2009). Deletion of N-type Ca^{2+} channel $Ca_V2.2$ results in hyperaggressive behaviors in mice. *The Journal of Biological Chemistry, 284*(5), 2738–2745. https://doi.org/10.1074/jbc.M807179200

Kim, Y. S., Kim, Y.-B., Kim, W. B., Yoon, B.-E., Shen, F.-Y., Lee, S. W., et al. (2015). Histamine resets the circadian clock in the suprachiasmatic nucleus through the H1R-$Ca_V1.3$-RyR pathway in the mouse. *The European Journal of Neuroscience, 42*(7), 2467–2477. https://doi.org/10.1111/ejn.13030

Kim, S.-H., Park, Y.-R., Lee, B., Choi, B., Kim, H., & Kim, C.-H. (2017). Reduction of $Ca_V1.3$ channels in dorsal hippocampus impairs the development of dentate gyrus newborn neurons and hippocampal-dependent memory tasks. *PLoS ONE, 12*(7), e0181138. https://doi.org/10.1371/journal.pone.0181138

Kisko, T. M., Braun, M. D., Michels, S., Witt, S. H., Rietschel, M., Culmsee, C., et al. (2018). *Cacna1c* haploinsufficiency leads to pro-social 50-kHz ultrasonic communication deficits in rats. *Disease Models & Mechanisms, 11*(6), dmm034116. https://doi.org/10.1242/dmm.034116

Kisko, T. M., Braun, M. D., Michels, S., Witt, S. H., Rietschel, M., Culmsee, C., et al. (2020). Sex-dependent effects of *Cacna1c* haploinsufficiency on juvenile social play behavior and pro-social 50-kHz ultrasonic communication in rats. *Genes, Brain, and Behavior, 19*(2), e12552. https://doi.org/10.1111/gbb.12552

Koppe, G., Mallien, A. S., Berger, S., Bartsch, D., Gass, P., Vollmayr, B., et al. (2017). *CACNA1C* gene regulates behavioral strategies in operant rule learning. *PLoS Biology, 15*(6), e2000936. https://doi.org/10.1371/journal.pbio.2000936

Koschak, A., Reimer, D., Huber, I., Grabner, M., Glossmann, H., Engel, J., et al. (2001). $\alpha1D$ ($Cav1.3$) subunits can form L-type Ca^{2+} channels activating at negative voltages. *The Journal of Biological Chemistry, 276*(25), 22100–22106. https://doi.org/10.1074/jbc.M101469200

Koschak, A., Obermair, G. J., Pivotto, F., Sinnegger-Brauns, M. J., Striessnig, J., & Pietrobon, D. (2007). Molecular nature of anomalous L-type calcium channels in mouse cerebellar granule cells. *The Journal of Neuroscience, 27*(14), 3855–3863. https://doi.org/10.1523/JNEUROSCI.4028-06.2007

Krey, J. F., Paşca, S. P., Shcheglovitov, A., Yazawa, M., Schwemberger, R., Rasmusson, R., et al. (2013). Timothy syndrome is associated with activity-dependent dendritic retraction in rodent and human neurons. *Nature Neuroscience, 16*(2), 201–209. https://doi.org/10.1038/nn.3307

Krueger, J. N., Moore, S. J., Parent, R., McKinney, B. C., Lee, A., & Murphy, G. G. (2017). A novel mouse model of the aged brain: Over-expression of the L-type voltage-gated calcium channel $Ca_V1.3$. *Behavioural Brain Research, 322*(Pt B), 241–249. https://doi.org/10.1016/j.bbr.2016.06.054

Kubota, M., Murakoshi, T., Saegusa, H., Kazuno, A., Zong, S., Hu, Q., et al. (2001). Intact LTP and fear memory but impaired spatial memory in mice lacking $Ca_V2.3$ (α_{1E}) channel. *Biochemical and Biophysical Research Communications, 282*(1), 242–248. https://doi.org/10.1006/bbrc.2001.4572

Kumar, D., Dedic, N., Flachskamm, C., Voulé, S., Deussing, J. M., & Kimura, M. (2015). Cacna1c (Cav1.2) modulates electroencephalographic rhythm and rapid eye movement sleep recovery. *Sleep, 38*(9), 1371–1380. https://doi.org/10.5665/sleep.4972

Ladera, C., Martín, R., Bartolomé-Martín, D., Torres, M., & Sánchez-Prieto, J. (2009). Partial compensation for N-type Ca^{2+} channel loss by P/Q-type Ca^{2+} channels underlines the differential release properties supported by these channels at cerebro-cortical nerve terminals. *The European Journal of Neuroscience, 29*(6), 1131–1140. https://doi.org/10.1111/j.1460-9568.2009.06675.x

Langwieser, N., Christel, C. J., Kleppisch, T., Hofmann, F., Wotjak, C. T., & Moosmang, S. (2010). Homeostatic switch in hebbian plasticity and fear learning after sustained loss of $Ca_V1.2$ calcium channels. *The Journal of Neuroscience, 30*(25), 8367–8375. https://doi.org/10.1523/JNEUROSCI.4164-08.2010

Le Quang, K., Naud, P., Qi, X.-Y., Duval, F., Shi, Y.-F., Gillis, M.-A., et al. (2011). Role of T-type calcium channel subunits in post-myocardial infarction remodelling probed with genetically engineered mice. *Cardiovascular Research, 91*(3), 420–428. https://doi.org/10.1093/cvr/cvr082

Lee, S.-C., Choi, S., Lee, T., Kim, H.-L., Chin, H., & Shin, H.-S. (2002). Molecular basis of R-type calcium channels in central amygdala neurons of the mouse. *Proceedings of the National Academy of Sciences of the United States of America, 99*(5), 3276–3281. https://doi.org/10.1073/pnas.052697799

Lee, J., Kim, D., & Shin, H.-S. (2004). Lack of delta waves and sleep disturbances during non-rapid eye movement sleep in mice lacking α_{1G}-subunit of T-type calcium channels. *Proceedings of the National Academy of Sciences of the United States of America,*

101(52), 18195–18199. https://doi.org/10.1073/pnas.0408089101

Lee, A. S., Ra, S., Rajadhyaksha, A. M., Britt, J. K., De Jesus-Cortes, H., Gonzales, K. L., et al. (2012). Forebrain elimination of *cacna1c* mediates anxiety-like behavior in mice. *Molecular Psychiatry, 17*(11), 1054–1055. https://doi.org/10.1038/mp.2012.71

Lee, J., Song, K., Lee, K., Hong, J., Lee, H., Chae, S., et al. (2013). Sleep spindles are generated in the absence of T-type calcium channel-mediated low-threshold burst firing of thalamocortical neurons. *Proceedings of the National Academy of Sciences of the United States of America, 110*(50), 20266–20271. https://doi.org/10.1073/pnas.1320572110

Lee, S. E., Lee, J., Latchoumane, C., Lee, B., Oh, S.-J., Saud, Z. A., et al. (2014). Rebound burst firing in the reticular thalamus is not essential for pharmacological absence seizures in mice. *Proceedings of the National Academy of Sciences of the United States of America, 111*(32), 11828–11833. https://doi.org/10.1073/pnas.1408609111

Lee, C. S., Dagnino-Acosta, A., Yarotskyy, V., Hanna, A., Lyfenko, A., Knoblauch, M., et al. (2015). Ca²⁺ permeation and/or binding to Ca$_V$1.1 fine-tunes skeletal muscle Ca²⁺ signaling to sustain muscle function. *Skeletal Muscle, 5*(1), 4. https://doi.org/10.1186/s13395-014-0027-1

Lee, A. S., Jesús-Cortés, H. D., Kabir, Z. D., Knobbe, W., Orr, M., Burgdorf, C., et al. (2016). The neuropsychiatric disease-associated gene *cacna1c* mediates survival of young hippocampal neurons. *eNeuro, 3*(2), ENEURO.0006-16.2016. https://doi.org/10.1523/ENEURO.0006-16.2016

Lei, D., Gao, X., Perez, P., Ohlemiller, K. K., Chen, C.-C., Campbell, K. P., et al. (2011). Anti-epileptic drugs delay age-related loss of spiral ganglion neurons via T-type calcium channel. *Hearing Research, 278*(1), 106–112. https://doi.org/10.1016/j.heares.2011.05.010

Lemke, T., Welling, A., Christel, C. J., Blaich, A., Bernhard, D., Lenhardt, P., et al. (2008). Unchanged β-adrenergic stimulation of cardiac L-type calcium channels in Ca$_V$1.2 phosphorylation site S1928A mutant mice. *The Journal of Biological Chemistry, 283*(50), 34738–34744. https://doi.org/10.1074/jbc.M804981200

Li, Y., & Bennett, D. J. (2003). Persistent sodium and calcium currents cause plateau potentials in motoneurons of chronic spinal rats. *Journal of Neurophysiology, 90*(2), 857–869. https://doi.org/10.1152/jn.00236.2003

Li, J., Zhao, L., Ferries, I. K., Jiang, L., Desta, M. Z., Yu, X., et al. (2010). Skeletal phenotype of mice with a null mutation in Ca$_V$1.3 L-type calcium channel. *Journal of Musculoskeletal & Neuronal Interactions, 10*(2), 180–187.

Li, Y., Wang, F., Zhang, X., Qi, Z., Tang, M., Szeto, C., et al. (2012). β-adrenergic stimulation increases Ca$_V$3.1 activity in cardiac myocytes through protein kinase A. *PLoS ONE, 7*(7), e39965. https://doi.org/10.1371/journal.pone.0039965

Li, G., Wang, J., Liao, P., Bartels, P., Zhang, H., Yu, D., et al. (2017). Exclusion of alternative exon 33 of Ca$_V$1.2 calcium channels in heart is proarrhythmogenic. *Proceedings of the National Academy of Sciences of the United States of America, 114*(21), E4288–E4295. https://doi.org/10.1073/pnas.1617205114

Li, C., Li, X., Bi, Z., Sugino, K., Wang, G., Zhu, T., et al. (2020). Comprehensive transcriptome analysis of cochlear spiral ganglion neurons at multiple ages. *eLife*, 9e50491. https://doi.org/10.7554/eLife.50491

Liao, P., Li, G., Yu, D. J., Yong, T. F., Wang, J. J., Wang, J., et al. (2009). Molecular alteration of Ca$_V$1.2 calcium channel in chronic myocardial infarction. *Pflügers Archiv – European Journal of Physiology, 458*(4), 701–711. https://doi.org/10.1007/s00424-009-0652-4

Lieb, A., Scharinger, A., Sartori, S., Sinnegger-Brauns, M. J., & Striessnig, J. (2012). Structural determinants of Ca$_V$1.3 L-type calcium channel gating. *Channels, 6*(3), 197–205. https://doi.org/10.4161/chan.21002

Lin, S.-S., Tzeng, B.-H., Lee, K.-R., Smith, R. J. H., Campbell, K. P., & Chen, C.-C. (2014). Ca$_V$3.2 T-type calcium channel is required for the NFAT-dependent Sox9 expression in tracheal cartilage. *Proceedings of the National Academy of Sciences of the United States of America, 111*(19), E1990–E1998. https://doi.org/10.1073/pnas.1323112111

Liss, B., & Striessnig, J. (2019). The potential of L-type calcium channels as a drug target for neuroprotective therapy in Parkinson's disease. *Annual Review of Pharmacology and Toxicology*, 59263–59289. https://doi.org/10.1146/annurev-pharmtox-010818-021214

Liu, Y., Harding, M., Dore, J., & Chen, X. (2017). Ca$_V$1.2, but not Ca$_V$1.3, L-type calcium channel subtype mediates nicotine-induced conditioned place preference in mice. *Progress in Neuro-Psychopharmacology & Biological Psychiatry*, 75176–75182. https://doi.org/10.1016/j.pnpbp.2017.02.004

Liu, G., Papa, A., Katchman, A. N., Zakharov, S. I., Roybal, D., Hennessey, J. A., et al. (2020). Mechanism of adrenergic Ca$_V$1.2 stimulation revealed by proximity proteomics. *Nature, 577*(7792), 695–700. https://doi.org/10.1038/s41586-020-1947-z

Llinás, R. R., Choi, S., Urbano, F. J., & Shin, H.-S. (2007). γ-Band deficiency and abnormal thalamocortical activity in P/Q-type channel mutant mice. *Proceedings of the National Academy of Sciences of the United States of America, 104*(45), 17819–17824. https://doi.org/10.1073/pnas.0707945104

Lu, Z.-J., Pereverzev, A., Liu, H.-L., Weiergräber, M., Henry, M., Krieger, A., et al. (2004). Arrhythmia in isolated prenatal hearts after ablation of the Ca$_V$2.3 (α$_{1E}$) subunit of voltage-gated Ca²⁺ channels. *Cellular Physiology and Biochemistry, 14*(1–2), 11–22. https://doi.org/10.1159/000076922

Lu, L., Zhang, Q., Timofeyev, V., Zhang, Z., Young, J. N., Shin, H.-S., et al. (2007). Molecular coupling of a Ca²⁺-activated K⁺ channel to L-type Ca²⁺ channels via α-Actinin2. *Circulation Research, 100*(1), 112–120. https://doi.org/10.1161/01.RES.0000253095.44186.72

Lübbert, M., Goral, R. O., Satterfield, R., Putzke, T., van den Maagdenberg, A. M., Kamasawa, N., et al. (2017). A novel region in the $Ca_V2.1$ α_1 subunit C-terminus regulates fast synaptic vesicle fusion and vesicle docking at the mammalian presynaptic active zone. *eLife*, 6e28412. https://doi.org/10.7554/eLife.28412

Lundt, A., Seidel, R., Soós, J., Henseler, C., Müller, R., Bakki, M., et al. (2019). $Ca_V3.2$ T-type calcium channels are physiologically mandatory for the auditory system. *Neuroscience*, 40981–40100. https://doi.org/10.1016/j.neuroscience.2019.04.024

Luvisetto, S., Marinelli, S., Panasiti, M. S., D'Amato, F. R., Fletcher, C. F., Pavone, F., et al. (2006). Pain sensitivity in mice lacking the $Ca_V2.1\alpha_1$ subunit of P/Q-type Ca^{2+} channels. *Neuroscience, 142*(3), 823–832. https://doi.org/10.1016/j.neuroscience.2006.06.049

Lv, P., Kim, H. J., Lee, J.-H., Sihn, C.-R., Gharaie, S. F., Mousavi-Nik, A., et al. (2014). Genetic, cellular, and functional evidence for Ca^{2+} inflow through $Ca_V1.2$ and $Ca_V1.3$ channels in murine spiral ganglion neurons. *The Journal of Neuroscience, 34*(21), 7383–7393. https://doi.org/10.1523/JNEUROSCI.5416-13.2014

M'Dahoma, S., Gadotti, V. M., Zhang, F.-X., Park, B., Nam, J. H., Onnis, V., et al. (2016). Effect of the T-type channel blocker KYS-05090S in mouse models of acute and neuropathic pain. *Pflügers Archiv – European Journal of Physiology, 468*(2), 193–199. https://doi.org/10.1007/s00424-015-1733-1

Maejima, T., Wollenweber, P., Teusner, L. U. C., Noebels, J. L., Herlitze, S., & Mark, M. D. (2013). Postnatal loss of P/Q-type channels confined to rhombic-lip-derived neurons alters synaptic transmission at the parallel fiber to purkinje cell synapse and replicates genomic *Cacna1a* mutation phenotype of ataxia and seizures in mice. *The Journal of Neuroscience, 33*(12), 5162–5174. https://doi.org/10.1523/JNEUROSCI.5442-12.2013

Makarenko, V. V., Peng, Y.-J., Yuan, G., Fox, A. P., Kumar, G. K., Nanduri, J., et al. (2015). CaV3.2 T-type Ca^{2+} channels in H_2S-mediated hypoxic response of the carotid body. *American Journal of Physiology. Cell Physiology, 308*(2), C146–C154. https://doi.org/10.1152/ajpcell.00141.2014

Makarenko, V. V., Ahmmed, G. U., Peng, Y.-J., Khan, S. A., Nanduri, J., Kumar, G. K., et al. (2016). $Ca_V3.2$ T-type Ca^{2+} channels mediate the augmented calcium influx in carotid body glomus cells by chronic intermittent hypoxia. *Journal of Neurophysiology, 115*(1), 345–354. https://doi.org/10.1152/jn.00775.2015

Mallmann, R. T., Elgueta, C., Sleman, F., Castonguay, J., Wilmes, T., van den Maagdenberg, A., et al. (2013). Ablation of $Ca_V2.1$ voltage-gated Ca^{2+} channels in mouse forebrain generates multiple cognitive impairments. *PLoS ONE, 8*(10), e78598. https://doi.org/10.1371/journal.pone.0078598

Mancarella, S., Yue, Y., Karnabi, E., Qu, Y., El-Sherif, N., & Boutjdir, M. (2008). Impaired Ca^{2+} homeostasis is associated with atrial fibrillation in the α_{1D} L-type Ca^{2+} channel KO mouse. *American Journal of Physiology.*

Heart and Circulatory Physiology, 295(5), H2017–H2024. https://doi.org/10.1152/ajpheart.00537.2008

Mangoni, M. E., Traboulsie, A., Leoni, A.-L., Couette, B., Marger, L., Le Quang, K., et al. (2006). Bradycardia and slowing of the atrioventricular conduction in mice lacking $Ca_V3.1/\alpha_{1G}$ T-type calcium channels. *Circulation Research, 98*(11), 1422–1430. https://doi.org/10.1161/01.RES.0000225862.14314.49

Mansergh, F., Orton, N. C., Vessey, J. P., Lalonde, M. R., Stell, W. K., Tremblay, F., et al. (2005). Mutation of the calcium channel gene *Cacna1f* disrupts calcium signaling, synaptic transmission and cellular organization in mouse retina. *Human Molecular Genetics, 14*(20), 3035–3046. https://doi.org/10.1093/hmg/ddi336

Marcantoni, A., Vandael, D. H. F., Mahapatra, S., Carabelli, V., Sinnegger-Brauns, M. J., Striessnig, J., et al. (2010). Loss of $Ca_V1.3$ channels reveals the critical role of L-type and BK channel coupling in pacemaking mouse adrenal chromaffin cells. *The Journal of Neuroscience, 30*(2), 491–504. https://doi.org/10.1523/JNEUROSCI.4961-09.2010

Marcantoni, M., Fuchs, A., Löw, P., Bartsch, D., Kiehn, O., & Bellardita, C. (2020). Early delivery and prolonged treatment with nimodipine prevents the development of spasticity after spinal cord injury in mice. *Science Translational Medicine*. https://doi.org/10.1126/scitranslmed.aay0167

Mark, M. D., Maejima, T., Kuckelsberg, D., Yoo, J. W., Hyde, R. A., Shah, V., et al. (2011). Delayed postnatal loss of P/Q-type calcium channels recapitulates the absence epilepsy, dyskinesia, and ataxia phenotypes of genomic Cacna1A mutations. *The Journal of Neuroscience, 31*(11), 4311–4326. https://doi.org/10.1523/JNEUROSCI.5342-10.2011

Mark, M. D., Krause, M., Boele, H.-J., Kruse, W., Pollok, S., Kuner, T., et al. (2015). Spinocerebellar ataxia type 6 protein aggregates cause deficits in motor learning and cerebellar plasticity. *The Journal of Neuroscience, 35*(23), 8882–8895. https://doi.org/10.1523/JNEUROSCI.0891-15.2015

Marschallinger, J., Sah, A., Schmuckermair, C., Unger, M., Rotheneichner, P., Kharitonova, M., et al. (2015). The L-type calcium channel $Ca_V1.3$ is required for proper hippocampal neurogenesis and cognitive functions. *Cell Calcium, 58*(6), 606–616. https://doi.org/10.1016/j.ceca.2015.09.007

Martín, R., Ladera, C., Bartolomé-Martín, D., Torres, M., & Sánchez-Prieto, J. (2008). The inhibition of release by mGlu7 receptors is independent of the Ca^{2+} channel type but associated to $GABA_B$ and adenosine A1 receptors. *Neuropharmacology, 55*(4), 464–473. https://doi.org/10.1016/j.neuropharm.2008.04.011

Martínez-Rivera, A., Hao, J., Tropea, T. F., Giordano, T. P., Kosovsky, M., Rice, R. C., et al. (2017). Enhancing VTA $Ca_V1.3$ L-type Ca^{2+} channel activity promotes cocaine and mood-related behaviors via overlapping AMPA receptor mechanisms in the nucleus accumbens. *Molecular Psychiatry, 22*(12), 1735–1745. https://doi.org/10.1038/mp.2017.9

Matsuda, Y., Saegusa, H., Zong, S., Noda, T., & Tanabe, T. (2001). Mice lacking Ca$_V$2.3 (α$_{1E}$) calcium channel exhibit hyperglycemia. *Biochemical and Biophysical Research Communications, 289*(4), 791–795. https://doi.org/10.1006/bbrc.2001.6051

Matsui, K., Tsubota, M., Fukushi, S., Koike, N., Masuda, H., Kasanami, Y., et al. (2019). Genetic deletion of Ca$_V$3.2 T-type calcium channels abolishes H$_2$S-dependent somatic and visceral pain signaling in C57BL/6 mice. *Journal of Pharmacological Sciences, 140*(3), 310–312. https://doi.org/10.1016/j.jphs.2019.07.010

McKinney, B. C., & Murphy, G. G. (2006). The L-Type voltage-gated calcium channel Ca$_V$1.3 mediates consolidation, but not extinction, of contextually conditioned fear in mice. *Learning & Memory, 13*(5), 584–589. https://doi.org/10.1101/lm.279006

McKinney, B. C., Sze, W., White, J. A., & Murphy, G. G. (2008). L-type voltage-gated calcium channels in conditioned fear: A genetic and pharmacological analysis. *Learning & Memory, 15*(5), 326–334. https://doi.org/10.1101/lm.893808

McKinney, B. C., Sze, W., Lee, B., & Murphy, G. G. (2009). Impaired long-term potentiation and enhanced neuronal excitability in the amygdala of Ca$_V$1.3 knockout mice. *Neurobiology of Learning and Memory, 92*(4), 519–528. https://doi.org/10.1016/j.nlm.2009.06.012

Mehregan, A., Ardestani, G., Akizawa, H., Carvacho, I., & Fissore, R. (2021). Deletion of TRPV3 and Ca$_V$3.2 T-type channels in mice undermines fertility and Ca²⁺ homeostasis in oocytes and eggs. *Journal of Cellular Science, 134*(13), jcs257956. https://doi.org/10.1242/jcs.257956

Mesirca, P., Torrente, A. G., & Mangoni, M. E. (2015). Functional role of voltage gated Ca²⁺ channels in heart automaticity. *Frontiers in Physiology*, 619. https://doi.org/10.3389/fphys.2015.00019

Miao, Q.-L., Herlitze, S., Mark, M. D., & Noebels, J. L. (2020). Adult loss of *Cacna1a* in mice recapitulates childhood absence epilepsy by distinct thalamic bursting mechanisms. *Brain, 143*(1), 161–174. https://doi.org/10.1093/brain/awz365

Mikkelsen, M. F., Björling, K., & Jensen, L. J. (2016). Age-dependent impact of Ca$_V$3.2 T-type calcium channel deletion on myogenic tone and flow-mediated vasodilatation in small arteries. *The Journal of Physiology, 594*(20), 5881–5898. https://doi.org/10.1113/JP271470

Miyazaki, T., Hashimoto, K., Shin, H.-S., Kano, M., & Watanabe, M. (2004). P/Q-type Ca²⁺ channel α$_{1A}$ regulates synaptic competition on developing cerebellar purkinje cells. *The Journal of Neuroscience, 24*(7), 1734–1743. https://doi.org/10.1523/JNEUROSCI.4208-03.2004

Miyazaki, T., Yamasaki, M., Hashimoto, K., Yamazaki, M., Abe, M., Usui, H., et al. (2012). Ca$_V$2.1 in cerebellar purkinje cells regulates competitive excitatory synaptic wiring, cell survival, and cerebellar biochemical compartmentalization. *The Journal of Neuroscience,* *32*(4), 1311–1328. https://doi.org/10.1523/JNEUROSCI.2755-11.2012

Moon, A. L., Brydges, N. M., Wilkinson, L. S., Hall, J., & Thomas, K. L. (2020). *Cacna1c* hemizygosity results in aberrant fear conditioning to neutral stimuli. *Schizophrenia Bulletin, 46*(5), 1231–1238. https://doi.org/10.1093/schbul/sbz127

Moore, S. J., & Murphy, G. G. (2020). The role of L-type calcium channels in neuronal excitability and aging. *Neurobiology of Learning and Memory*, 173107230. https://doi.org/10.1016/j.nlm.2020.107230

Moosmang, S., Schulla, V., Welling, A., Feil, R., Feil, S., Wegener, J. W., et al. (2003). Dominant role of smooth muscle L-type calcium channel Ca$_V$1.2 for blood pressure regulation. *The EMBO Journal, 22*(22), 6027–6034. https://doi.org/10.1093/emboj/cdg583

Moosmang, S., Haider, N., Klugbauer, N., Adelsberger, H., Langwieser, N., Müller, J., et al. (2005). Role of hippocampal Ca$_V$1.2 Ca²⁺ channels in NMDA receptor-independent synaptic plasticity and spatial memory. *The Journal of Neuroscience, 25*(43), 9883–9892. https://doi.org/10.1523/JNEUROSCI.1531-05.2005

Müller, C. S., Haupt, A., Bildl, W., Schindler, J., Knaus, H.-G., Meissner, M., et al. (2010). Quantitative proteomics of the Ca$_V$2 channel nano-environments in the mammalian brain. *Proceedings of the National Academy of Sciences of the United States of America, 107*(34), 14950–14957. https://doi.org/10.1073/pnas.1005940107

Murakami, M., Ohba, T., Wu, T.-W., Fujisawa, S., Suzuki, T., Takahashi, Y., et al. (2007). Modified sympathetic regulation in N-type calcium channel null-mouse. *Biochemical and Biophysical Research Communications, 354*(4), 1016–1020. https://doi.org/10.1016/j.bbrc.2007.01.087

Nakagawasai, O., Onogi, H., Mitazaki, S., Sato, A., Watanabe, K., Saito, H., et al. (2010). Behavioral and neurochemical characterization of mice deficient in the N-type Ca²⁺ channel α$_{1B}$ subunit. *Behavioural Brain Research, 208*(1), 224–230. https://doi.org/10.1016/j.bbr.2009.11.042

Nakayama, H., Bodi, I., Correll, R. N., Chen, X., Lorenz, J., Houser, S. R., et al. (2009). α1G-dependent T-type Ca²⁺ current antagonizes cardiac hypertrophy through a NOS3-dependent mechanism in mice. *The Journal of Clinical Investigation, 119*(12), 3787–3796. https://doi.org/10.1172/JCI39724

Namkung, Y., Skrypnyk, N., Jeong, M.-J., Lee, T., Lee, M.-S., Kim, H.-L., et al. (2001). Requirement for the L-type Ca²⁺ channel α$_{1D}$ subunit in postnatal pancreatic β cell generation. *The Journal of Clinical Investigation, 108*(7), 1015–1022. https://doi.org/10.1172/JCI13310

Nanou, E., & Catterall, W. A. (2018). Calcium channels, synaptic plasticity, and neuropsychiatric disease. *Neuron, 98*(3), 466–481. https://doi.org/10.1016/j.neuron.2018.03.017

Nanou, E., Scheuer, T., & Catterall, W. A. (2016a). Calcium sensor regulation of the Ca$_V$2.1 Ca²⁺ channel contributes to long-term potentiation and spatial learning. *Proceedings of the National Academy of Sciences of the United States of America, 113*(46), 13209–13214. https://doi.org/10.1073/pnas.1616206113

Nanou, E., Yan, J., Whitehead, N. P., Kim, M. J., Froehner, S. C., Scheuer, T., et al. (2016b). Altered short-term synaptic plasticity and reduced muscle strength in mice with impaired regulation of presynaptic $Ca_V2.1$ Ca^{2+} channels. *Proceedings of the National Academy of Sciences of the United States of America, 113*(4), 1068–1073. https://doi.org/10.1073/pnas.1524650113

Nelson, M. T., Woo, J., Kang, H.-W., Vitko, I., Barrett, P. Q., Perez-Reyes, E., et al. (2007). Reducing agents sensitize C-type nociceptors by relieving high-affinity zinc inhibition of T-type calcium channels. *The Journal of Neuroscience, 27*(31), 8250–8260. https://doi.org/10.1523/JNEUROSCI.1800-07.2007

Newton, P. M., Orr, C. J., Wallace, M. J., Kim, C., Shin, H.-S., & Messing, R. O. (2004). Deletion of N-type calcium channels alters ethanol reward and reduces ethanol consumption in mice. *The Journal of Neuroscience, 24*(44), 9862–9869. https://doi.org/10.1523/JNEUROSCI.3446-04.2004

Nishida, M., Ishikawa, T., Saiki, S., Sunggip, C., Aritomi, S., Harada, E., et al. (2013). Voltage-dependent N-type Ca^{2+} channels in endothelial cells contribute to oxidative stress-related endothelial dysfunction induced by angiotensin II in mice. *Biochemical and Biophysical Research Communications, 434*(2), 210–216. https://doi.org/10.1016/j.bbrc.2013.03.040

Nystoriak, M. A., Nieves-Cintrón, M., Patriarchi, T., Buonarati, O. R., Prada, M. P., Morotti, S., et al. (2017). Ser1928 phosphorylation by PKA stimulates the L-type Ca^{2+} channel $Ca_V1.2$ and vasoconstriction during acute hyperglycemia and diabetes. *Science Signaling*. https://doi.org/10.1126/scisignal.aaf9647

Ohn, T.-L., Rutherford, M. A., Jing, Z., Jung, S., Duque-Afonso, C. J., Hoch, G., et al. (2016). Hair cells use active zones with different voltage dependence of Ca^{2+} influx to decompose sounds into complementary neural codes. *Proceedings of the National Academy of Sciences of the United States of America, 113*(32), E4716–E4725. https://doi.org/10.1073/pnas.1605737113

Ohno, S., Yokoi, H., Mori, K., Kasahara, M., Kuwahara, K., Fujikura, J., et al. (2016). Ablation of the N-type calcium channel ameliorates diabetic nephropathy with improved glycemic control and reduced blood pressure. *Scientific Reports, 6*(1), 27192. https://doi.org/10.1038/srep27192

Olson, P. A., Tkatch, T., Hernandez-Lopez, S., Ulrich, S., Ilijic, E., Mugnaini, E., et al. (2005). G-protein-coupled receptor modulation of striatal $Ca_V1.3$ L-type Ca^{2+} channels is dependent on a shank-binding domain. *The Journal of Neuroscience, 25*(5), 1050–1062. https://doi.org/10.1523/JNEUROSCI.3327-04.2005

Omilusik, K., Priatel, J. J., Chen, X., Wang, Y. T., Xu, H., Choi, K. B., et al. (2011). The $Ca_V1.4$ calcium channel is a critical regulator of T cell receptor signaling and naive T cell homeostasis. *Immunity, 35*(3), 349–360. https://doi.org/10.1016/j.immuni.2011.07.011

Omilusik, K., Nohara, L., Stanwood, S., & Jefferies, W. (2013). Weft, warp, and weave: The intricate tapestry of calcium channels regulating T lymphocyte function. *Frontiers in Immunology*, 4164. https://doi.org/10.3389/fimmu.2013.00164

Orestes, P., Bojadzic, D., Chow, R. M., & Todorovic, S. M. (2009). Mechanisms and functional significance of inhibition of neuronal T-type calcium channels by isoflurane. *Molecular Pharmacology, 75*(3), 542–554. https://doi.org/10.1124/mol.108.051664

Ortner, N. J., Bock, G., Vandael, D. H. F., Mauersberger, R., Draheim, H. J., Gust, R., et al. (2014). Pyrimidine-2,4,6-triones are a new class of voltage-gated L-type Ca^{2+} channel activators. *Nature Communications, 5*(1), 3897. https://doi.org/10.1038/ncomms4897

Ortner, N. J., Bock, G., Dougalis, A., Kharitonova, M., Duda, J., Hess, S., et al. (2017). Lower affinity of isradipine for L-type Ca^{2+} channels during substantia nigra dopamine neuron-like activity: Implications for neuroprotection in Parkinson's disease. *The Journal of Neuroscience, 37*(28), 6761–6777. https://doi.org/10.1523/JNEUROSCI.2946-16.2017

Ortner, N. J., Kaserer, T., Copeland, J. N., & Striessnig, J. (2020). De novo CACAN1D Ca^{2+} channelopathies: Clinical phenotypes and molecular mechanism. *Pflügers Archiv – European Journal of Physiology, 472*(7), 755–773. https://doi.org/10.1007/s00424-020-02418-w

Osanai, M., Saegusa, H., Kazuno, A., Nagayama, S., Hu, Q., Zong, S., et al. (2006). Altered cerebellar function in mice lacking $Ca_V2.3$ Ca^{2+} channel. *Biochemical and Biophysical Research Communications, 344*(3), 920–925. https://doi.org/10.1016/j.bbrc.2006.03.206

Panagiotakos, G., Haveles, C., Arjun, A., Petrova, R., Rana, A., Portmann, T., et al. (2019). Aberrant calcium channel splicing drives defects in cortical differentiation in Timothy syndrome. *eLife*, 8e51037. https://doi.org/10.7554/eLife.51037

Parajuli, L. K., Nakajima, C., Kulik, A., Matsui, K., Schneider, T., Shigemoto, R., et al. (2012). Quantitative regional and ultrastructural localization of the $Ca_V2.3$ subunit of R-type calcium channel in mouse brain. *The Journal of Neuroscience, 32*(39), 13555–13567. https://doi.org/10.1523/JNEUROSCI.1142-12.2012

Park, C., Kim, J.-H., Yoon, B.-E., Choi, E.-J., Lee, C. J., & Shin, H.-S. (2010a). T-type channels control the opioidergic descending analgesia at the low threshold-spiking GABAergic neurons in the periaqueductal gray. *Proceedings of the National Academy of Sciences of the United States of America, 107*(33), 14857–14862. https://doi.org/10.1073/pnas.1009532107

Park, C. Y., Shcheglovitov, A., & Dolmetsch, R. (2010b). The CRAC channel activator STIM1 binds and inhibits L-type voltage-gated calcium channels. *Science, 330*(6000), 101–105. https://doi.org/10.1126/science.1191027

Park, Y.-G., Park, H.-Y., Lee, C. J., Choi, S., Jo, S., Choi, H., et al. (2010c). $Ca_V3.1$ is a tremor rhythm pacemaker in the inferior olive. *Proceedings of the National Academy of Sciences of the United States* of America, *107*(23), 10731–10736. https://doi.org/10.1073/pnas.1002995107

Paşca, S. P., Portmann, T., Voineagu, I., Yazawa, M., Shcheglovitov, A., Paşca, A. M., et al. (2011). Using iPSC-derived neurons to uncover cellular phenotypes associated with Timothy syndrome. *Nature Medicine, 17*(12), 1657–1662. https://doi.org/10.1038/nm.2576

Patriarchi, T., Qian, H., Di Biase, V., Malik, Z. A., Chowdhury, D., Price, J. L., et al. (2016). Phosphorylation of Ca$_V$1.2 on S1928 uncouples the L-type Ca2+ channel from the β2 adrenergic receptor. *The EMBO Journal, 35*(12), 1330–1345. https://doi.org/10.15252/embj.201593409

Pellegrini, C., Lecci, S., Lüthi, A., & Astori, S. (2016). Suppression of sleep spindle rhythmogenesis in mice with deletion of Ca$_V$3.2 and Ca$_V$3.3 T-type Ca²⁺ channels. *Sleep, 39*(4), 875–885. https://doi.org/10.5665/sleep.5646

Pereverzev, A., Mikhna, M., Vajna, R., Gissel, C., Henry, M., Weiergräber, M., et al. (2002). Disturbances in glucose-tolerance, insulin-release, and stress-induced hyperglycemia upon disruption of the Ca$_V$2.3 (α$_{1E}$) subunit of voltage-gated Ca²⁺ channels. *Molecular Endocrinology, 16*(4), 884–895. https://doi.org/10.1210/mend.16.4.0801

Petrenko, A. B., Tsujita, M., Kohno, T., Sakimura, K., & Baba, H. (2007). Mutation of α$_{1G}$ T-type calcium channels in mice does not change anesthetic requirements for loss of the righting reflex and minimum alveolar concentration but delays the onset of anesthetic induction. *Anesthesiology, 106*(6), 1177–1185. https://doi.org/10.1097/01.anes.0000267601.09764.e6

Picard, E., Carvalho, F. A., Agosti, F., Bourinet, E., Ardid, D., Eschalier, A., et al. (2019). Inhibition of Ca$_V$3.2 calcium channels: A new target for colonic hypersensitivity associated with low-grade inflammation. *British Journal of Pharmacology, 176*(7), 950–963. https://doi.org/10.1111/bph.14608

Pitman, K. A., Ricci, R., Gasperini, R., Beasley, S., Pavez, M., Charlesworth, J., et al. (2020). The voltage-gated calcium channel Ca$_V$1.2 promotes adult oligodendrocyte progenitor cell survival in the mouse corpus callosum but not motor cortex. *Glia, 68*(2), 376–392. https://doi.org/10.1002/glia.23723

Platzer, J., Engel, J., Schrott-Fischer, A., Stephan, K., Bova, S., Chen, H., et al. (2000). Congenital deafness and sinoatrial node dysfunction in mice lacking class D L-type Ca²⁺ channels. *Cell, 102*(1), 89–97. https://doi.org/10.1016/S0092-8674(00)00013-1

Poetschke, C., Dragicevic, E., Duda, J., Benkert, J., Dougalis, A., DeZio, R., et al. (2015). Compensatory T-type Ca²⁺ channel activity alters D2-autoreceptor responses of Substantia nigra dopamine neurons from Ca$_V$1.3 L-type Ca²⁺ channel KO mice. *Scientific Reports, 5*(1), 13688. https://doi.org/10.1038/srep13688

Poomvanicha, M., Wegener, J. W., Blaich, A., Fischer, S., Domes, K., Moosmang, S., et al. (2011). Facilitation and Ca²⁺-dependent inactivation are modified by mutation of the Ca$_V$1.2 channel IQ motif. *The Journal of*

Biological Chemistry, 286(30), 26702–26707. https://doi.org/10.1074/jbc.M111.247841

Poomvanicha, M., Matthes, J., Domes, K., Patrucco, E., Angermeier, E., Laugwitz, K.-L., et al. (2017). Beta-adrenergic regulation of the heart expressing the Ser1700A/Thr1704A mutated Ca$_V$1.2 channel. *Journal of Molecular and Cellular Cardiology*, 11110–11116. https://doi.org/10.1016/j.yjmcc.2017.07.119

Powell, J. A. (1990). Muscular dysgenesis: A model system for studying skeletal muscle development. *The FASEB Journal, 4*(10), 2798–2808. https://doi.org/10.1096/fasebj.4.10.2197156

Pudukulatham, Z., Zhang, F.-X., Gadotti, V. M., M'Dahoma, S., Swami, P., Tamboli, Y., et al. (2016). Synthesis and characterization of a disubstituted piperazine derivative with T-type channel blocking action and analgesic properties. *Molecular Pain*, 121744806916641678. https://doi.org/10.1177/1744806916641678

Qian, H., Patriarchi, T., Price, J. L., Matt, L., Lee, B., Nieves-Cintrón, M., et al. (2017). Phosphorylation of Ser1928 mediates the enhanced activity of the L-type Ca²⁺ channel Ca$_V$1.2 by the β2-adrenergic receptor in neurons. *Science Signaling, 10*(463), eaaf9659. https://doi.org/10.1126/scisignal.aaf9659

Rajakulendran, S., Kaski, D., & Hanna, M. G. (2012). Neuronal P/Q-type calcium channel dysfunction in inherited disorders of the CNS. *Nature Reviews. Neurology, 8*(2), 86–96. https://doi.org/10.1038/nrneurol.2011.228

Ramachandran, K. V., Hennessey, J. A., Barnett, A. S., Yin, X., Stadt, H. A., Foster, E., et al. (2013). Calcium influx through L-type Ca$_V$1.2 Ca²⁺ channels regulates mandibular development. *The Journal of Clinical Investigation, 123*(4), 1638–1646. https://doi.org/10.1172/JCI66903

Rebellato, P., Kaczynska, D., Kanatani, S., Rayyes, I. A., Zhang, S., Villaescusa, C., et al. (2019). The T-type Ca²⁺ channel Ca$_V$3.2 regulates differentiation of neural progenitor cells during cortical development via Caspase-3. *Neuroscience, 402*, 78–89. https://doi.org/10.1016/j.neuroscience.2019.01.015

Redecker, T. M., Kisko, T. M., Schwarting, R. K. W., & Wöhr, M. (2019). Effects of *Cacna1c* haploinsufficiency on social interaction behavior and 50-kHz ultrasonic vocalizations in adult female rats. *Behavioural Brain Research, 367*, 35–52. https://doi.org/10.1016/j.bbr.2019.03.032

Rosati, B., Yan, Q., Lee, M. S., Liou, S.-R., Ingalls, B., Foell, J., et al. (2011). Robust L-type calcium current expression following heterozygous knockout of the Ca$_V$1.2 gene in adult mouse heart. *The Journal of Physiology, 589*(13), 3275–3288. https://doi.org/10.1113/jphysiol.2011.210237

Rose, S. J., Kriener, L. H., Heinzer, A. K., Fan, X., Raike, R. S., van den Maagdenberg, A. M. J. M., et al. (2014). The first knockin mouse model of episodic ataxia type 2. *Experimental Neurology*, 261553–261562. https://doi.org/10.1016/j.expneurol.2014.08.001

Rossignol, E., Kruglikov, I., van den Maagdenberg, A. M. J. M., Rudy, B., & Fishell, G. (2013). Ca$_V$2.1 ablation in cortical interneurons selectively impairs fast-spiking basket cells and causes generalized seizures. *Annals of Neurology, 74*(2), 209–222. https://doi.org/10.1002/ana.23913

Saegusa, H., Kurihara, T., Zong, S., Minowa, O., Kazuno, A., Han, W., et al. (2000). Altered pain responses in mice lacking α$_{1E}$ subunit of the voltage-dependent Ca^{2+} channel. *Proceedings of the National Academy of Sciences of the United States of America, 97*(11), 6132–6137. https://doi.org/10.1073/pnas.100124197

Saegusa, H., Kurihara, T., Zong, S., Kazuno, A., Matsuda, Y., Nonaka, T., et al. (2001). Suppression of inflammatory and neuropathic pain symptoms in mice lacking the N-type Ca^{2+} channel. *The EMBO Journal, 20*(10), 2349–2356. https://doi.org/10.1093/emboj/20.10.2349

Saegusa, H., Wakamori, M., Matsuda, Y., Wang, J., Mori, Y., Zong, S., et al. (2007). Properties of human Ca$_V$2.1 channel with a spinocerebellar ataxia type 6 mutation expressed in Purkinje cells. *Molecular and Cellular Neurosciences, 34*(2), 261–270. https://doi.org/10.1016/j.mcn.2006.11.006

Sahu, G., Asmara, H., Zhang, F.-X., Zamponi, G. W., & Turner, R. W. (2017). Activity-dependent facilitation of Ca$_V$1.3 calcium channels promotes KCa3.1 activation in hippocampal neurons. *The Journal of Neuroscience, 37*(46), 11255–11270. https://doi.org/10.1523/JNEUROSCI.0967-17.2017

Sakata, Y., Saegusa, H., Zong, S., Osanai, M., Murakoshi, T., Shimizu, Y., et al. (2001). Analysis of Ca^{2+} currents in spermatocytes from mice lacking Ca$_V$2.3 (α$_{1E}$) Ca^{2+} channel. *Biochemical and Biophysical Research Communications, 288*(4), 1032–1036. https://doi.org/10.1006/bbrc.2001.5871

Sakata, Y., Saegusa, H., Zong, S., Osanai, M., Murakoshi, T., Shimizu, Y., et al. (2002). Ca$_V$2.3 (α$_{1E}$) Ca^{2+} channel participates in the control of sperm function. *FEBS Letters, 516*(1), 229–233. https://doi.org/10.1016/S0014-5793(02)02529-2

Sakkaki, S., Gangarossa, G., Lerat, B., Françon, D., Forichon, L., Chemin, J., et al. (2016). Blockade of T-type calcium channels prevents tonic-clonic seizures in a maximal electroshock seizure model. *Neuropharmacology, 101*, 320–329. https://doi.org/10.1016/j.neuropharm.2015.09.032

Santiago González, D. A., Cheli, V. T., Zamora, N. N., Lama, T. N., Spreuer, V., Murphy, G. G., et al. (2017). Conditional deletion of the L-type calcium channel Ca$_V$1.2 in NG2-positive cells impairs remyelination in mice. *The Journal of Neuroscience, 37*(42), 10038–10051. https://doi.org/10.1523/JNEUROSCI.1787-17.2017

Satheesh, S. V., Kunert, K., Rüttiger, L., Zuccotti, A., Schönig, K., Friauf, E., et al. (2012). Retrocochlear function of the peripheral deafness gene Cacna1d. *Human Molecular Genetics, 21*(17), 3896–3909. https://doi.org/10.1093/hmg/dds217

Scanzi, J., Accarie, A., Muller, E., Pereira, B., Aissouni, Y., Goutte, M., et al. (2016). Colonic overexpression of the T-type calcium channel Ca$_V$3.2 in a mouse model of visceral hypersensitivity and in irritable bowel syndrome patients. *Neurogastroenterology and Motility, 28*(11), 1632–1640. https://doi.org/10.1111/nmo.12860

Schafe, G. E. (2008). Rethinking the role of L-type voltage-gated calcium channels in fear memory extinction. *Learning & Memory, 15*(5), 324–325. https://doi.org/10.1101/lm.996908

Schampel, A., Volovitch, O., Koeniger, T., Scholz, C.-J., Jörg, S., Linker, R. A., et al. (2017). Nimodipine fosters remyelination in a mouse model of multiple sclerosis and induces microglia-specific apoptosis. *Proceedings of the National Academy of Sciences of the United States of America, 114*(16), E3295–E3304. https://doi.org/10.1073/pnas.1620052114

Scharinger, A., Eckrich, S., Vandael, D. H., Schönig, K., Koschak, A., Hecker, D., et al. (2015). Cell-type-specific tuning of Ca$_V$1.3 Ca^{2+}-channels by a C-terminal automodulatory domain. *Frontiers in Cellular Neuroscience, 9309*. https://doi.org/10.3389/fncel.2015.00309

Schierberl, K., Hao, J., Tropea, T. F., Ra, S., Giordano, T. P., Xu, Q., et al. (2011). Ca$_V$1.2 L-type Ca^{2+} channels mediate cocaine-induced GluA1 trafficking in the nucleus accumbens, a long-term adaptation dependent on ventral tegmental area Ca$_V$1.3 channels. *The Journal of Neuroscience, 31*(38), 13562–13575. https://doi.org/10.1523/JNEUROSCI.2315-11.2011

Schnütgen, F., Doerflinger, N., Calléja, C., Wendling, O., Chambon, P., & Ghyselinck, N. B. (2003). A directional strategy for monitoring Cre-mediated recombination at the cellular level in the mouse. *Nature Biotechnology, 21*(5), 562–565. https://doi.org/10.1038/nbt811

Schrauwen, I., Helfmann, S., Inagaki, A., Predoehl, F., Tabatabaiefar, M. A., Picher, M. M., et al. (2012). A mutation in CABP2, expressed in cochlear hair cells, causes autosomal-recessive hearing impairment. *American Journal of Human Genetics, 91*(4), 636–645. https://doi.org/10.1016/j.ajhg.2012.08.018

Schredelseker, J., Shrivastav, M., Dayal, A., & Grabner, M. (2010). Non–Ca^{2+}-conducting Ca^{2+} channels in fish skeletal muscle excitation-contraction coupling. *Proceedings of the National Academy of Sciences of the United States of America, 107*(12), 5658–5663. https://doi.org/10.1073/pnas.0912153107

Schulla, V., Renström, E., Feil, R., Feil, S., Franklin, I., Gjinovci, A., et al. (2003). Impaired insulin secretion and glucose tolerance in β cell-selective Ca$_V$1.2 Ca^{2+} channel null mice. *The EMBO Journal, 22*(15), 3844–3854. https://doi.org/10.1093/emboj/cdg389

Seidel, E., Schewe, J., Zhang, J., Dinh, H. A., Forslund, S. K., Markó, L., et al. (2021). Enhanced Ca^{2+} signaling, mild primary aldosteronism, and hypertension in a familial hyperaldosteronism mouse model (Cacna1hM1560V/+). *Proceedings of the National Academy of Sciences of the United States of America,*

118(17), e2014876118. https://doi.org/10.1073/pnas.2014876118

Seisenberger, C., Specht, V., Welling, A., Platzer, J., Pfeifer, A., Kühbandner, S., et al. (2000). Functional embryonic cardiomyocytes after disruption of the L-type α_{1C} (Ca$_V$1.2) calcium channel gene in the mouse. *The Journal of Biological Chemistry, 275*(50), 39193–39199. https://doi.org/10.1074/jbc.M006467200

Shao, Y., Alicknavitch, M., & Farach-Carson, M. C. (2005). Expression of voltage sensitive calcium channel (VSCC) L-type Ca$_V$1.2 (α_{1C}) and T-type Ca$_V$3.2 (α_{1H}) subunits during mouse bone development. *Developmental Dynamics, 234*(1), 54–62. https://doi.org/10.1002/dvdy.20517

Shi, L., Chang, J. Y.-A., Yu, F., Ko, M. L., & Ko, G. Y.-P. (2017). The contribution of L-type Ca$_V$1.3 channels to retinal light responses. *Frontiers in Molecular Neuroscience, 10*394. https://doi.org/10.3389/fnmol.2017.00394

Singh, A., Hamedinger, D., Hoda, J.-C., Gebhart, M., Koschak, A., Romanin, C., et al. (2006). C-terminal modulator controls Ca²⁺-dependent gating of Ca$_V$1.4 L-type Ca²⁺ channels. *Nature Neuroscience, 9*(9), 1108–1116. https://doi.org/10.1038/nn1751

Singh, A., Gebhart, M., Fritsch, R., Sinnegger-Brauns, M. J., Poggiani, C., Hoda, J.-C., et al. (2008). Modulation of voltage- and Ca²⁺-dependent gating of Ca$_V$1.3 L-type calcium channels by alternative splicing of a C-terminal regulatory domain. *The Journal of Biological Chemistry, 283*(30), 20733–20744. https://doi.org/10.1074/jbc.M802254200

Sinnegger-Brauns, M. J., Hetzenauer, A., Huber, I. G., Renström, E., Wietzorrek, G., Berjukov, S., et al. (2004). Isoform-specific regulation of mood behavior and pancreatic β cell and cardiovascular function by L-type Ca²⁺ channels. *The Journal of Clinical Investigation, 113*(10), 1430–1439. https://doi.org/10.1172/JCI20208

Sinnegger-Brauns, M. J., Huber, I. G., Koschak, A., Wild, C., Obermair, G. J., Einzinger, U., et al. (2009). Expression and 1,4-dihydropyridine-binding properties of brain L-type calcium channel isoforms. *Molecular Pharmacology, 75*(2), 407–414. https://doi.org/10.1124/mol.108.049981

Siwek, M. E., Müller, R., Henseler, C., Broich, K., Papazoglou, A., & Weiergräber, M. (2014). The Ca$_V$2.3 R-type voltage-gated Ca²⁺ channel in mouse sleep architecture. *Sleep, 37*(5), 881–892. https://doi.org/10.5665/sleep.3652

Sochivko, D., Pereverzev, A., Smyth, N., Gissel, C., Schneider, T., & Beck, H. (2002). The Ca$_V$2.3 Ca²⁺ channel subunit contributes to R-type Ca²⁺ currents in murine hippocampal and neocortical neurones. *The Journal of Physiology, 542*(3), 699–710. https://doi.org/10.1113/jphysiol.2002.020677

Song, I., Kim, D., Choi, S., Sun, M., Kim, Y., & Shin, H.-S. (2004). Role of the α_{1G} T-type calcium channel in spontaneous absence seizures in mutant mice. *The Journal of Neuroscience, 24*(22), 5249–5257. https://doi.org/10.1523/JNEUROSCI.5546-03.2004

Sridharan, P. S., Lu, Y., Rice, R. C., Pieper, A. A., & Rajadhyaksha, A. M. (2020). Loss of Ca$_V$1.2 channels impairs hippocampal theta burst stimulation-induced long-term potentiation. *Channels, 14*(1), 287–293. https://doi.org/10.1080/19336950.2020.1807851

Srinivasan, P. P., Parajuli, A., Price, C., Wang, L., Duncan, R. L., & Kirn-Safran, C. B. (2015). Inhibition of T-type voltage sensitive calcium channel reduces load-induced OA in mice and suppresses the catabolic effect of bone mechanical stress on chondrocytes. *PLoS ONE, 10*(5), e0127290. https://doi.org/10.1371/journal.pone.0127290

Srivastava, U., Aromolaran, A. S., Fabris, F., Lazaro, D., Kassotis, J., Qu, Y., et al. (2017). Novel function of α_{1D} L-type calcium channel in the atria. *Biochemical and Biophysical Research Communications, 482*(4), 771–776. https://doi.org/10.1016/j.bbrc.2016.11.109

Srivastava, U., Jennings-Charles, R., Qu, Y. S., Sossalla, S., Chahine, M., & Boutjdir, M. (2020). Novel re-expression of L-type calcium channel Cav1.3 in left ventricles of failing human heart. *Heart Rhythm, 17*(7), 1193–1197. https://doi.org/10.1016/j.hrthm.2020.02.025

Stamboulian, S., Kim, D., Shin, H.-S., Ronjat, M., De Waard, M., & Arnoult, C. (2004). Biophysical and pharmacological characterization of spermatogenic T-type calcium current in mice lacking the Ca$_V$3.1 (α_{1G}) calcium channel: Ca$_V$3.2 (α_{1H}) is the main functional calcium channel in wild-type spermatogenic cells. *Journal of Cellular Physiology, 200*(1), 116–124. https://doi.org/10.1002/jcp.10480

Striessnig, J. (2021). Voltage-gated Ca²⁺-channel α_1-subunit de novo Missense mutations: Gain or loss of function – implications for potential therapies. *Frontiers in Synaptic Neuroscience, 13*634760. https://doi.org/10.3389/fnsyn.2021.634760

Sultana, N., Dienes, B., Benedetti, A., Tuluc, P., Szentesi, P., Sztretye, M., et al. (2016). Restricting calcium currents is required for correct fiber type specification in skeletal muscle. *Development, 143*(9), 1547–1559. https://doi.org/10.1242/dev.129676

Surmeier, D. J., Guzman, J. N., Sanchez-Padilla, J., & Goldberg, J. A. (2011). The origins of oxidant stress in Parkinson's disease and therapeutic strategies. *Antioxidants & Redox Signaling, 14*(7), 1289–1301. https://doi.org/10.1089/ars.2010.3521

Svenningsen, P., & Hansen, P. B. L. (2016). The genetic background affects the vascular response in T-type calcium channels 3.2 deficient mice. *Acta Physiologica (Oxford, England), 217*(2), 101–102. https://doi.org/10.1111/apha.12655

Svenningsen, P., Andersen, K., Thuesen, A. D., Shin, H.-S., Vanhoutte, P. M., Skøtt, O., et al. (2014). T-type Ca²⁺ channels facilitate NO-formation, vasodilatation and NO-mediated modulation of blood pressure. *Pflügers Archiv – European Journal of Physiology, 466*(12), 2205–2214. https://doi.org/10.1007/s00424-014-1492-4

Sykes, L., Clifton, N. E., Hall, J., & Thomas, K. L. (2018). Regulation of the expression of the psychiatric risk gene Cacna1c during associative learning. *Molecular Neuropsychiatry, 4*(3), 149–157. https://doi.org/10.1159/000493917

Sykes, L., Haddon, J., Lancaster, T. M., Sykes, A., Azzouni, K., Ihssen, N., et al. (2019). Genetic variation in the psychiatric risk gene *CACNA1C* modulates reversal learning across species. *Schizophrenia Bulletin, 45*(5), 1024–1032. https://doi.org/10.1093/schbul/sby146

Takei, T., Saegusa, H., Zong, S., Murakoshi, T., Makita, K., & Tanabe, T. (2003). Anesthetic sensitivities to propofol and halothane in mice lacking the R-type (Ca$_V$2.3) Ca^{2+} channel. *Anesthesia and Analgesia, 97*(1), 96–103. https://doi.org/10.1213/01.ANE.0000065548.83253.5C

Takemura, H., Yasui, K., Opthof, T., Niwa, N., Horiba, M., Shimizu, A., et al. (2005). Subtype switching of L-type Ca^{2+} channel from Ca$_V$1.3 to Ca$_V$1.2 in embryonic murine ventricle. *Circulation Journal: Official Journal of Japanese Circulation Society, 69*(11), 1405–1411. https://doi.org/10.1253/circj.69.1405

Talley, E. M., Cribbs, L. L., Lee, J.-H., Daud, A., Perez-Reyes, E., & Bayliss, D. A. (1999). Differential distribution of three members of a gene family encoding low voltage-activated (T-type) calcium channels. *The Journal of Neuroscience, 19*(6), 1895–1911. https://doi.org/10.1523/JNEUROSCI.19-06-01895.1999

Tamura, K., Stecher, G., & Kumar, S. (2021). MEGA11: Molecular evolutionary genetics analysis version 11. *Molecular Biology and Evolution, 38*(7), 3022–3027. https://doi.org/10.1093/molbev/msab120

Tan, B. Z., Jiang, F., Tan, M. Y., Yu, D., Huang, H., Shen, Y., et al. (2011). Functional characterization of alternative splicing in the C terminus of L-type Ca$_V$1.3 channels. *The Journal of Biological Chemistry, 286*(49), 42725–42735. https://doi.org/10.1074/jbc.M111.265207

Tanabe, T., Beam, K. G., Powell, J. A., & Numa, S. (1988). Restoration of excitation-contraction coupling and slow calcium current in dysgenic muscle by dihydropyridine receptor complementary DNA. *Nature, 336*(6195), 134–139. https://doi.org/10.1038/336134a0

Tanabe, T., Adams, B. A., Numa, S., & Beam, K. G. (1991). Repeat I of the dihydropyridine receptor is critical in determining calcium channel activation kinetics. *Nature, 352*(6338), 800–803. https://doi.org/10.1038/352800a0

Tatsuki, F., Sunagawa, G. A., Shi, S., Susaki, E. A., Yukinaga, H., Perrin, D., et al. (2016). Involvement of Ca^{2+}-dependent hyperpolarization in sleep duration in mammals. *Neuron, 90*(1), 70–85. https://doi.org/10.1016/j.neuron.2016.02.032

Temme, S. J., & Murphy, G. G. (2017). The L-type voltage-gated calcium channel Ca$_V$1.2 mediates fear extinction and modulates synaptic tone in the lateral amygdala. *Learning & Memory, 24*(11), 580–588. https://doi.org/10.1101/lm.045773.117

Temme, S. J., Bell, R. Z., Fisher, G. L., & Murphy, G. G. (2016). Deletion of the mouse homolog of *CACNA1C* disrupts discrete forms of hippocampal-dependent memory and neurogenesis within the dentate gyrus. *eNeuro, 3*(6), ENEURO.0118-16.2016. https://doi.org/10.1523/ENEURO.0118-16.2016

Thuesen, A. D., Andersen, H., Cardel, M., Toft, A., Walter, S., Marcussen, N., et al. (2014). Differential effect of T-type voltage-gated Ca^{2+} channel disruption on renal plasma flow and glomerular filtration rate in vivo. *American Journal of Physiology. Renal Physiology, 307*(4), F445–F452. https://doi.org/10.1152/ajprenal.00016.2014

Thuesen, A. D., Andersen, K., Lyngsø, K. S., Burton, M., Brasch-Andersen, C., Vanhoutte, P. M., et al. (2018). Deletion of T-type calcium channels Ca$_V$3.1 or Ca$_V$3.2 attenuates endothelial dysfunction in aging mice. *Pflügers Archiv – European Journal of Physiology, 470*(2), 355–365. https://doi.org/10.1007/s00424-017-2068-x

Timic Stamenic, T., Feseha, S., Valdez, R., Zhao, W., Klawitter, J., & Todorovic, S. M. (2019). Alterations in oscillatory behavior of central medial thalamic neurons demonstrate a key role of Ca$_V$3.1 isoform of T-channels during isoflurane-induced anesthesia. *Cerebral Cortex, 29*(11), 4679–4696. https://doi.org/10.1093/cercor/bhz002

Tippens, A. L., Pare, J.-F., Langwieser, N., Moosmang, S., Milner, T. A., Smith, Y., et al. (2008). Ultrastructural evidence for pre- and postsynaptic localization of Ca$_V$1.2 L-type Ca^{2+} channels in the rat hippocampus. *Journal of Comparative Neurology, 506*(4), 569–583. https://doi.org/10.1002/cne.21567

To, K. H. T., Gui, P., Li, M., Zawieja, S. D., Castorena-Gonzalez, J. A., & Davis, M. J. (2020). T-type, but not L-type, voltage-gated calcium channels are dispensable for lymphatic pacemaking and spontaneous contractions. *Scientific Reports, 10*(1), 70. https://doi.org/10.1038/s41598-019-56953-3

Todorov, B., van de Ven, R. C. G., Kaja, S., Broos, L. A. M., Verbeek, S. J., Plomp, J. J., et al. (2006). Conditional inactivation of the *Cacna1a* gene in transgenic mice. *Genesis, 44*(12), 589–594. https://doi.org/10.1002/dvg.20255

Toriyama, H., Wang, L., Saegusa, H., Zong, S., Osanai, M., Murakoshi, T., et al. (2002). Role of Ca$_V$2.3 (α_{1E}) Ca^{2+} channel in ischemic neuronal injury. *NeuroReport, 13*(2), 261–265.

Torrente, A. G., Mesirca, P., Bidaud, I., & Mangoni, M. E. (2020). Channelopathies of voltage-gated L-type Ca$_V$1.3/α_{1D} and T-type Ca$_V$3.1/α_{1G} Ca^{2+} channels in dysfunction of heart automaticity. *Pflügers Archiv – European Journal of Physiology, 472*(7), 817–830. https://doi.org/10.1007/s00424-020-02421-1

Tottene, A., Conti, R., Fabbro, A., Vecchia, D., Shapovalova, M., Santello, M., et al. (2009). Enhanced excitatory transmission at cortical synapses as the basis for facilitated spreading depression in CaV2.1 knockin migraine mice. *Neuron, 61*(5), 762–773. https://doi.org/10.1016/j.neuron.2009.01.027

Toyoda, F., Mesirca, P., Dubel, S., Ding, W.-G., Striessnig, J., Mangoni, M. E., et al. (2017). Ca$_V$1.3 L-type Ca²⁺ channel contributes to the heartbeat by generating a dihydropyridine-sensitive persistent Na⁺ current. *Scientific Reports, 7*(1), 7869. https://doi.org/10.1038/s41598-017-08191-8

Tscherter, A., David, F., Ivanova, T., Deleuze, C., Renger, J. J., Uebele, V. N., et al. (2011). Minimal alterations in T-type calcium channel gating markedly modify physiological firing dynamics. *The Journal of Physiology, 589*(7), 1707–1724. https://doi.org/10.1113/jphysiol.2010.203836

Tuluc, P., Theiner, T., Jacobo-Piqueras, N., Geisler, S. M. (2004) Role of high voltage-gated Ca2+ channel subunits in pancreatic β-Cell insulin release. From structure to function. Cells. 2021 Aug 6;10(8):2004. doi: 10.3390/cells10082004. PMID: 34440773; PMCID: PMC8393260.

Tuluc, P., Molenda, N., Schlick, B., Obermair, G. J., Flucher, B. E., & Jurkat-Rott, K. (2009). A Ca$_V$1.1 Ca²⁺ channel splice variant with high conductance and voltage-sensitivity alters EC coupling in developing skeletal muscle. *Biophysical Journal, 96*(1), 35–44. https://doi.org/10.1016/j.bpj.2008.09.027

Tuluc, P., Yarov-Yarovoy, V., Benedetti, B., & Flucher, B. E. (2016). Molecular interactions in the voltage sensor controlling gating properties of Ca$_V$ calcium channels. *Structure, 24*(2), 261–271. https://doi.org/10.1016/j.str.2015.11.011

Uhrig, S., Vandael, D., Marcantoni, A., Dedic, N., Bilbao, A., Vogt, M. A., et al. (2017). Differential roles for L-type calcium channel subtypes in alcohol dependence. *Neuropsychopharmacology, 42*(5), 1058–1069. https://doi.org/10.1038/npp.2016.266

Urbano, F. J., Piedras-Rentería, E. S., Jun, K., Shin, H.-S., Uchitel, O. D., & Tsien, R. W. (2003). Altered properties of quantal neurotransmitter release at endplates of mice lacking P/Q-type Ca²⁺ channels. *Proceedings of the National Academy of Sciences of the United States of America, 100*(6), 3491–3496. https://doi.org/10.1073/pnas.0437991100

van den Maagdenberg, A. M. J. M., Pietrobon, D., Pizzorusso, T., Kaja, S., Broos, L. A. M., Cesetti, T., et al. (2004). A *Cacna1a* knockin migraine mouse model with increased susceptibility to cortical spreading depression. *Neuron, 41*(5), 701–710. https://doi.org/10.1016/S0896-6273(04)00085-6

van den Maagdenberg, A. M. J. M., Pizzorusso, T., Kaja, S., Terpolilli, N., Shapovalova, M., Hoebeek, F. E., et al. (2010). High cortical spreading depression susceptibility and migraine-associated symptoms in Ca$_V$2.1 S218L mice. *Annals of Neurology, 67*(1), 85–98. https://doi.org/10.1002/ana.21815

Voisin, T., Bourinet, E., & Lory, P. (2016). Genetic alteration of the metal/redox modulation of Ca$_V$3.2 T-type calcium channel reveals its role in neuronal excitability. *The Journal of Physiology, 594*(13), 3561–3574. https://doi.org/10.1113/JP271925

Völkening, B., Schönig, K., Kronenberg, G., Bartsch, D., & Weber, T. (2017). Deletion of psychiatric risk gene *Cacna1c* impairs hippocampal neurogenesis in cell-autonomous fashion. *Glia, 65*(5), 817–827. https://doi.org/10.1002/glia.23128

Waltereit, R., Mannhardt, S., Nescholta, S., Maser-Gluth, C., & Bartsch, D. (2008). Selective and protracted effect of nifedipine on fear memory extinction correlates with induced stress response. *Learning & Memory, 15*(5), 348–356. https://doi.org/10.1101/lm.808608

Wang, R., & Lewin, G. R. (2011). The Ca$_V$3.2 T-type calcium channel regulates temporal coding in mouse mechanoreceptors. *The Journal of Physiology, 589*(9), 2229–2243. https://doi.org/10.1113/jphysiol.2010.203463

Wang, Y., Deng, X., Mancarella, S., Hendron, E., Eguchi, S., Soboloff, J., et al. (2010). The calcium store sensor, STIM1, reciprocally controls Orai and Ca$_V$1.2 channels. *Science, 330*(6000), 105–109. https://doi.org/10.1126/science.1191086

Wang, G., Bochorishvili, G., Chen, Y., Salvati, K. A., Zhang, P., Dubel, S. J., et al. (2015). Ca$_V$3.2 calcium channels control NMDA receptor-mediated transmission: A new mechanism for absence epilepsy. *Genes & Development, 29*(14), 1535–1551. https://doi.org/10.1101/gad.260869.115

Wang, H., Zhang, X., Xue, L., Xing, J., Jouvin, M.-H., Putney, J. W., et al. (2016). Low-voltage-activated Ca$_V$3.1 calcium channels shape T helper cell cytokine profiles. *Immunity, 44*(4), 782–794. https://doi.org/10.1016/j.immuni.2016.01.015

Wang, X., Saegusa, H., Huntula, S., & Tanabe, T. (2019). Blockade of microglial Ca$_V$1.2 Ca²⁺ channel exacerbates the symptoms in a Parkinson's disease model. *Scientific Reports, 9*(1), 9138. https://doi.org/10.1038/s41598-019-45681-3

Watase, K., Barrett, C. F., Miyazaki, T., Ishiguro, T., Ishikawa, K., Hu, Y., et al. (2008). Spinocerebellar ataxia type 6 knockin mice develop a progressive neuronal dysfunction with age-dependent accumulation of mutant Ca$_V$2.1 channels. *Proceedings of the National Academy of Sciences of the United States of America, 105*(33), 11987–11992. https://doi.org/10.1073/pnas.0804350105

Wegener, J. W., Schulla, V., Lee, T.-S., Koller, A., Feil, S., Feil, R., et al. (2004). An essential role of Ca$_V$1.2 L-type calcium channel for urinary bladder function. *The FASEB Journal, 18*(10), 1159–1161. https://doi.org/10.1096/fj.04-1516fje

Wegener, J. W., Schulla, V., Koller, A., Klugbauer, N., Feil, R., & Hofmann, F. (2006). Control of intestinal motility by the Ca$_V$1.2 L-type calcium channel in mice. *FASEB, 20*(8), 1260–1262. https://doi.org/10.1096/fj.05-5292fje

Weiergräber, M., Henry, M., Südkamp, M., de Vivie, E.-R., Hescheler, J., & Schneider, T. (2005). Ablation of Ca$_V$2.3/E–type voltage–gatedcalcium channel results in cardiac arrhythmiaand altered autonomic control within themurine cardiovascular system. *Basic Research in Cardiology, 100*(1), 1–13. https://doi.org/10.1007/s00395-004-0488-1

Weiergräber, M., Henry, M., Krieger, A., Kamp, M., Radhakrishnan, K., Hescheler, J., et al. (2006). Altered seizure susceptibility in mice lacking the $Ca_V2.3$ E-type Ca^{2+} channel. *Epilepsia, 47*(5), 839–850. https://doi.org/10.1111/j.1528-1167.2006.00541.x

Weiergräber, M., Henry, M., Radhakrishnan, K., Hescheler, J., & Schneider, T. (2007). Hippocampal seizure resistance and reduced neuronal excitotoxicity in mice lacking the $Ca_V2.3$ E/R-type voltage-gated calcium channel. *Journal of Neurophysiology, 97*(5), 3660–3669. https://doi.org/10.1152/jn.01193.2006

Weiergräber, M., Henry, M., Ho, M. S. P., Struck, H., Hescheler, J., & Schneider, T. (2008). Altered thalamocortical rhythmicity in $Ca_V2.3$-deficient mice. *Molecular and Cellular Neurosciences, 39*(4), 605–618. https://doi.org/10.1016/j.mcn.2008.08.007

Weiss, J., Pyrski, M., Weissgerber, P., & Zufall, F. (2014). Altered synaptic transmission at olfactory and vomeronasal nerve terminals in mice lacking N-type calcium channel $Ca_V2.2$. *The European Journal of Neuroscience, 40*(10), 3422–3435. https://doi.org/10.1111/ejn.12713

Westenbroek, R. E., Bausch, S. B., Lin, R. C. S., Franck, J. E., Noebels, J. L., & Catterall, W. A. (1998). Upregulation of L-type Ca^{2+} channels in reactive astrocytes after brain injury, hypomyelination, and ischemia. *The Journal of Neuroscience, 18*(7), 2321–2334. https://doi.org/10.1523/JNEUROSCI.18-07-02321.1998

Weyrer, C., Turecek, J., Niday, Z., Liu, P. W., Nanou, E., Catterall, W. A., et al. (2019). The role of $Ca_V2.1$ channel facilitation in synaptic facilitation. *Cell Reports, 26*(9), 2289–2297.e3. https://doi.org/10.1016/j.celrep.2019.01.114

White, J. A., McKinney, B. C., John, M. C., Powers, P. A., Kamp, T. J., & Murphy, G. G. (2008). Conditional forebrain deletion of the L-type calcium channel $Ca_V1.2$ disrupts remote spatial memories in mice. *Learning & Memory, 15*(1), 1–5. https://doi.org/10.1101/lm.773208

Wild, A. R., Sinnen, B. L., Dittmer, P. J., Kennedy, M. J., Sather, W. A., & Dell'Acqua, M. L. (2019). Synapse-to-nucleus communication through NFAT is mediated by L-type Ca^{2+} channel Ca^{2+} spike propagation to the soma. *Cell Reports, 26*(13), 3537–3550.e4. https://doi.org/10.1016/j.celrep.2019.03.005

Wildburger, N. C., Lin-Ye, A., Baird, M. A., Lei, D., & Bao, J. (2009). Neuroprotective effects of blockers for T-type calcium channels. *Molecular Neurodegeneration, 4*(1), 44. https://doi.org/10.1186/1750-1326-4-44

Wilson, S. M., Toth, P. T., Oh, S. B., Gillard, S. E., Volsen, S., Ren, D., et al. (2000). The status of voltage-dependent calcium channels in α_{1E} knock-out mice. *The Journal of Neuroscience, 20*(23), 8566–8571. https://doi.org/10.1523/JNEUROSCI.20-23-08566.2000

Wöhr, M., Willadsen, M., Kisko, T. M., Schwarting, R. K. W., & Fendt, M. (2020). Sex-dependent effects of *Cacna1c* haploinsufficiency on behavioral inhibition evoked by conspecific alarm signals in rats. *Progress in Neuro-Psychopharmacology & Biological Psychiatry, 99109849*. https://doi.org/10.1016/j.pnpbp.2019.109849

Wu, J., Marmorstein, A. D., Striessnig, J., & Peachey, N. S. (2007). Voltage-dependent calcium channel $Ca_V1.3$ subunits regulate the light peak of the electroretinogram. *Journal of Neurophysiology, 97*(5), 3731–3735. https://doi.org/10.1152/jn.00146.2007

Wu, F., Mi, W., Hernández-Ochoa, E. O., Burns, D. K., Fu, Y., Gray, H. F., et al. (2012). A calcium channel mutant mouse model of hypokalemic periodic paralysis. *The Journal of Clinical Investigation, 122*(12), 4580–4591. https://doi.org/10.1172/JCI66091

Xu, M., Welling, A., Paparisto, S., Hofmann, F., & Klugbauer, N. (2003). Enhanced expression of L-type $Ca_V1.3$ calcium channels in murine embryonic hearts from $Ca_V1.2$-deficient mice. *The Journal of Biological Chemistry, 278*(42), 40837–40841. https://doi.org/10.1074/jbc.M307598200

Yamada, Y., Kinoshita, H., Kuwahara, K., Nakagawa, Y., Kuwabara, Y., Minami, T., et al. (2014). Inhibition of N-type Ca^{2+} channels ameliorates an imbalance in cardiac autonomic nerve activity and prevents lethal arrhythmias in mice with heart failure. *Cardiovascular Research, 104*(1), 183–193. https://doi.org/10.1093/cvr/cvu185

Yamazaki, K., Shigetomi, E., Ikeda, R., Nishida, M., Kiyonaka, S., Mori, Y., et al. (2006). Blocker-resistant presynaptic voltage-dependent Ca^{2+} channels underlying glutamate release in mice nucleus tractus solitarii. *Brain Research, 1104*(1), 103–113. https://doi.org/10.1016/j.brainres.2006.05.077

Yang, L., Dai, D.-F., Yuan, C., Westenbroek, R. E., Yu, H., West, N., et al. (2016a). Loss of β-adrenergic-stimulated phosphorylation of $Ca_V1.2$ channels on Ser1700 leads to heart failure. *Proceedings of the National Academy of Sciences of the United States of America, 113*(49), E7976–E7985. https://doi.org/10.1073/pnas.1617116113

Yang, S., Ben-Shalom, R., Ahn, M., Liptak, A. T., van Rijn, R. M., Whistler, J. L., et al. (2016b). β-arrestin-dependent dopaminergic regulation of calcium channel activity in the axon initial segment. *Cell Reports, 16*(6), 1518–1526. https://doi.org/10.1016/j.celrep.2016.06.098

Yokoyama, K., Kurihara, T., Saegusa, H., Zong, S., Makita, K., & Tanabe, T. (2004). Blocking the R-type ($Ca_V2.3$) Ca^{2+} channel enhanced morphine analgesia and reduced morphine tolerance. *The European Journal of Neuroscience, 20*(12), 3516–3519. https://doi.org/10.1111/j.1460-9568.2004.03810.x

Yue, D. T. (2004). The dawn of high-resolution structure for the queen of ion channels. *Neuron, 42*(3), 357–359. https://doi.org/10.1016/S0896-6273(04)00259-4

Zaman, T., Lee, K., Park, C., Paydar, A., Choi, J. H., Cheong, E., et al. (2011). $Ca_V2.3$ channels are critical for oscillatory burst discharges in the reticular thalamus and absence epilepsy. *Neuron, 70*(1), 95–108. https://doi.org/10.1016/j.neuron.2011.02.042

Zamora, N. N., Cheli, V. T., González, D. A. S., Wan, R., & Paez, P. M. (2020). Deletion of voltage-gated calcium channels in astrocytes during demyelination reduces brain inflammation and promotes myelin regeneration in mice. *The Journal of Neuroscience, 40*(17), 3332–3347. https://doi.org/10.1523/JNEUROSCI.1644-19.2020

Zamponi, G. W., Striessnig, J., Koschak, A., & Dolphin, A. C. (2015). The physiology, pathology, and pharmacology of voltage-gated calcium channels and their future therapeutic potential. *Pharmacological Reviews, 67*(4), 821–870. https://doi.org/10.1124/pr.114.009654

Zanos, P., Bhat, S., Terrillion, C. E., Smith, R. J., Tonelli, L. H., & Gould, T. D. (2015). Sex-dependent modulation of age-related cognitive decline by the L-type calcium channel gene *Cacna1c* (Ca$_V$1.2). *The European Journal of Neuroscience, 42*(8), 2499–2507. https://doi.org/10.1111/ejn.12952

Zhang, Z., He, Y., Tuteja, D., Xu, D., Timofeyev, V., Zhang, Q., et al. (2005). Functional roles of Ca$_V$1.3(α_{1D}) calcium channels in atria. *Circulation, 112*(13), 1936–1944. https://doi.org/10.1161/CIRCULATIONAHA.105.540070

Zhang, J., Berra-Romani, R., Sinnegger-Brauns, M. J., Striessnig, J., Blaustein, M. P., & Matteson, D. R. (2007). Role of Ca$_V$1.2 L-type Ca²⁺ channels in vascular tone: Effects of nifedipine and Mg²⁺. *American Journal of Physiology. Heart and Circulatory Physiology, 292*(1), H415–H425. https://doi.org/10.1152/ajpheart.01214.2005

Zhao, Y., Huang, G., Wu, J., Wu, Q., Gao, S., Yan, Z., et al. (2019). Molecular basis for ligand modulation of a mammalian voltage-gated Ca²⁺ channel. *Cell, 177*(6), 1495–1506.e12. https://doi.org/10.1016/j.cell.2019.04.043

Zheng, H., Park, K. S., Koh, S. D., & Sanders, K. M. (2014). Expression and function of a T-type Ca²⁺ conductance in interstitial cells of Cajal of the murine small intestine. *American Journal of Physiology. Cell Physiology, 306*(7), C705–C713. https://doi.org/10.1152/ajpcell.00390.2013

Zuccotti, A., Lee, S. C., Campanelli, D., Singer, W., Satheesh, S. V., Patriarchi, T., et al. (2013). L-type Ca$_V$1.2 deletion in the cochlea but not in the brainstem reduces noise vulnerability: Implication for Ca$_V$1.2-mediated control of cochlear BDNF expression. *Frontiers in Molecular Neuroscience, 6*, 20. https://doi.org/10.3389/fnmol.2013.00020

Zühlke, R. D., Pitt, G. S., Deisseroth, K., Tsien, R. W., & Reuter, H. (1999). Calmodulin supports both inactivation and facilitation of L-type calcium channels. *Nature, 399*(6732), 159–162. https://doi.org/10.1038/20200

Voltage-Gated Calcium Channels (VGCCs) and Synaptic Transmission

Rayan Saghian and Lu-Yang Wang

Abstract

The intrinsic characteristics of neurons, as well as their distinctive firing properties, are defined by the intricate interplay of different ion channels. Among these channels, voltage-gated calcium channels (VGCCs) have a pivotal role in conveying electrical impulses to chemical signals in the form of synaptic transmission between neurons. The functions of calcium were first established more than 50 years ago by Bernard Katz and his colleagues (Katz B, Miledi R, Nature 215(5101):651, 1967), and it is now well known that VGCCs transduce electrical activity to a calcium influx into nerve terminals, initiating the vesicular release of chemical neurotransmitters at highly specialized junctions or synapses (Augustine GJ, J Physiol 431:343–364, 1990). These transmitters in turn activate ligand-gated ion channels to excite or inhibit postsynaptic neurons. In this chapter, we will highlight recent progress toward understanding the VGCCs, with an emphasis on their extensive network of interactions with other presynaptic proteins in regulating the strength, fidelity, and short-term plasticity of synaptic transmission.

Keywords

Voltage-gated calcium channels · Neurotransmitter release · Active zone · Synapse · Active zone proteins · Synaptic transmission · SNAREs

All Figures are created with BioRender.com

R. Saghian · L.-Y. Wang (✉)
Program in Neuroscience and Mental Health, SickKids Research Institute & Department of Physiology, University of Toronto, Toronto, ON, Canada
e-mail: luyang.wang@utoronto.ca

Abbreviations

ATP	Adenosine triphosphate
AZ	Active zone
CaBP-1	Calcium-binding protein-1
cAMP	Cyclic adenosine monophosphate
CDF	Calcium-dependent facilitation
CDI	Calcium-dependent inactivation
EGTA	Ethylene glycol tetraacetic acid
GDP	Guanosine diphosphate
gSTED	Gated stimulated emission depletion
GTP	Guanosine triphosphate
hMFB	Hippocampal mossy fiber boutons
NCS-1	Neural calcium sensor-1
NSF	N-ethylmaleimide-sensitive factor
p	Release probability
RIM	Rab3-interacting molecules

© The Author(s), under exclusive license to Springer Nature Switzerland AG 2022
G. W. Zamponi, N. Weiss (eds.), *Voltage-Gated Calcium Channels*,
https://doi.org/10.1007/978-3-031-08881-0_12

RIM-BP	RIM-binding proteins
RRP	Readily release pool
RVP	Reserve vesicle pool
SAM	Sterile α-motif
SNAP	Synaptosomal-associated proteins
SNARE	Soluble N-ethylmaleimide-sensitive factor-attachment protein receptors
Stx	Syntaxins
SV	Synaptic vesicle
Syb	Synaptobrevin
t-SNAREs	Target membrane SNAREs
v-SNAREs	Vesicular SNAREs
VAMP	Vesicle-associated membrane protein
VGCC	Voltage-gated calcium channel
VILIP-2	Visinin-like protein 2

Introduction

VGCCs in Central Nerve Terminals

Upon the arrival of action potentials at the axonal terminals, VGCCs are opened by changes in membrane potential and allow an influx of Ca^{2+} to sharply increase its intracellular concentrations. The nature of transient Ca^{2+} rise is strongly influenced by the presynaptic spike waveform, density, and subtypes of different VGCCs at any given synapse (Yang & Wang, 2006). In general, there are two main groups of VGCCs, high- and low-threshold channels. The high-threshold VGCCs encompass P/Q-, N-, R-, and L-type calcium channels (Table 1) (Catterall, 2000), serving as pivotal entry points for Ca^{2+} into excitable cells. Pharmacological (Wheeler et al., 1994) and immunocytochemical labeling (Timmermann et al., 2002) of calcium channel distribution suggest that the N-type (also known as Cav2.2) and P/Q-type (also known as Cav2.1) VGCCs are the main subtypes expressed in presynaptic nerve terminals (Evans & Zamponi, 2006).

In nerve terminals, neurotransmitters are packed in synaptic vesicles (SVs), some of which are closely docked and/or tethered with presynaptic active zones (AZs) and ready to be released in response to Ca^{2+} influx. This subset of SVs is known as the readily releasable pool (RRP).

Upon the depletion of RRP-SVs, more distally located SVs in the reserve vesicle pool (RVP) will replenish the depleted release sites in a Ca^{2+}-dependent manner such that synaptic transmission can sustain the information flow between neurons (Wang & Kaczmarek, 1998). At any synapse, a single action potential typically leads to the fusion of only a small percentage ($\sim10\%$) of RRP-SVs (Schneggenburger et al., 2002), leaving a large fraction of RRP for subsequent action potentials. The key requirement for tight temporal regulation of calcium evoked release is that SVs must be placed at less than 100 nanometer from VGCCs to sense high "local" Ca^{2+} transients, while activity-dependent buildup of residual Ca^{2+} can increase the rate of replenishment from RVP to RRP (Wang & Kaczmarek, 1998; Neher & Brose, 2018).

RRP-SVs appear to be at different states in relationship to the presynaptic AZ; namely docked, primed, or fused (Bharat et al., 2014). At the docked state, SVs are located close to the membrane with partially zipped fusion proteins known as soluble N-ethylmaleimide-sensitive factor (NSF)-attachment protein receptors (SNAREs). At the primed stated, synaptic vesicles are fully zipped and prepared to be fused rapidly in response to a calcium influx (Neher & Brose, 2018). A sudden increase in cytoplasmic calcium by the opening of VGCCs leads to the fusion of docked vesicles, referred to as the fused state (Ge et al., 2019), where the vesicle membrane merges with the target membrane to open the fusion pore and unload neurotransmitters from within (Fig. 1).

To ensure this spatiotemporal control of the release (Neher & Brose, 2018), synaptic proteins are highly organized near AZs where a dense collection of vesicle-associated proteins and the cytomatrix streamline, the release and replenishment of RRP-SVs, and RVP-SVs by coupling/tethering them to VGCC channels, such that fast and synchronous fusion can be readily triggered by Ca^{2+} influx during action potentials. The strength and accuracy of synaptic transmission depend on the number of VGCCs clustered in the active zone and their physical proximity to SVs (Rozov et al., 2001; Holderith et al., 2012).

Table 1 Types of VGCCs (see also Part IV "Pharmacology of VGCCs")

Voltage-gated calcium channels	High threshold voltage-activated	Current type	Channel type	Gene name and human chromosome	Specific inhibitors	Function
		L (Long lasting)	Cav1.1 (α_{1S})	CACNA1S; 1q31-32	Dihydropyridines, Phenylalkylamines, Benzothiazepines	Excitation–contraction coupling
			Cav1.2 (α_{1C})	CACNA1C; 12p13.3	Dihydropyridines, Phenylalkylamines, Benzothiazepines	Excitation–contraction coupling, hormone release, regulation of transcription, synaptic integration
			Cav1.3 (α_{1D})	CACNA1D; 3p14.3	Dihydropyridines, Phenylalkylamines, Benzothiazepines	Hormone release, regulation of transcription, synaptic integration
			Cav1.4 (α_{1F})	CACNA1F; Xp11.23	–	Neurotransmitter release from rods and bipolar cells
		P/Q (Purkinje cells)	Cav2.1 (α_{1A})	CACNA1A; 19p13	Ω-agatoxin G IVA, Ω-agatoxin G IVB	Neurotransmitter release, dendritic calcium transients
		N (Neuronal type)	Cav2.2 (α_{1B})	CACNA1B; 9q34	Ω-Conotoxin-G VI A	Neurotransmitter release, dendritic calcium transients
		R (Resistance)	Cav2.3 (α_{1E})	CACNA1E; 1q25-31	SNX-482	Repetitive firing
	Low threshold voltage-activated	T (Transient current)	Cav3.1 (α_{1G})	CACNA1G; 17q22	–	Pacemaking, repetitive firing
			Cav3.2 (α_{1H})	CACNA1H; 16p13.3	–	Pacemaking, repetitive firing
			Cav3.3 (α_{1I})	CACNA1I; 22q12.3-13-2		Pacemaking, repetitive firing

VGCCs and Interactive Proteins in AZs

Five evolutionarily conserved protein families—RIM, Munc13, RIM-BP, α-liprin, and ELKS proteins—have been shown to form the core of AZs (Südhof, 2012) (Table 2). In addition, piccolo and bassoon are also associated with AZs in vertebrates (Tom Dieck et al., 1998; Limbach et al., 2011). Neurotransmitter release at the AZ scrupulously involves SNARE proteins that are present on both the plasma and vesicular membranes and form tight complexes to facilitate synaptic vesicle fusion. Exocytosis is mediated by synaptobrevin situated on synaptic vesicles, as well as syntaxin-1 and SNAP-25, located at the presynaptic plasma membrane (Fig. 2), but VGCCs have been shown to interact with a number of synaptic proteins including SNAREs themselves (Table 3).

RIMs

RIMs (Rab3-interacting molecules) is essential for synaptic vesicle docking and priming (Koushika et al., 2001; Kaeser et al., 2011), for recruiting VGCCs to active zones (Kaeser et al., 2011), as well as for the short-term plasticity of neurotransmitter release (Schoch et al., 2002). This protein is made of five domains: an N-terminal zinc finger domain, a central PDZ domain, and two C-terminal C2 domains (Kaeser et al., 2011). The RIMs N-terminal zinc finger binds to the C2A domain of Munc13 (Lu et al., 2006). This interaction links synaptic vesicles to active zones and keeps them in close proximity to the priming factor Munc13 (Kaeser et al., 2011). RIM proteins were found to be essential for short- and long-term synaptic plasticity by affecting the readily releasable pool of vesicles (Schoch et al., 2002). RIMs are also required for Ca^{2+}

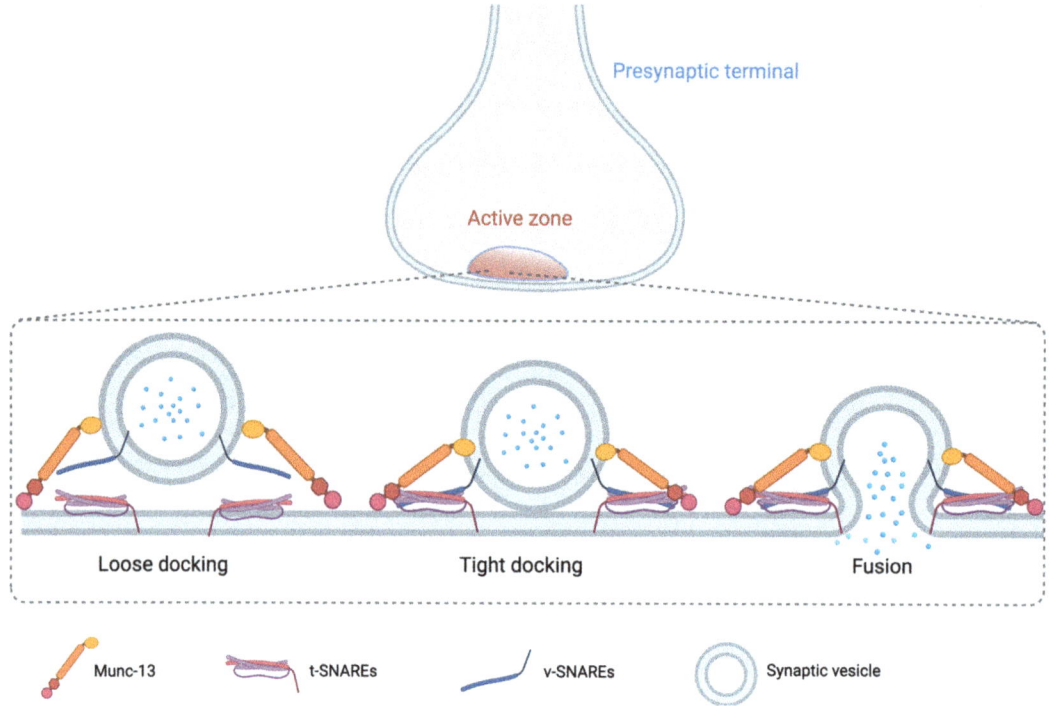

Fig. 1 Primed, docked, and fused synaptic vesicles

Table 2 Active zone proteins

Proteins	Function
Structural proteins	
Piccolo	Recruits synaptic vesicles to the active zone
Bassoon	Recruits synaptic vesicles to the active zone
RIMs	Recruitment of VGCCs to active zone
ELKs	Recruitment of synaptic vesicles
Docking and priming	
Munc-13	Primes the SNARE protein before fusion
α-liprins	Recruitment of synaptic vesicles
SNAREs	Mediates vesicular fusion
SNAP25	Targets membrane SNARE that binds to VAMP and syntaxin
Synaptobrevin/ VAMP	Vesicular SNARE
Syntaxin	Targets membrane SNARE
Calcium channels	
Voltage-gated calcium channels	Allows the rapid influx of calcium during an action potential

channel trafficking to presynaptic active zones (Han et al., 2011). In addition, RIM-binding proteins (RIM-BPs) form a complex with RIMs in the active zone (Wang et al., 2000) and also directly bind to VGCCs (Hibino et al., 2002).

Previous studies *in vitro* indicate a direct interaction of RIM-1 with $Ca_v2.2$ channels as well as with SNAP-25 and synaptotagmin via the C2 domains in a Ca^{2+}-dependent manner (Coppola et al., 2001). At low Ca^{2+} concentrations, RIM-1 preferentially binds SNAP-25, whereas an increase in Ca^{2+} concentration (>75 μM) favors its interaction with synaptotagmin-1 (Coppola et al., 2001). Further insight was gained by Kiyonaka et al. (2007), showing that β-subunit of VGCCs directly interacts with the RIM protein. By this interaction, RIM bridges between the synaptic vesicles and the VGCCs and holds them at near vicinity to each other (Kiyonaka et al., 2007). However, biochemical studies using synaptosome membranes failed to demonstrate any co-immunoprecipitation or co-labeling of $Ca_v2.2$ and RIM (Wong & Stanley, 2010). Kaeser et al. (2011) showed instead that RIM proteins directly interact

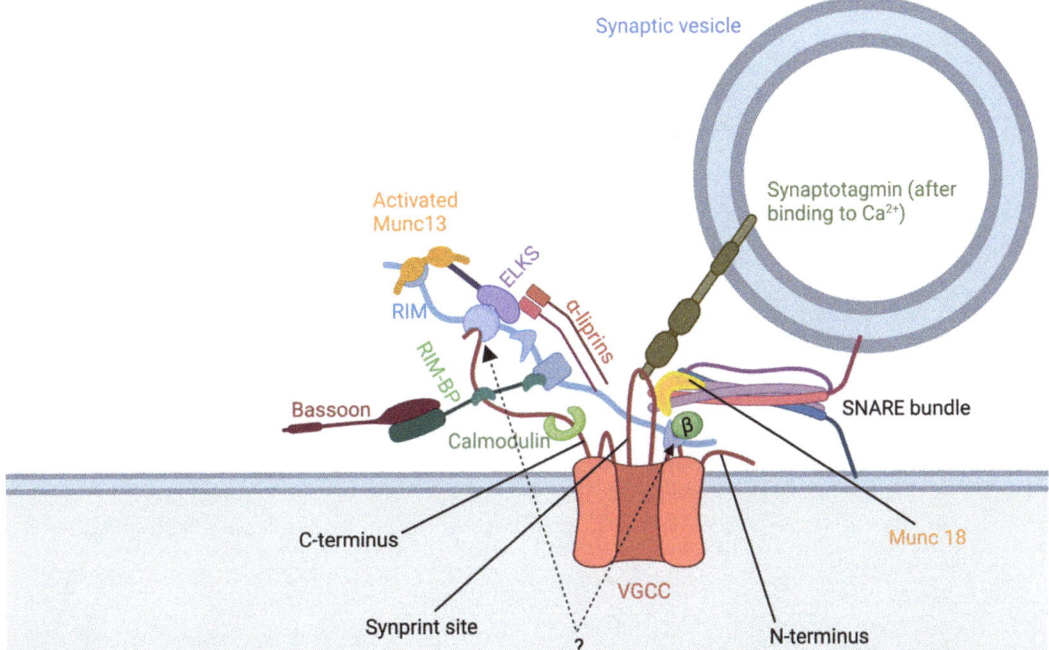

Fig. 2 Molecular model of the active zone proteins. Active zone proteins interactions with the VGCC can be direct or indirect. Different studies showed different interaction sites for RIM with the C-terminus of the VGCC or the β-subunit (indicated by the question mark). β-subunit directly interacts with the beta interacting domain (BID) of the intracellular loop between transmembrane domain I and II of the VGCC (see part1, Voltage-Gated Calcium Channel Auxiliary ß Subunits)

with the C-terminus of α1-subunits and facilitate their presynaptic localization as well as binding VGCCs to the synaptic vesicles. A possible explanation for these discrepancies might be due to redundancy in the interactions to the active zone proteins, such that disrupting one protein interaction can be compensated by allosteric interactions sites from other proteins within the complex so that readouts are insufficiently disrupted. Another possibility is that these interactions are not mutually exclusive and synergistically promote the association of RIM-1 and VGCCs. Alternatively, this might be related to diversity of splicing variants of VGCCs implicated in the synaptic heterogeneity (see below). The same neuronal active zones can show pleiotropic properties by rearranging their protein matrix.

Rab6-binding protein (Monier et al., 2002). Ohtsuka et al. (2002) showed its role as an active zone protein and renamed it as CAST. The name ELKS is derived from "protein rich in the amino acids E (Glutamic acid), L (leucine), K (Lysine), and S (Serine)" (Patwardhan et al., 2017). ELKS binding to the RIM PDZ domains led to the discovery of its role in the AZ (Wang et al., 2002). It also directly binds to α-liprins (Ko et al., 2003). Although ELKS is not an indispensable component of the *C. elegans* release machinery (Deken et al., 2005), its deletion in mice does reduce the presynaptic Ca^{2+} influx in the inhibitory synapses of the hippocampal neurons (Kaeser et al., 2009) and decrease the $Ca_v2.1$ clustering in the calyx of Held (Dong et al., 2018) synapse and at *Drosophila* neuromuscular junctions (Kittel et al., 2006).

ELKS

ELKS/RAB6-interacting/CAST family member1 was first discovered in 1999 (Knowles et al., 2006) and rediscovered in 2002 by the name

MUNC 13

The UNC-13 gene in *C. elegans* was discovered in 1991 (Maruyama & Brenner, 1991), but its precise functions were not well established until

Table 3 Active zone proteins interactions with VGCCs

Interactions with VGCCs		
Direct interactions	Proteins	Interaction site with VGCCs
	Syntaxin and SNAP-25	Synprint
	Synaptotagmin	Synprint (After binding to calcium)
	Munc-18	Synprint

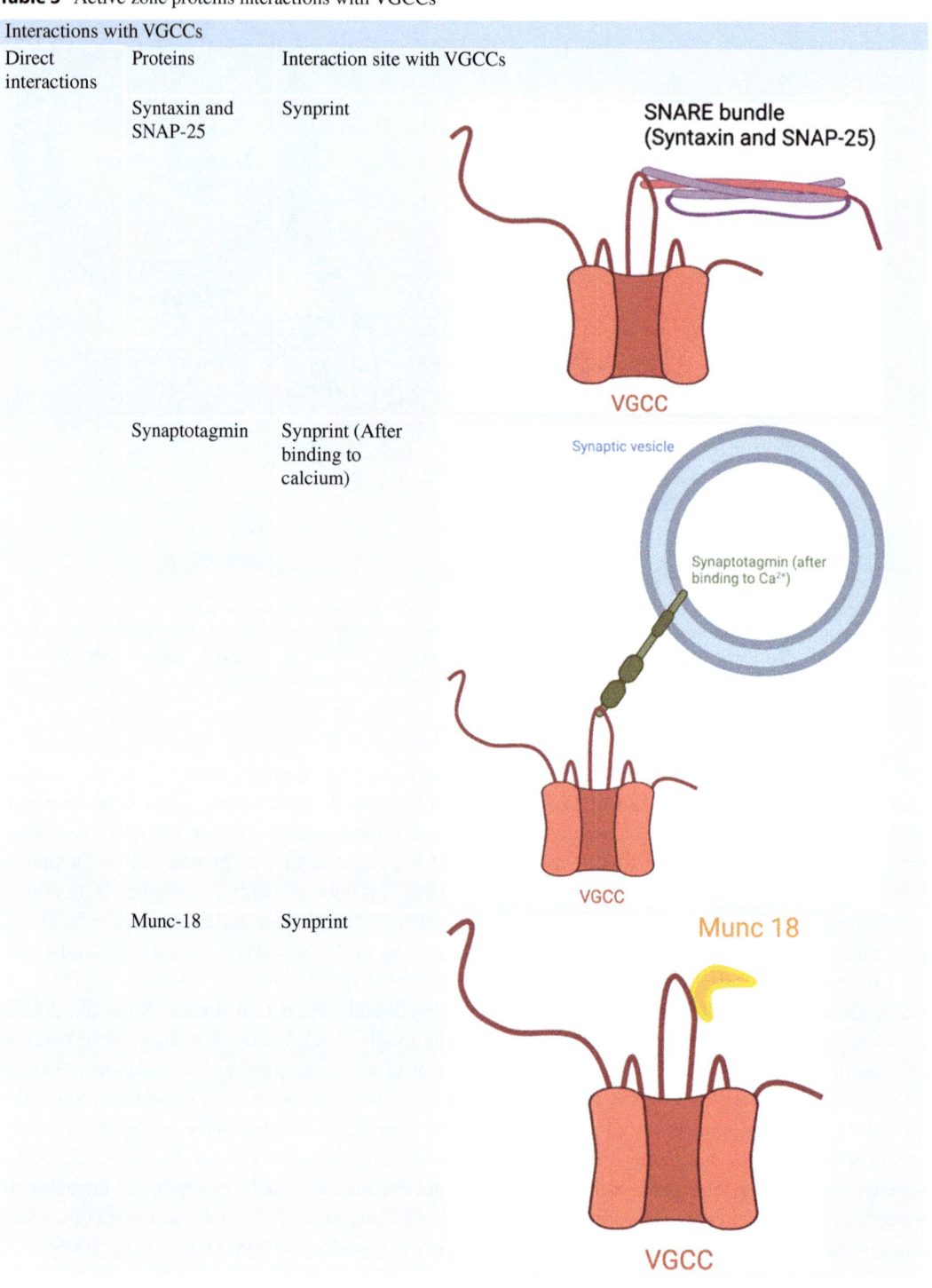

(continued)

Table 3 (continued)

	RIM[a]	C-terminus of VGCCs	

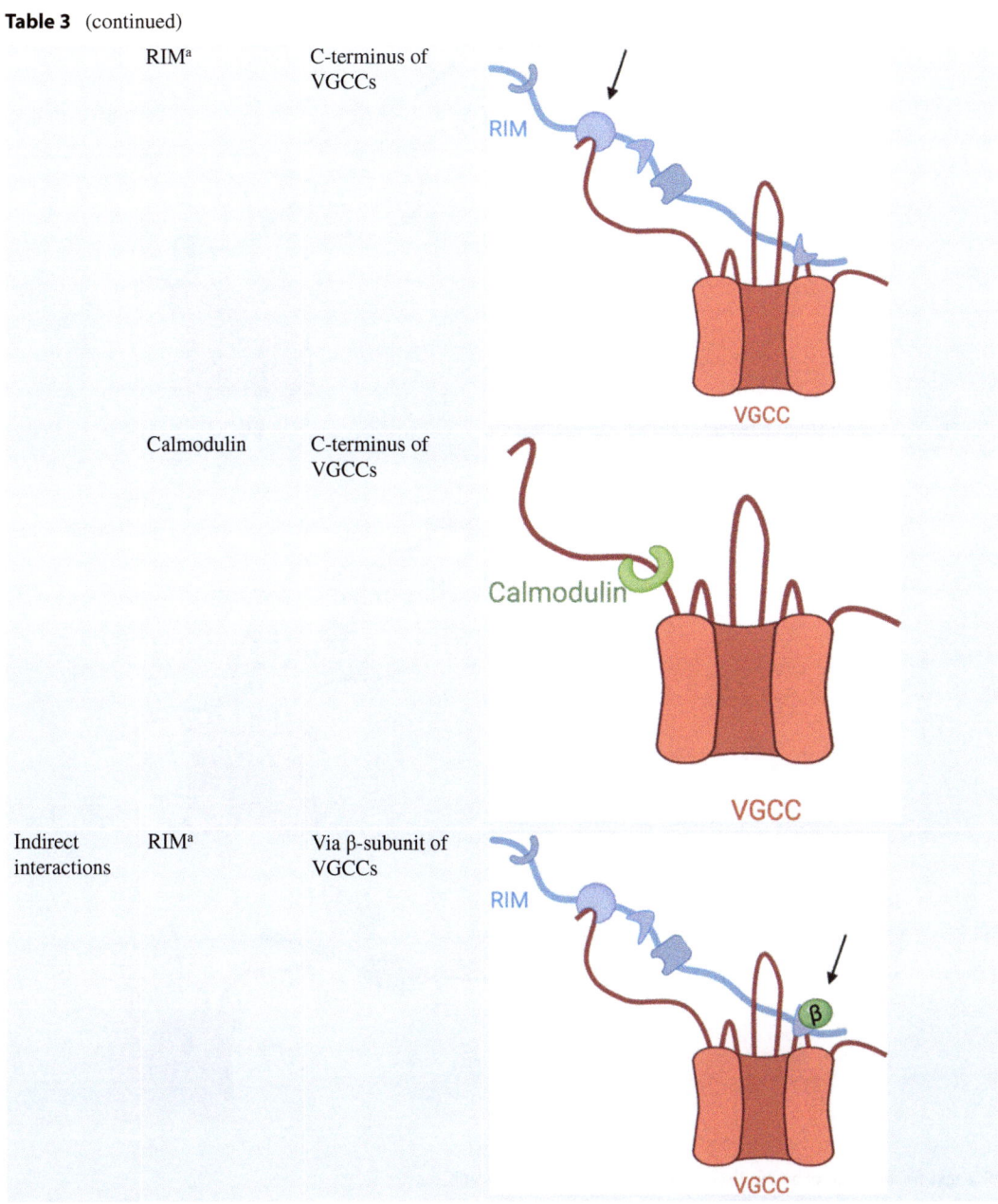

	Calmodulin	C-terminus of VGCCs	

Indirect interactions	RIM[a]	Via β-subunit of VGCCs	

(continued)

Table 3 (continued)

	ELKS	Via RIM and α-liprins and β-subunit	
	Munc-13	Via RIM and ELKS	
	Bassoon	Via RIM-BP	

[a]See text for both direct and indirect interactions of RIM with VGCCs

the characterization of the mammalian homologs (Brose et al., 1995), named Munc13. This protein is located in the AZ and is essential for synaptic vesicle priming (Augustin et al., 1999). While some isoforms of Munc13 associate with RIM, other isoforms can bind to ELKS and be recruited to the active zone at different synapses (Kawabe et al., 2017). These synapse-specific interactions can determine basal synaptic vesicle priming and short-term plasticity, and therefore lead to the molecular and functional heterogeneity of pre-synaptic active zones (Kawabe et al., 2017).

α- and β-liprins

α- and β-liprins are related proteins composed of a coiled-coil N-terminal, and three C-terminal sterile α-motif (SAM) domains (Serra-Pagès et al., 1995). The N-terminal of α-liprins binds to itself and form homodimers (Taru & Jin, 2011) to the RIM C2B domain (Schoch et al., 2002) and to ELKS (Dai et al., 2006). The SAM domains at the C-terminal bind to β-liprins to form heterodimers (Serra-Pagès et al., 1995). The role of α-liprins in the presynaptic AZs was first discovered when a loss-of-function mutation in *C. elegans* (Zhen & Jin, 1999) and *Drosophila* (Kaufmann et al., 2002). α-liprin was found to increase the AZ size and to interrupt docked SV accumulation (Wong et al., 2018). Together, it seems that α-liprin plays a role in AZ formation as well as the recruitment of synaptic vesicles.

Piccolo and Bassoon

Piccolo and bassoon are among the presynaptic proteins found in vertebrates and their major function is to guide synaptic vesicles to the AZ (Mukherjee et al., 2010; Hallermann et al., 2010). Loss of bassoon causes impaired neurotransmitter release, which can lead to spontaneous epileptic seizure (Altrock et al., 2003). Several studies on synapses lacking bassoon showed its role in the recruitment of SVs into the active zone as well as the positional priming of the VGCCs (Mukherjee et al., 2010; Jing et al., 2013). Similar to RIMs, bassoon also binds to RIM-BPs (Hibino et al., 2002) but unlike RIM that tethers various types of VGCCs (e.g., both $Ca_V2.2$, and $Ca_V2.1$) to the AZs (Han et al., 2011), bassoon selectively places P/Q-type channels adjacent to the release sites of hippocampal synapses (Davydova et al., 2014). A functional link between piccolo and VGCC localization at brain synapses is less studied. Only a small number of studies addressed this issue and showed that the C2A domain of piccolo binds calcium with low affinity, which causes conformational changes in this structure to impact its association with phospholipids (Gerber et al., 2001). However, no effect on synaptic transmission was observed if the C2A domain was disrupted (Mukherjee et al., 2010).

SNAREs

Soluble N-ethylmaleimide-sensitive factor attachment protein receptor (SNARE) proteins facilitate membrane fusion events in eukaryotic cells. SNAREs are categorized in three families depending on their subcellular localizations. Synaptosomal-associated proteins (SNAP) and syntaxins (Stx) are expressed on the target plasma membrane (t-SNAREs). Synaptobrevin (Syb) (also known as VAMP; vesicle-associated membrane protein) is vesicular SNAREs (v-SNARE) enriched on the vesicle membranes, respectively. Recent studies showed that several SNARE proteins are found on both vesicles and target membranes; therefore, v- and t-SNARE are sometimes also called R-and Q-SNAREs (Fasshauer et al., 1998). Often, R-SNAREs work as v-SNAREs and Q-SNAREs as t-SNAREs. The R-SNAREs contribute an arginine (R) residue and Q-SNAREs contribute a glutamine (Q) residue in the formation of the zero ionic layer during the assembly of the core SNARE complex.

In order for the vesicle and plasma membranes to fuse, cells have to overcome the energy barrier. To circumvent this energy barrier, SNAREs form high-affinity complexes that drive the docking and fusion of the vesicles to the target membrane. SNARE complex formation brings the two membranes in close proximity in an exothermic process that can supersede the energy barrier that self-repulses the poly cationic lipids found at the surface of vesicles and the inner face of the plasma membrane (Sudhof & Rothman, 2009). SNARE-dependent fusion is necessary for all hydro-soluble neurotransmitters (Sudhof & Rothman, 2009). The SNARE complex in the presynaptic region consists of a 25kDa synaptosomal-associated protein (SNAP25) as well as Stx-1 and Syb-2/VAMP-2 (Sudhof & Rothman, 2009; Jahn & Scheller, 2006; Rizo & Rosenmund, 2008). Calcium sensor proteins such as synaptotagmins are critical for completing calcium-dependent exocytosis in milliseconds (Sudhof & Rothman, 2009).

Overall, a cycle of ATP-independent association and ATP-dependent dissociation of the SNARE proteins is required for fusion to occur. In this cycle, the thermodynamically unfavored

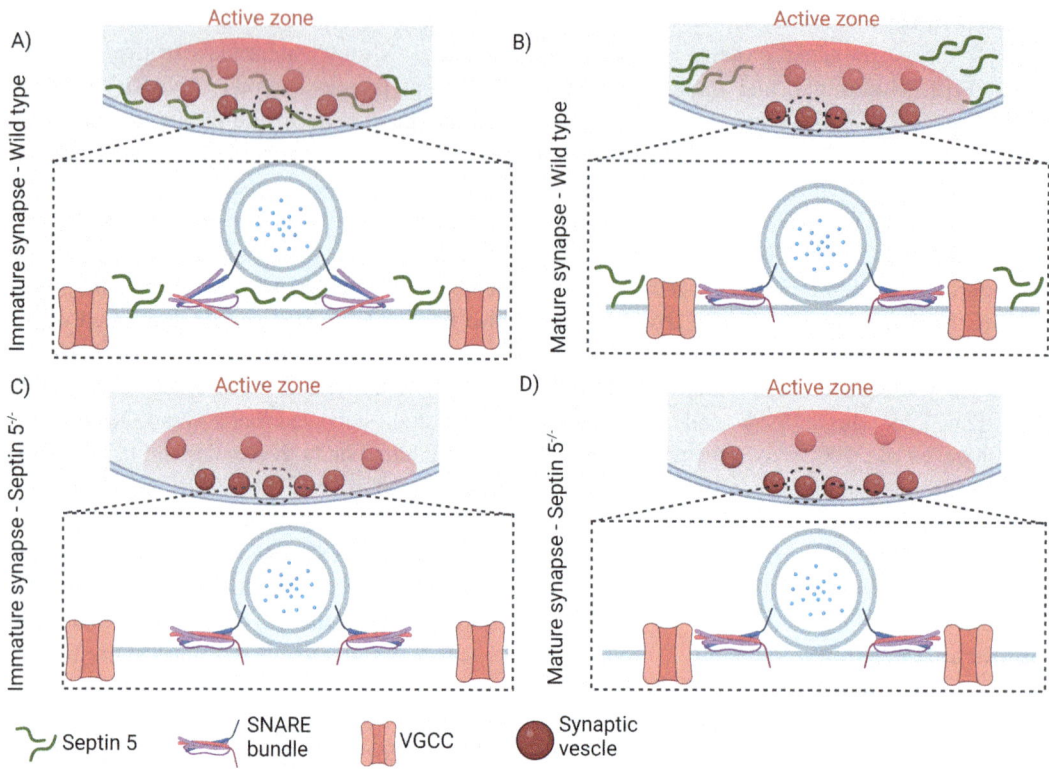

Fig. 3 Role of septin-5 filaments in the developing calyx of Held synapse. (**a**) At immature wild-type synapses, septin-5 filaments serve as a physical barrier between synaptic vesicles and their docking sites in the active zone, reducing the number of fully zippered SNARE complexes. This reflects a typical microdomain organization of release modality. (**b**) At mature wild-type synapses, septin-5 filaments are no longer present at the active zone, resulting in more docked synaptic vesicles, reflective of the nanodomain release modality. (**c**) Immature septin-5 $^{-/-}$ synapses display a nanodomain-like coupling because of the absence of the septin-5 physical barrier between synaptic vesicles and active zone. (**d**) Mature septin-5$^{-/-}$ synapses appear unaltered from their wild-type counterparts because in both septin-5 is absent at the active zone

reaction of the membrane–membrane fusion is coupled to an exergonic reaction of SNARE proteins folding, later followed by endergonic reaction of unfolding SNARE (Sudhof & Rothman, 2009). The unfolding relies on the energy from hydrolyzing ATP, facilitated by a specialized ATPase, NSF, that permits the recycling of SNARE proteins to their initial state for another round of vesicular fusion events. NSF is a hexamer that uses 3 to 6 ATP molecules with each catalytic cycle (~20–40 kcal/mole to disrupt the tightly bound SNARE complex) (Sudhof & Rothman, 2009).

In addition to their key role in synaptic transmission, syntaxin co-localizes with Ca$_v$2.1 and Ca$_v$2.2, the two key VGCCs in neurotransmission

(Cohen et al., 1991). These are direct interactions between the VGCCs and syntaxin, as well as SNAP-25. The interaction site is within the intracellular loop between domains II and III, called synaptic protein interaction site (synprint) of the Ca$_v$2.1 (Rettig et al., 1996) and Ca$_v$2.2 (Sheng et al., 1994). Syntaxin and SNAP-25 interactions are not competitive and they bind to two separate microdomains separated by a flexible linker within the synprint site (Rettig et al., 1996; Yokoyama et al., 2005). The importance of these interactions on neurotransmission was investigated either by the deletion of the synprint site on VGCCs or by inserting peptides with similar sequences to compete with synprint interactions. Results showed that disruption of the VGCCs–

SNARE coupling alters, and in some cases inhibits, synaptic transmission (Mochida et al., 1996; Rettig et al., 1997; Harkins et al., 2004; Keith et al., 2007). The functional impact of VGCCs–syntaxin interaction was shown by electrophysiological studies where syntaxin binding shifts the voltage dependence of inactivation toward more hyperpolarized membrane potentials and therefore reducing Ca^{2+} entry into synaptic terminals (Bezprozvanny et al., 1995; Zhong et al., 1999; Degtiar et al., 2000; Altrock et al., 2003). Syntaxin, in addition to its effect on voltage dependence, promotes G-protein modulation of the channels (Jarvis et al., 2000; Jarvis & Zamponi, 2001; Gerachshenko et al., 2005).

A similar alteration can be observed when SNAP-25 is co-expressed with $Ca_v2.1$ and $Ca_v2.2$ (Wiser et al., 1996; Zhong et al., 1999). The reverse modification was examined in glutamatergic neurons where silencing of SNAP-25 leads to an increase of Ca^{2+} currents carried by $Ca_v2.1$ channels (Condliffe et al., 2010). While interacting with the synprint site of the $Ca_v2.2$, syntaxin can also bind to the βγ-subunit of G-proteins, this interaction optimizes the tonic inhibition of the $Ca_v2.2$ by G-proteins (Jarvis et al., 2000; Zurawski et al., 2017) (Also see chapter "Modulation by GPCRs and second messengers"). However, it remains unknown if SNAP-25 directly affects the gating of VGCCs.

Two domains on synaptotagmin, C2A and C2B (Davletov & Südhof, 1993; Sutton et al., 1995), are negatively charged and bind Ca^{2+}. The conformational changes after Ca^{2+} binding allow the C2A domain to insert into the plasma membrane (Fernández-Chacón et al., 2001), whereas C2B binds to the synprint site of both $Ca_v2.1$ and $Ca_v2.2$ (Sheng et al., 1997). Although the binding of synaptotagmin does not alter the VGCCs gating properties, it reduces syntaxin-dependent inhibition of $Ca_v2.2$ channels (Wiser et al., 1997). This reversal inhibition might be due to competition for the synprint site in a calcium-dependent manner (Sheng et al., 1996).

Another protein that interacts with the synprint site is Munc-18. This protein binds to the II–III intracellular loop of $Ca_v2.2$ with a higher affinity than the I–II loop and does not bind to other channel intracellular domains (Chan et al., 2007). Studies on co-expression of Munc-18 and $Ca_v2.2$ showed that this interaction has no effects on the channel's properties (Gladycheva et al., 2004). Munc-18 also interacts with syntaxin (Dulubova et al., 1999). Therefore, although Munc-18 is not a direct modulator of VGCCs, it may have an important role in regulating Ca^{2+} influx into nerve terminals, in a syntaxin-dependent manner, to regulate transmitter release.

To better understand the importance of these interactions, Heck et al. (2019) showed how deletion of the C-terminus of VGCCs can alter SV exocytosis. In this study, a deletion of the exon 47 of $Ca_v2.1$ calcium channels ($Ca_v2.1\Delta ex47$), which encodes the C-terminus binding sites for a number of active zone proteins including RIM and RIM-BP, leads to less diffusibility and activity-dependent accumulation of $Ca_v2.1\Delta ex47$ in synaptic nanodomains, reducing neurotransmitter release at hippocampal synapses (Heck et al., 2019).

VGCCs and Developmental Plasticity of Neurotransmitter Release

In fast-spiking central synapses, pioneering studies showed a developmental switch from a mixture of N- and P/Q-type to exclusively P/Q-type VGCCs in mediating neurotransmitter release in the rat and mouse auditory brainstem synapses, thalamic and cerebellar synapses (Iwasaki et al., 2000). The β-subunits also show a marked developmental switch in isoforms, from β1 and β3 to β2 and β4 isoforms. These changes suggest that calcium channels change their subunit composition upon neuronal differentiation (Schlick et al., 2010). In hippocampus, the upregulation of β2 during the development is more noticeable than that of β4. Together these changes give rise to a dominant CaV2.3 and β2 expressions in the 8-week-old hippocampal neurons (Schlick et al., 2010). Given that both α- and β-subunits of VGCCs can interact with synaptic proteins such as RIM, such switches can impact the spatial couplings between VGCCs and SVs, leading to different release probabilities and RRP in different central synapses.

This is best exemplified at the developing calyx of Held synapse where the fidelity of synaptic transmission exhibits a strong age-dependent upregulation. There is a developmental transformation of neurotransmitter release modality from "microdomain" in immature synapses to "nanodomain" in mature synapses (Fedchyshyn & Wang, 2005). In microdomain modality, many "loosely coupled" N- and P/Q-VGCCs cooperatively drive the fusion of single SVs, and on the contrary, fewer "tightly coupled P/Q-VGCCs" are sufficient to trigger the release in nanodomain coupling. This developmental remodeling is achieved by shortening the spatial distance as evidenced by decreasing efficacy of the slow calcium chelator, EGTA, to reduce neurotransmitter release, independent of obvious age-dependent changes in voltage dependence, or gating kinetics of VGCCs, and the sensitivity of calcium sensor (Fedchyshyn & Wang, 2005; Wang et al., 2008). Different from fast-spiking sensory synapses like the calyx of Held synapse, hippocampal excitatory neurons also use both N- and P/Q-type VGCCs to mediate transmitter release but N-type appears to form "preferred slots" over P/Q-type VGCCs (Cao et al., 2004), although the concept on "preferred slots" is under challenge (Lübbert et al., 2017). These studies suggest different central synapses may strategize expressing their choice of VGCCs to optimize their individual functionalities.

Cytomatrix proteins appear to be a key player in determining the physical coupling distance between VGCCs and SVs. For example, Septin-5 is a core molecular substrate in differentiating distinct release modalities at the calyx of Held synapse (Yang et al., 2010). Septins are a conserved multigene family of filamentous, guanine nucleotide (GTP/GDP)-binding proteins that are best known for their roles in cytokinesis. A deletion of septin-5 in immature synapses led to precocious organization, reminiscent of functional phenotypes associated with mature synapses (Yang et al., 2010). More precisely, upon deletion of septin-5, synaptic vesicles were localized closer (or more clustered toward) to the active

zones, quantal output was elevated, readily releasable pool of vesicles was increased, and fewer VGCCs were required to induce single fusion events in immature synapses. Most importantly, acute disruption of septin-5 filament with septin-5 antibody transforms the coupling of VGCCs to synaptic vesicles from loose (microdomain) to tight (nanodomain) and potentiates quantal outputs in immature synapses. This supported the idea that clusters of septin-5 act as a physical, filamentous barrier between vesicles and their target docking sites on the terminal membrane. Upon removal of the septin-5 barrier, synaptic vesicles were able to dock in close proximity to VGCCs at the active zones (Fig. 3) (Yang et al., 2010). Humans have up to 14 septin isoforms (Cao et al., 2007). It remains unclear how septins regulate the distance between VGCCs and SVs, despite the fact that septin-5 is known to interact with syntaxin (Beites et al., 1999). However, there is no evidence thus far for their direct interactions with VGCCs. It should be noted that septin dysfunction or dysregulation leads to a variety of neurological disorders including Alzheimer's Disease (i.e., Sept1,2,3,4,7) (Kinoshita et al., 1998), Parkinson's Disease (Son et al., 2005) (i.e., Sept4,5), and Schizophrenia (i.e., Sept4,5,6,11) (Barr et al., 2004).

VGCCs and Short-Term Synaptic Plasticity

Functional synaptic connections are not rigid and can change in response to environmental stimuli. The best-known form of this adaptation is called synaptic plasticity, meaning that a synapse can strengthen or weaken over time, depending on an increase or decrease in the history of cell-to-cell interactions. These changes can be found in two forms: short- and long-term plasticity. Long-term refers to the synaptic changes that last for hours or longer (Nicoll, 2017), unlike short-term that acts on a timescale of tens of milliseconds to a few minutes. Plasticity can refer to facilitation or depression of the synaptic activity. The former is defined by an increase in probability of neurotransmitter release in response to presynaptic action potentials, while the latter is usually attributed to the depletion of the readily releasable vesicles and many other mechanisms (Zucker & Regehr, 2002).

VGCCs have central roles in synaptic plasticity. When an action potential reaches a presynaptic terminal and calcium channels open, calcium concentration around the channels, also known as "local calcium," rises to tens or even hundreds of micromolar and lasts as long as the calcium channels are open (Simon & Llinás, 1985; Augustine et al., 1991; Yamada & Zucker, 1992). Calcium concentration declines significantly with distance from an open channel due to diffusion and presence of calcium buffers (Neher, 1998). The local calcium can activate low-affinity calcium sensors such as synaptotagmin1/2, which lead to vesicle fusion. The diffused calcium gives rise to "residual calcium," around hundreds of nanomolar that lasts for tens to hundreds of milliseconds, until this calcium is largely cleared from the synaptic terminal. In 1968, Katz and Miledi proposed that synaptic facilitation could be a consequence of residual calcium that is available for the second release event when two stimulus are received within a short timescale (Katz & Miledi, 1968) which today is known as the residual calcium hypothesis of facilitation.

VGCCs in the nerve terminals or transfected cells exhibit paired-pulse facilitation or frequency-dependent facilitation during trains of depolarizing stimuli (Borst & Sakmann, 1998). This upregulation results from the residual Ca^{2+} binding to Ca^{2+} sensors (Burgoyne & Weiss, 2001). These calcium sensors, for example, calcium-modulated protein (calmodulin), can be different from those needed for exocytosis, e.g., synaptotagmin (Burgoyne & Weiss, 2001) (See chapter "Modulation by GPCRs and second messengers"). Calmodulin directly binds to the C-terminus of the VGCCs and has two high-affinity and two low-affinity Ca^{2+}-binding sites (EF hands) on its C-terminal and N-terminal, respectively (Gifford et al., 2007). Local increases in intracellular calcium are sensed by the high-affinity sites, whereas prolonged or repetitive activation of VGCCs causes the residual Ca^{2+} to

bind to the low-affinity sites which causes inactivation of the calcium current on the 100 ms to 1 s timescale (DeMaria et al., 2001). These dynamic changes are known as calcium-dependent facilitation or inactivation of calcium channels (CDF and CDI). CDF appears to be associated with P/Q- but not N-type channels. Prevention of calmodulin regulation by introducing the IM-AA mutation (Zühlke et al., 2000) in the calmodulin-binding site of the α_1-subunit of P/Q-type or CaV2.1 VGCCs blocks this facilitation (DeMaria et al., 2001).

Considering that calmodulin is ubiquitously expressed if CDF and CDI of $Ca_V2.1$ channels by calmodulin were mainly responsible for synaptic plasticity, then all synapses expressing $Ca_V2.1$ should have the same patterns of synaptic facilitation or depression. However, different synapses have different patterns of synaptic plasticity (Abbott & Regehr, 2004). So how can calcium channel regulation contribute to the diversity of synaptic function? Besides calmodulin, there is a large family of calcium sensor proteins that are differentially expressed, including neural calcium sensor-1 (NCS-1), calcium-binding protein-1 (CaBP-1), and visinin-like protein 2 (VILIP-2) (Girard et al., 2015). Injecting NCS-1 into the presynaptic nerve terminal at the calyx of Held increases $Ca_V2.1$ currents and synaptic facilitation (Tsujimoto et al., 2002), which is the different effect of calmodulin. The same pattern of increased CDF that opposes inactivation of $Ca_V2.1$ is observed in the presence of VILIP-2 (Lautermilch et al., 2005). CaBP-1, on the other hand, blocks CDF and enhances CDI (Lee et al., 2002). These results show that the diversity of the $Ca_V2.1$ modulation contributes to divergent forms of synaptic plasticity. Regulation of VGCC functions may occur at different levels including regulation of channel's mRNA splicing, trafficking to the membrane (Lipscombe et al., 2013), and G-Protein-mediated inhibition of presynaptic VGCCs (Tedford & Zamponi, 2006; Currie, 2010; Zamponi & Currie, 2013) (See chapter "Regulation of VGCCs").

Classical studies using electron microscopy on neuromuscular junction synapses showed that repeated stimulation for several minutes leads to physical depletion of RRP (Dickinson-Nelson & Reese, 1983). Can this rapid reduction of the RRP size be linked to the loss of physically docked synaptic vesicles or is it due to the CDI of VGCCs? This question was addressed by showing that even after complete deletion of the RRP, there are still a large pool of remaining synaptic vesicles that could also be released if intracellular calcium was increased rapidly and globally (Wadel et al., 2007). These results suggest that inactivation of presynaptic calcium channels after repetitive stimulations would cause less Ca^{2+} influx, which leads to nearby synaptic vesicles to be exposed to less than 10 μM Ca^{2+}, the physiological Ca^{2+} level for phasic neurotransmission (Schneggenburger & Neher, 2000). Another evidence that showed the rapid phase of synaptic depression is indeed due to CDI from the calyx of Held where CDI decreased release probability (Xu & Wu, 2005). Therefore, inactivation of presynaptic calcium channels contributes to the rapid synaptic depression in particular synapses.

Modulation of VGCC activities is often mediated by second messengers, such as cyclic adenosine monophosphate (cAMP)/protein kinase A and C pathways (Byrne & Kandel, 1996). Besides the changes in Ca^{2+} current during repeated stimulation, VGCC localization might also contribute to the synaptic plasticity. A recent study on hippocampal mossy fiber boutons (hMFB) showed that cAMP application enhances the neurotransmitter release in the range of 5–10 min (Fukaya et al., 2021). Although the peak of local Ca^{2+} concentrations and Ca^{2+} influx at the release sites were increased, this facilitation was independent of the changes in the Ca^{2+} sensitivity of synaptic vesicle release through cAMP application. Evidence with gated stimulated emission depletion (gSTED) microscopy implicates that the synaptic facilitation was due to the increase in the size of the P/Q type Ca^{2+} channel clusters near the release sites with no changes to the RRP size (Fukaya et al., 2021). However, it remains to be examined that whether the accumulation of Ca^{2+} channels is due to neighboring channels being added to each release size or the channels of each cluster get closer without altering their total number.

VGCCs and Synaptic Heterogeneity

As mentioned above, there are two main groups of VGCCs, low and high thresholds, and each has multiple subgroups (see Table 1). While some peripheral neurons, like superior cervical ganglion cells, are known to express only one presynaptic channel (Mochida et al., 2003), it is well known that a majority of brain regions and neurons express multiple VGCCs (Vacher et al., 2008). Considering the additional diversity of the auxiliary subunits and the fact that all α1-subunits seem to be capable of assembling with all β and α2δ isoforms, the complexity of possible subunit compositions becomes enormous. For example, 10 distinct presynaptic VGCCs $α_1$ isoforms plus three $α_2δ$ and four β-subunits already give 120 possible channel compositions, and that is without considering the splice variants that exist for all of the genes. Adding the divers VGCC interacting proteins (e.g., SNARE proteins), the synaptic diversity surges.

The simplest cellular mechanism to manage this diversity is to limit the number of isoforms expressed in a single cell at a given time (e.g., during development). For example, in skeletal muscle, CaV1.1/$β_{1a}$/$α_2δ$-1 (Arikkath & Campbell, 2003) are the dominant types, in cardiac myocytes CaV1.2/$β_2$/$α_2δ$-1 (Arikkath & Campbell, 2003), in retinal photoreceptor cells CaV1.4/$β_2$/$α_2δ$-4 (Wycisk et al., 2006; Ball et al., 2002), and in the cerebellum CaV2.1/$β_4$/$α_2δ$-2 (Ludwig et al., 1997; Brodbeck et al., 2002). In contrast, the cerebral cortex shows a more heterogeneous expression of VGCC isoforms (Ludwig et al., 1997; Klugbauer et al., 1999). Existing evidence indicates that these spatiotemporal expression patterns in cerebellum and cortex, respectively, are also reflected in the $α_1$-subunit expression of specific types of neurons, such as cerebellar granule cells and hippocampal pyramidal cells (Hell et al., 1993; Lorenzon & Foehring, 1995; Randall & Tsien, 1995; Westenbroek et al., 1995).

Spatiotemporal Heterogeneity of Diverse Types of VGCCs

The expression profiles of VGCCs in mouse brain and cultured hippocampal neurons showed that overall, L-type calcium channels have the lowest expression and among those, $Ca_V1.2$ and $Ca_V1.3$ are the only L-type calcium channels expressed in mouse brain. In the cerebellum, they are found at a ratio of 4:1 ($Ca_V1.2$: $Ca_V1.3$), whereas in the cortex and hippocampus, $Ca_V1.2$ and $Ca_V1.3$ are expressed equally (Schlick et al., 2010).

$Ca_V2.1$ has the highest expression of $\alpha1$-subunit in the cortex and cerebellum. $Ca_V2.3$ is the dominant type of VGCCs in the hippocampus (Schlick et al., 2010). Considering this diversity, it is important to know 1) if VGCCs $\alpha1$ isoforms prefer specific auxiliary isoforms (e.g., $\beta_{2 \text{ or }} \beta_4$), 2) if certain isoforms have co-regulated expression pattern (e.g., $\alpha1$ and $\beta_{2 \text{ or }} \alpha1$ and β_4), and 3) if this preference or co-regulation is persistent across the brain regions.

Studies on expression profiles of VGCCs and their auxiliary subunits showed that CaV2.1 has a specific preference for $\beta4$- and $\alpha2\delta$-2 subunits. This finding is based on their high expression profile in cerebellum and their co-upregulation during the development of hippocampal pyramidal cells (Schlick et al., 2010). Meanwhile, $Ca_V1.2$, $Ca_V2.2$, and $Ca_V2.3$ expressions correlate positively with $\beta1$, $\beta3$, $\alpha2\delta$-1, and $\alpha2\delta$-3, respectively. It is worth mentioning that these auxiliary subunits inversely correlate with the level of $Ca_V2.1$ (Schlick et al., 2010). The specificity of $Ca_V2.1/\beta_4/\alpha_2\delta$-2 combination contrast to a diverse association of $Ca_V2.2$ and $Ca_V2.3$ with any combination of β_1, β_3, $\alpha_2\delta$-1, and $\alpha_2\delta$-3 may well explain why the loss of $Ca_V2.1$ in double-mutant mice causes severe neurological diseases (Jun et al., 1999), while loss of $Ca_V2.2$ or 2.3 has little to weak neurological phenotype, indicating that they can be compensated (Saegusa et al., 2000; Ino et al., 2001).

Morphological Modules of VGCC Clusters in Synaptic Heterogeneity

Synapses exhibit vast diversity in their structure and function. Even in a homogeneous population of connections, large variability is found in their shapes and sizes as well as the functional properties such as the probability of the neurotransmission, the RRP size, and the short- and long-term plasticity of the synapses (Schikorski & Stevens, 1997; Shepherd & Harris, 1998; Atwood & Karunanithi, 2002).

The idea of structural and functional continuum in the synapses was examined by a study on calyx of Held and neurons of the medial nucleus of the trapezoid body (MNTB). The morphology of synaptic terminals contributes to different synaptic properties, such as changes in short-term plasticity (Grande & Wang, 2011). The calyx of Held is one of the biggest glutamatergic (excitatory) presynaptic terminals, located in the MNTB of the brainstem, and has a critical role in the binaural sound localization circuitry (Kuwabara et al., 1991; Smith et al., 1991; Forsythe, 1994). Therefore, different short-term plasticity, represented by different types of calyx of Held, defines the level of signals delivered to other brainstem centers for the computation of interaural timing and intensity differences required for sound localization (Grande & Wang, 2011). The mature calyx of Held can be broadly classified into three distinct types of morphologies based on their number of swellings (Fig. 4). It was shown that the Type I represents strong short-term depression after a high-frequency stimulation, while Type III shows short-term facilitation before going to depression, and Type II had intermediate characterizations of both extremes (Grande & Wang, 2011). The strong depression in type 1 calyces might be due to the smaller RRP size compared to the Type 3 ones (Grande & Wang, 2011). Another aspect of structural and functional continuum can be found in the correlations between the number of swellings and the residual calcium in the terminal as the narrowed geometry of the swellings dictate the diffusion of calcium (Grande & Wang, 2011).

VGCC distribution is locally organized into different clusters, whereas calcium sensors (e.g., synaptotagmin) are located at the perimeter of them (<30 nm). Each cluster may contain variable numbers of VGCCs which leads to different release probabilities (p); the more VGCCs per cluster, the higher the p (Nakamura et al., 2015).

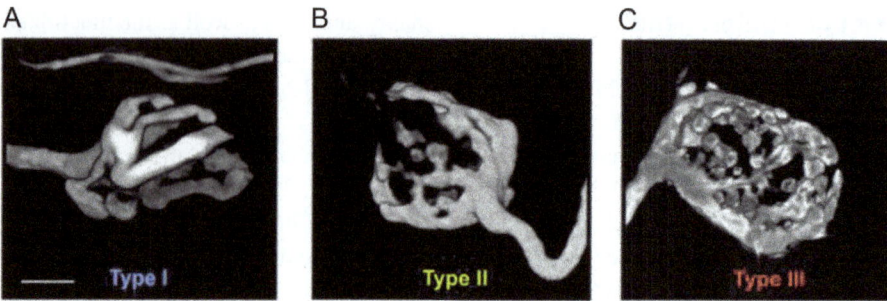

Fig. 4 The three different types of calyx of Held in the mature brain. (**a**) Type I with 0 to 5 swellings (also referred to as simple synapse). (**b**) Type II calyx with the intermediate number of swellings (8–12), and (**c**) Type III or complex calyx with more than 15 swellings. (Figure from Grande and Wang 2011)

It is worth mentioning that previous studies showed that the release probability of individual terminals as well as the total amount of Ca^{2+} influx have a linear correlation with the size of the active zone in hippocampal pyramidal cells (Holderith et al., 2012). In addition, $Ca_V2.1$ has two main patterns of distributions in the soma and the dendrites of Purkinje cells of the cerebellum, that is, the scattered and the clustered. The scattered pattern is non-uniform distributions of the channels from soma to distal dendrite with gradual increase in their density, and the clustered pattern is concentrated within intramembrane clusters in the soma and proximal dendrites (Indriati et al., 2013).

The release mechanisms at the active zones were better understood after the number of VGCCs per active zone was determined (Sheng et al., 2012). Each active zone includes 5–218 (mean, 42) of VGCCs, 1–10 (mean, 5) readily releasable synaptic vesicles, and after a 2-ms depolarization, a maximum five of these vesicles can be released (Sheng et al., 2012). An interesting finding here was the wide range of VGCCs in each AZ. Considering that each action potential can open up to 35 VGCCs, this diversity can lead to varied release strength by regulating the release probability and the RRP size (Sheng et al., 2012).

The idea of synaptic heterogeneity was further advanced by Fekete et al. (2019) when VGCCs in the calyx were clustered in two different types (simple versus complex). As the number of swelling increases, the probability of the neurotrans-mitter release decreases (Fekete et al., 2019). As mentioned above, this can be due to a smaller cluster size and longer coupling distance between the synaptic vesicles and VGCCs in swellings than that in stalks. Importantly, disrupting septin preferentially increases the synaptic strength in the complex (or type III) synapses (Fekete et al., 2019) (Fig. 5). However, it remains to be explored how different compartments segregate with different number and/or splicing variants of P/Q-type channels to diversify bimodal distribution of morphological ensembles.

Conclusion

In this chapter, we outlined the role of VGCCs in neurotransmitter release. VGCCs enable the communications between neurons by converting electrical signals into the release of neurotransmitters. The evolutionary conserved release machinery shows the importance of this phenomenon on brain functions (Südhof, 2012). Evolution tends to preserve those that can adapt to the changes in the environment as best exemplified by the brain, which manifests flexibility. As we discussed here, different types of VGCCs along with a range of auxiliary subunits and binding proteins dynamically interact in response to the environmental stimuli. The most noticeable changes are those occurring during development. The brain starts from a state of being erroneous and imprecise, having excessive numbers of synapses that are trimmed developmentally to tune

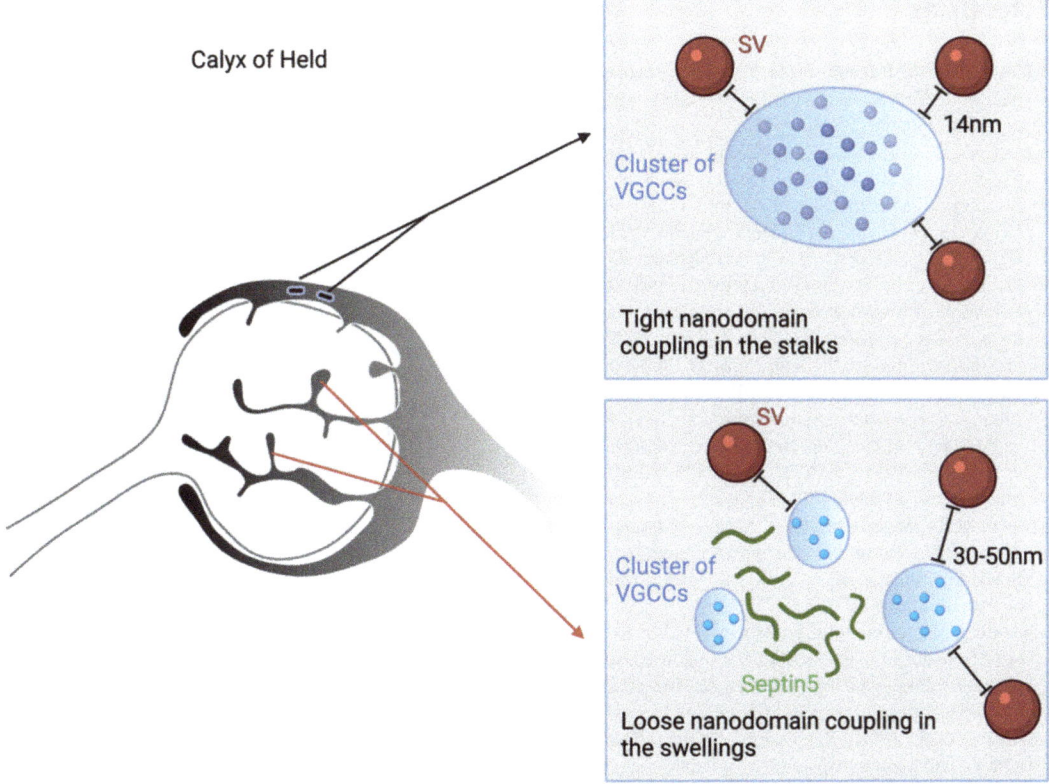

Fig. 5 Synaptic heterogeneity on the calyx of Held. In the complex (Type III) calyces, swellings have smaller VGCC cluster, larger coupling distance between the synaptic vesicles and the calcium channels, and less release probability compared to the stalks

the plasticity, fidelity, and accuracy that can adequately respond to in the changing environment. This process is linked to the changes in the protein matrix of each synapse which then gives rise to the circuit modifications in the whole brain. This chapter highlights a few examples of developmental changes in the VGCCs and their partners that form protein complexes.

Even after the brain has fully matured, modifications continue to take place in the form of synaptic plasticity. The greater the use of a certain neuronal pathway, the more plastic it becomes, deviating from the initial properties of that neuronal circuit. Not all neuronal connections are the same, synaptic heterogeneity from genetically homogeneous neurons assures diverse synaptic connections even within parallel circuits of the same brain region. All these diversities make each neuron unique, and, undoubtedly, complicates our understanding of the whole brain network seen from a limited, purely "connectomic" picture of the circuits.

> If the human brain were so simple that we could understand it, we would be so simple that we couldn't – Emerson M. Pugh.

Acknowledgments This work was supported by grants from Canadian Institutes of Health Research, the Natural Science and Engineering Research Council, Canada Research Chairs Program, and SickKids Research Institute and University of Toronto.

References

Abbott, L. F., & Regehr, W. G. (2004). Synaptic computation. *Nature, 431*(7010), 796–803. https://doi.org/10.1038/nature03010

Altrock, W. D., Tom Dieck, S., Sokolov, M., Meyer, A. C., Sigler, A., Brakebusch, C., Fässler, R., Richter, K., Boeckers, T. M., Potschka, H., Brandt, C., Löscher, W., Grimberg, D., Dresbach, T., Hempelmann, A., Hassan, H., Balschun, D., Frey, J. U., Brandstätter, J. H., … Gundelfinger, E. D. (2003). Functional inactivation of a fraction of excitatory synapses in mice deficient for the active zone protein bassoon. *Neuron, 37*(5), 787–800. https://doi.org/10.1016/s0896-6273(03)00088-6

Arikkath, J., & Campbell, K. P. (2003). Auxiliary subunits: Essential components of the voltage-gated calcium channel complex. *Current Opinion in Neurobiology, 13*(3), 298–307. https://doi.org/10.1016/s0959-4388(03)00066-7

Atwood, H. L., & Karunanithi, S. (2002). Diversification of synaptic strength: Presynaptic elements. *Nature Reviews. Neuroscience, 3*(7), 497–516. https://doi.org/10.1038/nrn876

Augustin, I., Rosenmund, C., Südhof, T. C., & Brose, N. (1999). Munc13-1 is essential for fusion competence of glutamatergic synaptic vesicles. *Nature, 400*(6743), 457–461. https://doi.org/10.1038/22768

Augustine, G. J. (1990). Regulation of transmitter release at the squid giant synapse by presynaptic delayed rectifier potassium current. *The Journal of Physiology, 431*, 343–364. https://doi.org/10.1113/jphysiol.1990.sp018333

Augustine, G. J., Adler, E. M., & Charlton, M. P. (1991). The calcium signal for transmitter secretion from presynaptic nerve terminals. *Annals of the New York Academy of Sciences, 635*, 365–381. https://doi.org/10.1111/j.1749-6632.1991.tb36505.x

Ball, S. L., Powers, P. A., Shin, H.-S., Morgans, C. W., Peachey, N. S., & Gregg, R. G. (2002). Role of the beta(2) subunit of voltage-dependent calcium channels in the retinal outer plexiform layer. *Investigative Ophthalmology & Visual Science, 43*(5), 1595–1603.

Barr, A. M., Young, C. E., Sawada, K., Trimble, W. S., Phillips, A. G., & Honer, W. G. (2004). Abnormalities of presynaptic protein CDCrel-1 in striatum of rats reared in social isolation: Relevance to neural connectivity in schizophrenia. *The European Journal of Neuroscience, 20*(1), 303–307. https://doi.org/10.1111/j.0953-816X.2004.03457.x

Beites, C. L., Xie, H., Bowser, R., & Trimble, W. S. (1999). The septin CDCrel-1 binds syntaxin and inhibits exocytosis. *Nature Neuroscience, 2*(5), 434–439. https://doi.org/10.1038/8100

Bezprozvanny, I., Scheller, R. H., & Tsien, R. W. (1995). Functional impact of syntaxin on gating of N-type and Q-type calcium channels. *Nature, 378*(6557), 623–626. https://doi.org/10.1038/378623a0

Bharat, T. A. M., Malsam, J., Hagen, W. J. H., Scheutzow, A., Söllner, T. H., & Briggs, J. A. G. (2014). SNARE and regulatory proteins induce local membrane protrusions to prime docked vesicles for fast calcium-triggered fusion. *EMBO Reports, 15*(3), 308–314. https://doi.org/10.1002/embr.201337807

Borst, J. G., & Sakmann, B. (1998). Facilitation of presynaptic calcium currents in the rat brainstem. *The Journal of Physiology, 513*(Pt 1), 149–155. https://doi.org/10.1111/j.1469-7793.1998.149by.x

Brodbeck, J., Davies, A., Courtney, J.-M., Meir, A., Balaguero, N., Canti, C., Moss, F. J., Page, K. M., Pratt, W. S., Hunt, S. P., Barclay, J., Rees, M., & Dolphin, A. C. (2002). The ducky mutation in Cacna2d2 results in altered Purkinje cell morphology and is associated with the expression of a truncated alpha 2 delta-2 protein with abnormal function. *The Journal of Biological Chemistry, 277*(10), 7684–7693. https://doi.org/10.1074/jbc.M109404200

Brose, N., Hofmann, K., Hata, Y., & Südhof, T. C. (1995). Mammalian homologues of Caenorhabditis elegans unc-13 gene define novel family of C2-domain proteins. *The Journal of Biological Chemistry, 270*(42), 25273–25280. https://doi.org/10.1074/jbc.270.42.25273

Burgoyne, R. D., & Weiss, J. L. (2001). The neuronal calcium sensor family of Ca2+-binding proteins. *The Biochemical Journal, 353*(Pt 1), 1–12.

Byrne, J. H., & Kandel, E. R. (1996). Presynaptic facilitation revisited: State and time dependence. *The Journal of Neuroscience: The Official Journal of the Society for Neuroscience, 16*(2), 425–435.

Cao, Y.-Q., Piedras-Rentería, E. S., Smith, G. B., Chen, G., Harata, N. C., & Tsien, R. W. (2004). Presynaptic Ca2+ channels compete for channel type-preferring slots in altered neurotransmission arising from Ca2+ channelopathy. *Neuron, 43*(3), 387–400. https://doi.org/10.1016/j.neuron.2004.07.014

Cao, L., Ding, X., Yu, W., Yang, X., Shen, S., & Yu, L. (2007). Phylogenetic and evolutionary analysis of the septin protein family in metazoan. *FEBS Letters, 581*(28), 5526–5532. https://doi.org/10.1016/j.febslet.2007.10.032

Catterall, W. A. (2000). Structure and regulation of voltage-gated Ca2+ channels. *Annual Review of Cell and Developmental Biology, 16*, 521–555. https://doi.org/10.1146/annurev.cellbio.16.1.521

Chan, A. W., Khanna, R., Li, Q., & Stanley, E. F. (2007). Munc18: A presynaptic transmitter release site N type (CaV2.2) calcium channel interacting protein. *Channels (Austin, Tex.), 1*(1), 11–20.

Cohen, M. W., Jones, O. T., & Angelides, K. J. (1991). Distribution of Ca2+ channels on frog motor nerve terminals revealed by fluorescent omega-conotoxin. *The Journal of Neuroscience: The Official Journal of the Society for Neuroscience, 11*(4), 1032–1039.

Condliffe, S. B., Corradini, I., Pozzi, D., Verderio, C., & Matteoli, M. (2010). Endogenous SNAP-25 regulates native voltage-gated calcium channels in glutamatergic neurons. *The Journal of Biological Chemistry,*

285(32), 24968–24976. https://doi.org/10.1074/jbc.M110.145813

Coppola, T., Magnin-Luthi, S., Perret-Menoud, V., Gattesco, S., Schiavo, G., & Regazzi, R. (2001). Direct interaction of the Rab3 effector RIM with Ca2+ channels, SNAP-25, and synaptotagmin. *The Journal of Biological Chemistry, 276*(35), 32756–32762. https://doi.org/10.1074/jbc.M100929200

Currie, K. P. M. (2010). G protein modulation of CaV2 voltage-gated calcium channels. *Channels (Austin, Tex.), 4*(6), 497–509. https://doi.org/10.4161/chan.4.6.12871

Dai, Y., Taru, H., Deken, S. L., Grill, B., Ackley, B., Nonet, M. L., & Jin, Y. (2006). SYD-2 Liprin-alpha organizes presynaptic active zone formation through ELKS. *Nature Neuroscience, 9*(12), 1479–1487. https://doi.org/10.1038/nn1808

Davletov, B. A., & Südhof, T. C. (1993). A single C2 domain from synaptotagmin I is sufficient for high affinity Ca^{2+}/phospholipid binding. *The Journal of Biological Chemistry, 268*(35), 26386–26390.

Davydova, D., Marini, C., King, C., Klueva, J., Bischof, F., Romorini, S., Montenegro-Venegas, C., Heine, M., Schneider, R., Schröder, M. S., Altrock, W. D., Henneberger, C., Rusakov, D. A., Gundelfinger, E. D., & Fejtova, A. (2014). Bassoon specifically controls presynaptic P/Q-type Ca(2+) channels via RIM-binding protein. *Neuron, 82*(1), 181–194. https://doi.org/10.1016/j.neuron.2014.02.012

Degtiar, V. E., Scheller, R. H., & Tsien, R. W. (2000). Syntaxin modulation of slow inactivation of N-type calcium channels. *The Journal of Neuroscience: The Official Journal of the Society for Neuroscience, 20*(12), 4355–4367.

Deken, S. L., Vincent, R., Hadwiger, G., Liu, Q., Wang, Z.-W., & Nonet, M. L. (2005). Redundant localization mechanisms of RIM and ELKS in Caenorhabditis elegans. *The Journal of Neuroscience: The Official Journal of the Society for Neuroscience, 25*(25), 5975–5983. https://doi.org/10.1523/JNEUROSCI.0804-05.2005

DeMaria, C. D., Soong, T. W., Alseikhan, B. A., Alvania, R. S., & Yue, D. T. (2001). Calmodulin bifurcates the local Ca2+ signal that modulates P/Q-type Ca2+ channels. *Nature, 411*(6836), 484–489. https://doi.org/10.1038/35078091

Dickinson-Nelson, A., & Reese, T. S. (1983). Structural changes during transmitter release at synapses in the frog sympathetic ganglion. *The Journal of Neuroscience: The Official Journal of the Society for Neuroscience, 3*(1), 42–52.

Dong, W., Radulovic, T., Goral, R. O., Thomas, C., Suarez Montesinos, M., Guerrero-Given, D., Hagiwara, A., Putzke, T., Hida, Y., Abe, M., Sakimura, K., Kamasawa, N., Ohtsuka, T., & Young, S. M. (2018). CAST/ELKS proteins control voltage-gated Ca2+ channel density and synaptic release probability at a mammalian central synapse. *Cell Reports, 24*(2), 284–293.e6. https://doi.org/10.1016/j.celrep.2018.06.024

Dulubova, I., Sugita, S., Hill, S., Hosaka, M., Fernandez, I., Südhof, T. C., & Rizo, J. (1999). A conformational switch in syntaxin during exocytosis: Role of munc18. *The EMBO Journal, 18*(16), 4372–4382. https://doi.org/10.1093/emboj/18.16.4372

Evans, R. M., & Zamponi, G. W. (2006). Presynaptic Ca2+ channels – Integration centers for neuronal signaling pathways. *Trends in Neurosciences, 29*(11), 617–624. https://doi.org/10.1016/j.tins.2006.08.006

Fasshauer, D., Sutton, R. B., Brunger, A. T., & Jahn, R. (1998). Conserved structural features of the synaptic fusion complex: SNARE proteins reclassified as Q- and R-SNAREs. *Proceedings of the National Academy of Sciences of the United States of America, 95*(26), 15781–15786. https://doi.org/10.1073/pnas.95.26.15781

Fedchyshyn, M. J., & Wang, L.-Y. (2005). Developmental transformation of the release modality at the calyx of Held synapse. *The Journal of Neuroscience: The Official Journal of the Society for Neuroscience, 25*(16), 4131–4140. https://doi.org/10.1523/JNEUROSCI.0350-05.2005

Fekete, A., Nakamura, Y., Yang, Y.-M., Herlitze, S., Mark, M. D., DiGregorio, D. A., & Wang, L.-Y. (2019). Underpinning heterogeneity in synaptic transmission by presynaptic ensembles of distinct morphological modules. *Nature Communications, 10*(1), 826. https://doi.org/10.1038/s41467-019-08452-2

Fernández-Chacón, R., Königstorfer, A., Gerber, S. H., García, J., Matos, M. F., Stevens, C. F., Brose, N., Rizo, J., Rosenmund, C., & Südhof, T. C. (2001). Synaptotagmin I functions as a calcium regulator of release probability. *Nature, 410*(6824), 41–49. https://doi.org/10.1038/35065004

Forsythe, I. D. (1994). Direct patch recording from identified presynaptic terminals mediating glutamatergic EPSCs in the rat CNS, in vitro. *The Journal of Physiology, 479*(Pt 3), 381–387. https://doi.org/10.1113/jphysiol.1994.sp020303

Fukaya, R., Maglione, M., Sigrist, S. J., & Sakaba, T. (2021). Rapid Ca2+ channel accumulation contributes to cAMP-mediated increase in transmission at hippocampal mossy fiber synapses. *Proceedings of the National Academy of Sciences of the United States of America, 118*(9), e2016754118. https://doi.org/10.1073/pnas.2016754118

Ge, D., Noakes, P. G., & Lavidis, N. A. (2019). What are neurotransmitter release sites and do they interact? *Neuroscience.* https://doi.org/10.1016/j.neuroscience.2019.11.017

Gerachshenko, T., Blackmer, T., Yoon, E. J., Bartleson, C., Hamm, H. E., & Alford, S. (2005). Gbetagamma acts at the C terminus of SNAP-25 to mediate presynaptic inhibition. *Nature Neuroscience, 8*(5), 597–605. https://doi.org/10.1038/nn1439

Gerber, S. H., Garcia, J., Rizo, J., & Südhof, T. C. (2001). An unusual C(2)-domain in the active-zone protein piccolo: Implications for Ca(2+) regulation of neurotransmitter release. *The EMBO Journal, 20*(7), 1605–1619. https://doi.org/10.1093/emboj/20.7.1605

Gifford, J. L., Walsh, M. P., & Vogel, H. J. (2007). Structures and metal-ion-binding properties of the

Ca2+-binding helix-loop-helix EF-hand motifs. *The Biochemical Journal, 405*(2), 199–221. https://doi.org/10.1042/BJ20070255

Girard, F., Venail, J., Schwaller, B., & Celio, M. R. (2015). The EF-hand Ca(2+)-binding protein super-family: A genome-wide analysis of gene expression patterns in the adult mouse brain. *Neuroscience, 294*, 116–155. https://doi.org/10.1016/j.neuroscience.2015.02.018

Gladycheva, S. E., Ho, C. S., Lee, Y. Y. F., & Stuenkel, E. L. (2004). Regulation of syntaxin1A-munc18 complex for SNARE pairing in HEK293 cells. *The Journal of Physiology, 558*(Pt 3), 857–871. https://doi.org/10.1113/jphysiol.2004.067249

Grande, G., & Wang, L.-Y. (2011). Morphological and functional continuum underlying heterogeneity in the spiking fidelity at the calyx of Held synapse in vitro. *The Journal of Neuroscience: The Official Journal of the Society for Neuroscience, 31*(38), 13386–13399. https://doi.org/10.1523/JNEUROSCI.0400-11.2011

Hallermann, S., Fejtova, A., Schmidt, H., Weyhersmüller, A., Silver, R. A., Gundelfinger, E. D., & Eilers, J. (2010). Bassoon speeds vesicle reloading at a central excitatory synapse. *Neuron, 68*(4), 710–723. https://doi.org/10.1016/j.neuron.2010.10.026

Han, Y., Kaeser, P. S., Südhof, T. C., & Schneggenburger, R. (2011). RIM determines Ca²+ channel density and vesicle docking at the presynaptic active zone. *Neuron, 69*(2), 304–316. https://doi.org/10.1016/j.neuron.2010.12.014

Harkins, A. B., Cahill, A. L., Powers, J. F., Tischler, A. S., & Fox, A. P. (2004). Deletion of the synaptic protein interaction site of the N-type (CaV2.2) calcium channel inhibits secretion in mouse pheochromocytoma cells. *Proceedings of the National Academy of Sciences of the United States of America, 101*(42), 15219–15224. https://doi.org/10.1073/pnas.04010011010401001101

Heck, J., Parutto, P., Ciuraszkiewicz, A., Bikbaev, A., Freund, R., Mitlöhner, J., Andres-Alonso, M., Fejtova, A., Holcman, D., & Heine, M. (2019). Transient confinement of CaV2.1 Ca2+-channel splice variants shapes synaptic short-term plasticity. *Neuron, 103*(1), 66–79.e12. https://doi.org/10.1016/j.neuron.2019.04.030

Hell, J. W., Westenbroek, R. E., Warner, C., Ahlijanian, M. K., Prystay, W., Gilbert, M. M., Snutch, T. P., & Catterall, W. A. (1993). Identification and differential subcellular localization of the neuronal class C and class D L-type calcium channel alpha 1 subunits. *The Journal of Cell Biology, 123*(4), 949–962. https://doi.org/10.1083/jcb.123.4.949

Hibino, H., Pironkova, R., Onwumere, O., Vologodskaia, M., Hudspeth, A. J., & Lesage, F. (2002). RIM binding proteins (RBPs) couple Rab3-interacting molecules (RIMs) to voltage-gated Ca(2+) channels. *Neuron, 34*(3), 411–423. https://doi.org/10.1016/s0896-6273(02)00667-0

Holderith, N., Lorincz, A., Katona, G., Rózsa, B., Kulik, A., Watanabe, M., & Nusser, Z. (2012). Release probability of hippocampal glutamatergic terminals scales with the size of the active zone. *Nature Neuroscience, 15*(7), 988–997. https://doi.org/10.1038/nn.3137

Indriati, D. W., Kamasawa, N., Matsui, K., Meredith, A. L., Watanabe, M., & Shigemoto, R. (2013). Quantitative localization of Cav2.1 (P/Q-type) voltage-dependent calcium channels in Purkinje cells: Somatodendritic gradient and distinct somatic coclustering with calcium-activated potassium channels. *The Journal of Neuroscience: The Official Journal of the Society for Neuroscience, 33*(8), 3668–3678. https://doi.org/10.1523/JNEUROSCI.2921-12.2013

Ino, M., Yoshinaga, T., Wakamori, M., Miyamoto, N., Takahashi, E., Sonoda, J., Kagaya, T., Oki, T., Nagasu, T., Nishizawa, Y., Tanaka, I., Imoto, K., Aizawa, S., Koch, S., Schwartz, A., Niidome, T., Sawada, K., & Mori, Y. (2001). Functional disorders of the sympathetic nervous system in mice lacking the alpha 1B subunit (Cav 2.2) of N-type calcium channels. *Proceedings of the National Academy of Sciences of the United States of America, 98*(9), 5323–5328. https://doi.org/10.1073/pnas.081089398

Iwasaki, S., Momiyama, A., Uchitel, O. D., & Takahashi, T. (2000). Developmental changes in calcium channel types mediating central synaptic transmission. *The Journal of Neuroscience: The Official Journal of the Society for Neuroscience, 20*(1), 59–65.

Jahn, R., & Scheller, R. H. (2006). SNAREs – Engines for membrane fusion. *Nature Reviews. Molecular Cell Biology, 7*(9), 631–643. https://doi.org/10.1038/nrm2002

Jarvis, S. E., & Zamponi, G. W. (2001). Distinct molecular determinants govern syntaxin 1A-mediated inactivation and G-protein inhibition of N-type calcium channels. *The Journal of Neuroscience: The Official Journal of the Society for Neuroscience, 21*(9), 2939–2948.

Jarvis, S. E., Magga, J. M., Beedle, A. M., Braun, J. E., & Zamponi, G. W. (2000). G protein modulation of N-type calcium channels is facilitated by physical interactions between syntaxin 1A and Gbetagamma. *The Journal of Biological Chemistry, 275*(9), 6388–6394. https://doi.org/10.1074/jbc.275.9.6388

Jing, Z., Rutherford, M. A., Takago, H., Frank, T., Fejtova, A., Khimich, D., Moser, T., & Strenzke, N. (2013). Disruption of the presynaptic cytomatrix protein bassoon degrades ribbon anchorage, multiquantal release, and sound encoding at the hair cell afferent synapse. *The Journal of Neuroscience: The Official Journal of the Society for Neuroscience, 33*(10), 4456–4467. https://doi.org/10.1523/JNEUROSCI.3491-12.2013

Jun, K., Piedras-Rentería, E. S., Smith, S. M., Wheeler, D. B., Lee, S. B., Lee, T. G., Chin, H., Adams, M. E., Scheller, R. H., Tsien, R. W., & Shin, H. S. (1999). Ablation of P/Q-type Ca(2+) channel currents, altered synaptic transmission, and progressive ataxia in mice lacking the alpha(1A)-subunit. *Proceedings of the National Academy of Sciences of the United States of America, 96*(26), 15245–15250. https://doi.org/10.1073/pnas.96.26.15245

Kaeser, P. S., Deng, L., Chávez, A. E., Liu, X., Castillo, P. E., & Südhof, T. C. (2009). ELKS2alpha/CAST deletion selectively increases neurotransmitter release at inhibitory synapses. *Neuron, 64*(2), 227–239. https://doi.org/10.1016/j.neuron.2009.09.019

Kaeser, P. S., Deng, L., Wang, Y., Dulubova, I., Liu, X., Rizo, J., & Südhof, T. C. (2011). RIM proteins tether Ca2+ channels to presynaptic active zones via a direct PDZ-domain interaction. *Cell, 144*(2), 282–295. https://doi.org/10.1016/j.cell.2010.12.029

Katz, B., & Miledi, R. (1967). Ionic requirements of synaptic transmitter release. *Nature, 215*(5101), 651. https://doi.org/10.1038/215651a0

Katz, B., & Miledi, R. (1968). The role of calcium in neuromuscular facilitation. *The Journal of Physiology, 195*(2), 481–492. https://doi.org/10.1113/jphysiol.1968.sp008469

Kaufmann, N., DeProto, J., Ranjan, R., Wan, H., & Van Vactor, D. (2002). Drosophila liprin-alpha and the receptor phosphatase Dlar control synapse morphogenesis. *Neuron, 34*(1), 27–38. https://doi.org/10.1016/s0896-6273(02)00643-8

Kawabe, H., Mitkovski, M., Kaeser, P. S., Hirrlinger, J., Opazo, F., Nestvogel, D., Kalla, S., Fejtova, A., Verrier, S. E., Bungers, S. R., Cooper, B. H., Varoqueaux, F., Wang, Y., Nehring, R. B., Gundelfinger, E. D., Rosenmund, C., Rizzoli, S. O., Südhof, T. C., Rhee, J.-S., & Brose, N. (2017). ELKS1 localizes the synaptic vesicle priming protein bMunc13-2 to a specific subset of active zones. *The Journal of Cell Biology, 216*(4), 1143–1161. https://doi.org/10.1083/jcb.201606086

Keith, R. K., Poage, R. E., Yokoyama, C. T., Catterall, W. A., & Meriney, S. D. (2007). Bidirectional modulation of transmitter release by calcium channel/syntaxin interactions in vivo. *The Journal of Neuroscience: The Official Journal of the Society for Neuroscience, 27*(2), 265–269. https://doi.org/10.1523/JNEUROSCI.4213-06.2007

Kinoshita, A., Kinoshita, M., Akiyama, H., Tomimoto, H., Akiguchi, I., Kumar, S., Noda, M., & Kimura, J. (1998). Identification of septins in neurofibrillary tangles in Alzheimer's disease. *The American Journal of Pathology, 153*(5), 1551–1560. https://doi.org/10.1016/S0002-9440(10)65743-4

Kittel, R. J., Wichmann, C., Rasse, T. M., Fouquet, W., Schmidt, M., Schmid, A., Wagh, D. A., Pawlu, C., Kellner, R. R., Willig, K. I., Hell, S. W., Buchner, E., Heckmann, M., & Sigrist, S. J. (2006). Bruchpilot promotes active zone assembly, Ca2+ channel clustering, and vesicle release. *Science (New York, N.Y.), 312*(5776), 1051–1054. https://doi.org/10.1126/science.1126308

Kiyonaka, S., Wakamori, M., Miki, T., Uriu, Y., Nonaka, M., Bito, H., Beedle, A. M., Mori, E., Hara, Y., De Waard, M., Kanagawa, M., Itakura, M., Takahashi, M., Campbell, K. P., & Mori, Y. (2007). RIM1 confers sustained activity and neurotransmitter vesicle anchoring to presynaptic Ca2+ channels. *Nature Neuroscience, 10*(6), 691–701. https://doi.org/10.1038/nn1904

Klugbauer, N., Lacinová, L., Marais, E., Hobom, M., & Hofmann, F. (1999). Molecular diversity of the calcium channel alpha2delta subunit. *The Journal of Neuroscience: The Official Journal of the Society for Neuroscience, 19*(2), 684–691.

Knowles, P. P., Murray-Rust, J., Kjaer, S., Scott, R. P., Hanrahan, S., Santoro, M., Ibáñez, C. F., & McDonald, N. Q. (2006). Structure and chemical inhibition of the RET tyrosine kinase domain. *The Journal of Biological Chemistry, 281*(44), 33577–33587. https://doi.org/10.1074/jbc.M605604200

Ko, J., Na, M., Kim, S., Lee, J.-R., & Kim, E. (2003). Interaction of the ERC family of RIM-binding proteins with the liprin-alpha family of multidomain proteins. *The Journal of Biological Chemistry, 278*(43), 42377–42385. https://doi.org/10.1074/jbc.M307561200

Koushika, S. P., Richmond, J. E., Hadwiger, G., Weimer, R. M., Jorgensen, E. M., & Nonet, M. L. (2001). A post-docking role for active zone protein Rim. *Nature Neuroscience, 4*(10), 997–1005. https://doi.org/10.1038/nn732

Kuwabara, N., DiCaprio, R. A., & Zook, J. M. (1991). Afferents to the medial nucleus of the trapezoid body and their collateral projections. *The Journal of Comparative Neurology, 314*(4), 684–706. https://doi.org/10.1002/cne.903140405

Lautermilch, N. J., Few, A. P., Scheuer, T., & Catterall, W. A. (2005). Modulation of CaV2.1 channels by the neuronal calcium-binding protein visinin-like protein-2. *The Journal of Neuroscience: The Official Journal of the Society for Neuroscience, 25*(30), 7062–7070. https://doi.org/10.1523/JNEUROSCI.0447-05.2005

Lee, A., Westenbroek, R. E., Haeseleer, F., Palczewski, K., Scheuer, T., & Catterall, W. A. (2002). Differential modulation of Ca(v)2.1 channels by calmodulin and Ca2+-binding protein 1. *Nature Neuroscience, 5*(3), 210–217. https://doi.org/10.1038/nn805

Limbach, C., Laue, M. M., Wang, X., Hu, B., Thiede, N., Hultqvist, G., & Kilimann, M. W. (2011). Molecular in situ topology of Aczonin/Piccolo and associated proteins at the mammalian neurotransmitter release site. *Proceedings of the National Academy of Sciences of the United States of America, 108*(31), E392-401. https://doi.org/10.1073/pnas.1101707108

Lipscombe, D., Allen, S. E., & Toro, C. P. (2013). Control of neuronal voltage-gated calcium ion channels from RNA to protein. *Trends in Neurosciences, 36*(10), 598–609. https://doi.org/10.1016/j.tins.2013.06.008

Lorenzon, N. M., & Foehring, R. C. (1995). Characterization of pharmacologically identified voltage-gated calcium channel currents in acutely isolated rat neocortical neurons. I. Adult neurons. *Journal of Neurophysiology, 73*(4), 1430–1442. https://doi.org/10.1152/jn.1995.73.4.1430

Lu, J., Machius, M., Dulubova, I., Dai, H., Südhof, T. C., Tomchick, D. R., & Rizo, J. (2006). Structural basis for a Munc13-1 homodimer to Munc13-1/RIM heterodimer switch. *PLoS Biology, 4*(7), e192. https://doi.org/10.1371/journal.pbio.0040192

Lübbert, M., Goral, R. O., Satterfield, R., Putzke, T., van den Maagdenberg, A. M., Kamasawa, N., & Young, S. M. (2017). A novel region in the CaV2.1 α1 subunit C-terminus regulates fast synaptic vesicle fusion and vesicle docking at the mammalian presynaptic active zone. *ELife, 6*, e28412. https://doi.org/10.7554/eLife.28412

Ludwig, A., Flockerzi, V., & Hofmann, F. (1997). Regional expression and cellular localization of the alpha1 and beta subunit of high voltage-activated calcium channels in rat brain. *The Journal of Neuroscience: The Official Journal of the Society for Neuroscience, 17*(4), 1339–1349.

Maruyama, I. N., & Brenner, S. (1991). A phorbol ester/diacylglycerol-binding protein encoded by the unc-13 gene of Caenorhabditis elegans. *Proceedings of the National Academy of Sciences of the United States of America, 88*(13), 5729–5733. https://doi.org/10.1073/pnas.88.13.5729

Mochida, S., Sheng, Z. H., Baker, C., Kobayashi, H., & Catterall, W. A. (1996). Inhibition of neurotransmission by peptides containing the synaptic protein interaction site of N-type Ca2+ channels. *Neuron, 17*(4), 781–788. https://doi.org/10.1016/s0896-6273(00)80209-3

Mochida, S., Westenbroek, R. E., Yokoyama, C. T., Itoh, K., & Catterall, W. A. (2003). Subtype-selective reconstitution of synaptic transmission in sympathetic ganglion neurons by expression of exogenous calcium channels. *Proceedings of the National Academy of Sciences of the United States of America, 100*(5), 2813–2818. https://doi.org/10.1073/pnas.262787299

Monier, S., Jollivet, F., Janoueix-Lerosey, I., Johannes, L., & Goud, B. (2002). Characterization of novel Rab6-interacting proteins involved in endosome-to-TGN transport. *Traffic (Copenhagen, Denmark), 3*(4), 289–297. https://doi.org/10.1034/j.1600-0854.2002.030406.x

Mukherjee, K., Yang, X., Gerber, S. H., Kwon, H.-B., Ho, A., Castillo, P. E., Liu, X., & Südhof, T. C. (2010). Piccolo and bassoon maintain synaptic vesicle clustering without directly participating in vesicle exocytosis. *Proceedings of the National Academy of Sciences of the United States of America, 107*(14), 6504–6509. https://doi.org/10.1073/pnas.1002307107

Nakamura, Y., Harada, H., Kamasawa, N., Matsui, K., Rothman, J. S., Shigemoto, R., Silver, R. A., DiGregorio, D. A., & Takahashi, T. (2015). Nanoscale distribution of presynaptic Ca(2+) channels and its impact on vesicular release during development. *Neuron, 85*(1), 145–158. https://doi.org/10.1016/j.neuron.2014.11.019

Neher, E. (1998). Usefulness and limitations of linear approximations to the understanding of Ca++ signals. *Cell Calcium, 24*(5–6), 345–357. https://doi.org/10.1016/s0143-4160(98)90058-6

Neher, E., & Brose, N. (2018). Dynamically primed synaptic vesicle states: Key to understand synaptic short-term plasticity. *Neuron, 100*(6), 1283–1291. https://doi.org/10.1016/j.neuron.2018.11.024

Nicoll, R. A. (2017). A brief history of long-term potentiation. *Neuron, 93*(2), 281–290. https://doi.org/10.1016/j.neuron.2016.12.015

Ohtsuka, T., Takao-Rikitsu, E., Inoue, E., Inoue, M., Takeuchi, M., Matsubara, K., Deguchi-Tawarada, M., Satoh, K., Morimoto, K., Nakanishi, H., & Takai, Y. (2002). Cast: A novel protein of the cytomatrix at the active zone of synapses that forms a ternary complex with RIM1 and munc13-1. *The Journal of Cell Biology, 158*(3), 577–590. https://doi.org/10.1083/jcb.200202083

Patwardhan, A., Bardin, S., Miserey-Lenkei, S., Larue, L., Goud, B., Raposo, G., & Delevoye, C. (2017). Routing of the RAB6 secretory pathway towards the lysosome related organelle of melanocytes. *Nature Communications, 8*, 15835. https://doi.org/10.1038/ncomms15835

Randall, A., & Tsien, R. W. (1995). Pharmacological dissection of multiple types of Ca2+ channel currents in rat cerebellar granule neurons. *The Journal of Neuroscience: The Official Journal of the Society for Neuroscience, 15*(4), 2995–3012.

Rettig, J., Sheng, Z. H., Kim, D. K., Hodson, C. D., Snutch, T. P., & Catterall, W. A. (1996). Isoform-specific interaction of the alpha1A subunits of brain Ca2+ channels with the presynaptic proteins syntaxin and SNAP-25. *Proceedings of the National Academy of Sciences of the United States of America, 93*(14), 7363–7368. https://doi.org/10.1073/pnas.93.14.7363

Rettig, J., Heinemann, C., Ashery, U., Sheng, Z. H., Yokoyama, C. T., Catterall, W. A., & Neher, E. (1997). Alteration of Ca^{2+} dependence of neurotransmitter release by disruption of Ca^{2+} channel/syntaxin interaction. *The Journal of Neuroscience: The Official Journal of the Society for Neuroscience, 17*(17), 6647–6656.

Rizo, J., & Rosenmund, C. (2008). Synaptic vesicle fusion. *Nature Structural & Molecular Biology, 15*(7), 665–674. https://doi.org/10.1038/nsmb.1450

Rozov, A., Burnashev, N., Sakmann, B., & Neher, E. (2001). Transmitter release modulation by intracellular Ca2+ buffers in facilitating and depressing nerve terminals of pyramidal cells in layer 2/3 of the rat neocortex indicates a target cell-specific difference in presynaptic calcium dynamics. *The Journal of Physiology, 531*(Pt 3), 807–826. https://doi.org/10.1111/j.1469-7793.2001.0807h.x

Saegusa, H., Kurihara, T., Zong, S., Minowa, O., Kazuno, A., Han, W., Matsuda, Y., Yamanaka, H., Osanai, M., Noda, T., & Tanabe, T. (2000). Altered pain responses in mice lacking alpha 1E subunit of the voltage-dependent Ca^{2+} channel. *Proceedings of the National Academy of Sciences of the United States of America, 97*(11), 6132–6137. https://doi.org/10.1073/pnas.100124197

Schikorski, T., & Stevens, C. F. (1997). Quantitative ultrastructural analysis of hippocampal excitatory synapses. *The Journal of Neuroscience: The Official Journal of the Society for Neuroscience, 17*(15), 5858–5867.

Schlick, B., Flucher, B. E., & Obermair, G. J. (2010). Voltage-activated calcium channel expression profiles in mouse brain and cultured hippocampal neurons. *Neuroscience, 167*(3), 786–798. https://doi.org/10.1016/j.neuroscience.2010.02.037

Schneggenburger, R., & Neher, E. (2000). Intracellular calcium dependence of transmitter release rates at a fast central synapse. *Nature, 406*(6798), 889–893. https://doi.org/10.1038/35022702

Schneggenburger, R., Sakaba, T., & Neher, E. (2002). Vesicle pools and short-term synaptic depression: Lessons from a large synapse. *Trends in Neurosciences, 25*(4), 206–212. https://doi.org/10.1016/s0166-2236(02)02139-2

Schoch, S., Castillo, P. E., Jo, T., Mukherjee, K., Geppert, M., Wang, Y., Schmitz, F., Malenka, R. C., & Südhof, T. C. (2002). RIM1alpha forms a protein scaffold for regulating neurotransmitter release at the active zone. *Nature, 415*(6869), 321–326. https://doi.org/10.1038/415321a

Serra-Pagès, C., Kedersha, N. L., Fazikas, L., Medley, Q., Debant, A., & Streuli, M. (1995). The LAR transmembrane protein tyrosine phosphatase and a coiled-coil LAR-interacting protein co-localize at focal adhesions. *The EMBO Journal, 14*(12), 2827–2838.

Sheng, Z. H., Rettig, J., Takahashi, M., & Catterall, W. A. (1994). Identification of a syntaxin-binding site on N-type calcium channels. *Neuron, 13*(6), 1303–1313. https://doi.org/10.1016/0896-6273(94)90417-0

Sheng, Z. H., Rettig, J., Cook, T., & Catterall, W. A. (1996). Calcium-dependent interaction of N-type calcium channels with the synaptic core complex. *Nature, 379*(6564), 451–454. https://doi.org/10.1038/379451a0

Sheng, Z. H., Yokoyama, C. T., & Catterall, W. A. (1997). Interaction of the synprint site of N-type Ca2+ channels with the C2B domain of synaptotagmin I. *Proceedings of the National Academy of Sciences of the United States of America, 94*(10), 5405–5410. https://doi.org/10.1073/pnas.94.10.5405

Sheng, J., He, L., Zheng, H., Xue, L., Luo, F., Shin, W., Sun, T., Kuner, T., Yue, D. T., & Wu, L.-G. (2012). Calcium-channel number critically influences synaptic strength and plasticity at the active zone. *Nature Neuroscience, 15*(7), 998–1006. https://doi.org/10.1038/nn.3129

Shepherd, G. M., & Harris, K. M. (1998). Three-dimensional structure and composition of CA3-->CA1 axons in rat hippocampal slices: Implications for presynaptic connectivity and compartmentalization. *The Journal of Neuroscience: The Official Journal of the Society for Neuroscience, 18*(20), 8300–8310.

Simon, S. M., & Llinás, R. R. (1985). Compartmentalization of the submembrane calcium activity during calcium influx and its significance in transmitter release.

Biophysical Journal, 48(3), 485–498. https://doi.org/10.1016/S0006-3495(85)83804-2

Smith, P. H., Joris, P. X., Carney, L. H., & Yin, T. C. (1991). Projections of physiologically characterized globular bushy cell axons from the cochlear nucleus of the cat. *The Journal of Comparative Neurology, 304*(3), 387–407. https://doi.org/10.1002/cne.903040305

Son, J. H., Kawamata, H., Yoo, M. S., Kim, D. J., Lee, Y. K., Kim, S., Dawson, T. M., Zhang, H., Sulzer, D., Yang, L., Beal, M. F., Degiorgio, L. A., Chun, H. S., Baker, H., & Peng, C. (2005). Neurotoxicity and behavioral deficits associated with Septin 5 accumulation in dopaminergic neurons. *Journal of Neurochemistry, 94*(4), 1040–1053. https://doi.org/10.1111/j.1471-4159.2005.03257.x

Südhof, T. C. (2012). The presynaptic active zone. *Neuron, 75*(1), 11–25. https://doi.org/10.1016/j.neuron.2012.06.012

Sudhof, T. C., & Rothman, J. E. (2009). Membrane fusion: Grappling with SNARE and SM proteins. *Science, 323*(5913), 474–477. https://doi.org/10.1126/science.1161748

Sutton, R. B., Davletov, B. A., Berghuis, A. M., Südhof, T. C., & Sprang, S. R. (1995). Structure of the first C2 domain of synaptotagmin I: A novel Ca2+/phospholipid-binding fold. *Cell, 80*(6), 929–938. https://doi.org/10.1016/0092-8674(95)90296-1

Taru, H., & Jin, Y. (2011). The Liprin homology domain is essential for the homomeric interaction of SYD-2/Liprin-α protein in presynaptic assembly. *The Journal of Neuroscience: The Official Journal of the Society for Neuroscience, 31*(45), 16261–16268. https://doi.org/10.1523/JNEUROSCI.0002-11.2011

Tedford, H. W., & Zamponi, G. W. (2006). Direct G protein modulation of Cav2 calcium channels. *Pharmacological Reviews, 58*(4), 837–862. https://doi.org/10.1124/pr.58.4.11

Timmermann, D. B., Westenbroek, R. E., Schousboe, A., & Catterall, W. A. (2002). Distribution of high-voltage-activated calcium channels in cultured gamma-aminobutyric acidergic neurons from mouse cerebral cortex. *Journal of Neuroscience Research, 67*(1), 48–61. https://doi.org/10.1002/jnr.10074

tom Dieck, S., Sanmartí-Vila, L., Langnaese, K., Richter, K., Kindler, S., Soyke, A., Wex, H., Smalla, K. H., Kämpf, U., Fränzer, J. T., Stumm, M., Garner, C. C., & Gundelfinger, E. D. (1998). Bassoon, a novel zinc-finger CAG/glutamine-repeat protein selectively localized at the active zone of presynaptic nerve terminals. *The Journal of Cell Biology, 142*(2), 499–509. https://doi.org/10.1083/jcb.142.2.499

Tsujimoto, T., Jeromin, A., Saitoh, N., Roder, J. C., & Takahashi, T. (2002). Neuronal calcium sensor 1 and activity-dependent facilitation of P/Q-type calcium currents at presynaptic nerve terminals. *Science (New York, N.Y.), 295*(5563), 2276–2279. https://doi.org/10.1126/science.1068278

Vacher, H., Mohapatra, D. P., & Trimmer, J. S. (2008). Localization and targeting of voltage-dependent ion channels in mammalian central neurons. *Physiological*

Reviews, 88(4), 1407–1447. https://doi.org/10.1152/physrev.00002.2008

Wadel, K., Neher, E., & Sakaba, T. (2007). The coupling between synaptic vesicles and Ca2+ channels determines fast neurotransmitter release. *Neuron, 53*(4), 563–575. https://doi.org/10.1016/j.neuron.2007.01.021

Wang, L. Y., & Kaczmarek, L. K. (1998). High-frequency firing helps replenish the readily releasable pool of synaptic vesicles. *Nature, 394*(6691), 384–388. https://doi.org/10.1038/28645

Wang, L.-Y., Neher, E., & Taschenberger, H. (2008). Synaptic vesicles in mature calyx of Held synapses sense higher nanodomain calcium concentrations during action potential-evoked glutamate release. *The Journal of Neuroscience: The Official Journal of the Society for Neuroscience, 28*(53), 14450–14458. https://doi.org/10.1523/JNEUROSCI.4245-08.2008

Wang, Y., Sugita, S., & Sudhof, T. C. (2000). The RIM/NIM family of neuronal C2 domain proteins. Interactions with Rab3 and a new class of Src homology 3 domain proteins. *The Journal of Biological Chemistry, 275*(26), 20033–20044. https://doi.org/10.1074/jbc.M909008199

Wang, Y., Liu, X., Biederer, T., & Südhof, T. C. (2002). A family of RIM-binding proteins regulated by alternative splicing: Implications for the genesis of synaptic active zones. *Proceedings of the National Academy of Sciences of the United States of America, 99*(22), 14464–14469. https://doi.org/10.1073/pnas.182532999

Westenbroek, R. E., Sakurai, T., Elliott, E. M., Hell, J. W., Starr, T. V., Snutch, T. P., & Catterall, W. A. (1995). Immunochemical identification and subcellular distribution of the alpha 1A subunits of brain calcium channels. *The Journal of Neuroscience: The Official Journal of the Society for Neuroscience, 15*(10), 6403–6418.

Wheeler, D. B., Randall, A., & Tsien, R. W. (1994). Roles of N-type and Q-type Ca2+ channels in supporting hippocampal synaptic transmission. *Science (New York, N.Y.), 264*(5155), 107–111. https://doi.org/10.1126/science.7832825

Wiser, O., Bennett, M. K., & Atlas, D. (1996). Functional interaction of syntaxin and SNAP-25 with voltage-sensitive L- and N-type Ca2+ channels. *The EMBO Journal, 15*(16), 4100–4110.

Wiser, O., Tobi, D., Trus, M., & Atlas, D. (1997). Synaptotagmin restores kinetic properties of a syntaxin-associated N-type voltage sensitive calcium channel. *FEBS Letters, 404*(2–3), 203–207. https://doi.org/10.1016/s0014-5793(97)00130-0

Wong, F. K., & Stanley, E. F. (2010). Rab3a interacting molecule (RIM) and the tethering of pre-synaptic transmitter release site-associated CaV2.2 calcium channels. *Journal of Neurochemistry, 112*(2), 463–473. https://doi.org/10.1111/j.1471-4159.2009.06466.x

Wong, M. Y., Liu, C., Wang, S. S. H., Roquas, A. C. F., Fowler, S. C., & Kaeser, P. S. (2018). Liprin-α3 controls vesicle docking and exocytosis at the active zone of hippocampal synapses. *Proceedings of the National Academy of Sciences of the United States of America, 115*(9), 2234–2239. https://doi.org/10.1073/pnas.1719012115

Wycisk, K. A., Budde, B., Feil, S., Skosyrski, S., Buzzi, F., Neidhardt, J., Glaus, E., Nürnberg, P., Ruether, K., & Berger, W. (2006). Structural and functional abnormalities of retinal ribbon synapses due to Cacna2d4 mutation. *Investigative Ophthalmology & Visual Science, 47*(8), 3523–3530. https://doi.org/10.1167/iovs.06-0271

Xu, J., & Wu, L.-G. (2005). The decrease in the presynaptic calcium current is a major cause of short-term depression at a calyx-type synapse. *Neuron, 46*(4), 633–645. https://doi.org/10.1016/j.neuron.2005.03.024

Yamada, W. M., & Zucker, R. S. (1992). Time course of transmitter release calculated from simulations of a calcium diffusion model. *Biophysical Journal, 61*(3), 671–682. https://doi.org/10.1016/S0006-3495(92)81872-6

Yang, Y.-M., & Wang, L.-Y. (2006). Amplitude and kinetics of action potential-evoked Ca2+ current and its efficacy in triggering transmitter release at the developing calyx of Held synapse. *The Journal of Neuroscience: The Official Journal of the Society for Neuroscience, 26*(21), 5698–5708. https://doi.org/10.1523/JNEUROSCI.4889-05.2006

Yang, Y.-M., Fedchyshyn, M. J., Grande, G., Aitoubah, J., Tsang, C. W., Xie, H., Ackerley, C. A., Trimble, W. S., & Wang, L.-Y. (2010). Septins regulate developmental switching from microdomain to nanodomain coupling of Ca(2+) influx to neurotransmitter release at a central synapse. *Neuron, 67*(1), 100–115. https://doi.org/10.1016/j.neuron.2010.06.003

Yokoyama, C. T., Myers, S. J., Fu, J., Mockus, S. M., Scheuer, T., & Catterall, W. A. (2005). Mechanism of SNARE protein binding and regulation of Cav2 channels by phosphorylation of the synaptic protein interaction site. *Molecular and Cellular Neurosciences, 28*(1), 1–17. https://doi.org/10.1016/j.mcn.2004.08.019

Zamponi, G. W., & Currie, K. P. M. (2013). Regulation of Ca(V)2 calcium channels by G protein coupled receptors. *Biochimica Et Biophysica Acta, 1828*(7), 1629–1643. https://doi.org/10.1016/j.bbamem.2012.10.004

Zhen, M., & Jin, Y. (1999). The liprin protein SYD-2 regulates the differentiation of presynaptic termini in C. elegans. *Nature, 401*(6751), 371–375. https://doi.org/10.1038/43886

Zhong, H., Yokoyama, C. T., Scheuer, T., & Catterall, W. A. (1999). Reciprocal regulation of P/Q-type Ca2+ channels by SNAP-25, syntaxin and synaptotagmin. *Nature Neuroscience, 2*(11), 939–941. https://doi.org/10.1038/14721

Zucker, R. S., & Regehr, W. G. (2002). Short-term synaptic plasticity. *Annual Review of Physiology, 64*, 355–405. https://doi.org/10.1146/annurev.physiol.64.092501.114547

Zühlke, R. D., Pitt, G. S., Tsien, R. W., & Reuter, H. (2000). Ca2+-sensitive inactivation and facilitation of

L-type Ca2+ channels both depend on specific amino acid residues in a consensus calmodulin-binding motif in the(alpha)1C subunit. *The Journal of Biological Chemistry, 275*(28), 21121–21129. https://doi.org/10.1074/jbc.M002986200

Zurawski, Z., Page, B., Chicka, M. C., Brindley, R. L., Wells, C. A., Preininger, A. M., Hyde, K., Gilbert, J. A., Cruz-Rodriguez, O., Currie, K. P. M., Chapman, E. R., Alford, S., & Hamm, H. E. (2017). Gβγ directly modulates vesicle fusion by competing with synaptotagmin for binding to neuronal SNARE proteins embedded in membranes. *The Journal of Biological Chemistry, 292*(29), 12165–12177. https://doi.org/10.1074/jbc.M116.773523

Functional Role and Plasticity of Voltage-Gated Calcium Channels in the Control of Heart Automaticity

Pietro Mesirca, Isabelle Bidaud, Eleonora Torre, Angelo G. Torrente, Alicia D'Souza, and Matteo E. Mangoni

Abstract

Voltage-gated calcium channels (VGCCs) control the heart rate and rhythm in everyday life. VGCCs importantly contribute to the generation of sinoatrial node (SAN) automaticity and are key mediators of sympathetic and parasympathetic regulation of heart rate. In this chapter, we will review and comment on recent evidence showing that the expression of SAN VGCCs is endowed with intrinsic plasticity under different physiologic situations such as chronic physical training, ageing, and cardiac disease. In particular, it has been shown that the expression of L-type $Ca_v1.3$ and T-type $Ca_v3.1$ channels is downregulated in the SAN of trained animals, which contributes to the adaptation of heart rate to endurance training. Similarly, training induces a reduction in the expression of L-type $Ca_v1.2$ channels in the atrioventricular node (AVN), to adapt the impulse conduction velocity to a slower basal SAN rate associated with physical activity. Finally, expression of both L-type (I_{CaL}) and T-type (I_{CaT}) calcium currents is diminished in the ageing SAN, resulting in age-related bradycardia. The complex molecular network involved in the plasticity of VGCCs is mostly unexplored and will provide an exciting new research avenue in the field of excitability of cardiac myocytes.

Keywords

Calcium channel · Sinoatrial node · Pacemaker activity · Training · Ageing · L-Type · T-Type · I_{KACh} · $Ca_v1.3$ · $Ca_v1.2$ · $Ca_v3.1$

P. Mesirca · I. Bidaud · E. Torre · A. G. Torrente
M. E. Mangoni (✉)
Institut de Génomique Fonctionnelle, Université de Montpellier, CNRS, INSERM, Montpellier, France

LabEx Ion Channels Science and Therapeutics (ICST), Montpellier, France
e-mail: matteo.mangoni@igf.cnrs.fr

A. D'Souza
Division of Cardiovascular Sciences, University of Manchester, Manchester, UK

Abbreviations

AVN	Atrioventricular node
CHB	congenital heart block
f-(HCN)	hyperpolarisation-activated "funny" channels
$I_{Cav1.2}$	$Ca_v1.2$-mediated I_{CaL}
$I_{Cav1.3}$	$Ca_v1.3$-mediated L-type calcium current (I_{CaL})
I_{KACh}	G protein activated potassium current

miR	micro RNA
NCX1	cardiac Na^+/Ca^{2+} exchanger
RyR2	type 2 ryanodine receptor
SAN	Sinoatrial node
SERCA2A	type 2A sarcoplasmic reticulum calcium pump
VGCCs	voltage-gated calcium channels

Introduction

The cardiac pacemaker mechanism is reliant on the ability of automatic myocytes in the sinoatrial node (SAN) to generate diastolic depolarisation. The cardiac pacemaker is an intrinsically complex structure, both morphologically and functionally (Mangoni & Nargeot, 2008). Recent work using single-cell RNA sequencing approaches demonstrated that pacemaker myocytes are embedded into a complex matrix of fibroblasts, proadypocytes, melanocytes, immune cells, and non-automatic atrial myocytes (Linscheid et al., 2019). Functional and paracrine interactions and intercellular cross-talk mechanisms between these cell types remain largely unexplored. Pacemaker myocytes are not phenotypically homogeneous, as they differ in size, morphology, and level of expression of critical ion channels involved in the generation and regulation of pacemaker activity (Boyett et al., 2000). Pacemaker myocytes do not either work altogether at the same time to generate SAN automaticity. Indeed, pacemaking is generated in a leading site that permanently shifts in different regions of the SAN area under the activation of sympathetic or parasympathetic branches of the autonomic nervous system (Boyett et al., 2000). To further complicate this scenario, it has been shown that a significant fraction of pacemaker cells in the SAN are silent, but can be recruited by neighbouring active myocytes when fast and robust pacemaking is needed, especially under adrenergic activation of the heartbeat (Bychkov et al., 2020). In conclusion, SAN functional structure is rather complex. Accordingly, pacemaking is an integrative, flexible, and robust process. As an example, the reader may ponder that to date, all attempts to suppress the positive chro-

notropic effect of catecholamines on pacemaker activity by blocking ion channels activity have failed.

Despite the intrinsic robustness of the pacemaker mechanism, the solid bulk of evidence exists showing that voltage-gated calcium channels (VGCCs) play a major role in the generation and regulation of SAN pacemaker activity. Even if most studies have been performed in animal models of SAN automaticity, genetic studies in families show that loss-of-function mutations in SAN VGCCs or autoimmune diseases underlie severe forms of primary (inheritable) SAN dysfunction (Torrente et al., 2020).

In this chapter, we will first discuss available evidence of the functional role of SAN VGCCs in the generation of the pacemaker mechanism. Then, we will review new studies showing that the expression of SAN VGCCs can be regulated on the long term by different pathophysiological conditions, including endurance training, ageing, and heart failure. Long-term regulation of SAN VGCCs constitutes a newly identified pathway to regulate heart automaticity and adapt it to the need of the organism. Finally, these studies open an entirely new field of research aiming at identification of signalling, transcriptional, and protein trafficking mechanisms underlying long-term regulation of SAN VGCCs.

Role of VGCCs in the Generation of Cardiac Pacemaker Activity

Automaticity in cardiac myocytes is due to diastolic depolarisation, a slow depolarising phase of the action potential cycle that drives the membrane potential from the end of repolarisation phase to the threshold of the following action potential (Mangoni & Nargeot, 2008). The sympathetic and parasympathetic branches of the autonomic nervous system regulate the slope of diastolic depolarisation in opposite ways. Sympathetic agonists increase the slope of diastolic depolarisation, while parasympathetic agonists and adenosine A1 receptors negatively regulate it (DiFrancesco, 1993; Irisawa et al., 1993). The ionic mechanisms underlying the

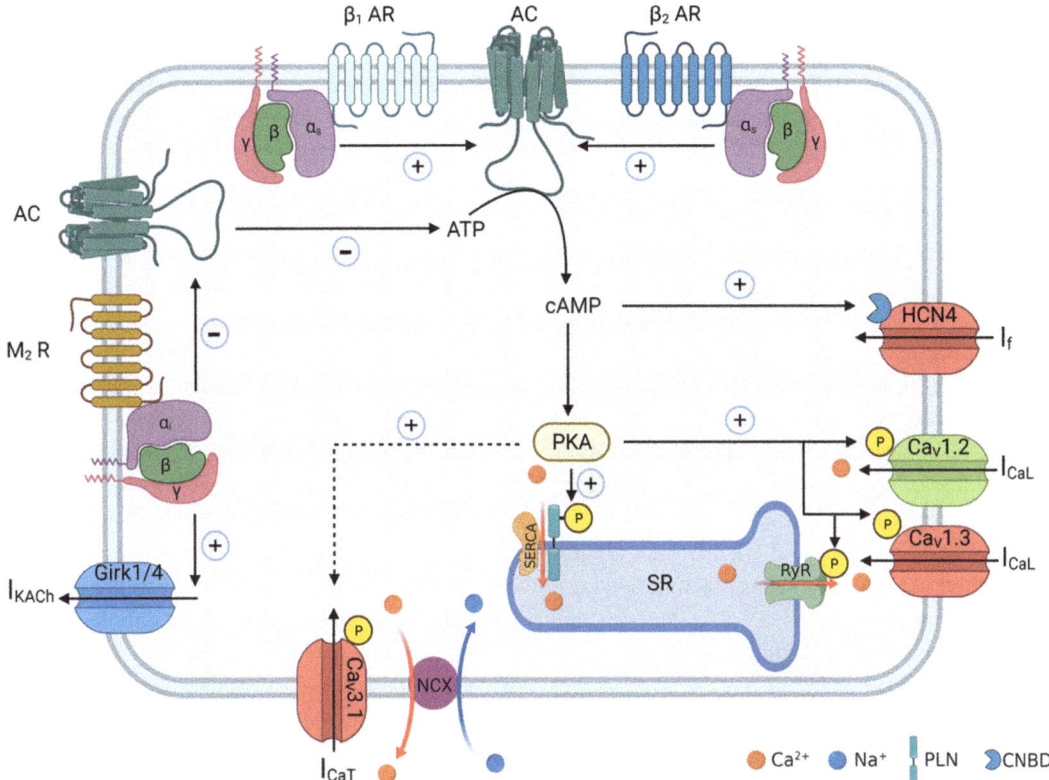

Fig. 1 Schematic drawing of a sinoatrial node (SAN) pacemaker myocyte. Represented are the principal receptors of the plasma membrane and ion channels. In addition, the sarcoplasmic reticulum (SR) together with RyRs, the SR calcium pump SERCA with its main regulator phospholamban (PLN) and the Na$^+$/Ca^{2+} exchanger (NCX) are shown. Main adrenergic (β1 AR and β2 AR) and cholinergic muscarinic (M$_2$R) receptors are displayed with the membrane delimited G-protein pathway control-ling the activity of adenylate cyclase (AC). The major cAMP-dependent protein kinase (PKA) involved in regulation of pacemaker activity by the autonomic nervous system is also shown. (P) indicates putative phosphorylation sites on ion channels and phospholamban. CNBD, indicates the cyclic nucleotide binding site on f-(HCN) channels. (Designed using BioRender software, with licence)

generation of diastolic depolarisation are still not fully understood and still stand as a hot topic in cardiovascular science. Since cardiac pacemaker activity is an electrical phenomenon, ion channels of the plasma membrane have been shown to play a key role in diastolic depolarisation. In addition, an important role of intracellular calcium release from ryanodine receptors (RyR) of the sarcoplasmic reticulum (SR) in pacemaker activity has been identified and investigated over the last two decades (Lakatta et al., 2010) (Fig. 1). In the late seventies, the generation of diastolic depolarisation and the positive chronotropic effect of catecholamines on SAN pacemaker activity have been explained in terms of the expression of hyperpolarisation-activated "funny" current (I_f) and the slow inward (I_{si}) current in SAN tissue (Brown et al., 1979). In this early study, adrenaline was shown to increase the amplitude of I_{si} and I_f suggesting an important role for these currents in the regulation of pacemaker activity. Importantly, the advent of the patch-clamp technique led to the reinterpretation of I_{si} in a composite current generated by the L-type calcium current (I_{CaL}) overlapping with "late" current carried by electrogenic Na$^+$/Ca^{2+} exchange mediated by the cardiac exchanger (NCX1) (Mangoni et al., 2006a). Since the reinterpretation of I_{si} into I_{CaL} and the molecular cloning of VGCC isoforms, important steps have been made in understanding the role of L- and T-type VGCCs in pacemaker activity.

The relevance of VGCCs in the generation and regulation of pacemaker activity has been often overlooked. In this respect, the reader is referred to review articles by DiFrancesco (1993) or Lakatta et al. (2010). Indeed, a common misconception in the field of heart physiology was that the "cardiac" I_{CaL} was uniform in expression and properties in cardiac tissue. From early 2000 it became progressively clear that pacemaker tissue differs from the working myocardium because of the co-expression of two separate L-type isoforms $Ca_v1.3$ and $Ca_v1.2$, together with a T-type isoform $Ca_v3.1$ (Mangoni et al., 2003, 2006a, b; Zhang et al., 2002).

In the adult heart, $Ca_v1.2$ channels are ubiquitously expressed in ventricular and supraventricular chambers, while $Ca_v1.3$ channels are expressed in the atria and the SAN (Mangoni et al., 2003; Marionneau et al., 2005; Chandler et al., 2009). The density of $Ca_v1.3$-mediated I_{CaL} ($I_{Cav1.3}$) is higher in SAN than in atrial myocytes (Zhang et al., 2005). In SAN myocytes, $Ca_v1.3$-mediated I_{CaL} accounts for 60–70% of total I_{CaL} (Mangoni et al., 2003; Baudot et al., 2020). Recent work from our group has shown that virtually all SAN myocytes co-express both $I_{Cav1.3}$ and $Ca_v1.2$-mediated I_{CaL} ($I_{Cav1.2}$) (Louradour et al. under revision). In contrast, in AVN myocytes, $Ca_v1.3$-mediated I_{CaL} can account for over 80% of total I_{CaL} (Marger et al., 2011).

$Ca_v1.3$ channels play a major role in pacemaker activity. Mice globally lacking $Ca_v1.3$ channels ($Ca_v1.3^{-/-}$) present with strong SAN bradycardia and second-degree AVN block (Platzer et al., 2000; Zhang et al., 2002; Marger et al., 2011; Mesirca et al., 2016a, b). In humans, loss of function in $Ca_v1.3$ channels due to a mutation in the *Cacna1D* gene underlies primary SAN bradycardia and AV block (Baig et al., 2011). A study in $Ca_v1.3^{-/-}$ mice has shown that expression of $Ca_v1.3$ is necessary for the pacemaker impulse to be initiated in the central SAN. Indeed, automaticity in $Ca_v1.3^{-/-}$ SAN-Atria preparations is predominantly initiated in the peripheral (caudal) SAN or in extra nodal sites in the coronary sinus, or in the right or left atrium (Baudot et al., 2020).

These observations obtained in vivo and ex vivo and in the intact SAN raise the question about the mechanistic impact of $Ca_v1.3$ channels in the pacemaker mechanism. In this regard, some important aspects of $I_{Cav1.3}$ are to be considered here. First, $I_{Cav1.3}$ activates at potentials spanning the whole range of diastolic depolarisation (−60 to −40 mV) in mouse SAN cells. Under β-adrenergic activation, the apparent threshold of $I_{Cav1.3}$ falls close to −55 mV after stepping from a holding potential of −60 mV (Mangoni et al., 2003). Second, we reported that when mouse SAN action potentials are applied as the voltage-clamp waveform to SAN myocytes, genetic ablation of $Ca_v1.3$ channels abolished a net $I_{Cav1.3}$ current component spanning the whole range of voltages corresponding to diastolic depolarisation (Torrente et al., 2016). This observation suggests also that measuring $I_{Cav1.3}$ under standard voltage clamp conditions may lead to underestimation of the contribution of this current to diastolic depolarisation. Finally, we have reported that, under the physiologic extracellular concentration of Na^+, $Ca_v1.3$ channels are necessary to generate the sustained inward Na^+ current (I_{st}) (Toyoda et al., 2017). The apparent threshold of I_{st} in mouse SAN cells falls around −70 mV and presents undistinguishable sensitivity to dihydropyridines and β-adrenergic agonists as $I_{Cav1.3}$ (Toyoda et al., 2017). While the exact mechanism by which $Ca_v1.3$ generates I_{st} beside $I_{Cav1.3}$ is still to be elucidated, the capability of $Ca_v1.3$ to mediate also a low voltage-activated Na^+ current can constitute an important mechanism contributing to diastolic depolarisation and pacemaker activity (Toyoda et al., 2017).

Another important role of $Ca_v1.3$ channels in the generation of cardiac pacemaking consists of its ability to regulate RyR-dependent Ca^{2+} release during spontaneous firing activity of SAN myocytes (Torrente et al., 2016; Baudot et al., 2020). Indeed, $Ca_v1.3$ channels have been shown to control the basal level of intracellular Ca^{2+}, the timing and synchronisation of RyR-dependent of local Ca^{2+} release units. Synchronisation of local Ca^{2+} release by $Ca_v1.3$ is predicted to stimulate the activity of NCX1 in the forward mode, generating an additional inward current in the diastolic phase (Baudot et al., 2020). In conclusion, available data indicate that $Ca_v1.3$ channels contribute to diastolic depolarisation and SAN automaticity by at least three mechanisms: (i) generation of inward Ca^{2+} current ($I_{Cav1.3}$), (ii) generation of sustained Na^+ current $I_{st,}$ and (iii) regulation of basal intracellular Ca^{2+} and synchronisation of local Ca^{2+} release events to stimulate electrogenic NCX1.

The specific role of $Ca_v1.2$ in pacemaker activity is less understood than that of $Ca_v1.3$ and it has not been investigated directly. Since SAN pacemaker myocytes co-express $I_{Cav1.3}$ and $I_{Cav1.2}$ and $Ca_v1.3^{-/-}$ myocytes show normal action potential upstroke phase and amplitude, we hypothesize that $Ca_v1.2$ contributes to the SAN action potential upstroke and duration in a similar way that in myocytes of the working myocardium. This hypothesis is supported by the observation that intracellular Ca^{2+} transients and sarcoplasmic reticulum Ca^{2+} load do not differ between wild-type and $Ca_v1.3^{-/-}$ SAN pacemaker myocytes, which indicates that $Ca_v1.3$ is not required for Ca^{2+} homeostasis and Ca^{2+} content in the sarcoplasmic reticulum. Future studies will clarify the role of $Ca_v1.2$ in relation to $Ca_v1.3$ channels in cardiac pacemaker tissue.

The predominant T-type cardiac isoform in the adult mouse heart is $Ca_v3.1$ (Mangoni et al., 2006b). mRNA encoding for $Ca_v3.2$ is consistently found in mouse and human SAN (Bohn et al., 2000; Marionneau et al., 2005; Tellez et al., 2006; Chandler et al., 2009). However, mice lacking $Ca_v3.1$ globally ($Ca_v3.1^{-/-}$) do not show residual I_{CaT} in pacemaker myocytes (Mangoni et al., 2006b; Baudot et al., 2020). Thus, it is possible that the translation of $Ca_v3.2$ mRNA is repressed in the adult cardiac conduction system and working myocardium at the postnatal stage. The relevance of T-type VGCCs in pacemaker activity has been investigated since the late eighties, in isolated rabbit SAN myocytes (Hagiwara et al., 1988) and using $Ca_v3.1^{-/-}$ mice (Mangoni et al., 2006b). Genetic ablation of $Ca_v3.1$ in mice reduces the basal heart rate and slows down atrioventricular conduction (Mangoni et al., 2006b).

$Ca_v3.1^{-/-}$ mice show 1st_degree atrioventricular block (prolongation of PR interval), with no isolated P waves. SAN bradycardia in $Ca_v3.1^{-/-}$ mice is observed at rest and at low heart rates in the daytime, without significant reduction of maximal heart rates recorded during periods of high activity at night (Mangoni et al., 2006b). Genetic ablation of $Ca_v3.1$ channels reduces the slope of diastolic depolarisation, despite the availability of I_{CaT} being low at diastolic voltages, which indicates that $Ca_v3.1$-mediated I_{CaT} contributes to determine the basal firing rate of pacemaker myocytes (Mangoni et al., 2006a, b).

New evidence about the role of SAN VGCCs in heart automaticity comes from the exploration of a double mutant mouse model lacking both $Ca_v1.3$ and $Ca_v3.1$ channels ($Ca_v1.3^{-/-}/Ca_v3.1^{-/-}$) (Baudot et al., 2020). These mice offer the interesting possibility to study the significance of co-expression of low-voltage activated L- $Ca_v1.3$ and T- $Ca_v3.1$ type channels in rhythmogenic centres and infer the role of other ion channels involved in pacemaking such as f-(HCN) and voltage-dependent Na^+ channels. Interestingly, the SAN phenotype of mice carrying concomitant ablation of $Ca_v1.3$ and $Ca_v3.1$ is not additive in comparison to the deletion of $Ca_v1.3$ alone. In contrast, double knockout mice present with partial heart block in vivo. It is possible that residual atrioventricular conduction in $Ca_v1.3^{-/-}/Ca_v3.1^{-/-}$ mice is due to elevated sympathetic tone because Langendorff-perfused $Ca_v1.3^{-/-}/Ca_v3.1^{-/-}$ mice have total heart block, dissociated rhythm and present long periods of ventricular tachycardia (Baudot et al., 2020). Finally, it has been shown that residual pacemaking in $Ca_v1.3^{-/-}/Ca_v3.1^{-/-}$ SAN-Atria preparations and pacemaker myocytes is predominantly supported, at low rates, by f-(HCN) and tetrodotoxin-sensitive Na^+ channels (Baudot et al., 2020). The phenotype of $Ca_v1.3^{-/-}/Ca_v3.1^{-/-}$ mice recapitulates primary SAN bradycardia and severe atrioventricular dysfunction observed in congenital heart block (CHB). CHB is an autoimmune disease in which pregnant women develop autoantibodies that target the fetus' cardiac conduction system (Lazzerini et al., 2019). Auto-antibodies against $Ca_v1.3$ and $Ca_v3.1$ have been isolated from maternal plasma.

These autoantibodies show potent neutralizing activity against SAN VGCC (Karnabi & Boutjdir, 2010). Even if CHB is a complex pathology involving loss of function of different proteins expressed in the conduction system, this pathology demonstrates the relevance of SAN VGCC in the generation of SAN pacemaker activity and impulse conduction in humans.

In conclusion, available evidence suggests that expression of SAN VGCC is important for the generation and regulation of pacemaker activity and that co-expression of $Ca_v1.3$ and $Ca_v3.1$ channels is essential for impulse conduction in the cardiac conduction system and for maintaining ventricular rhythmicity, roles that $Ca_v1.2$ and f-(HCN) channels could not sustain by themselves or in association. Because of their pivotal role in heart automaticity, the regulation of SAN VGCCs constitutes a powerful effector of pathophysiological conditions that modulate heart rate.

Plasticity of Sinus Node Calcium Channels in Training

The decrease in resting heart rate in athletes is considered a physiological and adaptive response of heart automaticity to endurance training. Clinical cardiologists have attributed a decrease in heart rate upon endurance training to elevated vagal tone with consequent slowing of SAN pacemaker activity (D'Souza et al., 2019). However, attempts to demonstrate that endurance athletes have increased vagal tone have consistently failed (see D'Souza et al. for review; D'Souza et al., 2019). Recent work has overturned this notion and new evidence indicates intrinsic remodelling in the expression of ion channels in the SAN and AVN as the downstream mechanism of training induced decrease in basal heart rate and atrioventricular conduction (D'Souza et al., 2014, 2017; Bidaud et al., 2020b; Mesirca et al., 2021b). The first indication that high vagal tone was not the primary mechanism in training-induced heart rate reduction came from a small cohort of trained mice and rats (D'Souza et al., 2014). It was shown that intrinsic SAN rate of trained mice was lower than that of

sedentary counterparts. A decrease in intrinsic SAN automaticity was accompanied by the decreased density of I_f, expression of Hcn4 mRNA and up-regulation of training-induced microRNAs (miRs) transcription signature (D'Souza et al., 2014). Targeting miR-423-5p reversed downregulation of Hcn4 mRNA and I_f, showing that miR antagonism could be a new way to rescue remodelling of SAN automaticity after exercise training (D'Souza et al., 2017). Exercise training-induced plasticity of intrinsic SAN automaticity and I_f is a new paradigm in the field of cardiac electrophysiology (Fig. 2). Recent work demonstrates that, together with f-(HCN), SAN VGCCs are also involved in remodelling heart automaticity in exercise training (Bidaud et al., 2020b). Similarly, remodelling of atrioventricular conduction upon physical training involves both I_f and I_{CaL} (Mesirca et al., 2021b). Standard 1.5 months exercise regimen in mice reduces I_{CaL} and I_{CaT} densities in mouse SAN (Fig. 3). However, contrary to *Hcn4*, mRNA expression of $Ca_v1.2$, $Ca_v1.3$, $Ca_v3.1$, and $Ca_v3.2$ α subunits is unaltered by exercise training (Bidaud et al., 2020b). Similarly, expression of L-type VGCC $α_2δ$ and β subunit isoforms is unaffected by training, indicating that the reduction in current density of I_{CaL} cannot be due to reduced expression of mRNA coding for auxiliary subunits. Interestingly, channel protein quantitation by western blotting shows significant training-induced reduction in the SAN expression levels of $Ca_v1.3$, but not of $Ca_v1.2$. This last observation is in line with the positive shift observed in SAN I_{CaL} in trained mice (Bidaud et al., 2020b). These data raise the interesting possibility of a differential pathway of regulation between f-(HCN) channels and VGCCs by training at post-transcriptional and post-translational levels in the SAN, respectively. In addition, other factors may control differentially the availability and trafficking of f-(HCN) channels and VGCCs under sedentary or exercise training conditions, such as protein kinases and endogenous factors that still need to be identified. A similar study involving a cohort of trained mice and racing horses has shown that downregulation of HCN4 and $Ca_v1.2$ channels underlie training-induced slowing of

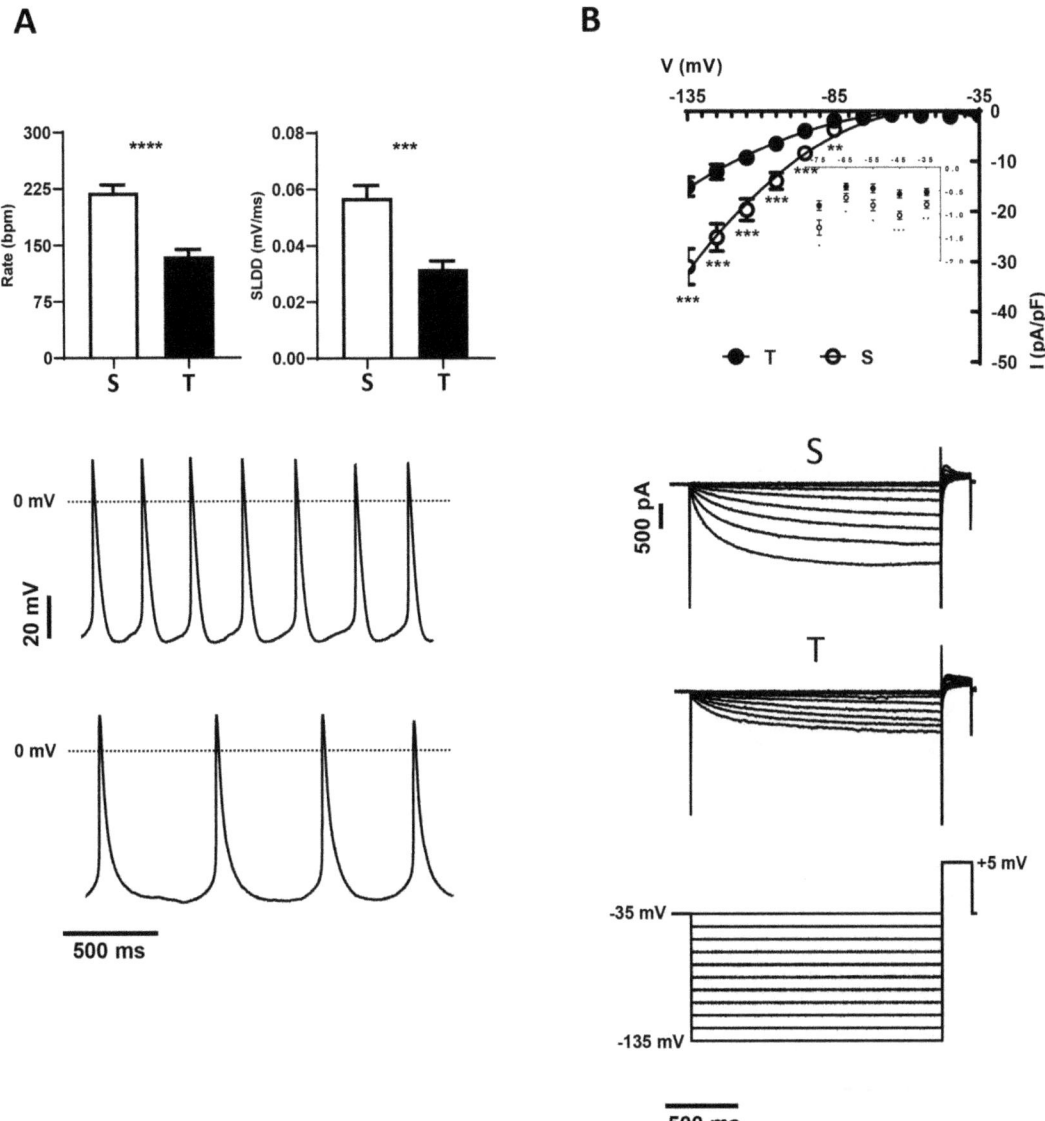

Fig. 2 (a) Training induced decrease of in vivo heart rate (upper panel, left), slope of linear diastolic depolarisation (right) recorded during spontaneous firing of SAN myocytes. Sample traces of spontaneous firing is shown below. (S) indicates data obtained from sedentary mice, (T) indicates data obtained from trained mice. Note that spontaneous firing of SAN myocytes isolated from trained mice is slower than that recorded in myocytes isolated from sedentary mice (upper panel). (b) I_f current-to-voltage relationship recorded in sedentary and trained mice (upper panel). Sample of I_f traces evoked by hyperpolarizing voltage-clamp pulses recorded in sedentary and trained mice (middle panel). Voltage-clamp protocol is depicted in the bottom panel. (Data from Bidaud et al., 2020b, with permission)

atrioventricular conduction. Different from the SAN, downregulation of AVN I_{CaL} involves miR-controlled expression of $Ca_v1.2$ mRNA. Western blots and immunofluorescence confirmed a reduction in $Ca_v1.2$ protein expression (Mesirca et al., 2021b). More generally, an interesting concept emerging from this work is that channels with distinct biophysical properties are jointly regulated to adapt heart automaticity to physiological challenges such as endurance physical

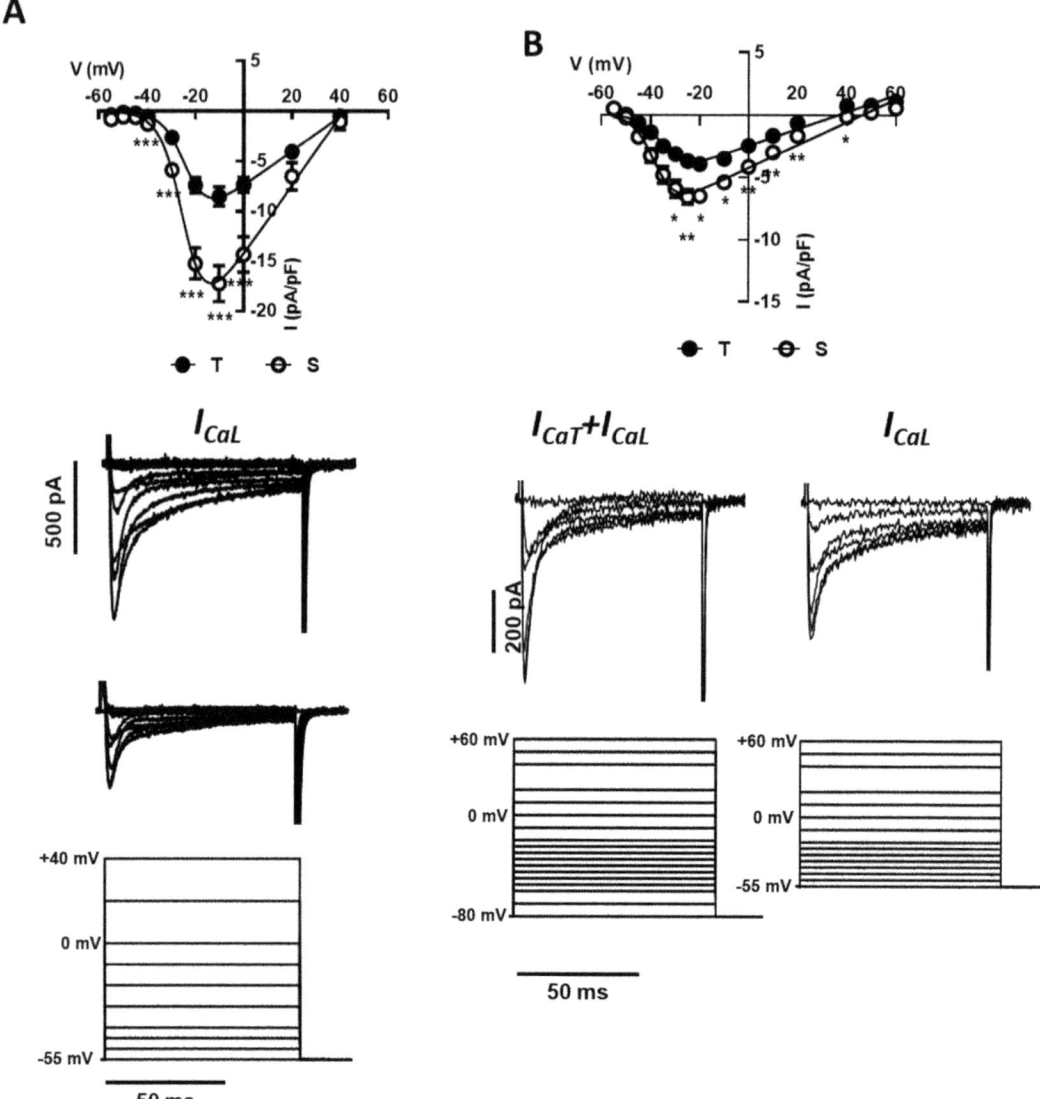

Fig. 3 Calcium currents in SAN myocytes isolated from sedentary and trained mice. (**a**) Top panel shows current-to-voltage relationship of I_{CaL} recorded in SAN myocytes from trained and sedentary mice. Middle panel show corresponding sample traces of I_{CaL}. Voltage-clamp protocol is displayed below. (**b**) Top panel shows current-to-voltage relationship of I_{CaT} recorded in SAN myocytes from trained and sedentary mice. Middle panels show I_{CaL} and I_{CaT} recorded from holding potential of −80 mV (left) and I_{CaL} traces used to deduce I_{CaT} by subtraction. Voltage-clamp protocol is depicted in the bottom panel. (Data from Bidaud et al., 2020b, with permission)

training. Interestingly, in the AVN, miR-211-5p, and miR-432 were found to be common regulators of $Ca_V1.2$ and HCN4. This observation indicates that the same repressive miR transcriptional program allows co-regulation of different ion channels, ensuring a phenotypic switch from "sedentary" to "trained" phenotype in the SAN and AVN. In this regard, numerical modelling of automaticity in mouse SAN pacemaker myocytes has shown that downregulation of I_f alone cannot fully explain SAN bradycardia induced by endurance training (Bidaud et al., 2020b). When only training induced changes in I_f are calculated, modelling predicts a 15% slowing of pacemaking which does not match experimental observations. However, when changes in I_f, I_{CaT} and I_{CaL} are concomitantly included in model parameters, a 38% slowing of pacemaker activity is calculated, which matches with 39% observed in SAN myocytes of trained mice. Interestingly, consistency between experimental observation and predicted pacemaking calculated from the model is obtained by attributing changes in I_{CaL} density and shift of half activation to $I_{Cav1.3}$ (+4 mV) rather than $I_{Cav1.2}$. When only changes in $I_{Cav1.3}$ are included in calculations, the diastolic interval is prolonged by 11%, which predicts diminished net inward diastolic current with consequent slowing of the computed pacemaker mechanism. As a comparison, modelling predicts 13% slowing of pacemaking when only the training-induced change in I_{CaT} amplitude in included in calculations. Thus, these calculations suggest two working hypotheses about how potential co-regulation and plasticity of SAN I_f and VGCCs allows tuning heart rate under different patho-physiological situations. First, calculations indicate the pacemaker mechanism to be no-linear in nature and, because of potential functional redundancy between I_f and $I_{Cav1.3}$, both currents need to be concomitantly downregulated to slow the basal heart rate to levels matching cardiovascular performance in endurance training. Second, it is notable that while endurance training downregulates the three major inward currents stimulating pacemaker activity, the G-protein activated K^+ current I_{KACh}, which constitutes the predominant brake of pacemaker activity (Wickman et al., 1998; Mesirca et al., 2013), is unaffected by the training transcriptional control program (Bidaud et al., 2020b). This factor also contributes to promote slowing of basal heart rate in endurance training, because we expect downregulation of inward currents exacerbates the tonic and phasic chronotropic effect of I_{KACh} activation. This hypothesis is supported by evidence showing that heart rate and rhythm in $Ca_v1.3^{-/-}$ (Mesirca et al., 2016a, b) and $Ca_v1.3^{-/-}/Ca_v3.1^{-/-}$ (Bidaud et al., 2020a) mice is rescued by genetic ablation or pharmacologic inhibition of I_{KACh}. Taken together, these studies show that the SAN pacemaker mechanism is endowed of significant functional

redundancy and that loss-of-function in VGCC can be rescued by inhibition of a tonic brake of pacemaker activity such as I_{KACh} (see (Mesirca et al., 2016b), for review). In the specific case of plasticity in SAN VGCC expression, it is noteworthy that endurance training downregulates the expression of $Ca_v1.3$, HCN4, and $Ca_v3.1$ channels via different pathways. This suggests that complex signalling pathways underlie phenotypic transition of the SAN from a "sedentary" to "trained" state.

While complexity generally ensures efficiency and flexibility to physiologic functions, it may lead to dysregulation with pathological consequences. In this respect, adaptation of SAN and AVN to training is a reversible phenomenon in model animals, de-training restores basal SAN rate and AVN conduction to values recorded before the training regimen (Mesirca et al., 2021b; D'Souza et al., 2014). However, even if athletes are considered among the healthiest individuals, clinical follow-up of cohorts of endurance athletes indicates this subpopulation to present a higher prevalence of bradyarrhythmia (Northcote et al., 1989a, b), AVN block (Stein et al., 2002) and atrial fibrillation (Andersen et al., 2013). These clinical studies suggest the existence of an irreversible component of post-transcriptional or post-translational regulation of VGCCs in SAN and AVN. It is possible that long-term endurance training induces irreversibility of SAN and AVN remodelling in some individuals. An interesting cue may come from observations of heart rhythm in $Ca_v1.3^{-/-}$ mice. As discussed above, these mice present with SAN bradycardia and atrioventricular block. However, they are also extremely prone to atrial fibrillation and flutter (Zhang et al., 2005; Mesirca et al., 2016a, b). Atrial arrhythmia in $Ca_v1.3^{-/-}$ mice can easily be recorded even with mild protocols of intracardiac pacing in vivo (Zhang et al., 2005; Mesirca et al., 2016a, b). Furthermore, spontaneous episodes of erratic atrial activity can be observed in telemetric recordings of ECGs in $Cav1.3^{-/-}$ mice, underscoring the fragility of atrial rhythm when $Ca_v1.3$ channels are downregulated. Because $Ca_v1.3$ is moderately expressed in the atria, genetic inactivation of *Cacna1D* in mice significantly shortens atrial action potential duration (Mesirca et al., 2016a, b). SAN dysfunction and shortening of atrial action potential duration following ablation of $Ca_v1.3$ could constitute a powerful pro-arrhythmic mechanism. Thus, in the case of long-term endurance training, concomitant downregulation of $Ca_v1.3$ and HCN4 may favour the incidence of atrial arrhythmias, by reducing SAN dominance in controlling atrial rhythmicity and by directly affecting atrial excitability, which favours mechanisms of impulse re-entry.

In conclusion, factors setting the boundary between physiologic plasticity of SAN VGCCs and irreversible downregulation of "pacemaker" channels that lead to pathology later in life are at present unknown. Nevertheless, irrespectively of lifestyle being sedentary, active, or even athletic, SAN automaticity is bound to decline with age. Consequently, it is possible that mechanisms operant in ageing of the SAN are also involved in pathological remodelling of VGCC and f-(HCN) channels in endurance training. Ageing of SAN VGCC is discussed in the next section.

SAN VGCC in Ageing of the Sino-atrial Node

Age-related SAN dysfunction is a growing disease in the population. It affects 1/600 individuals, especially in the age range of 75–85 years (Jensen et al., 2014). The incidence of age-related SAN dysfunction is forecasted to double in the next 50 years. Because bradycardia in the elderly is generally symptomatic, the incidence of SAN dysfunction constitutes a very reliable gauge of the necessity of implantation of permanent electronic pacemakers (Brignole et al., 2013). While electronic pacemakers are very performant, a consistent trend in the implantation of increasingly complex and expensive devices has been demonstrated in clinical studies (Jensen et al., 2014). There is thus a general interest in studying the mechanisms underlying age-related SAN dysfunction to develop new therapeutic strategies able to reduce the need for electronic pacemaker implantations (Mesirca et al., 2021a). Besides

symptomatic SAN dysfunction, heart automaticity and rate decline with age. Indeed, the maximal heart rate expected in standard cardiologic effort tests is corrected by the age of the subject. In has been demonstrated in humans that both basal and maximal heart rates decrease with age (Peters et al., 2020). In this section, we will discuss current available evidence of the role of SAN VGCC in the ageing process of heart automaticity.

SAN ageing has been studied using the SAN of rats, guinea pigs, and mice. Comparisons between 2 (young)- and 24–32 (aged) old months mice are currently used as a standard model of SAN ageing (Peters et al., 2020). Telemetric recordings of ECGs in young and aged mice demonstrated that ageing is correlated with decrease in both intrinsic heart rate – measured during pharmacologic blockade of autonomic input – and the maximal heart rate (Larson et al., 2013). Patch-clamp studies of SAN myocytes isolated from mice aged 24 to 32 months show that basal firing frequency of action potentials is slower than that recorded in young SAN myocytes. The slope of diastolic depolarisation is also negatively correlated with the age of the mouse donor of pacemaker myocytes. However, the relative degree of the positive chronotropic effect of isoproterenol on pacemaker activity of SAN myocytes does not differ between young and aged mice (Larson et al., 2013), demonstrating that direct stimulation of the cAMP-dependent pathway can still increase cellular automaticity. Recordings of calcium currents in pacemaker myocytes show a lower density of I_{CaL} and I_{CaT} in the SAN of aged mice in comparison to young counterparts (Mesirca et al. unpublished observation) (Larson et al., 2013). Reduction in densities of calcium current was reported to occur in the absence of a change in voltages for half activation of I_{CaL} and I_{CaT} (Larson et al., 2013). This observation suggests that either $I_{Cav1.2}$, or both $I_{Cav1.3}$ and $I_{Cav1.2}$ are diminished in the aged SAN, because downregulation of $I_{Cav1.3}$ alone would have resulted in a positive shift in activation of total I_{CaL}. In addition, the sensitivity of I_{CaL} to isoproterenol did not change with age. This observation is of high interest because it indicates that age-dependent decrease in I_{CaL} density is not secondary to a decrease in basal levels of cAMP but is potentially caused by altered intracellular signaling pathways controlling channel expression and trafficking. It has also been reported that a decrease in densities of I_{CaL} and I_{CaT} is paralleled by age-dependent hypertrophy of SAN myocytes (Larson et al., 2013). Similar to exercise training, downregulation of I_{CaL} and I_{CaT} is accompanied by a reduction of I_f density, suggesting that some signalling mechanisms that are operant in training-dependent downregulation of SAN ionic currents are at work in the aged SAN. Conflicting results exist as to age-dependent expression of Ca$_v$1.2. A study using aged rat SAN showed moderate augmentation in Ca$_v$1.2 mRNA. However, downregulation of Ca$_v$1.2 protein has been reported in the SAN of aged guinea pigs (Jones et al., 2007). The study by Tellez et al., did not find age-dependent differences in mRNA coding for Ca$_v$3.1 and Ca$_v$1.3 (Tellez et al., 2011). However, this does not exclude that Ca$_v$1.3 and/or Ca$_v$3.1 channel proteins are downregulated because exercise training reduces I_{CaL} density and Ca$_v$1.3 expression at the membrane, without significant downregulation of Ca$_v$1.3 mRNA (Bidaud et al., 2020b). Ageing in the SAN involves also downregulation of proteins involved in intracellular Ca^{2+} handling. Age-dependent downregulation of RyR2 mRNA has been reported in the aged rat SAN (Tellez et al., 2011). In a separate study, concomitant downregulation of RyR2, SERCA2A, and NCX1 proteins has been shown in the aged mouse SAN (Liu et al., 2014). Liu et al. have proposed that downregulation of intracellular Ca^{2+} handling proteins in aged SAN myocytes leads to a reduction in sarcoplasmic reticulum Ca^{2+} load with the consequent decrease in RyR2-dependent Ca^{2+} release and diminished diastolic inward current generated by NCX1 (Liu et al., 2014). Decreased NCX1 activity is expected to slow basal action potential firing rate, according to the coupled-clock model of pacemaking (Lakatta et al., 2010). However, since we have demonstrated that Ca$_v$1.3 channels control RyR2 -dependent Ca^{2+} release during pacemaker activity we expect that if Ca$_v$1.3 channels are altered by ageing the overall

effect on pacemaker activity will depend on the combined effects of reduced availability of RyR2-dependent Ca^{2+} release sites and reduced $I_{Cav1.3}$.

In conclusion, negative regulation of SAN VGCCs contributes to the decline of heart automaticity in ageing. It will be interesting to determine the relative contribution of Ca$_v$1.3 and Ca$_v$3.1 to this decline. Since $I_{Cav1.3}$ plays a major role in pacemaking, we expect that alteration in expression of Ca$_v$1.3 will have a major impact on pacemaking in the aged SAN. In addition, a central question will be the elucidation of the molecular mechanisms leading to decline in expression and activity of SAN VGCCs and particularly, if this mechanism shares some common aspects with those involved in plasticity of VGCC in training.

VGCCs in Sino-atrial Node Dysfunction Secondary to Heart Failure

SAN dysfunction can be associated with heart failure (Mesirca et al., 2021a). A clinical study has reported that SAN dysfunction is responsible for up to 40% of sudden death in hospitals in patients carrying heart failure (Gang et al., 2010). More generally, SAN dysfunction is associated with worsened patient outcomes. Previous works have attributed SAN dysfunction associated with heart failure to downregulation of I_f (Verkerk et al., 2003; Yanni et al., 2011; Boyett et al., 2021), or pacemaker myocytes apoptosis activated by oxidized calmodulin kinase II (Swaminathan et al., 2011). A new role for VGCCs in SAN dysfunction secondary to heart failure is nevertheless emerging. Indeed, new evidence indicates that SAN VGCC is downregulated in some forms of heart failure. In a recent study, Mesquita et al. have employed a rat model of heart failure with preserved ejection fraction (Mesquita et al., 2022). Rats carrying heart failure present with significant SAN bradycardia, and hallmarks of SAN dysfunction. Downregulation of I_{CaL}, I_{CaT} and I_f is observed in the SAN of failing rat hearts. Interestingly, both

Ca$_v$1.3 mRNA and protein are downregulated. Numerical modelling work using a model of human SAN automaticity suggests concomitant downregulation of $I_{Cav1.3}$ and I_f to be the predominant mechanism of SAN bradycardia in this rat model of heart failure (Mesquita et al., 2022). It is possible that discrepancies in the proposed role of ion channels indicated in studies of SAN bradycardia come from different models of heart failure used (e.g. aortic constriction, chronic administration of angiotensin II, salt diet....). Nevertheless, recent work indicates in Ca$_v$1.3 as a new interesting actor in SAN dysfunction secondary to heart failure.

Downregulation of I_{CaL}, I_{CaT} and I_f seems to be a common mechanism in SAN bradycardia secondary to endurance training, ageing and heart failure. It is thus tempting to hypothesize that at least part of the signalling pathways controlling the expression of Ca$_v$1.3, Ca$_v$3.1 and HCN4 channels is conserved in these pathophysiological situations. In this regard, a comparative analysis of human and atrial tissue has identified 48 miRs downregulated in the SAN in comparison to atrial tissue (Petkova et al., 2020). Among those, miR-486-3p is predicted to inhibit the expression of f-(HCN) channels HCN1 and HCN4, as well as Ca$_v$1.3 and Ca$_v$3.1. This result suggests the existence of miRs that can control the expression of different ion channels involved in SAN automaticity hence heart rate.

Conclusions

Heart automaticity is a highly integrated phenomenon that relies on the activity of several ion channel families. Among those, SAN VGCCs Ca$_v$1.3 and Ca$_v$3.1 play a highly relevant role in pacemaker activity. Over the last 15 years, extensive work demonstrated that Ca$_v$1.3 channels are a pivotal pacemaker mechanism to generate basal heart rate and is potentially involved in the positive chronotropic response of pacemaker activity to catecholamines. Recent work has also unraveled the plasticity of SAN VGCC expression to adapt heart automaticity to physiological challenges. In addition, SAN VGCCs are also involved in important pathophysiological processes such as ageing of heart automaticity or SAN dysfunction in heart failure. One striking finding is that expression of SAN VGCC is often co-regulated with that of f-(HCN) channels. Dissection of the pathways involved in the co-regulation of VGCC and f-channels will constitute an important avenue of research in the next decade.

References

Andersen, K., Farahmand, B., Ahlbom, A., Held, C., Ljunghall, S., Michaelsson, K., & Sundstrom, J. (2013). Risk of arrhythmias in 52 755 long-distance cross-country skiers: A Cohort study. *European Heart Journal, 34*, 3624–3631.

Baig, S. M., Koschak, A., Lieb, A., Gebhart, M., Dafinger, C., Nurnberg, G., Ali, A., Ahmad, I., Sinnegger-Brauns, M. J., Brandt, N., Engel, J., Mangoni, M. E., Farooq, M., Khan, H. U., Nurnberg, P., Striessnig, J., & Bolz, H. J. (2011). Loss of Ca(v)1.3 (CACNA1D) function in a human channelopathy with bradycardia and congenital deafness. *Nature Neuroscience, 14*, 77–84.

Baudot, M., Torre, E., Bidaud, I., Louradour, J., Torrente, A. G., Fossier, L., Talssi, L., Nargeot, J., Barrère-Lemaire, S., Mesirca, P., & Mangoni, M. E. (2020). Concomitant genetic ablation of L-type Cav1.3 (α1D) and T-type Cav3.1 (α1G) Ca2+ channels disrupts heart automaticity. *Scientific Reports, 10*, 18906.

Bidaud, I., Chong, A. C. Y., Carcouet, A., Waard, S., Charpentier, F., Ronjat, M., Waard, M., Isbrandt, D., Wickman, K., Vincent, A., Mangoni, M. E., & Mesirca, P. (2020a). Inhibition of G protein-gated K(+) channels by tertiapin-Q rescues sinus node dysfunction and atrioventricular conduction in mouse models of primary bradycardia. *Scientific Reports, 10*, 9835.

Bidaud, I., D'Souza, A., Forte, G., Torre, E., Greuet, D., Thirard, S., Anderson, C., Chung You Chong, A., Torrente, G., Roussel, J., Wickman, K., Boyett, M. R., Mangoni, M. E., & Mesirca, P. (2020b). Genetic ablation of G protein-gated inwardly rectifying K(+) channels prevents training-induced sinus bradycardia. *Frontiers in Physiology, 11*, 519382.

Bohn, G., Moosmang, S., Conrad, H., Ludwig, A., Hofmann, F., & Klugbauer, N. (2000). Expression of T- and L-type calcium channel mRNA in murine sino-atrial node. *FEBS Letters, 481*, 73–76.

Boyett, M. R., Honjo, H., & Kodama, I. (2000). The sinoatrial node, a heterogeneous pacemaker structure. *Cardiovascular Research, 47*, 658–687.

Boyett, M. R., Yanni, J., Tellez, J., Bucchi, A., Mesirca, P., Cai, X., Sjrj, L., Wilson, C., Anderson, C., Ariyaratnam, J., Stuart, L., Nakao, S., Abd Allah, E.,

Jones, S., Lancaster, M., Stephenson, R., Chandler, N., Smith, M., Bussey, C., ... D'Souza, A. (2021). Regulation of sinus node pacemaking and atrioventricular node conduction by HCN channels in health and disease. *Progress in Biophysics and Molecular Biology, 166*, 61–85.

Brignole, M., Auricchio, A., Baron-Esquivias, G., Bordachar, P., Boriani, G., Breithardt, O. A., Cleland, J., Deharo, J. C., Delgado, V., Elliott, P. M., Gorenek, B., Israel, C. W., Leclercq, C., Linde, C., Mont, L., Padeletti, L., Sutton, R., Vardas, P. E., E. S. C. Committee for Practice Guidelines, ... Wilson, C. M. (2013). 2013 ESC Guidelines on cardiac pacing and cardiac resynchronization therapy: The Task Force on cardiac pacing and resynchronization therapy of the European Society of Cardiology (ESC). Developed in collaboration with the European Heart Rhythm Association (EHRA). *European Heart Journal, 34*, 2281–2329.

Brown, H. F., DiFrancesco, D., & Noble, S. J. (1979). How does adrenaline accelerate the heart? *Nature, 280*, 235–236.

Bychkov, R., Juhaszova, M., Tsutsui, K., Coletta, C., Stern, M. D., Maltsev, V. A., & Lakatta, E. G. (2020). Synchronized cardiac impulses emerge from heterogeneous local calcium signals within and among cells of pacemaker tissue. *JACC: Clinical Electrophysiology, 6*, 907–931.

Chandler, N. J., Greener, I. D., Tellez, J. O., Inada, S., Musa, H., Molenaar, P., Difrancesco, D., Baruscotti, M., Longhi, R., Anderson, R. H., Billeter, R., Sharma, V., Sigg, D. C., Boyett, M. R., & Dobrzynski, H. (2009). Molecular architecture of the human sinus node: Insights into the function of the cardiac pacemaker. *Circulation, 119*, 1562–1575.

DiFrancesco, D. (1993). Pacemaker mechanisms in cardiac tissue. *Annual Review of Physiology, 55*, 455–472.

D'Souza, A., Bucchi, A., Johnsen, A. B., Logantha, S. J., Monfredi, O., Yanni, J., Prehar, S., Hart, G., Cartwright, E., Wisloff, U., Dobryznski, H., DiFrancesco, D., Morris, G. M., & Boyett, M. R. (2014). Exercise training reduces resting heart rate via downregulation of the funny channel HCN4. *Nature Communications, 5*, 3775.

D'Souza, A., Pearman, C. M., Wang, Y., Nakao, S., Logantha, S., Cox, C., Bennett, H., Zhang, Y., Johnsen, A. B., Linscheid, N., Poulsen, P. C., Elliott, J., Coulson, J., McPhee, J., Robertson, A., da Costa Martins, P. A., Kitmitto, A., Wisloff, U., Cartwright, E. J., ... Boyett, M. R. (2017). Targeting miR-423-5p reverses exercise training-induced HCN4 channel remodeling and sinus bradycardia. *Circulation Research, 121*, 1058–1068.

D'Souza, A., Trussell, T., Morris, G. M., Dobrzynski, H., & Boyett, M. R. (2019). Supraventricular arrhythmias in athletes: Basic mechanisms and new directions. *Physiology (Bethesda), 34*, 314–326.

Gang, U. J., Jons, C., Jorgensen, R. M., Abildstrom, S. Z., Haarbo, J., Messier, M. D., Huikuri, H. V., & Thomsen, P. E. (2010). Heart rhythm at the time of death documented by an implantable loop recorder. *Europace, 12*, 254–260.

Hagiwara, N., Irisawa, H., & Kameyama, M. (1988). Contribution of two types of calcium currents to the pacemaker potentials of rabbit sino-atrial node cells. *The Journal of Physiology, 395*, 233–253.

Irisawa, H., Brown, H. F., & Giles, W. (1993). Cardiac pacemaking in the sinoatrial node. *Physiological Reviews, 73*, 197–227.

Jensen, P. N., Gronroos, N. N., Chen, L. Y., Folsom, A. R., deFilippi, C., Heckbert, S. R., & Alonso, A. (2014). Incidence of and risk factors for sick sinus syndrome in the general population. *Journal of the American College of Cardiology, 64*, 531–538.

Jones, S. A., Boyett, M. R., & Lancaster, M. K. (2007). Declining into failure: The age-dependent loss of the L-type calcium channel within the sinoatrial node. *Circulation, 115*, 1183–1190.

Karnabi, E., & Boutjdir, M. (2010). Role of calcium channels in congenital heart block. *Scandinavian Journal of Immunology, 72*, 226–234.

Lakatta, E. G., Maltsev, V. A., & Vinogradova, T. M. (2010). A coupled SYSTEM of intracellular Ca2+ clocks and surface membrane voltage clocks controls the timekeeping mechanism of the heart's pacemaker. *Circulation Research, 106*, 659–673.

Larson, E. D., Clair, J. R. S., Sumner, W. A., Bannister, R. A., & Proenza, C. (2013). Depressed pacemaker activity of sinoatrial node myocytes contributes to the age-dependent decline in maximum heart rate. *Proceedings of the National Academy of Sciences of the United States of America, 110*, 18011–18016.

Lazzerini, P. E., Laghi-Pasini, F., Boutjdir, M., & Capecchi, P. L. (2019). Cardioimmunology of arrhythmias: The role of autoimmune and inflammatory cardiac channelopathies. *Nature Reviews. Immunology, 19*, 63–64.

Linscheid, N., Logantha, S., Poulsen, P. C., Zhang, S., Schrolkamp, M., Egerod, K. L., Thompson, J. J., Kitmitto, A., Galli, G., Humphries, M. J., Zhang, H., Pers, T. H., Olsen, J. V., Boyett, M., & Lundby, A. (2019). Quantitative proteomics and single-nucleus transcriptomics of the sinus node elucidates the foundation of cardiac pacemaking. *Nature Communications, 10*, 2889.

Liu, J., Sirenko, S., Juhaszova, M., Sollott, S. J., Shukla, S., Yaniv, Y., & Lakatta, E. G. (2014). Age-associated abnormalities of intrinsic automaticity of sinoatrial nodal cells are linked to deficient cAMP-PKA-Ca(2+) signaling. *American Journal of Physiology. Heart and Circulatory Physiology, 306*, H1385–H1397.

Mangoni, M. E., & Nargeot, J. (2008). Genesis and regulation of the heart automaticity. *Physiological Reviews, 88*, 919–982.

Mangoni, M. E., Couette, B., Bourinet, E., Platzer, J., Reimer, D., Striessnig, J., & Nargeot, J. (2003). Functional role of L-type Cav1.3 Ca2+ channels in cardiac pacemaker activity. *Proceedings of the National Academy of Sciences of the United States of America, 100*, 5543–5548.

Mangoni, M. E., Couette, B., Marger, L., Bourinet, E., Striessnig, J., & Nargeot, J. (2006a). Voltage-dependent calcium channels and cardiac pacemaker activity: From ionic currents to genes. *Progress in Biophysics and Molecular Biology, 90,* 38–63.

Mangoni, M. E., Traboulsie, A., Leoni, A. L., Couette, B., Marger, L., Le Quang, K., Kupfer, E., Cohen-Solal, A., Vilar, J., Shin, H. S., Escande, D., Charpentier, F., Nargeot, J., & Lory, P. (2006b). Bradycardia and slowing of the atrioventricular conduction in mice lacking CaV3.1/alpha1G T-type calcium channels. *Circulation Research, 98,* 1422–1430.

Marger, L., Mesirca, P., Alig, J., Torrente, A., Dubel, S., Engeland, B., Kanani, S., Fontanaud, P., Striessnig, J., Shin, H. S., Isbrandt, D., Ehmke, H., Nargeot, J., & Mangoni, M. E. (2011). Functional roles of Ca(v)1.3, Ca(v)3.1 and HCN channels in automaticity of mouse atrioventricular cells: Insights into the atrioventricular pacemaker mechanism. *Channels (Austin, Tex.), 5,* 251–261.

Marionneau, C., Couette, B., Liu, J., Li, H., Mangoni, M. E., Nargeot, J., Lei, M., Escande, D., & Demolombe, S. (2005). Specific pattern of ionic channel gene expression associated with pacemaker activity in the mouse heart. *The Journal of Physiology, 562,* 223–234.

Mesirca, P., Marger, L., Toyoda, F., Rizzetto, R., Audoubert, M., Dubel, S., Torrente, A. G., Difrancesco, M. L., Muller, J. C., Leoni, A. L., Couette, B., Nargeot, J., Clapham, D. E., Wickman, K., & Mangoni, M. E. (2013). The G-protein-gated K+ channel, IKACh, is required for regulation of pacemaker activity and recovery of resting heart rate after sympathetic stimulation. *The Journal of General Physiology, 142,* 113–126.

Mesirca, P., Bidaud, I., Briec, F., Evain, S., Torrente, A. G., Le Quang, K., Leoni, A. L., Baudot, M., Marger, L., Chong, A. C. Y., Nargeot, J., Striessnig, J., Wickman, K., Charpentier, F., & Mangoni, M. E. (2016a). G protein-gated IKACh channels as therapeutic targets for treatment of sick sinus syndrome and heart block. *Proceedings of the National Academy of Sciences of the United States of America, 113,* E932–E941.

Mesirca, P., Bidaud, I., & Mangoni, M. E. (2016b). Rescuing cardiac automaticity in L-type Cav1.3 channelopathies and beyond. *The Journal of Physiology, 594,* 5869–5879.

Mesirca, P., Fedorov, V. V., Hund, T. J., Torrente, A. G., Bidaud, I., Mohler, P. J., & Mangoni, M. E. (2021a). Pharmacologic approach to sinoatrial node dysfunction. *Annual Review of Pharmacology and Toxicology, 61,* 757–778.

Mesirca, P., Nakao, S., Nissen, S. D., Forte, G., Anderson, C., Trussell, T., Li, J., Cox, C., Zi, M., Logantha, S., Yaar, S., Cartensen, H., Bidaud, I., Stuart, L., Soattin, L., Morris, G. M., da Costa Martins, P. A., Cartwright, E. J., Oceandy, D., … D'Souza, A. (2021b). Intrinsic electrical remodeling underlies atrioventricular block in athletes. *Circulation Research, 129,* e1–e20.

Mesquita, T., Zhang, R., Cho, J. H., Zhang, R., Lin, Y. N., Sanchez, L., Goldhaber, J. I., Yu, J. K., Liang, J. A., Liu, W., Trayanova, N. A., & Cingolani, E. (2022). Mechanisms of sinoatrial node dysfunction in heart failure with preserved ejection fraction. *Circulation, 145,* 45–60.

Northcote, R. J., Canning, G. P., & Ballantyne, D. (1989a). Electrocardiographic findings in male veteran endurance athletes. *British Heart Journal, 61,* 155–160.

Northcote, R. J., Rankin, A. C., Scullion, R., & Logan, W. (1989b). Is severe bradycardia in veteran athletes an indication for a permanent pacemaker? *BMJ, 298,* 231–232.

Peters, C. H., Sharpe, E. J., & Proenza, C. (2020). Cardiac pacemaker activity and aging. *Annual Review of Physiology, 82,* 21–43.

Petkova, M., Atkinson, A. J., Yanni, J., Stuart, L., Aminu, A. J., Ivanova, A. D., Pustovit, K. B., Geragthy, C., Feather, A., Li, N., Zhang, Y., Oceandy, D., Perde, F., Molenaar, P., D'Souza, A., Fedorov, V. V., & Dobrzynski, H. (2020). Identification of key small non-coding microRNAs controlling pacemaker mechanisms in the human sinus node. *Journal of the American Heart Association, 9,* e016590.

Platzer, J., Engel, J., Schrott-Fischer, A., Stephan, K., Bova, S., Chen, H., Zheng, H., & Striessnig, J. (2000). Congenital deafness and sinoatrial node dysfunction in mice lacking class D L-type Ca2+ channels. *Cell, 102,* 89–97.

Stein, R., Medeiros, C. M., Rosito, G. A., Zimerman, L. I., & Ribeiro, J. P. (2002). Intrinsic sinus and atrio-ventricular node electrophysiologic adaptations in endurance athletes. *Journal of the American College of Cardiology, 39,* 1033–1038.

Swaminathan, P. D., Purohit, A., Soni, S., Voigt, N., Singh, M. V., Glukhov, A. V., Gao, Z., He, B. J., Luczak, E. D., Joiner, M. L., Kutschke, W., Yang, J., Donahue, J. K., Weiss, R. M., Grumbach, I. M., Ogawa, M., Chen, P. S., Efimov, I., Dobrev, D., … Anderson, M. E. (2011). Oxidized CaMKII causes cardiac sinus node dysfunction in mice. *The Journal of Clinical Investigation, 121,* 3277–3288.

Tellez, J. O., Dobrzynski, H., Greener, I. D., Graham, G. M., Laing, E., Honjo, H., Hubbard, S. J., Boyett, M. R., & Billeter, R. (2006). Differential expression of ion channel transcripts in atrial muscle and sinoatrial node in rabbit. *Circulation Research, 99,* 1384–1393.

Tellez, J. O., McZewski, M., Yanni, J., Sutyagin, P., Mackiewicz, U., Atkinson, A., Inada, S., Beresewicz, A., Billeter, R., Dobrzynski, H., & Boyett, M. R. (2011). Ageing-dependent remodelling of ion channel and Ca2+ clock genes underlying sino-atrial node pacemaking. *Experimental Physiology, 96,* 1163–1178.

Torrente, A. G., Mesirca, P., Neco, P., Rizzetto, R., Dubel, S., Barrere, C., Sinegger-Brauns, M., Striessnig, J., Richard, S., Nargeot, J., Gomez, A. M., & Mangoni, M. E. (2016). L-type Cav1.3 channels regulate ryanodine receptor-dependent Ca2+ release during sino-atrial node pacemaker activity. *Cardiovascular Research, 109*, 451–461.

Torrente, A. G., Mesirca, P., Bidaud, I., & Mangoni, M. E. (2020). Channelopathies of voltage-gated L-type Cav1.3/alpha1D and T-type Cav3.1/alpha1G Ca(2+) channels in dysfunction of heart automaticity. *Pflügers Archiv, 472*, 817–830.

Toyoda, F., Mesirca, P., Dubel, S., Ding, W. G., Striessnig, J., Mangoni, M. E., & Matsuura, H. (2017). CaV1.3 L-type Ca2+ channel contributes to the heartbeat by generating a dihydropyridine-sensitive persistent Na+ current. *Scientific Reports, 7*, 7869.

Verkerk, A. O., Wilders, R., Coronel, R., Ravesloot, J. H., & Verheijck, E. E. (2003). Ionic remodeling of sino-atrial node cells by heart failure. *Circulation, 108*, 760–766.

Wickman, K., Nemec, J., Gendler, S. J., & Clapham, D. E. (1998). Abnormal heart rate regulation in GIRK4 knockout mice. *Neuron, 20*, 103–114.

Yanni, J., Tellez, J. O., Maczewski, M., Mackiewicz, U., Beresewicz, A., Billeter, R., Dobrzynski, H., & Boyett, M. R. (2011). Changes in ion channel gene expression underlying heart failure-induced sinoatrial node dysfunction. *Circulation. Heart Failure, 4*, 496–508.

Zhang, Z., Xu, Y., Song, H., Rodriguez, J., Tuteja, D., Namkung, Y., Shin, H. S., & Chiamvimonvat, N. (2002). Functional roles of Ca(v)1.3 (alpha(1D)) calcium channel in sinoatrial nodes: Insight gained using gene-targeted null mutant mice. *Circulation Research, 90*, 981–987.

Zhang, Z., He, Y., Tuteja, D., Xu, D., Timofeyev, V., Zhang, Q., Glatter, K. A., Xu, Y., Shin, H. S., Low, R., & Chiamvimonvat, N. (2005). Functional roles of Cav1.3(alpha1D) calcium channels in atria: Insights gained from gene-targeted null mutant mice. *Circulation, 112*, 1936–1944.

An Integral View on Calcium Channels and Transporters Shaping Calcium and Exocytotic Signals in Chromaffin Cells

Ana Fernández, Antonio M. García-de Diego, Luis Gandía, Antonio G. García, and Jesús M. Hernandez-Guijo

Abstract

This chapter focuses on calcium movements and their impact on exocytosis in adrenal medullary chromaffin cells (CCs). Upon depolarization, Ca^{2+} enters the cell through open plasmalemmal voltage-dependent calcium channels (Ca_v channels) driven by a large electrochemical gradient. This gives rise to sub-plasmalemmal high-Ca^{2+} microdomains (HCMDs) near active exocytotic sites were a fraction of secretory catecholamine-storing chromaffin granules are docked, the so-called readily releasable vesicle pool (RRVP). These high transients of cytosolic Ca^{2+} concentrations ($[Ca^{2+}]_c$) are required for the fast exocytotic release of vesicle cargo to the extracellular space.

The HCMD is cleared by three regulatory mechanisms, namely the binding of Ca^{2+} to calcium-binding proteins (CBPs), the rapid and efficient Ca^{2+} uptake into mitochondria through the mitochondrial calcium uniporter (MICU), and the uptake of Ca^{2+} into the endoplasmic reticulum (ER) through the sarcoendoplasmic reticulum ATPase (SERCA). The subsequent release of Ca^{2+} back into the cytosol from mitochondria, through the mitochondrial Na^+/Ca^{2+} exchanger (MNCX) and from the ER via the inositol trisphosphate receptor (IP_3R) and the ryanodine receptor (RyR), generates low-Ca^{2+} microdomains (LCMDs) at cytosolic sites. This lower $[Ca^{2+}]_c$ is required for the transport of mature new chromaffin granules from a reserve vesicle pool (RVP) to the RRVP; this process serves to refill the RRVP with new vesicles and to secure new rounds of exocytosis. The Ca^{2+} efflux to the extracellular space through two calcium transporters (i.e., a Ca^{2+}-ATPase or Ca^{2+} pump and the plasmalemmal Na^+/Ca^{2+} exchanger (PNCX)), also participates in the regulation of exocytosis.

Finally, an integrative picture of Ca^{2+} movements and the balance of Ca^{2+} in the context of exocytosis is drawn. The review ends with a corollary on the vast amount of knowledge accumulated during the last 60 years since Douglas and Rubin demonstrated the absolute requirement of Ca^{2+} for the secretion of catecholamines from the perfused cat adrenal gland, triggered by acetylcholine. A prediction on the evolution of the

A. Fernández · A. M. García-de Diego · L. Gandía
A. G. García (✉) · J. M. Hernandez-Guijo
Instituto-Fundación Teófilo Hernando, IIS La Princesa, IRYCIS Department of Pharmacology, School of Medicine, Universidad Autónoma de Madrid, Madrid, Spain
e-mail: agg@uam.es; jesusmiguel.hernandez@uam.es

G. W. Zamponi, N. Weiss (eds.), *Voltage-Gated Calcium Channels*,
https://doi.org/10.1007/978-3-031-08881-0_14

stimulus-secretion coupling process and its impact on stress and disease is finally made.

Keywords

Chromaffin cells · Calcium channels · Calcium transporters · Calcium signaling · Exocytosis

Abbreviations

Cav	voltage-dependent calcium channels
CBPs	calcium-binding proteins
CCs	adrenal medullary chromaffin cells
CICR	Ca^{2+}-induced Ca^{2+} release mechanism
CNS	central nervous system
ER	endoplasmic reticulum
HCMDs	high-calcium microdomains
IP3R	inositoltrisphosphate receptor (IP_3R)
LCMDs	low-calcium microdomains
MICO	mitochondrial calcium uniporter
MNCX	mitochondrial Na^+/Ca^{2+} exchanger
NKA	Na^+/K^+-ATPase
PNCX	Plasmalemmal Na^+/Ca^{2+} exchanger
RRVP	ready-release vesicle pool
RVP	reserve vesicle pool
RyR	ryanodine receptor
SERCA	sarcoendoplasmic reticulum ATPase

Introduction

This review evaluates the circulation of calcium ions (Ca^{2+}) in mammalian adrenal medullary chromaffin cells and their role in the triggering of secretory responses. The adrenal medulla is a type of amplifying arm of the hypothalamus-sympathoadrenal axis; this has the critical function of activating the highly coordinated fight-or-flight response to stress to preserve body homeostasis and secure survival (Cannon, 1929). This explosive response is achieved by the sudden local release of norepinephrine from the sympathetic nerve terminals that innervate practically all organs of the body and by the release of norepinephrine and epinephrine into the systemic circulation from the adrenal gland.

The necessary and obligatory role of calcium ions (Ca^{2+}) in the triggering of catecholamine release from the perfused cat adrenal gland was demonstrated 60 years ago with two simple but elegant experiments: (1) the stimulation with acetylcholine (ACh), the physiological neurotransmitter at the splanchnic nerve-chromaffin cell synapse, triggered a healthy secretory response that was abolished in Ca^{2+}-deprived medium (Douglas & Rubin, 1961); and (2) ACh augmented the uptake of radiotracer $^{45}Ca^{2+}$ into the adrenal medulla CCs (Douglas & Poisner, 1961). In the following years, additional experiments on Ca^{2+} and secretion in various glands and neurons led William Douglas to formulate the general hypothesis on the stimulus-secretion coupling that implicated Ca^{2+} as a second messenger to trigger secretion (Douglas, 1968). This concept was extrapolated from the hypothesis of excitation-contraction coupling coined by Sandow in 1952 to stress the role of Ca^{2+} in muscle contraction (Sandow, 1952) and was inspired by previous elegant experiments on acetylcholine (ACh) release at the muscle end plate and the requirement of Ca^{2+} (del Castillo & Katz, 1956). Since then, much research done in laboratories all over the world has clarified many physiological, neurochemical, and pharmacological features of ion channels, cell excitability, Ca^{2+} signals, exocytosis, and endocytosis in CCs from different mammalian species, including humans.

But the interest in using CCs as a biological model grew much beyond physiology and pharmacology. This was due to the neuron-like phenotype of CCs, as shown in early experiments with growth factors (Aloe & Levi-Montalcini, 1979; Unsicker et al., 1978). Thus, CCs have been used to clarify the basic molecular mechanisms involved in cell excitability, synaptic transmission, regulated neurotransmitter release, and the mechanism of action of many drugs acting on the central nervous system (CNS) (Borges et al., 2018).

The CC has also been crucial for understanding the dynamics of Ca^{2+} movements, from Ca^{2+} entry through voltage-dependent calcium channels (Ca_v), Ca^{2+} redistribution into organelles, Ca^{2+} extrusion, and the overall handling of Ca^{2+} by the cell (García et al., 2006). Furthermore, most proteins of the exocytotic machinery were initially characterized in CCs (Tagaya et al., 1995; Hohne-Zell & Gratzl, 1996; Sorensen et al., 2002; Dhara et al., 2018). Additionally, CCs have been the major cell type employed in the development and refinement of patch-clamp and amperometric recording techniques (Hamill et al., 1981; Whitman et al., 1991). This led to the clarification of most of the molecular mechanisms that regulate cell excitability, Ca^{2+}-dependent secretion, and the protein machinery that regulates the fusion of vesicle and plasmalemma membranes during the exocytotic process. CCs have also been used in studies of growth factors and neural development (Aloe and Levy-Montalcini, 1979) and more recently, they also have been valuable in studies of altered Ca^{2+} handling and exocytotic fusion pore in transgenic mouse models of neurodegenerative diseases (de Diego & García, 2018).

Several reviews have focused on different aspects of CC biology during the last two decades. In 2006, a comprehensive review on Ca_v channels, Ca^{2+} signalling, and exocytosis included aspects related to the subtypes of Ca_v channels expressed by CCs of different mammalian species, the dynamics of Ca^{2+} entry and its subsequent redistribution into cell organelles, the shaping of cytosolic Ca^{2+} transients by Ca_v channels, the endoplasmic reticulum (ER) and mitochondria, the regulation by neurotransmitters of Ca_v currents, and how this complex machinery is involved in the regulation of Ca^{2+} homeostasis during cell stimulation and the regulation of pre- and exocytotic steps (García et al., 2006). In 2007 a review focused on CC excitability and the differential regulation of norepinephrine and epinephrine release by different CNS nuclei (de Diego et al., 2007). In 2012, three reviews appeared. Two of them emphasized aspects linked to intracellular Ca^{2+} handling by organelles and its implication in the regulation of exocytosis (García et al., 2012; García-Sancho et al., 2012). The third review dealt with Ca_v channels and their contribution to CC excitability, exocytosis, and endocytosis (Mahapatra et al., 2012). More recently, an extensive review on most aspects of CC biology in health and disease has appeared (Carbone et al., 2019).

Here we focus on three aspects of Ca^{2+} signalling and secretion: (i) the movements of Ca^{2+} linked to the various elements controlling the CC Ca^{2+} balance namely Ca_v channels, mitochondria, the ER, and plasmalemmal Ca^{2+} transporters; (ii) analysis of the individual implication of those elements on the exocytotic release of catecholamines; (iii) an integrative view of Ca^{2+} movements and exocytosis during CC activation, is finally presented.

Calcium Balance in Chromaffin Cells

The activation of CCs by physiological ACh gives rise to local transient increases of cytosolic calcium concentrations ($[Ca^{2+}]_c$). Two main Ca^{2+} signals occur upon CC stimulation: (i) high-calcium microdomains (HCMDs) occurring at subplasmalemmal sites as a result of sudden Ca^{2+} entry through Ca_v channels to trigger fast exocytosis; and (ii) the low-calcium microdomains (LCMDs) occurring at inner cytosolic sites to promote the transport of new secretory vesicles, to refill the secretory machinery with new vesicles to replace those that were used in the previous cycle of exocytosis. Of course, Ca^{2+} ions also play several other functions in CCs, such as the regulation of ion channels and cell excitability and the production of ATP by mitochondria, which is required for docking of the new vesicles to exocytotic subplasmalemmal sites and for storing catecholamines at secretory vesicles. This topic has been recently reviewed (Carbone et al., 2019) and hence, we will not discuss it further here.

To generate and maintain these various activities, the entry of Ca^{2+}, its ulterior redistribution into the endoplasmic reticulum (ER) and mitochondria, and its efflux through plasmalemmal calcium transporters secure the Ca^{2+} balance, cell

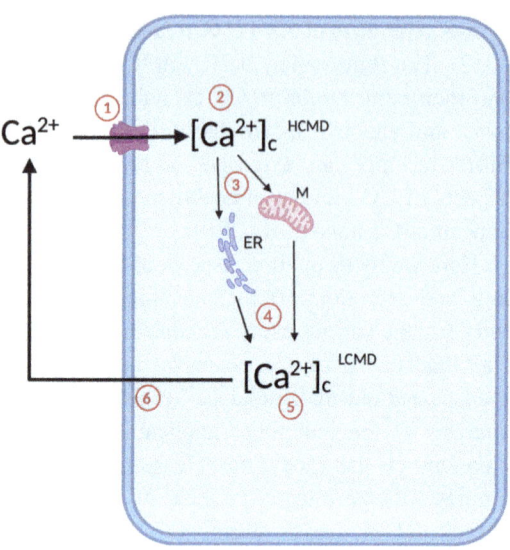

Ca^{2+} → [Ca^{2+}]$_c$ HCMD

M

ER

[Ca^{2+}]$_c$ LCMD

Fig. 1 The cycle of Ca^{2+} in a chromaffin cell. Cell depolarization opens Ca_v channels that allows Ca^{2+} entry (1) to form a sharp subplasmalemmal elevation of [Ca^{2+}]$_c$, the high-Ca^{2+} microdomain (2, HCMD), that is rapidly cleared by mitochondria and the endoplasmic reticulum (3), that release it back into the cytosol (4) giving rise to a low Ca^{2+} microdomain (5, LCMD) that finally is dissipated by plasmalemmal Ca^{2+} transporters that promote Ca^{2+} efflux (6) to restore the cell Ca^{2+} balance

functions, and cell viability. This is achieved by all elements that intervene in the circulation of Ca^{2+} in CCs, which are summarized in Fig. 1.

The main elements of this highly elaborated machinery to control Ca^{2+} movements in CCs will be briefly described next.

Calcium Influx Through Voltage-Dependent Calcium Channels

Practically all known subtypes of neuronal Ca_v channels are expressed in CCs of various mammalian species. This has been revealed by the combination of several methodological strategies. For instance, the combination of RT-PCR, in situ hybridization, and mRNA levels in bovine adrenal medulla tissue and CCs, revealed the expression of Ca_v1 (L-type), $Ca_v2.1$ (PQ-type), and $Ca_v2.2$ (N-type) channels (García-Palomero et al., 2001). Functionally, the various Ca_v channels have been characterized through the moni-

toring of whole-cell inward calcium currents (I_{ca}), $^{45}Ca^{2+}$ fluxes, and the use of drugs and toxins that block specific channel subtypes. Thus, Ca_v1 (L-type) and Ca_v2 (P/Q-, N-, and R-type) have been identified and characterized in CCs from cat (Albillos et al., 1994), cow (Gandía et al., 1993, 1997; Artalejo et al., 1994; Lukyanetz and Neher, 1999), mouse (Hernández-Guijo et al., 1998; Albillos et al., 2000; Aldea et al., 2002), rat (Gandía et al., 1995), and human (Pérez-Álvarez et al., 2008; Gandía et al., 1998). In rat and mouse CCs, $Ca_v3.2$ T-type channels are also expressed during immature stages and in response to stressful conditions where they preserve their particular low-voltage range of activation and fast-inactivating kinetics (Bournaud et al., 2001; Novara et al., 2004; Carabelli et al., 2007a, b; Carbone & Carabelli, 2009; Levitsky & Lopez-Barneo, 2009; Hill et al., 2011; Carbone et al., 2014).

Notable species differences concerning the relative fraction of the whole-cell Ca^{2+} current (I_{Ca}) carried by each Ca_v subtype in CCs have been demonstrated (Fig. 2).

Thus, in the cat (Albillos et al., 1994), mouse (Hernández-Guijo et al., 1998) and rat CCs (Gandía et al., 1995) L-type channels (Ca_v1) account for nearly 50% of the total calcium conductance. In contrast, as much as 80% of the calcium current is carried by N-type channels ($Ca_v2.2$) in pig CCs (Kitamura et al., 1997), and about 45% of the whole-cell current is carried by N-type channels in cat CCs (Albillos et al., 1994). A 30% fraction of the N-type current accounts for the whole Ca_v current in bovine (López et al., 1994b), rat (Gandía et al., 1995), mouse (Hernández-Guijo et al., 1998) and human CCs (Gandía et al., 1998).

Of note is the predominance of P/Q-type channels ($Ca_v2.1$) carrying 50% of the whole Ca_v current in bovine CCs (Albillos et al., 1996), and this fraction is even higher (60%) in human CCs (Gandía et al., 1998). In sharp contrast, PQ-type channels carry only 5% of the whole current in pig (Kitamura et al., 1997) and cat (Albillos et al., 1994) CCs. Finally, P/Q-type channels contribute to 20% of the current in rat CCs (Gandía et al., 1995) and 30% in mouse CCs (Hernández-

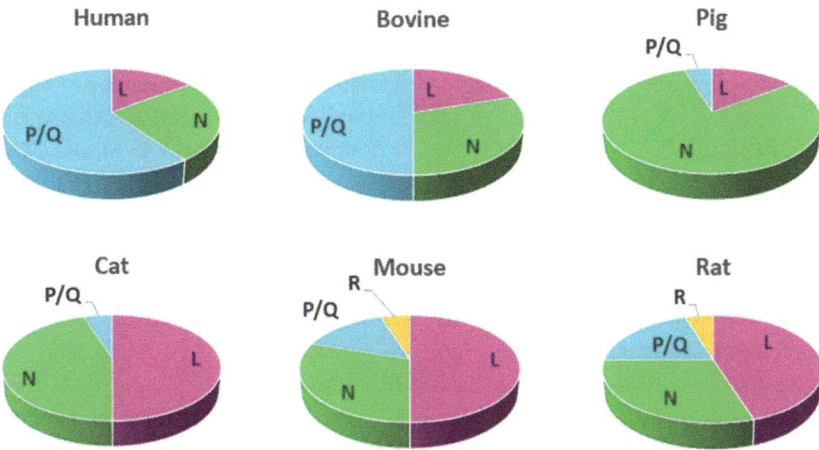

Fig. 2 Relative contribution of Ca_v subtypes to the whole-cell inward current in voltage-clamped chromaffin cells (CCs) from six mammalian species. (Adapted from García et al., 2006)

Guijo et al., 1998) although a latter study reported that P/Q-type channels accounted for only 15% of the whole current in mouse CCs (Aldea et al., 2002).

These drastic differences in the current density carried by Ca_v subtypes in CCs from six mammalian species must have some kind of teleological meaning, either from physiological or pathophysiological points of view. As not all Ca_v channel subtypes contribute equally to trigger and maintain a given secretory response of catecholamines in a stressful condition; and as Ca_v T-type channels carrying a tiny current may also contribute to exocytosis under certain "pathological" conditions, we may guess that any Ca^{2+} entering the cell during a depolarization may trigger exocytosis. This could question the hypothesis that some Ca_v subtypes could be specialized to regulate exocytosis and other relevant functions of CCs. Further discussion of this intriguing topic will be done later on, in sections where pre-exocytotic steps and exocytosis are discussed. This picture is further complicated by the observation that CCs also express plasmalemmal store-operated calcium channels and ligand-gated calcium channels; Ca^{2+} entry through these channels may also trigger exocytosis. We will not comment further on these channels (García et al., 2012).

Intracellular Calcium Movements

Calcium entering CCs upon their depolarization generates subplasmalemmal HCMDs that are dissipated by calcium-binding proteins (CBPs) (so-called cytosolic calcium buffers), Ca^{2+} sequestration by the ER and mitochondria, and its subsequent release back into the cytosol. We will comment on these coordinated sequential steps that critically maintain the balance of intracellular Ca^{2+} ion movements in CCs.

Cytosolic Calcium Buffers

The buffering of Ca^{2+} and its diffusion after transient depolarization (action potentials) or sustained depolarization have been extensively studied in Erwin Neher's laboratory, essentially in bovine CCs (Neher, 1998a). However, the identity of those immobile calcium buffers has been scarcely studied in CCs. For instance, in bovine CCs, parvalbumin contains Ca^{2+}/Mg^{2+} mixed sites exhibiting slow Ca^{2+}-binding kinetics. However, these CBPs do contribute to the clearance of large Ca^{2+} loads, thus extending the duration of the ($[Ca^{2+}]_c$) transient by transforming the monoexponential decay into a biexponential one (Lee et al., 2000). On the other hand, calbindin-D28K is homogeneously distributed in the cytosol of bovine CCs but it concentrates in submembrane areas in mouse CCs (Alés et al., 2002). This differently affects Ca^{2+} signalling and exocytosis in both cell types (Alés et al., 2005).

Endoplasmic Reticulum

The sarcoendoplasmic reticulum Mg^{2+}-and ATP-dependent calcium-ATPase (SERCA) takes up Ca^{2+} when elevated in the cytosol (Carafoli, 1987). These early observations led to the concept of intracellular calcium stores of practically all non-excitable and excitable cells (Berridge, 1998). The release of Ca^{2+} from the ER store back into the cytosol takes place via two channels, namely the inositol trisphosphate receptor channel (IP_3R) and the ryanodine receptor channel (RyR). The IP_3R channel is opened by IP_3 generated as a result of the activation of G-protein coupled membrane receptors. And RyR is activated by augmented $[Ca^{2+}]_c$ where Ca^{2+} binding to RyR opens the channel to release ER Ca^{2+} into the cytosol through the Ca^{2+}-induced Ca^{2+} release mechanism (CICR).

ER Ca^{2+} fluxes have been explored in various studies done in CCs. Thus, a parallel increase of IP_3 and $[Ca^{2+}]_c$ occurred upon challenging bovine CCs with histamine or angiotensin II (Stauderman & Pruss, 1990). Furthermore, histamine-elicited $[Ca^{2+}]_c$ increase was mimicked by IP_3 suggesting that stimulation of histamine receptors at the plasma membrane was coupled to IP_3 generation and the subsequent stimulation of IP_3Rs to cause ER Ca^{2+} release into the cytosol (Stauderman & Murawsky, 1991; Stauderman et al. 1991).

Concerning RyR channels and their role in releasing ER Ca^{2+}, various early studies established that bovine CCs express a caffeine-sensitive ER calcium store (Cheek & Thastrup, 1989). In these cells transfected with ER-targeted aequorins with different Ca^{2+} affinities, it was directly shown that the ER is a high-capacity Ca^{2+} store of as much as 500 µM. Furthermore, Ca^{2+} entry through Ca_v channels during cell depolarization elicited ER Ca^{2+} release through the CICR mechanism. In addition, the wave of Ca^{2+} from subplasmalemmal sites to the inner cell core, elicited by 100 ms pulses in voltage-champed bovine CCs and monitored with confocal microscopy, was delayed and reduced in magnitude in

ryanodine-treated cells. All these data suggest that the ER of bovine CCs is a single thapsigargin-sensitive Ca^{2+} store that can release Ca^{2+} both through IP_3Rs and RyRs (Alonso et al., 1999; Villalobos et al., 1992). In mouse CCs it was initially shown that they had a small or non-existent CICR (Rigual et al., 2002). However, a subsequent report demonstrated the expression of RyRs and the presence of a CICR mechanism in these cells (Wu et al., 2010).

Mitochondria

Elegant experiments done in the laboratory of Bertil Hille demonstrated that mitochondria acted as rapid and reversible Ca^{2+} buffers during stimulation of rat CCs (Park et al., 1996; Babcock et al., 1997). However, the levels of Ca^{2+} elevations at the mitochondrial matrix ($[Ca^{2+}]_m$) were in the low micromolar range (Babcock et al., 1997). This was likely due to underestimation by saturation of the measuring fluorescent Ca^{2+} probe. The use of mitochondrially targeted aequorins with low Ca^{2+} affinity in bovine CCs revealed a surprising rapid millimolar $[Ca^{2+}]_m$ elevation upon cell stimulation with ACh, caffeine or K^+ (Montero et al., 2000). This high-capacity mitochondrial Ca^{2+} uptake through the calcium uniporter (MICU) is due to the large driving force generated by the high transmembrane potential difference (near -180 mV) associated with the respiratory chain and ATP hydrolysis (Duchen, 2000). This accumulated Ca^{2+} is then released back into the cytosol by an electroneutral antiporter that exports Ca^{2+} from the matrix by exchanging one Ca^{2+} ion for two Na^+ ions, through the mitochondrial Na^+/Ca^{2+} exchanger (MNCX) (Carafoli, 1979).

Calcium Efflux

Plasma membrane calcium transporters (i.e. the Ca^{2+} pump and the plasma membrane Na^+/Ca^{2+} exchanger (PNCX)) maintain the long-term Ca^{2+} homeostasis through a well-balanced Ca^{2+} influx and Ca^{2+} efflux activities. The Ca^{2+} pump is a Ca^2-ATPase with high Ca^{2+} affinity for Ca^{2+} in the range of 10^{-7} M that operates as an electrogenic

Ca^{2+}/H^+ exchanger with a 1:1 stoichiometry (Salvador et al., 1998).

The PNCX is responsible for an electrogenic exchange of 3 Na^+ ions for 1 Ca^{2+} ion. Physiologically, the PNCX transports Na^+ into the cell and Ca^{2+} is extruded from the cytosol (the so-called forward mode) (Baker et al., 1969). Contrarily, during membrane depolarization and the opening of voltage-gated Na^+ channels the electrochemical gradient for Na^+ is reversed; thus, the NCX moves intracellular Na^+ out of the cell and extracellular Ca^{2+} into the cell, and the reversed mode of the PNCX (Blaustein & Lederer, 1999). Bovine CCs express the isoform NCX1 that mediates Na^+-dependent Ca^{2+} influx (Liu & Kao, 1990) or intracellular Ca^{2+} export to the extracellular space (Powis et al., 1991).

Calcium Movements and Exocytosis in Chromaffin Cells

As discussed above, during CC activation the Ca^{2+} balance is established with a high variety of Ca_v channel subtypes and calcium transporters located at the plasma membrane and at membranes of intracellular organelles, mainly the ER and mitochondria. The pharmacological inhibition or activation of each of these structures has been invaluable to define their participation of each one in the shaping of the LCMDs and the HCMDs required for the Ca^{2+}-dependent activation of pre-exocytotic steps (i.e. intracellular transport of secretory vesicles) and the last steps of exocytosis (i.e. membrane fusion, expansion of the fusion pore, and explosive catecholamine release of the vesicle cargo into the extracellular space) during CC stimulation. The pharmacological tools used to define the contribution of the different Ca_v channel subtypes and calcium transporters to exocytosis are summarized in Table 1.

Calcium Influx and Exocytosis

The subplasmalemmal HCMDs required for the explosive secretion of catecholamines during CC stimulation are primarily generated by Ca^{2+} entry

Table 1 Drugs, toxins and compounds are used to target the different ion channels and transporters that contribute to Ca^{2+} movements during stimulation of CCs, through the monitoring of Ca^{2+} currents, Ca^{2+} fluxes, or secretion

Targets	Blocker	Activator
Cav1 channel (L current)	Nifedipine	Bayk8644
Cav2.1 channel (P/Q current)	ω-conotoxin MVIIC	–
Cav2.2 channel (N current)	ω-conotoxin GVIA	–
Cav2.3 channel (R current)	SNX-482	–
Cav3 (T current)	Mibefradil	–
SERCA	Thapsigargin	–
Ryanodine receptor	Ryanodine	Caffeine
MICU	Ru360	–
MNCX	CGP37157	–
Ψm	FCCP	–
PNCX	SEA0400	–

SERCA Sarcoendoplasmic reticulum calcium ATPase, *MICU* mitochondrial calcium uniporter, *MNCX* mitochondrial sodium-calcium exchanger, *Ψm* mitochondrial membrane potential that is dissipated by protonophore FCCP

through Ca_v channels that open during depolarization of the plasma membrane (Fig. 3). The coexistence of various Ca_v channel subtypes, however, raises the question of whether each one of them could drive sufficient Ca^{2+} across the plasmalemma to trigger exocytosis. Another complication is that the relative contribution of each Cav cannel subtype to exocytosis varies with their different expression in the various animal species so far studied. A third problem is related to the observation that upon distinct stimuli applied to CCs (i.e. physiological ACh, action potentials, step depolarizations, depolarization with high K^+), the protagonist in triggering the secretion of one channel type may be favoured over another. The types of stimuli used to trigger secretion from CCs of various mammalian species and the effects of selective blockers and activators on such responses will be described in the ensuing sections.

Cat Chromaffin Cells

An early experiment in the perfused cat adrenal gland stimulated with mild K^+ depolarizing pulses demonstrated that BayK8644 elicited a

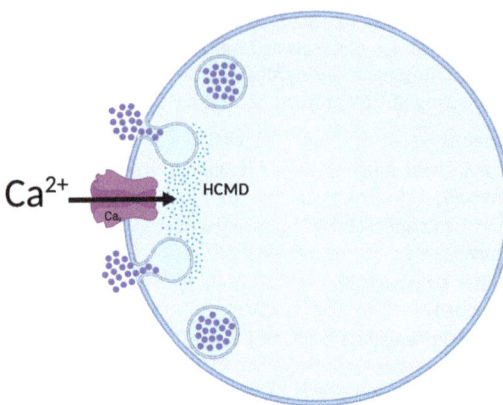

Fig. 3 High-calcium microdomain (HCMD) formed around the inner mouth of Ca$_v$ channels and docked and primed secretory vesicles within 300 nm of the Ca$_v$ channel, that undergoes rapid exocytosis. Docked vesicles away from the HCMD cannot undergo rapid exocytosis

drastic potentiation of secretory responses. As those enhanced responses were fully blocked by nifedipine, it was concluded that secretion was driven by L-type channels (García et al., 1984). Latter studies also showed that DHPs, verapamil and diltiazem fully blocked the secretion responses triggered by the nicotinic agonist dimethylphenylpiperazinium (DMPP) or by K$^+$ (Gandía et al., 1987; Cárdenas et al., 1988). As the whole-cell Ca^{2+} current of cat CCs was contributed 50% by L-type channels and 50% by N-type channels (Albillos et al., 1994), the effects on secretion by DHP furnidipine (to block L-type channels) and ω-conotoxin GVIA (to block N-type channels) were tested; catecholamine release elicited by 10s pulses of 70 mM K$^+$ was blocked by 95% with furnidipine and by only 25% with ω-conotoxin GVIA. These experiments were done in both the perfused gland and isolated CCs. Interestingly, furnidipine and ω-conotoxin GVIA equally blocked the ^{45}Ca^{2+} entry elicited by cell depolarization, meaning that the influx of Ca^{2+} through L-type channels had a predominant role in controlling secretion (López et al., 1994a). An explanation for this selectivity could be found in the type of stimulation used, 10s depolarizing pulses with high K$^+$ that could inactivate N-type channels; it could be that with

shorter stimuli, the contribution of N-type channels to secretion could be higher.

Bovine Chromaffin Cells

As discussed in section "Calcium influx through voltage-dependent calcium channels", bovine CCs express L-, N-, and PQ-type Ca$_v$ channels (García-Palomero et al., 2001). In spite of much research, the contribution of each channel type to exocytosis is still uncertain. Surprisingly, the blocker of N-type channels ω-conotoxin GVIA was ineffective or had a mild effect in blocking K$^+$-evoked secretion (Owen et al., 1989; Artalejo et al., 1991; Duarte et al., 1993; Jiménez et al., 1993; López et al., 1994b; Artalejo et al., 1994; Fernández et al., 1995). In contrast, DHPs exerted pronounced effects on functional parameters i.e. BayK8644 caused a notable augmentation of K$^+$-elicited ^{45}Ca^{2+} uptake in bovine CCs that was blocked by nifedipine (García et al., 1984), while nitrendipine completely inhibited the K$^+$-evoked secretion in those cells (Ceña et al., 1983). However, other studies found that DHPs did not block more than 40–50% of the K$^+$-induced secretory response (Owen et al., 1989; Gandía et al., 1990; Jiménez et al., 1993).

Blockers of PQ-type channels partially decreased the secretory responses triggered by K$^+$ in bovine CCs. Thus FTX, ω-agatoxin IVA, and ω-conotoxin MVIIC partially blocked secretion by 50–75%; when combined with a DHP, full blockade was achieved (Duarte et al., 1993; López et al., 1994b; Baltazar et al., 1997). Thus, it seems that K$^+$-evoked secretion is controlled by Ca^{2+} entering through L- and PQ-type Ca$_v$ channels. A latter study provided evidence indicating that secretion was more tightly coupled to P/Q-type channels (Lara et al., 1998).

A different picture emerged from electrophysiological experiments in voltage-clamped bovine CCS where Ca^{2+} entry (I$_{Ca}$) and capacitance increments (ΔCm) as an indication of exocytosis, were simultaneously monitored. Three studies concluded that ΔCm (exocytosis) triggered by single-step depolarizations was dependent on Ca^{2+} entry regardless of the subtype of Ca$_v$ channel (Engisch & Nowicky, 1996; Lukyanetz & Neher, 1999; Ulate et al., 2000).

Rat Chromaffin Cells

In the perfused rat adrenal gland, the DHP isradipine fully blocked the nicotine- or K^+ − elicited secretion. However, when secretion was triggered by field stimulation of splanchnic nerve terminals, DHPs only achieved a partial blockade, suggesting the dominance of L channels in controlling secretion evoked by nicotine or K^+, and the contribution of other Ca_v cannel subtypes when secretion was stimulated by a presynaptic input (López et al., 1992). In a latter study also done in the perfused adrenal, the high K^+-elicited secretion was strongly blocked by L-type channel inhibitors while the response to ACh was equally blocked by furnidipine or ω-conotoxin MVIIC. However, upon electrical field stimulation that evokes the presynaptic release of ACh and other cotransmitters (Wakade, 1981), the catecholamine secretion response may be contributed by N-type channels probably by a toxin action on splanchnic nerve terminals (Santana et al., 1999).

Upon measuring I_{Ca} and ΔCm in voltage-clamped single rat CCs, it was shown that nicardipine blocked secretion by 60% while ω-conotoxin GVIA caused a 40% blockade, suggesting the contribution of L- and N-type channels in evoking the whole capacitance increase as an indicator of exocytosis (Kim et al., 1995). Latter studies suggested that secretion was supported by all available Ca_v channels, including the T-type (Carabelli et al., 2002; Giancipoli et al., 2006).

Mouse Chromaffin Cells

In voltage-clamped mouse CCs with simultaneous monitoring of I_{Ca} and ΔCm, a study suggested that exocytosis was proportional to the relative current carried by L-type (40%), N-type (34%), P/Q-type (14%), and R-type (11%) Ca_v channels (Aldea et al., 2002). Upon deletion of the α_{1A} subunit of P/Q-type channels, the L-type component rose to 53%, N- and R-type components were similar and obviously, the P/Q component was absent; this suggests that Ca_v channel

subtypes do not co-localize with the secretory machinery, showing similar efficacy depending on the Ca^{2+} entering the cell when using the perforated-patch configuration of the patch-clamp technique and single 200-ms duration step-depolarizing pulses (Aldea et al., 2002). In contrast, another study in acutely isolated mouse adrenal slices led to different conclusions: although R-type channels accounted for only 22% of I_{Ca}, they controlled as much as 55% of rapid secretion, meaning that a close proximity of these channels to the secretory machinery existed (Albillos et al., 2000).

In mouse CCs there are studies trying to establish the role of L-channel subtypes ($Ca_v1.2$ and $Ca_v1.3$) in controlling exocytosis. Thus, deletion of $Ca_v1.3$ subunits decreased exocytosis at very negative potentials. In this manner the $Ca_v1.3$ channel seems to contribute to the low-threshold exocytosis in a way similar to the T-type $Ca_v3.2$ channels (Carabelli et al., 2003, 2007a, b; Giancippoli et al., 2006; Comunanza et al., 2010; Weiss et al., 2012).

Dog Chromaffin Cells

An in vivo study in the anaesthetized dog explored the release of catecholamines into the bloodstream, upon the electrical stimulation of the splanchnic nerve or the infusion of ACh. At the infusion rate of 0.4 μg/mL, ω-conotoxin GVIA reduced secretion elicited by electrical stimulation by 30%; neither nifedipine nor verapamil had any effect. However, upon ACh stimulation, secretion was reduced by 50% by either the toxin or nifedipine, suggesting that N- and L-type channels contributed to secretion (Kimura et al., 1994).

Intracellular Calcium Movements and Exocytosis

As discussed in section "Intracellular calcium movements", the bidirectional fluxes of Ca^{2+} ions into organelles play a notable role in dissipating and generating various types of Ca^{2+} microdomains at distinct sites in the cytosol. In this section, we will review how the Ca^{2+} movements principally occurring at the ER Ca^{2+} store and mitochondria, do also contribute to the shaping of secretory responses of CCs.

Calcium Fluxes at the ER Calcium Store and Exocytosis

Although the fast catecholamine release response from the adrenal gland is dominated by nicotinic receptors for ACh (nAChRs) and strictly depends on external Ca^{2+} (Douglas, 1968), the muscarinic receptors for ACh (mAChRs) do also mediate a healthy although slower secretory response; using agonists such as muscarine, pilocarpine and methacholine, the deletion of Ca^{2+} from the fluid perfusing the cat adrenal gland almost abolished such muscarinic response (Poisner & Douglas, 1966). This was corroborated later in the cat adrenal gland (Borges et al., 1987; Ballesta et al., 1989). In rat CCs the view somehow differed. In one report it was observed that muscarine-evoked secretion was present in Ca^{2+}-free solution in perfused rat adrenals (Harish et al., 1987). Another study in perfused guinea-pig adrenals showed that upon Ca^{2+} removal, the secretory responses to ACh and pilocarpine were still present; the response to the muscarinic agonist was only 10% of that obtained in the presence of Ca^{2+} and disappeared on challenging the gland for the second time in Ca^{2+}-free medium (Nakazato et al., 1988).

In the perfused cat adrenal gland, methacholine elicited substantial secretion that was blocked by atropine, indicating the mediation of mAChRs; as the response was markedly depressed by ER Ca^{2+} depletion with a combination of the calcium ionophore ionomycin and the calcium chelator EGTA, this study concluded that mobilization of Ca^{2+} from intracellular ER stores contributed to the muscarinic secretory response (Abad et al., 1992). However, such a response may also involve a contribution from external Ca^{2+} entering the cell through an ionophoric channel coupled to mAChRs (Borges et al., 1987).

The release of ER Ca^{2+} through the RyR is also able of triggering the release of catecholamines. Thus, in isolated bovine CCs, caffeine caused a mild secretory response (Cheek et al.,

1990) that was present even in the absence of extracellular Ca^{2+} (Teraoka et al., 1991; Lara et al., 1997). Caffeine also produced an increase of exoctytosis monitored with capacitance (ΔCm) in voltage-clamped bovine CCs (Heinemann et al., 1993).

An obvious approach to study the eventual role of the ER Ca^{2+} store in shaping the secretory responses evoked by cell depolarization is its depletion by the SERCA blocker thapsigargin. In voltage-clamped bovine CCs, no effect on ΔCm was shown upon ER Ca^{2+} depletion in one study (Mollard et al., 1995), but a rundown of exocytosis was demonstrated in another (Pan & Fox, 2000). Additionally, in intact bovine CCs stimulated with ACh, the severe ER Ca^{2+} depletion elicited by a mixture of caffeine, ryanodine and thapsigargin, halved the secretory response (Cuchillo-Ibáñez et al., 2002); this suggests that the contribution of ER and the CICR mechanism to the secretory response is quite visible when cells are stimulated with physiological ACh.

Mitochondrial Calcium Movements and Exocytosis

As discussed above, mitochondria are highly effective in clearing the HCMDs occurring at subplasmalemmal sites as a consequence of external Ca^{2+} entry through Ca_v channels. Hence, the blockade of mitochondrial Ca^{2+} uptake should deeply affect the shape of exocytotic responses. In line with this prediction is the result of one experiment in which the dissipation of the proton gradient by protonophores, which drastically decreased the Ca^{2+} buffering capacity of mitochondria (Montero et al., 2000; Yang and Kao, 2001), gave rise to a pronounced augmentation of exocytotic responses in voltage-clamped single bovine CCs, monitored as ΔCm (Giovannucci et al., 1999) and in populations of bovine CCs stimulated with ACh, K^+ or caffeine (Montero et al., 2000; Cuchillo-Ibáñez et al., 2002, 2004; Caricati-Neto et al., 2013). Upon K^+ stimulation of bovine CCs in the presence of L-type Ca_v activator FPL64176, the potentiation of secretory responses by protonophores was even more drastically augmented (Montero et al., 2001). The block of MICU also led to enhanced secretion in single bovine CCs (Yang & Kao, 2001). In mouse CCs, an unexpected decrease of K^+-elicited secretion in the presence of a protonophore was reported (Alés et al., 2005); this could be due to the different expression of Ca_v channel subtypes in bovine versus mouse CCs and/or the different rates of inactivation of such channels during the blockade by protonophores, of mitochondrial Ca^{2+} sequestration (Hernández-Guijo et al., 2001).

The rate of mitochondrial Ca^{2+} circulation also influences $[Ca^{2+}]_c$ signals; in fact, the pharmacological inhibition of the MNCX prolongs the duration of the $[Ca^{2+}]_c$ transient as the rate of mitochondrial Ca^{2+} release into the cytosol slows down (Montero et al., 2000). This gives rise to slower decay of quantal exocytotic responses to repeated brief ACh pulses, or as an augmentation of previously desensitized ACh secretory responses in mouse CCs. This has been interpreted to be "physiologically relevant in situations of prolonged stressful conflicts where sustained catecholamine release responses are regulated by changes in the rate of mitochondrial Ca^{2+} circulation through the MNCX, which couples respiration and ATP synthesis to long-term stimulation of CCs by endogenously released ACh" (López-Gil et al., 2017).

Calcium Efflux and Exocytosis

The efflux of Ca^{2+} ions from CCs into the extracellular space is carried out by plasmalemmal Ca^{2+} ATPase (the Ca^{2+} pump) and the PNCX. As discussed above, both transporters contribute to maintaining the long-term Ca^{2+} homeostasis through well-balanced Ca^{2+} influx and Ca^{2+} efflux activities. The high affinity/low capacity of Ca^{2+} transport precludes a major role of the Ca^{2+} pump in shaping Ca^{2+} and exocytotic signals. The reverse is true for the low affinity/high capacity for Ca^{2+} transport of the PNCX; its manipulation either by changes of Na^+ and Ca^{2+} gradients or by pharmacological ligands has revealed its relevant role in regulating the Ca^{2+}-exocytosis relationship.

By itself, the cardiac glycoside ouabain augmented the basal release of catecholamines in quiescent CCs (Banks, 1967; García et al., 1980; Aunis & García, 1981). This correlated well with [^{3}H]-ouabain binding to plasmalemmal Na^+/K^+-ATPase (NKA) (Aunis & García, 1981) and with K^+ movements, suggesting the inhibition of NKA as the cause of augmented secretion. Ouabain also increased the K^+-evoked catecholamine release from cat CCs (Esquerro et al., 1980) and bovine CCs (de Pascual & García, 2007; Milla et al., 2011). These effects were initially interpreted as a secondary activation of the reversal mode of the PNCX; ouabain upsets the Na^+ and K^+ gradients across the plasma membrane, thus collapsing the Na^+ gradient that can secondarily drive Ca^{2+} entry through the NCX to augment Ca^{2+} available to trigger exocytosis.

In the light of subsequent experiments, an alternative hypothesis emerged. It was based on the observation that ouabain caused the release of ER Ca^{2+} at the time the glycoside augmented the secretory responses to sequential K^+ pulses. Congruent with this, ER Ca^{2+} depletion prevented the enhanced secretion elicited by ouabain. Furthermore, ouabain contributed to the maintenance of the sequential secretory responses to K^+ pulses, which tended to fade off in the absence of the cardiotonic drug (Milla et al., 2011). As ouabain is present in the adrenal cortex (Li et al., 1998) and cortical hormones contribute to the regulation of different aspects of CC biology, it is plausible that the endogenous glycoside may have a physiological role in controlling Ca^{2+} and exocytotic signals, particularly under long-term stressful conflicts.

An Integrated View of Calcium and Exocytosis Signals in Chromaffin Cells

As for the investigation to clarify the functions of different organ systems, the stimulus-secretion coupling process occurring during stress at the adrenal medulla has been studied mostly in simple models such as acutely isolated or cultured CCs; some more physiological systems have scarcely been used, as the case is for adrenal slices or perfused intact adrenals. Some information on the process has also been obtained in intact animals subjected to stress.

The adrenal CCs in the intact animal or in the human are being continuously stimulated by the low-frequency discharge (0.1–0.2 Hz) of action potentials (APs) of the sympathoadrenal axis; this leads to the release of ACh quanta at the synapse formed by splanchnic nerve terminals with the surface of CCs, that releases low amounts of catecholamines into the blood stream. When a sudden stress occurs, the discharge of APs at sympathetic nerves drastically augments to 20–30 Hz; in a time second or less, a pronounced release of noradrenaline and adrenaline occurs thus preparing practically all the organs of the body for the fight-or-flight response (de Diego et al., 2007). The stress may last for a long time and so, the release from catecholamine-storing vesicles has to be maintained for minutes or even hours; in this manner, the body homeostasis and animal survival are warranted. This requires the augmented synthesis and storage of catecholamines and the continuous flow of new vesicles from a reserve vesicle pool (RVP) toward the subplasmalemmal secretory machinery (Carbone et al., 2019).

In the early 1990s, the laboratory of Erwin Neher defined various vesicle pools in voltage-clamped bovine CCs. After the application of a maximal depolarizing pulse, the depletion of a ready-release vesicle pool (RRVP) at exocytotic sites recovered within one to several minutes. The recovery was accelerated upon mild elevations of $[Ca^{2+}]_c$ that were subthreshold for exocytosis, elicited by ER Ca^{2+} release by histamine or with mild depolarizing pulses to −35 mV to augment Ca^{2+} entry through Ca_v channels (von Ruden & Neher, 1993). Thus, it seems that different $[Ca^{2+}]_c$ microdomains are required to trigger the last fast exocytotic steps (subplasmalemmal HCMDs) and preexocytotic steps involved in vesicle transport from the RVP to the RRVP (LCMD at inner cytosolic sites).

In CCs, the disposition of the Ca^{2+} signalling system favour the formation of HCMDs in areas close to Ca_v channels. This allows the regulation of distinct functions by the same triggering Ca^{2+} signals but at different spatial and temporal patterns. Those Ca^{2+} signals depend on Ca^{2+} fluxes at the plasma membrane and cytosolic organelles, as well as on Ca^{2+} diffusion and buffering by the cytosol, as discussed in section "Calcium balance in chromaffin cells".

Subplasmalemmal High-Calcium Microdomains Are Required to Trigger Fast Exocytosis from Docked Vesicles

Two approaches have been followed to inquire about the rate of Ca^{2+} entry through Ca_v channels upon stimulation of bovine CCs. In voltage-clamped cells, a 500 ms depolarizing pulse to 0 mV elicited a mean I_{Ca} of 250 pA. For a 15 μm cell diameter, this current is equivalent to a rate of Ca^{2+} entry of 700 μmol · l cells^{-1} · s^{-1} (Zhou & Neher, 1993). On the other hand, the monitoring of Ca^{2+} entry with the radiotracer $^{45}Ca^{2+}$ in populations of cultured bovine CCs depolarized with 59 mM K$^+$ during 1–5 s, estimated a linear Ca^{2+} entry of 400 μmol · l cells^{-1} · s^{-1} (Artalejo et al., 1987), a rate not far from the previous one. This high rate gives rise to the formation of HCMDs of 0.3 μM diameter and concentrations as high as 5–10 μM (Neher & Augustine, 1992; Von Rüden et al., 1993; Michelena et al., 1995; Chow et al., 1996; Klingauf & Neher, 1997).

Some studies indicate a certain degree of colocalization of Cav channels and exocytotic sites (Robinson et al., 1995; Zhou & Misler, 1995; Retting & Neher, 2002; Olivos & Artalejo, 2004). Furthermore, the simultaneous monitoring of $[Ca^{2+}]_c$ and exocytosis with evanescent microscopy showed fast ($T_{1/2}$, 100 ms) and localized (350 nm diameter) hot spots of high $[Ca^{2+}]_c$ (Becherer et al., 2003). However, only the docked secretory vesicles of the RRVP located at 300 nm away from a Ca_v cannel could undergo rapid exocytosis and only about 10% of those vesicles (the "primed" ones) entered in the last step of rapid exocytosis (Becherer et al., 2003; Olivos & Artalejo, 2004).

Termination of an Exocytotic Round by Dissipation of the High-Calcium Microdomain

The cytosolic Ca^{2+} buffering and Ca^{2+} clearance in bovine CCs have been amply studied in Erwin Neher laboratory (Zhou & Neher, 1993; Xu et al., 1997). In these reports it was found that the cytosolic CBPs are scarcely mobile, have a low Ca^{2+} affinity with a K_d around 100 µM, with an activity coefficient of 1/40. In another study, the two-dimensional diffusion coefficient was around 40 $µm^2/s$, with inhomogeneities at the plasma membrane and nuclear envelope (Naraghi et al., 1998). These properties could explain the rapid diffusion of $[Ca^{2+}]_c$ from the microdomain towards the surrounding cytosol, as well as the fact that HCMDs are highly restricted in time and space (Neher, 1998b). This is reinforced by the observation that the presence of mobile Ca^{2+} buffers accelerates the diffusion of Ca^{2+}, thus opposing the formation of HCMDs (Sala & Hernández-Cruz, 1990; Nowicky & Pinter, 1993; Alonso et al., 2003). This view was strengthened by an experiment showing that 50 µM of the Ca^{2+} probe fura-2 increased four times the apparent rate of Ca^{2+} diffusion in bovine CCs (Zhou & Neher, 1993).

A second highly relevant element that markedly contributes to the clearance of HCMDs is related to the ability of mitochondria to take up high amounts of $[Ca^{2+}]_c$. This is illustrated by three types of experiments. Cytosolic Ca^{2+} is rapidly taken up by nearby mitochondria through the low-affinity high-capacity MICU, following an exponential relationship with respect $[Ca^{2+}]_c$. Thus in rat CCs mitochondrial uptake rates (indirectly calculated from the changes of $[Ca^{2+}]_c$) of 0.4 to 0.7 µM/s at $[Ca^{2+}]_c$ of 0.5 to 2 µM occur (Herrington et al., 1996). In bovine CCs, at saturating $[Ca^{2+}]_c$ concentrations of MICU (200 µM) much larger rates of

Ca^{2+} uptake were found, i.e. 120 µM/s equivalent to 4800 $µmol \cdot l\, cells^{-1} \cdot s^{-1}$ (Xu et al., 1997). Using mitochondrial-targeted aequorins in bovine CCs, even higher rates of maximal Ca^{2+} uptake were reported i.e. 160 µM/s equivalent to 6000 $µmol \cdot l\, cells^{-1} \cdot s^{-1}$ (Villalobos et al., 1992). Of interest was the determination of the rate of mitochondrial Ca^{2+} uptake during maximal K^+ depolarization of bovine CCs loaded with mitochondrial-targeted aequorin, which amounted to around 1100 $µmol \cdot l\, cells^{-1} \cdot s^{-1}$ (Villalobos et al., 1992). These differences in the rates of Ca^{2+} uptake are consistent with the exponential $[Ca^{2+}]_c$ dependence of the activity of MICU (Montero et al., 2000, 2001).

The relative contribution of cytosolic calcium buffers and mitochondria to the clearance of HCMDs depends on the strength of CC stimulation. At low stimulation, $[Ca^{2+}]_c$ reaches 1 µM levels and the low-affinity MICU contributes little to the dissipation of the $[Ca^{2+}]_c$ transients; under these conditions, the cytosolic Ca^{2+} buffers, and the high-affinity Ca^{2+} pumps (plasma membrane and ER SERCA) dominate the clearance of $[Ca^{2+}]_c$ elevations. However, at high stimulation rates, the HCMD reaches $[Ca^{2+}]_c$ levels capable of activating mitochondrial Ca^{2+} uptake by MICU. In this situation, most Ca^{2+} entering the cell through Ca_V channels (i.e. around 500 $µmol \cdot l\, cells^{-1} \cdot s^{-1}$) is taken up by a subpopulation of mitochondria (M1) located nearby the Ca_V channels at subplasmalemmal sites that as mentioned above, take up Ca^{2+} at a rate substantially higher than the rate of Ca^{2+} entry through Ca_V channels (Herrington et al., 1996; Park et al., 1996; Xu et al., 1997; Montero et al., 2000; Villalobos et al., 1992) (Fig. 4).

As mitochondria rapidly take up most of the Ca^{2+} entering through Ca_V channels during cell depolarization, it was predicted that interruption of mitochondrial Ca^{2+} uptake could highly impact on exocytosis. The driving force for Ca^{2+} uptake into mitochondria through MICU is the large mitochondrial membrane potential of −150 to −180 mV; this will favour the Ca^{2+} accumulation at the mitochondrial matrix up to 5–6 orders of magnitude above the $[Ca^{2+}]_c$ (Bernardi, 1999). Such Ca^{2+} uptake may be blocked through two

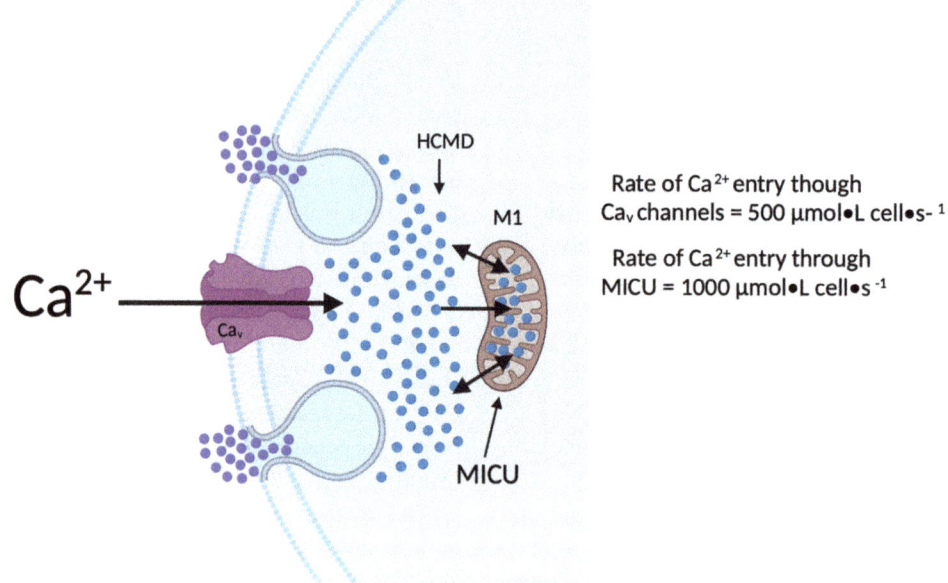

Fig. 4 Under conditions of high stimulation of CCs, the high-calcium microdomain (HCMD) underneath the plasma membrane, nearby Ca_v channels and exocytotic sites are mostly cleared by a subpopulation of nearby mitochondria

pharmacological strategies namely, the use of protonophores or the blockers of MICU.

Dissipation of the proton gradient by protonophores leads to mitochondrial depolarization and the loss of the driving force for Ca^{2+} entry, drastically decreasing the Ca^{2+} buffering capacity of mitochondria (Montero et al., 2000; Yang & Kao, 2001). This leads to enhanced exocytosis in voltage-clamped single bovine CCs (Giovanucci et al., 1999), in fast-perfused bovine CCs depolarized with ACh or K^+ (Montero et al., 2000; Cuchillo-Ibañez et al., 2002, 2004; Caricati-Neto et al., 2013), as well as in embryonic and adult rat CCs (Vestring et al., 2015). Blockade of MICU in bovine CCs (Yang & Kao, 2001) and mouse CCs (López-Gil et al., 2017) also led to augmented secretion. All these experiments emphasize the relevant role that mitochondria play in the shaping of Ca^{2+} and exocytotic signals at subplasmalemmal sites.

Vesicle Flow Is Catalyzed by Low-Calcium Microdomains

Reloading of the exhausted RRP after a round of exocytosis requires vesicle movements from the RVP from the inner cell core to subplasmalemmal exocytotic sites. As rapid exocytosis, vesicle transport is also a Ca^{2+} – dependent process; however, unlike exocytosis, this process requires much lower $[Ca^{2+}]_c$ elevations, the so-called LCMDs. Three Ca^{2+}-release pathways seem to contribute to these milder elevations of $[Ca^{2+}]_c$ at the inner cell core, namely IP_3R, RyRs and the MNCX.

Upon stimulation with histamine of bovine CCs, a mild cytosolic $[Ca^{2+}]_c$ elevation is produced as a result of ER Ca^{2+} release through IP_3R; (Alonso et al., 1999); this contributed to reload the RRVP to augment exocytosis monitored as ΔCm in voltage-clamped cells (Von Rüden & Neher, 1993). Histamine is a powerful secretagogue that elicits relevant catecholamine secretion usually during anaphylactic reactions; thus, it is unlikely that histamine plays a role in vesicle transport and reloading of the RRP under the subsequent exocytotic rounds, mediated by the firing of splanchnic nerves during stress. However, in the same study, Von Rüden and Neher also demonstrated that RRP reload was also achieved by mild cell depolarisations that only caused mild $[Ca^{2+}]_c$ elevations that were subthreshold for exocytosis (Von Rüden & Neher, 1993). This could

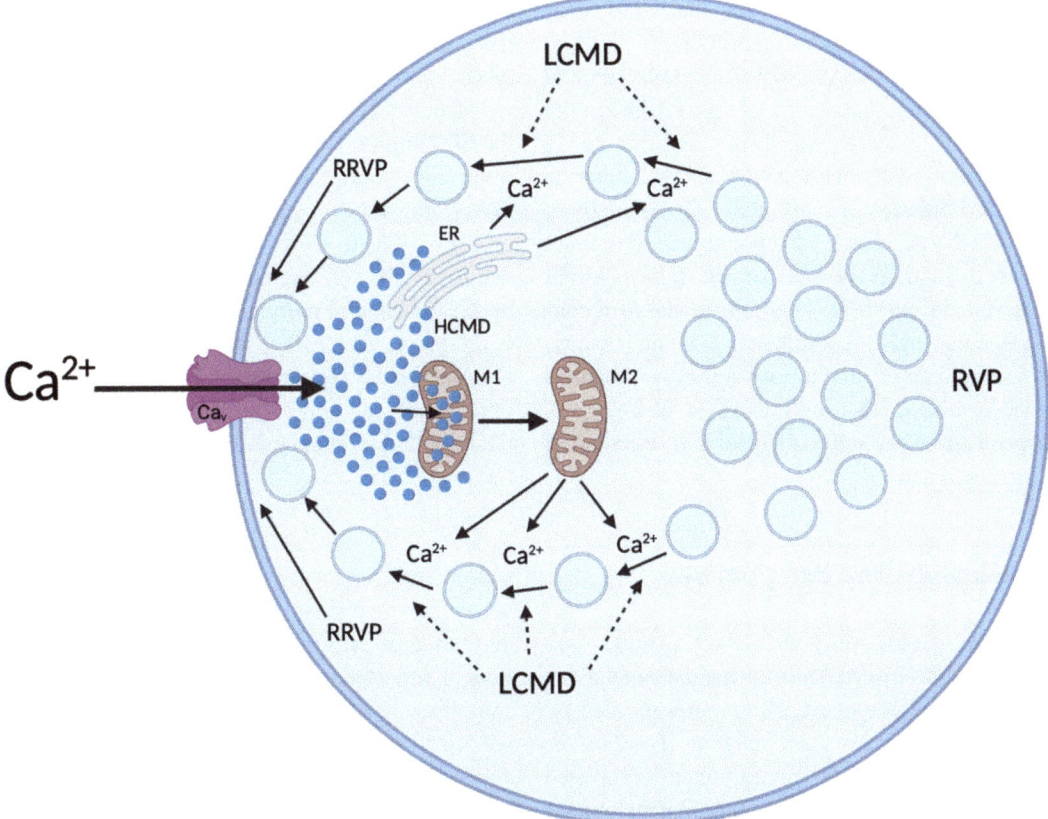

Fig. 5 Scheme showing an integral view of the high-Ca^{2+} microdomain (HCMD), the dissipation of such HCMD by Ca^{2+} redistribution into the endoplasmic reticulum (ER) and the pool M1 mitochondria, and the formation of the low-Ca^{2+} microdomain (LCMD), during Ca^{2+} entry through Ca_v channels occurring upon CC activation. Note also the role of the HCMD nearby the plasmalemma in the triggering of fast exocytosis, and of the LCMD in facilitating vesicle transport from a reserve vesicle subplasmalemmal RRVP, after exocytosis

be achieved by activation of ER Ca^{2+} release through RyR by Ca^{2+} entry via Cav channels (Alonso et al., 1999). In fact, using confocal microscopy in bovine CCs, a wave of $[Ca^{2+}]_c$ elevation elicited by short depolarizing stimuli, was shown to propagate from nearby plasmalemmal regions to the cell inner core; such Ca^{2+} wave was sustained by CICR (Alonso et al., 1999) (Fig. 5).

The third element contributing to the generation of LCMDs was the mitochondrial Ca^{2+} release through the MNCX. At the cell core, the mitochondrial pool (M2) takes up $[Ca^{2+}]_c$ at concentrations much lower ($\simeq 1$ μM) compared with the higher concentrations occurring at exocytotic sites during cell stimulation (10–30 μM or even more). Hence, M2 mitochondria releases Ca^{2+} through their MNCX generating LCMDs at inner cytosolic areas, to also favour vesicle transport and RRP replenishment.

An Integrated View of Calcium and Exocytosis in Adrenal Chromaffin Cells

Figure 5 shows an integrated view of LCMDs generated by Ca^{2+} release from the ER and mitochondria, to promote the transport of vesicles from the RVP towards the subplasmalemmal RRVP. The figure also shows the HCMDs formed at exocytotic sites as a result of Ca^{2+} entry through Ca_v channels during CC activation; this is required to trigger fast exocytosis of RRVP vesicles docked at the plasma membrane.

Considering that the RVP contains around 20,000 vesicles, that during CC activation around 200 vesicles undergo exocytosis, and that newly synthesized catecholamines occur as a function of CC stimulation, it seems that CCs are ready to provide the necessary catecholamines to prepare the body and its organs for the fight-or-flight response during short or prolonged stressful conflicts. This highly coordinated Ca^{2+} and exocytotic machinery is maintained active by enhanced ATP synthesis that curiously, is also tightly dependent on the mitochondrial Ca^{2+} circulation during cell activation.

Conclusions and Perspectives

We have learned much ever since Douglas and Rubin first discovered the requirements of Ca^{2+} to trigger the release of catecholamines elicited by ACh in the perfused cat adrenal (Douglas & Rubin, 1961). Such Ca^{2+} requirement comprises complex Ca^{2+} movements though plasmalemmal Cav channels and transporters as well as through intracellular ER and mitochondria ion channels and transporters. The nature and function of all these elements have been clarified and their contribution to the generation and shaping of high and low $[Ca^{2+}]_c$ microdomains; their implication in the last steps of exocytosis and in pre-exocytotic steps, have been clarified too. Hence, we are now in a position to offer an integral view of the dynamics of Ca^{2+} movements and exocytosis occurring during the activation of adrenal medullary CCs, to survive the various stressful conflicts taking place along the life of mammals.

However, several questions remain obscure and are needed for further investigation. They are as follows: (1) CCs express various subtypes of Ca_v channels; (2) their role in Ca^{2+} handling and exocytosis is only partially known; (3) there are drastic mammalian species differences in the relative density of expression of the different Ca_v

channel subtypes that require clarification; (4) the preferential location of a given Ca_v channel type with the exocytotic machinery is still a matter of debate; (5) the transfer of Ca^{2+} ions from the ER to mitochondria and vice versa, is a matter deserving study, as it can impinge on pre-exocytotic and exocytotic steps; (6) although the group of Emilio Carbone has nicely studied in detail the CC cell excitability and its regulation by Ca_v channels, particularly of the L subtype (Carbone et al., 2019), the projection of these findings to the regulation of $[Ca^{2+}]_c$ and exocytotic signals remains largely unexplored; (7) also the implication in exocytosis of the regulation of Ca_v channels by exogenous and endogenous neurotransmitters (an aspect not treated in this review), is needed of new experimental approaches; (8) most studies here reported have been performed in isolated and cultured CCs, that may undergo substantial changes in the expression of ion channels and transporters; although complex, new experimental approaches are needed to explore the role of Ca_v channel subtypes in low $[Ca^{2+}]_c$ signalling and exocytosis both in adrenal medullary slices and the intact adrenal gland; (9) studies on the changes in the expression and function of Ca_v channels upon chronic stress, and the implication of those changes for exocytosis are indicated; (10) profound changes in low $[Ca^{2+}]_c$ and exocytotic signals are being explored in transgenic mouse models of neurodegenerative diseases, hypertension and other diseases (reviewed by Carbone et al., 2019); and (11) plastic changes of catecholamine synthesis and release occur in the adrenal medulla of animal models of neuropathic pain (Arribas-Blázquez et al., 2020). Additionally, we guess that CCs will continue to serve as extremely useful models to decipher further intricacies of the protein interactions of the secretory machinery, taking place during exocytosis.

Acknowledgments Supported by PID2020-117127RB-I00 and EU Horizon 2020 Research and Innovation Program under Marie Skłodowska-Curie, Grant Agreement N. 766124.

References

Abad, F., Garrido, B., Lopez, M. G., & García, A. G. (1992). The source of calcium for muscarinic-mediated catecholamine release from cat adrenals. *The Journal of Physiology, 445*, 725–740. https://doi.org/10.1113/jphysiol.1992.sp018947

Albillos, A., Artalejo, A. R., Lopez, M. G., Gandia, L., Garcia, A. G., & Carbone, E. (1994). Calcium channel subtypes in cat chromaffin cells. *The Journal of Physiology, 477*, 197–213. https://doi.org/10.1113/jphysiol.1992.sp018947

Albillos, A., Garcia, A. G., Olivera, B., & Gandia, L. (1996). Re-evaluation of the P/Q Ca^{2+} channel components of Ba^{2+} currents in bovine chromaffin cells superfused with solutions containing low and high Ba^{2+} concentrations. *Pflügers Archiv / European Journal of Physiology, 432*, 1030–1038. https://doi.org/10.1007/s004240050231

Albillos, A., Neher, E., & Moser, T. (2000). R-Type Ca^{2+} channels are coupled to the rapid component of secretion in mouse adrenal slice chromaffin cells. *The Journal of Neuroscience, 20*, 8323–8330. https://doi.org/10.1523/JNEUROSCI.20-22-08323.2000

Aldea, M., Jun, K., Shin, H. S., Andres-Mateos, E., Solis-Garrido, L. M., Montiel, C., Garcia, A. G., & Albillos, A. (2002). A perforated patch-clamp study of calcium currents and exocytosis in chromaffin cells of wild-type and alpha(1A) knockout mice. *Journal of Neurochemistry, 81*, 911–921. https://doi.org/10.1046/j.1471-4159.2002.00845.x

Alés, E., Cano-Abad, M., García, A. G., & López, M. G. (2002). Different cellular distribution of calbindin D28k: Implications for the cytosolic Ca(2+) and exocytotic signals in single bovine & mouse chromaffin cells. *Annals of the New York Academy of Sciences, 971*, 168–170. https://doi.org/10.1111/j.1749-6632.2002.tb04458.x

Alés, E., Fuentealba, J., García, A. G., & López, M. G. (2005). Depolarization evokes different patterns of calcium signals and exocytosis in bovine and mouse chromaffin cells: The role of mitochondria. *The European Journal of Neuroscience, 21*(1), 142–150. https://doi.org/10.1111/j.1460-9568.2004.03861.x

Aloe, L., & Levi-Montalcini, R. (1979). Nerve growth factor-induced transformation of immature chromaffin cells in vivo into sympathetic neurons: Effect of antiserum to nerve growth factor. *Proceedings of the National Academy of Sciences of the United States of America, 76*, 1246–1250. https://doi.org/10.1073/pnas.76.3.1246

Alonso, M. T., Barrero, M. J., Michelena, P., Carnicero, E., Cuchillo, I., Garcia, A. G., Garcia-Sancho, J., Montero, M., & Alvarez, J. (1999). Ca^{2+}-induced Ca^{2+} release in chromaffin cells seen from inside the ER with targeted aequorin. *The Journal of Cell Biology, 144*, 241–254. https://doi.org/10.1083/jcb.144.2.241

Alonso, M. T., Chamero, P., Villalobos, C., & Garcia-Sancho, J. (2003). Fura-2 antagonises calcium-induced calcium release. *Cell Calcium, 33*(27–35), 2003. https://doi.org/10.1016/s0143-4160(02)00179-3

Arribas-Blázquez, M., Olivos-Oré, L. A., Barahona, M. V., Wojnicz, A., De Pascual, R., Sánchez de la Muela, M., García, A. G., & Artalejo, A. R. (2020). The adrenal medulla modulates mechanical allodynia in a rat model of neuropathic pain. *International Journal of Molecular Sciences, 21*(21), 8325. https://doi.org/10.3390/ijms21218325

Artalejo, C. R., Garcia, A. G., & Aunis, D. (1987). Chromaffin cell calcium channel kinetics measured isotopically through fast calcium, strontium, and barium fluxes. *The Journal of Biological Chemistry, 262*, 915–926. PMID: 2433271.

Artalejo, C. R., Mogul, D. J., Perlman, R. L., & Fox, A. P. (1991). Three types of bovine chromaffin cell Ca^{2+} channels: Facilitation increases the opening probability of a 27 pS channel. *The Journal of Physiology, 444*, 213–240. https://doi.org/10.1113/jphysiol.1991.sp018874

Artalejo, C. R., Adams, M. E., & Fox, A. P. (1994). Three types of Ca^{2+} channel trigger secretion with different efficacies in chromaffin cells. *Nature, 367*, 72–76. https://doi.org/10.1038/367072a0

Aunis, D., & Garcia, A. G. (1981). Correlation between catecholamine secretion from bovine isolated chromaffin cells and [3H]-ouabain binding to plasma membranes. *British Journal of Pharmacology, 72*, 31–40. https://doi.org/10.1111/j.1476-5381.1981.tb09101.x

Babcock, D. F., Herrington, J., Goodwin, P. C., Park, Y. B., & Hille, B. (1997). Mitochondrial participation in the intracellular Ca^{2+} network. *The Journal of Cell Biology, 136*, 833–844. https://doi.org/10.1083/jcb.136.4.833

Baker, P. F., Blaustein, M. P., Hodgkin, A. L., & Steinhardt, R. A. (1969). The influence of calcium on sodium efflux in squid axons. *The Journal of Physiology, 200*(2), 431–458. https://doi.org/10.1113/jphysiol.1969.sp008702

Ballesta, J. J., Borges, R., García, A. G., & Hidalgo, M. J. (1989). Secretory and radioligand binding studies on muscarinic receptors in bovine and feline chromaffin cells. *The Journal of Physiology, 418*, 411–426. https://doi.org/10.1113/jphysiol.1989.sp017849

Baltazar, G., Ladeira, I., Carvalho, A. P., & Duarte, E. P. (1997). Two types of omega-agatoxin IVA-sensitive Ca^{2+} channels are coupled to adrenaline and noradrenaline release in bovine adrenal chromaffin cells. *Pflügers Archiv/European Journal of Physiology, 434*(5), 592–598. https://doi.org/10.1007/s004240050441

Banks, P. (1967). The effect of ouabain on the secretion of catecholamines and on the intracellular concentration of potassium. *The Journal of Physiology, 93*(3), 631–637. https://doi.org/10.1113/jphysiol.1967.sp008383

Becherer, U., Moser, T., Stuhmer, W., & Oheim, M. (2003). Calcium regulates exocytosis at the level of single vesicles. *Nature Neuroscience, 6*, 846–853. https://doi.org/10.1038/nn1087

Bernardi, P. (1999). Mitochondrial transport of cations: Channels, exchangers, and permeability transition. *Physiological Reviews, 79*, 1127–1155. https://doi.org/10.1152/physrev.1999.79.4.1127

Berridge, M. J. (1998). Neuronal calcium signaling. *Neuron, 21*(1), 13–26. https://doi.org/10.1016/s0896-6273(00)80510-3

Blaustein, M. P., & Lederer, W. J. (1999). Sodium/calcium exchange: Its physiological implications. *Physiological Reviews, 79*(3), 763–854. https://doi.org/10.1152/physrev.1999.79.3.763

Borges, R., Ballesta, J. J., & García, A. G. (1987). M2 muscarinoceptor-associated ionophore at the cat adrenal medulla. *Biochemical and Biophysical Research Communications, 144*(2), 965–972. https://doi.org/10.1016/s0006-291x(87)80058-x

Borges, R., Gandía, L., & Carbone, E. (2018). Old and emerging concepts on adrenal chromaffin cell stimulus-secretion coupling. *Pflügers Archiv European Journal of Physiology, 470*(1), 1–6. https://doi.org/10.1007/s00424-017-2082-z

Bournaud, R., Hidalgo, J., Yu, H., Jaimovich, E., & Shimahara, T. (2001). Low threshold T- type calcium current in rat embryonic chromaffin cells. *The Journal of Physiology, 537*, 35–44. https://doi.org/10.1111/j.1469-7793.2001.0035k.x

Cannon, W. B. (1929). Organization for physiological homeostasis. *Physiological Reviews, 9*, 399–431. https://doi.org/10.1152/physrev.1929.9.3.399

Carabelli, V., D'Ascenzo, M., Carbone, E., & Grassi, C. (2002). Nitric oxide inhibits neuroendocrine Ca(V)1 L-channel gating via cGMP-dependent protein kinase in cell-attached patches of bovine chromaffin cells. *The Journal of Physiology, 541*, 351–366. https://doi.org/10.1113/jphysiol.2002.017749

Carabelli, V., Giancippoli, A., Baldelli, P., Carbone, E., & Artalejo, A. R. (2003). Distinct potentiation of L-type currents and secretion by cAMP in rat chromaffin cells. *Biophysical Journal, 85*, 1326–1337. https://doi.org/10.1016/S0006-3495(03)74567-6

Carabelli, V., Marcantoni, A., Comunanza, V., & Carbone, E. (2007a). Fast exocytosis mediated by T- and L-type channels in chromaffin cells: Distinct voltage-dependence but similar Ca^{2+}-dependence. *European Biophysics Journal with Biophysics Letters, 36*, 753–762. https://doi.org/10.1007/s00249-007-0138-2

Carabelli, V., Marcantoni, A., Comunanza, V., De Luca, A., Diaz, J., Borges, R., & Carbone, E. (2007b). Chronic hypoxia up-regulates alpha(1H) T-type channels and low-threshold catecholamine secretion in rat chromaffin cells. *The Journal of Physiology, 584*, 149–165. https://doi.org/10.1113/jphysiol.2007.132274

Carafoli, E. (1979). The calcium cycle of mitochondria. *FEBS Letters, 104*, 1–5. https://doi.org/10.1016/0014-5793(79)81073-x

Carafoli, E. (1987). Intracellular calcium homeostasis. *Annual Review of Biochemistry, 56*, 395–433. https://doi.org/10.1146/annurev.bi.56.070187.002143

Carbone, E., & Carabelli, V. (2009). O2 sensing in chromaffin cells: New duties for T-type channels. *The Journal of Physiology, 587*, 1859–1860. https://doi.org/10.1113/jphysiol.2009.172197

Carbone, E., Calorio, C., & Vandael, D. H. (2014). T-type channel-mediated neurotransmitter release. *Pflügers Archiv / European Journal of Physiology, 466*, 677–687. https://doi.org/10.1007/s00424-014-1489-z

Carbone, E., Borges, R., Eiden, L. E., García, A. G., & Hernández-Cruz, A. (2019). Chromaffin cells of the adrenal medulla: Physiology, pharmacology, and disease. *Comprehensive Physiology, 9*(4), 1443–1502. https://doi.org/10.1002/cphy.c190003

Cardenas, A. M., Montiel, C., Esteban, C., Borges, R., & Garcia, A. G. (1988). Secretion from adrenaline- and noradrenaline-storing adrenomedullary cells is regulated by a common dihydropyridine-sensitive calcium channel. *Brain Research, 456*, 364–366. https://doi.org/10.1016/0006-8993(88)90240-5

Caricati-Neto, A., Padin, J. F., Silva-Junior, E. D., Fernandez-Morales, J. C., de Diego, A. M., Jurkiewicz, A., & Garcia, A. G. (2013). Novel features on the regulation by mitochondria of calcium and secretion transients in chromaffin cells challenged with acetylcholine at 37°C. *Physiological Reports, 1*(e00182), 2013. https://doi.org/10.1002/phy2.182

Ceña, V., Nicolas, G. P., Sanchez-Garcia, P., Kirpekar, S. M., & Garcia, A. G. (1983). Pharmacological dissection of receptor-associated and voltage-sensitive ionic channels involved in catecholamine release. *Neuroscience, 10*, 1455–1462. https://doi.org/10.1016/0306-4522(83)90126-4

Cheek, T. R., & Thastrup, O. (1989). Internal Ca^{2+} mobilization & secretion in bovine adrenal chromaffin cells. *Cell Calcium, 10*, 213–221. https://doi.org/10.1016/0143-4160(89)90004-3

Cheek, T. R., O'Sullivan, A. J., Moreton, R. B., Berridge, M. J., & Burgoyne, R. D. (1990). The caffeine-sensitive Ca^{2+} store in bovine adrenal chromaffin cells; an examination of its role in triggering secretion and Ca^{2+} homeostasis. *FEBS Letters, 266*(1–2), 91–95. https://doi.org/10.1016/0014-5793(90)81514-o

Chow, R. H., Klingauf, J., Heinemann, C., Zucker, R. S., & Neher, E. (1996). Mechanisms determining the time course of secretion in neuroendocrine cells. *Neuron, 16*, 369–376. https://doi.org/10.1016/s0896-6273(00)80054-9

Comunanza, V., Marcantoni, A., Vandael, D. H., Mahapatra, S., Gavello, D., Carabelli, V., & Carbone, E. (2010). CaV1.3 as pacemaker channels in adrenal chromaffin cells: Specific role on exo- and endocytosis? *Channels (Austin, Tex.), 4*(6), 440–446. https://doi.org/10.4161/chan.4.6.12866

Cuchillo-Ibanez, I., Olivares, R., Aldea, M., Villarroya, M., Arroyo, G., Fuentealba, J., Garcia, A. G., & Albillos, A. (2002). Acetylcholine and potassium elicit different patterns of exocytosis in chromaffin cells when the intracellular calcium handling is disturbed. *Pflügers Archiv / European Journal of Physiology, 444*, 133–142. https://doi.org/10.1007/s00424-002-0810-4

Cuchillo-Ibanez, I., Lejen, T., Albillos, A., Rose, S. D., Olivares, R., Villarroya, M., Garcia, A. G., & Trifaro, J. M. (2004). Mitochondrial calcium sequestration protein kinase C cooperate in the regulation of cortical F-actin disassembly and secretion in bovine chromaffin cells. *The Journal of Physiology, 560*, 63–76. https://doi.org/10.1113/jphysiol.2004.064063

de Diego, A. M. G., & Garcia, A. G. (2018). Altered exocytosis in chromaffin cells from mouse models of neurodegenerative diseases. *Acta Physiologica, 224*, e13090. https://doi.org/10.1111/apha.13090

de Diego, A. M., Gandia, L., & Garcia, A. G. (2007). A physiological view of the central and peripheral mechanisms that regulate the release of catecholamines at the adrenal medulla. *Acta Physiologica, 192*, 287–301. https://doi.org/10.1111/j.1748-1716.2007.01807.x

de Pascual, R., & García, A. G. (2007). Ouabain augments and maintains the catecholamine release responses evoked by repetitive pulses of potassium, caffeine or histamine in perifused bovine chromaffin cells. *European Journal of Pharmacology, 568*(1–3), 99–105. https://doi.org/10.1016/j.ejphar.2007.04.050

del Castillo, J., & Katz, B. (1956). Biophysical aspects of neuro-muscular transmission. *Progress in Biophysics and Biophysical Chemistry, 6*, 121–170. PMID: 13420190.

Dhara, M., Mohrmann, R., & Bruns, D. (2018). v-SNARE function in chromaffin cells. *Pflügers Archiv / European Journal of Physiology, 470*, 169–180. https://doi.org/10.1007/s00424-017-2066-z

Douglas, W. W. (1968). Stimulus-secretion coupling: The concept and clues from chromaffin and other cells. *British Journal of Pharmacology, 34*, 453–474. https://doi.org/10.1111/j.1476-5381.1968.tb08474.x

Douglas, W. W., & Poisner, A. M. (1961). Stimulation of uptake of calcium-45 in the adrenal gland by acetylcholine. *Nature, 192*, 1299. https://doi.org/10.1038/1921299a0

Douglas, W. W., & Rubin, R. P. (1961). The role of calcium in the secretory response of the adrenal medulla to acetylcholine. *The Journal of Physiology, 159*, 40–57. https://doi.org/10.1113/jphysiol.1961.sp006791

Duarte, C. B., Rosario, L. M., Sena, C. M., & Carvalho, A. P. (1993). A toxin fraction (FTX) from the funnel-web spider poison inhibits dihydropyridine-insensitive Ca^{2+} channels coupled to catecholamine release in bovine adrenal chromaffin cells. *Journal of Neurochemistry, 60*, 908–913. https://doi.org/10.1111/j.1471-4159.1993.tb03236.x

Duchen, M. R. (2000). Mitochondria and calcium: From cell signalling to cell death. *The Journal of Physiology, 529*, 57–68. https://doi.org/10.1111/j.1469-7793.2000.00057.x

Engisch, K. L., & Nowycky, M. C. (1996). Calcium dependence of large dense-cored vesicle exocytosis evoked by calcium influx in bovine adrenal chromaffin cells.

The Journal of Neuroscience, 16, 1359–1369. https://doi.org/10.1523/JNEUROSCI.16-04-01359.1996

Esquerro, E., Garcia, A. G., Herandez, M., Kirpekar, S. M., & Prat, J. C. (1980). Catecholamine secretory response to calcium reintroduction in the perfused cat adrenal gland treated with ouabain. *Biochemical Pharmacology, 29*, 2669–2673. https://doi.org/10.1016/0006-2952(80)90084-2

Fernandez, J. M., Granja, R., Izaguirre, V., Gonzalez-Garcia, C., & Cena, V. (1995). Omega-Conotoxin GVIA blocks nicotine-induced catecholamine secretion by blocking the nicotinic receptor-activated inward currents in bovine chromaffin cells. *Neuroscience Letters, 191*, 59–62. https://doi.org/10.1016/0304-3940(95)11558-x

Gandia, L., Lopez, M. G., Fonteriz, R. I., Artalejo, C. R., & Garcia, A. G. (1987). Relative sensitivities of chromaffin cell calcium channels to organic and inorganic calcium antagonists. *Neuroscience Letters, 77*, 333–338. https://doi.org/10.1016/0304-3940(87)90523-4

Gandia, L., Michelena, P., de Pascual, R., Lopez, M. G., & Garcia, A. G. (1990). Different sensitivities to dihydropyridines of catecholamine release from cat and ox adrenals. *Neuroreport, 1*, 119–122. https://doi.org/10.1097/00001756-199010000-00009

Gandia, L., Albillos, A., & Garcia, A. G. (1993). Bovine chromaffin cells possess FTX- sensitive calcium channels. *Biochemical and Biophysical Research Communications, 194*, 671–676. https://doi.org/10.1006/bbrc.1993.1874

Gandia, L., Borges, R., Albillos, A., & Garcia, A. G. (1995). Multiple calcium channel subtypes in isolated rat chromaffin cells. *Pflügers Archiv / European Journal of Physiology, 430*, 55–63. https://doi.org/10.1007/BF00373839

Gandía, L., Lara, B., Imperial, J. S., Villarroya, M., Albillos, A., Maroto, R., García, A. G., & Olivera, B. M. (1997). Analogies and differences between omega-conotoxins MVIIC and MVIID: Binding sites and functions in bovine chromaffin cells. *Pflügers Archiv / European Journal of Physiology, 435*(1), 55–64. https://doi.org/10.1007/s004240050483

Gandia, L., Mayorgas, I., Michelena, P., Cuchillo, I., de Pascual, R., Abad, F., Novalbos, J. M., Larranaga, E., & Garcia, A. G. (1998). Human adrenal chromaffin cell calcium channels: Drastic current facilitation in cell clusters, but not in isolated cells. *Pflügers Archiv / European Journal of Physiology, 436*, 696–704. https://doi.org/10.1007/s004240050691

Garcia, A. G., Hernandez, M., Horga, J. F., & Sanchez-Garcia, P. (1980). On the release of catecholamines and dopamine-beta-hydroxylase evoked by ouabain in the perfused cat adrenal gland. *British Journal of Pharmacology, 68*(3), 571–583. https://doi.org/10.1111/j.1476-5381.1980.tb14573.x

Garcia, A. G., Sala, F., Reig, J. A., Viniegra, S., Frias, J., Fonteriz, R., & Gandia, L. (1984). Dihydropyridine BAY-K-8644 activates chromaffin cell calcium channels. *Nature, 309*, 69–71. https://doi.org/10.1038/309069a0

García, A. G., De-Diego, A. M. G., Gandía, L., Borges, R., & García-Sancho, J. (2006). Calcium signaling and exocytosis in adrenal chromaffin cells. *Physiological Reviews, 86*(4), 1093–1131. https://doi.org/10.1152/physrev.00039.2005

Garcia, A. G., Padin, F., Fernandez-Morales, J. C., Maroto, M., & Garcia-Sancho, J. (2012). Cytosolic organelles shape calcium signals and exo-endocytotic responses of chromaffin cells. *Cell Calcium, 51*, 309–320. https://doi.org/10.1016/j.ceca.2011.12.004

Garcia-Palomero, E., Renart, J., Andres-Mateos, E., Solis-Garrido, L. M., Matute, C., Herrero, C. J., Garcia, A. G., & Montiel, C. (2001). Differential expression of calcium channel subtypes in the bovine adrenal medulla. *Neuroendocrinology, 74*, 251–261. https://doi.org/10.1159/000054692

García-Sancho, J., de Diego, A. M. G., & García, A. G. (2012). Mitochondria and chromaffin cell function. *Pflügers Archiv / European Journal of Physiology, 464*(1), 33–41. https://doi.org/10.1007/s00424-012-1074-2

Giancippoli, A., Novara, M., de Luca, A., Baldelli, P., Marcantoni, A., Carbone, E., & Carabelli, V. (2006). Low-threshold exocytosis induced by cAMPrecruited Ca(V)3.2 (alpha(1H)) channels in rat chromaffin cells. *Biophysical Journal, 90*, 1830–1841. https://doi.org/10.1529/biophysj.105.071647

Giovannucci, D. R., Hlubek, M. D., & Stuenkel, E. L. (1999). Mitochondria regulate the Ca(2+)-exocytosis relationship of bovine adrenal chromaffin cells. *The Journal of Neuroscience, 19*, 9261–9270. https://doi.org/10.1523/jneurosci.19-21-09261.1999

Hamill, O. P., Marty, A., Neher, E., Sakmann, B., & Sigworth, F. J. (1981). Improved patch- clamp techniques for high-resolution current recording from cells and cell-free membrane patches. *Pflügers Archiv / European Journal of Physiology, 391*, 85–100. https://doi.org/10.1007/BF00656997

Harish, O. E., Kao, L. S., Raffaniello, R., Wakade, A. R., & Schneider, A. S. (1987). Calcium dependence of muscarinic receptor-mediated catecholamine secretion from the perfused rat adrenal medulla. *Journal of Neurochemistry, 48*(6), 1730–1735. https://doi.org/10.1111/j.1471-4159.1987.tb05730.x

Heinemann, C., von Ruden, L., Chow, R. H., & Neher, E. (1993). A two-step model of secretion control in neuroendocrine cells. *Pflügers Archiv / European Journal of Physiology, 424*, 105–112. https://doi.org/10.1007/BF00374600

Hernandez-Guijo, J. M., de Pascual, R., Garcia, A. G., & Gandia, L. (1998). Separation of calcium channel current components in mouse chromaffin cells superfused with low- and high-barium solutions. *Pflügers Archiv / European Journal of Physiology, 436*, 75–82. https://doi.org/10.1007/s004240050606

Hernandez-Guijo, J. M., Maneu-Flores, V. E., Ruiz-Nuño, A., Villarroya, M., Garcia, A. G., & Gandia, L. (2001). Calcium-dependent inhibition of L, N, and P/Q Ca^{2+} channels in chromaffin cells: Role of mitochondria.

The Journal of Neuroscience, 21, 2553–2560. https://doi.org/10.1523/JNEUROSCI.21-08-02553.2001

Herrington, J., Park, Y. B., Babcock, D. F., & Hille, B. (1996). Dominant role of mitochondria in clearance of large Ca^{2+} loads from rat adrenal chromaffin cells. *Neuron, 16*, 219–228. https://doi.org/10.1016/s0896-6273(00)80038-0

Hill, J., Chan, S. A., Kuri, B., & Smith, C. (2011). Pituitary adenylate cyclase-activating peptide (PACAP) recruits low voltage-activated T-type calcium influx under acute sympathetic stimulation in mouse adrenal chromaffin cells. *The Journal of Biological Chemistry, 286*, 42459–42469. https://doi.org/10.1074/jbc.M111.289389

Hohne-Zell, B., & Gratzl, M. (1996). Adrenal chromaffin cells contain functionally different SNAP-25 monomers and SNAP-25/syntaxin heterodimers. *FEBS Letters, 394*, 109–116. https://doi.org/10.1016/0014-5793(96)00931-3

Jimenez, R. R., Lopez, M. G., Sancho, C., Maroto, R., & Garcia, A. G. (1993). A component of the catecholamine secretory response in the bovine adrenal gland is resistant to dihydropyridines and omega-conotoxin. *Biochemical and Biophysical Research Communications, 191*, 1278–1283. https://doi.org/10.1006/bbrc.1993.1355

Kim, S. J., Lim, W., & Kim, J. (1995). Contribution of L- and N-type calcium currents to exocytosis in rat adrenal medullary chromaffin cells. *Brain Research, 675*(1–2), 289–296. https://doi.org/10.1016/0006-8993(95)00085-5

Kimura, T., Takeuchi, A., & Satoh, S. (1994). Inhibition by omega-conotoxin GVIA of adrenal catecholamine release in response to endogenous and exogenous acetylcholine. *European Journal of Pharmacology, 264*(2), 169–175. https://doi.org/10.1016/0014-2999(94)00459-5

Kitamura, N., Ohta, T., Ito, S., & Nakazato, Y. (1997). Calcium channel subtypes in porcine adrenal chromaffin cells. *Pflügers Archiv / European Journal of Physiology, 434*, 179–187. https://doi.org/10.1007/s004240050381

Klingauf, J., & Neher, E. (1997). Modeling buffered Ca^{2+} diffusion near the membrane: Implications for secretion in neuroendocrine cells. *Biophysical Journal, 72*, 674–690. https://doi.org/10.1016/s0006-3495(97)78704-6

Lara, B., López, M. G., Villarroya, M., Gandía, L., Cleeman, L., Morad, M., & García, A. G. (1997). A caffeine-sensitive Ca^{2+} store modulates K^+-evoked secretion in chromaffin cells. *The American Journal of Physiology, 272*(4 Pt 1), C1211–C1221. https://doi.org/10.1152/ajpcell.1997.272.4.C1211

Lara, B., Gandia, L., Martinez-Sierra, R., Torres, A., & Garcia, A. G. (1998). Q-type Ca^{2+} channels are located closer to secretory sites than L-type channels: Functional evidence in chromaffin cells. *Pflügers Archiv / European Journal of Physiology, 435*, 472–478. https://doi.org/10.1007/s004240050541

Lee, S. H., Rosenmund, C., Schwaller, B., & Neher, E. (2000). Differences in Ca²⁺ buffering properties between excitatory and inhibitory hippocampal neurons from the rat. *The Journal of Physiology, 525*(Pt 2), 405–418. https://doi.org/10.1111/j.1469-7793.2000.t01-3-00405.x

Levitsky, K. L., & Lopez-Barneo, J. (2009). Developmental change of T-type Ca²⁺ channel expression and its role in rat chromaffin cell responsiveness to acute hypoxia. *The Journal of Physiology, 587*, 1917–1929. https://doi.org/10.1113/jphysiol.2009.168989

Li, S., Eim, C., Kirch, U., Lang, R. E., & Schoner, W. (1998). Bovine adrenals and hypothalamus are a major source of proscillaridin A- and ouabain-immunoreactivities. *Life Sciences, 62*(11), 1023–1033. https://doi.org/10.1016/s0024-3205(98)00023-x

Liu, P. S., & Kao, L. S. (1990). Na(+)-dependent Ca²⁺ influx in bovine adrenal chromaffin cells. *Cell Calcium, 11*, 573–579. https://doi.org/10.1016/0143-4160(90)90011-i

Lopez, M. G., Shukla, R., Garcia, A. G., & Wakade, A. R. (1992). A dihydropyridine-resistant component in the rat adrenal secretory response to splanchnic nerve stimulation. *Journal of Neurochemistry, 58*, 2139–2144. https://doi.org/10.1111/j.1471-4159.1992.tb10956.x

Lopez, M. G., Albillos, A., de la Fuente, M. T., Borges, R., Gandia, L., Carbone, E., Garcia, A. G., & Artalejo, A. R. (1994a). Localized L-type calcium channels control exocytosis in cat chromaffin cells. *Pflügers Archiv / European Journal of Physiology, 427*, 348–354. https://doi.org/10.1007/BF00374544

Lopez, M. G., Villarroya, M., Lara, B., Martinez-Sierra, R., Albillos, A., Garcia, A. G., & Gandia, L. (1994b). Q- and L-type Ca²⁺ channels dominate the control of secretion in bovine chromaffin cells. *FEBS Letters, 349*, 331–337. https://doi.org/10.1016/0014-5793(94)00696-2

Lopez-Gil, A., Nanclares, C., Mendez-Lopez, I., Martinez-Ramirez, C., de los Rios, C., Padin-Nogueira, J. F., Montero, M., Gandia, L., & Garcia, A. G. (2017). The quantal catecholamine release from mouse chromaffin cells challenged with repeated ACh pulses is regulated by the mitochondrial Na(+)/Ca(2+) exchanger. *The Journal of Physiology, 595*, 2129–2146. https://doi.org/10.1113/JP273339

Lukyanetz, E. A., & Neher, E. (1999). Different types of calcium channels and secretion from bovine chromaffin cells. *The European Journal of Neuroscience, 11*, 2865–2873. https://doi.org/10.1046/j.1460-9568.1999.00707.x

Mahapatra, S., Calorio, C., Vandael, D. H. F., Marcantoni, A., Carabelli, V., & Carbone, E. (2012). Calcium channel types contributing to chromaffin cell excitability, exocytosis and endocytosis. *Cell Calcium, 51*, 321–330. https://doi.org/10.1016/j.ceca.2012.01.005

Michelena, P., Vega, T., Montiel, C., Lopez, M. G., Garcia-Perez, L. E., Gandia, L., & Garcia, A. G. (1995). Effects of tyramine & calcium on the kinetics of secretion in intact and electroporated chromaf-fin cells superfused at high speed. *Pflügers Archiv / European Journal of Physiology, 431*, 283–296. https://doi.org/10.1007/BF00410202

Milla, J., Montesinos, M. S., Machado, J. D., Borges, R., Alonso, E., Moreno-Ortega, A. J., Cano-Abad, M. F., Garcia, A. G., & Ruiz-Nuño, A. (2011). Ouabain enhances exocytosis through the regulation of calcium handling by theendoplasmic reticulum of chromaffin cells. *Cell Calcium, 50*, 332–342. https://doi.org/10.1016/j.ceca.2011.06.002

Mollard, P., Seward, E. P., & Nowycky, M. C. (1995). Activation of nicotinic receptors triggers exocytosis from bovine chromaffin cells in the absence of membrane depolarization. *Proceedings of the National Academy of Sciences of the United States of America, 92*, 3065–3069. https://doi.org/10.1073/pnas.92.7.3065

Montero, M., Alonso, M. T., Carnicero, E., Cuchillo-Ibanez, I., Albillos, A., Garcia, A. G., Garcia-Sancho, J., & Alvarez, J. (2000). Chromaffin-cell stimulation triggers fast millimolar mitochondrial Ca²⁺ transients that modulate secretion. *Nature Cell Biology, 2*, 57–61. https://doi.org/10.1038/35000001

Montero, M., Alonso, M. T., Albillos, A., Cuchillo-Ibáñez, I., Olivares, R., García, A. G., García-Sancho, J., & Alvarez, J. (2001). Control of secretion by mitochondria depends on the size of the local [Ca2+] after chromaffin cell stimulation. *The European Journal of Neuroscience, 13*(12), 2247–2254. https://doi.org/10.1046/j.0953-816x.2001.01602.x

Nakazato, Y., Ohga, A., Oleshansky, M., Tomita, U., & Yamada, Y. (1988). Voltage-independent catecholamine release mediated by the activation of muscarinic receptors in guinea-pig adrenal glands. *British Journal of Pharmacology, 93*(1), 101–109. https://doi.org/10.1111/j.1476-5381.1988.tb11410.x

Naraghi, M., Muller, T. H., & Neher, E. (1998). Two-dimensional determination of the cellular Ca²⁺ binding in bovine chromaffin cells. *Biophysical Journal, 75*, 1635–1647. https://doi.org/10.1016/S0006-3495(98)77606-4

Neher, E. (1998a). Usefulness & limitations of linear approximations to the understanding of Ca⁺⁺ signals. *Cell Calcium, 24*, 345–357. https://doi.org/10.1016/s0143-4160(98)90058-6

Neher, E. (1998b). Vesicle pools and Ca²⁺ microdomains: New tools for understanding their roles in neurotransmitter release. *Neuron, 20*, 389–399. https://doi.org/10.1016/s0896-6273(00)80983-6

Neher, E., & Augustine, G. J. (1992). Calcium gradients and buffers in bovine chromaffin cells. *The Journal of Physiology, 450*, 273–301. https://doi.org/10.1113/jphysiol.1992.sp019127

Novara, M., Baldelli, P., Cavallari, D., Carabelli, V., Giancippoli, A., & Carbone, E. (2004). Exposure to cAMP and beta-adrenergic stimulation recruits Ca(V)3 T-type channels in rat chromaffin cells through Epac cAMP-receptor proteins. *The Journal of Physiology, 558*, 433–449. https://doi.org/10.1113/jphysiol.2004.061184

Nowycky, M. C., & Pinter, M. J. (1993). Time courses of calcium and calcium-bound buffers following calcium influx in a model cell. *Biophysical Journal, 64*, 77–91. https://doi.org/10.1016/S0006-3495(93)81342-0

Olivos, O. L., & Artalejo, A. R. (2004). Intracellular Ca^{2+} microdomain-triggered exocytosis in neuroendocrine cells. *Trends in Neurosciences, 27*, 113–115. https://doi.org/10.1016/j.tins.2004.01.001

Owen, P. J., Marriott, D. B., & Boarder, M. R. (1989). Evidence for a dihydropyridine- sensitive and conotoxin-insensitive release of noradrenaline and uptake of calcium in adrenal chromaffin cells. *British Journal of Pharmacology, 97*, 133–138. https://doi.org/10.1111/j.1476-5381.1989.tb11933.x

Pan, C. Y., & Fox, A. P. (2000). Rundown of secretion after depletion of intracellular calcium stores in bovine adrenal chromaffin cells. *Journal of Neurochemistry, 75*, 1132–1139. https://doi.org/10.1046/j.1471-4159.2000.0751132.x

Park, Y. B., Herrington, J., Babcock, D. F., & Hille, B. (1996). Ca^{2+} clearance mechanisms in isolated rat adrenal chromaffin cells. *The Journal of Physiology, 492*, 329–346. https://doi.org/10.1113/jphysiol.1996.sp021312

Perez-Alvarez, A., Hernandez-Vivanco, A., Cano-Abad, M., & Albillos, A. (2008). Pharmacological and biophysical properties of Ca^{2+} channels and subtype distributions in human adrenal chromaffin cells. *Pflügers Archiv / European Journal of Physiology, 456*, 1149–1162. https://doi.org/10.1007/s00424-008-0492-7

Poisner, A. M., & Douglas, W. W. (1966). The need for calcium in adrenomedullary secretion evoked by biogenic amines, polypeptides, and muscarinic agents. *Proceedings of the Society for Experimental Biology and Medicine, 123*(1), 62–64. https://doi.org/10.3181/00379727-123-31402

Powis, D. A., O'Brien, K. J., & Von Grafenstein, H. R. (1991). Calcium export by sodium-calcium exchange in bovine chromaffin cells. *Cell Calcium, 12*, 493–504. https://doi.org/10.1016/0143-4160(91)90031-9

Rettig, J., & Neher, E. (2002). Emerging roles of presynaptic proteins in Ca++-triggered exocytosis. *Science, 298*, 781–785. https://doi.org/10.1126/science.1075375

Rigual, R., Montero, M., Rico, A. J., Prieto-Lloret, J., Alonso, M. T., & Alvarez, J. (2002). Modulation of secretion by the endoplasmic reticulum in mouse chromaffin cells. *The European Journal of Neuroscience, 16*, 1690–1696. https://doi.org/10.1046/j.1460-9568.11-2.02244.x

Robinson, I. M., Finnegan, J. M., Monck, J. R., Wightman, R. M., & Fernandez, J. M. (1995). Colocalization of calcium entry & exocytotic release sites in adrenal chromaffin cells. *Proceedings of the National Academy of Sciences of the United States of America, 92*, 2474–2478. https://doi.org/10.1073/pnas.92.7.2474

Sala, F., & Hernandez-Cruz, A. (1990). Calcium diffusion modeling in a spherical neuron. Relevance of buffering properties. *Biophysical Journal, 57*, 313–324. https://doi.org/10.1016/S0006-3495(90)82533-9

Salvador, J. M., Inesi, G., Rigaud, J. L., & Mata, A. M. (1998). Ca^{2+} transport by reconstituted synaptosomal ATPase is associated with H^+ countertransport and net charge displacement. *The Journal of Biological Chemistry, 273*, 18230–18234. https://doi.org/10.1074/jbc.273.29.18230

Sandow, A. (1952). Excitation-contraction coupling in muscular response. *The Yale Journal of Biology and Medicine, 25*, 176–201. PMCID: PMC2599245.

Santana, F., Michelena, P., Jaen, R., Garcia, A. G., & Borges, R. (1999). Calcium channel subtypes and exocytosis in chromaffin cells: A different view from the intact rat adrenal. *Naunyn-Schmiedeberg's Archives of Pharmacology, 360*, 33–37. https://doi.org/10.1007/s002109900041

Sorensen, J. B., Matti, U., Wei, S. H., Nehring, R. B., Voets, T., Ashery, U., Binz, T., Neher, E., & Rettig, J. (2002). The SNARE protein SNAP-25 is linked to fast calcium triggering of exocytosis. *Proceedings of the National Academy of Sciences of the United States of America, 99*, 1627–1632. https://doi.org/10.1073/pnas.251673298

Stauderman, K. A., & Murawsky, M. M. (1991). The inositol 1,4,5-trisphosphate-forming agonist histamine activates a ryanodine-sensitive Ca^{2+} release mechanism in bovine adrenal chromaffin cells. *The Journal of Biological Chemistry, 266*, 19150–19153. PMID: 1918031.

Stauderman, K. A., & Pruss, R. M. (1990). Different patterns of agonist-stimulatedincreases of 3H-inositol phosphate isomers and cytosolic Ca^{2+} in bovine chromaffin cells: Comparison of the effects of histamine and angiotensin II. *Journal of Neurochemistry, 54*, 946–953. https://doi.org/10.1111/j.1471-4159.1990.tb02342.x

Stauderman, K. A., McKinney, R. A., & Murawsky, M. M. (1991). The role of caffeine- sensitive Ca^{2+} stores in agonist- and inositol 1,4,5-trisphosphate-induced Ca^{2+} release from bovine adrenal chromaffin cells. *The Biochemical Journal, 278*, 643–650. https://doi.org/10.1042/bj2780643

Tagaya, M., Toyonaga, S., Takahashi, M., Yamamoto, A., Fujiwara, T., Akagawa, K., Moriyama, Y., & Mizushima, S. (1995). Syntaxin 1 (HPC-1) is associated with chromaffin granules. *The Journal of Biological Chemistry, 270*, 15930–15933. https://doi.org/10.1074/jbc.270.27.15930

Teraoka, H., Nakazato, Y., & Ohga, A. (1991). Ryanodine inhibits caffeine-evoked Ca^{2+} mobilization and catecholamine secretion from cultured bovine adrenal chromaffin cells. *Journal of Neurochemistry, 57*(6), 1884–1890. https://doi.org/10.1111/j.1471-4159.1991.tb06399.x

Ulate, G., Scott, S. R., Gonzalez, J., Gilabert, J. A., & Artalejo, A. R. (2000). Extracellular ATP regulates exocytosis in inhibiting multiple Ca(2+) channel types in bovine chromaffin cells. *Pflügers Archiv / European Journal of Physiology, 439*, 304–314. https://doi.org/10.1007/s004249900185

Unsicker, K., Habura-Fluh, O., & Zwarg, U. (1978). Different types of small granulecontaining cells and neurons in the guinea-pig adrenal medulla. *Cell and Tissue Research, 189*, 109–130. https://doi.org/10.1007/BF00223124

Vestring, S., Fernandez-Morales, J. C., Mendez-Lopez, I., de Diego, A. M., Padin, J. F., & Garcia, A. G. (2015). Tight mitochondrial control of calcium and exocytotic signals in chromaffin cells at embryonic life. *Pflügers Archiv / European Journal of Physiology, 467*, 2589–2601. https://doi.org/10.1007/s00424-015-1724-2

Villalobos, C., Fonteriz, R., Lopez, M. G., Garcia, A. G., & Garcia-Sancho, J. (1992). Inhibition of voltage-gated Ca^{2+} entry into GH3 and chromaffin cells by imidazole antimycotics and other cytochrome P450 blockers. *The FASEB Journal, 6*, 2742–2747. https://doi.org/10.1096/fasebj.6.9.1319362

von Ruden, L., & Neher, E. (1993). A Ca-dependent early step in the release of catecholamines from adrenal chromaffin cells. *Science, 262*, 1061–1065. https://doi.org/10.1126/science.8235626

von Ruden, L., Garcia, A. G., & Lopez, M. G. (1993). The mechanism of Ba(2+)-induced exocytosis from single chromaffin cells. *FEBS Letters, 336*, 48–52. https://doi.org/10.1016/0014-5793(93)81606-z

Wakade, A. R. (1981). Studies on secretion of catecholamines evoked by acetylcholine or transmural stimulation of the rat adrenal gland. *The Journal of Physiology, 313*, 463–480. https://doi.org/10.1113/jphysiol.1981.sp013676

Weiss, N., Hameed, H., Fernández-Fernández, J. M., Fablet, K., Karmazinova, M., Poillot, C., Proft, J., Chen, L., Bidaud, I., Monteil, A., Huc-Brandt, S., Lacinova, L., Lory, P., Zamponi, G. W., & De Waard, M. (2012). A Ca(v)3.2/syntaxin-1A signaling complex controls T-type channel activity and low-threshold exocytosis. *The Journal of Biological Chemistry, 287*(4), 2810–2818. https://doi.org/10.1074/jbc.M111.290882

Wightman, R. M., Jankowski, J. A., Kennedy, R. T., Kawagoe, K. T., Schroeder, T. J., Leszczyszyn, D. J., Near, J. A., Diliberto, E. J., Jr., & Viveros, O. H. (1991). Temporally resolved catecholamine spikes correspond to single vesicle release from individual chromaffin cells. *Proceedings of the National Academy of Sciences of the United States of America, 88*, 10754–10758. https://doi.org/10.1073/pnas.88.23.10754

Wu, P. C., Fann, M. J., & Kao, L. S. (2010). Characterization of Ca^{2+} signaling pathways in mouse adrenal medullary chromaffin cells. *Journal of Neurochemistry, 112*, 1210–1222. https://doi.org/10.1111/j.1471-4159.2009.06533.x

Xu, T., Naraghi, M., Kang, H., & Neher, E. (1997). Kinetic studies of Ca^{2+} binding and Ca^{2+} clearance in the cytosol of adrenal chromaffin cells. *Biophysical Journal, 73*, 532–545. https://doi.org/10.1016/S0006-3495(97)78091-3

Yang, D. M., & Kao, L. S. (2001). Relative contribution of the Na(+)/Ca(2+) exchanger, mitochondria and endoplasmic reticulum in the regulation of cytosolic Ca(2+) and catecholamine secretion of bovine adrenal chromaffin cells. *Journal of Neurochemistry, 76*, 210–216. https://doi.org/10.1046/j.1471-4159.2001.00055.x

Zhou, Z., & Misler, S. (1995). Action potential-induced quantal secretion of catecholamines from rat adrenal chromaffin cells. *The Journal of Biological Chemistry, 270*, 3498–3505. PMID: 7876083.

Zhou, Z., & Neher, E. (1993). Mobile and immobile calcium buffers in bovine adrenal chromaffin cells. *The Journal of Physiology, 469*, 245–273. https://doi.org/10.1113/jphysiol.1993.sp019813

Pancreatic β Cell Ca$_V$ Channels in Health and Disease

Shao-Nian Yang, Yue Shi, Kaixuan Zhao, Guang Yang, Jia Yu, and Per-Olof Berggren

Abstract

Multiple types of voltage-gated calcium (Ca$_V$) channels including Ca$_V$1.2, Ca$_V$1.3, Ca$_V$2.1, Ca$_V$2.2, Ca$_V$2.3, Ca$_V$3.1, and Ca$_V$3.2 operate in electrically excitable β cells under both physiological and pathophysiological conditions. Physiologically functioning β cell Ca$_V$ channels mediate spatiotemporally distinct Ca^{2+} entries to trigger exquisite glucose-stimulated insulin secretion (GSIS) and to sustain β cell maturity, viability, and physiological expansion, thereby maintaining blood glucose homeostasis. In addition, these channels also interconnect with exocytotic proteins, InsP$_3$ receptors, and PKC to form signalosomes for fine-tuning GSIS. Pathologically affected β cell Ca$_V$ channels, resulting from their phenotypic switch, mutation, altered expression and dysregulation, impair GSIS, and compensatory β cell expansion, induce β cell dedifferentiation, and even drive β cell death due to insufficient or excessive Ca^{2+} influx. In fact, abnormal upregulation of β cell Ca$_V$ channels couples diabetogenic serum components, like apolipoprotein CIII and IgGs in sera from patients with amyotrophic lateral sclerosis and type 2 diabetes, to Ca^{2+}-dependent β cell demise. Consequently, these detrimental Ca^{2+}-dependent events lead to insulin deficiency, hyperglycemia, and diabetes. The efficacy of curative treatments for diabetes critically relies on molecular target- and mechanism-based therapies. β Cell Ca$_V$ channels per se, their regulation mechanisms and signaling pathways most likely serve as druggable targets for medical intervention of Ca^{2+}-dependent β cell dysfunction, abnormal expansion, dedifferentiation, and death in association with diabetes.

Keywords

β cell · Calcium channel · Cell death · Cell dedifferentiation · Cell expansion · Cell maturity · Cell viability · Glucose-stimulated insulin secretion

Abbreviations

[Ca^{2+}]$_i$	cytoplasmic-free Ca^{2+} concentration
ALS	amyotrophic lateral sclerosis
ApoCIII	apolipoprotein CIII
BBW	Bio Bred/Worchester diabetic
Ca$_V$	voltage-gated calcium

S.-N. Yang (✉) · Y. Shi · K. Zhao · G. Yang
J. Yu · P.-O. Berggren
The Rolf Luft Research Center for Diabetes and Endocrinology, Karolinska Institutet, Stockholm, Sweden
e-mail: shao-nian.yang@ki.se;
per-olof.berggren@ki.se

© The Author(s), under exclusive license to Springer Nature Switzerland AG 2022
G. W. Zamponi, N. Weiss (eds.), *Voltage-Gated Calcium Channels*,
https://doi.org/10.1007/978-3-031-08881-0_15

GSIS	glucose-stimulated insulin secretion
PKA	protein kinase A
PKC	protein kinase C
SR-BI	scavenger receptor class B type I
T1D	type 1 diabetes
T2D	type 2 diabetes

Introduction

Nature has endowed the β cell with multiple types of voltage-gated calcium (Ca$_V$) channels including Ca$_V$1.2, Ca$_V$1.3, Ca$_V$2.1, Ca$_V$2.2, Ca$_V$2.3, Ca$_V$3.1, and Ca$_V$3.2 (Fig. 1). These Ca$_V$ channels share some similar features but possess distinct characteristics in their intrinsic properties and functionalities (Yang & Berggren, 2005a, b, 2006; Yang et al., 2014). As in other types of electrically excitable cells, β cell Ca$_V$ channels transit from closed pores to Ca^{2+}-conduits upon membrane depolarization, resulting from ATP-sensitive potassium (K$_{ATP}$) channel closure by ATP produced in glucose metabolism. This glucose-stimulated conformational transition of Ca$_V$ channels allows extracellular Ca^{2+} to rapidly enter the cytoplasm of β cells, initially generating electrical signals of the β cell and subsequently raising cytoplasmic-free Ca^{2+} concentration ([Ca^{2+}]$_i$) (Yang & Berggren, 2005a, b, 2006; Yang et al., 2014). [Ca^{2+}]$_i$ that serves as a versatile and ubiquitous second messenger and controls almost all known molecular and cellular events in the β cell, like glucose metabolism, insulin secretion, β cell endocytosis, differentiation, maturation, survival, growth, protein phosphorylation, gene expression, protein and organelle trafficking (Fig. 1) (Yang & Berggren, 2005a, b, 2006; Yang et al., 2014). Therefore, physiologically orchestrated β cell Ca$_V$ channels mediate appropriate Ca^{2+} influx to ensure adequate β cell function, identity, and viability, thereby maintaining satisfactory insulin release and glucose homeostasis in the body (Fig. 1) (Yang & Berggren, 2005a, b, 2006; Yang et al., 2014). Pathological disturbances in β cell Ca$_V$ channels, resulting from their mutation, altered expression, and

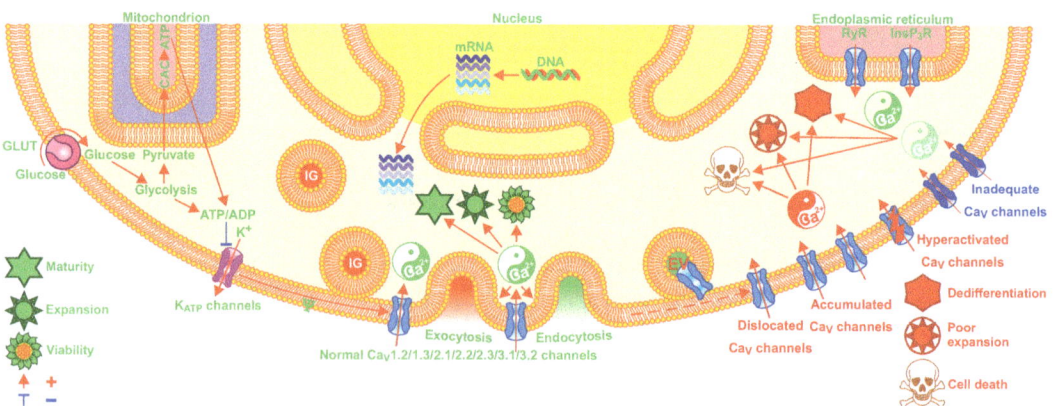

Fig. 1 A scheme illustrating pancreatic β cell Ca$_V$ channels in health and disease. The β cell is equipped with Ca$_V$1.2, Ca$_V$1.3, Ca$_V$2.1, Ca$_V$2.2, Ca$_V$2.3, Ca$_V$3.1, and Ca$_V$3.2 channels. It converts glucose into ATP via cytosolic glycolysis and mitochondrial citric acid cycle bringing about K$_{ATP}$ channel closure and plasma membrane depolarization to open Ca$_V$ channels. In physiological contexts, β cell Ca$_V$ channels take center stage in GSIS and a part in β cell maturity, viability, and expansion by mediating spatiotemporally distinct Ca^{2+} entries, which trigger Ca^{2+}-dependent protein-protein interactions and enzymatic responses and are involved in almost all known molecular and cellular events in the β cell. When suffering genetic mutations or undergoing phenotype switch, altered expression, and dysregulation under pathological conditions, β cell Ca$_V$ channels become inadequate, hyperactivated, accumulated, or dislocated resulting in impaired GSIS, β cell dedifferentiation, poor expansion, and death due to insufficient, excessive, misspaced or ill-timed Ca^{2+} influx. Consequently, insulin deficiency, hyperglycemia, and diabetes occur. *CAC* citric acid cycle, *Ca$_V$* voltage-gated calcium, *EV* endocytotic vesicles, *GLUT* glucose transporter, *IG* insulin-containing granule, *InsP$_3$R* InsP$_3$ receptor, *K$_{ATP}$* ATP-sensitive, *RyR* ryanodine receptor, *ψ* depolarization

dysregulation, impair insulin secretion and even drive β cell destruction due to insufficient or excessive Ca^{2+} influx (Fig. 1). Consequently, this brings about insulin deficiency, resultant hyperglycemia and diabetes (Yang & Berggren, 2005a, b, 2006; Yang et al., 2014). Effective treatments for diabetes depend on molecular target- and mechanism-based therapies. β Cell Ca$_V$ channels and their regulation mechanisms and signaling pathways most likely serve as druggable targets for medical intervention of Ca^{2+}-dependent β cell dedifferentiation and death associated with diabetes. Therefore, β cell Ca$_V$ channels, their regulation mechanisms and signaling pathways have intrigued numerous researchers and clinicians including us. A series of seminal findings have been gained from basic studies by combining the patch-clamp technique, fluorescence microscopy, biochemical methods, pharmacological approaches, cellular and molecular techniques, and clinical investigations in patients with diabetes (Ovalle et al., 2018; Yang & Berggren, 2005a, b, 2006; Yang et al., 2014).

This chapter focuses on the physiological and pathological aspects of Ca$_V$ channels in β cells.

β Cell Ca$_V$ Channels Under Healthy Circumstances

The role of β cell Ca$_V$ channels in the regulation of glucose-stimulated insulin secretion (GSIS) and the participation of Ca$_V$ channels in the control of β cell maturity, viability, and expansion as well as signaling networks of β cell Ca$_V$ channel subunits have thoroughly been characterized under healthy circumstances (Fig. 1) (Yang & Berggren, 2005a, b, 2006; Yang et al., 2014).

Ca$_V$ Channel Regulation of Glucose-Stimulated Insulin Secretion

β Cell Ca$_V$ channels serve as the pivot of a molecular network controlling GSIS where they exert insulinotropic action in a complex manner (Fig. 1) (Yang & Berggren, 2005a, b, 2006; Yang et al., 2014). They not only directly steer insulin-secretory granule trafficking and trigger its exocytosis by mediating Ca^{2+} influx, but also indirectly regulate these two processes by facilitating the generation of glucose metabolism-derived signals. These allosteric actions are the most important for β cell Ca$_V$ channel regulation of GSIS (Yang & Berggren, 2005a, b, 2006; Yang et al., 2014). In addition, the β cell Ca$_V$ channel-mediated Ca^{2+} influx in response to glucose stimulation also acts as an indispensable player in guaranteeing β cell maturity, growth and viability and ultimately adequate β cell mass and function by controlling β cell transcriptome, proteome, signalome, and metabolome (Fig. 1). This is well accounted for by the findings from a series of studies showing that Ca^{2+} passing through β cell Ca$_V$ channels takes an eminent part in β cell gene expression, protein phosphorylation, proliferation, differentiation, and metabolism (for details, see section "Involvement of Ca$_V$ channels in β cell maturity, viability, and expansion") (Benes et al., 1998; Berridge et al., 1998; Efrat et al., 1991; Jing et al., 2005; Namkung et al., 2001; Popiela & Moore, 1991; Sjoholm, 1995; Yu et al., 2020). Moreover, some of the β cell Ca$_V$ channel subunits are also endowed with non-channel functionality that regulates GSIS (Berggren et al., 2004; Rajagopal et al., 2014). In fact, Ca$_V$β$_3$ subunits represent a good paradigm for the non-channel functionality of Ca$_V$ channel subunits. Ca$_V$β$_3$ subunit-deficient β cells release significantly more insulin when subjected to glucose stimulation (Berggren et al., 2004; Rajagopal et al., 2014). This is attributed to the facts that the Ca$_V$β$_3$ subunit physically interacts with InsP$_3$ receptors and serves as a brake for intracellular Ca^{2+} mobilization from endoplasmic reticulum stores and that this subunit acts as a protein kinase C (PKC) inhibitor, thereby reducing GSIS (see section "Ca$_V$ channel-centered interaction networks in the β cell") (Belkacemi et al., 2018; Berggren et al., 2004). Differential regulation of insulin-secretory granule trafficking and exocytosis in GSIS by distinct types of β cell Ca$_V$ channels are discussed as follows.

In the process of GSIS, many steps such as glucose catabolism, insulin-secretory granule trafficking, docking and priming are involved in

$[Ca^{2+}]_i$ contributed from β cell Ca_V channel-mediated Ca^{2+} entry and importantly the final and key step, i.e., the fusion of insulin-secretory granule with the plasma membrane, is triggered by Ca^{2+} influx through Ca_V channels. This brings β cell Ca_V channels onto center stage in regulation of GSIS (Salinno et al., 2019; Yang & Berggren, 2005a, b, 2006; Yang et al., 2014). Of particular importance is that distinct types of β cell Ca_V channels are likely to mediate spatiotemporally dissimilar Ca^{2+} influxes to differentially regulate insulin-secretory granule trafficking and exocytosis in GSIS (Barg et al., 2002; MacDonald et al., 2005; Rorsman & Braun, 2013; Rorsman & Renstrom, 2003; Yang & Berggren, 2005a, b, 2006; Yang et al., 2014). In general, all known types of β cell Ca_V channels including $Ca_V1.2$, $Ca_V1.3$, $Ca_V2.1$, $Ca_V2.2$, $Ca_V2.3$, $Ca_V3.1$, and $Ca_V3.2$ channels are engaged in the regulation of GSIS by mediating Ca^{2+} influx (Lang, 1999; Yang & Berggren, 2005a).

The current consensus is that the Ca_V1 channel takes a predominant role over other types of Ca_V channels in Ca^{2+} triggered insulin exocytosis, the most distal step of GSIS (Yang & Berggren, 2005a). Studies with Ca_V channel blockers reveal that Ca^{2+} influx through the Ca_V1 channel brings about 60–80% of GSIS from mouse, rat, and human islets whose β cells possess various types of Ca_V channels (Davalli et al., 1996; Ohta et al., 1993; Schulla et al., 2003). Similarly, Ca_V1 channel-mediated Ca^{2+} entry also acts as a major player in triggering insulin exocytosis in insulin-secreting cell lines (Satin, 2000; Sher et al., 2003; Yang & Berggren, 2005a). Extensive investigations elucidated the role of Ca_V1 channels in dynamic insulin secretion following glucose stimulation (Davalli et al., 1996; Ohta et al., 1993; Schulla et al., 2003; Wollheim & Sharp, 1981). In an early study, the Ca_V1 channel blocker verapamil was shown to selectively inhibit the second phase of GSIS (Wollheim et al., 1978). However, accumulated experimental results verified that Ca_V1 channels regulate both phases of insulin secretion, predominantly triggering insulin release during the first phase in the mouse islet (Davalli et al., 1996; Ohta et al., 1993;

Schulla et al., 2003). Dynamic perifusion in combination with pharmacological manipulation showed that three Ca_V1 channel antagonists nifedipine, diltiazem, and verapamil significantly reduce both the first and second phase of GSIS from perifused rat islets (Ohta et al., 1993). The Ca_V1 channel agonist Bay K8644 prominently elevated GSIS at both phases in human perifused islets (Davalli et al., 1996). Furthermore, islets from $Ca_V1.2$ subunit knockout ($Ca_V1.2^{-/-}$) mice drastically reduced their first phase insulin secretion when perifused with stimulatory glucose. Capacitance measurements verified that the $Ca_V1.2^{-/-}$ resulted in a selective impairment in the initial rapid component of insulin exocytosis (Schulla et al., 2003).

Two subtypes of Ca_V1 channels, $Ca_V1.2$ and $Ca_V1.3$ channels, operate in the β cell (Namkung et al., 2001; Schulla et al., 2003; Wiser et al., 1999; Yang et al., 1999). They indeed distinctly contribute to GSIS, but the distinctive contribution has not been thoroughly characterized in different species and still stands as a controversial issue. Rat β cells express a higher level of $Ca_V1.3$ subunit mRNA that is 2.5 times greater than that of $Ca_V1.2$ subunit mRNA (Iwashima et al., 1993). In $Ca_V1.2^{-/-}$ mice, β cell Ca_V currents were reduced by about 45% and their first phase insulin secretion decreased by about 80%, while the remaining Ca_V channel currents and insulin release were insensitive to Ca_V1 channel blockers (Schulla et al., 2003). Another study showed that mouse islet β cells displayed negative $Ca_V1.3$ subunit immunoreactivity (Barg et al., 2001). Comparison of the data obtained from $Ca_V1.2^{-/-}$ mice and those acquired from $Ca_V1.2^{+/+}$ mice have given the impression that $Ca_V1.2$ channels alone mediate Ca^{2+} influx to couple glucose stimulation to insulin secretion in mouse islet β cells. However, other observations provide evidence that both $Ca_V1.3$ subunit mRNA and protein are present in islet β cells from some strains of mice (Namkung et al., 2001; Yang et al., 1999, 2015). $Ca_V1.3^{-/-}$ β cells compensatorily increase the expression of $Ca_V1.2$ subunit proteins (Namkung et al., 2001). Patch-clamp analysis witnessed that although both the total voltage-gated Ba^{2+} current density and L-type current density did not differ between Ca$_V$1.3$^{-/-}$ and Ca$_V$1.3$^{+/+}$ mouse islet β cells, the biophysical properties of L-type Ca^{2+} currents in Ca$_V$1.3$^{-/-}$ mouse islet β cells were markedly altered. Genetic ablation of the Ca$_V$1.3 subunit only resulted in a shift of about 10 mV towards more positive potentials at the lower voltage range (Namkung et al., 2001). Mutant islets released less insulin than wild-type islets when bathed in a buffer containing 3 mM glucose, but displayed insulin secretion similar to that of wild-type islets when exposed to 6 mM or higher concentrations of glucose. These observations suggest that compensatory upregulation of Ca$_V$1.2 channels occurs in Ca$_V$1.3$^{-/-}$ mouse islet β cells to maintain insulin secretion capacity (Namkung et al., 2001). The findings reflect that Ca$_V$1.3 subunits appear to mediate Ca^{2+} entry into wild-type mouse β cells for basal insulin secretion and stimulus-secretion coupling at the lower range of glucose concentrations (Namkung et al., 2001). The distinct contribution of Ca$_V$1.2 and Ca$_V$1.3 subtypes to insulin exocytosis remains to be further investigated.

Electrophysiological and pharmacological studies substantiate that Ca$_V$2.1 channels participate in GSIS from rat islet β cells (Ligon et al., 1998). Partial high voltage-activated Ca^{2+} currents disappeared in rat islet β cells exposed to the Ca$_V$2.1 channel blocker ω-Aga IVA. This pharmacological blockade inhibited the DHP-resistant component of GSIS by about 30% (Ligon et al., 1998). The involvement of Ca$_V$2.1 channels in GSIS is also true in rat insulin-secreting RINm5F cells (Sher et al., 2003). Moreover, Ca$_V$2.1 channels play a prominent role in the regulation of GSIS from human islet β cells. Human β cells exposed to ω-Aga IVA showed about a 65% reduction in GSIS (Sher et al., 2003). However, the participation of Ca$_V$2.1 channels in the regulation of GSIS from mouse islet β cells remains to be examined. The regulation of GSIS by Ca$_V$2.2 channels is still controversial. A study observed that the Ca$_V$2.2 channel blocker ω-CTX GVIA did not appreciably alter GSIS in human islets (Davalli et al., 1996). On the contrary, another observation showed that this blocker significantly inhibited GSIS in these islets (Sher et al., 2003). In addition, ω-CTX GVIA produced inhibitory effects on second phase GSIS from rat islets but did not affect first phase GSIS and high K$^+$-evoked insulin exocytosis. This inconsistency led to speculation that the ω-CTX GVIA-induced impairment of second phase GSIS occurred due to toxic effects rather than blockade of Ca$_V$2.2 channels (Komatsu et al., 1989). However, this speculation was made without considering the possibility that the β cell Ca$_V$2.2 channel-mediated Ca^{2+} influx does not directly trigger insulin exocytosis per se, but rather is engaged in signaling critical for second phase insulin secretion. Indeed, further work on the effect of ω-CTX GVIA on second phase insulin secretion verifies that ω-CTX GVIA significantly decreased the ATP/ADP ratio (Ohta et al., 1993). This suggests that Ca^{2+} entry through Ca$_V$2.2 channels participates in Ca^{2+}-dependent glucose metabolism, thereby promoting later phase production of ATP, which is critical for second phase insulin secretion (Ainscow & Rutter, 2002; Rorsman & Renstrom, 2003). It appears that an increase in glucose concentration can only result in a limited initial increase in cytosolic ATP, which then brings about the sequential events, K$_{ATP}$ channel closure, Ca$_V$1 channel opening and first phase insulin secretion. In addition, non-Ca$_V$1 channel-mediated Ca^{2+} influx induced by glucose stimulation is likely to promote mitochondrial metabolism resulting in later production of ATP, which activates the signaling pathway specific for second phase insulin secretion in a K$_{ATP}$ channel-independent manner (Ainscow & Rutter, 2002; Rorsman & Renstrom, 2003). Of note, Ca^{2+}-dependent regulation of mitochondrial metabolism is complex. As an example, the oxidative phosphorylation inhibitors carbonyl cyanide mchlorophenylhydrazone and sodium azide elevate [Ca^{2+}]$_i$, but reduce mitochondrial membrane potential and ATP production in pancreatic β cells (Rustenbeck et al., 1997). It is worthwhile to point out that Ca$_V$2.2 channel-mediated Ca^{2+} entry is likely to directly trigger insulin exocytosis in insulin-secreting RINm5F and INS-1 cells. Experimental evidence for this comes from an observation that shows that ω-CTX GVIA inhibits both Ca_V currents and insulin secretion in these insulin-secreting cell lines, the latter being stimulated not only by glucose but also by high K^+ (Sher et al., 1992, 2003).

The use of the $Ca_V2.3^{-/-}$ mouse model and the $Ca_V2.3$ channel selective peptide blocker SNX-482 greatly advanced our understanding of the role of $Ca_V2.3$ channels in the regulation of GSIS (Jing et al., 2005; Newcomb et al., 1998; Pereverzev et al., 2002a; Yang & Berggren, 2005b). It is clear that Ca^{2+} entry through $Ca_V2.3$ channels regulates insulin secretion from both the pancreatic β cell line INS-1 and primary mouse β cells (Jing et al., 2005; Pereverzev et al., 2002a, b; Vajna et al., 2001; Yang & Berggren, 2005b). An initial study showed that $Ca_V2.3^{-/-}$ does not influence β cell mass and insulin content. Importantly, however, $Ca_V2.3$-deficient mice exhibited impaired insulin secretion, glucose intolerance and hyperglycemia. Unfortunately, this study did not examine dynamic insulin granule exocytosis and phasic insulin secretion from $Ca_V2.3^{-/-}$ β cells (Pereverzev et al., 2002a). Later, capacitance analysis showed that SNX-482 significantly reduces the late component of depolarization-induced exocytotic events without appreciable influence on the exocytotic response to the first depolarization in mouse pancreatic β cells (Schulla et al., 2003). Akin to the pharmacological manipulation with SNX-482, the genetic ablation of the $Ca_V2.3$ subunit gene gives rise to selective suppression of the late component without altering the early component of the depolarization-induced capacitance responses (Jing et al., 2005). This β cell capacitance analysis offers an indication that $Ca_V2.3$ channels are likely to selectively control second phase insulin secretion. The indication has been verified by performing a phasic insulin secretion assay. It revealed that either genetic deletion of the $Ca_V2.3$ subunit gene or pharmacological block of the $Ca_V2.3$ channel with SNX-482 resulted in significant impairment in second phase GSIS without affecting first phase insulin secretion. These findings demonstrate that $Ca_V2.3$ channels mediate a specific form of Ca^{2+} entry whereby

they selectively regulate second phase GSIS (Jing et al., 2005; Schulla et al., 2003). Taken together, it is quite plausible to postulate that as distinct from the β cell Ca_V1 channels, which are physically associated with and functionally coupled to the exocytotic machinery, the $Ca_V2.3$ channel appears to be distant from exocytotic sites in the β cell. On the basis of this, the $Ca_V2.3$ channel-mediated Ca^{2+} influx is likely to specifically recruit insulin-containing granules from the reserve pool to the readily releasable pool/ immediately releasable pool to govern second phase GSIS (Jing et al., 2005; Yang & Berggren, 2005b). Of particular interest is the finding showing that $Ca_V2.3^{-/-}$ islets displayed a 20% decrease in integral $[Ca^{2+}]_i$, a 30% decrease in $[Ca^{2+}]_i$ oscillation frequency, and a 50% decrease in insulin secretion. This points out that both the amount of $[Ca^{2+}]_i$ and the $[Ca^{2+}]_i$ oscillation frequency serve to regulate insulin secretion, especially in the second phase. They do so by activating Ca^{2+}-dependent adenylyl cyclase (AC) and phospholipase C resulting in the specific generation of cAMP and diacylglycerol (DAG), important insulinotropic signals, in second phase insulin secretion in pancreatic β cells (Jing et al., 2005; Yang & Berggren, 2005b).

Ca_V3 channels are likely to be engaged in stimulus-secretion coupling in the β cell. The Ca_V3 channel blocker $NiCl_2$ reduced insulin secretion from INS-1 cells in a dose-dependent manner (Bhattacharjee et al., 1997). The nonselective Ca_V1 and Ca_V3 channel antagonist flunarizine inhibited both glucose- and K^+-induced insulin secretion in perifused rat islets (Ohta et al., 1993). Ca_V3 channels appear to serve as key players in pacemaking in human islet β cells. This is most strikingly seen at membrane potentials around the threshold for action potential firing (−55 mV) and particularly important for insulin secretion induced by glucose at its threshold concentration. Blockade of Ca_V3 channels with NNC 55-0396 results in a slowdown of action potential firing and reduced insulin secretion in response to stimulation with 6 mM glucose (Braun et al., 2008). The exact role of the Ca_V3 channel in insulin secretion from human islet β cells remains for further research.

Involvement of Ca$_V$ Channels in β Cell Maturity, Viability, and Expansion

Guaranteeing the adequate functional β cell mass under different metabolic conditions requires not only a reasonably high number of viable mature β cells, but also the sufficient capacity of β cell expansion including β cell proliferation, neogenesis, and hypertrophy (Bell & Polonsky, 2001; Mathis et al., 2001; Pipeleers et al., 2008; Rhodes, 2005; Wajchenberg, 2007). Naturally, β cells reach their full maturity through intrauterine differentiation and postnatal maturation (Liu & Hebrok, 2017; Salinno et al., 2019). Mature β cells are equipped with β cell-specific transcriptome, proteome, interactome, signalome and metabolome, and are characterized by exquisite glucose sensitivity and adequate insulin secretory capacity (Liu & Hebrok, 2017; Salinno et al., 2019). β Cells not only expand during intrauterine differentiation and postnatal maturation but also after postnatal development in some situations, e.g., pregnancy, where insulin demand increases (Liu & Hebrok, 2017; Salinno et al., 2019). Available evidence indicates that ion channels and in particular Ca$_V$ channels are involved in β cell maturity, viability, and expansion by altering membrane potentials and mediating Ca^{2+} influx (Fig. 1). This is because β cell maturity, viability, and expansion rely on a number of Ca^{2+}-dependent cellular and molecular events like protein phosphorylation, transcription, metabolism, cell cycle, proliferation, differentiation, and maturation (Fig. 1) (Berridge et al., 1998; Yang & Berggren, 2005a, 2006; Yang et al., 2014).

Genetic ablation of Ca$_V$α1 subunits significantly impairs the postnatal expansion of β cells (Jing et al., 2005; Namkung et al., 2001). The Ca$_V$1.3$^{-/-}$ mouse model shows that the Ca$_V$1.3 channel is indispensable for postnatal islet β cell expansion. Newborn Ca$_V$1.3$^{-/-}$ and Ca$_V$1.3$^{+/+}$ mice possess an equivalent number of islets when normalized to body weight although the former weighs less than the latter. In adulthood, however, differences between adult mutant and wild-type mice become obvious, the number and the size of islets in adult mutant mice drastically decline due to impaired β cell proliferation rather than β cell death (Namkung et al., 2001). Interestingly, Ca$_V$2.3 subunit deletion markedly impedes islet cell differentiation and maturation. This is suggested by an immunocytochemical observation showing that the majority of the glucagon-positive cells are also insulin-positive in Ca$_V$2.3$^{-/-}$ mouse islets. Such a type of polyhormonal cells has been considered immature fetal β cells (Hrvatin et al., 2014). This immunocytochemical phenotype indicates that the Ca$_V$2.3 channel-mediated Ca^{2+} influx is likely to drive the expression of some genes critical for β cell differentiation and maturation. Furthermore, pharmacological manipulation of Ca$_V$ channel opening and closure markedly influences β cell proliferation and viability (Popiela & Moore, 1991; Sjoholm, 1995). It has been demonstrated that depolarization and hyperpolarization of β cells with the selective K$_{ATP}$ channel blocker glibenclamide and opener diazoxide, which result in subsequent Ca$_V$ channel opening and closure, induce facilitation of DNA synthesis and an impediment in β cell proliferation, respectively (Sjoholm, 1995). Importantly, the Ca$_V$1 channel blockers D-600 and diltiazem can produce a similar effect, i.e., a striking inhibition of β cell proliferation, as glibenclamide (Popiela & Moore, 1991; Sjoholm, 1995).

Ca$_V$ channel-dependent regulation of β cell maturity, viability, and expansion is based on Ca^{2+}-dependent maintenance of normal transcriptome and proteome in the β cell. The expression of a range of genes in the β cell is associated with Ca^{2+} influx through Ca$_V$ channels (Efrat et al., 1991; German et al., 1990; Lee & Laychock, 2000; Macfarlane et al., 2000). For example, the insulin gene, the most specific β cell gene, is regulated by glucose. This glucose-stimulated regulation depends on Ca^{2+} influx through β cell Ca$_V$ channels. It is effectively abolished by the Ca$_V$ channel blockers, such as D-600 and verapamil (Efrat et al., 1991; German et al., 1990). This is also true for the islet amyloid polypeptide amylin. This peptide is co-localized and co-secreted with insulin and regulates glucose metabolism as well as insulin resistance. The β cell elevates amylin gene transcription when exposed to stimulatory glucose. Such glucose-induced transcription can no longer occur in the presence of the Ca_V1 channel blocker verapamil (Macfarlane et al., 2000). Inositol 1,4,5-trisphosphate ($InsP_3$) receptor gene expression is also under the influence of Ca_V channel-mediated Ca^{2+} influx. The Ca_V1 channel blocker nimodipine effectively ablates the expression of $InsP_3$ receptor type II and III genes induced by protein kinase A (PKA) activation (Lee & Laychock, 2000).

All factors that are able to alter β cell Ca_V1 channel activity or density may affect β cell maturity, viability, and expansion. Thus far, a limited number of factors have been examined. In fact, β cells fail with regard to maturity, viability, and expansion in a significant proportion of patients with diabetes. Understanding the special involvement of Ca_V channels in these processes may lead to the development of new preventive measures and curative treatments for diabetes.

Ca_V Channel-Centered Interaction Networks in the β Cell

Cellular proteins interact with each other in concert to constitute a plethora of cellular signaling networks (Bennett et al., 1992; Bezprozvanny et al., 1995; Ji et al., 2002; Kang et al., 2002; Leveque et al., 1992; Mochida et al., 1996; Rettig et al., 1996; Sheng et al., 1994, 1996, 1997; Wiser et al., 1999; Yang et al., 1999). As a typical example, Ca_V channel subunits interact with many other proteins to create complex signaling networks in addition to forming Ca^{2+} conducting pores in the plasma membrane (Belkacemi et al., 2018; Yang & Berggren, 2005a, 2006; Yang et al., 2014). In a Ca_V channel subunit complex, Ca_V channels no longer sense and respond only to voltage depolarization, but are also modulated by their interacting partners (for additional information, please see chapter "Modulation of VGCCs by G-protein Coupled Receptors and Their Second Messengers" by Mark and colleagues). They signal not only by altering membrane potentials and conducting Ca^{2+}, but also by physically acting on their interacting partners.

Exceptionally, Ca_V channel subunits can even stay away from the Ca_V channel subunit complex to function as non-channel proteins and crosstalk with other signaling molecules (Belkacemi et al., 2018; Berggren et al., 2004; Hibino et al., 2003; Yang & Berggren, 2005a, 2006; Yang et al., 1999, 2014). This is well exemplified by the fact that a short splice variant of the $Ca_V\beta_4$ subunit enters the nucleus where it directly acts on the nuclear protein chromobox protein 2 and controls gene silencing (Hibino et al., 2003). Here we discuss some Ca_V channel-centered interaction networks in the β cell including Ca_V1 channel-exocytotic proteins, $Ca_V\beta_3$ subunit-$InsP_3$ receptor, and $Ca_V\beta$ subunit-PKC signaling pathways.

Ca_V1 Channel-Exocytotic Protein Interaction Networks

The first evidence for interactions between Ca_V channels and exocytotic proteins results from observations of the pull-down of ω-CTX GVIA binding proteins by antibodies against syntaxin or synaptotagmin (Bennett et al., 1992; Leveque et al., 1992). Soon after that, Ca_V channels were found to physically associate with exocytotic proteins and thereby forming an exocytotic signalosome. This exocytotic signalosome in combination with their upstream and downstream molecules makes up a complex signaling network whereby Ca_V channels and exocytotic proteins interact with each other to efficiently and precisely orchestrate complex cellular and molecular processes (Bezprozvanny et al., 1995; Mochida et al., 1996; Rettig et al., 1996; Sheng et al., 1994, 1996, 1997). In distinct cell types, specific types of Ca_V subunits selectively interact with exocytotic proteins likely due to their subcellular localization. In neurons, $Ca_V2.1$ or $Ca_V2.2$ subunits are not only colocalized, but also physically associated with exocytotic proteins in active zones, but Ca_V1 subunits are not (Sheng et al., 1998). By contrast, a Ca_V1 subunit-exocytotic protein signaling network indeed operates in pancreatic β cells (Ji et al., 2002; Kang et al., 2002; Wiser et al., 1999; Yang et al.,

1999). Both Ca$_V$1.2 and Ca$_V$1.3 subunits have similar interaction with the exocytotic machinery as the neuronal Ca$_V$2.1 and Ca$_V$2.2 subunits (Ji et al., 2002; Kang et al., 2002; Wiser et al., 1999; Yang et al., 1999).

The interaction of Ca$_V$1.3 subunits with syntaxin 1A in the pancreatic β cell has been experimentally verified (Yang et al., 1999). This interaction indeed has clear functional significance. Interrupting the formation of a syntaxin 1A/Ca$_V$1.3 subunit complex not only induces the rundown of β cell Ca$_V$1 channel activity but also impairs insulin exocytosis independently of this rundown. These findings corroborate that the interaction between the β cell Ca$_V$1 channel and syntaxin 1A is required for a proper β cell function (Yang et al., 1999). The β cell Ca$_V$1.2 subunit also interacts with exocytotic proteins (Wiser et al., 1999). Its II-III loop is physically associated with syntaxin 1A, SNAP-25 and synatotagmin. Among these interaction partners, syntaxin 1A dampens the amplitude of Ca$_V$1.2 subunit-mediated currents. This inhibitory effect is partially counteracted by synaptotagmin (Wiser et al., 1999). Disruption of this Ca$_V$1.2 subunit-exocytotic protein complex almost completely abolishes depolarization-evoked exocytosis (Kang et al., 2002; Wiser et al., 1999). β Cell Ca$_V$1 channels functionally interact with distinct domains within SNAP-25 (Ji et al., 2002). Interestingly, the long N-terminus (SNAP-25$_{(1-197)}$) and the short C-terminus of this exocytotic protein (SNAP-25$_{(198-206)}$) produce opposite effects on Ca$_V$1 channel activity, namely stimulation and inhibition, respectively. The stimulatory effect of SNAP-25$_{(1-197)}$ is weaker than the inhibitory effect of SNAP-25$_{(198-206)}$ (Ji et al., 2002).

In conclusion, the β cell Ca$_V$1 channel physically associates with the exocytotic machinery to constitute Ca$_V$1 channel-exocytotic protein interaction networks. Such networks serve to anchor β cell Ca$_V$1 channels and exocytotic machinery in close proximity to each other within the site of insulin exocytosis and to fine-tune β cell Ca$_V$1 channel function. This allows them to gain the most effective use of Ca^{2+} influx for insulin exocytosis and avoid adverse effects of unnecessary excessive Ca^{2+} influx.

Ca$_V$β$_3$ Subunit-InsP$_3$ Receptor Signaling Pathway

Both the plasma membrane Ca$_V$ channels and intracellular InsP$_3$ receptors serve to create the exquisitely fine-tuned dynamics of [Ca^{2+}]$_i$, namely, the precise spatiotemporal arrangement of [Ca^{2+}]$_i$, by mediating Ca^{2+} influx from the extracellular space and Ca^{2+} mobilization from the intracellular store (Catterall, 2000; Taylor et al., 2004). The dynamics of [Ca^{2+}]$_i$ act as a ubiquitous second messenger encoding a rich assortment of information to control Ca^{2+}-dependent protein-protein interactions and enzymatic responses in the cell (Berridge et al., 2000, 2003).

Unexpectedly, the Ca$_V$β$_3$ subunit does not behave as a required building block of β cell Ca$_V$ channels. Instead, it promiscuously crosstalks with the intracellular Ca^{2+} release machinery to establish a Ca$_V$β$_3$ subunit-intracellular Ca^{2+} store network (Berggren et al., 2004). This is strongly supported by the following findings: genetic ablation of Ca$_V$β$_3$ subunits (β$_3$$^{-/-}$) did not alter the activity and gating properties of Ca$_V$ channels at both the single channel and whole-cell level in the β cell, but prominently enhanced InsP$_3$-induced Ca^{2+} release and markedly increased the frequency of glucose-induced [Ca^{2+}]$_i$ oscillations in islets. Intact Ca$_V$β$_3$$^{-/-}$ islets significantly increased their GSIS, whereas permeabilized Ca$_V$β$_3$$^{-/-}$ islets showed unaltered Ca^{2+}-evoked insulin secretion in comparison to wild-type islets. Restoration of the Ca$_V$β$_3$ subunit effectively counteracted the enhancement of GSIS induced by the deletion of this subunit. Importantly, mutant mice displayed a better glucose tolerance and more efficient glucose homeostasis. Furthermore, ectopic expression of the Ca$_V$β$_3$ subunit in COS-7 cells, where there are no endogenous Ca$_V$ channels, significantly dampened Ca^{2+} release from InsP$_3$-sensitive stores. The endogenous Ca$_V$β$_3$ subunit in pancreatic β cells and ectopically expressed Ca$_V$β$_3$ subunit in COS-7 were mainly localized in intracellular compartments resembling ER where InsP$_3$ receptors reside. In addition to the above findings, the large cytoplasmic region of the InsP$_3$ receptor subunit carries numerous recognition sites for a range of small molecules and proteins like $InsP_3$, Ca^{2+}, nucleotides, protein kinases and phosphatases, calmodulin, apoptotic proteins, transient receptor potential channels, and G-protein $\beta\gamma$ subunits. (Taylor et al., 2004). Such a molecular organization opens the possibility for $InsP_3$ receptors to interact with $Ca_V\beta_3$ subunits. This prompted us to propose that physical and functional interactions of the $Ca_V\beta_3$ subunit with $InsP_3$ receptors operate in the pancreatic β cell to act as a brake on $InsP_3$ receptor-mediated Ca^{2+} mobilization from intracellular stores (Berggren et al., 2004).

Recently, the aforementioned proposal, by taking advantage of $Ca_V\beta_3^{-/-}$ mouse model and heterologous expression systems, was closely examined and substantiated (Belkacemi et al., 2018). Co-immunoprecipitation and immunoblot assay revealed that antibodies against $Ca_V\beta_3$ subunits or $InsP_3$ receptor subtypes 1–3 effectively co-immunoprecipitated $InsP_3$ receptor subtypes 1–3 or $Ca_V\beta_3$ subunits, but were unable to do so when the SH3-HOOK domain of $Ca_V\beta_3$ subunits was deleted. $[Ca^{2+}]_i$ measurements showed that the presence of $Ca_V\beta_3$ subunits resulted in inhibition of Ca^{2+} mobilization from $InsP_3$-sensitive stores, whereas the absence of $Ca_V\beta_3$ subunits gave rise to sensitization of $InsP_3$-sensitive Ca^{2+} stores to low $InsP_3$ levels. The inhibitory effect required the $Ca_V\beta_3$ SH3-HOOK domain binding to $InsP_3$ receptors. Furthermore, radioligand binding assays revealed that $InsP_3$ receptors reduced their sensitivity to $InsP_3$ in the presence of $Ca_V\beta_3$ subunits. Overall, these findings demonstrate that the $Ca_V\beta_3$ subunit indeed physically and functionally interacts with $InsP_3$ receptors and acts as a brake on $InsP_3$ receptor-mediated Ca^{2+} mobilization from intracellular stores (Belkacemi et al., 2018). Releasing the brake by ablating the $Ca_V\beta_3$ subunit is beneficial for glucose homeostasis and skin wound healing, which is relevant for patients with diabetes (Belkacemi et al., 2018; Berggren et al., 2004). However, the physiological significance of the interaction between $Ca_V\beta_3$ subunits and $InsP_3$ receptors remains enigmatic and interesting.

$Ca_V\beta$ Subunit-PKC Signaling Pathway

Interestingly, a promiscuous interaction of $Ca_V\beta_2$ or $Ca_V\beta_3$ subunits with PKCα has been found to limit insulin secretion (Rajagopal et al., 2014). Following injection of $Ca_V\beta_2$ or $Ca_V\beta_3$ siRNAs into the tail vein, mice showed elevated serum insulin levels and improved glucose homeostasis. The same did not hold true for mice subjected to tail vein injection of $Ca_V\beta_1$ or β_4 siRNAs. A clear $Ca_V\beta_2$- or $Ca_V\beta_3$-immunoreactive band appeared in cytoplasmic fractions of islets isolated from control mice, but not from mice treated with $Ca_V\beta_2$ or $Ca_V\beta_3$ siRNAs. The treatment indeed effectively silenced $Ca_V\beta_2$ or $Ca_V\beta_3$ subunit expression. Enhanced GSIS occurred in islets and islet cells from mice treated with $Ca_V\beta_2$ or $Ca_V\beta_3$ siRNAs, but not in those from controls (Rajagopal et al., 2014). Furthermore, there was a clear translocation of PKCα, PKCβII, and PKCθ, to the plasma membrane of islet β cells obtained from $Ca_V\beta_2$- or β_3 siRNA-treated mice. This reflects released $Ca\nu\beta_2$ and β_3 subunit inhibition on PKCα, PKCβII, and PKCθ. Moreover, mice treated with PKCα, but not with PKCβII or PKCθ siRNAs displayed a reduction in serum insulin levels and an elevation in blood glucose concentrations. This raises the possibility that PKCβII and PKCθ are not as important as PKCα in the regulation of GSIS. Taken together, these findings demonstrate that $Ca_V\beta_2$ or $Ca_V\beta_3$ subunits interact with PKCα inhibiting GSIS (Rajagopal et al., 2014).

β Cell Ca_V Channels in Pathology

Massive efforts have been made to understand β cell Ca_V channels in pathology (Yang & Berggren, 2005a, 2006; Yang et al., 2014). The resulting findings provide strong evidence that β cell Ca_V channels suffer deterioration in several aspects under unhealthy circumstances and especially in diabetes (Fig. 1). Here we discuss pathological alterations of β cell Ca_V channel phenotypes, associations of Ca_V channel gene mutation and polymorphism with abnormal insulin secretion and diabetes, engagement of Ca$_V$ channels in β cell dedifferentiation and death as well as Ca$_V$ channel-mediated coupling of diabetogenic serum components to Ca^{2+}-dependent β cell demise.

Pathological Alterations of β Cell Ca$_V$ Channel Phenotypes

A series of papers verify that the phenotypic switch of β cell Ca$_V$ channels happens in a diabetic milieu although β cell Ca$_V$ channelome is genetically inherited. In fact, T-type Ca^{2+} currents are not detectable in the normal mouse pancreatic β cell (Ashcroft & Rorsman, 1989; Satin, 2000; Schulla et al., 2003; Sher et al., 2003; Yang & Berggren, 2005a). However, such Ca^{2+} currents appear in NOD mouse pancreatic β cells (Wang et al., 1996). Importantly, the newly appeared T-type Ca^{2+} currents elevate basal [Ca^{2+}]$_i$ and thereby driving the Ca^{2+}-dependent apoptosis of NOD mouse β cells (Wang et al., 1996, 1999). This is undoubtedly involved in the development of type 1 diabetes (T1D) in the NOD mice, an animal model of human T1D characterized by leukocytic infiltration in islets and β cell death (Solomon & Sarvetnick, 2004; Yoshida & Kikutani, 2000). Experimental evidence shows that inflammatory cytokines, produced by the immune cells in the peri-islet and intra-islet infiltrate, induce abnormal expression of Ca$_V$3 channels, resultant excessive [Ca^{2+}]$_i$ and consequent Ca^{2+}-dependent apoptosis in β cells, but not in a glucagon-secreting α cell line (Wang et al., 1999). The above findings clearly pinpoint cytokines as the main culprit causing the phenotypic switch of β cell Ca$_V$ channels and Ca^{2+}-dependent β cell death, thereby playing an important role in the development of T1D. However, some deep-seated problems remain to be addressed. For example, what are the mechanistic details whereby cytokines selectively initiate the appearance of β cell Ca$_V$3 channels in the T1D situation? Do cytokines turn on de novo expression of β cell Ca$_V$3 channel genes or just activate functionally concealed Ca$_V$3 channels in β cells?

Association of Ca$_V$ Channel Gene Mutations and Polymorphisms with Abnormal Insulin Secretion and Diabetes

Type 2 diabetes (T2D) is a complex polygenic trait that is phenotypically characterized by a progressive loss and/or impairment in functional β cell mass (American Diabetes Association, 2014; Mathis et al., 2001; Saltiel, 2001; van Belle et al., 2011). It is rational to deem genes critical for GSIS and/or β cell viability, as exemplified by Ca$_V$ channel genes, as potential diabetes risk genes (American Diabetes Association, 2014; Ashcroft & Rorsman, 2004; Holmkvist et al., 2007; Mathis et al., 2001; Reinbothe et al., 2013; Saltiel, 2001; van Belle et al., 2011). Efforts have been put into research understanding the association of Ca$_V$ channel gene mutation and polymorphism with abnormal insulin secretion and diabetes. Some interesting findings on the association of Ca$_V$ channel gene mutations and polymorphisms with abnormal insulin secretion and diabetes are discussed below.

It has been clearly verified that Ca$_V$1.2 missense mutations are the cause of the monogenetic disorder Timothy syndrome. This inherited disease manifests with a wide clinical spectrum that involves the cardiac, endocrine, immune and central nervous systems (Splawski et al., 2004). This complex clinical spectrum stems from a single nucleotide G > A transition at position 1216 or substitution of glycine with arginine at residue 406 (G406R) in the Ca$_V$1.2 subunit gene. Such mutant Ca$_V$1.2 subunits almost completely lose voltage-dependent inactivation. The loss of voltage-dependent inactivation in Ca$_V$1.2 channels should enhance insulin secretion in patients with Timothy syndrome. In fact, 36% of the patients suffered from episodic hypoglycemia, reflecting that excessive insulin secretion occurs due to the exaggerated Ca^{2+} influx through the mutant Ca$_V$1.2 channels in the pancreatic β cell. Moreover, such episodic hypoglycemia caused by inadequate insulin secretion has led to the decease of some affected patients (Splawski et al., 2004).

Investigation of human $Ca_V1.3$ subunit gene variation in 918 Japanese type 2 diabetics and 336 control subjects identified an ATG repeat expansion of this gene in type 2 diabetics (Yamada et al., 2001). However, this mutation appears at low frequency and is not closely associated with the development of common T2D. Nevertheless, it may give rise to a subgroup of this polygenic disease (Yamada et al., 2001; Yang & Berggren, 2005a, 2006). Furthermore, the association of the single nucleotide polymorphisms (SNPs) rs312480, rs312486 and rs9841978 in the $Ca_V1.3$ subunit gene in humans with impaired insulin secretion and T2D has been characterized in 8987 non-diabetic Finnish and Swedish people and 2830 Finnish and Swedish patients with T2D. The obtained data show that the SNP rs312480 is associated with decreased mRNA expression of the $Ca_V1.3$ subunit gene and impaired GSIS and that the SNPs rs312486 and rs9841978 are linked to T2D (Reinbothe et al., 2013). Moreover, the SNP rs673391 in the human $Ca_V2.3$ subunit has also been identified by the Botnia Study Group in collaboration with others (Holmkvist et al., 2007). It is closely associated with T2D and impaired insulin secretion and in particular reduced second-phase insulin secretion (Holmkvist et al., 2007).

The association of $Ca_V2.1$ subunit gene mutation-caused spinocerebellar ataxia type 6 with T2D has been found in a Japanese family (Takiyama et al., 1998; Yang & Berggren, 2005a, 2006). In this five-generation family, thirteen members are diagnosed with $Ca_V2.1$ subunit mutation and suffer from spinocerebellar ataxia type 6. Three out of the five patients examined also have overt T2D. This $Ca_V2.1$ subunit mutant carries abnormal CAG repeat expansion encoding a polyglutamine tract (Takiyama et al., 1998; Yang & Berggren, 2005a, 2006). Characterization of biophysical properties and surface expression of this mutant shows contradictory results from two groups (Piedras-Renteria et al., 2001; Toru et al., 2000; Yang & Berggren, 2005a, 2006). One group reveals that the polyglutamine-containing $Ca_V2.1$ subunit expressed in HEK 293 cells displayed a negative shift of voltage-dependent inactivation resulting in reduced Ca^{2+} influx (Toru

et al., 2000; Yang & Berggren, 2005a, 2006). In contrast, the other witnesses that the mutant $Ca_V2.1$ subunit does not alter its biophysical properties, but instead is expressed more abundantly in the plasma membrane in HEK 293 cells, giving rise to higher Ca_V channel current density as compared to the wild-type $Ca_V2.1$ subunit (Piedras-Renteria et al., 2001; Yang & Berggren, 2005a, 2006). It remains to be seen how the mutant $Ca_V2.1$ channel behaves in the β cell to induce T2D concurrent with spinocerebellar ataxia type 6 (Takiyama et al., 1998; Yang & Berggren, 2005a, 2006).

Interestingly, a genomewide linkage analysis has found a novel neonatal diabetes locus mapped to chromosome 10p12.1-p13 in a large consanguineous family with autosomal recessively inherited neonatal diabetes (Sellick et al., 2003; Yang & Berggren, 2005a, 2006). This region contains the $Ca_V\beta_2$ subunit gene, which is predominantly expressed in the β cell (Berggren et al., 2004). All affected individuals in this family phenotypically manifested low or even undetectable levels of circulating insulin indicating inadequate β cell mass and impaired insulin synthesis and secretion (Sellick et al., 2003; Yang & Berggren, 2005a, 2006). These findings indicate that the $Ca_V\beta_2$ subunit gene should be considered as one of the potential susceptibility genes for neonatal diabetes (Sellick et al., 2003; Yang & Berggren, 2005a, 2006).

Engagement of Ca_V Channels in β Cell Dedifferentiation and Death

Correctly-localized and properly-functioning Ca_V channels with an appropriate density serve as central players in the creation and orchestration of dynamics, homeostasis, and remodeling of $[Ca^{2+}]_i$ in the β cell (Yang & Berggren, 2005a, 2006; Yang et al., 2014). This is critical for β cell maturity, viability, and expansion, as discussed earlier. Conversely, in pathological contexts, β cell Ca_V channels undergo hypoactivation or hyperactivation, become abnormally scarce or abundant, or are out of position (Fig. 1). These pathological disturbances make β cell Ca_V

channels mediate either deleteriously high or harmfully low Ca^{2+} influx, resulting in exaggerated and minimized $[Ca^{2+}]_i$, respectively, or have them conduct Ca^{2+} into wrong subcellular locations. This leads to activating detrimental signaling pathways driving β cell dedifferentiation or immaturity and β cell death (Fig. 1) (Yang & Berggren, 2005a, 2006; Yang et al., 2014).

Ca_V Channels in β Cell Dedifferentiation

Strong evidence has accumulated to support that abnormal Ca_V channels can contribute to β cell dedifferentiation (Fig. 1). In islets of GK rats, a non-obese, T2D model, pathological hyperactivation of β cell Ca_V3 and Ca_V1 channels appear coincidentally with lower levels of insulin and exocytotic proteins, including vesicle-associated membrane protein-2 (VAMP-2), synaptotagmin III, cysteine string protein (CSP), the mammalian homolog of the unc-18 gene (Munc-18), α-soluble N-ethylmaleimide-sensitive attachment protein (α-SNAP), N-ethylmaleimide-sensitive factor (NSF) and synaptosomal-associated protein of 25 kDa (SNAP-25) (Kato et al., 1996; Rose et al., 2007; Salinno et al., 2019; Yang & Berggren, 2005a, b, 2006; Yang et al., 2014; Zhang et al., 2002). Furthermore, GK rats suffer impairment of GSIS, which has thus far been considered the most important and most specific parameter representing the metabolic and functional maturity of the β cell (Kato et al., 1996; Rose et al., 2007; Salinno et al., 2019; Yang & Berggren, 2005a, b, 2006; Yang et al., 2014). It is most likely that pathological Ca^{2+} influx through hyperactivated β cell Ca_V3 and Ca_V1 channels drives β cell dedifferentiation by decreasing expression of exocytotic proteins, resulting in impaired GSIS of GK islets. There is also a possibility that such a GK rat β cell dedifferentiation occurs due to hyperactivated Ca_V channels being redistributed apart from exocytotic sites to instead serve as a Ca^{2+}-dependent β cell dedifferentiation signal (Yang & Berggren, 2005a, 2006; Yang et al., 2014).

A recent study demonstrates that enhanced expression of β cell Ca_V3.1 channels drive the dedifferentiation of these cells, as evidenced by impaired GSIS and aberrant glucose homeostasis (Yu et al., 2020). Detailed analysis shows that de novo expression of a recombinant adenovirus encoding enhanced green fluorescent protein-Ca_V3.1 subunit (Ad-EGFP-Ca_V3.1) in the β cell increases typical T-type Ca^{2+} currents and basal $[Ca^{2+}]_i$. Islets transduced with Ad-EGFP-Ca_V3.1 secrete less insulin under both basal and first phases in response to glucose stimulation and display the inability to normalize hyperglycemia in diabetic rats. Furthermore, Ad-EGFP-Ca_V3.1-transduced β cells exhibit decreased phosphorylated FoxO1 in their cytoplasm, increased FoxO1 nuclear retention and reduced syntaxin 1A, SNAP-25 and synaptotagmin III. Importantly, these disturbances are reduced by inhibiting Ca_V3.1 channels or the Ca^{2+}-dependent phosphatase calcineurin. These findings corroborate that enhanced expression of β cell Ca_V3.1 channels induces β cell dedifferentiation via a signaling pathway consisting of initial excessive Ca^{2+} influx, subsequent activation of calcineurin, consequent dephosphorylation, and nuclear retention of FoxO1 and eventual FoxO1-mediated downregulation of β cell exocytotic proteins. This Ca^{2+}-dependent β cell dedifferentiation is strong enough to interfere with glucose homeostasis and hence being a pathogenic mechanism in diabetes (Yu et al., 2020).

Ca_V Channels and β Cell Death

A set of molecules responsible for cell survival can no longer function and another set of molecules being in charge of cell demise becomes activated when $[Ca^{2+}]_i$ goes below or over a physiological range or becomes spatiotemporally impaired owing to the pathological conductivity, abnormal density or dislocation of β cell Ca_V channels. Such pathologically-disturbed $[Ca^{2+}]_i$ drives β cells to death (Fig. 1) (Chang et al., 2004; Juntti-Berggren et al., 1993, 2004; Wang et al., 1999).

Pathologically exaggerated $[Ca^{2+}]_i$ mediated by hyperactivated β cell Ca_V channels can drive apoptotic and necrotic β cell death by activating various Ca^{2+}-sensitive enzymes, e.g., calcineurin, endonucleases, transglutaminase, and calpains (Chang et al., 2004; Efanova et al., 1998; Huo et al., 2003; Juntti-Berggren et al., 1993, 2004; Wang et al., 1999; Zaitsev et al., 2001). The serine- and threonine-specific protein phosphatase (PP) calcineurin is conserved in all eukaryotes and undergo activation in a Ca^{2+}- and calmodulin-dependent manner. It acts as a key player in cell death signaling. Activated calcineurin catalyzes dephosphorylation of the Bcl-2 family member Bad to play its pro-apoptotic role (Orrenius et al., 2003). It has been shown that calcineurin orchestrates β cell Ca_V1 channel-mediated Ca^{2+} influx and the cytokine signal transduction cascade in the process of β cell death. Treatment with interleukin 1β induces pancreatic β cell apoptosis. This apoptotic event is largely counteracted by either the Ca_V1 channel inhibitor D-600 or the calcineurin inhibitor deltamethrin (Zaitsev et al., 2001). Furthermore, hyperactivated Ca_V channels can activate calpain by overloading insulin-secreting MIN6N8 cells with Ca^{2+} and subsequently calpain initiates activation of calcineurin. Upon activation, calcineurin directly dephosphorylates Bad bringing about apoptosis of the insulin-secreting cells (Chang et al., 2004).

Ca^{2+}-dependent endonucleases induce programmed cell death i.e., apoptosis by catalyzing the cleavage of chromosomal DNA into oligonucleosomal size fragments, a biochemical hallmark of apoptosis (Orrenius et al., 2003). This happens in pancreatic β cells where these DNA cutting enzymes become activated to evoke Ca^{2+}-dependent β cell apoptosis as pathologically excessive Ca^{2+} ions enter β cells subsequent to Ca_V channel activation (Efanova et al., 1998; Yang & Berggren, 2006). This is well exemplified by the fact that high glucose exposure induces typical Ca^{2+}-dependent apoptosis which is mimicked by the K_{ATP} channel blocker tolbutamide, abolished by either the K_{ATP} channel opener diazoxide or the Ca_V1 channel blocker D-600, and prevented by the endonuclease inhibitor aurintricarboxylic acid (Efanova et al., 1998; Yang & Berggren, 2006).

Transglutaminases are widely distributed in various organs, tissues, and body fluids to cata-

lyze Ca^{2+}-dependent post-translational modification of proteins. Transglutaminase 2, also known as tissue transglutaminase, is ubiquitously expressed in mammalian cells including β cells where it extensively cross-links various cytoskeletal proteins and nucleosomal histones as the distal steps of apoptosis (Ballestar et al., 1996; Esposito & Caputo, 2005; Yang & Berggren, 2006). Transglutaminase 2 is activated by Ca^{2+} and inhibited by GTP. The activated enzyme extensively cross-links various cytoskeletal proteins, such as microtubule protein tau, β-tubulin, actin, myosin, spectrin, thymosin β, troponin T, and vimentin. This Ca^{2+}-dependent cross-linking of cytoskeletal proteins brings the apoptotic process into its final steps (Esposito & Caputo, 2005). The activated transglutaminase 2 also regulates post-translational modification of nuclear proteins, like core histones. The resultant histone cross-linking is speculated to mediate the chromatin condensation in apoptosis (Ballestar et al., 1996). Importantly, activated transglutaminase 2 also induces apoptotic cell death in a Ca^{2+}-dependent fashion (Orrenius et al., 2003). Experimental data show that insulin-secreting HIT-T15 cells subjected to depletion of GTP with mycophenolic acid display increased transglutaminase 2 activity and in turn undergo apoptosis. Of particular interest is that β cell apoptosis induced by activated transglutaminase 2 is effectively counteracted by reducing extracellular Ca^{2+} concentrations (Huo et al., 2003). Most likely transglutaminase 2 acts as a potential decoder of excessive Ca^{2+} entry through hyperactivated Ca_V channels in β cell apoptosis.

Calpains are Ca^{2+}-activated neutral cysteine proteases. Among them, the two best characterized isoforms are μ- and m-calpain, which are so named because they are activated in vitro by concentrations of Ca^{2+} in μM and mM ranges, respectively. The activated μ- and m-calpains proteolyze a number of cytoskeletal, membrane-associated, and regulatory proteins involved in a range of cellular processes including necrosis and apoptosis (Goll et al., 2003; Liu et al., 2004). An interesting study shows that apoptotic signaling cascade proceeds in a sequence of cytokine, PKC, Ca_V channels, calpain, calcineurin, Bad, cytochrome c, and caspases in insulin-secreting MIN6N8 cells (Chang et al., 2004). Co-treatment with interferon-γ and TNF-α significantly enhances high Ca$_V$ channel activity, prominently increases [Ca^{2+}]$_i$ and consequently induces apoptotic death of these insulin-secreting cells. The effects are effectively abolished by the PKC inhibitor chelerythrine. It is likely that PKC phosphorylation results in the enhanced activity of high Ca$_V$ channels through which excessive Ca^{2+} influx occurs triggering Ca^{2+}-dependent apoptosis. In fact, the excessive Ca^{2+} influx first activates the Ca^{2+}-activated protease calpain. Then, the activated calpain induces activation of calcineurin, which catalyzes the dephosphorylation of Bad at S112. Subsequently, Bad dephosphorylation leads to the release of cytochrome c from the mitochondria to the cytoplasm. Eventually, caspases 9, 3, and 7 are cleaved and activated resulting in apoptosis (Chang et al., 2004).

In general, there are two ways for Ca$_V$ channels to be engaged in β cell apoptosis (Fig. 1). On one hand, Ca$_V$ channels undergo upregulation of their conductivity and/or density in the β cell. The upregulated Ca$_V$ channels mediate excessive Ca^{2+} influx into the β cell resulting in abnormally exaggerated [Ca^{2+}]$_i$ which in turn initiates apoptotic cascades in the β cell. This is exemplified by the case of calpain-mediated β cell apoptosis triggered by hyperactivated Ca$_V$ channel-mediated Ca^{2+} entry (Fig. 1) (Chang et al., 2004). On the other hand, physiological Ca^{2+} influx through β cell Ca$_V$ channels can serve as a permissive condition for initiation of apoptosis under some conditions. This is well illustrated by the fact that interleukin 1β cannot induce β cell apoptosis without Ca^{2+} influx through Ca$_V$1 channels that are not directly affected by this cytokine (Zaitsev et al., 2001).

The loss of β cells occurs in both T1D and T2D. The former is characterized by the absolute loss of pancreatic β cells, whereas the latter is defined by not only the progressive loss of β cell function but also increased β cell apoptosis (Mathis et al., 2001). As illustrated above, hyperactivation of β cell Ca$_V$ channels plays an important role in β cell apoptosis (Fig. 1) (Chang et al.,

2004; Juntti-Berggren et al., 1993, 2004; Wang et al., 1999). These findings pinpoint β cell Ca$_V$ channels as potential therapeutic targets for the prevention of β cell loss during the development of diabetes. In fact, it has been demonstrated that systemic application of the Ca$_V$1 channel blocker verapmil ameliorates and even prevents low-dose streptozotocin-induced progressive diabetes in mice through reduction of β cell apoptosis and promotion of β cell survival and function (Xu et al., 2012).

Ca$_V$ Channel-Mediated Coupling of Diabetogenic Serum Components to Ca^{2+}-Dependent β Cell Demise

T1D sera contain diabetogenic components that hyperactivate β cell Ca$_V$1 channels to provoke Ca^{2+}-dependent β cell death (Fig. 1) (Juntti-Berggren et al., 1993, 2004; Shi et al., 2014). In fact, diabetogenic components are also present in sera from patients with neurodegenerative disorders like amyotrophic lateral sclerosis (ALS) (Shi et al., 2019).

Originally, T1D serum has been verified to hyperactivate β cell Ca$_V$1 channels driving pathological Ca^{2+} influx into β cells and consequent Ca^{2+}-dependent β cell apoptosis (Juntti-Berggren et al., 1993). In cell-attached single-channel recordings, unitary Ca$_V$1 currents display more frequent openings in β cells exposed to T1D serum. Consistent with single Ca$_V$1 channel analysis, whole-cell patch-clamp recordings reveal that Ca$_V$1 currents are massively increased in T1D serum-treated β cells. As a direct consequence of the hyperactivation of β cell Ca$_V$1 channels, excessive Ca^{2+} ions enter β cells resulting in pathologically exaggerated [Ca^{2+}]$_i$ in these cells. The Ca^{2+} overload eventually causes β cell apoptosis as verified by the appearance of typical apoptotic DNA "ladder". Interestingly, the T1D serum-induced β cell apoptosis disappears upon pharmacological ablation of Ca$_V$1 channels. This demonstrates that hyperactivated Ca$_V$1 channels serve to convey apoptotic cues from T1D serum to the intrinsic apoptotic machinery in β cells (Juntti-Berggren et al., 1993).

More recently, the subtypes of Ca_V1 channels hyperactivated by T1D serum and the biophysical mechanisms responsible for T1D serum-induced hyperactivation of β cell Ca_V1 channels have been clarified by using $Ca_V1.2$ and $Ca_V1.3$ knockout mice in combination with patch-clamp recordings and single-cell RT-PCR analysis (Yang et al., 2015). Interestingly, a subgroup of islet β cells from $Ca_V1.2$ knockout mice express functional $Ca_V1.3$ channels. Importantly, T1D sera increase whole-cell Ca_V currents in islet β cells from $Ca_V1.2$- or $Ca_V1.3$-deficient mice. They boost the open probability and number of functional unitary Ca_V1 channels in $Ca_V1.2$- and $Ca_V1.3$-deficient β cells. It is clear that T1D sera hyperactivate both $Ca_V1.2$ and $Ca_V1.3$ channels by increasing their conductivity and number (Yang et al., 2015). These findings pinpoint $Ca_V1.2$ and $Ca_V1.3$ channels as potential targets for anti-diabetes therapy.

To nail down the actual factor(s), i.e. molecular component(s), responsible for T1D serum-induced hyperactivation of β cell Ca_V1 channels, high-performance liquid chromatography of T1D sera has been carried out. The obtained data show that T1D sera contain significantly higher levels of apolipoprotein CIII (ApoCIII). Interestingly, patch-clamp recordings show that this apolipoprotein is capable of mimicking T1D serum to hyperactivate β cell Ca_V channels. Like T1D sera, ApoCIII markedly increases whole-cell Ca_V channel currents and $[Ca^{2+}]_i$ in the β cell and prominently provokes β cell apoptosis. Neutralization of ApoCIII with anti-ApoCIII antibodies effectively abolishes both T1D serum- and ApoCIII-induced elevation in $[Ca^{2+}]_i$ and apoptosis (Juntti-Berggren et al., 2004). The findings provide promising evidence that ApoCIII is likely to act as the actual factor in T1D sera to hyperactivate β cell Ca_V channels.

Unfortunately, the above findings cannot reveal which Ca_V channel type(s) are targeted by ApoCIII and whether Ca_V channel conductivity, density, or both are influenced by ApoCIII treatment? These two issues have been addressed by combining patch-clamp techniques and pharmacological manipulation (Shi et al., 2014). Cell-attached single-channel recordings reveal that

ApoCIII increases both the open probability and density of unitary Ca_V1 currents. Whole-cell patch-clamp analysis visualizes that ApoCIII enhances whole-cell Ca^{2+} currents and the enhancement can no longer occur in the presence of the Ca_V1 channel blocker nimodipine. The findings corroborate that ApoCIII selectively targets β cell Ca_V1 channels and increases their conductivity and density (Shi et al., 2014).

The key issue of what the signaling mechanisms are whereby ApoCIII hyperactivates β cell Ca_V channels has drawn great attention. The findings obtained in other cell types suggest that plasma membrane-standing scavenger receptor class B type I (SR-BI) can directly sense extracellular ApoCIII and subsequently signals downstream to the SR-BI interaction partner β1 integrin that activates PKA, PKC, and Src kinase to phosphorylate their substrates like Ca_V1 channels. Indeed, SR-BI physically associates with ApoCIII in a CHO heterologous expression system and also interacts with β1 integrin in microglial cells (Bamberger et al., 2003; Xu et al., 1997). ApoCIII activates PKC through indirect interaction with β1 integrin in monocytic cells (Kawakami et al., 2006). Furthermore, β1 integrin activation leads to increased Ca_V1 channel activity in neurons, ventricular myocytes, and vascular smooth muscle cells through stimulation of PKA, PKC, and Src kinase (Gui et al., 2006; Rueckschloss & Isenberg, 2004; Waitkus-Edwards et al., 2002; Wu et al., 2001). In fact, all these components are expressed in β cells (Bosco et al., 2000; Kantengwa et al., 1997; Mukai et al., 2011; Nikolova et al., 2006; Yang & Berggren, 2006). Hyperactivation of Ca_V1 channels by ApoCIII is hardly influenced by individual inhibition of PKA, PKC, or Src, but significantly abolished by a cocktail of PKA, PKC, and Src or by coinhibition of PKA and Src. Moreover, knockdown of β1 integrin or SR-BI effectively prevents ApoCIII from hyperactivating β cell Ca_V channels. These findings demonstrate that ApoCIII hyperactivates β cell Ca_V1 channels through SR-BI/β1 integrin-dependent coactivation of PKA and Src (Shi et al., 2014).

Moreover, additional factors in T1D sera may signal upstream of β cell Ca_V channels to drive β cell apoptosis as suggested by a study performed in neuroblastoma cells (Pittenger et al., 1997). This study shows that Fas-specific antibodies in T1D sera induce neuroblastoma cell apoptosis similar to T1D-induced β cell apoptosis. This indicates that Fas-specific antibodies in T1D sera may act as another candidate factor involved in the hyperactivation of β cell Ca$_V$ channels and Ca^{2+}-dependent β cell death (Pittenger et al., 1997). As a matter of fact, N1E-115 murine neuroblastoma cells show a gradual increase in [Ca^{2+}]$_i$ followed by Fas-mediated apoptosis characterized by the appearance of condensed chromatin, shrunken cytoplasm, and DNA fragmentation when exposed to T1D sera (Pittenger et al., 1997). Given that β cell apoptosis happens following activation of the Fas signaling pathway or exposure to T1D sera, it is intriguing to clarify if Fas-specific antibodies in T1D sera drive β cell apoptosis through hyperactivation of β cell Ca$_V$1 channels (Maedler et al., 2001).

Possible involvement of inhibitory G proteins in T1D serum-induced hyperactivation of β cell Ca$_V$ channels is suggested by studies with serum from the T1D animal model, Bio Bred/Worchester diabetic (BBW) rat. BBW rat serum-induced hyperactivation of neuronal Ca$_V$ channels is very similar to human T1D-induced hyperactivation of β cell Ca$_V$ channels. Nondiabetic rat dorsal root ganglion neurons display a drastic increase in Ca$_V$ channel activity following treatment with the BBW rat serum. This occurs most likely due to impaired regulation of the inhibitory G protein-Ca$_V$ channel complex (Ristic et al., 1998). Interestingly, both high voltage-activated and low voltage-activated Ca^{2+} channels in dorsal root ganglion neurons from the BBW rats undergo an enhancement due to a decrease in opiate-mediated inhibition of PTX-sensitive, G protein-coupled Ca$_V$ channels (Hall et al., 1995, 1996, 2001). The β cell also accommodates inhibitory G proteins that down-regulate β cell Ca$_V$ channel activity (Ammala et al., 1992; Robertson et al., 1991). There is a high likelihood that human T1D serum-induced hyperactivation of β cell Ca$_V$ channels involves inhibitory G proteins.

T1D serum has been found to promote the expression of Ca$_V$3 channels in a particular type of neurons with triangular soma in cerebellar granule cell cultures (Chandra et al., 2001). This is interesting because T1D serum does not affect the Ca$_V$3 channel in RINm5F cells, an insulin-secreting cell line (Juntti-Berggren et al., 1993; Shi et al., 2014). Maybe this is due to differences in signaling pathways upstream of Ca$_V$3 channels between nonproliferating primary cells like cerebellar granule neurons and islet β cells and the highly proliferative RINm5F cell line. As a matter of fact, current available data on islet β cells cannot fully rule out the possibility that T1D sera also attack β cell Ca$_V$3 channels to provoke Ca^{2+}-dependent β cell death.

Indeed, a T-lymphocyte-mediated autoimmune attack is crucial for β cell death in T1D (Mathis et al., 2001). In addition, the above-discussed findings add another layer of complexity to the mechanism of β cell death in T1D and suggest that factors such as ApoCIII and Fas-specific antibodies in T1D sera can attack β cell-cell Ca$_V$ channels and initiate cell death (Juntti-Berggren et al., 2004; Pittenger et al., 1997; Shi et al., 2014). The complete repertoire of diabetogenic factors in T1D sera and their involvement in β cell Ca$_V$ channels are of therapeutic interest and need further investigation.

In addition to T1D sera, sera from patients with amyotrophic lateral sclerosis (ALS) and T2D (ALS-T2D) also express diabetogenic factors (Shi et al., 2019). About 60% ALS-T2D patients have positive sera that exaggerate [Ca^{2+}]$_i$ responses induced by K$^+$ depolarization in mouse islet cells. The component responsible for the effect has been identified as pathogenic immunoglobulin Gs (IgGs). These pathogenic IgGs not only immunocapture Ca$_V$α$_2$δ1 subunits in the β cell plasma membrane, but also hyperactivate β cell Ca$_V$1 channels. The hyperactivated Ca$_V$1 channels conduct excessive Ca^{2+} entry leading to pathologically exaggerated increases in [Ca^{2+}]$_i$ that activate multiple Ca^{2+}-dependent events detrimental to β cell function and viability. These detrimental events include impaired mitochondrial function, disturbed [Ca^{2+}]$_i$ dynamics, reduced GSIS, and nonviable β cells. The findings verify that cytotoxic ALS-T2D-IgG autoantibodies appear in sera from ALS-T2D patients, behaving as a causal link between two closely-associated disorders ALS and T2D by immunoattacking $Ca_V\alpha_2\delta1$ subunits. Interestingly, this study suggests that in a fraction of patients T2D is likely to result from immune destruction of β cells by cytotoxic ALS-T2D-IgG autoantibodies reflecting altered humoral immunity. This may suggest a new pharmacological treatment strategy for patients suffering from a combination of ALS and T2D.

Conclusions

Over decades, persistent attempts and continuous efforts to understand pancreatic β cell Ca_V channels in health and disease have led to significant progress (Yang & Berggren, 2005a, b, 2006; Yang et al., 2014). Electrically excitable β cells accommodate at least seven $Ca_V\alpha1$ subunits including $Ca_V1.2$, $Ca_V1.3$, $Ca_V2.1$, $Ca_V2.2$, $Ca_V2.3$, $Ca_V3.1$, and $Ca_V3.2$. These pore-forming subunits and certain Ca_V auxiliary subunits are noncovalently assembled into different types of Ca_V channels that conduct L-, P/Q-, N-, R-, and T-type Ca_V currents, respectively (Braun et al., 2008; Yang & Berggren, 2005a, b, 2006; Yang et al., 2014). As reflected by the name of these channels, they undergo conformational transitions from an impermeable state to a highly permeable pore upon membrane depolarization initiated by the glucose-induced closure of K_{ATP} channels, mediating rapid Ca^{2+} entry into β cells (Yang & Berggren, 2005a, b, 2006; Yang et al., 2014). Such a Ca^{2+} entry not only generates electrical signals in the β cell plasma membrane but also the versatile intracellular messenger $[Ca^{2+}]_i$ (Yang & Berggren, 2005a, b, 2006; Yang et al., 2014). This makes Ca_V channels different from other ion channels in promoting Ca^{2+}-dependent molecular and cellular events besides producing electrical signals (Catterall, 2000).

In physiological contexts, β cell Ca_V channels take center stage in GSIS and are engaged in the control of β cell maturity, viability, and expansion, thereby acting as a multifaceted player in β cell signaling (Fig. 1) (Yang & Berggren, 2005a, b, 2006; Yang et al., 2014). Impressively, β cell Ca_V channels interplay with other non-channel proteins to create complex signaling networks in addition to promoting Ca^{2+} influx over the plasma membrane (Belkacemi et al., 2018; Berggren et al., 2004; Ji et al., 2002; Rajagopal et al., 2014; Wiser et al., 1999; Yang & Berggren, 2005a, 2006; Yang et al., 1999, 2014). This renders β cell Ca_V channels versatile, efficient, and even specific in β cell signaling. β Cell Ca_V1 subunits and exocytotic proteins specifically interact at the site of exocytosis to drive insulin release, thus preventing detrimental actions of unnecessary excessive Ca^{2+} influx (Ji et al., 2002; Wiser et al., 1999; Yang & Berggren, 2005a, 2006; Yang et al., 1999). Of interest is that promiscuous $Ca_V\beta$ subunits are networking with $InsP_3$ receptors and PKC to limit GSIS (Belkacemi et al., 2018; Berggren et al., 2004; Rajagopal et al., 2014; Yang & Berggren, 2005a, 2006; Yang et al., 2014). It seems difficult to reconcile the evolutionary conservation of $Ca_V\beta$ subunits and their negative action on GSIS, but it becomes understandable why these signaling networks are conserved in β cells if adding pathological roles of insulin hypersecretion into the context. They likely function as protective mechanisms against insulin hypersecretion and its obesogenic and even carcinogenic effects (Erion & Corkey, 2017; Gallagher & LeRoith, 2020). Surprisingly, $Ca_V\beta_3$ subunits do not appear as a required building block for β cell Ca_V channels, but instead an interaction partner of $InsP_3$ receptors to serve as a brake on Ca^{2+} mobilization from $InsP_3$-sensitive stores (Belkacemi et al., 2018; Berggren et al., 2004; Yang & Berggren, 2005a, 2006; Yang et al., 1999, 2014).

β Cell Ca_V channels suffering from inherited defects or acquired impairments conduct pathological Ca^{2+} influx, being insufficient, excessive, or spatiotemporally disorganized, causing β cell dysfunction, dedifferentiation, destruction, aberrant expansion, and eventual diabetes (Fig. 1) (Yang & Berggren, 2005a, 2006; Yang et al., 2014). They undergo either hyperactivation or hypoactivation under diabetic conditions (Yang & Berggren, 2005a, 2006; Yang et al., 2014). Excessively-expressed Ca$_V$β$_3$ subunits drive diabetic phenotypes in obese mice, whereas Ca$_V$β$_3$ gene silencing reverses these phenotypes (Lee et al., 2018). Ca$_V$3 channels show no functional operation in healthy mouse β cells, but become active in NOD mouse β cells (Yang & Berggren, 2005a, 2006; Yang et al., 2014). Ca$_V$ channel gene mutation and polymorphism are closely associated with abnormal insulin secretion and diabetes (Holmkvist et al., 2007; Splawski et al., 2004). For example, the Ca$_V$1.2^{G406R} mutant in patients with Timothy syndrome displays little voltage-dependent inactivation and mediates exaggerated Ca^{2+} influx that enhances insulin secretion resulting in lethal hypoglycemia (Splawski et al., 2004). The close association of the SNPs rs312480, rs312486, and rs9841978 in the Ca$_V$1.3 subunit gene with impaired insulin secretion and T2D is verified in Finnish and Swedish patients (Holmkvist et al., 2007). Interestingly, Ca$_V$3.1 channels overexpressed in β cells activate detrimental signaling pathways driving β cell dedifferentiation or maturity loss through FoxO1-mediated downregulation of β cell exocytotic proteins (Yu et al., 2020). In worst-case scenarios, β cell Ca$_V$ channels mediate exaggerated Ca^{2+} entry resulting in excessive [Ca^{2+}]$_i$ and consequent Ca^{2+}-dependent β cell death by activating various Ca^{2+}-sensitive enzymes, e.g., calcineurin, endonucleases, transglutaminase and calpains (Chang et al., 2004; Efanova et al., 1998; Huo et al., 2003; Juntti-Berggren et al., 1993, 2004; Wang et al., 1999; Zaitsev et al., 2001). In general, β cell Ca$_V$ channels serve to couple diabetogenic factors, such as ApoCIII, Ca$_V$α$_1$δ2-specific, and Fas-specific antibodies in sera from patients with T1D or ALS-T1D to Ca^{2+}-dependent β cell demise (Juntti-Berggren et al., 1993, 2004; Shi et al., 2014, 2019; Yang et al., 2014, 2015; Yang & Berggren, 2005a, 2006). More specifically, Ca$_V$1 channels increase their density and activity in the β cell plasma membrane in response to ApoCIII

through SR-BI/β1 integrin-dependent coactivation of PKA and Src kinase (Shi et al., 2014). Cytotoxic IgG autoantibodies in sera from ALS-T2D patients immunocapture Ca$_V$α$_2$δ1 subunits and hyperactivate β cell Ca$_V$ channels leading to impairments in [Ca^{2+}]$_i$ dynamics, mitochondrial function, insulin secretion, and survival of β cells, thereby serving as a causal link between ALS and T2D (Shi et al., 2019).

The current knowledge on β cell Ca$_V$ channels have been gained mostly from studies with rodent β cells, but seldom from those with human ones (Yang & Berggren, 2005a, 2006; Yang et al., 2014). Caution should be exercised in generalizing such knowledge to human β cells. Human β cell Ca$_V$ channelome, physiology, pathology, regulation mechanisms, and therapeutic potentials should be subjected to further high-priority research. The findings from studies with rodent β cells support that β cell Ca$_V$1 and Ca$_V$3 channels and their associated interplayers ApoCIII, InsP$_3$ receptors, PKC and calcineurin most likely serve as druggable targets for clinical intervention of Ca^{2+}-dependent β cell dedifferentiation and death associated with diabetes (Belkacemi et al., 2018; Berggren et al., 2004; Juntti-Berggren et al., 2004; Rajagopal et al., 2014; Shi et al., 2014, 2019; Yang & Berggren, 2005a, 2006; Yang et al., 2014; Yu et al., 2020). Of particular interest are clinical trials of Ca$_V$3 channel blockers for diabetes treatment. Thus far, the visualization of the ultra-high resolution structure of Ca$_V$ channels has indeed been turned into reality with the development of advanced technologies, like x-ray crystallography, cryo-electron microscopy, stimulated emission depletion microscopy, and single-molecule microscopy. However, vivid super-ultra high-resolution imaging of instantaneous switch in Ca$_V$ channel conformation and dynamic redistribution of Ca$_V$ channels in the plasma membrane is still infeasible. This has created a wide range of physiological enigmas and medical dilemmas and prevented in-depth understanding of β cell Ca$_V$ channels in health and disease. To resolve these issues, one has to develop and apply super-ultra high spatiotemporal resolution technologies.

References

Ainscow, E. K., & Rutter, G. A. (2002). Glucose-stimulated oscillations in free cytosolic ATP concentration imaged in single islet β-cells: Evidence for a Ca^{2+}-dependent mechanism. *Diabetes, 51*, S162–S170.

American Diabetes Association. (2014). Diagnosis and classification of diabetes mellitus. *Diabetes Care, 37*, S81–S90.

Ammala, C., Berggren, P. O., Bokvist, K., & Rorsman, P. (1992). Inhibition of L-type calcium channels by internal GTPγS in mouse pancreatic β cells. *Pflügers Archiv, 420*, 72–77.

Ashcroft, F. M., & Rorsman, P. (1989). Electrophysiology of the pancreatic β-cell. *Progress in Biophysics and Molecular Biology, 54*, 87–143.

Ashcroft, F., & Rorsman, P. (2004). Type 2 diabetes mellitus: Not quite exciting enough? *Human Molecular Genetics, 13*, R21–R31.

Ballestar, E., Abad, C., & Franco, L. (1996). Core histones are glutaminyl substrates for tissue transglutaminase. *The Journal of Biological Chemistry, 271*, 18817–18824.

Bamberger, M. E., Harris, M. E., McDonald, D. R., Husemann, J., & Landreth, G. E. (2003). A cell surface receptor complex for fibrillar beta-amyloid mediates microglial activation. *The Journal of Neuroscience, 23*, 2665–2674.

Barg, S., Ma, X., Eliasson, L., Galvanovskis, J., Gopel, S. O., Obermuller, S., Platzer, J., Renstrom, E., Trus, M., Atlas, D., et al. (2001). Fast exocytosis with few Ca^{2+} channels in insulin-secreting mouse pancreatic β cells. *Biophysical Journal, 81*, 3308–3323.

Barg, S., Eliasson, L., Renstrom, E., & Rorsman, P. (2002). A subset of 50 secretory granules in close contact with L-type Ca^{2+} channels accounts for first-phase insulin secretion in mouse β-cells. *Diabetes, 51*, S74–S82.

Belkacemi, A., Hui, X., Wardas, B., Laschke, M. W., Wissenbach, U., Menger, M. D., Lipp, P., Beck, A., & Flockerzi, V. (2018). IP_3 receptor-dependent cytoplasmic Ca^{2+} signals are tightly controlled by $Ca_V beta3$. *Cell Reports, 22*, 1339–1349.

Bell, G. I., & Polonsky, K. S. (2001). Diabetes mellitus and genetically programmed defects in β-cell function. *Nature, 414*, 788–791.

Benes, C., Roisin, M. P., Van Tan, H., Creuzet, C., Miyazaki, J., & Fagard, R. (1998). Rapid activation and nuclear translocation of mitogen-activated protein kinases in response to physiological concentration of glucose in the MIN6 pancreatic β cell line. *The Journal of Biological Chemistry, 273*, 15507–15513.

Bennett, M. K., Calakos, N., & Scheller, R. H. (1992). Syntaxin: A synaptic protein implicated in docking of synaptic vesicles at presynaptic active zones. *Science, 257*, 255–259.

Berggren, P. O., Yang, S. N., Murakami, M., Efanov, A. M., Uhles, S., Kohler, M., Moede, T., Fernstrom, A., Appelskog, I. B., Aspinwall, C. A., et al. (2004).

Removal of Ca^{2+} channel β_3 subunit enhances Ca^{2+} oscillation frequency and insulin exocytosis. *Cell, 119*, 273–284.

Berridge, M. J., Bootman, M. D., & Lipp, P. (1998). Calcium-a life and death signal. *Nature, 395*, 645–648.

Berridge, M. J., Lipp, P., & Bootman, M. D. (2000). The versatility and universality of calcium signalling. *Nature Reviews. Molecular Cell Biology, 1*, 11–21.

Berridge, M. J., Bootman, M. D., & Roderick, H. L. (2003). Calcium signalling: Dynamics, homeostasis and remodelling. *Nature Reviews. Molecular Cell Biology, 4*, 517–529.

Bezprozvanny, I., Scheller, R. H., & Tsien, R. W. (1995). Functional impact of syntaxin on gating of N-type and Q-type calcium channels. *Nature, 378*, 623–626.

Bhattacharjee, A., Whitehurst, R. M., Jr., Zhang, M., Wang, L., & Li, M. (1997). T-type calcium channels facilitate insulin secretion by enhancing general excitability in the insulin-secreting β-cell line, INS-1. *Endocrinology, 138*, 3735–3740.

Bosco, D., Meda, P., Halban, P. A., & Rouiller, D. G. (2000). Importance of cell-matrix interactions in rat islet β-cell secretion in vitro: Role of α6β1 integrin. *Diabetes, 49*, 233–243.

Braun, M., Ramracheya, R., Bengtsson, M., Zhang, Q., Karanauskaite, J., Partridge, C., Johnson, P. R., & Rorsman, P. (2008). Voltage-gated ion channels in human pancreatic β-cells: Electrophysiological characterization and role in insulin secretion. *Diabetes, 57*, 1618–1628.

Catterall, W. A. (2000). Structure and regulation of voltage-gated Ca^{2+} channels. *Annual Review of Cell and Developmental Biology, 16*, 521–555.

Chandra, J., Yang, S. N., Kohler, M., Zaitsev, S., Juntti-Berggren, L., Berggren, P. O., Zhivotovsky, B., & Orrenius, S. (2001). Effects of serum from patients with type 1 diabetes on primary cerebellar granule cells. *Diabetes, 50*, S77–S81.

Chang, I., Cho, N., Kim, S., Kim, J. Y., Kim, E., Woo, J. E., Nam, J. H., Kim, S. J., & Lee, M. S. (2004). Role of calcium in pancreatic islet cell death by IFN-γ/TNF-α. *Journal of Immunology, 172*, 7008–7014.

Davalli, A. M., Biancardi, E., Pollo, A., Socci, C., Pontiroli, A. E., Pozza, G., Clementi, F., Sher, E., & Carbone, E. (1996). Dihydropyridine-sensitive and -insensitive voltage-operated calcium channels participate in the control of glucose-induced insulin release from human pancreatic β cells. *The Journal of Endocrinology, 150*, 195–203.

Efanova, I. B., Zaitsev, S. V., Zhivotovsky, B., Kohler, M., Efendic, S., Orrenius, S., & Berggren, P. O. (1998). Glucose and tolbutamide induce apoptosis in pancreatic β-cells. A process dependent on intracellular Ca^{2+} concentration. *The Journal of Biological Chemistry, 273*, 33501–33507.

Efrat, S., Surana, M., & Fleischer, N. (1991). Glucose induces insulin gene transcription in a murine pancreatic β-cell line. *The Journal of Biological Chemistry, 266*, 11141–11143.

Erion, K. A., & Corkey, B. E. (2017). Hyperinsulinemia: A cause of obesity? *Current Obesity Reports, 6*, 178–186.

Esposito, C., & Caputo, I. (2005). Mammalian transglutaminases. Identification of substrates as a key to physiological function and physiopathological relevance. *The FEBS Journal, 272*, 615–631.

Gallagher, E. J., & LeRoith, D. (2020). Hyperinsulinaemia in cancer. *Nature Reviews. Cancer, 20*, 629–644.

German, M. S., Moss, L. G., & Rutter, W. J. (1990). Regulation of insulin gene expression by glucose and calcium in transfected primary islet cultures. *The Journal of Biological Chemistry, 265*, 22063–22066.

Goll, D. E., Thompson, V. F., Li, H., Wei, W., & Cong, J. (2003). The calpain system. *Physiological Reviews, 83*, 731–801.

Gui, P., Wu, X., Ling, S., Stotz, S. C., Winkfein, R. J., Wilson, E., Davis, G. E., Braun, A. P., Zamponi, G. W., & Davis, M. J. (2006). Integrin receptor activation triggers converging regulation of Cav1.2 calcium channels by c-Src and protein kinase A pathways. *The Journal of Biological Chemistry, 281*, 14015–14025.

Hall, K. E., Sima, A. A., & Wiley, J. W. (1995). Voltage-dependent calcium currents are enhanced in dorsal root ganglion neurones from the Bio Bred/Worchester diabetic rat. *The Journal of Physiology, 486*, 313–322.

Hall, K. E., Sima, A. A., & Wiley, J. W. (1996). Opiate-mediated inhibition of calcium signaling is decreased in dorsal root ganglion neurons from the diabetic BB/W rat. *The Journal of Clinical Investigation, 97*, 1165–1172.

Hall, K. E., Liu, J., Sima, A. A., & Wiley, J. W. (2001). Impaired inhibitory G-protein function contributes to increased calcium currents in rats with diabetic neuropathy. *Journal of Neurophysiology, 86*, 760–770.

Hibino, H., Pironkova, R., Onwumere, O., Rousset, M., Charnet, P., Hudspeth, A. J., & Lesage, F. (2003). Direct interaction with a nuclear protein and regulation of gene silencing by a variant of the Ca^{2+}-channel β$_4$ subunit. *Proceedings of the National Academy of Sciences of the United States of America, 100*, 307–312.

Holmkvist, J., Tojjar, D., Almgren, P., Lyssenko, V., Lindgren, C. M., Isomaa, B., Tuomi, T., Berglund, G., Renstrom, E., & Groop, L. (2007). Polymorphisms in the gene encoding the voltage-dependent Ca^{2+} channel Ca$_V$2.3 (CACNA1E) are associated with type 2 diabetes and impaired insulin secretion. *Diabetologia, 50*, 2467–2475.

Hrvatin, S., O'Donnell, C. W., Deng, F., Millman, J. R., Pagliuca, F. W., DiIorio, P., Rezania, A., Gifford, D. K., & Melton, D. A. (2014). Differentiated human stem cells resemble fetal, not adult, beta cells. *Proceedings of the National Academy of Sciences of the United States of America, 111*, 3038–3043.

Huo, J., Metz, S. A., & Li, G. (2003). Role of tissue transglutaminase in GTP depletion-induced apoptosis of insulin-secreting (HIT-T15) cells. *Biochemical Pharmacology, 66*, 213–223.

Iwashima, Y., Pugh, W., Depaoli, A. M., Takeda, J., Seino, S., Bell, G. I., & Polonsky, K. S. (1993). Expression of calcium channel mRNAs in rat pancreatic islets and downregulation after glucose infusion. *Diabetes, 42*, 948–955.

Ji, J., Yang, S. N., Huang, X., Li, X., Sheu, L., Diamant, N., Berggren, P. O., & Gaisano, H. Y. (2002). Modulation of L-type Ca^{2+} channels by distinct domains within SNAP-25. *Diabetes, 51*, 1425–1436.

Jing, X., Li, D. Q., Olofsson, C. S., Salehi, A., Surve, V. V., Caballero, J., Ivarsson, R., Lundquist, I., Pereverzev, A., Schneider, T., et al. (2005). Ca$_V$2.3 calcium channels control second-phase insulin release. *The Journal of Clinical Investigation, 115*, 146–154.

Juntti-Berggren, L., Larsson, O., Rorsman, P., Ammala, C., Bokvist, K., Wahlander, K., Nicotera, P., Dypbukt, J., Orrenius, S., Hallberg, A., et al. (1993). Increased activity of L-Type Ca^{2+} channels exposed to serum from patients with type I diabetes. *Science, 261*, 86–90.

Juntti-Berggren, L., Refai, E., Appelskog, I., Andersson, M., Imreh, G., Dekki, N., Uhles, S., Yu, L., Griffiths, W. J., Zaitsev, S., et al. (2004). Apolipoprotein CIII promotes Ca^{2+}-dependent β cell death in type 1 diabetes. *Proceedings of the National Academy of Sciences of the United States of America, 101*, 10090–10094.

Kang, Y., Huang, X., Pasyk, E. A., Ji, J., Holz, G. G., Wheeler, M. B., Tsushima, R. G., & Gaisano, H. Y. (2002). Syntaxin-3 and syntaxin-1A inhibit L-type calcium channel activity, insulin biosynthesis and exocytosis in β-cell lines. *Diabetologia, 45*, 231–241.

Kantengwa, S., Baetens, D., Sadoul, K., Buck, C. A., Halban, P. A., & Rouiller, D. G. (1997). Identification and characterization of α3β1 integrin on primary and transformed rat islet cells. *Experimental Cell Research, 237*, 394–402.

Kato, S., Ishida, H., Tsuura, Y., Tsuji, K., Nishimura, M., Horie, M., Taminato, T., Ikehara, S., Odaka, H., Ikeda, H., et al. (1996). Alterations in basal and glucose-stimulated voltage-dependent Ca^{2+} channel activities in pancreatic β cells of non-insulin-dependent diabetes mellitus GK rats. *The Journal of Clinical Investigation, 97*, 2417–2425.

Kawakami, A., Aikawa, M., Libby, P., Alcaide, P., Luscinskas, F. W., & Sacks, F. M. (2006). Apolipoprotein CIII in apolipoprotein B lipoproteins enhances the adhesion of human monocytic cells to endothelial cells. *Circulation, 113*, 691–700.

Komatsu, M., Yokokawa, N., Takeda, T., Nagasawa, Y., Aizawa, T., & Yamada, T. (1989). Pharmacological characterization of the voltage-dependent calcium channel of pancreatic β-cell. *Endocrinology, 125*, 2008–2014.

Lang, J. (1999). Molecular mechanisms and regulation of insulin exocytosis as a paradigm of endocrine secretion. *European Journal of Biochemistry, 259*, 3–17.

Lee, B., & Laychock, S. G. (2000). Regulation of inositol trisphosphate receptor isoform expression in glucose-desensitized rat pancreatic islets: Role of cyclic adenosine 3′,5′-monophosphate and calcium. *Endocrinology, 141*, 1394–1402.

Lee, K., Kim, J., Kohler, M., Yu, J., Shi, Y., Yang, S. N., Ryu, S. H., & Berggren, P. O. (2018). Blocking Ca^{2+} channel β₃ subunit reverses diabetes. *Cell Reports, 24*, 922–934.

Leveque, C., Hoshino, T., David, P., Shoji-Kasai, Y., Leys, K., Omori, A., Lang, B., El Far, O., Sato, K., & Martin-Moutot, N. (1992). The synaptic vesicle protein synaptotagmin associates with calcium channels and is a putative Lambert-Eaton myasthenic syndrome antigen. *Proceedings of the National Academy of Sciences of the United States of America, 89*, 3625–3629.

Ligon, B., Boyd, A. E., & Dunlap, K. (1998). Class A calcium channel variants in pancreatic islets and their role in insulin secretion. *The Journal of Biological Chemistry, 273*, 13905–13911.

Liu, J. S., & Hebrok, M. (2017). All mixed up: Defining roles for β-cell subtypes in mature islets. *Genes & Development, 31*, 228–240.

Liu, X., Van Vleet, T., & Schnellmann, R. G. (2004). The role of calpain in oncotic cell death. *Annual Review of Pharmacology and Toxicology, 44*, 349–370.

MacDonald, P. E., Joseph, J. W., & Rorsman, P. (2005). Glucose-sensing mechanisms in pancreatic β-cells. *Philosophical Transactions of the Royal Society of London. Series B, Biological Sciences, 360*, 2211–2225.

Macfarlane, W. M., Campbell, S. C., Elrick, L. J., Oates, V., Bermano, G., Lindley, K. J., Aynsley-Green, A., Dunne, M. J., James, R. F., & Docherty, K. (2000). Glucose regulates islet amyloid polypeptide gene transcription in a PDX1- and calcium-dependent manner. *The Journal of Biological Chemistry, 275*, 15330–15335.

Maedler, K., Spinas, G. A., Lehmann, R., Sergeev, P., Weber, M., Fontana, A., Kaiser, N., & Donath, M. Y. (2001). Glucose induces β-cell apoptosis via upregulation of the Fas receptor in human islets. *Diabetes, 50*, 1683–1690.

Mathis, D., Vence, L., & Benoist, C. (2001). β-Cell death during progression to diabetes. *Nature, 414*, 792–798.

Mochida, S., Sheng, Z. H., Baker, C., Kobayashi, H., & Catterall, W. A. (1996). Inhibition of neurotransmission by peptides containing the synaptic protein interaction site of N-type Ca^{2+} channels. *Neuron, 17*, 781–788.

Mukai, E., Fujimoto, S., Sato, H., Oneyama, C., Kominato, R., Sato, Y., Sasaki, M., Nishi, Y., Okada, M., & Inagaki, N. (2011). Exendin-4 suppresses Src activation and reactive oxygen species production in diabetic Goto-Kakizaki rat islets in an Epac-dependent manner. *Diabetes, 60*, 218–226.

Namkung, Y., Skrypnyk, N., Jeong, M. J., Lee, T., Lee, M. S., Kim, H. L., Chin, H., Suh, P. G., Kim, S. S., & Shin, H. S. (2001). Requirement for the L-type Ca^{2+} channel α₁D subunit in postnatal pancreatic β cell generation. *The Journal of Clinical Investigation, 108*, 1015–1022.

Newcomb, R., Szoke, B., Palma, A., Wang, G., Chen, X., Hopkins, W., Cong, R., Miller, J., Urge, L., Tarczy-Hornoch, K., et al. (1998). Selective peptide antagonist of the class E calcium channel from the venom of the tarantula Hysterocrates gigas. *Biochemistry, 37*, 15353–15362.

Nikolova, G., Jabs, N., Konstantinova, I., Domogatskaya, A., Tryggvason, K., Sorokin, L., Fassler, R., Gu, G., Gerber, H. P., Ferrara, N., et al. (2006). The vascular basement membrane: A niche for insulin gene expression and β cell proliferation. *Developmental Cell, 10*, 397–405.

Ohta, M., Nelson, J., Nelson, D., Meglasson, M. D., & Erecinska, M. (1993). Effect of Ca^{++} channel blockers on energy level and stimulated insulin secretion in isolated rat islets of Langerhans. *The Journal of Pharmacology and Experimental Therapeutics, 264*, 35–40.

Orrenius, S., Zhivotovsky, B., & Nicotera, P. (2003). Regulation of cell death: The calcium-apoptosis link. *Nature Reviews. Molecular Cell Biology, 4*, 552–565.

Ovalle, F., Grimes, T., Xu, G., Patel, A. J., Grayson, T. B., Thielen, L. A., Li, P., & Shalev, A. (2018). Verapamil and beta cell function in adults with recent-onset type 1 diabetes. *Nature Medicine, 24*, 1108–1112.

Pereverzev, A., Mikhna, M., Vajna, R., Gissel, C., Henry, M., Weiergraber, M., Hescheler, J., Smyth, N., & Schneider, T. (2002a). Disturbances in glucose-tolerance, insulin-release, and stress-induced hyperglycemia upon disruption of the Ca$_V$2.3 (α₁E) subunit of voltage-gated Ca^{2+} channels. *Molecular Endocrinology, 16*, 884–895.

Pereverzev, A., Vajna, R., Pfitzer, G., Hescheler, J., Klockner, U., & Schneider, T. (2002b). Reduction of insulin secretion in the insulinoma cell line INS-1 by overexpression of a Ca$_V$2.3 (α1E) calcium channel antisense cassette. *European Journal of Endocrinology, 146*, 881–889.

Piedras-Renteria, E. S., Watase, K., Harata, N., Zhuchenko, O., Zoghbi, H. Y., Lee, C. C., & Tsien, R. W. (2001). Increased expression of α₁A Ca^{2+} channel currents arising from expanded trinucleotide repeats in spinocerebellar ataxia type 6. *The Journal of Neuroscience, 21*, 9185–9193.

Pipeleers, D., Chintinne, M., Denys, B., Martens, G., Keymeulen, B., & Gorus, F. (2008). Restoring a functional β-cell mass in diabetes. *Diabetes, Obesity & Metabolism, 10*(Suppl 4), 54–62.

Pittenger, G. L., Liu, D., & Vinik, A. I. (1997). The apoptotic death of neuroblastoma cells caused by serum from patients with insulin-dependent diabetes and neuropathy may be Fas-mediated. *Journal of Neuroimmunology, 76*, 153–160.

Popiela, H., & Moore, W. (1991). Tolbutamide stimulates proliferation of pancreatic beta cells in culture. *Pancreas, 6*, 464–469.

Rajagopal, S., Fields, B. L., & Kamatchi, G. L. (2014). Contribution of protein kinase Calpha in the stimulation of insulin by the down-regulation of Cavbeta subunits. *Endocrine, 47*, 463–471.

Reinbothe, T. M., Alkayyali, S., Ahlqvist, E., Tuomi, T., Isomaa, B., Lyssenko, V., & Renstrom, E. (2013). The human L-type calcium channel Ca$_V$1.3 regulates insulin release and polymorphisms in CACNA1D associate with type 2 diabetes. *Diabetologia, 56*, 340–349.

Rettig, J., Sheng, Z. H., Kim, D. K., Hodson, C. D., Snutch, T. P., & Catterall, W. A. (1996). Isoform-specific interaction of the α$_{1A}$ subunits of brain Ca^{2+} channels with the presynaptic proteins syntaxin and SNAP-25. *Proceedings of the National Academy of Sciences of the United States of America, 93*, 7363–7368.

Rhodes, C. J. (2005). Type 2 diabetes-a matter of β-cell life and death? *Science, 307*, 380–384.

Ristic, H., Srinivasan, S., Hall, K. E., Sima, A. A., & Wiley, J. W. (1998). Serum from diabetic BB/W rats enhances calcium currents in primary sensory neurons. *Journal of Neurophysiology, 80*, 1236–1244.

Robertson, R. P., Seaquist, E. R., & Walseth, T. F. (1991). G proteins and modulation of insulin secretion. *Diabetes, 40*, 1–6.

Rorsman, P., & Braun, M. (2013). Regulation of insulin secretion in human pancreatic islets. *Annual Review of Physiology, 75*, 155–179.

Rorsman, P., & Renstrom, E. (2003). Insulin granule dynamics in pancreatic beta cells. *Diabetologia, 46*, 1029–1045.

Rose, T., Efendic, S., & Rupnik, M. (2007). Ca^{2+}-secretion coupling is impaired in diabetic Goto Kakizaki rats. *The Journal of General Physiology, 129*, 493–508.

Rueckschloss, U., & Isenberg, G. (2004). Contraction augments L-type Ca^{2+} currents in adherent guinea-pig cardiomyocytes. *The Journal of Physiology, 560*, 403–411.

Rustenbeck, I., Herrmann, C., & Grimmsmann, T. (1997). Energetic requirement of insulin secretion distal to calcium influx. *Diabetes, 46*, 1305–1311.

Salinno, C., Cota, P., Bastidas-Ponce, A., Tarquis-Medina, M., Lickert, H., & Bakhti, M. (2019). β-Cell maturation and identity in health and disease. *The International Journal of Molecular Sciences, 20*, 5417.

Saltiel, A. R. (2001). New perspectives into the molecular pathogenesis and treatment of type 2 diabetes. *Cell, 104*, 517–529.

Satin, L. S. (2000). Localized calcium influx in pancreatic β-cells: Its significance for Ca^{2+}-dependent insulin secretion from the islets of Langerhans. *Endocrine, 13*, 251–262.

Schulla, V., Renstrom, E., Feil, R., Feil, S., Franklin, I., Gjinovci, A., Jing, X. J., Laux, D., Lundquist, I., Magnuson, M. A., et al. (2003). Impaired insulin secretion and glucose tolerance in β cell-selective Ca$_V$1.2 Ca^{2+} channel null mice. *The EMBO Journal, 22*, 3844–3854.

Sellick, G. S., Garrett, C., & Houlston, R. S. (2003). A novel gene for neonatal diabetes maps to chromosome 10p12.1-p13. *Diabetes, 52*, 2636–2638.

Sheng, Z. H., Rettig, J., Takahashi, M., & Catterall, W. A. (1994). Identification of a syntaxin-binding site on N-type calcium channels. *Neuron, 13*, 1303–1313.

Sheng, Z. H., Rettig, J., Cook, T., & Catterall, W. A. (1996). Calcium-dependent interaction of N-type calcium channels with the synaptic core complex. *Nature, 379*, 451–454.

Sheng, Z. H., Yokoyama, C. T., & Catterall, W. A. (1997). Interaction of the synprint site of N-type Ca^{2+} channels with the C2B domain of synaptotagmin I. *Proceedings of the National Academy of Sciences of the United States of America, 94*, 5405–5410.

Sheng, Z. H., Westenbroek, R. E., & Catterall, W. A. (1998). Physical link and functional coupling of presynaptic calcium channels and the synaptic vesicle docking/fusion machinery. *Journal of Bioenergetics and Biomembranes, 30*, 335–345.

Sher, E., Biancardi, E., Pollo, A., Carbone, E., Li, G., Wollheim, C. B., & Clementi, F. (1992). ω-Conotoxin-sensitive, voltage-operated Ca^{2+} channels in insulin-secreting cells. *European Journal of Pharmacology, 216*, 407–414.

Sher, E., Giovannini, F., Codignola, A., Passafaro, M., Giorgi-Rossi, P., Volsen, S., Craig, P., Davalli, A., & Carrera, P. (2003). Voltage-operated calcium channel heterogeneity in pancreatic β cells: Physiopathological implications. *Journal of Bioenergetics and Biomembranes, 35*, 687–696.

Shi, Y., Yang, G., Yu, J., Yu, L., Westenbroek, R., Catterall, W. A., Juntti-Berggren, L., Berggren, P. O., & Yang, S. N. (2014). Apolipoprotein CIII hyperactivates β cell Ca$_V$1 channels through SR-BI/β1 integrin-dependent coactivation of PKA and Src. *Cellular and Molecular Life Sciences, 71*, 1289–1303.

Shi, Y., Park, K. S., Kim, S. H., Yu, J., Zhao, K., Yu, L., Oh, K. W., Lee, K., Kim, J., Chaggar, K., et al. (2019). IgGs from patients with amyotrophic lateral sclerosis and diabetes target Ca$_V$α2δ1 subunits impairing islet cell function and survival. *Proceedings of the National Academy of Sciences of the United States of America, 116*(52), 26816–26822.

Sjoholm, A. (1995). Regulation of insulinoma cell proliferation and insulin accumulation by peptides and second messengers. *Upsala Journal of Medical Sciences, 100*, 201–216.

Solomon, M., & Sarvetnick, N. (2004). The pathogenesis of diabetes in the NOD mouse. *Advances in Immunology, 84*, 239–264.

Splawski, I., Timothy, K. W., Sharpe, L. M., Decher, N., Kumar, P., Bloise, R., Napolitano, C., Schwartz, P. J., Joseph, R. M., Condouris, K., et al. (2004). Ca$_V$1.2 calcium channel dysfunction causes a multisystem disorder including arrhythmia and autism. *Cell, 119*, 19–31.

Takiyama, Y., Sakoe, K., Namekawa, M., Soutome, M., Esumi, E., Ogawa, T., Ishikawa, K., Mizusawa, H., Nakano, I., & Nishizawa, M. (1998). A Japanese family with spinocerebellar ataxia type 6 which includes three individuals homozygous for an expanded CAG repeat in the SCA6/CACNL1A4 gene. *Journal of the Neurological Sciences, 158*, 141–147.

Taylor, C. W., da Fonseca, P. C., & Morris, E. P. (2004). IP$_3$ receptors: The search for structure. *Trends in Biochemical Sciences, 29*, 210–219.

Toru, S., Murakoshi, T., Ishikawa, K., Saegusa, H., Fujigasaki, H., Uchihara, T., Nagayama, S., Osanai, M., Mizusawa, H., & Tanabe, T. (2000). Spinocerebellar ataxia type 6 mutation alters P-type calcium channel function. *The Journal of Biological Chemistry, 275*, 10893–10898.

Vajna, R., Klockner, U., Pereverzev, A., Weiergraber, M., Chen, X., Miljanich, G., Klugbauer, N., Hescheler, J., Perez-Reyes, E., & Schneider, T. (2001). Functional coupling between 'R-type' Ca^{2+} channels and insulin secretion in the insulinoma cell line INS-1. *European Journal of Biochemistry, 268*, 1066–1075.

van Belle, T. L., Coppieters, K. T., & von Herrath, M. G. (2011). Type 1 diabetes: Etiology, immunology, and therapeutic strategies. *Physiological Reviews, 91*, 79–118.

Waitkus-Edwards, K. R., Martinez-Lemus, L. A., Wu, X., Trzeciakowski, J. P., Davis, M. J., Davis, G. E., & Meininger, G. A. (2002). $\alpha_4\beta_1$ Integrin activation of L-type calcium channels in vascular smooth muscle causes arteriole vasoconstriction. *Circulation Research, 90*, 473–480.

Wajchenberg, B. L. (2007). β-Cell failure in diabetes and preservation by clinical treatment. *Endocrine Reviews, 28*, 187–218.

Wang, L., Bhattacharjee, A., Fu, J., & Li, M. (1996). Abnormally expressed low-voltage-activated calcium channels in β-cells from NOD mice and a related clonal cell line. *Diabetes, 45*, 1678–1683.

Wang, L., Bhattacharjee, A., Zuo, Z., Hu, F., Honkanen, R. E., Berggren, P. O., & Li, M. (1999). A low voltage-activated Ca^{2+} current mediates cytokine-induced pancreatic β-cell death. *Endocrinology, 140*, 1200–1204.

Wiser, O., Trus, M., Hernandez, A., Renstrom, E., Barg, S., Rorsman, P., & Atlas, D. (1999). The voltage sensitive L$_C$-type Ca^{2+} channel is functionally coupled to the exocytotic machinery. *Proceedings of the National Academy of Sciences of the United States of America, 96*, 248–253.

Wollheim, C. B., & Sharp, G. W. (1981). Regulation of insulin release by calcium. *Physiological Reviews, 61*, 914–973.

Wollheim, C. B., Kikuchi, M., Renold, A. E., & Sharp, G. W. (1978). The roles of intracellular and extracellular Ca^{++} in glucose-stimulated biphasic insulin release by rat islets. *The Journal of Clinical Investigation, 62*, 451–458.

Wu, X., Davis, G. E., Meininger, G. A., Wilson, E., & Davis, M. J. (2001). Regulation of the L-type calcium channel by $\alpha_5\beta_1$ integrin requires signaling between focal adhesion proteins. *The Journal of Biological Chemistry, 276*, 30285–30292.

Xu, S., Laccotripe, M., Huang, X., Rigotti, A., Zannis, V. I., & Krieger, M. (1997). Apolipoproteins of HDL can directly mediate binding to the scavenger receptor SR-BI, an HDL receptor that mediates selective lipid uptake. *Journal of Lipid Research, 38*, 1289–1298.

Xu, G., Chen, J., Jing, G., & Shalev, A. (2012). Preventing β-cell loss and diabetes with calcium channel blockers. *Diabetes, 61*, 848–856.

Yamada, Y., Kuroe, A., Li, Q., Someya, Y., Kubota, A., Ihara, Y., Tsuura, Y., & Seino, Y. (2001). Genomic variation in pancreatic ion channel genes in Japanese type 2 diabetic patients. *Diabetes/Metabolism Research and Reviews, 17*, 213–216.

Yang, S. N., & Berggren, P. O. (2005a). β-Cell Ca$_V$ channel regulation in physiology and pathophysiology. *American Journal of Physiology. Endocrinology and Metabolism, 288*, E16–E28.

Yang, S. N., & Berggren, P. O. (2005b). Ca$_V$2.3 channel and PKCλ: New players in insulin secretion. *The Journal of Clinical Investigation, 115*, 16–20.

Yang, S. N., & Berggren, P. O. (2006). The role of voltage-gated calcium channels in pancreatic β-cell physiology and pathophysiology. *Endocrine Reviews, 27*, 621–676.

Yang, S. N., Larsson, O., Branstrom, R., Bertorello, A. M., Leibiger, B., Leibiger, I. B., Moede, T., Kohler, M., Meister, B., & Berggren, P. O. (1999). Syntaxin 1 interacts with the L$_D$ subtype of voltage-gated Ca^{2+} channels in pancreatic β cells. *Proceedings of the National Academy of Sciences of the United States of America, 96*, 10164–10169.

Yang, S. N., Shi, Y., Yang, G., Li, Y., Yu, J., & Berggren, P. O. (2014). Ionic mechanisms in pancreatic β cell signaling. *Cellular and Molecular Life Sciences, 71*, 4149–4177.

Yang, G., Shi, Y., Yu, J., Li, Y., Yu, L., Welling, A., Hofmann, F., Striessnig, J., Juntti-Berggren, L., Berggren, P. O., et al. (2015). Ca$_V$1.2 and Ca$_V$1.3 channel hyperactivation in mouse islet β cells exposed to type 1 diabetic serum. *Cellular and Molecular Life Sciences, 72*, 1197–1207.

Yoshida, K., & Kikutani, H. (2000). Genetic and immunological basis of autoimmune diabetes in the NOD mouse. *Reviews in Immunogenetics, 2*, 140–146.

Yu, J., Shi, Y., Zhao, K., Yang, G., Yu, L., Li, Y., Andersson, E. M., Ammala, C., Yang, S. N., & Berggren, P. O. (2020). Enhanced expression of β cell Ca$_V$3.1 channels impairs insulin release and glucose homeostasis. *Proceedings of the National Academy of Sciences of the United States of America, 117*, 448–453.

Zaitsev, S. V., Appelskog, I. B., Kapelioukh, I. L., Yang, S. N., Kohler, M., Efendic, S., & Berggren, P. O. (2001). Imidazoline compounds protect against interleukin 1β-induced β-cell apoptosis. *Diabetes, 50*, S70–S76.

Zhang, W., Khan, A., Ostenson, C. G., Berggren, P. O., Efendic, S., & Meister, B. (2002). Down-regulated expression of exocytotic proteins in pancreatic islets of diabetic GK rats. *Biochemical and Biophysical Research Communications, 291*, 1038–1044.

The Skeletal Muscle Calcium Channel

Bernhard E. Flucher and Kurt G. Beam

Abstract

Ca$_V$1.1 is the voltage-gated calcium channel isoform uniquely expressed in skeletal muscle. Its primary function is that of the voltage sensor for excitation–contraction (EC) coupling. Activation of Ca$_V$1.1 voltage sensors upon membrane depolarization directly leads to opening of the ryanodine receptor (RyR1) and calcium release from intracellular stores, without the participation of calcium currents through Ca$_V$1.1. This function requires the precise organization of Ca$_V$1.1 in the skeletal muscle triads and its interactions with a strictly defined set of protein partners, comprising the auxiliary channel subunits $\alpha_2\delta$-1, β_{1a}, and γ_1;

the scaffold protein STAC3; and the RyR1 calcium release channel. Conversely, all these interaction partners define the unique properties of Ca$_V$1.1's L-type calcium currents (LTCC). Whereas calcium influx through Ca$_V$1.1 is expendable for EC coupling, it counteracts depletion of calcium stores and maintains contractile force during prolonged activation. Genetic variants of Ca$_V$1.1 altering calcium homeostasis cause a variety of skeletal muscle diseases.

Keywords

Voltage-gated calcium channel · Ca$_V$1.1 · Skeletal muscle · Excitation–contraction coupling · Voltage sensor

We dedicate this chapter to the memory of Jeanne Powell (1933–2002), whose insightful work revealed that the link between depolarization and calcium release was absent in the skeletal muscle of *dysgenic* mice, and who thus implemented an experimental model permitting homologous expression and analysis of Ca$_V$1.1 that greatly benefited the field, including the authors.

B. E. Flucher (✉)
Institute of Physiology, Medical University Innsbruck, Innsbruck, Austria
e-mail: bernhard.e.flucher@i-med.ac.at

K. G. Beam
Department of Physiology and Biophysics, University of Colorado, Anschutz Medical Campus, Aurora, CO, USA
e-mail: kurt.beam@cuanschutz.edu

Introduction

Ca$_V$1.1 is the best of all channels, and Ca$_V$1.1 the worst of all channels. For the experimentalist this certainly holds true. Biochemists love Ca$_V$1.1 because it is highly and exclusively expressed in the largest tissue of the vertebrate body, in skeletal muscle, and because it has a high-affinity drug binding site—both ideal properties for its biochemical purification and characterization. Consequently, the skeletal muscle Ca$_V$1.1 was the first calcium channel to be isolated (Curtis & Catterall, 1984), the first to be

© The Author(s), under exclusive license to Springer Nature Switzerland AG 2022
G. W. Zamponi, N. Weiss (eds.), *Voltage-Gated Calcium Channels*,
https://doi.org/10.1007/978-3-031-08881-0_16

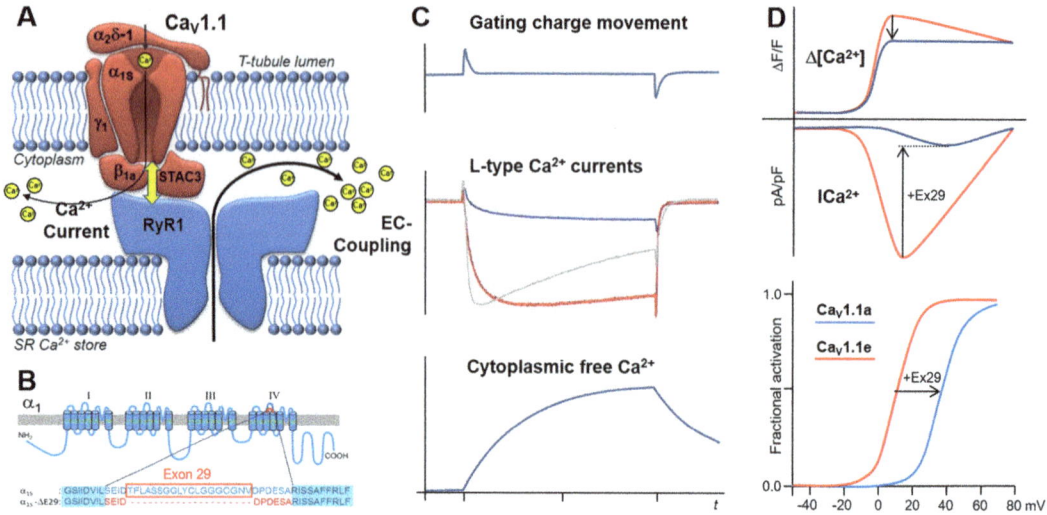

Fig. 1 The macromolecular $Ca_V1.1$ complex and its functions as calcium channel and voltage sensor of skeletal muscle EC coupling. (**a**) In the skeletal muscle triad, $Ca_V1.1$ functions depend on its association with a defined set of auxiliary proteins and the ryanodine receptor (RyR1). Conformational coupling of $Ca_V1.1$ and RyR1 (yellow double arrow) is independent of calcium influx but requires the β_{1a} subunit and STAC3. (**b**) Alternative splicing of exon 29 in the extracellular IVS3–S4 linker results in two functionally distinct channels. (**c**) $Ca_V1.1$'s action as voltage sensor can be assessed by measuring On- and Off-gating charges (top). Depolarization triggers slowly activating calcium currents (middle) and a massive release of calcium from sarcoplasmic reticulum (SR) calcium stores (bottom). (**d**) Inclusion of exon 29 reduces the current amplitude and shifts the voltage-dependence of activation to more positive potentials. Color code of traces: blue, $Ca_V1.1a$ (including exon 29); red, $Ca_V1.1e$ (lacking exon 29); and gray, $Ca_V1.2$ for comparison

cloned and sequenced (Tanabe et al., 1987), and the first member of the mammalian Ca_V/Na_V family for which the protein structure has been solved at atomic resolution (Wu et al., 2015, 2016). On the other hand, electrophysiologists hated this channel, because, contrary to the other nine members of the Ca_V family, $Ca_V1.1$ resisted heterologous expression and characterization in *Xenopus* oocytes and in mammalian cells. Whereas a handful of such studies were successful (Perez-Reyes et al., 1989; Lacerda et al., 1991; Varadi et al., 1991; Lory et al., 1992; Johnson et al., 1997; Ren & Hall, 1997), the general verdict after 30 years of calcium channel research was: $Ca_V1.1$ does not express well in heterologous cells. (But note that recently this has changed, as will be discussed in detail below.) Instead, functional analysis of $Ca_V1.1$ was either performed by measuring the intrinsic currents in muscle fibers of amphibians (Sanchez & Stefani, 1978) and mammals (Donaldson & Beam, 1983; Lamb & Walsh, 1987) or in cul-

tured myotubes (Beam et al., 1986; Cognard et al., 1986; Sipos et al., 1997). Analyses of recombinant $Ca_V1.1$ required its expression in myotubes of $Ca_V1.1$-null mutant (*dysgenic*) mice (Tanabe et al., 1988; Neuhuber et al., 1998). Naturally, this limited functional analysis of this channel isoform to a small number of labs specializing on skeletal muscle. Whereas the difficulties with heterologous expression of $Ca_V1.1$ complicated and even hampered electrophysiological characterization of this channel isoform, as for example simple co-expression studies were not easily possible, the fact that $Ca_V1.1$ has largely been characterized in the native environment of skeletal myotubes is also of great advantage. Particularly, because $Ca_V1.1$'s characteristic channel properties are shaped by its interaction partners in the skeletal muscle triad and because its primary physiological function is that of the voltage sensor for excitation–contraction (EC) coupling, wherein it closely interacts with other triad proteins.

The ambivalent nature of $Ca_V1.1$ is manifested not only in its conflicting accessibility to different experimental approaches, but equally in its dual functions as calcium channel and voltage sensor for EC coupling (Melzer et al., 1995) (Fig. 1). Of all Ca_V channels, the classical $Ca_V1.1$ isoform requires the strongest depolarization for activation and shows the slowest activation kinetics. In fact, it is such a poor channel that it probably opens very little during physiological activation of skeletal muscle and its physiological importance as calcium channel is disputable (Bannister & Beam, 2013a). On the other hand, $Ca_V1.1$ is a vital signaling protein, essential for activation of skeletal muscle contraction in response to action potentials. Muscles lacking $Ca_V1.1$ are paralyzed and loss of function, as in $Ca_V1.1$-null mice, leads to death at birth from respiratory failure (Powell, 1990). The function of $Ca_V1.1$ in skeletal muscle EC coupling relies on its voltage sensors and its physical communication with the calcium release channel (type 1 ryanodine receptor—RyR1), but does not require its channel function (Rios & Brum, 1987; Melzer et al., 1995). Actually, in fish skeletal muscle $Ca_V1.1s$ have lost the ability to conduct calcium entirely (Schredelseker et al., 2010). Nevertheless, mammalian $Ca_V1.1$ can also form a proper calcium channel. Alternative splicing resulting in the exclusion of exon 29 normalizes the gating properties of $Ca_V1.1$ such that it conducts currents very similar to those of the neuronal/cardiac $Ca_V1.2$ (Tuluc et al., 2009). This process is developmentally regulated, so that alternative splicing of a single exon switches $Ca_V1.1$ from a perfect calcium channel into a channel with poor gating properties at the period around birth. Interestingly, the voltage-sensor function of $Ca_V1.1$ is not affected by this isoform switch, as both the embryonic $Ca_V1.1e$ and the adult $Ca_V1.1a$ isoforms equally support EC coupling (Tuluc & Flucher, 2011).

Therefore, in one sense $Ca_V1.1$ is prototypical for its family, in that much of what we know about voltage-gated calcium channels has first been detected in the skeletal muscle channel. At the same time, $Ca_V1.1$ is atypical in that its gating properties make it a poor channel and its channel function is non-essential for its primary physiological role in skeletal muscle EC coupling. For the experimentalist, this ambivalent nature holds specific challenges as well as unique opportunities.

$Ca_V1.1$ Voltage Sensing

Both physiological functions of the skeletal muscle $Ca_V1.1$, channel gating and activation of EC coupling, depend on its ability to sense changes in the membrane potential. Voltage sensing is a fundamental property of all voltage-gated ion channels. In their classical work, Hodgkin and Huxley (Hodgkin & Huxley, 1952) predicted that voltage-dependent processes (i.e., channel gating or activation of EC coupling) should be controlled by mobile charges trapped in the plane of the membrane, and that, upon changes in the transmembrane potential, displacement of these charges should give rise to measurable currents. The experimental validation of this prediction was achieved only about 20 years later. Using a three-electrode voltage-clamp configuration in frog skeletal muscle fibers and a pulse protocol designed to isolate the non-linear component of the capacitative currents, Schneider and Chandler (Schneider & Chandler, 1973) succeeded in measuring such charge-displacement currents on depolarization and repolarization of muscle fibers. Today, we know that these gating currents correspond to the activation and deactivation of the $Ca_V1.1$ voltage sensors. So even before its molecular identity had been discovered, $Ca_V1.1$ was the first voltage-gated ion channel for which this characteristic property has been recorded.[1]

The charge-displacement current recorded in skeletal muscle fulfilled the criteria expected for ion channel gating currents. It moved outward upon membrane depolarization (ON-gating current) and inward upon repolarization (OFF-gating current), the amount of moved charge was equal

---

[1]Note that within a month of this publication, the seminal work of Armstrong and Bezanilla on gating currents of sodium channels in squid axons was reported (Armstrong & Bezanilla, 1973).

in both directions (charge conservation), and it increased in a sigmoidal manner (i.e., it saturated) in response to increasing depolarizations. All these properties are consistent with the back-and-forth movement of a fixed number of charges trapped in the plane of the membrane. Subsequent studies demonstrated that the properties of the gating currents measured in skeletal muscle fibers closely matched the activation properties of contractions and myoplasmic calcium signals. Over a wide range of test pulses, the amount of moved charge and the maximal rate of calcium release from the sarcoplasmic reticulum (SR) increased linearly with each other (Melzer et al., 1986; Lamb & Walsh, 1987). Thus, the gating currents in skeletal muscle fibers represent the action of the voltage sensor of EC coupling (recently reviewed in Hernández-Ochoa & Schneider, 2018).

In the late 1980s, several lines of research converged to reveal the molecular identity of the voltage sensor of EC coupling. Rios and Brum (Rios & Brum, 1987) demonstrated that charge movements and calcium release from the sarcoplasmic reticulum are inhibited in parallel by low concentrations of dihydropyridines, known inhibitors of L-type calcium currents. The electrophysiological characterization of a mutant mouse model with failing skeletal muscle EC coupling, revealed that these *dysgenic* mice specifically lack L-type calcium currents in skeletal muscle (Klaus et al., 1983; Beam et al., 1986). Finally, cloning and sequencing of the skeletal muscle dihydropyridine receptor showed that its domain structure closely resembled that of the previously cloned voltage-gated sodium channel (Tanabe et al., 1987). Together, these findings strongly indicated that the voltage sensor for EC coupling, the skeletal muscle dihydropyridine receptor, and the skeletal muscle L-type calcium channel are all the same molecule. This notion was ultimately confirmed when all these functions could be restored in *dysgenic* ($Ca_V1.1$-null) myotubes, by injecting the cDNA of the newly cloned $Ca_V1.1$ (Tanabe et al., 1988; Adams et al., 1990). Thus, the skeletal muscle dihydropyridine receptor (i.e., $Ca_V1.1$) was

identified as the molecular entity constituting the voltage sensor of EC coupling and of skeletal muscle L-type calcium currents.

Based on structural evidence (discussed below) it is likely that four independently functioning $Ca_V1.1$ channels are associated with one RyR1 calcium release channel. Accordingly, a quantitative allosteric model has been proposed according to which the independent action of each of the four $Ca_V1.1$ (voltage sensors) equally contributes to increasing the open probability of the calcium release channel (Ríos et al., 1993). This model accurately describes the relationship of voltage sensing and calcium release experimentally determined in muscle fibers. However, considering the known structure of voltage-gated calcium channels, with four independently functioning voltage-sensing domains (VSDs) each, this mechanistic model needs to be reassessed. The assumptions of the original model are fulfilled only, if either only a single VSD of each of the four $Ca_V1.1$ controls RyR1 opening, or if all VSDs in a single $Ca_V1.1$ that are functionally coupled to the RyR1 move in concert and thus act like one. Yet, molecularly and functionally the four VSDs of $Ca_V1.1$ are rather diverse (Fernández-Quintero et al., 2021; El Ghaleb et al., 2021). The question as to how many and which of the four VSDs of $Ca_V1.1$ are involved in the regulation of EC coupling is subject to current investigations in several laboratories (Flucher & Tuluc, 2017; Banks et al., 2021; Savalli et al., 2021a) and is discussed further below.

$Ca_V1.1$ in Skeletal Muscle Triads

Before discussing the function of $Ca_V1.1$ in EC coupling, it is important to briefly address its unique localization and molecular interactions in skeletal muscle triads. Triads are junctions between a transverse (T-) tubule and two terminal cisternae of the sarcoplasmic reticulum (SR). T-tubules are tubular extensions of the plasma membrane forming the tracks for the rapid propagation of action potentials into the depth of the muscle fiber. Terminal SR cisternae are the intracellular calcium storage and release organelles,

Skeletal muscle triad - junctional membranes

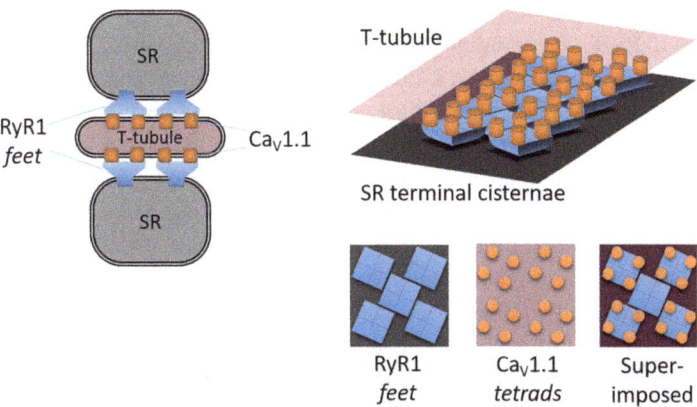

Fig. 2 $Ca_V1.1$ is localized in the T-tubule membrane of skeletal muscle triads (left), which are junctions between one T-tubule, functioning as track for the propagation of action potentials into the muscle fiber, and two terminal SR cisternae, functioning as calcium storage and release organelles. In electron microscopy preparations, the cytoplasmic domains of RyR1, called feet, display a tilted checker-board array, and $Ca_V1.1$ assemble in groups of four, called tetrads, matching the RyR1 arrays in size and orientation (right). This coordinated arrangement supports conformational coupling between four $Ca_V1.1$ (voltage sensors) and each RyR1 homo-tetramer (calcium release channel)

containing the calcium-binding protein calsequestrin in its lumen and the calcium release channel (RyR1) in its membrane. $Ca_V1.1$ is highly concentrated in the junctional face of the T-tubules,[2] in direct opposition to RyR1 in the junctional SR membrane (Fig. 2). Remarkably, its ultrastructural arrangement closely matches that of the RyR1 (Franzini-Armstrong, 2018). The large homo-tetrameric calcium release channels are organized in parallel, laterally shifted rows, extending their large cytoplasmic domains (called feet) toward the T-tubule. In freeze-fracture preparations of junctional T-tubules groups of four integral membrane particles (called tetrads) are observed, which match the RyR1 arrays in shape and disposition (Block et al., 1988). These integral membrane particles

---

[2]Note that particularly in immature muscles functionally analogous junctions are formed between a single SR cisternae and T-tubule (called dyads) or the plasma membrane (called peripheral junctions). These junctions resemble triads with respect to their molecular composition and function in EC coupling. For simplicity reasons we use the term triad inclusively for all of these junctions.

are lacking in *dysgenic* muscles, but re-appear when such $Ca_V1.1$-null myotubes are reconstituted with recombinant $Ca_V1.1$—thus, identifying them as $Ca_V1.1$ (Franzini-Armstrong et al., 1991; Takekura et al., 1995). Importantly, this matching ultrastructural arrangement of $Ca_V1.1$ and RyR1 is specific for skeletal muscle. Whereas in heart muscle the analogous isoforms, $Ca_V1.2$ and RyR2, are also concentrated in the opposing membranes of the analogous plasma membrane–SR junctions, $Ca_V1.2$ lacks the organization in tetrads (Franzini-Armstrong, 2018). Consequently, the unique array of $Ca_V1.1$ tetrads in skeletal muscle has been interpreted as the structural basis for a physical/conformational coupling between the voltage sensor and calcium release channel necessary for skeletal muscle-type EC coupling. This intimate structural and functional coupling of $Ca_V1.1$ with the skeletal muscle RyR1 is further supported by findings showing that pharmacologically blocking RyR1 with high concentrations of ryanodine reduced the spacing of $Ca_V1.1$ particles in the tetrads and shifts the voltage-dependence of voltage-sensing and calcium current activation to more hyperpolarizing potentials (Balog & Gallant, 1999; Paolini et al., 2004; Bannister & Beam, 2009).

Both $Ca_V1.1$ and RyR1 accumulate in the triads in the absence of the respective other channel (Flucher et al., 1993; Takekura et al., 1995), indicating that their incorporation in the T-tubule–SR junction depends on other triad proteins. Junctophilins are primary candidates for this targeting function, as their knockdown in cultured myotubes inhibits $Ca_V1.1$ clustering in the triads (Nakada et al., 2018) and co-expression of junctophilin 2 is essential for the reconstitution of junctional EC coupling complexes in heterologous cells (Perni et al., 2017). Notably, the carboxy-terminus of $Ca_V1.1$ contains sequences necessary for triad targeting as well as for binding to junctophilins (Flucher et al., 2000; Nakada et al., 2018).

While RyR1 is not necessary for triad targeting of $Ca_V1.1$, it is required for tetrad formation. $Ca_V1.1$ tetrads are absent from triads in RyR1 knockout mice (Takekura et al., 1995), indicative of important role of RyR1 in establishing a physical interaction with the voltage sensor. Yet, $Ca_V1.1$ similarly contributes to this interaction. While expression of $Ca_V1.1$ in *dysgenic* myotubes restores tetrad formation and skeletal muscle EC coupling, expression of the closely related $Ca_V1.2$ isoform does not (Takekura et al., 1994; Nakai et al., 1998b). Apparently, specific sequences in the skeletal muscle isoforms of both channels are required for their characteristic structural and functional interaction. This is further highlighted by the finding that both, skeletal muscle-type EC coupling and the organization of $Ca_V1.1$ in tetrads, require the same "critical" sequences in a cytoplasmic loop of $Ca_V1.1$, as well as the same isoform of the auxiliary β subunit (Takekura et al., 2004; Schredelseker et al., 2005). Together, these results demonstrate that the function of $Ca_V1.1$ as voltage sensor of EC coupling depends on its highly specific structural organization within the skeletal muscle triads and on its specific interactions with a unique set of triad proteins.

Skeletal Muscle Excitation–Contraction Coupling

As already mentioned, the most important role of $Ca_V1.1$ in skeletal muscle is to control the release of calcium from the SR via RyR1 (that is, skeletal-type EC coupling calcium release). Unlike cardiac muscle, SR calcium release in skeletal muscle does not depend upon the entry of extracellular calcium, which has been demonstrated by removal of extracellular calcium (Armstrong et al., 1972) and by the ability of $Ca_V1.1$ to elicit such release when it contains mutations that eliminate calcium permeability of the pore (Dirksen & Beam, 1999; Dayal et al., 2010, 2017). Furthermore, with wild-type $Ca_V1.1$ the amplitude of the cytoplasmic calcium transient elicited by voltage-clamp steps increases sigmoidally with potential, reaches a plateau, and does not decline even for strong depolarizations beyond the calcium reversal potential at which entry of extracellular calcium should be abolished (García & Beam, 1994). Of course, the sigmoidal voltage-dependence of SR calcium release does not eliminate another possibility, which is that $Ca_V1.1$ activated a second messenger pathway, which then caused the activation of RyR1. However, even at 16 °C, the activation and deactivation of RyR1 in response to an action potential occur within a few milliseconds in mammalian fast twitch fibers (Hollingworth & Baylor, 2013). Thus, the rates of generation and breakdown of such a hypothetical second messenger would have to be rapid, even at temperatures far below physiological conditions, and also coordinately regulated to produce a sigmoidal response amplitude. Accordingly, most investigators believe that there is some sort of conformational coupling between $Ca_V1.1$ and RyR1—although the exact molecular mechanism still remains elusive.

Conformational coupling between the voltage sensor in the T-tubules and the calcium release channel in the SR could be direct or indirect and encompass a number of possible mechanisms. For example, one or more parts of the $Ca_V1.1$ complex could push or pull at domains of the RyR1, thereby gating its channel. Another possibility is that depolarization could cause domains of the $Ca_V1.1$ complex to adopt a conformation that functions as a RyR1-activating ligand. Thus, conformational coupling of components of the $Ca_V1.1$ complex with RyR1 does not necessarily require biochemical binding. It could just as well be accomplished by repulsive or state-dependent interactions, which might explain the persistent failure to identify mutual binding sites. In any case, either of the mechanisms mentioned above would require that the $Ca_V1.1$ complex be precisely oriented with respect to RyR1 in both the resting and activated states. The tetradic arrangement of $Ca_V1.1$, described earlier, indicates that this requirement is met in skeletal muscle triads and that $Ca_V1.1$ is physically linked to RyR1 (Block et al., 1988; Paolini et al., 2004). A considerable amount of work has been devoted to trying to identify the links between cytoplasmic domains of the $Ca_V1.1$ complex and RyR1.

Early work focused on $Ca_V1.1$ itself and involved the expression of Ca_V chimeric constructs in *dysgenic* myotubes, null for endogenous $Ca_V1.1$. These chimeras were based on $Ca_V1.1$ and either $Ca_V1.2$ (the cardiac L-type calcium channel) or α_{1M} (the fly muscle calcium channel), because neither of these latter two channels support skeletal-type calcium release. These experiments demonstrated that skeletal-type calcium release depended on the presence of a segment of the cytoplasmic loop connecting repeats II and III of $Ca_V1.1$ (Nakai et al., 1998b; Wilkens et al., 2001; Kugler et al., 2004b) (Fig. 3). In the earliest study (Nakai et al., 1998b), residues 720–765 (the "critical domain") were found to be sufficient for full restoration of skeletal-type calcium release, but the apparent boundaries of this segment depend somewhat on the chimeric construct, most likely because there is structural conservation between some regions of $Ca_V1.1$ and either $Ca_V1.2$ or α_{1M} (Kugler et al., 2004b). Based on the expression of similar constructs in *dysgenic* myotubes, the critical domain was also found to be important for the organization of $Ca_V1.1$ into tetrads (Takekura et al., 1994). Interestingly, a construct ($Ca_V1.1$ with a II–III loop replaced by that of α_{1M}) formed tetrads but did not support skeletal-type calcium release, indicating that although tetrads may be necessary

for skeletal-type calcium release, they are not sufficient.

Like $Ca_V1.1$, the β_{1a} auxiliary subunit is essential for both skeletal-type calcium release and tetrad formation (Coronado et al., 2004). Thus, $Ca_V1.1$ is moderately expressed in relaxed (β_1-null) zebrafish, but does not support calcium release or form tetrads (Schredelseker et al., 2005). The C-terminus of β_{1a} appears to be particularly important for both features as demonstrated for calcium release in β_1-null mouse myotubes (Beurg et al., 1999) and for calcium release and tetrads in relaxed (β_1-null) zebrafish (Dayal et al., 2013). In the latter work, it was also found that the neuronal β_3 isoform could restore tetrads but not calcium release. Like $Ca_V1.1$ and β_{1a}, the accessory protein STAC3 is also essential for EC coupling calcium release, which is absent in both myotubes from mice null for STAC3 (Nelson et al., 2013; Polster et al., 2016) and in Stac3-null zebrafish (Horstick et al., 2013; Linsley et al., 2016). Because its role was only discovered quite recently, our knowledge is still limited about the domains of STAC3 that are important for signaling from $Ca_V1.1$ to RyR1. Intriguingly, the highest affinity interaction appears to occur between the first of the two SH3 domains in STAC3 and residues 749–758 of the "critical domain" of $Ca_V1.1$ (Wong King Yuen et al., 2017; Polster et al., 2018b). However, other regions of STAC3 are also involved as indicated by the reduction in calcium release as a result of substituting either STAC1 or STAC2 for STAC3 (Polster et al., 2018b) and of mutations in the second SH3 domain (Rufenach et al., 2020). Additionally, regions of $Ca_V1.1$ other than the II–III loop are likely to be involved in interactions with STAC3 (Niu et al., 2018).

Together, these findings indicate that conformational coupling of $Ca_V1.1$ and RyR1 encompasses complex interactions of three proteins: $Ca_V1.1$, β_{1a}, and STAC3. Because these proteins, plus RyR1 and junctophilin 2, were sufficient to reconstitute skeletal muscle-type EC coupling in HEK cells, we can be confident that no other muscle-specific proteins are critically involved in this interaction (Perni et al., 2017). Although the kinds of experiments described

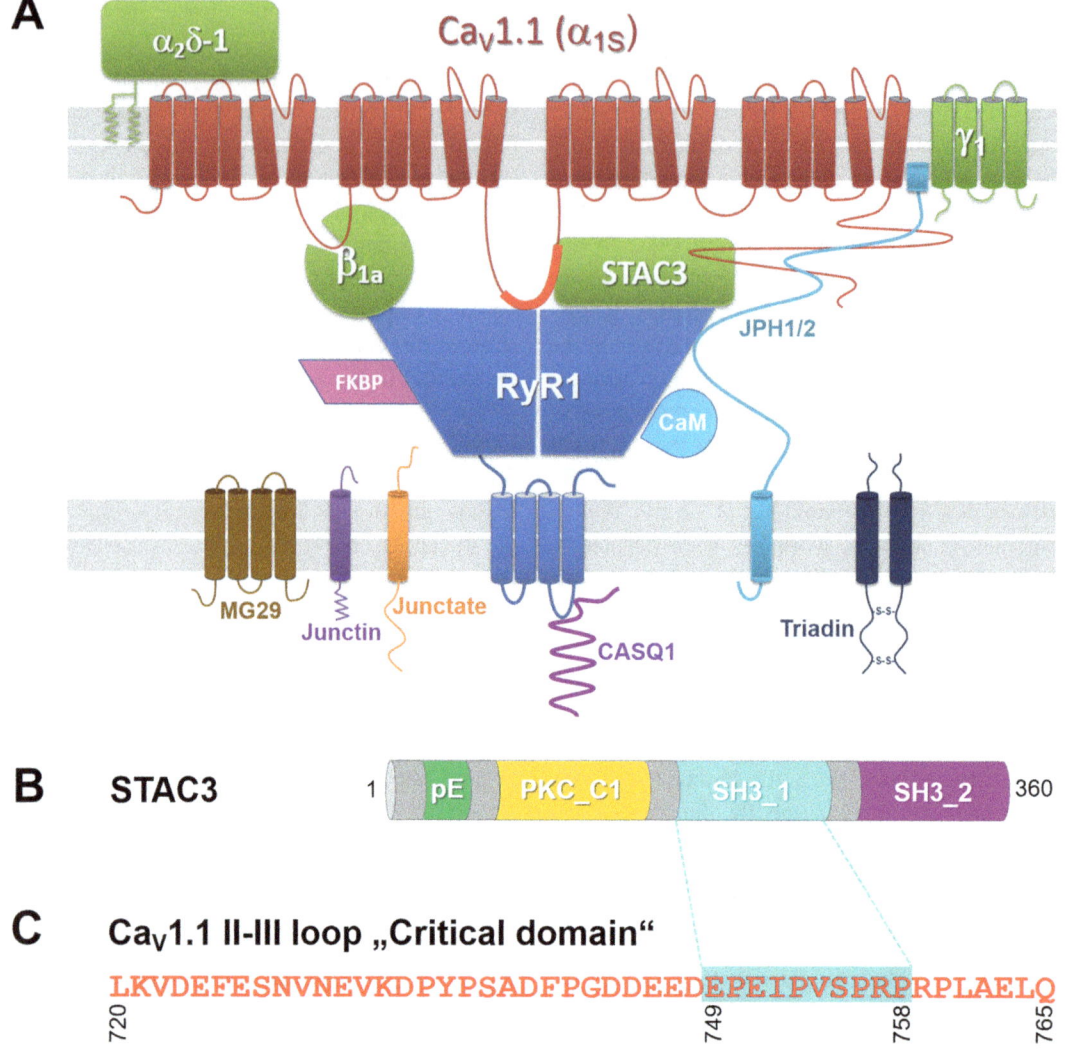

B **STAC3**

C **Ca$_V$1.1 II-III loop „Critical domain"**

LKVDEFESNVNEVKDPYPSADFPGDDEEDEPEIPVSPRPRPLAELQ

720   749   758   765

Fig. 3 Triad junction proteins co-localizing with Ca$_V$1.1. (a) Schematic illustration of some of the proteins that localize at triad junctions in skeletal muscle (MG29, mitsugumin29; CASQ1, calsequestrin1; JPH1/2, junctophilins 1 and 2; FKBP12, FK506 binding protein12; CaM, calmodulin). Although these proteins are important for regulating EC coupling calcium release, only five proteins are sufficient to couple depolarization to calcium release from the sarcoplasmic reticulum: Ca$_V$1.1, β_{1a}, STAC3, JPH2, and RyR1. In the diagram, β_{1a}, the "critical domain" of the cytoplasmic loop connecting repeats II and III of Ca$_V$1.1 (shown in red), and STAC3 are all shown as contacting RyR, because they have all been shown to have a direct role in conformational coupling between Ca$_V$1.1 and RyR1, but whether this is the case remains unknown. Interactions between the junctophilins and both Ca$_V$1.1 and RyR1 have also been described; it is not known whether these are directly involved in conformational coupling. (b) Domain structure of STAC3. (c) Sequence of the "critical domain" of the cytoplasmic loop connecting repeats II and III of Ca$_V$1.1. The binding site for the SH3-1 domain of STAC3 is highlighted

above are invaluable for identifying sites of importance in Ca$_V$1.1, β_{1a}, and STAC3, these sites may, or may not, be in direct contact with RyR1. Clearly, to identify sites of contact, it will be necessary to obtain structure of the key proteins assembled as a functioning unit. It is to be hoped that such structure will be forthcoming in the near future.

The (Enigmatic) Role of LTCC in Skeletal Muscle

Compared to the current properties of the other members of the Ca_V family of calcium channels, $Ca_V1.1$ makes a poor channel. Even when expressed at high levels in skeletal muscle cells, its current density remains small, due to a very low open probability (Dirksen & Beam, 1995, 1996). It is the Ca_V channel requiring the most positive potentials for current activation. In fact, currents through $Ca_V1.1$ are activated only at about 30 mV more positive potentials than EC coupling (Melzer et al., 1995). $Ca_V1.1$ currents activate very slowly and show little inactivation. Altogether, these gating properties raised doubt as to whether $Ca_V1.1$ channels would open at all during the brief depolarizations of skeletal muscle action potentials. Notably, however, $Ca_V1.1$ comes in two splice-variants with strikingly different gating properties. Exclusion of the alternatively spliced exon 29 in the embryonic splice variant ($Ca_V1.1e$) substantially left-shifts the voltage-dependence of activation and increases the current amplitude so that the resulting whole cell currents are similar to those of the cardiac/neuronal $Ca_V1.2$ (as will be discussed in detail below) (Tuluc et al., 2009; Tuluc & Flucher, 2011). Altogether, the physiological importance of $Ca_V1.1$ as voltage-gated calcium channel has been and still is a highly controversial issue.

As mentioned before, calcium currents through $Ca_V1.1$ are expandable for skeletal muscle EC coupling in most vertebrates. Whereas EC coupling in heart and smooth muscle requires extracellular calcium and its influx via LTCCs into the muscle cell (Bers, 2002), the seminal experiment of Armstrong, Bezanilla, and Horowicz (Armstrong et al., 1972) for the first time demonstrated in frog skeletal muscle that it continues to contract for up to 20 min in the absence of extracellular free calcium. Subsequent studies, aimed at unraveling the mechanism of skeletal muscle EC coupling and using various experimental paradigms and biological preparations, corroborated these findings and ascertained that skeletal muscle EC coupling is independent of calcium currents through $Ca_V1.1$ (Melzer et al., 1995).

Almost five decades and many studies later, knock-in mouse models provided what could be considered the ultimate proof of this notion. Two laboratories independently generated mice in which the pore of $Ca_V1.1$ was made non-conductive for calcium. In the mice from the Texan group, a mutation was introduced in $Ca_V1.1$ that had previously been shown to reduce the affinity of the selectivity filter for calcium and shift its ion selectivity to monovalent ions (Georgiou et al., 2015). In the Austrian mice a mutation in the selectivity filter was introduced, which increased the affinity for calcium, resulting in a complete pore block of the $Ca_V1.1$ channel (Dayal et al., 2017). Importantly, in both mice skeletal muscle EC coupling was normal. Also, the overall development was normal in both mouse models. Only the Texan mice showed a decrease in calcium-dependent enzyme activity, an increased proportion of fast IIb muscle fiber types, and increased fatigability (Georgiou et al., 2015). Thus, calcium currents through $Ca_V1.1$ are expandable not only for the process of EC coupling, but also for normal development of skeletal muscle and for long-term maintenance of motor activity in mice.

Phylogenetically, the role of calcium currents in skeletal muscle function shows at least three distinct qualities. In invertebrates, the mechanism of skeletal muscle EC coupling is calcium-induced calcium release. Similar to mammalian cardiac muscles, these skeletal muscles require calcium influx through the voltage-gated calcium channel to trigger further SR calcium release and thus activate muscle contraction. As explained in detail above, in mammalian skeletal muscles, EC coupling is independent of calcium influx. $Ca_V1.1$ currents exist, but are small and ineffective in triggering contractions. Finally, in teleost fish skeletal muscles, genetic variations in the selectivity filter of $Ca_V1.1$ render the calcium channel non-conductive, turning $Ca_V1.1$ into a dedicated voltage sensor of skeletal muscle EC coupling. In fact, fish possess gene duplications of $Ca_V1.1$, and both copies contain current-blocking amino acid substitutions in their selectivity filter, albeit distinct ones (Di Biase & Franzini-Armstrong, 2005; Schredelseker et al., 2010). This phylogenetic progression indicates that, in parallel to the increasing differentiation and specialization of skeletal muscle, $Ca_V1.1$ exchanged its calcium channel function for an exclusive voltage-sensor function. Apparently, calcium influx not only is expendable for skeletal muscle function, but there may even be an advantage of shutting down the calcium currents and exclusively relying on intracellular cycling of calcium for muscle contraction. So, are the calcium currents in mammalian muscle just an "evolutionary remnant" half-way to perfection of skeletal muscle EC coupling in fish muscles, or do they play some role in normal physiology, even if not an essential one?

There are several arguments for a physiological role of $Ca_V1.1$ currents in mammalian skeletal muscle. The conviction, that things in nature do not exist without reason, is one of them, although not easily accessible to experimental testing. Yet many researchers observed that, although contractions in mammalian muscles can be elicited in the absence of extracellular calcium or when currents have been blocked by pharmacological agents or blocking divalent cations, such contractile activity cannot be sustained for long (Spiecker et al., 1979; Brum et al., 1988; Dulhunty & Gage, 1988). Upon prolonged membrane depolarizations or during repetitive trains of action potentials, the contractile force and the amplitude of the calcium transients dwindle. Apparently, efficient refilling of the SR calcium stores requires some influx from the extracellular compartment and LTCCs may contribute to this aspect of calcium homeostasis (Robin & Allard, 2013). In fact, a $Ca_V1.1$-dependent calcium influx associated with prolonged depolarization or pulse trains has been described that was called excitation-coupled calcium entry (ECCE) (Bannister et al., 2009; Bannister & Beam, 2013a). ECCE is dependent on the presence of both $Ca_V1.1$ and RyR1 and it shows a broad pharmacological profile for drugs affecting store-operated currents, RyR1-dependent calcium release, as well as L-type calcium currents. However, all these drugs were shown to similarly affect $Ca_V1.1$ currents, indicating that ECCE represents the physiological role of $Ca_V1.1$ during prolonged activity of skeletal muscle (Bannister et al., 2009).

Independently, the existence of a small but significant calcium influx has been characterized using the highly sensitive manganese-quenching technique in isolated muscle fibers subjected to continuous action potential firing (Robin & Allard, 2015). This calcium influx could be blocked with lanthanum or the L-type channel blocker nifedipine, and this block resulted in an accelerated decline of the calcium signals. Together, these experiments support a role of $Ca_V1.1$ currents in normal physiology of mammalian skeletal muscles important for maintaining sufficient SR calcium loading and contractile force during prolonged activity.

The apparently contradictory observation that in the knock-in mice carrying non-conducting $Ca_V1.1$ channels skeletal, muscle function, and particularly endurance are unaltered compared to normal controls (Dayal et al., 2017) could be explained by compensatory mechanisms. First, the skeletal muscle is a highly plastic tissue that adjusts its size and composition to the need. We grow just as much muscle as our regular activity level requires. Thus, reduced strength or endurance of muscles, due to inefficient calcium store refilling, could be compensated by additional muscle growth and adjusted fiber composition, although an altered fiber-type composition has only been noted in one of the two non-conducting $Ca_V1.1$ mouse models (Georgiou et al., 2015; Dayal et al., 2017). Alternatively, compensation at the molecular level, like strengthening of the store-operated calcium entry, could explain the normal phenotype in the non-conducting $Ca_V1.1$ mice. In any case, experimental evidence supports a function of the small LTCCs in sustaining calcium filling of the SR stores and thus sustaining the force during prolonged activation of mammalian skeletal muscle.

Of equal importance to the question as to the physiological function of $Ca_V1.1$ currents is the question as to whether and why calcium currents through $Ca_V1.1$ need to be curtailed for proper skeletal muscle function (Flucher & Tuluc, 2017). Intuitively, one would assume that for a process like muscle contraction that depends on the regulation by cytoplasmic calcium signals, any additional calcium flux into the cytoplasm should be welcome. Nevertheless, it appears as if calcium influx through $Ca_V1.1$ is actively minimized, both phylogenetically as muscle becomes increasingly specialized, and ontogenetically by developmentally regulated alternative splicing.

The Structural Basis of $Ca_V1.1$ Function

Associated Proteins

The skeletal muscle voltage-gated calcium channel and voltage sensor for EC coupling is comprised of multiple proteins, all of which contribute to its specific functional properties (Fig. 4). In addition to the $Ca_V1.1$ α_{1S} subunit, which contains the voltage sensors and the channel pore, the functional channel complex contains the classical auxiliary channel subunits $\alpha_2\delta$-1, β_{1a}, and γ_1; the muscle-specific scaffold protein STAC3; and the SR calcium release channel RyR1. In variance to other excitable cells, there is no evidence that any other isoforms of the pore-forming or auxiliary Ca_V subunits are expressed in skeletal muscle and functionally contribute to its function. This rigid subunit composition is well in line with the concept that in skeletal muscle, $Ca_V1.1$ is part of a single-purpose molecular machine that does not require or even tolerate much molecular diversity.

Also consistent with this concept, β_{1a}, STAC3, and RyR1 are all essential for $Ca_V1.1$'s primary function in skeletal muscle EC coupling. Muscle-specific knockout of any one of these proteins results in non-contractile muscles and the homozygous knockout animals die at birth of respiratory failure (Takeshima et al., 1996; Gregg et al., 1996; Horstick et al., 2013; Nelson et al., 2013). Conversely, β_{1a}, STAC3, and RyR1 are sufficient to restore skeletal muscle EC coupling in non-muscle cells (plus junctophilin 2 needed for establishing plasma membrane SR junctions) (Perni et al., 2017). Thus, the minimal complement of the $Ca_V1.1$ EC coupling machine is $Ca_V1.1$, β_{1a}, STAC3, and RyR1. The molecular requirements for the $Ca_V1.1$ channel function are similar, but not identical. In skeletal muscle,

knockout of β_{1a}, STAC3, and RyR1 reduces $Ca_V1.1$ calcium currents to different degrees but does not completely abolish them. For β_{1a} and STAC3, this most likely reflects their important roles in membrane targeting of the channel. In the case of the RyR1, the reduced current density and shifted voltage-dependence observed in RyR1 knockout myotubes reflect a direct-current promoting effect of RyR1 on $Ca_V1.1$, termed retrograde coupling (Nakai et al., 1996; Avila et al., 2001). Conversely, functional expression of $Ca_V1.1$ currents in non-muscle cells requires β_{1a} and STAC3, but not RyR1 (Polster et al., 2015).

As to the specific functions of the associated proteins, the cytoplasmic β_{1a} subunit promotes functional membrane expression of $Ca_V1.1$ (Arikkath & Campbell, 2003; Buraei & Yang, 2010). In skeletal muscle, β_{1a} requires association with $Ca_V1.1$ for its own incorporation into the triads (Neuhuber et al., 1998), where it functions in the structural organization of $Ca_V1.1$ in the tetrads opposite RyR1 (Schredelseker et al., 2005). As tetrad formation is a prerequisite for skeletal muscle EC coupling, this defect in itself explains the failure of EC coupling in β_{1a} knockout mice. A direct function of β_{1a} in linking the voltage sensor ($Ca_V1.1$) with the calcium release channel (RyR1) has been proposed (Coronado et al., 2004), but lacks experimental confirmation.

Co-expression of STAC3 and β_{1a} turned out to be the key for functional expression of $Ca_V1.1$ in heterologous cells (Polster et al., 2015). Also, in muscle cells, STAC3 promotes membrane expression of $Ca_V1.1$. However, STAC proteins interact with two functionally important domains of Ca_V1 channels (Wong King Yuen et al., 2017; Campiglio et al., 2018; Polster et al., 2018b; Niu et al., 2018). Structure–function analysis demonstrated an interaction of STAC3 with sequences in the cytoplasmic loop connecting repeats II and III of $Ca_V1.1$ that previously have been shown to be essential for skeletal muscle-specific EC coupling (Wong King Yuen et al., 2017). Mutations on either side of this interaction site perturb depolarization-induced calcium release in skeletal myotubes (Polster et al., 2016; Wong King Yuen et al., 2017; Rufenach et al., 2020), thus indicating a central function of STAC3 in the sig-

Ca$_V$1.1 functional domains

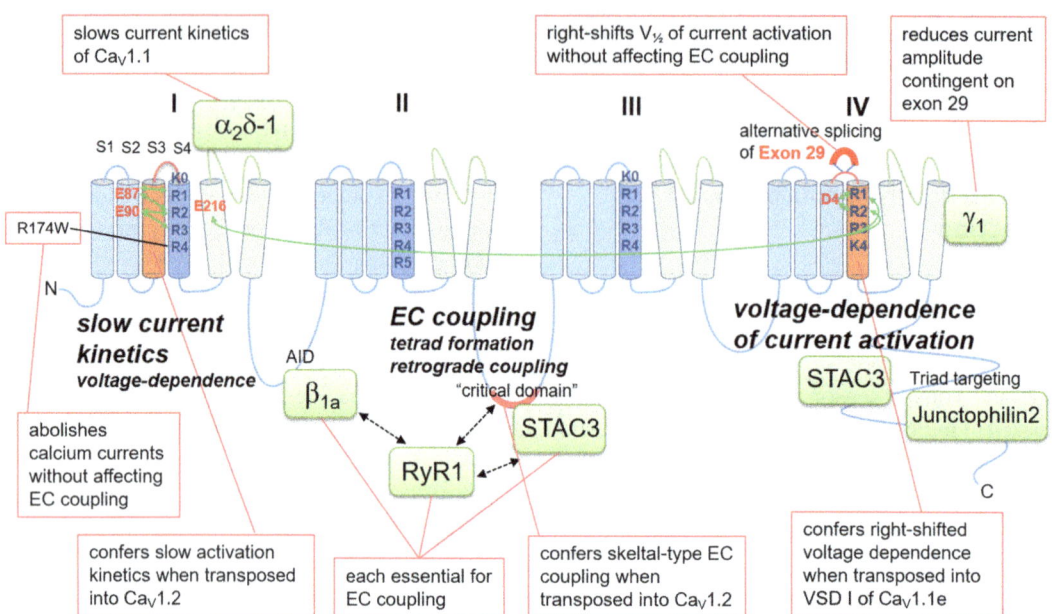

Fig. 4 Domain structure of Ca$_V$1.1 consisting of four homologous repeats (I–IV), each with a separate VSD (S1–S4, blue/orange) and a PD (S4–S5). The S4 segments of the four VSDs contain different numbers of positive gating charges (R/K). These form ion-pairs (green double-arrows) with negative countercharges (E/D) in the S2 and S3 helices within the respective VSD and, in the case of VSD IV, with a glutamate (E216) in the adjacent PD of repeat I. These ionic interactions are essential for determining the kinetics and voltage-dependence of current activation. The minimal essential sequences determining these gating properties comprise the S3–S4 linker and either S3 or S4 of VSD I and IV, respectively (orange). Auxiliary subunits and associated proteins bind at particular intracellular or extracellular loops and thus regulate Ca$_V$1.1 current properties and skeletal muscle EC coupling. Existing data (given in boxes) indicate that VSDs I and IV determine specific aspects of current gating, whereas VSD II and/or III might control EC coupling

naling between Ca$_V$1.1 and RyR1 (Flucher & Campiglio, 2019). Whether STAC3 also directly interacts with RyR1, and thus establishes a direct link between the voltage sensor and effector of EC coupling, remains to be demonstrated. In addition, STAC3 interacts with the C-terminus of Ca$_V$1.1, and this interaction is important for membrane targeting of Ca$_V$1.1 in heterologous cells (Niu et al., 2018). In Ca$_V$1.2 and Ca$_V$1.3 channels, a corresponding C-terminal interaction site of STAC proteins has been localized in the proximal C-terminus, in the vicinity of the calmodulin binding site (EF hands and IQ motif) (Campiglio et al., 2018; Niu et al., 2018; Polster et al., 2018a). In these channel isoforms, the association of STAC interfered with the known function of calmodulin in calcium-dependent current inactivation. However, sequence differences in the IQ motif of Ca$_V$1.1, compared to Ca$_V$1.2, ablate calmodulin binding and, therefore, can by themselves account for the lack of calcium-dependent inactivation in skeletal muscle (Ohrtman et al., 2008). Whether the interaction of STAC3 with the C-terminus of Ca$_V$1.1 serves functions other than anchoring STAC3 to Ca$_V$1.1 and promoting its membrane targeting is still unknown.

The α$_2$δ-1 subunit is not essential for the function of Ca$_V$1.1 as channel or in EC coupling, but stabilizes its normal channel properties (Flucher et al., 2005; Fuller-Bicer et al., 2009). In skeletal muscle, the membrane-anchored extracellular

$\alpha_2\delta$-1 is targeted into the membrane independently, but requires $Ca_V1.1$ to become incorporated into skeletal muscle triads (Flucher et al., 1991; Obermair et al., 2005). Its major functional effect on $Ca_V1.1$ is in promoting its characteristic slow current activation (Obermair et al., 2005). Interestingly, slowing current kinetics is not the intrinsic activity of $\alpha_2\delta$-1. Instead, it accentuates the intrinsic properties of Ca_V channels, in that the association of $\alpha_2\delta$-1 makes $Ca_V1.1$ slower, while making $Ca_V1.2$ faster (Tuluc et al., 2007; Obermair et al., 2008).

The γ_1 subunit is exclusively expressed in skeletal muscle where it tightly associates with the $Ca_V1.1$ complex, but γ_1 is not essential for EC coupling (Melzer et al., 2006). This integral membrane protein belongs to the family of AMPA receptor regulatory proteins in the brain, where they function in anchoring glutamate receptors in the postsynaptic membrane (Straub & Tomita, 2012). In non-muscle cells, γ_1 supported membrane expression of $Ca_V1.1$, but at the same time inhibited the voltage-sensor function and currents of $Ca_V1.1$ (Polster et al., 2016). In skeletal muscles, γ_1 suppresses both calcium currents and depolarization-induced calcium release by shifting their steady-state availability to more positive potentials (Freise et al., 2000; Ahern et al., 2001; Ursu et al., 2001, 2004). Thus, the action of γ_1 is to stabilize the inactivated state of $Ca_V1.1$, similar to the pharmacological action of calcium channel antagonists (Andronache et al., 2007).

The RyR1 calcium release channel is the effector of $Ca_V1.1$ in skeletal muscle EC coupling (i.e., anterograde coupling of the voltage sensor to the release channel), but also the chief structural organizer of the EC coupling complex, necessary for the tetrad arrangement of $Ca_V1.1$, and a modulator of $Ca_V1.1$ channel function (i.e., retrograde coupling). In myotubes of RyR1 knockout mice the calcium current size is dramatically decreased without a parallel reduction of gating charge movements, and the speed of current activation is increased (Nakai et al., 1996; Avila & Dirksen, 2000). Furthermore, the sensitivity of $Ca_V1.1$ to the channel activator BayK 8644 is increased in RyR1-null myotubes, and, conversely, pharmacological block of the RyR1 in normal myotubes shifts the voltage-dependence of current activation and inactivation (Bannister & Beam, 2009). In $Ca_V1.1$, the retrograde actions of the RyR1 depend on the same sequence domains that are also critical for EC coupling (Grabner et al., 1999). However, on the side of the RyR1 these actions can be dissociated to different structural domains (Nakai et al., 1998a; Sheridan et al., 2006; Bannister et al., 2016). The multitude of effects of RyR1 on $Ca_V1.1$ currents highlight their intimate interactions within a common signaling complex (Flucher, 2016).

Together, the structural and functional interactions between $Ca_V1.1$ and all these associated proteins demonstrate an important aspect for understanding $Ca_V1.1$ function in skeletal muscle. The two channels and the associated proteins are highly specialized components of a molecular machine optimized for skeletal muscle EC coupling. As such, the specific properties of each component are important—in several cases essential—for the bi-directional function of the machine. This is different from the notion of auxiliary subunits being dynamic channel modulators, functioning by association-dissociation or isoform-switching, as has been commonly invoked because of their "modulatory" actions upon heterologous co-expression. Accordingly, mature skeletal muscle expresses only a single isoform of each of these proteins, and, for the most part, they cannot be replaced by heterologous isoforms without compromising $Ca_V1.1$'s EC coupling or channel function.

Today, we probably know all the essential components of this machine. However, the list of the non-essential components will likely continue to grow. For example, recent discoveries in heart muscle identified the negative regulator of $Ca_V1.2$ currents, Rad, as the target of protein kinase A in the fight-or-flight response (Liu et al., 2020). Under basal conditions, the small G-protein Rad forms a tripartite complex with the β subunit and the $Ca_V1.2$ I–II loop, thus inhibiting the calcium currents. In response to β-adrenergic stimulation, Rad dissociates from β

and releases the inhibition of $Ca_V1.2$ (Papa et al., 2021). Also, in skeletal muscle, β-adrenergic stimulation increases calcium currents and contractility and the molecular mechanism is incompletely understood (Johnson et al., 1997; Emrick et al., 2010). Therefore, protein kinase A, AKAPs, or even a member of the RGK family of small G-proteins may well be associated with the $Ca_V1.1$–RyR1 signaling machine (Miranda et al., 2021).

Functionally Distinct $Ca_V1.1$ Isoforms

There is one notable exception to the notion of a relatively rigid molecular machine with non-replaceable components: the pore-forming $α_1$ subunit of $Ca_V1.1$ itself. As described above, the $Ca_V1.1$ channel properties refer to the classical splice variant expressed in mature skeletal muscle. During fetal development and in diseased muscle, another $Ca_V1.1$ splice variant is expressed that has dramatically different channel-gating properties (Tuluc & Flucher, 2011). For easy distinction, we named the classical splice variant $Ca_V1.1a$ (for adult) and the alternatively spliced variant expressed in embryonic muscle and cultured myotubes $Ca_V1.1e$ (for embryonic) (Tuluc et al., 2009). In $Ca_V1.1e$, exon 29 is excluded, which codes for a 19 amino acid sequence in the extracellular loop connecting the S3 and S4 helices in the fourth repeat. Compared to the classical splice variant, $Ca_V1.1e$ activates at about 30 mV less depolarized potentials, its activation kinetics are about twice as fast, and its open probability is increased resulting in an approximately fivefold higher current density. Interestingly, these altered channel-gating properties are not reflected in its voltage sensing and EC coupling functions, as both the voltage-dependence of the gating currents and EC coupling are unaltered by the exclusion of exon 29 (Tuluc et al., 2009). Accordingly, $Ca_V1.1e$ is properly incorporated into triads.

Therefore, the embryonic splice variant $Ca_V1.1e$ is both, a fully functional calcium channel and voltage sensor for EC coupling—although the extra calcium influx component is reflected in a steeper increase and higher peak of the curve reflecting the voltage-dependence of the myoplasmic calcium signals (Tuluc et al., 2009; Flucher & Tuluc, 2011). At present, the physiological relevance of this developmental switch from a proper to a poor calcium channel in mammalian muscle is unclear. The $Ca_V1.1e$ calcium currents may serve developmental functions. For example, in the absence of EC coupling in RyR1 knock-out mice, this calcium influx is sufficient to support important calcium-dependent processes in neuro-muscular junction development (Kaplan et al., 2018; Kaplan & Flucher, 2019). What is clear, however, is that curtailing the calcium influx by developmentally regulated alternative splicing of exon 29 is important for proper muscle development and function. In a mouse model that does not undergo this splicing event and consequently solely expresses $Ca_V1.1e$ throughout life, fiber-type composition is amiss and muscles experience severe mitochondrial damage (Sultana et al., 2016). Apparently, the additional calcium influx through $Ca_V1.1e$ has little effect on EC coupling but alters other calcium-signaling processes like fiber-type specification and eventually causes muscle disease. In fact, this mechanism probably contributes to the dystrophic phenotype in myotonic dystrophy type 1 and type 2, where $Ca_V1.1e$ is aberrantly expressed in adult patients (Tang et al., 2012).

Structural Determinants of $Ca_V1.1$ Functions

The Structure of $Ca_V1.1$

The $Ca_V1.1$ channel complex was the first member of the Ca_V family for which the protein structure has been solved at high resolution using cryo-electron microscopy (Wu et al., 2015, 2016). Like the previously solved structures of homo-tetrameric potassium channels or bacterial sodium channels, $Ca_V1.1$ showed the typical fourfold symmetry, with a central pore composed of the S5–S6 segments of all four repeats, and four separate voltage-sensing domains (VSDs) positioned around it in a domain-swapped arrangement; i.e., each VSD (S1–S4) is positioned adjacent to the pore domain (S5–S6) of the next repeat. Because the membrane potential breaks down in the course of the protein preparation, the S4 helices of the voltage sensors were in the up-position, and the gate was closed, probably corresponding to an inactivated or relaxed state. The extracellular $\alpha_2\delta$-1 subunit was associated with the extracellular parts of the pore loops of repeats I, II, and III, plus an interaction with the IS1–S2 linker of the first VSD. The four helix bundles of the γ_1 subunit were associated with the fourth VSD.

While the overall fold closely resembled that of the previously known homo-tetrameric channel structures, the structure of the pseudo-tetrameric $Ca_V1.1$ deviated in several notable aspects, owing to the fact that its four repeats are structurally highly conserved but not identical to each other. This affected the conduction path with the selectivity filter and the channel gate, which in the bacterial Na_VAb are highly symmetrical (Payandeh et al., 2011), while in the $Ca_V1.1$ structure this is not the case (Wu et al., 2016). Also the structures of the four VSDs differ from each other. While overall they display the fold of the well-characterized VSDs of Na_VAb, they differ in the number and position of the gating charges in the S4 helices, as well as in their interactions with counter-charges in the adjacent helices. This is expected considering the distinct primary structure of the four VSDs, but this structural diversity is also highly relevant considering the functional diversity of Ca_V channels in general and of $Ca_V1.1$ in particular. It suggests the possibility that the specific gating properties of mammalian Ca_V and Na_V channels are encoded in the distinct structures of their four VSDs; and further that the four VSDs may work independently of each other and differentially contribute to the specific gating properties of the respective channel.

Structure–Function Differences of $Ca_V1.1$ VSDs

From the mammalian sodium channel, it is known that VSDs I–III move rapidly and in concert, and are sufficient to activate pore opening (Ahern et al., 2016). VSD IV moves more slowly and its up-state is prerequisite for the channel to enter into the inactivated state (Jiang et al., 2020). Whereas in quantitative models of EC coupling the four VSDs of $Ca_V1.1$ have been regarded as a single unit switching between two states (Ríos et al., 1993; Hernández-Ochoa & Schneider, 2018), structure–function analysis of $Ca_V1.1$ indicated distinct behaviors and functions of its four VSDs (Fig. 4).

Analyses of chimeras of $Ca_V1.1$ and $Ca_V1.2$ indicated that the first VSD of $Ca_V1.1$ determines its characteristically slow activation kinetics. Despite a high degree of similarity, the sequence of IS3 plus the IS3–S4 linker was sufficient to transfer slow skeletal muscle-like activation onto $Ca_V1.2$, and vice versa (Nakai et al., 1994; Tuluc et al., 2016a). Interestingly, neither this nor more extensive VSD I sequences were capable of restoring slow activation when moved from the first to the fourth VSD of $Ca_V1.1$. This indicated that a single VSD (VSD I) determines slow activation kinetics and that the similarities between VSDs I of $Ca_V1.1$ and $Ca_V1.2$ are greater than those between VSDs I and IV of the same channel ($Ca_V1.1$) (Tuluc et al., 2016a). Recent mutagenesis studies elucidated the molecular mechanism by which VSD I regulates activation kinetics. Specific ion-pair interaction between outer gating charges and countercharges in IS3 slows down the sequential transitions as IS4 moves from the resting up to the activated state, thereby delaying the activation of the channel gate (Fernández-Quintero et al., 2021). These experiments also indicate that VSD I is rate-limiting for the activation of the ion channel. The conclusions from the mutagenesis study are in line with recent voltage-clamp fluorometry experiments, which monitor the voltage-dependence and kinetics of each VSD one at a time (Savalli et al., 2021b). The data are consistent with an allosteric gating model and suggest that VSD I is almost exclusively responsible for gating the $Ca_V1.1a$ channel pore.

The dramatic effects of alternative splicing of exon 29 on the voltage-dependence of current activation clearly demonstrate the special role of VSD IV in regulation of $Ca_V1.1$ voltage sensitivity (Tuluc et al., 2009; Tuluc & Flucher, 2011). A series of recent structure–function studies revealed the molecular mechanism by which inclusion of exon 29 shifts the voltage-dependence of activation out of the physiological range. The IVS4 helix differs from IS4 in that its gating charges form scarce ionic interactions in the resting states, allowing it to move rapidly up upon membrane depolarization (Fernández-Quintero et al., 2021). However, in the activated state, IVS4 of the embryonic $Ca_V1.1e$ isoform is stabilized by multiple interactions with two ion-pair partners: one in the fourth VSD itself (IVS3), the other in the pore domain of the first repeat (IS5) (Tuluc et al., 2016b; El Ghaleb et al., 2021). Insertion of the 19 amino acid exon 29 into the IVS3–S4 linker laterally displaces the participating helices and breaks the ion-bonds with the two countercharges. This results in the destabilization of the activated state and the associated 30 mV shift of activation to more depolarizing potentials. Similar mechanisms to regulate voltage-dependence of activation by alternative splicing in the IVS3–S4 loop are found in $Ca_V1.2$ and $Ca_V1.3$. However, there the magnitude of the effects are much smaller (Liu et al., 2016; Costé de Bagneaux et al., 2018; El Ghaleb et al., 2021).

This modulatory mechanism highlights the importance of the fourth VSD in determining the specific voltage-dependence of activation. In $Ca_V1.1a$, loss of stabilization of the activated state clearly becomes limiting for channel gating. Together with the role of VSD I in the regulation of activation kinetics, this finding further demonstrates how controlling distinct aspects of the gating properties can be divided between different VSDs of a pseudo-tetrameric channel. However, this does not preclude contributions of other VSDs to a specific gating property. Whereas the mechanism regulating current kinetics could not be transferred from VSD I to VSD IV, regulation

of the voltage-dependence of activation could be transferred from VSD IV to VSDI by moving the IVS3–S4 linker and IVS4 to that repeat (Tuluc et al., 2016a). A possible contribution of VSD I to setting the voltage-dependence of activation is also evident in the effects of mutating VSD I gating charges and binding partners (El Ghaleb et al., 2019; Fernández-Quintero et al., 2021). Probably, stabilization of any VSD in the activated state can contribute to some degree to this gating property.

Overall, these recent findings highlight the importance of the structural differences between the four VSDs of $Ca_V1.1$, and probably of all voltage-gated calcium channels, for shaping their characteristic gating properties and modulating them. It particularly shows the importance of the molecular interactions guiding and stabilizing the gating charges as S4 moves up and down upon activation and deactivation. This specialization among the four VSDs also enables modulation of specific properties by interactions with auxiliary subunits. Note that $\alpha_2\delta$-1, which stabilizes slow activation of $Ca_V1.1$, possesses a unique interaction with VSD I, and γ_1, which shows exon 29-dependent suppression of calcium currents, associates with VSD IV (Wu et al., 2016).

These unique effects of VSDs I and IV on gating properties also demonstrate that at least two of the four VSDs of $Ca_V1.1$ are critically involved in channel gating. At this point, we lack details about the roles of VSD II and VSD III in channel activation. Also unresolved is the central question, as to which VSDs regulate the activation of the RyR1 calcium release channel. The data discussed above allow us to conclude that VSDs I and IV are not important for skeletal muscle EC coupling, because activation of EC coupling is faster than VSD I's current activation and EC coupling activates at substantially more negative membrane potentials than $Ca_V1.1a$ currents and it is unaffected by the alternative splicing in VSD IV. The non-involvement of VSD I in skeletal muscle EC coupling is further evidenced by a disease mutation of the innermost gating charge (R174W), which abolished the calcium currents without affecting depolarization-induced calcium release (Eltit et al., 2012; Bannister & Beam, 2013b). This leaves VSDs II and III for EC coupling.

A role of VSD II and/or VSD III in regulating EC coupling is consistent with the importance of the cytoplasmic loop connecting these two VSDs in functional and structural coupling of $Ca_V1.1$ and the RyR1. Sequences in this loop have been shown to be critical for skeletal muscle-type EC coupling (anterograde coupling of $Ca_V1.1$ and RyR1) and for the augmentation of $Ca_V1.1$ calcium currents by RyR1 (retrograde coupling) (Nakai et al., 1998b; Grabner et al., 1999; Kugler et al., 2004a). Moreover, STAC3, which is an essential component of the EC coupling signaling complex, associates with the II–III loop of $Ca_V1.1$ (Wong King Yuen et al., 2017). Therefore, a role of one or both of the adjacent VSDs in initiating the conformational changes in the II–III loop that accomplish this mechanical signaling process is plausible. Yet, since the same II–III loop sequences are also important in $Ca_V1.1$ tetrad formation and the β_{1a} subunit also contributes to this structural prerequisite of skeletal muscle EC coupling (Takekura et al., 1994; Schredelseker et al., 2005), activation of the RyR1 may well involve much more complex and widespread interactions between the two proteins. Unfortunately, the structures of the cytoplasmic loops are only partially resolved and, as of today, structures in complex with RyR1 are lacking. Therefore, the structural basis of skeletal muscle EC coupling remains elusive.

Recent advances with voltage-clamp fluorometry of $Ca_V1.1$ are beginning to provide answers to this pertinent question (Banks et al., 2021; Savalli et al., 2021b). This technique uses $Ca_V1.1$ constructs in which individual S4 segments are fluorescently tagged, and monitored upon membrane depolarization. In the work of Banks et al., the constructs were expressed in differentiated muscle fibers, which allowed comparisons between the fluorescence signals, gating charge movements, and calcium transients. This study represents the first time a voltage-gated channel ($Ca_V1.1$) has been analyzed by voltage-clamp fluorometry in its native environment. They char-

acterized the fluorescence signals from each of the four S4 labels in response to an action potential waveform and found that all were too slow to correspond to gating charge movement (Banks et al., 2021), meaning that additional work will be needed to determine their relationship both to S4 movements and their role in EC coupling. After expression in *Xenopus* oocytes, Savalli et al. found that, in response to step changes in voltage, there were rapid changes in fluorescence associated with IIS4, IIIS4, and IVS4, and a slow change in IS4 (Savalli et al., 2021b). In the aggregate, these fluorescence signals had kinetics and voltage-dependence consistent with gating charge movement, and the kinetics of the IS4 signal were similar to those of Ca^{2+} current activation. On the other hand, the voltage-dependence and kinetics of the signals from IIS4 and IIIS4 were consistent with the possibility that they control EC coupling calcium release, but this could not be directly tested in the oocyte system, which lacked the interaction with RyR1.

$Ca_V1.1$ Channelopathies

Genetic variants of $Ca_V1.1$ have been associated with several muscle diseases, including myotonic dystrophy, malignant hyperthermia susceptibility, dystrophic myopathies, hypokalemic periodic paralysis (hypoPP) and normokalemic periodic paralysis (normoPP; Fig. 5).

The physiological function of $Ca_V1.1$ as voltage sensor of EC coupling and its curtailed calcium currents also make $Ca_V1.1$ unique with respect to its involvement in genetically caused diseases. Because its calcium currents are intrinsically small and activate only upon strong and long-lasting depolarizations, genetic variants that typically affect the current via Ca_V channels are of little consequence in $Ca_V1.1$ (Flucher, 2020). On the other hand, reversing the channel properties to those of the embryonic isoform, thus causing a gain of channel function, can produce disease, as exemplified by the splicing disorder myotonic dystrophy (Tang et al., 2012).

However, its intimate association with the RyR1 makes $Ca_V1.1$ especially susceptible to functional

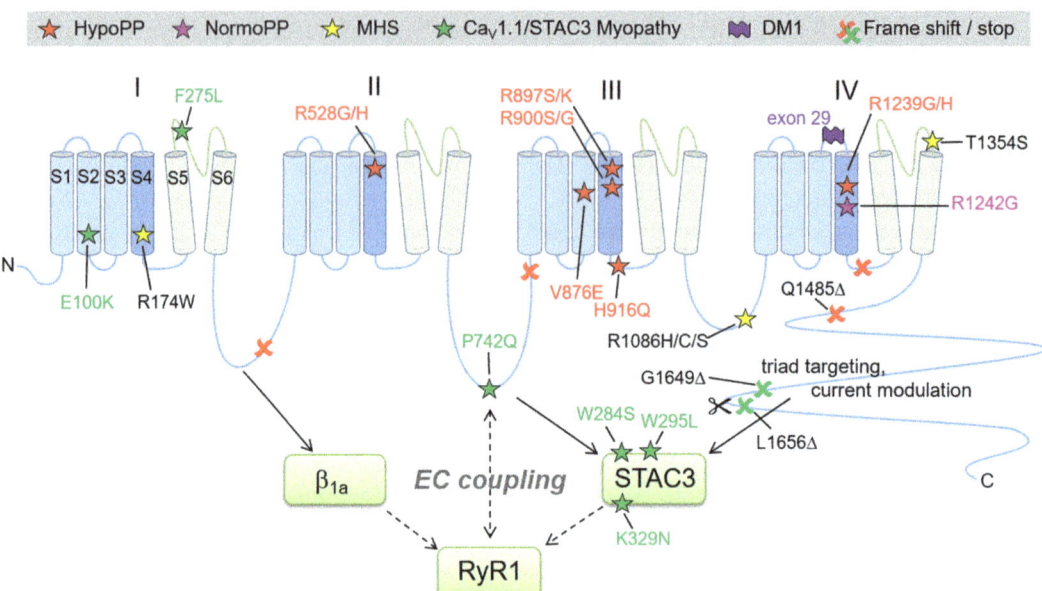

Fig. 5 CACNA1S variants associated with various muscle diseases. Amino acid substitutions of the S4 gating charges resulting in gating pore currents cause periodic paralysis (HypoPP, NormoPP), whereas splicing defects and mutations resulting in increased calcium influx through $Ca_V1.1$ or in altered calcium handling by RyR1 result in malignant hyperthermia susceptibility (MHS) and/or various forms of myopathies, respectively. (Modified from Flucher, 2020)

alterations, which ultimately affect the function of the SR calcium release channel. For example, genetic $Ca_V1.1$ variants have been reported that sensitize the RyR1 to caffeine-induced calcium release and cause malignant hyperthermia susceptibility (Monnier et al., 1997; Weiss et al., 2004; Pirone et al., 2010; Toppin et al., 2010; Eltit et al., 2012). This disease typically results from mutations in the RyR1 itself that render it more sensitive or even leaky at rest (Dowling et al., 2014; Treves et al., 2017). One of the $Ca_V1.1$ malignant hyperthermia susceptibility mutations resides in a voltage-sensing domain, which would be consistent with an effect on the voltage-sensor function of $Ca_V1.1$. However, biophysical characterization revealed that the amino acid exchange negatively affected the calcium current but not EC coupling (Eltit et al., 2012; Bannister & Beam, 2013b). Hence, the pathomechanisms of these disease variants remain unclear at present.

Other $Ca_V1.1$ variants have been associated with different forms of muscle dystrophies. It is likely that the disease phenotypes are caused by an altered calcium homeostasis. However, the different nature and variable locations of the mutations in $Ca_V1.1$ do not reveal a cohesive pathomechanism. Most likely, mutations affecting the $Ca_V1.1$ channel function and such affecting the RyR1 channel lead to similar disease phenotypes (Jungbluth et al., 2018; Flucher, 2020).

The best understood $Ca_V1.1$ channelopathies are hypokalemic periodic paralysis (HypoPP) and normokalemic periodic paralysis (NormoPP) (Jurkat-Rott et al., 2012). The underlying disease variants affect the gating charges of the voltage-sensing domains and probably share a common pathomechanism with similar HypoPP and NormoPP variants of the skeletal muscle voltage-gated sodium channel (Sokolov et al., 2007; Struyk & Cannon, 2007). The exchange of the positive amino acid in the voltage sensor creates a hydrophilic pore through the voltage-sensing domain, allowing the flow of protons or monovalent cations through this non-canonical conduit (Monteleone et al., 2017). Thus, gating pore currents (also termed omega currents) are small, state-dependent leak currents. In mutations affecting the outer gating charges (R1, R2), the omega pore opens at hyperpolarized membrane potentials, at which these amino acids reside in the hydrophobic constriction site of the VSD. Consequently, these leak currents occur at resting membrane potentials and cause the long-lasting membrane depolarizations observed in HypoPP. Channel variants with mutations of the inner gating charges (R3) generate omega currents in the activated state, leading to muscle weakness associated with normal potassium levels (NormoPP). In $Ca_V1.1$, several HypoPP mutations of R1 and R2, and one NormoPP mutation have been reported (Lapie et al., 1996; Jurkat-Rott et al., 1998; Morrill et al., 1998; Morrill & Cannon, 1999; Fan et al., 2013; Fuster et al., 2017a). In addition, a non-canonical HypoPP mutation of an uncharged residue in IIIS3 has been identified (Fuster et al., 2017b), for which the pathomechanism is not known.

Since $Ca_V1.1$ functions in the context of a complex molecular machine and multiple associated proteins participate in its functions or modulate them, genetic variations in any of them may also cause $Ca_V1.1$-related channelopathies. A notable example is the recently identified STAC3 protein, genetic variants of which are linked to a muscle disease named Native American myopathy (NAM) (Horstick et al., 2013; Nelson et al., 2013). Consistent with its important function in EC coupling, the examined NAM mutations lowered the affinity of STAC3 to the II–III loop of $Ca_V1.1$ and decreased the efficiency of depolarization-induced calcium signals in skeletal myotubes (Polster et al., 2016; Wong King Yuen et al., 2017; Rufenach et al., 2020).

Conclusion and Outlook

Whereas the skeletal muscle $Ca_V1.1$ often has been treated like the exception among the members of the Ca_V family of ion channels, it actually is more appropriately considered the prototype of this family. The often bemoaned circumstance that it requires the environment of the skeletal muscle triad for functional expression turns out to be a great asset. From the very beginning, it forced researchers to study and consider this channel in the context of its role as an integral component of a sophisticated signaling complex, rather than an isolated functional unit. The interdependence of $Ca_V1.1$ with the other constituents of the skeletal muscle triad is evident in its primary role in EC coupling, in its channel properties, and in its pathophysiology. None of this could be understood by solely studying the isolated $Ca_V1.1$ in heterologous cells. While considerable progress has been made in understanding $Ca_V1.1$ functions and multiple interactions in skeletal muscle, many of the big questions remain incompletely understood. How is $Ca_V1.1$ structurally organized within the triad? What is the complete complement of its interaction partners and what are the interaction sites? And most of all, how is $Ca_V1.1$'s voltage-sensing function mechanically coupled to gating of the RyR1 calcium release channel? These are the challenges for the continuing research efforts concerning $Ca_V1.1$. The recent advances in structural biology combined with a revival of classical biophysical analyses have spurred astonishing progress in the field and renewed the hope that finding answers to these long-standing questions about the physiological role of $Ca_V1.1$ in skeletal muscle is within reach.

References

Adams, B. A., Tanabe, T., Mikami, A., Numa, S., & Beam, K. G. (1990). Intramembrane charge movement restored in dysgenic skeletal muscle by injection of dihydropyridine receptor cDNAs. *Nature, 346,* 569–572.

Ahern, C. A., Powers, P. A., Biddlecome, G. H., Roethe, L., Vallejo, P., Mortenson, L., Strube, C., Campbell, K. P., Coronado, R., & Gregg, R. G. (2001). Modulation of L-type Ca2+ current but not activation of Ca2+ release by the gamma1 subunit of the dihydropyridine receptor of skeletal muscle. *BMC Physiology, 1*, 8.

Ahern, C. A., Payandeh, J., Bosmans, F., & Chanda, B. (2016). The hitchhiker's guide to the voltage-gated sodium channel galaxy. *The Journal of General Physiology, 147*, 1–24.

Andronache, Z., Ursu, D., Lehnert, S., Freichel, M., Flockerzi, V., & Melzer, W. (2007). The auxiliary subunit gamma 1 of the skeletal muscle L-type Ca2+ channel is an endogenous Ca2+ antagonist. *Proceedings of the National Academy of Sciences of the United States of America, 104*, 17885–17890.

Arikkath, J., & Campbell, K. P. (2003). Auxiliary subunits: Essential components of the voltage-gated calcium channel complex. *Current Opinion in Neurobiology, 13*, 298–307.

Armstrong, C. M., & Bezanilla, F. (1973). Currents related to movement of the gating particles of the sodium channels. *Nature, 242*, 459–461.

Armstrong, C. M., Bezanilla, F. M., & Horowicz, P. (1972). Twitches in the presence of ethylene glycol bis(−aminoethyl ether)-N,N'-tetracetic acid. *Biochimica et Biophysica Acta, 267*, 605–608.

Avila, G., & Dirksen, R. T. (2000). Functional impact of the ryanodine receptor on the skeletal muscle L-type Ca(2+) channel. *The Journal of General Physiology, 115*, 467–480.

Avila, G., O'Connell, K. M., Groom, L. A., & Dirksen, R. T. (2001). Ca2+ release through ryanodine receptors regulates skeletal muscle L-type Ca2+ channel expression. *The Journal of Biological Chemistry, 276*, 17732–17738.

Balog, E. M., & Gallant, E. M. (1999). Modulation of the sarcolemmal L-type current by alteration in SR Ca2+ release. *The American Journal of Physiology, 276*, C128–C135.

Banks, Q., Bibollet, H., Contreras, M., Bennett, D. F., & Bannister, R. A. (2021). Voltage sensor movements of CaV1.1 during an action potential in skeletal muscle fibers. *Proceedings of the National Academy of Sciences of the United States of America, 118*, 1–7.

Bannister, R. A., & Beam, K. G. (2009). Ryanodine modification of RyR1 retrogradely affects L-type Ca(2+) channel gating in skeletal muscle. *Journal of Muscle Research and Cell Motility, 30*, 217–223.

Bannister, R. A., & Beam, K. G. (2013a). CaV1.1: The atypical prototypical voltage-gated Ca2+ channel. *Biochimica et Biophysica Acta, Biomembranes, 1828*, 1587–1597.

Bannister, R. A., & Beam, K. G. (2013b). Impaired gating of an L-type Ca2+ channel carrying a mutation linked to malignant hyperthermia. *Biophysical Journal, 104*, 1917–1922.

Bannister, R. A., Pessah, I. N., & Beam, K. G. (2009). The skeletal L-type Ca(2+) current is a major contributor to excitation-coupled Ca(2+) entry. *The Journal of General Physiology, 133*, 79–91.

Bannister, R. A., Sheridan, D. C., & Beam, K. G. (2016). Distinct components of retrograde CaV1.1-RyR1 coupling revealed by a lethal mutation in RyR1. *Biophysical Journal, 110*, 912–921.

Beam, K. G., Knudson, C. M., & Powell, J. A. (1986). A lethal mutation in mice eliminates the slow calcium current in skeletal muscle cells. *Nature, 320*, 168–170.

Bers, D. M. (2002). Cardiac excitation–contraction coupling. *Nature, 415*, 198–205.

Beurg, M., Ahern, C. A., Vallejo, P., Conklin, M. W., Powers, P. A., Gregg, R. G., & Coronado, R. (1999). Involvement of the carboxy-terminus region of the dihydropyridine receptor beta1a subunit in excitation-contraction coupling of skeletal muscle. *Biophysical Journal, 77*, 2953–2967.

Block, B. A., Imagawa, T., Campbell, K. P., & Franzini-Armstrong, C. (1988). Structural evidence for direct interaction between the molecular components of the transverse tubule/sarcoplasmic reticulum junction in skeletal muscle. *The Journal of Cell Biology, 107*, 2587–2600.

Brum, G., Fitts, R., Pizarro, G., & Ríos, E. (1988). Voltage sensors of the frog skeletal muscle membrane require calcium to function in excitation-contraction coupling. *The Journal of Physiology, 398*, 475–505.

Buraei, Z., & Yang, J. (2010). The beta subunit of voltage-gated calcium channels. *Physiological Reviews, 90*, 1461–1506.

Campiglio, M., Costé de Bagneaux, P., Ortner, N. J., Tuluc, P., Van Petegem, F., & Flucher, B. E. (2018). STAC proteins associate to the IQ domain of CaV1.2 and inhibit calcium-dependent inactivation. *Proceedings of the National Academy of Sciences, 115*, 1376–1381.

Cognard, C., Romey, G., Galizzi, J. P., Fosset, M., & Lazdunski, M. (1986). Dihydropyridine-sensitive Ca2+ channels in mammalian skeletal muscle cells in culture: Electrophysiological properties and interactions with Ca2+ channel activator (Bay K8644) and inhibitor (PN 200-110). *Proceedings of the National Academy of Sciences of the United States of America, 83*, 1518–1522.

Coronado, R., Ahern, C. A., Sheridan, D. C., Cheng, W., Carbonneau, L., & Bhattacharya, D. (2004). Functional equivalence of dihydropyridine receptor a1S and b1a subunits in triggering excitation-contraction coupling in skeletal muscle. *Biological Research, 37*, 1–15.

Costé de Bagneaux, P., Campiglio, M., Benedetti, B., Tuluc, P., & Flucher, B. E. (2018). Role of putative voltage-sensor countercharge D4 in regulating gating properties of Ca V 1.2 and Ca V 1.3 calcium channels. *Channels, 12*, 249–261.

Curtis, B. M., & Catterall, W. A. (1984). Purification of the calcium antagonist receptor of the voltage-sensitive calcium channel from skeletal muscle transverse tubules. *Biochemistry, 23*, 2113–2118.

Dayal, A., Schredelseker, J., Franzini-Armstrong, C., & Grabner, M. (2010). Skeletal muscle excitation-contraction coupling is independent of a conserved heptad repeat motif in the C-terminus of the DHPR??1a subunit. *Cell Calcium, 47*, 500–506.

Dayal, A., Bhat, V., Franzini-Armstrong, C., & Grabner, M. (2013). Domain cooperativity in the β1a subunit is essential for dihydropyridine receptor voltage sensing in skeletal muscle. *PNAS, 110*, 7488–7493.

Dayal, A., Schrötter, K., Pan, Y., Föhr, K., Melzer, W., & Grabner, M. (2017). The Ca2+influx through the mammalian skeletal muscle dihydropyridine receptor is irrelevant for muscle performance. *Nature Communications, 8*, 1–14.

Di Biase, V., & Franzini-Armstrong, C. (2005). Evolution of skeletal type e–c coupling. *The Journal of Cell Biology, 171*, 695–704.

Dirksen, R. T., & Beam, K. G. (1995). Single calcium channel behavior in native skeletal muscle. *The Journal of General Physiology, 105*, 227–247.

Dirksen, R. T., & Beam, K. G. (1996). Unitary behavior of skeletal, cardiac, and chimeric L-type Ca2+ channels expressed in dysgenic myotubes. *The Journal of General Physiology, 107*, 731–742.

Dirksen, R. T., & Beam, K. G. (1999). Role of calcium permeation in dihydropyridine receptor function. Insights into channel gating and excitation-contraction coupling. *The Journal of General Physiology, 114*, 393–403.

Donaldson, P. L., & Beam, K. G. (1983). Calcium currents in a fast-twitch skeletal muscle of the rat. *The Journal of General Physiology, 82*, 449–468.

Dowling, J. J., Lawlor, M. W., & Dirksen, R. T. (2014). Triadopathies: An emerging class of skeletal muscle diseases. *Neurotherapeutics, 11*, 773–785.

Dulhunty, A. F., & Gage, P. W. (1988). Effects of extracellular calcium concentration and dihydropyridines on contraction in mammalian skeletal muscle. *The Journal of Physiology, 399*, 63–80.

El Ghaleb, Y., Campiglio, M., & Flucher, B. E. (2019). Correcting the R165K substitution in the first voltage-sensor of CaV1.1 right-shifts the voltage-dependence of skeletal muscle calcium channel activation. *Channels (Austin, Tex.), 13*, 62–71.

El Ghaleb, Y., Fernández-Quintero, M. L., Monteleone, S., Tuluc, P., Campiglio, M., Liedl, K. R., & Flucher, B. E. (2021). Ion-pair interactions between voltage-sensing domain IV and pore domain I regulate CaV1.1 gating. *Biophysical Journal, 120*(20), 4429–4441. https://doi.org/10.1016/J.BPJ.2021.09.004

Eltit, J. M., Bannister, R. A., Moua, O., Altamirano, F., Hopkins, P. M., Pessah, I. N., Molinski, T. F., López, J. R., Beam, K. G., & Allen, P. D. (2012). Malignant hyperthermia susceptibility arising from altered resting coupling between the skeletal muscle L-type Ca2+ channel and the type 1 ryanodine receptor. *Proceedings of the National Academy of Sciences of the United States of America, 109*, 7923–7928.

Emrick, M. A., Sadilek, M., Konoki, K., & Catterall, W. A. (2010). Beta-adrenergic-regulated phosphorylation of the skeletal muscle Ca(V)1.1 channel in the fight-or-flight response. *Proceedings of the National*

Academy of Sciences of the United States of America, 107, 18712–18717.

Fan, C., Lehmann-Horn, F., Weber, M. A., Bednarz, M., Groome, J. R., Jonsson, M. K. B., & Jurkat-Rott, K. (2013). Transient compartment-like syndrome and normokalaemic periodic paralysis due to a Cav1.1 mutation. *Brain, 136*, 3775–3786.

Fernández-Quintero, M. L., El Ghaleb, Y., Tuluc, P., Campiglio, M., Liedl, K. R., & Flucher, B. E. (2021). Structural determinants of voltage-gating properties in calcium channels. *eLife, 10*, e64087. https://doi.org/10.7554/eLife.64087

Flucher, B. E. (2016). Retrograde coupling: Muscle's orphan signaling pathway? *Biophysical Journal, 110*, 870–871.

Flucher, B. E. (2020). Skeletal muscle CaV1.1 channelopathies. *Pflügers Archiv / European Journal of Physiology, 472*, 739–754.

Flucher, B. E., & Campiglio, M. (2019). STAC proteins: The missing link in skeletal muscle EC coupling and new regulators of calcium channel function. *Biochimica et Biophysica Acta (BBA)—Molecular Cell Research, 1866*, 1101–1110.

Flucher, B. E., & Tuluc, P. (2011). A new L-type calcium channel isoform required for normal patterning of the developing neuromuscular junction. *Channels, 5*, 1–7.

Flucher, B. E., & Tuluc, P. (2017). How and why are calcium currents curtailed in the skeletal muscle voltage-gated calcium channels? *The Journal of Physiology, 595*, 1451–1463.

Flucher, B. E., Phillips, J. L., & Powell, J. A. (1991). Dihydropyridine receptor α subunits in normal and dysgenic muscle in vitro: Expression of alpha 1 is required for proper targeting and distribution of alpha 2. *The Journal of Cell Biology, 115*(5), 1345–1356. https://doi.org/10.1083/jcb.115.5.1345

Flucher, B. E., Andrews, S. B., Fleischer, S., Marks, A. R., Caswell, A., & Powell, J. A. (1993). Triad formation: Organization and function of the sarcoplasmic reticulum calcium release channel and triadin in normal and dysgenic muscle in vitro. *The Journal of Cell Biology, 123*, 1161–1174.

Flucher, B. E., Kasielke, N., & Grabner, M. (2000). The triad targeting signal of the skeletal muscle calcium channel is localized in the COOH terminus of the ??(1S) subunit. *The Journal of Cell Biology, 151*, 467–477.

Flucher, B. E., Obermair, G. J., Tuluc, P., Schredelseker, J., Kern, G., & Grabner, M. (2005). The role of auxiliary dihydropyridine receptor subunits in muscle. *Journal of Muscle Research and Cell Motility, 26*(1), 1–6. https://doi.org/10.1007/s10974-005-9000-2

Franzini-Armstrong, C. (2018). The relationship between form and function throughout the history of excitation–contraction coupling. *The Journal of General Physiology, 150*, 189–210.

Franzini-Armstrong, C., Pincon-Raymond, M., & Rieger, F. (1991). Muscle fibers from dysgenic mouse in vivo lack a surface component of peripheral couplings. *Developmental Biology, 146*, 364–376.

Freise, D., Held, B., Wissenbach, U., Pfeifer, A., Trost, C., Himmerkus, N., Schweig, U., Freichel, M., Biel, M., Hofmann, F., Hoth, M., & Flockerzi, V. (2000). Absence of the gamma subunit of the skeletal muscle dihydropyridine receptor increases L-type Ca2+ currents and alters channel inactivation properties. *The Journal of Biological Chemistry, 275*, 14476–14481.

Fuller-Bicer, G. A., Varadi, G., Koch, S. E., Ishii, M., Bodi, I., Kadeer, N., Muth, J. N., Mikala, G., Petrashevskaya, N. N., Jordan, M. A., Zhang, S.-P., Qin, N., Flores, C. M., Isaacsohn, I., Varadi, M., Mori, Y., Jones, W. K., & Schwartz, A. (2009). Targeted disruption of the voltage-dependent calcium channel alpha2/delta-1-subunit. *American Journal of Physiology. Heart and Circulatory Physiology, 297*, H117–H124.

Fuster, C., Perrot, J., Berthier, C., Jacquemond, V., & Allard, B. (2017a). Elevated resting H+ current in the R1239H type 1 hypokalaemic periodic paralysis mutated Ca2+ channel. *The Journal of Physiology, 595*, 6417–6428.

Fuster, C., Perrot, J., Berthier, C., Jacquemond, V., Charnet, P., & Allard, B. (2017b). Na leak with gating pore properties in hypokalemic periodic paralysis V876E mutant muscle Ca channel. *The Journal of General Physiology, 149*, 1139–1148.

García, J., & Beam, K. G. (1994). Measurement of calcium transients and slow calcium current in myotubes. *The Journal of General Physiology, 103*, 107–123.

Georgiou, D. K., Dagnino-Acosta, A., Lee, C. S., Griffin, D. M., Wang, H., Lagor, W. R., Pautler, R. G., Dirksen, R. T., & Hamilton, S. L. (2015). Ca2+ binding/permeation via calcium channel, Ca V 1.1, regulates the intracellular distribution of the fatty acid transport protein, CD36, and fatty acid metabolism. *The Journal of Biological Chemistry, 290*, 23751–23765.

Grabner, M., Dirksen, R. T., Suda, N., & Beam, K. G. (1999). The II-III loop of the skeletal muscle dihydropyridine receptor is responsible for the Bi-directional coupling with the ryanodine receptor. *The Journal of Biological Chemistry, 274*, 21913–21919.

Gregg, R. G., Messing, A., Strube, C., Beurg, M., Moss, R., Behan, M., Sukhareva, M., Haynes, S., Powell, J. A., Coronado, R., & Powers, P. A. (1996). Absence of the beta subunit (cchb1) of the skeletal muscle dihydropyridine receptor alters expression of the alpha 1 subunit and eliminates excitation-contraction coupling. *Proceedings of the National Academy of Sciences of the United States of America, 93*, 13961–13966.

Hernández-Ochoa, E. O., & Schneider, M. F. (2018). Voltage sensing mechanism in skeletal muscle excitation-contraction coupling: Coming of age or midlife crisis? *Skeletal Muscle, 8*, 22.

Hodgkin, A. L., & Huxley, A. F. (1952). A quantitative description of membrane current and its application to conduction and excitation in nerve. *The Journal of Physiology, 117*, 500–544.

Hollingworth, S., & Baylor, S. M. (2013). Comparison of myoplasmic calcium movements during excitation-contraction coupling in frog twitch and mouse fast-twitch muscle fibers. *The Journal of General Physiology, 141*, 567–583.

Horstick, E. J., Linsley, J. W., Dowling, J. J., Hauser, M. A., McDonald, K. K., Ashley-Koch, A., Saint-Amant, L., Satish, A., Cui, W. W., Zhou, W., Sprague, S. M., Stamm, D. S., Powell, C. M., Speer, M. C., Franzini-Armstrong, C., Hirata, H., & Kuwada, J. Y. (2013). Stac3 is a component of the excitation-contraction coupling machinery and mutated in Native American myopathy. *Nature Communications, 4*, 1952.

Jiang, D., Shi, H., Tonggu, L., El-din, T. M. G., Lenaeus, M. J., Zhao, Y., Yoshioka, C., Zheng, N., & Catterall, W. A. (2020). Structure of the cardiac sodium channel. *Cell, 180*, 1–13.

Johnson, B. D., Brousal, J. P., Peterson, B. Z., Gallombardo, P. A., Hockerman, G. H., Lai, Y., Scheuer, T., & Catterall, W. A. (1997). Modulation of the cloned skeletal muscle L-type Ca2+ channel by anchored cAMP-dependent protein kinase. *The Journal of Neuroscience, 17*, 1243–1255.

Jungbluth, H., Treves, S., Zorzato, F., Sarkozy, A., Ochala, J., Sewry, C., Phadke, R., Gautel, M., & Muntoni, F. (2018). Congenital myopathies: Disorders of excitation-contraction coupling and muscle contraction. *Nature Reviews. Neurology, 14*, 151–167.

Jurkat-Rott, K., Uetz, U., Pika-Hartlaub, U., Powell, J., Fontaine, B., Melzer, W., & Lehmann-Horn, F. (1998). Calcium currents and transients of native and heterologously expressed mutant skeletal muscle DHP receptor 1/2 subunits (R528H). *FEBS Letters, 423*, 198–204.

Jurkat-Rott, K., Groome, J., & Lehmann-Horn, F. (2012). Pathophysiological role of omega pore current in channelopathies. *Frontiers in Pharmacology, 3*, 1–19.

Kaplan, M. M., & Flucher, B. E. (2019). Postsynaptic CaV1.1-driven calcium signaling coordinates presynaptic differentiation at the developing neuromuscular junction. *Scientific Reports, 9*, 18450.

Kaplan, M. M., Sultana, N., Benedetti, A., Obermair, G. J., Linde, N. F., Papadopoulos, S., Dayal, A., Grabner, M., & Flucher, B. E. (2018). Calcium influx and release cooperatively regulate AChR patterning and motor axon outgrowth during neuromuscular junction formation. *Cell Reports, 23*, 3891–3904.

Klaus, M. M., Scordilis, S. P., Rapalus, J. M., Briggs, R. T., & Powell, J. A. (1983). Evidence for dysfunction in the regulation of cytosolic Ca2+ in excitation-contraction uncoupled dysgenic muscle. *Developmental Biology, 99*, 152–165.

Kugler, G., Grabner, M., Platzer, J., Striessnig, J., & Flucher, B. E. (2004a). The monoclonal antibody mAB 1A binds to the excitation-contraction coupling domain in the II-III loop of the skeletal muscle calcium channel α1S subunit. *Archives of Biochemistry and Biophysics, 427*, 91–100.

Kugler, G., Weiss, R. G., Flucher, B. E., & Grabner, M. (2004b). Structural requirements of the dihydropyridine receptor alpha1S II-III Loop for skeletal-type excitation-contraction coupling. *The Journal of Biological Chemistry, 279*, 4721–4728.

Lacerda, A. E., Kim, H. S., Ruth, P., Perez-Reyes, E., Flockerzi, V., Hofmann, F., Birnbaumer, L., & Brown, A. M. (1991). Normalization of current kinetics by interaction between the alpha 1 and beta subunits of the skeletal muscle dihydropyridine-sensitive Ca2+ channel. *Nature, 352*, 527–530.

Lamb, G. D., & Walsh, T. (1987). Calcium currents, charge movement and dihydropyridine binding in fast- and slow-twitch muscles of rat and rabbit. *The Journal of Physiology, 393*, 595–617.

Lapie, P., Goudet, C., Nargeot, J., Fontaine, B., & Lory, P. (1996). Electrophysiological properties of the hypokalaemic periodic paralysis mutation (R528H) of the skeletal muscle alpha 1s subunit as expressed in mouse L cells. *FEBS Letters, 382*, 244–248.

Linsley, J. W., Hsu, I.-U., Groom, L., Yarotskyy, V., Lavorato, M., Horstick, E. J., Linsley, D., Wang, W., Franzini-Armstrong, C., Dirksen, R. T., & Kuwada, J. Y. (2016). Congenital myopathy results from misregulation of a muscle Ca2+ channel by mutant Stac3. *Proceedings of the National Academy of Sciences, 114*, E228–E236.

Liu, N., Liu, Y., Yang, Y., & Liu, X. (2016). Linker flexibility of IVS3-S4 loops modulates voltage-dependent activation of L-type Ca 2+ channels. *Channels, 6950*, 00–00.

Liu, G., Papa, A., Katchman, A. N., Zakharov, S. I., Roybal, D., Hennessey, J. A., Kushner, J., Yang, L., Chen, B. X., Kushnir, A., Dangas, K., Gygi, S. P., Pitt, G. S., Colecraft, H. M., Ben-Johny, M., Kalocsay, M., & Marx, S. O. (2020). Mechanism of adrenergic CaV1.2 stimulation revealed by proximity proteomics. *Nature, 577*, 695–700.

Lory, P., Varadi, G., & Schwartz, A. (1992). The beta subunit controls the gating and dihydropyridine sensitivity of the skeletal muscle Ca2+ channel. *Biophysical Journal, 63*, 1421–1424.

Melzer, W., Schneider, M. F., Simon, B. J., & Szucs, G. (1986). Intramembrane charge movement and calcium release in frog skeletal muscle. *The Journal of Physiology, 373*, 481–511.

Melzer, W., Herrmann-Frank, A., & Lüttgau, H.-C. (1995). The role of Ca2++ ions in excitation-contraction coupling of skeletal muscle fibres. *Biochimica et Biophysica Acta, 1241*, 59–116.

Melzer, W., Andronache, Z., & Ursu, D. (2006). Functional roles of the gamma subunit of the skeletal muscle DHP-receptor. *Journal of Muscle Research and Cell Motility, 27*, 307–314.

Miranda, D. R., Voss, A. A., & Bannister, R. A. (2021). Into the spotlight: RGK proteins in skeletal muscle. *Cell Calcium, 98*, 102439.

Monnier, N., Procaccio, V., Stieglitz, P., & Lunardi, J. (1997). Malignant-hyperthermia susceptibility is associated with a mutation of the alpha 1-subunit of the human dihydropyridine-sensitive L-type voltage-dependent calcium-channel receptor in skeletal muscle. *American Journal of Human Genetics, 60*, 1316–1325.

Monteleone, S., Lieb, A., Pinggera, A., Negro, G., Fuchs, J. E., Hofer, F., Striessnig, J., Tuluc, P., & Liedl, K. R. (2017). Mechanisms responsible for ω-pore currents in Cav calcium channel voltage-sensing domains. *Biophysical Journal, 113*, 1485–1495.

Morrill, J. A., & Cannon, S. C. (1999). Effects of mutations causing hypokalaemic periodic paralysis on the skeletal muscle L-type Ca2+ channel expressed in Xenopus laevis oocytes. *The Journal of Physiology, 520*(Pt 2), 321–336.

Morrill, J. A., Brown, R. H., & Cannon, S. C. (1998). Gating of the L-type Ca channel in human skeletal myotubes: An activation defect caused by the hypokalemic periodic paralysis mutation R528H. *The Journal of Neuroscience, 18*, 10320–10334.

Nakada, T., Kashihara, T., Komatsu, M., Kojima, K., Takeshita, T., & Yamada, M. (2018). Physical interaction of junctophilin and the CaV1.1 C terminus is crucial for skeletal muscle contraction. *Proceedings of the National Academy of Sciences of the United States of America, 115*, 4507–4512.

Nakai, J., Adams, B. A., Imoto, K., & Beam, K. G. (1994). Critical roles of the S3 segment and S3-S4 linker of repeat I in activation of L-type calcium channels. *Proceedings of the National Academy of Sciences, 91*, 1014–1018.

Nakai, J., Dirksen, R. T., Nguyen, H. T., Pessah, I. N., Beam, K. G., & Allen, P. D. (1996). Enhanced dihydropyridine receptor channel activity in the presence of ryanodine receptor. *Nature, 380*, 72–75.

Nakai, J., Sekiguchi, N., Rando, T. A., Allen, P. D., & Beam, K. G. (1998a). Two regions of the ryanodine receptor involved in coupling with L-type Ca2+ channels. *The Journal of Biological Chemistry, 273*, 13403–13406.

Nakai, J., Tanabe, T., Konno, T., Adams, B., & Beam, K. G. (1998b). Localization in the II-III loop of the dihydropyridine receptor of a sequence critical for excitation-contraction coupling. *The Journal of Biological Chemistry, 273*, 24983–24986.

Nelson, B. R., Wu, F., Liu, Y., Anderson, D. M., McAnally, J., Lin, W., Cannon, S. C., Bassel-Duby, R., & Olson, E. N. (2013). Skeletal muscle-specific T-tubule protein STAC3 mediates voltage-induced Ca2+ release and contractility. *Proceedings of the National Academy of Sciences of the United States of America, 110*, 11881–11886.

Neuhuber, B., Gerster, U., Döring, F., Glossmann, H., Tanabe, T., & Flucher, B. E. (1998). Association of calcium channel alpha1S and beta1a subunits is required for the targeting of beta1a but not of alpha1S into skeletal muscle triads. *Proceedings of the National Academy of Sciences of the United States of America, 95*, 5015–5020.

Niu, J., Yang, W., Yue, D. T., Inoue, T., & Ben-Johny, M. (2018). Duplex signaling by CaM and Stac3 enhances CaV1.1 function and provides insights into congenital myopathy. *The Journal of General Physiology, 150,* 1145–1161.

Obermair, G. J., Kugler, G., Baumgartner, S., Tuluc, P., Grabner, M., & Flucher, B. E. (2005). The Ca2+ channel alpha2delta-1 subunit determines Ca2+ current kinetics in skeletal muscle but not targeting of alpha1S or excitation-contraction coupling. *The Journal of Biological Chemistry, 280,* 2229–2237.

Obermair, G. J., Tuluc, P., & Flucher, B. E. (2008). Auxiliary Ca2+ channel subunits: Lessons learned from muscle. *Current Opinion in Pharmacology, 8,* 311–318.

Ohrtman, J., Ritter, B., Polster, A., Beam, K. G., & Papadopoulos, S. (2008). Sequence differences in the IQ motifs of CaV1.1 and Ca V1.2 strongly impact calmodulin binding and calcium-dependent inactivation. *The Journal of Biological Chemistry, 283,* 29301–29311.

Paolini, C., Fessenden, J. D., Pessah, I. N., & Franzini-Armstrong, C. (2004). Evidence for conformational coupling between two calcium channels. *Proceedings of the National Academy of Sciences of the United States of America, 101,* 12748–12752.

Papa, A., et al. (2021). Adrenergic CaV1.2 activation via rad phosphorylation converges at α1CI-II loop. *Circulation Research, 128,* 76–88.

Payandeh, J., Scheuer, T., Zheng, N., & Catterall, W. A. (2011). The crystal structure of a voltage-gated sodium channel. *Nature, 475,* 353–358.

Perez-Reyes, E., Kim, H. S., Lacerda, A. E., Horne, W., Wei, X. Y., Rampe, D., Campbell, K. P., Brown, A. M., & Birnbaumer, L. (1989). Induction of calcium currents by the expression of the alpha 1-subunit of the dihydropyridine receptor from skeletal muscle. *Nature, 340,* 233–236.

Perni, S., Lavorato, M., & Beam, K. G. (2017). De novo reconstitution reveals the proteins required for skeletal muscle voltage-induced Ca2+ release. *Proceedings of the National Academy of Sciences, 114,* 13822–13827.

Pirone, A., Schredelseker, J., Tuluc, P., Gravino, E., Fortunato, G., Flucher, B. E., Carsana, A., Salvatore, F., & Grabner, M. (2010). Identification and functional characterization of malignant hyperthermia mutation T1354S in the outer pore of the Cavalpha1S-subunit. *American Journal of Physiology. Cell Physiology, 299,* C1345–C1354.

Polster, A., Perni, S., Bichraoui, H., & Beam, K. G. (2015). Stac adaptor proteins regulate trafficking and function of muscle and neuronal L-type Ca2+ channels. *Proceedings of the National Academy of Sciences of the United States of America, 112,* 602–606.

Polster, A., Nelson, B. R., Olson, E. N., & Beam, K. G. (2016). Stac3 has a direct role in skeletal muscle-type excitation—Contraction coupling that is disrupted by a myopathy-causing mutation. *Proceedings of the National Academy of Sciences, 113,* 10986–10991.

Polster, A., Dittmer, P. J., Perni, S., Bichraoui, H., Sather, W. A., & Beam, K. G. (2018a). Stac proteins suppress Ca2+−dependent inactivation of neuronal L-type

Ca2+ channels. *The Journal of Neuroscience, 38,* 9215–9227.

Polster, A., Nelson, B. R., Papadopoulos, S., Olson, E. N., & Beam, K. G. (2018b). Stac proteins associate with the critical domain for excitation–contraction coupling in the II–III loop of CaV1.1. *The Journal of General Physiology, 150,* 613–624.

Powell, J. A. (1990). Muscular dysgenesis: A model system for studying skeletal muscle development. *The FASEB Journal, 4,* 2798–2808.

Ren, D., & Hall, L. M. (1997). Functional expression and characterization of skeletal muscle dihydropyridine receptors in Xenopus oocytes. *The Journal of Biological Chemistry, 272,* 22393–22396.

Rios, E., & Brum, G. (1987). Involvement of dihydropyridine receptors in excitation–contraction coupling in skeletal muscle. *Nature, 325,* 717–720.

Ríos, E., Karhanek, M., Ma, J., & González, A. (1993). An allosteric model of the molecular interactions of excitation-contraction coupling in skeletal muscle. *The Journal of General Physiology, 102,* 449–481.

Robin, G., & Allard, B. (2013). Major contribution of sarcoplasmic reticulum Ca(2+) depletion during long-lasting activation of skeletal muscle. *The Journal of General Physiology, 141,* 557–565.

Robin, G., & Allard, B. (2015). Voltage-gated Ca2+ influx through L-type channels contributes to sarcoplasmic reticulum Ca2+ loading in skeletal muscle. *The Journal of Physiology, 539,* 4781–4797.

Rufenach, B., Christy, D., Flucher, B. E., Bui, J. M., Gsponer, J., Campiglio, M., & Van Petegem, F. (2020). Multiple sequence variants in STAC3 affect interactions with CaV1.1 and excitation-contraction coupling. *Structure, 28*(8), 922–932.

Sanchez, J. A., & Stefani, E. (1978). *Inward calcium current in twitch muscle fibres of the frog.* Available at https://physoc.onlinelibrary.wiley.com/doi/abs/10.1113/jphysiol.1978.sp012496?sid=nlm%3Apubmed

Savalli, N., Angelini, M., Steccanella, F., Neely, A., & Olcese, R. (2021a). Voltage sensor operation in the embryonic splice variant of skeletal CaV1.1 channels. *Biophysical Journal, 120,* 155A.

Savalli, N., Angelini, M., Steccanella, F., Wier, J., Wu, F., Quinonez, M., Difranco, M., Neely, A., Cannon, S. C., & Olcese, R. (2021b). The distinct role of the four voltage sensors of the skeletal CaV1.1 channel in voltage-dependent activation. *The Journal of General Physiology, 153*(11), e202112915. https://doi.org/10.1085/jgp.202112915

Schneider, M. F., & Chandler, W. K. (1973). Voltage dependent charge movement of skeletal muscle: A possible step in excitation-contraction coupling. *Nature, 242,* 244–246.

Schredelseker, J., Di Biase, V., Obermair, G. J., Felder, E. T., Flucher, B. E., Franzini-Armstrong, C., & Grabner, M. (2005). The beta 1a subunit is essential for the assembly of dihydropyridine-receptor arrays in skeletal muscle. *Proceedings of the National Academy of Sciences of the United States of America, 102,* 17219–17224.

Schredelseker, J., Shrivastav, M., Dayal, A., & Grabner, M. (2010). Non-Ca2+−conducting Ca2+ channels in fish skeletal muscle excitation-contraction coupling. *Proceedings of the National Academy of Sciences of the United States of America, 107*, 5658–5663.

Sheridan, D. C., Takekura, H., Franzini-Armstrong, C., Beam, K. G., Allen, P. D., & Perez, C. F. (2006). Bidirectional signaling between calcium channels of skeletal muscle requires multiple direct and indirect interactions. *Proceedings of the National Academy of Sciences of the United States of America, 103*, 19760–19765.

Sipos, I., Harasztosi, C., & Melzer, W. (1997). L-type calcium current activation in cultured human myotubes. *Journal of Muscle Research and Cell Motility, 18*, 353–367.

Sokolov, S., Scheuer, T., & Catterall, W. A. (2007). Gating pore current in an inherited ion channelopathy. *Nature, 446*, 76–78.

Spiecker, W., Melzer, W., & Lüttgau, H. C. (1979). Extracellular Ca2+ and excitation-contraction coupling. *Nature, 280*, 158–160.

Straub, C., & Tomita, S. (2012). The regulation of glutamate receptor trafficking and function by TARPs and other transmembrane auxiliary subunits. *Current Opinion in Neurobiology, 22*, 488–495.

Struyk, A. F., & Cannon, S. C. (2007). A Na+ channel mutation linked to hypokalemic periodic paralysis exposes a proton-selective gating pore. *The Journal of General Physiology, 130*, 11–20.

Sultana, N., Dienes, B., Benedetti, A., Tuluc, P., Szentesi, P., Sztretye, M., Rainer, J., Hess, M. W., Schwarzer, C., Obermair, G. J., Csernoch, L., & Flucher, B. E. (2016). Restricting calcium currents is required for correct fiber type specification in skeletal muscle. *Development, 143*, 1547–1559.

Takekura, H., Bennett, L., Tanabe, T., Beam, K. G., & Franzini-Armstrong, C. (1994). Restoration of junctional tetrads in dysgenic myotubes by dihydropyridine receptor cDNA. *Biophysical Journal, 67*, 793–803.

Takekura, H., Nishi, M., Noda, T., Takeshima, H., & Franzini-Armstrong, C. (1995). Abnormal junctions between surface membrane and sarcoplasmic reticulum in skeletal muscle with a mutation targeted to the ryanodine receptor. *Proceedings of the National Academy of Sciences of the United States of America, 92*, 3381–3385.

Takekura, H., Paolini, C., Franzini-Armstrong, C., Kugler, G., Grabner, M., & Flucher, B. E. (2004). Differential contribution of skeletal and cardiac II-III loop sequences to the assembly of dihydropyridine- receptor arrays in skeletal muscle. *Molecular Biology of the Cell, 15*, 5408–5419.

Takeshima, H., Ikemoto, T., Nishi, M., Nishiyama, N., Shimuta, M., Sugitani, Y., Kuno, J., Saito, I., Saito, H., Endo, M., Iino, M., & Noda, T. (1996). Generation and characterization of mutant mice lacking ryanodine

receptor type 3. *The Journal of Biological Chemistry, 271*, 19649–19652.

Tanabe, T., Takeshima, H., Mikami, A., Flockerzi, V., Takahashi, H., Kangawa, K., Kojima, M., Matsuo, H., Hirose, T., & Numa, S. (1987). Primary structure of the receptor for calcium channel blockers from skeletal muscle. *Nature, 328*, 313–318.

Tanabe, T., Beam, K. G., Powell, J. A., & Numa, S. (1988). Restoration of excitation-contraction coupling and slow calcium current in dysgenic muscle by dihydropyridine receptor complementary DNA. *Nature, 336*, 134–139.

Tang, Z. Z., Yarotskyy, V., Wei, L., Sobczak, K., Nakamori, M., Eichinger, K., Moxley, R. T., Dirksen, R. T., & Thornton, C. A. (2012). Muscle weakness in myotonic dystrophy associated with misregulated splicing and altered gating of Ca v1.1 calcium channel. *Human Molecular Genetics, 21*, 1312–1324.

Toppin, P. J., Chandy, T. T., Ghanekar, A., Kraeva, N., Beattie, W. S., & Riazi, S. (2010). A report of fulminant malignant hyperthermia in a patient with a novel mutation of the CACNA1S gene. *Canadian Journal of Anaesthesia, 57*, 689–693.

Treves, S., Jungbluth, H., Voermans, N., Muntoni, F., & Zorzato, F. (2017). Ca2+ handling abnormalities in early-onset muscle diseases: Novel concepts and perspectives. *Seminars in Cell & Developmental Biology, 64*, 201–212. https://doi.org/10.1016/j.semcdb.2016.07.017

Tuluc, P., & Flucher, B. E. (2011). Divergent biophysical properties, gating mechanisms, and possible functions of the two skeletal muscle CaV1.1 calcium channel splice variants. *Journal of Muscle Research and Cell Motility, 32*, 249–256.

Tuluc, P., Kern, G., Obermair, G. J., & Flucher, B. E. (2007). Computer modeling of siRNA knockdown effects indicates an essential role of the Ca2+ channel α2δ-1 subunit in cardiac excitation-contraction coupling. *Proceedings of the National Academy of Sciences of the United States of America, 104*, 11091–11096. https://doi.org/10.1073/pnas.0700577104

Tuluc, P., Molenda, N., Schlick, B., Obermair, G. J., Flucher, B. E., & Jurkat-Rott, K. (2009). A Cav1.1 Ca2+ channel splice variant with high conductance and voltage-sensitivity alters EC coupling in developing skeletal muscle. *Biophysical Journal, 96*, 35–44.

Tuluc, P., Benedetti, B., Coste De Bagneaux, P., Grabner, M., & Flucher, B. E. (2016a). Two distinct voltage-sensing domains control voltage sensitivity and kinetics of current activation in CaV1.1 calcium channels. *The Journal of General Physiology, 147*, 437–449.

Tuluc, P., Yarov-Yarovoy, V., Benedetti, B., & Flucher, B. E. (2016b). Molecular interactions in the voltage sensor controlling gating properties of CaV calcium channels. *Structure, 24*, 261–271.

Ursu, D., Sebille, S., Dietze, B., Freise, B., Flockerzi, V., Melzer, W., Freise, D., Flockerzi, V., & Melzer, W. (2001). Excitation-contraction coupling in skeletal muscle of a mouse lacking the dihydropyridine receptor subunit gamma1. *The Journal of Physiology, 533*, 367–377.

Ursu, D., Schuhmeier, R. P., Freichel, M., Flockerzi, V., & Melzer, W. (2004). Altered inactivation of Ca2+ current and Ca2+ release in mouse muscle fibers deficient in the DHP receptor gamma1 subunit. *The Journal of General Physiology, 124*, 605–618.

Varadi, G., Lory, P., Schultz, D., Varadi, M., & Schwartz, A. (1991). *Acceleration of activation and inactivation by the beta subunit of the skeletal muscle calcium channel.* Available at https://www.nature.com/articles/352159a0

Weiss, R. G., O'Connell, K. M. S., Flucher, B. E., Allen, P. D., Grabner, M., & Dirksen, R. T. (2004). Functional analysis of the R1086H malignant hyperthermia mutation in the DHPR reveals an unexpected influence of the III-IV loop on skeletal muscle EC coupling. *American Journal of Physiology. Cell Physiology, 287*, C1094–C1102.

Wilkens, C. M., Kasielke, N., Flucher, B. E., Beam, K. G., & Grabner, M. (2001). Excitation-contraction coupling is unaffected by drastic alteration of the sequence surrounding residues L720-L764 of the alpha 1S II-III loop. *Proceedings of the National Academy of Sciences of the United States of America, 98*, 5892–5897.

Wong King Yuen, S. M., Campiglio, M., Tung, C.-C., Flucher, B. E., & Van Petegem, F. (2017). Structural insights into binding of STAC proteins to voltage-gated calcium channels. *Proceedings of the National Academy of Sciences, 114*, E9520–E9528.

Wu, J., Yan, Z., Li, Z., Yan, C., Lu, S., Dong, M., & Yan, N. (2015). Structure of the voltage-gated calcium channel Cav1.1 complex. *Science (80-), 350*, aad2395-1–9.

Wu, J., Yan, Z., Li, Z., Qian, X., Lu, S., Dong, M., Zhou, Q., & Yan, N. (2016). Structure of the voltage-gated calcium channel Cav1.1 at 3.6 Å resolution. *Nature, 537*, 191–196.

Ca$_v$1 L-Type Calcium Channels in the Auditory and Visual Systems

Alexandra Koschak and Amy Lee

Abstract

In sensory neurons, presynaptic voltage-gated L-type calcium channels (Ca$_v$1) mediate calcium influx as a first important step in synaptic transmission. This review illuminates the role of Ca$_v$1 channels in the auditory and visual systems. Specifically, we describe the functional properties of Ca$_v$ channels in sensory receptor cells of the retina and the inner ear considering also their regulation by alternative splicing and interaction with calcium-binding proteins. Further, we pay particular attention to the role of Ca$_v$1 channels in the maturation and development of the visual and hearing systems. Finally, we review Ca$_v$1 channelopathies that affect the visual and hearing systems describing how dysregulation of these channels leads to disorders of vision and hearing in humans.

Keywords

Voltage-gated calcium channel · Channelopathy · Ribbon synapse · Hair cell · Photoreceptor

A. Koschak (✉)
Pharmacology and Toxicology, Institute of Pharmacy, University of Innsbruck, Innsbruck, Austria
e-mail: alexandra.koschak@uibk.ac.at

A. Lee (✉)
Department of Neuroscience, University of Texas-Austin, Austin, TX, USA
e-mail: amy.lee1@austin.utexas.edu

Introduction

The transmission of visual and auditory information begins at synapses formed by photoreceptors and bipolar cells in the retina and hair cells in the inner ear. A functional specialization of these synapses is the synaptic ribbon, an organelle that primes synaptic vesicles for release and supports continuous exocytosis, generally at high rates (for review, see Sterling & Matthews, 2005). Unlike fast phasic neurotransmitter release evoked by action potentials, transmission at ribbon synapses occurs in response to graded changes in the membrane potential. A second hallmark of ribbon synapses is a reliance on Ca$_v$1 L-type Ca^{2+} channels (e.g., reviewed in Joiner & Lee 2015; Pangrsic et al., 2018), in contrast to Ca$_v$2 channels that predominate at conventional synapses (e.g., reviewed in Mochida, 2018; Dolphin & Lee, 2020). Arranged in clusters in the presynaptic density beneath the synaptic ribbon, Ca$_v$1 channels mediate the influx of Ca^{2+} ions that enable vesicular fusion with the plasma membrane and subsequent glutamate release into the synaptic cleft. Compared to their expression in other cell-types (e.g., myocytes), Ca$_v$1 channels at ribbon synapses exhibit unique properties that are well tailored to support the requirements for analog transmission. These properties arise from unique molecular determinants within the Ca$_v$1 protein as well as interactions of/with various proteins. Mutations in genes affecting the function of Ca_v1 channels have deleterious consequences for hearing, balance, or vision. In this review, we will discuss the structural and functional specialization of Ca_v1 channels that enable sensory information to be encoded reliably at ribbon synapses.

Role of Ca_v1 Channels in Hearing and Vision

Ca_v1 Channel Family

Macromolecular protein complexes like Ca_v channels comprise a number of protein subunits: the pore-forming Ca^{2+}-channel $Ca_v\alpha_1$-subunit is formed by one of the ten isoforms; accessory subunits contribute to the channel complex and include one of the four different $Ca_v\beta$ and $Ca_v\alpha_2\delta$ subunits (for review, see Buraei & Yang, 2010; Dolphin, 2012. See also chapters "Subunit Architecture and Atomic Structure of Voltage Gated Ca2+ Channels" by Dr. Catterall, "Voltage-Gated Calcium Channel Auxiliary β Subunits" by Dr. Colecraft, and "Regulation of Calcium Channels and Synaptic Function by Auxiliary α2δ Subunits" by Drs. Dolphin and Obermair). Most of the channel's biophysical and pharmacological properties are conferred by the $Ca_v\alpha_1$ subunit; however, accessory subunits enhance $Ca_v\alpha_1$ trafficking and functional expression at the plasma membrane and influence the channels' biophysical properties. According to their functional characteristics, Ca_v channels are grouped as L-type (Ca_v1 family), P/Q-, N-, and R-type (Ca_v2 family), and T-type (Ca_v3 family) (for review, see Zamponi et al., 2015).

In this chapter, we focus on Ca_v1 channels, which mediate L-type currents and neurotransmitter release at ribbon synapses upon changes in graded potentials. The Ca_v1 gating properties allow for fast responses to fluctuations in the membrane potential caused by changes in the intensity of a sensory stimulus. The Ca_v1 channel family is composed of four isoforms: $Ca_v1.1$, $Ca_v1.2$, $Ca_v1.3$, and $Ca_v1.4$. Both $Ca_v1.3$ and $Ca_v1.4$ channels activate at relatively hyperpolarized potentials that are in the range of sensory

cell potentials, and inactivate slowly (Koschak et al., 2001, 2003; Xu & Lipscombe, 2001; Baumann et al., 2004). Thus, they are highly suited to regulate tonic transmission at visual or auditory sensory synapses (for review, see Pangrsic et al., 2018). The pharmacology of Ca_v1 channels is well described. Among many clinically relevant Ca_v1 Ca^{2+} channel modulators, dihydropyridines (DHPs) like amlodipine, felodipine, or nifedipine are widely used for the treatment of hypertension (for review, see Zamponi et al., 2015). As demonstrated in heterologous expression systems, recombinant $Ca_v1.3$ and $Ca_v1.4$ exhibit about five- to tenfold lower sensitivity to DHPs compared to $Ca_v1.2$ (Koschak et al., 2001, 2003; Xu & Lipscombe, 2001; Zanetti et al., 2021). Therefore, DHPs can't be considered as being able to definitively distinguish between Ca_v1 subtypes, making the pharmacological distinction of native Ca_v1 channels arduous.

Ca_v1 Channels in Hearing

Hearing relies on the ability of inner hair cells (IHCs) to transduce mechanical stimuli into an electrochemical signal. Sound waves entering the cochlea cause the deflection of hair bundles in the apical membrane of IHCs, which modulates the opening of mechanoelectrical transduction channels. This results in a graded change in the IHC membrane potential, which controls the opening of Ca_v1 channels clustered near the presynaptic ribbon. Ca^{2+} influx through these channels promotes the release of glutamate sensed by postsynaptic spiral ganglion neuron (SGN) afferents, which transmit auditory information into the central nervous system via the auditory nerve. IHC synapses are tonically active and support spontaneous firing above 100 Hz (Liberman, 1978). IHCs encode sound intensity with an increase in afferent firing rates and communicate the temporal features of sound by phase-locking transmitter release to frequencies up to 5 kHz (Goutman, 2012).

Based on human genetics and animal studies, $Ca_v1.3$ is the predominant Ca_v subtype in IHCs and outer hair cells (OHCs), which mediate sound amplification. In IHCs and OHCs of Ca$_v$1.3 (KO) knock out mice, Ca^{2+} currents are nearly abolished (Platzer et al., 2000; Brandt et al., 2003; Michna et al., 2003). Ca$_v$1.3 KO mice are deaf and lack sound-evoked auditory brainstem responses (ABR; Platzer et al., 2000; Brandt et al., 2003). Similarly, a loss-of function mutation in the *CACNA1D* gene encoding Ca$_v$1.3 causes congenital deafness in humans (Baig et al., 2011; see also, section "Calcium channelopathies in hair cells"). Compared to other Ca$_v$ subtypes, Ca$_v$1.3 activates quite rapidly with time constants for activation and latency to first opening in the sub-millisecond range (Zampini et al., 2010, 2013; Inagaki & Lee, 2013; Schnee & Ricci, 2003). Fast activation and deactivation of Ca$_v$1.3 is likely to be important for triggering auditory nerve firing to reflect sound frequency (i.e., phase locking) and for minimizing delays in synaptic transmission relevant for binaural sound localization (Grothe et al., 2010). Ca$_v$1.3 is also distinguished by its activation at relatively negative voltages. For example, in immature mouse IHCs, the threshold for Ca^{2+} current activation was found to be ~ -70 mV (Zampini et al., 2010). Since the resting membrane potential of these IHCs is ~ -60 mV (Marcotti et al., 2003), a fraction of activated Ca$_v$1.3 channels is expected to support tonic transmitter release at rest and increased channel opening with depolarization promotes greater rates of transmitter release thus signaling an increase in sound intensity. Compared to their counterparts in heterologous expression systems and in the heart, Ca$_v$1.3 channels in IHCs exhibit little Ca^{2+}- or voltage-dependent inactivation (CDI, VDI). CDI depends on the EF-hand Ca^{2+}-binding protein, calmodulin (CaM), which interacts with a consensus site (i.e., IQ domain) in the cytoplasmic C-terminal region of Ca$_v$1.3. CDI is evidenced by a faster decay of currents carried by Ca^{2+} than by Ba^{2+} as Ba^{2+} binds poorly to CaM (Ben-Johny & Yue, 2014). The prolonged opening of Ca$_v$1.3 due to limited CDI is thought to support sustained transmitter release that is characteristic of the IHC ribbon synapses. Despite decades of research, the mechanism underlying the limited CDI of Ca$_v$1.3 in IHCs has not been definitively estab-

lished. Possibilities include alternative splicing of the IQ domain (Shen et al., 2006; see also section on "Alternative splicing" and chapter "Splicing and Editing to Fine-Tune Activity of High Voltage-Activated Calcium Channels" by Dr. Huang and colleagues) and interactions with various proteins including Rab3-interacting molecule 2 (RIM2; Gebhart et al., 2010; Jung et al., 2015) and CaBPs, a family of Ca^{2+}-binding proteins related to CaM (see section "Ca$_v$1.3 and Ca$_v$1.4 channel interaction with Ca^{2+} binding proteins").

Ca$_v$1 Channels in Vision

There have been some controversies surrounding the identity of Ca$_v$1 channels at retinal ribbon synapses. Immunohistochemistry and in situ hybridization suggest the presence of all Ca$_v$1 subunits at mouse retinal ribbon synapses (Berntson et al., 2003; Morgans et al., 2005; Xiao et al., 2007; Busquet et al., 2010; Kersten et al., 2010; Hasan et al., 2016; Seitter & Koschak, 2018). However, most studies suggest the predominant expression of Ca$_v$1.4 at photoreceptor ribbon synapses where they mediate sustained Ca^{2+} entry needed for continuous neurotransmitter release (for review, see Pangrsic et al., 2018; Koschak et al., 2021). For tonic release, a few Ca$_v$1 channels that activate rapidly at relatively negative voltages (Bartoletti et al., 2011; Heidelberger & Matthews, 1992; von Gersdorff & Matthews, 1996; Bartoletti & Thoreson, 2011) and inactivate slowly are sufficient (Bartoletti et al., 2011). Consistent with these observations in invertebrate retinas, heterologous expression studies showed that recombinant Ca$_v$1.4 currents indeed activate at membrane potentials negative to -40 mV and show fast activation (Koschak et al., 2001, 2003; Baumann et al., 2004). Moreover, their VDI is slow (Koschak et al., 2001, 2003; Baumann et al., 2004; McRory et al., 2004) and CDI is limited (Singh et al., 2006; Wahl-Schott et al., 2006). Although the kinetics of inactivation of native Ca$_v$1.4 currents is faster, the general properties of recombinant Ca$_v$1.4 established in heterologous systems lead to a negative window of activation preserved under physiological temperatures (Peloquin et al., 2008).

In contrast to $Ca_v1.3$ (see section "Ca_v1 channels in hearing"), the lack of CDI in $Ca_v1.4$ is well explained by an inhibitory domain in their C-terminus that actively suppresses CDI (Singh et al., 2006; Wahl-Schott et al., 2006). Termed as an inhibitor of CDI (ICDI; Wahl-Schott et al., 2006), this domain has also been referred to as a C-terminal modulator (CTM) because other gating properties were also shown to be affected (Singh et al., 2006). Similar to $Ca_v1.2$ channels, the modulation conferred by the C-terminus of $Ca_v1.4$ (and also $Ca_v1.3$) channels is due to the interaction between a proximal (PCRD) and a distal (DCRD) C-terminal regulatory domain (Singh et al., 2006, 2008) competing with calmodulin (CaM) binding (Liu et al., 2010). CDI is therefore absent ($Ca_v1.4$) or weaker ($Ca_v1.3$; Burtscher et al., 2014; Singh et al., 2006, 2008). Of note, the CTM also affects the activation gating (Singh et al., 2006, 2008; Bock et al., 2011; Hulme et al., 2005, 2006; Burtscher et al., 2014) and/or the open probability of Ca_v1 channels (Singh et al., 2008; Bock et al., 2011; Hulme et al., 2005, 2006).

Regulation of $Ca_v1.3$ and $Ca_v1.4$ Functional Properties

Alternative Splicing

Alternative splicing represents a key mechanism for the regulation of the functional properties of Ca_v1 channels, which—among Ca_v1 channels—is well described in the C-terminus of $Ca_v1.3$ (Singh et al., 2008; Bock et al., 2011; Tan et al., 2011; Hofer et al., 2021) and $Ca_v1.4$ subunits (Tan et al., 2012; Haeseleer et al., 2016). Alternative splicing can affect the DCRD domain (Singh et al., 2008; Bock et al., 2011) or the CaM-binding IQ domain of $Ca_v1.3$ (Tan et al., 2011) thereby changing the channels' biophysical properties. Alternative splicing of the $Ca_v1.3$ C-terminus results in "long" ($Ca_v1.3_L$ also termed $Ca_v1.3_{42}$) and "short" ($Ca_v1.3_{42A}$ or $Ca_v1.3_{43S}$)

channels likely with different functions in different tissues (Singh et al., 2008; Bock et al., 2011; Tan et al., 2011; for review, see Seitter & Koschak, 2018; see also, chapter "Splicing and Editing to Fine-Tune Activity of High Voltage-Activated Calcium Channels" by Huang and colleagues). In mouse IHCs, alternative splicing in the C-terminus of $Ca_v1.3$ may contribute to functional heterogeneity (Vincent et al., 2017) whereas in OHCs, the expression of $Ca_v1.3$ channels lacking the IQ motif ($Ca_v1.3_{IQ\Delta}$) might explain the almost complete absence of CDI in these cells (Shen et al., 2006). In addition to the $Ca_v1.3_{IQ\Delta}$ splice variant, studies have also identified and characterized an isoform lacking exon 41 ($Ca_v1.3_{\Delta41}$) thereby truncating the C-terminus beyond the EF-hand. $Ca_v1.3_{\Delta41}$ channels showed only a slight depolarizing shift in the voltage-dependence of activation and attenuated CDI, as expected from the removal of the entire IQ motif, but also enhanced VDI (Tan et al., 2011). Interestingly, analyses of C-terminal $Ca_v1.3$ splice variants revealed that pharmacological properties can be modulated by alternative splicing (Huang et al., 2013). Splice variants like $Ca_v1.3_{42a}$ and $Ca_v1.3_{43s}$ with intact CaM–IQ domain interaction but fast inactivation showed consistently lower DHP sensitivity compared to "long" $Ca_v1.3_{42}$ channels. In comparison $Ca_v1.3_{\Delta41}$ showed greater DHP sensitivity, which was explained by the loss of CaM binding to the IQ motif rather than by differences in CDI properties (Huang et al., 2013).

As can be inferred from $Ca_v1.3$ channels (Singh et al., 2008; Bock et al., 2011), alternative splicing in the C-terminus of $Ca_v1.4$ leads to pronounced CDI and a marked hyperpolarizing shift of activation voltage due to substantial deletions including the CTM (Ex43* and ΔEx47/ $Ca_v1.4_{\Delta ex47}$; see also section "CaBPs in hair cells and photoreceptors") (Williams et al., 2018). A variant that lacks only a portion of exon 45 (Δex p45) and thereby results in a minor shortening of the C-terminus only, has no effect on the current–voltage relationship or CDI properties (Haeseleer et al., 2016). Of note, alternative splicing in the IVS3–S4 linker is also common in the Ca_v1 channel family and likewise confers functional diversity in terms of modulating the influence of the S4 voltage sensors. Correspondingly, skipping of exon 32 in the IVS3–S4 linker (Ca$_v$1.4$_{\Delta ex32}$) that shortens the linker is likely to alter channel activation (Tan et al., 2012). So far, the ΔEx32 variant has only been studied in a chimera (α_{1D}E32$_{1D\Delta}$ in Liu et al., 2017), which exhibits a significant shift of the activation curve toward more depolarized potentials.

Ca$_v$1.3 and Ca$_v$1.4 Channel Interaction with Ca^{2+}-Binding Proteins

In addition to their association with Ca$_v$β and Ca$_v$α2δ subunits, Ca$_v$1 channels interact with a variety of proteins that regulate the function and localization, and therefore the synaptic contribution of these channels in the retina and inner ear (Pangrsic et al., 2018). Among such Ca$_v$-interacting proteins, members of the superfamily of EF-hand Ca^{2+}-binding proteins have emerged as key regulators of Ca$_v$1 channels and potentially other effectors at ribbon synapses.

Calmodulin (CaM)

CaM is the primordial member of a superfamily of EF-hand Ca^{2+}-binding proteins that have emerged as major regulators of Ca$_v$ channels (reviewed in Hardie & Lee, 2016; Ben-Johny et al., 2015). The structural and functional highlights of this "calmodulation" have been reviewed recently (Dick et al., 2016; see also, chapter "Calmodulin Regulation of Voltage-Gated Calcium Channels" by Drs. Dick and Ben Johny). Permeating Ca^{2+} ions bind to Ca^{2+}-free (apo)CaM that is tethered to the IQ-domain and potentially other sites within the channel protein. Ca^{2+}-bound CaM then promotes conformational changes that favor CDI. The N- and C-terminal lobes of CaM each contain a pair of EF-hand Ca^{2+}-binding motifs that have different affinities for Ca^{2+} (~6-times higher for the C- as compared to the N-lobe; Faas et al., 2011). The CaM C-lobe binds Ca^{2+} and releases it slowly, thus displaying "local Ca^{2+}

selectivity" (Tadross et al., 2008) near open channels. Because of this property, CDI of Ca$_v$1 channels is not blunted by high concentrations of the Ca^{2+} chelator BAPTA in intracellular recording solutions (Thomas & Lee, 2016). In contrast, the CaM N-lobe binds and releases Ca^{2+} rapidly (Faas et al., 2011), mediating a "global Ca^{2+} selectivity" for Ca^{2+} signals at relatively distant sites. However, CaM binding to the N-terminal spatial Ca^{2+}-transforming element (NSCaTE), which is only found in Ca$_v$1.2 and Ca$_v$1.3, causes the CaM N-lobe to switch to a local Ca^{2+} selectivity causing a form of CDI that can be blunted by BAPTA (Dick et al., 2008). In IHCs, Ca$_v$1 channels exhibit CDI that is sensitive to BAPTA or EGTA (Johnson & Marcotti, 2008; Grant & Fuchs, 2008; Inagaki & Lee, 2013; von Gersdorff & Matthews, 1996; Corey et al., 1984). Mechanisms that favor NSCaTE-mediated CDI could be important for restricting Ca$_v$1 channels to specific patterns of neurotransmitter release mediated by more global elevations in intracellular Ca^{2+}. For example, Ca^{2+}-induced Ca^{2+} release mediated by ryanodine receptors supports slow, non-ribbon-mediated exocytosis in rods that is independent of Ca$_v$1.4 (Chen et al., 2014).

CaBPs in Hair Cells and Photoreceptors

CaBPs are EF-hand Ca^{2+}-binding proteins with ~50% homology with CaM. Like CaM, CaBPs possess N- and C-lobes with pairs of EF-hand Ca^{2+}-binding motifs at least one of which is non-functional (for review, see Haeseleer et al., 2002). Unlike CaM, CaBPs are expressed primarily in neuronal cell-types (CaBP1/caldendrin) (Kim et al., 2014) throughout the central nervous system, retina (CaBP1, CaBP2, CaBP4, CaBP5) (Haeseleer et al., 2004; Rieke et al., 2008; Sinha et al., 2016), and the inner ear (CaBP1, CaBP2, CaBP4, CaBP5) (Cui et al., 2007; Yang et al., 2016). Because CaBPs bind to sites including the IQ-domain of Ca$_v$1 channels, competition with CaM is thought to underlie limited CDI upon co-expression of CaBPs with Ca$_v$1 channels. However, since they also bind to other cytoplasmic domains of the Ca_v1 protein, CaBPs may also regulate CDI by a non-competitive mechanism (Oz et al., 2013; Yang et al., 2014).

Because they temper CDI of recombinant $Ca_v1.3$ in heterologous expression systems, CaBPs are logical candidates for mediating the slowly inactivating properties of $Ca_v1.3$ in IHCs. By quantitative PCR, CaBP1 and CaBP2 are the major CaBPs expressed in the cochlea (Yang et al., 2016), with immunolabeling suggesting the presence of CaBP1, CaBP2, CaBP4, and CaBP5 in IHCs (Cui et al., 2007; Yang et al., 2006). The presence of multiple CaBPs in IHCs may account for the limited impact of singly knocking out CaBP1, CaBP2, and CaBP4 on CDI of the IHC Ca^{2+} current (Cui et al., 2007; Yang et al., 2018b; Picher et al., 2017). The faster inactivation of Ca^{2+} currents in IHCs of CaBP2 KO mice seems to be due to an effect on VDI rather than on CDI and correlates with abnormal ABR thresholds (Picher et al., 2017). In adult mice, CaBP2 is also expressed in OHCs where it could similarly regulate $Ca_v1.3$. In support of this possibility, distortion product otoacoustic emissions (DPOAEs), which reflect OHC function, are abnormal in CaBP2 KO mice (Yang et al., 2018b).

Within the retina, CaBP4 is the predominant CaBP expressed in rods and cones (Haeseleer et al., 2004; Maeda et al., 2005). Within the synaptic terminals of these cells, CaBP4 interacts with $Ca_v1.4$ and shifts the voltage-dependence of activation of the channel to more hyperpolarized potentials (Haeseleer et al., 2004; Lee et al., 2015). This effect of CaBP4 is only observed for $Ca_v1.4$ channels containing the CTM (Shaltiel et al., 2012). As seen for $Ca_v1.3$ (Cui et al., 2007; Lee et al., 2007), CaBP4 can suppress CDI of $Ca_v1.4$ but only for splice variants that lack a functional CTM. For example, $Cav1.4_{\Delta ex47}$ is a human splice variant that lacks a portion of the CTM and exhibits robust CDI and a hyperpolarized activation range (Haeseleer et al., 2016; Williams et al., 2018). Co-expression of CaBP4 with $Cav1.4_{\Delta ex47}$ suppresses CDI without additional effect on the activation properties (Haeseleer et al., 2016). CaBP4 may ensure prolonged opening of splice variants lacking a por-

tion or all of the CTM (i.e., $Ca_v1.4$ ex 43[*]; Tan et al., 2012) in order to support sustained neurotransmission from photoreceptors in darkness.

CaBPs in Other Retinal/Cochlear Cell-Types

In addition to photoreceptors and IHCs, CaBPs are expressed in other neurons of the retinal and cochlear circuitries. Within the cochlea, CaBP1 is most highly expressed in SGNs (Yang et al., 2016), where it regulates CDI of Ca_v1 channels and their contribution to neurite outgrowth (Yang et al., 2018a). Increased CDI of Ca_v1 channels correlates with heightened excitability of SGNs in CaBP1 KO mice, perhaps due to decreased coupling of Ca_v1 channels to Ca^{2+}-activated K^+ channels (Yang et al., 2018b). This cellular phenotype could underlie the enhanced synchrony of auditory nerve fibers as suggested by abnormal amplitudes and kinetics of wave I in the ABRs of CaBP1 KO mice (Yang et al., 2018b).

In the inner retina, CaBP1 and CaBP2 are expressed in OFF type 1 and type 2 cone bipolar cells (Menger et al., 1999; Haverkamp & Wassle, 2000; Haverkamp et al., 2003b; Sinha et al., 2016), and CaBP2 is additionally expressed in ON type 6 cone bipolar cells (Sinha et al., 2016). In CaBP2 KO mice, ON alpha-ganglion cells exhibit reduced excitatory postsynaptic current (EPSC) amplitudes, which could result from increased CDI of presynaptic Ca_v1 channels in CaBP2 KO ON cone bipolar cells (Sinha et al., 2016). In contrast, EPSC amplitudes of OFF T-ganglion cells are significantly increased in CaBP1 KO mice (Sinha et al., 2016), which could stem from altered function of Ca_v channels or other targets in amacrine cells—inhibitory interneurons in which CaBP1 is expressed (Sinha et al., 2016). CaBP5 is expressed in rod bipolar cells as well as in ON type 5 and OFF type 3 cone bipolar cells (Haverkamp et al., 2003a; Ghosh et al., 2004). CaBP5 interacts with and suppresses CDI of recombinant $Ca_v1.2$ in a heterologous expression system, and has been reported to co-localize with $Ca_v1.2$ in rod bipolar cells (Rieke et al., 2008). Thus, increased CDI could contribute to the reduced sensitivity of rod-mediated ganglion cell light responses in CaBP5 KO mice (Rieke et al., 2008). However, CaBP5 may also affect aspects of synaptic vesicle cycling via interactions with Munc18–1 and myosin VI (Sokal & Haeseleer, 2011).

Role of Ca$_v$1 Channels for the Development of the Visual and Hearing System

Ca$_v$1.3 Channels in Inner Hair Cell Maturation

In addition to their roles in triggering exocytosis at mature IHC synapses, Ca$_v$1.3 channels are critical for the maturation of IHCs. Immature IHCs exhibit spontaneous Ca^{2+}-dependent action potentials (Kros et al., 1998; Tritsch et al., 2007). The decline in this pace-making activity toward hearing onset is attributed to the combined effects of a decline in Ca^{2+} currents and an increase in K$^+$ currents (Marcotti et al., 2003). IHCs from Ca$_v$1.3 KO mice lack pre-sensory spiking as well as other aspects of IHC maturation that include the expression of BK K$^+$ channels and the disappearance of direct olivocochlear innervation of IHCs (Brandt et al., 2003; Nemzou et al., 2006). The developmental reduction in Ca$_v$1.3 current density is mediated by thyroid hormone signaling and is critical for improved Ca^{2+}-efficiency of exocytosis in mature IHCs (Sendin et al., 2007; Brandt et al., 2007). Although independent of a reduction in *Cacna1d* mRNA (Sendin et al., 2007), the diminished Ca$_v$1.3 current density in mature IHCs could involve the PDZ scaffold harmonin, which binds to Ca$_v$1.3 and promotes ubiquitination and proteasomal degradation of the channel protein (Gregory et al., 2011). Interestingly, extrasynaptic Ca$_v$1.3 channels (i.e., those outside of ribbon sites) are removed during maturation, while synaptic Ca$_v$1.3 channels undergo a change in organization from clusters from round to stripe-like (Wong et al., 2014).

While required for the maturation of IHC synapses, Ca$_v$1.3 is dispensable for ribbon formation and overall molecular assembly and structure of these synapses (Brandt et al., 2003; Nemzou et al., 2006). This is in stark contrast to the absolute requirement for Ca$_v$1.4 expression for the integrity of photoreceptor synapses, as outlined below.

Ca$_v$1.4 Channels in Photoreceptor Synapse Development

Ca$_v$1.4 channels are crucial for the fate of retinal photoreceptors because their development depends on their proper function and expression. In the absence of functional Ca$_v$1.4 channels, photoreceptor ribbon synapses remain mostly immature as reported in studies of different Ca$_v$1.4 mutant animal models: Ca$_v$1.4-deficient mice (Ca$_v$1.4 KO: Mansergh et al., 2005; Raven et al., 2008; Specht et al., 2009; Zabouri & Haverkamp, 2013; Liu et al., 2013; Knoflach et al., 2013), mice carrying a naturally occurring mutation resulting in strongly reduced Ca$_v$1.4 expression (nob2: Chang et al., 2006), a gain-of-function point mutation (Ca$_v$1.4-IT: Liu et al., 2013; Knoflach et al., 2013; Regus-Leidig et al., 2014; Michalakis et al., 2014), or a non-conducting Ca$_v$1.4 mutation (Cav1.4-G369i: Maddox et al., 2020). The structural integrity of proteins forming the ribbon or the arciform density/plasma membrane compartment is affected in Ca$_v$1.4 mutant retinas making this channel an important synapse organizer (Mansergh et al., 2005; Raven et al., 2008; Specht et al., 2009; Zabouri & Haverkamp, 2013; Liu et al., 2013; Knoflach et al., 2013; Chang et al., 2006; Maddox et al., 2020).

Whereas the initial stages of ribbon formation in photoreceptor are largely unaffected, ribbon structures appear elongated and/or spherical and are free-floating rather than anchored to the membrane after postnatal day (P) 8 in Ca$_v$1.4-deficient (Ca$_v$1.4-KO) mice. These synaptic alterations are observed later (after P13) in retinas with dysfunctional Ca$_v$1.4 channels (Ca$_v$1.4-IT; Liu et al., 2013). This finding suggests a role for Ca$_v$1.4 in different stages of ribbon development: the initial assembly might be independent of Ca$_v$1.4 activity because ribbons develop normally through P13 whereas maturation depends on the proper $Ca_v1.4$ activation properties. A recent report using $Ca_v1.4$-G369i mice in which a non-conducting mutant form of $Ca_v1.4$ is expressed shows that adult rod ribbons still form in the absence of $Ca_v1.4$-mediated Ca^{2+} signals. Mutant $Ca_v1.4$ channels are properly expressed and targeted in photoreceptors but do not mediate voltage-dependent Ca^{2+} current (Maddox et al., 2020).

Postsynaptic elements in the signaling complex of rod bipolar cells can also be assembled in the absence of $Ca_v1.4$ Ca^{2+} signals, although the subcellular distribution of the postsynaptic proteins was partially obscured (Maddox et al., 2020). Resulting from presynaptic impairment of rod and cone photoreceptors, $Ca_v1.4$ dysfunction leads to dendritic sprouting of second-order neurons—rod and cone bipolar and horizontal cells—which can form ectopic synapses in the retinal outer nuclear layer (ONL). Ectopic ribbons and other presynaptic proteins therefore typically appear in the ONL (Mansergh et al., 2005; Chang et al., 2006; Liu et al., 2013; Knoflach et al., 2013, 2015; Regus-Leidig et al., 2014; Waldner et al., 2018; Maddox et al., 2020; Zanetti et al., 2021). Consistent with the more severe presynaptic defects in $Ca_v1.4$-deficient mice, postsynaptic remodeling starts earlier in $Ca_v1.4$-KO (around P11) than in $Ca_v1.4$-IT (Liu et al., 2013; no data for $Ca_v1.4$-G369i are available).

Channelopathies Affecting the Visual and Hearing System

Calcium Channelopathies in Hair Cells

Based on the requirement for $Ca_v1.3$ in IHC exocytosis, it is not surprising that mutations impacting the function of $Ca_v1.3$ cause various forms of hearing impairment in humans. For example, a mutation that disrupts Ca^{2+} conductance through $Ca_v1.3$ channels causes sinoatrial node dysfunction and deafness (*SANDD*, OMIM#614896; Baig et al., 2011); of note, also the cardiac pheno-

type of affected individuals is similar to that in $Ca_v1.3$ KO mice, and stems from the critical role of $Ca_v1.3$ in pace-making by sinoatrial nodal cells (Platzer et al., 2000; Liaqat et al., 2019).

Mutations in genes encoding $Ca_v1.3$-interacting proteins are also linked to hearing impairment. In particular, mutations in the *CABP2* gene lead to autosomal recessive moderate to severe non-syndromic hearing impairment (DFNB93; Schrauwen et al., 2012; Picher et al., 2017; Sheyanth et al., 2021; Koohiyan et al., 2019). One such *CABP2* mutation causes premature truncation of the protein after the second EF-hand domain and limits the ability of CaBP2 to suppress CDI of $Ca_v1.3$ (Schrauwen et al., 2012). Audiograms of affected individuals indicate an auditory synaptopathy phenotype consistent with a loss-of-function of $Ca_v1.3$ (Schrauwen et al., 2012). The ability of virally mediated expression of CaBP2 to restore $Ca_v1.3$ function as well as normal hearing in CaBP2 KO mice suggests gene therapy as a viable approach to restore hearing in individuals harboring genetic variants in *CABP2* (Oestreicher et al., 2021).

Calcium Channelopathies in the Retina

The important role of $Ca_v1.4$ in the retina is supported by the observation that mutations in the encoding *CACNA1F* gene cause several forms of human retinal diseases (OMIM #300071, #300476, #300600). The majority of $Ca_v1.4$ mutations were identified in humans with congenital stationary night blindness type 2 (CSNB2; Zeitz et al., 2015). $Ca_v1.4$ mutations can also affect proteins co-clustering with $Ca_v1.4$ in mature photoreceptor synapses (see section "$Ca_v1.4$ channels in photoreceptor synapse development") thereby altering ON and OFF bipolar signaling pathways. For precise functional phenotyping and diagnosis of CSNB2 patients, electroretinograms (ERGs) of both scotopic and photopic responses are performed (Zeitz et al., 2015). CSNB2 patients present with an abnormal dim scotopic ERG, and a typical negative bright-flash ERG shows severely reduced β-waves with α-waves preserved. Moreover, oscillatory potentials are missing. The reduction in the β-wave in both scotopic and photopic ERGs points toward a transmission problem from photoreceptors to bipolar cells (for review, see Zeitz et al., 2015).

CSNB2 phenotypes have been reported also for mutations in the Ca$_v$α$_2$δ$_4$ subunit (*CACNA2D4* gene; Wycisk et al., 2006) and CaBP4 (Zeitz et al., 2006)—both proteins that interact with Ca$_v$1.4 (see section "CaBPs in hair cells and photoreceptors"). Similarities in the functional and morphological phenotypes have recently been reviewed in detail (Zeitz et al., 2015). CSNB2 patients show variable levels of night blindness together with myopia, nystagmus, strabismus, and low visual acuity (Bijveld et al., 2013). Also, daylight and color vision as well as visual acuity can be affected, and more than half of the patients suffer from photophobia (Bijveld et al., 2013). Different mutation phenotypes that cause different channel defects—reported in various heterologous expression systems (McRory et al., 2004; Hemara-Wahanui et al., 2005; Hoda et al., 2005, 2006; Singh et al., 2006; Peloquin et al., 2008; Burtscher et al., 2014)—might explain some of the variable clinical manifestations of the disease. The spectrum of Ca$_v$1.4 dysfunction comprises severe structural changes due to missense mutations such that functional channels are unlikely to assemble or show strongly reduced or abolished Ca$_v$1.4 activity (for details and review, see Stockner & Koschak, 2013; Koschak et al., 2021).

Various mouse models helped to better understand the pathology of CSNB2. The first important conclusions on the role of the channel were drawn from two Ca$_v$1.4 KO models (Mansergh et al., 2005; Specht et al., 2009). Still, their functional phenotype is different than in human patients because Ca$_v$1.4 KO mice are functionally blind, lacking ganglion cell responses to physiological light stimuli (Knoflach et al., 2015) and visually evoked cortical activation (Mansergh et al., 2005). A naturally occurring mouse model, nob-2 (Chang et al., 2006), in which the full-length Ca$_v$1.4 protein is expressed at significantly reduced levels due to alternative splicing in exon 2, is more similar to the human CSNB2 phenotype (Doering et al., 2008). These mice show reduced ERG responses; however, optokinetic responses (spatial contrast sensitivity) are similar to wild-type mice (Doering et al., 2008).

The mentioned Ca$_v$1.4-IT missense mutation (Ca$_v$1.4-I745T: Hope et al., 2005; and mouse homolog Ca$_v$1.4-I756T: Specht et al., 2009), found in a New Zealand family, is of particular interest because the human ERG phenotype is well reflected in a mouse model carrying the respective mutation (Knoflach et al., 2013). However, this mouse strain shows progressive retinal degeneration (reduction in the size of the ONL; Knoflach et al., 2013; Regus-Leidig et al., 2014), a finding not reported in the New Zealand family, but long-term clinical investigations like ERG testing and/or optical coherence tomography were not undertaken. In mouse Ca$_v$1.4-IT retinas, the reduction in the size of the ONL is rather due to a loss of rods because cones do not undergo significant cell death (Zanetti et al., 2021). In contrast to rods, which were retracting into the ONL, cones retained their synaptic terminals in the (OPL) outer plexiform layer although their morphology is altered (Zanetti et al., 2021).

Of note, heterologously expressed Ca$_v$1.4-IT channels show a 30 mV hyperpolarizing shift of the voltage-dependence of activation and significantly slower inactivation kinetics (Hemara-Wahanui et al., 2005). Although Ca^{2+} currents have not been directly measured electrophysiologically in photoreceptors, Ca^{2+} imaging experiments showed increased intracellular Ca^{2+} concentration at the ribbon synapse (Regus-Leidig et al., 2014) in line with the pronounced hyperpolarizing shift seen in the voltage-dependence of Ca$_v$1.4-IT channels (Hemara-Wahanui et al., 2005). The question then arises as why the same mutation affects rods and cones differently? This finding possibly relates to specific internal Ca^{2+} modulation because cone pedicles possess the machinery to remove intra-terminal free Ca^{2+} more rapidly during light adaptation compared to rod spherules (Johnson et al., 2007) and might therefore be less susceptible to Ca^{2+}-induced toxicity. Still, patients with CSNB2 show a reduced cone function in the

ERG and quite severe daylight symptoms (Zeitz et al., 2015). Although in Ca$_V$1.4-IT retinas normal glutamate release from cones suggested functional cones, pathways involving cone-to-cone bipolar cell transmission including the secondary pathway are strongly impaired in this mouse model (Zanetti et al., 2021).

In adult Ca$_V$1.4-deficient mice, Waldner and colleagues reported that cone photoreceptor structure and photopic visual function—measured by immunohistochemistry, optokinetic response, and electroretinography—deteriorates progressively with age (Waldner et al., 2018). Moreover, cone axons can form ectopic, synapse-like contacts with rod bipolar cell dendrites in the ONL (Waldner et al., 2018), which they normally do not contact in wild-type retinas—a finding that has not been reported in other investigations of Ca$_V$1.4-KO, Ca$_V$1.4-IT, or Ca$_V$1.4-G369i retinas.

Conclusion

Studies of Ca$_V$1 channels in heterologous expression systems as well as in animal models have yielded critical insights into how the properties of these channels may help shape the initial transmission of visual and auditory information at ribbon synapses. However, we still know little about the mechanisms that help traffic these channels to their sites of action and maintain their clustering near synaptic ribbons. Despite the increasing number of splice variants that have been identified for Ca$_V$1.3 and Ca$_V$1.4, the physiological significance of these variants for synaptic signaling remains to be determined. Finally, we still lack a comprehensive view of how and when post-translational modifications as well as protein interactions impact the functions of Ca$_V$1 channels as exocytotic triggers. Addressing these questions will benefit from exogenous expression of modified Ca$_V$1 proteins into retinal and inner ear cell-types, and is an important challenge for future studies.

Acknowledgments The work was supported by the (NIH) National Institutes of Health (R03NS115653 and R01EY026817 to AL), the (FWF) Austrian Science Funds (P32747 and Doc30 to AK), the University of Innsbruck, and the Center for Molecular Biosciences Innsbruck (CMBI).

References

Baig, S. M., Koschak, A., Lieb, A., Gebhart, M., Dafinger, C., Nurnberg, G., Ali, A., Ahmad, I., Sinnegger-Brauns, M. J., Brandt, N., Engel, J., Mangoni, M. E., Farooq, M., Khan, H. U., Nurnberg, P., Striessnig, J., & Bolz, H. J. (2011). Loss of Ca(v)1.3 (CACNA1D) function in a human channelopathy with bradycardia and congenital deafness. *Nature Neuroscience, 14*, 77–84.

Bartoletti, T. M., & Thoreson, W. B. (2011). Quantal amplitude at the cone ribbon synapse can be adjusted by changes in cytosolic glutamate. *Molecular Vision, 17*, 920–931.

Bartoletti, T. M., Jackman, S. L., Babai, N., Mercer, A. J., Kramer, R. H., & Thoreson, W. B. (2011). Release from the cone ribbon synapse under bright light conditions can be controlled by the opening of only a few Ca(2+) channels. *Journal of Neurophysiology, 106*, 2922–2935.

Baumann, L., Gerstner, A., Zong, X., Biel, M., & Wahl-Schott, C. (2004). Functional characterization of the L-type Ca^{2+} channel Ca$_V$1.4α1 from mouse retina. *Investigative Ophthalmology & Visual Science, 45*, 708–713.

Ben-Johny, M., & Yue, D. T. (2014). Calmodulin regulation (calmodulation) of voltage-gated calcium channels. *The Journal of General Physiology, 143*, 679–692.

Ben-Johny, M., Dick, I. E., Sang, L., Limpitikul, W. B., Kang, P. W., Niu, J., Banerjee, R., Yang, W., Babich, J. S., Issa, J. B., Lee, S. R., Namkung, H., Li, J., Zhang, M., Yang, P. S., Bazzazi, H., Adams, P. J., Joshi-Mukherjee, R., Yue, D. N., & Yue, D. T. (2015). Towards a unified theory of calmodulin regulation (calmodulation) of voltage-gated calcium and sodium channels. *Current Molecular Pharmacology, 8*, 188–205.

Berntson, A., Taylor, W. R., & Morgans, C. W. (2003). Molecular identity, synaptic localization, and physiology of calcium channels in retinal bipolar cells. *Journal of Neuroscience Research, 71*, 146–151.

Bijveld, M. M., Florijn, R. J., Bergen, A. A., van den Born, L. I., Kamermans, M., Prick, L., Riemslag, F. C., van Schooneveld, M. J., Kappers, A. M., & van Genderen, M. M. (2013). Genotype and phenotype of 101 dutch patients with congenital stationary night blindness. *Ophthalmology, 120*, 2072–2081.

Bock, G., Gebhart, M., Scharinger, A., Jangsangthong, W., Busquet, P., Poggiani, C., Sartori, S., Mangoni, M. E., Sinnegger-Brauns, M. J., Herzig, S., Striessnig, J., & Koschak, A. (2011). Functional properties of a newly identified C-terminal splice variant of Cav1.3 L-type

Ca2+ channels. *The Journal of Biological Chemistry, 286*, 42736–42748.

Brandt, A., Striessnig, J., & Moser, T. (2003). CaV1.3 channels are essential for development and presynaptic activity of cochlear inner hair cells. *The Journal of Neuroscience, 23*, 10832–10840.

Brandt, N., Kuhn, S., Munkner, S., Braig, C., Winter, H., Blin, N., Vonthein, R., Knipper, M., & Engel, J. (2007). Thyroid hormone deficiency affects postnatal spiking activity and expression of Ca2+ and K+ channels in rodent inner hair cells. *The Journal of Neuroscience, 27*, 3174–3186.

Buraei, Z., & Yang, J. (2010). The ss subunit of voltage-gated Ca2+ channels. *Physiological Reviews, 90*, 1461–1506.

Burtscher, V., Schicker, K., Novikova, E., Pohn, B., Stockner, T., Kugler, C., Singh, A., Zeitz, C., Lancelot, M. E., Audo, I., Leroy, B. P., Freissmuth, M., Herzig, S., Matthes, J., & Koschak, A. (2014). Spectrum of Cav1.4 dysfunction in congenital stationary night blindness type 2. *Biochimica et Biophysica Acta, 1838*, 2053–2065.

Busquet, P., Nguyen, N. K., Schmid, E., Tanimoto, N., Seeliger, M. W., Ben-Yosef, T., Mizuno, F., Akopian, A., Striessnig, J., & Singewald, N. (2010). CaV1.3 L-type Ca2+ channels modulate depression-like behaviour in mice independent of deaf phenotype. *The International Journal of Neuropsychopharmacology, 13*, 499–513.

Chang, B., Heckenlively, J. R., Bayley, P. R., Brecha, N. C., Davisson, M. T., Hawes, N. L., Hirano, A. A., Hurd, R. E., Ikeda, A., Johnson, B. A., McCall, M. A., Morgans, C. W., Nusinowitz, S., Peachey, N. S., Rice, D. S., Vessey, K. A., & Gregg, R. G. (2006). The nob2 mouse, a null mutation in Cacna1f: Anatomical and functional abnormalities in the outer retina and their consequences on ganglion cell visual responses. *Visual Neuroscience, 23*, 11–24.

Chen, M., Krizaj, D., & Thoreson, W. B. (2014). Intracellular calcium stores drive slow non-ribbon vesicle release from rod photoreceptors. *Frontiers in Cellular Neuroscience, 8*, 20.

Corey, D. P., Dubinsky, J. M., & Schwartz, E. A. (1984). The calcium current in inner segments of rods from the salamander (Ambystoma tigrinum) retina. *The Journal of Physiology, 354*, 557–575.

Cui, G., Meyer, A. C., Calin-Jageman, I., Neef, J., Haeseleer, F., Moser, T., & Lee, A. (2007). Ca2+-binding proteins tune Ca2+-feedback to Cav1.3 channels in mouse auditory hair cells. *The Journal of Physiology, 585*, 791–803.

Dick, I. E., Tadross, M. R., Liang, H., Tay, L. H., Yang, W., & Yue, D. T. (2008). A modular switch for spatial Ca2+ selectivity in the calmodulin regulation of CaV channels. *Nature, 451*, 830–834.

Dick, I. E., Limpitikul, W. B., Niu, J., Banerjee, R., Issa, J. B., Ben-Johny, M., Adams, P. J., Kang, P. W., Lee, S. R., Sang, L., Yang, W., Babich, J., Zhang, M., Bazazzi, H., Yue, N. C., & Tomaselli, G. F. (2016). A rendezvous with the queen of ion channels: Three

decades of ion channel research by David T Yue and his Calcium Signals Laboratory. *Channels (Austin, Tex.), 10*, 20–32.

Doering, C. J., Rehak, R., Bonfield, S., Peloquin, J. B., Stell, W. K., Mema, S. C., Sauve, Y., & McRory, J. E. (2008). Modified Ca(v)1.4 expression in the Cacna1f(nob2) mouse due to alternative splicing of an ETn inserted in exon 2. *PLoS One, 3*, e2538.

Dolphin, A. C. (2012). Calcium channel auxiliary alph-a2delta and beta subunits: Trafficking and one step beyond. *Nature Reviews. Neuroscience, 13*, 542–555.

Dolphin, A. C., & Lee, A. (2020). Presynaptic calcium channels: Specialized control of synaptic neurotransmitter release. *Nature Reviews. Neuroscience, 21*, 213–229.

Faas, G. C., Raghavachari, S., Lisman, J. E., & Mody, I. (2011). Calmodulin as a direct detector of Ca2+ signals. *Nature Neuroscience, 14*, 301–304.

Gebhart, M., Juhasz-Vedres, G., Zuccotti, A., Brandt, N., Engel, J., Trockenbacher, A., Kaur, G., Obermair, G. J., Knipper, M., Koschak, A., & Striessnig, J. (2010). Modulation of Cav1.3 Ca2+ channel gating by Rab3 interacting molecule. *Molecular and Cellular Neurosciences, 44*, 246–259.

Ghosh, K. K., Bujan, S., Haverkamp, S., Feigenspan, A., & Wassle, H. (2004). Types of bipolar cells in the mouse retina. *The Journal of Comparative Neurology, 469*, 70–82.

Goutman, J. D. (2012). Transmitter release from cochlear hair cells is phase locked to cyclic stimuli of different intensities and frequencies. *The Journal of Neuroscience, 32*, 17025–1735a.

Grant, L., & Fuchs, P. (2008). Calcium- and calmodulin-dependent inactivation of calcium channels in inner hair cells of the rat cochlea. *Journal of Neurophysiology, 99*, 2183–2193.

Gregory, F. D., Bryan, K. E., Pangrsic, T., Calin-Jageman, I. E., Moser, T., & Lee, A. (2011). Harmonin inhibits presynaptic Cav1.3 Ca(2)(+) channels in mouse inner hair cells. *Nature Neuroscience, 14*, 1109–1111.

Grothe, B., Pecka, M., & McAlpine, D. (2010). Mechanisms of sound localization in mammals. *Physiological Reviews, 90*, 983–1012.

Haeseleer, F., Imanishi, Y., Sokal, I., Filipek, S., & Palczewski, K. (2002). Calcium-binding proteins: Intracellular sensors from the calmodulin superfamily. *Biochemical and Biophysical Research Communications, 290*, 615–623.

Haeseleer, F., Imanishi, Y., Maeda, T., Possin, D. E., Maeda, A., Lee, A., Rieke, F., & Palczewski, K. (2004). Essential role of Ca2+-binding protein 4, a Cav1.4 channel regulator, in photoreceptor synaptic function. *Nature Neuroscience, 7*, 1079–1087.

Haeseleer, F., Williams, B., & Lee, A. (2016). Characterization of C-terminal splice variants of Cav1.4 Ca2+ channels in human retina. *The Journal of Biological Chemistry, 291*, 15663–15673.

Hardie, J., & Lee, A. (2016). Decalmodulation of Cav1 channels by CaBPs. *Channels (Austin, Tex.), 10*, 33–37.

Hasan, N., Ray, T. A., & Gregg, R. G. (2016). CACNA1S expression in mouse retina: Novel isoforms and antibody cross-reactivity with GPR179. *Visual Neuroscience, 33*, E009.

Haverkamp, S., & Wassle, H. (2000). Immunocytochemical analysis of the mouse retina. *The Journal of Comparative Neurology, 424*, 1–23.

Haverkamp, S., Ghosh, K. K., Hirano, A. A., & Wassle, H. (2003a). Immunocytochemical description of five bipolar cell types of the mouse retina. *The Journal of Comparative Neurology, 455*, 463–476.

Haverkamp, S., Haeseleer, F., & Hendrickson, A. (2003b). A comparison of immunocytochemical markers to identify bipolar cell types in human and monkey retina. *Visual Neuroscience, 20*, 589–600.

Heidelberger, R., & Matthews, G. (1992). Calcium influx and calcium current in single synaptic terminals of goldfish retinal bipolar neurons. *The Journal of Physiology, 447*, 235–256.

Hemara-Wahanui, A., Berjukow, S., Hope, C. I., Dearden, P. K., Wu, S. B., Wilson-Wheeler, J., Sharp, D. M., Lundon-Treweek, P., Clover, G. M., Hoda, J. C., Striessnig, J., Marksteiner, R., Hering, S., & Maw, M. A. (2005). A CACNA1F mutation identified in an X-linked retinal disorder shifts the voltage dependence of Cav1.4 channel activation. *Proceedings of the National Academy of Sciences of the United States of America, 102*, 7553–7558.

Hoda, J. C., Zaghetto, F., Koschak, A., & Striessnig, J. (2005). Congenital stationary night blindness type 2 mutations S229P, G369D, L1068P, and W1440X alter channel gating or functional expression of Ca(v)1.4 L-type Ca2+ channels. *The Journal of Neuroscience, 25*, 252–259.

Hoda, J. C., Zaghetto, F., Singh, A., Koschak, A., & Striessnig, J. (2006). Effects of congenital stationary night blindness type 2 mutations R508Q and L1364H on Cav1.4 L-type Ca2+ channel function and expression. *Journal of Neurochemistry, 96*, 1648–1658.

Hofer, N. T., Pinggera, A., Nikonishyna, Y. V., Tuluc, P., Fritz, E. M., Obermair, G. J., & Striessnig, J. (2021). Stabilization of negative activation voltages of Cav1.3 L-Type Ca(2+)-channels by alternative splicing. *Channels (Austin, Tex.), 15*, 38–52.

Hope, C. I., Sharp, D. M., Hemara-Wahanui, A., Sissingh, J. I., Lundon, P., Mitchell, E. A., Maw, M. A., & Clover, G. M. (2005). Clinical manifestations of a unique X-linked retinal disorder in a large New Zealand family with a novel mutation in CACNA1F, the gene responsible for CSNB2. *Clinical & Experimental Ophthalmology, 33*, 129–136.

Huang, H., Yu, D., & Soong, T. W. (2013). C-terminal alternative splicing of CaV1.3 channels distinctively modulates their dihydropyridine sensitivity. *Molecular Pharmacology, 84*, 643–653.

Hulme, J. T., Konoki, K., Lin, T. W., Gritsenko, M. A., Camp, D. G., 2nd, Bigelow, D. J., & Catterall, W. A. (2005). Sites of proteolytic processing and noncovalent association of the distal C-terminal domain of CaV1.1 channels in skeletal muscle. *Proceedings of the National Academy of Sciences of the United States of America, 102*, 5274–5279.

Hulme, J. T., Yarov-Yarovoy, V., Lin, T. W., Scheuer, T., & Catterall, W. A. (2006). Autoinhibitory control of the Ca,1.2 channel by its proteolytically processed distal C-terminal domain. *The Journal of Physiology, 576*(Pt 1), 87–102.

Inagaki, A., & Lee, A. (2013). Developmental alterations in the biophysical properties of Ca(v) 1.3 Ca(2+) channels in mouse inner hair cells. *Channels (Austin, Tex.), 7*, 171–181.

Johnson, J. E., Jr., Perkins, G. A., Giddabasappa, A., Chaney, S., Xiao, W., White, A. D., Brown, J. M., Waggoner, J., Ellisman, M. H., & Fox, D. A. (2007). Spatiotemporal regulation of ATP and Ca2+ dynamics in vertebrate rod and cone ribbon synapses. *Molecular Vision, 13*, 887–919.

Johnson, S. L., & Marcotti, W. (2008). Biophysical properties of CaV1.3 calcium channels in gerbil inner hair cells. *The Journal of Physiology, 586*, 1029–1042.

Joiner, M. L., & Lee, A. (2015). Voltage-gated Cav1 channels in disorders of vision and hearing. *Current Molecular Pharmacology, 8*, 143–148.

Jung, S., Oshima-Takago, T., Chakrabarti, R., Wong, A. B., Jing, Z., Yamanbaeva, G., Picher, M. M., Wojcik, S. M., Gottfert, F., Predoehl, F., Michel, K., Hell, S. W., Schoch, S., Strenzke, N., Wichmann, C., & Moser, T. (2015). Rab3-interacting molecules 2alpha and 2beta promote the abundance of voltage-gated CaV1.3 Ca2+ channels at hair cell active zones. *Proceedings of the National Academy of Sciences of the United States of America, 112*, E3141–E3149.

Kersten, F. F., van Wijk, E., van Reeuwijk, J., van der Zwaag, B., Marker, T., Peters, T. A., Katsanis, N., Wolfrum, U., Keunen, J. E., Roepman, R., & Kremer, H. (2010). Association of whirlin with Cav1.3 (alpha1D) channels in photoreceptors, defining a novel member of the usher protein network. *Investigative Ophthalmology & Visual Science, 51*, 2338–2346.

Kim, K. Y., Scholl, E. S., Liu, X., Shepherd, A., Haeseleer, F., & Lee, A. (2014). Localization and expression of CaBP1/caldendrin in the mouse brain. *Neuroscience, 268*, 33–47.

Knoflach, D., Kerov, V., Sartori, S. B., Obermair, G. J., Schmuckermair, C., Liu, X., Sothilingam, V., Garrido, M. G., Baker, S. A., Glosmann, M., Schicker, K., Seeliger, M., Lee, A., & Koschak, A. (2013). Cav1.4 IT mouse as model for vision impairment in human congenital stationary night blindness type 2. *Channels (Austin, Tex.), 7*, 503–513.

Knoflach, D., Schicker, K., Glosmann, M., & Koschak, A. (2015). Gain-of-function nature of Cav1.4 L-type calcium channels alters firing properties of mouse retinal ganglion cells. *Channels (Austin, Tex.), 9*, 298–306.

Koohiyan, M., Noori-Daloii, M. R., Hashemzadeh-Chaleshtori, M., Salehi, M., Abtahi, H., & Tabatabaiefar, M. A. (2019). A novel pathogenic variant in the CABP2 gene causes severe nonsyndromic hearing loss in a consanguineous Iranian family. *Audiology & Neuro-Otology, 24*, 258–263.

Koschak, A., Reimer, D., Huber, I., Grabner, M., Glossmann, H., Engel, J., & Striessnig, J. (2001). alpha 1D (Cav1.3) subunits can form l-type Ca2+ channels activating at negative voltages. *The Journal of Biological Chemistry, 276*, 22100–22106.

Koschak, A., Reimer, D., Walter, D., Hoda, J. C., Heinzle, T., Grabner, M., & Striessnig, J. (2003). Ca$_v$1.4α1 subunits can form slowly inactivating dihydropyridine-sensitive L-type Ca^{2+} channels lacking Ca^{2+}-dependent inactivation. *The Journal of Neuroscience, 23*, 6041–6049.

Koschak, A., Fernandez-Quintero, M. L., Heigl, T., Ruzza, M., Seitter, H., & Zanetti, L. (2021). Cav1.4 dysfunction and congenital stationary night blindness type 2. *Pflügers Archiv, 473*, 1437–1454.

Kros, C. J., Ruppersberg, J. P., & Rusch, A. (1998). Expression of a potassium current in inner hair cells during development of hearing in mice. *Nature, 394*, 281–284.

Lee, A., Jimenez, A., Cui, G., & Haeseleer, F. (2007). Phosphorylation of the Ca2+−binding protein CaBP4 by protein kinase C zeta in photoreceptors. *The Journal of Neuroscience, 27*, 12743–12754.

Lee, A., Wang, S., Williams, B., Hagen, J., Scheetz, T. E., & Haeseleer, F. (2015). Characterization of Cav1.4 complexes (alpha11.4, beta2, and alpha2delta4) in HEK293T cells and in the retina. *The Journal of Biological Chemistry, 290*, 1505–1521.

Liaqat, K., Schrauwen, I., Raza, S. I., Lee, K., Hussain, S., Chakchouk, I., Nasir, A., Acharya, A., Abbe, I., Umair, M., Ansar, M., Ullah, I., Shah, K., G. University of Washington Center for Mendelian, Bamshad, M. J., Nickerson, D. A., Ahmad, W., & Leal, S. M. (2019). Identification of CACNA1D variants associated with sinoatrial node dysfunction and deafness in additional Pakistani families reveals a clinical significance. *Journal of Human Genetics, 64*, 153–160.

Liberman, M. C. (1978). Auditory-nerve response from cats raised in a low-noise chamber. *The Journal of the Acoustical Society of America, 63*, 442–455.

Liu, X., Yang, P. S., Yang, W., & Yue, D. T. (2010). Enzyme-inhibitor-like tuning of Ca(2+) channel connectivity with calmodulin. *Nature, 463*, 968–972.

Liu, X., Kerov, V., Haeseleer, F., Majumder, A., Artemyev, N., Baker, S. A., & Lee, A. (2013). Dysregulation of Ca(v)1.4 channels disrupts the maturation of photoreceptor synaptic ribbons in congenital stationary night blindness type 2. *Channels (Austin, Tex.), 7*, 514–523.

Liu, N., Liu, Y., Yang, Y., & Liu, X. (2017). Linker flexibility of IVS3-S4 loops modulates voltage-dependent activation of L-type Ca(2+) channels. *Channels (Austin, Tex.), 11*, 34–45.

Maddox, J. W., Randall, K. L., Yadav, R. P., Williams, B., Hagen, J., Derr, P. J., Kerov, V., Della Santina, L., Baker, S. A., Artemyev, N., Hoon, M., & Lee, A. (2020). A dual role for Cav1.4 Ca(2+) channels in the molecular and structural organization of the rod photoreceptor synapse. *eLife, 9*, e62184.

Maeda, T., Lem, J., Palczewski, K., & Haeseleer, F. (2005). A critical role of CaBP4 in the cone synapse.

Investigative Ophthalmology & Visual Science, 46, 4320–4327.

Mansergh, F., Orton, N. C., Vessey, J. P., Lalonde, M. R., Stell, W. K., Tremblay, F., Barnes, S., Rancourt, D. E., & Bech-Hansen, N. T. (2005). Mutation of the calcium channel gene Cacna1f disrupts calcium signaling, synaptic transmission and cellular organization in mouse retina. *Human Molecular Genetics, 14*, 3035–3046.

Marcotti, W., Johnson, S. L., Holley, M. C., & Kros, C. J. (2003). Developmental changes in the expression of potassium currents of embryonic, neonatal and mature mouse inner hair cells. *The Journal of Physiology, 548*, 383–400.

McRory, J. E., Hamid, J., Doering, C. J., Garcia, E., Parker, R., Hamming, K., Chen, L., Hildebrand, M., Beedle, A. M., Feldcamp, L., Zamponi, G. W., & Snutch, T. P. (2004). The CACNA1F gene encodes an L-type calcium channel with unique biophysical properties and tissue distribution. *The Journal of Neuroscience, 24*, 1707–1718.

Menger, N., Seidenbecher, C. I., Gundelfinger, E. D., & Kreutz, M. R. (1999). The cytoskeleton-associated neuronal calcium-binding protein caldendrin is expressed in a subset of amacrine, bipolar and ganglion cells of the rat retina. *Cell and Tissue Research, 298*, 21–32.

Michalakis, S., Shaltiel, L., Sothilingam, V., Koch, S., Schludi, V., Krause, S., Zeitz, C., Audo, I., Lancelot, M. E., Hamel, C., Meunier, I., Preising, M. N., Friedburg, C., Lorenz, B., Zabouri, N., Haverkamp, S., Garcia Garrido, M., Tanimoto, N., Seeliger, M. W., … Wahl-Schott, C. A. (2014). Mosaic synaptopathy and functional defects in Cav1.4 heterozygous mice and human carriers of CSNB2. *Human Molecular Genetics, 23*, 1538–1550.

Michna, M., Knirsch, M., Hoda, J. C., Muenkner, S., Langer, P., Platzer, J., Striessnig, J., & Engel, J. (2003). Ca$_v$1.3 (α1D) Ca^{2+} currents in neonatal outer hair cells of mice. *The Journal of Physiology, 553*, 747–758.

Mochida, S. (2018). Presynaptic calcium channels. *Neuroscience Research, 127*, 33–44.

Morgans, C. W., Bayley, P. R., Oesch, N. W., Ren, G., Akileswaran, L., & Taylor, W. R. (2005). Photoreceptor calcium channels: Insight from night blindness. *Visual Neuroscience, 22*, 561–568.

Nemzou, N. R., Bulankina, A. V., Khimich, D., Giese, A., & Moser, T. (2006). Synaptic organization in cochlear inner hair cells deficient for the CaV1.3 (alpha1D) subunit of L-type Ca2+ channels. *Neuroscience, 141*, 1849–1860.

Oestreicher, D., Picher, M. M., Rankovic, V., Moser, T., & Pangrsic, T. (2021). Cabp2-gene therapy restores inner hair cell calcium currents and improves hearing in a DFNB93 mouse model. *Frontiers in Molecular Neuroscience, 14*, 689415.

Oz, S., Benmocha, A., Sasson, Y., Sachyani, D., Almagor, L., Lee, A., Hirsch, J. A., & Dascal, N. (2013). Competitive and non-competitive regulation of calcium-dependent inactivation in CaV1.2 L-type Ca2+ channels by calmodulin and Ca2+-binding protein 1. *The Journal of Biological Chemistry, 288,* 12680–12691.

Pangrsic, T., Singer, J. H., & Koschak, A. (2018). Voltage-gated calcium channels: Key players in sensory coding in the retina and the inner ear. *Physiological Reviews, 98,* 2063–2096.

Peloquin, J. B., Doering, C. J., Rehak, R., & McRory, J. E. (2008). Temperature dependence of Cav1.4 calcium channel gating. *Neuroscience, 151,* 1066–1083.

Picher, M. M., Gehrt, A., Meese, S., Ivanovic, A., Predoehl, F., Jung, S., Schrauwen, I., Dragonetti, A. G., Colombo, R., Van Camp, G., Strenzke, N., & Moser, T. (2017). Ca(2+)-binding protein 2 inhibits Ca(2+)-channel inactivation in mouse inner hair cells. *Proceedings of the National Academy of Sciences of the United States of America, 114,* E1717–E1726.

Platzer, J., Engel, J., Schrott-Fischer, A., Stephan, K., Bova, S., Chen, H., Zheng, H., & Striessnig, J. (2000). Congenital deafness and sinoatrial node dysfunction in mice lacking class D L-type Ca^{2+} channels. *Cell, 102,* 89–97.

Raven, M. A., Orton, N. C., Nassar, H., Williams, G. A., Stell, W. K., Jacobs, G. H., Bech-Hansen, N. T., & Reese, B. E. (2008). Early afferent signaling in the outer plexiform layer regulates development of horizontal cell morphology. *The Journal of Comparative Neurology, 506,* 745–758.

Regus-Leidig, H., Atorf, J., Feigenspan, A., Kremers, J., Maw, M. A., & Brandstatter, J. H. (2014). Photoreceptor degeneration in two mouse models for congenital stationary night blindness type 2. *PLoS One, 9,* e86769.

Rieke, F., Lee, A., & Haeseleer, F. (2008). Characterization of Ca2+−binding protein 5 knockout mouse retina. *Investigative Ophthalmology & Visual Science, 49,* 5126–5135.

Schnee, M. E., & Ricci, A. J. (2003). Biophysical and pharmacological characterization of voltage-gated calcium currents in turtle auditory hair cells. *The Journal of Physiology, 549,* 697–717.

Schrauwen, I., Helfmann, S., Inagaki, A., Predoehl, F., Tabatabaiefar, M. A., Picher, M. M., Sommen, M., Zazo Seco, C., Oostrik, J., Kremer, H., Dheedene, A., Claes, C., Fransen, E., Chaleshtori, M. H., Coucke, P., Lee, A., Moser, T., & Van Camp, G. (2012). A mutation in CABP2, expressed in cochlear hair cells, causes autosomal-recessive hearing impairment. *American Journal of Human Genetics, 91,* 636–645.

Seitter, H., & Koschak, A. (2018). Relevance of tissue specific subunit expression in channelopathies. *Neuropharmacology, 132,* 58–70.

Sendin, G., Bulankina, A. V., Riedel, D., & Moser, T. (2007). Maturation of ribbon synapses in hair cells is driven by thyroid hormone. *The Journal of Neuroscience, 27,* 3163–3173.

Shaltiel, L., Paparizos, C., Fenske, S., Hassan, S., Gruner, C., Rotzer, K., Biel, M., & Wahl-Schott, C. A. (2012). Complex regulation of voltage-dependent activation and inactivation properties of retinal voltage-gated Cav1.4 L-type Ca2+ channels by Ca2+−binding pro-

tein 4 (CaBP4). *The Journal of Biological Chemistry, 287,* 36312–36321.

Shen, Y., Yu, D., Hiel, H., Liao, P., Yue, D. T., Fuchs, P. A., & Soong, T. W. (2006). Alternative splicing of the $Ca_v1.3$ channel IQ domain, a molecular switch for Ca^{2+}-dependent inactivation within auditory hair cells. *The Journal of Neuroscience, 26,* 10690–10699.

Sheyanth, I. N., Hojland, A. T., Okkels, H., Lolas, I., Thorup, C., & Petersen, M. B. (2021). First reported CABP2-related non-syndromic hearing loss in Northern Europe. *Molecular Genetics & Genomic Medicine, 9,* e1639.

Singh, A., Hamedinger, D., Hoda, J. C., Gebhart, M., Koschak, A., Romanin, C., & Striessnig, J. (2006). C-terminal modulator controls Ca2+-dependent gating of Ca(v)1.4 L-type Ca2+ channels. *Nature Neuroscience, 9,* 1108–1116.

Singh, A., Gebhart, M., Fritsch, R., Sinnegger-Brauns, M. J., Poggiani, C., Hoda, J. C., Engel, J., Romanin, C., Striessnig, J., & Koschak, A. (2008). Modulation of voltage- and Ca2+-dependent gating of CaV1.3 L-type calcium channels by alternative splicing of a C-terminal regulatory domain. *The Journal of Biological Chemistry, 283,* 20733–20744.

Sinha, R., Lee, A., Rieke, F., & Haeseleer, F. (2016). Lack of CaBP1/caldendrin or CaBP2 leads to altered ganglion cell responses. *eNeuro, 3,* ENEURO.0099-16.2016.

Sokal, I., & Haeseleer, F. (2011). Insight into the role of Ca2+-binding protein 5 in vesicle exocytosis. *Investigative Ophthalmology & Visual Science, 52,* 9131–9141.

Specht, D., Wu, S. B., Turner, P., Dearden, P., Koentgen, F., Wolfrum, U., Maw, M., Brandstatter, J. H., & tom Dieck, S. (2009). Effects of presynaptic mutations on a postsynaptic Cacna1s calcium channel colocalized with mGluR6 at mouse photoreceptor ribbon synapses. *Investigative Ophthalmology & Visual Science, 50,* 505–515.

Sterling, P., & Matthews, G. (2005). Structure and function of ribbon synapses. *Trends in Neurosciences, 28,* 20–29.

Stockner, T., & Koschak, A. (2013). What can naturally occurring mutations tell us about Ca(v)1.x channel function? *Biochimica et Biophysica Acta, 1828,* 1598–1607.

Tadross, M. R., Dick, I. E., & Yue, D. T. (2008). Mechanism of local and global Ca2+ sensing by calmodulin in complex with a Ca2+ channel. *Cell, 133,* 1228–1240.

Tan, B. Z., Jiang, F., Tan, M. Y., Yu, D., Huang, H., Shen, Y., & Soong, T. W. (2011). Functional characterization of alternative splicing in the C terminus of L-type CaV1.3 channels. *The Journal of Biological Chemistry, 286,* 42725–42735.

Tan, G. M., Yu, D., Wang, J., & Soong, T. W. (2012). Alternative splicing at C terminus of Ca(V)1.4 calcium channel modulates calcium-dependent inactivation, activation potential, and current density. *The Journal of Biological Chemistry, 287,* 832–847.

Thomas, J. R., & Lee, A. (2016). Measuring Ca2+-dependent modulation of voltage-gated Ca2+ channels in HEK-293T cells. *Cold Spring Harbor Protocols, 2016*(9), pdb.prot087213.

Tritsch, N. X., Yi, E., Gale, J. E., Glowatzki, E., & Bergles, D. E. (2007). The origin of spontaneous activity in the developing auditory system. *Nature, 450*, 50–55.

Vincent, P. F., Bouleau, Y., Charpentier, G., Emptoz, A., Safieddine, S., Petit, C., & Dulon, D. (2017). Different CaV1.3 channel isoforms control distinct components of the synaptic vesicle cycle in auditory inner hair cells. *The Journal of Neuroscience, 37*, 2960–2975.

von Gersdorff, H., & Matthews, G. (1996). Calcium-dependent inactivation of calcium current in synaptic terminals of retinal bipolar neurons. *The Journal of Neuroscience, 16*, 115–122.

Wahl-Schott, C., Baumann, L., Cuny, H., Eckert, C., Griessmeier, K., & Biel, M. (2006). Switching off calcium-dependent inactivation in L-type calcium channels by an autoinhibitory domain. *Proceedings of the National Academy of Sciences of the United States of America, 103*, 15657–15662.

Waldner, D. M., Giraldo Sierra, N. C., Bonfield, S., Nguyen, L., Dimopoulos, I. S., Sauve, Y., Stell, W. K., & Bech-Hansen, N. T. (2018). Cone dystrophy and ectopic synaptogenesis in a Cacna1f loss of function model of congenital stationary night blindness (CSNB2A). *Channels (Austin, Tex.), 12*, 17–33.

Williams, B., Haeseleer, F., & Lee, A. (2018). Splicing of an automodulatory domain in Cav1.4 Ca(2+) channels confers distinct regulation by calmodulin. *The Journal of General Physiology, 150*, 1676–1687.

Wong, A. B., Rutherford, M. A., Gabrielaitis, M., Pangrsic, T., Gottfert, F., Frank, T., Michanski, S., Hell, S., Wolf, F., Wichmann, C., & Moser, T. (2014). Developmental refinement of hair cell synapses tightens the coupling of Ca2+ influx to exocytosis. *The EMBO Journal, 33*, 247–264.

Wycisk, K. A., Zeitz, C., Feil, S., Wittmer, M., Forster, U., Neidhardt, J., Wissinger, B., Zrenner, E., Wilke, R., Kohl, S., & Berger, W. (2006). Mutation in the auxiliary calcium-channel subunit CACNA2D4 causes autosomal recessive cone dystrophy. *American Journal of Human Genetics, 79*, 973–977.

Xiao, H., Chen, X., & Steele, E. C., Jr. (2007). Abundant L-type calcium channel Ca(v)1.3 (alpha1D) subunit mRNA is detected in rod photoreceptors of the mouse retina via in situ hybridization. *Molecular Vision, 13*, 764–771.

Xu, W., & Lipscombe, D. (2001). Neuronal Ca$_v$1.3α1 L-type channels activate at relatively hyperpolarized membrane potentials and are incompletely inhibited by dihydropyridines. *The Journal of Neuroscience, 21*, 5944–5951.

Yang, P. S., Alseikhan, B. A., Hiel, H., Grant, L., Mori, M. X., Yang, W., Fuchs, P. A., & Yue, D. T. (2006). Switching of Ca^{2+}-dependent inactivation of Ca$_v$1.3 channels by calcium binding proteins of auditory hair cells. *The Journal of Neuroscience, 26*, 10677–10689.

Yang, P. S., Johny, M. B., & Yue, D. T. (2014). Allostery in Ca(2)(+) channel modulation by calcium-binding proteins. *Nature Chemical Biology, 10*, 231–238.

Yang, T., Scholl, E. S., Pan, N., Fritzsch, B., Haeseleer, F., & Lee, A. (2016). Expression and localization of CaBP Ca2+ binding proteins in the mouse cochlea. *PLoS One, 11*, e0147495.

Yang, T., Choi, J. E., Soh, D., Tobin, K., Joiner, M. L., Hansen, M., & Lee, A. (2018a). CaBP1 regulates Cav1 L-type Ca(2+) channels and their coupling to neurite growth and gene transcription in mouse spiral ganglion neurons. *Molecular and Cellular Neurosciences, 88*, 342–352.

Yang, T., Hu, N., Pangrsic, T., Green, S., Hansen, M., & Lee, A. (2018b). Functions of CaBP1 and CaBP2 in the peripheral auditory system. *Hearing Research, 364*, 48–58.

Zabouri, N., & Haverkamp, S. (2013). Calcium channel-dependent molecular maturation of photoreceptor synapses. *PLoS One, 8*, e63853.

Zampini, V., Johnson, S. L., Franz, C., Lawrence, N. D., Munkner, S., Engel, J., Knipper, M., Magistretti, J., Masetto, S., & Marcotti, W. (2010). Elementary properties of CaV1.3 Ca(2+) channels expressed in mouse cochlear inner hair cells. *The Journal of Physiology, 588*, 187–199.

Zampini, V., Johnson, S. L., Franz, C., Knipper, M., Holley, M. C., Magistretti, J., Masetto, S., & Marcotti, W. (2013). Burst activity and ultrafast activation kinetics of CaV1.3 Ca(2)(+) channels support presynaptic activity in adult gerbil hair cell ribbon synapses. *The Journal of Physiology, 591*, 3811–3820.

Zamponi, G. W., Striessnig, J., Koschak, A., & Dolphin, A. C. (2015). The physiology, pathology, and pharmacology of voltage-gated calcium channels and their future therapeutic potential. *Pharmacological Reviews, 67*, 821–870.

Zanetti, L., Kilicarslan, I., Netzer, M., Babai, N., Seitter, H., & Koschak, A. (2021). Function of cone and cone-related pathways in CaV1.4 IT mice. *Scientific Reports, 11*, 2732.

Zeitz, C., Kloeckener-Gruissem, B., Forster, U., Kohl, S., Magyar, I., Wissinger, B., Matyas, G., Borruat, F. X., Schorderet, D. F., Zrenner, E., Munier, F. L., & Berger, W. (2006). Mutations in CABP4, the gene encoding the Ca2+-binding protein 4, cause autosomal recessive night blindness. *American Journal of Human Genetics, 79*, 657–667.

Zeitz, C., Robson, A. G., & Audo, I. (2015). Congenital stationary night blindness: An analysis and update of genotype-phenotype correlations and pathogenic mechanisms. *Progress in Retinal and Eye Research, 45*, 58–110.

Voltage-Gated Calcium Channels in the Afferent Pain Pathway

Laurent Ferron and Gerald W. Zamponi

Abstract

Voltage-gated calcium channels are important mediators of signal transduction in the primary afferent pain pathway. In rodents, T-type calcium channels regulate the excitability of afferent fibers and spinal cord interneurons, and contribute to synaptic release in dorsal horn synapses. N-type calcium channels are the primary driver of neurotransmission in afferent fiber terminals in the superficial dorsal horn. N-type calcium channel activity in chronic pain states is splice isoform dependent and is regulated by ancillary Cavα2δ subunits. During chronic pain, T-type calcium channel activity can be dysregulated by post-translational modification such as glycosylation and ubiquitination. Both N-type and T-type calcium channels are potential drug targets for treating pain, with the former also being a target of G-protein-coupled receptors such as opioid receptors. R-type calcium channels may also play a possible role in afferent nociceptive signaling.

Keywords

Calcium channel · Pain · Nociception · T-type · N-type · Cav3.2 · Cav2.2 · Opioid receptors · Gabapentinoids · USP5 · Cavα2δ

Abbreviations

Cdk5	cyclin-dependent kinase 5
CFA	Complete Freund's Adjuvant
CGRP	calcitonin gene-related peptide
CRMP2	collapsin response mediator protein 2
DRG	Dorsal root ganglia
GPCR	G-protein-coupled receptor
MrgpR	Mas-related G-protein-coupled receptor
NeuN	neuronal nuclear protein
NF200	neurofilament protein 200
NMDA	N-methyl-D-aspartate
OR	opioid receptor
PKC	protein kinase C
TRKA	tropomyosin receptor kinase A
TRKB	tropomyosin receptor kinase B
TRPV1	transient receptor potential cation channel subfamily V member 1
vGLut3	vesicular glutamate transporter 3

L. Ferron · G. W. Zamponi (✉)
University of Calgary, Calgary, AB, Canada
e-mail: zamponi@ucalgary.ca

Introduction

Nociception is a key physiological process that has a critical protective function across many different species (Dubin & Patapoutian, 2010). The same neuroanatomical pathways that support nociceptive processing can become dysregulated from an adaptive to a maladaptive state and result in chronic hypersensitivity that serves no useful biological purpose (Kuner & Flor, 2017). Chronic pain is a debilitating condition that affects more than 20% of adult North Americans, and treatment of chronic pain represents one of the most serious health challenges facing society (Schopflocher et al., 2011; Reitsma et al., 2012). It is a major cause of lost productivity, and leads to a substantially reduced quality of life in both adults and children, along with staggering health care and economic costs (Gaskin & Richard, 2012; Mayer et al., 2019).

Painful stimuli are first detected by a set of diverse and specialized nociceptors that have their nerve endings in the skin or internal organs, and their cell bodies in either the dorsal root ganglia (DRG) or trigeminal ganglia (Fig. 1). In response to noxious stimuli such as heat, mechanical pressure, or chemicals, action potentials are initiated in the distal nerve endings, which then propagate along the afferent axons to nerve terminals located in the superficial layers of the spinal dorsal horn and brain stem (for cephalic neurons of the trigeminal ganglia) (Dubin &

Patapoutian, 2010; Basbaum et al., 2009). Synaptic transmission then leads to the initiation of action potentials in second-order neurons that project to various regions in the brain, where they are processed to ultimately give rise to the perception of an unpleasant sensation. The brain extends descending projections to the spinal cord, which modulate ascending nociceptive inputs through activation of excitatory and inhibitory spinal interneurons. Afferent sensory fibers are heterogeneous and do not just convey nociceptive inputs, but also other input modalities such as touch, proprioception, and itch, and ascending spinal dorsal columns are similarly organized along modalities (Niu et al., 2013). Nociceptive fibers are classified into (1) non-myelinated, slowly conducting peptidergic and non-peptidergic C-fibers, and (2) myelinated, more rapidly conducting A-fibers (Basbaum et al., 2009). These can be further subdivided based on their diameter and more precisely, based on a number of different markers that in many cases show overlapping expression patterns (e.g., tropomyosin receptor kinase A [TRKA], Ret, calcitonin gene-related peptide [CGRP], substance P, islocetin B4, transient receptor potential cation channel subfamily V member 1 [TRPV1], Mas-related G-protein-coupled receptor [MrgpR] among others). Furthermore, different subtypes of afferent fibers terminate in different laminae in the spinal cord (Molliver et al., 1995; Li et al., 2011). A more detailed description of afferent

Fig. 1 Schematic representation of the afferent pain pathway, illustrating the roles of Cav3.2 and Cav2.2 calcium channels

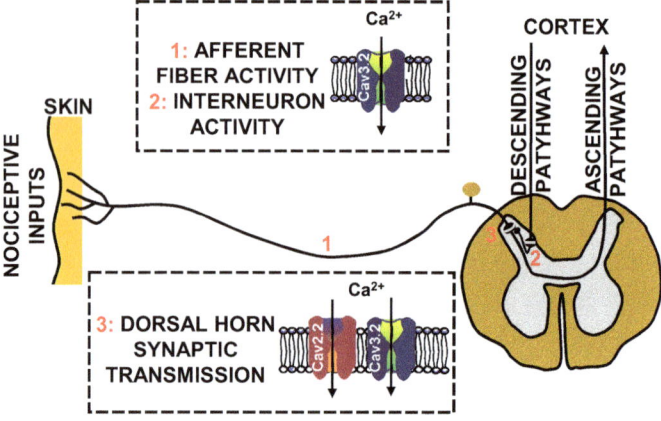

fiber types is beyond the scope of this chapter but can be found in recent transcriptomic analyses and comprehensive reviews (Finnerup et al., 2021; Sharma et al., 2020; Kupari et al., 2021). Equally complex is the organization of the spinal dorsal horn and its role in processing nociceptive information (Peirs et al., 2015; Gangadharan & Kuner, 2015; Harding et al., 2020). Suffice it to say that dorsal horn neurons are responsible for the integration of nociceptive peripheral inputs, and the activity of a heterogeneous set of spinal interneurons, and their inputs from descending pathways.

Voltage-gated calcium channels play important roles in these processes (Zamponi, 2016). Cav3.2 (T-type) calcium channels regulate the firing of certain types of afferent sensory neurons and contribute to synaptic transmission in dorsal horn synapses. They also appear to be involved in shaping the firing properties of spinal interneurons. Cav2.2 (N-type) channels are the major calcium channel subtype that is involved in synaptic transmission between nociceptive fibers and second-order neurons in the spinal cord (Fig. 1). In this chapter, we focus primarily on the roles of these two calcium channel isoforms in afferent nociceptive signaling.

N-Type Calcium Channels in the Afferent Pain Pathway

N-type (Cav2.2) channels exhibit a mainly neuronal distribution and they are enriched at presynaptic terminals where they coordinate the synchronous release of neurotransmitters (Nowycky et al., 1985; Westenbroek et al., 1992; Wheeler et al., 1994). In terminals of primary afferent neurons, calcium entry via Cav2.2 channels triggers the release of neurotransmitters such as glutamate, substance P, and CGRP (Gruner & Silva, 1994; Holz et al., 1988; Maggi et al., 1990; Santicioli et al., 1992). Specific Cav2.2 channel blockers such as ω-conotoxin GVIA and MVIIA (Olivera et al., 1994) have been used to demonstrate the involvement of this channel type in the afferent pain pathway (basal thermal and mechan-

ical nociception), and they have also highlighted the potential of N-type blockers as potent analgesics (Chaplan et al., 1994; Malmberg & Yaksh, 1994; Bowersox et al., 1996; Diaz & Dickenson, 1997; Sluka, 1998). The participation of Cav2.2 channels in the pain pathway was later confirmed by the phenotype of Cav2.2 channel knockout mice. Indeed, these mice are hyposensitive to pain (Saegusa et al., 2001; Kim et al., 2001; Hatakeyama et al., 2001), and they only present a relatively mild central nervous system (CNS) phenotype that includes reduced anxiety levels and reduced alcohol withdrawal symptoms (Saegusa et al., 2001; Newton et al., 2004).

The *CACNA1B* gene, which encodes the Cav2.2α1 subunit, is subject to extensive alternative splicing (Gray et al., 2007). The chapter by T.W. Soong (chapter "Splicing and Editing to Fine-Tune Activity of High Voltage-Activated Calcium Channels") deals in detail with this regulatory mechanism for voltage-gated calcium channels. In this section, we focus on splice variants that are relevant to the afferent pain pathway. Exon 37 encodes a short section (14 amino acids) of the cytoplasmic carboxy-terminal tail of the channel and can generate two mutually exclusive variants: E37a and E37b (Lipscombe et al., 2002; Bell et al., 2004). In heterologous expression systems, Cav2.2 E37a produces larger calcium currents than its counterpart Cav2.2 E37b, which is probably due to the fact that Cav2.2 E37a is both more resistant to degradation by the proteasome and more efficiently trafficked to the plasma membrane (Marangoudakis et al., 2012; Macabuag & Dolphin, 2015). It is noteworthy that Cav2.2 E37a is selectively expressed in peripheral nociceptive neurons that are positive for both TRPV1 and Nav1.8 channels, and its expression is critical for pain signaling (Bell et al., 2004; Altier et al., 2007). Indeed, using an small interfering RNA (siRNA) knockdown strategy, it was shown that Cav2.2 channels containing E37a mediate basal nociception and inflammatory pain (Altier et al., 2007).

In a chronic pain context, an upregulation of Cav2.2 channel expression has been shown in dorsal root ganglia and in the dorsal horn of the spinal cord (Cizkova et al., 2002; Yokoyama et al., 2003). However, it is still unclear whether the increase of Cav2.2 channels in the dorsal horn results from an increase of channels in central neurons or from an accumulation of channels in primary afferent terminals or a combination of both. Interestingly, while several studies have investigated the transcription level of Cav2.2 channels in chronic pain models, none has shown a correlation between level of Cav2.2 channel proteins and level of Cav2.2 channel mRNAs (messenger RNA), suggesting that chronic pain-induced changes in Cav2.2 channel expression occur at a translational and/or post-translational level (Altier et al., 2007; Umeda et al., 2006). Cav2.2-channel-deficient mice subjected to inflammatory and neuropathic insults exhibit reduced pain hypersensitivity (Hatakeyama et al., 2001; Saegusa et al., 2001; Kim et al., 2001). Moreover, whereas thermal hyperalgesia induced by sciatic nerve ligation relies mainly on the expression of E37a, tactile allodynia depends on both E37a and E37b splice variants (Altier et al., 2007). E37a was also shown to play a critical role in morphine-dependent analgesia (Andrade et al., 2010). Finally, in a model of chronic inflammatory pain, the expression of a Cav2.2 channel splice variant that lacks exon 18a was shown to be increased in DRG neurons (Asadi et al., 2009). Exon 18 encodes a region of the channel that corresponds to the II–III linker, which plays a critical role in presynaptic targeting (Zamponi et al., 2015). Although the physiological relevance of this variant switch is still unclear, it has been suggested that it could constitute a regulatory mechanism to attenuate excitability and reduce synaptic transmission (Asadi et al., 2009).

In the central nervous system, excitatory synaptic transmission usually depends on the dual activity of Cav2.2 and Cav2.1 channels. However, using a multi-electrode array recording system, a study has shown that in the anterior cingulate cortex (ACC), Cav2.2 channels are the only voltage-gated calcium channels involved in glutamatergic excitatory synaptic transmission (Kang et al., 2013). Interestingly, the ACC plays important roles in pain perception and chronic pain (Xu et al., 2008; Li et al., 2010).

Interaction Between CRMP2 and Cav2.2: A Therapeutic Target for Chronic Pain

For more than a decade, the interaction between Cav2.2 channels and collapsin response mediator protein 2 (CRMP2) has attracted increasing interest (Brittain et al., 2009). Initially identified as a protein involved in axonal growth, CRMP2 was then shown to directly interact with Cav2.2 channels and to promote their forward trafficking to the plasma membrane (Brittain et al., 2009). The CRMP2/Cav2.2 interaction was shown to modulate vesicular release in hippocampal neurons (Brittain et al., 2009; Chew & Khanna, 2018). In DRG neurons, Cav2.2-dependent secretion of CGRP was enhanced by the over-expression of CRMP2, which pointed to its involvement in the pain pathway (Chi et al., 2009). These initial studies led to the identification of interaction domains between the two proteins and the development of a short 14 amino-acid peptide (CBD3) derived from CRMP2 that can interfere with the binding of CRMP2 to Cav2.2 (Brittain et al., 2011). Conjugated to the HIV1 TAT protein and injected in vivo, CBD3 peptide is able to suppress inflammatory, neuropathic, postoperative, diabetic, migraine, and HIV-related pain (Brittain et al., 2011; Wilson et al., 2011; François-Moutal et al., 2015; Xie et al., 2016; Piekarz et al., 2012; Ripsch et al., 2012).

CRMP2/Cav2.2 interactions can be modulated by several mechanisms (Moutal et al., 2019). Phosphorylation of CRMP2 by cyclin-dependent kinase 5 (Cdk5) has been shown to reinforce its interaction with Cav2.2 channels (Brittain et al., 2012) and thus potentially enhance calcium currents. Interestingly, activation of Cdk5 was increased in animal models of neuropathic pain and inhibiting Cdk5 activity attenuated mechanical allodynia (Li et al., 2014; Yang et al., 2014;

Brittain et al., 2012; Moutal et al., 2016a, b). Cdk5 has multiple cellular targets, and thus inhibiting its phosphorylation activity likely disrupts other pathways that (in addition to reducing CRMP2/Cav2.2 interaction) can account for the improvement of the pain phenotype (Yang et al., 2014).

SUMOylation, the post-translational addition of small ubiquitin-related modifier (SUMO) peptide, can modulate CRMP2 function (Dustrude et al., 2013). A putative SUMO-interaction motif in CRMP2 has been linked to its role in regulating calcium influx via voltage-gated calcium channels (Ju et al., 2013). However, there is still no evidence showing involvement of CRMP2 SUMOylation and Cav2.2 channels in pain pathways.

Finally, the tumor suppressor protein neurofibromin has been shown to interact with CRMP2 via its C-terminal domain, inhibiting its function on Cav2.2 channels (Moutal et al., 2017a, b, 2019). Neurofibromatosis type 1 patients present mutations in the *NF1* gene, which encodes neurofibromin. These mutations often result in truncated proteins and afflicted patients suffer from idiopathic chronic pain (Gutmann et al., 2017; Esposito et al., 2015). In mutant mice heterozygous for *Nf1* ($Nf1^{+/-}$), N-type current densities are increased in DRG neurons and vesicular release is enhanced compared with wild-type neurons (Duan et al., 2014). Of note, the use of the CBD3 peptide to treat these $Nf1^{+/-}$ mice restores calcium currents and vesicular release to control levels indicating that neurofibromin contributes to the same regulatory pathway as CRMP2–Cav2.2 (Wilson et al., 2012). Moreover, a clustered regularly (CRISPR)/Cas9 strategy was used to truncate the C-terminal domain of neurofibromin in rats and the effect on calcium channel function was examined (Moutal et al., 2017b). This latter study showed an increase in voltage-gated calcium currents in DRG neurons and the animals developed thermal hyperalgesia. Further investigation of the interaction between neurofibromin and CRMP2 led to the design of a 15 amino-acid peptide (CNRP1) derived from CRMP2 C-terminal sequence, which, when coupled to a TAT peptide, was able to mimic the negative regulation of neurofibromin on CRMP2–Cav2.2 interaction in vivo (Moutal et al., 2017a). Indeed, intrathecal injection of TAT-CNRP1 alleviated nociceptive responses in animal models of inflammatory, post-surgical, and neuropathic pain (Moutal et al., 2017a). Interestingly, CNRP1 was also used in a proteomic analysis of a synaptic membrane library to reveal a novel interaction between CRMP2 and the presynaptic protein syntaxin 1A (Moutal et al., 2017a). Syntaxin 1A, which interacts with the synprint (Synaptic Protein Interaction) domain of Cav2.2 and Cav2.1, plays an important role in trafficking the channels to presynaptic terminals and promotes synaptic vesicle docking (Sheng et al., 1994; Gandini & Zamponi, 2021; Bennett et al., 1992). Altogether, these data reveal the importance of CRMP2 as a presynaptic interaction hub regulating Cav2.2 trafficking and highlight its potential as therapeutic target for pain treatment (Khanna et al., 2020).

Cavα2δ Subunits and Neuropathic Pain

Cavα2δ are critical auxiliary subunits that are required for Cav1.X and Cav2.X channel complexes to be trafficked to the plasma membrane, and for Cav2.X to be targeted to presynaptic terminals (Dolphin, 2016; Ferron et al., 2021). A detailed account on the structure and functions of Cavα2δs is provided in the chapter by A.C. Dolphin (chapter "Regulation of Calcium Channels and Synaptic Function by Auxiliary α2δ Subunits"). In this section, we will focus on Cavα2δ subunits, and particularly Cavα2δ-1, in a context of chronic pain. Cavα2δ-1 mRNA expression is enhanced in DRG neurons in animal models of chronic pain (Newton et al., 2001; Wang et al., 2002; Bauer et al., 2009; Lana et al., 2014). This transcriptional effect is associated with an increase in protein expression in the soma of DRG neurons and also along the axons and in the terminals within the spinal cord (Bauer et al., 2009; Luo et al., 2001). Interestingly, in vivo overexpression of Cavα2δ-1 is sufficient to induce tactile allodynia and hyperalgesia (Li et al., 2006). In addition, mice lacking the *Cacna2d1* gene that encodes Cavα2δ-1 have a reduced sensitivity to mechanical and cold stimuli, and delayed mechanical hypersensitivity following sciatic nerve ligation (Patel et al., 2013). Altogether, these data support a key role of Cavα2δ-1 in controlling DRG neuron excitability and its involvement in the development of chronic pain.

Cavα2δ-1 is the target of gabapentinoids (gabapentin and pregabalin), a class of drugs prescribed to treat chronic pain in humans (Rosenberg et al., 1997; Field et al., 2006, 2007; Taylor et al., 2007). Chronic application of gabapentinoids to DRG neurons in culture inhibits synaptic transmission by reducing the trafficking of Cavα2δ-1, and consequently Cav2.2, to the plasma membrane (Hendrich et al., 2008, 2012; Cassidy et al., 2014). In animal models of neuropathic pain, chronic treatment with pregabalin partially prevents the increase of presynaptic Cavα2δ-1 in the dorsal horn of the spinal cord and alleviates allodynia (Bauer et al., 2009).

Besides their role as auxiliary subunits for Cav channels, Cavα2δs have been involved in direct interactions with other ion channels and have been linked to other biological functions. Indeed, the N-terminal domain of KCa1.1 (BK) potassium channels competes with Cav2.2 channels for the binding of Cavα2δ-1 and reduces Cav2.2 surface expression and calcium current density (Zhang et al., 2018). Importantly, over-expressing a membrane-bound BK channel N-terminus peptide has an analgesic effect in mouse models of inflammatory and neuropathic pain. In addition, Cavα2δ-1 has been shown to interact with N-methyl-D-aspartate (NMDA) receptors and to increase their delivery at dorsal horn synapses during neuropathic pain development (Chen et al., 2018). Finally, Cavα2δs have been shown to play a role in synaptogenesis via extracellular matrix protein thrombospondins and thrombospondin-4 has been associated with the development of neuropathic pain (Eroglu

et al., 2009; Yu et al., 2018; Kim et al., 2012). Altogether, these studies reinforce the interest of Cavα2δ-1 as a therapeutic target for the treatment of chronic pain, and they also highlight the fact that the benefits of gabapentinoid treatments in vivo are not solely attributable to the normalization of Cav2.2-dependent synaptic transmission in the spinal cord.

N-Type Channel Blockers as Pain Therapeutics

In addition to the gabapentinoids discussed in the preceding section, a number of blockers of Cav2.2 channels have been shown to mediate analgesia in preclinical models, and some in a clinical setting. The chapter by Lewis and Adams (chapter "Pharmacology and Structure-Function of Venom Peptide Inhibitors of N-Type (Cav2.2) Calcium Channels") discusses in detail the inhibition of voltage-gated calcium channels by peptide toxins. Briefly, let us recall that ω-conotoxins are among the most selective inhibitors of N-type calcium channels. Both ω-conotoxins GVIA (*Conus geographus*) and MVIIA (*Conus magus*) potently inhibit native and transiently expressed Cav2.2 channels in an almost irreversible manner by occluding the pore of the channel (Olivera et al., 1984; Reynolds et al., 1986; Feng et al., 2001). The selective action of MVIIA, the analgesic effects in preclinical models (Chaplan et al., 1994; Bowersox & Luther, 1998), and the fact that it can be synthesized in vitro ultimately led this molecule to being developed as a potential therapeutic for pain. Under the commercial name Prialt, MVIIA passed phase III clinical trials and was ultimately approved for use in patients with severe cancer pain (Atanassoff et al., 2000; Miljanich, 2004). A major limitation of Prialt is that it does not cross the blood-brain barrier and therefore has to be delivered intrathecally via an implanted pump. Another is the fact that the therapeutic window is relatively narrow, and surprising side effects such as memory problems, hypotension, and vision problems have been described (Penn & Paice, 2000; Staats et al., 2004). Several additional conotoxins derived from other species such as *Conus fulmen* and *Conus catus* have been explored as possible analgesics (Adams et al., 2003; Lee et al., 2010; Sadeghi et al., 2013), with ω-conotoxin CVID having been tested in clinical trials, but ultimately not pursued further (Schroeder et al., 2006). Despite the limitations of MVIIA as a human pain therapeutic, it did serve to validate N-type calcium channels as a possible therapeutic target from pain.

Small organic blockers of N-type calcium channels that can penetrate the blood-brain barrier would be able to circumvent the need for intrathecal delivery. These include several distinct classes of compounds, such as amino-acid derivatives, N-triazoles, piperazines, piperidines, and even some dihydropyridines, that are normally thought to act on L-type channels (for review, see Schroeder et al., 2006). Many of these derivatives are not strictly specific for Cav2.2 channels, such as TROX-1, and N-triazole derivative that acts on all Cav2 channel isoforms but mediates analgesia in rodent models of osteoarthritis pain (Rahman et al., 2015; Abbadie et al., 2010; Swensen et al., 2012). L-cysteine (Seko et al., 2002) and amino-piperidine derivatives (Teodori et al., 2004) are N-type channel-blocking molecules with analgesic properties. The mixed L-/N-type channel-blocking dihydropyridine cilnidipine (Uneyama et al., 1997; Koganei et al., 2009; Yamamoto et al., 2016) has shown promise as an analgesic in preclinical models. A new N-type channel-blocking chemical scaffold was developed based on the structure of piperazine-based antipsychotics (Tytgat et al., 1991) with N-type channel-blocking activity (Zamponi et al., 2009; Pajouhesh et al., 2010). This includes a lead compound termed Z160 that was remarkably efficacious in a number of rodent pain models, but unfortunately failed phase II human trials due to lack of efficacy. The discrepancy between the preclinical and clinical data remains unresolved. To our knowledge, there are no small organic direct blockers of N-type calcium channels approved for use in humans.

Opioid and GABA-B Receptor Regulation of N-Type Calcium Channels

N-type calcium channels are subject to powerful modulation by a wide array of G-protein-coupled receptors (GPCRs) (for review, see Tedford & Zamponi, 2006). A detailed description of GPCR modulation of voltage-gated calcium channels is provided in the chapter by Herlitze et al. (chapter "Modulation of VGCCs by G-protein Coupled Receptors and Their Second Messengers"). In a nutshell, N-type calcium channels are modulated in a voltage-dependent manner through the direct interaction of G-protein βγ subunits with binding sites in the N-terminus and domain I–II linker regions of the Cav2.2 α1 subunit (Herlitze et al., 1996; Zamponi et al., 1997; Agler et al., 2005), leading to an inhibition of voltage-dependent activation of the channels that can be reversed by strong membrane depolarizations. In addition, there are also voltage-independent pathways that involve the activation of classical messenger cascades such as protein kinases (Luebke & Dunlap, 1994). In the context of nociceptive signaling, N-type calcium channel modulation by the family of opioid receptors (ORs) is particularly important. ORs comprise a family of four different receptor subtypes (μ-, δ-, κ-opioid receptors [MOR, DOR, KOR], and opioid receptor like receptors [NOP]; Waldhoer et al., 2004; McDonald & Lambert, 2005), that can undergo alternative splicing to create additional diversity (Gavériaux-Ruff et al., 1997; Pasternak, 2018; Piltonen et al., 2019). The different opioid receptor family members share a common ability to inhibit N-type calcium channels, and can heteromerize to form complexes with altered signaling properties (Evans et al., 2010). Different types of ORs are expressed in different types of afferent fibers, with MORs and DORs modulating different pain modalities (Scherrer et al., 2009). Activation of these receptors by their ligands leads to an activation of G-protein-coupled inwardly rectifying potassium channels (Blanchet & Lüscher, 2002; Marker et al., 2004), and direct voltage-dependent inhibition of presynaptic N-type calcium channels. In the afferent pain pathway, this leads to reduced neuronal excitability and reduced synaptic transmission in the spinal dorsal horn, leading to analgesia (Kondo et al., 2005; Beaudry et al., 2011). There appears to be tonic opioid control over nociceptive inputs such that DOR null mice show tactile allodynia (Nozaki et al., 2012). MORs are the pharmacological target of clinically active opioids such as morphine and fentanyl, whereas the pentazocine is a clinically approved KOR agonist (Goldstein, 1985). Action of opioids on MOR and DOR activation can lead to respiratory depression (Field et al., 1999), and repeated opioid use can lead to addiction (Darcq & Kieffer, 2018). Interestingly, KORs do not affect respiration, which would be an advantage in the development of OR agonists for pain. Several DOR agonists have been explored in clinical trials (Spahn & Stein, 2017), and bivalent ligands of OR subtypes are being explored as possibly less addictive pain medications (Vardanyan et al., 2017).

MOR modulation of Cav2.2 channels expressed in tA-201 cells has been shown to occur in an OR splice isoform-dependent manner (Gandini et al., 2019). Furthermore, MORs couple differentially to Cav2.2 channels with alternatively spliced exon 37 sequences (Gandini et al., 2019; Raingo et al., 2007). This highlights an immense potential for functional diversity in OR coupling to N-type calcium channels, and thus fine tuning of synaptic activity. Further complexity is generated by the formation of signaling complexes between NOP receptors and Cav2.2 channels that leads to (1) agonist-independent modulation of Cav2.2 channel activity, and (2) receptor-mediated regulation of forward trafficking and internalization of the channels (Beedle et al., 2004; Altier et al., 2006). Furthermore, co-expression with NOP can recruit other OR subtypes into Cav2.2 complexes (Evans et al., 2010). How these signaling complexes affect pain modulation remains to be determined.

GABA-B receptors have also been shown to modulate pain transmission via actions on N-type channels in rodents (Terrence et al., 1985), but the clinical utility of agonists such as baclofen has not been borne out due to CNS side effects

(Bortolato et al., 2010). The α-conotoxin Vc1.1, which has GABA-B receptor agonist activity (Callaghan et al., 2008), has been explored as a possible pain therapeutic, but despite promising preclinical results (Castro et al., 2017, 2018), has failed clinical development (Carstens et al., 2011). This molecule mediates voltage-independent modulation rather than the classical Gβγ-mediated voltage-dependent inhibition of Cav2.2 activity (Callaghan et al., 2008). In addition to OR and GABA-B receptors, many other types of GPCRs are known to modulate pain transmission (for review, see Pan et al., 2008), including cannabinoid, and adrenergic and muscarinic receptors. Although these receptors all modulate N-type channels, it is unclear if and how their analgesic actions can be directly attributed to the modulation of these channels in the afferent pathway and/or brain.

Other Types of Cav2 Calcium Channels

P/Q-type (Cav2.1) calcium channels are critical mediators of certain types of congenital migraines. Their role in headache is described in chapter "Voltage-Gated Calcium Channels and Migraine" by Dr. Pietrobon and will not be discussed here further. R-type (Cav2.3) calcium channels have emerged as potential players in afferent pain signaling. Due to their hyperpolarized activation range, Cav2.3 channels are well suited toward supporting neuronal firing (e.g., Park et al., 2010; Zaman et al., 2011; Gutzmann et al., 2019), and furthermore, they have been shown to contribute to neurotransmitter release at a subset of CNS synapses (Myoga & Regehr, 2011; Dietrich et al., 2003). They are expressed in peripheral and colonic sensory neurons (Yusaf et al., 2001; Fang et al., 2007; Qian et al., 2013). Mice lacking Cav2.3 channels show resistance to chemically induced seizures (in agreement with the role of these channels in neuronal excitability; Weiergräber et al., 2007; Dibué-Adjei et al., 2017), and partial resistance to inflammatory pain (Saegusa et al., 2000, 2002). Cav2.3 channels can be inhibited by ORs (Berecki et al.,

2016), and direct inhibitors of Cav2.3 channels have been shown to mediate antinociception (Murakami et al., 2004, 2007), possibly by acting on Cav2.3 channels in the spinal dorsal horn. A potential role of Cav2.3 channels in afferent pain signaling is also supported by the notion that natural mixed blockers of Cav2.3 and Cav2.2 channels are able to reverse neuropathic pain in rodents (Shan et al., 2019). Given the paucity of selective small organic Cav2.3 channel blockers, the precise role of Cav2.3 in pain signaling is not yet fully understood. In this context we note that gain-of-function mutations in Cav2.3 channels have been linked to seizure disorders in children, without any associated pain hypersensitivity (Helbig et al., 2018). On the other hand, the R-type channel blocker topiramate has been shown to mediate analgesia in patients with polyneuropathy (Nazarbaghi et al., 2017; but see Wiffen et al., 2013). Hence, the jury is out on whether Cav2.3 is a suitable target for treating pain.

Role of T-Type Calcium Channels in Afferent Pain Signaling

T-type calcium channels are uniquely suited toward regulating cellular excitability (Fig. 2). They activate at hyperpolarized membrane potentials and display a hyperpolarized half inactivation potential (Nowycky et al., 1985; Perez-Reyes, 2003). The overlap between steady-state activation and inactivation curves supports a large window current that allows these channels to be active near typical neuronal resting potentials (Iftinca & Zamponi, 2009; Fig. 2a). At rest, a large fraction of T-type calcium channels is tonically inactivated. Upon membrane hyperpolarization, as occurs, for example, during a

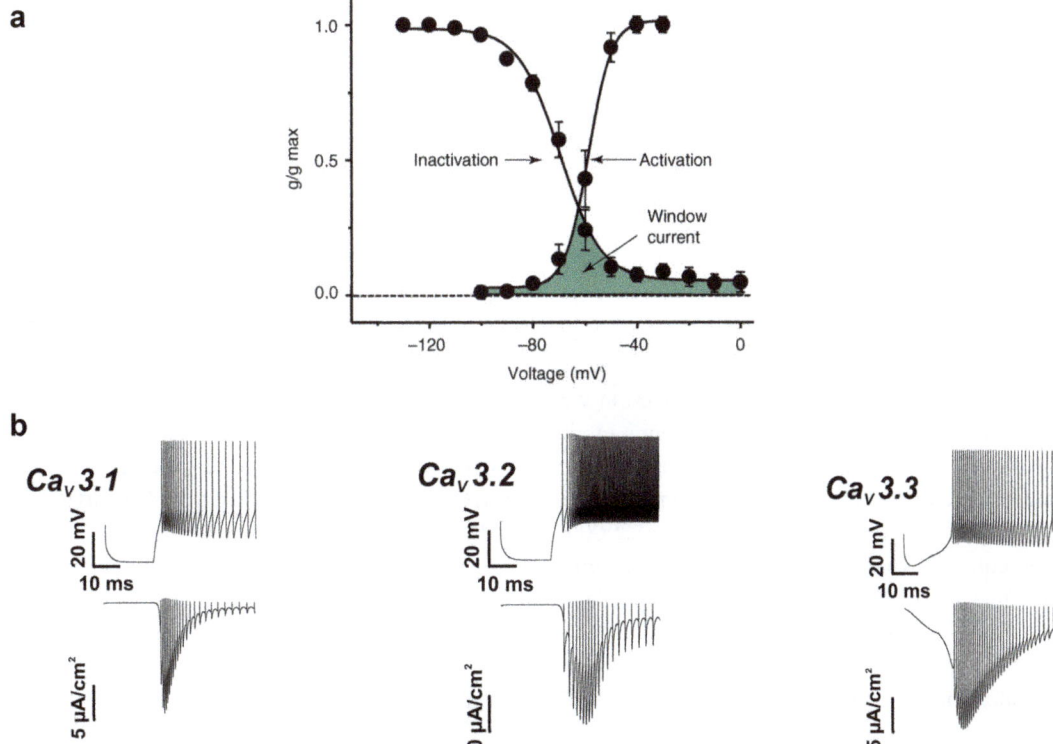

Fig. 2 (**a**) Schematic representation of the steady-state activation and inactivation curves for T-type calcium channels, highlighting the large window current. (Adapted with permission from Iftinca & Zamponi, 2009). (**b**) Effects of I_{CaT} in a model neuron on rebound depolarizations. Neuronal output (*top traces*) and corresponding inward I_{CaT} current trajectories (*lower traces*) are shown for the three Ca$_v$3 isoforms. (Fernandez et al., 2021)

GABAergic synaptic input, T-type channels recover from inactivation, giving rise to a greatly increased T-type current upon a subsequent membrane depolarization. This in turn triggers the opening of voltage-gated sodium channels and supports a rapid activation discharge, termed a rebound burst (Coulter et al., 1989; Huguenard & Prince, 1992; Fig. 2b). These collective biophysical properties allow T-type calcium channels to regulate the excitability of primary afferent nociceptive fibers, as well as the firing of spinal interneurons (Candelas et al., 2019). T-type calcium channels also associate with the synaptic vesicle release protein syntaxin 1A to support low-threshold exocytosis (Weiss et al., 2012). This is relevant for synaptic transmission in the dorsal horn of the spinal cord (Jacus et al., 2012; García-Caballero et al., 2014). A role of T-type calcium channels in afferent pain signaling in rodents was originally supported by two lines of evidence. First, delivery of known blockers of T-type calcium channels such as ethosuximide mediated analgesic effects in rodent pain models (Matthews & Dickenson, 2001; Dogrul et al., 2003; Flatters & Bennett, 2004). Second, in an elegant study, Bourinet and coworkers delivered siRNA against the three known T-type calcium channel isoforms, and found that either pan knockdown of Cav3 or specific knockdown of the Cav3.2 isoform resulted in potent analgesia in rat neuropathic pain models, thus identifying a single specific Cav3 channel isoform as a key mediator of afferent pain signaling (Bourinet et al., 2005). Along these lines, similar protection was observed upon knockdown of Cav3.2 in a model of visceral pain induced by intracolonic delivery of butyric acid and in streptozotocin-induced diabetic neuropathy in rats (Messinger et al., 2009). Subsequent experiments in Cav3.2 null mice revealed a more complex picture, with mice lacking Cav3.2 showing reduced sensitivity to intraplantar formalin, but relatively normal response to intraplantar injection of Complete Freund's Adjuvant (CFA) and in neuropathic pain models (Choi et al., 2007; García-Caballero et al., 2014). This is most likely due to developmental compensatory mechanisms.

In rodents, Cav3.2 channels are expressed in specific subtypes of primary afferent fibers including fibers positive for peripherin and vesicular glutamate transporter 3 (vGlut3), which mark a subset of nociceptive C-fibers, and neurofilament protein 200 (NF200) and tropomyosin receptor kinase B (TRKB), which are found in D-hair mechanoreceptors (François et al., 2015). In the spinal cord, Cav3.2 has been shown to overlap with the markers for neuronal nuclear protein (NeuN), Tlx3, Pax2, calretinin, protein kinase C gamma (PKCγ), calbindin, nNos, and parvalbumin, indicating expression in both glutamatergic and GABAergic interneurons (Candelas et al., 2019). The precise function of Cav3.2 channel in spinal cord neurons needs to be explored further, but there is clear evidence that these channels are able to shape the firing properties of lamina II cells (Candelas et al., 2019).

There is also emerging evidence of Cav3.2 channel mutations that are linked to chronic pain. Two heterozygous mutations (P769L and A1059S) in different alleles were identified in a pediatric chronic pain patient (Souza et al., 2019). When introduced and expressed in tsA-201 cells, the mutations mediated slight gain-of-function effects; they were however dependent on the particular recording conditions used for electrophysiological measurements. A number of Cav3.2 channel mutations were also identified in patients with trigeminal neuralgia (Dong et al., 2020); however, to date, they have not been functionally characterized.

Dysregulation of Cav3.2 Channel Expression and Function in Chronic Pain States

An important role of Cav3.2 channels is further underscored by the observation that these channels are upregulated in sensory neurons and/or spinal cord in a number of chronic pain conditions in rodents. This includes peripheral nerve injury (Jagodic et al., 2008; Wen et al., 2010; Feng et al., 2019; Tomita et al., 2020), spinal cord injury (Lauzadis et al., 2020), chronic diabetic conditions (Jagodic et al., 2007), visceral and CFA-induced inflammation (Marger et al., 2011; García-Caballero et al., 2014), osteoarthritis (Shin et al., 2020), certain types of chemotherapy-induced neuropathies (Li et al., 2017; Tomita et al., 2019), and models of post-surgical pain (Joksimovic et al., 2018). Two different molecular mechanisms that may contribute to this enhanced Cav3.2 activity have been identified. First, N-linked glycosylation of a set of four different asparagine residues has been shown to lead to an increase in Cav3.2 channel surface expression (Weiss et al., 2013; Orestes et al., 2013). Importantly, exposing cells to high glucose levels promoted an increase in Cav3.2 surface expression, whereas neuraminidase inhibited Cav3.2 channel membrane expression (Orestes et al., 2013). Thus, glycosylation may at least in part contribute to the upregulation of T-type channel activity during diabetic pain states. Second, Cdk5- and PKC-mediated phosphorylation has been associated with an increase in Cav3.2 channel activity in different models of neuropathic pain (Gomez et al., 2020; Gaifullina et al., 2019). Third, Cav3.2 channels are regulated by ubiquitination by the ubiquitin ligase WWP1 and the deubiquitinase USP5 (García-Caballero et al., 2014). The intracellular linker region between domains III and IV of Cav3.2 contains a ubiquitin ligase consensus motif as well as two lysine residues that are ubiquitinated. Importantly, this region also associates with USP5, an enzyme that is upregulated in the dorsal root ganglia and spinal cord in several mouse models of chronic pain (Gadotti et al., 2015b). This includes models of inflammatory pain induced by intraplantar delivery of CFA, neuropathic pain induced by sciatic nerve injury, visceral pain, chronic diabetes (Gadotti et al., 2015b), and surgery (Joksimovic et al., 2018). Importantly, depletion of USP5 or preventing the association of USP5 with Cav3.2 by delivery of decoy peptides mediated analgesia in these conditions in both male and female mice (García-Caballero et al., 2014, 2016; Gadotti & Zamponi, 2018). Small organic molecules targeting the USP5–Cav3.2 interaction also mediated analgesia (Gadotti et al., 2015b), overall suggest-

ing this molecular interaction as a potential drug target for treating pain. Interestingly, USP5 upregulation could also be observed in response to transcutaneous non-invasive optogenetic activation of C-fibers (Stemkowski et al., 2016). This led to a transient sensitization of the stimulated paw that subsided after 24 h along with a decrease in USP5 levels back to baseline. This then suggests that activity-dependent upregulation of USP5 and an associated increase in Cav3.2 channel activity may have originally evolved as an adaptive (i.e., protective) response that can become maladaptive under certain circumstances. Altogether, there are at least two molecular mechanisms that may contribute to the aberrant upregulation of Cav3.2 channels in chronic pain states, and it is possible that there are others. This could include the Cavα2δ subunit, which has been shown to promote cell surface expression of Cav3.2 channels in expression system even though this does not appear to occur via a physical interaction (Dubel et al., 2004).

We note that the trigeminal system appears to be different from other peripheral afferents in that trigeminal neuropathic pain has been associated with an upregulation of Cav3.3 rather than Cav3.2, and inhibiting these channels mediates analgesia (Montera et al., 2021) suggesting that trigeminal neuralgia may involve different T-type calcium channel signaling than peripheral nerve neuropathy. Finally, T-type calcium channels do not only contribute to nociceptive signaling in the afferent pain pathway, but also there is evidence that they do so in the brain. Cav3.1 null mice have been reported to show increased visceral pain sensitivity, and this was attributed to alterations in function of ventroposterolateral thalamic neurons (Kim et al., 2003). Along these lines, mice lacking Cav3.1 showed a reduction in trigeminal neuropathic pain, due to alterations in thalamic signaling (Choi et al., 2016). Direct delivery of the T-type channel inhibitor NCC-55-0396 into the anterior cingulate cortex was shown to inhibit neuropathic pain in rats (Shen et al., 2015). From a therapeutic point of view, it is thus important to consider all types of Cav3 channels, both peripherally and in the CNS.

T-Type Calcium Channel Blockers as Possible Pain Therapeutics

There is considerable evidence from preclinical models that inhibitors of T-type calcium channels mediate analgesia in a number of different chronic pain models, including diabetic, visceral, inflammatory, and neuropathic pain arising from physical nerve injury. Several detailed reviews of different classes of T-type calcium channel blockers with analgesic properties in rodent pain models were recently published (Snutch & Zamponi, 2018; Weiss & Zamponi, 2019a, b) and we refer the reader to these publications for additional detail. It has been known for some time that certain dihydropyridines can effectively block T-type calcium channels. For example, nimodipine blocks Cav3 channels in the micromolar range, although it should be noted that nimodipine has a much higher affinity for L-type calcium channels, especially at depolarized membrane potentials. Nonetheless, hexahydroquinoline–dihydropyridine derivatives have been shown to block T-type calcium channel preferentially over L-types (Bladen et al., 2014), with several derivatives of this compound series mediating analgesia in mouse models of both neuropathic and inflammatory pain (Bladen et al., 2015a; Gadotti et al., 2015a). Another class of T-type channel inhibitors with analgesic properties is derived from cannabinoids and endocannabinoids. The cannabinoid receptor ligands anandamide (Chemin et al., 2001), as well as N-arachnidonylyglycine (Barbara et al., 2009; Ross et al., 2009) were both shown to potently block Cav3 calcium channels. Importantly, the analgesic effects of these endocannabinoids were abolished in Cav3.2 null mice (Barbara et al., 2009), indicating that they inhibit pain via this calcium channel subtype. Subsequently, a number of compounds related to cannabinoid receptor agonists were developed and shown to not only inhibit Cav3.2 calcium channels but also mediate analgesia by blocking this channel subtype (You et al., 2011; Berger et al., 2014). Derivatives of this compound series that do not act on cannabinoid receptors were subsequently developed, and shown to mediate potent T-type channel inhibition along with analgesic effects (Bladen et al., 2015b), as well as the ability to inhibit acute itch in mice (Gadotti et al., 2020). Piperidine- and piperazine-based compounds such as penfluridol and flunarizine have been known to block Cav3-type calcium channels for some time (Santi et al., 2002). A number of related compounds have since been developed, including TTA-P2 and Z944, and shown to inhibit pain (Choe et al., 2011; Harding et al., 2021). Of note, Z944 has also been shown to mediate analgesic CNS effects by affecting thalamocortical connectivity in rats with neuropathic pain (LeBlanc et al., 2016).

Only a limited number of clinical pain studies with T-type calcium channels have been reported. ABT-639, a potent T-type calcium channel inhibitor, has failed multiple phase II clinical trials (Wallace et al., 2016), including for diabetic pain (Serra et al., 2015; Ziegler et al., 2015). Recently, ethosuximide was tested as a potential pain drug, but failed a randomized controlled trial for neuropathic pain (Kerckhove et al., 2018). Z944 on the other hand successfully completed a phase 1 trial; however, to date no data on efficacy in phase II are available.

The availability of a cryo-electron microscopy (EM) structure for Cav3.1 channels (Zhao et al., 2019) in complex with Z944 has opened new avenues for rational design of T-type calcium channel blockers. Homology models for Cav3.2 and Cav3.3 based on the Cav3.1 structure will aid the development of compounds with specificity for individual T-type calcium channel isoforms, opening new therapeutic options for pain as well as new means for probing the physiological roles of T-type calcium channels in a broader context.

Conclusions

In summary, Cav2.2 and Cav3.2 calcium channels are the predominant calcium channel isoforms involved in the afferent pain pathways, with a possible contribution from Cav2.3. The only clinically approved calcium channel inhibitors for treating chronic pain are the gabapentinoids, which target the ancillary Cavα2δ subunit of Cav2.2 channels; Prialt, which specifically target Cav2.2; and opioids, which inhibit Cav2.2 channel activity via activation of MORs. There remains a paucity of direct small organic inhibitors of Cav2.2 and Cav3.2 channels for clinical use in chronic pain.

References

Abbadie, C., McManus, O. B., Sun, S. Y., Bugianesi, R. M., Dai, G., Haedo, R. J., et al. (2010). Analgesic effects of a substituted N-triazole oxindole (TROX-1), a state-dependent, voltage-gated calcium channel 2 blocker. *The Journal of Pharmacology and Experimental Therapeutics, 334*(2), 545–555. https://doi.org/10.1124/jpet.110.166363

Adams, D. J., Smith, A. B., Schroeder, C. I., Yasuda, T., & Lewis, R. J. (2003). Omega-conotoxin CVID inhibits a pharmacologically distinct voltage-sensitive calcium channel associated with transmitter release from preganglionic nerve terminals. *The Journal of Biological Chemistry, 278*(6), 4057–4062. https://doi.org/10.1074/jbc.M209969200

Agler, H. L., Evans, J., Tay, L. H., Anderson, M. J., Colecraft, H. M., & Yue, D. T. (2005). G protein-gated inhibitory module of N-type (ca(v)2.2) ca2+ channels. *Neuron, 46*(6), 891–904. https://doi.org/10.1016/j.neuron.2005.05.011

Altier, C., Khosravani, H., Evans, R. M., Hameed, S., Peloquin, J. B., Vartian, B. A., et al. (2006). ORL1 receptor-mediated internalization of N-type calcium channels. *Nature Neuroscience, 9*(1), 31–40. https://doi.org/10.1038/nn1605

Altier, C., Dale, C. S., Kisilevsky, A. E., Chapman, K., Castiglioni, A. J., Matthews, E. A., et al. (2007). Differential role of N-type calcium channel splice isoforms in pain. *The Journal of Neuroscience, 27*(24), 6363–6373. https://doi.org/10.1523/JNEUROSCI.0307-07.2007

Andrade, A., Denome, S., Jiang, Y. Q., Marangoudakis, S., & Lipscombe, D. (2010). Opioid inhibition of N-type Ca2+ channels and spinal analgesia couple to alternative splicing. *Nature Neuroscience, 13*(10), 1249–1256. https://doi.org/10.1038/nn.2643

Asadi, S., Javan, M., Ahmadiani, A., & Sanati, M. H. (2009). Alternative splicing in the synaptic protein interaction site of rat Ca(v)2.2 (alpha (1B)) calcium channels: Changes induced by chronic inflammatory pain. *Journal of Molecular Neuroscience, 39*(1–2), 40–48. https://doi.org/10.1007/s12031-008-9159-2

Atanassoff, P. G., Hartmannsgruber, M. W., Thrasher, J., Wermeling, D., Longton, W., Gaeta, R., et al. (2000).

Ziconotide, a new N-type calcium channel blocker, administered intrathecally for acute postoperative pain. *Regional Anesthesia and Pain Medicine, 25*(3), 274–278. https://doi.org/10.1016/s1098-7339(00)90010-5

Barbara, G., Alloui, A., Nargeot, J., Lory, P., Eschalier, A., Bourinet, E., et al. (2009). T-type calcium channel inhibition underlies the analgesic effects of the endogenous lipoamino acids. *The Journal of Neuroscience, 29*(42), 13106–13114. https://doi.org/10.1523/JNEUROSCI.2919-09.2009

Basbaum, A. I., Bautista, D. M., Scherrer, G., & Julius, D. (2009). Cellular and molecular mechanisms of pain. *Cell, 139*(2), 267–284. https://doi.org/10.1016/j.cell.2009.09.028

Bauer, C. S., Nieto-Rostro, M., Rahman, W., Tran-Van-Minh, A., Ferron, L., Douglas, L., et al. (2009). The increased trafficking of the calcium channel subunit alpha2delta-1 to presynaptic terminals in neuropathic pain is inhibited by the alpha2delta ligand pregabalin. *The Journal of Neuroscience, 29*(13), 4076–4088. https://doi.org/10.1523/JNEUROSCI.0356-09.2009

Beaudry, H., Dubois, D., & Gendron, L. (2011). Activation of spinal mu- and delta-opioid receptors potently inhibits substance P release induced by peripheral noxious stimuli. *The Journal of Neuroscience, 31*(37), 13068–13077. https://doi.org/10.1523/JNEUROSCI.1817-11.2011

Beedle, A. M., McRory, J. E., Poirot, O., Doering, C. J., Altier, C., Barrere, C., et al. (2004). Agonist-independent modulation of N-type calcium channels by ORL1 receptors. *Nature Neuroscience, 7*(2), 118–125. https://doi.org/10.1038/nn1180

Bell, T. J., Thaler, C., Castiglioni, A. J., Helton, T. D., & Lipscombe, D. (2004). Cell-specific alternative splicing increases calcium channel current density in the pain pathway. *Neuron, 41*(1), 127–138. https://doi.org/10.1016/s0896-6273(03)00801-8

Bennett, M. K., Calakos, N., & Scheller, R. H. (1992). Syntaxin: A synaptic protein implicated in docking of synaptic vesicles at presynaptic active zones. *Science, 257*(5067), 255–259. https://doi.org/10.1126/science.1321498

Berecki, G., Motin, L., & Adams, D. J. (2016). Voltage-gated R-type calcium channel inhibition via human μ-, δ-, and κ-opioid receptors is voltage-independently mediated by Gβγ protein subunits. *Molecular Pharmacology, 89*(1), 187–196. https://doi.org/10.1124/mol.115.101154

Berger, N. D., Gadotti, V. M., Petrov, R. R., Chapman, K., Diaz, P., & Zamponi, G. W. (2014). NMP-7 inhibits chronic inflammatory and neuropathic pain via block of Cav3.2 T-type calcium channels and activation of CB2 receptors. *Molecular Pain, 10*, 77. https://doi.org/10.1186/1744-8069-10-77

Bladen, C., Gündüz, M. G., Şimşek, R., Şafak, C., & Zamponi, G. W. (2014). Synthesis and evaluation of 1,4-dihydropyridine derivatives with calcium channel blocking activity. *Pflügers Archiv, 466*(7), 1355–1363. https://doi.org/10.1007/s00424-013-1376-z

Bladen, C., Gadotti, V. M., Gündüz, M. G., Berger, N. D., Şimşek, R., Şafak, C., et al. (2015a). 1,4-Dihydropyridine derivatives with T-type calcium channel blocking activity attenuate inflammatory and neuropathic pain. *Pflügers Archiv, 467*(6), 1237–1247. https://doi.org/10.1007/s00424-014-1566-3

Bladen, C., McDaniel, S. W., Gadotti, V. M., Petrov, R. R., Berger, N. D., Diaz, P., et al. (2015b). Characterization of novel cannabinoid based T-type calcium channel blockers with analgesic effects. *ACS Chemical Neuroscience, 6*(2), 277–287. https://doi.org/10.1021/cn500206a

Blanchet, C., & Lüscher, C. (2002). Desensitization of mu-opioid receptor-evoked potassium currents: Initiation at the receptor, expression at the effector. *Proceedings of the National Academy of Sciences of the United States of America, 99*(7), 4674–4679. https://doi.org/10.1073/pnas.072075399

Bortolato, M., Frau, R., Orrù, M., Fà, M., Dessì, C., Puligheddu, M., et al. (2010). GABAB receptor activation exacerbates spontaneous spike-and-wave discharges in DBA/2J mice. *Seizure, 19*(4), 226–231. https://doi.org/10.1016/j.seizure.2010.02.007

Bourinet, E., Alloui, A., Monteil, A., Barrère, C., Couette, B., Poirot, O., et al. (2005). Silencing of the Cav3.2 T-type calcium channel gene in sensory neurons demonstrates its major role in nociception. *The EMBO Journal, 24*(2), 315–324. https://doi.org/10.1038/sj.emboj.7600515

Bowersox, S. S., & Luther, R. (1998). Pharmacotherapeutic potential of omega-conotoxin MVIIA (SNX-111), an N-type neuronal calcium channel blocker found in the venom of Conus magus. *Toxicon, 36*(11), 1651–1658. https://doi.org/10.1016/s0041-0101(98)00158-5

Bowersox, S. S., Gadbois, T., Singh, T., Pettus, M., Wang, Y. X., & Luther, R. R. (1996). Selective N-type neuronal voltage-sensitive calcium channel blocker, SNX-111, produces spinal antinociception in rat models of acute, persistent and neuropathic pain. *The Journal of Pharmacology and Experimental Therapeutics, 279*(3), 1243–1249.

Brittain, J. M., Piekarz, A. D., Wang, Y., Kondo, T., Cummins, T. R., & Khanna, R. (2009). An atypical role for collapsin response mediator protein 2 (CRMP-2) in neurotransmitter release via interaction with presynaptic voltage-gated calcium channels. *The Journal of Biological Chemistry, 284*(45), 31375–31390. https://doi.org/10.1074/jbc.M109.009951

Brittain, J. M., Duarte, D. B., Wilson, S. M., Zhu, W., Ballard, C., Johnson, P. L., et al. (2011). Suppression of inflammatory and neuropathic pain by uncoupling CRMP-2 from the presynaptic Ca²⁺ channel complex. *Nature Medicine, 17*(7), 822–829. https://doi.org/10.1038/nm.2345

Brittain, J. M., Wang, Y., Eruvwetere, O., & Khanna, R. (2012). Cdk5-mediated phosphorylation of CRMP-2 enhances its interaction with CaV2.2. *FEBS Letters, 586*(21), 3813–3818. https://doi.org/10.1016/j.febslet.2012.09.022

Callaghan, B., Haythornthwaite, A., Berecki, G., Clark, R. J., Craik, D. J., & Adams, D. J. (2008). Analgesic alpha-conotoxins Vc1.1 and Rg1A inhibit N-type calcium channels in rat sensory neurons via GABAB receptor activation. *The Journal of Neuroscience, 28*(43), 10943–10951. https://doi.org/10.1523/JNEUROSCI.3594-08.2008

Candelas, M., Reynders, A., Arango-Lievano, M., Neumayer, C., Fruquière, A., Demes, E., et al. (2019). Cav3.2 T-type calcium channels shape electrical firing in mouse Lamina II neurons. *Scientific Reports, 9*(1), 3112. https://doi.org/10.1038/s41598-019-39703-3

Carstens, B. B., Clark, R. J., Daly, N. L., Harvey, P. J., Kaas, Q., & Craik, D. J. (2011). Engineering of conotoxins for the treatment of pain. *Current Pharmaceutical Design, 17*(38), 4242–4253. https://doi.org/10.2174/138161211798999401

Cassidy, J. S., Ferron, L., Kadurin, I., Pratt, W. S., & Dolphin, A. C. (2014). Functional exofacially tagged N-type calcium channels elucidate the interaction with auxiliary α2δ-1 subunits. *Proceedings of the National Academy of Sciences of the United States of America, 111*(24), 8979–8984. https://doi.org/10.1073/pnas.1403731111

Castro, A., Raver, C., Li, Y., Uddin, O., Rubin, D., et al. (2017). Cortical regulation of nociception of the trigeminal nucleus caudalis. *The Journal of Neuroscience, 37*(47), 11431–11440. https://doi.org/10.1523/JNEUROSCI.3897-16.2017

Castro, J., Grundy, L., Deiteren, A., Harrington, A. M., O'Donnell, T., et al. (2018). Cyclic analogues of α-conotoxin Vc1.1 inhibit colonic nociceptors and provide analgesia in a mouse model of chronic abdominal pain. *British Journal of Pharmacology, 175*(12), 2384–2398. https://doi.org/10.1111/bph.14115

Chaplan, S. R., Pogrel, J. W., & Yaksh, T. L. (1994). Role of voltage-dependent calcium channel subtypes in experimental tactile allodynia. *The Journal of Pharmacology and Experimental Therapeutics, 269*(3), 1117–1123.

Chemin, J., Monteil, A., Perez-Reyes, E., Nargeot, J., & Lory, P. (2001). Direct inhibition of T-type calcium channels by the endogenous cannabinoid anandamide. *The EMBO Journal, 20*(24), 7033–7040. https://doi.org/10.1093/emboj/20.24.7033

Chen, J., Li, L., Chen, S. R., Chen, H., Xie, J. D., Sirrieh, R. E., et al. (2018). The α2δ-1-NMDA receptor complex is critically involved in neuropathic pain development and gabapentin therapeutic actions. *Cell Reports, 22*(9), 2307–2321. https://doi.org/10.1016/j.celrep.2018.02.021

Chew, L. A., & Khanna, R. (2018). CRMP2 and voltage-gated ion channels: Potential roles in neuropathic pain. *Neuro-Signals, 2*(1), NS20170220. https://doi.org/10.1042/NS20170220

Chi, X. X., Schmutzler, B. S., Brittain, J. M., Wang, Y., Hingtgen, C. M., Nicol, G. D., et al. (2009). Regulation of N-type voltage-gated calcium channels (Cav2.2) and transmitter release by collapsin response mediator protein-2 (CRMP-2) in sensory neurons. *Journal* of Cell Science, 122(Pt 23), 4351–4362. https://doi.org/10.1242/jcs.053280

Choe, W., Messinger, R. B., Leach, E., Eckle, V. S., Obradovic, A., Salajegheh, R., et al. (2011). TTA-P2 is a potent and selective blocker of T-type calcium channels in rat sensory neurons and a novel antinociceptive agent. Molecular Pharmacology, 80(5), 900–910. https://doi.org/10.1124/mol.111.073205

Choi, S., Na, H. S., Kim, J., Lee, J., Lee, S., Kim, D., et al. (2007). Attenuated pain responses in mice lacking Ca(V)3.2 T-type channels. Genes, Brain, and Behavior, 6(5), 425–431. https://doi.org/10.1111/j.1601-183X.2006.00268.x

Choi, S., Yu, E., Hwang, E., & Llinás, R. R. (2016). Pathophysiological implication of CaV3.1 T-type Ca2+ channels in trigeminal neuropathic pain. Proceedings of the National Academy of Sciences of the United States of America, 113(8), 2270–2275. https://doi.org/10.1073/pnas.1600418113

Cizkova, D., Marsala, J., Lukacova, N., Marsala, M., Jergova, S., Orendacova, J., et al. (2002). Localization of N-type Ca2+ channels in the rat spinal cord following chronic constrictive nerve injury. Experimental Brain Research, 147(4), 456–463. https://doi.org/10.1007/s00221-002-1217-3

Coulter, D. A., Huguenard, J. R., & Prince, D. A. (1989). Calcium currents in rat thalamocortical relay neurones: Kinetic properties of the transient, low-threshold current. The Journal of Physiology, 414, 587–604. https://doi.org/10.1113/jphysiol.1989.sp017705

Darcq, E., & Kieffer, B. L. (2018). Opioid receptors: Drivers to addiction? Nature Reviews. Neuroscience, 19(8), 499–514. https://doi.org/10.1038/s41583-018-0028-x

Diaz, A., & Dickenson, A. H. (1997). Blockade of spinal N- and P-type, but not L-type, calcium channels inhibits the excitability of rat dorsal horn neurones produced by subcutaneous formalin inflammation. Pain, 69(1–2), 93–100. https://doi.org/10.1016/s0304-3959(96)03271-x

Dibué-Adjei, M., Kamp, M. A., Alpdogan, S., Tevoufouet, E. E., Neiss, W. F., Hescheler, J., et al. (2017). Cav2.3 (R-type) calcium channels are critical for mediating anticonvulsive and neuroprotective properties of lamotrigine in vivo. Cellular Physiology and Biochemistry, 44(3), 935–947. https://doi.org/10.1159/000485361

Dietrich, D., Kirschstein, T., Kukley, M., Pereverzev, A., von der Brelie, C., Schneider, T., et al. (2003). Functional specialization of presynaptic Cav2.3 Ca2+ channels. Neuron, 39(3), 483–496. https://doi.org/10.1016/s0896-6273(03)00430-6

Dogrul, A., Gardell, L. R., Ossipov, M. H., Tulunay, F. C., Lai, J., & Porreca, F. (2003). Reversal of experimental neuropathic pain by T-type calcium channel blockers. Pain, 105(1–2), 159–168. https://doi.org/10.1016/s0304-3959(03)00177-5

Dolphin, A. C. (2016). Voltage-gated calcium channels and their auxiliary subunits: Physiology and pathophysiology and pharmacology. The Journal

of Physiology, 594(19), 5369–5390. https://doi.org/10.1113/JP272262

Dong, W., Jin, S. C., Allocco, A., Zeng, X., Sheth, A. H., Panchagnula, S., et al. (2020). Exome sequencing implicates impaired GABA signaling and neuronal ion transport in trigeminal neuralgia. iScience, 23(10), 101552. https://doi.org/10.1016/j.isci.2020.101552

Duan, J. H., Hodgdon, K. E., Hingtgen, C. M., & Nicol, G. D. (2014). N-type calcium current, Cav2.2, is enhanced in small-diameter sensory neurons isolated from Nf1+/− mice. Neuroscience, 270, 192–202. https://doi.org/10.1016/j.neuroscience.2014.04.021

Dubel, S. J., Altier, C., Chaumont, S., Lory, P., Bourinet, E., & Nargeot, J. (2004). Plasma membrane expression of T-type calcium channel alpha(1) subunits is modulated by high voltage-activated auxiliary subunits. The Journal of Biological Chemistry, 279(28), 29263–29269. https://doi.org/10.1074/jbc.M313450200

Dubin, A. E., & Patapoutian, A. (2010). Nociceptors: The sensors of the pain pathway. The Journal of Clinical Investigation, 120(11), 3760–3772. https://doi.org/10.1172/JCI42843

Dustrude, E. T., Wilson, S. M., Ju, W., Xiao, Y., & Khanna, R. (2013). CRMP2 protein SUMOylation modulates NaV1.7 channel trafficking. The Journal of Biological Chemistry, 288(34), 24316–24331. https://doi.org/10.1074/jbc.M113.474924

Eroglu, C., Allen, N. J., Susman, M. W., O'Rourke, N. A., Park, C. Y., Ozkan, E., et al. (2009). Gabapentin receptor alpha2delta-1 is a neuronal thrombospondin receptor responsible for excitatory CNS synaptogenesis. Cell, 139(2), 380–392. https://doi.org/10.1016/j.cell.2009.09.025

Esposito, T., Piluso, G., Saracino, D., Uccello, R., Schettino, C., Dato, C., et al. (2015). A novel diagnostic method to detect truncated neurofibromin in neurofibromatosis 1. Journal of Neurochemistry, 135(6), 1123–1128. https://doi.org/10.1111/jnc.13396

Evans, R. M., You, H., Hameed, S., Altier, C., Mezghrani, A., Bourinet, E., et al. (2010). Heterodimerization of ORL1 and opioid receptors and its consequences for N-type calcium channel regulation. The Journal of Biological Chemistry, 285(2), 1032–1040. https://doi.org/10.1074/jbc.M109.040634

Fang, Z., Park, C. K., Li, H. Y., Kim, H. Y., Park, S. H., Jung, S. J., et al. (2007). Molecular basis of Ca(v)2.3 calcium channels in rat nociceptive neurons. The Journal of Biological Chemistry, 282(7), 4757–4764. https://doi.org/10.1074/jbc.M605248200

Feng, Z. P., Hamid, J., Doering, C., Bosey, G. M., Snutch, T. P., & Zamponi, G. W. (2001). Residue Gly1326 of the N-type calcium channel alpha 1B subunit controls reversibility of omega-conotoxin GVIA and MVIIA block. The Journal of Biological Chemistry, 276(19), 15728–15735. https://doi.org/10.1074/jbc.M100406200

Feng, J. X., Ma, L. X., Jiao, C., Kuang, H. X., Zeng, F., et al. (2019). Nerve injury elevates functional Cav3.2 channels in superficial spinal dorsal horn.

Molecular Pain, 15, 1744806919836569. https://doi.org/10.1177/1744806919836569

Fernandez, F. R., Iftinca, M. C., Zamponi, G. W., & Turner, R. W. (2021). Modeling temperature- and Cav3 subtype-dependent alterations in T-type calcium mediated burst firing. *Molecular Brain, 14*, 115.

Ferron, L., Koshti, S., & Zamponi, G. W. (2021). The life cycle of voltage-gated Ca^{2+} channels in neurons: An update on the trafficking of neuronal calcium channels. *Neuro-Signals, 5*(1), NS20200095. https://doi.org/10.1042/NS20200095

Field, M. J., Carnell, A. J., Gonzalez, M. I., McCleary, S., Oles, R. J., Smith, R., et al. (1999). Enadoline, a selective kappa-opioid receptor agonist shows potent antihyperalgesic and antiallodynic actions in a rat model of surgical pain. *Pain, 80*(1–2), 383–389. https://doi.org/10.1016/s0304-3959(98)00237-1

Field, M. J., Cox, P. J., Stott, E., Melrose, H., Offord, J., Su, T. Z., et al. (2006). Identification of the alpha2-delta-1 subunit of voltage-dependent calcium channels as a molecular target for pain mediating the analgesic actions of pregabalin. *Proceedings of the National Academy of Sciences of the United States of America, 103*(46), 17537–17542. https://doi.org/10.1073/pnas.0409066103

Field, M. J., Li, Z., & Schwarz, J. B. (2007). Ca2+ channel alpha2-delta ligands for the treatment of neuropathic pain. *Journal of Medicinal Chemistry, 50*(11), 2569–2575. https://doi.org/10.1021/jm060650z

Finnerup, N. B., Kuner, R., & Jensen, T. S. (2021). Neuropathic pain: From mechanisms to treatment. *Physiological Reviews, 101*(1), 259–301. https://doi.org/10.1152/physrev.00045.2019

Flatters, S. J., & Bennett, G. J. (2004). Ethosuximide reverses paclitaxel- and vincristine-induced painful peripheral neuropathy. *Pain, 109*(1–2), 150–161. https://doi.org/10.1016/j.pain.2004.01.029

François, A., Schüetter, N., Laffray, S., Sanguesa, J., Pizzoccaro, A., Dubel, S., et al. (2015). The low-threshold calcium channel Cav3.2 determines low-threshold mechanoreceptor function. *Cell Reports, 10*(3), 370–382. https://doi.org/10.1016/j.celrep.2014.12.042

François-Moutal, L., Wang, Y., Moutal, A., Cottier, K. E., Melemedjian, O. K., Yang, X., et al. (2015). A membrane-delimited N-myristoylated CRMP2 peptide aptamer inhibits CaV2.2 trafficking and reverses inflammatory and postoperative pain behaviors. *Pain, 156*(7), 1247–1264. https://doi.org/10.1097/j.pain.0000000000000147

Gadotti, V. M., & Zamponi, G. W. (2018). Disrupting USP5/Cav3.2 interactions protects female mice from mechanical hypersensitivity during peripheral inflammation. *Molecular Brain, 11*(1), 60. https://doi.org/10.1186/s13041-018-0405-4

Gadotti, V. M., Bladen, C., Zhang, F. X., Chen, L., Gündüz, M. G., Şimşek, R., et al. (2015a). Analgesic effect of a broad-spectrum dihydropyridine inhibitor of voltage-gated calcium channels. *Pflügers*

Archiv, 467(12), 2485–2493. https://doi.org/10.1007/s00424-015-1725-1

Gadotti, V. M., Caballero, A. G., Berger, N. D., Gladding, C. M., Chen, L., Pfeifer, T. A., et al. (2015b). Small organic molecule disruptors of Cav3.2 – USP5 interactions reverse inflammatory and neuropathic pain. *Molecular Pain, 11*, 12. https://doi.org/10.1186/s12990-015-0011-8

Gadotti, V. M., Kreitinger, J. M., Wageling, N. B., Budke, D., Diaz, P., & Zamponi, G. W. (2020). Cav3.2 T-type calcium channels control acute itch in mice. *Molecular Brain, 13*(1), 119. https://doi.org/10.1186/s13041-020-00663-9

Gaifullina, A. S., Lazniewska, J., Gerasimova, E. V., Burkhanova, G. F., Rzhepetskyy, Y., Tomin, A., et al. (2019). A potential role for T-type calcium channels in homocysteinemia-induced peripheral neuropathy. *Pain, 160*(12), 2798–2810. https://doi.org/10.1097/j.pain.0000000000001669

Gandini, M. A., & Zamponi, G. W. (2021). Voltage-gated calcium channel nanodomains: Molecular composition and function. *The FEBS Journal, 289*(3), 614–633. https://doi.org/10.1111/febs.15759

Gandini, M. A., Souza, I. A., Raval, D., Xu, J., Pan, Y. X., & Zamponi, G. W. (2019). Differential regulation of Cav2.2 channel exon 37 variants by alternatively spliced µ-opioid receptors. *Molecular Brain, 12*(1), 98. https://doi.org/10.1186/s13041-019-0524-6

Gangadharan, V., & Kuner, R. (2015). Unravelling spinal circuits of pain and mechanical allodynia. *Neuron, 87*(4), 673–675. https://doi.org/10.1016/j.neuron.2015.08.013

García-Caballero, A., Gadotti, V. M., Stemkowski, P., Weiss, N., Souza, I. A., Hodgkinson, V., et al. (2014). The deubiquitinating enzyme USP5 modulates neuropathic and inflammatory pain by enhancing Cav3.2 channel activity. *Neuron, 83*(5), 1144–1158. https://doi.org/10.1016/j.neuron.2014.07.036

Garcia-Caballero, A., Gadotti, V. M., Chen, L., & Zamponi, G. W. (2016). A cell-permeant peptide corresponding to the cUBP domain of USP5 reverses inflammatory and neuropathic pain. *Molecular Pain, 12*, 1744806916642444. https://doi.org/10.1177/1744806916642444

Gaskin, D. J., & Richard, P. (2012). The economic costs of pain in the United States. *The Journal of Pain, 13*(8), 715–724. https://doi.org/10.1016/j.jpain.2012.03.009

Gavériaux-Ruff, C., Peluso, J., Befort, K., Simonin, F., Zilliox, C., & Kieffer, B. L. (1997). Detection of opioid receptor mRNA by RT-PCR reveals alternative splicing for the delta- and kappa-opioid receptors. *Brain Research. Molecular Brain Research, 48*(2), 298–304. https://doi.org/10.1016/s0169-328x(97)00109-5

Goldstein, G. (1985). Pentazocine. *Drug and Alcohol Dependence, 14*(3–4), 313–323. https://doi.org/10.1016/0376-8716(85)90064-x

Gomez, K., Vallecillo, T. G. M., Moutal, A., Perez-Miller, S., Delgado-Lezama, R., Felix, R., et al. (2020). The role of cyclin-dependent kinase 5 in neuropathic pain.

Pain, 161(12), 2674–2689. https://doi.org/10.1097/j.pain.0000000000002027

Gray, A. C., Raingo, J., & Lipscombe, D. (2007). Neuronal calcium channels: Splicing for optimal performance. *Cell Calcium, 42*(4–5), 409–417. https://doi.org/10.1016/j.ceca.2007.04.003

Gruner, W., & Silva, L. R. (1994). Omega-conotoxin sensitivity and presynaptic inhibition of glutamatergic sensory neurotransmission in vitro. *The Journal of Neuroscience, 14*(5 Pt 1), 2800–2808.

Gutmann, D. H., Ferner, R. E., Listernick, R. H., Korf, B. R., Wolters, P. L., & Johnson, K. J. (2017). Neurofibromatosis type 1. *Nature Reviews. Disease Primers, 3*, 17004. https://doi.org/10.1038/nrdp.2017.4

Gutzmann, J. J., Lin, L., & Hoffman, D. A. (2019). Functional coupling of Cav2.3 and BK potassium channels regulates action potential repolarization and short-term plasticity in the mouse hippocampus. *Frontiers in Cellular Neuroscience, 13*, 27. https://doi.org/10.3389/fncel.2019.00027

Harding, E. K., Boivin, B., & Salter, M. W. (2020). Intracellular calcium responses encode action potential firing in spinal cord lamina I neurons. *The Journal of Neuroscience, 40*(23), 4439–4456. https://doi.org/10.1523/JNEUROSCI.0206-20.2020

Harding, E. K., Dedek, A., Bonin, R. P., Salter, M. W., Snutch, T. P., & Hildebrand, M. E. (2021). The T-type calcium channel antagonist, Z944, reduces spinal excitability and pain hypersensitivity. *British Journal of Pharmacology, 178*(17), 3517–3532. https://doi.org/10.1111/bph.15498

Hatakeyama, S., Wakamori, M., Ino, M., Miyamoto, N., Takahashi, E., Yoshinaga, T., et al. (2001). Differential nociceptive responses in mice lacking the alpha(1B) subunit of N-type Ca(2+) channels. *Neuroreport, 12*(11), 2423–2427. https://doi.org/10.1097/00001756-200108080-00027

Helbig, K. L., Lauerer, R. J., Bahr, J. C., Souza, I. A., Myers, C. T., Uysal, B., et al. (2018). De novo pathogenic variants in CACNA1E cause developmental and epileptic encephalopathy with contractures, macrocephaly, and dyskinesias. *American Journal of Human Genetics, 103*(5), 666–678. https://doi.org/10.1016/j.ajhg.2018.09.006

Hendrich, J., Van Minh, A. T., Heblich, F., Nieto-Rostro, M., Watschinger, K., Striessnig, J., et al. (2008). Pharmacological disruption of calcium channel trafficking by the alpha2delta ligand gabapentin. *Proceedings of the National Academy of Sciences of the United States of America, 105*(9), 3628–3633. https://doi.org/10.1073/pnas.0708930105

Hendrich, J., Bauer, C. S., & Dolphin, A. C. (2012). Chronic pregabalin inhibits synaptic transmission between rat dorsal root ganglion and dorsal horn neurons in culture. *Channels (Austin, Tex.), 6*(2), 124–132. https://doi.org/10.4161/chan.19805

Herlitze, S., Garcia, D. E., Mackie, K., Hille, B., Scheuer, T., & Catterall, W. A. (1996). Modulation of Ca2+ channels by G-protein beta gamma subunits. *Nature, 380*(6571), 258–262. https://doi.org/10.1038/380258a0

Holz, G. G., Dunlap, K., & Kream, R. M. (1988). Characterization of the electrically evoked release of substance P from dorsal root ganglion neurons: Methods and dihydropyridine sensitivity. *The Journal of Neuroscience, 8*(2), 463–471.

Huguenard, J. R., & Prince, D. A. (1992). A novel T-type current underlies prolonged Ca(2+)-dependent burst firing in GABAergic neurons of rat thalamic reticular nucleus. *The Journal of Neuroscience, 12*(10), 3804–3817.

Iftinca, M. C., & Zamponi, G. W. (2009). Regulation of neuronal T-type calcium channels. *Trends in Pharmacological Sciences, 30*(1), 32–40. https://doi.org/10.1016/j.tips.2008.10.004

Jacus, M. O., Uebele, V. N., Renger, J. J., & Todorovic, S. M. (2012). Presynaptic Cav3.2 channels regulate excitatory neurotransmission in nociceptive dorsal horn neurons. *The Journal of Neuroscience, 32*(27), 9374–9382. https://doi.org/10.1523/JNEUROSCI.0068-12.2012

Jagodic, M. M., Pathirathna, S., Nelson, M. T., Mancuso, S., Joksovic, P. M., Rosenberg, E. R., et al. (2007). Cell-specific alterations of T-type calcium current in painful diabetic neuropathy enhance excitability of sensory neurons. *The Journal of Neuroscience, 27*(12), 3305–3316. https://doi.org/10.1523/JNEUROSCI.4866-06.2007

Jagodic, M. M., Pathirathna, S., Joksovic, P. M., Lee, W., Nelson, M. T., Naik, A. K., et al. (2008). Upregulation of the T-type calcium current in small rat sensory neurons after chronic constrictive injury of the sciatic nerve. *Journal of Neurophysiology, 99*(6), 3151–3156. https://doi.org/10.1152/jn.01031.2007

Joksimovic, S. L., Joksimovic, S. M., Tesic, V., García-Caballero, A., Feseha, S., Zamponi, G. W., et al. (2018). Selective inhibition of $Ca_V3.2$ channels reverses hyperexcitability of peripheral nociceptors and alleviates postsurgical pain. *Science Signaling, 11*(545), eaao4425. https://doi.org/10.1126/scisignal.aao4425

Ju, W., Li, Q., Wilson, S. M., Brittain, J. M., Meroueh, L., & Khanna, R. (2013). SUMOylation alters CRMP2 regulation of calcium influx in sensory neurons. *Channels (Austin, Tex.), 7*(3), 153–159. https://doi.org/10.4161/chan.24224

Kang, S. J., Liu, M. G., Shi, T. Y., Zhao, M. G., Kaang, B. K., & Zhuo, M. (2013). N-type voltage gated calcium channels mediate excitatory synaptic transmission in the anterior cingulate cortex of adult mice. *Molecular Pain, 9*, 58. https://doi.org/10.1186/1744-8069-9-58

Kerckhove, N., Pereira, B., Soriot-Thomas, S., Alchaar, H., Deleens, R., Hieng, V. S., et al. (2018). Efficacy and safety of a T-type calcium channel blocker in patients with neuropathic pain: A proof-of-concept, randomized, double-blind and controlled trial. *European Journal of Pain, 22*(7), 1321–1330. https://doi.org/10.1002/ejp.1221

Khanna, R., Moutal, A., Perez-Miller, S., Chefdeville, A., Boinon, L., & Patek, M. (2020). Druggability of CRMP2 for neurodegenerative diseases. *ACS Chemical Neuroscience, 11*(17), 2492–2505. https://doi.org/10.1021/acschemneuro.0c00307

Kim, C., Jun, K., Lee, T., Kim, S. S., McEnery, M. W., et al. (2001). Altered nociceptive response in mice deficient in the alpha(1B) subunit of the voltage-dependent calcium channel. *Molecular and Cellular Neurosciences, 18*(2), 235–245. https://doi.org/10.1006/mcne.2001.1013

Kim, D., Park, D., Choi, S., Lee, S., Sun, M., et al. (2003). Thalamic control of visceral nociception mediated by T-type Ca2+ channels. *Science, 302*(5642), 117–119. https://doi.org/10.1126/science.1088886

Kim, S. D., Li, K. W., Boroujerdi, A., Peter Yu, Y., Zhou, C. Y., et al. (2012). Thrombospondin-4 contributes to spinal sensitization and neuropathic pain states. *The Journal of Neuroscience, 32*(26), 8977–8987. https://doi.org/10.1523/JNEUROSCI.6494-11.2012

Koganei, H., Shoji, M., & Iwata, S. (2009). Suppression of formalin-induced nociception by cilnidipine, a voltage-dependent calcium channel blocker. *Biological & Pharmaceutical Bulletin, 32*(10), 1695–1700. https://doi.org/10.1248/bpb.32.1695

Kondo, I., Marvizon, J. C., Song, B., Salgado, F., Codeluppi, S., Hua, X. Y., et al. (2005). Inhibition by spinal mu- and delta-opioid agonists of afferent-evoked substance P release. *The Journal of Neuroscience, 25*(14), 3651–3660. https://doi.org/10.1523/JNEUROSCI.0252-05.2005

Kuner, R., & Flor, H. (2017). Structural plasticity and reorganisation in chronic pain. *Nature Reviews. Neuroscience, 18*(2), 113. https://doi.org/10.1038/nrn.2017.5

Kupari, J., Usoskin, D., Parisien, M., Lou, D., Hu, Y., Fatt, M., et al. (2021). Single cell transcriptomics of primate sensory neurons identifies cell types associated with chronic pain. *Nature Communications, 12*(1), 1510. https://doi.org/10.1038/s41467-021-21725-z

Lana, B., Schlick, B., Martin, S., Pratt, W. S., Page, K. M., Goncalves, L., et al. (2014). Differential upregulation in DRG neurons of an α2δ-1 splice variant with a lower affinity for gabapentin after peripheral sensory nerve injury. *Pain, 155*(3), 522–533. https://doi.org/10.1016/j.pain.2013.12.001

Lauzadis, J., Liu, H., Lu, Y., Rebecchi, M. J., Kaczocha, M., & Puopolo, M. (2020). Contribution of T-type calcium channels to spinal cord injury-induced hyperexcitability of nociceptors. *The Journal of Neuroscience, 40*(38), 7229–7240. https://doi.org/10.1523/JNEUROSCI.0517-20.2020

LeBlanc, B. W., Lii, T. R., Huang, J. J., Chao, Y. C., Bowary, P. M., Cross, B. S., et al. (2016). T-type calcium channel blocker Z944 restores cortical synchrony and thalamocortical connectivity in a rat model of neuropathic pain. *Pain, 157*(1), 255–263. https://doi.org/10.1097/j.pain.0000000000000362

Lee, S., Kim, Y., Back, S. K., Choi, H. W., Lee, J. Y., Jung, H. H., et al. (2010). Analgesic effect of highly reversible ω-conotoxin FVIA on N type Ca2+ channels. *Molecular Pain, 6*, 97. https://doi.org/10.1186/1744-8069-6-97

Li, Y. C., Zhang, X. L., Matthews, E. A., Li, K. W., Kurwa, A., et al. (2006). Calcium channel alpha2delta1 subunit mediates spinal hyperexcitability in pain modulation. *Pain, 125*(1–2), 20–34. https://doi.org/10.1016/j.pain.2006.04.022

Li, Y. X., Ko, H. G., Chen, T., Descalzi, G., Koga, K., et al. (2010). Alleviating neuropathic pain hypersensitivity by inhibiting PKMzeta in the anterior cingulate cortex. *Science, 330*(6009), 1400–1404. https://doi.org/10.1126/science.1191792

Li, L., Rutlin, M., Abraira, V. E., Cassidy, C., Kus, L., et al. (2011). The functional organization of cutaneous low-threshold mechanosensory neurons. *Cell, 147*(7), 1615–1627. https://doi.org/10.1016/j.cell.2011.11.027

Li, K., Zhao, G. Q., Li, L. Y., Wu, G. Z., & Cui, S. S. (2014). Epigenetic upregulation of Cdk5 in the dorsal horn contributes to neuropathic pain in rats. *Neuroreport, 25*(14), 1116–1121. https://doi.org/10.1097/WNR.0000000000000237

Li, Y., Tatsui, C. E., Rhines, L. D., North, R. Y., Harrison, D. S., et al. (2017). Dorsal root ganglion neurons become hyperexcitable and increase expression of voltage-gated T-type calcium channels (Cav3.2) in paclitaxel-induced peripheral neuropathy. *Pain, 158*(3), 417–429. https://doi.org/10.1097/j.pain.0000000000000774

Lipscombe, D., Pan, J. Q., & Gray, A. C. (2002). Functional diversity in neuronal voltage-gated calcium channels by alternative splicing of Ca(v)alpha1. *Molecular Neurobiology, 26*(1), 21–44. https://doi.org/10.1385/MN:26:1:021

Luebke, J. I., & Dunlap, K. (1994). Sensory neuron N-type calcium currents are inhibited by both voltage-dependent and -independent mechanisms. *Pflügers Archiv, 428*(5–6), 499–507. https://doi.org/10.1007/BF00374571

Luo, Z. D., Chaplan, S. R., Higuera, E. S., Sorkin, L. S., Stauderman, K. A., Williams, M. E., et al. (2001). Upregulation of dorsal root ganglion (alpha)2(delta) calcium channel subunit and its correlation with allodynia in spinal nerve-injured rats. *The Journal of Neuroscience, 21*(6), 1868–1875.

Macabuag, N., & Dolphin, A. C. (2015). Alternative splicing in Ca(V)2.2 regulates neuronal trafficking via adaptor protein complex-1 adaptor protein motifs. *The Journal of Neuroscience, 35*(43), 14636–14652. https://doi.org/10.1523/JNEUROSCI.3034-15.2015

Maggi, C. A., Tramontana, M., Cecconi, R., & Santicioli, P. (1990). Neurochemical evidence for the involvement of N-type calcium channels in transmitter secretion from peripheral endings of sensory nerves in guinea pigs. *Neuroscience Letters, 114*(2), 203–206. https://doi.org/10.1016/0304-3940(90)90072-h

Malmberg, A. B., & Yaksh, T. L. (1994). Voltage-sensitive calcium channels in spinal nociceptive processing: Blockade of N- and P-type channels inhibits formalin-induced nociception. *The Journal of Neuroscience, 14*(8), 4882–4890.

Marangoudakis, S., Andrade, A., Helton, T. D., Denome, S., Castiglioni, A. J., & Lipscombe, D. (2012). Differential ubiquitination and proteasome regulation of Ca(V)2.2 N-type channel splice isoforms. *The Journal of Neuroscience, 32*(30), 10365–10369. https://doi.org/10.1523/JNEUROSCI.0851-11.2012

Marger, F., Gelot, A., Alloui, A., Matricon, J., Ferrer, J. F., Barrère, C., et al. (2011). T-type calcium channels contribute to colonic hypersensitivity in a rat model of irritable bowel syndrome. *Proceedings of the National Academy of Sciences of the United States of America, 108*(27), 11268–11273. https://doi.org/10.1073/pnas.1100869108

Marker, C. L., Stoffel, M., & Wickman, K. (2004). Spinal G-protein-gated K+ channels formed by GIRK1 and GIRK2 subunits modulate thermal nociception and contribute to morphine analgesia. *The Journal of Neuroscience, 24*(11), 2806–2812. https://doi.org/10.1523/JNEUROSCI.5251-03.2004

Matthews, E. A., & Dickenson, A. H. (2001). Effects of ethosuximide, a T-type Ca(2+) channel blocker, on dorsal horn neuronal responses in rats. *European Journal of Pharmacology, 415*(2–3), 141–149. https://doi.org/10.1016/s0014-2999(01)00812-3

Matthews, E. A., Bee, L. A., Stephens, G. J., & Dickenson, A. H. (2007). The Cav2.3 calcium channel antagonist SNX-482 reduces dorsal horn neuronal responses in a rat model of chronic neuropathic pain. *The European Journal of Neuroscience, 25*(12), 3561–3569. https://doi.org/10.1111/j.1460-9568.2007.05605.x

Mayer, S., Spickschen, J., Stein, K. V., Crevenna, R., Dorner, T. E., & Simon, J. (2019). The societal costs of chronic pain and its determinants: The case of Austria. *PLoS One, 14*(3), e0213889. https://doi.org/10.1371/journal.pone.0213889

McDonald, J., & Lambert, D. G. (2005). Opioid receptors. *Continuing Education in Anaesthesia, Critical Care & Pain, 5*(1), 22–25. https://doi.org/10.1093/bjaceaccp/mki004

Messinger, R. B., Naik, A. K., Jagodic, M. M., Nelson, M. T., Lee, W. Y., Choe, W. J., et al. (2009). In vivo silencing of the Ca(V)3.2 T-type calcium channels in sensory neurons alleviates hyperalgesia in rats with streptozocin-induced diabetic neuropathy. *Pain, 145*(1–2), 184–195. https://doi.org/10.1016/j.pain.2009.06.012

Miljanich, G. P. (2004). Ziconotide: Neuronal calcium channel blocker for treating severe chronic pain. *Current Medicinal Chemistry, 11*(23), 3029–3040. https://doi.org/10.2174/0929867043363884

Molliver, D. C., Radeke, M. J., Feinstein, S. C., & Snider, W. D. (1995). Presence or absence of TrkA protein distinguishes subsets of small sensory neurons with unique cytochemical characteristics and dorsal horn projections. *The Journal of Comparative Neurology, 361*(3), 404–416. https://doi.org/10.1002/cne.903610305

Montera, M., Goins, A., Cmarko, L., Weiss, N., Westlund, K. N., & Alles, S. R. A. (2021). Trigeminal neuropathic pain is alleviated by inhibition of Cav3.3 T-type calcium channels in mice. *Channels (Austin, Tex.), 15*(1), 31–37. https://doi.org/10.1080/19336950.2020.1859248

Moutal, A., Chew, L. A., Yang, X., Wang, Y., Yeon, S. K., Telemi, E., et al. (2016a). (S)-lacosamide inhibition of CRMP2 phosphorylation reduces postoperative and neuropathic pain behaviors through distinct classes of sensory neurons identified by constellation pharmacology. *Pain, 157*(7), 1448–1463. https://doi.org/10.1097/j.pain.0000000000000555

Moutal, A., François-Moutal, L., Perez-Miller, S., Cottier, K., Chew, L. A., Yeon, S. K., et al. (2016b). (S)-lacosamide binding to collapsin response mediator protein 2 (CRMP2) regulates CaV2.2 activity by subverting its phosphorylation by Cdk5. *Molecular Neurobiology, 53*(3), 1959–1976. https://doi.org/10.1007/s12035-015-9141-2

Moutal, A., Wang, Y., Yang, X., Ji, Y., Luo, S., Dorame, A., et al. (2017a). Dissecting the role of the CRMP2-neurofibromin complex on pain behaviors. *Pain, 158*(11), 2203–2221. https://doi.org/10.1097/j.pain.0000000000001026

Moutal, A., Yang, X., Li, W., Gilbraith, K. B., Luo, S., Cai, S., et al. (2017b). CRISPR/Cas9 editing of Nf1 gene identifies CRMP2 as a therapeutic target in neurofibromatosis type 1-related pain that is reversed by (S)-Lacosamide. *Pain, 158*(12), 2301–2319. https://doi.org/10.1097/j.pain.0000000000001002

Moutal, A., White, K. A., Chefdeville, A., Laufmann, R. N., Vitiello, P. F., Feinstein, D., et al. (2019). Dysregulation of CRMP2 post-translational modifications drive its pathological functions. *Molecular Neurobiology, 56*(10), 6736–6755. https://doi.org/10.1007/s12035-019-1568-4

Murakami, M., Nakagawasai, O., Suzuki, T., Mobarakeh, I. I., Sakurada, Y., Murata, A., et al. (2004). Antinociceptive effect of different types of calcium channel inhibitors and the distribution of various calcium channel alpha 1 subunits in the dorsal horn of spinal cord in mice. *Brain Research, 1024*(1–2), 122–129. https://doi.org/10.1016/j.brainres.2004.07.066

Myoga, M. H., & Regehr, W. G. (2011). Calcium microdomains near R-type calcium channels control the induction of presynaptic long-term potentiation at parallel fiber to purkinje cell synapses. *The Journal of Neuroscience, 31*(14), 5235–5243. https://doi.org/10.1523/JNEUROSCI.5252-10.2011

Nazarbaghi, S., Amiri-Nikpour, M. R., Eghbal, A. F., & Valizadeh, R. (2017). Comparison of the effect of topiramate versus gabapentin on neuropathic pain in patients with polyneuropathy: A randomized clinical trial. *Electronic Physician, 9*(10), 5617–5622. https://doi.org/10.19082/5617

Newton, A. R., Bingham, S., Case, P. C., Sanger, G. J., & Lawson, S. N. (2001). Dorsal root ganglion neurons show increased expression of the calcium channel alpha2delta-1 subunit following partial sciatic nerve injury. *Brain Research. Molecular Brain Research, 95*(1–2), 1–8. https://doi.org/10.1016/s0169-328x(01)00188-7

Newton, M. P., Orr, C. J., Wallace, M. J., Kim, C., Shin, H. S., et al. (2004). Deletion of N-type calcium channels alters ethanol reward and reduces ethanol consumption in mice. *The Journal of Neuroscience, 24*(44), 9862–9869. https://doi.org/10.1523/JNEUROSCI.3446-04.2004

Niu, J., Ding, L., Li, J. J., Kim, H., Liu, J., Li, H., et al. (2013). Modality-based organization of ascending somatosensory axons in the direct dorsal column pathway. *The Journal of Neuroscience, 33*(45), 17691–17709. https://doi.org/10.1523/JNEUROSCI.3429-13.2013

Nowycky, M. C., Fox, A. P., & Tsien, R. W. (1985). Three types of neuronal calcium channel with different calcium agonist sensitivity. *Nature, 316*(6027), 440–443. https://doi.org/10.1038/316440a0

Nozaki, C., Le Bourdonnec, B., Reiss, D., Windh, R. T., Little, P. J., Dolle, R. E., et al. (2012). δ-Opioid mechanisms for ADL5747 and ADL5859 effects in mice: Analgesia, locomotion, and receptor internalization. *The Journal of Pharmacology and Experimental Therapeutics, 342*(3), 799–807. https://doi.org/10.1124/jpet.111.188987

Olivera, B. M., McIntosh, J. M., Cruz, L. J., Luque, F. A., & Gray, W. R. (1984). Purification and sequence of a presynaptic peptide toxin from Conus geographus venom. *Biochemistry, 23*(22), 5087–5090. https://doi.org/10.1021/bi00317a001

Olivera, B. M., Miljanich, G. P., Ramachandran, J., & Adams, M. E. (1994). Calcium channel diversity and neurotransmitter release: The omega-conotoxins and omega-agatoxins. *Annual Review of Biochemistry, 63*, 823–867. https://doi.org/10.1146/annurev.bi.63.070194.004135

Orestes, P., Osuru, H. P., McIntire, W. E., Jacus, M. O., Salajegheh, R., Jagodic, M. M., et al. (2013). Reversal of neuropathic pain in diabetes by targeting glycosylation of Ca(V)3.2 T-type calcium channels. *Diabetes, 62*(11), 3828–3838. https://doi.org/10.2337/db13-0813

Pajouhesh, H., Feng, Z. P., Ding, Y., Zhang, L., Morrison, J. L., Belardetti, F., et al. (2010). Structure-activity relationships of diphenylpiperazine N-type calcium channel inhibitors. *Bioorganic & Medicinal Chemistry Letters, 20*(4), 1378–1383. https://doi.org/10.1016/j.bmcl.2010.01.008

Pan, H. L., Wu, Z. Z., Zhou, H. Y., Chen, S. R., Zhang, H. M., & Li, D. P. (2008). Modulation of pain transmission by G-protein-coupled receptors. *Pharmacology & Therapeutics, 117*(1), 141–161. https://doi.org/10.1016/j.pharmthera.2007.09.003

Park, J. Y., Remy, S., Varela, J., Cooper, D. C., Chung, S., Kang, H. W., et al. (2010). A post-burst after depolarization is mediated by group i metabotropic glutamate receptor-dependent upregulation of Ca(v)2.3 R-type calcium channels in CA1 pyramidal neurons. *PLoS Biology, 8*(11), e1000534. https://doi.org/10.1371/journal.pbio.1000534

Pasternak, G. W. (2018). Mu opioid pharmacology: 40 years to the promised land. *Advances in Pharmacology, 82*, 261–291. https://doi.org/10.1016/bs.apha.2017.09.006

Patel, R., Bauer, C. S., Nieto-Rostro, M., Margas, W., Ferron, L., Chaggar, K., et al. (2013). α2δ-1 gene deletion affects somatosensory neuron function and delays mechanical hypersensitivity in response to peripheral nerve damage. *The Journal of Neuroscience, 33*(42), 16412–16426. https://doi.org/10.1523/JNEUROSCI.1026-13.2013

Peirs, C., Williams, S. P., Zhao, X., Walsh, C. E., Gedeon, J. Y., Cagle, N. E., et al. (2015). Dorsal horn circuits for persistent mechanical pain. *Neuron, 87*(4), 797–812. https://doi.org/10.1016/j.neuron.2015.07.029

Penn, R. D., & Paice, J. A. (2000). Adverse effects associated with the intrathecal administration of ziconotide. *Pain, 85*(1–2), 291–296. https://doi.org/10.1016/s0304-3959(99)00254-7

Perez-Reyes, E. (2003). Molecular physiology of low-voltage-activated t-type calcium channels. *Physiological Reviews, 83*(1), 117–161. https://doi.org/10.1152/physrev.00018.2002

Piekarz, A. D., Due, M. R., Khanna, M., Wang, B., Ripsch, M. S., Wang, R., et al. (2012). CRMP-2 peptide mediated decrease of high and low voltage-activated calcium channels, attenuation of nociceptor excitability, and anti-nociception in a model of AIDS therapy-induced painful peripheral neuropathy. *Molecular Pain, 8*, 54. https://doi.org/10.1186/1744-8069-8-54

Piltonen, M., Parisien, M., Grégoire, S., Chabot-Doré, A. J., Jafarnejad, S. M., Bérubé, P., et al. (2019). Alternative splicing of the delta-opioid receptor gene suggests existence of new functional isoforms. *Molecular Neurobiology, 56*(4), 2855–2869. https://doi.org/10.1007/s12035-018-1253-z

Qian, A., Song, D., Li, Y., Liu, X., Tang, D., Yao, W., et al. (2013). Role of voltage gated Ca2+ channels in rat visceral hypersensitivity change induced by 2,4,6-trinitrobenzene sulfonic acid. *Molecular Pain, 9*, 15. https://doi.org/10.1186/1744-8069-9-15

Rahman, W., Patel, R., & Dickenson, A. H. (2015). Electrophysiological evidence for voltage-gated calcium channel 2 (Cav2) modulation of mechano- and thermosensitive spinal neuronal responses in a rat model of osteoarthritis. *Neuroscience, 305*, 76–85. https://doi.org/10.1016/j.neuroscience.2015.07.073

Raingo, J., Castiglioni, A. J., & Lipscombe, D. (2007). Alternative splicing controls G protein-dependent inhibition of N-type calcium channels in nociceptors. *Nature Neuroscience, 10*(3), 285–292. https://doi.org/10.1038/nn1848

Reitsma, M. I., Tranmer, J. E., Buchanan, D. M., & VanDenKerhof, E. G. (2012). The epidemiology of chronic pain in Canadian men and women between 1994 and 2007: Results from the longitudinal component of the National Population Health Survey. *Pain Research & Management, 17*(3), 166–172.

Reynolds, I. J., Wagner, J. A., Snyder, S. H., Thayer, S. A., Olivera, B. M., & Miller, R. J. (1986). Brain voltage-sensitive calcium channel subtypes differentiated by omega-conotoxin fraction GVIA. *Proceedings of the National Academy of Sciences of the United States of America, 83*(22), 8804–8807. https://doi.org/10.1073/pnas.83.22.8804

Ripsch, M. S., Ballard, C. J., Khanna, M., Hurley, J. H., White, F. A., & Khanna, R. (2012). A peptide uncoupling CRMP-2 from the presynaptic Ca2+ channel complex demonstrates efficacity in animal models of migraine and AIDS therapy-induced neuropathy. *Translational Neuroscience, 3*(1), 1–8. https://doi.org/10.2478/s13380-012-0002-4

Rosenberg, J. M., Harrell, C., Ristic, H., Werner, R. A., & de Rosayro, A. M. (1997). The effect of gabapentin on neuropathic pain. *The Clinical Journal of Pain, 13*(3), 251–255. https://doi.org/10.1097/00002508-199709000-00011

Ross, H. R., Gilmore, A. J., & Connor, M. (2009). Inhibition of human recombinant T-type calcium channels by the endocannabinoid N-arachidonoyl dopamine. *British Journal of Pharmacology, 156*(5), 740–750. https://doi.org/10.1111/j.1476-5381.2008.00072.x

Sadeghi, M., Murali, S. S., Lewis, R. J., Alewood, P. F., Mohammadi, S., & Christie, M. J. (2013). Novel ω-conotoxins from C. catus reverse signs of mouse inflammatory pain after systemic administration. *Molecular Pain, 9*, 51. https://doi.org/10.1186/1744-8069-9-51

Saegusa, H., Kurihara, T., Zong, S., Minowa, O., Kazuno, A., Han, W., et al. (2000). Altered pain responses in mice lacking alpha 1E subunit of the voltage-dependent Ca2+ channel. *Proceedings of the National Academy of Sciences of the United States of America, 97*(11), 6132–6137. https://doi.org/10.1073/pnas.100124197

Saegusa, H., Kurihara, T., Zong, S., Kazuno, A., Matsuda, Y., Nonaka, T., et al. (2001). Suppression of inflammatory and neuropathic pain symptoms in mice lacking the N-type Ca2+ channel. *The EMBO Journal, 20*(10), 2349–2356. https://doi.org/10.1093/emboj/20.10.2349

Saegusa, H., Matsuda, Y., & Tanabe, T. (2002). Effects of ablation of N- and R-type Ca(2+) channels on pain transmission. *Neuroscience Research, 43*(1), 1–7. https://doi.org/10.1016/s0168-0102(02)00017-2

Santi, C. M., Cayabyab, F. S., Sutton, K. G., McRory, J. E., Mezeyova, J., Hamming, K. S., et al. (2002). Differential inhibition of T-type calcium channels by neuroleptics. *The Journal of Neuroscience, 22*(2), 396–403.

Santicioli, P., Del Bianco, E., Tramontana, M., Geppetti, P., & Maggi, C. A. (1992). Release of calcitonin gene-related peptide like-immunoreactivity induced by electrical field stimulation from rat spinal afferents is mediated by conotoxin-sensitive calcium channels. *Neuroscience Letters, 136*(2), 161–164. https://doi.org/10.1016/0304-3940(92)90039-a

Scherrer, G., Imamachi, N., Cao, Y. Q., Contet, C., Mennicken, F., O'Donnell, D., et al. (2009). Dissociation of the opioid receptor mechanisms that control mechanical and heat pain. *Cell, 137*(6), 1148–1159. https://doi.org/10.1016/j.cell.2009.04.019

Schopflocher, D., Taenzer, P., & Jovey, R. (2011). The prevalence of chronic pain in Canada. *Pain Research & Management, 16*(6), 445–450. https://doi.org/10.1155/2011/876306

Schroeder, C. I., Doering, C. J., Zamponi, G. W., & Lewis, R. J. (2006). N-type calcium channel blockers: Novel therapeutics for the treatment of pain. *Medicinal Chemistry, 2*(5), 535–543. https://doi.org/10.2174/157340606778250216

Seko, T., Kato, M., Kohno, H., Ono, S., Hashimura, K., Takenobu, Y., et al. (2002). L-cysteine based N-type calcium channel blockers: Structure-activity relationships of the C-terminal lipophilic moiety, and oral analgesic efficacy in rat pain models. *Bioorganic & Medicinal Chemistry Letters, 12*(17), 2267–2269. https://doi.org/10.1016/s0960-894x(02)00456-0

Serra, J., Duan, W. R., Locke, C., Solà, R., Liu, W., & Nothaft, W. (2015). Effects of a T-type calcium channel blocker, ABT-639, on spontaneous activity in C-nociceptors in patients with painful diabetic neuropathy: A randomized controlled trial. *Pain, 156*(11), 2175–2183. https://doi.org/10.1097/j.pain.0000000000000249

Shan, Z., Cai, S., Yu, J., Zhang, Z., Vallecillo, T. G. M., Serafini, M. J., et al. (2019). Reversal of peripheral neuropathic pain by the small-molecule natural product physalin F via block of CaV2.3 (R-type) and CaV2.2 (N-type) voltage-gated calcium channels. *ACS Chemical Neuroscience, 10*(6), 2939–2955. https://doi.org/10.1021/acschemneuro.9b00166

Sharma, N., Flaherty, K., Lezgiyeva, K., Wagner, D. E., Klein, A. M., & Ginty, D. D. (2020). The emergence of transcriptional identity in somatosensory neurons. *Nature, 577*(7790), 392–398. https://doi.org/10.1038/s41586-019-1900-1

Shen, F. Y., Chen, Z. Y., Zhong, W., Ma, L. Q., Chen, C., Yang, Z. J., et al. (2015). Alleviation of neuropathic pain by regulating T-type calcium channels in rat anterior cingulate cortex. *Molecular Pain, 11*, 7. https://doi.org/10.1186/s12990-015-0008-3

Sheng, Z. H., Rettig, J., Takahashi, M., & Catterall, W. A. (1994). Identification of a syntaxin-binding site on N-type calcium channels. *Neuron, 13*(6), 1303–1313. https://doi.org/10.1016/0896-6273(94)90417-0

Shin, S. M., Cai, Y., Itson-Zoske, B., Qiu, C., Hao, X., Xiang, H., et al. (2020). Enhanced T-type calcium channel 3.2 activity in sensory neurons contributes to neuropathic-like pain of monosodium iodoacetate-induced knee osteoarthritis. *Molecular Pain, 16*, 1744806920963807. https://doi.org/10.1177/1744806920963807

Sluka, K. A. (1998). Blockade of N- and P/Q-type calcium channels reduces the secondary heat hyperalgesia induced by acute inflammation. *The Journal of Pharmacology and Experimental Therapeutics, 287*(1), 232–237.

Snutch, T. P., & Zamponi, G. W. (2018). Recent advances in the development of T-type calcium channel blockers for pain intervention. *British Journal of Pharmacology, 175*(12), 2375–2383. https://doi.org/10.1111/bph.13906

Souza, I. A., Gandini, M. A., Zhang, F. X., Mitchell, W. G., Matsumoto, J., Lerner, J., et al. (2019). Pathogenic Cav3.2 channel mutation in a child with primary generalized epilepsy. *Molecular Brain, 12*(1), 86. https://doi.org/10.1186/s13041-019-0509-5

Spahn, V., & Stein, C. (2017). Targeting delta opioid receptors for pain treatment: Drugs in phase I and II clinical development. *Expert Opinion on Investigational Drugs, 26*(2), 155–160. https://doi.org/10.1080/13543784.2017.1275562

Staats, P. S., Yearwood, T., Charapata, S. G., Presley, R. W., Wallace, M. S., Byas-Smith, M., et al. (2004). Intrathecal ziconotide in the treatment of refractory pain in patients with cancer or AIDS: A randomized controlled trial. *JAMA, 291*(1), 63–70. https://doi.org/10.1001/jama.291.1.63

Stemkowski, P., García-Caballero, A., Gadotti, V. M., M'Dahoma, S., Huang, S., Black, S. A. G., et al. (2016). TRPV1 nociceptor activity initiates USP5/T-type channel-mediated plasticity. *Cell Reports, 17*(11), 2901–2912. https://doi.org/10.1016/j.celrep.2016.11.047

Swensen, A. M., Herrington, J., Bugianesi, R. M., Dai, G., Haedo, R. J., Ratliff, K. S., et al. (2012). Characterization of the substituted N-triazole oxindole TROX-1, a small-molecule, state-dependent inhibitor of Ca(V)2 calcium channels. *Molecular Pharmacology, 81*(3), 488–497. https://doi.org/10.1124/mol.111.075226

Taylor, C. P., Angelotti, T., & Fauman, E. (2007). Pharmacology and mechanism of action of pregabalin: The calcium channel alpha2-delta (alpha2-delta) subunit as a target for antiepileptic drug discovery. *Epilepsy Research, 73*(2), 137–150. https://doi.org/10.1016/j.eplepsyres.2006.09.008

Tedford, H. W., & Zamponi, G. W. (2006). Direct G protein modulation of Cav2 calcium channels. *Pharmacological Reviews, 58*(4), 837–862. https://doi.org/10.1124/pr.58.4.11

Teodori, E., Baldi, E., Dei, S., Gualtieri, F., Romanelli, M. N., Scapecchi, S., et al. (2004). Design, synthesis, and preliminary pharmacological evaluation of 4-aminopiperidine derivatives as N-type calcium channel blockers active on pain and neuropathic pain. *Journal of Medicinal Chemistry, 47*(24), 6070–6081. https://doi.org/10.1021/jm0499231

Terrence, C. F., Fromm, G. H., & Tenicela, R. (1985). Baclofen as an analgesic in chronic peripheral nerve disease. *European Neurology, 24*(6), 380–385. https://doi.org/10.1159/000115830

Tomita, S., Sekiguchi, F., Deguchi, T., Miyazaki, T., Ikeda, Y., Tsubota, M., et al. (2019). Critical role of Ca$_v$3.2 T-type calcium channels in the peripheral neuropathy induced by bortezomib, a proteasome-inhibiting chemotherapeutic agent, in mice. *Toxicology, 413*, 33–39. https://doi.org/10.1016/j.tox.2018.12.003

Tomita, S., Sekiguchi, F., Kasanami, Y., Naoe, K., Tsubota, M., Wake, H., et al. (2020). Ca$_v$3.2 overexpression in L4 dorsal root ganglion neurons after L5 spinal nerve cutting involves Egr-1, USP5 and HMGB1 in rats: An emerging signaling pathway for neuropathic pain. *European Journal of Pharmacology, 888*, 173587. https://doi.org/10.1016/j.ejphar.2020.173587

Tytgat, J., Pauwels, P. J., Vereecke, J., & Carmeliet, E. (1991). Flunarizine inhibits a high-threshold inactivating calcium channel (N-type) in isolated hippocampal neurons. *Brain Research, 549*(1), 112–117. https://doi.org/10.1016/0006-8993(91)90606-v

Umeda, M., Ohkubo, T., Ono, J., Fukuizumi, T., & Kitamura, K. (2006). Molecular and immunohistochemical studies in expression of voltage-dependent Ca2+ channels in dorsal root ganglia from streptozotocin-induced diabetic mice. *Life Sciences, 79*(21), 1995–2000. https://doi.org/10.1016/j.lfs.2006.06.039

Uneyama, H., Takahara, A., Dohmoto, H., Yoshimoto, R., Inoue, K., & Akaike, N. (1997). Blockade of N-type Ca2+ current by cilnidipine (FRC-8653) in acutely dissociated rat sympathetic neurones. *British Journal of Pharmacology, 122*(1), 37–42. https://doi.org/10.1038/sj.bjp.0701342

Vardanyan, R. S., Cain, J. P., Haghighi, S. M., Kumirov, V. K., McIntosh, M. I., Sandweiss, A. J., et al. (2017). Synthesis and investigation of mixed μ-opioid and δ-opioid agonists as possible bivalent ligands for treatment of pain. *Journal of Heterocyclic Chemistry, 54*(2), 1228–1235. https://doi.org/10.1002/jhet.2696

Waldhoer, M., Bartlett, S. E., & Whistler, J. L. (2004). Opioid receptors. *Annual Review of Biochemistry, 73*, 953–990. https://doi.org/10.1146/annurev.biochem.73.011303.073940

Wallace, M., Duan, R., Liu, W., Locke, C., & Nothaft, W. (2016). A randomized, double-blind, placebo-controlled, crossover study of the T-type calcium channel blocker ABT-639 in an intradermal capsaicin experimental pain model in healthy adults. *Pain Medicine, 17*(3), 551–560. https://doi.org/10.1093/pm/pnv068

Wang, H., Sun, H., Della Penna, K., Benz, R. J., Xu, J., Gerhold, D. L., et al. (2002). Chronic neuropathic pain is accompanied by global changes in gene expression and shares pathobiology with neurodegenerative diseases. *Neuroscience, 114*(3), 529–546. https://doi.org/10.1016/s0306-4522(02)00341-x

Weiergräber, M., Henry, M., Radhakrishnan, K., Hescheler, J., & Schneider, T. (2007). Hippocampal seizure resistance and reduced neuronal excitotoxicity in mice lacking the Cav2.3 E/R-type voltage-gated calcium channel. *Journal of Neurophysiology, 97*(5), 3660–3669. https://doi.org/10.1152/jn.01193.2006

Weiss, N., & Zamponi, G. W. (2019a). T-type calcium channels: From molecule to therapeutic opportunities. *The International Journal of Biochemistry &*

Cell Biology, 108, 34–39. https://doi.org/10.1016/j.biocel.2019.01.008

Weiss, N., & Zamponi, G. W. (2019b). T-type channel druggability at a crossroads. *ACS Chemical Neuroscience, 10*(3), 1124–1126. https://doi.org/10.1021/acschemneuro.9b00031

Weiss, N., Hameed, S., Fernández-Fernández, J. M., Fablet, K., Karmazinova, M., Poillot, C., et al. (2012). A Ca(v)3.2/syntaxin-1A signaling complex controls T-type channel activity and low-threshold exocytosis. *The Journal of Biological Chemistry, 287*(4), 2810–2818. https://doi.org/10.1074/jbc.M111.290882

Weiss, N., Black, S. A., Bladen, C., Chen, L., & Zamponi, G. W. (2013). Surface expression and function of Cav3.2 T-type calcium channels are controlled by asparagine-linked glycosylation. *Pflügers Archiv, 465*(8), 1159–1170. https://doi.org/10.1007/s00424-013-1259-3

Wen, X. J., Xu, S. Y., Chen, Z. X., Yang, C. X., Liang, H., & Li, H. (2010). The roles of T-type calcium channel in the development of neuropathic pain following chronic compression of rat dorsal root ganglia. *Pharmacology, 85*(5), 295–300. https://doi.org/10.1159/000276981

Westenbroek, R. E., Hell, J. W., Warner, C., Dubel, S. J., Snutch, T. P., & Catterall, W. A. (1992). Biochemical properties and subcellular distribution of an N-type calcium channel alpha 1 subunit. *Neuron, 9*(6), 1099–1115. https://doi.org/10.1016/0896-6273(92)90069-p

Wheeler, D. B., Randall, A., & Tsien, R. W. (1994). Roles of N-type and Q-type Ca2+ channels in supporting hippocampal synaptic transmission. *Science, 264*(5155), 107–111. https://doi.org/10.1126/science.7832825

Wiffen, P. J., Derry, S., Lunn, M. P., & Moore, R. A. (2013). Topiramate for neuropathic pain and fibromyalgia in adults. *Cochrane Database of Systematic Reviews, 8*, CD008314. https://doi.org/10.1002/14651858.CD008314.pub3

Wilson, S. M., Brittain, J. M., Piekarz, A. D., Ballard, C. J., Ripsch, M. S., Cummins, T. R., et al. (2011). Further insights into the antinociceptive potential of a peptide disrupting the N-type calcium channel-CRMP-2 signaling complex. *Channels (Austin, Tex.), 5*(5), 449–456. https://doi.org/10.4161/chan.5.5.17363

Wilson, S. M., Schmutzler, B. S., Brittain, J. M., Dustrude, E. T., Ripsch, M. S., Pellman, J. J., et al. (2012). Inhibition of transmitter release and attenuation of anti-retroviral-associated and tibial nerve injury-related painful peripheral neuropathy by novel synthetic Ca2+ channel peptides. *The Journal of Biological Chemistry, 287*(42), 35065–35077. https://doi.org/10.1074/jbc.M112.378695

Xie, J. Y., Chew, L. A., Yang, X., Wang, Y., Qu, C., Federici, L. M., et al. (2016). Sustained relief of ongoing experimental neuropathic pain by a CRMP2 peptide aptamer with low abuse potential. *Pain, 157*(9), 2124–2140. https://doi.org/10.1097/j.pain.0000000000000628

Xu, H., Wu, L. J., Wang, H., Zhang, X., Vadakkan, K. I., Kim, S. S., et al. (2008). Presynaptic and postsynaptic amplifications of neuropathic pain in the anterior cingulate cortex. *The Journal of Neuroscience,* 28(29), 7445–7453. https://doi.org/10.1523/JNEUROSCI.1812-08.2008

Yamamoto, S., Suzuki, Y., Ono, H., Kume, K., & Ohsawa, M. (2016). N- and L-type calcium channels blocker cilnidipine ameliorates neuropathic pain. *European Journal of Pharmacology, 793*, 66–75. https://doi.org/10.1016/j.ejphar.2016.11.001

Yang, L., Gu, X., Zhang, W., Zhang, J., & Ma, Z. (2014). Cdk5 inhibitor roscovitine alleviates neuropathic pain in the dorsal root ganglia by downregulating N-methyl-D-aspartate receptor subunit 2A. *Neurological Sciences, 35*(9), 1365–1371. https://doi.org/10.1007/s10072-014-1713-9

Yokoyama, K., Kurihara, T., Makita, K., & Tanabe, T. (2003). Plastic change of N-type Ca channel expression after preconditioning is responsible for prostaglandin E2-induced long-lasting allodynia. *Anesthesiology, 99*(6), 1364–1370. https://doi.org/10.1097/00000542-200312000-00019

You, H., Gadotti, V. M., Petrov, R. R., Zamponi, G. W., & Diaz, P. (2011). Functional characterization and analgesic effects of mixed cannabinoid receptor/T-type channel ligands. *Molecular Pain, 7*, 89. https://doi.org/10.1186/1744-8069-7-89

Yu, P. Y., Gong, N., Kweon, T. D., Vo, B., & Luo, Z. D. (2018). Gabapentin prevents synaptogenesis between sensory and spinal cord neurons induced by thrombospondin-4 acting on pre-synaptic $Ca_v\alpha_2\delta_1$ subunits and involving T-type Ca^{2+} channels. *British Journal of Pharmacology, 175*(12), 2348–2361. https://doi.org/10.1111/bph.14149

Yusaf, S. P., Goodman, J., Pinnock, R. D., Dixon, A. K., & Lee, K. (2001). Expression of voltage-gated calcium channel subunits in rat dorsal root ganglion neurons. *Neuroscience Letters, 311*(2), 137–141. https://doi.org/10.1016/s0304-3940(01)02038-9

Zaman, T., Lee, K., Park, C., Paydar, A., Choi, J. H., Cheong, E., et al. (2011). Cav2.3 channels are critical for oscillatory burst discharges in the reticular thalamus and absence epilepsy. *Neuron, 70*(1), 95–108. https://doi.org/10.1016/j.neuron.2011.02.042

Zamponi, G. W. (2016). Targeting voltage-gated calcium channels in neurological and psychiatric diseases. *Nature Reviews. Drug Discovery, 15*(1), 19–34. https://doi.org/10.1038/nrd.2015.5

Zamponi, G. W., Bourinet, E., Nelson, D., Nargeot, J., & Snutch, T. P. (1997). Crosstalk between G proteins and protein kinase C mediated by the calcium channel alpha1 subunit. *Nature, 385*(6615), 442–446. https://doi.org/10.1038/385442a0

Zamponi, G. W., Feng, Z. P., Zhang, L., Pajouhesh, H., Ding, Y., Belardetti, F., et al. (2009). Scaffold-based design and synthesis of potent N-type calcium channel blockers. *Bioorganic & Medicinal Chemistry Letters, 19*(22), 6467–6472. https://doi.org/10.1016/j.bmcl.2009.09.008

Zamponi, G. W., Striessnig, J., Koschak, A., & Dolphin, A. C. (2015). The physiology, pathology, and pharmacology of voltage-gated calcium channels and their future therapeutic potential. *Pharmacological*

Reviews, 67(4), 821–870. https://doi.org/10.1124/pr.114.009654

Zhang, F. X., Gadotti, V. M., Souza, I. A., Chen, L., & Zamponi, G. W. (2018). BK potassium channels suppress Cavα2δ subunit function to reduce inflammatory and neuropathic pain. *Cell Reports, 22*(8), 1956–1964. https://doi.org/10.1016/j.celrep.2018.01.073

Zhao, Y., Huang, G., Wu, Q., Wu, K., Li, R., Lei, J., et al. (2019). Cryo-EM structures of apo and antagonist-bound human Ca$_v$3.1. *Nature, 576*(7787), 492–497. https://doi.org/10.1038/s41586-019-1801-3

Ziegler, D., Duan, W. R., An, G., Thomas, J. W., & Nothaft, W. (2015). A randomized double-blind, placebo-, and active-controlled study of T-type calcium channel blocker ABT-639 in patients with diabetic peripheral neuropathic pain. *Pain, 156*(10), 2013–2020. https://doi.org/10.1097/j.pain.0000000000000263

Voltage-Gated Calcium Channels and Migraine

Daniela Pietrobon

Abstract

Missense mutations in *CACNA1A*, the gene encoding the pore-forming subunit of the human voltage-gated $Ca_V2.1$ (P/Q-type calcium) channel, cause a rare form of migraine with aura (familial hemiplegic migraine type 1: FHM1). $Ca_V2.1$ channels play a dominant role in initiating synaptic transmission at brain excitatory and inhibitory synapses. In this chapter, we first summarize current understanding of the pathophysiological mechanisms underlying migraine headache and migraine aura and then describe the functional consequences of FHM1 mutations. After a brief discussion of the alterations produced by the mutations in the $Ca_V2.1$ channel's biophysical properties, we focus on functional studies in knock-in mouse models carrying an FHM1 mutation in the orthologous gene. We first discuss the migraine-relevant phenotypes in FHM1 mice, in particular their increased susceptibility to cortical spreading depression (CSD), the phenomenon that underlies migraine aura and may trigger the headache mechanisms. Then we discuss the alterations in synaptic transmission in the cerebral cortex of FHM1 mice and the mechanisms underlying their enhanced susceptibility to CSD. In the last part, we discuss the alterations in the trigeminovascular pain pathway.

Keywords

Calcium channel · $Ca_V2.1$ · Migraine · Spreading depression · Synaptic transmission · Cerebral cortex · Trigeminovascular · Knock-in mouse model

Abbreviations

$[K]_e$	extracellular concentration of K^+ ions
Ca	calcium
CBF	cerebral blood flow
CGRP	calcitonin gene-related peptide
CSD	cortical spreading depression
E/I	excitatory–inhibitory balance
FHM	familial hemiplegic migraine
FHM1	familial hemiplegic migraine type 1
FHM2	familial hemiplegic migraine type 2
FHM3	familial hemiplegic migraine type 3
FS	fast-spiking
GPCR	G-protein-coupled receptor
GWAS	genome-wide association studies

D. Pietrobon (✉)
Department of Biomedical Sciences, Padova
Neuroscience Center, CNR Institute of Neuroscience,
University of Padova, Padova, Italy
e-mail: daniela.pietrobon@unipd.it

mEPSC   miniature   excitatory   postsynaptic
         current
P2X3R   P2X3 receptor
TG   trigeminal ganglion
TNC   trigeminal nucleus caudalis
VGCC   voltage-gated calcium channel

Introduction

Migraine is a common, disabling, and costly brain disorder, characterized by attacks of typically throbbing, unilateral headache with certain associated features such as nausea and amplification of percepts from one or more senses (photophobia, phonophobia, osmophobia, allodynia) (Pietrobon & Moskowitz, 2013; Goadsby et al., 2017; Brennan & Pietrobon, 2018). The headache phase of migraine begins with the activation and sensitization of trigeminal sensory afferents that innervate cranial tissues, in particular the meninges, and subsequent activation and sensitization of second-order neurons in the trigeminal nucleus caudalis (TNC) and dorsal horn of the first cervical segments; the TNC projects directly or indirectly to the thalamus and different areas of the brainstem and forebrain that are involved in different aspects of pain and other aspects of the complex migraine symptomatology (Pietrobon & Moskowitz, 2013; Burstein et al., 2015; Goadsby et al., 2017; Brennan & Pietrobon, 2018). Whereas the properties of pial afferents remain largely unknown, the dural afferents have properties typical of nociceptors in other tissues (Pietrobon & Moskowitz, 2013; Levy et al., 2019). Sterile meningeal inflammation is considered to be a key mechanism that may underlie the sustained activation and sensitization of meningeal nociceptors during migraine attacks (Pietrobon & Moskowitz, 2013; Levy et al., 2019). The first direct evidence of meningeal inflammation was recently obtained in migraine with aura patients (Hadjikhani et al., 2020).

Migraine is much more than an episodic headache and a pain disorder. It is a complex brain disorder characterized by a global dysfunction in multisensory information processing and integration. Indeed, hypersensitivity to sensory stimuli

may persists in the interictal period, during which the brains of migraineurs show altered processing of noxious and non-noxious sensory information. Interestingly, the magnitude of some of these alterations increases in the interictal period toward the next attack and becomes maximal the day before the attack in temporal coincidence with prodromal symptoms (such as difficulty with speech, reading, and concentration; increased emotionality; irritability; sensory hypersensitivity) that in many migraineurs are highly predictive of the attack. The nature of the prodromal symptoms as well as that of typical migraine triggers (e.g., stress, sleep deprivation, oversleeping, hunger, strong sensory stimulation) point to involvement of different brain areas in migraine onset (Pietrobon & Moskowitz, 2013; Burstein et al., 2015; Schwedt et al., 2015; Goadsby et al., 2017; Brennan & Pietrobon, 2018).

In a third of patients, the headache is preceded by transient sensory (most frequently visual) disturbances, the so-called migraine aura, whose neurophysiological correlate is recognized to be cortical spreading depression (CSD) (Pietrobon & Moskowitz, 2014; Ayata & Lauritzen, 2015). CSD is a self-sustaining, slowly propagating (2–5 mm/min) wave of nearly complete depolarization of a sizable population of brain cells that lasts about 1 min and silences electrical activity for several minutes. CSD can be induced in healthy brain tissue by intense depolarizing stimuli that increase the extracellular concentration of K^+ ions, $[K]_e$, above a critical threshold. Although the mechanisms of initiation and propagation of experimental CSD are still incompletely understood, initiation of a positive feedback cycle mediating a self-regenerative neuronal depolarization and a regenerative increase in $[K]_e$, as a consequence of activation of specific neuronal cation channels and generation of net inward current, appears required for ignition of CSD (Pietrobon & Moskowitz, 2014).

There is evidence from animal studies that CSD can activate and sensitize the trigeminovascular pain pathway and hence can initiate the headache mechanism. In fact, a single experimentally induced CSD can lead to delayed sustained increases in dural blood flow and in ongoing activity of rat dural nociceptors and TNC trigeminovascular neurons as well as delayed sensitization of these neurons (Bolay et al., 2002; Zhang et al., 2010, 2011; Zhao & Levy, 2015, 2016; Melo-Carrillo et al., 2017a, b). According to an interesting study, the trigeminal activation may result from CSD-induced release of pro-inflammatory molecules in the meninges as a consequence of parenchymal inflammation, initiated by CSD-induced opening of pannexin1 channels and inflammasome activation (Karatas et al., 2013). Rapid pro-inflammatory response in the cerebral cortex was also triggered by CSD when the latter was induced through the intact skull using optogenetics (Takizawa et al., 2020). Moreover CSD leads to a closure of paravascular space within the cortex and an impairment of glymphatic flow, which may compromise the clearance of a variety of excitatory and inflammatory chemicals known to increase after a CSD (Schain et al., 2017). CSD-induced pial and dural macrophage activation has been uncovered in another study (Schain et al., 2018). Activation of peptidergic meningeal nociceptors and consequent release of pro-inflammatory neuropeptides, including calcitonin gene-related peptide (CGRP), from their peripheral nerve endings might then further promote meningeal inflammation (Pietrobon & Moskowitz, 2013; Levy et al., 2019). In awake animals, CSD induces behavioral alterations associated with pain perception and periorbital tactile allodynia, which are reversed by drugs efficacious in migraine headache treatment (sumatriptan or a CGRP receptor antagonist) (Filiz et al., 2019; Harriott et al., 2021). Intravenous administration of monoclonal antibodies against CGRP, which are efficacious in migraine treatment (Russo, 2015), inhibits the CSD-induced activation of A-δ meningeal nociceptors and the CSD-induced activation and sensitization of high-threshold TNC neurons, which receive input from A-δ fibers (Melo-Carrillo et al., 2017a, b).

The neurobiological mechanisms of the primary brain dysfunctions underlying the onset of a migraine attack, susceptibility to CSD, and the global dysfunction in multisensory information processing remain largely unknown and are key unanswered questions in migraine neurobiology.

Migraine is a complex polygenic genetic disorder, with heritability estimates as high as 50% (Ferrari et al., 2015; Sutherland & Griffiths, 2017). Although genome-wide association studies (GWAS) are providing increasing insights into the common genetic variants associated with migraine (Gormley et al., 2016; Hautakangas et al., 2021), the study of the functional consequences of GWAS hits is very difficult, if not impossible, given also the fact that they generally lie in intronic or intergenic regions. In contrast, rare monogenic forms of migraine are caused by mutations that directly affect protein function, and the functional consequences of the disease-causing mutations can be studied in genetic mouse models of the disease, which provide a unique opportunity to study the cellular and circuit mechanisms of the primary brain dysfunctions causing a migraine disorder and thus tackle the above questions.

Familial hemiplegic migraine is a rare monogenic form of migraine with aura. Apart from the motor weakness or hemiplegia during aura (and the possible longer duration of aura), typical FHM attacks resemble "common" migraine with aura attacks and both types of attacks may alternate in patients and co-occur within families (Pietrobon, 2007; Russell & Ducros, 2011; Ferrari et al., 2015; Di Stefano et al., 2020). Thus, FHM and migraine with aura are considered to be part of the same spectrum and may share pathogenetic mechanisms. Indeed, a very recent GWAS of more than 100,000 migraine cases identified risk variants specific for migraine with aura in *CACNA1A*, and provided the first gene-based support for the increased sharing of common variants between monogenic and polygenic forms of migraine with aura (Hautakangas et al., 2021). Some FHM mutations give rise to "atypical" severe FHM attacks and the patients carrying these mutations show additional ictal and/or permanent neurological features such as epilepsy, loss of consciousness, ataxia, and cognitive impairment (Pietrobon, 2007; Russell & Ducros, 2011; Ferrari et al., 2015; Di Stefano et al., 2020).

Thus far, three FHM causative genes, all encoding ion channels or transporters, have been identified: *CACNA1A* (FHM1), *ATP1A2* (FHM2), and *SCN1A* (FHM3). *CACNA1A* and *SCN1A* encode the pore-forming subunits of the voltage-gated ion channels $Ca_V2.1$ and $Na_V1.1$, while *ATP1A2* encodes the α2 Na/K-ATPase (Ophoff et al., 1996; De Fusco et al., 2003; Dichgans et al., 2005). Knock-in mouse models for FHM1, FHM2, and FHM3 were generated by introduction of human mutations in the orthologous genes (van den Maagdenberg et al., 2004, 2010; Leo et al., 2011; Bøttger et al., 2016; Jansen et al., 2020; Auffenberg et al., 2021).

After a brief discussion of the functional consequences of FHM1 mutations on recombinant $Ca_V2.1$ channels, here, I will focus on the functional studies in FHM1 knock-in mice (FHM1 mice) carrying the mutation R192Q, which causes typical FHM attacks in humans (pure FHM) (Ophoff et al., 1996; van den Maagdenberg et al., 2004), and on the insights into migraine pathophysiology and the mechanisms underlying the susceptibility to CSD obtained from these studies. I will only briefly discuss the different findings in the S218L FHM1 mouse model, which may give insights into the additional clinical features of the severe syndrome caused by the S218L mutation, which include slowly progressive cerebellar ataxia and atrophy, epileptic seizures, coma or profound stupor, and severe, sometimes fatal, cerebral edema triggered by even a trivial head trauma (Kors et al., 2001; van den Maagdenberg et al., 2010).

Ca$_V$2.1 Channels and Functional Consequences of FHM1 Mutations on Channel Biophysical Properties

$Ca_V2.1$ (P/Q-type calcium) channels are widely expressed in the nervous system, including all structures implicated in the pathogenesis of migraine and/or migraine pain, and play a dominant role in controlling neurotransmitter release, particularly at excitatory and inhibitory central synapses (Pietrobon, 2013; see chapter "Voltage Gated Calcium Channels (VGCCs) and Synaptic

Transmission" by Wang discussing voltage-gated calcium channels [VGCCs] and synaptic transmission). Among the presynaptic Ca channels, $Ca_V2.1$ channels are unique in their capacity of interacting with and being modulated in a complex manner by a number of Ca-binding proteins, and may exhibit both Ca-dependent inactivation and Ca-dependent facilitation (Catterall & Few, 2008; see chapter "Calmodulin Regulation of Voltage-gated Calcium Channels" by Dick and Ben Johnny discussing calcium regulation of VGCCs). Ca-dependent regulation of presynaptic $Ca_V2.1$ channels may play a crucial role in short-term synaptic plasticity (Catterall et al., 2013; see also chapters "Calmodulin Regulation of Voltage-gated Calcium Channels" by Dick and Ben Johnny and "Voltage Gated Calcium Channels (VGCCs) and Synaptic Transmission" by Wang). Their somatodendritic localization points to additional postsynaptic roles, e.g., in regulation of neuronal intrinsic excitability and gene transcription (Pietrobon, 2013; see also chapter "Cav3 Calcium Channel Interactions with Potassium Channels" by Turner, discussing calcium channel complexes with other channels).

Numerous different mutations have been found in *CACNA1A* of FHM1 patients, which all produce substitutions of conserved amino acids in important functional regions of the $Ca_V2.1$ channel, including the pore lining and the voltage sensor (Fig. 1; see also Pietrobon, 2013; Izquierdo-Serra et al., 2020 for review and references). Analysis of the single channel properties of mutant human recombinant $Ca_V2.1$ channels (with eight different FHM1 mutations) and of the $Ca_V2.1$ current in neurons of FHM1 mice revealed that the FHM1 mutations produce a gain-of-function, mainly due to increased channel open probability and channel activation at lower voltages (Tottene et al., 2002, 2005, 2009; van den Maagdenberg et al., 2004, 2010; Catterall et al., 2008; Inchauspe et al., 2010; Fioretti et al., 2011; Di Guilmi et al., 2014). Analysis of the whole-cell current carried by mutant recombinant $Ca_V2.1$ channels (with 17 different FHM1 or sporadic hemiplegic migraine mutations) also revealed consistent shifts to lower voltages of

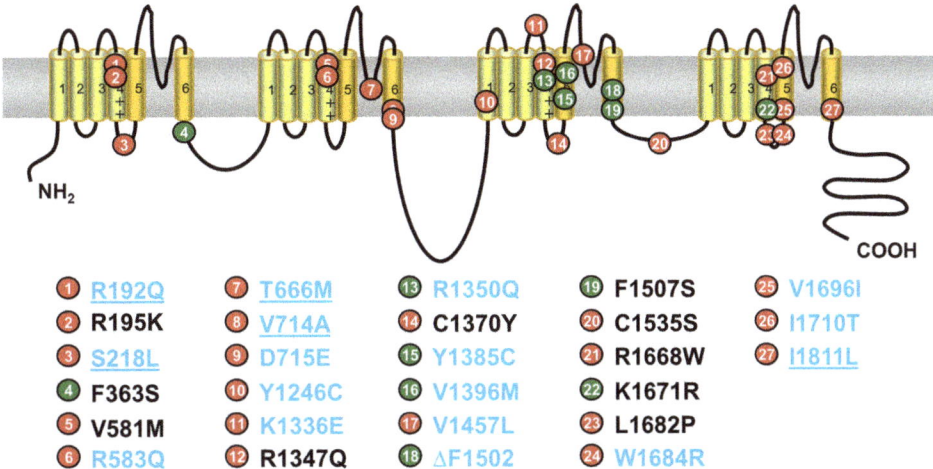

① R192Q	⑦ T666M	⑬ R1350Q	⑲ F1507S	㉕ V1696I
② R195K	⑧ V714A	⑭ C1370Y	⑳ C1535S	㉖ I1710T
③ S218L	⑨ D715E	⑮ Y1385C	㉑ R1668W	㉗ I1811L
④ F363S	⑩ Y1246C	⑯ V1396M	㉒ K1671R	
⑤ V581M	⑪ K1336E	⑰ V1457L	㉓ L1682P	
⑥ R583Q	⑫ R1347Q	⑱ ΔF1502	㉔ W1684R	

Fig. 1 Location of FHM1 (red) and some sporadic hemiplegic migraine mutations (green) in the secondary structure of the $Ca_V2.1\alpha1$ subunit. Reference sequence Genbank Acc N. X998897. The mutations written in blue are those whose effects on the biophysical properties of recombinant $Ca_V2.1$ channels were studied in heterologous expression systems; the mutations underlined are those whose effects were studied also in transfected neurons from $Ca_V2.1-/-$ mice

channel activation, although the interpretation of the data for several mutations is confused by a decrease in maximal current, due in most cases to a decrease in expression of functional channels linked to the heterologous expression system and not observed in neurons of FHM1 mice (see Pietrobon, 2010, 2013 for a more thorough discussion of this point and references) (Izquierdo-Serra et al., 2020; Gandini et al., 2021). In agreement with the lower activation threshold of human S218L $Ca_V2.1$ channels compared to R192Q $Ca_V2.1$ channels (Tottene et al., 2005), the gain-of-function of the P/Q-type calcium current was larger in neurons of S218L compared to R192Q FHM1 mice (van den Maagdenberg et al., 2004, 2010). The specific gating alterations produced by FHM1 mutations on recombinant channels, including the magnitude (or even presence) of the shift to lower voltages of channel activation, may be dependent on the specific $Ca_V2.1$ splice variant and/or auxiliary subunit (Mullner et al., 2004; Adams et al., 2009). Indeed, neuron-specific effects have been uncovered in FHM1 mice, likely due to expression of specific $Ca_V2.1$ splice variants and/or auxiliary subunits (Tottene et al., 2009; Fioretti et al., 2011; Vecchia et al., 2014), which may help to explain why a mutation

in a channel widely expressed in the nervous system (Westenbroek et al., 1995) produces the specific neuronal dysfunctions leading to migraine (see below).

FHM1 mutations may also affect Ca-dependent and G-protein-dependent modulation of $Ca_V2.1$ channels. Ca-dependent facilitation of recombinant channels was largely reduced by the R192Q mutation and almost eliminated by the S218L mutation (Adams et al., 2010), suggesting that the increase in the open probability of mutant channels partially or completely occludes further facilitation. However, the findings for native neuronal $Ca_V2.1$ channels were somewhat different, since Ca-dependent facilitation was unaltered in Purkinje cells (Adams et al., 2010) and Calyx of Held synaptic terminals (Inchauspe et al., 2010) of R192Q FHM1 mice, and it was reduced in Purkinje cells of S218L FHM1 mice (Adams et al., 2010), suggesting that also the effect of FHM1 mutations on Ca-dependent regulation may be $Ca_V2.1$ variant-dependent.

G-protein-mediated inhibitory modulation of recombinant $Ca_V2.1$ channels (see chapter "Modulation of VGCCs by G-protein Coupled Receptors and Their Second Messengers" by Herlitze, Mark, and Schwitalia, discussing modulation of VGCCs by GPCRs) was also reduced by FHM1 mutations (Melliti et al., 2003; Weiss et al., 2008; Serra et al., 2009; Garza-Lopez et al., 2012), an effect that may lead to further gain-of-function of Ca influx through mutant channels during neuromodulation. Investigation of voltage-dependent G-protein modulation using action potential-like stimuli in cells overexpressing Gβγ subunits revealed a reduced Gβγ-mediated inhibition (which was dependent on the duration of the action potential) and a reduced facilitation of the Ca current carried by mutant channels after a train of short depolarizing pulses (Serra et al., 2009). Indirect evidence of reduced G-protein-mediated inhibition of native neuronal $Ca_V2.1$ channels in R192Q FHM1 mice was obtained in a study investigating the effect of the FHM1 mutation on sleep regulation by adenosine receptors in vivo, since FHM1 mice showed reduced responsiveness to either inhibition or enhancement of adenosine receptor activation (Deboer et al., 2013).

Migraine-Relevant Phenotypes in FHM1 Mouse Models

Homozygous R192Q knock-in mice (R192Q FHM1 mice) do not show an overt phenotype (van den Maagdenberg et al., 2004), but they show signs of photophobia in a modified elevated plus maze in which the safe closed arms are brightly illuminated and they show behavioral changes suggestive of unilateral head pain when subjected to novelty or restrain stress (Chanda et al., 2013). Interestingly, systemic administration of the antimigraine drug rizatriptan normalized these pain measures. Given the higher female prevalence in migraine, another interesting finding is that the increase in some of these behavioral pain measures was larger in female mice (Chanda et al., 2013). Homozygous S218L knock-in mice (S218L FHM1 mice) exhibit the main features of the human S218L clinical syndrome, including attacks of hemiparesis, attacks of generalized seizures, mild cerebellar ataxia, and brain edema after mild head impact. These features were not observed in het-

erozygous S218L mice (van den Maagdenberg et al., 2010).

A key migraine-relevant phenotype of FHM1 mice is increased susceptibility to experimentally induced CSD, as revealed by a lower stimulation threshold for CSD induction and an increased rate of CSD propagation in vivo (van den Maagdenberg et al., 2004, 2010) or a higher frequency of CSDs elicited by prolonged epidural high KCl (Eikermann-Haerter et al., 2009b). In agreement with the higher female prevalence in migraine, the velocity of propagation and the frequency of CSDs induced by prolonged KCl were larger in female than in male FHM1 mice; the sex difference was abrogated by ovariectomy and enhanced by orchiectomy, suggesting that female and male gonadal hormones exert reciprocal effects on CSD susceptibility (Eikermann-Haerter et al., 2009a, b).

Many migraineurs, including FHM patients, report that stress or relief after stress trigger an attack (Hansen et al., 2011). In male R192Q FHM1 mice, the frequency of CSDs induced by prolonged KCl increased after administration of the stress hormone corticosterone, but not after acute restrain stress; neither the stress hormone nor restrain stress affected CSD frequency in male wild-type mice (Shyti et al., 2015). Even chronic stress did not affect CSD frequency and velocity in male wild-type mice (Kaufmann & Brennan, 2018; Yapici-Eser et al., 2018; Balkaya et al., 2019), while contradictory findings were reported in wild-type mice regarding the effect of stress on the threshold for CSD induction, being this either lower (Yapici-Eser et al., 2018) or unaltered (Balkaya et al., 2019) after acute or chronic stress. In the latter study, the CSD threshold was unaltered after acute or chronic stress even in heterozygous S218L mice. Possibly, strain differences in stress response may have contributed to the discrepant findings (Balkaya et al., 2019). However, in heterozygous S218L (but not wild type) mice, chronic stress followed by relief for 36 h reduced the electrical CSD threshold and increased the frequency of KCl-induced CSDs, without affecting their propagation velocity (Balkaya et al., 2019).

In contrast with wild-type mice, a single CSD caused prolonged hemiplegia with leaning and circling in FHM1 mice, and CSD readily propagated into the striatum in the majority of FHM1 but not wild-type mice, suggesting corticostriatal CSD propagation as a likely explanation for the more severe motor deficits in FHM1 mutants (Eikermann-Haerter et al., 2009b, 2011; Cain et al., 2017). Moreover, the typical reduction in cerebral blood flow (CBF) after CSD was more prolonged in R192Q FHM1 mice (Khennouf et al., 2016) and more severe in heterozygous S218L FHM1 mice (Eikermann-Haerter et al., 2015). During CSD, the increase in $[Ca^{2+}]_{in}$ in cortical neurons and neuropil was larger in both FHM1 mouse models compared to wild-type mice (Eikermann-Haerter et al., 2015; Khennouf et al., 2016), in agreement with the gain-of-function of $Ca_V2.1$ channels in cortical pyramidal cells from both mutants (Tottene et al., 2009; Vecchia et al., 2015). The metabolic oxygen consumption was larger in FHM1 mice and resulted in a larger decrease in tissue oxygenation during CSD compared to wild-type mice, due to a larger mismatch between oxygen consumption and oxygen supply from the CBF, being CBF similar in the two genotypes during CSD (Khennouf et al., 2016; but cf. Eikermann-Haerter et al., 2015).

In good correlation with the larger $Ca_V2.1$ gain-of-function produced by the S218L compared to the R192Q mutation (Tottene et al., 2002, 2005; van den Maagdenberg et al., 2004, 2010), the strength of CSD facilitation as well as the severity of the post-CSD neurological motor deficits and the propensity of CSD to propagate into subcortical structures were larger in S218L compared to R192Q FHM1 mice (Eikermann-Haerter et al., 2009b, 2011; van den Maagdenberg et al., 2010; Cain et al., 2017). Moreover, unlike the R192Q mice, the S218L mice developed multiple CSDs after a single CSD-inducing stimulus, which were more frequent in homozygotes compared to heterozygotes (van den Maagdenberg et al., 2010). Furthermore, unlike the R192Q mice (and the heterozygous S218L mice), the homozygous S218L mutants developed generalized seizure after about 1 h from a single CSD (Eikermann-Haerter et al., 2009b). In some cases the generalized seizure was fatal, and in this case it was associated with brainstem spreading depolarization observed in the pons and medulla; the appearance of seizure-related spreading depolarization in the brainstem correlated with respiratory arrest that was followed by cardiac arrest and death (Loonen et al., 2019).

Migraine-Relevant Alterations in the Cerebral Cortex of FHM1 Mouse Models

Direct measurement of synaptic transmission at different cortical excitatory synapses in R192Q FHM1 mice (including layer 2/3 pyramidal cell synapses onto both fast-spiking [FS] and somatostatin-expressing interneurons in slices of somatosensory cortex) revealed enhanced excitatory neurotransmission, due to enhanced action-potential (AP)-evoked Ca^{2+} influx through mutant presynaptic $Ca_V2.1$ channels and enhanced probability of glutamate release at cortical pyramidal cell synapses (Tottene et al., 2009; Marchionni et al., 2022). Congruently, short-term synaptic depression during trains of action potentials was also enhanced at layer 2/3 pyramidal cell synapses onto both pyramidal cells and FS interneurons (Tottene et al., 2009; Marchionni et al., 2022). Excitatory synaptic transmission was also enhanced at thalamocortical synapses onto both layer 4 regular-spiking principal neurons and layer 4 FS interneurons in R192Q FHM1 mice (Tottene et al., 2019). As expected, if an increased probability of glutamate release underlies the gain-of-function of thalamocortical synaptic transmission, short-term depression at thalamocortical synapses on layer 4 principal neurons was larger in FHM1 compared to wild-type mice; in striking contrast, short-term depression was unaltered at thalamocortical synapses on layer 4 FS interneurons (Tottene et al., 2019). Since a single thalamocortical axon can impinge on both layer 4 principal neurons and FS interneurons (Gabernet et al., 2005; Inoue & Imoto, 2006), these findings indicate postsynaptic target cell-dependent plasticity at thalamocortical synapses (Tottene et al., 2019).

Indirect evidence for gain-of-function of excitatory neurotransmission in FHM1 mice was also obtained at excitatory synapses onto dorsal suprachiasmatic nucleus neurons (van Oosterhout et al., 2008), at parallel fiber–Purkinje cell synapses (Adams et al., 2010), and at stratum radiatum CA1 synapses (Dilekoz et al., 2015). Long-term potentiation at these hippocampal synapses was also enhanced in R192Q mutants, while long-term depression was unaltered; paradoxically, spatial learning in contextual fear conditioning and Morris water maze tests was impaired (Dilekoz et al., 2015).

In striking contrast with the enhanced glutamatergic transmission, inhibitory GABAergic transmission at cortical interneuron synapses, including fast-spiking and somatostatin-expressing interneuron synapses, was unaltered in R192Q FHM1 mice, despite being initiated by $Ca_V2.1$ channels (Tottene et al., 2009; Vecchia et al., 2014; Marchionni et al., 2022). Expression of inhibitory interneuron-specific $Ca_V2.1$ channels whose gating properties are barely affected by the FHM1 mutation likely underlies this unexpected finding (Vecchia et al., 2014). Both the lack of effect on inhibitory neurotransmission and the gain-of-function effect on excitatory transmission appear to be a common feature of FHM1 mutations since they were shown also for the severe S218L mutation (Vecchia et al., 2015).

CSD rescue experiments have shown that there is a causal link between increased excitatory transmission at cortical synapses and facilitation of experimental CSD in R192Q FHM1 mice. In fact, the facilitation of initiation and propagation of CSD in mutant cortical slices was completely eliminated when action-potential-evoked glutamate release at pyramidal cell synapses was brought back to wild-type values by partially inhibiting the $Ca_V2.1$ channels (Tottene et al., 2009). The finding that propagation of CSD to subcortical structures in R192Q FHM1 mice was eliminated by systemic treatment with pregabalin, a drug that reduced excitatory transmission in R192Q hippocampal slices, suggests that the increased propensity of CSD to propagate into subcortical structures is also linked to increased

excitatory synaptic transmission (Cain et al., 2017).

The important role of excessive glutamatergic transmission in migraine mechanisms, in particular CSD susceptibility, is underscored and supported by the functional studies in heterozygous FHM2 knock-in mice (FHM2 mice) carrying a loss-of-function mutation in the astrocytic α2 Na/K-ATPase, which in humans causes pure FHM (Leo et al., 2011). FHM2 mice show reduced rate of synaptic glutamate clearance by cortical astrocytes during neuronal activity and reduced density of glutamate transporters GLT-1a at perisynaptic astrocytic processes, which mirrors the reduced expression of the α2 Na/K-ATPase (Capuani et al., 2016; Crivellaro et al., 2021; Parker et al., 2021). Imaging of glutamate with a genetically encoded glutamate sensor in awake, head-fixed FHM2 mice revealed spontaneous, focal, high-amplitude glutamate transients in layer 1 (glutamatergic plumes), which are not present in wild-type mice and arise as a consequence of impaired astrocyte clearance of synaptically released glutamate (Parker et al., 2021). The impaired rate of glutamate clearance by astrocytes and the consequent specific activation of diheteromeric GluN1-N2B N-methyl-D-aspartate (NMDA) receptors in apical dendrites of pyramidal cells in FHM2 mice can account for most of the facilitation of CSD induction and propagation in FHM2 cortical slices (Capuani et al., 2016; Crivellaro et al., 2021). Moreover, imaging of glutamate at the site of induction of CSD in awake mice revealed that an increase in plumes' frequency and in basal glutamate precedes and may predict CSD initiation (Parker et al., 2021).

Together with pharmacological data in wild-type mice providing strong support for a key role of glutamate NMDA receptors and $Ca_V2.1$ channels in initiation and/or propagation of experimental CSD (Tottene et al., 2011; Pietrobon & Moskowitz, 2014; Chung et al., 2018), the findings in FHM1 and FHM2 mice support a model of CSD initiation in which $Ca_V2.1$-dependent glutamate release and activation of NMDA receptors are key elements for generation of the net self-sustaining inward current necessary to initiate the positive feedback cycle that ignites a propagating CSD when the removal of K⁺ and glutamate from the interstitium does not keep pace with their release (Pietrobon & Moskowitz, 2014).

In FHM1 mice carrying the severe S218L mutation, gain-of-function of additional $Ca_V2.1$-dependent processes, besides enhanced glutamatergic synaptic transmission, likely underlies the particularly high susceptibility to CSD and high propensity of CSD to spread into subcortical structures as well as some of the unique features of CSD, including its recurrence, which were not observed in R192Q FHM1 mice (Eikermann-Haerter et al., 2009b, 2011; van den Maagdenberg et al., 2010; Vecchia et al., 2015; Cain et al., 2017). A specific feature of S218L FHM1 mice is the presence of a fraction of mutant $Ca_V2.1$ channels that is open at the resting potential, as revealed by the reduced frequency of miniature excitatory postsynaptic currents (mEPSCs) after blockade of $Ca_V2.1$ channels in cortical slices from both heterozygous and homozygous S218L mutants (Vecchia et al., 2015), and by increased baseline $[Ca^{2+}]_{in}$ in layer 2/3 axonal boutons and shafts in heterozygous S218L mice in vivo (Eikermann-Haerter et al., 2015). Both an increase in basal $[Ca^{2+}]_{in}$ in synaptic terminals and a reduction in mEPSCs' frequency after blocking $Ca_V2.1$ channels were measured at Calyx of Held synapses in brainstem slices from S218L mice (Di Guilmi et al., 2014). In contrast, the mEPSCs' frequency at cortical and brainstem synapses was not altered in R192Q FHM1 mice, indicating that presynaptic $Ca_V2.1$ channels carrying the R192Q mutation are closed at resting potential in brain slices (Tottene et al., 2009; Inchauspe et al., 2012). Probably as a consequence of the increase in baseline $[Ca^{2+}]_{in}$, heterozygous S218L mice showed some alterations in axonal and dendritic morphology in the resting state, including slightly larger boutons (Eikermann-Haerter et al., 2015). It remains to be seen whether this is a specific functional consequence of severe FHM1 mutations, which contributes to some of the additional clinical features associated with them.

The differential effect of FHM1 mutations on cortical excitatory and inhibitory synaptic transmissions (and on short-term synaptic plasticity at different synapses) suggested the hypothesis that impaired regulation of the excitatory–inhibitory (E/I) balance in specific neuronal circuits may be a key pathogenic mechanism in FHM and possibly migraine (Vecchia & Pietrobon, 2012; Pietrobon & Brennan, 2019). Investigation of a key microcircuit mediating feedforward inhibition, the thalamocortical disynaptic feedforward inhibitory microcircuit resulting from the divergence of thalamocortical fibers onto layer 4 principal neurons and FS interneurons making inhibitory synapses onto layer 4 principal neurons, revealed a dysregulation of the E/I balance in layer 4 principal neurons in R192Q FHM1 mice during repetitive thalamic stimulation (Tottene et al., 2019). Despite the fact that thalamocortical excitatory transmission is enhanced in FHM1 mice throughout the stimulation, the dysregulation of the E/I balance does not skew the balance toward excitation and does not lead to cortical overexcitation; on the contrary, it relatively skews the balance toward inhibition and reduces cortical recurrent network activity induced by repetitive thalamic firing (Tottene et al., 2019). This unexpected finding is due to the differential effect of the FHM1 mutation on short-term synaptic plasticity at thalamocortical synapses on layer 4 principal neurons and FS interneurons mentioned above, whereby, during repetitive thalamic stimulation, the increase in the disynaptic feedforward inhibition of principal neurons consequent to enhanced recruitment of FS interneurons (due to the gain-of-function of excitatory transmission at the thalamocortical synapses on FS interneurons) in FHM1 mice becomes larger than the increase in thalamocortical monosynaptic excitation of principal neurons (Tottene et al., 2019). This could, at least in part, explain the unexpected reduction in neuronal calcium responses to prolonged repeated whisker stimulation reported in anesthetized R192Q FHM1 mice (Khennouf et al., 2016) and the reduced amplitude of visual-evoked potentials recorded in response to light flashes in the visual cortex of freely behaving

R192Q FHM1 mice using epidural electroencephalography (EEG) (Perenboom et al., 2021).

Interestingly, in contrast with the findings in the FHM1 mouse model, a skewing toward excitation of the E/I balance of thalamocortical-evoked responses in layer 4 principal neurons has been reported in some animal models of epilepsy, and this dysregulation has been proposed to contribute to the unconstrained cortical recurrent excitation and the seizures observed in these animals (Sun et al., 2005; Sasaki et al., 2006; Paz & Huguenard, 2015). These animal models include the tottering mice, which carry a loss-of-function mutation in the $Ca_V2.1$ channel (Pietrobon, 2005; Sasaki et al., 2006) and are resistant to induction of experimental CSD (Ayata et al., 2000), i.e., show the opposite $Ca_V2.1$ and CSD phenotype than the FHM1 mice. Thus, a dysregulation of the E/I balance in the thalamocortical feedforward inhibitory microcircuit does not seem to affect in an obvious manner CSD susceptibility. Much work remains to be done in the FHM mouse models to identify the relevant dysfunctional cortical circuits and establish whether indeed dysfunctional regulation of the E/I balance in these circuits may favor ignition of a "spontaneous" CSD (and identify the specific conditions in which this may occur).

On the other hand, the alterations in the thalamocortical feedforward inhibitory microcircuit uncovered in R192Q FHM1 mice may lead to alterations in sensory processing, which might contribute to the interictal dysfunctions in sensory processing shown by migraineurs. In fact, the thalamocortical feedforward inhibitory microcircuit is critical in gating information flow to the cortex (including trigeminovascular nociceptive information) and in controlling the time window for integration of thalamic inputs and hence the temporal precision of cortical sensory responses; it also participates in sensory gain modulation and other important features of sensory processing, including spatial focusing of cortical sensory responses and rapid sensory adaptation (Moore, 2004; Isaacson & Scanziani, 2011; Tremblay et al., 2016; Bruno, 2011; Whitmire & Stanley, 2016). Possibly, the alterations in the thalamocortical feedforward inhibi-

tory circuit in R192Q FHM1 mice (Tottene et al., 2019) could contribute to explain the enhanced photic drive in the EEG beta-gamma band in response to frequency-dependent visual stimulation paradigms uncovered in freely behaving R192Q FHM1 mice (Perenboom et al., 2021).

Alterations in the Trigeminovascular Pain Pathway in FHM1 Mouse Models

The function of the trigeminovascular pain pathway is expected to be altered in FHM1 mice because $Ca_V2.1$ channels are involved in controlling neurotransmitter release at different levels in the trigeminovascular system, e.g., CGRP release from capsaicin-sensitive perivascular terminals of meningeal nociceptors as well as release at central synapses onto TNC neurons, including synapses of the descending inhibitory and facilitatory pathways that regulate trigeminovascular pain transmission (Pietrobon, 2013, and references therein). However, relatively few studies investigated the function of the trigeminovascular pain network in FHM1 mice; moreover, most of these studies were in vitro and focused on the peripheral part of the network.

Measurements of CGRP release from dura mater in fluid-filled hemisected skulls revealed that neither basal nor K^+-evoked CGRP release were significantly altered in adult R192Q FHM1 mice (Fioretti et al., 2011; Chan et al., 2019). Since a large fraction of peptidergic dural trigeminal ganglion (TG) afferents are capsaicin-sensitive (Pietrobon & Moskowitz, 2013), these findings are consistent with the unaltered $Ca_V2.1$ current found in small capsaicin-sensitive TG neurons from adult R192Q FHM1 mice; according to retrograde labeling from the dura, these neurons constitute the majority of small dural afferents (Fioretti et al., 2011). This may also help explain the finding that dural artery vasodilation induced in vivo by systemic capsaicin was not enhanced in R192Q FHM1 mice. In fact, vasodilation induced by both systemic capsaicin and CGRP was decreased (Chan et al., 2019), suggesting downregulation and/or desensitization of blood vessel CGRP receptors, perhaps as a compensatory mechanism. The lower fraction of CGRP-expressing neurons uncovered in trigeminal ganglia of R192Q FHM1 mice (Mathew et al., 2011) might be an additional compensatory mechanism, which would contribute to the unaltered CGRP release from dura mater in the FHM1 mutants. These findings argue against the idea that increased CGRP release from perivascular TG fibers at the dura and consequent increased vasodilation and mast cell degranulation facilitate the development of neurogenic inflammation in FHM1 compared to wild-type mice following activation of meningeal nociceptors, e.g., by CSD. This further underscores the importance of neuron subtype-specific effects of FHM1 mutations in the pathophysiology of FHM1.

Judging from the finding of unaltered *c-fos* expression in the TNC after in vivo electrical stimulation of the dura in FHM1 mice (Park et al., 2014), also synaptic transmission at the central terminals of dural TG afferents might be unaffected by the FHM1 mutation. This would be consistent with unaltered $Ca_V2.1$ current in most dural TG afferents of FHM1 mice, as was shown in small capsaicin-sensitive afferents (Fioretti et al., 2011). On the other hand, the finding of increased *c-fos* expression in several thalamic nuclei after in vivo electrical stimulation of the dura (Park et al., 2014) is consistent with increased synaptic transmission at TNC–thalamus excitatory synapses as a consequence of gain-of-function of $Ca_V2.1$ channels located at thalamic synaptic terminals of TNC neurons. As described in the previous section, thalamocortical synaptic transmission was shown to be enhanced in R192Q FHM1 mice (Tottene et al., 2019). The increased strength of neurotransmission at both TNC–thalamus and thalamocortical synapses in FHM1 mice would contribute to increase the gain of the trigeminovascular pain pathway.

Depending on the study, K+-evoked CGRP release from isolated trigeminal ganglia was either increased (Fioretti et al., 2011) or unaltered (Chan et al., 2019) in adult R192Q FHM1 mice; in the latter study, also CGRP release from TNC was unaltered in the mutants (but note that, since the FHM1 mutations shift to lower voltage

channel activation without affecting maximal open probability [Pietrobon, 2010], the K+ concentration in Chan et al. [2019] was likely too high to be able to reveal increased CGRP release). Enhanced K+-evoked CGRP release from trigeminal ganglia implies gain-of-function of $Ca_V2.1$ channels in some TG neurons in R192Q FHM1 mice; this was indeed shown in a subpopulation of small capsaicin-insensitive neurons, which do not innervate the dura (Fioretti et al., 2011). Given that in these neurons the action-potential-evoked $Ca_V2.1$ current is larger in FHM1 mice (Fioretti et al., 2011), one predicts enhanced transmitter release upon their activation. However, the function, transmitters, and possible involvement in migraine pain of this subpopulation of small TG neurons remain unknown.

In cultured TG neurons from R192Q FHM1 mice pups, also basal (besides K+-evoked) CGRP release was increased, suggesting opening of mutant $Ca_V2.1$ channels at resting potential (Ceruti et al., 2011). Congruently, these cultured TG neurons show interesting $Ca_V2.1$-dependent alterations such as loss of constitutive inhibition of ATP-gated P2X3 receptors (P2X3R) by brain natriuretic peptide receptors, which leads to increased P2X3R current and enhanced excitability in response to ATP in FHM1 mice (Marchenkova et al., 2016a, b). The neuronal upregulation of the P2X3R function, as well as the upregulation of P2X7 receptors function uncovered in satellite glial cells and macrophages, was eliminated after inhibition of the CGRP receptors (Ceruti et al., 2011; Hullugundi et al., 2013; Nowodworska et al., 2017). This is consistent with the idea that the increased basal release of CGRP promotes sensitization of P2X3R-expressing TG neurons, cross-talk between neurons and satellite glial cells and macrophages, resulting in a local persistent inflammatory environment in the FHM1 TG (Ceruti et al., 2011; Hullugundi et al., 2013; Nowodworska et al., 2017). However, basal release of CGRP was not increased in trigeminal ganglia from adult FHM1 mice (Fioretti et al., 2011; Chan et al., 2019), warranting caution in drawing conclusions regarding migraine pain mechanisms from findings in pups' TG cultures. Whether the adult TG shows a basal inflammatory phenotype in FHM1 mutants remains unclear. In FHM1 ganglia the number of active macrophages was increased (in all divisions), but the protein level of the pro-inflammatory cytokines IL1beta, IL6, and TNF alpha was unaltered (Franceschini et al., 2013). Interestingly, a larger fraction of TG neurons was immunoreactive for active phosphorylated CaMKII in FHM1 compared to wild-type ganglia; the difference in amount of phosphoprotein between the two genotypes was eliminated after blockade of $Ca_V2.1$ channels, suggesting facilitation of basal $Ca_V2.1$-dependent Ca signaling in FHM1 TG neurons (Nair et al., 2010).

Conclusions

The mutations in *CACNA1A* causing FHM1, a rare form of migraine with aura, produce gain-of-function of the $Ca_V2.1$ channel, mainly due to increased channel open probability and channel activation at lower voltages. A key migraine-relevant phenotype of knock-in mice carrying FHM1 mutations is increased susceptibility to experimentally induced CSD, the phenomenon that underlies migraine aura and may trigger the headache mechanisms. The FHM1 mouse models show increased excitatory synaptic transmission at intracortical and thalamocortical synapses and unaltered cortical inhibitory neurotransmission. There is a causal link between enhanced glutamatergic neurotransmission and enhanced susceptibility to CSD. Impaired regulation of the excitatory–inhibitory balance in specific neuronal circuits may be a key pathogenic mechanism in FHM and possibly migraine.

References

Adams, P. J., Garcia, E., David, L. S., Mulatz, K. J., Spacey, S. D., & Snutch, T. P. (2009). Ca(V)2.1 P/Q-type calcium channel alternative splicing affects functional impact of familial hemiplegic migraine mutations: Implications for calcium channelopathies. *Channels (Austin, Tex.), 3*, 110–121.

Adams, P. J., Rungta, R. L., Garcia, E., van den Maagdenberg, A. M., MacVicar, B. A., & Snutch, T. P.

(2010). Contribution of calcium-dependent facilitation to synaptic plasticity revealed by migraine mutations in the P/Q-type calcium channel. *Proceedings of the National Academy of Sciences of the United States of America, 107*, 18694–18699.

Auffenberg, E., Hedrich, U. B., Barbieri, R., Miely, D., Groschup, B., Wuttke, T. V., Vogel, N., Lührs, P., Zanardi, I., Bertelli, S., Spielmann, N., Gailus-Durner, V., Fuchs, H., Hrabě de Angelis, M., Pusch, M., Dichgans, M., Lerche, H., Gavazzo, P., Plesnila, N., & Freilinger, T. (2021). Hyperexcitable interneurons trigger cortical spreading depression in an Scn1a migraine model. *The Journal of Clinical Investigation, 131*, e142202.

Ayata, C., & Lauritzen, M. (2015). Spreading depression, spreading depolarizations, and the cerebral vasculature. *Physiological Reviews, 95*, 953–993.

Ayata, C., Shimizu-Sasamata, M., Lo, E. H., Noebels, J. L., & Moskowitz. (2000). Impaired neurotransmitter release and elevated threshold for cortical spreading depression in mice with mutations in the a_{1A} subunit of P/Q type calcium channels. *Neuroscience, 95*, 639–645.

Balkaya, M., Seidel, J. L., Sadeghian, H., Qin, T., Chung, D. Y., Eikermann-Haerter, K., van den Maagdenberg, A., Ferrari, M. D., & Ayata, C. (2019). Relief following chronic stress augments spreading depolarization susceptibility in familial hemiplegic migraine mice. *Neuroscience, 415*, 1–9.

Bolay, H., Reuter, U., Dunn, A. K., Huang, Z., Boas, D. A., & Moskowitz, M. A. (2002). Intrinsic brain activity triggers trigeminal meningeal afferents in a migraine model. *Nature Medicine, 8*, 136–142.

Bøttger, P., Glerup, S., Gesslein, B., Illarionova, N. B., Isaksen, T. J., Heuck, A., Clausen, B. H., Füchtbauer, E.-M., Gramsbergen, J. B., Gunnarson, E., Aperia, A., Lauritzen, M., Lambertsen, K. L., Nissen, P., & Lykke-Hartmann, K. (2016). Glutamate-system defects behind psychiatric manifestations in a familial hemiplegic migraine type 2 disease-mutation mouse model. *Scientific Reports, 6*, 22047.

Brennan, K. C., & Pietrobon, D. (2018). A systems neuroscience approach to migraine. *Neuron, 97*, 1004–1021.

Bruno, R. M. (2011). Synchrony in sensation. *Current Opinion in Neurobiology, 21*, 701–708.

Burstein, R., Noseda, R., & Borsook, D. (2015). Migraine: Multiple processes, complex pathophysiology. *The Journal of Neuroscience, 35*, 6619–6629.

Cain, S. M., Bohnet, B., LeDue, J., Yung, A. C., Garcia, E., Tyson, J. R., Alles, S. R. A., Han, H., van den Maagdenberg, A. M. J. M., Kozlowski, P., MacVicar, B. A., & Snutch, T. P. (2017). In vivo imaging reveals that pregabalin inhibits cortical spreading depression and propagation to subcortical brain structures. *Proceedings of the National Academy of Sciences, 114*, 2401–2406.

Capuani, C., Melone, M., Tottene, A., Bragina, L., Crivellaro, G., Santello, M., Casari, G., Conti, F., & Pietrobon, D. (2016). Defective glutamate and K^+ clearance by cortical astrocytes in familial hemiplegic migraine type 2. *EMBO Molecular Medicine, 8*, 967–986.

Catterall, W. A., & Few, A. P. (2008). Calcium channel regulation and presynaptic plasticity. *Neuron, 59*, 882–901.

Catterall, W. A., Dib-Haji, M. H., Meisler, M. H., & Pietrobon, D. (2008). Inherited neuronal ion channelopathies: New windows on complex neurological disorders. *The Journal of Neuroscience, 28*, 11768–11777.

Catterall, W. A., Leal, K., & Nanou, E. (2013). Calcium channels and short-term synaptic plasticity. *The Journal of Biological Chemistry, 288*, 10742–10749.

Ceruti, S., Villa, G., Fumagalli, M., Colombo, L., Magni, G., Zanardelli, M., Fabbretti, E., Verderio, C., van den Maagdenberg, A. M., Nistri, A., & Abbracchio, M. P. (2011). Calcitonin gene-related peptide-mediated enhancement of purinergic neuron/glia communication by the algogenic factor bradykinin in mouse trigeminal ganglia from wild-type and R192Q Cav2.1 Knock-in mice: Implications for basic mechanisms of migraine pain. *The Journal of Neuroscience, 31*, 3638–3649.

Chan, K. Y., Labastida-Ramirez, A., Ramirez-Rosas, M. B., Labruijere, S., Garrelds, I. M., Danser, A. H., van den Maagdenberg, A. M., & MaassenVanDenBrink, A. (2019). Trigeminovascular calcitonin gene-related peptide function in Cacna1a R192Q-mutated knock-in mice. *Journal of Cerebral Blood Flow and Metabolism, 39*, 718–729.

Chanda, M. L., Tuttle, A. H., Baran, I., Atlin, C., Guindi, D., Hathaway, G., Israelian, N., Levenstadt, J., Low, D., Macrae, L., O'Shea, L., Silver, A., Zendegui, E., Mariette Lenselink, A., Spijker, S., Ferrari, M. D., van den Maagdenberg, A. M., & Mogil, J. S. (2013). Behavioral evidence for photophobia and stress-related ipsilateral head pain in transgenic Cacna1a mutant mice. *Pain, 154*, 1254–1262.

Chung, D. Y., Sugimoto, K., Fischer, P., Bohm, M., Takizawa, T., Sadeghian, H., Morais, A., Harriott, A., Oka, F., Qin, T., Henninger, N., Yaseen, M. A., Sakadzic, S., & Ayata, C. (2018). Real-time noninvasive in vivo visible light detection of cortical spreading depolarizations in mice. *Journal of Neuroscience Methods, 309*, 143–146.

Crivellaro, G., Tottene, A., Vitale, M., Melone, M., Casari, G., Conti, F., Santello, M., & Pietrobon, D. (2021). Specific activation of GluN1-N2B NMDA receptors underlies facilitation of cortical spreading depression in a genetic mouse model of migraine with reduced astrocytic glutamate clearance. *Neurobiology of Disease, 156*, 105419.

De Fusco, M., Marconi, R., Silvestri, L., Atorino, L., Rampoldi, L., Morgante, L., Ballabio, A., Aridon, P., & Casari, G. (2003). Haploinsufficiency of ATP1A2 encoding the Na+/K+ pump alpha2 subunit associated with familial hemiplegic migraine type 2. *Nature Genetics, 33*, 192–196.

Deboer, T., van Diepen, H. C., Ferrari, M. D., Van den Maagdenberg, A. M., & Meijer, J. H. (2013). Reduced sleep and low adenosinergic sensitivity in cacna1a R192Q mutant mice. *Sleep, 36*, 127–136.

Di Guilmi, M. N., Wang, T., Inchauspe, C. G., Forsythe, I. D., Ferrari, M. D., van den Maagdenberg, A. M., Borst, J. G., & Uchitel, O. D. (2014). Synaptic gain-of-function effects of mutant Cav2.1 channels in a mouse model of familial hemiplegic migraine are due to increased basal [Ca2+]i. *The Journal of Neuroscience, 34*, 7047–7058.

Di Stefano, V., Rispoli, M. G., Pellegrino, N., Graziosi, A., Rotondo, E., Napoli, C., Pietrobon, D., Brighina, F., & Parisi, P. (2020). Diagnostic and therapeutic aspects of hemiplegic migraine. *Journal of Neurology, Neurosurgery, and Psychiatry, 91*, 764–771.

Dichgans, M., Freilinger, T., Eckstein, G., Babini, E., Lorenz-Depiereux, B., Biskup, S., Ferrari, M. D., Herzog, J., van den Maagdenberg, A. M., Pusch, M., & Strom, T. M. (2005). Mutation in the neuronal voltage-gated sodium channel SCN1A in familial hemiplegic migraine. *Lancet, 366*, 371–377.

Dilekoz, E., Houben, T., Eikermann-Haerter, K., Balkaya, M., Lenselink, A. M., Whalen, M. J., Spijker, S., Ferrari, M. D., van den Maagdenberg, A. M. J. M., & Ayata, C. (2015). Migraine mutations impair hippocampal learning despite enhanced long-term potentiation. *The Journal of Neuroscience, 35*, 3397–3402.

Eikermann-Haerter, K., Baum, M. J., Ferrari, M. D., van den Maagdenberg, A. M., Moskowitz, M. A., & Ayata, C. (2009a). Androgenic suppression of spreading depression in familial hemiplegic migraine type 1 mutant mice. *Annals of Neurology, 66*, 564–568.

Eikermann-Haerter, K., Dilekoz, E., Kudo, C., Savitz, S. I., Waeber, C., Baum, M. J., Ferrari, M. D., van den Maagdenberg, A. M. J. M., Moskowitz, M. A., & Ayata, C. (2009b). Genetic and hormonal factors modulate spreading depression and transient hemiparesis in mouse models of familial hemiplegic migraine type 1. *Journal of Clinical Investigation, 119*, 99–109.

Eikermann-Haerter, K., Yuzawa, I., Qin, T., Wang, Y., Baek, K., Kim, Y. R., Hoffmann, U., Dilekoz, E., Waeber, C., Ferrari, M. D., van den Maagdenberg, A. M., Moskowitz, M. A., & Ayata, C. (2011). Enhanced subcortical spreading depression in familial hemiplegic migraine type 1 mutant mice. *The Journal of Neuroscience, 31*, 5755–5763.

Eikermann-Haerter, K., Arbel-Ornath, M., Yalcin, N., Yu, E. S., Kuchibhotla, K. V., Yuzawa, I., Hudry, E., Willard, C. R., Climov, M., Keles, F., Belcher, A. M., Sengul, B., Negro, A., Rosen, I. A., Arreguin, A., Ferrari, M. D., van den Maagdenberg, A. M. J. M., Bacskai, B. J., & Ayata, C. (2015). Abnormal synaptic Ca2+ homeostasis and morphology in cortical neurons of familial hemiplegic migraine type 1 mutant mice. *Annals of Neurology, 78*, 193–210.

Ferrari, M. D., Klever, R. R., Terwindt, G. M., Ayata, C., & van den Maagdenberg, A. M. J. M. (2015). Migraine pathophysiology: Lessons from mouse models and human genetics. *The Lancet Neurology, 14*, 65–80.

Filiz, A., Tepe, N., Eftekhari, S., Boran, H. E., Dilekoz, E., Edvinsson, L., & Bolay, H. (2019). CGRP receptor antagonist MK-8825 attenuates cortical spreading depression induced pain behavior. *Cephalalgia, 39,* 354–365.

Fioretti, B., Catacuzzeno, L., Sforna, L., Gerke-Duncan, M. B., van den Maagdenberg, A. M. J. M., Franciolini, F., Connor, M., & Pietrobon, D. (2011). Trigeminal ganglion neuron subtype-specific alterations of CaV2.1 calcium current and excitability in a Cacna1a mouse model of migraine. *The Journal of Physiology, 589,* 5879–5895.

Franceschini, A., Vilotti, S., Ferrari, M. D., van den Maagdenberg, A. M., Nistri, A., & Fabbretti, E. (2013). TNFalpha levels and macrophages expression reflect an inflammatory potential of trigeminal ganglia in a mouse model of familial hemiplegic migraine. *PLoS One, 8,* e52394.

Gabernet, L., Jadhav, S. P., Feldman, D. E., Carandini, M., & Scanziani, M. (2005). Somatosensory integration controlled by dynamic thalamocortical feed-forward inhibition. *Neuron, 48,* 315–327.

Gandini, M. A., Souza, I. A., Ferron, L., Michell-Innes, A., & Zamponi, G. W. (2021). The de novo CACNA1A pathogenic variant Y1384C associated with hemiplegic migraine, early onset cerebellar atrophy and developmental delay leads to a loss of $Ca_V2.1$ channel function. *Molecular Brain, 14,* 27.

Garza-Lopez, E., Sandoval, A., Gonzalez-Ramirez, R., Gandini, M. A., Van den Maagdenberg, A., De Waard, M., & Felix, R. (2012). Familial hemiplegic migraine type 1 mutations W1684R and V1696I alter G protein-mediated regulation of Ca(V)2.1 voltage-gated calcium channels. *Biochimica et Biophysica Acta, 1822,* 1238–1246.

GBD 2015 Neurological Disorders Collaborator Group (2017). Global, regional, and national burden of neurological disorders during 1990–2015: A systematic analysis for the Global Burden of Disease Study 2015. *Lancet Neurology, 16,* 877–897.

Goadsby, P. J., Holland, P. R., Martins-Oliveira, M., Hoffmann, J., Schankin, C., & Akerman, S. (2017). Pathophysiology of migraine: A disorder of sensory processing. *Physiological Reviews, 97,* 553–622.

Gormley, P., Anttila, V., Winsvold, B. S., Palta, P., Esko, T., Pers, T. H., Farh, K.-H., Cuenca-Leon, E., Muona, M., Furlotte, N. A., Kurth, T., Ingason, A., McMahon, G., Ligthart, L., Terwindt, G. M., Kallela, M., Freilinger, T. M., Ran, C., Gordon, S. G., … Palotie, A. (2016). Meta-analysis of 375,000 individuals identifies 38 susceptibility loci for migraine. *Nature Genetics, 48,* 856.

Hadjikhani, N., Albrecht, D. S., Mainero, C., Ichijo, E., Ward, N., Granziera, C., Zurcher, N. R., Akeju, O., Bonnier, G., Price, J., Hooker, J. M., Napadow, V., Nahrendorf, M., Loggia, M. L., & Moskowitz, M. A. (2020). Extra-axial inflammatory signal in parameninges in migraine with visual aura. *Annals of Neurology, 87,* 939–949.

Hansen, J. M., Hauge, A. W., Ashina, M., & Olesen, J. (2011). Trigger factors for familial hemiplegic migraine. *Cephalalgia, 31,* 1274–1281.

Harriott, A. M., Chung, D. Y., Uner, A., Bozdayi, R. O., Morais, A., Takizawa, T., Qin, T., & Ayata, C. (2021). Optogenetic spreading depression elicits trigeminal pain and anxiety behavior. *Annals of Neurology, 89,* 99–110.

Hautakangas, H., Winsvold, B. S., Ruotsalainen, S. E., Bjornsdottir, G., Harder, A. V. E., Kogelman, L. J. A., Thomas, L. F., Noordam, R., Benner, C., Gormley, P., Artto, V., Banasik, K., Bjornsdottir, A., Boomsma, D. I., Brumpton, B. M., Sølvsten Burgdorf, K., Buring, J. E., Chalmer, M. A., de Boer, I., … Pirinen, M. (2021). Genome-wide analysis of 102,084 migraine cases identifies 123 risk loci and subtype-specific risk alleles. *medRxiv 2021.2001.2020.21249647.*

Hullugundi, S. K., Ferrari, M. D., van den Maagdenberg, A. M. J. M., & Nistri, A. (2013). The mechanism of functional up-regulation of P2X3 receptors in trigeminal sensory neurons in a genetic mouse model of familial hemiplegic migraine type 1 (FHM-1). *PLoS One, 8,* e60677.

Inchauspe, C. G., Urbano, F. J., Di Guilmi, M. N., Forsythe, I. D., Ferrari, M. D., van den Maagdenberg, A. M., & Uchitel, O. D. (2010). Gain of function in FHM-1 Ca(V)2.1 knock-in mice is related to the shape of the action potential. *Journal of Neurophysiology, 104,* 291–299.

Inchauspe, C. G., Urbano, F. J., Di Guilmi, M. N., Ferrari, M. D., van den Maagdenberg, A. M., Forsythe, I. D., & Uchitel, O. D. (2012). Presynaptic CaV2.1 calcium channels carrying familial hemiplegic migraine mutation R192Q allow faster recovery from synaptic depression in mouse calyx of Held. *Journal of Neurophysiology, 108,* 2967–2976.

Inoue, T., & Imoto, K. (2006). Feedforward inhibitory connections from multiple thalamic cells to multiple regular-spiking cells in layer 4 of the somatosensory cortex. *Journal of Neurophysiology, 96,* 1746–1754.

Isaacson, J. S., & Scanziani, M. (2011). How inhibition shapes cortical activity. *Neuron, 72,* 231–243.

Izquierdo-Serra, M., Fernandez-Fernandez, J. M., & Serrano, M. (2020). Rare CACNA1A mutations leading to congenital ataxia. *European Journal of Physiology, 472,* 791–809.

Jansen, N. A., Dehghani, A., Linssen, M. M. L., Breukel, C., Tolner, E. A., & van den Maagdenberg, A. M. J. M. (2020). First FHM3 mouse model shows spontaneous cortical spreading depolarizations. *Annals of Clinical Translational Neurology, 7,* 132–138.

Karatas, H., Erdener, S. E., Gursoy-Ozdemir, Y., Lule, S., Eren-Kocak, E., Sen, Z. D., & Dalkara, T. (2013). Spreading depression triggers headache by activating neuronal Panx1 channels. *Science, 339,* 1092–1095.

Kaufmann, D., & Brennan, K. C. (2018). The effects of chronic stress on migraine relevant phenotypes in male mice. *Frontiers in Cellular Neuroscience, 12,* 294.

Khennouf, L., Gesslein, B., Lind, B. L., van den Maagdenberg, A. M. J. M., & Lauritzen, M. (2016). Activity-dependent calcium, oxygen, and vascular responses in a mouse model of familial hemiplegic migraine type 1. *Annals of Neurology, 80,* 219–232.

Kors, E. E., Terwindt, G. M., Vermeulen, F. L., Fitzsimons, R. B., Jardine, P. E., Heywood, P., Love, S., van den Maagdenberg, A. M., Haan, J., Frants, R. R., & Ferrari, M. D. (2001). Delayed cerebral edema and fatal coma after minor head trauma: Role of the CACNA1A calcium channel subunit gene and relationship with familial hemiplegic migraine. *Annals of Neurology, 49*, 753–760.

Leo, L., Gherardini, L., Barone, V., De Fusco, M., Pietrobon, D., Pizzorusso, T., & Casari, G. (2011). Increased susceptibility to cortical spreading depression in the mouse model of familial hemiplegic migraine type 2. *PLoS Genetics, 7*, e1002129.

Levy, D., Labastida-Ramirez, A., & MaassenVanDenBrink, A. (2019). Current understanding of meningeal and cerebral vascular function underlying migraine headache. *Cephalalgia, 39*, 1606–1622.

Loonen, I. C. M., Jansen, N. A., Cain, S. M., Schenke, M., Voskuyl, R. A., Yung, A. C., Bohnet, B., Kozlowski, P., Thijs, R. D., Ferrari, M. D., Snutch, T. P., van den Maagdenberg, A., & Tolner, E. A. (2019). Brainstem spreading depolarization and cortical dynamics during fatal seizures in Cacna1a S218L mice. *Brain, 142*, 412–425.

Marchenkova, A., van den Maagdenberg, A. M. J. M., & Nistri, A. (2016a). Loss of inhibition by brain natriuretic peptide over P2X3 receptors contributes to enhanced spike firing of trigeminal ganglion neurons in a mouse model of familial hemiplegic migraine type-1. *Neuroscience, 331*, 197–205.

Marchenkova, A., Vilotti, S., Ntamati, N., Maagdenberg, A. M. V. D., & Nistri, A. (2016b). Inefficient constitutive inhibition of P2X3 receptors by brain natriuretic peptide system contributes to sensitization of trigeminal sensory neurons in a genetic mouse model of familial hemiplegic migraine. *Molecular Pain, 12*, 1744806916646110.

Marchionni, I., Pilati, N., Forli, A., Sessolo, M., Tottene, A., & Pietrobon, D. (2022). Enhanced feedback inhibition due to increased recruitment of somatostatin-expressing interneurons and enhanced cortical recurrent excitation in a genetic mouse model of migraine. *The Journal of Neuroscience. 10.1523/JNEUROSCI.0228-22.2022*

Mathew, R., Andreou, A. P., Chami, L., Bergerot, A., van den Maagdenberg, A. M., Ferrari, M. D., & Goadsby, P. J. (2011). Immunohistochemical characterization of calcitonin gene-related peptide in the trigeminal system of the familial hemiplegic migraine 1 knock-in mouse. *Cephalalgia, 31*, 1368–1380.

Melliti, K., Grabner, M., & Seabrook, G. R. (2003). The familial hemiplegic migraine mutation R192Q reduces G-protein-mediated inhibition of P/Q-type (Ca(V)2.1) calcium channels expressed in human embryonic kidney cells. *The Journal of Physiology, 546*, 337–347.

Melo-Carrillo, A., Noseda, R., Nir, R.-R., Schain, A. J., Stratton, J., Strassman, A. M., & Burstein, R. (2017a). Selective inhibition of trigeminovascular neurons by fremanezumab: A humanized monoclonal anti-

CGRP antibody. *The Journal of Neuroscience, 37*, 7149–7163.

Melo-Carrillo, A., Strassman, A. M., Nir, R.-R., Schain, A. J., Noseda, R., Stratton, J., & Burstein, R. (2017b). Fremanezumab—A humanized monoclonal anti-CGRP antibody—Inhibits thinly myelinated (Aδ) but not unmyelinated (C) meningeal nociceptors. *The Journal of Neuroscience, 37*, 10587–10596.

Moore, C. I. (2004). Frequency-dependent processing in the vibrissa sensory system. *Journal of Neurophysiology, 91*, 2390–2399.

Mullner, C., Broos, L. A., van den Maagdenberg, A. M., & Striessnig, J. (2004). Familial hemiplegic migraine type 1 mutations K1336E, W1684R, and V1696I Alter Cav2.1 Ca2+ channel gating: Evidence for beta-subunit isoform-specific effects. *The Journal of Biological Chemistry, 279*, 51844–51850.

Nair, A., Simonetti, M., Birsa, N., Ferrari, M. D., van den Maagdenberg, A. M., Giniatullin, R., Nistri, A., & Fabbretti, E. (2010). Familial hemiplegic migraine Ca(v)2.1 channel mutation R192Q enhances ATP-gated P2X3 receptor activity of mouse sensory ganglion neurons mediating trigeminal pain. *Molecular Pain, 6*, 48.

Nowodworska, A., van den Maagdenberg, A. M. J. M., Nistri, A., & Fabbretti, E. (2017). In situ imaging reveals properties of purinergic signalling in trigeminal sensory ganglia in vitro. *Purinergic Signalling, 13*, 511–520.

Ophoff, R. A., Terwindt, G. M., Vergouwe, M. N., van Eijk, R., Oefner, P. J., Hoffman, S. M. G., Lamerdin, J. E., Mohrenweiser, H. W., Bulman, D. E., Ferrari, M., Haan, J., Lindhout, D., van Hommen, G.-J. B., Hofker, M. H., Ferrari, M. D., & Frants, R. R. (1996). Familial hemiplegic migraine and episodic ataxia type-2 are caused by mutations in the Ca²⁺ channel gene CACNL1A4. *Cell, 87*, 543–552.

Park, J., Moon, H., Akerman, S., Holland, P. R., Lasalandra, M. P., Andreou, A. P., Ferrari, M. D., van den Maagdenberg, A. M. J. M., & Goadsby, P. J. (2014). Differential trigeminovascular nociceptive responses in the thalamus in the familial hemiplegic migraine 1 knock-in mouse: A Fos protein study. *Neurobiology of Disease, 64*, 1–7.

Parker, P. D., Suryavanshi, P., Melone, M., Sawant-Pokam, P. A., Reinhart, K. M., Kaufmann, D., Theriot, J. J., Pugliese, A., Conti, F., Shuttleworth, C. W., Pietrobon, D., & Brennan, K. C. (2021). Non-canonical glutamate signaling in a genetic model of migraine with aura. *Neuron, 109*, 611–628.e618.

Paz, J. T., & Huguenard, J. R. (2015). Microcircuits and their interactions in epilepsy: Is the focus out of focus? *Nature Neuroscience, 18*, 351–359.

Perenboom, M. J. L., Schenke, M., Ferrari, M. D., Terwindt, G. M., van den Maagdenberg, A. M. J. M., & Tolner, E. A. (2021). Responsivity to light in familial hemiplegic migraine type 1 mutant mice reveals frequency-dependent enhancement of visual network excitability. *The European Journal of Neuroscience, 53*, 1672–1686.

Pietrobon, D. (2005). Function and dysfunction of synaptic calcium channels: Insights from mouse models. *Current Opinion in Neurobiology, 15*, 257–265.

Pietrobon, D. (2007). Familial hemiplegic migraine. *Neurotherapeutics, 4*, 274–284.

Pietrobon, D. (2010). CaV2.1 channelopathies. *Pflügers Archiv, 460*, 375–393.

Pietrobon, D. (2013). Calcium channels and migraine. *Biochimica et Biophysica Acta (BBA) – Biomembranes, 1828*, 1655–1665.

Pietrobon, D., & Brennan, K. C. (2019). Genetic mouse models of migraine. *The Journal of Headache and Pain, 20*, 79.

Pietrobon, D., & Moskowitz, M. A. (2013). Pathophysiology of migraine. *Annual Review of Physiology, 75*, 365–391.

Pietrobon, D., & Moskowitz, M. A. (2014). Chaos and commotion in the wake of cortical spreading depression and spreading depolarizations. *Nature Reviews. Neuroscience, 15*, 379–393.

Russell, M. B., & Ducros, A. (2011). Sporadic and familial hemiplegic migraine: Pathophysiological mechanisms, clinical characteristics, diagnosis, and management. *Lancet Neurology, 10*, 457–470.

Russo, A. F. (2015). Calcitonin gene-related peptide (CGRP): A new target for migraine. *Annual Review of Pharmacology and Toxicology, 55*, 533–552.

Sasaki, S., Huda, K., Inoue, T., Miyata, M., & Imoto, K. (2006). Impaired feedforward inhibition of the thalamocortical projection in epileptic Ca2+ channel mutant mice, tottering. *The Journal of Neuroscience, 26*, 3056–3065.

Schain, A. J., Melo-Carrillo, A., Strassman, A. M., & Burstein, R. (2017). Cortical spreading depression closes paravascular space and impairs glymphatic flow: Implications for migraine headache. *The Journal of Neuroscience, 37*, 2904–2915.

Schain, A. J., Melo-Carrillo, A., Borsook, D., Grutzendler, J., Strassman, A. M., & Burstein, R. (2018). Activation of pial and dural macrophages and dendritic cells by cortical spreading depression. *Annals of Neurology, 83*, 508–521.

Schwedt, T. J., Chiang, C. C., Chong, C. D., & Dodick, D. W. (2015). Functional MRI of migraine. *Lancet Neurology, 14*, 81–91.

Serra, S. A., Fernandez-Castillo, N., Macaya, A., Cormand, B., Valverde, M. A., & Fernandez-Fernandez, J. M. (2009). The hemiplegic migraine-associated Y1245C mutation in CACNA1A results in a gain of channel function due to its effect on the voltage sensor and G-protein-mediated inhibition. *Pflügers Archiv, 458*, 489–502.

Shyti, R., Eikermann-Haerter, K., van Heiningen, S. H., Meijer, O. C., Ayata, C., Joëls, M., Ferrari, M. D., van den Maagdenberg, A. M. J. M., & Tolner, E. A. (2015). Stress hormone corticosterone enhances susceptibility to cortical spreading depression in familial hemiplegic migraine type 1 mutant mice. *Experimental Neurology, 263*, 214–220.

Sun, Q. Q., Huguenard, J. R., & Prince, D. A. (2005). Reorganization of barrel circuits leads to thalamically-evoked cortical epileptiform activity. *Thalamus & Related Systems, 3*, 261–273.

Sutherland, H. G., & Griffiths, L. R. (2017). Genetics of migraine: Insights into the molecular basis of migraine disorders. *Headache: The Journal of Head and Face Pain, 57*, 537–569.

Takizawa, T., Qin, T., Lopes de Morais, A., Sugimoto, K., Chung, J. Y., Morsett, L., Mulder, I., Fischer, P., Suzuki, T., Anzabi, M., Böhm, M., Qu, W. S., Yanagisawa, T., Hickman, S., Khoury, J. E., Whalen, M. J., Harriott, A. M., Chung, D. Y., & Ayata, C. (2020). Non-invasively triggered spreading depolarizations induce a rapid pro-inflammatory response in cerebral cortex. *Journal of Cerebral Blood Flow and Metabolism, 40*, 1117–1131.

Tottene, A., Fellin, T., Pagnutti, S., Luvisetto, S., Striessnig, J., Fletcher, C., & Pietrobon, D. (2002). Familial hemiplegic migraine mutations increase Ca(2+) influx through single human CaV2.1 channels and decrease maximal CaV2.1 current density in neurons. *Proceedings of the National Academy of Sciences of the United States of America, 99*, 13284–13289.

Tottene, A., Pivotto, F., Fellin, T., Cesetti, T., van den Maagdenberg, A. M., & Pietrobon, D. (2005). Specific kinetic alterations of human CaV2.1 calcium channels produced by mutation S218L causing familial hemiplegic migraine and delayed cerebral edema and coma after minor head trauma. *The Journal of Biological Chemistry, 280*, 17678–17686.

Tottene, A., Conti, R., Fabbro, A., Vecchia, D., Shapovalova, M., Santello, M., van den Maagdenberg, A. M. J. M., Ferrari, M. D., & Pietrobon, D. (2009). Enhanced excitatory transmission at cortical synapses as the basis for facilitated spreading depression in ca(v)2.1 knockin migraine mice. *Neuron, 61*, 762–773.

Tottene, A., Urbani, A., & Pietrobon, D. (2011). Role of different voltage-gated Ca2+ channels in cortical spreading depression: Specific requirement of P/Q-type Ca2+ channels. *Channels (Austin, Tex.), 5*, 110–114.

Tottene, A., Favero, M., & Pietrobon, D. (2019). Enhanced thalamocortical synaptic transmission and dysregulation of the excitatory-inhibitory balance at the thalamocortical feedforward inhibitory microcircuit in a genetic mouse model of migraine. *The Journal of Neuroscience, 39*, 9841–9851.

Tremblay, R., Lee, S., & Rudy, B. (2016). GABAergic interneurons in the neocortex: From cellular properties to circuits. *Neuron, 91*, 260–292.

van den Maagdenberg, A. M., Pietrobon, D., Pizzorusso, T., Kaja, S., Broos, L. A., Cesetti, T., van de Ven, R. C., Tottene, A., van der Kaa, J., Plomp, J. J., Frants, R. R., & Ferrari, M. D. (2004). A Cacna1a knockin migraine mouse model with increased susceptibility to cortical spreading depression. *Neuron, 41*, 701–710.

van den Maagdenberg, A. M., Pizzorusso, T., Kaja, S., Terpolilli, N., Shapovalova, M., Hoebeek, F. E., Barrett, C. F., Gherardini, L., van de Ven, R. C.,

Todorov, B., Broos, L. A., Tottene, A., Gao, Z., Fodor, M., De Zeeuw, C. I., Frants, R. R., Plesnila, N., Plomp, J. J., Pietrobon, D., & Ferrari, M. D. (2010). High cortical spreading depression susceptibility and migraine-associated symptoms in Ca(v)2.1 S218L mice. *Annals of Neurology, 67*, 85–98.

van Oosterhout, F., Michel, S., Deboer, T., Houben, T., van de Ven, R. C. G., Albus, H., Westerhout, J., Vansteensel, M. J., Ferrari, M. D., van den Maagdenberg, A. M. J. M., & Meijer, J. H. (2008). Enhanced circadian phase resetting in R192Q Cav2.1 calcium channel migraine mice. *Annals of Neurology, 64*, 315–324.

Vecchia, D., & Pietrobon, D. (2012). Migraine: A disorder of brain excitatory–inhibitory balance? *Trends in Neurosciences, 35*, 507–520.

Vecchia, D., Tottene, A., van den Maagdenberg, A. M. J. M., & Pietrobon, D. (2014). Mechanism underlying unaltered cortical inhibitory synaptic transmission in contrast with enhanced excitatory transmission in CaV2.1 knockin migraine mice. *Neurobiology of Disease, 69*, 225–234.

Vecchia, D., Tottene, A., van den Maagdenberg, A. M. J. M., & Pietrobon, D. (2015). Abnormal cortical synaptic transmission in CaV2.1 knockin mice with the S218L missense mutation which causes a severe familial hemiplegic migraine syndrome in humans. *Frontiers in Cellular Neuroscience, 9*, 8.

Weiss, N., Sandoval, A., Felix, R., Van den Maagdenberg, A., & De Waard, M. (2008). The S218L familial hemiplegic migraine mutation promotes deinhibition of Ca(v)2.1 calcium channels during direct G-protein regulation. *Pflügers Archiv, 457*, 315–326.

Westenbroek, R. E., Sakurai, T., Elliott, E. M., Hell, J. W., Starr, T. V. B., Snutch, T. P., & Catterall, W. A. (1995). Immunochemical identification and subcellular distribution of the a_{1A} subunits of brain calcium channels. *The Journal of Neuroscience, 15*, 6403–6418.

Whitmire, C. J., & Stanley, G. B. (2016). Rapid sensory adaptation redux: A circuit perspective. *Neuron, 92*, 298–315.

Yapici-Eser, H., Donmez-Demir, B., Kilic, K., Eren-Kocak, E., & Dalkara, T. (2018). Stress modulates cortical excitability via alpha-2 adrenergic and glucocorticoid receptors: As assessed by spreading depression. *Experimental Neurology, 307*, 45–51.

Zhang, X., Levy, D., Noseda, R., Kainz, V., Jakubowski, M., & Burstein, R. (2010). Activation of meningeal nociceptors by cortical spreading depression: Implications for migraine with aura. *The Journal of Neuroscience, 30*, 8807–8814.

Zhang, X., Levy, D., Kainz, V., Noseda, R., Jakubowski, M., & Burstein, R. (2011). Activation of central trigeminovascular neurons by cortical spreading depression. *Annals of Neurology, 69*, 855–865.

Zhao, J., & Levy, D. (2015). Modulation of intracranial meningeal nociceptor activity by cortical spreading depression: A reassessment. *Journal of Neurophysiology, 113*, 2778–2785.

Zhao, J., & Levy, D. (2016). Cortical spreading depression promotes persistent mechanical sensitization of intracranial meningeal afferents: Implications for the intracranial mechanosensitivity of migraine. *eNeuro, 3*(6), ENEURO.0287-16.2016.

T-Type Calcium Channels in Epilepsy

Anne-Sophie Sack and Terrance P. Snutch

Abstract

Epilepsy is a complex neurological disorder commonly thought to involve hyperexcitable neurons that recruit neuronal populations within circuits to generate aberrant synchronous activity. T-type calcium channels are expressed in key epileptic brain regions where they shape neuronal firing patterns and excitability. Accordingly, altered expression, localization, and/or biophysical properties of T-type channels can have important consequences concerning the activity of circuits implicated in epileptic seizures. Epileptic seizures are spontaneous and episodic, making it difficult to predict when and how often they will occur. Evidence from acquired and genetic epilepsy models provides crucial information concerning how T-type calcium channels contribute to both the progression of epilepsy and seizure activity itself. Understanding the involvement of T-type calcium channels in the transition from normal activity to epileptic activity, as well as in the maintenance of seizures, is key to developing therapeutic strategies.

Keywords

T-type calcium channel · Temporal lobe epilepsy · Absence epilepsy · Burst-firing · Therapeutics · $Ca_v3.1$ · $Ca_v3.2$ · $Ca_v3.3$

Abbreviations

ADP	afterdepolarization
DG	dentate gyrus
GAERS	genetic absence epilepsy rats from Strasbourg
KO	knockout
LTP	long-term potentiation
LTS	low-threshold calcium spike
MTF1	metal-regulatory transcription factor 1
RTN	reticular thalamic nucleus
SE	status epilepticus
SNPs	single nucleotide polymorphisms
SWD	spike wave discharge
TLE	temporal lobe epilepsy
VB	ventrobasal thalamic groups
WAG/Rij	Wistar Albino Glaxo from Rijswijk

A.-S. Sack
Michael Smith Laboratories, University of British Columbia, Vancouver, BC, Canada

T. P. Snutch (✉)
Djavad Mowafaghian Centre for Brain Health, University of British Columbia, Vancouver, BC, Canada
e-mail: snutch@msl.ubc.ca

Introduction

Epilepsy is a complex neurological disorder with a multitude of clinical presentations and risk factors generally characterized by recurrent and unprovoked seizures. Seizures occur as a result of abnormal neuronal discharges and can have symptoms ranging from brief staring spells to tonic-clonic convulsions. The behavioral manifestation of seizures depends on several factors including which regions of the brain are involved (Stafstrom, 2006; Velíšková & Velíšek, 2017), and in this chapter we take the approach of separating epilepsies based on the circuits involved in seizures. A primary classification of seizures is by the onset of abnormal activity: focal or generalized; where a focal onset originates in one or more localized brain regions, while generalized onset begins with widespread distribution over both hemispheres (Devinsky et al., 2018). Seizures do not result from abnormal activity in single neurons, but instead involve coordination and propagation of neuronal activity. This is particularly well established in absence epilepsy, where generalized seizures require coordination of thalamus and cortex, but also has been suggested in temporal lobe epilepsy (TLE), where focal seizures occur (Bertram, 2009). In both forms of epilepsy, studies have implicated burst-firing of neurons in relevant circuits as key to both initiating seizures and driving their propagation. T-type calcium channels are particularly well suited to mediate such burst-firing.

T-type calcium channels (see chapter "A Lived History of Early Calcium Channel Discoveries Over the Past Half-Century" by Emilio Carbone) are low-threshold channels that activate in response to small depolarizations from hyperpolarized resting membrane potentials, generating an initial inward current that leads to the threshold for activation of voltage-gated sodium channels, and are consequently referred to as "first responders" (Cain & Snutch, 2010). The contribution of low-threshold calcium currents to action potential firing in mammalian neurons was initially characterized in inferior olivary neurons where hyperpolarization of the membrane uncovered a low-threshold calcium spike (LTS; Llinás

& Yarom, 1981a, b). Since then, LTSs have been identified in neurons in multiple brain regions (for example, see Jahnsen & Llinás, 1984; Greene et al., 1986; Wilcox et al., 1988).

T-type channels participate in the generation of bursts due to their unique biophysical properties. The gating properties of T-type channels have been reviewed thoroughly by others (Huguenard, 1996; Perez-Reyes, 2003; Talavera & Nilius, 2006), but will be briefly described here with respect to bursting. T-type channels activate in response to small membrane depolarizations (between -75 and -60 mV). Notably, the rate of activation is voltage dependent, with the time constant of activation significantly slower near resting membrane potentials than at maximal activation (Fig. 1a–c). This impacts the frequency of action potentials generated in a burst, a phenomenon nicely illustrated in thalamic neurons where the maximal activation of T-type currents is more depolarized in reticular thalamic neurons (~ -10 mV) compared to thalamocortical neurons (~ -30 mV) (Huguenard & Prince, 1992). Accordingly, as reticular thalamic neurons are depolarized near the LTS threshold (~ -50 mV), the rate of T-type channel activation is slower and suggested to underlie the increase in action potential frequency on the rising phase of the LTS (Fig. 1g) (Huguenard, 1996).

Upon activation, T-type channels inactivate rapidly thus generating a transient inward current. Inactivation is also voltage dependent, with the time course of inactivation slow at the threshold and reducing with further depolarization until reaching a plateau (Fig. 1d). This characteristic also helps shape burst-firing; for instance, in reticular thalamic cells the slower inactivation has been proposed to underlie the longer burst duration compared to that for thalamocortical neurons (Fig. 1h) (Huguenard & Prince, 1992).

The rate of T-type channel recovery from inactivation (de-inactivation) following repolarization determines the availability of channels that can be subsequently recruited and consequently impacts LTSs and burst generation (Fig. 1e, i). For example, in deep cerebellar nuclei cells, Purkinje-cell-evoked GABAergic inhibition hyperpolarizes the membrane potential

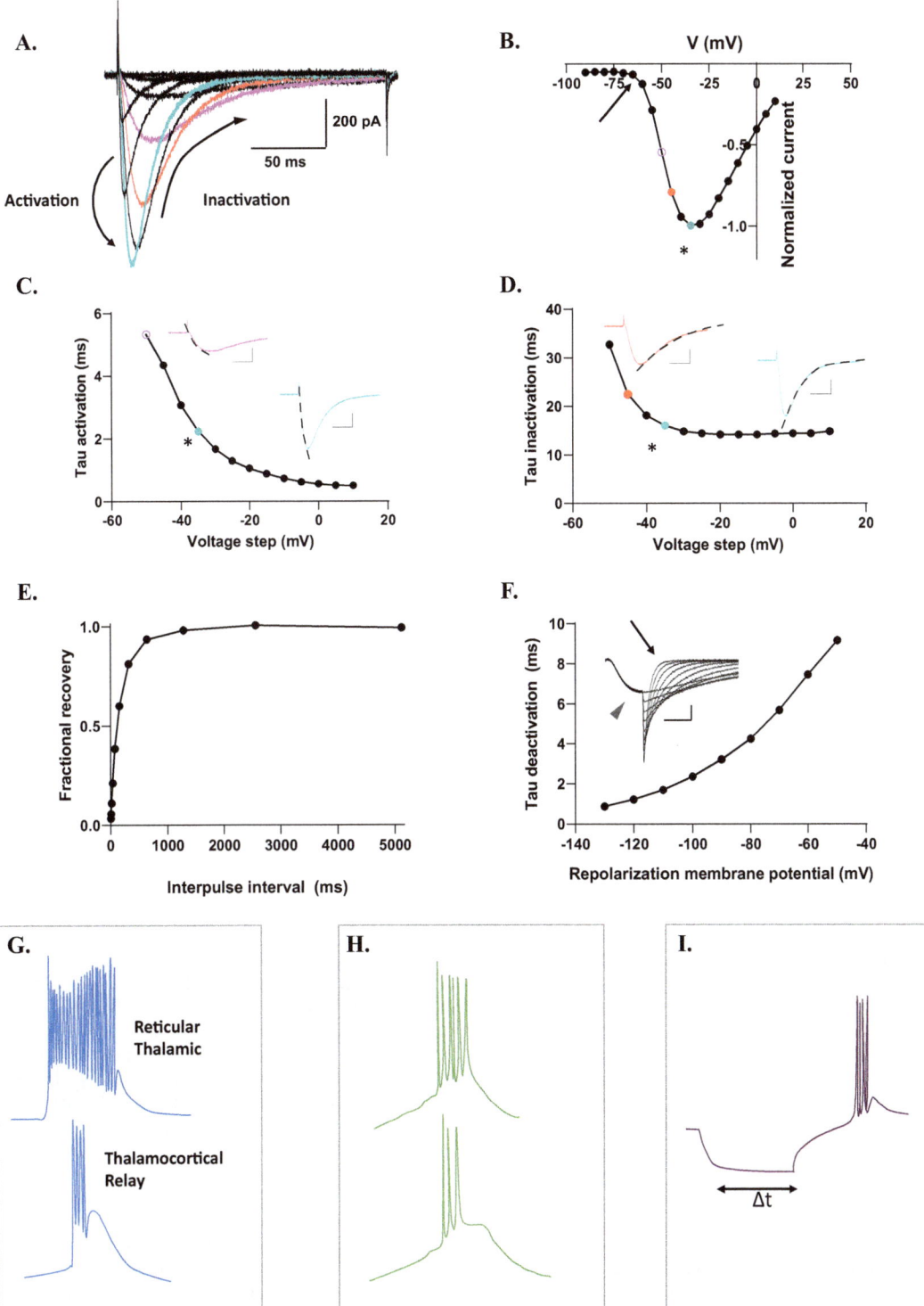

Fig. 1 T-type channel gating properties affect bursting behaviors. (**a**) At the whole-cell level, T-type calcium channels activate in response to membrane depolarization to generate rapidly inactivating currents. In order to illustrate the time course of T-type currents at different membrane potentials, the traces highlighted in blue, purple,

sufficiently to allow a proportion of T-type channels to recover from inactivation leading to the generation of rebound bursts (Engbers et al., 2011).

Lastly of relevance, T-type channels close slowly with return to resting membrane potential through a process called deactivation (Fig. 1f). The slow deactivation of T-type channels compared to high-voltage-activated calcium channels is important concerning the repolarization of action potentials where the membrane potential becomes increasingly negative and as T-type channels deactivate they mediate a significant Ca^{2+} influx owing to the greater driving force (McCobb & Beam, 1991).

In addition to regulating bursting patterns, the biophysical properties of T-type channels allow individual neurons to exhibit distinct voltage-dependent firing patterns. For instance, thalamic neurons can fire either tonic or bursts of action potentials depending upon the resting membrane potential (Llinás & Jahnsen, 1982). Similarly, medial habenula neurons display a range of firing patterns based on the initial membrane potential;

burst-firing occurs from hyperpolarized membrane potentials while tonic firing and action potential doublets are observed at depolarized membrane potentials where T-type channels are inactivated (Vickstrom et al., 2020). Hence, although T-type channels generate relatively tiny and transient currents, they contribute profoundly to influencing action potential firing patterns that impact neuronal circuits.

Mammalian genomes possess three T-type calcium channel genes: *CACNA1G*, *CACNA1H* and *CACAN1I*, which encode the pore subunits α1G, α1H, and α1I, most recently referred to as $Ca_v3.1$, $Ca_v3.2$, and $Ca_v3.3$, respectively. Each T-type channel subtype exhibits distinct biophysical properties and consequently is predicted to contribute differentially to firing patterns. $Ca_v3.3$ channels activate and inactivate at a slower rate than $Ca_v3.1$ and $Ca_v3.2$, where $Ca_v3.1$ has the fastest kinetics. Further, $Ca_v3.3$ channels activate and inactivate at the most hyperpolarized membrane potentials followed by $Ca_v3.1$ and $Ca_v3.2$ at more positive potentials (McRory et al., 2001). Accordingly, in terms of burst-firing, $Ca_v3.1$

Fig. 1 (continued) and red are used to show their corresponding tau values plotted in the activation and inactivation graphs (panels C and D). (**b**) The relationship between T-type channel currents and voltage is shown as an IV curve, where the maximal activation is labeled with an asterisk (*). As T-type channels activate close to resting membrane potential (arrow), they are called low-threshold activated. (**c**) The rate of activation is voltage dependent, such that time constant of activation (tau activation) is slower near resting membrane potential (shown in purple trace with corresponding value on graph) and faster around maximal activation of T-type currents (shown in blue; max current labeled with * from IV). (**d**) The rate of inactivation is also voltage dependent with the time constant of inactivation (tau inactivation) slower near resting membrane potentials (shown in red) and faster around maximal activation of T-type currents (asterisk and shown in blue). (**e**) T-type channels recover from inactivation following repolarization in a time-dependent manner. The plot shows the fraction of channels that have recovered from inactivation following increasing duration between successive pulses. (**f**) T-type channels also deactivate with return to negative membrane potentials. This produces a tail current (arrow) with an amplitude and kinetics that changes with voltage. The plot shows the rate of tail current decay (tau deactivation) produced when the cell is repolarized to various potentials after a depolarizing voltage step (arrowhead). Lastly, how these biophysical properties influence burst-firing in distinct neurons is illustrated in panels G–I. (**g**) The impact of the rate of activation on bursting is seen in thalamocortical relay and reticular thalamic neurons. Here, the slow onset of T-type current in reticular thalamic neurons (upper) is suggested to contribute to the increase in frequency on the rising phase of the low-threshold calcium spike (LTS) shown in the upper trace compared to that for the thalamocortical relay neuron (lower). (**h**) The contribution of inactivation kinetics to burst-firing is shown in model neurons. A slower inactivation could lead to an increase in the number of action potentials per burst (upper trace) compared to a neuron with faster inactivation (lower trace). (**i**) T-type channel recovery from inactivation can impact the availability of channels that can be subsequently recruited. This is well illustrated with rebound burst-firing. Here, a neuron is hyperpolarized sufficiently long enough for T-type channels to fully recover from inactivation, resulting in rebound burst-firing. (T-type current recordings from recombinant $Ca_v3.1$ [Snutch lab, unpublished]. Scale bar in panels C and D: $y = 200$ pA; $x = 20$ ms. Scale bar in panel F: $y = 500$ pA; $x = 5$ ms). (Thalamocortical and reticular thalamic neuron burst-firing examples drawn based upon Huguenard & Prince, 1992; Huguenard, 1996) channels are thought to contribute toward short rapid bursts, $Ca_v3.2$ to longer bursts, and $Ca_v3.3$ the long-lasting sustained bursts (Cain & Snutch, 2010; Chemin et al., 2002). Consequently, the subtype expression of T-type channels in distinct neuronal populations is likely to contribute to a range of burst patterns across neuronal subtypes and circuitries.

Further diversity in neuronal functional activity mediated by T-type channels arises from their subcellular localizations (Lambert et al., 2014). For example, T-type channels located in distal dendrites of thalamic reticular neurons can amplify afferent inputs suggesting a role in synaptic integration (Crandall et al., 2010; Lambert et al., 2014). Conversely, T-type channels located in axons are predicted to contribute, via a cholinergic mechanism, toward setting the threshold for action potential initiation in hippocampal granule cells (Martinello et al., 2015). Here, synaptically released acetylcholine shifts T-type channel activation toward hyperpolarization, in turn increasing Ca^{2+} levels that inhibit Kv7/M channels located at the axon initial segment (Martinello et al., 2015). This area of study is only beginning to uncover the numerous roles of T-type channels in modulating activity in distinct circuits beyond the generation of bursts based on their localization.

In general, epilepsy research focuses on a combination of changes in excitatory synapses, inhibitory synapses, and intrinsic neuronal excitability (Yaari & Beck, 2006). Among hypotheses, burst-firing of neurons in epileptic regions is intriguing and is included in many models and at various stages of epilepsy. The transformation of a non-epileptic brain into a brain capable of generating unprovoked seizures can involve changes in expression and/or trafficking of T-type channels triggered by pathological events. Further, mutations that modify the biophysical properties of T-type channels can also alter neuronal firing patterns that lead to aberrant activity. Given the importance of T-type channels in burst-firing and network excitability, the present chapter will review the involvement of T-type channels in epilepsy, with an emphasis on TLE and absence epilepsy.

T-Type Calcium Channels and Temporal Lobe Epilepsy

TLE is a common form of epilepsy with focal seizures that include temporal lobe structures, such as the hippocampus. Mesial temporal lobe epilepsy is the most prevalent and often presents with hippocampal sclerosis (Engel & Salamon, 2015). Risk factors for TLE include damage to the limbic structures resulting from febrile seizures, head trauma, infection, tumors, malformations, and genetic factors (Tatum, 2012). Human studies and animal models have shown that TLE is not limited to the hippocampus but extends to regions including the amygdala, entorhinal cortex, thalamus, and temporopolar and lateral temporal neocortices (Bernhardt et al., 2013). However, substantial research in TLE has focused on the hippocampus, and so this region serves as a good model for understanding epilepsy-related changes.

The hippocampus formation consists of subregions that include the dentate gyrus (DG); Cornu Ammonis area CA3, CA2, and CA1; and the subiculum. The entorhinal cortex provides a major source of cortical input to the hippocampus, forming the tri-synaptic hippocampus circuit (Fig. 2a). This well-characterized circuit involves projections from the entorhinal cortex via the perforant path to DG granule cells, which then excite CA3 pyramidal cells via the mossy fiber pathway, in turn exciting CA1 pyramidal cells via the Schaffer collateral pathway. The hippocampus also receives inputs from other regions including direct projections from the entorhinal cortex to the CA1 (temporoammonic pathway), midline thalamic nucleus reuniens, amygdala, and others (Witter & Amaral, 2004). Excitation can be further spread throughout subregions through synapses between neighboring cells as shown for the recurrent collaterals established in the CA3 (MacVicar & Dudek, 1980), and associational connections along the CA1 longitudinal axis (Yang et al., 2014).

All three T-type calcium channel subtypes are expressed to varying degrees in hippocampal subfields. In CA3 and CA1 pyramidal neurons, immunolabeling showed that $Ca_v3.1$ is expressed

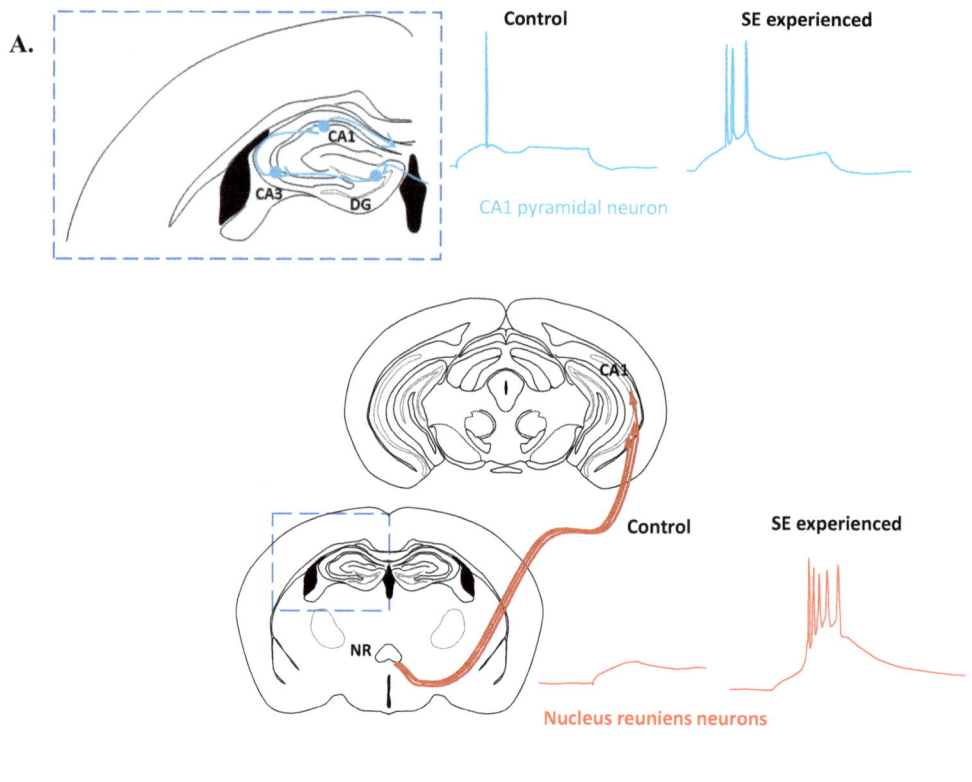

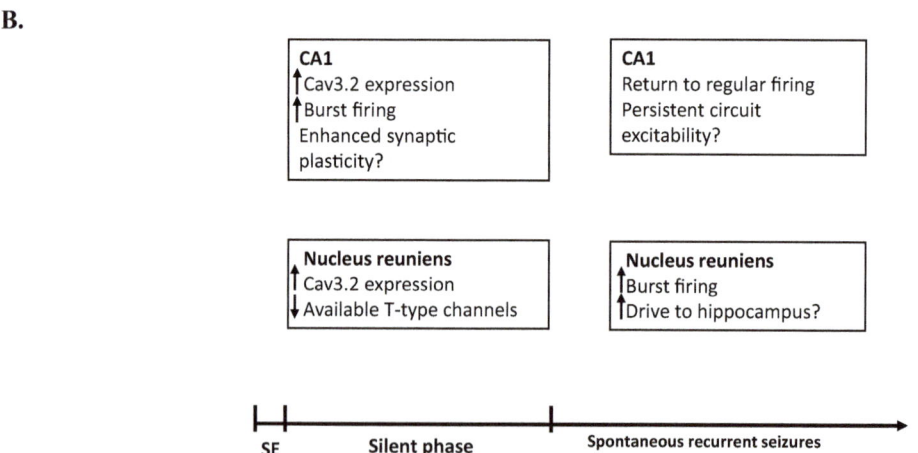

Fig. 2 Potential circuits involved in pilocarpine-induced temporal lobe epilepsy (TLE). (**a**) Tri-synaptic hippocampal circuit (upper panel) illustrated in a coronal section from rodent brain. Afferent excitation from the entorhinal cortex (not shown) excites dentate gyrus (DG) granule cells via the perforant pathway. Granule cell axons (mossy fiber) in turn excite CA3 pyramidal cells, which then synapse onto CA1 pyramidal cells via Schaffer collaterals. Following pilocarpine-induced status epilepticus (SE), $Ca_v3.2$ expression is transiently increased in the CA1 fol-

lowed by an increase in T-type currents and burst-firing (blue traces show just suprathreshold current step in control $Ca_v3.2$ +/+ mice). The midline thalamus nuclei, nucleus reuniens (NR), project to CA1, bypassing the tri-synaptic circuit (lower panel). In midline thalamus nuclei, $Ca_v3.2$ expression is increased both at 10 days (silent phase) and 31 days (spontaneous recurrent seizures phase) post pilocarpine-induced SE. Burst-firing is increased during the spontaneous recurrent seizures phase (red traces show burst-firing in midline thalamic neurons at a predominantly in the soma and proximal dendrites, $Ca_v3.2$ in the soma and proximal dendrites with labeling extending to mid-dendritic regions, and $Ca_v3.3$ further into distal dendrites (McKay et al., 2006). Conversely, in the DG, T-type channels are mostly expressed at the soma of granule cells, with $Ca_v3.2$ expression being the highest, followed by $Ca_v3.3$, and $Ca_v3.1$ the lowest (McKay et al., 2006). T-type currents have been observed in CA1 pyramidal neurons (Takahashi et al., 1991), and burst-firing in these neurons can be evoked by blocking K^+ channels in the distal apical dendrites, allowing backpropagating spikes to activate dendritic Ni^{2+}-sensitive Ca^{2+} channels (Magee & Carruth, 1999). Although this current was subsequently suggested to be mediated by R-type calcium channels (Metz et al., 2005), the robust expression of T-type calcium channels in hippocampal subfields suggests that an increase in T-type channel activity associated with TLE pathophysiology could alter action potential firing and excitability.

TLE animal models include those that induce status epilepticus (SE) using chemoconvulsants such as kainic acid or pilocarpine triggering epileptogenesis and kindling models that involve repeated electrical stimulations resulting in increased seizure susceptibility. These models have shown roles for T-type channels in both epileptogenesis and the generation of spontaneous recurrent seizures. The following section reviews the evidence for T-type channels in these TLE animal models.

Pilocarpine Model

The pilocarpine model of TLE was first described almost 40 years ago by Turski's group and involves an injection (typically intraperitoneal [i.p.]) of pilocarpine, a muscarinic receptor agonist, to induce SE (Turski et al., 1983). Three characteristic stages follow pilocarpine injection: an acute phase (24 h) that progressively leads to status epilepticus, a silent phase (4–44 days) with normal EEG and behavior, and lasting chronic phase where animals display spontaneous recurrent seizures (Cavalheiro, 1995). In the hippocampus, pilocarpine-induced convulsions increase the release of neurotransmitters including glutamate and involve NMDA receptors in the maintenance of seizures (Smolders et al., 1997). This model is widely used to induce SE and reproduces several features of human TLE including aberrant DG axon sprouting known as mossy fiber sprouting (Curia et al., 2008). Studies using the pilocarpine model have found a pivotal role for T-type channels in mediating changes in intrinsic excitability in key brain regions.

The majority of CA1 pyramidal cells under normal conditions are regular spiking neurons, namely, they fire a train of independent action potentials with suprathreshold stimulation (Jensen et al., 1994). However, following pilocarpine-induced SE, a large proportion of CA1 pyramidal neurons undergo changes in intrinsic excitability to fire bursts of action potentials (Sanabria et al., 2001). Several lines of evi-

---

Fig. 2 (continued) holding potential around −65 mV with a −50 pA current injection). (**b**) Timeline of T-type-mediated effects in two key regions after pilocarpine-induced SE. In the CA1, T-type channel-mediated burst-firing is increased early after SE during the silent phase. This burst-firing might evoke alterations in the circuit, such as the strengthening of synapses via long-term potentiation (LTP). During spontaneous recurrent seizures, CA1 pyramidal neurons return to regular firing. At this point, it is possible that transient burst-firing during the silent phase has altered the hyperexcitable circuit such that cells can be recruited more easily. In the midline thalamic nucleus (nucleus reuniens), $Ca_v3.2$ expression is increased during the silent phase as well; however, a hyperpolarizing shift in the steady state inactivation and activation curves results in a decrease in available T-type channels near resting membrane potentials. During the spontaneous recurrent seizure phase, burst-firing is increased, which has been suggested to increase the drive both onto the hippocampus CA1 neurons and entorhinal cortex. Overall, these data highlight both the regional and temporal specificities associated with the SE-induced changes. (CA1 burst-firing redrawn from Becker et al., 2008; midline thalamic neurons redrawn from Graef et al., 2009; nucleus reuniens projections and coordinates based upon: Vertes, 2015; Dolleman-van der Weel et al., 2017; Scheel et al., 2020)

dence have implicated T-type channels, and more specifically $Ca_v3.2$ located in the distal apical dendrites, in the bursting mode of firing. These include that bursting is blocked by Ni^{2+} (Sanabria et al., 2001), that CA1 pyramidal neuron T-type currents are increased (Su et al., 2002), as is $Ca_v3.2$ expression early after SE, and finally that bursting is significantly attenuated in $Ca_v3.2$ knockout (KO) mice (Becker et al., 2008). In further support, $Ca_v3.2$ KO mice display reduced neuronal damage, mossy fiber sprouting, as well as decreased frequency and severity of seizures (Becker et al., 2008). The mechanism underlying bursting in these neurons has been suggested to involve the spike afterdepolarization (ADP), which normally is subthreshold, however after SE is enhanced (Su et al., 2002). Accordingly, with an increased T-type current in the distal apical dendrites, a backpropagating action potential from the soma into the dendrites, which normally would not evoke a large Ca^{2+} current, now causes a large depolarization via a T-type calcium-channel-mediated current, spreading back into the axo-soma leading to suprathreshold ADP and triggering spiking (Yaari et al., 2007). This process repeats itself until the cell is repolarized via K^+-channel-mediated currents, and is termed a "ping-pong mechanism" (Yaari et al., 2007). How these neurons contribute to the generation of seizures is still unclear; however, one possibility is that a small percentage of CA1 pyramidal neurons that generate spontaneous bursts are key to initiating the epileptiform burst serving as potential pacemakers (Sanabria et al., 2001).

Interestingly, the increase in $Ca_v3.2$ expression and burst-firing is transient, with the proportion of bursting CA1 pyramidal neurons peaking early after SE then declining in the following weeks (Becker et al., 2008; Chen et al., 2011). How then can a transient effect of $Ca_v3.2$ on CA1 pyramidal neurons profoundly impact seizures? Strengthening of synapses within the circuit via long-term potentiation (LTP) has been proposed (Becker et al., 2008). LTP at the Schaffer collateral-commissural fibers in the CA1 has been suggested to increase early (1 day) after SE and decrease in the subsequent weeks in a similar TLE model (Zhang et al., 2010), although this

would be predicted to occur before the emergence of bursting. Another interesting possibility involves strengthening of neighboring CA1 neurons along the interlamellar CA1 network (Yang et al., 2014; Sun et al., 2018). In line with this notion, recurrent excitatory synapses are increased in the CA1 after treatment with chemoconvulsants (Esclapez et al., 1999; Shao & Dudek, 2004). Future studies will be needed to understand the consequences of this transient burst-firing on the hippocampal circuit.

Similarly, neurons in the midline thalamus have been shown to increase $Ca_v3.2$-mediated burst-firing after pilocarpine treatment (Graef et al., 2009). Here, the gating properties of T-type calcium channels are altered such that a higher proportion of channels are available near resting membrane potential and neurons can generate a larger number of action potentials per burst (Fig. 2a). Notably, this happens during the spontaneous recurrent seizure period. The midline thalamus, in particular the nucleus reuniens, has been shown to project directly to the CA1 (Dolleman-van Der Weel & Witter, 1996), providing strong excitation that bypasses the trisynaptic hippocampal circuit (Bertram & Zhang, 1999). Additionally, enhanced excitability of the nucleus reuniens could further target CA1 indirectly via projections to the entorhinal cortex, resulting in aberrant excitation (Graef et al., 2009; Graef & Godwin, 2010).

Lastly, a possible mechanism underlying the increase in $Ca_v3.2$ expression after SE has been suggested to involve zinc ions (Zn^{2+}). Zn^{2+} is found in glutamatergic terminals in several regions of the brain including the hippocampus (Frederickson et al., 2000). Following pilocarpine-induced SE, Zn^{2+} is released from the presynaptic boutons to postsynaptic neurons (Suh et al., 2001). While Zn^{2+} acutely inhibits $Ca_v3.2$ channels, long-term exposure to Zn^{2+} upregulates $Ca_v3.2$ channels in CA1 pyramidal cells (Ekstein et al., 2012). In SE-experienced animals, the rise in Zn^{2+} has been suggested to increase $Ca_v3.2$ transcription by activating the metal-regulatory transcription factor 1 (MTF1) that then binds to the metal-responsive elements in the $Ca_v3.2$ gene promoter (Van Loo et al.,

2015). Together, these suggest a possible Zn^{2+}-dependent transcriptional pathway driving increased $Ca_v3.2$-mediated currents that lead to enhanced burst-firing after SE.

Kainic Acid Model

Another common chemoconvulsant model of TLE employs kainic acid, a glutamate analog that binds to kainate-sensitive glutamate receptor subunits (GluK1-5) (Rusina et al., 2021). In rodents, kainic acid injection either intracerebrally or systemically induces SE similar to that for pilocarpine and leads to the occurrence of spontaneous seizures after a latent period (Lévesque & Avoli, 2013). Early studies using i.p. injection of kainic acid showed that both T-type currents and high-voltage-activated calcium currents were increased in DG granule cells (Beck et al., 1998). $Ca_v3.1$ has been implicated in the acute phase of kainic-acid-induced epilepsy as $Ca_v3.1$ knockout mice showed reduced seizure duration but increased frequency (Kim, 2015). T-type channel blockade has been suggested to be protective against several aspects of TLE induced by kainic acid (Casillas-Espinosa et al., 2019). Here, continuous subcutaneous infusion of the T-type channel antagonist Z944 after SE had anti-epileptogenetic effects and further decreased psychiatric and cognitive impairments. In addition, this chronic T-type antagonist treatment decreased expression of $Ca_v3.1$ and $Ca_v3.2$ in the hippocampus. Hence, the involvement of T-type channels in a second chemoconvulsant model strengthens the notion that T-type channels are involved in TLE and that selective channel blockade is potentially beneficial for the treatment of epilepsy.

Kindling Model

The kindling model of epilepsy involves a progressive increase in seizure susceptibility induced by repeated electrical stimulations (McNamara et al., 1980). This is most commonly achieved by stimulation of limbic brain regions, such as the amygdala or hippocampus, which initially produces little behavioral responses but with continued stimulation progresses to bilateral clonic convulsions (Goddard, 1967; Goddard et al., 1969). This response is considered permanent as it can be elicited after several weeks of rest from stimulation (Goddard et al., 1969) and suggests that kindling induces persistent changes that render the neuronal circuits susceptible to seizures. The kindling model of epilepsy is similar to certain human epilepsies producing similar behavioral patterns, pharmacological profiles, and EEG recordings such as complex partial seizures found in TLE (Sato et al., 1990).

Early experiments using the kindling model stimulating the Schaffer collaterals showed that low-voltage-activated calcium currents in the CA1 were increased immediately after kindling and persisted 6 weeks later, albeit with partial recovery (Faas et al., 1996). Similar to kainic acid model, the T-type channel blocker Z944 notably delayed the progression of amygdala kindling, however, with no anti-seizure effect once rats were fully kindled (Casillas-Espinosa et al., 2015). Specifically, rats treated with Z944 required more stimulations to reach class III, IV, and V seizures, and most did not reach the fully kindled state within 30 stimulations. This suggests that T-type channels are involved early in epileptogenesis, although their exact roles remain to be discovered. It is possible that T-type channels play a role in burst-firing seen in CA1 pyramidal neurons (Yamada & Bilkey, 1991), similar to that observed in the pilocarpine model.

Overall, the various animal models of TLE provide evidence for T-type channel involvement in the development of epilepsy. T-type channels seem to play a crucial role in epileptogenesis early after status epilepticus, where increased channel expression might enhance the proportion of burst-firing neurons. The enhanced burst-firing may have long-term consequences that lead to a hyperexcitable circuit, which persists even after firing returns to basal. The distinct temporal windows for the increased T-type channel currents and associated burst-firing in CA1 neurons compared to the nucleus reuniens neurons suggest that T-type channels may contribute separate roles in brain regions that initiate, propagate, and modulate seizure activity in epilepsies (Fig. 2b).

T-Type Channels and Absence Epilepsy

Absence seizures are a type of generalized seizures that involve periods of brief unconsciousness with loss of voluntary movements and a characteristic 2.5–4 Hz (5–11 Hz in animals) spike wave discharge (SWD) in EEG recordings (Crunelli et al., 2020). These include childhood absence epilepsy, juvenile absence epilepsy, and absence status epilepticus (Jallon & Latour, 2005). The thalamocortical network has been shown to be crucial in absence seizures, and includes thalamocortical relay neurons in the ventrobasal thalamic groups (VB), corticothalamic pyramidal neurons in layers V–VI of the sensory cerebral cortex, and neurons in the reticular thalamic nucleus (RTN; Cain & Snutch, 2012). Corticothalamic pyramidal neurons send excitatory projections to the thalamocortical relay neurons, which in turn send excitatory projections back to the cortex. In this loop, axons from thalamocortical relay neurons and cortical pyramidal neurons also pass through the RTN innervating GABAergic neurons. The RTN provides GABAergic inhibition to the thalamocortical relay neurons (Fig. 3). The network model for absence seizures involves T-type channel-mediated burst-firing in RTN and thalamocortical relay neurons. Dependent upon membrane potential, thalamic neurons fire in two modes: burst-firing or tonic firing. As shown by Llinás and Jahnsen, when the membrane is relatively depolarized, thalamic neurons fire tonic spikes as T-type channels are in their inactivated state (Llinás & Jahnsen, 1982). Conversely, if the membrane is hyperpolarized sufficiently that T-type channels are de-inactivated, for instance in response to GABA-mediated hyperpolarization, neurons can fire bursts of action potentials, including a rebound burst following hyperpolarization. During an absence seizure, the thalamocortical network is proposed to be in a hypersynchronous state (Fogerson & Huguenard,

2016). Namely, that RTN cells fire bursts of action potentials, inhibiting thalamocortical relay neurons. The hyperpolarization de-inactivates T-type channels in thalamocortical relay neurons and results in rebound burst-firing following excitation, activating the cortex and RTN. The cortex then sends glutamatergic projections to RTN cells and thalamocortical relay neurons, continuing the loop (Fig. 3). The following section will review the involvement of T-type channels in two animal models of absence epilepsy: the genetic absence epilepsy rats from Strasbourg (GAERS) and Wistar Albino Glaxo from Rijswijk (WAG/Rij). Studies implicating T-type channels in these models have been reviewed previously (Khosravani & Zamponi, 2006; Cain & Snutch, 2012). Here, we incorporate recent research that expands on the contributing mechanism of T-type channels in the main thalamic nuclei.

GAERS

GAERS are a well-established rat model of absence epilepsy that originated from inbred rats that spontaneously developed SWDs with behavioral arrest similar to human absence epilepsy (Vergnes et al., 1982; Marescaux & Vergnes, 1995). Specific T-type channels have been implicated in the main thalamic nuclei involved in the thalamocortical network associated with absence seizures. The RTN is composed of GABAergic neurons that receive projections from both the corticothalamic pyramidal neurons and thalamocortical neurons (Pinault, 2004). Pioneering studies implicated Ca^{2+} channels in the RTN as a key mediator in SWDs where injection of Cd^{2+} into the RTN suppressed SWDs (Avanzini et al., 1993). Low-voltage-activated Ca^{2+} currents were subsequently found to be increased in the RTN of juvenile GAERS, with no change in VB thalamocortical relay neurons (Tsakiridou et al., 1995). In line with this, the expression of $Ca_v3.2$ channels was found to be higher in the RTN of juvenile GAERS, while in adult GAERS $Ca_v3.2$ and $Ca_v3.1$ were increased in the RTN and ventral posterior thalamic relay nuclei, respectively (Talley et al., 2000). Notably, while GAERS is

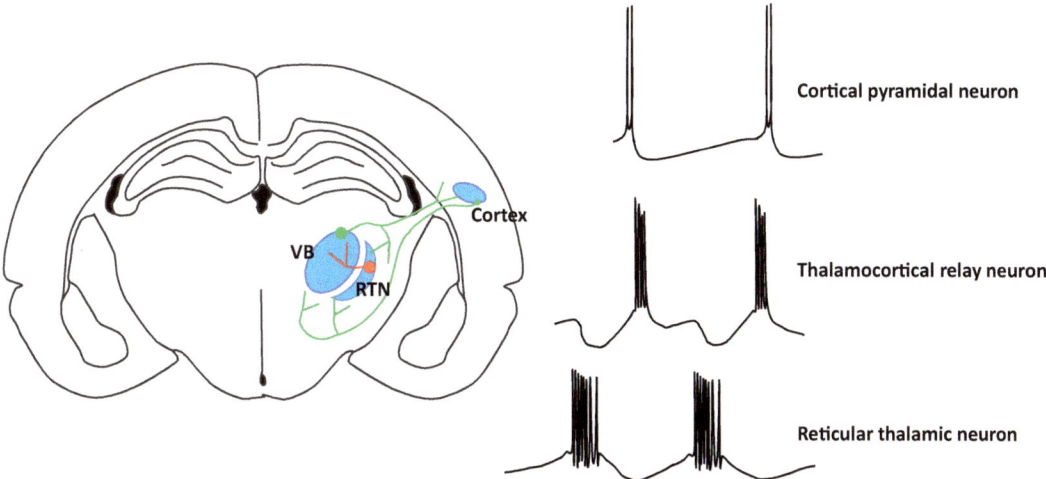

Fig. 3 Model of thalamocortical network involved in absence seizures. Coronal section from rodent brain illustrating the three main brain regions thought to be involved in absence seizures: layer V–VI of the sensory cerebral cortex, ventrobasalposterior thalamus (VB), and reticular thalamic nucleus (RTN). During absence seizures, T-type channels contribute to creating a self-propagating oscillatory loop within the circuit. Here, RTN cells send inhibitory projections (shown as T-type channel-mediated bursts in the lower trace) to thalamocortical relay neurons, which de-inactivates T-type channels to produce a rebound burst (middle trace), and activate RTN and cortical pyramidal neurons (upper trace). Cortical pyramidal neurons send projections to both RTN and thalamocortical relay neurons, repeating the loop. (Firing patterns based on Huguenard, 2019; Avanzini & Franceschetti, 2003)

polygenetic, a single nucleotide mutation in the *CACNA1H* gene in GAERS (causing an R1584P substitution) was shown to genetically associate with a proportion of seizure activity (Powell et al., 2009). The effect of this mutation on $Ca_v3.2$ channel function is splice variant specific, such that in the presence of exon 25, the channels recovered from inactivation faster, resulting in more channels being available during subsequent depolarizations and leading to larger charge transference carried by Ca^{2+} through the channel during bursts (Powell et al., 2009). Neurons in the RTN from GAERS animals exhibit enhanced burst-firing in the SWD frequency range, an effect that increases during development with a concurrent developmental increase in the ratio of $Ca_v3.2(+25)/Ca_v3.2(-)$ splice variants (Cain et al., 2018). Enhanced burst-firing is prevented with either $Ca_v3.2$ channel blockade (Tringham et al., 2012) or selective thalamic $Ca_v3.2$ knockdown (Cain et al., 2018). Notably, the selective thalamic knockdown of $Ca_v3.2$ also reduced the GAERS seizure phenotype (Cain et al., 2018). Interestingly, the GAERS mutation has been suggested to increase $Ca_v3.2$ surface expression by altering interaction with calnexin, which normally increases endoplasmic reticulum (ER) retention of $Ca_v3.2$ and consequently reduces surface expression (Proft et al., 2017).

However, the relationship between SWDs and burst-firing is complicated by the fact that tonic firing has also been shown to generate SWDs in knockout mice (Lee et al., 2014). Here, $Ca_v3.3$ knockout mice, adeno-associated virus (AAV) mediated knockdown of $Ca_v3.3$ in the RTN, and double knockout mice of $Ca_v3.2$ and $Ca_v3.3$ surprisingly show enhanced γ-butyrolactone-induced SWDs. Although RTN burst-firing was reduced in both knockout mouse models, tonic firing was increased suggesting that enhanced excitability without burst-firing may also be able to trigger the pathological activity required for absence seizures.

In contrast, in thalamocortical relay neurons, $Ca_v3.1$ has been suggested to be important in absence epilepsy. As mentioned, $Ca_v3.1$ expression is increased in the ventral posterior thalamic relay nuclei of adult GAERS (Talley et al., 2000). Evidence for the role of $Ca_v3.1$ in absence epilepsy comes from $Ca_v3.1$ knockout mice, which show reduced generation of GABA$_B$ receptor-mediated SWDs likely due to a lack of burst-firing in thalamocortical relay neurons seen in these mice (Kim et al., 2001). In line with this, a transgenic mouse line with increased Ca$_v$3.1 expression showed rhythmic cortical SWDs associated with periods of behavioral arrest consistent with a role of this channel in absence epilepsy (Ernst et al., 2009). Others have recently suggested that T-type channels in VB thalamocortical relay neurons are not directly involved in absence seizures, but instead that the seizure properties of thalamocortical neurons are driven by cortical excitation and feedforward inhibition by RTN, which depends on T-type channels (McCafferty et al., 2018; Crunelli et al., 2020). Overall, the current models of absence epilepsy are evolving as new methodological approaches are used to tackle relevant research questions (Huguenard, 2019).

Lastly, it is important to note that high-voltage-activated calcium channels are also implicated in absence epilepsy and impact T-type channels. Of particular relevance, selective knockout of *CACNA1A*, encoding for P/Q (Ca$_v$2.1) channels, in layer VI corticothalamic pyramidal neurons showed increased T-type currents in both thalamocortical relay neurons and RTN neurons as well as increased rebound bursting following hyperpolarization in RTN neurons (Bomben et al., 2016).

Clinical manifestations of epilepsy vary depending upon the circuitry involved. Coexistence of generalized and focal epilepsy is rarely observed in patients (Nicolson et al., 2004). Interestingly, a role for T-type channels in limiting overlap between epilepsy subtypes has been proposed. For example, GAERS are resistant to amygdala kindling such that they do not develop stage V seizures but instead remain at stage II (Eşkazan et al., 2002). Further, amygdala kindling leads to reduced firing frequency and increased burst-firing in thalamic reticular nucleus of non-epileptic control rats but is not observed in GAERS (Çarçak et al., 2014). The resistance to kindling-induced changes in firing patterns was attributed to the Ca$_v$3.2 R1584P mutation; however, the R1584P mutation alone

did not confer the resistance to amygdala kindling progression seen in GAERS (Çarçak et al., 2019). Thalamic structures were proposed to limit seizure propagation from the limbic region stimulation in GAERS (Çarçak et al., 2014), an effect that was previously established wherein a co-stimulation of thalamic reticular nucleus during hippocampus kindling suppressed seizure development (Nanobashvili et al., 2003). Overall, it is possible that abnormal activity in thalamocortical networks in GAERS might limit subsequent recruitment of circuits required for TLE.

WAG/Rij

Another rat model of absence epilepsy is Wistar Albino Glaxo from Rijswijk (WAG/Rij), an inbred model that exhibits SWDs and seizures (Coenen et al., 1992). The expression pattern of T-type channels varies between thalamic nuclei and is higher in WAG/Rij, where the lateral geniculate nucleus and centrolateral nucleus express Ca$_v$3.1 and Ca$_v$3.3, and the reticular thalamic nucleus expresses Ca$_v$3.2 and Ca$_v$3.3 (Broicher et al., 2008). Further, within the dorsal part of the lateral geniculate nucleus both thalamocortical relay cells and GABAergic interneurons express T-type calcium channels, with Ca$_v$3.1 expressed in thalamocortical cells and Ca$_v$3.1 and Ca$_v$3.3 expressed in interneurons (Broicher et al., 2007). Interestingly, a splice variant in exons 25 and 26 of *CACNA1G* has been identified with thalamocortical relay neurons primarily expressing a splice variant missing the exon 26 (referred to as Iso1 α1G-a), and interneurons expressing an intermediated splice variant that lacked the 3′ terminus of exon 25 (referred to as Iso2 α1G-bc) (Broicher et al., 2007). WAG/Rij also have increased T-type current densities as well as differences between some activation and inactivation kinetic parameters within the three thalamic nuclei, suggested to result in more robust LTS generation (Broicher et al., 2008).

Together, these two absence epilepsy models highlight crucial roles for T-type channels in the pathological activity seen in the thalamocortical network during absence seizures. In support, clinically relevant antiepileptic drugs used in childhood absence epilepsy patients such as ethosuximide and valproate block T-type channels among other targets (Powell et al., 2014). A proportion of patients with childhood absence epilepsy are pharmaco-resistant and many exhibit neuropsychiatric comorbidities (Crunelli et al., 2020). The T-type channel blocker Z944 has been shown to attenuate seizure activity and RTN burst-firing in GAERS (Tringham et al., 2012) as well as reduce behavioral comorbidities (Marks et al., 2016, 2019; Henbid et al., 2017). Further, another T-type channel blocker ACT-709478 has also been shown to significantly reduce seizures in WAG/Rij (Bezençon et al., 2017). These agents along with another T-type channel blocker, CX-8998, are currently in clinical trials for epilepsy (Steriade et al., 2020). Overall, while subtype specificity remains an ongoing concern, the recent advances in the development of T-type channel blockers are promising for treatment of absence epilepsy.

T-Type Channel Mutations in Epilepsy

A factor that highlights the relationship between T-type calcium channels and epilepsy is mutations observed in patients. T-type channelopathies have previously been linked to several disorders including epilepsy (Adams & Snutch, 2012; Weiss & Zamponi, 2020). Variants in *CACNA1H* gene have been reported for absence epilepsy and temporal lobe epilepsy (Heron et al., 2007; Chourasia et al., 2019; Souza et al., 2019; Wei et al., 2021). Some of these variants have been tested for their effects on the biophysical properties of recombinant $Ca_v3.2$ expressed in heterologous systems (Khosravani et al., 2004; Vitko, 2005). A number of variants resulted in a gain of function characteristics, consistent with a role of $Ca_v3.2$ in seizures; however, some variants did not have any detectable effect or even had opposing effects. One example is childhood absence epilepsy, a polygenic disorder where missense mutations in the *CACNA1H* gene have been reported in patient populations (Chen et al.,

2003; Heron et al., 2007). Many of these were shown to alter the biophysical properties of $Ca_v3.2$ albeit varying in their impact (Khosravani et al., 2004) and with some predicted to increase neuronal firing in modeling studies (Vitko, 2005). Single nucleotide polymorphisms (SNPs) located within the I–II loop were found to increase the surface expression of $Ca_v3.2$ channels (Vitko et al., 2007). One SNP (C456S) studied further was shown to reduce the threshold for LTS generation by increasing T-type calcium currents as well as impact downstream changes in gene expression via an increase in the Ca^{2+}-regulated transcription factor calcium-responsive transactivator (CREST) activity (Eckle et al., 2014). However, mutations (or SNPs) in *CACNA1H* have also been identified in the general population, highlighting the complex polygenic nature of the epilepsies (Calhoun et al., 2020) and highlighting that a proportion of SNPs/mutations are likely part of the many risk factors that contribute to increasing epilepsy susceptibility.

Variants in *CACNA1G* have also been reported in patients with idiopathic generalized epilepsy such as juvenile myoclonic epilepsy (Singh et al., 2007). However, these variants did not significantly impact the electrophysiological properties when expressed in HEK293 cells. Further, *CACNA1G* has been suggested as a genetic modifier of epilepsy in voltage-gated sodium channel transgenic mouse models of focal epilepsy and Dravet syndrome (Calhoun et al., 2016, 2017). Overall, further highlighting the complexity of the various factors involved in epilepsy where T-type channels appear to be one of the multiple players capable of increasing the susceptibility and risk of seizures.

Conclusions

Multiple lines of evidence from rodent studies to genetic mutations in patients have identified roles for T-type channels in the epilepsies. Owing to their biophysical properties, T-type channels are able to shape firing patterns by modulating action potential threshold and firing frequency as well as other less explored roles such as synaptic integration and synaptic transmission (Lambert et al., 2014). The differential expression patterns and biophysical properties of T-type channel subtypes contribute to the diversity of firing patterns of neuronal populations. Both types of epilepsy discussed here, TLE and absence epilepsy, highlight the different roles T-type channels can play in two distinct circuits that underlie seizure activity (Fig. 2).

In TLE, rodent models have shown that increased neuronal T-type current following status epilepticus can transform regular firing neurons to burst-firing. Here, burst-firing has been suggested to involve dendritic T-type channels where backpropagation of action potentials triggers T-type calcium currents, which propagate to the soma generating an ADP and subsequent burst-firing (Yaari et al., 2007). The transient nature of the T-type-mediated burst-firing may have lasting implications that strengthen synapses to generate circuitry that is hyperexcitable, producing sustained and recurrent excitation.

Conversely, in the thalamocortical network circuitry involved in absence epilepsy, the neuronal membrane potential dictates firing patterns. A commonly accepted model suggests that during absence seizures, RTN cells fire bursts of action potentials inhibiting thalamocortical relay neurons via GABAergic transmission. Thalamocortical relay neurons generate a T-type channel-mediated rebound bursts following inhibition, which in turn excite corticothalamic pyramidal neurons and RTN cells. Cortical neurons send excitatory projections back to RTN and thalamocortical relay neurons creating a feedback loop. The burst-firing in thalamocortical relay neurons is suggested to filter out incoming activity as LTSs backpropagate into the dendritic tree, leading to a self-sustaining oscillatory loop (Cain & Snutch, 2013).

In summary, T-type channels operate in a narrow voltage range to mediate rapidly inactivating currents that contribute to abnormal activity within circuits that can manifest as seizures and highlight their importance in neuronal firing generation and propagation. Besides their contribution to LTSs and burst-firing, new findings concerning T-type channel modulation are expected to expand our understanding of the mechanisms mediated by these channels in a variety of epileptic circuits. For example, transcriptional regulation-dependent effects, such as the Zn^{2+}-dependent MTF1 pathway discussed in the TLE section, add additional complexity in the way the channels impact neuronal circuits (van Loo & Becker, 2020). Notably, the epigenetic modulation of T-type channels has recently been shown in response to treatment with antiepileptic drug valproic acid (Kim et al., 2020), providing an additional mechanism that impacts the progression of epilepsy. As new technological advances continue to emerge, such as the characterization of the drug binding site of a selective T-type channel blocker by cryo-EM (Zhao et al., 2019), potentials for discovery of new antiepileptic drugs will likely follow.

References

Adams, P. J., & Snutch, T. P. (2012). Calcium channelopathies: Voltage-gated calcium channels. *Sub-Cellular Biochemistry, 45*, 215–251. https://doi.org/10.1007/978-1-4020-6191-2_8

Avanzini, G., & Franceschetti, S. (2003). Cellular biology of epileptogenesis. *Lancet Neurology, 2*(1), 33–42. https://doi.org/10.1016/S1474-4422(03)00265-5

Avanzini, G., Vergnes, M., Spreafico, R., & Marescaux, C. (1993). Calcium-dependent regulation of genetically determined spike and waves by the reticular thalamic nucleus of rats. *Epilepsia, 34*(1), 1–7. https://doi.org/10.1111/j.1528-1157.1993.tb02369.x

Beck, H., Steffens, R., Elger, C. E., & Heinemann, U. (1998). Voltage-dependent Ca2+ currents in epilepsy. *Epilepsy Research, 32*(1–2), 321–332. https://doi.org/10.1016/S0920-1211(98)00062-X

Becker, A. J., Pitsch, J., Sochivko, D., Opitz, T., Staniek, M., Chen, C. C., Campbell, K. P., Schoch, S., Yaari, Y., & Beck, H. (2008). Transcriptional upregulation of Cav3.2 mediates epileptogenesis in the pilocarpine model of epilepsy. *The Journal of Neuroscience, 28*(49), 13341–13353. https://doi.org/10.1523/JNEUROSCI.1421-08.2008

Bernhardt, B. C., Hong, S.-J., Bernasconi, A., & Bernasconi, N. (2013). Imaging structural and functional brain networks in temporal lobe epilepsy. *Frontiers in Human Neuroscience, 7*, 624. https://doi.org/10.3389/fnhum.2013.00624

Bertram, E. H. (2009). Temporal lobe epilepsy: Where do the seizures really begin? *Epilepsy & Behavior, 14*(Suppl 1), 32–37. https://doi.org/10.1016/j.yebeh.2008.09.017

Bertram, E. H., & Zhang, D. X. (1999). Thalamic excitation of hippocampal CA1 neurons: A comparison with the effects of CA3 stimulation. *Neuroscience, 92*(1), 15–26. https://doi.org/10.1016/S0306-4522(98)00712-X

Bezençon, O., et al. (2017). Discovery of a potent, selective T-type calcium channel blocker as a drug candidate for the treatment of generalized epilepsies. *Journal of Medicinal Chemistry, 60*(23), 9769–9789. https://doi.org/10.1021/ACS.JMEDCHEM.7B01236/SUPPL_FILE/JM7B01236_SI_002.CSV

Bomben, V. C., Aiba, I., Qian, J., Mark, M. D., Herlitze, S., & Noebels, J. L. (2016). Isolated P/Q calcium channel deletion in layer VI corticothalamic neurons generates absence epilepsy. *The Journal of Neuroscience, 36*(2), 405–418. https://doi.org/10.1523/JNEUROSCI.2555-15.2016

Broicher, T., Kanyshkova, T., Landgraf, P., Rankovic, V., Meuth, P., Meuth, S. G., Pape, H.-C., & Budde, T. (2007). Specific expression of low-voltage-activated calcium channel isoforms and splice variants in thalamic local circuit interneurons. *Molecular and Cellular Neurosciences, 36*(2), 132–145. https://doi.org/10.1016/j.mcn.2007.05.013

Broicher, T., Kanyshkova, T., Meuth, P., Pape, H.-C., & Budde, T. (2008). Correlation of T-channel coding gene expression, IT, and the low threshold Ca2+ spike in the thalamus of a rat model of absence epilepsy. *Molecular and Cellular Neurosciences, 39*(3), 384–399. https://doi.org/10.1016/j.mcn.2008.07.012

Cain, S. M., & Snutch, T. P. (2010). Contributions of T-type calcium channel isoforms to neuronal firing. *Channels, 4*(6), 475–482. https://doi.org/10.4161/chan.4.6.14106

Cain, S. M., & Snutch, T. P. (2012). Voltage-gated calcium channels in epilepsy. In J. L. Noebels, M. Avoli, M. A. Rogawski, R. W. Olsen, & A. V. Delgado-Escueta (Eds.), *Jasper's basic mechanisms of the epilepsies* (4th ed., pp. 66–84). Oxford University Press.

Cain, S. M., & Snutch, T. P. (2013). T-type calcium channels in burst-firing, network synchrony, and epilepsy. *Biochimica et Biophysica Acta, Biomembranes, 1828*(7), 1572–1578. https://doi.org/10.1016/j.bbamem.2012.07.028

Cain, S. M., Tyson, J. R., Choi, H.-B., Ko, R., Lin, P. J. C., LeDue, J. M., Powell, K. L., Bernier, L.-P., Rungta, R. L., Yang, Y., Cullis, P. R., O'Brien, T. J., MacVicar, B. A., & Snutch, T. P. (2018). CaV3.2 drives sustained burst-firing, which is critical for absence seizure propagation in reticular thalamic neurons. *Epilepsia, 59*(4), 778–791. https://doi.org/10.1111/EPI.14018

Calhoun, J. D., Hawkins, N. A., Zachwieja, N. J., & Kearney, J. A. (2016). Cacna1g is a genetic modifier of epilepsy caused by mutation of voltage-gated sodium channel Scn2a. *Epilepsia, 57*(6), e103–e107. https://doi.org/10.1111/epi.13390

Calhoun, J. D., Hawkins, N. A., Zachwieja, N. J., & Kearney, J. A. (2017). Cacna1g is a genetic modifier of epilepsy in a mouse model of Dravet syndrome.

Epilepsia, 58(8), e111–e115. https://doi.org/10.1111/epi.13811

Calhoun, J. D., Huffman, A. M., Bellinski, I., Kinsley, L., Bachman, E., Gerard, E., Kearney, J. A., & Carvill, G. L. (2020). CACNA1H variants are not a cause of monogenic epilepsy. *Human Mutation, 41*(6), 1138–1144. https://doi.org/10.1002/humu.24017

Çarçak, N., Zheng, T., Ali, I., Abdullah, A., French, C., Powell, K. L., Jones, N. C., Van Raay, L., Rind, G., Onat, F., & O'Brien, T. J. (2014). The effect of amygdala kindling on neuronal firing patterns in the lateral thalamus in the GAERS model of absence epilepsy. *Epilepsia, 55*(5), 654–665. https://doi.org/10.1111/epi.12592

Çarçak, N., Ali, I., Powell, K., Zheng, T., Onat, F., & O'Brien, T. J. (2019). Ca v 3.2 T-type calcium channel mutation influences kindling-induced thalamic neuronal firing patterns in Genetic Absence Epilepsy Rats from Strasbourg. *Epilepsia, 60*(7), 1378–1386. https://doi.org/10.1111/epi.16076

Casillas-Espinosa, P. M., Hicks, A., Jeffreys, A., Snutch, T. P., O'Brien, T. J., Powell, K. L., & Spafford, J. D. (2015). Z944, a novel selective t-type calcium channel antagonist delays the progression of seizures in the amygdala kindling model. *PLoS One, 10*(8), e0130012. https://doi.org/10.1371/journal.pone.0130012

Casillas-Espinosa, P. M., Shultz, S. R., Braine, E. L., Jones, N. C., Snutch, T. P., Powell, K. L., & O'Brien, T. J. (2019). Disease-modifying effects of a novel T-type calcium channel antagonist, Z944, in a model of temporal lobe epilepsy. *Progress in Neurobiology, 182*, 101677. https://doi.org/10.1016/j.pneurobio.2019.101677

Cavalheiro, E. A. (1995). The pilocarpine model of epilepsy. *Italian Journal of Neurological Sciences, 16*(1–2), 33–37. https://doi.org/10.1007/BF02229072

Chemin, J., Monteil, A., Perez-Reyes, E., Bourinet, E., Nargeot, J., & Lory, P. (2002). Specific contribution of human T-type calcium channel isotypes (α1G, α1H and α1I) to neuronal excitability. *The Journal of Physiology, 540*(1), 3–14. https://doi.org/10.1113/jphysiol.2001.013269

Chen, Y., Lu, J., Pan, H., Zhang, Y., Wu, H., Xu, K., Liu, X., Jiang, Y., Bao, X., Yao, Z., Ding, K., Lo, W. H. Y., Qiang, B., Chan, P., Shen, Y., & Wu, X. (2003). Association between genetic variation of CACNA1H and childhood absence epilepsy. *Annals of Neurology, 54*(2), 239–243. https://doi.org/10.1002/ana.10607

Chen, S., Su, H., Yue, C., Remy, S., Royeck, M., Sochivko, D., Opitz, T., Beck, H., & Yaari, Y. (2011). An increase in persistent sodium current contributes to intrinsic neuronal bursting after status epilepticus. *Journal of Neurophysiology, 105*(1), 117–129. https://doi.org/10.1152/jn.00184.2010

Chourasia, N., Ossó-Rivera, H., Ghosh, A., Von Allmen, G., & Koenig, M. K. (2019). Expanding the phenotypic spectrum of CACNA1H mutations. *Pediatric Neurology, 93*, 50–55. https://doi.org/10.1016/j.pediatrneurol.2018.11.017

Coenen, A. M. L., Drinkenburg, W. H. I. M., Inoue, M., & van Luijtelaar, E. L. J. M. (1992). Genetic models of absence epilepsy, with emphasis on the WAG/Rij strain of rats. *Epilepsy Research, 12*(2), 75–86. https://doi.org/10.1016/0920-1211(92)90029-S

Crandall, S. R., Govindaiah, G., & Cox, C. L. (2010). Low-threshold Ca2+ current amplifies distal dendritic signaling in thalamic reticular neurons. *The Journal of Neuroscience, 30*(46), 15419–15429. https://doi.org/10.1523/JNEUROSCI.3636-10.2010

Crunelli, V., Lorincz, M. L., McCafferty, C., Lambert, R. C., Leresche, N., Di Giovanni, G., & David, F. (2020). Clinical and experimental insight into pathophysiology, comorbidity and therapy of absence seizures. *Brain, 143*(8), 2341–2368. https://doi.org/10.1093/BRAIN/AWAA072

Curia, G., Longo, D., Biagini, G., Jones, R. S. G., & Avoli, M. (2008). The pilocarpine model of temporal lobe epilepsy. *Journal of Neuroscience Methods, 172*(2), 143–157. https://doi.org/10.1016/j.jneumeth.2008.04.019

Devinsky, O., Vezzani, A., O'Brien, T. J., Jette, N., Scheffer, I. E., De Curtis, M., & Perucca, P. (2018). Epilepsy. *Nature Reviews. Disease Primers, 4*(1), 1–24. https://doi.org/10.1038/nrdp.2018.24

Dolleman-van Der Weel, M. J., & Witter, M. P. (1996). Projections from the nucleus reuniens thalami to the entorhinal cortex, hippocampal field CA1, and the subiculum in the rat arise from different populations of neurons. *The Journal of Comparative Neurology, 364*(4), 637–650. https://doi.org/10.1002/(SICI)1096-9861(19960122)364:4<637::AID-CNE3>3.0.CO;2-4

Dolleman-van der Weel, M. J., Lopes da Silva, F. H., & Witter, M. P. (2017). Interaction of nucleus reuniens and entorhinal cortex projections in hippocampal field CA1 of the rat. *Brain Structure & Function, 222*(5), 2421–2438. https://doi.org/10.1007/s00429-016-1350-6

Eckle, V. S., Shcheglovitov, A., Vitko, I., Dey, D., Yap, C. C., Winckler, B., & Perez-Reyes, E. (2014). Mechanisms by which a CACNA1H mutation in epilepsy patients increases seizure susceptibility. *The Journal of Physiology, 592*(4), 795–809. https://doi.org/10.1113/jphysiol.2013.264176

Ekstein, D., Benninger, F., Daninos, M., Pitsch, J., van Loo, K. M. J., Becker, A. J., & Yaari, Y. (2012). Zinc induces long-term upregulation of T-type calcium current in hippocampal neurons in vivo. *The Journal of Physiology, 590*(22), 5895–5905. https://doi.org/10.1113/jphysiol.2012.242537

Engbers, J. D. T., Anderson, D., Tadayonnejad, R., Mehaffey, W. H., Molineux, M. L., & Turner, R. W. (2011). Distinct roles for I T and I H in controlling the frequency and timing of rebound spike responses. *The Journal of Physiology, 589*(22), 5391–5413. https://doi.org/10.1113/jphysiol.2011.215632

Engel, J., & Salamon, N. (2015). Temporal lobe epilepsy. In A. W. Toga (Ed.), *Brain mapping: An encyclopedic reference* (pp. 853–860). Academic Press.

Ernst, W. L., Zhang, Y., Yoo, J. W., Ernst, S. J., & Noebels, J. L. (2009). Genetic enhancement of thalamocortical network activity by elevating α1G-mediated low-voltage-activated calcium current induces pure absence epilepsy. *The Journal of Neuroscience, 29*(6), 1615–1625. https://doi.org/10.1523/JNEUROSCI.2081-08.2009

Esclapez, M., Hirsch, J. C., Ben-Ari, Y., & Bernard, C. (1999). Newly formed excitatory pathways provide a substrate for hyperexcitability in experimental temporal lobe epilepsy. *The Journal of Comparative Neurology, 408*(4), 449–460. https://doi.org/10.1002/(SICI)1096-9861(19990614)408:4<449::AID-CNE1>3.0.CO;2-R

Eşkazan, E., Onat, F. Y., Aker, R., & Öner, G. (2002). Resistance to propagation of amygdaloid kindling seizures in rats with genetic absence epilepsy. *Epilepsia, 43*(10), 1115–1119. https://doi.org/10.1046/J.1528-1157.2002.35601.X

Faas, G. C., Vreugdenhil, M., & Wadman, W. J. (1996). Calcium currents in pyramidal CA1 neurons in vitro after kindling epileptogenesis in the hippocampus of the rat. *Neuroscience, 75*(1), 57–67. https://doi.org/10.1016/0306-4522(96)00254-0

Fogerson, P. M., & Huguenard, J. R. (2016). Tapping the brakes: Cellular and synaptic mechanisms that regulate thalamic oscillations. *Neuron, 92*(4), 687–704. https://doi.org/10.1016/j.neuron.2016.10.024

Frederickson, C. J., Suh, S. W., Silva, D., Frederickson, C. J., & Thompson, R. B. (2000). Importance of zinc in the central nervous system: The zinc-containing neuron. *The Journal of Nutrition, 130*(5), 1471S–1483S. https://doi.org/10.1093/jn/130.5.1471S

Goddard, G. V. (1967). Development of epileptic seizures through brain stimulation at low intensity. *Nature, 214*(5092), 1020–1021. https://doi.org/10.1038/2141020a0

Goddard, G. V., McIntyre, D. C., & Leech, C. K. (1969). A permanent change in brain function resulting from daily electrical stimulation. *Experimental Neurology, 25*(3), 295–330. https://doi.org/10.1016/0014-4886(69)90128-9

Graef, J. D., & Godwin, D. W. (2010). Intrinsic plasticity in acquired epilepsy: Too much of a good thing? *The Neuroscientist, 16*(5), 487–495. https://doi.org/10.1177/1073858409358776

Graef, J. D., Nordskog, B. K., Wiggins, W. F., & Godwin, D. W. (2009). An acquired channelopathy involving thalamic T-type Ca 2+ channels after status epilepticus. *The Journal of Neuroscience, 29*(14), 4430–4441. https://doi.org/10.1523/JNEUROSCI.0198-09.2009

Greene, R. W., Haas, H. L., McCarley, R. W., & Haas, H. L. (1986). A low threshold calcium spike mediates firing pattern alterations in pontine reticular neurons. *Science, 234*(4777), 738–740. https://doi.org/10.1126/SCIENCE.3775364

Henbid, M. T., Marks, W. N., Collins, M. J., Cain, S. M., Snutch, T. P., & Howland, J. G. (2017). Sociability impairments in Genetic Absence Epilepsy Rats from Strasbourg: Reversal by the T-type calcium channel antagonist Z944. *Experimental*

Neurology, 296, 16–22. https://doi.org/10.1016/j.expneurol.2017.06.022

Heron, S. E., Khosravani, H., Varela, D., Bladen, C., Williams, T. C., Newman, M. R., Scheffer, I. E., Berkovic, S. F., Mulley, J. C., & Zamponi, G. W. (2007). Extended spectrum of idiopathic generalized epilepsies associated with CACNA1H functional variants. *Annals of Neurology, 62*(6), 560–568. https://doi.org/10.1002/ANA.21169

Huguenard, J. R. (1996). Low-threshold calcium currents in central nervous system neurons. *Annual Review of Physiology, 58*, 329–348. https://doi.org/10.1146/ANNUREV.PH.58.030196.001553

Huguenard, J. (2019). Current controversy: Spikes, bursts, and synchrony in generalized absence epilepsy: Unresolved questions regarding thalamocortical synchrony in absence epilepsy. *Epilepsy Currents, 19*(2), 105–111. https://doi.org/10.1177/1535759719835355

Huguenard, J. R., & Prince, D. A. (1992). A novel T-type current underlies prolonged Ca(2+)-dependent burst firing in GABAergic neurons of rat thalamic reticular nucleus. *The Journal of Neuroscience, 12*(10), 3804–3817. https://doi.org/10.1523/JNEUROSCI.12-10-03804.1992

Jahnsen, H., & Llinás, R. (1984). Ionic basis for the electro-responsiveness and oscillatory properties of guinea-pig thalamic neurones in vitro. *The Journal of Physiology, 349*(1), 227–247. https://doi.org/10.1113/JPHYSIOL.1984.SP015154

Jallon, P., & Latour, P. (2005). Epidemiology of idiopathic generalized epilepsies. *Epilepsia, 46*(Suppl. 9), 10–14. https://doi.org/10.1111/J.1528-1167.2005.00309.X

Jensen, M. S., Azouz, R., & Yaari, Y. (1994). Variant firing patterns in rat hippocampal pyramidal cells modulated by extracellular potassium. *Journal of Neurophysiology, 71*(3), 831–839. https://doi.org/10.1152/jn.1994.71.3.831

Khosravani, H., & Zamponi, G. W. (2006). Voltage-gated calcium channels and idiopathic generalized epilepsies. *Physiological Reviews, 86*(3), 941–966. https://doi.org/10.1152/physrev.00002.2006

Khosravani, H., Altier, C., Simms, B., Hamming, K. S., Snutchll, T. P., Mezeyova, J., McRory, J. E., & Zamponi, G. W. (2004). Gating effects of mutations in the Cav3.2 T-type calcium channel associated with childhood absence epilepsy. *The Journal of Biological Chemistry, 279*(11), 9681–9684. https://doi.org/10.1074/jbc.C400006200

Kim, C.-H. (2015). Cav3.1 T-type calcium channel modulates the epileptogenicity of hippocampal seizures in the kainic acid-induced temporal lobe epilepsy model. *Brain Research, 1622*, 204–216. https://doi.org/10.1016/j.brainres.2015.06.015

Kim, D., Song, I., Keum, S., Lee, T., Jeong, M.-J., Kim, S.-S., McEnery, M. W., & Shin, H.-S. (2001). Lack of the burst firing of Thalamocortical relay neurons and resistance to absence seizures in mice lacking α1G T-type Ca2+ channels. *Neuron, 31*(1), 35–45. https://doi.org/10.1016/S0896-6273(01)00343-9

Kim, J. W., Oh, H. A., Kim, S. R., Ko, M. J., Seung, H., Lee, S. H., & Shin, C. Y. (2020). Epigenetically upregulated t-type calcium channels contribute to abnormal proliferation of embryonic neural progenitor cells exposed to valproic acid. *Biomolecules & Therapeutics, 28*(5), 389–396. https://doi.org/10.4062/biomolther.2020.027

Lambert, R. C., Bessaïh, T., Crunelli, V., & Leresche, N. (2014). The many faces of T-type calcium channels. *Pflügers Archiv – European Journal of Physiology, 466*(3), 415–423. https://doi.org/10.1007/s00424-013-1353-6

Lee, S. E., Lee, J., Latchoumane, C., Lee, B., Oh, S. J., Saud, Z. A., Park, C., Sun, N., Cheong, E., Chen, C. C., Choi, E. J., Lee, C. J., & Shin, H. S. (2014). Rebound burst firing in the reticular thalamus is not essential for pharmacological absence seizures in mice. *Proceedings of the National Academy of Sciences of the United States of America, 111*(32), 11828–11833. https://doi.org/10.1073/pnas.1408609111

Lévesque, M., & Avoli, M. (2013). The kainic acid model of temporal lobe epilepsy. *Neuroscience and Biobehavioral Reviews, 37*(10), 2887–2899. https://doi.org/10.1016/j.neubiorev.2013.10.011

Llinás, R., & Jahnsen, H. (1982). Electrophysiology of mammalian thalamic neurones in vitro. *Nature, 297*(5865), 406–408. https://doi.org/10.1038/297406a0

Llinás, R., & Yarom, Y. (1981a). Electrophysiology of mammalian inferior olivary neurones in vitro. Different types of voltage-dependent ionic conductances. *The Journal of Physiology, 315*(1), 549–567. https://doi.org/10.1113/JPHYSIOL.1981.SP013763

Llinás, R., & Yarom, Y. (1981b). Properties and distribution of ionic conductances generating electroresponsiveness of mammalian inferior olivary neurones in vitro. *The Journal of Physiology, 315*(1), 569–584. https://doi.org/10.1113/JPHYSIOL.1981.SP013764

MacVicar, B. A., & Dudek, F. E. (1980). Local synaptic circuits in rat hippocampus: Interactions between pyramidal cells. *Brain Research, 184*(1), 220–223. https://doi.org/10.1016/0006-8993(80)90602-2

Magee, J. C., & Carruth, M. (1999). Dendritic voltage-gated ion channels regulate the action potential firing mode of hippocampal CA1 pyramidal neurons. *Journal of Neurophysiology, 82*(4), 1895–1901. https://doi.org/10.1152/jn.1999.82.4.1895

Marescaux, C., & Vergnes, M. (1995). Genetic Absence Epilepsy in Rats from Strasbourg (GAERS). *Italian Journal of Neurological Sciences, 16*(1–2), 113–118. https://doi.org/10.1007/BF02229083

Marks, W. N., Cain, S. M., Snutch, T. P., & Howland, J. G. (2016). The T-type calcium channel antagonist Z944 rescues impairments in crossmodal and visual recognition memory in Genetic Absence Epilepsy Rats from Strasbourg. *Neurobiology of Disease, 94*, 106–115. https://doi.org/10.1016/j.nbd.2016.06.001

Marks, W. N., Zabder, N. K., Greba, Q., Cain, S. M., Snutch, T. P., & Howland, J. G. (2019). The T-type calcium channel blocker Z944 reduces conditioned fear in Genetic Absence Epilepsy Rats from Strasbourg and the non-epileptic control strain. *The European Journal of Neuroscience, 50*(6), 3046–3059. https://doi.org/10.1111/EJN.14406

Martinello, K., Huang, Z., Lujan, R., Tran, B., Watanabe, M., Cooper, E. C., Brown, D. A., & Shah, M. M. (2015). Cholinergic afferent stimulation induces axonal function plasticity in adult hippocampal granule cells. *Neuron, 85*(2), 346–363. https://doi.org/10.1016/j.neuron.2014.12.030

McCafferty, C., David, F., Venzi, M., Lorincz, M. L., Delicata, F., Atherton, Z., Recchia, G., Orban, G., Lambert, R. C., Di Giovanni, G., Leresche, N., & Crunelli, V. (2018). Cortical drive and thalamic feed-forward inhibition control thalamic output synchrony during absence seizures. *Nature Neuroscience, 21*(5), 744–756. https://doi.org/10.1038/s41593-018-0130-4

McCobb, D. P., & Beam, K. G. (1991). Action potential waveform voltage-clamp commands reveal striking differences in calcium entry via low and high voltage-activated calcium channels. *Neuron, 7*(1), 119–127. https://doi.org/10.1016/0896-6273(91)90080-J

McKay, B. E., McRory, J. E., Molineux, M. L., Hamid, J., Snutch, T. P., Zamponi, G. W., & Turner, R. W. (2006). CaV3 T-type calcium channel isoforms differentially distribute to somatic and dendritic compartments in rat central neurons. *The European Journal of Neuroscience, 24*(9), 2581–2594. https://doi.org/10.1111/j.1460-9568.2006.05136.x

McNamara, J. O., Constant Byrne, M., Dasheiff, R. M., & Gregory Fitz, J. (1980). The kindling model of epilepsy: A review. *Progress in Neurobiology, 15*(2), 139–159. https://doi.org/10.1016/0301-0082(80)90006-4

McRory, J. E., Santi, C. M., Hamming, K. S. C., Mezeyova, J., Sutton, K. G., Baillie, D. L., Stea, A., & Snutch, T. P. (2001). Molecular and functional characterization of a family of rat brain T-type calcium channels. *The Journal of Biological Chemistry, 276*(6), 3999–4011. https://doi.org/10.1074/jbc.M008215200

Metz, A. E., Jarsky, T., Martina, M., & Spruston, N. (2005). R-type calcium channels contribute to after-depolarization and bursting in hippocampal CA1 pyramidal neurons. *The Journal of Neuroscience, 25*(24), 5763–5773. https://doi.org/10.1523/JNEUROSCI.0624-05.2005

Nanobashvili, Z., Chachua, T., Nanobashvili, A., Bilanishvili, I., Lindvall, O., & Kokaia, Z. (2003). Suppression of limbic motor seizures by electrical stimulation in thalamic reticular nucleus. *Experimental Neurology, 181*(2), 224–230. https://doi.org/10.1016/S0014-4886(03)00045-1

Nicolson, A., Chadwick, D. W., & Smith, D. F. (2004). The coexistence of idiopathic generalized epilepsy and partial epilepsy. *Epilepsia, 45*(6), 682–685. https://doi.org/10.1111/J.0013-9580.2004.45003.X

Perez-Reyes, E. (2003). Molecular physiology of low-voltage-activated T-type calcium channels. *Physiological Reviews, 83*(1), 117–161. https://doi.org/10.1152/physrev.00018.2002

Pinault, D. (2004). The thalamic reticular nucleus: Structure, function and concept. *Brain Research Reviews, 46*(1), 1–31. https://doi.org/10.1016/j.brainresrev.2004.04.008

Powell, K. L., Cain, S. M., Ng, C., Sirdesai, S., David, L. S., Kyi, M., Garcia, E., Tyson, J. R., Reid, C. A., Bahlo, M., Foote, S. J., Snutch, T. P., & O'Brien, T. J. (2009). A Cav3.2 T-type calcium channel point mutation has splice-variant-specific effects on function and segregates with seizure expression in a polygenic rat model of absence epilepsy. *The Journal of Neuroscience, 29*(2), 371–380. https://doi.org/10.1523/JNEUROSCI.5295-08.2009

Powell, K. L., Cain, S. M., Snutch, T. P., & O'Brien, T. J. (2014). Low threshold T-type calcium channels as targets for novel epilepsy treatments. *British Journal of Clinical Pharmacology, 77*(5), 729–739. https://doi.org/10.1111/BCP.12205

Proft, J., Rzhepetskyy, Y., Lazniewska, J., Zhang, F. X., Cain, S. M., Snutch, T. P., Zamponi, G. W., & Weiss, N. (2017). The Cacna1h mutation in the GAERS model of absence epilepsy enhances T-type Ca2+ currents by altering calnexin-dependent trafficking of Cav3.2 channels. *Scientific Reports, 7*(1). https://doi.org/10.1038/s41598-017-11591-5

Rusina, E., Bernard, C., & Williamson, A. (2021). The kainic acid models of temporal lobe epilepsy. *eNeuro, 8*(2), ENEURO.0337-20.2021. https://doi.org/10.1523/ENEURO.0337-20.2021

Sanabria, E. R. G., Su, H., & Yaari, Y. (2001). Initiation of network bursts by Ca2+-dependent intrinsic bursting in the rat pilocarpine model of temporal lobe epilepsy. *The Journal of Physiology, 532*(1), 205–216. https://doi.org/10.1111/j.1469-7793.2001.0205g.x

Sato, M., Racine, R. J., & McIntyre, D. C. (1990). Kindling: Basic mechanisms and clinical validity. *Electroencephalography and Clinical Neurophysiology, 76*(5), 459–472. https://doi.org/10.1016/0013-4694(90)90099-6

Scheel, N., Wulff, P., & Mooij-van Malsen, J. G. (2020). Afferent connections of the thalamic nucleus reuniens in the mouse. *The Journal of Comparative Neurology, 528*(7), 1189–1202. https://doi.org/10.1002/cne.24811

Shao, L. R., & Dudek, F. E. (2004). Increased excitatory synaptic activity and local connectivity of hippocampal CA1 pyramidal cells in rats with kainate-induced epilepsy. *Journal of Neurophysiology, 92*(3), 1366–1373. https://doi.org/10.1152/jn.00131.2004

Singh, B., Monteil, A., Bidaud, I., Sugimoto, Y., Suzuki, T., Hamano, S., Oguni, H., Osawa, M., Alonso, M. E., Delgado-Escueta, A. V., Inoue, Y., Yasui-Furukori, N., Kaneko, S., Lory, P., & Yamakawa, K. (2007). Mutational analysis of CACNA1G in idiopathic generalized epilepsy. *Human Mutation, 28*(5), 524–525. https://doi.org/10.1002/humu.9491

Smolders, I., Khan, G. M., Manil, J., Ebinger, G., & Michotte, Y. (1997). NMDA receptor-mediated pilocarpine-induced seizures: Characterization in freely moving rats by microdialysis. *British Journal* of *Pharmacology, 121*(6), 1171–1179. https://doi.org/10.1038/sj.bjp.0701231

Souza, I. A., Gandini, M. A., Zhang, F. X., Mitchell, W. G., Matsumoto, J., Lerner, J., Pierson, T. M., & Zamponi, G. W. (2019). Pathogenic Cav3.2 channel mutation in a child with primary generalized epilepsy. *Molecular Brain, 12*(1), 1–6. https://doi.org/10.1186/s13041-019-0509-5

Stafstrom, C. E. (2006). Epilepsy: A review of selected clinical syndromes and advances in basic science. *Journal of Cerebral Blood Flow and Metabolism, 26*(8), 983–1004. https://doi.org/10.1038/sj.jcbfm.9600265

Steriade, C., French, J., & Devinsky, O. (2020). Epilepsy: Key experimental therapeutics in early clinical development. *Expert Opinion on Investigational Drugs, 29*(4), 373–383. https://doi.org/10.1080/13543784.2020.1743678

Su, H., Sochivko, D., Becker, A., Chen, J., Jiang, Y., Yaari, Y., & Beck, H. (2002). Upregulation of a T-type Ca2+ channel causes a long-lasting modification of neuronal firing mode after status epilepticus. *The Journal of Neuroscience, 22*(9), 3645–3655. https://doi.org/10.1523/jneurosci.22-09-03645.2002

Suh, S. W., Thompson, R. B., & Frederickson, C. J. (2001). Loss of vesicular zinc and appearance of perikaryal zinc after seizures induced by pilocarpine. *Neuroreport, 12*(7), 1523–1525. https://doi.org/10.1097/00001756-200105250-00044

Sun, D. G., Kang, H., Tetteh, H., Su, J., Lee, J., Park, S. W., He, J., Jo, J., Yang, S., & Yang, S. (2018). Long term potentiation, but not depression, in interlamellar hippocampus CA1. *Scientific Reports, 8*(1), 1–9. https://doi.org/10.1038/s41598-018-23369-4

Takahashi, K., Ueno, S., & Akaike, N. (1991). Kinetic properties of T-type Ca2+ currents in isolated rat hippocampal CA1 pyramidal neurons. *Journal of Neurophysiology, 65*(1), 148–155. https://doi.org/10.1152/jn.1991.65.1.148

Talavera, K., & Nilius, B. (2006). Biophysics and structure-function relationship of T-type Ca2+ channels. *Cell Calcium, 40*(2), 97–114. https://doi.org/10.1016/J.CECA.2006.04.013

Talley, E. M., Solórzano, G., Depaulis, A., Perez-Reyes, E., & Bayliss, D. A. (2000). Low-voltage-activated calcium channel subunit expression in a genetic model of absence epilepsy in the rat. *Molecular Brain Research, 75*(1), 159–165. https://doi.org/10.1016/S0169-328X(99)00307-1

Tatum, W. O. (2012). Mesial temporal lobe epilepsy. *Journal of Clinical Neurophysiology, 29*(5), 356–365. https://doi.org/10.1097/WNP.0b013e31826b3ab7

Tringham, E., Powell, K. L., Cain, S. M., Kuplast, K., Mezeyova, J., Weerapura, M., Eduljee, C., Jiang, X., Smith, P., Morrison, J. L., Jones, N. C., Braine, E., Rind, G., Fee-Maki, M., Parker, D., Pajouhesh, H., Parmar, M., O'Brien, T. J., & Snutch, T. P. (2012). T-type calcium channel blockers that attenuate thalamic burst firing and suppress absence seizures.

Science Translational Medicine, 4(121), 121ra19. https://doi.org/10.1126/scitranslmed.3003120

Tsakiridou, E., Bertollini, L., De Curtis, M., Avanzini, G., & Pape, H. C. (1995). Selective increase in T-type calcium conductance of reticular thalamic neurons in a rat model of absence epilepsy. *The Journal of Neuroscience, 15*(4), 3110–3117. https://doi.org/10.1523/jneurosci.15-04-03110.1995

Turski, W. A., Cavalheiro, E. A., Schwarz, M., Czuczwar, S. J., Kleinrok, Z., & Turski, L. (1983). Limbic seizures produced by pilocarpine in rats: Behavioural, electroencephalographic and neuropathological study. *Behavioural Brain Research, 9*(3), 315–335. https://doi.org/10.1016/0166-4328(83)90136-5

van Loo, K. M. J., & Becker, A. J. (2020). Transcriptional regulation of channelopathies in genetic and acquired epilepsies. *Frontiers in Cellular Neuroscience, 13*, 587. https://doi.org/10.3389/fncel.2019.00587

Van Loo, K. M. J., Schaub, C., Pitsch, J., Kulbida, R., Opitz, T., Ekstein, D., Dalal, A., Urbach, H., Beck, H., Yaari, Y., Schoch, S., & Becker, A. J. (2015). Zinc regulates a key transcriptional pathway for epileptogenesis via metal-regulatory transcription factor 1. *Nature Communications, 6*, 8688. https://doi.org/10.1038/ncomms9688

Velíšková, J., & Velíšek, L. (2017). Behavioral characterization and scoring of seizures in rodents. In A. Pitkänen, P. S. Buckmaster, A. S. Galanopoulou, & S. L. B. T. Moshé (Eds.), *Models of seizures and epilepsy: Second edition* (pp. 111–123). Academic Press.

Vergnes, M., Marescaux, C., Micheletti, G., Reis, J., Depaulis, A., Rumbach, L., & Warter, J. M. (1982). Spontaneous paroxysmal electroclinical patterns in rat: A model of generalized non-convulsive epilepsy. *Neuroscience Letters, 33*(1), 97–101. https://doi.org/10.1016/0304-3940(82)90136-7

Vertes, R. P. (2015). Major diencephalic inputs to the hippocampus: Supramammillary nucleus and nucleus reuniens. Circuitry and function. *Progress in Brain Research, 219*, 121–144. https://doi.org/10.1016/BS.PBR.2015.03.008

Vickstrom, C. R., Liu, X., Zhang, Y., Mu, L., Kelly, T. J., Yan, X., Hu, M. M., Snarrenberg, S. T., & Liu, Q. S. (2020). T-type calcium channels contribute to burst firing in a subpopulation of medial habenula neurons. *eNeuro, 7*(4), 1–13. https://doi.org/10.1523/ENEURO.0201-20.2020

Vitko, I. (2005). Functional characterization and neuronal modeling of the effects of childhood absence epilepsy variants of CACNA1H, a T-type calcium channel. *The Journal of Neuroscience, 25*(19), 4844–4855. https://doi.org/10.1523/JNEUROSCI.0847-05.2005

Vitko, I., Bidaud, I., Arias, J. M., Mezghrani, A., Lory, P., & Perez-Reyes, E. (2007). The I-II loop controls plasma membrane expression and gating of Cav3.2 T-type Ca2+ channels: A paradigm for childhood absence epilepsy mutations. *The Journal of Neuroscience, 27*(2), 322–330. https://doi.org/10.1523/JNEUROSCI.1817-06.2007

Wei, Z., Liu, C., Wu, Z., Cao, M., Qiao, X., Han, T., Zhang, Y., Liu, Y., & Deng, Y. (2021). The prognosis of epilepsy patients with CACNA1H missense variants: A longitudinal cohort study. *Seizure, 91*, 52–59. https://doi.org/10.1016/j.seizure.2021.05.019

Weiss, N., & Zamponi, G. W. (2020). Genetic T-type calcium channelopathies. *Journal of Medical Genetics, 57*(1), 1–10. https://doi.org/10.1136/JMEDGENET-2019-106163

Wilcox, K. S., Gutnick, M. J., & Christoph, G. R. (1988). Electrophysiological properties of neurons in the lateral habenula nucleus: An in vitro study. *Journal of Neurophysiology, 59*(1), 212–225. https://doi.org/10.1152/jn.1988.59.1.212

Witter, M. P., & Amaral, D. G. (2004). Hippocampal formation. In G. Paxinos (Ed.), *The rat nervous system* (3rd ed., pp. 635–704). Academic Press.

Yaari, Y., & Beck, H. (2006). "Epileptic neurons" in temporal lobe epilepsy. *Brain Pathology, 12*(2), 234–239. https://doi.org/10.1111/j.1750-3639.2002.tb00438.x

Yaari, Y., Yue, C., & Su, H. (2007). Recruitment of apical dendritic T-type Ca 2 + channels by backpropagating spikes underlies de novo intrinsic bursting in hippocampal epileptogenesis. *The Journal of Physiology, 580*(2), 435–450. https://doi.org/10.1113/jphysiol.2007.127670

Yamada, N., & Bilkey, D. K. (1991). Kindling-induced persistent alterations in the membrane and synaptic properties of CA1 pyramidal neurons. *Brain Research, 561*(2), 324–331. https://doi.org/10.1016/0006-8993(91)91611-4

Yang, S., Yang, S., Moreira, T., Hoffman, G., Carlson, G. C., Bender, K. J., Alger, B. E., & Tang, C.-M. (2014). Interlamellar CA1 network in the hippocampus. *Proceedings of the National Academy of Sciences, 111*(35), 12919–12924. https://doi.org/10.1073/PNAS.1405468111

Zhang, Y., Cai, G.-E., Yang, Q., Lu, Q.-C., Li, S.-T., & Ju, G. (2010). Time-dependent changes in learning ability and induction of long-term potentiation in the lithium–pilocarpine-induced epileptic mouse model. *Epilepsy & Behavior, 17*(4), 448–454. https://doi.org/10.1016/j.yebeh.2010.02.008

Zhao, Y., Huang, G., Wu, Q., Wu, K., Li, R., Lei, J., Pan, X., & Yan, N. (2019). Cryo-EM structures of apo and antagonist-bound human Cav3.1. *Nature, 576*(7787), 492–497. https://doi.org/10.1038/s41586-019-1801-3

Novel Insights into the Role of Voltage-Gated Calcium Channel Genes in Psychiatric Disorders

Camryn Berry, Herie Sun, Vladimir Tkachev,
Anjali M. Rajadhyaksha, and Arturo Andrade

Abstract

Calcium entering via voltage-gated calcium channels is a second messenger key for cellular functions including gene expression, neuronal excitability, neurogenesis, neuronal differentiation, and neurotransmitter release. Alterations in these calcium-dependent processes have been observed in psychiatric disorders. Furthermore, genetic studies have identified associations of risk genetic variants of voltage-gated calcium channel genes (*CACNA1A-I*) with psychiatric disorders. Thus, these channels are becoming promising targets to treat these pathologies. In this chapter, we will discuss evidence linking calcium to psychiatric disorders, and then we will review key genetic studies that have found strong associations among voltage-gated calcium channel genes and psychiatric disorders. Next, we will examine the role of voltage-gated calcium channels in neurobiological mechanisms linked to psychiatric disorders. We will analyze evidence from animal models that link voltage-gated to behavioral endophenotypes observed in psychiatric disorders. Finally, we will discuss the current view and challenges to target these channels to treat psychiatric disorders.

Keywords

Voltage-gated calcium channels · Psychiatric disorders · Calcium · Behavior · Neurodevelopmental disorders · Endophenotypes · Neurotransmitter release · Neurogenesis · Excitation/inhibition balance · Polymorphism · Risk genetic variants

C. Berry · V. Tkachev
Department of Biological Sciences, College of Life Sciences and Agriculture, University of New Hampshire, Durham, NH, USA

H. Sun · A. M. Rajadhyaksha (✉)
Department of Pediatric Neurology, Pediatrics, Feil Family Brain and Mind and Research Institute, Weill Cornell Autism Research Project, Weill Cornell Medicine, New York, NY, USA
e-mail: amr2011@med.cornell.edu

A. Andrade (✉)
Robert J. and Nancy D. Carney Institute for Brain Science, Brown University, Providence, RI, USA
e-mail: Arturo_Andrade@brown.edu

Abbreviations

ADHD	attention deficit and hyperactivity disorder
ASD	autism spectrum disorder
BD	bipolar disorder
BDNF	brain-derived neurotrophic factor
CaMKII	Ca^2/calmodulin-dependent protein kinase II

G. W. Zamponi, N. Weiss (eds.), *Voltage-Gated Calcium Channels*,
https://doi.org/10.1007/978-3-031-08881-0_21

CCK cholecystokinin
CPu caudate putamen
CREB cAMP response-element binding protein
EA2 episodic ataxia type 2
eIF2α eukaryotic initiation factor 2α
FHM1 familial hemiplegic ataxia type 1
FST forced-swim test
GWAS genome-wide association study
MDD major depressive disorder
mTOR mammalian target of rapamycin
NAc nucleus accumbens
PV parvalbumin
RDoC research domain criteria
SCZ schizophrenia
SNc substantia nigra pars compacta
SNP single nucleotide polymorphism
TS timothy syndrome
TST tail suspension test
VGCC voltage-gated calcium channel
VTA ventral tegmental area

Introduction

Psychiatric disorders affect a large percentage of the global population, treatments are scarce, and little is known about their neurobiology (Health, 2019). Psychiatric disorders are pathologies with main outcomes as altered states of behavior, thus psychiatric disorders are also known as behavioral disorders (Association, 2000; Health, 2019). Schizophrenia (SCZ), Bipolar Disorder (BD), Autism Spectrum Disorder (ASD), Attention-Deficit and Hyperactivity Disorder (ADHD), and Major Depression Disorder (MDD) share common genetic risk factors (Cross-Disorder et al., 2013). Among the common risk factors in these disorders are various genes for voltage-gated calcium channels (VGCCs) including *CACNA1C*, which encodes for L-type channel $Ca_V 1.2$ (Cross-Disorder et al., 2013). Thus a "calcium signaling hypothesis" to explain the physiopathology of psychiatric disorders has been gaining traction in recent years. This is further supported by the expression pattern and the important function of voltage-gated calcium channels in areas of the brain that are linked to psychiatric disorders.

L-type channels, particularly $Ca_V 1.2$ and $Ca_V 1.3$, are expressed in dopamine neurons of the ventral tegmental area (VTA) and glutamatergic neurons of the forebrain where they control gene expression and firing patterns (Lee et al., 2012; Liu et al., 2014; Terrillion et al., 2017; Martínez-Rivera et al., 2017). These areas have been linked to schizophrenia (Rajarajan et al., 2018; Karlsgodt et al., 2010; Howes et al., 2017; Sonnenschein et al., 2020). The presynaptic calcium channels $Ca_V 2.1$ (P/Q-type), $Ca_V 2.2$ (N-type), and $Ca_V 2.3$ (R-type) control transmitter release in most synapses in the brain affecting many neurotransmitter systems (Dolphin & Lee, 2020). For example, $Ca_V 2.2$ channels dominate in the release of dopamine in synapses from the nucleus accumbens (NAc) and caudate putamen (CPu) (Brimblecombe et al., 2015). Finally, $Ca_V 3.1$–3.3 (T-type) channels control the firing frequency of thalamic neurons, which have been linked to attention, cognitive function, and schizophrenia (Astori et al., 2011; Ferrarelli & Tononi, 2011).

In this chapter, we will discuss current evidence pointing at the role of VGCCs in psychiatric disorders including but not limited to ASD, SCZ, BD, ADHD, and MDD. First, we will provide a brief description of psychiatric disorders. Next, we will discuss how altered calcium signaling is thought to contribute to the physiopathology of psychiatric disorders. Then, we will summarize key genetic studies linking genes for VGCCs with psychiatric disorders, and how these studies have helped to support the overarching idea that alterations in calcium signaling contribute to the neurobiological mechanisms of psychiatric disorders. We will review important studies in both human subjects and animal models that are providing important cellular and molecular cues to understand psychiatric disorders. Finally, we will provide a perspective on targeting VGCCs to treat psychiatric disorders.

Overview of Psychiatric Disorders

Psychiatric disorders or mental illnesses are defined as mental, behavioral, or emotional disorders (Association, 2000). It is estimated that nearly 19% of the US population in 2021 had a type of psychiatric disorder (Health, 2019). Psychiatric disorders represent a burden not only for patients but also for their families and caregivers. Thus, understanding the cellular and molecular mechanisms of these pathologies is becoming a public health priority.

Psychiatric disorders are genetically complex. Due to the capability to sequence and analyze whole genomes of patients and their families, we have been able to identify genetic risk factors that are linked to psychiatric disorders (Lee et al., 2021; O'Donovan & Owen, 2016). These findings have provided cues for the development of robust animal models to dissect neurobiological mechanisms (Harrison et al., 2018). The combination of these studies with computational tools is allowing a comprehensive modeling of the behavioral consequences of genetic risk factors—and their interaction with the environment—in the context of psychiatric disorders (Huys et al., 2016; Burmeister et al., 2008; Robinson & Bergen, 2021; Tsuang et al., 2004).

Psychiatric disorders comprise a complex array of conditions that have common genetic risk factors and affect overlapping brain circuits, as a result underlying signs and symptoms are shared among psychiatric disorders (Smail et al., 2021). Some of the most common risk genetic factors identified in genome-wide association studies are genes for VGCCs. These findings across multiple studies have implicated alterations in calcium signaling as a major player in the pathology of psychiatric disorders.

Implications of Calcium Signaling in Psychiatric Disorders

The links between calcium signaling and psychiatric disorders are thought to result from the key role of calcium in controlling crucial neuronal processes including gene expression, synaptic plasticity, transmitter release, and excitability (Berridge, 2014; Casamassima et al., 2010; Harrison et al., 2018, 2019). Initial studies showed elevated intracellular calcium in platelets and lymphocytes of patients with BD (Dubovsky

et al., 1994; Emamghoreishi et al., 1997). Furthermore, calcium levels in cerebrospinal fluid were found to correlate with mood state of patients with BD (reduced in manic states, increased in depressed states) (Carman & Wyatt, 1979; Jimerson et al., 1979; Levine et al., 1999). These changes are responsive to lithium, which is one of the few treatments available for BD (Harrison et al., 2018). Furthermore, drugs that target calcium-dependent processes like pregabalin and lamotrigine have shown some promise to treat bipolar disorder (Reid et al., 2013; Takenaka et al., 1989).

Alterations in intracellular calcium and in the expression/localization of calcium-binding proteins have been observed in patients with schizophrenia (Bojarski et al., 2010). And this disruption in calcium signaling leads to alterations of neurotransmitters including dopamine and glutamate (Bojarski et al., 2010). In ASD, alterations in calcium signaling have been observed in genetic disorders like Timothy Syndrome (TS; a cardiac disorder with autistic traits). Here, an increase in calcium entry via $Ca_V1.2$ leads to arrhythmias and autistic traits (Splawski et al., 2004, 2005). In addition, psychotropic drugs have shown to alter calcium signaling (Dubovsky & Franks, 1983; Glen, 1985). This collective body of evidence strongly suggests the involvement of calcium signaling in the physiopathology of psychiatric disorders.

Genetic Studies Linking VGCC Genes to Psychiatric Disorders

In this section, we will review the critical studies that led to the discovery of mutations and risk variants associated with psychiatric disorders for genes that encode L-type channels (*CACNA1C* and *CACNA1D*), presynaptic calcium channels P/Q-, N-, and R-type (*CACNA1A*, *CACNA1B*, *CACNA1E*, respectively), and T-type channels (*CACNA1G*, *CACNA1H*, and *CACNA1I*). We will not provide an exhaustive list of all the genetic studies that have associated VGCC risk variants with psychiatric disorders, but rather we will describe studies that have been replicated in several genetic screenings and that have led to the description of molecular mechanisms associated with psychiatric disorders (described in the following section). For a comprehensive review of the genetic risk variants that have been identified in VGCC genes see (Andrade et al., 2019; Heyes et al., 2015).

L-Type Channel Genes (*CACNA1C* and *CACNA1D*)

One of the most solid pieces of evidence linking VGCCs, particularly *CACNA1C*, to psychiatric disorders arose from studies in TS. In TS, a mutation in $Ca_V1.2$, G406R, results in a gain-of-function, due to a dramatic reduction in the channel's ability to undergo inactivation (Splawski et al., 2004, 2005). TS is a multisystem disorder accompanied by cardiac deficits, and abnormalities of the hand, foot, and face, notably ~80% of patients with TS also exhibit traits of ASD (Splawski et al., 2004, 2005). Furthermore, the single nucleotide polymorphism (SNP) rs 1006737 in the intronic region between exon 3 and 4 of *CACNA1C* is a genetic risk variant that has been replicated in multiple genetic studies of BD. The SNP rs 1,006,737 was originally linked to BD (Ferreira et al., 2008; Stahl et al., 2019). However recent studies have linked this SNP, among others in *CACNA1C*, to numerous psychiatric conditions including SCZ, drug dependence associated with BD, MDD, and ASD (Heyes et al., 2015; Green et al., 2010; Hamshere et al., 2013; Moon et al., 2018; Ripke et al., 2013; Schizophrenia, 2014; Li et al., 2015).

Evidence exists linking genetic risk variants for *CACNA1D* to BD, SCZ, ADHD, MDD, and ASD (Ortner et al., 2020). However, there is more evidence linking *CACNA1D* to ASD. Interestingly, the G407R risk variant in *CACNA1D* associated with ASD is similar to the G406R mutation in $Ca_V1.2$ that causes TS (Pinggera et al., 2015, 2017). This mutation together with A749G results in $Ca_V1.3$ gain-of-function and has been confirmed in two independent whole-exome sequencing studies (Pinggera et al., 2015, 2017; De Rubeis et al., 2014).

Presynaptic Calcium Channel Genes (*CACNA1A, CACNA1B, CACNA1E*)

More in-depth studies of clinical cases of patients with genetic mutations in the *CACNA1A* gene ($Ca_V2.1$) like those that cause familial hemiplegic ataxia type 1 (FHM1) and episodic ataxia type 2 (EA2) have revealed strong links of $Ca_V2.1$ to psychiatric disorders in particular neurodevelopment disorders. Patients with EA2 and FHM1 exhibit learning deficits, and traits of ADHD and ASD (Indelicato & Boesch, 2021; Lupien-Meilleur et al., 2021). More recently, the $Ca_V2.1$ mutation R1673P has been linked to a severe disorder characterized by ataxia, hypotonia, cerebellar atrophy, and neurodevelopmental defects (Luo et al., 2017; Tyagi et al., 2019). Finally, genome-wide association studies (GWAS) have linked variants in *CACNA1A* with SCZ in a Xhosa population of South Africa (Gulsuner et al., 2020; Hidalgo et al., 2021).

Some GWAS studies have unveiled associations of the *CACNA1B* ($Ca_V2.2$) gene with SCZ and BD (Moskvina et al., 2009). Whole-exome sequencing has shown the presence of large deletions and variants of *CACNA1B* as genetic risks for SCZ (Ament et al., 2015; Purcell et al., 2014). Furthermore, allelic variants of *CACNA1B* are linked to treatment outcomes in the initial manifestation of SCZ in South African cohorts (O'Connell et al., 2019). Several studies have reported that *CACNA1B* is linked to ASD, MDD, and BD. For example, a monogenic duplication in *CACNA1B* has been linked to Asperger syndrome (Yatsenko et al., 2012). *CACNA1B*, together with *CACNA1C*, has also been associated with suicide risk in patients with MDD (Tombácz et al., 2017).

Little evidence exists of risk variants for *CACNA1E* ($Ca_V2.3$) being associated with psychiatric disorders; however, one study pointed at de novo mutations in *CACNA1E* as risk factors for ASD (O'Roak et al., 2012). Another de novo synonymous variant of *CACNA1E,* predicted to affect the regulation of an exonic splice site, was identified in a patient with ASD (Takata et al., 2016). SNPs in *CACNA1E* have been linked to neuroticism and subjective well-being—both traits related to MDD—, as well as, SSRI response (Lin et al., 2018; Okbay et al., 2016). *CACNA1E* risk variants are also linked to cortical and cerebellar defects in working memory, a condition observed in SCZ patients (Heck et al., 2014).

T-Type Calcium Channel Genes (*CACNA1G, CACNA1H, CACNA1I*)

CACNA1G (Ca$_V$3.1) was originally identified as a part of the chromosomic region 17q11–121, which is a genetic risk for ASD (Cantor et al., 2005). Later *CACNA1G* was confirmed as a novel genetic risk candidate for ASD because SNPs within intron 9 had the strongest association with ASD relative to other genes present in the 17q11–21 region (Strom et al., 2010).

There is some evidence supporting the role of *CACNA1H* (Ca$_V$3.2) variants in the physiopathology of psychiatric disorders. For example, four missense mutations (R212C, R902W, W962C, and A1874V) located throughout the Ca$_V$3.2 channel including voltage sensors of DI-II, the DII P-loop, and C-terminus, respectively, were identified in ASD probands. Although these mutations have been shown to modify Ca$_V$3.2 function, they have low penetrance (Chourasia et al., 2019; Splawski et al., 2006). Using ultradeep sequencing of 78 ASD candidate genes in both cerebellar and cortical tissues, *CACNA1H* variations were found in ASD cases (D'Gama et al., 2015). A few more genetic studies using whole-exome sequencing identified de novo missense variations in *CACNA1H* that are predicted to be disruptive in ASD probands (Takata et al., 2018). All of these studies support that *CACNA1H* is a susceptibility gene for ASD. *CACNA1H* has also been linked to SCZ. One study found two disruptive deletions in Ca$_V$3.2 in patients with SCZ (Purcell et al., 2014), and a second GWAS coupled to meta-analysis in a Swedish population found an association of *CACNA1H* with SCZ (Li et al., 2018).

Several studies have strongly suggested variations in *CACNA1I* (Ca$_V$3.3) as a risk factor for SCZ (El Ghaleb et al., 2021; Irish, 2012; Manoach et al., 2016; Schizophrenia, 2014; Xie et al., 2018; Xu et al., 2018). Two variations (R1346H

and T797M) in the P-loop of Ca$_V$3.3 were identified by exome sequencing of trio samples that included probands, parents, and unaffected siblings when available (Gulsuner et al., 2013). R1346H impairs N-glycosylation of Ca$_V$3.3 channels preventing membrane surface expression and thereby reducing overall calcium currents (Andrade et al., 2016). In addition, noncoding variations in *CACNA1I* have been also identified as risk factors for SCZ in various GWAS studies (Goes et al., 2015; Xie et al., 2018; Xu et al., 2018; Zhang et al., 2018). *CACNA1I* has been associated with cognitive ability and sleep spindle activity, defects in both traits have been observed in patients with SCZ (Lam et al., 2017; Merikanto et al., 2019). Finally, *CACNA1I* has also been linked to neurodevelopmental disorders including ADHD and other conditions with strong cognitive impairment (Dalley & Robbins, 2017; El Ghaleb et al., 2021; Elia et al., 2011; Sanchez-Roige et al., 2019).

Calcium-Dependent Molecular Mechanisms in the Pathology of Psychiatric Disorders

Calcium entering through VGCCs is a well-known second messenger capable of controlling a multitude of cellular processes in neurons, thus the links between VGCC function and psychiatric disorders are becoming clear. In this section, we will discuss the current evidence supporting detailed neurobiological mechanisms involving VGCCs in the pathophysiology of psychiatric disorders. We will focus on studies involving animal models, human-derived cells, and human studies that have used genetic screenings as a foundation.

L-Type Calcium Channels

Role of L-Type Channels in Altered Gene Expression in Psychiatric Disorders

Calcium entering via L-type channels, including Ca$_V$1.2 and Ca$_V$1.3, has been long associated with activity-dependent gene expression, particularly via the cAMP response-element binding protein, CREB (Moosmang et al., 2005; Deisseroth et al., 1998; Dolmetsch et al., 2001). This calcium-dependent regulation of gene expression leads to long-term adaptations to neuronal activity. Over the years, new calcium-dependent pathways that regulate gene expression have been discovered, for example, the mammalian target of the rapamycin (mTOR) pathway (Costa-Mattioli & Monteggia, 2013; Nandagopal & Roux, 2015) and the eukaryotic initiation factor 2α (eIF2α) signaling pathway (Wek & Cavener, 2007). Interestingly, these two pathways have also been involved in the dysregulation of mRNA translation and protein synthesis in dendrites, which have been hypothesized to be an important mechanism underlying behavioral impairments associated with psychiatric disorders (Forrest et al., 2018; Iasevoli et al., 2013; Penzes et al., 2011). More specifically, selective ablation of $Ca_V1.2$ in forebrain glutamatergic neurons lowers protein synthesis concomitant with reduced expression of mTOR complex 1 (mTORC1) (Kabir et al., 2017). In addition, this selective ablation results in increased phosphorylation of eIF2a that leads to repression of mRNA translation (Kabir et al., 2017).

Control of Neuronal Excitability by L-Type Channels in Psychiatric Disorders

L-type channels control neuronal activity by various mechanisms including regulation of firing frequency and synaptic plasticity. L-type channels couple to calcium-dependent potassium channels in different types of central neurons including midbrain dopamine neurons, excitatory hippocampal neurons, and excitatory forebrain neurons to control firing frequency (Marrion & Tavalin, 1998; Durante et al., 2004; Iyer et al., 2017; Puopolo et al., 2007). In addition, calcium entering via L-type channels controls long-term synaptic plasticity in excitatory and inhibitory synapses via their ability to regulate gene expression (Voglis & Tavernarakis, 2006). Thus, L-type channels can have a robust impact on the activity of individual neurons, as well as circuits.

Because of the presence of L-type channels in both excitatory and inhibitory neurons, the activity of these channels can impact the excitatory/inhibitory balance. Small perturbations in the excitatory/inhibitory balance can have strong downstream effects at the molecular, cellular, circuit, and behavioral levels (Sohal & Rubenstein, 2019; Tatti et al., 2017). Evidence supporting the role of L-type channels in controlling excitation/inhibition balance in brain areas that are thought to be affected in psychiatric disorders (e.g., cortex) came from ex vivo studies using L-type channel blockers (Gong et al., 2007). Here, L-type channel blockage resulted in synaptic scaling of forebrain cortical neurons by an increase in excitatory postsynaptic currents (Gong et al., 2007). Furthermore, L-type channel blockage also leads to a reduction in membrane receptors that mediate inhibitory neurotransmission including $GABA_A$ (Saliba et al., 2009). In line with this, glutamatergic neuron-specific forebrain knockout of $Ca_V1.2$ is associated with increased excitation/inhibition balance indicative of increased cortical excitation that underlies behavioral cortical deficits (Kabir et al., 2017).

Role of L-Type Channels in Neurogenesis in the Context of Psychiatric Disorders

Both $Ca_V1.2$ and $Ca_V1.3$ channels have been linked to the generation and/or survival of new neurons in the adult hippocampus of rodents. L-type channels regulate the bidirectional transition between neuronal precursors and adult hippocampal neurons via mechanisms involving CREB signaling, Ca^2/calmodulin-dependent protein kinase II (CaMKII), and brain-derived neurotrophic factor (BDNF) (Deisseroth et al., 1998; Lee et al., 2016; Moosmang et al., 2005; Striessnig et al., 2006; Völkening et al., 2017; Moon et al., 2018). Loss of $Ca_V1.2$ results in reduced survival of newborn hippocampal neurons, whereas loss of $Ca_V1.3$ leads to defects in both proliferation of neuronal progenitor cells as well as survival of newborn hippocampal neurons (Lee et al., 2016; Marschallinger et al., 2015; Bavley et al., 2021; Moon et al., 2020; Temme et al., 2016; Tigaret et al., 2021).

Hippocampal adult neurogenesis has been shown to be key for cognitive function including the formation of spatial memories (Deng et al., 2010; Dupret et al., 2008; Yau et al., 2015), thus defects in adult neurogenesis have been hypothesized to underlie cognitive deficits associated with psychiatric disorders such as SCZ (Apple et al., 2017). Further evidence to support a link between adult neurogenesis and psychiatric diseases arises from the observations that antidepressive drugs such as SSRIs, tricyclic antidepressants, and SNRIs lead to increased neurogenesis in rodent models (Malberg, 2004). All this supports the intriguing hypothesis that L-type channel activity leads to increased hippocampal neurogenesis to compensate for cognitive deficits observed in psychiatric disorders.

Presynaptic Calcium Channels

The predominant localization of P/Q-, N-, and R-type channels in presynaptic terminals in both excitatory and inhibitory synapses in the brain suggests that these channels play critical roles in the excitation/inhibition balance (Dolphin & Lee, 2020). Furthermore, presynaptic calcium channels are primary targets for G-protein coupled receptors that are activated by psychotropic drugs. This together with the identification of genetic risk variants in presynaptic calcium channel genes opens the possibility of targeting these channels to design new treatments for psychiatric disorders.

Presynaptic Control of Transmitter Release and Psychiatric Disorders

$Ca_V2.1$ channels are dominant throughout the central nervous system. $Ca_V2.1$ is the major calcium channels involved in the release of glutamate in central synapses and is the dominant calcium channels in controlling GABA release in specific synapses of the brain like those that originate from parvalbumin-expressing (PV+) interneurons (Takahashi & Momiyama, 1993; Zaitsev et al., 2007). PV+ interneurons are fast-spiking GABAergic neurons that provide strong inhibitory control of cortical and subcortical circuits

(Kawaguchi et al., 1987; Kawaguchi & Kubota, 1997). This interneuron population has attracted a lot of attention because it has been linked to SCZ, ASD, depression, and stress (Ferguson & Gao, 2018; Lewis et al., 2012; Nahar et al., 2021). This evidence has been found in both animal models and humans (Kaar et al., 2019; Lewis et al., 2012; Mukherjee et al., 2019; Filice et al., 2020; Sagi et al., 2020). PV+ interneurons provide a strong regulatory control of the inhibitory/excitatory balance of glutamatergic neurons in cortical and subcortical areas of the brain by preventing feedforward excitation of other excitatory neurons (Ferguson & Gao, 2018). As described above defects in this balance are associated with psychiatric disorders. The strategic localization of $Ca_V2.1$ channels in presynaptic terminals of PV+ interneurons supports a potential role of these channels in psychiatric disorders.

It is well known that $Ca_V2.2$ channels are the dominant presynaptic calcium channels controlling transmitter release in peripheral synapses (Dolphin & Lee, 2020). However, $Ca_V2.2$ channels in the central nervous system are dominant in a few but important neuronal populations for psychiatric disorders. Among these neuronal populations are cholecystokinin-expressing interneurons, dopaminergic, serotoninergic, and adrenergic neurons, thus $Ca_V2.2$ channels effectively control multiple neurotransmitter systems including GABA, dopamine, serotonin, and noradrenaline ones (Brimblecombe et al., 2015; Beuckmann et al., 2003; Lipscombe et al., 1989).

$Ca_V2.2$ channels dominate GABA release from neurons expressing the peptide cholecystokinin (CCK+) in the hippocampus and amygdala (Blazon et al., 2021; Bunda et al., 2019a; Pelkey et al., 2017; Hefft & Jonas, 2005). Interestingly, these neurons have been linked to emotional processing and cognitive functions (Nguyen et al., 2020). CCK+ interneurons are also a major cellular target for cannabinoids and SSRIs (Medrihan et al., 2017). $Ca_V2.2$ channels also control the release of dopamine in synapses from the substantia nigra pars compacta (SNc) to CPu and VTA to NAc (Brimblecombe et al., 2015), areas that have been linked to psychiatric disorders including SCZ and ADHD. All this evidence further aligns with a role of $Ca_V2.2$ channels in psychiatric disorders.

$Ca_V2.3$ channels are activated by smaller depolarizations compared to $Ca_V2.1$ and $Ca_V2.2$ channels (Metz et al., 2005; Randall & Tsien, 1997). $Ca_V2.3$ channels are rarely the dominant presynaptic calcium channels controlling transmitter release, except for specific cholinergic synapses of the habenula (Zhang et al., 2016). Here $Ca_V2.3$ channels are important to regulate fear memory expression by potentiating the release of glutamate, acetylcholine, and neurokinin B thereby exciting interpeduncular neurons (Zhang et al., 2016). $Ca_V2.3$ channels also regulate asynchronous release from small hippocampal boutons and play a role in long-term potentiation (Dietrich et al., 2003; Ermolyuk et al., 2013).

Modulation by GPCRs

The calcium that enters via presynaptic calcium channels, particularly $Ca_V2.1$ and $Ca_V2.2$, is subject to regulation by GPCRs including dopamine receptors 1 and 2 (D1R and D2R), serotonin receptors 2A/1B (5HT2A/B), and adrenergic receptors 2A (A2A) (Albert & Vahid-Ansari, 2019; Boehm & Huck, 1996; Cardozo & Bean, 1995; Kisilevsky et al., 2008; Kisilevsky & Zamponi, 2008; Kolaj & Renaud, 2001; Komatsu, 2015; Lu et al., 2018; Murat et al., 2019; Stępnicki et al., 2018). Interestingly, drugs that target D2R like haloperidol have robust psychotropic effects (Fox et al., 1994), and recently it was observed that the early response to antidepressants is mediated via the control of GABA release by 5-HT1B receptors in synapses of hippocampal interneurons (Medrihan et al., 2017). However, a specific role in the modulation of presynaptic calcium channels by GPCRs in psychiatric disorders has not been established.

T-Type Channels and Control of Firing Frequency

T-type channels open at lower voltages and inactivate faster compared to Ca_V1 and Ca_V2 subfamilies (Perez-Reyes, 2003). Because of these properties, Ca_V3 channels play a major role in controlling the rhythmic firing of action potentials (Lambert et al., 2014; Zamponi et al., 2015). Of the T-type channels, $Ca_V3.3$ controls the rhythmic bursts of action potential in relay neurons in the thalamus (Suzuki & Rogawski, 1989). The rhythmic activity of thalamic relay neurons is thought to originate from sleep spindles (waves of electrical activity observed during sleep) (Steriade et al., 1985; Thankachan et al., 2019; Halassa et al., 2011). These spindles have been shown to be altered in patients with SCZ (Ferrarelli & Tononi, 2011; Ferrarelli et al., 2007; Manoach et al., 2016; Manoach & Stickgold, 2019). More broadly, defects in sleep have been observed across multiple psychiatric disorders (Benca et al., 1992; Krystal, 2012; Riemann et al., 2020). Thus, T-type channels are becoming an increasingly attractive target to treat these conditions.

Behavioral Roles of VGCCS in the Context of Psychiatric Disorders

The major consequences of psychiatric disorders are altered patterns of behavior that can result in impairments to carry out vital functions. A few of these patterns of behavior can be characterized as endophenotypes, which are quantitative neurobehavioral traits associated with a disorder, are reasonably heritable, cosegregate with the disease, and are independent of the clinical status of the disorder (Flint & Munafò, 2007). For example, deficits in working memory have been observed in patients with SCZ (Huang et al., 2019; Lett et al., 2014). Many of the endophenotypes associated with psychiatric disorders can be extremely difficult to dissect at the cellular and molecular levels in humans. Thus, the development of animal models capable of emulating defined behavioral traits associated with psychiatric disorders, with the concomitant careful considerations of their limitations, can provide robust insights into the neurobiology and treatment for psychiatric disorders (Baker et al., 2020; Kokras & Dalla, 2014; Nestler & Hyman, 2010).

Based on research domain criteria (RDoC), the four main constructs used to study behavioral models in the context of psychiatric disorders are negative valence systems, positive valence systems, cognitive systems, and social processes. In this section, we will discuss human studies that have linked genetic risk variants in VGCCs with endophenotypes. Next, we will review relevant work related to the design of animal models with their corresponding behavioral characterization. These models have provided insights into the neurobiological basis of psychiatric disorders.

VGCCs in Human Endophenotypes Associated with Psychiatric Disorders

The SNP rs1006737 in between exons 3 and 4 of *CACNA1C* has been identified in multiple studies as a genetic risk variant for BD (Ferreira et al., 2008; Bigos et al., 2010). Furthermore, studies of human carriers of rs1006737 have shown its association with the ability to recognize emotions based on facial expressions (facial emotion recognition) in both healthy individuals and patients with BD (Nieratschker et al., 2015; Soeiro-de-Souza et al., 2012). Facial emotion recognition endophenotypes have been observed in BD patients (Ruihua et al., 2021). Two other endophenotypes have been associated with rs1006737, baseline and affective startle, and reversal learning (Pasparakis et al., 2015; Sykes et al., 2019). In startle, individuals with the risk A allele for rs1006737 showed heightened contextual sensitivity, affective reactivity, and inefficient emotional appraisal (Pasparakis et al., 2015; Sykes et al., 2019). These behavioral observations align with increased anxiety and depression in BD patients. Individuals with rs100637 also showed impaired reversal learning (suppression of a learned association between reward and stimulus, but also the acquisition of a new opposite association) (Sykes et al., 2019).

Alterations in sleep have been strongly linked to SCZ, BD, and MDD. Some of these sleep-related endophenotypes have been linked to VGCCs. For example, SNPs in *CACNA1C* have been associated with narcolepsy as well as sleep latency and duration (Jansen et al., 2019; Shimada et al., 2010). Furthermore, variants of $Ca_V3.3$ have been associated with sleep spindles (Andrade et al., 2016; Ghoshal et al., 2020). The rs 4,652,676 in *CACNA1E* has been linked to neuroticism and subjective well-being, both endophenotypes associated with MDD (Okbay et al., 2016). Finally, the SNP rs 199,694,726 in *CACNA1I* has been linked to impulsive behavior under extreme negative emotions, which is commonly associated with ADHD (Dalley & Robbins, 2017).

VGCCs in Behavioral Assessments of Animal Models

VGCCs in Negative Valence Systems

Negative valence systems are responsible for the behavioral responses to aversive situations including fear, anxiety, and depression. Altered fear responses and anxiety have been widely observed in psychiatric disorders (Kaufman & Charney, 2000; Mineka et al., 1998). Animal models for fear and anxiety are limited, however, they are extremely useful to initiate the investigation of the behavioral consequences of genetic risk variants for psychiatric disorders. Among the tests that are most commonly used to model negative valence systems in rodents are those based on the reluctance of rodents to explore new/open environments (e.g., open field, elevated plus maze, light-dark conflict test, and novelty-induced hypophagia test).

In recent years, scientists have developed sophisticated models to establish links between VGCC function and behavior. For example, heterozygous for $Ca_V1.2$ knockout mice shows reduced exploratory behavior in various tests (Dao et al., 2010; Dedic et al., 2018; Jaric et al., 2019; Kabir et al., 2017; Kabitzke et al., 2018; Lee et al., 2012). Furthermore, glutamatergic-specific ablation of $Ca_V1.2$ in the forebrain and prefrontal cortex showed similar results, suggesting that $Ca_V1.2$ channels in these sites are key for exploratory behavior, and likely anxiety (Kabir et al., 2017; Lee et al., 2012).

In addition to Ca$_V$1.2, other VGCCs have been studied at the behavioral level. Deletion of Ca$_V$2.1 in the forebrain principal results in increased exploratory behavior (Mallmann et al., 2013). Global knockout of Ca$_V$2.2 channels leads to increased exploratory behavior, reduced startle, and enhanced aggression (Beuckmann et al., 2003; Nakagawasai et al., 2010; Saegusa et al., 2001). Furthermore, the specific ablation of Ca$_V$2.2 splice variants leads to enhanced exploratory behavior in both baseline and aversive conditions (Bunda et al., 2019b). Similar results have been observed in global Ca$_V$2.3 and Ca$_V$3.2 knockout mice (Kubota et al., 2001). All these studies combined with genetic screenings in patients with psychiatric disorders support a potential role of Ca$_V$2.1, Ca$_V$2.2, Ca$_V$2.3, and Ca$_V$3.2 in negative valence systems related to the processing of threats.

Similar to anxiety, depression is often observed in patients with psychiatric disorders including but not limited to BD and MDD (Buckley et al., 2009; Gorman, 1996). Depression is linked to altered hedonic responses and impaired responses under life-threatening conditions. In rodents, these responses have been modeled with a sucrose preference test (hedonic response), tail suspension test (TST), and forced-swim test (FST; survival responses). VGCCs have been linked to this type of behavioral response in rodents. Loss of Ca$_V$1.2 and Ca$_V$1.3 channels results in reduced immobility in both tail suspension and forced swim tests (antidepressive related behavior) (Busquet et al., 2010; Dao et al., 2010; Kabir et al., 2017). More specifically, deletion of Ca$_V$1.2 in glutamatergic neurons of the forebrain resulted in similar observations to the ones in Ca$_V$1.2 heterozygous mice (Kabir et al., 2017). Selective stimulation of Ca$_V$1.3 led to increased immobility in the forced swim test (Sinnegger-Brauns et al., 2004). The role of Ca$_V$2.2 channels has been also tested in TST and FST by lateral ventricle infusion of the specific inhibitor, ω-conotoxin GVIA (ω-ctx GVIA). Here, mice infused with ω-ctx GVIA showed increased immobility in both TST and FST, suggesting a role of these channels in behavioral responses necessary for survival in rodents (Zhou et al., 2015).

VGCCs in Positive Valence Systems

Positive valence systems are related to behavioral responses to positive situations like reward seeking, consummatory behavior, and reward/habit formation. Alterations in positive valence systems have been observed in psychiatric disorders. For example, substance use disorders (addiction) are a comorbidity with BD and MDD (Kessler et al., 2009; Post & Kalivas, 2013). There is a strong body of evidence suggesting a link between VGCCs and altered behavioral responses to addictive substances. L-type channel blockers modify the responses of rodents to addictive substances including alcohol, cocaine, morphine, and nicotine, these responses include decreased self-administration, sensitization, and reduced withdrawal symptoms following chronic intake (Biała & Langwiński, 1996; Gatch, 2002; Kuzmin et al., 1992; Pierce et al., 1998; Reimer & Martin-Iverson, 1994; Uhrig et al., 2017; Whittington et al., 1991; Zharkovsky et al., 1993). These observations nicely align with the association of the SNP rs1034936 *CACNA1C* with alcohol and lifetime cocaine abuse in patients with BD (Mosheva et al., 2020). Furthermore, Ca$_V$1.2 channels in the prefrontal cortex circuits are key for cocaine- and stress-primed reinstatement of cocaine-conditioned place preference (Bavley et al., 2020). Ca$_V$1.3 channels also play roles in cocaine dependence (Martínez-Rivera et al., 2017). Activation of Ca$_V$1.3 channels in dopaminergic neurons of VTA is key for acquisition of cocaine conditioned place preference and cocaine sensitization (Martínez-Rivera et al., 2017; Schierberl et al., 2011) and drive depressive-like behaviors (Martínez-Rivera et al., 2017). Despite the clear dominance of N-type channels in regulating dopamine release in NAc (Brimblecombe et al., 2015), very few studies exist that link these presynaptic calcium channels with addictive behaviors. However, reduced alcohol consumption and development of place preference at low ethanol doses were observed in Ca$_V$2.2 knockout mice (Newton et al., 2004).

These studies suggest that L-type and N-type channels might play a role in addiction, which might be related to their functional role in circuits involved in addictive behavior in both humans and rodents.

VGCCs in Systems for Social Processes

Systems for social processes mediate responses to interpersonal settings including perception and interpretation of others' actions. Alterations in systems for social processes are extensively observed in psychiatric disorders particularly ASD (Kennedy & Adolphs, 2012). L-type channels have been linked to social behavior in rodents. Genetic mouse models with a gain-of-function mutation in $Ca_V1.2$ that causes TS result in reduced interaction with a social object and reduced social vocalizations (Bader et al., 2011). Rats with *Cacna1c* haploinsufficiency show deficits in prosocial ultrasonic communication (Kisko et al., 2018, 2020; Redecker et al., 2019). Ultrasonic communication is an important modality of social communication in rodents (Engelhardt et al., 2018; Knutson et al., 1998). $Ca_V1.2$ channels in glutamatergic forebrain neurons also play key roles in social behavioral tasks in mice (Kabir et al., 2017). Pharmacological activation of $Ca_V1.3$ induces social deficits in mice (Martínez-Rivera et al., 2017). These observations align with recent findings linking gain-of-function mutations in *CACNA1D* with ASD (Hofer et al., 2020; Limpitikul et al., 2016; Pinggera et al., 2015, 2017, 2018). Few studies exist linking Ca_V2 or Ca_V3 channels to social behavior. However, $Ca_V2.2$ knockout mice exhibit a strong aggressive phenotype (Kim et al., 2009), a trait that can be linked to psychiatric disorders (Nelson & Trainor, 2007).

VGCCs in Cognitive Systems

Cognitive systems are widely diverse and are composed of those responsible for attention, perception, declarative memory, language, cognitive control, and working memory. Interestingly, impairments in attention, executive function, and working memory have been observed in psychiatric disorders, particularly MDD, ADHD, and

SCZ (Heinrichs, 2005; Lieberman et al., 2001). Several behavioral tasks have been developed to assess cognitive processes in rodents, these include the Morris Water Maze test, Y-mase test, and fear conditioning tasks. Various genetic models for VGCCs have been tested on these behavioral tasks. Neuronal-specific loss of $Ca_V1.2$ leads to deficits in hippocampal-dependent context discrimination and acquisition of spatial memory (Moosmang et al., 2005; Temme et al., 2016). Furthermore, mice harboring gain-of-function $Ca_V1.2$ mutations that cause TS showed defects in reversal learning in the Morris Water Maze Test (Bader et al., 2011). Interestingly, the behavioral deficits are not observed during the acquisition phase, but rather are prompted when the experimental conditions change. These changes required certain cognitive flexibility, and a lack of cognitive flexibility has been observed in patients with SCZ and ASD (D'Cruz et al., 2013; Geurts et al., 2009; Waltz, 2017).

In addition to $Ca_V1.2$ channels, $Ca_V2.1$, $Ca_V2.3$, and $Ca_V3.2$ have been linked to cognitive function. Neuron-specific forebrain deletion of $Ca_V2.1$ channels results in deficits in spatial learning and reference memory, and reduced recognition memory (Mallmann et al., 2013). Global $Ca_V2.3$ knockout mice show impaired spatial memory in the MWM test (Kubota et al., 2001). Finally, $Ca_V3.2$ knockout mice show defects in recognition memory in the novel object recognition test (Gangarossa et al., 2014). $Ca_V3.2$ knockout mice also showed impairments in context-cued trace fear conditioning, step-down, and step-through passive avoidance tasks, suggesting defects in hippocampus-dependent memory (Gangarossa et al., 2014). All these studies support a role of these VGCCs in cognitive processes.

Targeting VGCCs to Treat Psychiatric Disorders

Understanding the role of VGCCs in the neurobiology of psychiatric disorders is providing insights into novel treatments. For example, repurposing of drugs approved by the Federal

Drug Administration that target VGCCs for the treatment of psychiatric disorders is promising. In this section, we will review recent advancements in the development of treatments for psychiatric disorders that focus on VGCCs. Next, we will discuss some of the challenges in targeting VGCCs and promising strategies.

L-type channel blockers like dihydropyridines (e.g., nifedipine, amlodipine, isradipine, nicardipine) and nondihydropyridines (e.g., verapamil, diltiazem) have been widely used to treat hypertension and cardiac arrhythmias because these channels control the contraction of vascular and cardiac smooth muscles. However, given the strong genetic evidence linking L-type channels to psychiatric disorders, the idea of repurposing L-type channel blockers has gained popularity. Drug repurposing is commonly used in pharmaceutics for drugs with known mechanisms. This offers the advantages of higher rates of success, less years to impact, lower costs, and better-known safety profiles when compared to novel compounds (Oprea & Mestres, 2012; Pushpakom et al., 2019).

L-type channel blockers in the context of psychiatric disorders were initially tested to alleviate symptoms of BD. For example, verapamil, which also targets T-type channels (Bergson et al., 2011), was evaluated to attenuate manic episodes but produced mixed effects (Cipriani et al., 2016; Gitlin & Weiss, 1984; Goodnick, 1996; Höschl & Kozený, 1989; Lenzi et al., 1995; Pazzaglia et al., 1998). However, L-type channel blockers have been shown to be effective in combination with mood stabilizers and antipsychotic drugs (Grunze et al., 1996; Lenzi et al., 1995; Mallinger et al., 2008). In SCZ, verapamil demonstrated promising results in reducing psychosis, and in improving positive and negative symptoms, anxiety, and depression (Bartko et al., 1991; Price, 1987; Price & Pascarzi, 1987). After these mixed results, interest in L-type channel blockers dwindled, but due to a better understanding of the neurobiology of psychiatric disorders, potential therapies that focus on L-type channel blockers are gaining interest. Regarding the other VGCCs involved in psychiatric disorders (Ca_V2s and Ca_V3s), their pharmacology is still in its infancy compared to

the pharmacological tools available for L-type channels.

A clear challenge in the utilization of drugs that target VGCCs to treat psychiatric disorders is the broad expression of these channels. In addition to the brain, L-type channels are also expressed in the cardiovascular system. Thus, drugs that target L-type channels to treat psychiatric disorders would need to avoid L-type channels that are expressed in the heart and smooth muscle. A possibility to overcome this barrier lies in alternative splicing. This cell-specific process allows for the differential expression of splice variants in the brain and peripheral tissues, thus presenting an opportunity for the design of drugs that target VGCCs in the brain (Clark et al., 2020; Hu et al., 2017; Lipscombe et al., 2013a, b; Lipscombe & Andrade, 2015; Clark et al., 2020; Hall & Tunbridge, 2021). However, our understanding of alternative splicing is still nascent; nonetheless, progress has been made to functionally characterize splice variants for VGCC genes in multiple cells and tissues.

Conclusions

In conclusion, there is a strong body of evidence arising from genetic studies, animal models, and human behavioral studies that link VGCC genes to psychiatric disorders. Therefore, VGCCs represent promising molecules to study neurobiological mechanisms of disease and to develop novel treatments for psychiatric disorders. The discovery of risk genetic variants in VGCCs in humans via genome-wide association studies has allowed the creation of robust animal models to dissect molecular mechanisms associated with psychiatric disorders. Furthermore, the availability of well-known drugs that target VGCCs makes these channels attractive substrates for future clinical studies. Despite the amount of evidence linking VGCCs to psychiatric disorders, there is still progress to be made to target specifically the VGCCs in the brain without affecting the ones located in the periphery. However, as sequencing, biochemistry and pharmacological tools evolve, we will be able to create specific drugs to target VGCCs in specific areas of the nervous system.

References

Albert, P. R., & Vahid-Ansari, F. (2019). The 5-HT1A receptor: Signaling to behavior. *Biochimie, 161,* 34–45.

Ament, S. A., Szelinger, S., Glusman, G., Ashworth, J., Hou, L., Akula, N., et al. (2015). Rare variants in neuronal excitability genes influence risk for bipolar disorder. *Proceedings of the National Academy of Sciences of the United States of America, 112,* 3576–3581.

Andrade, A., Hope, J., Allen, A., Yorgan, V., Lipscombe, D., & Pan, J. Q. (2016). A rare schizophrenia risk variant of CACNA1I disrupts $Ca_V3.3$ channel activity. *Scientific Reports, 6,* 34233.

Andrade, A., Brennecke, A., Mallat, S., Brown, J., Gomez-Rivadeneira, J., Czepiel, N., et al. (2019). Genetic associations between voltage-gated calcium channels and psychiatric disorders. *International Journal of Molecular Sciences, 20,* E3537.

Apple, D. M., Fonseca, R. S., & Kokovay, E. (2017). The role of adult neurogenesis in psychiatric and cognitive disorders. *Brain Research, 1655,* 270–276.

Association, A. P. (2000). *Diagnostic and statistical manual of mental disorders.*

Astori, S., Wimmer, R. D., Prosser, H. M., Corti, C., Corsi, M., Liaudet, N., et al. (2011). The Ca(V)3.3 calcium channel is the major sleep spindle pacemaker in thalamus. *Proceedings of the National Academy of Sciences of the United States of America, 108,* 13823–13828.

Bader, P. L., Faizi, M., Kim, L. H., Owen, S. F., Tadross, M. R., Alfa, R. W., et al. (2011). Mouse model of Timothy syndrome recapitulates triad of autistic traits. *Proceedings of the National Academy of Sciences of the United States of America, 108,* 15432–15437.

Baker, M., Hong, S. I., Kang, S., & Choi, D. S. (2020). Rodent models for psychiatric disorders: Problems and promises. *Laboratory Animal Research, 36,* 9.

Bartko, G., Horvath, S., Zador, G., & Frecska, E. (1991). Effects of adjunctive verapamil administration in chronic schizophrenic patients. *Progress in Neuro-Psychopharmacology & Biological Psychiatry, 15,* 343–349.

Bavley, C. C., Fetcho, R. N., Burgdorf, C. E., Walsh, A. P., Fischer, D. K., Hall, B. S., et al. (2020). Cocaine- and stress-primed reinstatement of drug-associated memories elicit differential behavioral and frontostriatal circuit activity patterns via recruitment of L-type Ca^{2+} channels. *Molecular Psychiatry, 25,* 2373–2391.

Bavley, C. C., Kabir, Z. D., Walsh, A. P., Kosovsky, M., Hackett, J., Sun, H., et al. (2021). Dopamine D1R-neuron cacna1c deficiency: A new model of extinction therapy-resistant post-traumatic stress. *Molecular Psychiatry, 26,* 2286–2298.

Benca, R. M., Obermeyer, W. H., Thisted, R. A., & Gillin, J. C. (1992). Sleep and psychiatric disorders. A meta-analysis. *Archives of General Psychiatry, 49,* 651–668; discussion 669.

Bergson, P., Lipkind, G., Lee, S. P., Duban, M. E., & Hanck, D. A. (2011). Verapamil block of T-type calcium channels. *Molecular Pharmacology, 79,* 411–419.

Berridge, M. J. (2014). Calcium signalling and psychiatric disease: Bipolar disorder and schizophrenia. *Cell and Tissue Research, 357,* 477–492.

Beuckmann, C. T., Sinton, C. M., Miyamoto, N., Ino, M., & Yanagisawa, M. (2003). N-type calcium channel alpha1B subunit (Cav2.2) knock-out mice display hyperactivity and vigilance state differences. *The Journal of Neuroscience, 23,* 6793–6797.

Biała, G., & Langwiński, R. (1996). Effects of calcium channel antagonists on the reinforcing properties of morphine, ethanol and cocaine as measured by place conditioning. *Journal of Physiology and Pharmacology, 47,* 497–502.

Bigos, K. L., Mattay, V. S., Callicott, J. H., Straub, R. E., Vakkalanka, R., Kolachana, B., et al. (2010). Genetic variation in CACNA1C affects brain circuitries related to mental illness. *Archives of General Psychiatry, 67,* 939–945.

Blazon, M., LaCarubba, B., Bunda, A., Czepiel, N., Mallat, S., Londrigan, L., et al. (2021). N-type calcium channels control GABAergic transmission in brain areas related to fear and anxiety. *OBM Neurobiology, 5.* https://doi.org/10.21926/obm.neurobiol.2101083

Boehm, S., & Huck, S. (1996). Inhibition of N-type calcium channels: The only mechanism by which presynaptic alpha 2-autoreceptors control sympathetic transmitter release. *The European Journal of Neuroscience, 8,* 1924–1931.

Bojarski, L., Debowska, K., & Wojda, U. (2010). In vitro findings of alterations in intracellular calcium homeostasis in schizophrenia. *Progress in Neuro-Psychopharmacology & Biological Psychiatry, 34,* 1367–1374.

Brimblecombe, K. R., Gracie, C. J., Platt, N. J., & Cragg, S. J. (2015). Gating of dopamine transmission by calcium and axonal N-, Q-, T- and L-type voltage-gated calcium channels differs between striatal domains. *The Journal of Physiology, 593,* 929–946.

Buckley, P. F., Miller, B. J., Lehrer, D. S., & Castle, D. J. (2009). Psychiatric comorbidities and schizophrenia. *Schizophrenia Bulletin, 35,* 383–402.

Bunda, A., LaCarubba, B., Akiki, M., & Andrade, A. (2019a). Tissue- and cell-specific expression of a splice variant in the II-III cytoplasmic loop of Cacna1b. *FEBS Open Bio, 9,* 1603–1616.

Bunda, A., LaCarubba, B., Bertolino, M., Akiki, M., Bath, K., Lopez-Soto, J., et al. (2019b). Cacna1b alternative splicing impacts excitatory neurotransmission and is linked to behavioral responses to aversive stimuli. *Molecular Brain, 12,* 81.

Burmeister, M., McInnis, M. G., & Zöllner, S. (2008). Psychiatric genetics: Progress amid controversy. *Nature Reviews. Genetics, 9*, 527–540.

Busquet, P., Nguyen, N. K., Schmid, E., Tanimoto, N., Seeliger, M. W., Ben-Yosef, T., et al. (2010). CaV1.3 L-type Ca2+ channels modulate depression-like behaviour in mice independent of deaf phenotype. *The International Journal of Neuropsychopharmacology, 13*, 499–513.

Cantor, R. M., Kono, N., Duvall, J. A., Alvarez-Retuerto, A., Stone, J. L., Alarcón, M., et al. (2005). Replication of autism linkage: Fine-mapping peak at 17q21. *American Journal of Human Genetics, 76*, 1050–1056.

Cardozo, D. L., & Bean, B. P. (1995). Voltage-dependent calcium channels in rat midbrain dopamine neurons: Modulation by dopamine and GABAB receptors. *Journal of Neurophysiology, 74*, 1137–1148.

Carman, J. S., & Wyatt, R. J. (1979). Calcium: Bivalent cation in the bivalent psychoses. *Biological Psychiatry, 14*, 295–336.

Casamassima, F., Hay, A. C., Benedetti, A., Lattanzi, L., Cassano, G. B., & Perlis, R. H. (2010). L-type calcium channels and psychiatric disorders: A brief review. *American Journal of Medical Genetics. Part B, Neuropsychiatric Genetics, 153B*, 1373–1390.

Chourasia, N., Ossó-Rivera, H., Ghosh, A., Von Allmen, G., & Koenig, M. K. (2019). Expanding the phenotypic spectrum of CACNA1H mutations. *Pediatric Neurology, 93*, 50–55.

Cipriani, A., Saunders, K., Attenburrow, M. J., Stefaniak, J., Panchal, P., Stockton, S., et al. (2016). A systematic review of calcium channel antagonists in bipolar disorder and some considerations for their future development. *Molecular Psychiatry, 21*, 1324–1332.

Clark, M. B., Wrzesinski, T., Garcia, A. B., Hall, N. A. L., Kleinman, J. E., Hyde, T., et al. (2020). Long-read sequencing reveals the complex splicing profile of the psychiatric risk gene CACNA1C in human brain. *Molecular Psychiatry, 25*, 37–47.

Costa-Mattioli, M., & Monteggia, L. M. (2013). mTOR complexes in neurodevelopmental and neuropsychiatric disorders. *Nature Neuroscience, 16*, 1537–1543.

Cross-Disorder, G. O. T. P. G. C., Lee, S. H., Ripke, S., Neale, B. M., Faraone, S. V., Purcell, S. M., et al. (2013). Genetic relationship between five psychiatric disorders estimated from genome-wide SNPs. *Nature Genetics, 45*, 984–994.

D'Cruz, A. M., Ragozzino, M. E., Mosconi, M. W., Shrestha, S., Cook, E. H., & Sweeney, J. A. (2013). Reduced behavioral flexibility in autism spectrum disorders. *Neuropsychology, 27*, 152–160.

D'Gama, A. M., Pochareddy, S., Li, M., Jamuar, S. S., Reiff, R. E., Lam, A. N., et al. (2015). Targeted DNA sequencing from autism spectrum disorder brains implicates multiple genetic mechanisms. *Neuron, 88*, 910–917.

Dalley, J. W., & Robbins, T. W. (2017). Fractionating impulsivity: Neuropsychiatric implications. *Nature Reviews. Neuroscience, 18*, 158–171.

Dao, D. T., Mahon, P. B., Cai, X., Kovacsics, C. E., Blackwell, R. A., Arad, M., et al. (2010). Mood disorder susceptibility gene CACNA1C modifies mood-related behaviors in mice and interacts with sex to influence behavior in mice and diagnosis in humans. *Biological Psychiatry, 68*, 801–810.

De Rubeis, S., He, X., Goldberg, A. P., Poultney, C. S., Samocha, K., Cicek, A. E., et al. (2014). Synaptic, transcriptional and chromatin genes disrupted in autism. *Nature, 515*, 209–215.

Dedic, N., Pöhlmann, M. L., Richter, J. S., Mehta, D., Czamara, D., Metzger, M. W., et al. (2018). Cross-disorder risk gene CACNA1C differentially modulates susceptibility to psychiatric disorders during development and adulthood. *Molecular Psychiatry, 23*, 533–543.

Deisseroth, K., Heist, E. K., & Tsien, R. W. (1998). Translocation of calmodulin to the nucleus supports CREB phosphorylation in hippocampal neurons. *Nature, 392*, 198–202.

Deng, W., Aimone, J. B., & Gage, F. H. (2010). New neurons and new memories: How does adult hippocampal neurogenesis affect learning and memory. *Nature Reviews. Neuroscience, 11*, 339–350.

Dietrich, D., Kirschstein, T., Kukley, M., Pereverzev, A., von der Brelie, C., Schneider, T., et al. (2003). Functional specialization of presynaptic Cav2.3 Ca2+ channels. *Neuron, 39*, 483–496.

Dolmetsch, R. E., Pajvani, U., Fife, K., Spotts, J. M., & Greenberg, M. E. (2001). Signaling to the nucleus by an L-type calcium channel-calmodulin complex through the MAP kinase pathway. *Science, 294*, 333–339.

Dolphin, A. C., & Lee, A. (2020). Presynaptic calcium channels: Specialized control of synaptic neurotransmitter release. *Nature Reviews. Neuroscience, 21*, 213–229.

Dubovsky, S. L., & Franks, R. D. (1983). Intracellular calcium ions in affective disorders: A review and an hypothesis. *Biological Psychiatry, 18*, 781–797.

Dubovsky, S. L., Thomas, M., Hijazi, A., & Murphy, J. (1994). Intracellular calcium signalling in peripheral cells of patients with bipolar affective disorder. *European Archives of Psychiatry and Clinical Neuroscience, 243*, 229–234.

Dupret, D., Revest, J. M., Koehl, M., Ichas, F., De Giorgi, F., Costet, P., et al. (2008). Spatial relational memory requires hippocampal adult neurogenesis. *PLoS One, 3*, e1959.

Durante, P., Cardenas, C. G., Whittaker, J. A., Kitai, S. T., & Scroggs, R. S. (2004). Low-threshold L-type calcium channels in rat dopamine neurons. *Journal of Neurophysiology, 91*, 1450–1454.

El Ghaleb, Y., Schneeberger, P. E., Fernández-Quintero, M. L., Geisler, S. M., Pelizzari, S., Polstra, A. M., et al. (2021). CACNA1I gain-of-function mutations differentially affect channel gating and cause neurodevelopmental disorders. *Brain, 144*, 2092–2106.

Elia, J., Glessner, J. T., Wang, K., Takahashi, N., Shtir, C. J., Hadley, D., et al. (2011). Genome-wide copy number variation study associates metabotropic glutamate receptor gene networks with attention deficit hyperactivity disorder. *Nature Genetics, 44*, 78–84.

Emamghoreishi, M., Schlichter, L., Li, P. P., Parikh, S., Sen, J., Kamble, A., et al. (1997). High intracellular calcium concentrations in transformed lymphoblasts from subjects with bipolar I disorder. *The American Journal of Psychiatry, 154*, 976–982.

Engelhardt, K., Schwarting, R. K. W., & Wöhr, M. (2018). Mapping trait-like socio-affective phenotypes in rats through 50-kHz ultrasonic vocalizations. *Psychopharmacology, 235*, 83–98.

Ermolyuk, Y. S., Alder, F. G., Surges, R., Pavlov, I. Y., Timofeeva, Y., Kullmann, D. M., et al. (2013). Differential triggering of spontaneous glutamate release by P/Q-, N- and R-type Ca2+ channels. *Nature Neuroscience, 16*, 1754–1763.

Ferguson, B. R., & Gao, W. J. (2018). PV interneurons: Critical regulators of E/I balance for prefrontal cortex-dependent behavior and psychiatric disorders. *Frontiers in Neural Circuits, 12*, 37.

Ferrarelli, F., & Tononi, G. (2011). The thalamic reticular nucleus and schizophrenia. *Schizophrenia Bulletin, 37*, 306–315.

Ferrarelli, F., Huber, R., Peterson, M. J., Massimini, M., Murphy, M., Riedner, B. A., et al. (2007). Reduced sleep spindle activity in schizophrenia patients. *The American Journal of Psychiatry, 164*, 483–492.

Ferreira, M. A., O'Donovan, M. C., Meng, Y. A., Jones, I. R., Ruderfer, D. M., Jones, L., et al. (2008). Collaborative genome-wide association analysis supports a role for ANK3 and CACNA1C in bipolar disorder. *Nature Genetics, 40*, 1056–1058.

Filice, F., Janickova, L., Henzi, T., Bilella, A., & Schwaller, B. (2020). The parvalbumin hypothesis of autism spectrum disorder. *Frontiers in Cellular Neuroscience, 14*, 577525.

Flint, J., & Munafò, M. R. (2007). The endophenotype concept in psychiatric genetics. *Psychological Medicine, 37*, 163–180.

Forrest, M. P., Parnell, E., & Penzes, P. (2018). Dendritic structural plasticity and neuropsychiatric disease. *Nature Reviews. Neuroscience, 19*, 215–234.

Fox, C. A., Mansour, A., & Watson, S. J. (1994). The effects of haloperidol on dopamine receptor gene expression. *Experimental Neurology, 130*, 288–303.

Gangarossa, G., Laffray, S., Bourinet, E., & Valjent, E. (2014). T-type calcium channel Cav3.2 deficient mice show elevated anxiety, impaired memory and reduced sensitivity to psychostimulants. *Frontiers in Behavioral Neuroscience, 8*, 92.

Gatch, M. B. (2002). Nitrendipine blocks the nociceptive effects of chronically administered ethanol. *Alcoholism, Clinical and Experimental Research, 26*, 1181–1187.

Geurts, H. M., Corbett, B., & Solomon, M. (2009). The paradox of cognitive flexibility in autism. *Trends in Cognitive Sciences, 13*, 74–82.

Ghoshal, A., Uygun, D. S., Yang, L., McNally, J. M., Lopez-Huerta, V. G., Arias-Garcia, M. A., et al.

(2020). Effects of a patient-derived de novo coding alteration of CACNA1I in mice connect a schizophrenia risk gene with sleep spindle deficits. *Translational Psychiatry, 10*, 29.

Gitlin, M. J., & Weiss, J. (1984). Verapamil as maintenance treatment in bipolar illness: A case report. *Journal of Clinical Psychopharmacology, 4*, 341–343.

Glen, A. I. (1985). Lithium prophylaxis of recurrent affective disorders. *Journal of Affective Disorders, 8*, 259–265.

Goes, F. S., McGrath, J., Avramopoulos, D., Wolyniec, P., Pirooznia, M., Ruczinski, I., et al. (2015). Genome-wide association study of schizophrenia in Ashkenazi Jews. *American Journal of Medical Genetics. Part B, Neuropsychiatric Genetics, 168*, 649–659.

Gong, B., Wang, H., Gu, S., Heximer, S. P., & Zhuo, M. (2007). Genetic evidence for the requirement of adenylyl cyclase 1 in synaptic scaling of forebrain cortical neurons. *The European Journal of Neuroscience, 26*, 275–288.

Goodnick, P. J. (1996). Treatment of mania: Relationship between response to verapamil and changes in plasma calcium and magnesium levels. *Southern Medical Journal, 89*, 225–226.

Gorman, J. M. (1996). Comorbid depression and anxiety spectrum disorders. *Depression and Anxiety, 4*, 160–168.

Green, E. K., Grozeva, D., Jones, I., Jones, L., Kirov, G., Caesar, S., et al. (2010). The bipolar disorder risk allele at CACNA1C also confers risk of recurrent major depression and of schizophrenia. *Molecular Psychiatry, 15*, 1016–1022.

Grunze, H., Walden, J., Wolf, R., & Berger, M. (1996). Combined treatment with lithium and nimodipine in a bipolar I manic syndrome. *Progress in Neuro-Psychopharmacology & Biological Psychiatry, 20*, 419–426.

Gulsuner, S., Walsh, T., Watts, A. C., Lee, M. K., Thornton, A. M., Casadei, S., et al. (2013). Spatial and temporal mapping of de novo mutations in schizophrenia to a fetal prefrontal cortical network. *Cell, 154*, 518–529.

Gulsuner, S., Stein, D. J., Susser, E. S., Sibeko, G., Pretorius, A., Walsh, T., et al. (2020). Genetics of schizophrenia in the South African Xhosa. *Science, 367*, 569–573.

Halassa, M. M., Siegle, J. H., Ritt, J. T., Ting, J. T., Feng, G., & Moore, C. I. (2011). Selective optical drive of thalamic reticular nucleus generates thalamic bursts and cortical spindles. *Nature Neuroscience, 14*, 1118–1120.

Hall, N. A. L., & Tunbridge, E. M. (2021). Brain-enriched CACNA1C isoforms as novel, selective targets for psychiatric indications. *Neuropsychopharmacology, 47*(1), 393–394.

Hamshere, M. L., Walters, J. T., Smith, R., Richards, A. L., Green, E., Grozeva, D., et al. (2013). Genome-wide significant associations in schizophrenia to ITIH3/4, CACNA1C and SDCCAG8, and extensive replication of associations reported by the Schizophrenia PGC. *Molecular Psychiatry, 18*, 708–712.

Harrison, P. J., Geddes, J. R., & Tunbridge, E. M. (2018). The emerging neurobiology of bipolar disorder. *Trends in Neurosciences, 41*, 18–30.

Harrison, P. J., Hall, N., Mould, A., Al-Juffali, N., & Tunbridge, E. M. (2019). Cellular calcium in bipolar disorder: Systematic review and meta-analysis. *Molecular Psychiatry, 26*(8), 4106–4116.

Health, N. S. O. D. U. A. (2019). *Substance abuse and mental service administration.*

Heck, A., Fastenrath, M., Ackermann, S., Auschra, B., Bickel, H., Coynel, D., et al. (2014). Converging genetic and functional brain imaging evidence links neuronal excitability to working memory, psychiatric disease, and brain activity. *Neuron, 81*, 1203–1213.

Hefft, S., & Jonas, P. (2005). Asynchronous GABA release generates long-lasting inhibition at a hippocampal interneuron-principal neuron synapse. *Nature Neuroscience, 8*, 1319–1328.

Heinrichs, R. W. (2005). The primacy of cognition in schizophrenia. *The American Psychologist, 60*, 229–242.

Heyes, S., Pratt, W. S., Rees, E., Dahimene, S., Ferron, L., Owen, M. J., et al. (2015). Genetic disruption of voltage-gated calcium channels in psychiatric and neurological disorders. *Progress in Neurobiology, 134*, 36–54.

Hidalgo, S., Campusano, J. M., & Hodge, J. J. L. (2021). The Drosophila ortholog of the schizophrenia-associated CACNA1A and CACNA1B voltage-gated calcium channels regulate memory, sleep and circadian rhythms. *Neurobiology of Disease, 155*, 105394.

Hofer, N. T., Tuluc, P., Ortner, N. J., Nikonishyna, Y. V., Fernándes-Quintero, M. L., Liedl, K. R., et al. (2020). Biophysical classification of a CACNA1D de novo mutation as a high-risk mutation for a severe neurodevelopmental disorder. *Molecular Autism, 11*, 4.

Höschl, C., & Kozený, J. (1989). Verapamil in affective disorders: A controlled, double-blind study. *Biological Psychiatry, 25*, 128–140.

Howes, O. D., McCutcheon, R., Owen, M. J., & Murray, R. M. (2017). The role of genes, stress, and dopamine in the development of schizophrenia. *Biological Psychiatry, 81*, 9–20.

Hu, Z., Liang, M. C., & Soong, T. W. (2017). Alternative splicing of L-type $Ca_v1.2$ calcium channels: Implications in cardiovascular diseases. *Genes (Basel), 8*, E344.

Huang, A. S., Rogers, B. P., Anticevic, A., Blackford, J. U., Heckers, S., & Woodward, N. D. (2019). Brain function during stages of working memory in schizophrenia and psychotic bipolar disorder. *Neuropsychopharmacology, 44*, 2136–2142.

Huys, Q. J., Maia, T. V., & Frank, M. J. (2016). Computational psychiatry as a bridge from neuroscience to clinical applications. *Nature Neuroscience, 19*, 404–413.

Iasevoli, F., Tomasetti, C., & de Bartolomeis, A. (2013). Scaffolding proteins of the post-synaptic density contribute to synaptic plasticity by regulating receptor localization and distribution: Relevance for neuropsychiatric diseases. *Neurochemical Research, 38*, 1–22.

Indelicato, E., & Boesch, S. (2021). From genotype to phenotype: Expanding the clinical spectrum of CACNA1A variants in the era of next generation sequencing. *Frontiers in Neurology, 12*, 639994.

Irish, S. G. C. A. T. W. T. C. C. C. (2012). Genome-wide association study implicates HLA-C*01:02 as a risk factor at the major histocompatibility complex locus in schizophrenia. *Biological Psychiatry, 72*, 620–628.

Iyer, R., Ungless, M. A., & Faisal, A. A. (2017). Calcium-activated SK channels control firing regularity by modulating sodium channel availability in midbrain dopamine neurons. *Scientific Reports, 7*, 5248.

Jansen, P. R., Watanabe, K., Stringer, S., Skene, N., Bryois, J., Hammerschlag, A. R., et al. (2019). Genome-wide analysis of insomnia in 1,331,010 individuals identifies new risk loci and functional pathways. *Nature Genetics, 51*, 394–403.

Jaric, I., Rocks, D., Cham, H., Herchek, A., & Kundakovic, M. (2019). Sex and estrous cycle effects on anxiety- and depression-related phenotypes in a two-hit developmental stress model. *Frontiers in Molecular Neuroscience, 12*, 74.

Jimerson, D. C., Post, R. M., Carman, J. S., van Kammen, D. P., Wood, J. H., Goodwin, F. K., et al. (1979). CSF calcium: Clinical correlates in affective illness and schizophrenia. *Biological Psychiatry, 14*, 37–51.

Kaar, S. J., Angelescu, I., Marques, T. R., & Howes, O. D. (2019). Pre-frontal parvalbumin interneurons in schizophrenia: A meta-analysis of post-mortem studies. *Journal of Neural Transmission (Vienna), 126*, 1637–1651.

Kabir, Z. D., Che, A., Fischer, D. K., Rice, R. C., Rizzo, B. K., Byrne, M., et al. (2017). Rescue of impaired sociability and anxiety-like behavior in adult cacna1c-deficient mice by pharmacologically targeting eIF2α. *Molecular Psychiatry, 22*, 1096–1109.

Kabitzke, P. A., Brunner, D., He, D., Fazio, P. A., Cox, K., Sutphen, J., et al. (2018). Comprehensive analysis of two Shank3 and the Cacna1c mouse models of autism spectrum disorder. *Genes, Brain, and Behavior, 17*, 4–22.

Karlsgodt, K. H., Sun, D., & Cannon, T. D. (2010). Structural and functional brain abnormalities in schizophrenia. *Current Directions in Psychological Science, 19*, 226–231.

Kaufman, J., & Charney, D. (2000). Comorbidity of mood and anxiety disorders. *Depression and Anxiety, 12*(Suppl 1), 69–76.

Kawaguchi, Y., & Kubota, Y. (1997). GABAergic cell subtypes and their synaptic connections in rat frontal cortex. *Cerebral Cortex, 7*, 476–486.

Kawaguchi, Y., Katsumaru, H., Kosaka, T., Heizmann, C. W., & Hama, K. (1987). Fast spiking cells in rat hippocampus (CA1 region) contain the calcium-binding protein parvalbumin. *Brain Research, 416*, 369–374.

Kennedy, D. P., & Adolphs, R. (2012). The social brain in psychiatric and neurological disorders. *Trends in Cognitive Sciences, 16*, 559–572.

Kessler, R. C., Aguilar-Gaxiola, S., Alonso, J., Chatterji, S., Lee, S., Ormel, J., et al. (2009). The global burden of mental disorders: An update from the WHO World Mental Health (WMH) surveys. *Epidemiologia e Psichiatria Sociale, 18*, 23–33.

Kim, C., Jeon, D., Kim, Y. H., Lee, C. J., Kim, H., & Shin, H. S. (2009). Deletion of N-type Ca(2+) channel Ca(v)2.2 results in hyperaggressive behaviors in mice. *The Journal of Biological Chemistry, 284*, 2738–2745.

Kisilevsky, A. E., & Zamponi, G. W. (2008). D2 dopamine receptors interact directly with N-type calcium channels and regulate channel surface expression levels. *Channels (Austin, Tex.), 2*, 269–277.

Kisilevsky, A. E., Mulligan, S. J., Altier, C., Iftinca, M. C., Varela, D., Tai, C., et al. (2008). D1 receptors physically interact with N-type calcium channels to regulate channel distribution and dendritic calcium entry. *Neuron, 58*, 557–570.

Kisko, T. M., Braun, M. D., Michels, S., Witt, S. H., Rietschel, M., Culmsee, C., et al. (2018). Cacna1c haploinsufficiency leads to pro-social 50-kHz ultrasonic communication deficits in rats. *Disease Models & Mechanisms, 11*, dmm034116.

Kisko, T. M., Braun, M. D., Michels, S., Witt, S. H., Rietschel, M., Culmsee, C., et al. (2020). Sex-dependent effects of Cacna1c haploinsufficiency on juvenile social play behavior and pro-social 50-kHz ultrasonic communication in rats. *Genes, Brain, and Behavior, 19*, e12552.

Knutson, B., Burgdorf, J., & Panksepp, J. (1998). Anticipation of play elicits high-frequency ultrasonic vocalizations in young rats. *Journal of Comparative Psychology, 112*, 65–73.

Kokras, N., & Dalla, C. (2014). Sex differences in animal models of psychiatric disorders. *British Journal of Pharmacology, 171*, 4595–4619.

Kolaj, M., & Renaud, L. P. (2001). Norepinephrine acts via alpha(2) adrenergic receptors to suppress N-type calcium channels in dissociated rat median preoptic nucleus neurons. *Neuropharmacology, 41*, 472–479.

Komatsu, H. (2015). Novel therapeutic GPCRs for psychiatric disorders. *International Journal of Molecular Sciences, 16*, 14109–14121.

Krystal, A. D. (2012). Psychiatric disorders and sleep. *Neurologic Clinics, 30*, 1389–1413.

Kubota, M., Murakoshi, T., Saegusa, H., Kazuno, A., Zong, S., Hu, Q., et al. (2001). Intact LTP and fear memory but impaired spatial memory in mice lacking Ca(v)2.3 (alpha(IE)) channel. *Biochemical and Biophysical Research Communications, 282*, 242–248.

Kuzmin, A., Zvartau, E., Gessa, G. L., Martellotta, M. C., & Fratta, W. (1992). Calcium antagonists isradipine and nimodipine suppress cocaine and morphine intravenous self-administration in drug-naive mice. *Pharmacology, Biochemistry, and Behavior, 41*, 497–500.

Lam, M., Trampush, J. W., Yu, J., Knowles, E., Davies, G., Liewald, D. C., et al. (2017). Large-scale cognitive GWAS meta-analysis reveals tissue-specific neural expression and potential nootropic drug targets. *Cell Reports, 21*, 2597–2613.

Lambert, R. C., Bessaïh, T., Crunelli, V., & Leresche, N. (2014). The many faces of T-type calcium channels. *Pflügers Archiv, 466*, 415–423.

Lee, A. S., Ra, S., Rajadhyaksha, A. M., Britt, J. K., De Jesus-Cortes, H., Gonzales, K. L., et al. (2012). Forebrain elimination of cacna1c mediates anxiety-like behavior in mice. *Molecular Psychiatry, 17*, 1054–1055.

Lee, A. S., De Jesús-Cortés, H., Kabir, Z. D., Knobbe, W., Orr, M., Burgdorf, C., et al. (2016). The neuropsychiatric disease-associated gene cacna1c mediates survival of young hippocampal neurons. *eNeuro, 3*, ENEURO.0006–16.2016.

Lee, P. H., Feng, Y. A., & Smoller, J. W. (2021). Pleiotropy and cross-disorder genetics among psychiatric disorders. *Biological Psychiatry, 89*, 20–31.

Lenzi, A., Marazziti, D., Raffaelli, S., & Cassano, G. B. (1995). Effectiveness of the combination verapamil and chlorpromazine in the treatment of severe manic or mixed patients. *Progress in Neuro-Psychopharmacology & Biological Psychiatry, 19*, 519–528.

Lett, T. A., Voineskos, A. N., Kennedy, J. L., Levine, B., & Daskalakis, Z. J. (2014). Treating working memory deficits in schizophrenia: A review of the neurobiology. *Biological Psychiatry, 75*, 361–370.

Levine, J., Stein, D., Rapoport, A., & Kurtzman, L. (1999). High serum and cerebrospinal fluid Ca/Mg ratio in recently hospitalized acutely depressed patients. *Neuropsychobiology, 39*, 63–70.

Lewis, D. A., Curley, A. A., Glausier, J. R., & Volk, D. W. (2012). Cortical parvalbumin interneurons and cognitive dysfunction in schizophrenia. *Trends in Neurosciences, 35*, 57–67.

Li, J., Zhao, L., You, Y., Lu, T., Jia, M., Yu, H., et al. (2015). Schizophrenia related variants in CACNA1C also confer risk of autism. *PLoS One, 10*, e0133247.

Li, W., Fan, C. C., Mäki-Marttunen, T., Thompson, W. K., Schork, A. J., Bettella, F., et al. (2018). A molecule-based genetic association approach implicates a range of voltage-gated calcium channels associated with schizophrenia. *American Journal of Medical Genetics. Part B, Neuropsychiatric Genetics, 177*, 454–467.

Lieberman, J. A., Perkins, D., Belger, A., Chakos, M., Jarskog, F., Boteva, K., et al. (2001). The early stages of schizophrenia: Speculations on pathogenesis, pathophysiology, and therapeutic approaches. *Biological Psychiatry, 50*, 884–897.

Limpitikul, W. B., Dick, I. E., Ben-Johny, M., & Yue, D. T. (2016). An autism-associated mutation in CaV1.3 channels has opposing effects on voltage- and Ca(2+)-dependent regulation. *Scientific Reports, 6*, 27235.

Lin, E., Kuo, P. H., Liu, Y. L., Yu, Y. W., Yang, A. C., & Tsai, S. J. (2018). A deep learning approach for predicting antidepressant response in major depression using clinical and genetic biomarkers. *Frontiers in Psychiatry, 9*, 290.

Lipscombe, D., & Andrade, A. (2015). Calcium channel CaVα₁ splice isoforms—Tissue specificity and drug action. *Current Molecular Pharmacology, 8*, 22–31.

Lipscombe, D., Kongsamut, S., & Tsien, R. W. (1989). Alpha-adrenergic inhibition of sympathetic neurotransmitter release mediated by modulation of N-type calcium-channel gating. *Nature, 340*, 639–642.

Lipscombe, D., Allen, S. E., & Toro, C. P. (2013a). Control of neuronal voltage-gated calcium ion channels from RNA to protein. *Trends in Neurosciences, 36*, 598–609.

Lipscombe, D., Andrade, A., & Allen, S. E. (2013b). Alternative splicing: Functional diversity among voltage-gated calcium channels and behavioral consequences. *Biochimica et Biophysica Acta, 1828*, 1522–1529.

Liu, Y., Harding, M., Pittman, A., Dore, J., Striessnig, J., Rajadhyaksha, A., et al. (2014). Cav1.2 and Cav1.3 L-type calcium channels regulate dopaminergic firing activity in the mouse ventral tegmental area. *Journal of Neurophysiology, 112*, 1119–1130.

Lu, C. W., Lin, T. Y., Huang, S. K., & Wang, S. J. (2018). 5-HT₁B receptor agonist CGS12066 presynaptically inhibits glutamate release in rat hippocampus. *Progress in Neuro-Psychopharmacology & Biological Psychiatry, 86*, 122–130.

Luo, X., Rosenfeld, J. A., Yamamoto, S., Harel, T., Zuo, Z., Hall, M., et al. (2017). Clinically severe CACNA1A alleles affect synaptic function and neurodegeneration differentially. *PLoS Genetics, 13*, e1006905.

Lupien-Meilleur, A., Jiang, X., Lachance, M., Taschereau-Dumouchel, V., Gagnon, L., Vanasse, C., et al. (2021). Reversing frontal disinhibition rescues behavioural deficits in models of CACNA1A-associated neurodevelopment disorders. *Molecular Psychiatry, 26*(12), 7225–7246.

Malberg, J. E. (2004). Implications of adult hippocampal neurogenesis in antidepressant action. *Journal of Psychiatry & Neuroscience, 29*, 196–205.

Mallinger, A. G., Thase, M. E., Haskett, R., Buttenfield, J., Luckenbaugh, D. A., Frank, E., et al. (2008). Verapamil augmentation of lithium treatment improves outcome in mania unresponsive to lithium alone: Preliminary findings and a discussion of therapeutic mechanisms. *Bipolar Disorders, 10*, 856–866.

Mallmann, R. T., Elgueta, C., Sleman, F., Castonguay, J., Wilmes, T., van den Maagdenberg, A., et al. (2013). Ablation of Ca(V)2.1 voltage-gated Ca²⁺ channels in mouse forebrain generates multiple cognitive impairments. *PLoS One, 8*, e78598.

Manoach, D. S., & Stickgold, R. (2019). Abnormal sleep spindles, memory consolidation, and schizophrenia. *Annual Review of Clinical Psychology, 15*, 451–479.

Manoach, D. S., Pan, J. Q., Purcell, S. M., & Stickgold, R. (2016). Reduced sleep spindles in schizophrenia: A treatable endophenotype that links risk genes to impaired cognition. *Biological Psychiatry, 80*, 599–608.

Marrion, N. V., & Tavalin, S. J. (1998). Selective activation of Ca2+−activated K+ channels by co-localized

Ca2+ channels in hippocampal neurons. *Nature, 395*, 900–905.

Marschallinger, J., Sah, A., Schmuckermair, C., Unger, M., Rotheneichner, P., Kharitonova, M., et al. (2015). The L-type calcium channel Cav1.3 is required for proper hippocampal neurogenesis and cognitive functions. *Cell Calcium, 58*, 606–616.

Martínez-Rivera, A., Hao, J., Tropea, T. F., Giordano, T. P., Kosovsky, M., Rice, R. C., et al. (2017). Enhancing VTA Cav1.3 L-type Ca²⁺ channel activity promotes cocaine and mood-related behaviors via overlapping AMPA receptor mechanisms in the nucleus accumbens. *Molecular Psychiatry, 22*, 1735–1745.

Medrihan, L., Sagi, Y., Inde, Z., Krupa, O., Daniels, C., Peyrache, A., et al. (2017). Initiation of behavioral response to antidepressants by cholecystokinin neurons of the dentate gyrus. *Neuron, 95*, 564–576.e4.

Merikanto, I., Utge, S., Lahti, J., Kuula, L., Makkonen, T., Lahti-Pulkkinen, M., et al. (2019). Genetic risk factors for schizophrenia associate with sleep spindle activity in healthy adolescents. *Journal of Sleep Research, 28*, e12762.

Metz, A. E., Jarsky, T., Martina, M., & Spruston, N. (2005). R-type calcium channels contribute to afterdepolarization and bursting in hippocampal CA1 pyramidal neurons. *The Journal of Neuroscience, 25*, 5763–5773.

Mineka, S., Watson, D., & Clark, L. A. (1998). Comorbidity of anxiety and unipolar mood disorders. *Annual Review of Psychology, 49*, 377–412.

Moon, A. L., Haan, N., Wilkinson, L. S., Thomas, K. L., & Hall, J. (2018). CACNA1C: Association with psychiatric disorders, behavior, and neurogenesis. *Schizophrenia Bulletin, 44*, 958–965.

Moon, A. L., Brydges, N. M., Wilkinson, L. S., Hall, J., & Thomas, K. L. (2020). Cacna1c hemizygosity results in aberrant fear conditioning to neutral stimuli. *Schizophrenia Bulletin, 46*, sbz127.

Moosmang, S., Haider, N., Klugbauer, N., Adelsberger, H., Langwieser, N., Müller, J., et al. (2005). Role of hippocampal Cav1.2 Ca2+ channels in NMDA receptor-independent synaptic plasticity and spatial memory. *The Journal of Neuroscience, 25*, 9883–9892.

Mosheva, M., Serretti, A., Stukalin, Y., Fabbri, C., Hagin, M., Horev, S., et al. (2020). Association between CANCA1C gene rs1034936 polymorphism and alcohol dependence in bipolar disorder. *Journal of Affective Disorders, 261*, 181–186.

Moskvina, V., Craddock, N., Holmans, P., Nikolov, I., Pahwa, J. S., Green, E., et al. (2009). Gene-wide analyses of genome-wide association data sets: Evidence for multiple common risk alleles for schizophrenia and bipolar disorder and for overlap in genetic risk. *Molecular Psychiatry, 14*, 252–260.

Mukherjee, A., Carvalho, F., Eliez, S., & Caroni, P. (2019). Long-lasting rescue of network and cognitive dysfunction in a genetic schizophrenia model. *Cell, 178*, 1387–1402.e14.

Murat, S., Bigot, M., Chapron, J., König, G. M., Kostenis, E., Battaglia, G., et al. (2019). 5-HT₂A receptor-dependent phosphorylation of mGlu$_2$ receptor at Serine 843 promotes mGlu$_2$ receptor-operated G$_{i/o}$ signaling. *Molecular Psychiatry, 24*, 1610–1626.

Nahar, L., Delacroix, B. M., & Nam, H. W. (2021). The role of parvalbumin interneurons in neurotransmitter balance and neurological disease. *Frontiers in Psychiatry, 12*, 679960.

Nakagawasai, O., Onogi, H., Mitazaki, S., Sato, A., Watanabe, K., Saito, H., et al. (2010). Behavioral and neurochemical characterization of mice deficient in the N-type Ca2+ channel alpha1B subunit. *Behavioural Brain Research, 208*, 224–230.

Nandagopal, N., & Roux, P. P. (2015). Regulation of global and specific mRNA translation by the mTOR signaling pathway. *Translation (Austin), 3*, e983402.

Nelson, R. J., & Trainor, B. C. (2007). Neural mechanisms of aggression. *Nature Reviews. Neuroscience, 8*, 536–546.

Nestler, E. J., & Hyman, S. E. (2010). Animal models of neuropsychiatric disorders. *Nature Neuroscience, 13*, 1161–1169.

Newton, P. M., Orr, C. J., Wallace, M. J., Kim, C., Shin, H. S., & Messing, R. O. (2004). Deletion of N-type calcium channels alters ethanol reward and reduces ethanol consumption in mice. *The Journal of Neuroscience, 24*, 9862–9869.

Nguyen, R., Venkatesan, S., Binko, M., Bang, J. Y., Cajanding, J. D., Briggs, C., et al. (2020). Cholecystokinin-expressing interneurons of the medial prefrontal cortex mediate working memory retrieval. *The Journal of Neuroscience, 40*, 2314–2331.

Nieratschker, V., Brückmann, C., & Plewnia, C. (2015). CACNA1C risk variant affects facial emotion recognition in healthy individuals. *Scientific Reports, 5*, 17349.

O'Connell, K. S., McGregor, N. W., Malhotra, A., Lencz, T., Emsley, R., & Warnich, L. (2019). Variation within voltage-gated calcium channel genes and antipsychotic treatment response in a South African first episode schizophrenia cohort. *The Pharmacogenomics Journal, 19*, 109–114.

O'Donovan, M. C., & Owen, M. J. (2016). The implications of the shared genetics of psychiatric disorders. *Nature Medicine, 22*, 1214–1219.

O'Roak, B. J., Vives, L., Girirajan, S., Karakoc, E., Krumm, N., Coe, B. P., et al. (2012). Sporadic autism exomes reveal a highly interconnected protein network of de novo mutations. *Nature, 485*, 246–250.

Okbay, A., Baselmans, B. M., De Neve, J. E., Turley, P., Nivard, M. G., Fontana, M. A., et al. (2016). Genetic variants associated with subjective well-being, depressive symptoms, and neuroticism identified through genome-wide analyses. *Nature Genetics, 48*, 624–633.

Oprea, T. I., & Mestres, J. (2012). Drug repurposing: Far beyond new targets for old drugs. *The AAPS Journal, 14*, 759–763.

Ortner, N. J., Kaserer, T., Copeland, J. N., & Striessnig, J. (2020). De novo CACNA1D Ca^{2+} channelopathies: Clinical phenotypes and molecular mechanism. *Pflügers Archiv, 472*, 755–773.

Pasparakis, E., Koiliari, E., Zouraraki, C., Tsapakis, E. M., Roussos, P., Giakoumaki, S. G., et al. (2015). The effects of the CACNA1C rs1006737 A/G on affective startle modulation in healthy males. *European Psychiatry, 30*, 492–498.

Pazzaglia, P. J., Post, R. M., Ketter, T. A., Callahan, A. M., Marangell, L. B., Frye, M. A., et al. (1998). Nimodipine monotherapy and carbamazepine augmentation in patients with refractory recurrent affective illness. *Journal of Clinical Psychopharmacology, 18*, 404–413.

Pelkey, K. A., Chittajallu, R., Craig, M. T., Tricoire, L., Wester, J. C., & McBain, C. J. (2017). Hippocampal GABAergic inhibitory interneurons. *Physiological Reviews, 97*, 1619–1747.

Penzes, P., Cahill, M. E., Jones, K. A., VanLeeuwen, J. E., & Woolfrey, K. M. (2011). Dendritic spine pathology in neuropsychiatric disorders. *Nature Neuroscience, 14*, 285–293.

Perez-Reyes, E. (2003). Molecular physiology of low-voltage-activated t-type calcium channels. *Physiological Reviews, 83*, 117–161.

Pierce, R. C., Quick, E. A., Reeder, D. C., Morgan, Z. R., & Kalivas, P. W. (1998). Calcium-mediated second messengers modulate the expression of behavioral sensitization to cocaine. *The Journal of Pharmacology and Experimental Therapeutics, 286*, 1171–1176.

Pinggera, A., Lieb, A., Benedetti, B., Lampert, M., Monteleone, S., Liedl, K. R., et al. (2015). CACNA1D de novo mutations in autism spectrum disorders activate Cav1.3 L-type calcium channels. *Biological Psychiatry, 77*, 816–822.

Pinggera, A., Mackenroth, L., Rump, A., Schallner, J., Beleggia, F., Wollnik, B., et al. (2017). New gain-of-function mutation shows CACNA1D as recurrently mutated gene in autism spectrum disorders and epilepsy. *Human Molecular Genetics, 26*, 2923–2932.

Pinggera, A., Negro, G., Tuluc, P., Brown, M. J., Lieb, A., & Striessnig, J. (2018). Gating defects of disease-causing de novo mutations in Ca$_v$1.3 Ca^{2+} channels. *Channels (Austin, Tex.), 12*, 388–402.

Post, R. M., & Kalivas, P. (2013). Bipolar disorder and substance misuse: Pathological and therapeutic implications of their comorbidity and cross-sensitisation. *The British Journal of Psychiatry, 202*, 172–176.

Price, W. A. (1987). Antipsychotic effects of verapamil in schizophrenia. *The Hillside Journal of Clinical Psychiatry, 9*, 225–230.

Price, W. A., & Pascarzi, G. A. (1987). Use of verapamil to treat negative symptoms in schizophrenia. *Journal of Clinical Psychopharmacology, 7*, 357.

Puopolo, M., Raviola, E., & Bean, B. P. (2007). Roles of subthreshold calcium current and sodium current in spontaneous firing of mouse midbrain dopamine neurons. *The Journal of Neuroscience, 27*, 645–656.

Purcell, S. M., Moran, J. L., Fromer, M., Ruderfer, D., Solovieff, N., Roussos, P., et al. (2014). A polygenic burden of rare disruptive mutations in schizophrenia. *Nature, 506*, 185–190.

Pushpakom, S., Iorio, F., Eyers, P. A., Escott, K. J., Hopper, S., Wells, A., et al. (2019). Drug repurposing: Progress, challenges and recommendations. *Nature Reviews. Drug Discovery, 18*, 41–58.

Rajarajan, P., Borrman, T., Liao, W., Schrode, N., Flaherty, E., Casiño, C., et al. (2018). Neuron-specific signatures in the chromosomal connectome associated with schizophrenia risk. *Science, 362*, eaat4311.

Randall, A. D., & Tsien, R. W. (1997). Contrasting biophysical and pharmacological properties of T-type and R-type calcium channels. *Neuropharmacology, 36*, 879–893.

Redecker, T. M., Kisko, T. M., Schwarting, R. K. W., & Wöhr, M. (2019). Effects of Cacna1c haploinsufficiency on social interaction behavior and 50-kHz ultrasonic vocalizations in adult female rats. *Behavioural Brain Research, 367*, 35–52.

Reid, J. G., Gitlin, M. J., & Altshuler, L. L. (2013). Lamotrigine in psychiatric disorders. *The Journal of Clinical Psychiatry, 74*, 675–684.

Reimer, A. R., & Martin-Iverson, M. T. (1994). Nimodipine and haloperidol attenuate behavioural sensitization to cocaine but only nimodipine blocks the establishment of conditioned locomotion induced by cocaine. *Psychopharmacology, 113*, 404–410.

Riemann, D., Krone, L. B., Wulff, K., & Nissen, C. (2020). Sleep, insomnia, and depression. *Neuropsychopharmacology, 45*, 74–89.

Ripke, S., O'Dushlaine, C., Chambert, K., Moran, J. L., Kähler, A. K., Akterin, S., et al. (2013). Genome-wide association analysis identifies 13 new risk loci for schizophrenia. *Nature Genetics, 45*, 1150–1159.

Robinson, N., & Bergen, S. E. (2021). Environmental risk factors for schizophrenia and bipolar disorder and their relationship to genetic risk: Current knowledge and future directions. *Frontiers in Genetics, 12*, 686666.

Ruihua, M., Meng, Z., Nan, C., Panqi, L., Hua, G., Sijia, L., et al. (2021). Differences in facial expression recognition between unipolar and bipolar depression. *Frontiers in Psychology, 12*, 619368.

Saegusa, H., Kurihara, T., Zong, S., Kazuno, A., Matsuda, Y., Nonaka, T., et al. (2001). Suppression of inflammatory and neuropathic pain symptoms in mice lacking the N-type Ca2+ channel. *The EMBO Journal, 20*, 2349–2356.

Sagi, Y., Medrihan, L., George, K., Barney, M., McCabe, K. A., & Greengard, P. (2020). Emergence of 5-HT5A signaling in parvalbumin neurons mediates delayed antidepressant action. *Molecular Psychiatry, 25*, 1191–1201.

Saliba, R. S., Gu, Z., Yan, Z., & Moss, S. J. (2009). Blocking L-type voltage-gated Ca2+ channels with dihydropyridines reduces gamma-aminobutyric acid type A receptor expression and synaptic inhibition. *The Journal of Biological Chemistry, 284*, 32544–32550.

Sanchez-Roige, S., Fontanillas, P., Elson, S. L., Gray, J. C., de Wit, H., MacKillop, J., et al. (2019). Genomewide Association studies of impulsive personality traits (BIS-11 and UPPS-P) and drug experimentation in up to 22,861 adult research participants identify loci in the CACNA1I and CADM2 genes. *The Journal of Neuroscience, 39*, 2562–2572.

Schierberl, K., Hao, J., Tropea, T. F., Ra, S., Giordano, T. P., Xu, Q., et al. (2011). Cav1.2 L-type Ca2+ channels mediate cocaine-induced GluA1 trafficking in the nucleus accumbens, a long-term adaptation dependent on ventral tegmental area Ca(v)1.3 channels. *The Journal of Neuroscience, 31*, 13562–13575.

Schizophrenia, W. G. O. T. P. G. C. (2014). Biological insights from 108 schizophrenia-associated genetic loci. *Nature, 511*, 421–427.

Shimada, M., Miyagawa, T., Kawashima, M., Tanaka, S., Honda, Y., Honda, M., et al. (2010). An approach based on a genome-wide association study reveals candidate loci for narcolepsy. *Human Genetics, 128*, 433–441.

Sinnegger-Brauns, M. J., Hetzenauer, A., Huber, I. G., Renström, E., Wietzorrek, G., Berjukov, S., et al. (2004). Isoform-specific regulation of mood behavior and pancreatic beta cell and cardiovascular function by L-type Ca 2+ channels. *The Journal of Clinical Investigation, 113*, 1430–1439.

Smail, M. A., Wu, X., Henkel, N. D., Eby, H. M., Herman, J. P., McCullumsmith, R. E., et al. (2021). Similarities and dissimilarities between psychiatric cluster disorders. *Molecular Psychiatry, 26*(9), 4853–4863.

Soeiro-de-Souza, M. G., Otaduy, M. C., Dias, C. Z., Bio, D. S., Machado-Vieira, R., & Moreno, R. A. (2012). The impact of the CACNA1C risk allele on limbic structures and facial emotions recognition in bipolar disorder subjects and healthy controls. *Journal of Affective Disorders, 141*, 94–101.

Sohal, V. S., & Rubenstein, J. L. R. (2019). Excitation-inhibition balance as a framework for investigating mechanisms in neuropsychiatric disorders. *Molecular Psychiatry, 24*, 1248–1257.

Sonnenschein, S. F., Gomes, F. V., & Grace, A. A. (2020). Dysregulation of midbrain dopamine system and the pathophysiology of schizophrenia. *Frontiers in Psychiatry, 11*, 613.

Splawski, I., Timothy, K. W., Sharpe, L. M., Decher, N., Kumar, P., Bloise, R., et al. (2004). Ca(V)1.2 calcium channel dysfunction causes a multisystem disorder including arrhythmia and autism. *Cell, 119*, 19–31.

Splawski, I., Timothy, K. W., Decher, N., Kumar, P., Sachse, F. B., Beggs, A. H., et al. (2005). Severe arrhythmia disorder caused by cardiac L-type calcium channel mutations. *Proceedings of the National Academy of Sciences of the United States of America, 102*, 8089–8096. discussion 8086.

Splawski, I., Yoo, D. S., Stotz, S. C., Cherry, A., Clapham, D. E., & Keating, M. T. (2006). CACNA1H mutations in autism spectrum disorders. *The Journal of Biological Chemistry, 281*, 22085–22091.

Stahl, E. A., Breen, G., Forstner, A. J., McQuillin, A., Ripke, S., Trubetskoy, V., et al. (2019). Genome-wide association study identifies 30 loci associated with bipolar disorder. *Nature Genetics, 51*, 793–803.

Stępnicki, P., Kondej, M., & Kaczor, A. A. (2018). Current concepts and treatments of schizophrenia. *Molecules, 23*, E2087.

Steriade, M., Deschênes, M., Domich, L., & Mulle, C. (1985). Abolition of spindle oscillations in thalamic neurons disconnected from nucleus reticularis thalami. *Journal of Neurophysiology, 54*, 1473–1497.

Striessnig, J., Koschak, A., Sinnegger-Brauns, M. J., Hetzenauer, A., Nguyen, N. K., Busquet, P., et al. (2006). Role of voltage-gated L-type Ca2+ channel isoforms for brain function. *Biochemical Society Transactions, 34*, 903–909.

Strom, S. P., Stone, J. L., Ten Bosch, J. R., Merriman, B., Cantor, R. M., Geschwind, D. H., et al. (2010). High-density SNP association study of the 17q21 chromosomal region linked to autism identifies CACNA1G as a novel candidate gene. *Molecular Psychiatry, 15*, 996–1005.

Suzuki, S., & Rogawski, M. A. (1989). T-type calcium channels mediate the transition between tonic and phasic firing in thalamic neurons. *Proceedings of the National Academy of Sciences of the United States of America, 86*, 7228–7232.

Sykes, L., Haddon, J., Lancaster, T. M., Sykes, A., Azzouni, K., Ihssen, N., et al. (2019). Genetic variation in the psychiatric risk gene CACNA1C modulates reversal learning across species. *Schizophrenia Bulletin, 45*, 1024–1032.

Takahashi, T., & Momiyama, A. (1993). Different types of calcium channels mediate central synaptic transmission. *Nature, 366*, 156–158.

Takata, A., Ionita-Laza, I., Gogos, J. A., Xu, B., & Karayiorgou, M. (2016). De novo synonymous mutations in regulatory elements contribute to the genetic etiology of autism and schizophrenia. *Neuron, 89*, 940–947.

Takata, A., Miyake, N., Tsurusaki, Y., Fukai, R., Miyatake, S., Koshimizu, E., et al. (2018). Integrative analyses of de novo mutations provide deeper biological insights into autism spectrum disorder. *Cell Reports, 22*, 734–747.

Takenaka, S., Sera, N., Tokiwa, H., Hirohata, I., & Hirohata, T. (1989). Identification of mutagens in Japanese pickles. *Mutation Research, 223*, 35–40.

Tatti, R., Haley, M. S., Swanson, O. K., Tselha, T., & Maffei, A. (2017). Neurophysiology and regulation of the balance between excitation and inhibition in neocortical circuits. *Biological Psychiatry, 81*, 821–831.

Temme, S. J., Bell, R. Z., Fisher, G. L., & Murphy, G. G. (2016). Deletion of the mouse homolog of CACNA1C disrupts discrete forms of hippocampal-dependent memory and neurogenesis within the dentate gyrus. *eNeuro, 3*, ENEURO.0118–16.2016.

Terrillion, C. E., Dao, D. T., Cachope, R., Lobo, M. K., Puche, A. C., Cheer, J. F., et al. (2017). Reduced levels of Cacna1c attenuate mesolimbic dopamine system function. *Genes, Brain, and Behavior, 16*, 495–505.

Thankachan, S., Katsuki, F., McKenna, J. T., Yang, C., Shukla, C., Deisseroth, K., et al. (2019). Thalamic reticular nucleus parvalbumin neurons regulate sleep spindles and electrophysiological aspects of schizophrenia in mice. *Scientific Reports, 9*, 3607.

Tigaret, C. M., Lin, T. E., Morrell, E. R., Sykes, L., Moon, A. L., O'Donovan, M. C., et al. (2021). Neurotrophin receptor activation rescues cognitive and synaptic abnormalities caused by hemizygosity of the psychiatric risk gene Cacna1c. *Molecular Psychiatry, 26*, 1748–1760.

Tombácz, D., Maróti, Z., Kalmár, T., Csabai, Z., Balázs, Z., Takahashi, S., et al. (2017). High-coverage whole-exome sequencing identifies candidate genes for suicide in victims with major depressive disorder. *Scientific Reports, 7*, 7106.

Tsuang, M. T., Bar, J. L., Stone, W. S., & Faraone, S. V. (2004). Gene-environment interactions in mental disorders. *World Psychiatry, 3*, 73–83.

Tyagi, S., Bendrick, T. R., Filipova, D., Papadopoulos, S., & Bannister, R. A. (2019). A mutation in Ca$_V$2.1 linked to a severe neurodevelopmental disorder impairs channel gating. *The Journal of General Physiology, 151*, 850–859.

Uhrig, S., Vandael, D., Marcantoni, A., Dedic, N., Bilbao, A., Vogt, M. A., et al. (2017). Differential roles for L-type calcium channel subtypes in alcohol dependence. *Neuropsychopharmacology, 42*, 1058–1069.

Voglis, G., & Tavernarakis, N. (2006). The role of synaptic ion channels in synaptic plasticity. *EMBO Reports, 7*, 1104–1110.

Völkening, B., Schönig, K., Kronenberg, G., Bartsch, D., & Weber, T. (2017). Deletion of psychiatric risk gene Cacna1c impairs hippocampal neurogenesis in cell-autonomous fashion. *Glia, 65*, 817–827.

Waltz, J. A. (2017). The neural underpinnings of cognitive flexibility and their disruption in psychotic illness. *Neuroscience, 345*, 203–217.

Wek, R. C., & Cavener, D. R. (2007). Translational control and the unfolded protein response. *Antioxidants & Redox Signaling, 9*, 2357–2371.

Whittington, M. A., Dolin, S. J., Patch, T. L., Siarey, R. J., Butterworth, A. R., & Little, H. J. (1991). Chronic dihydropyridine treatment can reverse the behavioural consequences of and prevent adaptations to, chronic ethanol treatment. *British Journal of Pharmacology, 103*, 1669–1676.

Xie, Y., Huang, D., Wei, L., & Luo, X. J. (2018). Further evidence for the genetic association between CACNA1I and schizophrenia. *Hereditas, 155*, 16.

Xu, W., Liu, Y., Chen, J., Guo, Q., Liu, K., Wen, Z., et al. (2018). Genetic risk between the CACNA1I gene and schizophrenia in Chinese Uygur population. *Hereditas, 155*, 5.

Yatsenko, S. A., Hixson, P., Roney, E. K., Scott, D. A., Schaaf, C. P., Ng, Y. T., et al. (2012). Human subtelomeric copy number gains suggest a DNA replication mechanism for formation: Beyond breakage-fusion-bridge for telomere stabilization. *Human Genetics, 131*, 1895–1910.

Yau, S. Y., Li, A., & So, K. F. (2015). Involvement of adult hippocampal neurogenesis in learning and forgetting. *Neural Plasticity, 2015*, 717958.

Zaitsev, A. V., Povysheva, N. V., Lewis, D. A., & Krimer, L. S. (2007). P/Q-type, but not N-type, calcium channels mediate GABA release from fast-spiking interneurons to pyramidal cells in rat prefrontal cortex. *Journal of Neurophysiology, 97*, 3567–3573.

Zamponi, G. W., Striessnig, J., Koschak, A., & Dolphin, A. C. (2015). The physiology, pathology, and pharmacology of voltage-gated calcium channels and their future therapeutic potential. *Pharmacological Reviews, 67*, 821–870.

Zhang, J., Tan, L., Ren, Y., Liang, J., Lin, R., Feng, Q., et al. (2016). Presynaptic excitation via GABAB receptors in habenula cholinergic neurons regulates fear memory expression. *Cell, 166*, 716–728.

Zhang, T., Zhu, L., Ni, T., Liu, D., Chen, G., Yan, Z., et al. (2018). Voltage-gated calcium channel activity and complex related genes and schizophrenia: A systematic investigation based on Han Chinese population. *Journal of Psychiatric Research, 106*, 99–105.

Zharkovsky, A., Tötterman, A. M., Moisio, J., & Ahtee, L. (1993). Concurrent nimodipine attenuates the withdrawal signs and the increase of cerebral dihydropyridine binding after chronic morphine treatment in rats. *Naunyn-Schmiedeberg's Archives of Pharmacology, 347*, 483–486.

Zhou, Y., Niimi, K., Li, W., & Takahashi, E. (2015). Role of Cav2. 2-mediated signaling in depressive behaviors. *Integrative Molecular Medicine, 2*, 369–372.

Calcium Channels and Selective Neuronal Vulnerability in Parkinson's Disease

Birgit Liss and D. James Surmeier

Abstract

Ca²⁺ entry through voltage-dependent ion channels (Ca$_v$) in the plasma membrane provides a signal coupling neuronal activity to a wide array of intracellular processes ranging from control of other ion channels to regulation of metabolism and gene expression. In Parkinson's disease (PD), the degeneration of substantia nigra pars compacta (SNc) dopaminergic neurons, causing the cardinal symptoms, has been tied to the prominent engagement of Ca$_v$ channels that modulate repetitive spiking, transmitter-release, and mitochondrial function. Here, we summarize the literature underlying this connection. We focus on Ca$_v$1 L-type Ca$_v$ channels, as epidemiological studies indicate that inhibition of these channels reduces the risk of developing PD. In addition, we discuss the translational implications of this literature and the prospect of selective Ca$_v$ channel modulators as disease-modifying drugs for early-stage PD.

Keywords

Ca²⁺ channel · Parkinson's disease · Dopamine · Substantia nigra · L-type · R-type · T-type · D2 dopamine receptor · Neuronal calcium sensors (NCS) · Mitochondria · Bioenergetics · Neurodegeneration · Homeostasis

B. Liss (✉)
Institute of Applied Physiology, Medical Faculty, Ulm University, Ulm, Germany

Linacre & New College, Oxford University, Oxford, UK
e-mail: birgit.liss@uni-ulm.de

D. J. Surmeier (✉)
Departments of Neuroscience, Feinberg School of Medicine, Northwestern University, Chicago, IL, USA
e-mail: j-surmeier@northwestern.edu

Parkinson's Disease: A Multifactorial Disease

Parkinson's disease (PD) is the second most common neurodegenerative disease, affecting 2–3% of the population over the age of 65 years (Bloem et al., 2021). With improvements in healthcare and population 'graying', the incidence of PD is expected to steadily rise (Poewe et al., 2017; GBD 2016 PD Collaborators, 2018). The defining features of PD are slowness of movement (bradykinesia) and muscle rigidity, commonly with resting tremor, and later with postural instability. There are a variety of non-motor symptoms that frequently become comorbid with PD movement deficits, like mood, sleep, and autonomous dysfunctions (Poewe et al., 2017; Berg et al., 2021; Bloem et al., 2021).

The pathogenic mechanisms driving PD have yet to be definitively identified. That said, there are many lines of evidence that suggest it is a multifactorial disease. Besides age, environmental toxins, and mutations in so-called PARK-genes that lead to familial forms of PD, there are a variety of other stressors that have been linked to disease risk, including inflammation, viral infections, trauma, and gut bacteria (Goldman, 2014; Caggiu et al., 2019; Menozzi et al., 2021; Romano et al., 2021). Many of these stressors challenge mitochondrial, proteasomal, and/or lysosomal function - all of which are important to PD pathogenesis (Michel et al., 2016; Bloem et al., 2021; Smeyne et al., 2021).

Another clue about PD mechanisms comes from the pattern and type of pathology that the disease produces. The pathological hallmarks of PD are (1) a distributed loss of neurons, primarily in the mesencephalon and brainstem, and (2) a broader, but largely overlapping distribution of intracellular inclusion-bodies composed of aggregated proteins, lipids, and organelles, referred to as Lewy pathology (LP) (Mahul-Mellier et al., 2020). What causes these two features of the disease and their relationship to PD pathogenesis continues to be debated, but there are two widely held theories. One is that aging-related mitochondrial dysfunction drives pathogenesis, leading to a bioenergetic crisis, proteostatic dysfunction, and ultimately death of vulnerable neurons. The other theory is that propagated forms of misfolded alpha-synuclein (αSYN), which is a major component of LP, drive pathogenesis in synaptically coupled networks (Zhang et al., 2018; Merchant et al., 2019). Mutations in the αSYN gene *SNCA* (*PARK1*), as well as duplications or triplications of *SNCA* (*PARK4*) are inextricably linked to some familial forms of PD (Bloem et al., 2021; Oliveira et al., 2021). Both theories have wide-ranging experimental and clinical support that has been reviewed elsewhere (Bloem et al., 2021; Day & Mullin, 2021; Vazquez-Velez & Zoghbi, 2021).

We will not wade into this scientific debate. Rather, we will advance the hypothesis that the neuronal phenotype and the way in which vulnerable neurons handle Ca^{2+} puts them at risk for both, mitochondrial dysfunction and αSYN pathology. Our focus will be on the functions of voltage-gated Ca^{2+} channels (Ca_v) in those dopaminergic neurons that are at greatest risk in PD. This literature is of translational interest as epidemiological studies have consistently found that the use of negative allosteric modulators of Ca_v1 channels (dihydropyridines) is associated with a reduced risk of developing PD (Becker et al., 2008; Ritz et al., 2010; Pasternak et al., 2012; Lee et al., 2014; Gudala et al., 2015; Tseng et al., 2021).

Selective Neuronal Vulnerability in Parkinson's

The loss of neurons in PD is very selective. As mentioned above, the cardinal, defining feature of PD is a disability in initiating and sustaining fluid purposeful movement, that is responsive to treatment with levodopa, a blood-brain-barrier permissive precursor of dopamine (Abbott, 2010). These motor deficits are attributable to the dysfunction and degeneration of dopaminergic midbrain neurons, particularly in the ventral tier of the SNc that innervate the striatum, a part of the basal ganglia (Damier et al., 1999). Although there is some debate about the extent of dopaminergic neuron loss that is necessary for the emergence of PD motor symptoms, it is worth considering that in their entirety, these neurons constitute a tiny fraction of the roughly 80–90 billion neurons in the human brain (~800,000 or ~0.01%) (Pakkenberg et al., 1991; Brichta & Greengard, 2014). In fact, the number of vulnerable neurons is likely to be even smaller. For example, within the mesencephalic dopaminergic neuron population, some subpopulations are largely resilient in PD, like those expressing the Ca^{2+} binding protein calbindin (CB) in the dorsal tier of the SNc, and those in the ventral tegmental area (VTA) (German et al., 1992; Damier et al., 1999). Even within the most vulnerable ventral tier of the SNc, there is heterogeneity that might influence vulnerability (Poulin et al., 2020).

Although the precise timing of neuronal loss and its association with PD symptoms is difficult to rigorously determine, SNc dopaminergic neuronal loss begins in the prodromal period of the disease and precedes that in other regions (Surmeier et al., 2017b). Later in the course of PD, there is a neuronal loss in several other brain regions as well, including the locus coeruleus (LC), pedunculopontine nucleus (PPN), raphe nuclei (RN), intralaminar thalamus, basal forebrain (BF), and dorsal motor nucleus of the vagus (DMV) (Surmeier et al., 2017b; Giguere et al., 2018).

The relationship between LP and neurodegeneration in PD is still debated. LP has been hypothesized to spread from a peripheral seeding site to the caudal medulla or olfactory system and then to other brain regions (Braak et al., 2003). But roughly half of PD cases do not conform to this model (Beach et al., 2009; Rietdijk et al., 2017). In these other cases, a central, brain seeding site has been suggested (Borghammer et al., 2021). Nevertheless, most of the nuclei that eventually display LP (e.g., SNc, RN, DMV, LC, PPN) are the ones in which there is some degree of neuronal loss (Surmeier et al., 2017b). What is less clear is the temporal (and potentially causal) relationship between neuronal loss and LP. As noted above, in the SNc neuronal loss precedes the appearance of LP (Surmeier et al., 2017b). Moreover, the severity of LP is not strongly correlated with neuronal loss or disease symptoms (Parkkinen et al., 2005). Postmortem analysis of brains taken from familial cases of PD has often found little or no LP, despite neuronal loss and levodopa-responsive motor disability (Wszolek et al., 2004; Ross et al., 2006; Hasegawa et al., 2009; Poulopoulos et al., 2012; Doherty & Hardy, 2013). What these studies suggest is that neuronal loss/dysfunction and PD can occur in the absence of LP. However, there is evidence that LP can promote neurodegeneration and that it is not merely a PD tombstone (Mahul-Mellier et al., 2020). What is more difficult to determine from this literature is how αSYN pathology begins and how it disrupts neuronal function (Eschbach et al., 2015; Kiechle et al., 2019; Oliveira et al., 2021). As with other potential pathological agents, it is critical to distinguish between biologically meaningful events and those that are merely pharmacological artifacts. Few studies have attempted to do the sort of rigorous quantitative analysis that would distinguish between these two possibilities (Abdelmotilib et al., 2017).

Neuronal Phenotype and Ca²⁺ Signaling

Genetic studies have failed to find clear evidence in favor of the hypothesis that neurons preferentially express genes linked to PD, which would have provided a simple explanation for pathogenesis (Reynolds et al., 2019). However, neurons in the brain vary enormously in their phenotype and this diversity may obscure a relationship when bulk profiling methods are used. More refined profiling of vulnerable neuronal subpopulations (Agarwal et al., 2020; Poulin et al., 2020; Gonzalez-Rodriguez et al., 2021) may find a connection between the expression level of genes, like *SCNA*, and the risk of neurodegeneration in PD.

Despite this gap, SNc dopaminergic neurons have a set of anatomical and physiological traits that are likely to contribute to their vulnerability. Many of these traits also are found in other vulnerable neurons, like noradrenergic LC neurons, serotonergic RN neurons, cholinergic PPN, DMV, and BF neurons. One of the traits that can be readily linked to PD risk is an extended terminal axonal arbor that is invested with a large number of transmitter release sites (Wong et al., 2019). In rodents, the terminal axonal arbor of SNc dopaminergic neurons in the striatum has a scale of millimeters and has been estimated to have more than 200,000 transmitter release sites (Bolam & Pissadaki, 2012). In humans, with the expansion of the striatum in parallel with the cerebral cortex, the size of the arbor in SNc dopaminergic neurons appears much larger and the number of release sites is estimated to 1–2 million for one individual axon (Pissadaki & Bolam, 2013; Diederich et al., 2019). In neighboring VTA dopaminergic neurons that are less vulnerable in PD, the axonal arbor appears to be much smaller (Matsuda et al., 2009). Pointing to the importance of the axonal arbor for defining neuronal vulnerability, decreasing its size in culture significantly reduces the vulnerability of SNc dopaminergic neurons to PD-linked stressors (Pacelli et al., 2015). The large SNc neuron axon arborization comes with a corresponding bioenergetic and proteostatic burden to support transmitter release and regenerative activity (Pissadaki & Bolam, 2013; Wong et al., 2019), which could render these neurons more vulnerable to PD stressors.

The somatodendritic physiology of SNc dopaminergic neurons also is distinctive. Unlike the vast majority of neurons in the brain, SNc dopaminergic neurons are autonomous pacemakers that spike at a slow and regular rate in the absence of synaptic stimulation (Fig. 1a). In vivo, synaptic input modulates this basal activity leading to pauses, as well as irregular or burst firing (Paladini & Roeper, 2014; Otomo et al., 2020).

There are two hypotheses that have been advanced to explain why SNc dopaminergic neurons are pacemakers. The most widely embraced idea is that pacemaking, which drives axonal spiking and transmitter release, is necessary for bidirectional control of striatal DA levels and goal-directed movement (Gerfen & Surmeier, 2011; Yttri & Dudman, 2016). Some variants of this hypothesis are at the heart of the classical theory of basal ganglia-linked movement disorders (Albin et al., 1989). While there is a local control of striatal DA release, primarily by striatal cholinergic interneurons (Ding et al., 2010; Threlfell et al., 2012; Shin et al., 2015; Mohebi et al., 2019), pacemaking provides a means of maintaining a tonic dopaminergic tone; not only in the striatum but throughout the basal ganglia, including the substantia nigra (Rice & Patel, 2015; Wichmann, 2019), as SNc neurons release dopamine not only from their axons but also from their somatodendritic region (Cheramy et al., 1981; Ludwig et al., 2016). Given the therapeutic efficacy of tonic dopaminergic signaling in PD patients (LeWitt, 2016), the importance of this tonic signaling cannot be underestimated.

Yet, there are features of the pacemaking and action potential phenotype of SNc dopaminergic neurons that cannot be accounted for by the need for tonic dopamine release in the striatum. The most prominent of these is associated with Ca^{2+} handling (Zampese & Surmeier, 2020). In particular, SNc dopaminergic neurons have (1) very broad spikes, which promote Ca^{2+} entry through plasma membrane Ca_v channels (Grace & Bunney, 1980; Bean, 2007; Hage & Khaliq, 2015), (2) large cytosolic Ca^{2+} transients associated with spikes, which reflect a combination of enhanced Ca^{2+} entry trough Ca_v channels, weak intrinsic Ca^{2+} buffering and release of Ca^{2+} from intracellular stores (Foehring et al., 2009; Guzman et al., 2009, 2018), and (3) voltage oscillations, even in the absence of spiking, that are dependent upon Ca_v channel activity (Nedergaard et al., 1993; Mercuri et al., 1994; Chan et al., 2007). These observations point to another role for Ca^{2+} beyond just helping to drive pacemaking, as this could be readily accomplished with monovalent cation channels.

Like most neurons in the brain, SNc dopaminergic neurons and other at-risk neurons, appear to express all major classes of voltage-dependent Ca^{2+} channels. As described in chapter "Subunit Architecture and Atomic Structure of Voltage Gated Ca^{2+} Channels" by Dr. Catterall, these Ca_v channels are diverse in their composition of pore-forming and regulatory subunits. Pore-forming $Ca_v\alpha_1$ subunits come from one of three classes (Cav1, Cav2 and Cav3, corresponding to L-, P/Q-, N-, R-, and T-type channels). Their associated regulatory $Ca_v\beta$ and $Ca_v\alpha_2\delta$ subunits also are heterogeneous (Dolphin, 2018). However, the exact subunit and splice variant composition and the physiological functions of distinct Ca_v channel complexes in axonal and somatodendritic regions of dopaminergic neurons are not fully defined. What might differentiate vulnerable and resistant neurons are the relative expression levels of these channels, their subcellular distribution, and their roles for intracellular Ca^{2+} signaling. On this score, it should be recognized that the functional significance of a channel type cannot be readily gleaned from its expression level or from conventional patch-clamp measurements from the soma, which constitutes only a small fraction of the total membrane surface area of the cell. A major step forward in our effort to

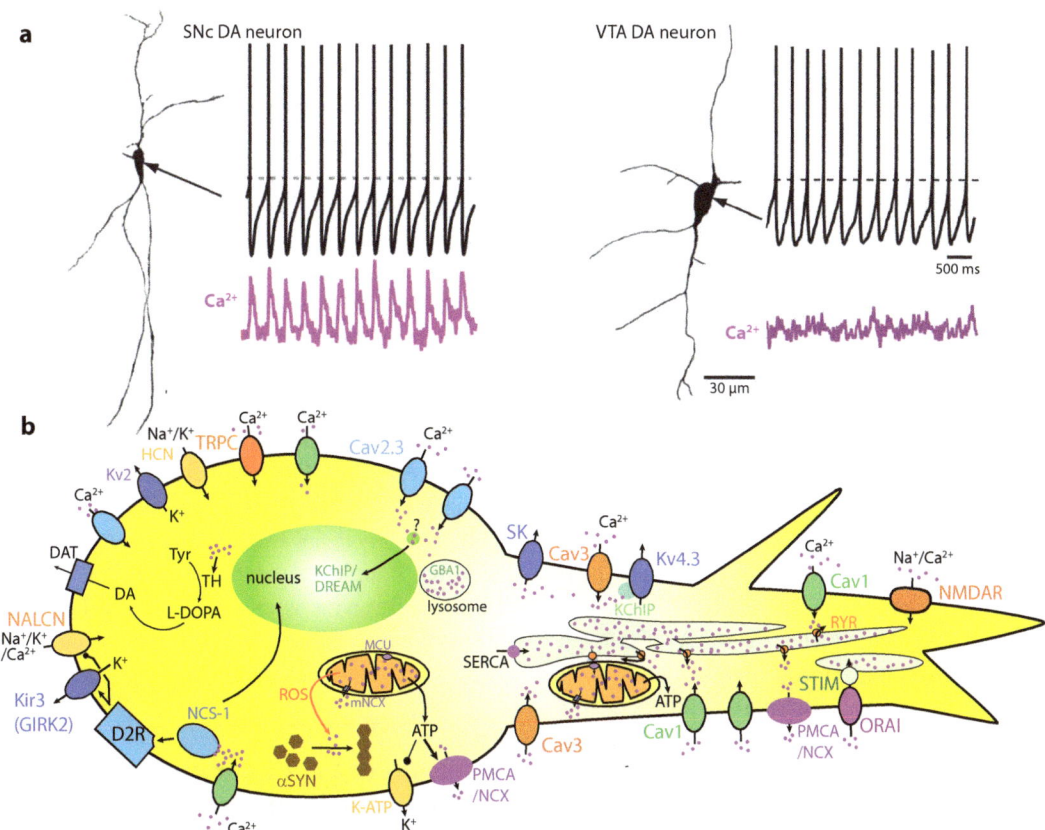

Fig. 1 Activity dependent Ca^{2+} signalling in somatodendrites if dopaminergic midbrain neurons. (**a**) Simultaneous whole-cell current clamp brain slice recordings of pacemaker activity (black trace, mV) from mouse dopaminergic neurons in SN and VTA, and of corresponding somatodendritic Ca^{2+} oscillations (purple traces, Ca^{2+} imaging). Note the prominent activity-related Ca^{2+} oscillations only in SN dopaminergic neurons. (**b**) Cartoon depicts a selection of somatodendritic ion channels, receptors, and transporters that generate or modulate the electrical activity patterns of dopaminergic midbrain neurons, cytosolic Ca^{2+} levels (small purple dots), and Ca^{2+} related signaling pathways. Note that functional expression of distinct Ca_v channels differs in vulnerable SN and resistant VTA dopaminergic neurons. For details, please see the text. (Figure adapted from Guzman et al., 2010; Liss & Striessnig, 2019). Abbreviations: αSYN α-synuclein, Ca_v voltage-gated Ca^{2+} channel, $D2R$ dopamine D2 autoreceptor, DA dopamine, DAT dopamine-transporter, $DREAM$ Downstream regulatory element antagonistic modulator, ER endoplasmatic reticulum, $GBA1$ glucocerebrosidase, HCN hyperpolarization-activated cyclic nucleotide gated cation channel, $K\text{-}ATP$ ATP-sensitive K^+ channel, $KChip3$ K_v channel interacting protein (K_v4 β-subunit), $Kir3/GIRK2$ G-protein coupled inwardly rectifying K^+ channel, K_v voltage-gated K^+ channel, K_v4 voltage and Ca^{2+} regulated A-type K^+ channel, $L\text{-}DOPA$ L-3,4-dihydroxyphenylalanin, MCU mitochondrial Ca^{2+} transporter, $mNCX$ mitochondrial Na^+/Ca^{2+} exchanger, $NCS\text{-}1$ neuronal Ca^{2+} sensor 1, NCX Na^+/Ca^{2+} exchanger, $NMDAR$ N-methyl-D-aspartate glutamate receptor, $ORA1$ Ca^{2+} release activated Ca^{2+} channel modulator protein 1, P phosphate, $PMCA$ Plasma membrane Ca^{2+} ATPase, ROS reactive oxygen species, RYR ryanodine receptor, $SERCA$ sarcoplasmic/endoplasmic reticulum Ca^{2+} ATPase, SK small conductance Ca^{2+} sensitive K^+ channel, SNc substantia nigra pars compacta, $STIM$ stromal interaction molecule 1, $TRPC$ transient receptor potential channel subtype C, Tyr tyrosine, VTA ventral tegmental area

better understand Ca^{2+} dynamics in SNc dopaminergic (and other at-risk) neurons has been to combine somatic patch-clamp techniques with quantitative Ca^{2+} imaging in somatic and dendritic compartments, using two-photon laser scanning microscopy (2PLSM) and ratiometric dyes, like Fura2 (Guzman et al., 2018).

Studies using these and complementary approaches have shown that the relative importance of distinct Ca^{2+} channel subtypes is dependent upon subcellular location (Fig. 1a). In the somatic region, spikes in SNc dopaminergic neurons are accompanied by a transient elevation in cytosolic Ca^{2+} that is driven in large part by high-voltage-activated (HVA) R-type $Ca_v2.3$ channels (Benkert et al., 2019). Alternative splicing of auxiliary $Ca_v\beta_2$ subunits in dopaminergic neurons enhances $Ca_v2.3$ channel currents during pacemaking and can change their engagement with the neuronal calcium sensor NCS-1 (Benkert et al., 2019; Siller et al., 2022). High-voltage-activated (HVA) $Ca_v1.3$ L-type channels also are likely contributors to somatic Ca^{2+} signaling in the somatic region of dopaminergic SNc neurons, as in a variety of other neurons (Ortner et al., 2017; Guzman et al., 2018). In proximal dendrites, low-voltage-activated (LVA) Ca_v3 Ca^{2+} channels make an important contribution to the cytosolic Ca^{2+} signal (Wolfart & Roeper, 2002; Evans et al., 2017; Guzman et al., 2018). In both proximal and distal dendrites, $Ca_v1.3$ L-type channels, which activate at more negative membrane potentials than $Ca_v1.2$ and Ca_v2 channels, are prominent determinants of activity-related Ca^{2+} transients in SNc neurons and other vulnerable neurons that have been studied, like LC and DMV neurons (Goldberg et al., 2012; Sanchez-Padilla et al., 2014; Guzman et al., 2018). It is worth mentioning that the effort to dissect the contribution of these channels has been hampered by the lack of selective antibodies, drugs, and genetic tools, like virally-delivered shRNA constructs (Guzman et al., 2018), or transgenic mice with mutated $Ca_v1.2$ channels (Sinnegger-Brauns et al., 2004), essential to move the field forward.

In both somatic and dendritic regions, Ca^{2+} entering ventral tier SNc dopaminergic neurons during the broad action potential is only weakly buffered by cytoplasmic proteins (Foehring et al., 2009), allowing Ca^{2+} to readily diffuse and interact with other proteins. In contrast, neighboring dorsal tier SNc and VTA dopaminergic neurons robustly express the Ca^{2+} binding protein calbindin, which helps to clamp cytosolic Ca^{2+} concen-

trations (Devine & Kittler, 2018). Interestingly, calbindin also appears to dampen axonal dopamine release (Brimblecombe et al., 2019), but how this precisely happens is unclear. In ventral tier SNc dopaminergic neurons, (and vulnerable LC neurons), the somatodendritic cytosolic Ca^{2+} oscillation reflects Ca^{2+} influx through plasma membrane Ca_v channels, weak buffering, and release of Ca^{2+} from intracellular stores (Sanchez-Padilla et al., 2014). As in other excitable cells, plasma membrane Ca_v1 channels appear to be juxtaposed to ryanodine receptors (RYRs) in the endoplasmic reticulum (ER), allowing plasma membrane Ca^{2+} influx to trigger Ca^{2+}-induced Ca^{2+} release (CICR) from the ER (Zampese & Surmeier, 2020). In this way, spiking becomes linked not only to ER Ca^{2+} signaling but to organelles that couple to the ER, like mitochondria (Fig. 1b). But what is the physiological function of Ca_v channel signaling in SNc dopaminergic and other at-risk neurons?

Functions of Ca_v Channels in Vulnerable Neurons

In axon terminals, Ca_v channels are particularly important for triggering neurotransmitter release (Sulzer et al., 2016). As outlined above, one of the key determinants of neuronal vulnerability in PD appears to be the size of the axonal arbor. Understanding the physiological roles of specific classes of Ca_v channels in this arbor, and how they might contribute to PD is a major challenge facing the field. Fortunately, the rapid evolution of genetically encoded optical sensors for Ca^{2+}, dopamine, and related targets is making this fertile ground for study.

What we do know is illustrated in Fig. 2. Using fast-scan cyclic voltammetry (FSCV), Cragg and her colleagues have shown that $Ca_v2.1/2.2$ channels (P/Q- and N-type) play expected roles in activity-dependent dopamine release (Brimblecombe et al., 2015). Interestingly, a recent study identified lower levels of these Ca_v-subtypes in dopamine neurons from adult PARK1 mice, indicating altered axonal dopamine release, in line with a described impairment

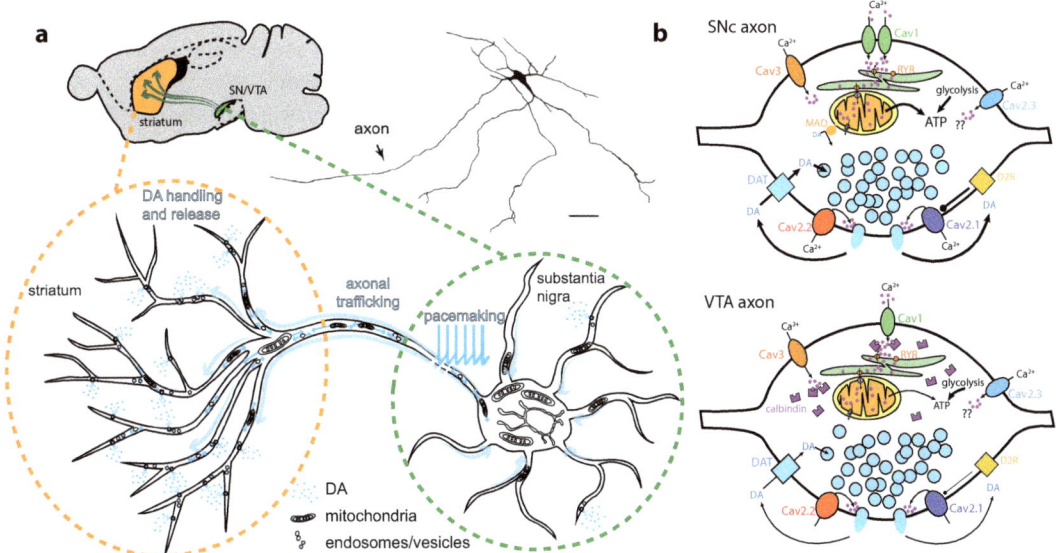

Fig. 2 Distinct roles of Ca$_v$ channels in axons of dopaminergic midbrain neurons. (**a**) Upper left: Schematic sagittal section of a mouse brain, illustrating the anatomical location of the substantia nigra and the ventral tegmental area (green) as well as their striatal axonal projections (red). Upper right: drawing of a typical rat dopaminergic midbrain neuron after Lucifer yellow injection (scale bar: 30 μm, Modified from Grace & Bunney, 1983). Lower: Schematic representation of a dopaminergic substania nigra neuron and its axonal projections. (**b**) Speculative schematic representation of en passant axon DA (blue circles) release sites for SNc and VTA dopaminergic neurons (projecting to the dorsal and the ventral striatum, respectively). The cartoon and thickness of arrows suggest a differential Cav channel regulation of dopamine release, terminal bioenergetics, DAT, and D2-autoreceptors. Note that the role of Ca$_v$2.3 (R-type) Ca$_v$ for axonal DA-release is currently unclear. For details, please see the text (Figure adapted from Brimblecombe et al., 2015; Zampese & Surmeier, 2020). Abbreviations: *CB* Ca^{2+} buffering protein calbindin, *Ca$_v$* voltage-gated Ca^{2+} channel, *D2R* dopamine D2 autoreceptor, *DA* dopamine, *DAT* dopamine-transporter, *RYR* ryanodine receptor, *SNc* substantia nigra pars compacta, *VTA* ventral tegmental areal

of striatal dopamine signaling in these mice (Kurz et al., 2010; Zhong et al., 2021).

While a role of Ca$_v$2.3 R-type channels for axonal dopamine release has not yet been addressed, Ca$_v$3 and Ca$_v$1 channels do participate in regulating axonal dopamine release in the highly arborized axons projecting to the dorsal striatum, but to a much lesser extent in those projecting to the ventral striatum (Brimblecombe et al., 2015). Precisely why there is a difference remains unclear, but the effects of Ca$_v$1 and Ca$_v$3 channel inhibitors were most evident with repetitive stimulation (Brimblecombe et al., 2015). In other systems where it has been studied in-depth, transmitter release in response to repetitive stimulation depends upon mitochondrial ATP production stimulated by Ca^{2+} entry (Rangaraju et al., 2014; Devine & Kittler, 2018). Given what we know about the coupling of Ca$_v$1 channels to ER

CICR and mitochondria in the somatodendritic region, it is tempting to speculate that their axon-terminal involvement (and possibly that of Ca$_v$3 channels) stems from their regulation of bioenergetics.

Besides participating in dopamine release, what Ca$_v$ channels do in SNc dopaminergic neurons has been controversial and is still not completely clear. Although early work suggested that Ca$_v$1 channels were necessary for pacemaking (Nedergaard et al., 1993; Mercuri et al., 1994), subsequent work, e.g. using channel-specific concentrations of Ca$_v$1 channel inhibitors and 2PLSM Ca^{2+} imaging to monitor channel function, revealed that although they made pacemaking more robust, they were not necessary (Guzman et al., 2009; Dragicevic et al., 2014; Shin et al., 2022). Ca$_v$2.3 and Ca$_v$3 channels also are not necessary for pacemaking, but rather modulate it (Wolfart & Roeper, 2002; Poetschke et al., 2015; Benkert et al., 2019; Evans et al., 2017). This makes good sense, as channels that are selective for monovalent cations (e.g., NALCN sodium leak or HCN cation channels), which are less expensive from a bioenergetic standpoint, are more commonly used to drive autonomous activity in neurons (Guzman et al., 2009; Khaliq & Bean, 2010; Lutas et al., 2016; Um et al., 2021).

As plasma membrane Ca^{2+} influx through Ca_v1 channels is not necessary for electrogenesis, what physiological function do they play? One key role is to control neuronal bioenergetics. The CICR, triggered by the opening of plasma membrane Ca_v1 channels, promotes Ca^{2+} loading of mitochondria (Fig. 1b), at specialized junctions (mitochondria-associated membranes or MAMs). This loading stimulates oxidative phosphorylation (OXPHOS) and ATP production by activating the malate-aspartate shuttle (MAS) and disinhibiting matrix tricarboxylic acid cycle (TCA) enzymes (Zampese & Surmeier, 2020; Diaz-Garcia et al., 2021). As in muscle (Viola & Hool, 2010; Diaz-Vegas et al., 2018), this pathway serves as a feed-forward control mechanism for mitochondrial ATP production that anticipates demand, rather than waiting for ATP levels to fall and disinhibit mitochondrial complex V (Zampese & Surmeier, 2020). This signaling mechanism provides a beautiful way of scaling somatodendritic bioenergetic production to activity. The feed-forward design also should help to ensure that in times of stress or elevated bioenergetic demand (e.g. prey-predator situations), these critical neurons, which modulate large swaths of motor and cognitive circuitry, do not shut down (Dragicevic et al., 2015; Duda et al., 2016; Surmeier et al., 2017b).

This control mechanism also could be scaled to fit the bioenergetic needs of the axonal arbor. While a great deal has been made of the local bioenergetic needs associated with regenerative activity and transmitter release (Bolam & Pissadaki, 2012; Pacelli et al., 2015; Sheng, 2017), the anabolic and catabolic demands associated with the arbor rest largely on the somatodendritic region where most of the protein and vesicular production/degradation machinery reside (Fig. 2). These processes are bioenergetically demanding and should scale with the dimensions of the axon.

However, feedforward control of mitochondrial OXPHOS comes at a cost. The electron flux necessary for OXPHOS generates reactive oxygen species (ROS, Fig. 1b), particularly at times when the stimulation precedes demand and the high mitochondrial membrane potential causes electrons to "stall" - increasing the probability of them moving to molecular oxygen and generating ROS (Murphy, 2009). Although redox signaling is important to a variety of cellular functions, excessive ROS generation can damage mitochondrial DNA, proteins, and lipids. Thus, all cells have a variety of oxidant defense systems localized to distinct subcellular regions (Murphy, 2009; Trist et al., 2019), but these systems are imperfect. Using genetically encoded sensors to monitor mitochondrial matrix redox status in ex vivo brain slices, it has been revealed that in pacemaking SNc dopaminergic neurons, mitochondrial matrix oxidant stress is high, particularly in dendritic and axonal mitochondria (Pacelli et al., 2015; Graves et al., 2020). A similar situation is found in vulnerable LC, DMV, and PPN neurons (Goldberg et al., 2012; Sanchez-Padilla et al., 2014; unpublished observations). In the SNc, elevated ROS generation is associated with mitochondrial DNA deletions and LP (Bender et al., 2006; Sanders et al., 2014; Dolle et al., 2016), and reduced mitochondrial complex I (MCI) function (Schapira et al., 1990; Reeve et al., 2014). Dopamine itself also increases mitochondrial oxidant stress, particularly in the massive axonal arbor of SNc neurons (Graves et al., 2020; Pacelli et al., 2015). This oxidant stress may be exacerbated by iron accumulation and the Fenton reaction (Berg & Youdim, 2006; Trist et al., 2019). Together, these factors cause progressive mitochondrial damage, further increasing metabolic stress.

In addition to damaging mitochondrial proteins and DNA, mitochondrial oxidant stress compromises lysosomal function (Burbulla et al., 2017) and may contribute to the disruption of chaperone-mediated autophagy (Martinez-Vicente & Cuervo, 2007), both of which are necessary for proper proteostasis. For instance, lysosomes degrade αSYN (Minakaki et al., 2020). Moreover, variants in the lysosomal glucocerebrosidase (coded by the *GBA-1* gene) are a major genetic risk factor for PD and can produce not only deficits in lysosomal function but also in intracellular Ca^{2+} homeostasis (Schondorf et al., 2014; Avenali et al., 2020). This could be particularly problematic in SNc dopaminergic neurons where activity-associated Ca^{2+} oscillations normally reach the micromolar concentrations, leading to the activation of calcineurin and the protease calpain, both of which promote protein aggregation (Dufty et al., 2007; Dehay et al., 2010; Diepenbroek et al., 2014; Guzman et al., 2018).

Thus, at-risk neurons - like SNc dopaminergic neurons - may begin to falter with aging, as the accumulated damage to mitochondria further compromises their function in a vicious cycle. Indeed, there is roughly a 5–10% decrease in the apparent density of SNc dopaminergic neurons in healthy individuals and a common hypothesis sees Parkinson's as a form of accelerated aging (Fearnley & Lees, 1991; Collier et al., 2011).

As illustrated in Fig. 1b, Ca_v channels regulate SNc dopaminergic neurons in a variety of other ways that can influence electrical activity and metabolic stress. For example, in SNc dopaminergic neurons, Ca^{2+} entry through Ca_v channels activates small conductance K^+ (SK) channels that give rise to a spike after-hyperpolarization (AHP) that controls the rate and the regularity of spiking and the switch to burst activity (Wolfart et al., 2001; Wolfart & Roeper, 2002; Iyer et al., 2017). In concert with K_v2 channels, big conductance Ca^{2+}-activated K^+ channels (BK) also regulate action potential repolarization in SNc dopaminergic neurons (Kimm et al., 2015); however, the role of specific Ca^{2+} channels in gating BK channels has not been determined.

Another spike-regulating mechanism found in SNc dopaminergic neurons involves voltage-activated A-type K_v4 K^+ channels. These channels are functionally coupled to Ca_v channels (Anderson et al., 2013) and are composed of pore-forming K_v4 subunits and regulatory DPPx and KChIP subunits (Rhodes et al., 2004; Anderson et al., 2013; Bahring, 2018; Carrillo-

Reid et al., 2019) (see also chapter "Cav3 Calcium Channel Interactions with Potassium Channels" by Dr. Turner). Ca_v channels positively modulate K_v4 gating through the Ca^{2+}-binding KChip1-4 subunits, members of the neuronal calcium sensor (NCS) superfamily (Mundhenk et al., 2019). This serves as a negative feedback system to modulate pacemaker activity and dendritic excitability of neurons (Liss et al., 2001; Kim et al., 2007; Carrillo-Reid et al., 2019). In SNc dopaminergic neurons, K_v4 channels are composed primarily of $K_v4.3$ and KChIP3 subunits, while in neighboring VTA dopaminergic neurons, K_v4 channels are also composed of $K_v4.2$ and KChIP4 subunits (Liss et al., 2001; Subramaniam & Roeper, 2016; Haddjeri-Hopkins et al., 2021). This subunit combination endows K_v4 channels in VTA dopaminergic neurons with slower inactivation kinetics (Holmqvist et al., 2002), which, along with lower expression of hyperpolarization and cAMP-activated cation (HCN) channels (Franz et al., 2000; Neuhoff et al., 2002; Lammel et al., 2008), enhances the ability of inhibitory signaling to pause spiking (Paladini & Roeper, 2014; Tarfa et al., 2017).

Yet another way in which Ca_v channels and intracellular Ca^{2+} regulate K^+ channel function in dopaminergic neurons is through NCS-1, another member of the NCS superfamily (Mundhenk et al., 2019). D2 dopamine receptors (D2Rs) act as autoreceptors at both axonal and dendritic dopamine release sites (Ford, 2014; Hikima et al., 2021). Like other autoreceptors, they are positively coupled to Kir3 channels (mainly GIRK2) through $G_{i/o}$ proteins, but a variety of additional D2-autoreceptor (D2-AR) signaling mechanisms are described, including inhibition of NALCN Na^+ channels (Ford, 2014; Philippart & Khaliq, 2018; Chen et al., 2020; Um et al., 2021).

In vulnerable SNc dopaminergic neurons, where Ca^{2+} is weakly buffered, somatodendritic D2-AR signaling is enhanced by NCS-1, in a Ca^{2+} and Ca_v channel-dependent fashion (Catoni et al., 2019). Ca^{2+} influx through somatodendritic $Ca_v1.3$ and also Ca_v3 channels effectively activates NCS-1 to blunt the desensitization of D2Rs

(Dragicevic et al., 2014; Poetschke et al., 2015). This is most likely accomplished by Ca^{2+} stimulated NCS-1 binding to the third intracellular loop of the D2R, which diminishes receptor internalization (Kabbani et al., 2002; Catoni et al., 2019; Mannal et al., 2021). In this way, NCS-1 helps sustain D2-receptor activation of Kir3 channels, increasing autoreceptor efficacy in negative feedback control of spiking and dopamine release. D2R coupling to Kir3 channels also enables high extracellular DA concentrations to hyperpolarize SNc dopaminergic neurons sufficiently to de-inactivate Ca_v3 T-type Ca^{2+} channels and promote rebound burst spiking (Evans et al., 2017). Thus, D2-receptor signaling may diminish stress on SNc dopaminergic neurons and be protective (Catoni et al., 2019).

An additional level of complexity is created by considering the role of Ca_v channels in regulating gene expression. Ca^{2+} entry into the somatodendritic region is a well-known means of coupling neuronal activity to gene expression (Lu et al., 2015; Heck et al., 2021). Although there are several signaling pathways that have been reported to be involved in this coupling, one that may be particularly important in SNc dopaminergic neurons is KChIP3, the Ca^{2+} sensing subunit of K_v4 A-type channels, which is also known as downstream regulatory element antagonist modulator (DREAM) (Buxbaum, 2004). DREAM can shuttle from the plasma membrane to the nucleus to act as a Ca^{2+}-dependent transcription-repressor (Zaidi et al., 2004; Naranjo & Mellstrom, 2012; Burgoyne et al., 2019). Among others, DREAM regulates the expression of Ca_v1 channel genes (Ronkainen et al., 2011) and has been implicated in modulating levodopa-induced dyskinesia, a common side-effect of levodopa treatment in late-stage PD patients (Ruiz-DeDiego et al., 2015).

In addition to DREAM, there are a variety of other ways in which activity-dependent Ca^{2+} entry may be coupled to gene expression in dopaminergic neurons. For example, NCS-1, which at the plasma-membrane can modulate Ca_v, K_v4, and D2-AR function, is also found at the nucleus and appears to control the expression of genes encoding for mitochondrial OXPHOS proteins

(Naranjo & Mellstrom, 2012; Burgoyne et al., 2019; Nakamura et al., 2019; Simons et al., 2019). Ca_v channels also may regulate neuronal gene expression through histone deacetylases (HDACs) (Mazzocchi et al., 2020). Class IIa HDACs (HDAC 4, 5, 7 & 9) are all expressed in ventral tier SNc dopaminergic neurons and can modulate Ca^{2+} and K^+ channels (Lugenbiel et al., 2018; Urbano et al., 2018), DREAM function, and the expression of neuronal Ca^{2+} transporters (Formisano et al., 2020). Particularly, the activity- and Ca^{2+}-dependent relocation of class IIa HDACs from the cytoplasm to the nucleus seems to be detrimental for ventral tier SNc neurons (Mazzocchi et al., 2020; Shukla & Tekwani, 2020). HDAC4 location was shown to be enriched in the nucleus in human iPSC-derived dopaminergic neurons from GBA-1 and some sporadic PD-patients, and pharmacological modulation of nuclear HDAC4 location rescued the pathological phenotype of these human neurons (Lang et al., 2019). Also for other vulnerable neurons, roles for HDACs are emerging (Corradini et al., 2014; Jiang & Zsombok, 2014).

The Linkage Between Ca_v Channels, Ca^{2+} Signaling and PD

Although there are strong reasons for linking Ca_v channels and Ca^{2+} signaling in dopaminergic neurons to their differential vulnerability in PD, there is a substantial amount of controversy swirling around this topic. One of the major limitations in the field has been the lack of models of PD pathogenesis that have reasonable predictive validity. The vast majority of work attempting to directly link Ca^{2+} channels to pathogenesis is based upon systemically administered or locally injected toxins, like 1-methyl-4-phenyl-1,2,3,6-tetrahydropyridine (MPTP) or 6-hydroxydopamine (6-OHDA) (Surmeier et al., 2017a; Liss & Striessnig, 2019). While these toxins induce preferential degeneration of SNc dopaminergic neurons, compounds that ameliorate this damage have consistently failed in clinical trials (Surmeier et al., 2017b). A plausible explanation for this disconnect is that these PD-toxins kill neurons through mechanisms that have little to do with PD (Surmeier et al., 2017b; Liss & Striessnig, 2019; Ortner, 2021). As a consequence, it is very difficult to draw firm conclusions about the role of Ca_v channels and related Ca^{2+} signaling in pathogenesis from this literature (to which both authors have contributed). Mice harboring PD-linked genetic mutations have consistently failed to produce levodopa-responsive parkinsonism, making them of limited utility in testing hypotheses about pathogenesis (Dawson et al., 2010; Guzman et al., 2010; Surmeier et al., 2017b). Because of this gap, the theoretical linkage between Ca^{2+} signaling and PD pathogenesis relies heavily upon two intermediates that have a well-established causal relationship to the disease: αSYN and mitochondria.

There are several lines of study linking Ca_v channels and Ca^{2+} to αSYN pathology and neurodegeneration (Leandrou et al., 2019). Elevated Ca^{2+} can promote aggregation of αSYN in the test tube, presumably by charge-shielding the COOH terminal tails; in a cellular environment, Ca^{2+} can act this way as well as by activating the Ca^{2+}-dependent protease calpain, which cleaves this tail region to promote aggregation (Dufty et al., 2007; Nath et al., 2011; Diepenbroek et al., 2014; Rcom-H'cheo-Gauthier et al., 2016). ROS can also promote αSYN misfolding and aggregation (Nath et al., 2011; Follett et al., 2013). As misfolded forms of αSYN increase cytosolic and mitochondrial oxidant stress (Hsu et al., 2000; Vila et al., 2000; Norris et al., 2003; Dryanovski et al., 2013; Hu et al., 2019), and the Ca^{2+} stimulation of mitochondrial OXPHOS increases the generation of ROS, there is a potential positive feedback loop created that could promote PD-pathology. Ca^{2+}, mitochondrial stress, and misfolded αSYN also disrupt lysosomal and autophagic function, necessary for Ca^{2+} homeostasis, αSYN degradation, and mitochondrial quality control (Mosharov et al., 2009; Burbulla et al., 2017). Lastly, as αSYN is a synaptic protein, its expression should be relatively high in neurons with a large number of axonal transmitter release sites, further increasing the proaggregate environment found in vulnerable SNc neurons (Volpicelli-Daley et al., 2011; Henderson et al., 2017, 2019; Courte et al., 2020).

Another Ca^{2+}-linked driver of pathology may be redox regulation of plasma membrane ion channels. In SNc dopaminergic neurons from mice over-expressing the A53T mutant form of αSYN (PARK1), elevated oxidant stress diminished K_v4 channel function, leading to an acceleration in pacemaking rate in vivo (Subramaniam et al., 2014). This sets the stage for a "death spiral" involving Ca^{2+} and αSYN, described above. In contrast, in response to nominally the same A53T aSYN overexpression, less vulnerabe cholinergic DMV neurons seem to downregulate Ca^{2+} channel expression to keep cellular redox levels within the normal range, potentially explaining their relative resistance to αSYN pathology (Lasser-Katz et al., 2017). Human SNc dopaminergic neurons that survive in PD display higher levels of $K_v4.3$ mRNA (Dragicevic et al., 2015), suggesting that slowing activity makes them less vulnerable to PD pathogenesis or that slowing activity is an adaptation to disease progression.

As outlined Ca^{2+} entry through Ca_v channels in vulnerable neurons stimulates feedforward control of mitochondrial OXPHOS, resulting in increased oxidant stress and mitochondrial damage. Elevated mitochondrial DNA damage and reduced MCI function are hallmarks of the SNc in PD patients (Schapira et al., 1990). Despite its plausibility, whether damage to mitochondrial OXPHOS capacity in vulnerable neurons is a factor in PD is somewhat controversial (Chen et al., 2019). For example, partial disruption of MCI function by deleting its NADH-ubiquinone oxidoreductase subunit S4 (Ndufs4) in dopaminergic neurons does not result in a parkinsonian phenotype, nor does it alter the sensitivity to environmental toxins linked to PD, suggesting that MCI dysfunction might just be a PD tombstone (Kim et al., 2015). On the other hand, the absence of a phenotype could simply reflect the ability of neurons to compensate for a modest impairment.

A more complete and progressive disruption of mitochondrial function in dopaminergic neurons induced by knocking out the mitochondrial transcription factor Tfam leads to a parkinsonian phenotype in the so-called MitoPark model (Ekstrand & Galter, 2009; Ricke et al., 2020).

Recent work in this model suggests that Ca^{2+} entry through Ca_v1 channels accelerates pathogenesis (Ricke et al., 2020), consistent with the earlier work (Guzman et al., 2010). Ricke et al. found that despite mitochondrial insufficiency in all DA neurons, only ventral tier SNc dopamine neurons showed signs of mitochondrial oxidant stress - that was attenuated by isradipine. But given the broad mitochondrial dysfunction induced in the MitoPark model, it does not resolve the question of whether mitochondrial deficits seen specifically in PD patients are causative.

To directly assess the consequences of selective disruption of MCI function, the gene coding for its catalytic subunit, the NADH-ubichinone oxidoreductase subunit S2 (Ndufs2), was deleted in dopaminergic neurons using intersectional genomics (Gonzalez-Rodriguez et al., 2021). This deletion led to near-complete loss in the ability of mitochondria to produce ATP in the resulting MCI-Park mice. In fact, roughly a month after deletion, mitochondria became net consumers of ATP to maintain their membrane potential, by running complex V in reverse. In line, SNc dopaminergic neurons upregulated the expression of genes associated with glycolysis and downregulated those associated with mitochondrial ATP production, effectively undergoing a Warburg shift (Garber, 2004). In parallel, there was a shift away from the pentose phosphate shunt, decreasing the glutathione redox couple. Despite these adaptations, within about a month mice began to manifest parkinsonian deficits; these deficits - including reduced movement speed, fragmentation of sequential motor acts, and impaired exploratory rearing - progressively worsened and were alleviated by systemic levodopa treatment. The progressive parkinsonian behavioral phenotype was attributable to an axon-first, progressive loss of SNc dopaminergic neuron function, mirroring that seen in humans (Kordower & Burke, 2018). At about 12–15 weeks of age, the mice displayed deficits in gait and stance, mirroring those in PD patients. Importantly, these deficits appeared without concomitant Lewy pathology or inflammation. Thus, loss of MCI function alone in dopaminergic neurons is sufficient to cause a progressive, human-like form of parkinsonism.

In addition, loss of MCI function induced a profound remodeling of cellular physiology, particularly features related to Ca^{2+} handling: (1) the broad spikes typical of SNc dopaminergic neurons narrowed by roughly half; (2) pacemaking slowed or stopped (but spikes could be readily evoked by somatic current injection or dendritic glutamate release), and (3) dendritic spike-associated Ca^{2+} transients were dramatically reduced, due to downregulation of functional $Ca_v1.3$ channels.

The suppression of pacemaking was not caused by the activation of ATP-sensitive potassium (K-ATP) channels, as previously shown in SNc dopamine neurons, in response to toxin treatment and acute inhibition of MCI (Liss et al., 2005; Schiemann et al., 2011), or to αSYN injections (Hill et al., 2021). Rather, there was a downregulation in HCN channels, as well as an upregulation in an array of K^+ channels that are sensitive to metabolic state, like TWIK K^+ channels (Gonzalez-Rodriguez et al., 2021). These changes all point to the conclusion that these features in wild-type mice are in place to control mitochondrial function and when this target is no longer responsive to Ca^{2+} signaling, a more bioenergetically economical and less stressful electrical phenotype is adopted if neurons are given the time to do so. Because of the long half-life of many mitochondria proteins for weeks (Fornasiero et al., 2018), Ndufs2 deletion appears to give neurons this opportunity. In principle, this slower kinetics should create a more faithful representation of the processes engaged in PD, where pathology develops over decades.

Are Ca_v Channels Viable Therapeutic Targets for Disease Modification in PD?

In agreement with the preclinical evidence, linking Ca_v channels to differential neuronal vulnerability and PD pathogenesis, epidemiological studies pointed to a particular role of Ca_v1 L-type channels, as they have consistently found that use of blood-brain-barrier permissive DHPs to treat hypertension is associated with reduced risk of developing PD by ~30% (Becker et al., 2008; Ritz et al., 2010; Tseng et al., 2021). Recent epidemiological work found this relationship to be dependent upon DHP dose, with the reduction in risk at the highest doses being profound (hazard ratio ~ 0.37) (Tseng et al., 2021).

Based upon the preclinical and epidemiological evidence, both Phase 2 and 3 clinical trials were conducted with the DHP isradipine (ParkinsonStudyGroup, 2013). Isradipine was chosen because of its safety, brain bioavailability, and because of all commonly used DHPs, it has the highest relative affinity for $Ca_v1.3$ L-type channels, although it is more selective for $Ca_v1.2$ channels (Sinnegger-Brauns et al., 2009; Surmeier et al., 2017b). In the Phase 2 study, 10 mg/day of a controlled release (CR) format isradipine was deemed the maximum tolerable dose, as 20 mg/day induced a variety of side effects including peripheral edema (ParkinsonStudyGroup, 2013). With this tolerability data in hand, a Phase 3 clinical trial with recently diagnosed, early-stage PD participants was initiated in North America (Biglan et al., 2017). However, because the CR formulation of isradipine was discontinued commercially just prior to trial launch, participants were given 5 mg immediate release (IR) tablets twice a day. Recently diagnosed PD patients were treated for 36 months and were then evaluated for disability using the Unified Parkinson's Disease Rating Scale (UPDRS). All of the participants initiated levodopa therapy during the trial and were well managed, leading to a very small change in the so-called on-state UPDRS score during the trial. In an attempt to correct for symptomatic medication treatment and uncover any potential difference in disease progression with isradipine treatment, UPDRS measurements during the on-state were "adjusted" using a linear regression model (ParkinsonStudyGroup, 2020). However, the phase 3 clinical trial failed to demonstrate a PD-protective effect, as the isradipine treatment group did not differ significantly from the placebo group in their adjusted UPDRS score, suggesting that isradipine did not modify disease course (ParkinsonStudyGroup, 2020; Venuto et al., 2021).

Although there are many reasons why such a trial might fail (Liss & Striessnig, 2019; Maiti & Perlmutter, 2020; Ortner, 2021), the usual refrain in this situation is "too little, too late." Two independent lines of study suggest that "too little" is at least part of the story. Dosing with isradipine was limited by peripheral side-effects associated with inhibition of $Ca_v1.2$ channels in the cardiovascular system, leading to relaxation of smooth muscle (Surmeier et al., 2017b). The pharmacokinetic (PK) analysis of patient plasma isradipine concentration after dosing with the IR formulation revealed that concentration peaked rapidly and then declined over the next few hours. With the 5 mg dose, this resulted in plasma isradipine concentrations that were very low (<5 nM) for most of the day, which is below the presumed EC_{50} for inhibiting human $Ca_v1.3$ channels in SNc dopaminergic neurons (Xu & Lipscombe, 2001; Ilijic et al., 2011; Ortner et al., 2017). Thus, likely very little inhibition of $Ca_v1.3$ channel function was achieved, except right after dosing. Interestingly, the PK profile of the participant pool varied considerably, with some participants showing relatively slow clearance of isradipine and, hence, greater exposure (Venuto et al., 2021). In those patients, there was a clear trend towards disease slowing, as the time to onset of symptomatic therapy was delayed, the aggregate dose of symptomatic medication at the end of the trial was reduced and non-tremor motor disability was less severe. In addition, patients with higher isradipine exposure had a decreased risk of needing antiparkinson medication during the 3-year trial (Venuto et al., 2021). This analysis also avoided reliance upon the UPDRS adjustment for symptomatic medication performed in the main trial.

This post-hoc analysis of STEADY-PD III suggested that a more sustained elevation in plasma isradipine concentration, even a suboptimal one, may have modified disease progression. To test this hypothesis, the STEADY-PD II trial data in which patients were given the extended-release CR isradipine formulation were re-examined (Surmeier et al., 2021). The initial analysis of this 12-month trial did not find evidence of disease modification using a standard analysis of variance with placebo, 5, 10, and 20 mg/day treatment groups. However, re-examination of the tablet logs found that the 20 mg/day treatment group was not protocol compliant because of side effects, and they were excluded from the analysis. The other decision made in the initial analysis was to use the "last observation carried forward" method for dealing with patients initiating symptomatic medication during the trial. In this method, the patient's UPDRS score when they started levodopa therapy was "carried forward" (LOCF) to the end of the trial and used as the outcome measure. As many patients started therapy early in the trial, the method is problematic when considering the impact of a drug on a slowly progressing disease. This protocol is not typically followed for trials that are looking for longitudinal changes and was abandoned in the reanalysis. Re-examination of the cohorts found no significant differences between them (e.g., age, disease severity at trial start, time since diagnosis). When UPDRS scores at the end of the trial were used (abandoning the LOCF protocol) and the analysis was restricted to patients taking 0, 5, and 10 mg/day of isradipine, a statistically significant reduction in total and motor UPDRS progression in the 10 mg/day group emerged. Using the adjustment equation used in the STEADY-PD III trial (which has uncertain validity) added variance to the data but only marginally diminished the P values associated with treatment.

Another point of interest in this reanalysis was the potential impact of rasagiline treatment on assessments of disease progression. Rasagiline is an irreversible inhibitor of monoamine oxidase B (MAO-B) that degrades dopamine. It is commonly used as an early symptomatic treatment for PD because it slows the degradation of dopamine (Dezsi & Vecsei, 2017; Alborghetti & Nicoletti, 2019). Moreover, there are a handful of studies that suggest that MAO-B inhibitors are neuroprotective (Tabakman et al., 2004). Recent work by our group has shown that MAO-B (and MAO-A) anchored to the outer membrane of mitochondria in dopaminergic neurons not only metabolizes cytosolic dopamine but also shuttles electrons generated by this metabolism to the electron transport chain to help support ATP production needed for dopamine release (Graves et al., 2020). Hence, MAO-B seems to be part of a feedforward control system for axonal mitochondria, just like $Ca_v1.3$ channels in the cell body. However, as in the somatodendritic region, the MAO-B pathway can increase mitochondrial oxidant stress, and amphetamin-induced degeneration of dopaminergic neurons has been shown to significantly attenuated by rasagiline treatment in vivo (Graves et al., 2020). Thus, both isradipine and MAO-B blockers (rasagiline or selegiline) lower mitochondrial oxidant stress in dopaminergic neurons. This insight raises the possibility that the beneficial effects of isradipine were occluded in patients taking MAO-B blockers. Indeed, restricting the analysis to just those patients taking CR isradipine but not rasagiline or selegiline increased the apparent disease slowing effects of the drug, making them resistant to "adjustment" for levodopa and DA agonist therapy (Surmeier et al., 2021).

Taken together, the reanalysis of the two clinical trials with isradipine suggests that keeping isradipine even modestly elevated in the plasma may slow PD progression. However, isradipine is off-patent and is unlikely to be used in another Phase 3 trial in any currently available format. The costs associated with running such a trial might not be recouped by entering the marketplace with a drug that could not be protected.

Conclusions

Despite the failed phase 3 isradipine clinical trial, there are many compelling reasons for targeting Ca_v channels to slow the progression of PD. One path forward in this effort would be to develop a selective inhibitor of $Ca_v1.3$ channels. These channels have a more restricted expression in the brain and periphery (Striessnig et al., 2014) and they are likely directly implicated in pathogenesis (as outlined above). Our group has initiated this effort to identify a class of compounds that preferentially bind to the dihydropyridine binding site of the pore-forming subunit of $Ca_v1.3$ channels (Cooper et al., 2020). Like DHPs, these compounds are negative allosteric modulators, which make them ideal drug candidates because of their potentially limited side-effect profile and preferential targeting of cells with relatively depolarized membrane potentials, like pacemaking SNc dopaminergic neurons. However, at this point in time, these compounds do not have a sufficient affinity or selectivity to be clinically viable (Huang et al., 2014; Ortner et al., 2014; Cooper et al., 2020; Ye et al., 2020). A medicinal chemistry effort is underway to correct this situation. Another strategy might be to target a subset of Ca_v channels implicated in pathogenesis, ideally without inhibiting striatal dopamine release. For example, it might be possible to inhibit $Ca_v2.3$ or Ca_v3 channels alone or in combination with Ca_v1 channels to slow progression (Benkert et al., 2019; Tabata et al., 2018).

Even with a suitable Ca^{2+} channel inhibitor in hand, it is possible that by the time patients are diagnosed with PD, it is too late for significant disease modification. At this point in time, about ~50–70% of SN DA neurons have lost their phenotype or have died (Lang & Lozano, 1998; Dauer & Przedborski, 2003; Collier et al., 2011). As consequence, there is the need to identify biomarkers that can reliably identify prodromal PD and forecast phenoconversion 5–10 years later. Patients with rapid eye movement (REM) sleep disorder phenoconvert to PD at high rates, but may represent a very distinctive type of disease with different determinants of progression (Berg et al., 2021). Essentially all of the correlates of the prodromal state (e.g., hyposmia, constipation) have very low specificity, making them less than ideal, Although many strategies have been tried, none have yet been validated.

While our focus has been on the potential benefit of Ca_v1 channel inhibitors in PD, subtype specific inhibitors of these channels are likely to be of value in other diseases, e.g. autism spectrum disorder (Pinggera et al., 2015; Zamponi et al., 2015; Harrison et al., 2020). Thus, there are compelling reasons to continue the effort to develop new drugs to target Ca_v1 channels.

Acknowledgments BL is funded by the German DFG, the Austrian FWF, the Alfried Krupp Foundation, the Boehringer Ingelheim-Ulm University (BIU) Center, the Wellcome Trust, and a Research Fellowship by the Hamburg Institute for Advanced Study, Hamburg University. DJS is funded by the JPB Foundation, MJFF, and NIH.

References

Abbott, A. (2010). Levodopa: The story so far. *Nature, 466*, S6–S7.

Abdelmotilib, H., Maltbie, T., Delic, V., Liu, Z., Hu, X., Fraser, K. B., Moehle, M. S., Stoyka, L., Anabtawi, N., Krendelchtchikova, V., Volpicelli-Daley, L. A., & West, A. (2017). alpha-synuclein fibril-induced inclusion spread in rats and mice correlates with dopaminergic Neurodegeneration. *Neurobiology of Disease, 105*, 84–98.

Agarwal, D., Sandor, C., Volpato, V., Caffrey, T. M., Monzon-Sandoval, J., Bowden, R., Alegre-Abarrategui, J., Wade-Martins, R., & Webber, C. (2020). A single-cell atlas of the human substantia nigra reveals cell-specific pathways associated with neurological disorders. *Nature Communications, 11*, 4183.

Albin, R. L., Young, A. B., & Penney, J. B. (1989). The functional anatomy of basal ganglia disorders. *Trends in Neurosciences, 12*, 366–375.

Alborghetti, M., & Nicoletti, F. (2019). Different generations of type-B monoamine oxidase inhibitors in Parkinson's disease: From bench to bedside. *Current Neuropharmacology, 17*, 861–873.

Anderson, D., Engbers, J. D., Heath, N. C., Bartoletti, T. M., Mehaffey, W. H., Zamponi, G. W., & Turner, R. W. (2013). The Cav3-Kv4 complex acts as a calcium sensor to maintain inhibitory charge transfer during extracellular calcium fluctuations. *The Journal of Neuroscience, 33*, 7811–7824.

Avenali, M., Blandini, F., & Cerri, S. (2020). Glucocerebrosidase defects as a major risk factor for Parkinson's disease. *Frontiers in Aging Neuroscience, 12*, 97.

Bahring, R. (2018). Kv channel-interacting proteins as neuronal and non-neuronal calcium sensors. *Channels (Austin, Tex.), 12*, 187–200.

Beach, T. G., Adler, C. H., Lue, L., Sue, L. I., Bachalakuri, J., Henry-Watson, J., Sasse, J., Boyer, S., Shirohi, S., Brooks, R., Eschbacher, J., White, C. L., 3rd, Akiyama, H., Caviness, J., Shill, H. A., Connor, D. J., Sabbagh, M. N., & Walker, D. G. (2009). Unified staging system for Lewy body disorders: Correlation with nigrostriatal degeneration, cognitive impairment and motor dysfunction. *Acta Neuropathologica, 117*, 613–634.

Bean, B. P. (2007). The action potential in mammalian central neurons. *Nature Reviews Neuroscience, 8*, 451–465.

Becker, C., Jick, S. S., & Meier, C. R. (2008). Use of antihypertensives and the risk of Parkinson disease. *Neurology, 70*, 1438–1444.

Bender, A., Krishnan, K. J., Morris, C. M., Taylor, G. A., Reeve, A. K., Perry, R. H., Jaros, E., Hersheson, J. S., Betts, J., Klopstock, T., Taylor, R. W., & Turnbull, D. M. (2006). High levels of mitochondrial DNA deletions in substantia nigra neurons in aging and Parkinson disease. *Nature Genetics, 38*, 515–517.

Benkert, J., et al. (2019). Cav2.3 channels contribute to dopaminergic neuron loss in a model of Parkinson's disease. *Nature Communications, 10*, 5094.

Berg, D., & Youdim, M. B. (2006). Role of iron in neurodegenerative disorders. *Topics in Magnetic Resonance Imaging, 17*, 5–17.

Berg, D., Borghammer, P., Fereshtehnejad, S. M., Heinzel, S., Horsager, J., Schaeffer, E., & Postuma, R. B. (2021). Prodromal Parkinson disease subtypes—Key to understanding heterogeneity. *Nature Reviews. Neurology, 17*, 349–361.

Biglan, K. M., Oakes, D., Lang, A. E., Hauser, R. A., Hodgeman, K., Greco, B., Lowell, J., Rockhill, R., Shoulson, I., Venuto, C., Young, D., Simuni, T., & Parkinson Study Group S-PDIII. (2017). A novel design of a Phase III trial of isradipine in early Parkinson disease (STEADY-PD III). *Annals of Clinical Translational Neurology, 4*, 360–368.

Bloem, B. R., Okun, M. S., & Klein, C. (2021). Parkinson's disease. *Lancet, 397*, 218.

Bolam, J. P., & Pissadaki, E. K. (2012). Living on the edge with too many mouths to feed: Why dopamine neurons die. *Movement Disorders, 27*, 1478–1483.

Borghammer, P., Horsager, J., Andersen, K., Van Den Berge, N., Raunio, A., Murayama, S., Parkkinen, L., & Myllykangas, L. (2021). Neuropathological evidence of body-first vs. brain-first Lewy body disease. *Neurobiology of Disease, 161*, 105557.

Braak, H., Del Tredici, K., Rub, U., de Vos, R. A., Jansen Steur, E. N., & Braak, E. (2003). Staging of brain pathology related to sporadic Parkinson's disease. *Neurobiology of Aging, 24*, 197–211.

Brichta, L., & Greengard, P. (2014). Molecular determinants of selective dopaminergic vulnerability in Parkinson's disease: An update. *Frontiers in Neuroanatomy, 8*, 152.

Brimblecombe, K. R., Gracie, C. J., Platt, N. J., & Cragg, S. J. (2015). Gating of dopamine transmission by calcium and axonal N-, Q-, T- and L-type voltage-gated calcium channels differs between striatal domains. *The Journal of Physiology, 593*, 929–946.

Brimblecombe, K. R., Vietti-Michelina, S., Platt, N. J., Kastli, R., Hnieno, A., Gracie, C. J., & Cragg, S. J. (2019). Calbindin-D28K limits dopamine release in ventral but not dorsal striatum by regulating Ca(2+) availability and dopamine transporter function. *ACS Chemical Neuroscience, 10*, 3419–3426.

Burbulla, L. F., Song, P., Mazzulli, J. R., Zampese, E., Wong, Y. C., Jeon, S., Santos, D. P., Blanz, J., Obermaier, C. D., Strojny, C., Savas, J. N., Kiskinis, E., Zhuang, X., Kruger, R., Surmeier, D. J., & Krainc, D. (2017). Dopamine oxidation mediates mitochondrial and lysosomal dysfunction in Parkinson's disease. *Science, 357*, 1255–1261.

Burgoyne, R. D., Helassa, N., McCue, H. V., & Haynes, L. P. (2019). Ca2+ sensors in neuronal function and dysfunction. *Cold Spring Harbor Perspectives in Biology, 11*, a035154.

Buxbaum, J. D. (2004). A role for calsenilin and related proteins in multiple aspects of neuronal function. *Biochemical and Biophysical Research Communications, 322*, 1140–1144.

Caggiu, E., Arru, G., Hosseini, S., Niegowska, M., Sechi, G., Zarbo, I. R., & Sechi, L. A. (2019). Inflammation, infectious triggers, and Parkinson's disease. *Frontiers in Neurology, 10*, 122.

Carrillo-Reid, L., Day, M., Xie, Z., Melendez, A. E., Kondapalli, J., Plotkin, J. L., Wokosin, D. L., Chen, Y., Kress, G. J., Kaplitt, M., Ilijic, E., Guzman, J. N., Chan, C. S., & Surmeier, D. J. (2019). Mutant huntingtin enhances activation of dendritic Kv4 K(+) channels in striatal spiny projection neurons. *eLife, 8*, e40818.

Catoni, C., Cali, T., & Brini, M. (2019). Calcium, dopamine and neuronal calcium sensor 1: Their contribution to Parkinson's disease. *Frontiers in Molecular Neuroscience, 12*, 55.

Chan, C. S., Guzman, J. N., Ilijic, E., Mercer, J. N., Rick, C., Tkatch, T., Meredith, G. E., & Surmeier, D. J. (2007). 'Rejuvenation' protects neurons in mouse models of Parkinson's disease. *Nature, 447*, 1081–1086.

Chen, C., Turnbull, D. M., & Reeve, A. K. (2019). Mitochondrial dysfunction in Parkinson's disease-cause or consequence? *Biology (Basel), 8*, 38.

Chen, R., Ferris, M. J., & Wang, S. (2020). Dopamine D2 autoreceptor interactome: Targeting the receptor complex as a strategy for treatment of substance use disorder. *Pharmacology & Therapeutics, 213*, 107583.

Cheramy, A., Leviel, V., & Glowinski, J. (1981). Dendritic release of dopamine in the substantia nigra. *Nature, 289*, 537–542.

Collaborators GBDPsD. (2018). Global, regional, and national burden of Parkinson's disease, 1990-2016: A systematic analysis for the Global Burden of Disease Study 2016. *The Lancet Neurology, 17*, 939–953.

Collier, T. J., Kanaan, N. M., & Kordower, J. H. (2011). Ageing as a primary risk factor for Parkinson's disease: Evidence from studies of non-human primates. *Nature Reviews Neuroscience, 12*, 359–366.

Cooper, G., Kang, S., Perez-Rosello, T., Guzman, J. N., Galtieri, D., Xie, Z., Kondapalli, J., Mordell, J., Silverman, R. B., & Surmeier, D. J. (2020). A single amino acid determines the selectivity and efficacy of selective negative allosteric modulators of CaV1.3 L-type calcium channels. *ACS Chemical Biology, 15*, 2539–2550.

Corradini, B. R., Iamashita, P., Tampellini, E., Farfel, J. M., Grinberg, L. T., & Moreira-Filho, C. A. (2014). Complex network-driven view of genomic mechanisms underlying Parkinson's disease: Analyses in dorsal motor vagal nucleus, locus coeruleus, and substantia nigra. *BioMed Research International, 2014*, 543673.

Courte, J., Bousset, L., Boxberg, Y. V., Villard, C., Melki, R., & Peyrin, J. M. (2020). The expression level of alpha-synuclein in different neuronal populations is the primary determinant of its prion-like seeding. *Scientific Reports, 10*, 4895.

Damier, P., Hirsch, E. C., Agid, Y., & Graybiel, A. M. (1999). The substantia nigra of the human brain. II. Patterns of loss of dopamine-containing neurons in Parkinson's disease. *Brain, 122*(Pt 8), 1437–1448.

Dauer, W., & Przedborski, S. (2003). Parkinson's disease: Mechanisms and models. *Neuron, 39*, 889–909.

Dawson, T. M., Ko, H. S., & Dawson, V. L. (2010). Genetic animal models of Parkinson's disease. *Neuron, 66*, 646–661.

Day, J. O., & Mullin, S. (2021). The genetics of Parkinson's disease and implications for clinical practice. *Genes (Basel), 12*, 1006.

Dehay, B., Bove, J., Rodriguez-Muela, N., Perier, C., Recasens, A., Boya, P., & Vila, M. (2010). Pathogenic lysosomal depletion in Parkinson's disease. *The Journal of Neuroscience, 30*, 12535–12544.

Devine, M. J., & Kittler, J. T. (2018). Mitochondria at the neuronal presynapse in health and disease. *Nature Reviews Neuroscience, 19*, 63–80.

Dezsi, L., & Vecsei, L. (2017). Monoamine oxidase B inhibitors in Parkinson's disease. *CNS & Neurological Disorders Drug Targets, 16*, 425–439.

Diaz-Garcia, C. M., Meyer, D. J., Nathwani, N., Rahman, M., Martinez-Francois, J. R., & Yellen, G. (2021). The distinct roles of calcium in rapid control of neuronal glycolysis and the tricarboxylic acid cycle. *eLife, 10*, e64821.

Diaz-Vegas, A. R., Cordova, A., Valladares, D., Llanos, P., Hidalgo, C., Gherardi, G., De Stefani, D., Mammucari, C., Rizzuto, R., Contreras-Ferrat, A., & Jaimovich, E. (2018). Mitochondrial calcium increase induced by RyR1 and IP3R channel activation after membrane depolarization regulates skeletal muscle metabolism. *Frontiers in Physiology, 9*, 791.

Diederich, N. J., James Surmeier, D., Uchihara, T., Grillner, S., & Goetz, C. G. (2019). Parkinson's disease: Is it a consequence of human brain evolution? *Movement Disorders, 34*, 453–459.

Diepenbroek, M., Casadei, N., Esmer, H., Saido, T. C., Takano, J., Kahle, P. J., Nixon, R. A., Rao, M. V., Melki, R., Pieri, L., Helling, S., Marcus, K., Krueger, R., Masliah, E., Riess, O., & Nuber, S. (2014). Overexpression of the calpain-specific inhibitor calpastatin reduces human alpha-synuclein processing, aggregation and synaptic impairment in [A30P]alphaSyn transgenic mice. *Human Molecular Genetics, 23*, 3975–3989.

Ding, J. B., Guzman, J. N., Peterson, J. D., Goldberg, J. A., & Surmeier, D. J. (2010). Thalamic gating of corticostriatal signaling by cholinergic interneurons. *Neuron, 67*, 294–307.

Doherty, K. M., & Hardy, J. (2013). Parkin disease and the Lewy body conundrum. *Movement Disorders, 28*, 702–704.

Dolle, C., Flones, I., Nido, G. S., Miletic, H., Osuagwu, N., Kristoffersen, S., Lilleng, P. K., Larsen, J. P., Tysnes, O. B., Haugarvoll, K., Bindoff, L. A., & Tzoulis, C. (2016). Defective mitochondrial DNA homeostasis in the substantia nigra in Parkinson disease. *Nature Communications, 7*, 13548.

Dolphin, A. C. (2018). Voltage-gated calcium channels: Their discovery, function and importance as drug targets. *Brain and Neuroscience Advances, 2*, 2398212818794805.

Dragicevic, E., Poetschke, C., Duda, J., Schlaudraff, F., Lammel, S., Schiemann, J., Fauler, M., Hetzel, A., Watanabe, M., Lujan, R., Malenka, R. C., Striessnig, J., & Liss, B. (2014). Cav1.3 channels control D2-autoreceptor responses via NCS-1 in substantia nigra dopamine neurons. *Brain, 137*, 2287–2302.

Dragicevic, E., Schiemann, J., & Liss, B. (2015). Dopamine midbrain neurons in health and Parkinson's disease: Emerging roles of voltage-gated calcium channels and ATP-sensitive potassium channels. *Neuroscience, 284*, 798–814.

Dryanovski, D. I., Guzman, J. N., Xie, Z., Galteri, D. J., Volpicelli-Daley, L. A., Lee, V. M., Miller, R. J., Schumacker, P. T., & Surmeier, D. J. (2013). Calcium entry and alpha-synuclein inclusions elevate dendritic mitochondrial oxidant stress in dopaminergic neurons. *The Journal of Neuroscience, 33*, 10154–10164.

Duda, J., Potschke, C., & Liss, B. (2016). Converging roles of ion channels, calcium, metabolic stress, and activity pattern of Substantia nigra dopaminergic neurons in health and Parkinson's disease. *Journal of Neurochemistry, 139*, 156–178.

Dufty, B. M., Warner, L. R., Hou, S. T., Jiang, S. X., Gomez-Isla, T., Leenhouts, K. M., Oxford, J. T., Feany, M. B., Masliah, E., & Rohn, T. T. (2007). Calpain-cleavage of alpha-synuclein: Connecting proteolytic processing to disease-linked aggregation. *The American Journal of Pathology, 170*, 1725–1738.

Ekstrand, M. I., & Galter, D. (2009). The MitoPark mouse—An animal model of Parkinson's disease with impaired respiratory chain function in dopamine neurons. *Parkinsonism & Related Disorders, 15*(Suppl 3), S185–S188.

Eschbach, J., von Einem, B., Muller, K., Bayer, H., Scheffold, A., Morrison, B. E., Rudolph, K. L., Thal, D. R., Witting, A., Weydt, P., Otto, M., Fauler, M., Liss, B., McLean, P. J., Spada, A. R., Ludolph, A. C., Weishaupt, J. H., & Danzer, K. M. (2015). Mutual exacerbation of peroxisome proliferator-activated receptor gamma coactivator 1alpha deregulation and alpha-synuclein oligomerization. *Annals of Neurology, 77*, 15–32.

Evans, R. C., Zhu, M., & Khaliq, Z. M. (2017). Dopamine inhibition differentially controls excitability of substantia nigra dopamine neuron subpopulations through T-type calcium channels. *The Journal of Neuroscience, 37*, 3704–3720.

Fearnley, J. M., & Lees, A. J. (1991). Ageing and Parkinson's disease: Substantia nigra regional selectivity. *Brain, 114*(Pt 5), 2283–2301.

Foehring, R. C., Zhang, X. F., Lee, J. C., & Callaway, J. C. (2009). Endogenous calcium buffering capacity of substantia nigral dopamine neurons. *Journal of Neurophysiology, 102*, 2326–2333.

Follett, J., Darlow, B., Wong, M. B., Goodwin, J., & Pountney, D. L. (2013). Potassium depolarization and raised calcium induces alpha-synuclein aggregates. *Neurotoxicity Research, 23*, 378–392.

Ford, C. P. (2014). The role of D2-autoreceptors in regulating dopamine neuron activity and transmission. *Neuroscience, 282*, 13–22.

Formisano, L., Laudati, G., Guida, N., Mascolo, L., Serani, A., Cuomo, O., Cantile, M., Boscia, F., Molinaro, P., Anzilotti, S., Pizzorusso, V., Di Renzo, G., Pignataro, G., & Annunziato, L. (2020). HDAC4 and HDAC5 form a complex with DREAM that epigenetically down-regulates NCX3 gene and its pharmacological inhibition reduces neuronal stroke damage. *Journal of Cerebral Blood Flow and Metabolism, 40*, 2081–2097.

Fornasiero, E. F., et al. (2018). Precisely measured protein lifetimes in the mouse brain reveal differences across tissues and subcellular fractions. *Nature Communications, 9*, 4230.

Franz, O., Liss, B., Neu, A., & Roeper, J. (2000). Single-cell mRNA expression of HCN1 correlates with a fast gating phenotype of hyperpolarization-activated cyclic nucleotide-gated ion channels (Ih) in central neurons. *The European Journal of Neuroscience, 12*, 2685–2693.

Garber, K. (2004). Energy boost: The Warburg effect returns in a new theory of cancer. *Journal of the National Cancer Institute, 96*, 1805–1806.

Gerfen, C. R., & Surmeier, D. J. (2011). Modulation of striatal projection systems by dopamine. *Annual Review of Neuroscience, 34*, 441–466.

German, D. C., Manaye, K. F., Sonsalla, P. K., & Brooks, B. A. (1992). Midbrain dopaminergic cell loss in Parkinson's disease and MPTP-induced parkinsonism: Sparing of calbindin-D28k-containing cells. *Annals of the New York Academy of Sciences, 648*, 42–62.

Giguere, N., Burke Nanni, S., & Trudeau, L. E. (2018). On cell loss and selective vulnerability of neuronal populations in Parkinson's disease. *Frontiers in Neurology, 9*, 455.

Goldberg, J. A., Guzman, J. N., Estep, C. M., Ilijic, E., Kondapalli, J., Sanchez-Padilla, J., & Surmeier, D. J. (2012). Calcium entry induces mitochondrial oxidant stress in vagal neurons at risk in Parkinson's disease. *Nature Neuroscience, 15*, 1414–1421.

Goldman, S. M. (2014). Environmental toxins and Parkinson's disease. *Annual Review of Pharmacology and Toxicology, 54*, 141–164.

Gonzalez-Rodriguez, P., Zampese, E., Stout, K. A., Guzman, J. N., Ilijic, E., Yang, B., Tkatch, T., Stavarache, M. A., Wokosin, D. L., Gao, L., Kaplitt, M. G., Lopez-Barneo, J., Schumacker, P. T., & Surmeier, D. J. (2021). Disruption of mitochondrial complex I induces progressive parkinsonism. *Nature, 599*, 650–656.

Grace, A. A., & Bunney, B. S. (1980). Nigral dopamine neurons: Intracellular recording and identification with L-dopa injection and histofluorescence. *Science, 210*, 654–656.

Grace, A. A., & Bunney, B. S. (1983). Intracellular and extracellular electrophysiology of nigral dopaminergic neurons—2. Action potential generating mechanisms and morphological correlates. *Neuroscience, 10*, 317–331.

Graves, S. M., Xie, Z., Stout, K. A., Zampese, E., Burbulla, L. F., Shih, J. C., Kondapalli, J., Patriarchi, T., Tian, L., Brichta, L., Greengard, P., Krainc, D., Schumacker, P. T., & Surmeier, D. J. (2020). Dopamine metabolism by a monoamine oxidase mitochondrial shuttle activates the electron transport chain. *Nature Neuroscience, 23*, 15–20.

Gudala, K., Kanukula, R., & Bansal, D. (2015). Reduced risk of Parkinson's disease in users of calcium channel blockers: A meta-analysis. *International Journal of Chronic Diseases, 2015*, 697404.

Guzman, J. N., Sanchez-Padilla, J., Chan, C. S., & Surmeier, D. J. (2009). Robust pacemaking in substantia nigra dopaminergic neurons. *The Journal of Neuroscience, 29*, 11011–11019.

Guzman, J. N., Sanchez-Padilla, J., Wokosin, D., Kondapalli, J., Ilijic, E., Schumacker, P. T., & Surmeier, D. J. (2010). Oxidant stress evoked by pacemaking in dopaminergic neurons is attenuated by DJ-1. *Nature, 468*, 696–700.

Guzman, J. N., Ilijic, E., Yang, B., Sanchez-Padilla, J., Wokosin, D., Galtieri, D., Kondapalli, J., Schumacker, P. T., & Surmeier, D. J. (2018). Systemic isradipine treatment diminishes calcium-dependent mitochondrial oxidant stress. *The Journal of Clinical Investigation, 128*, 2266–2280.

Haddjeri-Hopkins, A., Tapia, M., Ramirez-Franco, J., Tell, F., Marqueze-Pouey, B., Amalric, M., & Goaillard, J. M. (2021). Refining the identity and role of Kv4 channels in mouse substantia nigra dopaminergic neurons. *eNeuro, 8*, ENEURO.0207-21.2021.

Hage, T. A., & Khaliq, Z. M. (2015). Tonic firing rate controls dendritic Ca2+ signaling and synaptic gain in substantia nigra dopamine neurons. *The Journal of Neuroscience, 35*, 5823–5836.

Harrison, P. J., Tunbridge, E. M., Dolphin, A. C., & Hall, J. (2020). Voltage-gated calcium channel blockers for psychiatric disorders: Genomic reappraisal. *The British Journal of Psychiatry, 216*, 250–253.

Hasegawa, K., Stoessl, A. J., Yokoyama, T., Kowa, H., Wszolek, Z. K., & Yagishita, S. (2009). Familial parkinsonism: Study of original Sagamihara PARK8 (I2020T) kindred with variable clinicopathologic outcomes. *Parkinsonism & Related Disorders, 15*, 300–306.

Heck, J., Palmeira Do Amaral, A. C., Weissbach, S., El Khallouqi, A., Bikbaev, A., & Heine, M. (2021). More than a pore: How voltage-gated calcium channels act on different levels of neuronal communication regulation. *Channels (Austin, Tex.), 15*, 322–338.

Henderson, M. X., Chung, C. H., Riddle, D. M., Zhang, B., Gathagan, R. J., Seeholzer, S. H., Trojanowski, J. Q., & Lee, V. M. Y. (2017). Unbiased proteomics of early Lewy body formation model implicates active microtubule affinity-regulating kinases (MARKs) in synucleinopathies. *The Journal of Neuroscience, 37*, 5870–5884.

Henderson, M. X., Cornblath, E. J., Darwich, A., Zhang, B., Brown, H., Gathagan, R. J., Sandler, R. M., Bassett, D. S., Trojanowski, J. Q., & Lee, V. M. Y. (2019). Spread of alpha-synuclein pathology through the brain connectome is modulated by selective vulnerability and predicted by network analysis. *Nature Neuroscience, 22*, 1248–1257.

Hikima, T., Lee, C. R., Witkovsky, P., Chesler, J., Ichtchenko, K., & Rice, M. E. (2021). Activity-dependent somatodendritic dopamine release in the substantia nigra autoinhibits the releasing neuron. *Cell Reports, 35*, 108951.

Hill, E., Gowers, R., Richardson, M. J. E., & Wall, M. J. (2021). alpha-synuclein aggregates increase the conductance of substantia nigra dopamine neurons, an effect partly reversed by the KATP channel inhibitor glibenclamide. *eNeuro, 8*, ENEURO.0330-20.2020.

Holmqvist, M. H., Cao, J., Hernandez-Pineda, R., Jacobson, M. D., Carroll, K. I., Sung, M. A., Betty, M., Ge, P., Gilbride, K. J., Brown, M. E., Jurman, M. E., Lawson, D., Silos-Santiago, I., Xie, Y., Covarrubias, M., Rhodes, K. J., Distefano, P. S., & An, W. F. (2002). Elimination of fast inactivation in Kv4 A-type potassium channels by an auxiliary subunit domain. *Proceedings of the National Academy of Sciences of the United States of America, 99*, 1035–1040.

Hsu, L. J., Sagara, Y., Arroyo, A., Rockenstein, E., Sisk, A., Mallory, M., Wong, J., Takenouchi, T., Hashimoto, M., & Masliah, E. (2000). alpha-synuclein promotes mitochondrial deficit and oxidative stress. *The American Journal of Pathology, 157*, 401–410.

Hu, D., Sun, X., Liao, X., Zhang, X., Zarabi, S., Schimmer, A., Hong, Y., Ford, C., Luo, Y., & Qi, X. (2019). Alpha-synuclein suppresses mitochondrial protease ClpP to trigger mitochondrial oxidative damage and neurotoxicity. *Acta Neuropathologica, 137*, 939–960.

Huang, H., Ng, C. Y., Yu, D., Zhai, J., Lam, Y., & Soong, T. W. (2014). Modest CaV1.342-selective inhibition by compound 8 is beta-subunit dependent. *Nature Communications, 5*, 4481.

Ilijic, E., Guzman, J. N., & Surmeier, D. J. (2011). The L-type channel antagonist isradipine is neuroprotective in a mouse model of Parkinson's disease. *Neurobiology of Disease, 43*, 364–371.

Iyer, R., Ungless, M. A., & Faisal, A. A. (2017). Calcium-activated SK channels control firing regularity by modulating sodium channel availability in midbrain dopamine neurons. *Scientific Reports, 7*, 5248.

Jiang, Y., & Zsombok, A. (2014). Regulation of neurons in the dorsal motor nucleus of the vagus by SIRT1. *Frontiers in Neuroscience, 7*, 270.

Kabbani, N., Negyessy, L., Lin, R., Goldman-Rakic, P., & Levenson, R. (2002). Interaction with neuronal calcium sensor NCS-1 mediates desensitization of the D2 dopamine receptor. *The Journal of Neuroscience, 22*, 8476–8486.

Khaliq, Z. M., & Bean, B. P. (2010). Pacemaking in dopaminergic ventral tegmental area neurons: Depolarizing drive from background and voltage-dependent sodium conductances. *The Journal of Neuroscience, 30*, 7401–7413.

Kiechle, M., et al. (2019). In vivo protein complementation demonstrates presynaptic alpha-synuclein oligomerization and age-dependent accumulation of 8-16-mer oligomer species. *Cell Reports, 29*, 2862–2874 e2869.

Kim, J., Jung, S. C., Clemens, A. M., Petralia, R. S., & Hoffman, D. A. (2007). Regulation of dendritic excitability by activity-dependent trafficking of the A-type K+ channel subunit Kv4.2 in hippocampal neurons. *Neuron, 54*, 933–947.

Kim, H. W., Choi, W. S., Sorscher, N., Park, H. J., Tronche, F., Palmiter, R. D., & Xia, Z. (2015). Genetic reduction of mitochondrial complex I function does not lead to loss of dopamine neurons in vivo. *Neurobiology of Aging, 36*, 2617–2627.

Kimm, T., Khaliq, Z. M., & Bean, B. P. (2015). Differential regulation of action potential shape and burst-frequency firing by BK and Kv2 channels in substantia nigra dopaminergic neurons. *The Journal of Neuroscience, 35*, 16404–16417.

Kordower, J. H., & Burke, R. E. (2018). Disease modification for Parkinson's disease: Axonal regeneration and trophic factors. *Movement Disorders, 33*, 678–683.

Kurz, A., Double, K. L., Lastres-Becker, I., Tozzi, A., Tantucci, M., Bockhart, V., Bonin, M., Garcia-Arencibia, M., Nuber, S., Schlaudraff, F., Liss, B., Fernandez-Ruiz, J., Gerlach, M., Wullner, U., Luddens, H., Calabresi, P., Auburger, G., & Gispert, S. (2010). A53T-alpha-synuclein overexpression impairs dopamine signaling and striatal synaptic plasticity in old mice. *PLoS One, 5*, e11464.

Lammel, S., Hetzel, A., Hackel, O., Jones, I., Liss, B., & Roeper, J. (2008). Unique properties of mesoprefrontal neurons within a dual mesocorticolimbic dopamine system. *Neuron, 57*, 760–773.

Lang, A. E., & Lozano, A. M. (1998). Parkinson's disease. First of two parts. *The New England Journal of Medicine, 339*, 1044–1053.

Lang, C., Campbell, K. R., Ryan, B. J., Carling, P., Attar, M., Vowles, J., Perestenko, O. V., Bowden, R., Baig, F., Kasten, M., Hu, M. T., Cowley, S. A., Webber, C., & Wade-Martins, R. (2019). Single-cell sequencing of iPSC-dopamine neurons reconstructs disease progression and identifies HDAC4 as a regulator of Parkinson cell phenotypes. *Cell Stem Cell, 24*, 93–106.

Lasser-Katz, E., Simchovitz, A., Chiu, W. H., Oertel, W. H., Sharon, R., Soreq, H., Roeper, J., & Goldberg, J. A. (2017). Mutant alpha-synuclein overexpression induces stressless pacemaking in vagal motoneurons at risk in

Parkinson's disease. *The Journal of Neuroscience, 37,* 47–57.

Leandrou, E., Emmanouilidou, E., & Vekrellis, K. (2019). Voltage-gated calcium channels and alpha-synuclein: Implications in Parkinson's disease. *Frontiers in Molecular Neuroscience, 12,* 237.

Lee, Y. C., Lin, C. H., Wu, R. M., Lin, J. W., Chang, C. H., & Lai, M. S. (2014). Antihypertensive agents and risk of Parkinson's disease: A nationwide cohort study. *PLoS One, 9,* e98961.

LeWitt, P. A. (2016). New levodopa therapeutic strategies. *Parkinsonism & Related Disorders, 22*(Suppl 1), S37–S40.

Liss, B., & Striessnig, J. (2019). The potential of L-type calcium channels as a drug target for neuroprotective therapy in Parkinson's disease. *Annual Review of Pharmacology, 59,* 263–289.

Liss, B., Franz, O., Sewing, S., Bruns, R., Neuhoff, H., & Roeper, J. (2001). Tuning pacemaker frequency of individual dopaminergic neurons by Kv4.3L and KChip3.1 transcription. *The EMBO Journal, 20,* 5715–5724.

Liss, B., Haeckel, O., Wildmann, J., Miki, T., Seino, S., & Roeper, J. (2005). K-ATP channels promote the differential degeneration of dopaminergic midbrain neurons. *Nature Neuroscience, 8,* 1742–1751.

Lu, L., Sirish, P., Zhang, Z., Woltz, R. L., Li, N., Timofeyev, V., Knowlton, A. A., Zhang, X. D., Yamoah, E. N., & Chiamvimonvat, N. (2015). Regulation of gene transcription by voltage-gated L-type calcium channel, Cav1.3. *The Journal of Biological Chemistry, 290,* 4663–4676.

Ludwig, M., Apps, D., Menzies, J., Patel, J. C., & Rice, M. E. (2016). Dendritic release of neurotransmitters. *Comprehensive Physiology, 7,* 235–252.

Lugenbiel, P., Govorov, K., Rahm, A. K., Wieder, T., Gramlich, D., Syren, P., Weiberg, N., Seyler, C., Katus, H. A., & Thomas, D. (2018). Inhibition of histone deacetylases induces K+ channel remodeling and action potential prolongation in HL-1 atrial cardiomyocytes. *Cellular Physiology and Biochemistry, 49,* 65–77.

Lutas, A., Lahmann, C., Soumillon, M., & Yellen, G. (2016). The leak channel NALCN controls tonic firing and glycolytic sensitivity of substantia nigra pars reticulata neurons. *eLife, 5,* e15271.

Mahul-Mellier, A. L., Burtscher, J., Maharjan, N., Weerens, L., Croisier, M., Kuttler, F., Leleu, M., Knott, G. W., & Lashuel, H. A. (2020). The process of Lewy body formation, rather than simply alpha-synuclein fibrillization, is one of the major drivers of neurodegeneration. *Proceedings of the National Academy of Sciences of the United States of America, 117,* 4971–4982.

Maiti, B., & Perlmutter, J. S. (2020). A clinical trial of isradipine: What went wrong? *Annals of Internal Medicine, 172,* 625–626.

Mannal, N.; Kleiner, K.; Fauler, M.; Dougalis, A.; Poetschke, C.;& Liss, B. (2021). Multi-Electrode Array Analysis Identifies Complex Dopamine

Responses and Glucose Sensing Properties of Substantia Nigra Neurons in Mouse Brain Slices. Front. *Synaptic Neurosci, 13,* 635050.

Martinez-Vicente, M., & Cuervo, A. M. (2007). Autophagy and neurodegeneration: When the cleaning crew goes on strike. *Lancet Neurology, 6,* 352–361.

Matsuda, W., Furuta, T., Nakamura, K. C., Hioki, H., Fujiyama, F., Arai, R., & Kaneko, T. (2009). Single nigrostriatal dopaminergic neurons form widely spread and highly dense axonal arborizations in the neostriatum. *The Journal of Neuroscience, 29,* 444–453.

Mazzocchi, M., Collins, L. M., Sullivan, A. M., & O'Keeffe, G. W. (2020). The class II histone deacetylases as therapeutic targets for Parkinson's disease. *Neuronal Signal, 4,* NS20200001.

Menozzi, E., Macnaughtan, J., & Schapira, A. H. V. (2021). The gut-brain axis and Parkinson disease: Clinical and pathogenetic relevance. *Annals of Medicine, 53,* 611–625.

Merchant, K. M., Cedarbaum, J. M., Brundin, P., Dave, K. D., Eberling, J., Espay, A. J., Hutten, S. J., Javidnia, M., Luthman, J., Maetzler, W., Menalled, L., Reimer, A. N., Stoessl, A. J., Weiner, D. M., & The Michael JFFASCPWG. (2019). A proposed roadmap for Parkinson's disease proof of concept clinical trials investigating compounds targeting alpha-synuclein. *Journal of Parkinson's Disease, 9,* 31–61.

Mercuri, N. B., Bonci, A., Calabresi, P., Stratta, F., Stefani, A., & Bernardi, G. (1994). Effects of dihydropyridine calcium antagonists on rat midbrain dopaminergic neurones. *British Journal of Pharmacology, 113,* 831–838.

Michel, P. P., Hirsch, E. C., & Hunot, S. (2016). Understanding dopaminergic cell death pathways in Parkinson disease. *Neuron, 90,* 675–691.

Minakaki, G., Krainc, D., & Burbulla, L. F. (2020). The convergence of alpha-synuclein, mitochondrial, and lysosomal pathways in vulnerability of midbrain dopaminergic neurons in Parkinson's disease. *Frontiers in Cell and Development Biology, 8,* 580634.

Mohebi, A., Pettibone, J. R., Hamid, A. A., Wong, J. T., Vinson, L. T., Patriarchi, T., Tian, L., Kennedy, R. T., & Berke, J. D. (2019). Dissociable dopamine dynamics for learning and motivation. *Nature, 570,* 65–70.

Mosharov, E. V., Larsen, K. E., Kanter, E., Phillips, K. A., Wilson, K., Schmitz, Y., Krantz, D. E., Kobayashi, K., Edwards, R. H., & Sulzer, D. (2009). Interplay between cytosolic dopamine, calcium, and alpha-synuclein causes selective death of substantia nigra neurons. *Neuron, 62,* 218–229.

Mundhenk, J., Fusi, C., & Kreutz, M. R. (2019). Caldendrin and calneurons-EF-hand CaM-like calcium sensors with unique features and specialized neuronal functions. *Frontiers in Molecular Neuroscience, 12,* 16.

Murphy, M. P. (2009). How mitochondria produce reactive oxygen species. *The Biochemical Journal, 417,* 1–13.

Nakamura, T. Y., Nakao, S., & Wakabayashi, S. (2019). Emerging roles of neuronal Ca(2+) sensor-1 in cardiac and neuronal tissues: A mini review. *Frontiers in Molecular Neuroscience, 12*, 56.

Naranjo, J. R., & Mellstrom, B. (2012). Ca^{2+}-dependent transcriptional control of Ca2+ homeostasis. *The Journal of Biological Chemistry, 287*, 31674–31680.

Nath, S., Goodwin, J., Engelborghs, Y., & Pountney, D. L. (2011). Raised calcium promotes alpha-synuclein aggregate formation. *Molecular and Cellular Neurosciences, 46*, 516–526.

Nedergaard, S., Flatman, J. A., & Engberg, I. (1993). Nifedipine- and omega-conotoxin-sensitive Ca2+ conductances in guinea-pig substantia nigra pars compacta neurones. *The Journal of Physiology, 466*, 727–747.

Neuhoff, H., Neu, A., Liss, B., & Roeper, J. (2002). I(h) channels contribute to the different functional properties of identified dopaminergic subpopulations in the midbrain. *The Journal of Neuroscience, 22*, 1290–1302.

Norris, E. H., Giasson, B. I., Ischiropoulos, H., & Lee, V. M. (2003). Effects of oxidative and nitrative challenges on alpha-synuclein fibrillogenesis involve distinct mechanisms of protein modifications. *The Journal of Biological Chemistry, 278*, 27230–27240.

Oliveira, L. M. A., Gasser, T., Edwards, R., Zweckstetter, M., Melki, R., Stefanis, L., Lashuel, H. A., Sulzer, D., Vekrellis, K., Halliday, G. M., Tomlinson, J. J., Schlossmacher, M., Jensen, P. H., Schulze-Hentrich, J., Riess, O., Hirst, W. D., El-Agnaf, O., Mollenhauer, B., Lansbury, P., & Outeiro, T. F. (2021). Alpha-synuclein research: Defining strategic moves in the battle against Parkinson's disease. *NPJ Parkinson's Disease, 7*, 65.

Ortner, N. J. (2021). Voltage-gated Ca(2+) channels in dopaminergic substantia nigra neurons: Therapeutic targets for neuroprotection in Parkinson's disease? *Frontiers in Synaptic Neuroscience, 13*, 636103.

Ortner, N. J., Bock, G., Vandael, D. H., Mauersberger, R., Draheim, H. J., Gust, R., Carbone, E., Tuluc, P., & Striessnig, J. (2014). Pyrimidine-2,4,6-triones are a new class of voltage-gated L-type Ca2+ channel activators. *Nature Communications, 5*, 3897.

Ortner, N. J., Bock, G., Dougalis, A., Kharitonova, M., Duda, J., Hess, S., Tuluc, P., Pomberger, T., Stefanova, N., Pitterl, F., Ciossek, T., Oberacher, H., Draheim, H. J., Kloppenburg, P., Liss, B., & Striessnig, J. (2017). Lower affinity of isradipine for L-type Ca(2+) channels during substantia nigra dopamine neuron-like activity: Implications for neuroprotection in Parkinson's disease. *The Journal of Neuroscience, 37*, 6761–6777.

Otomo, K., Perkins, J., Kulkarni, A., Stojanovic, S., Roeper, J., & Paladini, C. A. (2020). In vivo patch-clamp recordings reveal distinct subthreshold signatures and threshold dynamics of midbrain dopamine neurons. *Nature Communications, 11*, 6286.

Pacelli, C., Giguere, N., Bourque, M. J., Levesque, M., Slack, R. S., & Trudeau, L. E. (2015). Elevated mitochondrial bioenergetics and axonal arborization size

are key contributors to the vulnerability of dopamine neurons. *Current Biology, 25*, 2349–2360.

Pakkenberg, B., Moller, A., Gundersen, H. J., Mouritzen Dam, A., & Pakkenberg, H. (1991). The absolute number of nerve cells in substantia nigra in normal subjects and in patients with Parkinson's disease estimated with an unbiased stereological method. *Journal of Neurology, Neurosurgery, and Psychiatry, 54*, 30–33.

Paladini, C. A., & Roeper, J. (2014). Generating bursts (and pauses) in the dopamine midbrain neurons. *Neuroscience, 282*, 109–121.

ParkinsonStudyGroup. (2013). Phase II safety, tolerability, and dose selection study of isradipine as a potential disease-modifying intervention in early Parkinson's disease (STEADY-PD). *Movement Disorders, 28*, 1823–1831.

ParkinsonStudyGroup. (2020). Isradipine versus placebo in early Parkinson disease: A randomized trial. *Annals of Internal Medicine, 172*, 591–598.

Parkkinen, L., Kauppinen, T., Pirttila, T., Autere, J. M., & Alafuzoff, I. (2005). Alpha-synuclein pathology does not predict extrapyramidal symptoms or dementia. *Annals of Neurology, 57*, 82–91.

Pasternak, B., Svanstrom, H., Nielsen, N. M., Fugger, L., Melbye, M., & Hviid, A. (2012). Use of calcium channel blockers and Parkinson's disease. *American Journal of Epidemiology, 175*, 627–635.

Philippart, F., & Khaliq, Z. M. (2018). Gi/o protein-coupled receptors in dopamine neurons inhibit the sodium leak channel NALCN. *eLife, 7*, e40984.

Pinggera, A., Lieb, A., Benedetti, B., Lampert, M., Monteleone, S., Liedl, K. R., Tuluc, P., & Striessnig, J. (2015). CACNA1D de novo mutations in autism spectrum disorders activate Cav1.3 L-type calcium channels. *Biological Psychiatry, 77*, 816–822.

Pissadaki, E. K., & Bolam, J. P. (2013). The energy cost of action potential propagation in dopamine neurons: Clues to susceptibility in Parkinson's disease. *Frontiers in Computational Neuroscience, 7*, 13.

Poetschke, C., Dragicevic, E., Duda, J., Benkert, J., Dougalis, A., DeZio, R., Snutch, T. P., Striessnig, J., & Liss, B. (2015). Compensatory T-type Ca2+ channel activity alters D2-autoreceptor responses of Substantia nigra dopamine neurons from Cav1.3 L-type Ca2+ channel KO mice. *Scientific Reports, 5*, 13688.

Poewe, W., Seppi, K., Tanner, C. M., Halliday, G. M., Brundin, P., Volkmann, J., Schrag, A. E., & Lang, A. E. (2017). Parkinson disease. *Nature Reviews. Disease Primers, 3*, 17013.

Poulin, J. F., Gaertner, Z., Moreno-Ramos, O. A., & Awatramani, R. (2020). Classification of midbrain dopamine neurons using single-cell gene expression profiling approaches. *Trends in Neurosciences, 43*, 155–169.

Poulopoulos, M., Levy, O. A., & Alcalay, R. N. (2012). The neuropathology of genetic Parkinson's disease. *Movement Disorders, 27*, 831–842.

Rangaraju, V., Calloway, N., & Ryan, T. A. (2014). Activity-driven local ATP synthesis is required for synaptic function. *Cell, 156*, 825–835.

Rcom-H'cheo-Gauthier, A. N., Osborne, S. L., Meedeniya, A. C., & Pountney, D. L. (2016). Calcium: Alpha-synuclein interactions in alpha-synucleinopathies. *Frontiers in Neuroscience, 10*, 570.

Reeve, A., Simcox, E., & Turnbull, D. (2014). Ageing and Parkinson's disease: Why is advancing age the biggest risk factor? *Ageing Research Reviews, 14*, 19–30.

Reynolds, R. H., Botia, J., Nalls, M. A., International Parkinson's Disease Genomics C, System Genomics of Parkinson's D, Hardy, J., Gagliano Taliun, S. A., & Ryten, M. (2019). Moving beyond neurons: The role of cell type-specific gene regulation in Parkinson's disease heritability. *NPJ Parkinson's Disease, 5*, 6.

Rhodes, K. J., Carroll, K. I., Sung, M. A., Doliveira, L. C., Monaghan, M. M., Burke, S. L., Strassle, B. W., Buchwalder, L., Menegola, M., Cao, J., An, W. F., & Trimmer, J. S. (2004). KChIPs and Kv4 alpha subunits as integral components of A-type potassium channels in mammalian brain. *The Journal of Neuroscience, 24*, 7903–7915.

Rice, M. E., & Patel, J. C. (2015). Somatodendritic dopamine release: Recent mechanistic insights. *Philosophical Transactions of the Royal Society of London. Series B, Biological Sciences, 370*, 20140185.

Ricke, K. M., Pass, T., Kimoloi, S., Fahrmann, K., Jungst, C., Schauss, A., Baris, O. R., Aradjanski, M., Trifunovic, A., Eriksson Faelker, T. M., Bergami, M., & Wiesner, R. J. (2020). Mitochondrial dysfunction combined with high calcium load leads to impaired antioxidant defense underlying the selective loss of nigral dopaminergic neurons. *The Journal of Neuroscience, 40*, 1975–1986.

Rietdijk, C. D., Perez-Pardo, P., Garssen, J., van Wezel, R. J., & Kraneveld, A. D. (2017). Exploring Braak's hypothesis of Parkinson's disease. *Frontiers in Neurology, 8*, 37.

Ritz, B., Rhodes, S. L., Qian, L., Schernhammer, E., Olsen, J. H., & Friis, S. (2010). L-type calcium channel blockers and Parkinson disease in Denmark. *Annals of Neurology, 67*, 600–606.

Romano, S., Savva, G. M., Bedarf, J. R., Charles, I. G., Hildebrand, F., & Narbad, A. (2021). Meta-analysis of the Parkinson's disease gut microbiome suggests alterations linked to intestinal inflammation. *NPJ Parkinson's Disease, 7*, 27.

Ronkainen, J. J., Hanninen, S. L., Korhonen, T., Koivumaki, J. T., Skoumal, R., Rautio, S., Ronkainen, V. P., & Tavi, P. (2011). Ca2+-calmodulin-dependent protein kinase II represses cardiac transcription of the L-type calcium channel alpha(1C)-subunit gene (Cacna1c) by DREAM translocation. *The Journal of Physiology, 589*, 2669–2686.

Ross, O. A., Toft, M., Whittle, A. J., Johnson, J. L., Papapetropoulos, S., Mash, D. C., Litvan, I., Gordon, M. F., Wszolek, Z. K., Farrer, M. J., & Dickson, D. W. (2006). Lrrk2 and Lewy body disease. *Annals of Neurology, 59*, 388–393.

Ruiz-DeDiego, I., Mellstrom, B., Vallejo, M., Naranjo, J. R., & Moratalla, R. (2015). Activation of DREAM (downstream regulatory element antagonistic modulator), a calcium-binding protein, reduces L-DOPA-induced dyskinesias in mice. *Biological Psychiatry, 77*, 95–105.

Sanchez-Padilla, J., Guzman, J. N., Ilijic, E., Kondapalli, J., Galtieri, D. J., Yang, B., Schieber, S., Oertel, W., Wokosin, D., Schumacker, P. T., & Surmeier, D. J. (2014). Mitochondrial oxidant stress in locus coeruleus is regulated by activity and nitric oxide synthase. *Nature Neuroscience, 17*, 832–840.

Sanders, L. H., McCoy, J., Hu, X., Mastroberardino, P. G., Dickinson, B. C., Chang, C. J., Chu, C. T., Van Houten, B., & Greenamyre, J. T. (2014). Mitochondrial DNA damage: Molecular marker of vulnerable nigral neurons in Parkinson's disease. *Neurobiology of Disease, 70*, 214–223.

Schapira, A. H., Cooper, J. M., Dexter, D., Clark, J. B., Jenner, P., & Marsden, C. D. (1990). Mitochondrial complex I deficiency in Parkinson's disease. *Journal of Neurochemistry, 54*, 823–827.

Schiemann, A., Hadzidiakos, D., & Spies, C. (2011). Managing ICU delirium. *Current Opinion in Critical Care, 17*, 131–140.

Schondorf, D. C., Aureli, M., McAllister, F. E., Hindley, C. J., Mayer, F., Schmid, B., Sardi, S. P., Valsecchi, M., Hoffmann, S., Schwarz, L. K., Hedrich, U., Berg, D., Shihabuddin, L. S., Hu, J., Pruszak, J., Gygi, S. P., Sonnino, S., Gasser, T., & Deleidi, M. (2014). iPSC-derived neurons from GBA1-associated Parkinson's disease patients show autophagic defects and impaired calcium homeostasis. *Nature Communications, 5*, 4028.

Sheng, Z. H. (2017). The interplay of axonal energy homeostasis and mitochondrial trafficking and anchoring. *Trends in Cell Biology, 27*, 403–416.

Shin, J. H., Adrover, M. F., Wess, J., & Alvarez, V. A. (2015). Muscarinic regulation of dopamine and glutamate transmission in the nucleus accumbens. *Proceedings of the National Academy of Sciences of the United States of America, 112*, 8124–8129.

Shin J, Kovacheva L, Thomas D, Stojanovic S, Knowlton CJ, Mankel J, Boehm J, Farassat N, Paladini C, Striessnig J, Canavier CC, Geisslinger G, & Roeper J. (2022). Cav1.3 calcium channels are full-range linear amplifiers of firing frequencies in lateral DA SN neurons. *Science Advances, 8*, eabm4560.

Shukla, S., & Tekwani, B. L. (2020). Histone deacetylases inhibitors in neurodegenerative diseases, neuroprotection and neuronal differentiation. *Frontiers in Pharmacology, 11*, 537.

Siller, A., Hofer, N. T., Tomagra, G., Wiederspohn, N., Hess, S., Benkert, J., Gaifullina, A., Spaich, D., Duda, J., Pötschke, C., Vilusic, K., Fritz, E. M., Schneider, T., Kloppenburg, P., Liss, B., Carabelli, V., Carbone, E., Ortner, N. J., & Striessnig, J. (2022). β2-subunit alternative splicing stabilizes Cav2.3 Ca2+ channel activity during continuous midbrain dopamine neuron-like activity. *Elife, 11*, e67464.

Simons, C., Benkert, J., Deuter, N., Poetschke, C., Pongs, O., Schneider, T., Duda, J., & Liss, B. (2019). NCS-1 deficiency affects mRNA levels of genes involved in regulation of ATP synthesis and mitochondrial stress in highly vulnerable substantia nigra dopaminergic neurons. *Frontiers in Molecular Neuroscience, 12,* 252.

Sinnegger-Brauns, M. J., Hetzenauer, A., Huber, I. G., Renstrom, E., Wietzorrek, G., Berjukov, S., Cavalli, M., Walter, D., Koschak, A., Waldschutz, R., Hering, S., Bova, S., Rorsman, P., Pongs, O., Singewald, N., & Striessnig, J. (2004). Isoform-specific regulation of mood behavior and pancreatic beta cell and cardiovascular function by L-type Ca 2+ channels. *The Journal of Clinical Investigation, 113,* 1430–1439.

Sinnegger-Brauns, M. J., Huber, I. G., Koschak, A., Wild, C., Obermair, G. J., Einzinger, U., Hoda, J. C., Sartori, S. B., & Striessnig, J. (2009). Expression and 1,4-dihydropyridine-binding properties of brain L-type calcium channel isoforms. *Molecular Pharmacology, 75,* 407–414.

Smeyne, R. J., Noyce, A. J., Byrne, M., Savica, R., & Marras, C. (2021). Infection and risk of Parkinson's disease. *Journal of Parkinson's Disease, 11,* 31–43.

Striessnig, J., Pinggera, A., Kaur, G., Bock, G., & Tuluc, P. (2014). L-type Ca(2+) channels in heart and brain. *Wiley Interdisciplinary Reviews Membrane Transport and Signaling, 3,* 15–38.

Subramaniam, M., & Roeper, J. (2016). Subtypes of midbrain dopamine neurons. *Handbook of Behavioral Neuroscience, 24,* 317–334.

Subramaniam, M., Althof, D., Gispert, S., Schwenk, J., Auburger, G., Kulik, A., Fakler, B., & Roeper, J. (2014). Mutant alpha-synuclein enhances firing frequencies in dopamine substantia nigra neurons by oxidative impairment of A-type potassium channels. *The Journal of Neuroscience, 34,* 13586–13599.

Sulzer, D., Cragg, S. J., & Rice, M. E. (2016). Striatal dopamine neurotransmission: Regulation of release and uptake. *Basal Ganglia, 6,* 123–148.

Surmeier, D. J., Halliday, G. M., & Simuni, T. (2017a). Calcium, mitochondrial dysfunction and slowing the progression of Parkinson's disease. *Experimental Neurology, 298,* 202–209.

Surmeier, D. J., Obeso, J. A., & Halliday, G. M. (2017b). Selective neuronal vulnerability in Parkinson disease. *Nature Reviews. Neuroscience, 18,* 101–113.

Surmeier, D. J., Nguyen, J. T., Lancki, N., Venuto, C. S., Oakes, D., Simuni, T., & Wyse, R. K. (2021). Re-analysis of the STEADY-PD II trial-evidence for slowing the progression of Parkinson's disease. *Movement Disorders, 37*(2), 334–342.

Tabakman, R., Lecht, S., & Lazarovici, P. (2004). Neuroprotection by monoamine oxidase B inhibitors: A therapeutic strategy for Parkinson's disease? *BioEssays, 26,* 80–90.

Tabata Y, Imaizumi Y, Sugawara M, Andoh-Noda T, Banno S, Chai M, Sone T, Yamazaki K, Ito M, Tsukahara K, Saya H, Hattori N, Kohyama J, & Okano H. (2018). T-type Calcium Channels Determine the Vulnerability of Dopaminergic Neurons to Mitochondrial Stress in Familial Parkinson Disease. *Stem Cell Reports, 11,* 1171–1184.

Tarfa, R. A., Evans, R. C., & Khaliq, Z. M. (2017). Enhanced sensitivity to hyperpolarizing inhibition in mesoaccumbal relative to nigrostriatal dopamine neuron subpopulations. *The Journal of Neuroscience, 37,* 3311–3330.

Threlfell, S., Lalic, T., Platt, N. J., Jennings, K. A., Deisseroth, K., & Cragg, S. J. (2012). Striatal dopamine release is triggered by synchronized activity in cholinergic interneurons. *Neuron, 75,* 58–64.

Trist, B. G., Hare, D. J., & Double, K. L. (2019). Oxidative stress in the aging substantia nigra and the etiology of Parkinson's disease. *Aging Cell, 18,* e13031.

Tseng, Y. F., Lin, H. C., Chao, J. C., Hsu, C. Y., & Lin, H. L. (2021). Calcium channel blockers are associated with reduced risk of Parkinson's disease in patients with hypertension: A population-based retrospective Cohort study. *Journal of Neuroscience, 424,* 117412.

Um, K. B., Hahn, S., Kim, S. W., Lee, Y. J., Birnbaumer, L., Kim, H. J., & Park, M. K. (2021). TRPC3 and NALCN channels drive pacemaking in substantia nigra dopaminergic neurons. *eLife, 10,* e70920.

Urbano, F. J., Bisagno, V., Mahaffey, S., Lee, S. H., & Garcia-Rill, E. (2018). Class II histone deacetylases require P/Q-type Ca(2+) channels and CaMKII to maintain gamma oscillations in the pedunculopontine nucleus. *Scientific Reports, 8,* 13156.

Vazquez-Velez, G. E., & Zoghbi, H. Y. (2021). Parkinson's disease genetics and pathophysiology. *Annual Review of Neuroscience, 44,* 87–108.

Venuto, C. S., Yang, L., Javidnia, M., Oakes, D., James Surmeier, D., & Simuni, T. (2021). Isradipine plasma pharmacokinetics and exposure-response in early Parkinson's disease. *Annals of Clinical Translational Neurology, 8,* 603–612.

Vila, M., Vukosavic, S., Jackson-Lewis, V., Neystat, M., Jakowec, M., & Przedborski, S. (2000). Alpha-synuclein up-regulation in substantia nigra dopaminergic neurons following administration of the parkinsonian toxin MPTP. *Journal of Neurochemistry, 74,* 721–729.

Viola, H. M., & Hool, L. C. (2010). Cross-talk between L-type Ca2+ channels and mitochondria. *Clinical and Experimental Pharmacology & Physiology, 37,* 229–235.

Volpicelli-Daley, L. A., Luk, K. C., Patel, T. P., Tanik, S. A., Riddle, D. M., Stieber, A., Meaney, D. F., Trojanowski, J. Q., & Lee, V. M. (2011). Exogenous alpha-synuclein fibrils induce Lewy body pathology leading to synaptic dysfunction and neuron death. *Neuron, 72,* 57–71.

Wichmann, T. (2019). Changing views of the pathophysiology of Parkinsonism. *Movement Disorders, 34,* 1130–1143.

Wolfart, J., & Roeper, J. (2002). Selective coupling of T-type calcium channels to SK potassium channels prevents intrinsic bursting in dopaminergic midbrain neurons. *The Journal of Neuroscience, 22,* 3404–3413.

Wolfart, J., Neuhoff, H., Franz, O., & Roeper, J. (2001). Differential expression of the small-conductance, calcium-activated potassium channel SK3 is critical for pacemaker control in dopaminergic midbrain neurons. *The Journal of Neuroscience, 21*, 3443–3456.

Wong, Y. C., Luk, K., Purtell, K., Burke Nanni, S., Stoessl, A. J., Trudeau, L. E., Yue, Z., Krainc, D., Oertel, W., Obeso, J. A., & Volpicelli-Daley, L. A. (2019). Neuronal vulnerability in Parkinson disease: Should the focus be on axons and synaptic terminals? *Movement Disorders, 34*, 1406–1422.

Wszolek, Z. K., Pfeiffer, R. F., Tsuboi, Y., Uitti, R. J., McComb, R. D., Stoessl, A. J., Strongosky, A. J., Zimprich, A., Muller-Myhsok, B., Farrer, M. J., Gasser, T., Calne, D. B., & Dickson, D. W. (2004). Autosomal dominant parkinsonism associated with variable synuclein and tau pathology. *Neurology, 62*, 1619–1622.

Xu, W., & Lipscombe, D. (2001). Neuronal Ca(V)1.3alpha(1) L-type channels activate at relatively hyperpolarized membrane potentials and are incompletely inhibited by dihydropyridines. *The Journal of Neuroscience, 21*, 5944–5951.

Ye, Q., Zhang, Z., Zhang, W., Ding, Y., Zhao, F., Zhang, J., & Song, Y. (2020). Investigation of the selectivity of L-type voltage-gated calcium channels 1.3 for pyrimidine-2,4,6-triones derivatives based on molecular dynamics simulation. *Molecules, 25*, 5440.

Yttri, E. A., & Dudman, J. T. (2016). Opponent and bidirectional control of movement velocity in the basal ganglia. *Nature, 533*, 402–406.

Zaidi, N. F., Thomson, E. E., Choi, E. K., Buxbaum, J. D., & Wasco, W. (2004). Intracellular calcium modulates the nuclear translocation of calsenilin. *Journal of Neurochemistry, 89*, 593–601.

Zampese, E., & Surmeier, D. J. (2020). Calcium, bioenergetics and Parkinson's disease. *Cells, 9*, 2045.

Zamponi, G. W., Striessnig, J., Koschak, A., & Dolphin, A. C. (2015). The physiology, pathology, and pharmacology of voltage-gated calcium channels and their future therapeutic potential. *Pharmacological Reviews, 67*, 821–870.

Zhang, Z., Nie, S., & Chen, L. (2018). Targeting prion-like protein spreading in neurodegenerative diseases. *Neural Regeneration Research, 13*, 1875–1878.

Zhong, J., Tang, G., Zhu, J., Wu, W., Li, G., Lin, X., Liang, L., Chai, C., Zeng, Y., Wang, F., Luo, L., Li, J., Chen, F., Huang, Z., Zhang, X., Zhang, Y., Liu, H., Qiu, X., Tang, S., & Chen, D. (2021). Single-cell brain atlas of Parkinson's disease mouse model. *Journal of Genetics and Genomics, 48*, 277–288.

Voltage-Dependent Calcium Channels (Ca$_V$s) and CatSper in Spermatogenic and Sperm Cells

Alberto Darszon, Juan J. Ferreira,
Ignacio López-González, Gerardo Orta,
Claudia L. Treviño, and Celia M. Santi

Abstract

As in most cell types, intracellular Ca^{2+} regulates many fundamental physiological functions in sperm and its progenitors. Ca^{2+} permeable and voltage-dependent channels in their plasma membrane deeply influence the intracellular Ca^{2+} concentration. Notably, the role of traditional voltage-dependent Ca^{2+} channels (Ca$_V$s) in spermatogenesis, sperm maturation, motility, and preparation for egg fusion is not well defined. Sperm possesses a unique and very complex Ca^{2+} channel regulated by intracellular pH, Ca^{2+}, and voltage called CatSper that is key for fertilization and influences motility, and possibly the acrosome reaction (AR). This exocytotic singular event encompasses the fusion of the acrosome vesicle in the posterior region of the head to the plasma membrane preparing this cell to fuse and fertilize the egg. Though Ca$_V$s and CatSper may participate in this reaction, it is still a pending question. In this chapter, we examine the characteristics of these Ca^{2+} channels and their involvement in sperm physiology.

A. Darszon (✉) · I. López-González · G. Orta
C. L. Treviño
Genética del Desarrollo y Fisiología Molecular,
Instituto de Biotecnología, Universidad Nacional
Autónoma de México, Cuernavaca, México
e-mail: darszon@ibt.unam.mx;
ignacio.lopez@ibt.unam.mx;
gerardo.orta@ibt.unam.mx;
claudia.trevino@ibt.unam.mx

J. J. Ferreira
Department of Obstetrics and Gynecology,
Washington University School of Medicine,
Saint Louis, MO, USA
e-mail: JuanFerreira@wustl.edu

C. M. Santi
Genética del Desarrollo y Fisiología Molecular,
Instituto de Biotecnología, Cuernavaca, México

Department of Obstetrics and Gynecology and
Department of Neurosciences, Washington University
School of Medicine, Saint Louis, MO, USA
e-mail: santic@wustl.edu

Keywords

CatSper channels · Mammalian sperm ·
Voltage-dependent Ca^{2+} channels ·
Spermatogenesis · CatSper regulation ·
Chemotaxis · Hyperactivation · Rheotaxis ·
Acrosome reaction · Hormonal regulation

Introduction

Sperm cells are small, morphologically polarized, and specialized cells that originate from their precursor cells in the testis during the process of spermatogenesis. They are constituted by a head that is connected to the flagellum, in many instances such

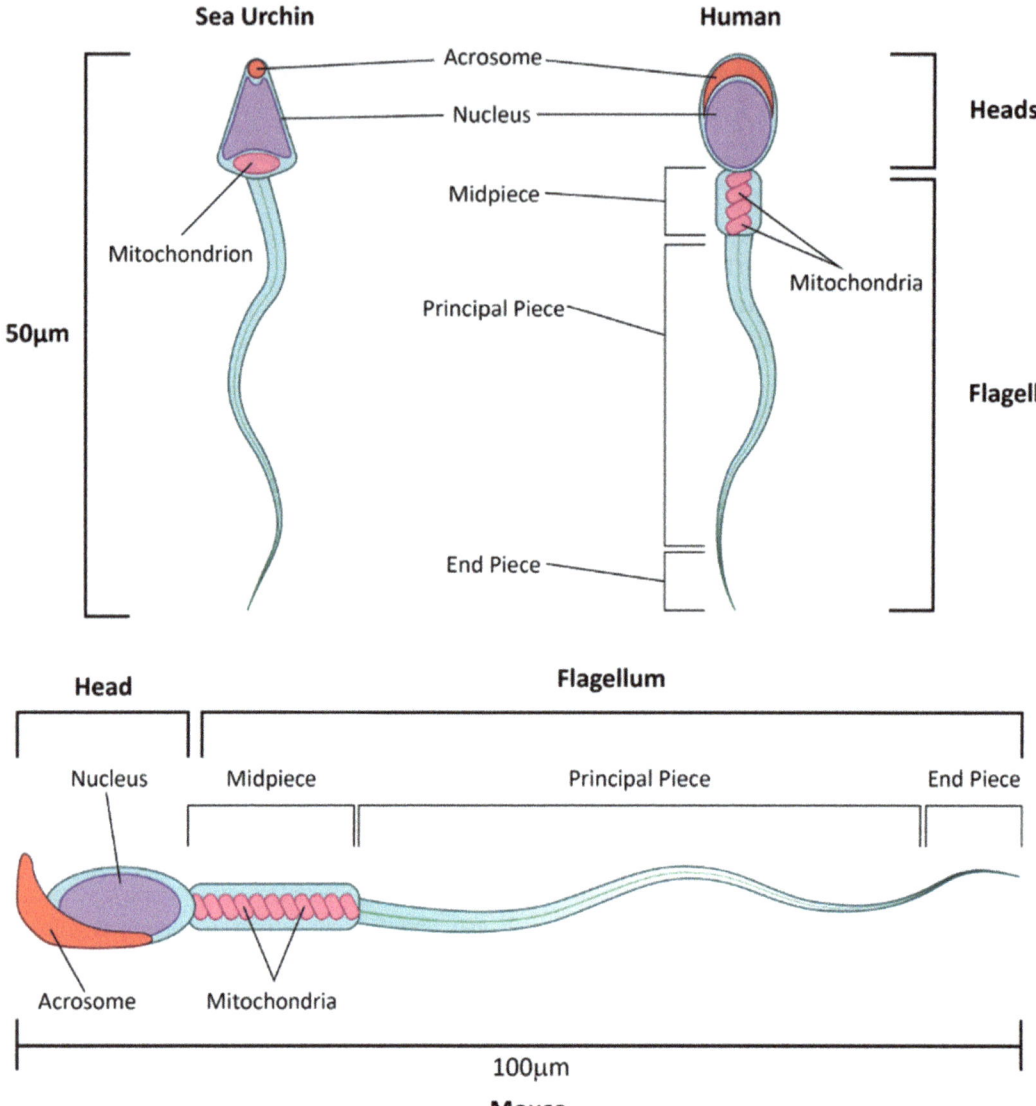

Fig. 1 Scheme of sea urchin, human, and mouse spermatozoa

as in mammals, through a neck or midpiece which is followed by the principal flagellar piece (Fig. 1). The head contains the haploid genetic material in the nucleus and in some species like the sea urchin it also contains a mitochondrion. The head also compraises the acrosome, a membranous and acidic organelle located over the anterior part of the sperm nucleus that is highly conserved throughout evolution. The exocytosis of the organelle is a required event for the sperm to fertilize the egg.

In mammals, the midpiece mainly contains the mitochondria, and the principal piece encloses the motor structures that allow the sperm to swim (Yanagimachi, 1994). Calcium influx through Ca^{2+} channels and the subsequent increase in $[Ca^{2+}]_i$ are key to processes that occur in each of these sperm regions: the exocytosis of the acrosome in the head, the activation of mitochondria in the midpiece, and the acquisition of motility in the flagellum (Darszon et al., 2011; Carafoli & Krebs, 2016; Cai et al.,

2015; Lishko et al., 2012; Wang et al., 2021; Vyklicka & Lishko, 2020; Brown et al., 2019). In addition, Ca^{2+} influx through Ca^{2+} channels influences membrane potential (V$_m$) affecting all other voltage-dependent ion transporters (Nau, 2008).

Sperm originate in the testis during the process of spermatogenesis that involves the multiplication of spermatogonia stem cells and the differentiation of these cells into spermatocytes, spermatids, and spermatozoa. Calcium influx might also play a relevant role in these processes. Mammalian sperm cells mature in their journey through the epididymis and the female reproductive tract in a process known as capacitation. As a consequence of capacitation, sperm acquire the ability to respond to physiological acrosome reaction (AR) inducers (Yanagimachi, 1994; Kornbluth & Fissore, 2015), and change the pattern of motility. Sperm flagellar beat changes from a symmetric mode called active, to a more asymmetric and vigorous one known as hyperactivated (Stival et al., 2016). Calcium entry and the increase in [Ca^{2+}]$_i$ play a fundamental role in sperm capacitation

The study of ion channels in spermatozoa has been challenging due to the structural characteristics of this cell. Overall, the sperm plasma membrane is rigid, a characteristic that precluded for a long-time studying sperm ion channels electrophysiologically *in situ* (Kirichok & Lishko, 2011). These difficulties were partially sidestepped by functionally studying sperm Ca^{2+} channels and their regulation *in vivo* using fluorescent dyes sensitive to Ca^{2+}, V$_m$, and intracellular pH (pH$_i$); incorporating ion channels into planar lipid bilayers; and exploring their presence using RT-PCR and specific antibodies (Darszon et al., 2011; Beltran et al., 2016; Correia et al., 2015). On the other hand, because sperm are basically unable to synthesize proteins, most of their ion channels are synthesized by their progenitor cells during spermatogenesis, mostly during the more mature pachytene, spermatocytes, and round spermatid stages. The inability of sperm to synthesize proteins impedes the use of many of the powerful molecular biology strategies, including interference RNA (iRNA).

Spermatogenic cells in addition to being transcriptionally active have the advantage of being amenable to traditional electrophysiology. The combination of these strategies provided the initial inklings of the ion channels present in these cells, some of which will prevail up to the mature sperm (Hagiwara & Kawa, 1984; Liévano et al., 1996; Santi et al., 1998; Jagannathan et al., 2002a, b; Xiao et al., 2006).

The seminal finding by Kirichok et al. (2006) that the cytoplasmic droplet in mouse sperm possesses a more fluid plasma membrane than the rest of the cell allowing the formation of gigaseals in this zone paved the way for the electrophysiological study of ion channels in these cells (Darszon et al., 2011; Wang et al., 2021; Vyklicka & Lishko, 2020).

There are many types of Ca^{2+} permeable channels but this book deals mainly with those that are classified within the voltage-dependent category or Ca$_V$s. Though several types of Ca$_V$s have been reported to be present in sperm, so far none of them have been recorded electrophysiologically in the mature male gamete (Darszon et al., 2011; Wang et al., 2021). Since CatSper, a Ca^{2+} selective, sperm-specific channel, mildly voltage dependent and pH dependent, is the main Ca^{2+} uptake channel of this cell that has been electrophysiologically characterized, we decided to include it in this channel in the chapter. Therefore, our main subjects will be Ca$_V$s and CatSper channels, their expression and function in spermatogenic cells and sperm. We apologize at the outset for possibly missing some valuable information, and for arbitrarily choosing a few species. There are comprehensive reviews dealing with sperm ion channels and transporters, including a broader species spectrum, where interested readers may complement their knowledge regarding this interesting field (Darszon et al., 2011; Lishko et al., 2012; Wang et al., 2021; Vyklicka & Lishko, 2020; Brown et al., 2019; Beltran et al., 2016; Boulais et al., 2019; Wudick et al., 2018; Wachten et al., 2017; Santi et al., 2013; Singaravelu & Singson, 2013).

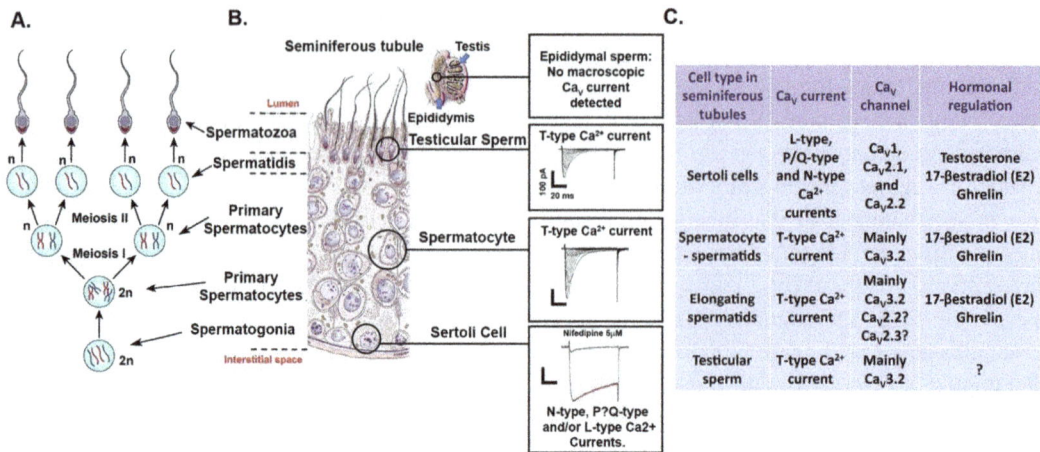

Fig. 2 Voltage-gated Ca²⁺ channels in the testis. (**a**) Morphological mature sperm are originated from germ cells contained in the seminiferous tubules of the testis, in a process known as spermatogenesis. At the basal compartment, mitotic divisions of spermatogonia produce primary spermatocytes (diploid cells) which start with the first meiotic division producing secondary spermatocytes (haploid cells) located at the adluminal compartment. After the second meiotic division, round spermatids are produced from secondary spermatocytes. Round spermatid s differentiate into mature sperm through morphological changes during spermiogenesis, eliminating part of their cytoplasmic content and producing flagella. Mature sperm are released into the lumen of the seminiferous tubules. Low voltage-gated Ca²⁺ currents (T-type) have been recorded from spermatogenic cells and testicular sperm, where their current amplitude decreases in the most advanced spermatogenic stages. Within the seminiferous tubules, spermatogenic cells are always accompanied by Sertoli cells which provide structural and metabolic support to the germ cells. Sertoli cells possess high voltage-gated Ca²⁺ currents constituted by N-, P/Q- and L-type channels. (**b**) Both spermatogenic cells' LVA and Sertoli cells' HVA Ca²⁺ currents are differentially modulated by sexual and metabolic hormones relevant for spermatogenesis. Therefore, controlling Sertoli cells protein secretion, nutritional support of spermatogenesis, expression of cyclins, and regulating spermatogenic cells apoptosis via Ca²⁺ influxes

Ca$_V$s in Spermatogenic Cells and Sperm

Spermatogenesis

Sperm originates from germ cells during spermatogenesis, a proliferating and differentiating process that takes place in the seminiferous tubules of the testis. Then, the morphological mature sperm travel to the epididymis where they are stored until ejaculation. Figure 2a schematically shows the process of spermatogenesis in mammals. This process starts with the mitotic division of spermatogonia stem cells, located at the basal membrane of the seminiferous tubules. Some of these cells differentiate into primary spermatocytes (diploid cells). Subsequently, the primary spermatocyte divides meiotically (meiosis I) into two secondary spermatocytes (haploid cells); then each secondary spermatocyte divides by meiosis II

into two equal haploid spermatids. The spermatids are differentiated into spermatozoa (sperm) in a process known as spermiogenesis. Cell division from a spermatogonium to a spermatid is incomplete; cells remain connected to one another by bridges of cytoplasm to allow synchronous development. At all stages of differentiation, the spermatogenic cells are in close contact with Sertoli cells which extend from the basal membrane to the lumen of the seminiferous tubule and provide structural and metabolic support to the developing sperm (Fig. 2) (Alberts et al., 2008).

As in other tissues, Ca²⁺ influx through the corresponding Ca2+ channels could play a crucial role in regulating these consecutive cellular divisions, potentially modulating intracellular signaling to facilitate morphological changes as germ cells progress through spermatogenesis. Unfortunately, there is very scant information regarding the role of Ca$_V$s and CatSper in spermatogenesis in marine organisms, therefore we will only discuss this matter in mammals.

Calcium Channel Expression Within Sertoli and Spermatogenic Cells

Ca$_V$s in Sertoli Cells

Sertoli cells reside in the seminiferous tubules where they play structural and secretory functions, nourishing spermatogenic cells through their differentiation and specialization processes to become morphologically mature sperm. Previous studies have demonstrated the presence of P/Q- and N-type Ca^{2+} currents (Ca$_V$2.1 and Ca$_V$2.2 Ca^{2+} channels, respectively) at the basal compartment of the seminiferous epithelium, where they were immunodetected in contact membranes between Sertoli cells and spermatogonia or early spermatocytes. Pharmacological experiments performed on Sertoli cell cultures in the presence of specific toxins indicate that both N- and P/Q-type Ca^{2+} channels could be involved in the regulation of protein secretion, which is important for germ cell maturation (Fragale et al., 2000).

A recent report has also suggested that Sertoli cells functionally express L-type Ca^{2+} channels (Ca$_V$1). The activation of L-selectin, by fucoidan and sialyl-Lewis-a, in Sertoli cells induces Ca^{2+} influx. This agonist-induced Ca^{2+} influx was inhibited by different L-type voltage-operated Ca^{2+} channel antagonists, such as nifedipine, diltiazem, or verapamil, indicating their participation in the fucoidan or sialyl Lewis-a-induced signaling cascade. On the other hand, cells treated with mibefradil, a T-type Ca$_V$3 antagonist, still present an L-selectin-induced [Ca^{2+}]i increase. These results indicate that the L-selectin activation of Sertoli cells induces Ca^{2+} influx, as a consequence of the L-type Ca$_V$ activation (Kao & Millette, 2007). Another study demonstrated that stimulation with specific L-selectin ligands activates the Rho GTPase and up-regulates the expression of Rho and Rac1 proteins. Consistent with previous observations, addition of nifedipine, but not mibefradil, abolished the activation of both Rho and Rac1. This led to the conclusion that

L-selectin activation induces Ca^{2+} influx through the L-type Ca^{2+} channels, and subsequently enhances Rho and Rac1 protein expression in both Sertoli cells and ASC-17D cell line (Limanjaya et al., 2020).

T-type Ca^{2+} Channel in Spermatogenic Cells

The iconic report by Hagiwara and Kawa (1984) provided a first characterization of ion currents present throughout rat spermatogenesis. They described a transient inward current followed by a slowly developing outward current produced when the membrane potential was depolarized to values more positive than −60 mV (Hagiwara & Kawa, 1984). The inward currents were identified as a Ca^{2+}-carried current, since replacement of external Ca^{2+} with Mn^{2+} reversibly diminished the current amplitude. According to Hagiwara and Kawa, during spermatogenesis, from spermatogonia to early spermatids, the density of Ca^{2+} current increased while the K$^+$ current density decreased significantly. However, in this report the molecular identity of the recorded Ca^{2+} inward current was not clearly established.

More recently, electrophysiological techniques were applied to mouse spermatogenic cells. In this species, the detected currents were identified as low voltage-activated (LVA) Ca^{2+} currents (Liévano et al., 1996; Arnoult et al., 1996; Santi et al., 1996). Male rodent germ cells express LVA Ca^{2+} channels (T-type) that are regulated by membrane potential and are responsible for the primary Ca^{2+} influx mechanism in the male germ cell line. These currents were evoked from a holding potential of −80 mV with voltage steps between −70 and +10 mV. Like somatic T-type Ca^{2+} currents, spermatogenic cell Ca^{2+} currents were transient, with both activation and inactivation kinetics becoming faster at stronger depolarizations showing the T-type typical crisscross current pattern. Coherent with LVA channels, deactivation of these Ca^{2+} currents displayed slow kinetics (Fig. 2a). Current–voltage (I-V) relationships obtained by averaging data from individual spermatogenic cells showed that the threshold for inward current activation was –60 mV, and the peak amplitude occurred between–20 and –30 mV, typical features for T-type channels. There was no indication of a second Ca^{2+} current component activated at more positive membrane potential values in these cells. Altogether, these biophysical parameters are sufficient experimental evidence to classify these currents as low-voltage activated Ca^{2+} currents and as the sole voltage-dependent Ca^{2+} current present in these cells. Because mature sperm cells are unable to synthesize new proteins, it was originally proposed that T-type Ca^{2+} channels could be also present in mature sperm, where they could contribute to the Ca^{2+} influx required for fertilization (Arnoult et al., 1996; Santi et al., 1996).

Regarding human spermatogenic cells, small T-type-like Ca^{2+} currents have been recorded from round spermatids freshly isolated from an orchidectomy (Jagannathan et al., 2002b). However, even when the activation threshold for this inward current was close to –60 mV, the I-V curve peak reported for human germ cells was right shifted, close to –10 mV, compared to the I-V curve peak observed in mouse spermatogenic cells.

Complementing these electrophysiological data, different reports contributed to the molecular identification of the $Ca_v\alpha 1$ subunit gene responsible for encoding the spermatogenic cell T-type Ca^{2+} currents. Using RT-PCR or PCR, transcripts for *Cacna1g, Cacna1h,* and *Cacna1i* genes ($Ca_V 3.1$, $Ca_V 3.2$ and $Ca_V 3.3$, respectively) were identified in mouse germ cells (Espinosa et al., 1999; Treviño et al., 2004); $Ca_v 3.1$ and $Ca_v 3.2$ subunit transcripts were detected in human testis/male germ cells (Jagannathan et al., 2002a, b); and a partial sequence of $Ca_v 3.1$ transcript from rat testis (only domain IV and C-terminus) was reported (Jagannathan et al., 2002a). Additionally, immune detection using confocal microscopy revealed the presence of the three $Ca_V 3$ channel isoforms in mouse spermatogenic cells (Treviño et al., 2004). Based on the biophysical properties of the mouse spermatogenic cell's T-type Ca^{2+} currents described above, the most likely molecular candidates to be responsible for the T-type Ca^{2+} currents present in

spermatogenic cells were the $Ca_v 3.1$ and/or $Ca_v 3.2$ channels.

Experiments done in the $Ca_V 3.1$ null and $Ca_V 3.2$ null mice demonstrate that $Ca_V 3.2$ is the main contributor to the T-type Ca^{2+} currents in spermatogenic cells and that $Ca_V 3.1$ contributes to a very negligible extent to the T-type Ca^{2+} currents (Treviño et al., 2004; Stamboulian et al., 2004; Escoffier et al., 2007).

Despite that initial report by Hagiwara and Kawa describing that the density of the Ca^{2+} current increased during spermatogenesis in rat (Hagiwara & Kawa, 1984), it was later shown that mouse spermatogenic cell T-type Ca^{2+} currents gradually decreased in amplitude as sperm mature, and were not detectable as a macroscopic Ca^{2+} current in epididymal sperm (Fig. 2b) (Xia & Ren, 2009), even when they were present in testicular sperm (Darszon et al., 2006). Certainly, the high Zn^{2+} concentration in the epididymis could contribute to reduce the T-type Ca^{2+} current amplitude, since Zn^{2+} reduces the number of available Ca^{2+} channels in sperm (Lopez-Gonzalez et al., 2016).

Pharmacology of the Low Voltage Activated Ca^{2+} Current in Spermatogenic Cells

Nickel (Ni^{2+}) at 200 µM and amiloride at 500 µM reduced the amplitude of these currents by 75% and 62%, respectively. Unexpectedly, 10 µM nifedipine also reduced peak currents by 53%, and significant inhibition of the Ca^{2+} current occurred at concentrations as low as 2 µM (Santi et al., 1996).

In addition to the classical blockers of T-channels described above, evidence indicates that anion channel blockers also affect the T-type Ca^{2+} current in mouse spermatogenic cells. Niflumic acid (NA), 1.9-dideoxyforskolin, and 5 nitro-2-(3-phenylpropylamine) benzoic acid block the current in a concentration- and voltage-dependent manner (IC_{50} at –20 mV: 43, 28 and 15 µM, respectively) (Espinosa et al., 1999). Additionally, NA blockade is more potent in the open and in the inactivated state than in the closed state of T-type channels and slows the recovery from inactivation (Balderas et al., 2012). Toxins from the scorpion *Parabuthus granulatus* also inhibit T-type Ca^{2+} currents in mouse spermatogenic cells (López-González et al., 2003). The modulation of T-type channels by endogenous or exogenous agents is crucial for intracellular Ca^{2+} homeostasis and it is relevant for spermatogenesis. Unfortunately, the expression of Ca$_V$ channels in mature sperm based on pharmacology is not conclusive because blockers of Ca$_V$ channels also inhibit the sperm-specific CatSper Ca^{2+} channel.

Other Ca$_V$s Channels in Spermatogenic Cells

Despite electrophysiological evidence that LVA Ca^{2+} currents (T-type, Ca$_V$3) are basically the only Ca$_V$s functionally expressed in mammalian spermatogenic cells, there are reports of the presence of HVA channels in rodent and human germ cell lines that remain controversial (Jagannathan et al., 2002a).

First, molecular biology strategies allowed identification of different HVA Ca^{2+} channel transcripts within testis and/or the male germ cell line. In mouse, transcripts for Ca$_V$2.1, Ca$_V$2.2, Ca$_V$1.2, Ca$_V$1.3, and Ca$_V$2.3 subunits were detected by RT-PCR (Liévano et al., 1996; Espinosa et al., 1999); whereas in rat, the transcript for a testis-specific Ca$_V$1.2 subunit was detected by in situ RT-PCR-PCR (Goodwin et al., 1997). Germ cell immunostaining demonstrated the expression of Ca$_V$2.1 and Ca$_V$1.2 subunits in mouse (Serrano et al., 1999) and Ca$_V$1.1 in rat (Goodwin et al., 2000).

Second, in elongated spermatids, new components of the Ca^{2+} currents were reported. Wennemuth and collaborators observed a 38% inhibition of the Ca^{2+} currents by ω-conotoxin GVIA, a toxin considered to be specific for Ca$_V$2.2 channels (Wennemuth et al., 2000). The remaining fraction of this current was almost completely blocked by an extra addition of Ni^{2+} (100 μM). Interestingly, this dose of Ni^{2+} inhibits 50–70% of non-toxin-pretreated Ca^{2+} currents (Treviño et al., 2004; Wennemuth et al., 2000). Even when previous reports indicate the expression of T-type like Ca^{2+} currents in mouse elongating spermatids, authors proposed that their recorded Ca^{2+} currents contained two components with similar kinetics but different pharmacology, one of which is Ni^{2+}-sensitive (probably R-type Ca^{2+} currents or Ca$_V$2.3), and another one sensitive to ω-conotoxin GVIA, therefore pharmacologically consistent with N-type Ca^{2+} channels (Ca$_V$2.2). A very small third component of the Ca^{2+} currents was resistant to both Ni^{2+} and ω-conotoxin GVIA (Wennemuth et al., 2000).

Interestingly, CatSper, the sperm-specific Ca^{2+} channel (see section "CatSper channels"), which is a unique sperm cation channel essential for sperm motility and male fertility, is exclusively expressed in meiotic and post-meiotic spermatogenic cells (Yu et al., 2015). A functional CatSper channel requires pore formation by four transmembrane subunits: CatSper1 to CatSper4. At least in mouse testis, CatSper2 mRNA is found before meiosis (as early as day 8 post-natal, probably in primary spermatocytes) whereas very low levels of CatSper1 mRNA are usually detected after meiosis (day 18 and 21 post-natal, from secondary spermatocytes to sperm); CatSper3 and CatSper4 mRNAs have similar expression profiles during mouse post-natal development, initiated on day 15 (after meiosis) (Li et al., 2007). Considering this information, it is quite possible that one of the new components observed in the elongating spermatids Ca^{2+} currents could be mediated by CatSper channel, but this hypothesis needs to be explored.

Voltage-Gated Calcium Channel Regulation in Mammalian Germ Cell Line

The correct function of mammalian testis depends on a plethora of sexual hormones, which act through different pathways (endocrine, paracrine, and autocrine). In addition to sexual hormones, a different group of these molecular messengers acts as metabolic hormones that play a relevant role in the maintenance of cellular energy balance during spermatogenesis. For instance, in testis the most important messengers are gonadotropins, follicle stimulating FSH, luteinizing hormones and androgens. Whereas gonadotropins maintain proper functioning of Leydig, myoid, and Sertoli cells, testosterone and FSH act via the Sertoli cells to regulate germ cell differentiation (Holdcraft & Braun, 2004). Different groups contributed to elucidate at least in part the hormonal-dependent regulation of voltage-gated Ca^{2+} channels in the seminiferous tubules (Espinosa et al., 2000; Liu et al., 2011; Sanchez-Cardenas et al., 2012; Martins et al., 2016; Zanatta et al., 2021). For a broader description of calcium channel modulation by second messengers, please refer to chapter "Modulation of VGCCs by G-Protein Coupled Receptors and Their Second Messengers" by Dr. Mark and colleagues.

Estradiol and Serum Albumin

Testosterone and its metabolites, dihydrotestosterone (DHT) and estradiol (E2), are the main sexual hormones due to their preponderant role in the regulation of gonadal and germ cell development in both genders as well as in the sexual differentiation of males (Holdcraft & Braun, 2004). In mouse spermatogenic cells, 17β-estradiol at nM concentrations produces a small inhibition (~10%) of the T-type Ca^{2+} currents which becomes significant at μM concentrations. A hormonally inactive isomer of 17β-estradiol (17α-estradiol) causes less reduction in T-type channel activity, indicating some specificity in the observed regulation. This inhibition seems too fast to be attributed to genomic activation, so it is likely due to a direct inhibition of the channel. Unlike for what has been reported for HVA channels, this 17β-estradiol downregulation probably does not involve activation of a G protein signaling pathway, since spermatogenic T-type Ca^{2+} channels do not seem to be modulated through G-protein activation (Fig. 2c) (Espinosa et al., 2000).

Activation of Growth Hormone Secretagogue Type 1a Receptor

In mammals, metabolic hormones, including ghrelin, leptin, and nesfatin-1, play essential regulatory roles in many biological processes, including morphogenesis, growth, and reproduction. In humans, ghrelin is a growth hormone-releasing peptide synthesized in the brain and gut that has been suggested to interfere with spermatogenesis, via the growth hormone secretagogue receptor expressed in Sertoli cells. Besides its regulatory role in Sertoli cells, a novel functional role of ghrelin is the modulation of T-type Ca^{2+} currents in mouse spermatogenic cells. Ghrelin inhibits T- currents in a dose-dependent manner. This inhibitory effect is blocked by D-Lys3-GHRP-6, a selective growth hormone secretagogue receptor 1a (GHS-R1a) antagonist, indicating that ghrelin acts through GSH-R1a inhibiting T-type Ca^{2+} currents via a PTX-sensitive novel PKC pathway (Fig. 2c) (Liu et al., 2011).

Arachidonic Acid

Phospholipids and fatty acids play important roles both as structural components of the Sertoli and spermatogenic cell plasma membranes and as molecular messengers that trigger the first meiotic division within the male germ cell line. Arachidonic acid (AA) is one of the main molecules among these components that is released from Sertoli cells in response to follicle-stimulating hormone. Compounds like AA could participate in the modulation of Ca_Vs in spermatogenic cells. In fact, AA inhibits the spermatogenic cell T-type Ca^{2+} current at micromolar concentrations ($IC_{50}= 4.7$ μM). Interestingly pathophysiological conditions leading to AA spontaneous oxidation enhance T-type Ca^{2+} current inhibition. Oxidized AA (AAox) is 25-fold more potent than AA at inhibiting T-type Ca^{2+} currents ($IC_{50}= 0.19$ μM) and it left shifts the I-V curve peak of both activation and steady-state inactivation curves. In addition, AAox slows T-type Ca^{2+} current deactivation kinetics and their recovery from inactivation; as a consequence, the fraction of inactivated Ca^{2+} channels is increased by this fatty acid (Bondarenko et al., 2019).

Calmodulin

Pharmacological studies focusing on environmental toxicity have provided experimental evidence on the physiological relevance of second messengers involved in male reproduction. Fenvalerate, a pyrethroid pesticide initially considered as a low-toxicity agent, causes male reproductive toxicity. This pyrethroid significantly increases calmodulin (CaM) expression in mouse testis. Enhanced CaM expression causes dilated seminiferous tubules and disturbs the arrangement of spermatogenic cells (Gao et al., 2012). CaM has been reported to activate the voltage-dependent T-type Ca^{2+} currents in mouse spermatogenic cells. Addition of CaM to the patch pipette solution increases the Ca^{2+} current density, and W7 and trifluoperazine, two CaM antagonists, decrease these currents in a concentration-dependent manner with IC$_{50}$s of 10 μM and 12 μM, respectively. W7 alters the channels' voltage dependence of activation and slows both activation and inactivation kinetics. It also induces inactivation at voltages at which T-channels are not activated, suggesting inactivation from the closed state. CaM-mediated up-regulation of these Ca^{2+} currents seems to be direct, because inhibitors of CaM-dependent kinase II and protein kinase A, as well as a CaM-activated phosphatase, have no effect on T-type Ca^{2+} currents in spermatogenic cells (Lopez-Gonzalez et al., 2001). These observations suggest that it is likely that the fenvalerate-induced overexpression of CaM increases T-type Ca^{2+} current density, contributing to the pathological phenotype produced by this pyrethroid in the seminiferous tubules.

Tyrosine Phosphorylation

In spermatogenic cells, T-type Ca^{2+} currents were reported to display a facilitation either by strong depolarizations or by antagonists of protein tyrosine kinases. Inhibitors of tyrosine phosphatases blocked voltage-dependent facilitation of these Ca^{2+} currents (Arnoult et al., 1997). Therefore, tyrosine phosphorylation diminishes the activity of male germ cell Ca$_V$3 channels while tyrosine phosphatases enhance it. This voltage-dependent facilitation seems to be also dependent on [Ca^{2+}]$_i$ as decreasing it below its resting level enhances this voltage-dependent facilitation (Stamboulian et al., 2002). Based on these results, the authors proposed a functional relationship between the T-type Ca^{2+} channels and some proteins whose activity is related to endoplasmic Ca^{2+} concentra-

tion (presumably TRP channels and/or inositol trisphosphate receptors). Though a massive tyrosine phosphorylation occurs during mammalian sperm capacitation, it is an open question if a similar process occurs in response to an agonist during spermatogenesis.

Functional Roles of Ca$_V$s

The physiological relevance of Ca$_V$s in the male germ cell line is not completely understood. As previously mentioned, a functional mammalian testis depends on a balance between sexual and metabolic hormones that trigger and/or regulate Ca^{2+} influxes necessary for spermatogenesis maintenance and progression.

Different experimental strategies have been used to study the role of Ca^{2+} influx on spermatogenesis. For instance, mouse testis Ca^{2+} oscillations have been studied using acute slice preparations of seminiferous tubules that preserve tissue architecture and intercellular communications. In this study, Ca^{2+} imaging revealed that a subpopulation of male germ cells displays spontaneous [Ca^{2+}]$_i$ fluctuations resulting from Ca^{2+} entry possibly throughout Ca$_V$3 channels. These spontaneous Ca^{2+} fluctuations frequently occur in a synchronous manner in spermatogenic cell subpopulations, probably mediated by intercellular communication via gap junctions and/or intercellular bridges. Furthermore, this *in vitro* cellular model preserved the biological capacity to respond to hormones like testosterone. For instance, addition of testosterone elevated [Ca^{2+}]$_i$ suggesting the presence of functional Sertoli cells (Sanchez-Cardenas et al., 2012). As mentioned earlier, three different voltage-gated Ca^{2+} channels could be involved in the testosterone-induced Sertoli cell response considering their expression in this cell type, namely Ca$_V$1, Ca$_V$2.1, or Ca$_V$2.2 channels (Fragale et al., 2000; Kao & Millette, 2007; Limanjaya et al., 2020). This [Ca^{2+}]$_i$ increase could regulate protein secretion during germ cell maturation, a matter that requires further evaluation.

The hormonal control of Ca^{2+} influx by 17-βestradiol (E2) was studied in rat testis and seminiferous tubule cultures, to evaluate how this hormone mediates its effect on the first spermatogenic wave. Addition of E2 (1 µM) increased Ca^{2+} influx (37%) in both whole testis and seminiferous tubules, upon the activation of ESR1/2 estrogen receptors and nifedipine-sensitive Ca_V1 Ca^{2+} channels expressed in Sertoli cells (Fig. 2c). This is the first time that E2 has been shown to stimulate a Ca^{2+} uptake in immature testis. Additionally, in this report E2 diminished cyclin A1 mRNA expression, did not affect cyclin B1 mRNA levels, but increased the protein level of both cyclins, which are proteins related to cell cycle regulation (Zanatta et al., 2021). As previously mentioned, 17-βestradiol inhibits the spermatogenic cell T-type Ca^{2+} currents in a concentration-dependent manner (Espinosa et al., 2000), indicating the differential effect of E2 on Sertoli and germinal cells within the seminiferous tubules.

The germ cell's Ca^{2+} influx also regulates the expression of the Bcl-2 protein family, composed of both cell death agonists and antagonists that differ in their structural features. Bcl-x exists in two isoforms, the anti-apoptotic form Bcl-xL and the proapoptotic form Bcl-xS. The critical balance between the two forms appears to be important for cell survival. In rat spermatogenic cells, the activation of T-type Ca^{2+} channels by the reactive oxygen species-induced cascade causes the increase of Bcl-xS expression while Bcl-xL is decreased. Consistently, addition of pimozide, a T-type Ca^{2+} channel blocker (Santi et al., 2002), resulted in a decrease in Bcl-xS and an increase in Bcl-xL expression, favoring cell survival due to prevention of mitochondrial potential loss, reduction of caspase-3 activity, and inhibition of DNA fragmentation (Mishra et al., 2006). This study provided a new insight into the possible role of Ca^{2+} influx in regulating spermatogenic cell´s apoptosis.

On the other hand, *in vivo* studies have contributed to evaluate the effect of Ca_V inhibitors on mammalian testis. Prepubertal male mice fed with low concentrations of nifedipine and ethosuximide (L-type and T-type Ca^{2+} channel blockers, respectively) for 20 days showed a significant reduction in body weight, testis size/weight, and sperm production in a dose-dependent fashion. These defects are associated with a spermatogenic cell arrest at the elongating spermatid stage. In addition, the number of Leydig cells, and consequently the serum testosterone level and the content of StAR protein mRNA, were also drastically reduced. Both T- and L-type Ca^{2+} channel blockers play an adverse role in normal spermatogenesis and steroidogenesis partly by blocking post-meiotic germ cell maturation, contributing to this pharmacologically induced male sterility (Lee et al., 2011). The interpretation of these results is complicated due to the complex pathological phenotypes observed and to the cross-pharmacological effect of the Ca^{2+} channel inhibitors used. For instance, nifedipine could inhibit both L and T-type Ca^{2+} channels since T-type Ca^{2+} channels in spermatogenic cells are sensitive to this drug (Santi et al., 1996; Arnoult et al., 1998). Therefore, at least in mammalian testis, nifedipine could inhibit both the Sertoli cell L-type Ca^{2+} current and the spermatogenic cell T-type Ca^{2+} current to cause abnormal spermatogenesis.

Chemotaxis

Multiple observations have led in recent years to the realization that sperm, depending on the species, develop different signaling modes to achieve fertilization (Kaupp & Strunker, 2017; Darszon et al., 2020). These decoding strategies contend with specific reproductive requirements emanating from distinct morphological characteristics of the female reproductive tract in internal fertilizers and from the varying properties of the surrounding media in internal and external fertilizers. Chemotaxis is an example, as in external and internal fertilizers sperm must find the egg under very different circumstances.

Chemotaxis is broadly defined as the movement of cells in the direction of a chemoattractant gradient, the chemoattractant being a peptide or any other chemical that attracts specific cells by chemotaxis (Eisenbach & Giojalas, 2006). Sperm from many species display chemotaxis toward the egg which enhances their fertilization probability (Yoshida & Yoshida, 2011; Guerrero et al., 2010; Kaupp et al., 2008). This fundamental process for many sperm species is dependent on orchestrated Ca^{2+} permeability changes involving Ca^{2+} channels (Wachten et al., 2017; Mata-Martinez et al., 2021). We will separately discuss the participation of Ca$_V$s and CatSper in this process starting with Ca$_V$s in sea urchin sperm, probably the best-known chemotactic model in these cells.

Do Ca$_V$s Participate in Sea Urchin Sperm Chemotaxis?

Diffusible sperm activating peptides (SAPs) from the outer envelope of the egg regulate sea urchin sperm motility (Suzuki, 1995). SAP binding to its receptor, which regulates the activity of a membrane Guanylate Cyclase, elevates intracellular cGMP that after a delay leads to the increase of cAMP levels. The cGMP elevation triggers fast changes of V_m, pH_i, and $[Ca^{2+}]_i$ (Darszon et al., 2011; Kaupp et al., 2008). The increase in cGMP first opens a K^+ selective channel (KCNG) (Strunker et al., 2006; Galindo et al., 2007) that hyperpolarizes sperm. Because the uptake of external Ca^{2+} induced by SAPs and egg jelly were found to be inhibited by Ca$_V$ blockers, such as dihydropyridines (Kaupp et al., 2003; Nishigaki et al., 2004; Wood et al., 2003), it was suggested that this V_m change could remove inactivation of Ca$_V$s. Transcripts of Ca$_V$1.2 and Ca$_V$2.3 were found in testis, and mammalian Ca$_V$ antibodies detected these two subunits in sea urchin sperm (Granados-Gonzalez et al., 2005). Despite this, all-further efforts from several groups have been unable to corroborate the presence of Ca$_V$s by proteomic strategies. In addition, modeling strategies using an experimentally backed logical network (Espinal-Enriquez et al., 2017) and a modular model of the speract (*S. purpuratus* and *L. pictus* SAP), signaling cascade have indicated that if present, Ca$_V$s may fine-tune flagellar regulation and that CatSper is the most important Ca^{2+} channel in this process (Priego-Espinosa et al., 2020).

Acrosome Reaction (AR)

Prior to fertilization, sperm must undergo the AR, an exocytotic process that liberates the content of this lysosome originated organelle. This process allows sperm penetration of the egg investments to finally reach and fuse with the egg. Regardless of the exact site where the AR occurs and the factors that trigger the AR, its occurrence in sperm strictly depends on Ca^{2+} (Yanagimachi & Usui, 1974). Therefore, the Ca^{2+} channels involved in the AR have been a matter of intense study.

T-type channels in spermatogenic cells and testicular sperm inactivate at potentials more depolarized than ~−50 mV. Therefore, at the resting potential of mammalian sperm (~−40 mV) these channels would be inactivated, and a hyperpolarization of the sperm plasma membrane should be produced to remove this inactivation and render these channels available to activate. Because membrane hyperpolarization is a hallmark of sperm capacitation, a model was built around this feature and the belief that T-type Ca^{2+} channels played a key role in the AR. During sperm capacitation in the female tract, hyperpolarization takes place and removes the inactivation from T-type channels. As a result, there is a fast and transient Ca^{2+} increase which is followed by a second and sustained Ca^{2+} increase, probably mediated by SOC channels (Arnoult et al., 1999). These independent but related Ca^{2+} increases were believed to be required for the induction of the AR. The evidence supporting this model was based on electrophysiology in spermatogenic cells, immunodetection and Ca^{2+} imaging studies in mature sperm, combined with pharmacological tools. However, in the era of sperm patch clamping, few Ca^{2+} channels have been registered in the mature cell, and T-type Ca^{2+} channels were not among them.

On the other hand, mouse knockout models have been used as an alternative strategy to determine the role of different Ca$_V$ channels. For example, studies using sperm from KO mice for Ca$_V$3.1 and Ca$_V$3.2 channels suggested that Ca$_V$3.2 was responsible for the Ca^{2+} influx induced by a potassium stimulus depolarization. However, parameters such as the AR and motility remained intact in Ca$_V$3.2 KO sperm (Escoffier et al., 2007). Ca$_V$3.1 KO mice are also fertile and the residual currents in spermatogenic cells from these mice differ from somatic cell T-type currents (Stamboulian et al., 2004), suggesting the presence of additional Ca$_v$ channels.

The role of Ca$_V$2.3 was also investigated using a KO model (Cohen et al., 2014). These authors reported that sperm lacking Ca$_V$2.3 are subfertile, display altered Ca^{2+} dynamics, and experience the AR only when induced with a Ca^{2+} ionophore but not with progesterone and zona pellucida (ZP). Additionally, they performed *in vitro* fertilization studies, and concluded that Ca$_V$2.3 channels were involved in early egg–sperm interactions since fertility was rescued when ZP nude oocytes were used.

CatSper Channels

Expression in Different Mammalian Species

Ca^{2+} signaling is one of the most important mechanisms for signal transduction from bacteria (Dominguez, 2004), plants (Kudla et al., 2010), fungi (Zelter et al., 2004) to animals (Berridge et al., 2003; Clapham, 2007). As animals evolved into more complex organisms, Ca^{2+} signaling became more extensive and versatile. The sperm-specific Ca^{2+} channel CatSper is one of the most complex ion channels known. Calcium influx through CatSper plays an essential role in mammalian sperm maturation, motility, and male fertility (Lishko et al., 2012; Darszon et al., 2006; Jimenez-Gonzalez et al., 2006; Whitaker, 2006). CatSper expression is conserved among mammals, but these channels are also expressed widely and heterogeneously in all eukaryotic lineages (Cai et al., 2014, 2015; Romero & Nishigaki, 2019; Cai & Clapham, 2008, 2012). They are present in sea urchins and tunicates (Strunker et al., 2006; Cai & Clapham, 2008), but they were not found in other organisms, like some birds (chickens and zebra finch) where CatSper genes degenerated into pseudogenes

(Cai & Clapham, 2008). This expression heterogeneity has brought up questions about the evolution and origins of CatSper and the signaling complex (Thomson & Wishart, 1991).

It is suspected that flagella and cilia arise from all eukaryotes' last common ancestor (Mitchell, 2007). Studies have shown that flagella and primary cilia have different Ca^{2+} signaling molecules, CatSper and TRPP, respectively (Ren et al., 2001; DeCaen et al., 2013; Delling et al., 2013). To examine the origin of the CatSper complex, Cai and collaborators analyzed the genome of the apusozoan protist *T. trahens*, a unicellular progenitor of Opisthokonta, a common unicellular ancestor of animals and fungi (Cai et al., 2015; Ruiz-Trillo et al., 2007), and hypothesized that the Ca^{2+} signaling mechanism from animals and fungi could be found in them (Cai et al., 2015). CatSper was present in *T. trahens* and fungal components of Ca^{2+} signaling (Cai et al., 2015; Cai & Clapham, 2012). This suggested that this complex had a role regulating cell motility that precedes the divergence into animal and fungi kingdom. Interestingly, CatSper homologs were not found previously in fungal genomes (Cai & Clapham, 2008). However, in *A. macrogynus*, a pheromone released by females induces Ca^{2+} increases similar to those induced in human sperm by progesterone (Pommerville et al., 1990), and not surprisingly, CatSper complexes were found in *A. Macrogynus* (Cai & Clapham, 2012).

CatSper was not found in bikont protists, algae, and plants. However, a recent search revealed: (1) the presence of the same CatSper complex components in *Aurantiochytrium limacinum* than in *T. trahens* (Cai et al., 2014), and (2) the presence of CatSper in one of the alga precursors of the lineage leading to green plants *Cyanophora paradoxa* (Price et al., 2012). The conservation of CatSpers in two bikont species or more suggests that the CatSper channel complex forms a Ca^{2+} signaling complex before the divergence into Unikonta and Bikonta. Alternatively, the presence in bikonts could be due to horizontal gene transfer. However, this is unlikely since all the proteins in the CatSper complex are necessary for expression.

In 2008, Cai and Clapham mentioned that sNHE showed a similar lineage-specific gene loss as that found for CatSper (Romero & Nishigaki, 2019; Cai & Clapham, 2008). Most recently, and with the hypothesis that CatSper, sNHE, and sAC are functionally coupled and regulate sperm motility, Romero and Nishigaki used comparative genomic analysis to explore phylogenetic relationships among CatSper, sNHE, and sAC in eukaryotes. They found that nearly all species of Metazoa with CatSper also preserved sNHE and sAC, like mammals, reptiles, the coelacanth, cartilaginous fishes, tunicates, amphioxus, Echinodermata, Brachiopoda, Cnidaria, and Ctenophora. Romero et al. concluded that a correlated distribution of the three proteins suggests coevolution in the Metazoa.

Molecular Composition, Biophysical Properties, Regulation, and Pharmacology

The CatSper channel is a multi-subunit complex located in the plasma membrane, in a quadrilateral column arrangement in the principal piece of the sperm flagella. This channel, one of the most complex yet described, permeates Ca^{2+} and is activated by voltage- and intracellular alkalinization (Kirichok et al., 2006; Chung et al., 2014). The CatSper complex includes the pore-forming proteins named CatSperα1-4 (forming a functional tetramer) and different auxiliary subunits, that comprise CatSperβ, γ, δ, ε, ζ, EFCAB9, η, and τ (Ren et al., 2001; Quill et al., 2001; Lobley et al., 2003; Liu et al., 2007; Wang et al., 2009; Chung et al., 2011, 2017; Hwang et al., 2019, 2021a, b; Lin et al., 2021) (Fig. 3a). The CatSper α1-4 subunits have six transmembrane (TM) segments enclosing the voltage-sensing domain (VSD:S1-S4) and the pore domain (PD:S5-S6). In the fourth segment (S4) of α1 and α2 subunits, four or five positively charged residues are found every three amino acids, which is a typical feature of the S4 segment of voltage-gated channels. In contrast, α3 and α4 have only two suggesting a correlation with the moderate voltage dependence of this channel. All α subunits have a pore-forming loop with the canonical Ca^{2+} selective channel motif. Notably, CatSper channels possess only aspartic acids in this motif (D536, D295, D227, and D237) which suggests a different cation selectivity of CatSper compared with typical voltage-gated Ca^{2+} channels (Senatore et al., 2013; Garza-Lopez et al., 2016; Li et al., 2021).

The auxiliary subunits CatSperβ, γ, δ, and ε, all contain a TM segment and a large extracellular domain which form a canopy-like structure that stabilizes the full complex through interactions with CatSper subunits α4,1,3 and 2, respectively (Lin et al., 2021). The ζ and EFCAB9 are two small cytosolic subunits and their mutual interaction is essential for the pH$_i$-dependence and Ca^{2+}-sensitive activation of the CatSper channel (Hwang et al., 2019).

Catsperτ (tau, CatSper Targeting subunit), the C2cd6-encoded (C2 Ca^{2+}-dependent Domain-containing protein 6) membrane associating C2 domain protein, independently migrates to flagella and serves as a major targeting component of the CatSper channel complex. The testis-specific *C2CD6* gene encodes two isoforms, one long (>200 kDa) and one short (70 kDa) in mouse and human. CatSperτ KO mouse sperm still form functional CatSper channels, albeit in significantly decreased quantities, but sperm fail to hyperactivate, rendering the male infertile (Hwang et al., 2021b; Yang et al., 2021).

Recently, Cryo-EM analysis of the CatSper complex allowed the identification of two new proteins associated with CatSper channels: TMEM262, a scramblase also denominated as CatSperη subunit and SLCO6C1 (solute carrier organic anion transporter family, member 6c1) an organic anion transporter. The C-terminal loop of CatSperη together with the loop of TM1 and TM2 constitute a platform that supports the stem domain of CatSperβ. The TM2 and TM3 segments of CatSperη form hydrophobic interactions with the TM domain of CatSperβ stabilizing the CatSper complex. SLCO6C1 is a testis-specific organic anion transporter protein (belonging to the OATP6 family) that in humans is closely related to its homologue SLCO6A1, predominantly expressed in testis

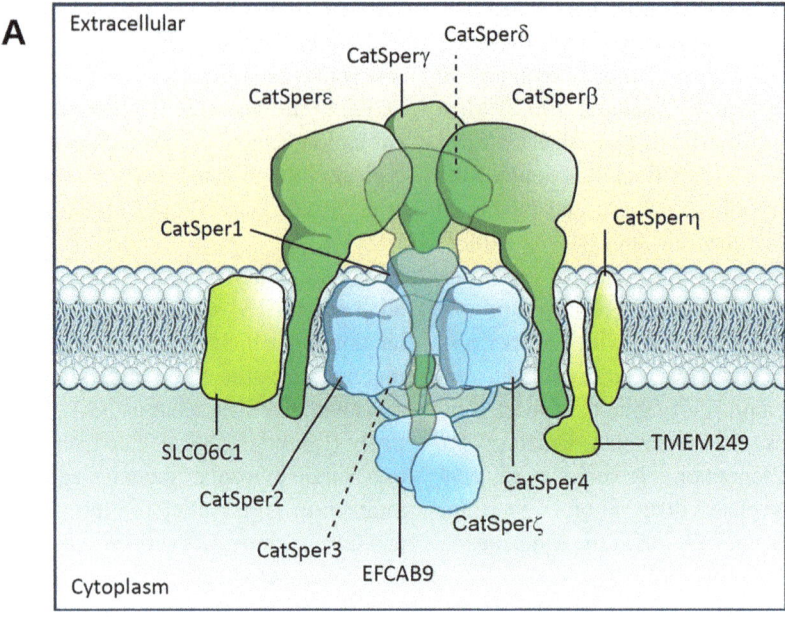

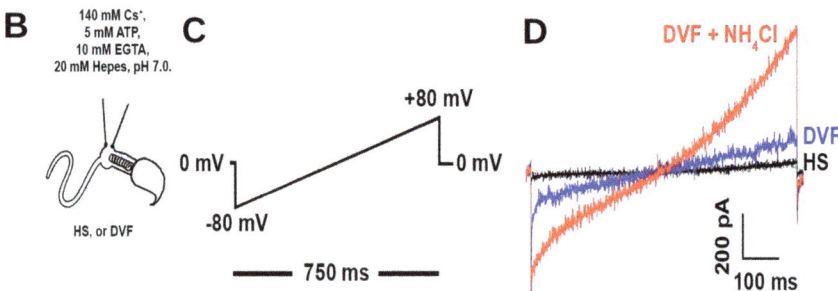

Fig. 3 (**a**) Structure of CatSper channels. (**b–d**) Whole-cell patch-clamp recordings from sperm. (**b**) A high resistance seal (>5 GOhms) was established between the pipette and the membrane of the cytoplasmic sperm droplet. (**c**) Voltage-ramp protocol used to stimulate CatSper currents. (**d**) Representative CatSper currents were obtained in the same cells under three different conditions, in physiological solution (HS), divalent free solution (DVF), and DVF + 10 mM NH_4Cl solutions. Solutions: **HS** (in mM): 135 NaCl, 5 KCl, 2 $CaCl_2$, 1 $MgSO_4$, 10 lactic acid, 1 pyruvic acid, 5 glucose, 20 Hepes, pH 7.4. **DVF** (in mM): 150 sodium gluconate, 2 Na2EDTA, 2 EGTA, 20 Hepes, pH 7.4

(Hagenbuch & Stieger, 2013; Fietz et al., 2013). The structure of SLCO6C1 contains the N-domain, the C-domain, and the extracellular Kazal domain between TM9 and TM10 of the C domain (Lin et al., 2021). SLCO6C1 interacts with CatSperε through two interfaces: in the TM9 with the TM domain of CatSperε, through hydrophobic interactions and in the extracellular region, the Kazal domain of SLCO6C1 associates with the stem domain of CatSperε through hydrophobic interactions. The glyco-sylation site in SLCO6C1 and the residues that are involved in its interactions with CatSperε are conserved among mouse, rat, and human (Lin et al., 2021). Additionally, TMEM249, another protein of the TMEM family, is likely to be a component of the CatSper complex. TMEM249 is also localized to the principal piece of sperm and contributes to the assembly of the CatSper complex by interacting with CatSperη and VSD4 in the transmembrane and cytosolic domains (Lin et al., 2021).

All together principal and auxiliary subunits constitute the macromolecular CatSper channel complex that forms quadrilinear nanodomains. Studies using cryo-electron tomograms revealed continuous rows with repeating units in a zigzag arrangement of ~25 nm in width, demonstrating that the CatSper complex is repeated within the rows. Extracellular structures of CatSper form canopy tents that connect pore-forming channels as beads on a zigzag string in mouse and human sperm flagella (Zhao et al., 2021).

All the evidence shows that the CatSper channel is a complex that functions as a critical structural and functional organizer to form four lines in quadrilateral arranged nanodomains that organize signaling proteins throughout the flagellum. This structural arrangement is decisive for conducting tyrosine phosphorylation in time and space during the capacitation process to acquire the ability to fertilize. When the sperm loses the CatSper channel (CatSper1-KO), the capacitation-dependent protein tyrosine phosphorylation is strongly affected and the Catsper regulatory protein, the ABHD2 enzyme, also loses its quadrilateral distribution (Chung et al., 2014; Miller et al., 2018). All these structural and functional alterations include the loss of hyperactivated mobility and show that CatSper is essential to constitute key Ca^{2+} domains for the correct migration of the sperm to the ovule and to achieve fertilization.

CatSper Regulation

CatSper channels are considered polymodal sensors due to their modulation by multiple chemical and physical cues (Brenker et al., 2012). On their journey to meet the egg, sperm encounter media with different pH, ionic, and hormone concentrations, different chemical and physical cues that CatSper channels integrate to generate specific spatio-temporal Ca^{2+} signals, which trigger specific flagellar motility patterns. CatSper channels share at least two forms of regulation in different species: voltage and pH regulation. Other forms of regulation, as we will discuss below, vary between different species.

V$_m$ Regulation

Though Neher et al. first recorded ion channel currents using the patch clamp technique in the late seventies (Neher et al., 1978), almost 30 years passed till the first whole-sperm recording of CatSper currents in mouse sperm (Kirichok et al., 2006). The sperm morphology, motility, small size and low cytoplasmic volume and tight links between its plasma membrane and rigid intracellular structures made it very difficult to obtain giga-seals along most of its membrane surface. These difficulties were surmounted when a region of the sperm plasma membrane known as the cytoplasmic droplet that is loosely attached to the rigid intracellular structures was identified. A tight giga-ohm seal can be formed with the patch pipette in this region. Then, in 2006 and 2011, Kirichok and Lishko recorded for the first time whole-cell CatSper currents in mice and human sperm, respectively (Kirichok et al., 2006; Lishko et al., 2011). CatSper currents have been so far the only constitutively active Ca^{2+} conductance recorded using the whole-cell patch clamp technique (Fig. 3b–d).

CatSper channels from human and mouse sperm are weakly voltage-dependent and Ca^{2+} selective but pass monovalent ions (Cs$^+$ or Na$^+$) under divalent-free conditions (Lishko et al., 2011, 2012; Kirichok et al., 2006). Because monovalent CatSper currents are significantly larger, most studies report CatSper currents measured under divalent-free conditions (Fig. 3b–d). However, tail currents in high divalent (Ba^{2+}) concentrations were used to obtain the conductance–voltage relationship for these channels. Human CatSper channels recorded in 50 mM Ba^{2+} showed a stronger voltage dependency than the mouse orthologue, with a steeper G-V curve (k values of 20 and 30, respectively). The V$_{1/2}$ of human CatSper activation is +85 mV versus +11 mV for mouse CatSper at pH$_i$ = 7.5 (Kirichok et al., 2006; Lishko et al., 2011), which brings up the question of how human CatSper channels activate at more physiological membrane potentials. The voltage dependence of CatSper channels of sea urchin sperm has not been established since whole-cell patch-clamp recordings from these cells have not been attainable yet. Studies done in CatSper channels from *Arbacia Punctulata* sea urchin sperm using voltage-sensitive fluorescent dyes reported a threshold for activation of CatSper channels slightly more positive than the resting membrane potential of these cells of around −50 mV (Seifert et al., 2015).

pH Regulation

In mammals and marine invertebrates, CatSper channels are activated by intracellular alkalinization (Kirichok et al., 2006; Lishko et al., 2011; Seifert et al., 2015; Miller et al., 2015; Strunker et al., 2011). In mammals, the encounter of sperm with high bicarbonate concentrations in semen and in the female genital tract, plus the increase in pH in the female genital tract at the time of ovulation, likely stimulates these channels. In mouse sperm, CatSper currents increase approximately sevenfold when pH_i is elevated from 6.0 to 7.0 (Kirichok et al., 2006), corresponding to a (leftward) shift in the G-V curve of −70 mV. Therefore, the pH_i alkalinization that occurs during capacitation might be an important trigger for mouse CatSper activation. In humans, several reports showed that CatSper is less pH_i sensitive than the mouse channel (Lishko et al., 2011), but how pH_i affects its GV curve has not been detailed. These findings indicate that in humans, other factors in conjunction with pH_i are important for the physiological activation of CatSper channels.

Sea urchin CatSper channels from *Arbacia punctulata* are also activated by intracellular alkalinization. Studies from Seifert (Seifert et al., 2015) using voltage, pH, and Ca^{2+} sensitive dyes estimated that intracellular alkalinization shifts the voltage dependence of the sea urchin CatSper channels to more negative values, comparable to the shift reported for mouse CatSper. Whole-cell patch clamp studies from sea urchin sperm are needed to confirm these data and determine the precise activation curves of these channels at different intracellular pH_i. Recently, on cell CatSper like currents were reported both in *S. purpuratus* and *L. pictus* sperm. These currents displayed a

pharmacology and pH_i dependence consistent with the known properties of CatSper (Loyo-Celis et al., 2021).

We still do not fully understand how pH activates the CatSper channel. A histidine-rich amino terminus of the CatSper1 subunit was initially proposed to sense pH_i in mouse and human CatSper (Kirichok et al., 2006; Ren et al., 2001). However, mouse and human CatSper channels differ in their pH sensitivity despite having similar histidine-rich CatSper1 N-termini (Lishko et al., 2011; Miller et al., 2015), and intracellular alkalinization seems to be enough to activate sea urchin CatSper channels (Seifert et al., 2015) despite their lack of a histidine-enriched N-terminus. These results suggest that other molecular mechanisms might be responsible for the pH modulation of CatSper channels. Hwang et al. recently showed that the CatSper zeta ζ- and EFCAB9 are responsible for the pH_i and Ca^{2+} sensitivity of the channel (Hwang et al., 2019). The authors proposed that in the non-capacitated state, CatSperζ/EFCAB9 complex localizes near the cytoplasmic mouth of the pore, inhibiting the channel gating and maintaining the channel closed. During capacitation, pH_i increases partially causing EFCAB9 to dissociate from CatSperζ, therefore releasing gate inhibition and opening the pore. Calcium flow through CatSper binds to EFCAB9 which undergoes a conformational change that prolongs CatSper channel opening (Hwang et al., 2019). Differences in amino acid sequence of the EFCAB9 subunit could explain species variations in pH_i sensitivity of CatSper channels. Notably, the absence of EFCAB9 and/or CatSperζ does not eliminate the pH_i dependence nor the requirement of cytosolic Ca^{2+} for CatSper activation, suggesting that additional molecular mechanisms participate in the pH_i activation of CatSper channels (Hwang et al., 2019).

HCO_3^- Regulation

One of the initial capacitation events occurs when sperm encounter an elevated HCO_3^- concentration in the semen. Work from many laboratories has shown that HCO_3^- modulates Ca^{2+} entry in mouse and human sperm (Buck et al., 1999) but did not define the role of CatSper in this signaling pathway. Recently, Orta et al. (2018) showed that HCO$_3^-$ activates mouse CatSper channels by activating a soluble adenylyl cyclase (sAC) present in sperm and PKA. In contrast, this mechanism was thought not to apply to human CatSper (Wang et al., 2020). Work from Chung's lab suggested that PKA does not directly phosphorylate CatSper channels (Chung et al., 2014, 2017). Therefore, the mechanism through which PKA activates the CatSper channel is unclear. Recent work provided evidence that in both mouse and human sperm, exposure to HCO$_3^-$ activates sAC/PKA pathway and Ca^{2+} entry through CatSper channels (Ferreira et al., 2021a). This work supports the initial findings reported by Orta et al. (2018) in mice (Orta et al., 2018). Also, it provides evidence that an increase in pH$_i$, induced by PKA activity, mediates the activation of CatSper channels. This increase in pH$_i$ sensitizes CatSper to increase [Ca^{2+}]$_i$ in response to membrane depolarization.

Hormone Regulation

In their journey to meet the egg, sperm must travel through the male and female reproductive tract, where they are exposed to various hormones, such as progesterone, prostaglandins, testosterone, estrogen, and even cortisol. These hormones can modify CatSper channels activity and its function.

Progesterone

The cumulus cells surrounding the egg produce the steroid hormone progesterone, which stimulates an immediate increase in [Ca^{2+}]$_i$ in mammalian sperm reported to be involved in sperm capacitation (Sumigama et al., 2015), chemotaxis (Eisenbach & Giojalas, 2006), hyperactivation (Alasmari et al., 2013a, b; Rennhack et al., 2018), and AR (Rennhack et al., 2018). The nuclear progesterone receptor does not mediate this rapid effect, instead, in 2011 two independent groups reported that CatSper channels mediate the non-

genomic rapid increases in [Ca^{2+}]$_i$ induced by progesterone in human sperm (Lishko et al., 2011; Strunker et al., 2011). These studies contributed complementary information on the action of progesterone on CatSper currents (Lishko et al., 2011) and [Ca^{2+}]$_i$ changes in human sperm (Strunker et al., 2011). Lishko's group showed that progesterone activated monovalent CatSper currents with a half-maximum effective concentration (IC$_{50}$) of 7.76 nM (Lishko et al., 2011). Strunker's group reported a 66 nM IC$_{50}$ of progesterone to activate Ba^{2+} CatSper tail currents and progesterone induced [Ca^{2+}]$_i$ elevations (Strunker et al., 2011).

Progesterone shifts CatSper voltage dependence toward more physiological, negative membrane potentials, while it does not significantly affect the channel conductance. At pH$_i$ 7.4, adding 500 nM progesterone produced a negative shift in CatSper half-activation voltage of ~30 mV (Lishko et al., 2011). By negatively shifting the G/V curve, progesterone allows human CatSper to achieve a similar activation at physiological voltages as the mouse CatSper channels by pH$_i$.

The progesterone-binding site associated with human CatSper stimulation is external as membrane-impermeable conjugated progesterone activates similar currents as progesterone (Lishko et al., 2011). Regarding the mechanism by which progesterone activates CatSper channels, Miller et al. proposed that the receptor alpha/beta hydrolase domain-containing protein 2 (ABHD2) is the mediator of the progesterone effect (Miller et al., 2016). In their model at rest, CatSper is inhibited by the endocannabinoid 2-arachidonoylglycerol (2-AG) in the flagellar membrane. Upon progesterone binding, ABHD2 degrades 2-AG relieving CatSper from inhibition (Miller et al., 2016).

CatSper channels from rhesus macaque sperm are also sensitive to progesterone (500 nM) (Sumigama et al., 2015), while CatSper currents from rat sperm are insensitive to 1 mM progesterone (1mM) (Sumigama et al., 2015). There are still mysteries regarding CatSper activation by progesterone in mouse. Concentrations up to 10 μM progesterone do not activate mouse CatSper channels (Lishko et al., 2011). However, 50–100 μM progesterone can trigger $[Ca^{2+}]_i$ increases absent in the CatSper KO mouse (unpublished observations). The initial CatSper report by Ren et al. (Ren et al., 2001) showed that, though diminished and altered, Ca^{2+} increases triggered by progesterone remained in CatSper KO mice. However, the authors used a very high progesterone concentration (1 mM) in those experiments, which could have had non-specific effects. In addition, these experiments were performed in non-backcrossed mice. Carlson et al. showed that backcrossing the CatSper heterozygous mice for several generations eliminated the defect in progressive motility initially reported in the CatSper knock-out mice (Carlson et al., 2003).

Other Steroid Hormones: Testosterone, Estrogen, and Cortisol

The regulation of human CatSper by estradiol, testosterone and hydrocortisone is controversial. Studies from Lishko's laboratory reported that pregnenolone-sulfate activates CatSper channels similarly to progesterone, probably by binding to the same site. Her laboratory also showed that testosterone and cortisol did not affect basal CatSper activity but prevented the activation of CatSper channels by progesterone and the plant triterpenoids, lupeol, and pristimerin (Mannowetz et al., 2017). On the other hand, Brenker et al. reported similar pregnenolone sulfate responses on human sperm CatSper currents but they did not find any antagonistic action of testosterone, estradiol, and hydrocortisone on the currents or in the Ca^{2+} signals (Brenker et al., 2018a). In contrast this group observed that testosterone, estradiol and hydrocortisone increased CatSper-currents in these cells and failed to induce Ca^{2+}-signals in sperm from a patient with deafness-infertility syndrome lacking functional CatSper-channels. They also did not see inhibition of both the progesterone-induced Ca^{2+}-signals and the increase in CatSper-currents by

the plant triterpenoids lupeol and pristimerin. A more recent study by Anders Rehfeld (Rehfeld, 2020) mostly supported Brenker's observations. It showed that testosterone, estradiol, and hydrocortisone induced Ca^{2+}-signals in human sperm with IC_{50} values in the μM range. Testosterone was the most potent (IC_{50} = 1.274 μM), followed by estradiol (IC_{50} = 1.510 μM) and hydrocortisone (IC_{50} = 12.45 μM). The steroids produced a dose-dependent inhibition of the subsequent progesterone-induced Ca^{2+}-signals with μM IC_{50s}. In contrast with Mannowetz's report (2017), these authors did not see inhibition of the progesterone response by the plant tripenoids lupeol and pristimerin (Mannowetz et al., 2017). Differences in experimental methods used could partly explain these controversial results. Lishko's group mainly based their conclusions on electrophysiological measurements of CatSper currents in human sperm. In contrast, Brenker and Rehfeld mostly used $[Ca^{2+}]_i$ measurements with Ca^{2+} sensitive dyes in sperm populations, in the whole sperm and not only in the principal piece where CatSper is located (Brenker et al., 2018a, b; Rehfeld, 2020; Mannowetz et al., 2018). In any case, more experiments are needed to clarify these discrepancies.

Prostaglandins

Prostaglandins are abundant in the seminal plasma (Mann & Lutwak-Mann, 1981) and are secreted by the oviduct and cumulus cells surrounding the oocyte (Espey & Richards, 2014). Prostaglandins have been reported to induce $[Ca^{2+}]_i$ increases in human sperm (Aitken et al., 1986; Shimizu et al., 1998; Schaefer et al., 1998) by unknown mechanisms, until the studies by Lishko's group showed that CatSper channels mediate these Ca^{2+} increases (Lishko et al., 2011).

The relative potency of the human CatSper activators is as follows: progesterone > PGF1 ≥ PGE1 > PGA1 > PGE2 > PGD2 (Lishko et al., 2011). Prostaglandin and progesterone effects are additive, suggesting that they have different binding sites (Lishko et al., 2011; Strunker et al., 2011; Jeschke et al., 2021). In contrast to the human CatSper channel, prostaglandins do not activate the mouse channel (Lishko et al., 2011). In the oviduct and the cumulus, sperm might be exposed to steroids and prostaglandins simultaneously (Ogra et al., 1974; Vastik-Fernandez et al., 1975; Libersky & Boatman, 1995a, b; Schuetz & Dubin, 1981). Brenker et al. (Brenker et al., 2018a) showed that steroids and prostaglandins synergically activate CatSper channels and elevate [Ca^{2+}]$_i$ to levels that are not reached by each ligand alone.

Other CatSper Regulatory Factors from Seminal Fluid and Female Reproductive Tract

Zinc

Zinc (Zn^{2+}) is present in the seminal plasma at high concentrations (Saaranen et al., 1987) and seems to be an important regulator of the CatSper channel. However, only a few studies reported their interaction. Jeschke et al. (Jeschke et al., 2021) showed that Zn^{2+} inhibits human CatSper activation by steroids and prostaglandins. However, Zn^{2+} interaction with CatSper has not been established yet. Inhibition of CatSper activation by Zn^{2+} might be relevant to prevent premature activation of the channel in the seminal plasma. Dilution of the seminal plasma in the female tract might release the Zn^{2+} inhibition allowing the subsequent CatSper activation.

Glycoproteins

Zona Pellucida (ZP), a glycoprotein layer surrounding the plasma membrane of mammalian oocytes, induces Ca^{2+} entry through CatSper channels in mouse sperm (Xia & Ren, 2009; De La Vega-Beltran et al., 2012; Chavez et al., 2014). ZP has an essential role in species-specific sperm-egg recognition and fertilization (Avella et al., 2013), preventing polyspermy and inducing *in vitro* acrosomal exocytosis in some species (Bleil & Wassarman, 1983; Wassarman et al., 1986; Wassarman, 1988; Florman et al., 1989; Florman & Storey, 1982). A recent report from Balbach et al. (2020) provided evidence that ZP-mediated activation of mouse CatSper is indirect through a change in pH$_i$.

Cysteine-Rich Secretory Proteins (CRISP)

CRISP are a group of glycoproteins with a role in the functioning of the mammalian reproductive system. These proteins are expressed in both male and female reproductive systems (Weigel Munoz et al., 2019). Work from Ernesto et al. showed that one of the protein family members, CRISP1, is also present in the female reproductive tract and in particular, is expressed by the cumulus cells (Ernesto et al., 2015). The authors reported that CRISP1 is an effective blocker of mouse CatSper and proposed that the inhibition of this channel by CRISP1 might cause temporary inhibition of hyperactivation resulting in more linear oriented swimming. The alternation between hyperactivated and linear swimming could be a novel mechanism that allows efficient sperm penetration of the egg's outer layers and successful fertilization (Ernesto et al., 2015).

Environmental Factors

Increasing amounts of endocrine disruptors (EDCs) in the environment in recent times could be partly responsible for the decline in human semen quality and the rise in infertility cases (Levine et al., 2017; Minguez-Alarcon et al., 2018; Mascarenhas et al., 2012). Work from several laboratories showed that EDCs activate CatSper *in vitro*, elevate [Ca^{2+}]$_i$, and trigger acrosomal exocytosis similarly to progesterone (Brenker et al., 2018a; Rehfeld et al., 2016; Schiffer et al., 2014; Tavares et al., 2013; Zou et al., 2017). Schiffer et al. and Rehfeld et al. also showed that EDCs compete either with progesterone or prostaglandins to activate CatSper and desensitize sperm for the respective physiological ligand (Rehfeld et al., 2016; Schiffer et al., 2014). However, application of low dose mixtures cooperatively elevates [Ca^{2+}]$_i$ in human sperm. Supporting these data, Brenker et al. showed that the EDCs BCSA and α-Zearalenol synergistically activate Ca^{2+} influx via CatSper in human sperm (Brenker et al., 2018b). It was also reported that bisphenols could affect Ca^{2+} signaling in human sperm by activating CatSper (Rehfeld et al., 2020; Wang et al., 2016). These findings suggest that even low doses of EDC mixtures in reproductive fluids might affect human CatSper *in vivo* and impact male fertility.

Cyclic Nucleotides

An increase in intracellular cAMP is a key factor and an early event in mammalian sperm capacitation (Visconti et al., 1995; Esposito et al., 2004; Buffone et al., 2014; Balbach et al., 2018; Wennemuth et al., 2003). In mammalian sperm, the HCO_3^- present in seminal fluid and the female reproductive tract activates sAC, which triggers the increase in intracellular cAMP and PKA activation. The activation of this HCO_3^-/sAC/cAMP/ PKA signaling pathway induces an increase in sperm flagellar beat frequency and sperm capacitation (Wennemuth et al., 2003; Hess et al., 2005; Nolan et al., 2004).

Membrane-permeable analogs of cyclic nucleotides trigger an increase of $[Ca^{2+}]_i$ in both mouse and human sperm (Xia & Ren, 2009; Ren et al., 2001; Brenker et al., 2012; Orta et al., 2018; Wang et al., 2020; Wennemuth et al., 2003; Wiesner et al., 1998; Kobori et al., 2000; Xia et al., 2007; Machado-Oliveira et al., 2008). These changes in $[Ca^{2+}]_i$ are not detected either in the CatSper KO mice (Ren et al., 2001) or in sperm from infertile patients that lack the *CatSper 2* gene (Wang et al., 2020), suggesting that CatSper channel is the mediator of the Ca^{2+} increase induced by cyclic nucleotides.

The mechanism of regulation of CatSper by cyclic nucleotides is a matter of intensive debate. None of the CatSper subunits contain consensus sequences for cyclic nucleotide-binding domains (Ren et al., 2001); therefore, the effect of cyclic nucleotides is likely to be indirect. In agreement with this, data from several laboratories support the hypothesis that cAMP activates CatSper channels through the activation of a kinase (Cisneros-Mejorado & Sanchez Herrera, 2012; Cisneros-Mejorado et al., 2014). More recently Orta et al. (2018) provided evidence that strongly suggests that CatSper channels are activated by cAMP through PKA activation. Whole-cell recordings from mouse CatSper channels revealed that CatSper monovalent currents are activated by adding HCO_3^- to the bath and by intracellular cAMP applied inside the patch-clamp pipette. The activation by HCO_3^- and cAMP was also blocked by different PKA inhibitors: PKI, H89, Rp-cAMPS. Neither HCO_3^- nor the intracellular application of cAMP affects sperm from the CatSper knock-out mice. These findings are consistent with the hypothesis that HCO_3^- enters sperm, activates sAC, and increases cAMP, which stimulates PKA. Once activated, PKA could directly phosphorylate the channel or activate it indirectly. In agreement with these findings, Ferreira et al. found that the HCO_3^-/ sAC/cAMP/PKA pathway activates both human and mouse CatSper channels (Ferreira et al., 2021a). The authors suggested that the effect of PKA on CatSper is indirect and mediated by an increase in pH_i. In contrast to the work from Orta and others, Kirichok et al. reported no effect of intracellular cyclic nucleotides nor with 8-Br-cAMP and 8-Br-cGMP on mouse CatSper currents (Kirichok et al., 2006). Different experimental conditions could explain the discrepancy of the results. Indeed, Kirichok et al. recorded divalent currents at either pH_i 6 or 8 where CatSper channels are either closed or fully active, while Orta et al. recorded monovalent CatSper currents at pH 7.0 (Kirichok et al., 2006; Orta et al., 2018).

In human sperm, the effect of permeable analogs of cAMP and cGMP on $[Ca^{2+}]i$ increases has been attributed to the binding of these molecules to an unidentified extracellular site of the CatSper channel (Brenker et al., 2012; Wang et al., 2020). The authors showed that infusion of cGMP or cAMP inside the sperm through a patch-clamp pipette fails to elevate $[Ca^{2+}]i$. The external activation of CatSper channels by permeable cyclic nucleotides does not seem to occur in mouse sperm. In addition, Wang et al. reported that the inhibitors of PKA, PKI 14–22 and H89 directly inhibit human CatSper channels while KT 5779 activates them (Wang et al., 2020). In contrast to these studies, data from Ferreira et al. showed that HCO_3^-, through the activation of sAC/cAMP and PKA, facilitates CatSper-dependent Ca^{2+} responses to membrane depolarization in human sperm (Ferreira et al., 2021a). These results suggest that PKA directly or indirectly regulates human CatSper activity. Ferreira et al. results suggest that CatSper activation by PKA is indirect through an increase in pH$_i$ while Orta's data discarded pH$_i$ as the mediator of PKA activation (Orta et al., 2018; Ferreira et al., 2021a).

Several studies showed that odorants stimulate Ca^{2+} influx and sperm motility through increases in intracellular cyclic nucleotides adding more complexity to the field. Work from Spehr et al. and Veitinger et al. proposed that odorants might serve as chemoattractants for sperm, activate olfactory receptors, stimulate the synthesis of cAMP by transmembrane adenylyl cyclases (tmACs), and open unknown Ca^{2+} channels (Spehr et al., 2004; Veitinger et al., 2011). The odorant-induced Ca^{2+} influx is impaired by inhibitors of tmACs and antagonists of odorant receptors (Spehr et al., 2003, 2004; Veitinger et al., 2011). Contesting these findings, Brenker et al. found that structurally diverse odorants directly activate the human CatSper channel by direct binding to an unidentified extracellular site of the channel and without involving metabotropic receptors and cAMP signaling (Brenker et al., 2012). The site of action of these odorants as well as cyclic nucleotides seem to be different from the sites where progesterone and prostaglandins bind to the channel (Brenker et al., 2012) Further studies are needed to elucidate the regulation of cAMP and PKA on CatSper channels in different species.

CatSper Channel Blockers

Since CatSper channels are essential for fertilization, special attention has been put in finding small molecules that block them, intending to develop an effective non-hormonal contraceptive agent. Several drugs block CatSper channels, but so far, no specific inhibitor has been identified. CatSper channel's complexity and the impossibility of expressing it in a heterologous system have made the task very difficult.

The T-type Ca^{2+} channel blockers mibefradil and NNC-0396 inhibit mouse and human CatSper channels in the micromolar range concentration (Brenker et al., 2012; Lishko et al., 2011).

MDL12330A also abolishes CatSper currents (Brenker et al., 2012; Lishko et al., 2011; Strunker et al., 2011). However, none of these drugs are selective for CatSper since they inhibit the mouse sperm-specific K$^+$ channel SLO3 (Zeng et al., 2011; Santi et al., 2010) and the human K$^+$ channel (Brenker et al., 2014) with similar potency (Carlson et al., 2003; Brenker et al., 2014; Mansell et al., 2014). In addition, NNC-0396, mibefradil, and MDL12330A trigger increases in intracellular [Ca^{2+}]i in human and mouse sperm (Strunker et al., 2011; Brenker et al., 2014; Chavez et al., 2018). These drugs also impair sperm vitality and motility (Tamburrino et al., 2014). HC-056456 significantly reduces mouse CatSper currents (Carlson et al., 2009). Low μM concentrations of HC-056456 inhibit >50% of [Ca^{2+}]i increases induced by alkaline depolarization and CatSper currents from mouse and human sperm. Unfortunately, the drug also blocks the K$^+$ channels present in sperm, with an IC$_{50}$ of 40 μM (Carlson et al., 2009). The steroid RU1968 inhibits CatSper current from invertebrates and mammalian sperm (Rennhack et al., 2018; Loyo-Celis et al., 2021). This drug did not block mouse SLO3 currents but inhibited the human CatSper channel with about 15-fold higher potency than the human SLO3 K$^+$ channel. In human sperm, RU1968 suppressed the motility responses evoked by progesterone and replicated the phenotype of CatSper-deficient sperm. Also, RU1968 abolishes CatSper-mediated chemotactic behavior in sea urchin sperm (Rennhack et al., 2018). RU1968 likely acts as a pore channel blocker, as shown for voltage-dependent Ca^{2+} channels. The binding site of the drug on the CatSper channel is not known. Its rapid effect on blocking Ca^{2+} influx through CatSper suggests that the compound binds to an unknown site on the extracellular side of the channel (Rennhack et al., 2018).

In conclusion, novel, non-hormonal, potent, and selective CatSper inhibitors are still needed. This class of inhibitors would help to reveal specific functions of CatSper channels and could be developed as novel contraceptive agents.

Chemotaxis

Sea urchin sperm, an excellent model to explore chemotaxis, express all the CatSper subunits described in mammalian sperm but ζ (Seifert et al., 2015; Loyo-Celis et al., 2021; Trotschel et al., 2020). As indicated in section 4.1. and summarizing work from several groups, the V_m hyperpolarization caused by chemoattractant (resact or speract) binding increases pH_i by activating a sperm-specific, voltage-dependent, Na^+/H^+ exchanger (sNHE). Due to its pH_i dependence, upon an alkalinization CatSper can open at more hyperpolarized potentials, resulting in Ca^{2+} influx and thereafter a depolarization (Kaupp et al., 2008; Darszon et al., 2008). The unique regulation of CatSper by pH_i, Vm and $[Ca^{2+}]_i$ endow it with the ability to contribute to the generation of $[Ca^{2+}]_i$ oscillatory responses. As sperm depolarize sNHE is inhibited, decreasing pH_i and diminishing CatSper's voltage sensitivity. These changes tip the V_m balance toward the hyperpolarizing direction due to KCNG channels and the cycle restarts (Priego-Espinosa et al., 2020). As briefly stated, CatSper plays a key role in sea urchin sperm and in other marine organisms.

Hyperactivation

In order to fertilize the oocyte, sperm motility must switch from linear and progressive to hyperactivated, during their transit in the female reproductive tract or when incubated *in vitro* under capacitated conditions (Overstreet et al., 1980; Suarez & Osman, 1987). Progressive motility is characterized by symmetrical, low-amplitude flagellar bending that drives the sperm in a straight-line-like pattern (Yanagimachi & Usui, 1974; Suarez & Osman, 1987; Yanagimachi, 1970; Turner, 2006). Hyperactivated motility in aqueous media is characterized by high amplitude and asymmetrical flagellar bending, which pushes the sperm in a trajectory that resembles a circular- or figure-eight pattern (Yanagimachi & Usui, 1974; Suarez & Osman, 1987; Yanagimachi, 1970; Ishijima et al., 2002). Hyperactivated motility facilitates sperm navigation in the viscous fluids

of the female reproductive tract, helps sperm detach from the oviductal walls, and penetrate the zona pellucida (Suarez & Pacey, 2006; Suarez et al., 1991; Suarez & Dai, 1992). Calcium influx through CatSper channels is the main trigger of hyperactivated motility in mice and human sperm (Kirichok et al., 2006; Ren et al., 2001; Carlson et al., 2003). Mouse sperm lacking CatSper-1, -2, -3, or -4 cannot fertilize the egg either *in vitro* due to a failure in hyperactivation and penetration of the zona pellucida (Kirichok et al., 2006; Ren et al., 2001; Quill et al., 2001; Carlson et al., 2003; Qi et al., 2007); or *in vivo* due to failure to ascend beyond the oviductal reservoir (Chung et al., 2011; Ho et al., 2009). CatSper seems also to be needed for both the initiation and the maintenance of mouse hyperactivated motility (Carlson et al., 2009).

In mouse sperm, Ca^{2+} influx through CatSper channels is responsible for high-amplitude pro-hook bends of the flagellum (bends in the direction of the hook on the head), whereas release of Ca^{2+} from intracellular stores seem to produce anti-hook flagellar bending. Chang et al suggested that the alternation of these two types of flagellar bending could serve to redirect sperm toward oocytes (Chang & Suarez, 2011). Chung et al. showed that the CatSperζ subunit is relevant to maintain the pro-hook flagellar bending. The authors reported that the anti-hook beating predominates in the CatSperζ null sperm disrupting sperm hyperactivation and *in vivo* migration in the female tract. These mice have a marked subfertility (Chung et al., 2017).

Recently, Ferreira and collaborators indicated that CatSper channels are also important for mitochondrial Ca^{2+} increases during capacitation (Ferreira et al., 2021b). This increase in mitochondrial Ca^{2+} contributes to mitochondrial efficiency and sperm hyperactivation (Ferreira et al., 2021b).

Calcium influx through CatSper channels is also key to human sperm hyperactivation and fertility. Humans with a deficiency in CatSper display a loss of sperm hyperactivation and male infertility (Brown et al., 2018, 2019; Jeschke et al., 2021; Avenarius et al., 2009; Schiffer et al., 2020; Smith et al., 2013; Williams et al., 2015).

The factors that trigger the activation of CatSper and sperm hyperactivation in the female genital tract are not completely understood. Most of the data we presented come from *in vitro* experiments where some factors have shown the capability of activating CatSper channels and induce sperm hyperactivation. A key activator of CatSper in both mouse and human is the increase in pH$_i$ that takes place during capacitation (Parrish et al., 1989; Vredenburgh-Wilberg & Parrish, 1995). In mouse sperm, CatSper activation, the subsequent [Ca^{2+}]$_i$ increase, and hyperactivation are associated with intracellular alkalinization and the activation of the sperm-specific K$^+$ channel SLO3 (Chavez et al., 2014; Marquez et al., 2007; Nishigaki et al., 2014a, b; Suarez, 2008). Initially it was proposed that alkalinization activates SLO3 channels hyperpolarizing the sperm, and at the same time shifts activation of CatSper to the hyperpolarized range increasing the driving force for Ca^{2+} entry (Navarro et al., 2007). Later, Chavez et al. proposed that the activation of the sperm-specific K$^+$ channel SLO3 promotes membrane hyperpolarization and activation of the sperm-specific and voltage-dependent sNHE (Woo et al., 2002; Liu et al., 2010; Wang et al., 2003), which further increases alkalinization triggering CatSper activation (Chavez et al., 2014; Woo et al., 2002; Liu et al., 2010; Wang et al., 2003). In human sperm, both the sNHE and the H$^+$ channel Hv1 could contribute to sperm alkalinization (Garcia & Meizel, 1999; Matamoros-Volante & Trevino, 2020) and CatSper activation. However, the association with the sperm-specific SLO3 channels is not completely clear and needs further investigation. Interestingly, the human SLO3 K channel is sensitive to pH$_i$ and intracellular Ca^{2+}, opening the possibility of SLO3 channels working downstream of CatSper channel activation (Brenker et al., 2014; Geng et al., 2017).

The activation of human CatSper could be triggered by prostaglandins, and/or steroids present in the seminal and follicular fluids respectively (Wang et al., 2021; Brown et al., 2019; Jeschke et al., 2021; Tamburrino et al., 2014). As we mentioned before, progesterone and prostaglandins activate CatSper channels *in vitro*, inducing a rapid Ca^{2+} increase associated with sperm hyperactivation (Lishko et al., 2011; Strunker et al., 2011; Jeschke et al., 2021). This, together with the high concentrations of Zn^{2+} in the seminal fluid (Zn^{2+} inhibits CatSper) and other factors secreted by the egg, could be part of the chemotactic system that guides human sperm (Alasmari et al., 2013a, b; Schiffer et al., 2014; Tamburrino et al., 2014; Aitken & Kelly, 1985; Baldi et al., 2009; Bedu-Addo et al., 2007; Oren-Benaroya et al., 2008; Publicover et al., 2008; Sanchez-Cardenas et al., 2014; Servin-Vences et al., 2012; Tamburrino et al., 2020).

Rheotaxis

Where to go? Sperm do not reach the fertilization site by swimming randomly in the female tract (Croxatto, 2002). However, how does sperm find its way to the oocyte? The presence of specific flows in the female reproductive tract, such as a cilia-created flow toward the distal oviduct and an opposite central lumen flow, has propped the idea that sperm can use this flow as a guidance cue. Rheotaxis, particularly positive rheotaxis, the ability of the sperm to swim against the flow, was reported decades ago, and recent studies showed rheotactic response for a variety of sperm and its importance in fertilization (Bukatin et al., 2015; El-Sherry et al., 2014; Kantsler et al., 2013, 2014a; Miki & Clapham, 2013; Tung et al., 2014). For several years it has been debated if the specific orientation of the sperm with the flow is an active process or a passive biomechanical and hydrodynamic process. In 1961, Bretherton et al. showed that human and bull sperm responses to vertical velocity differ between live and dead sperm and that both human and bull live sperm exhibit positive rheotaxis (Bretherton, 1997).

The ability of the sperm to rotate around its longitudinal axis and reorienting itself against the flow, called rolling, is an essential factor for the sperm's rheotactic capacity (Bukatin et al., 2015; Kantsler et al., 2013, 2014a; Miki & Clapham, 2013). Rolling enables a sperm to swim progressively even if the flagellum beats asymmetrically.

The factors that contribute to the rolling phenomenon are still not completely understood. On one hand, the contribution of physical factors as steric surface alignment (Bukatin et al., 2015; Kantsler et al., 2013; Denissenko et al., 2012), the weathervane effect (Kantsler et al., 2014a, b; Tung et al., 2014; Mathijssen et al., 2019), and influence of the sperm morphology could play a role in rheotactic orientation (Friedrich et al., 2010), but this needs experimental confirmation in conditions simulating the female reproductive tract. On the other hand, the active contribution of the flagellar beating to the rolling behavior in response to the flow is a topic of debate in the field. In most species, $[Ca^{2+}]_i$ changes carried by CatSper channels regulate sperm flagellar beating (Kirichok et al., 2006; Kaupp et al., 2003; Ren et al., 2001; Quill et al., 2001; Lishko et al., 2011; Seifert et al., 2015; Strunker et al., 2011; Publicover et al., 2008; Fechner et al., 2015; Lishko & Kirichok, 2010; Loux et al., 2013) and it has been suggested that the regulation of $[Ca^{2+}]_i$ by CatSper is essential for rolling and reorientation of the sperm. In 2013, Miki and Clapham showed that Catsper1 null mouse sperm do not show rolling and rheotaxis behavior (Miki & Clapham, 2013). In the case of human sperm, Miller et al proposed that rolling is at least in part the result of an asymmetrical Ca^{2+} influx via CatSper, driven by changes in pH_i that follow the local organization of CatSper and Hv1 channels along the flagellum (Miller et al., 2018). However, the theory that CatSper channels are essential for rolling and reorienting the sperm is debated. First, evidence showed that the rolling of mouse sperm does not require extracellular Ca^{2+} (Babcock et al., 2014; Muschol et al., 2018), and that exposure of human sperm to gradients of flow velocities does not evoke measurable changes in $[Ca^{2+}]_i$ (Zhang et al., 2016). Second, Schiffer et al. 2019 recently showed that rolling and rheotaxis are still present in human sperm from CatSper-deficient patients and in the absence of external Ca^{2+} (Schiffer et al., 2020). Third, it has been proposed that the quadrilateral organization of CatSper and associated signaling components can provide the ultrastructure necessary for rolling and rheotaxis (Miller et al., 2018).

Schiffer and collaborators contested this hypothesis by comparing the sperm navigation properties between *CatSper2* null and *CatSper1* null. *CatSper2* null lacks functional CatSper channels (*CatSper2* null) but conserves the special quadrilateral arrangement of the channels and the associated complex, while CatSper 1 null sperm lack the CatSper channels entirely together with the spatial arrangement. Interestingly the authors found that both *CatSper1* and *CatSper2* null mice sperm showed longitudinal rolling and rheotaxis (Schiffer et al., 2020).

Thus, it remains unclear whether rolling and rheotaxis are passive biomechanical and hydrodynamic processes or an active process guided by Ca^{2+} and H^+ signaling. Subtle changes in the rheotactic performance of the sperm could be fundamental for sperm fertilization capacity and need to be explored. Further studies of CatSper-deficient sperm in a broader range of physiological flow velocities and viscosities similar to the conditions encountered in the oviduct can elucidate the importance of $[Ca^{2+}]_i$ and CatSper function in rheotaxis. It will also be essential to evaluate the rolling and rheotaxis efficiency in wild-type and CatSper-deficient sperm under different stages of activation, for example, under capacitated conditions or stimulated by different factors such as prostaglandins and progesterone in human sperm.

Acrosomal Reaction (AR)

Although the importance of CatSper in mouse and human fertility is undoubted, its role during the AR is not that clear. Initially, sperm from CatSper KO mice were reported to undergo $[Ca^{2+}]_i$ increases in response to high progesterone doses (1 mM) (Ren et al., 2001). Later work reported that a HCO_3^- depolarization/alkalization stimuli, failed to elicit a Ca^{2+} rise in sperm-lacking CatSper (Carlson et al., 2003). However, the ability of these sperm to undergo the AR was not explored until 2007 (Xia et al., 2007). This group reported that the levels of AR induced by a depolarization/alkalization stimulus as well as ionophore, were indistinguishable from those obtained in wild type sperm (Xia et al., 2007). In 2009, Xia and Ren reported that sperm from CatSper KO mice failed to elevate Ca^{2+} in response to ZP stimulus (although 18% of the cells did show a delayed Ca^{2+} increase) (Xia & Ren, 2009).

Progesterone has been known for more than 30 years to have profound effects in sperm physiology (Tamburrino et al., 2020). This hormone induces a complex Ca^{2+} increase, the AR and influences sperm motility. As mentioned earlier, progesterone in human sperm is a physiological ligand that indirectly activates CatSper (Lishko et al., 2011; Strunker et al., 2011) and causes a robust Ca^{2+} increase that most likely includes the activation of other Ca^{2+} entry/release routes from internal stores. Interestingly, nearly 100% of human capacitated sperm stimulated with progesterone respond with a transient followed by a sustained [Ca^{2+}]$_i$ increase, yet only 7–12% of these cells acrosome react. It is known that only a fraction of these cells is indeed capacitated. Why do only 10–20% of mammalian sperm fully capacitate and undergo AR is far from clear. Although progesterone does not activate CatSper in mouse sperm, at high concentrations (50–100 μM), it does induce the AR (Romarowski et al., 2016).

Genetic models to eliminate a specific protein readily available in mouse models are not accessible in humans. Therefore, a few pharmacological studies have explored the role of CatSper during the AR in human sperm. It was noted earlier that CatSper channels are pharmacologically very promiscuous, with some compounds inducing the AR themselves (Brenker et al., 2012; Chavez et al., 2014; Barratt & Publicover, 2012). Recently, Rahban et al. reported that sertraline, an antidepressant widely prescribed, inhibits CatSper (Rahban et al., 2021). This drug reduced the AR induced by progesterone as well as human CatSper currents. Further research is needed to clarify the role in the AR of CatSper channels in different species.

Conclusions and Perspectives

It is surprising that at the present time Ca$_V$ participation in sperm physiology is still ill-defined. It must be stressed that sperm are tiny, polarized cells with a very minute cytoplasmic volume; therefore a few Ca^{2+} channels can play a pivotal role in a specific function such as the AR. In light of this, it is also worth pointing out that (1) detection of only a few channels by proteomic and immunological strategies can be very difficult, (2) sperm electrophysiological recordings are only performed in a handful of labs in the world, and (3) many of the potent strategies of molecular biology are not applicable to sperm. These matters attest to the sometimes-paused pace at which the field advances. In contrast, since CatSper is the most prevalent sperm Ca^{2+} channel and is present in significant numbers, its importance and characterization have progressed enormously. We now know it has at least 14 subunits and is arranged in an extravagant manner, so it can have a functional and structural influence in how the sperm flagella beat, and thus in fertilization.

The unprecedented developments in imaging strategies are allowing the experimenter to look at local changes in sperm signaling processes with unimaginable time and structural resolution. Certainly, cell biology is undergoing an amazing revolution, and we can expect our understanding of sperm physiology and reproduction to be significantly enriched.

Acknowledgements Contract grant sponsor: Consejo Nacional de Ciencia y Tecnología (CONACyT-Mexico). Contract grant numbers: 71 to AD; 319694 to ILG Contract grant sponsor: Dirección General de Asuntos del Personal Académico/ Universidad Nacional Autónoma de México (DGAPA/ UNAM). Contract grant numbers: IN207122 to CT; IN200919 to AD; IN205518 and IN208321 to ILG. Contract grant sponsor: National Institutes of Health (NIH) Contract grant numbers: NICHD R01 grant # R01HD069631 to CMS.

References

Aitken, R. J., & Kelly, R. W. (1985). Analysis of the direct effects of prostaglandins on human sperm function. *Journal of Reproduction and Fertility, 73*(1), 139–146. https://doi.org/10.1530/jrf.0.0730139

Aitken, R. J., Irvine, S., & Kelly, R. W. (1986). Significance of intracellular calcium and cyclic adenosine 3′,5′-monophosphate in the mechanisms by which prostaglandins influence human sperm function. *Journal of Reproduction and Fertility, 77*(2), 451–462. https://doi.org/10.1530/jrf.0.0770451.

Alasmari, W., Barratt, C. L., Publicover, S. J., Whalley, K. M., Foster, E., Kay, V., et al. (2013a). The clinical significance of calcium-signalling pathways mediating human sperm hyperactivation. *Human Reproduction, 28*(4), 866–876. https://doi.org/10.1093/humrep/des467

Alasmari, W., Costello, S., Correia, J., Oxenham, S. K., Morris, J., Fernandes, L., et al. (2013b). Ca2+ signals generated by CatSper and Ca2+ stores regulate different behaviors in human sperm. *The Journal of Biological Chemistry, 288*(9), 6248–6258. https://doi.org/10.1074/jbc.M112.439356

Alberts, B., Johnson, A., Lewis, J., Morgan, D., Raff, M., Roberts, K., et al. (2008). Sexual reproduction: Meiosis, germ cells, and fertilization. In M. A. S. Granum (Ed.), *Molecular biology of the cell* (5th ed., pp. 1269–1304). Garland Science.

Arnoult, C., Cardullo, R. A., Lemos, J. R., & Florman, H. M. (1996). Activation of mouse sperm T-type Ca2+ channels by adhesion to the egg zona pellucida. *Proceedings of the National Academy of Sciences of the United States of America, 93*(23), 13004–13009. https://doi.org/10.1073/pnas.93.23.13004

Arnoult, C., Lemos, J. R., & Florman, H. M. (1997). Voltage-dependent modulation of T-type calcium channels by protein tyrosine phosphorylation. *The EMBO Journal, 16*(7), 1593–1599. https://doi.org/10.1093/emboj/16.7.1593

Arnoult, C., Villaz, M., & Florman, H. M. (1998). Pharmacological properties of the T-type Ca2+ current of mouse spermatogenic cells. *Molecular Pharmacology*, 1104–1111.

Arnoult, C., Kazam, I. G., Visconti, P. E., Kopf, G. S., Villaz, M., & Florman, H. M. (1999). Control of the low voltage-activated calcium channel of mouse sperm by egg ZP3 and by membrane hyperpolarization during capacitation. *Proceedings of the National Academy of Sciences of the United States of America, 96*(12), 6757–6762. https://doi.org/10.1073/pnas.96.12.6757

Avella, M. A., Xiong, B., & Dean, J. (2013). The molecular basis of gamete recognition in mice and humans. *Molecular Human Reproduction, 19*(5), 279–289. https://doi.org/10.1093/molehr/gat004

Avenarius, M. R., Hildebrand, M. S., Zhang, Y., Meyer, N. C., Smith, L. L., Kahrizi, K., et al. (2009). Human male infertility caused by mutations in the CATSPER1 channel protein. *American Journal of Human Genetics, 84*(4), 505–510. https://doi.org/10.1016/j.ajhg.2009.03.004

Babcock, D. F., Wandernoth, P. M., & Wennemuth, G. (2014). Episodic rolling and transient attachments create diversity in sperm swimming behavior. *BMC Biology, 12*, 67. https://doi.org/10.1186/s12915-014-0067-3

Balbach, M., Beckert, V., Hansen, J. N., & Wachten, D. (2018). Shedding light on the role of cAMP in mammalian sperm physiology. *Molecular and Cellular Endocrinology, 468*, 111–120. https://doi.org/10.1016/j.mce.2017.11.008

Balbach, M., Hamzeh, H., Jikeli, J. F., Brenker, C., Schiffer, C., Hansen, J. N., et al. (2020). Molecular mechanism underlying the action of zona-pellucida glycoproteins on mouse sperm. *Frontiers in Cell and Development Biology, 8*, 572735. https://doi.org/10.3389/fcell.2020.572735

Balderas, E., Ateaga-Tlecuitl, R., Rivera, M., Gomora, J. C., & Darszon, A. (2012). Niflumic acid blocks native and recombinant T-type channels. *Journal of Cellular Physiology, 227*(6), 2542–2555. https://doi.org/10.1002/jcp.22992

Baldi, E., Luconi, M., Muratori, M., Marchiani, S., Tamburrino, L., & Forti, G. (2009). Nongenomic activation of spermatozoa by steroid hormones: facts and fictions. *Molecular and Cellular Endocrinology, 308*(1-2), 39–46. https://doi.org/10.1016/j.mce.2009.02.006

Barratt, C. L., & Publicover, S. J. (2012). Sperm are promiscuous and CatSper is to blame. *The EMBO Journal, 31*(7), 1624–1626. https://doi.org/10.1038/emboj.2012.62

Bedu-Addo, K., Barratt, C. L., Kirkman-Brown, J. C., & Publicover, S. J. (2007). Patterns of [Ca2+](i) mobilization and cell response in human spermatozoa exposed to progesterone. *Developmental Biology, 302*(1), 324–332. https://doi.org/10.1016/j.ydbio.2006.09.040

Beltran, C., Trevino, C. L., Mata-Martinez, E., Chavez, J. C., Sanchez-Cardenas, C., Baker, M., et al. (2016). Role of ion channels in the sperm acrosome reaction. *Advances in Anatomy, Embryology, and Cell Biology, 220*, 35–69. https://doi.org/10.1007/978-3-319-30567-7_3

Berridge, M. J., Bootman, M. D., & Roderick, H. L. (2003). Calcium signalling: Dynamics, homeostasis and remodelling. *Nature Reviews Molecular Cell Biology, 4*(7), 517–529. https://doi.org/10.1038/nrm1155

Bleil, J. D., & Wassarman, P. M. (1983). Sperm-egg interactions in the mouse: Sequence of events and induction of the acrosome reaction by a zona pellucida glycoprotein. *Developmental Biology, 95*(2), 317–324. https://doi.org/10.1016/0012-1606(83)90032-5

Bondarenko, O., Corzo, G., Santana, F. L., Del Rio-Portilla, F., Darszon, A., & Lopez-Gonzalez, I. (2019). Nonenzymatically oxidized arachidonic acid regulates T-type Ca(2+) currents in mouse spermatogenic cells. *FEBS Letters, 593*(14), 1735–1750. https://doi.org/10.1002/1873-3468.13448

Boulais, M., Demoy-Schneider, M., Alavi, S. M. H., & Cosson, J. (2019). Spermatozoa motility in bivalves: Signaling, flagellar beating behavior, and energetics.

Theriogenology, 136, 15–27. https://doi.org/10.1016/j.theriogenology.2019.06.025

Brenker, C., Goodwin, N., Weyand, I., Kashikar, N. D., Naruse, M., Krahling, M., et al. (2012). The CatSper channel: A polymodal chemosensor in human sperm. *The EMBO Journal, 31*(7), 1654–1665. https://doi.org/10.1038/emboj.2012.30

Brenker, C., Zhou, Y., Muller, A., Echeverry, F. A., Trotschel, C., Poetsch, A., et al. (2014). The Ca2+-activated K+ current of human sperm is mediated by Slo3. *eLife, 3*, e01438. https://doi.org/10.7554/eLife.01438

Brenker, C., Schiffer, C., Wagner, I. V., Tuttelmann, F., Ropke, A., Rennhack, A., et al. (2018a). Action of steroids and plant triterpenoids on CatSper Ca(2+) channels in human sperm. *Proceedings of the National Academy of Sciences of the United States of America, 115*(3), E344–E3E6. https://doi.org/10.1073/pnas.1717929115

Brenker, C., Rehfeld, A., Schiffer, C., Kierzek, M., Kaupp, U. B., Skakkebaek, N. E., et al. (2018b). Synergistic activation of CatSper Ca2+ channels in human sperm by oviductal ligands and endocrine disrupting chemicals. *Human Reproduction, 33*(10), 1915–1923. https://doi.org/10.1093/humrep/dey275

Bretherton, F. P. (1997). Rothschild NMV. Rheotaxis of spermatozoa. *Proceedings of the Royal Society of London, Series B: Biological Sciences, 153*(953), 490–502. https://doi.org/10.1098/rspb.1961.0014

Brown, S. G., Miller, M. R., Lishko, P. V., Lester, D. H., Publicover, S. J., Barratt, C. L. R., et al. (2018). Homozygous in-frame deletion in CATSPERE in a man producing spermatozoa with loss of CatSper function and compromised fertilizing capacity. *Human Reproduction, 33*(10), 1812–1816. https://doi.org/10.1093/humrep/dey278

Brown, S. G., Publicover, S. J., Barratt, C. L. R., & Martins da Silva, S. J. (2019). Human sperm ion channel (dys)function: Implications for fertilization. *Human Reproduction Update, 25*(6), 758–776. https://doi.org/10.1093/humupd/dmz032

Buck, J., Sinclair, M. L., Schapal, L., Cann, M. J., & Levin, L. R. (1999). Cytosolic adenylyl cyclase defines a unique signaling molecule in mammals. *Proceedings of the National Academy of Sciences of the United States of America, 96*(1), 79–84. https://doi.org/10.1073/pnas.96.1.79

Buffone, M. G., Wertheimer, E. V., Visconti, P. E., & Krapf, D. (2014). Central role of soluble adenylyl cyclase and cAMP in sperm physiology. *Biochimica et Biophysica Acta, 1842*(12 Pt B), 2610–2620. https://doi.org/10.1016/j.bbadis.2014.07.013

Bukatin, A., Kukhtevich, I., Stoop, N., Dunkel, J., & Kantsler, V. (2015). Bimodal rheotactic behavior reflects flagellar beat asymmetry in human sperm cells. *Proceedings of the National Academy of Sciences of the United States of America, 112*(52), 15904–15909. https://doi.org/10.1073/pnas.1515159112

Cai, X., & Clapham, D. E. (2008). Evolutionary genomics reveals lineage-specific gene loss and rapid evolution of a sperm-specific ion channel complex: CatSpers and CatSperbeta. *PLoS One, 3*(10), e3569. https://doi.org/10.1371/journal.pone.0003569

Cai, X., & Clapham, D. E. (2012). Ancestral Ca2+ signaling machinery in early animal and fungal evolution. *Molecular Biology and Evolution, 29*(1), 91–100. https://doi.org/10.1093/molbev/msr149

Cai, X., Wang, X., & Clapham, D. E. (2014). Early evolution of the eukaryotic Ca2+ signaling machinery: conservation of the CatSper channel complex. *Molecular Biology and Evolution, 31*(10), 2735–2740. https://doi.org/10.1093/molbev/msu218

Cai, X., Wang, X., Patel, S., & Clapham, D. E. (2015). Insights into the early evolution of animal calcium signaling machinery: A unicellular point of view. *Cell Calcium, 57*(3), 166–173. https://doi.org/10.1016/j.ceca.2014.11.007

Carafoli, E., & Krebs, J. (2016). Why calcium? How calcium became the best communicator. *The Journal of Biological Chemistry, 291*(40), 20849–20857. https://doi.org/10.1074/jbc.R116.735894

Carlson, A. E., Westenbroek, R. E., Quill, T., Ren, D., Clapham, D. E., Hille, B., et al. (2003). CatSper1 required for evoked Ca2+ entry and control of flagellar function in sperm. *Proceedings of the National Academy of Sciences of the United States of America, 100*(25), 14864–14868. https://doi.org/10.1073/pnas.2536658100

Carlson, A. E., Burnett, L. A., del Camino, D., Quill, T. A., Hille, B., Chong, J. A., et al. (2009). Pharmacological targeting of native CatSper channels reveals a required role in maintenance of sperm hyperactivation. *PLoS One, 4*(8), e6844. https://doi.org/10.1371/journal.pone.0006844

Chang, H., & Suarez, S. S. (2011). Two distinct Ca(2+) signaling pathways modulate sperm flagellar beating patterns in mice. *Biology of Reproduction, 85*(2), 296–305. https://doi.org/10.1095/biolreprod.110.089789

Chavez, J. C., Ferreira, J. J., Butler, A., De La Vega Beltran, J. L., Trevino, C. L., Darszon, A., et al. (2014). SLO3 K+ channels control calcium entry through CATSPER channels in sperm. *The Journal of biological chemistry., 289*(46), 32266–32275. https://doi.org/10.1074/jbc.M114.607556

Chavez, J. C., De la Vega-Beltran, J. L., Jose, O., Torres, P., Nishigaki, T., Trevino, C. L., et al. (2018). Acrosomal alkalization triggers Ca(2+) release and acrosome reaction in mammalian spermatozoa. *Journal of Cellular Physiology, 233*(6), 4735–4747. https://doi.org/10.1002/jcp.26262

Chung, J. J., Navarro, B., Krapivinsky, G., Krapivinsky, L., & Clapham, D. E. (2011). A novel gene required for male fertility and functional CATSPER channel formation in spermatozoa. *Nature Communications, 2*, 153. https://doi.org/10.1038/ncomms1153

Chung, J. J., Shim, S. H., Everley, R. A., Gygi, S. P., Zhuang, X., & Clapham, D. E. (2014). Structurally distinct Ca(2+) signaling domains of sperm flagella orchestrate tyrosine phosphorylation and motility. *Cell, 157*(4), 808–822. https://doi.org/10.1016/j.cell.2014.02.056

Chung, J. J., Miki, K., Kim, D., Shim, S. H., Shi, H. F., Hwang, J. Y., et al. (2017). CatSperzeta regulates the structural continuity of sperm Ca(2+) signaling domains and is required for normal fertility. *eLife, 6.* https://doi.org/10.7554/eLife.23082

Cisneros-Mejorado, A., & Sanchez Herrera, D. P. (2012). cGMP and cyclic nucleotide-gated channels participate in mouse sperm capacitation. *FEBS Letters, 586*(2), 149–153. https://doi.org/10.1016/j.febslet.2011.12.006

Cisneros-Mejorado, A., Hernandez-Soberanis, L., Islas-Carbajal, M. C., & Sanchez, D. (2014). Capacitation and Ca(2+) influx in spermatozoa: Role of CNG channels and protein kinase G. *Andrology, 2*(1), 145–154. https://doi.org/10.1111/j.2047-2927.2013.00169.x

Clapham, D. E. (2007). Calcium signaling. *Cell, 131*(6), 1047–1058. https://doi.org/10.1016/j.cell.2007.11.028

Cohen, R., Buttke, D. E., Asano, A., Mukai, C., Nelson, J. L., Ren, D., et al. (2014). Lipid modulation of calcium flux through CaV2.3 regulates acrosome exocytosis and fertilization. *Developmental Cell, 28*(3), 310–321. https://doi.org/10.1016/j.devcel.2014.01.005

Correia, J., Michelangeli, F., & Publicover, S. (2015). Regulation and roles of Ca2+ stores in human sperm. *Reproduction, 150*(2), R65–R76. https://doi.org/10.1530/REP-15-0102

Croxatto, H. B. (2002). Physiology of gamete and embryo transport through the Fallopian tube. *Reproductive BioMedicine Online, 4*(2), 160–169. https://doi.org/10.1016/s1472-6483(10)61935-9

Darszon, A., Lopez-Martinez, P., Acevedo, J. J., Hernandez-Cruz, A., & Trevino, C. L. (2006). T-type Ca2+ channels in sperm function. *Cell Calcium, 40*(2), 241–252. https://doi.org/10.1016/j.ceca.2006.04.028

Darszon, A., Guerrero, A., Galindo, B. E., Nishigaki, T., & Wood, C. D. (2008). Sperm-activating peptides in the regulation of ion fluxes, signal transduction and motility. *The International Journal of Developmental Biology, 52*(5-6), 595–606. https://doi.org/10.1387/ijdb.072550ad

Darszon, A., Nishigaki, T., Beltran, C., & Trevino, C. L. (2011). Calcium channels in the development, maturation, and function of spermatozoa. *Physiological Reviews, 91*(4), 1305–1355. https://doi.org/10.1152/physrev.00028.2010

Darszon, A., Nishigaki, T., Lopez-Gonzalez, I., Visconti, P. E., & Trevino, C. L. (2020). Differences and similarities: The richness of comparative sperm physiology. *Physiology (Bethesda, Md.), 35*(3), 196–208. https://doi.org/10.1152/physiol.00033.2019

De La Vega-Beltran, J. L., Sanchez-Cardenas, C., Krapf, D., Hernandez-Gonzalez, E. O., Wertheimer, E., Trevino, C. L., et al. (2012). Mouse sperm membrane potential hyperpolarization is necessary and sufficient to prepare sperm for the acrosome reaction. *The Journal of biological chemistry., 287*(53), 44384–44393. https://doi.org/10.1074/jbc.M112.393488

DeCaen, P. G., Delling, M., Vien, T. N., & Clapham, D. E. (2013). Direct recording and molecular identification of the calcium channel of primary cilia. *Nature, 504*(7479), 315–318. https://doi.org/10.1038/nature12832

Delling, M., DeCaen, P. G., Doerner, J. F., Febvay, S., & Clapham, D. E. (2013). Primary cilia are specialized calcium signalling organelles. *Nature, 504*(7479), 311–314. https://doi.org/10.1038/nature12833

Denissenko, P., Kantsler, V., Smith, D. J., & Kirkman-Brown, J. (2012). Human spermatozoa migration in microchannels reveals boundary-following navigation. *Proceedings of the National Academy of Sciences of the United States of America, 109*(21), 8007–8010. https://doi.org/10.1073/pnas.1202934109

Dominguez, D. C. (2004). Calcium signalling in bacteria. *Molecular Microbiology, 54*(2), 291–297. https://doi.org/10.1111/j.1365-2958.2004.04276.x

Eisenbach, M., & Giojalas, L. C. (2006). Sperm guidance in mammals – An unpaved road to the egg. Nature Reviews. *Molecular Cell Biology, 7*(4), 276–285. https://doi.org/10.1038/nrm1893

El-Sherry, T. M., Elsayed, M., Abdelhafez, H. K., & Abdelgawad, M. (2014). Characterization of rheotaxis of bull sperm using microfluidics. *Integrative Biology: Quantitative Biosciences from Nano to Macro, 6*(12), 1111–1121. https://doi.org/10.1039/c4ib00196f

Ernesto, J. I., Weigel Munoz, M., Battistone, M. A., Vasen, G., Martinez-Lopez, P., Orta, G., et al. (2015). CRISP1 as a novel CatSper regulator that modulates sperm motility and orientation during fertilization. *The Journal of Cell Biology., 210*(7), 1213–1224. https://doi.org/10.1083/jcb.201412041

Escoffier, J., Boisseau, S., Serres, C., Chen, C. C., Kim, D., Stamboulian, S., et al. (2007). Expression, localization and functions in acrosome reaction and sperm motility of Ca(V)3.1 and Ca(V)3.2 channels in sperm cells: an evaluation from Ca(V)3.1 and Ca(V)3.2 deficient mice. *Journal of Cellular Physiology, 212*(3), 753–763. https://doi.org/10.1002/jcp.21075

Espey, L. L., & Richards, J. S. (2014). *Knobil and Neill's physiology of reproduction* (4th ed.). Academic. 15th November 2014.

Espinal-Enriquez, J., Priego-Espinosa, D. A., Darszon, A., Beltran, C., & Martinez-Mekler, G. (2017). Network model predicts that CatSper is the main Ca(2+) channel in the regulation of sea urchin sperm motility. *Scientific Reports, 7*(1), 4236. https://doi.org/10.1038/s41598-017-03857-9

Espinosa, F., López-González, I., Serrano, C. J., Gasque, G., de la Vega-Beltrán, J. L., Treviño, C. L., Darszon, A., et al. (1999). Anion channel blockers differentially affect t-type Ca2+ currents of mouse spermatogenic cells, ?1E currents expressed inXenopus oocytes and the sperm acrosome reaction. *Developmental Genetics, 25*(2), 103–114. https://doi.org/10.1002/(sici)1520-6408(1999)25:2<103::aid-dvg4>3.0.co;2-b

Espinosa, F., López-González, I., Muñoz-Garay, C., Felix, R., De la Vega-Beltrán, J. L., Kopf, G. S., et al. (2000). Dual regulation of the T-type Ca2+current by serum albumin and β-estradiol in mammalian spermatogenic cells. *FEBS Letters, 475*(3), 251–256. https://doi.org/10.1016/s0014-5793(00)01688-4

Esposito, G., Jaiswal, B. S., Xie, F., Krajnc-Franken, M. A., Robben, T. J., Strik, A. M., et al. (2004). Mice deficient for soluble adenylyl cyclase are infertile because of a severe sperm-motility defect. *Proceedings of the National Academy of Sciences of* the United States of America, 101(9), 2993–2998. https://doi.org/10.1073/pnas.0400050101

Fechner, S., Alvarez, L., Bonigk, W., Muller, A., Berger, T. K., Pascal, R., et al. (2015). A K(+)-selective CNG channel orchestrates Ca(2+) signalling in zebrafish sperm. eLife, 4. https://doi.org/10.7554/eLife.07624

Ferreira, J. J., Lybaert, P., Puga-Molina, L. C., & Santi, C. M. (2021a). Conserved mechanism of bicarbonate-induced sensitization of CatSper channels in human and mouse sperm. Frontiers in Cell and Development Biology, 9, 733653. https://doi.org/10.3389/fcell.2021.733653

Ferreira, J. J., Cassina, A., Irigoyen, P., Ford, M., Pietroroia, S., Peramsetty, N., et al. (2021b). Increased mitochondrial activity upon CatSper channel activation is required for mouse sperm capacitation. Redox Biology, 48, 102176. https://doi.org/10.1016/j.redox.2021.102176

Fietz, D., Bakhaus, K., Wapelhorst, B., Grosser, G., Gunther, S., Alber, J., et al. (2013). Membrane transporters for sulfated steroids in the human testis--cellular localization, expression pattern and functional analysis. PLoS One, 8(5), e62638. https://doi.org/10.1371/journal.pone.0062638

Florman, H. M., & Storey, B. T. (1982). Mouse gamete interactions: The zona pellucida is the site of the acrosome reaction leading to fertilization in vitro. Developmental Biology, 91(1), 121–130. https://doi.org/10.1016/0012-1606(82)90015-x

Florman, H. M., Tombes, R. M., First, N. L., & Babcock, D. F. (1989). An adhesion-associated agonist from the zona pellucida activates G protein-promoted elevations of internal Ca2+ and pH that mediate mammalian sperm acrosomal exocytosis. Developmental Biology, 135(1), 133–146. https://doi.org/10.1016/0012-1606(89)90164-4

Fragale, A., Aguanno, S., Kemp, M., Reeves, M., Price, K., Beattie, R., et al. (2000). Identification and cellular localisation of voltage-operated calcium channels in immature rat testis. Molecular and Cellular Endocrinology, 162(1-2), 25–33. https://doi.org/10.1016/s0303-7207(00)00213-6

Friedrich, B. M., Riedel-Kruse, I. H., Howard, J., & Julicher, F. (2010). High-precision tracking of sperm swimming fine structure provides strong test of resistive force theory. The Journal of Experimental Biology, 213(Pt 8), 1226–1234. https://doi.org/10.1242/jeb.039800

Galindo, B. E., de la Vega-Beltran, J. L., Labarca, P., Vacquier, V. D., & Darszon, A. (2007). Sp-tetraKCNG: A novel cyclic nucleotide gated K(+) channel. Biochemical and Biophysical Research Communications, 354(3), 668–675. https://doi.org/10.1016/j.bbrc.2007.01.035

Gao, X., Wang, Q., Wang, J., Wang, C., Lu, L., Gao, R., et al. (2012). Expression of calmodulin in germ cells is associated with fenvalerate-induced male reproductive toxicity. Archives of Toxicology, 86(9), 1443–1451. https://doi.org/10.1007/s00204-012-0825-3

Garcia, M. A., & Meizel, S. (1999). Regulation of intracellular pH in capacitated human spermatozoa

by a Na+/H+ exchanger. Molecular Reproduction and Development, 52(2), 189–195. https://doi.org/10.1002/(sici)1098-2795(199902)52:2<189::aid-mrd10>3.0.co;2-d

Garza-Lopez, E., Chavez, J. C., Santana-Calvo, C., Lopez-Gonzalez, I., & Nishigaki, T. (2016). Cd(2+) sensitivity and permeability of a low voltage-activated Ca(2+) channel with CatSper-like selectivity filter. Cell Calcium, 60(1), 41–50. https://doi.org/10.1016/j.ceca.2016.03.011

Geng, Y., Ferreira, J. J., Dzikunu, V., Butler, A., Lybaert, P., Yuan, P., et al. (2017). A genetic variant of the sperm-specific SLO3 K(+) channel has altered pH and Ca(2+) sensitivities. The Journal of Biological Chemistry, 292(21), 8978–8987. https://doi.org/10.1074/jbc.M117.776013

Goodwin, L. O., Leeds, N. B., Hurley, I., Mandel, F. S., Pergolizzi, R. G., & Benoff, S. (1997). Isolation and characterization of the primary structure of testis-specific L-type calcium channel: Implications for contraception. Molecular Human Reproduction, 3(3), 255–268. https://doi.org/10.1093/molehr/3.3.255

Goodwin, L. O., Karabinus, D. S., Pergolizzi, R. G., & Benoff, S. (2000). L-type voltage-dependent calcium channel alpha-1C subunit mRNA is present in ejaculated human spermatozoa. Molecular Human Reproduction, 6(2), 127–136. https://doi.org/10.1093/molehr/6.2.127

Granados-Gonzalez, G., Mendoza-Lujambio, I., Rodriguez, E., Galindo, B. E., Beltran, C., & Darszon, A. (2005). Identification of voltage-dependent Ca2+ channels in sea urchin sperm. FEBS Letters, 579(29), 6667–6672. https://doi.org/10.1016/j.febslet.2005.10.035

Guerrero, A., Wood, C. D., Nishigaki, T., Carneiro, J., & Darszon, A. (2010). Tuning sperm chemotaxis. Biochemical Society Transactions, 38(5), 1270–1274. https://doi.org/10.1042/BST0381270

Hagenbuch, B., & Stieger, B. (2013). The SLCO (former SLC21) superfamily of transporters. Molecular Aspects of Medicine, 34(2-3), 396–412. https://doi.org/10.1016/j.mam.2012.10.009

Hagiwara, S., & Kawa, K. (1984). Calcium and potassium currents in spermatogenic cells dissociated from rat seminiferous tubules. The Journal of Physiology., 356, 135–149. https://doi.org/10.1113/jphysiol.1984.sp015457

Hess, K. C., Jones, B. H., Marquez, B., Chen, Y., Ord, T. S., Kamenetsky, M., et al. (2005). The "soluble" adenylyl cyclase in sperm mediates multiple signaling events required for fertilization. Developmental Cell, 9(2), 249–259. https://doi.org/10.1016/j.devcel.2005.06.007

Ho, K., Wolff, C. A., & Suarez, S. S. (2009). CatSper-null mutant spermatozoa are unable to ascend beyond the oviductal reservoir. Reproduction, Fertility, and Development, 21(2), 345–350. https://doi.org/10.1071/rd08183

Holdcraft, R. W., & Braun, R. E. (2004). Hormonal regulation of spermatogenesis. International Journal of Andrology, 27(6), 335–342. https://doi.org/10.1111/j.1365-2605.2004.00502.x

Hwang, J. Y., Mannowetz, N., Zhang, Y., Everley, R. A., Gygi, S. P., Bewersdorf, J., et al. (2019). Dual sensing of physiologic pH and calcium by EFCAB9 regulates sperm motility. *Cell, 177*(6), 1480–1494. e19. https://doi.org/10.1016/j.cell.2019.03.047

Hwang, J. Y., Maziarz, J., Wagner, G. P., & Chung, J. J. (2021a). Molecular evolution of CatSper in mammals and function of sperm hyperactivation in gray short-tailed opossum. *Cells, 10*(5). https://doi.org/10.3390/cells10051047

Hwang, J. Y., Wang, H., Lu, Y., Ikawa, M., & Chung, J.-J. (2021b). *C2cd6-encoded CatSperτ targets sperm calcium channel to Ca2+ signaling domains in the flagellar membrane*. https://doi.org/10.1101/2021.08.16.456347

Ishijima, S., Baba, S. A., Mohri, H., & Suarez, S. S. (2002). Quantitative analysis of flagellar movement in hyperactivated and acrosome-reacted golden hamster spermatozoa. *Molecular Reproduction and Development, 61*(3), 376–384. https://doi.org/10.1002/mrd.10017

Jagannathan, S., Publicover, S. J., & Barratt, C. L. (2002a). Voltage-operated calcium channels in male germ cells. *Reproduction, 123*(2), 203–215. https://doi.org/10.1530/rep.0.1230203

Jagannathan, S., Punt, E. L., Gu, Y., Arnoult, C., Sakkas, D., Barratt, C. L., et al. (2002b). Identification and localization of T-type voltage-operated calcium channel subunits in human male germ cells. Expression of multiple isoforms. *The Journal of Biological Chemistry, 277*(10), 8449–8456. https://doi.org/10.1074/jbc.M105345200

Jeschke, J. K., Biagioni, C., Schierling, T., Wagner, I. V., Borgel, F., Schepmann, D., et al. (2021). The action of reproductive fluids and contained steroids, prostaglandins, and Zn(2+) on CatSper Ca(2+) channels in human sperm. *Frontiers in Cell and Development Biology, 9*, 699554. https://doi.org/10.3389/fcell.2021.699554

Jimenez-Gonzalez, C., Michelangeli, F., Harper, C. V., Barratt, C. L., & Publicover, S. J. (2006). Calcium signalling in human spermatozoa: A specialized 'toolkit' of channels, transporters and stores. *Human Reproduction Update, 12*(3), 253–267. https://doi.org/10.1093/humupd/dmi050

Kantsler, V., Dunkel, J., Polin, M., & Goldstein, R. E. (2013). Ciliary contact interactions dominate surface scattering of swimming eukaryotes. *Proceedings of the National Academy of Sciences of the United States of America, 110*(4), 1187–1192. https://doi.org/10.1073/pnas.1210548110

Kantsler, V., Dunkel, J., Blayney, M., & Goldstein, R. E. (2014a). Correction: Rheotaxis facilitates upstream navigation of mammalian sperm cells. *eLife, 3*, e03521. https://doi.org/10.7554/eLife.03521

Kantsler, V., Dunkel, J., Blayney, M., & Goldstein, R. E. (2014b). Rheotaxis facilitates upstream navigation of mammalian sperm cells. *eLife, 3*, e02403. https://doi.org/10.7554/eLife.02403

Kao, T. J., & Millette, C. F. (2007). L-type voltage-operated Ca(+2) channels modulate transient Ca(+2) influx triggered by activation of Sertoli cell surface L-selectin. *Journal of Cellular Biochemistry, 101*(4), 1023–1037. https://doi.org/10.1002/jcb.21135

Kaupp, U. B., & Strunker, T. (2017). Signaling in sperm: More different than similar. *Trends in Cell Biology, 27*(2), 101–109. https://doi.org/10.1016/j.tcb.2016.10.002

Kaupp, U. B., Solzin, J., Hildebrand, E., Brown, J. E., Helbig, A., Hagen, V., et al. (2003). The signal flow and motor response controling chemotaxis of sea urchin sperm. *Nature Cell Biology, 5*(2), 109–117. https://doi.org/10.1038/ncb915

Kaupp, U. B., Kashikar, N. D., & Weyand, I. (2008). Mechanisms of sperm chemotaxis. *Annual Review of Physiology, 70*, 93–117. https://doi.org/10.1146/annurev.physiol.70.113006.100654

Kirichok, Y., & Lishko, P. V. (2011). Rediscovering sperm ion channels with the patch-clamp technique. *Molecular Human Reproduction, 17*(8), 478–499. https://doi.org/10.1093/molehr/gar044

Kirichok, Y., Navarro, B., & Clapham, D. E. (2006). Whole-cell patch-clamp measurements of spermatozoa reveal an alkaline-activated Ca2+ channel. *Nature, 439*(7077), 737–740. https://doi.org/10.1038/nature04417

Kobori, H., Miyazaki, S., & Kuwabara, Y. (2000). Characterization of intracellular Ca(2+) increase in response to progesterone and cyclic nucleotides in mouse spermatozoa. *Biology of Reproduction, 63*(1), 113–120. https://doi.org/10.1095/biolreprod63.1.113

Kornbluth, S., & Fissore, R. (2015). Vertebrate reproduction. *Cold Spring Harbor Perspectives in Biology, 7*(10), a006064. https://doi.org/10.1101/cshperspect.a006064

Kudla, J., Batistic, O., & Hashimoto, K. (2010). Calcium signals: The lead currency of plant information processing. *The Plant Cell, 22*(3), 541–563. https://doi.org/10.1105/tpc.109.072686

Lee, J. H., Ahn, H. J., Lee, S. J., Gye, M. C., & Min, C. K. (2011). Effects of L- and T-type Ca(2)(+) channel blockers on spermatogenesis and steroidogenesis in the prepubertal mouse testis. *Journal of Assisted Reproduction and Genetics, 28*(1), 23–30. https://doi.org/10.1007/s10815-010-9480-x

Levine, H., Jorgensen, N., Martino-Andrade, A., Mendiola, J., Weksler-Derri, D., Mindlis, I., et al. (2017). Temporal trends in sperm count: a systematic review and meta-regression analysis. *Human Reproduction Update, 23*(6), 646–659. https://doi.org/10.1093/humupd/dmx022

Li, H. G., Ding, X. F., Liao, A. H., Kong, X. B., & Xiong, C. L. (2007). Expression of CatSper family transcripts in the mouse testis during post-natal development and human ejaculated spermatozoa: Relationship to sperm motility. *Molecular Human Reproduction, 13*(5), 299–306. https://doi.org/10.1093/molehr/gam009

Li, Y., Mi, P., Chen, X., Wu, J., Qin, W., Shen, Y., et al. (2021). Dynamic profiles and transcriptional preferences of histone modifications during spermiogenesis. *Endocrinology, 162*(1). https://doi.org/10.1210/endocr/bqaa210

Libersky, E. A., & Boatman, D. E. (1995a). Progesterone concentrations in serum, follicular fluid, and oviductal fluid of the golden hamster during the periovulatory period. *Biology of Reproduction, 53*(3), 477–482. https://doi.org/10.1095/biolreprod53.3.477

Libersky, E. A., & Boatman, D. E. (1995b). Effects of progesterone on in vitro sperm capacitation and egg penetration in the golden hamster. *Biology of Reproduction, 53*(3), 483–487. https://doi.org/10.1095/biolreprod53.3.483

Liévano, A., Santi, C. M., Serrano, C. J., Treviño, C. L., Bellvé, A. R., Hernández-Cruz, A., et al. (1996). T-type Ca2+channels andα1Eexpression in spermatogenic cells, and their possible relevance to the sperm acrosome reaction. *FEBS Letters, 388*(2-3), 150–154. https://doi.org/10.1016/0014-5793(96)00515-7

Limanjaya, I., Hsu, T. I., Chuang, J. Y., & Kao, T. J. (2020). L-selectin activation regulates Rho GTPase activity via Ca(+2) influx in Sertoli cell line, ASC-17D cells. *Biochemical and Biophysical Research Communications, 525*(4), 1011–1017. https://doi.org/10.1016/j.bbrc.2020.03.011

Lin, S., Ke, M., Zhang, Y., Yan, Z., & Wu, J. (2021). Structure of a mammalian sperm cation channel complex. *Nature, 595*(7869), 746–750. https://doi.org/10.1038/s41586-021-03742-6

Lishko, P. V., & Kirichok, Y. (2010). The role of Hv1 and CatSper channels in sperm activation. *The Journal of Physiology, 588*(Pt 23), 4667–4672. https://doi.org/10.1113/jphysiol.2010.194142

Lishko, P. V., Botchkina, I. L., & Kirichok, Y. (2011). Progesterone activates the principal Ca2+ channel of human sperm. *Nature, 471*(7338), 387–391. https://doi.org/10.1038/nature09767

Lishko, P. V., Kirichok, Y., Ren, D., Navarro, B., Chung, J. J., & Clapham, D. E. (2012). The control of male fertility by spermatozoan ion channels. *Annual Review of Physiology, 74*, 453–475. https://doi.org/10.1146/annurev-physiol-020911-153258

Liu, J., Xia, J., Cho, K. H., Clapham, D. E., & Ren, D. (2007). CatSperbeta, a novel transmembrane protein in the CatSper channel complex. *The Journal of Biological Chemistry, 282*(26), 18945–18952. https://doi.org/10.1074/jbc.M701083200

Liu, T., Huang, J. C., Zuo, W. L., Lu, C. L., Chen, M., Zhang, X. S., et al. (2010). A novel testis-specific Na+/H+ exchanger is involved in sperm motility and fertility. *Frontiers in Bioscience, 2*, 566–581. https://doi.org/10.2741/e115

Liu, K., Jiang, D., Zhang, T., Tao, J., Shen, L., & Sun, X. (2011). Activation of growth hormone secretagogue type 1a receptor inhibits T-type Ca2+ channel currents through pertussis toxin-sensitive novel protein kinase C pathway in mouse spermatogenic cells. *Cellular Physiology and Biochemistry: International Journal of Experimental Cellular Physiology, Biochemistry, and Pharmacology, 27*(5), 613–624. https://doi.org/10.1159/000329983

Lobley, A., Pierron, V., Reynolds, L., Allen, L., & Michalovich, D. (2003). Identification of human and mouse CatSper3 and CatSper4 genes: Characterisation of a common interaction domain and evidence for expression in testis. *Reproductive Biology and Endocrinology: RB & E., 1*, 53. https://doi.org/10.1186/1477-7827-1-53

Lopez-Gonzalez, I., De La Vega-Beltran, J. L., Santi, C. M., Florman, H. M., Felix, R., & Darszon, A. (2001). Calmodulin antagonists inhibit T-type Ca(2+) currents in mouse spermatogenic cells and the zona pellucida-induced sperm acrosome reaction. *Developmental Biology, 236*(1), 210–219. https://doi.org/10.1006/dbio.2001.0314

López-González, I., Olamendi-Portugal, T., De La Vega-Beltrán, J. L., Van der Walt, J., Dyason, K., Possani, L. D., et al. (2003). Scorpion toxins that block T-type Ca2+ channels in spermatogenic cells inhibit the sperm acrosome reaction. *Biochemical and Biophysical Research Communications, 300*(2), 408–414. https://doi.org/10.1016/s0006-291x(02)02859-0

Lopez-Gonzalez, I., Trevino, C. L., & Darszon, A. (2016). Regulation of spermatogenic cell T-type Ca(2+) currents by Zn(2+): Implications in male reproductive physiology. *Journal of Cellular Physiology, 231*(3), 659–667. https://doi.org/10.1002/jcp.25112

Loux, S. C., Crawford, K. R., Ing, N. H., Gonzalez-Fernandez, L., Macias-Garcia, B., Love, C. C., et al. (2013). CatSper and the relationship of hyperactivated motility to intracellular calcium and pH kinetics in equine sperm. *Biology of Reproduction, 89*(5), 123. https://doi.org/10.1095/biolreprod.113.111708

Loyo-Celis, V., Orta, G., Beltran, C., & Darszon, A. (2021). CatSper channels in sea urchin sperm. *Cell Calcium, 99*, 102466. https://doi.org/10.1016/j.ceca.2021.102466

Machado-Oliveira, G., Lefievre, L., Ford, C., Herrero, M. B., Barratt, C., Connolly, T. J., et al. (2008). Mobilisation of Ca2+ stores and flagellar regulation in human sperm by S-nitrosylation: a role for NO synthesised in the female reproductive tract. *Development, 135*(22), 3677–3686. https://doi.org/10.1242/dev.024521

Mann, T., & Lutwak-Mann, C. (1981). *Male reproductive function and semen: Themes and trends in physiology, biochemistry, and investigative andrology.* Springer.

Mannowetz, N., Miller, M. R., & Lishko, P. V. (2017). Regulation of the sperm calcium channel CatSper by endogenous steroids and plant triterpenoids. *Proceedings of the National Academy of Sciences of the United States of America, 114*(22), 5743–5748. https://doi.org/10.1073/pnas.1700367114

Mannowetz, N., Mundt, N., & Lishko, P. V. (2018). Reply to Brenker et al.: The plant triterpenoid pristimerin inhibits calcium influx into human spermatozoa via CatSper. *Proceedings of the National Academy of Sciences of the United States of America, 115*(3), E347–E3E8. https://doi.org/10.1073/pnas.1719673115

Mansell, S. A., Publicover, S. J., Barratt, C. L., & Wilson, S. M. (2014). Patch clamp studies of human sperm under physiological ionic conditions reveal three functionally and pharmacologically distinct cation channels. *Molecular Human Reproduction, 20*(5), 392–408. https://doi.org/10.1093/molehr/gau003

Marquez, B., Ignotz, G., & Suarez, S. S. (2007). Contributions of extracellular and intracellular Ca2+ to regulation of sperm motility: Release of intracellular stores can hyperactivate CatSper1 and CatSper2 null sperm. *Developmental Biology, 303*(1), 214–221. https://doi.org/10.1016/j.ydbio.2006.11.007

Martins, A. D., Sa, R., Monteiro, M. P., Barros, A., Sousa, M., Carvalho, R. A., et al. (2016). Ghrelin acts as energy status sensor of male reproduction by modulating Sertoli cells glycolytic metabolism and mitochondrial bioenergetics. *Molecular and Cellular Endocrinology, 434*, 199–209. https://doi.org/10.1016/j.mce.2016.07.008

Mascarenhas, M. N., Flaxman, S. R., Boerma, T., Vanderpoel, S., & Stevens, G. A. (2012). National, regional, and global trends in infertility prevalence since 1990: A systematic analysis of 277 health surveys. *PLoS Medicine, 9*(12), e1001356. https://doi.org/10.1371/journal.pmed.1001356

Mata-Martinez, E., Sanchez-Cardenas, C., Chavez, J. C., Guerrero, A., Trevino, C. L., Corkidi, G., et al. (2021). Role of calcium oscillations in sperm physiology. *Bio Systems, 209*, 104524. https://doi.org/10.1016/j.biosystems.2021.104524

Matamoros-Volante, A., & Trevino, C. L. (2020). Capacitation-associated alkalization in human sperm is differentially controlled at the subcellular level. *Journal of Cell Science, 133*(2). https://doi.org/10.1242/jcs.238816

Mathijssen, A., Figueroa-Morales, N., Junot, G., Clement, E., Lindner, A., & Zottl, A. (2019). Oscillatory surface rheotaxis of swimming E. coli bacteria. Nature communications, *10*(1), 3434. https://doi.org/10.1038/s41467-019-11360-0

Miki, K., & Clapham, D. E. (2013). Rheotaxis guides mammalian sperm. *Current Biology: CB, 23*(6), 443–452. https://doi.org/10.1016/j.cub.2013.02.007

Miller, M. R., Mansell, S. A., Meyers, S. A., & Lishko, P. V. (2015). Flagellar ion channels of sperm: Similarities and differences between species. *Cell Calcium, 58*(1), 105–113. https://doi.org/10.1016/j.ceca.2014.10.009

Miller, M. R., Mannowetz, N., Iavarone, A. T., Safavi, R., Gracheva, E. O., Smith, J. F., et al. (2016). Unconventional endocannabinoid signaling governs sperm activation via the sex hormone progesterone. *Science, 352*(6285), 555–559. https://doi.org/10.1126/science.aad6887

Miller, M. R., Kenny, S. J., Mannowetz, N., Mansell, S. A., Wojcik, M., Mendoza, S., et al. (2018). Asymmetrically positioned flagellar control units regulate human sperm rotation. *Cell Reports, 24*(10), 2606–2613. https://doi.org/10.1016/j.celrep.2018.08.016

Minguez-Alarcon, L., Williams, P. L., Chiu, Y. H., Gaskins, A. J., Nassan, F. L., Dadd, R., et al. (2018). Secular trends in semen parameters among men attending a fertility center between 2000 and 2017: Identifying potential predictors. *Environment International, 121*(Pt 2), 1297–1303. https://doi.org/10.1016/j.envint.2018.10.052

Mishra, D. P., Pal, R., & Shaha, C. (2006). Changes in cytosolic Ca2+ levels regulate Bcl-xS and Bcl-xL expression in spermatogenic cells during apoptotic death. *The Journal of Biological Chemistry, 281*(4), 2133–2143. https://doi.org/10.1074/jbc.M508648200

Mitchell, D. R. (2007). The evolution of eukaryotic cilia and flagella as motile and sensory organelles. *Adv. Exp. Med. Biol., 607*, 130–140. https://doi.org/10.1007/978-0-387-74021-8_11

Muschol, M., Wenders, C., & Wennemuth, G. (2018). Four-dimensional analysis by high-speed holographic imaging reveals a chiral memory of sperm flagella. *PLoS One, 13*(6), e0199678. https://doi.org/10.1371/journal.pone.0199678

Nau, C. (2008). Voltage-gated ion channels. *Handbook of Experimental Pharmacology, 182*, 85–92. https://doi.org/10.1007/978-3-540-74806-9_4

Navarro, B., Kirichok, Y., & Clapham, D. E. (2007). KSper, a pH-sensitive K+ current that controls sperm membrane potential. *Proceedings of the National Academy of Sciences of the United States of America, 104*(18), 7688–7692. https://doi.org/10.1073/pnas.0702018104

Neher, E., Sakmann, B., & Steinbach, J. H. (1978). The extracellular patch clamp: A method for resolving currents through individual open channels in biological membranes. *Pflügers Archiv – European Journal of Physiology, 375*(2), 219–228. https://doi.org/10.1007/BF00584247

Nishigaki, T., Wood, C. D., Tatsu, Y., Yumoto, N., Furuta, T., Elias, D., et al. (2004). A sea urchin egg jelly peptide induces a cGMP-mediated decrease in sperm intracellular Ca 2+ before its increase. *Developmental Biology, 272*(2), 376–388. https://doi.org/10.1016/j.ydbio.2004.04.035

Nishigaki, T., González-Cota, A. L., & Orta Salazar, G. J. (2014a). *CatSper in Male Infertility*, 713–728. https://doi.org/10.1007/978-3-642-40282-1_34

Nishigaki, T., Jose, O., Gonzalez-Cota, A. L., Romero, F., Trevino, C. L., & Darszon, A. (2014b). Intracellular pH in sperm physiology. *Biochemical and Biophysical Research Communications, 450*(3), 1149–1158. https://doi.org/10.1016/j.bbrc.2014.05.100

Nolan, M. A., Babcock, D. F., Wennemuth, G., Brown, W., Burton, K. A., & McKnight, G. S. (2004). Sperm-specific protein kinase A catalytic subunit Calpha2 orchestrates cAMP signaling for male fertility. *Proceedings of the National Academy of Sciences of the United States of America, 101*(37), 13483–13488. https://doi.org/10.1073/pnas.0405580101

Ogra, S. S., Kirton, K. T., Tomasi, T. B., Jr., & Lippes, J. (1974). Prostaglandins in the human fallopian tube. *Fertil Steril, 25*, 250.

Oren-Benaroya, R., Orvieto, R., Gakamsky, A., Pinchasov, M., & Eisenbach, M. (2008). The sperm chemoattractant secreted from human cumulus cells is progesterone. *Human Reproduction, 23*(10), 2339–2345. https://doi.org/10.1093/humrep/den265

Orta, G., de la Vega-Beltran, J. L., Martin-Hidalgo, D., Santi, C. M., Visconti, P. E., & Darszon, A. (2018). CatSper channels are regulated by protein kinase A. *The Journal of Biological Chemistry, 293*(43), 16830–16841. https://doi.org/10.1074/jbc.RA117.001566

Overstreet, J. W., Yanagimachi, R., Katz, D. F., Hayashi, K., & Hanson, F. W. (1980). Penetration of human spermatozoa into the human zona pellucida and the zona-free hamster egg: A study of fertile donors and infertile patients**supported in part by a grant from the international planned parenthood federation (to

R. Y.). *Fertility and Sterility, 33*(5), 534–542. https://doi.org/10.1016/s0015-0282(16)44720-5

Parrish, J. J., Susko-Parrish, J. L., & First, N. L. (1989). Capacitation of bovine sperm by heparin: Inhibitory effect of glucose and role of intracellular pH. *Biology of Reproduction, 41*(4), 683–699. https://doi.org/10.1095/biolreprod41.4.683

Pommerville, J. C., Strickland, J. B., & Harding, K. E. (1990). Pheromone interactions and ionic communication in gametes of aquatic fungus Allomyces macrogynus. *Journal of Chemical Ecology, 16*(1), 121–131. https://doi.org/10.1007/BF01021274

Price, D. C., Chan, C. X., Yoon, H. S., Yang, E. C., Qiu, H., Weber, A. P., et al. (2012). Cyanophora paradoxa genome elucidates origin of photosynthesis in algae and plants. *Science, 335*(6070), 843–847. https://doi.org/10.1126/science.1213561

Priego-Espinosa, D. A., Darszon, A., Guerrero, A., Gonzalez-Cota, A. L., Nishigaki, T., Martinez-Mekler, G., et al. (2020). Modular analysis of the control of flagellar Ca2+-spike trains produced by CatSper and CaV channels in sea urchin sperm. *PLoS Computational Biology, 16*(3), e1007605. https://doi.org/10.1371/journal.pcbi.1007605

Publicover, S. J., Giojalas, L. C., Teves, M. E., de Oliveira, G. S., Garcia, A. A., Barratt, C. L., et al. (2008). Ca2+ signalling in the control of motility and guidance in mammalian sperm. *Frontiers in Bioscience: A Journal and Virtual Library, 13*, 5623–5637. https://doi.org/10.2741/3105

Qi, H., Moran, M. M., Navarro, B., Chong, J. A., Krapivinsky, G., Krapivinsky, L., et al. (2007). All four CatSper ion channel proteins are required for male fertility and sperm cell hyperactivated motility. *Proceedings of the National Academy of Sciences of the United States of America, 104*(4), 1219–1223. https://doi.org/10.1073/pnas.0610286104

Quill, T. A., Ren, D., Clapham, D. E., & Garbers, D. L. (2001). A voltage-gated ion channel expressed specifically in spermatozoa. *Proceedings of the National Academy of Sciences of the United States of America, 98*(22), 12527–12531. https://doi.org/10.1073/pnas.221454998

Rahban, R., Rehfeld, A., Schiffer, C., Brenker, C., Egeberg Palme, D. L., Wang, T., et al. (2021). The antidepressant Sertraline inhibits CatSper Ca2+ channels in human sperm. *Human Reproduction, 36*(10), 2638–2648. https://doi.org/10.1093/humrep/deab190

Rehfeld, A. (2020). Revisiting the action of steroids and triterpenoids on the human sperm Ca2+ channel CatSper. *Molecular Human Reproduction, 26*(11), 816–824. https://doi.org/10.1093/molehr/gaaa062

Rehfeld, A., Dissing, S., & Skakkebaek, N. E. (2016). Chemical UV filters mimic the effect of progesterone on Ca(2+) signaling in human sperm cells. *Endocrinology, 157*(11), 4297–4308. https://doi.org/10.1210/en.2016-1473

Rehfeld, A., Andersson, A. M., & Skakkebaek, N. E. (2020). Bisphenol A diglycidyl ether (BADGE) and bisphenol analogs, but not bisphenol A (BPA), activate the CatSper Ca(2+) channel in human sperm. *Frontiers in Endocrinology, 11*, 324. https://doi.org/10.3389/fendo.2020.00324

Ren, D., Navarro, B., Perez, G., Jackson, A. C., Hsu, S., Shi, Q., et al. (2001). A sperm ion channel required for sperm motility and male fertility. *Nature, 413*(6856), 603–609. https://doi.org/10.1038/35098027

Rennhack, A., Schiffer, C., Brenker, C., Fridman, D., Nitao, E. T., Cheng, Y. M., et al. (2018). A novel cross-species inhibitor to study the function of CatSper Ca(2+) channels in sperm. *British Journal of Pharmacology, 175*(15), 3144–3161. https://doi.org/10.1111/bph.14355

Romarowski, A., Sanchez-Cardenas, C., Ramirez-Gomez, H. V., Puga Molina Ldel, C., Trevino, C. L., Hernandez-Cruz, A., et al. (2016). A specific transitory increase in intracellular calcium induced by progesterone promotes acrosomal exocytosis in mouse sperm. *Biology of Reproduction, 94*(3), 63. https://doi.org/10.1095/biolreprod.115.136085

Romero, F., & Nishigaki, T. (2019). Comparative genomic analysis suggests that the sperm-specific sodium/proton exchanger and soluble adenylyl cyclase are key regulators of CatSper among the Metazoa. *Zoological Letters, 5*, 25. https://doi.org/10.1186/s40851-019-0141-3

Ruiz-Trillo, I., Burger, G., Holland, P. W., King, N., Lang, B. F., Roger, A. J., et al. (2007). The origins of multicellularity: A multi-taxon genome initiative. *Trends in Genetics: TIG., 23*(3), 113–118. https://doi.org/10.1016/j.tig.2007.01.005

Saaranen, M., Suistomaa, U., Kantola, M., Saarikoski, S., & Vanha-Perttula, T. (1987). Lead, magnesium, selenium and zinc in human seminal fluid: Comparison with semen parameters and fertility. *Human Reproduction, 2*(6), 475–479. https://doi.org/10.1093/oxfordjournals.humrep.a136573

Sanchez-Cardenas, C., Guerrero, A., Trevino, C. L., Hernandez-Cruz, A., & Darszon, A. (2012). Acute slices of mice testis seminiferous tubules unveil spontaneous and synchronous Ca2+ oscillations in germ cell clusters. *Biology of Reproduction, 87*(4), 92. https://doi.org/10.1095/biolreprod.112.100255

Sanchez-Cardenas, C., Servin-Vences, M. R., Jose, O., Trevino, C. L., Hernandez-Cruz, A., & Darszon, A. (2014). Acrosome reaction and Ca(2)(+) imaging in single human spermatozoa: New regulatory roles of [Ca(2)(+)]i. *Biology of Reproduction, 91*(3), 67. https://doi.org/10.1095/biolreprod.114.119768

Santi, C. M., Darszon, A., & Hernandez-Cruz, A. (1996). A dihydropyridine-sensitive T-type Ca2+ current is the main Ca2+ current carrier in mouse primary spermatocytes. *The American Journal of Physiology., 271*(5 Pt 1), C1583–C1593. https://doi.org/10.1152/ajpcell.1996.271.5.C1583

Santi, C. M., Santos, T., Hernandez-Cruz, A., & Darszon, A. (1998). Properties of a novel pH-dependent Ca2+ permeation pathway present in male germ cells with possible roles in spermatogenesis and mature sperm function. *The Journal of General Physiology, 112*(1), 33–53. https://doi.org/10.1085/jgp.112.1.33

Santi, C. M., Cayabyab, F. S., Sutton, K. G., McRory, J. E., Mezeyova, J., Hamming, K. S., et al. (2002). Differential inhibition of T-type calcium channels by neuroleptics. *The Journal of Neuroscience,*

22(2), 396–403. https://doi.org/10.1523/jneurosci.22-02-00396.2002

Santi, C. M., Martinez-Lopez, P., de la Vega-Beltran, J. L., Butler, A., Alisio, A., Darszon, A., et al. (2010). The SLO3 sperm-specific potassium channel plays a vital role in male fertility. *FEBS Letters, 584*(5), 1041–1046. https://doi.org/10.1016/j.febslet.2010.02.005

Santi, C. M., Orta, G., Salkoff, L., Visconti, P. E., Darszon, A., & Trevino, C. L. (2013). K+ and Cl- channels and transporters in sperm function. *Current Topics in Developmental Biology, 102*, 385–421. https://doi.org/10.1016/B978-0-12-416024-8.00014-3

Schaefer, M., Hofmann, T., Schultz, G., & Gudermann, T. (1998). A new prostaglandin E receptor mediates calcium influx and acrosome reaction in human spermatozoa. *Proceedings of the National Academy of Sciences of the United States of America, 95*(6), 3008–3013. https://doi.org/10.1073/pnas.95.6.3008

Schiffer, C., Muller, A., Egeberg, D. L., Alvarez, L., Brenker, C., Rehfeld, A., et al. (2014). Direct action of endocrine disrupting chemicals on human sperm. *EMBO Reports, 15*(7), 758–765. https://doi.org/10.15252/embr.201438869

Schiffer, C., Rieger, S., Brenker, C., Young, S., Hamzeh, H., Wachten, D., et al. (2020). Rotational motion and rheotaxis of human sperm do not require functional CatSper channels and transmembrane Ca(2+) signaling. *The EMBO Journal, 39*(4), e102363. https://doi.org/10.15252/embj.2019102363

Schuetz, A. W., & Dubin, N. H. (1981). Progesterone and prostaglandin secretion by ovulated rat cumulus cell-oocyte complexes. *Endocrinology, 108*(2), 457–463. https://doi.org/10.1210/endo-108-2-457

Seifert, R., Flick, M., Bonigk, W., Alvarez, L., Trotschel, C., Poetsch, A., et al. (2015). The CatSper channel controls chemosensation in sea urchin sperm. *The EMBO Journal, 34*(3), 379–392. https://doi.org/10.15252/embj.201489376

Senatore, A., Monteil, A., van Minnen, J., Smit, A. B., & Spafford, J. D. (2013). NALCN ion channels have alternative selectivity filters resembling calcium channels or sodium channels. *PLoS One, 8*(1), e55088. https://doi.org/10.1371/journal.pone.0055088

Serrano, C. J., Treviño, C. L., Felix, R., & Darszon, A. (1999). Voltage-dependent Ca2+channel subunit expression and immunolocalization in mouse spermatogenic cells and sperm. *FEBS Letters, 462*(1-2), 171–176. https://doi.org/10.1016/s0014-5793(99)01518-5

Servin-Vences, M. R., Tatsu, Y., Ando, H., Guerrero, A., Yumoto, N., Darszon, A., et al. (2012). A caged progesterone analog alters intracellular Ca2+ and flagellar bending in human sperm. *Reproduction, 144*(1), 101–109. https://doi.org/10.1530/REP-11-0268

Shimizu, Y., Yorimitsu, A., Maruyama, Y., Kubota, T., Aso, T., & Bronson, R. A. (1998). Prostaglandins induce calcium influx in human spermatozoa. *Molecular Human Reproduction, 4*(6), 555–561. https://doi.org/10.1093/molehr/4.6.555

Singaravelu, G., & Singson, A. (2013). Calcium signaling surrounding fertilization in the nematode

Caenorhabditis elegans. *Cell Calcium, 53*(1), 2–9. https://doi.org/10.1016/j.ceca.2012.11.009

Smith, J. F., Syritsyna, O., Fellous, M., Serres, C., Mannowetz, N., Kirichok, Y., et al. (2013). Disruption of the principal, progesterone-activated sperm Ca2+ channel in a CatSper2-deficient infertile patient. *Proceedings of the National Academy of Sciences of the United States of America, 110*(17), 6823–6828. https://doi.org/10.1073/pnas.1216588110

Spehr, M., Gisselmann, G., Poplawski, A., Riffell, J. A., Wetzel, C. H., Zimmer, R. K., et al. (2003). Identification of a testicular odorant receptor mediating human sperm chemotaxis. *Science, 299*(5615), 2054–2058. https://doi.org/10.1126/science.1080376

Spehr, M., Schwane, K., Riffell, J. A., Barbour, J., Zimmer, R. K., Neuhaus, E. M., et al. (2004). Particulate adenylate cyclase plays a key role in human sperm olfactory receptor-mediated chemotaxis. *The Journal of Biological Chemistry, 279*(38), 40194–40203. https://doi.org/10.1074/jbc.M403913200

Stamboulian, S., De Waard, M., Villaz, M., & Arnoult, C. (2002). Functional interaction between mouse spermatogenic LVA and thapsigargin-modulated calcium channels. *Developmental Biology, 252*(1), 72–83. https://doi.org/10.1006/dbio.2002.0844

Stamboulian, S., Kim, D., Shin, H. S., Ronjat, M., De Waard, M., & Arnoult, C. (2004). Biophysical and pharmacological characterization of spermatogenic T-type calcium current in mice lacking the CaV3.1 (alpha1G) calcium channel: CaV3.2 (alpha1H) is the main functional calcium channel in wild-type spermatogenic cells. *Journal of Cellular Physiology, 200*(1), 116–124. https://doi.org/10.1002/jcp.10480

Stival, C., Puga Molina Ldel, C., Paudel, B., Buffone, M. G., Visconti, P. E., & Krapf, D. (2016). Sperm capacitation and acrosome reaction in mammalian sperm. *Advances in Anatomy, Embryology, and Cell Biology, 220*, 93–106. https://doi.org/10.1007/978-3-319-30567-7_5

Strunker, T., Weyand, I., Bonigk, W., Van, Q., Loogen, A., Brown, J. E., et al. (2006). A K+-selective cGMP-gated ion channel controls chemosensation of sperm. *Nature Cell Biology, 8*(10), 1149–1154. https://doi.org/10.1038/ncb1473

Strunker, T., Goodwin, N., Brenker, C., Kashikar, N. D., Weyand, I., Seifert, R., et al. (2011). The CatSper channel mediates progesterone-induced Ca2+ influx in human sperm. *Nature, 471*(7338), 382–386. https://doi.org/10.1038/nature09769

Suarez, S. S. (2008). Control of hyperactivation in sperm. *Human Reproduction Update, 14*(6), 647–657. https://doi.org/10.1093/humupd/dmn029

Suarez, S. S., & Dai, X. (1992). Hyperactivation enhances mouse sperm capacity for penetrating viscoelastic media. *Biology of Reproduction, 46*(4), 686–691. https://doi.org/10.1095/biolreprod46.4.686

Suarez, S. S., & Osman, R. A. (1987). Initiation of hyperactivated flagellar bending in mouse sperm within the female reproductive tract. *Biology of Reproduction, 36*(5), 1191–1198. https://doi.org/10.1095/biolreprod36.5.1191

Suarez, S. S., & Pacey, A. A. (2006). Sperm transport in the female reproductive tract. *Human Reproduction Update, 12*(1), 23–37. https://doi.org/10.1093/humupd/dmi047

Suarez, S. S., Katz, D. F., Owen, D. H., Andrew, J. B., & Powell, R. L. (1991). Evidence for the function of hyperactivated motility in sperm. *Biology of Reproduction, 44*(2), 375–381. https://doi.org/10.1095/biolreprod44.2.375

Sumigama, S., Mansell, S., Miller, M., Lishko, P. V., Cherr, G. N., Meyers, S. A., et al. (2015). Progesterone accelerates the completion of sperm capacitation and activates CatSper channel in spermatozoa from the rhesus macaque. *Biology of Reproduction, 93*(6), 130. https://doi.org/10.1095/biolreprod.115.129783

Suzuki, N. (1995). Structure, function and biosynthesis of sperm-activating peptides and fucose sulfate glycoconjugate in the extracellular coat of sea urchin eggs. *Zoological Science, 12*(1), 13–27. https://doi.org/10.2108/zsj.12.13

Tamburrino, L., Marchiani, S., Minetti, F., Forti, G., Muratori, M., & Baldi, E. (2014). The CatSper calcium channel in human sperm: Relation with motility and involvement in progesterone-induced acrosome reaction. *Human Reproduction, 29*(3), 418–428. https://doi.org/10.1093/humrep/det454

Tamburrino, L., Marchiani, S., Muratori, M., Luconi, M., & Baldi, E. (2020). Progesterone, spermatozoa and reproduction: An updated review. *Molecular and Cellular Endocrinology, 516*, 110952. https://doi.org/10.1016/j.mce.2020.110952

Tavares, R. S., Mansell, S., Barratt, C. L., Wilson, S. M., Publicover, S. J., & Ramalho-Santos, J. (2013). p,p′-DDE activates CatSper and compromises human sperm function at environmentally relevant concentrations. *Human Reproduction, 28*(12), 3167-3177. https://doi.org/10.1093/humrep/det372.

Thomson, M. F., & Wishart, G. J. (1991). Temperature-mediated regulation of calcium flux and motility in fowl spermatozoa. *Journal of Reproduction and Fertility, 93*(2), 385–391. https://doi.org/10.1530/jrf.0.0930385

Treviño, C. L., Felix, R., Castellano, L. E., Gutiérrez, C., Rodríguez, D., Pacheco, J., et al. (2004). Expression and differential cell distribution of low-threshold Ca2+ channels in mammalian male germ cells and sperm. *FEBS Letters, 563*(1–3), 87–92. https://doi.org/10.1016/s0014-5793(04)00257-1

Trotschel, C., Hamzeh, H., Alvarez, L., Pascal, R., Lavryk, F., Bonigk, W., et al. (2020). Absolute proteomic quantification reveals design principles of sperm flagellar chemosensation. *The EMBO Journal, 39*(4), e102723. https://doi.org/10.15252/embj.2019102723

Tung, C. K., Ardon, F., Fiore, A. G., Suarez, S. S., & Wu, M. (2014). Cooperative roles of biological flow and surface topography in guiding sperm migration revealed by a microfluidic model. *Lab on a Chip, 14*(7), 1348–1356. https://doi.org/10.1039/c3lc51297e

Turner, R. M. (2006). Moving to the beat: A review of mammalian sperm motility regulation. *Reproduction, Fertility, and Development, 18*(1–2), 25–38. https://doi.org/10.1071/rd05120

Vastik-Fernandez, J., Gimeno, M. F., Lima, F., & Gimeno, A. L. (1975). Spontaneous motility and distribution of prostaglandins in different segments of human Fallopian tubes. *American Journal of Obstetrics and Gynecology, 122*(5), 663–668. https://doi.org/10.1016/0002-9378(75)90067-8

Veitinger, T., Riffell, J. R., Veitinger, S., Nascimento, J. M., Triller, A., Chandsawangbhuwana, C., et al. (2011). Chemosensory Ca2+ dynamics correlate with diverse behavioral phenotypes in human sperm. *The Journal of Biological Chemistry, 286*(19), 17311–17325. https://doi.org/10.1074/jbc.M110.211524

Visconti, P. E., Moore, G. D., Bailey, J. L., Leclerc, P., Connors, S. A., Pan, D., et al. (1995). Capacitation of mouse spermatozoa. II. Protein tyrosine phosphorylation and capacitation are regulated by a cAMP-dependent pathway. *Development, 121*(4), 1139–1150.

Vredenburgh-Wilberg, W. L., & Parrish, J. J. (1995). Intracellular pH of bovine sperm increases during capacitation. *Molecular Reproduction and Development, 40*(4), 490–502. https://doi.org/10.1002/mrd.1080400413

Vyklicka, L., & Lishko, P. V. (2020). Dissecting the signaling pathways involved in the function of sperm flagellum. *Current Opinion in Cell Biology, 63*, 154–161. https://doi.org/10.1016/j.ceb.2020.01.015

Wachten, D., Jikeli, J. F., & Kaupp, U. B. (2017). Sperm sensory signaling. *Cold Spring Harbor Perspectives in Biology, 9*(7). https://doi.org/10.1101/cshperspect.a028225

Wang, D., King, S. M., Quill, T. A., Doolittle, L. K., & Garbers, D. L. (2003). A new sperm-specific Na+/H+ exchanger required for sperm motility and fertility. *Nature Cell Biology, 5*(12), 1117–1122. https://doi.org/10.1038/ncb1072

Wang, H., Liu, J., Cho, K. H., & Ren, D. (2009). A novel, single, transmembrane protein CATSPERG is associated with CATSPER1 channel protein. *Biology of Reproduction, 81*(3), 539–544. https://doi.org/10.1095/biolreprod.109.077107

Wang, H. F., Liu, M., Li, N., Luo, T., Zheng, L. P., & Zeng, X. H. (2016). Bisphenol A impairs mature sperm functions by a CatSper-relevant mechanism. *Toxicological Sciences: An Official Journal of the Society of Toxicology, 152*(1), 145–154. https://doi.org/10.1093/toxsci/kfw070

Wang, T., Young, S., Krenz, H., Tuttelmann, F., Ropke, A., Krallmann, C., et al. (2020). The Ca(2+) channel CatSper is not activated by cAMP/PKA signaling but directly affected by chemicals used to probe the action of cAMP and PKA. *The Journal of Biological Chemistry, 295*(38), 13181–13193. https://doi.org/10.1074/jbc.RA120.013218

Wang, H., McGoldrick, L. L., & Chung, J. J. (2021). Sperm ion channels and transporters in male fertility and infertility. *Nature Reviews Urology, 18*(1), 46–66. https://doi.org/10.1038/s41585-020-00390-9

Wassarman, P. M. (1988). Zona pellucida glycoproteins. *Annual Review of Biochemistry, 57*, 415–442. https://doi.org/10.1146/annurev.bi.57.070188.002215

Wassarman, P. M., Bleil, J. D., Florman, H. M., Greve, J. M., Roller, R. J., & Salzmann, G. S. (1986). Nature of the mouse egg's receptor for sperm. *Advances in Experimental Medicine and Biology, 207*, 55–77. https://doi.org/10.1007/978-1-4613-2255-9_5

Weigel Munoz, M., Carvajal, G., Curci, L., Gonzalez, S. N., & Cuasnicu, P. S. (2019). Relevance of CRISP proteins for epididymal physiology, fertilization, and fertility. *Andrology, 7*(5), 610–617. https://doi.org/10.1111/andr.12638

Wennemuth, G., Westenbroek, R. E., Xu, T., Hille, B., & Babcock, D. F. (2000). CaV2.2 and CaV2.3 (N- and R-type) Ca2+ channels in depolarization-evoked entry of Ca2+ into mouse sperm. *The Journal of Biological Chemistry, 275*(28), 21210–21217. https://doi.org/10.1074/jbc.M002068200

Wennemuth, G., Carlson, A. E., Harper, A. J., & Babcock, D. F. (2003). Bicarbonate actions on flagellar and Ca2+ -channel responses: Initial events in sperm activation. *Development, 130*(7), 1317–1326. https://doi.org/10.1242/dev.00353

Whitaker, M. (2006). Calcium at fertilization and in early development. *Physiological Reviews, 86*(1), 25–88. https://doi.org/10.1152/physrev.00023.2005

Wiesner, B., Weiner, J., Middendorff, R., Hagen, V., Kaupp, U. B., & Weyand, I. (1998). Cyclic nucleotide-gated channels on the flagellum control Ca2+ entry into sperm. *The Journal of Cell Biology., 142*(2), 473–484. https://doi.org/10.1083/jcb.142.2.473

Williams, H. L., Mansell, S., Alasmari, W., Brown, S. G., Wilson, S. M., Sutton, K. A., et al. (2015). Specific loss of CatSper function is sufficient to compromise fertilizing capacity of human spermatozoa. *Human Reproduction, 30*(12), 2737–2746. https://doi.org/10.1093/humrep/dev243

Woo, A. L., James, P. F., & Lingrel, J. B. (2002). Roles of the Na,K-ATPase alpha4 isoform and the Na+/H+ exchanger in sperm motility. *Molecular Reproduction and Development, 62*(3), 348–356. https://doi.org/10.1002/mrd.90002

Wood, C. D., Darszon, A., & Whitaker, M. (2003). Speract induces calcium oscillations in the sperm tail. *The Journal of Cell Biology., 161*(1), 89–101. https://doi.org/10.1083/jcb.200212053

Wudick, M. M., Michard, E., Oliveira Nunes, C., & Feijo, J. A. (2018). Comparing plant and animal glutamate receptors: Common traits but different fates? *Journal of Experimental Botany.* https://doi.org/10.1093/jxb/ery153

Xia, J., & Ren, D. (2009). Egg coat proteins activate calcium entry into mouse sperm via CATSPER channels. *Biology of Reproduction, 80*(6), 1092–1098. https://doi.org/10.1095/biolreprod.108.074039

Xia, J., Reigada, D., Mitchell, C. H., & Ren, D. (2007). CATSPER channel-mediated Ca2+ entry into mouse sperm triggers a tail-to-head propagation. *Biology of Reproduction, 77*(3), 551–559. https://doi.org/10.1095/biolreprod.107.061358

Xiao, H., Zhang, X. C., Zhang, L., Dai, X. Q., Gong, W., Cheng, J., et al. (2006). Fenvalerate modifies T-type Ca2+ channels in mouse spermatogenic cells. *Reproductive Toxicology, 21*(1), 48–53. https://doi.org/10.1016/j.reprotox.2005.05.007

Yanagimachi, R. (1970). In vitro capacitation of golden hamster spermatozoa by homologous and heterologous blood sera. *Biology of Reproduction, 3*(2), 147–153. https://doi.org/10.1093/biolreprod/3.2.147

Yanagimachi, R. (1994). Fertility of mammalian spermatozoa: Its development and relativity. *Zygote, 2*(4), 371–372. https://doi.org/10.1017/s0967199400002240

Yanagimachi, R., & Usui, N. (1974). Calcium dependence of the acrosome reaction and activation of guinea pig spermatozoa. *Experimental Cell Research, 89*(1), 161–174. https://doi.org/10.1016/0014-4827(74)90199-2

Yang, F., Gervasi, M. G., Leu, N. A., Tourzani, D. A., Ruthel, G., Visconti, P. E., et al. (2021). *C2CD6 is required for assembly of the CatSper calcium channel complex and fertilization.* https://doi.org/10.1101/2021.07.06.451342

Yoshida, M., & Yoshida, K. (2011). Sperm chemotaxis and regulation of flagellar movement by Ca2+. *Molecular Human Reproduction, 17*(8), 457–465. https://doi.org/10.1093/molehr/gar041

Yu, Q., Mei, X. Q., Ding, X. F., Dong, T. T., Dong, W. W., & Li, H. G. (2015). Construction of a catsper1 DNA vaccine and its antifertility effect on male mice. *PLoS One, 10*(5), e0127508. https://doi.org/10.1371/journal.pone.0127508

Zanatta, A. P., Goncalves, R., Ourique da Silva, F., Pedrosa, R. C., Zanatta, L., Bouraima-Lelong, H., et al. (2021). Estradiol and 1alpha,25(OH)2 vitamin D3 share plasma membrane downstream signal transduction through calcium influx and genomic activation in immature rat testis. *Theriogenology, 172*, 36–46. https://doi.org/10.1016/j.theriogenology.2021.05.030

Zelter, A., Bencina, M., Bowman, B. J., Yarden, O., & Read, N. D. (2004). A comparative genomic analysis of the calcium signaling machinery in Neurospora crassa, Magnaporthe grisea, and Saccharomyces cerevisiae. *Fungal Genetics and Biology: FG & B, 41*(9), 827–841. https://doi.org/10.1016/j.fgb.2004.05.001

Zeng, X. H., Yang, C., Kim, S. T., Lingle, C. J., & Xia, X. M. (2011). Deletion of the Slo3 gene abolishes alkalization-activated K+ current in mouse spermatozoa. *Proceedings of the National Academy of Sciences of the United States of America, 108*(14), 5879–5884. https://doi.org/10.1073/pnas.1100240108

Zhang, Z., Liu, J., Meriano, J., Ru, C., Xie, S., Luo, J., et al. (2016). Human sperm rheotaxis: A passive physical process. *Scientific Reports, 6*, 23553. https://doi.org/10.1038/srep23553

Zhao, Y., Wang, H., Wiesehoefer, C., Shah, N. B., Reetz, E., Hwang, J. Y., et al. (2021). *3D structure and in situ arrangements of CatSper channel in the sperm flagellum.* https://doi.org/10.1101/2021.06.19.448910

Zou, Q. X., Peng, Z., Zhao, Q., Chen, H. Y., Cheng, Y. M., Liu, Q., et al. (2017). Diethylstilbestrol activates CatSper and disturbs progesterone actions in human spermatozoa. *Human Reproduction, 32*(2), 290–298. https://doi.org/10.1093/humrep/dew332

Voltage-Gated Calcium Channels as Key Regulators of Cancer Progression

Andrew D. James and William J. Brackenbury

Abstract

Calcium (Ca^{2+}) is a universal and versatile second messenger regulating everything from cell growth to cell death. As a result, Ca^{2+} regulates many processes that are hallmarks of cancer, such as proliferation, angiogenesis, invasion and metastasis. One key part of the Ca^{2+} signalling toolkit is the voltage-gated calcium channel (VGCC) family; VGCCs are aberrantly expressed in many cancers, and emerging evidence implicates VGCCs in numerous key cancer processes. VGCCs are thus an attractive putative oncological target, especially since they are readily druggable and existing clinically available VGCC-targeting drugs could be repurposed to cancer. This chapter will explore the evidence for altered VGCC expression in cancer, its impact on cancer cell function and patient outcome, and avenues for future pharmacological intervention.

Keywords

Calcium signalling · Cancer · Chemotherapy · Ion channels · Tumour growth · Voltage-gated calcium channels

A. D. James · W. J. Brackenbury (✉)
Department of Biology, York Biomedical Research Institute, University of York, York, UK
e-mail: andrew.james@york.ac.uk;
william.brackenbury@york.ac.uk

Abbreviations

APA	aldosterone-producing adenoma
BCF	radiofrequency electromagnetic fields at breast cancer-specific frequencies
BK channels	KCa1.1 calcium-activated potassium channels
Ca^{2+}	calcium
CDK2	cyclin-dependent kinase 2
CXCL11	C-X-C motif chemokine 11
DHT	5α-dihydrotestosterone
EMT	epithelial-mesenchymal transition
ER+	estrogen receptor positive
ERK1/2	extracellular signal-regulated protein kinase 1/2
HCC TICs	hepatocellular carcinoma tumour-initiating cells
HER2+	human epidermal growth factor receptor 2 positive
HIF-1	hypoxia inducible factor 1
IL-6	interleukin 6
MAPK	mitogen-activated protein kinase
NED	neuroendocrine differentiation
NF-kB	nuclear factor kappa-light-chain-enhancer of activated B cells
NLK	nemo-like kinase
PI3K	phosphoinositide 3-kinase
PKC-β	protein kinase C β
PTEN	phosphatase and tensin homolog deleted on chromosome 10

VGCC   voltage-gated Ca^{2+} channel
γH2AX   γ H2A histone family member X

Introduction

Changes in cytosolic calcium (Ca^{2+}) are among the most important signals in cell biology, regulating diverse processes at all stages of a cell's life including gene expression, apoptosis, cell motility and exocytosis. The diversity of messages conveyed by Ca^{2+} signals is achieved by a multiplicity of channels, transporters, pumps (ATPases) and binding proteins, known collectively as the Ca^{2+} signalling toolkit. These proteins act in concert to shape intracellular Ca^{2+} signals, which typically manifest as oscillations in intracellular $[Ca^{2+}]$ that are spatially and temporally distinct. It is this spatiotemporal diversity that underlies the versatility of Ca^{2+} signalling and enables such a range of intracellular messages to be conveyed by this single species of ion (Berridge et al., 2000).

Key regulators of such Ca^{2+} signals in excitable cells (such as neurons and cardiomyocytes) are voltage-gated Ca^{2+} channels (VGCCs). Upon activation by membrane depolarisation, VGCCs open to allow Ca^{2+} in the extracellular space to enter the cytoplasm, thereby increasing intracellular $[Ca^{2+}]$. VGCCs are particularly important for excitation-contraction coupling, neuronal excitability and synaptic transmission, which has led to the development of VGCC blockers indicated for the treatment of hypertension and epilepsy (Catterall et al., 2020). VGCCs are also expressed in certain nonexcitable cells such as osteoblasts (Liu et al., 2017), osteoclasts (Grössinger et al., 2018) and immune cells (Vig & Kinet, 2009; Davenport et al., 2015). Ca^{2+} conductance via VGCCs is facilitated by the pore-forming α_1 subunit ($Ca_v1–3.x$). Moreover, with the exception of some $Ca_v3.x$ channels, the α1 subunit is typically associated with multiple auxiliary subunits ($\alpha_2\delta_{1–4}$, $\beta_{1–4}$, $\gamma_{1–8}$) that regulate its function; functional diversity therefore is achieved via α1 subtypes, splice variants and auxiliary subunits (Catterall et al., 2005). However, many VGCCs and their auxiliary sub-

units are aberrantly expressed in numerous cancers and an emerging body of evidence implicates VGCCs in malignancy and tumour progression (Buchanan & McCloskey, 2016; Haworth & Brackenbury, 2019).

Malignant tumours are characterised by key cancer hallmarks including uncontrolled cell proliferation and insensitivity to anti-growth signals, neoangiogenesis, resistance to apoptosis and metastatic invasion (Hanahan & Weinberg, 2011). Importantly, Ca^{2+} signalling via VGCCs is a key regulator of many of the mechanisms underlying these hallmark processes. For example, in healthy smooth muscle cells, cell-cycle progression and cell proliferation are regulated by Ca^{2+} signals mediated by VGCCs, which can act both directly and indirectly to induce associated gene expression changes (Kuga et al., 1996; Barbado et al., 2009; Borowiec et al., 2014). Moreover, emerging evidence implicates numerous members of the Ca^{2+} signalling toolkit, including VGCCs, in malignancy and metastasis (Monteith et al., 2017; Bruce & James, 2020). This chapter will explore the current evidence for aberrant expression of VGCCs in cancer and the pathophysiological role they play in tumour progression and patient outcome.

Altered VGCC Expression in Cancer

In addition to excitable cells, VGCCs are often expressed in proliferating embryonic tissues, and their expression decreases in the latter after birth and is absent in adult tissues (Lory et al., 2006). Moreover, many aggressive cancers exhibit gene expression signatures similar to those of embryonic stem cells (Ben-Porath et al., 2008; Wong et al., 2008; Somervaille et al., 2009; Mathieu et al., 2011) and neuronal cells (Onganer et al., 2005; Jung et al., 2020) There is now a substantial body of evidence indicating altered gene and protein expression of VGCCs in many cancer cell lines and patient tissue samples; a recent meta-analysis of microarray data publicly available on the ONCOMINE platform identified altered VGCC gene signatures within patient samples from a diverse range of cancers, with particular isoforms associated with certain cancer types (Wang et al., 2015). It is also clear that some cancer types exhibit altered expression of multiple Ca$_v$ family subtypes and auxiliary subunits. For example, numerous studies have reported functional expression of Ca$_v$1, Ca$_v$2 and Ca$_v$3 channels in various neuroblastoma cell lines (Reuveny & Narahashi, 1993; Kito et al., 1999; Neelands et al., 2000; Grassi et al., 2004; Chiou, 2006; Park et al., 2010; Sousa et al., 2013). Of particular clinical relevance, VGCC expression in tumours is often associated with more aggressive phenotypes and poor prognosis, suggesting that VGCCs might be a histological biomarker with which to stratify patients. Tables 1 and 2 summarise the altered expression in Ca$_v$ α$_1$ and auxiliary subunits, respectively, across a range of cancers. The following section will consider in turn the evidence for altered expression of the three major pore-forming Ca$_v$ α$_1$-subunit families and their subunits (α$_2$δ, β and γ) in cancer.

Ca$_v$1.X Family

Numerous studies have identified expression changes in Ca$_v$1.2 and Ca$_v$1.3 in various cancers. *CACNA1C* (encoding Ca$_v$1.2) is overexpressed in colon cancer (Wang et al., 2000), and copy number amplifications in this gene have been reported in ovarian cancer (Davis et al., 2015). Moreover, overexpression of the *CACNA1C* gene is associated with differentiation in oesophageal squamous cell carcinoma (Shi et al., 2015). Interestingly, *CACNA1C* is expressed in MCF-7 breast cancer cells, where its expression is regulated by androgen signalling (Marques et al., 2015).

CACNA1D (encoding Ca$_v$1.3) is expressed in Hec-1A endometrial carcinoma cells (Bao et al., 2012). Moreover, Ca$_v$1.3 is upregulated in atypical hyperplasia and endometrial carcinoma compared with healthy endometrial tissue (Hao et al., 2015). *CACNA1D* expression levels are also significantly elevated in prostate cancer tissues relative to healthy tissues; since *CACNA1D* is a target gene of v-ets erythroblastosis virus E26 onco-

gene homolog (*ERG*) (Paulo et al., 2012), *CACNA1D* expression is particularly elevated in prostate cancers harbouring *ERG* overexpression (Setlur et al., 2008; Jhavar et al., 2009; Boormans et al., 2013) or the urinary transmembrane protease, serine 2:ERG (*TMPRSS2-ERG*) gene fusion (Chen et al., 2014c; Geybels et al., 2015). Moreover, *CACNA1D* expression is correlated with more aggressive prostate cancer (Zhu et al., 2015). Interestingly, the treatment of LNCaP cells with 5α-dihydrotestosterone (DHT) evoked increases in intracellular Ca^{2+} that were abolished by treatment with nifedipine, verapamil and diltiazem, supporting the notion that functional Ca$_v$1.x are expressed in prostate cancer (Sun, 2006).

In contrast to Ca$_v$1.2 and Ca$_v$1.3, only limited evidence is available for the involvement of Ca$_v$1.1 and Ca$_v$1.4 in cancer. Interestingly, *CACNA1S* and *CACNA1C* polymorphisms are associated with bladder cancer risk (Chen et al., 2016). *CACNA1F* (encoding Ca$_v$1.4) overexpression has been observed in testicular teratoma, and *CACNA1S* (encoding Ca$_v$1.1) overexpression has been observed in brain tumours and leukaemia (Wang et al., 2015), but their functional expression and role in cancer progression have yet to be explored.

Ca$_v$2.X Family

CACNA1A (encoding Ca$_v$2.1) overexpression has been identified in ovarian, brain, uterine, ovarian lung and cervical cancers, leukaemia and sarcoma (Wang et al., 2015). Similarly, *CACNA1B* (encoding Ca$_v$2.2) overexpression has been observed in prostate and breast cancers, and *CACNA1E* (encoding Ca$_v$2.3) overexpression in oesophageal, uterine (Wang et al., 2015) and squamous cell bladder cancers (Zhang et al., 2016). *CACNA1A* is hypermethylated in ovarian clear cell adenocarcinoma, and promoter hypermethylation correlates with worse prognosis (Ho et al., 2012). Moreover, *CACNA1A* and *CACNA1G* hypermethylation has been observed in lung cancer cell lines and tissues (Castro et al.,

Table 1 Altered expression of VGCC a_1 subunits in cancer

Ca$_v$ number	Gene name	Cancer type	Gene expression	Protein expression	References
Ca$_v$1.1	*CACNA1S*	Brain	++		Wang et al. (2015)
		Leukaemia	++		Wang et al. (2015)
Ca$_v$1.2	*CACNA1C*	Brain	++		Wang et al. (2015)
		Breast	++		Marques et al. (2015) and Wang et al. (2015)
		Colorectal	++		Wang et al. (2000, 2015)
		Gastric	++		Wang et al. (2015)
		Leukaemia	++		Wang et al. (2015)
		Pancreatic	++		Wang et al. (2015)
		Prostate	++		Wang et al. (2015)
		Sarcoma	++		Wang et al. (2015)
		Skin	++		Wang et al. (2015)
		Uterine	++		Wang et al. (2015)
		Oesophageal	++		Shi et al. (2015) and Wang et al. (2015)
		Ovarian	++		Davis et al. (2015)
Ca$_v$1.3	*CACNA1D*	Neuroblastoma	+		Chiou (2006) and Sousa et al. (2013)
		Endometrial	++	++	Bao et al. (2012) and Hao et al. (2015)
		Prostate	++	++	Setlur et al. (2008), Jhavar et al. (2009), Boormans et al. (2013), Chen et al. (2014c), and Geybels et al. (2015)
Ca$_v$1.4	*CACNA1F*	Neuroblastoma	+		Park et al. (2010)
		Testicular	++		Wang et al. (2015)
Ca$_v$2.1	*CACNA1A*	Leukaemia	++		Wang et al. (2015)
		Ovarian	++		Wang et al. (2015)
		Sarcoma	++		Wang et al. (2015)
		Brain	++		Wang et al. (2015)
		Cervical	++		Wang et al. (2015)
		Lung	++		Wang et al. (2015)
		Uterine	++		Wang et al. (2015)
Ca$_v$2.2	*CACNA1B*	Neuroblastoma	+		Chiou (2006) and Sousa et al. (2013)
		Prostate	++		Wang et al. (2015)
		Breast	++		Wang et al. (2015)
Ca$_v$2.3	*CACNA1E*	Neuroblastoma	+		Chiou (2006)
		Pancreatic neuroendocrine	+		Mergler et al. (2005)
		Oesophageal	++		Wang et al. (2015)
		Uterine	++		Wang et al. (2015)
		Wilms' nephroblastoma tumours	++	++	Natrajan et al. (2006)
		Bladder squamous cell carcinoma	++		Zhang et al. (2016)
Ca$_v$3.1	*CACNA1G*	Neuroblastoma	+		Sousa et al. (2013)
		Sarcoma	++		Wang et al. (2015)
		Colorectal	++		Wang et al. (2015)
		Uterus	++		Wang et al. (2015)
		Lung	++		Wang et al. (2015)
		Prostate	++		Wang et al. (2015)
		Breast	++	+	Taylor et al. (2008), Squecco et al. (2015), and Wang et al. (2015)

(continued)

Table 1 (continued)

Ca$_v$ number	Gene name	Cancer type	Gene expression	Protein expression	References
		Oesophageal	++		Lu et al. (2008)
		Glioblastoma	+	+	Latour et al. (2004) and Zhang et al. (2012)
		Neuroblastoma	+	+	Panner et al. (2005)
		Acute lymphocytic leukaemia		++	Huang et al. (2015)
		Non-small cell lung carcinoma		++	Suo et al. (2018)
		Hepatocellular carcinoma	+		Li et al. (2009)
		Melanoma	++		Das et al. (2012) and Maiques et al. (2017)
Ca$_v$3.2	*CACNA1H*	Oesophageal	++		Lu et al. (2008)
		Prostate neuroendocrine	++	++	Mariot et al. (2002), Gray et al. (2004), Gackière et al. (2008), and Weaver et al. (2015a)
		Renal	++		Wang et al. (2015)
		Sarcoma	++		Wang et al. (2015)
		Gastric	++		Wang et al. (2015)
		Breast	+	+	Gray et al. (2004), Taylor et al. (2008), Squecco et al. (2015), and Sharma et al. (2019)
		Glioma	+	+	Panner et al. (2005)
		Neuroblastoma	+	+	Gray et al. (2004) and Panner et al. (2005)
		Acute lymphocytic leukaemia		++	Huang et al. (2015)
		T-cell	+		Gray et al. (2004)
		Glioblastoma	+	+	Zhang et al. (2012, 2017)
		Hepatocellular carcinoma	+		Li et al. (2009) and Jimenez et al. (2019)
		Melanoma	++		Das et al. (2012) and Maiques et al. (2017)
		Bladder squamous cell carcinoma	++		Zhang et al. (2016)
Ca$_v$3.3	*CACNA1I*	Oesophageal	++		Lu et al. (2008)
		Breast	++		Wang et al. (2015)
		Glioblastoma	+	+	Zhang et al. (2012)
		Sarcoma	++		Wang et al. (2015)
		Hepatocellular carcinoma	+		Li et al. (2009)
		Melanoma	++		Das et al. (2012)

+, expressed; ++, overexpressed

2010). Functional Ca$_v$2.3 channels appear to be expressed within BON pancreatic carcinoid cells, where they couple to IGF-1 regulation of chromogranin A secretion (Mergler et al., 2005). *CACNA1E* is overexpressed in relapsing Wilms' nephroblastoma tumours, and its nuclear localisation correlates with reduced relapse-free survival (Natrajan et al., 2006). Interestingly, *CACNA1E* mutation rates are significantly higher in non-small cell lung cancer tumours from patients exposed to severe air pollution relative to patients from regions without poor air quality (Yu et al., 2015), implicating VGCC mutations in cancers arising from environmental triggers.

Table 2 Altered expression of VGCC auxiliary subunits in cancer

Cav number	Gene name	Cancer type	Gene expression	Protein expression	References
$\alpha_2\delta_1$	CACNA2D1	Ovarian		++	Yu et al. (2016))
		Hepatocellular		++	Amhimmid Badr et al. (2018)
		Small cell lung carcinoma		++	Yu et al. (2018)
		Non-small cell lung carcinoma		+	Sui et al. (2018)
		Gastric cancer stem cells	++		Shiozaki et al. (2021)
$\alpha_2\delta_2$	CACNA2D2	Prostate	++	++	Warnier et al. (2015)
		Pancreatic	++		Cromer et al. (2015)
		Lung	−		Lerman and Minna (2000), Huang et al. (2014), and Lindskog et al. (2014)
		Uterine	−		Mitra et al. (2012)
		Head/neck squamous cell	−		Ghosh et al. (2008)
$\alpha_2\delta_3$	CACNA2D3	Oesophageal squamous cell carcinoma	−	−	Qin et al. (2008) and Li et al. (2013b)
		Nasopharyngeal carcinoma	−		Wong et al. (2013)
		Glioma	−		Jin et al. (2017)
		Neuroblastoma	−		De Preter et al. (2006) and Thorell et al. (2009)
$\alpha_2\delta_4$	CACNA2D4	Pituitary adenomas	++		Taniguchi-Ponciano et al. (2020)
β_1	CACNB1	Colon	++		Gao et al. (2018)
		HER2+ gastric	++		Xiang et al. (2018)
β_2	CACNB2	Head/neck squamous cell carcinoma	−		Oshima et al. (2021)
		Gastric	++		Liu et al. (2020)
β_3	CACNB3	Non-small cell lung	++		Mitra et al. (2011) and Chen et al. (2014b)
β_4	CACNB4	Gastric cancer stem cells	++		Shiozaki et al. (2021)
γ_1	CACNG1	HER2+ metastatic breast cancer	−		Omarini et al. (2018)
γ_4	CACNG4	Bladder squamous cell carcinoma	++		Zhang et al. (2016)
		Breast cancer	++	+	Kanwar et al. (2020)

+, expressed; ++, overexpressed; −, downregulated

Ca$_v$3.X Family

Three isoforms of the Ca$_v$3 α_1 subunit have been identified, encoded by *CACNA1G*, *CACNA1H* and *CACNA1I*. Of the Ca$_v$3.x subfamily, *CACNA1G* (encoding Ca$_v$3.1) overexpression has been observed in sarcoma, colorectal, uterus, lung, prostate and breast tumours (Wang et al., 2015), glioma (Latour et al., 2004) and in SH-SY5Y human neuroblastoma cells (Sousa et al., 2013). Ca$_v$3.1 protein is overexpressed in some non-small cell lung carcinomas where it is associated with poor prognosis (Suo et al., 2018). The *CACNA1G* gene is also aberrantly methylated in lung cancer (Castro et al., 2010) colon cancer, gastric cancer and acute myelogenous leukaemia (Toyota et al., 1999), suggesting that inactivation of *CACNA1G* may be involved in the progression of these cancers.

CACNA1H (encoding $Ca_v3.2$) is overexpressed in neuroendocrine prostate cancer (Mariot et al., 2002; Gackière et al., 2008), renal cell carcinoma, sarcoma, gastric cancer (Wang et al., 2015) and squamous cell bladder cancer (Zhang et al., 2016). *CACNA1H* is also expressed in human leukaemia, neuroblastoma and breast and prostate cancer cell lines (Gray et al., 2004). Interestingly, functional $Ca_v3.2$ expression in androgen-dependent LNCaP prostate cancer cells is upregulated following interleukin 6 (IL-6) stimulation (Weaver et al., 2015a) and following differentiation into a neuroendocrine, androgen-refractory state (Mariot et al., 2002; Weaver et al., 2015a), suggesting that $Ca_v3.2$ expression responds to cytokine signalling within the tumour microenvironment to facilitate progression towards a more aggressive phenotype. MCF-7 breast cancer cells express both *CACNA1G* and *CACNA1H* mRNA (Taylor et al., 2008) and elicit $Ca_v3.1/Ca_v3.2$ currents (Squecco et al., 2015). *CACNA1I* (encoding $Ca_v3.3$) is overexpressed in breast cancer and sarcoma (Wang et al., 2015). Moreover, all three major $Ca_v3.x$ genes (*CACNA1G*, *CACNA1H* and *CACNA1I*) are overexpressed in human oesophageal cancer (Lu et al., 2008) and melanoma (Das et al., 2012; Maiques et al., 2017), with increased expression correlated with poor prognosis.

$\alpha_2\delta$ Subunits

In terms of the role of Ca_v auxiliary subunits in cancer, the most is known about the role of the $\alpha_2\delta$ subunit. $\alpha_2\delta$ subunits are upregulated in some cancers but downregulated in others, suggesting that they have juxtaposing roles dependent on cancer type. Increased serum levels of $\alpha_2\delta_1$ protein have been observed in hepatocellular carcinoma patients (Amhimmid Badr et al., 2018). Moreover, $\alpha_2\delta_1$ protein expression has been identified in hepatocellular tumour-initiating cells, with the presence of $\alpha_2\delta_1$-expressing cells at the tumour margins following hepatectomy predicting rapid recurrence (Zhao et al., 2013). Similarly, $\alpha_2\delta_1$ expression has been identified in small cell lung carcinoma cell lines and clinical samples

where it is associated with a cancer stem cell-like phenotype (Yu et al., 2018). $\alpha_2\delta_1$ is also expressed in non-small cell lung carcinoma cells (Sui et al., 2018) and in epithelial ovarian cancer, where high expression is associated with poor prognosis (Yu et al., 2016). Various miRNAs that are downregulated in cancer target *CACNA2D1* expression, including in medulloblastoma and chronic myeloid leukaemia (Ruan et al., 2015; Zhang et al., 2018), suggesting that the loss of these miRNAs may lead to overexpression of $\alpha_2\delta_1$ in cancer cells.

In contrast to *CACNA2D1*, *CACNA2D2* is downregulated, mutated or deleted in numerous cancers, and has therefore been suggested as a potential tumour suppressor. *CACNA2D2* lies within chromosome 3p (at 3p21.3) (Wanajo et al., 2008), which is commonly deleted in numerous cancers including breast and lung (Hung et al., 1995; Martinez et al., 2001). Early studies identified reduced mRNA levels of *CACNA2D2* in many lung cancer cell lines, particularly NSCLCs, relative to normal lung (Lerman & Minna, 2000), positioning *CACNA2D2* as a potential tumour suppressor gene. More recent studies have since correlated *CACNA2D2* expression with improved survival in patients with lung cancer (Lindskog et al., 2014). Interestingly, *CACNA2D2* is also a predicted target of miR-205, leading to its downregulation in squamous cell lung carcinoma with high miR-205 expression (Huang et al., 2014). In other cancers, *CACNA2D2* is downregulated via promoter hypermethylation. In uterine cervical carcinoma, the *CACNA2D2* gene is commonly deleted or methylated, with alterations in *CACNA2D2* corresponding with poor prognosis (Mitra et al., 2012). Similarly, *CACNA2D2* promoter hypermethylation and reduced expression have been identified in head and neck squamous cell carcinoma (Ghosh et al., 2008). Single nucleotide polymorphisms in *CACNA2D2* have been identified in small-cell lung cancer (Angeloni et al., 2001). In contrast, *CACNA2D2* gene and protein expression are increased in prostate tumours (Warnier et al., 2015) and in insulin-secreting pancreatic adenomas (Cromer et al., 2015) where expression appears to be tumourigenic. Thus, despite early reports suggesting *CACNA2D2* functions as a tumour suppressor, the involvement of *CACNA2D2* in cancer is clearly complex and context-dependent.

Similar to *CACNA2D2*, *CACNA2D3* also lies on chromosome 3p (at 3p21.1). Due to 3p deletion, *CACNA2D3* is frequently downregulated or absent in oesophageal squamous cell carcinoma (Qin et al., 2008; Li et al., 2013b), nasopharyngeal carcinoma (Wong et al., 2013), endometrial cancer (Kong et al., 2020), neuroblastoma (De Preter et al., 2006; Thorell et al., 2009) and glioma (Jin et al., 2017), with downregulation associated with lymph node metastasis and poor prognosis. Similarly, the frequent methylation of *CACNA2D3* correlates with poor prognosis in gastric (Wanajo et al., 2008) and breast cancers (Palmieri et al., 2012). It is therefore tempting to speculate that *CACNA2D3* acts as a tumour suppressor and that its loss leads to remodelling of Ca^{2+} signalling that facilitates cancer progression. In contrast, little is known about *CACNA2D4* in cancer beyond occasional mutations in hepatocellular carcinoma (Kan et al., 2013; Xiong et al., 2019) and upregulated expression in non-functioning pituitary adenomas (Taniguchi-Ponciano et al., 2020).

β Subunits

$Ca_v\beta$ subunit overexpression is correlated with poor survival in numerous cancers. *CACNB1* is overexpressed in human epidermal growth factor receptor 2 positive (HER2+) gastric cancer (Xiang et al., 2018) and colon cancer (Gao et al., 2018), with expression levels associated with prognostic risk in the latter. *CACNB2* polymorphisms are associated with prognostic risk in bladder cancer (Chen et al., 2016). Similarly, higher expression and corresponding decreased methylation of *CACNB2* corresponds with worse overall survival in patients with *H. pylori*-induced gastric cancer (Liu et al., 2020). Somatic mutations in *CACNB2* have also been linked to lung cancer (Tomoshige et al., 2015). Similarly, *CACNB3* is significantly and consistently overexpressed in human non-small cell lung cancer

(Chen et al., 2014b), and increased *CACNB3* expression has been observed in patients with recurrent disease relative to recurrence-free patients (Mitra et al., 2011). On the other hand, evidence also suggests that the $Ca_v\beta$ subunits may act as tumour suppressors. *CACNB2* expression is regulated by miR-31 in head and neck squamous cell carcinoma, with low expression in clinical specimens an independent predictor of worse patient outcome (Oshima et al., 2021). Relatively little is known about *CACNB4* expression in cancer, though evidence suggests that *CACNB4* is upregulated in gastric cancer stem cells (Shiozaki et al., 2021).

γ Subunits

Eight $Ca_v\gamma$ subunits exist, encoded by genes *CACNG1* to *CACNG8*, though evidence for altered expression or gene promoter methylation in cancer is limited to *CACNG1*, *CACNG4* and *CACNG7*. *CACNG1* expression is downregulated in HER2+ metastatic breast cancer cases where disease progression occurred within a year of diagnosis (Omarini et al., 2018). Alterations in *CACNG7* gene methylation have been identified in endometrial cancer (Makabe et al., 2019) and renal cell carcinoma (Slater et al., 2013), suggesting that *CACNG7* downregulation is associated with cancer progression. On the other hand, *CACNG4* is upregulated in bladder squamous cell carcinoma (Zhang et al., 2016) and breast cancer (Kanwar et al., 2020), and is associated with lymph node metastasis in the latter. While expression of the microRNA miR-197 is downregulated in leiomyoma relative to healthy myometrium, *CACNG7* is downregulated when miR-197 is overexpressed experimentally; this suggests that miR-197 loss might lead to *CACNG7* upregulation in uterine cancers (Ling et al., 2015). Interestingly, machine learning approaches using a pan-cancer dataset have also predicted *CACNG3* as a potential oncogene (Kumar et al., 2015). Further research is required to resolve these contrasting findings regarding γ subunit expression in cancer.

Functional Consequences of VGCC Expression in Cancer

Tumours are complex tissue compartments that comprise multiple heterogeneous and distinct cell types that interact with one another and the surrounding tumour microenvironment. Within this microenvironment, tumours exhibit distinctive 'hallmark' properties and behaviours that facilitate growth of the tumour, its invasion into surrounding healthy tissues and ultimately metastatic spread. In 2000, Hanahan and Weinberg proposed six hallmark properties of cancer cells, including resistance to cell death, sustained proliferative signalling, evasion of growth suppressors, induction of angiogenesis, activating invasion and metastasis, and enabling of replicative immortality (Hanahan & Weinberg, 2000). This has since been expanded to include two additional hallmarks: deregulation of cellular energetics and evasion of immune surveillance (Hanahan & Weinberg, 2011). As an incipient premalignant lesion progresses to malignant transformation, tumorigenic cells gradually develop these hallmarks as they transition towards a neoplastic and malignant state. At the transcriptional level, this is commonly driven by the cumulative mutation or loss of tumour suppressor genes (Bieging et al., 2014) alongside the abnormal expression of oncogenes or gain-of-function mutations within proto-oncogenic genes (Cooper, 2000; Liu et al., 2019).

Due to its ubiquitous role in many intracellular signalling pathways, Ca^{2+}-mediated signalling plays an important role in the processes underlying these hallmarks (Monteith et al., 2017; Bruce & James, 2020). Moreover, a considerable body of evidence indicates that both the α1 subunit (Buchanan & McCloskey, 2016) and the auxiliary subunits (Haworth & Brackenbury, 2019) of VGCCs act to regulate these hallmark processes, thereby regulating tumour cell differentiation, growth and proliferation, neoangiogenesis, resistance to apoptosis and chemotherapy, migration and invasion. Moreover, it is becoming clear that VGCCs can have both oncogenic and tumour-suppressive roles depending on the cancer type and the VGCC subunit involved. The following section will explore the evidence supporting the functional role of VGCCs in cancer progression. Figure 1 summarises the Ca_v isoforms and auxiliary subunits putatively involved in these key cancer hallmarks.

Tumour Growth and Proliferation

Relative to healthy cells, cancer cells exhibit independence from exogenous mitogenic growth signals, thereby acquiring an active, self-sustained proliferating state. Furthermore, a large body of evidence now suggests that aberrant VGCC activity influences cell-cycle progression and proliferation in cancer cells to facilitate this tumour growth, particularly in hormone-driven tumours, melanoma and cancers of epithelial origin. $Ca_v1.x$ channels appear to play an important role in regulating proliferation in hormone-driven cancers. In prostate cancer cells, Ca^{2+} influx is stimulated by the synthetic androgen R1881; this was abolished by nifedipine treatment or shRNA knockdown of *CACNA1D* gene expression, thereby suppressing androgen receptor transactivation and cell growth (Chen et al., 2014c). Similarly, 17β-oestradiol rapidly induced expression of $Ca_v1.3$ (within 30 min) in endometrial cancer Ishikawa cells via GPR30, and $Ca_v1.3$ knockdown inhibited 17β-oestradiol-induced Ca^{2+} influx, migration and proliferation (Hao et al., 2015). Inward Ca^{2+} currents have been identified in oestrogen receptor positive (ER+) MCF-7 breast cancer cells (Ca_v1 and Ca_v3), and inhibition with nifedipine and Ni^{2+} inhibits cell proliferation (Squecco et al., 2015). Nifedipine has also been shown to increase autophagy and inhibit proliferation and migration of Hec-1A endometrial carcinoma cells (Bao et al., 2012). Together, these studies suggest an intimate link between hormone signalling and upregulation of $Ca_v1.3$ expression and activity, and implicate $Ca_v1.3$ as a key mediator of proliferation in hormone-driven cancers.

While hormones influence Ca_v regulation of proliferation in some cancers, in others, aberrant Ca_v function may also regulate proliferation by stimulating dysregulated hormone secretion.

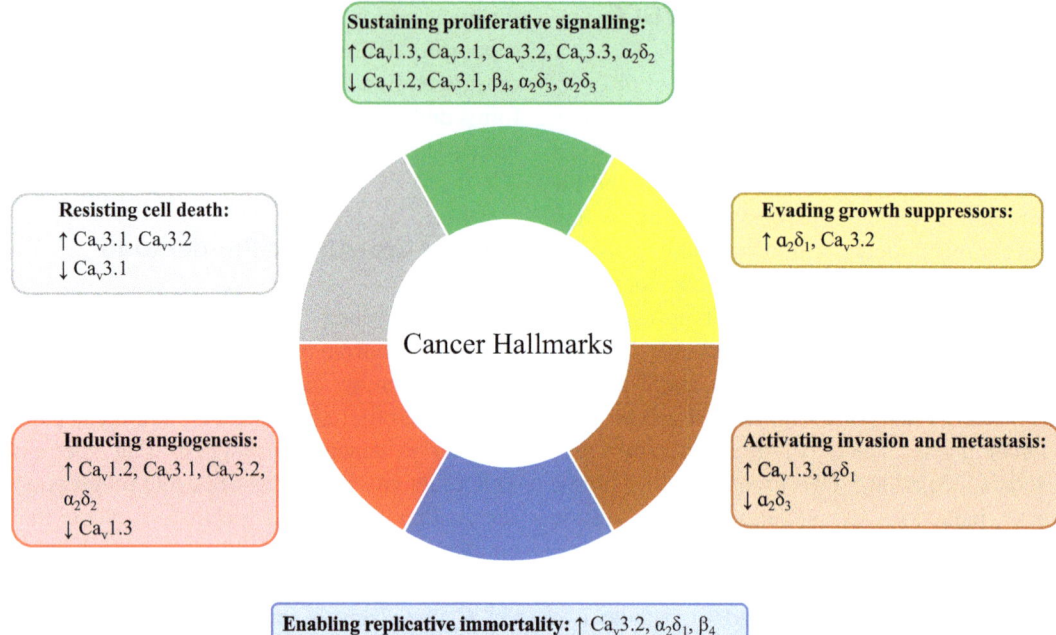

Fig. 1 The hallmarks of cancer and putative involvement of VGCCs and auxiliary subunits. Arrows denote positive or negative regulation by the indicated VGCC subunit. (Adapted from Hanahan and Weinberg 2011)

Scholl et al. identified five somatic and germline *CACNA1D* mutations in aldosterone-producing adenomas (APAs) which altered residues within the channel pore, resulting in channel activation at less depolarised potentials and impaired channel inactivation (Scholl et al., 2013). These findings suggest that $Ca_v1.3$ mediates Ca^{2+} influx in APAs. It is interesting to note that mutations in the *KCNJ5* K^+ channel, which are associated with ~40% of APAs, lead to activation of VGCCs via membrane depolarisation, thereby facilitating aldosterone production and proliferation (Choi et al., 2011; Scholl et al., 2012). It is therefore tempting to speculate that gain-of-function *CACNA1D* mutations can play a similar role in APAs.

A large body of evidence links $Ca_v3.x$ channels with the regulation of proliferation in numerous cancers, including melanoma, glioma and cancers of epithelial origin (Panner & Wurster, 2006), with $Ca_v3.x$ expression typically associated with an increased proliferative rate. The $Ca_v3.x$ inhibitors mibefradil and NNC-55-0396 both inhibit the proliferation of MOLT4 leukaemia cells by halting G1-S progression, corresponding to a decrease in extracellular signal-regulated protein kinase (ERK1/2) phosphorylation (Huang et al., 2015). NNC-55-0396 treatment or siRNA knockdown of *CACNA1G* and *CACNA1H* slows proliferation of MCF-7 cells, but has no effect on the non-cancerous breast epithelial cell line MCF-10A (Taylor et al., 2008). Interestingly, in this study, *CACNA1G* and *CACNA1H* mRNA expression was only observed in subconfluent, proliferating MCF-7 cells, and was not observed in confluent cells (Taylor et al., 2008); similar results have been observed for $Ca_v1.2$ channels in colon cancer (Wang et al., 2000), suggesting that Ca_v expression changes depend on the proliferation state of cancer cells. Similarly, *CACNA1G* and *CACNA1H* expression in U87MG glioma cells and N1E-115 neuroblastoma cells decreases under conditions conferring slowed proliferation (increased expression of cyclin D1) and serum starvation. Moreover, treatment with mibefradil and knockdown of α1 subunits slowed proliferation, while α1 overexpression accelerated it (Panner et al., 2005). Tumour xenograft studies have also highlighted the role of $Ca_v3.x$ in tumour proliferation in vivo.

Selective pharmacological inhibition of Ca_v3 channels using 3,4-dihydroquinazoline 1 or KYS05090 inhibit A549 xenograft tumour growth in BALB/c nude mice (Jung et al., 2010; Kang et al., 2012), while $Ca_v3.1$ knockdown decreases U87-MG xenograft tumour growth in vivo (Visa et al., 2019).

Further studies have provided mechanistic insights into the role of $Ca_v3.x$ channels in tumour growth and proliferation. In melanoma cells, the $Ca_v3.x$ blocker kurtoxin increased the number of cells in G1 phase and reduced the number in the proliferative S phase, indicating cell-cycle arrest (Das et al., 2012). In SNU449 hepatocellular carcinoma cells exhibiting functional $Ca_v3.x$ currents, mibefradil induced an increase of phosphorylated ERK1/2 and downregulated genes associated with cell cycle and cell proliferation (Li et al., 2009). Inhibition of $Ca_v3.1$ and $Ca_v3.2$ channels by the collagen XVIII proteolytic fragment endostatin (a broad-spectrum angiogenesis inhibitor) slows proliferation of U87 glioblastoma cells (Zhang et al., 2012), suggesting that tumour matrix remodelling may regulate tumour cell proliferation via VGCCs. In TE8 oesophageal cancer cells, treatment with mibefradil and *CACNA1G* knockdown both inhibited proliferation, potentially via upregulation of the cyclin-dependent kinase p21(CIP1) (Lu et al., 2008). Interestingly, this upregulation of p21(CIP1) via VGCC blockade was almost completely abolished by p53 silencing, suggesting that $Ca_v3.1$ regulates proliferation via a p53-dependent p21(CIP1) pathway. Evidence suggests that $Ca_v3.2$ channels regulate the growth of LNCaP cells (Gackière et al., 2013); $Ca_v3.2$ channels formed macromolecular complexes with $K_{Ca}1.1$ calcium-activated potassium (BK) channels in these cells, with BK channels setting the resting membrane potential to promote constitutive Ca^{2+} entry via $Ca_v3.2$. $Ca_v3.3$ channels also regulate proliferation via mechanosensitive mechanisms; elevated interstitial pressure due to tumour growth against the constraining extracellular matrix stimulates cancer cell proliferation through $Ca_v3.3$-mediated Ca^{2+} influx and subsequent Protein Kinase C β (PKC-β) and nuclear factor kappa-light-chain-enhancer of activated B cells (NF-kB) signalling (Basson et al., 2015). It is clear that VGCCs have complex regulatory roles in the proliferation of numerous cancers, and further research is needed to fully elucidate these mechanisms.

On the other hand, there is contrasting evidence that VGCC expression may also inversely correlate with increased proliferation under certain conditions. Upregulation of $Ca_v1.2$ expression in MCF-7 cells by the androgen 5α-dihydrotestosterone correlated with decreased rather than increased tumour growth, suggesting that the role of $Ca_v1.x$ channels in regulating proliferation in hormone-driven cancers is complex (Marques et al., 2015). Interestingly, $Ca_v3.1$ knockdown or inhibition with ProTx-I increased MCF-7 cell proliferation, while $Ca_v3.1$ overexpression decreased proliferation, whereas altering $Ca_v3.2$ expression had no effect (Ohkubo, 2012). Thus, despite the large body of evidence supporting the pro-proliferative effects of Ca_v expression in tumours, there may be circumstances where they in fact have the opposite effect. Furthermore, VGCC blockers may inhibit cancer cell proliferation via mechanisms other than blockade of VGCC-mediated Ca^{2+} conductance. Although amlodipine, diltiazem and verapamil slowed proliferation of HT-39 breast cancer cells and inhibited tumour growth in vivo, no depolarisation-induced Ca^{2+} fluxes were detected in these cells (Taylor & Simpson, 1992). Therefore, targeting the relationship between VGCCs and cancer cell proliferation may require specific approaches tailored to individual cancer types.

One interesting and novel approach to targeting VGCCs in cancer is via amplitude-modulated radiofrequency electromagnetic fields. Tumour-specific modulation frequencies have been identified in patients with various forms of cancer (Barbault et al., 2009), and administration of radiofrequency electromagnetic fields at breast cancer-specific frequencies (BCF) inhibited the growth of brain-metastatic breast cancer cells and two brain-metastatic patient derived xenografts (Sharma et al., 2019). Importantly, the pan-$Ca_v3.x$ inhibitor ethosuximide and *CACNA1H* knockdown both attenuated the growth suppression induced by BCF, which was shown to be mediated by the activation of *CACNA1H*, increased Ca^{2+} influx and p38/mitogen-activated protein kinase (MAPK) activation. Similar results have been reported in hepatocellular carcinoma (Jimenez et al., 2019). These findings raise the possibility of targeting VGCC-dependent regulation of cancer cell proliferation noninvasively without the administration of small molecule inhibitors.

Regulation of tumour cell proliferation is not limited to the Ca$_v$α$_1$ subunit; auxiliary subunits also play a role. Interestingly, there is evidence that β subunits can regulate proliferation via intracellular regulation of the Wnt/β-catenin signalling pathway. This key signalling pathway controls embryonic development and tissue regeneration in healthy cells, but when dysregulated, it can drive cancer cell proliferation and tumour progression (Klaus & Birchmeier, 2008). In hepatoma cell lines, VGCC β$_4$ subunits translocate to the nucleus and interact with the TCF4 transcription factor, thereby inhibiting binding of β-catenin to TCF4 and suppressing gene transcription downstream of Wnt signalling (Rima et al., 2017). Thus, the β4 subunit may act as a tumour suppressor in hepatoma that opposes dysregulated cell proliferation.

In addition to the β subunits, studies have identified a role for α$_2$δ in tumour growth and proliferation. Exogenous expression of *CACNA2D2* in 3p21.3-deficient non-small-cell lung cancer cells inhibits cell proliferation, whereas overexpression leads to increased intracellular [Ca^{2+}] and induces apoptosis (Carboni et al., 2003). These findings suggest that deletion of 3p21.3 (and hence loss of *CACNA2D2* expression) facilitates lung tumour growth and positions α$_2$δ$_2$ as a tumour suppressor in lung cancer. In contrast, α$_2$δ$_2$ subunit overexpression in LNCaP prostate cancer cells increases cell proliferation and tumorigenicity in vivo, and treatment with the α2δ ligand gabapentin reduces tumour development in xenograft models (Warnier et al., 2015). These contrasting results indicate that α$_2$δ$_2$ can both promote and inhibit proliferation in different cancer types. Further work is required to resolve these differences and delineate the competing mechanisms involved.

The α$_2$δ$_3$ subunit also appears to regulate proliferation in cancer. *CACNA2D3* overexpression in Ishikawa endometrial cancer cells inhibited in vitro proliferation and prevented growth of tumour xenografts in nude mice (Kong et al., 2020). In this study, progesterone induced *CACNA2D3* expression and inhibited tumour growth in vivo; knockdown of *CACNA2D3* blocked the tumour-suppressive effects of progesterone on the p38/MAPK, phosphatase and tensin homolog deleted on chromosome 10 (PTEN, a tumour suppressor) and phosphoinositide 3-kinase(PI3K)/p-AKT pathways. In addition, overexpression of *CACNA2D3* in KYSE30 and KYSE510 oesophageal squamous cell carcinoma cells slowed cell proliferation and tumour growth in vivo. This was accompanied by upregulated p53 and p21 expression, downregulated expression of cyclin-dependent kinase 2 (CDK2, which has a role in the G1/S transition) and cell-cycle arrest at G1/S phase (Li et al., 2013b). Taken together, these data suggest that *CACNA2D3* can act as a tumour suppressor to oppose tumour growth.

Hypoxia and Angiogenesis

Sustained tumour growth requires the supporting formation of new blood vessels to provide rapidly proliferating tumour cells with oxygen and metabolites. Pathological neoangiogenesis is thus a key mediator of tumour progression (Palma et al., 2017). Tumour angiogenesis is commonly mediated by the transcription factor hypoxia inducible factor 1 (HIF-1), which stabilises under hypoxic conditions. Interestingly, the Ca$_v$3.x inhibitor NNC-55-0396 has been identified as a regulator of HIF-1 (Kim et al., 2015). In this study, NNC-55-0396 inhibited tumour-associated angiogenesis both in vitro and in vivo by inhibiting HIF-1 protein stabilisation, thus suppressing HIF-1 signal transduction. Hypoxia also induces upregulation of Ca$_v$1.2 and Ca$_v$3.1 (and to a lesser extent Ca$_v$3.2) in JG and M28 melanoma cells while downregulating $Ca_v1.3$ expression (Das et al., 2012). Interestingly, $\alpha_2\delta_2$-overexpressing LNCaP cells form tumours in vivo with increased vasculature and elevated expression of CD31 (a vascular marker), indicating that the $\alpha_2\delta$ subunit may also have a role in increasing tumour angiogenesis (Warnier et al., 2015). These studies suggest a positive reciprocal regulation between hypoxic signalling and VGCC expression, and further research is required to fully characterise how these factors interplay in the context of angiogenesis and cancer progression.

Apoptosis and Chemotherapy Resistance

Apoptosis is a biological process of programmed cell death that enables removal of dying or dysfunctional cells in a regulated manner. Many cancer treatments aim to eliminate tumours by inducing apoptosis and other related cell death pathways in malignant cells. However, dysregulated apoptotic signalling pathways enable cancer cells to avoid programmed cell death, thereby facilitating cancer cell proliferation, resistance to chemotherapy and tumour recurrence (Hou et al., 2018). VGCCs have been implicated in the regulation of programmed cell death in cancer cells, and their blockade has been shown to induce apoptosis. For example, the $Ca_v3.x$ channel antagonist BK10040 induced apoptosis in human lung (A549) and pancreatic (Mia PaCa-2) cancer cell lines (Choi et al., 2014). Similar results were obtained with the $Ca_v3.x$ channel antagonist KYS05090 in A549 cells (Rim et al., 2014) and in A2780 and A2780-T ovarian cancer cells (Jang et al., 2013). The $Ca_v3.x$ blocker kurtoxin significantly reduced the viability of melanoma cell lines while having no effect on control melanocytes; in contrast, blockers of other Ca_v subtypes had limited or no effect (Das et al., 2012). $Ca_v3.x$ blockade was also shown to inhibit autophagy and promote apoptosis via caspase activation in malignant melanoma, potentially via ER stress (Das et al., 2013). Similarly, selective inhibition or knockdown of $Ca_v3.1$ channels slowed proliferation and induced apoptosis in HCT116 cells expressing p53 (Dziegielewska et al., 2014). Interestingly, these antiproliferative effects were shown to be due to phosphorylation of p53 and activation of the p38/MAPK cascade, suggesting that overexpression of $Ca_v3.1$ channels acts to dampen the p53 tumour suppressor pathway. Collectively, these studies position the $Ca_v3.x$ subfamily as regulators of apoptosis in various cancers.

VGCCs have also been shown to regulate the sensitivity of cancer cells to chemotherapy. Knockdown of *CACNA1G* or *CACNA1H* (but not *CACNA1C*) or treatment with mibefradil induces apoptosis in glioblastoma cell lines and sensitised cells to ionising radiation. This effect was ascribed to inhibition of the mTORC2/Akt signalling pathway, specifically via decreased phosphorylation of Akt, Rictor and Bad, ultimately resulting in activation of caspases (Valerie et al., 2013). Disruption of G1/S transition via mibefradil inhibition of Ca^{2+} entry increases the efficacy of temozolomide in xenograft models of glioblastoma multiforme (Keir et al., 2013). Inhibition of $Ca_v3.2$ in tumour initiating glioblastoma stem-like cells leads to increased sensitivity to chemotherapy, suppression of AKT/mTOR pathways and oncogene expression, stimulation of proapoptotic pathways, and expression of tumour suppressor genes (Zhang et al., 2017). Similar results have been observed when combining mibefradil with carboplatin treatment in models of ovarian cancer (Dziegielewska et al., 2016). Interestingly, $Ca_v3.1$ expression and autophagy are increased in temozolomide-resistant glioblastoma multiforme cells (Visa et al., 2019). Furthermore, autophagy- and apoptosis-mediated cell death in A549 lung cancer cells was induced by the $Ca_v3.x$ inhibitor KYS05090, although this was not directly related to changes in intracellular $[Ca^{2+}]$ (Rim et al., 2014). In contrast, $Ca_v3.1$ knockdown was shown to inhibit apoptosis induced by cyclophosphamide (Ohkubo, 2012). Similar to their role in proliferation, VGCCs therefore likely have cancer subtype-specific roles in the regulation of apoptosis resistance and may act via mechanisms other than Ca^{2+} conductance.

The $\alpha_2\delta$ subunit also regulates cell death and resistance to therapy in lung cancer. Small cell lung carcinoma cells expressing $\alpha_2\delta_1$ demonstrate cancer stem cell-like properties (self-renewal, tumorigenicity, differentiation potential, increased expression of cancer stem cell-related genes), and an increased proportion of $\alpha_2\delta_1$-positive cells corresponds to increased resistance to chemotherapy (Yu et al., 2018). $\alpha_2\delta_1$ also regulates resistance to radiotherapy in non-small cell lung cancer. *CACNA2D1* is highly expressed by A549 cells, and its knockdown enhances sensitivity to radiotherapy; the opposite effect was observed when *CACNA2D1* was overexpressed in PC9 and H1975 cells that normally have low or no *CACNA2D1* expression. The authors attributed this to $a_2\delta_1$ conferring enhanced capacity for DNA damage repair; cells overexpressing *CACNA2D1* had higher levels of activated ATM serine/threonine kinase (a DNA damage repair initiator) and exhibited faster decreases in γ H2A histone family member X (γH2AX) foci (a marker of DNA double-strand breaks) following treatment with radiation (Sui et al., 2018). On the other hand, low *CACNA2D3* expression is associated with poor response to platinum-based therapy in oesophageal squamous cell carcinoma; overexpression of *CACNA2D3* increased sensitivity to cisplatin by increasing intracellular $[Ca^{2+}]$ and inhibiting PI3K and Akt phosphorylation, while *CACNA2D3* knockdown induced resistance (Nie et al., 2019). Based on these results, it is tempting to speculate that $\alpha_2\delta$ subunit expression levels may be a predictive biomarker of treatment response for certain cancers.

Differentiation

Dedifferentiation of cancer cells to an immature phenotype bearing little resemblance to their tissue of origin is a hallmark of many tumours, with poor differentiation associated with aggressive tumour cell behaviour. Similarly, transdifferentiation is a process by which tumour cells take on the features of different cell lineages, which can influence tumour progression. For example, neuroendocrine differentiation (NED)

in prostate cancer describes a process where tumour cells differentiate to a neuroendocrine-like phenotype that lacks androgen receptor expression (Hu et al., 2015). Early-stage prostate cancer is dependent on androgen signalling; however, later stages associated with NED exhibit a switch to an androgen-refractory state that is correlated with poor prognosis. Evidence suggests that VGCCs, in particular $Ca_v3.2$, might play a role in the morphological changes associated with NED and may contribute to the associated poor prognosis in prostate cancer. Androgen-sensitive LNCaP prostate cancer cells overexpress functional $Ca_v3.2$ following NED induced by treatment with IL-6 (Weaver et al., 2015a), membrane permeable cAMP analogues or steroid deprivation (Mariot et al., 2002). $Ca_v3.2$ activity following NED results in a phenotypic change that alters secretion from prostate cancer cells, potentially promoting the release of mitogenic factors (Gackière et al., 2008; Fukami et al., 2015). Furthermore, treatment with the $Ca_v3.x$ inhibitor NCC 55–0396 blocked the IL-6/cAMP analogue-induced neurite-like outgrowth typically consistent with NED (Weaver et al., 2015a). Treatment with sodium butyrate, a naturally occurring by-product of dietary fibre fermentation and histone deacetylase inhibitor, induces differentiation and $Ca_v3.2$ expression in LNCaP cells (Weaver et al., 2015b). In this study, blockade of $Ca_v3.2$ with Ni^{2+} attenuated outgrowth of neurite-like processes associated with differentiation and decreased cell viability. These studies implicate $Ca_v3.2$ in prostate cancer progression following differentiation to an androgen-independent stage. More work needs to be done to evaluate whether similar Ca_v-dependent mechanisms operate in other tumour types.

Migration and Invasion

Metastatic spread is the major cause of cancer-related mortality, and is characterised by breakdown of extracellular matrix, epithelial-mesenchymal transition (EMT, whereby epithelial cells transdifferentiate to a mesenchymal-like phenotype characterised by loss of cell polarity and cell adhesion and a gain of migratory behaviours), invasion of surrounding healthy tissues and ultimately metastasis to distant sites (Ribatti et al., 2020). Numerous studies have linked VGCCs to invasion in various cancer subtypes. For example, $Ca_v1.3$ knockdown and nifedipine treatment inhibited migration in Ishikawa (Hao et al., 2015) and Hec-1A (Bao et al., 2012) endometrial carcinoma cells, respectively. In non-small cell lung cancer cells, $\alpha_2\delta_1$ regulates in vitro spheroid-forming capacity, which is a measure of self-renewal ability (Sui et al., 2018). Interestingly, $\alpha_2\delta_1$ is expressed specifically within the infiltrating front of advanced-stage epithelial ovarian tumours (Yu et al., 2016). Similarly, $\alpha_2\delta_1$ regulates Ca^{2+} influx in hepatocellular carcinoma tumour-initiating cells (HCC TICs) at the tumour margins, where it correlates with poor outcome (Zhao et al., 2013). In this study, *CACNA2D1* knockdown inhibited xenograft tumour formation, while its overexpression enhanced sphere formation, and targeting $\alpha_2\delta_1$ with a monoclonal antibody eliminated TICs in vivo. HCC TICs also exhibited upregulation of inflammation-related cytokine C-X-C motif chemokine 11 (CXCL11), which induced expression of stem cell-related genes and promoted the acquisition of tumorigenic and self-renewal properties and chemoresistance (Zhang et al., 2019). These findings suggest that $\alpha_2\delta_1$ may have a role in invasive behaviour and seeding of metastatic sites.

On the other hand, $\alpha_2\delta_3$ appears to act as a metastasis suppressor in nasopharyngeal carcinoma by opposing invasive behaviour. In nasopharyngeal carcinoma cells, *CACNA2D3* overexpression mediates an increase in intracellular $[Ca^{2+}]$ and, via Nemo-like kinase (NLK) activation, antagonises Wnt signalling-mediated proliferation (both in vitro and in vivo), snail-mediated EMT and matrix metallopeptidase 7-mediated invasion (Wong et al., 2013). Moreover, in this study, silencing of *CACNA2D3* by siRNA inhibited its tumour-suppressive effects. Similar results have since been reported in glioma cell lines (Jin et al., 2017).

Targeting VGCCs in Cancer: Evidence from Clinical Studies

The existence of clinically available drugs that target the $\alpha1$ subunit (such as dihydropyridine drugs) or auxiliary subunits (such as the gabapentinoid $\alpha_2\delta$ ligands) makes VGCCs attractive therapeutic candidates for either novel therapeutics or drug repurposing in cancer. In particular, drugs blocking $Ca_v3.x$ hold promise due to their ability to inhibit proliferation and induce apoptosis, cell-cycle arrest and sensitivity to chemotherapy (Barceló et al., 2020). The use of the $Ca_v3.x$ inhibitor mibefradil to inhibit G1/S phase transition in gliomas, thereby increasing sensitivity to chemotherapy and radiotherapy, has shown efficacy in in vivo models (Dziegielewska et al., 2016), and combination treatment with mibefradil and either temozolomide or radiotherapy has been found to be well tolerated in Phase I studies (Holdhoff et al., 2017; Lester-Coll et al., 2018). Thus, utilising $Ca_v3.x$ blockers as chemotherapy/radiotherapy sensitisers is an attractive therapeutic strategy. However, further research is needed to determine the clinical efficacy of this approach in glioma.

Several epidemiological studies have also explored whether treatment with VGCC inhibitors confers improved outcome in patients with cancer, though these have yielded conflicting results. While numerous studies have linked Ca^{2+} channel blocker use with a decreased risk of advanced-stage prostate cancer (Fitzpatrick et al., 2001; Debes et al., 2004; Kemppainen et al., 2011; Geybels et al., 2017), other studies found no association with prostate cancer risk, aggressiveness or survival (Rodriguez et al., 2009; Poch et al., 2013). Similarly equivocal results have been observed in breast cancer, with conflicting studies separately reporting that VGCC inhibitors confer an increase (Saltzman et al., 2013), reduction (Fitzpatrick et al., 1997; Fryzek et al., 2006) or no difference (Michels et al., 1998; Li et al., 2013a; Bergman et al., 2014; Chen et al., 2014a; Devore et al., 2015) in breast cancer risk. Furthermore, studies in patients with colorectal cancer revealed no associations between VGCC-targeting drugs and cancer risk (Michels et al.,

1998; Boudreau et al., 2008). Further studies are required to reconcile these conflicting results, particularly in light of the promising results from both in vitro and translational in vivo studies described in the previous sections. However, as discussed above, VGCC α_1 and auxiliary subunit expression is commonly associated with certain cancer subtypes and stages. Since cancer is a heterogeneous disease, stratification of patients and subsequent personalised targeting of medication could reveal benefits in subpopulations of patients within clinical cohorts. A further limitation of retrospective epidemiological studies is that patients taking VGCC-targeting medications present with comorbidities and thus are likely to be unrepresentative of the wider patient population. Importantly, future clinical research should assess whether VGCC blocker use is associated with reduced cancer risk or improved patient outcome in other cancers beyond those focused on in the studies described above.

Concluding Remarks

It is clear that VGCC expression is significantly altered in cancer, and that functional expression of VGCCs has diverse effects on cancer formation, progression and resistance to chemotherapy. However, in many cases, the precise molecular mechanisms involved, and whether these are regulated by Ca^{2+} flux via the $Ca_v\alpha_1$ subunit or via non-canonical mechanisms such as C-terminus cleavage (Gomez-Ospina et al., 2006) remain to be determined. Similarly, VGCC auxiliary subunits also appear to play important and diverse roles in tumour progression, though it is not yet clear whether they exert these effects via regulation of the α_1 subunit or by secondary mechanisms. Little is known about how the combined expression of different subunits influences cancer progression and which subunits take prominence in this case. Nevertheless, it seems that the specific role of both $Ca_v\alpha_1$ and auxiliary subunits in cancer is dependent on the tumour subtype and the stage of malignancy. Future research should aim to further delineate the mechanisms underlying these observations and to explore the role of

VGCCs in other cancers where this evidence is lacking. Moreover, the potential of VGCC expression in cancer as a means to risk-stratify patients and of VGCCs as druggable targets should be thoroughly investigated. Given that VGCC-targeting drugs are clinically available, the potential for repurposing existing medications is a promising area for future research.

References

Amhimmid Badr, S., Waheeb Fahmi, M., Mahmoud Nomir, M., & Mohammad El-Shishtawy, M. (2018). Calcium channel α2δ1 subunit as a novel biomarker for diagnosis of hepatocellular carcinoma. *Cancer Biology & Medicine, 15*, 52–60.

Angeloni, D., Duh, F. M., Wei, M. F., Johnson, B. E., & Lerman, M. I. (2001). A G-to-A single nucleotide polymorphism in intron 2 of the human CACNA2D2 gene that maps at 3p21.3. *Molecular and Cellular Probes, 15*, 125–127.

Bao, X.-X., Xie, B.-S., Li, Q., Li, X.-P., Wei, L.-H., & Wang, J.-L. (2012). Nifedipine induced autophagy through Beclin1 and mTOR pathway in endometrial carcinoma cells. *Chinese Medical Journal, 125*, 3120–3126.

Barbado M, Fablet K, Ronjat M, De Waard M (2009) Gene regulation by voltage-dependent calcium channels. Biochimica et Biophysica Acta (BBA) - Molecular Cell Research 1793:1096–1104 https://doi.org/10.1016/j.bbamcr.2009.02.004.

Barbault, A., Costa, F. P., Bottger, B., Munden, R. F., Bomholt, F., Kuster, N., & Pasche, B. (2009). Amplitude-modulated electromagnetic fields for the treatment of cancer: Discovery of tumor-specific frequencies and assessment of a novel therapeutic approach. *Journal of Experimental & Clinical Cancer Research, 28*, 51.

Barceló C, Sisó P, Maiques O, de la Rosa I, Martí RM, Macià A (2020) T-type calcium channels: A potential novel target in melanoma Cancers 12:391 https://doi.org/10.3390/cancers12020391.

Basson, M. D., Zeng, B., Downey, C., Sirivelu, M. P., & Tepe, J. J. (2015). Increased extracellular pressure stimulates tumor proliferation by a mechanosensitive calcium channel and PKC-β. *Molecular Oncology, 9*, 513–526.

Ben-Porath, I., Thomson, M. W., Carey, V. J., Ge, R., Bell, G. W., Regev, A., & Weinberg, R. A. (2008). An embryonic stem cell-like gene expression signature in poorly differentiated aggressive human tumors. *Nature Genetics, 40*, 499–507.

Bergman, G. J., Khan, S., Danielsson, B., & Borg, N. (2014). Breast cancer risk and use of calcium channel blockers using Swedish population registries. *JAMA Internal Medicine, 174*, 1700–1701.

Berridge, M. J., Lipp, P., & Bootman, M. D. (2000). The versatility and universality of calcium signalling. *Nature Reviews. Molecular Cell Biology, 1*, 11–21.

Bieging, K. T., Mello, S. S., & Attardi, L. D. (2014). Unravelling mechanisms of p53-mediated tumour suppression. *Nature Reviews. Cancer, 14*, 359–370.

Boormans, J. L., Korsten, H., Ziel-van der Made, A. J. C., van Leenders, G. J. L. H., de Vos, C. V., Jenster, G., & Trapman, J. (2013). Identification of TDRD1 as a direct target gene of ERG in primary prostate cancer. *International Journal of Cancer, 133*, 335–345.

Borowiec, A.-S., Bidaux, G., Pigat, N., Goffin, V., Bernichtein, S., & Capiod, T. (2014). Calcium channels, external calcium concentration and cell proliferation. *European Journal of Pharmacology, 739*, 19–25.

Boudreau DM, Koehler E, Rulyak SJ, Haneuse S, Harrison R, Mandelson MT (2008) Cardiovascular medication use and risk for colorectal cancer. Cancer Epidemiology, Biomarkers & Prevention 17:3076–3080 https://doi.org/10.1158/1055-9965.epi-08-0095.

Bruce JIE, James AD (2020) Targeting the calcium signalling machinery in cancer Cancers 12:2351 https://doi.org/10.3390/cancers12092351.

Buchanan, P. J., & McCloskey, K. D. (2016). Ca channels and cancer: Canonical functions indicate benefits of repurposed drugs as cancer therapeutics. *European Biophysics Journal, 45*, 621–633.

Carboni, G. L., Gao, B., Nishizaki, M., Xu, K., Minna, J. D., Roth, J. A., & Ji, L. (2003). CACNA2D2-mediated apoptosis in NSCLC cells is associated with alterations of the intracellular calcium signaling and disruption of mitochondria membrane integrity. *Oncogene, 22*, 615–626.

Castro, M., Grau, L., Puerta, P., Gimenez, L., Venditti, J., Quadrelli, S., & Sánchez-Carbayo, M. (2010). Multiplexed methylation profiles of tumor suppressor genes and clinical outcome in lung cancer. *Journal of Translational Medicine, 8*, 86.

Catterall, W. A., Perez-Reyes, E., Snutch, T. P., & Striessnig, J. (2005). International Union of Pharmacology. XLVIII. Nomenclature and structure-function relationships of voltage-gated calcium channels. *Pharmacological Reviews, 57*, 411–425.

Catterall, W. A., Lenaeus, M. J., & Gamal El-Din, T. M. (2020). Structure and pharmacology of voltage-gated sodium and calcium channels. *Annual Review of Pharmacology and Toxicology, 60*, 133–154.

Chen, Q., Zhang, Q., Zhong, F., Guo, S., Jin, Z., Shi, W., Chen, C., & He, J. (2014a). Association between calcium channel blockers and breast cancer: A meta-analysis of observational studies. *Pharmacoepidemiology and Drug Safety, 23*, 711–718.

Chen, R., Khatri, P., Mazur, P. K., Polin, M., Zheng, Y., Vaka, D., Hoang, C. D., Shrager, J., Xu, Y., Vicent, S., Butte, A. J., & Sweet-Cordero, E. A. (2014b). A meta-analysis of lung cancer gene expression identifies PTK7 as a survival gene in lung adenocarcinoma. *Cancer Research, 74*, 2892–2902.

Chen, R., Zeng, X., Zhang, R., Huang, J., Kuang, X., Yang, J., Liu, J., Tawfik, O., Thrasher, J. B., & Li, B. (2014c). Cav1.3 channel α1D protein is overexpressed and modulates androgen receptor transactivation in prostate cancers. *Urologic Oncology, 32*, 524–536.

Chen, M., Rothman, N., Ye, Y., Gu, J., Scheet, P. A., Huang, M., Chang, D. W., Dinney, C. P., Silverman, D. T., Figueroa, J. D., Chanock, S. J., & Wu, X. (2016). Pathway analysis of bladder cancer genome-wide association study identifies novel pathways involved in bladder cancer development. *Genes & Cancer, 7*, 229–239.

Chiou, W.-F. (2006). Effect of Abeta exposure on the mRNA expression patterns of voltage-sensitive calcium channel alpha 1 subunits (alpha 1A-alpha 1D) in human SK-N-SH neuroblastoma. *Neurochemistry International, 49*, 256–261.

Choi M et al. (2011) K channel mutations in adrenal aldosterone-producing adenomas and hereditary hypertension. Science 331:768–772 https://doi.org/10.1126/science.1198785.

Choi, D. L., Jang, S. J., Cho, S., Choi, H.-E., Rim, H.-K., Lee, K.-T., & Lee, J. Y. (2014). Inhibition of cellular proliferation and induction of apoptosis in human lung adenocarcinoma A549 cells by T-type calcium channel antagonist. *Bioorganic & Medicinal Chemistry Letters, 24*, 1565–1570.

Cooper, G. M. (2000). *The cell: A molecular approach.* Sinauer Associates.

Cromer, M. K., Choi, M., Nelson-Williams, C., Fonseca, A. L., Kunstman, J. W., Korah, R. M., Overton, J. D., Mane, S., Kenney, B., Malchoff, C. D., Stalberg, P., Akerström, G., Westin, G., Hellman, P., Carling, T., Björklund, P., & Lifton, R. P. (2015). Neomorphic effects of recurrent somatic mutations in Yin Yang 1 in insulin-producing adenomas. *Proceedings of the National Academy of Sciences of the United States of America, 112*, 4062–4067.

Das, A., Pushparaj, C., Bahí, N., Sorolla, A., Herreros, J., Pamplona, R., Vilella, R., Matias-Guiu, X., Martí, R. M., & Cantí, C. (2012). Functional expression of voltage-gated calcium channels in human melanoma. *Pigment Cell & Melanoma Research, 25*, 200–212.

Das, A., Pushparaj, C., Herreros, J., Nager, M., Vilella, R., Portero, M., Pamplona, R., Matias-Guiu, X., Martí, R. M., & Cantí, C. (2013). T-type calcium channel blockers inhibit autophagy and promote apoptosis of malignant melanoma cells. *Pigment Cell & Melanoma Research, 26*, 874–885.

Davenport, B., Li, Y., Heizer, J. W., Schmitz, C., & Perraud, A.-L. (2015). Signature channels of excitability no more: L-type channels in immune cells. *Frontiers in Immunology, 6*, 375.

Davis, S. J., et al. (2015). Enhanced GAB2 expression is associated with improved survival in high-grade serous ovarian cancer and sensitivity to PI3K inhibition. *Molecular Cancer Therapeutics, 14*, 1495–1503.

De Preter, K., Vandesompele, J., Heimann, P., Yigit, N., Beckman, S., Schramm, A., Eggert, A., Stallings, R. L., Benoit, Y., Renard, M., De Paepe, A., Laureys,

G., Påhlman, S., & Speleman, F. (2006). Human fetal neuroblast and neuroblastoma transcriptome analysis confirms neuroblast origin and highlights neuroblastoma candidate genes. *Genome Biology, 7*, R84.

Debes, J. D., Roberts, R. O., Jacobson, D. J., Girman, C. J., Lieber, M. M., Tindall, D. J., & Jacobsen, S. J. (2004). Inverse association between prostate cancer and the use of calcium channel blockers. *Cancer Epidemiology, Biomarkers & Prevention, 13*, 255–259.

Devore, E. E., Kim, S., Ramin, C. A., Wegrzyn, L. R., Massa, J., Holmes, M. D., Michels, K. B., Tamimi, R. M., Forman, J. P., & Schernhammer, E. S. (2015). Antihypertensive medication use and incident breast cancer in women. *Breast Cancer Research and Treatment, 150*, 219–229.

Dziegielewska, B., Brautigan, D. L., Larner, J. M., & Dziegielewski, J. (2014). T-type Ca²⁺ channel inhibition induces p53-dependent cell growth arrest and apoptosis through activation of p38-MAPK in colon cancer cells. *Molecular Cancer Research, 12*, 348–358.

Dziegielewska B, Casarez EV, Yang WZ, Gray LS, Dziegielewski J, Slack-Davis JK (2016) T-type Ca2 channel inhibition sensitizes ovarian cancer to carboplatin. Molecular Cancer Therapeutics 15:460–470 https://doi.org/10.1158/1535-7163.mct-15-0456.

Fitzpatrick, A. L., Daling, J. R., Furberg, C. D., Kronmal, R. A., & Weissfeld, J. L. (1997). Use of calcium channel blockers and breast carcinoma risk in postmenopausal women. *Cancer, 80*, 1438–1447.

Fitzpatrick, A. L., Daling, J. R., Furberg, C. D., Kronmal, R. A., & Weissfeld, J. L. (2001). Hypertension, heart rate, use of antihypertensives, and incident prostate cancer. *Annals of Epidemiology, 11*, 534–542.

Fryzek, J. P., Poulsen, A. H., Lipworth, L., Pedersen, L., Nørgaard, M., McLaughlin, J. K., & Friis, S. (2006). A cohort study of antihypertensive medication use and breast cancer among Danish women. *Breast Cancer Research and Treatment, 97*, 231–236.

Fukami, K., Sekiguchi, F., Yasukawa, M., Asano, E., Kasamatsu, R., Ueda, M., Yoshida, S., & Kawabata, A. (2015). Functional upregulation of the H2S/Cav3.2 channel pathway accelerates secretory function in neuroendocrine-differentiated human prostate cancer cells. *Biochemical Pharmacology, 97*, 300–309.

Gackière, F., Bidaux, G., Delcourt, P., Van Coppenolle, F., Katsogiannou, M., Dewailly, E., Bavencoffe, A., Van Chuoï-Mariot, M. T., Mauroy, B., Prevarskaya, N., & Mariot, P. (2008). CaV3.2 T-type calcium channels are involved in calcium-dependent secretion of neuroendocrine prostate cancer cells. *The Journal of Biological Chemistry, 283*, 10162–10173.

Gackière, F., Warnier, M., Katsogiannou, M., Derouiche, S., Delcourt, P., Dewailly, E., Slomianny, C., Humez, S., Prevarskaya, N., Roudbaraki, M., & Mariot, P. (2013). Functional coupling between large-conductance potassium channels and Cav3.2 voltage-dependent calcium channels participates in prostate cancer cell growth. *Biology Open, 2*, 941–951.

Gao, P., He, M., Zhang, C., & Geng, C. (2018). Integrated analysis of gene expression signatures associated with colon cancer from three datasets. *Gene, 654*, 95–102.

Geybels, M. S., Alumkal, J. J., Luedeke, M., Rinckleb, A., Zhao, S., Shui, I. M., Bibikova, M., Klotzle, B., van den Brandt, P. A., Ostrander, E. A., Fan, J.-B., Feng, Z., Maier, C., & Stanford, J. L. (2015). Epigenomic profiling of prostate cancer identifies differentially methylated genes in TMPRSS2:ERG fusion-positive versus fusion-negative tumors. *Clinical Epigenetics, 7*, 128.

Geybels, M. S., McCloskey, K. D., Mills, I. G., & Stanford, J. L. (2017). Calcium channel blocker use and risk of prostate cancer by TMPRSS2:ERG gene fusion status. *Prostate, 77*, 282–290.

Ghosh, S., Ghosh, A., Maiti, G. P., Alam, N., Roy, A., Roy, B., Roychoudhury, S., & Panda, C. K. (2008). Alterations of 3p21.31 tumor suppressor genes in head and neck squamous cell carcinoma: Correlation with progression and prognosis. *International Journal of Cancer, 123*, 2594–2604.

Gomez-Ospina N, Tsuruta F, Barreto-Chang O, Hu L, Dolmetsch R (2006) The C terminus of the L-type voltage-gated calcium channel CaV1.2 encodes a transcription factor. Cell 127:591–606 https://doi.org/10.1016/j.cell.2006.10.017.

Grassi, C., D'Ascenzo, M., Torsello, A., Martinotti, G., Wolf, F., Cittadini, A., & Azzena, G. B. (2004). Effects of 50 Hz electromagnetic fields on voltage-gated Ca²⁺ channels and their role in modulation of neuroendocrine cell proliferation and death. *Cell Calcium, 35*, 307–315.

Gray, L. S., Perez-Reyes, E., Gomora, J. C., Haverstick, D. M., Shattock, M., McLatchie, L., Harper, J., Brooks, G., Heady, T., & Macdonald, T. L. (2004). The role of voltage gated T-type Ca²⁺ channel isoforms in mediating "capacitative" Ca²⁺ entry in cancer cells. *Cell Calcium, 36*, 489–497.

Grössinger, E. M., Kang, M., Bouchareychas, L., Sarin, R., Haudenschild, D. R., Borodinsky, L. N., & Adamopoulos, I. E. (2018). Ca-dependent regulation of NFATc1 via KCa3.1 in inflammatory osteoclastogenesis. *Journal of Immunology, 200*, 749–757.

Hanahan D, Weinberg RA (2000) The hallmarks of cancer Cell 100:57–70 https://doi.org/10.1016/s0092-8674(00)81683-9.

Hanahan D, Weinberg RA (2011) Hallmarks of cancer: The next generation Cell 144:646–674 https://doi.org/10.1016/j.cell.2011.02.013.

Hao J, Bao X, Jin B, Wang X, Mao Z, Li X, Wei L, Shen D, Wang J (2015) Ca 2 channel subunit α 1D promotes proliferation and migration of endometrial cancer cells mediated by 17β-estradiol via the G protein-coupled estrogen receptor. The FASEB Journal 29:2883–2893 https://doi.org/10.1096/fj.14-265603.

Haworth, A. S., & Brackenbury, W. J. (2019). Emerging roles for multifunctional ion channel auxiliary subunits in cancer. *Cell Calcium, 80*, 125–140.

Ho, C.-M., Huang, C.-J., Huang, C.-Y., Wu, Y.-Y., Chang, S.-F., & Cheng, W.-F. (2012). Promoter methylation status of HIN-1 associated with outcomes of ovarian clear cell adenocarcinoma. *Molecular Cancer, 11*, 53.

Holdhoff, M., Ye, X., Supko, J. G., Nabors, L. B., Desai, A. S., Walbert, T., Lesser, G. J., Read, W. L., Lieberman, F. S., Lodge, M. A., Leal, J., Fisher, J. D., Desideri, S., Grossman, S. A., Wahl, R. L., & Schiff, D. (2017). Timed sequential therapy of the selective T-type calcium channel blocker mibefradil and temozolomide in patients with recurrent high-grade gliomas. *Neuro-Oncology, 19*, 845–852.

Hou Z-Y, Tong X-P, Peng Y-B, Zhang B-K, Yan M (2018) Broad targeting of triptolide to resistance and sensitization for cancer therapy. Biomedicine & Pharmacotherapy 104:771–780 https://doi.org/10.1016/j.biopha.2018.05.088.

Hu, C.-D., Choo, R., & Huang, J. (2015). Neuroendocrine differentiation in prostate cancer: A mechanism of radioresistance and treatment failure. *Frontiers in Oncology, 5*, 90.

Huang, W., Jin, Y., Yuan, Y., Bai, C., Wu, Y., Zhu, H., & Lu, S. (2014). Validation and target gene screening of hsa-miR-205 in lung squamous cell carcinoma. *Chinese Medical Journal, 127*, 272–278.

Huang, W., Lu, C., Wu, Y., Ouyang, S., & Chen, Y. (2015). T-type calcium channel antagonists, mibefradil and NNC-55-0396 inhibit cell proliferation and induce cell apoptosis in leukemia cell lines. *Journal of Experimental & Clinical Cancer Research, 34*, 54.

Hung, J., Kishimoto, Y., Sugio, K., Virmani, A., McIntire, D. D., Minna, J. D., & Gazdar, A. F. (1995). Allele-specific chromosome 3p deletions occur at an early stage in the pathogenesis of lung carcinoma. *JAMA, 273*, 1908.

Jang, S. J., Choi, H. W., Choi, D. L., Cho, S., Rim, H.-K., Choi, H.-E., Kim, K.-S., Huang, M., Rhim, H., Lee, K.-T., & Lee, J. Y. (2013). In vitro cytotoxicity on human ovarian cancer cells by T-type calcium channel blockers. *Bioorganic & Medicinal Chemistry Letters, 23*, 6656–6662.

Jhavar, S., et al. (2009). Integration of ERG gene mapping and gene-expression profiling identifies distinct categories of human prostate cancer. *BJU International, 103*, 1256–1269.

Jimenez H et al. (2019) Tumour-specific amplitude-modulated radiofrequency electromagnetic fields induce differentiation of hepatocellular carcinoma via targeting Cav3.2 T-type voltage-gated calcium channels and Ca2 influx. eBioMedicine 44:209–224 https://doi.org/10.1016/j.ebiom.2019.05.034.

Jin, Y., Cui, D., Ren, J., Wang, K., Zeng, T., & Gao, L. (2017). CACNA2D3 is downregulated in gliomas and functions as a tumor suppressor. *Molecular Carcinogenesis, 56*, 945–959.

Jung, S. Y., Lee, S. H., Kang, H. B., Park, H. A., Chang, S. K., Kim, J., Choo, D. J., Oh, C. R., Kim, Y. D., Seo, J. H., Lee, K.-T., & Lee, J. Y. (2010). Antitumor activity of 3,4-dihydroquinazoline dihydrochloride in A549 xenograft nude mice. *Bioorganic & Medicinal Chemistry Letters, 20*, 6633–6636.

Jung, E., Alfonso, J., Monyer, H., Wick, W., & Winkler, F. (2020). Neuronal signatures in cancer. *International Journal of Cancer, 147*, 3281–3291.

Kan, Z., et al. (2013). Whole-genome sequencing identifies recurrent mutations in hepatocellular carcinoma. *Genome Research, 23*, 1422–1433.

Kang, H. B., Rim, H.-K., Park, J. Y., Choi, H. W., Choi, D. L., Seo, J.-H., Chung, K.-S., Huh, G., Kim, J., Choo, D. J., Lee, K.-T., & Lee, J. Y. (2012). In vivo evaluation of oral anti-tumoral effect of 3,4-dihydroquinazoline derivative on solid tumor. *Bioorganic & Medicinal Chemistry Letters, 22*, 1198–1201.

Kanwar, N., Carmine-Simmen, K., Nair, R., Wang, C., Moghadas-Jafari, S., Blaser, H., Tran-Thanh, D., Wang, D., Wang, P., Wang, J., Pasculescu, A., Datti, A., Mak, T., Lewis, J. D., & Done, S. J. (2020). Amplification of a calcium channel subunit CACNG4 increases breast cancer metastasis. *eBioMedicine, 52*, 102646.

Keir ST, Friedman HS, Reardon DA, Bigner DD, Gray LA (2013) Mibefradil, a novel therapy for glioblastoma multiforme: cell cycle synchronization and interlaced therapy in a murine model. Journal of Neuro-Oncology 111:97–102 https://doi.org/10.1007/s11060-012-0995-0.

Kemppainen KJ, Tammela TLJ, Auvinen A, Murtola TJ (2011) The association between antihypertensive drug use and incidence of prostate cancer in Finland: A population-based case–control study. Cancer Causes & Control 22:1445–1452 https://doi.org/10.1007/s10552-011-9819-3.

Kim, K. H., Kim, D., Park, J. Y., Jung, H. J., Cho, Y.-H., Kim, H. K., Han, J., Choi, K.-Y., & Kwon, H. J. (2015). NNC 55-0396, a T-type Ca²⁺ channel inhibitor, inhibits angiogenesis via suppression of hypoxia-inducible factor-1α signal transduction. *Journal of Molecular Medicine, 93*, 499–509.

Kito, M., Maehara, M., & Watanabe, K. (1999). Three types of voltage-dependent calcium currents developing in cultured human neuroblastoma cells. *Nagoya Journal of Medical Science, 62*, 39–45.

Klaus A, Birchmeier W (2008) Wnt signalling and its impact on development and cancer. Nature Reviews Cancer 8:387–398 https://doi.org/10.1038/nrc2389.

Kong, X., Li, M., Shao, K., Yang, Y., Wang, Q., & Cai, M. (2020). Progesterone induces cell apoptosis via the CACNA2D3/Ca²⁺/p38 MAPK pathway in endometrial cancer. *Oncology Reports, 43*, 121–132.

Kuga, T., Kobayashi, S., Hirakawa, Y., Kanaide, H., & Takeshita, A. (1996). Cell cycle – Dependent expression of L- and T-type Ca²⁺ currents in rat aortic smooth muscle cells in primary culture. *Circulation Research, 79*, 14–19.

Kumar, R. D., Searleman, A. C., Swamidass, S. J., Griffith, O. L., & Bose, R. (2015). Statistically identifying tumor suppressors and oncogenes from pan-cancer genome-sequencing data. *Bioinformatics, 31*, 3561–3568.

Latour, I., Louw, D. F., Beedle, A. M., Hamid, J., Sutherland, G. R., & Zamponi, G. W. (2004). Expression of T-type calcium channel splice variants in human glioma. *Glia, 48*, 112–119.

Lerman, M. I., & Minna, J. D. (2000). The 630-kb lung cancer homozygous deletion region on human chromosome 3p21.3: Identification and evaluation of the resident candidate tumor suppressor genes. The International Lung Cancer Chromosome 3p21.3 Tumor Suppressor Gene Consortium. *Cancer Research, 60*, 6116–6133.

Lester-Coll NH, Supko JG, Kluytenaar J, Pavlik KF, Yu JB, Moliterno J, Piepmeier J, Becker K, Baehring JM, Huttner A, Vortmeyer A, Contessa JN, Ramani R, Lampert R, Yao X, Bindra R (2018) Mibefradil dihydrochloride with hypofractionated radiation for recurrent glioblastoma: A phase I dose expansion trial. Journal of Clinical Oncology 36:e14046–e14046 https://doi.org/10.1200/jco.2018.36.15_suppl.e14046.

Li, Y., Liu, S., Lu, F., Zhang, T., Chen, H., Wu, S., & Zhuang, H. (2009). A role of functional T-type Ca²⁺ channel in hepatocellular carcinoma cell proliferation. *Oncology Reports, 22*, 1229–1235.

Li, C. I., Daling, J. R., Tang, M.-T. C., Haugen, K. L., Porter, P. L., & Malone, K. E. (2013a). Use of antihypertensive medications and breast cancer risk among women aged 55 to 74 years. *JAMA Internal Medicine, 173*, 1629–1637.

Li, Y., Zhu, C.-L., Nie, C.-J., Li, J.-C., Zeng, T.-T., Zhou, J., Chen, J., Chen, K., Fu, L., Liu, H., Qin, Y., & Guan, X.-Y. (2013b). Investigation of tumor suppressing function of CACNA2D3 in esophageal squamous cell carcinoma. *PLoS One, 8*, e60027.

Lindskog, C., Fagerberg, L., Hallström, B., Edlund, K., Hellwig, B., Rahnenführer, J., Kampf, C., Uhlén, M., Pontén, F., & Micke, P. (2014). The lung-specific proteome defined by integration of transcriptomics and antibody-based profiling. *The FASEB Journal, 28*, 5184–5196.

Ling, J., Wu, X., Fu, Z., Tan, J., & Xu, Q. (2015). Systematic analysis of gene expression pattern in has-miR-197 over-expressed human uterine leiomyoma cells. *Biomedicine & Pharmacotherapy, 75*, 226–233.

Liu L, Li H, Cui Y, Li R, Meng F, Ye Z, Zhang X (2017) Calcium channel opening rather than the release of ATP causes the apoptosis of osteoblasts induced by overloaded mechanical stimulation. Cellular Physiology and Biochemistry 42:441–454 https://doi.org/10.1159/000477592.

Liu, P., Wang, Y., & Li, X. (2019). Targeting the untargetable KRAS in cancer therapy. *Acta Pharmaceutica Sinica B, 9*, 871–879.

Liu, D., Ma, X., Yang, F., Xiao, D., Jia, Y., & Wang, Y. (2020). Discovery and validation of methylated-differentially expressed genes in Helicobacter pylori-induced gastric cancer. *Cancer Gene Therapy, 27*, 473–485.

Lory, P., Bidaud, I., & Chemin, J. (2006). T-type calcium channels in differentiation and proliferation. *Cell Calcium, 40*, 135–146.

Lu, F., Chen, H., Zhou, C., Liu, S., Guo, M., Chen, P., Zhuang, H., Xie, D., & Wu, S. (2008). T-type Ca²⁺ channel expression in human esophageal carcinomas: A functional role in proliferation. *Cell Calcium, 43*, 49–58.

Maiques O, Macià A, Moreno S, Barceló C, Santacana M, Vea A, Herreros J, Gatius S, Ortega E, Valls J, Chen BJ, Llobet-Navas D, Matias-Guiu X, Cantí C, Marti RM (2017) Immunohistochemical analysis of T-type calcium channels in acquired melanocytic naevi and melanoma. British Journal of Dermatology 176:1247–1258 https://doi.org/10.1111/bjd.15121.

Makabe, T., Arai, E., Hirano, T., Ito, N., Fukamachi, Y., Takahashi, Y., Hirasawa, A., Yamagami, W., Susumu, N., Aoki, D., & Kanai, Y. (2019). Genome-wide DNA methylation profile of early-onset endometrial cancer: Its correlation with genetic aberrations and comparison with late-onset endometrial cancer. *Carcinogenesis, 40*, 611–623.

Mariot, P., Vanoverberghe, K., Lalevee, N., Rossier, M. F., & Prevarskaya, N. (2002). Overexpression of an alpha 1H (Cav3.2) T-type calcium channel during neuroendocrine differentiation of human prostate cancer cells. *The Journal of Biological Chemistry, 277*, 10824–10833.

Marques, R., Peres, C. G., Vaz, C. V., Gomes, I. M., Figueira, M. I., Cairrão, E., Verde, I., Maia, C. J., & Socorro, S. (2015). 5α-Dihydrotestosterone regulates the expression of L-type calcium channels and calcium-binding protein regucalcin in human breast cancer cells with suppression of cell growth. *Medical Oncology, 32*, 228.

Martinez, A., Walker, R. A., Shaw, J. A., Dearing, S. J., Maher, E. R., & Latif, F. (2001). Chromosome 3p allele loss in early invasive breast cancer: Detailed mapping and association with clinicopathological features. *Molecular Pathology, 54*, 300–306.

Mathieu, J., et al. (2011). HIF induces human embryonic stem cell markers in cancer cells. *Cancer Research, 71*, 4640–4652.

Mergler, S., Strauss, O., Strowski, M., Prada, J., Drost, A., Langrehr, J., Neuhaus, P., Wiedenmann, B., & Ploeckinger, U. (2005). Insulin-like growth factor-1 increases intracellular calcium concentration in human primary neuroendocrine pancreatic tumor cells and a pancreatic neuroendocrine tumor cell line (BON-1) via R-type Ca² channels and regulates chromogranin A secretion in BON-1 cells. *Neuroendocrinology, 82*, 87–102. https://doi.org/10.1159/000091008

Michels KB, Rosner BA, Walker AM, Stampfer MJ, Manson JE, Colditz GA, Hennekens CH, Willett WC (1998) Calcium channel blockers, cancer incidence, and cancer mortality in a cohort of U.S. Women. Cancer 83:2003–2007 https://doi.org/10.1002/(sici)1097-0142(19981101)83:9<2003::aid-cncr17>3.0.co;2-3.

Mitra R et al. (2011) Prediction of postoperative recurrence-free survival in non–small cell lung cancer by using an internationally validated gene expression model. Clinical Cancer Research 17:2934–2946 https://doi.org/10.1158/1078-0432.ccr-10-1803.

Mitra, S., Mazumder Indra, D., Basu, P. S., Mondal, R. K., Roy, A., Roychoudhury, S., & Panda, C. K. (2012). Alterations of RASSF1A in premalignant cervical lesions: Clinical and prognostic significance. *Molecular Carcinogenesis, 51*, 723–733.

Monteith GR, Prevarskaya N, Roberts-Thomson SJ (2017) The calcium–cancer signalling nexus. Nature Reviews Cancer 17:373–380 https://doi.org/10.1038/nrc.2017.18.

Natrajan, R., Little, S. E., Reis-Filho, J. S., Hing, L., Messahel, B., Grundy, P. E., Dome, J. S., Schneider, T., Vujanic, G. M., Pritchard-Jones, K., & Jones, C. (2006). Amplification and overexpression of CACNA1E correlates with relapse in favorable histology Wilms tumors. *Clinical Cancer Research, 12*, 7284–7293.

Neelands TR, King APJ, Macdonald RL (2000) Functional expression of L-, N-, P/Q-, and R-type calcium channels in the human NT2-N cell line. Journal of Neurophysiology 84:2933–2944 https://doi.org/10.1152/jn.2000.84.6.2933.

Nie, C., Qin, X., Li, X., Tian, B., Zhao, Y., Jin, Y., Li, Y., Wang, Q., Zeng, D., Hong, A., & Chen, X. (2019). CACNA2D3 enhances the chemosensitivity of esophageal squamous cell carcinoma to cisplatin via inducing Ca-mediated apoptosis and suppressing PI3K/Akt pathways. *Frontiers in Oncology, 9*, 185.

Ohkubo, T. (2012). T-type voltage-activated calcium channel Cav3.1, but not Cav3.2, is involved in the inhibition of proliferation and apoptosis in MCF-7 human breast cancer cells. *International Journal of Oncology.* https://doi.org/10.3892/ijo.2012.1422.

Omarini, C., Bettelli, S., Caprera, C., Manfredini, S., Caggia, F., Guaitoli, G., Moscetti, L., Toss, A., Cortesi, L., Kaleci, S., Maiorana, A., Cascinu, S., Conte, P. F., & Piacentini, F. (2018). Clinical and molecular predictors of long-term response in HER2 positive metastatic breast cancer patients. *Cancer Biology & Therapy, 19*, 879–886.

Onganer, P. U., Seckl, M. J., & Djamgoz, M. B. A. (2005). Neuronal characteristics of small-cell lung cancer. *British Journal of Cancer, 93*, 1197–1201.

Oshima S, Asai S, Seki N, Minemura C, Kinoshita T, Goto Y, Kikkawa N, Moriya S, Kasamatsu A, Hanazawa T, Uzawa K (2021) Identification of tumor suppressive genes regulated by miR-31-5p and miR-31-3p in head and neck squamous cell carcinoma. International Journal of Molecular Sciences 22:6199 https://doi.org/10.3390/ijms22126199.

Palma MD, De Palma M, Biziato D, Petrova TV (2017) Microenvironmental regulation of tumour angiogenesis. Nature Reviews Cancer 17:457–474 https://doi.org/10.1038/nrc.2017.51.

Palmieri, C., Rudraraju, B., Monteverde, M., Lattanzio, L., Gojis, O., Brizio, R., Garrone, O., Merlano, M., Syed, N., Lo Nigro, C., & Crook, T. (2012). Methylation of the calcium channel regulatory subunit α2δ-3 (CACNA2D3) predicts site-specific relapse in oestrogen receptor-positive primary breast carcinomas. *British Journal of Cancer, 107*, 375–381.

Panner, A., & Wurster, R. D. (2006). T-type calcium channels and tumor proliferation. *Cell Calcium, 40*, 253–259.

Panner A, Cribbs LL, Zainelli GM, Origitano TC, Singh S, Wurster RD (2005) Variation of T-type calcium channel protein expression affects cell division of cultured tumor cells. Cell Calcium 37:105–119 https://doi.org/10.1016/j.ceca.2004.07.002.

Park, J. H., Park, S. J., Chung, M. K., Jung, K. H., Choi, M. R., Kim, Y., Chai, Y. G., Kim, S. J., & Park, K. S. (2010). High expression of large-conductance Ca^{2+}-activated K+ channel in the CD133+ subpopulation of SH-SY5Y neuroblastoma cells. *Biochemical and Biophysical Research Communications, 396*, 637–642.

Paulo, P., Ribeiro, F. R., Santos, J., Mesquita, D., Almeida, M., Barros-Silva, J. D., Itkonen, H., Henrique, R., Jerónimo, C., Sveen, A., Mills, I. G., Skotheim, R. I., Lothe, R. A., & Teixeira, M. R. (2012). Molecular subtyping of primary prostate cancer reveals specific and shared target genes of different ETS rearrangements. *Neoplasia, 14*, 600–611.

Poch, M. A., Mehedint, D., Green, D. J., Payne-Ondracek, R., Fontham, E. T. H., Bensen, J. T., Attwood, K., Wilding, G. E., Guru, K. A., Underwood, W., Mohler, J. L., & Heemers, H. V. (2013). The association between calcium channel blocker use and prostate cancer outcome. *Prostate, 73*, 865–872.

Qin, Y. R., Fu, L., Sham, P. C., Kwong, D. L. W., Zhu, C. L., Chu, K. K. W., Li, Y., & Guan, X.-Y. (2008). Single-nucleotide polymorphism-mass array reveals commonly deleted regions at 3p22 and 3p14.2 associate with poor clinical outcome in esophageal squamous cell carcinoma. *International Journal of Cancer, 123*, 826–830.

Reuveny E, Narahashi T (1993) Two types of high voltage-activated calcium channels in SH-SY5Y human neuroblastoma cells. Brain Research 603:64–73 https://doi.org/10.1016/0006-8993(93)91300-h.

Ribatti, D., Tamma, R., & Annese, T. (2020). Epithelial-mesenchymal transition in cancer: A historical overview. *Translational Oncology, 13*, 100773.

Rim, H.-K., Cho, S., Shin, D.-H., Chung, K.-S., Cho, Y.-W., Choi, J.-H., Lee, J. Y., & Lee, K.-T. (2014). T-type Ca^{2+} channel blocker, KYS05090 induces autophagy and apoptosis in A549 cells through inhibiting glucose uptake. *Molecules, 19*, 9864–9875.

Rima, M., Daghsni, M., Lopez, A., Fajloun, Z., Lefrancois, L., Dunach, M., Mori, Y., Merle, P., Brusés, J. L., De Waard, M., & Ronjat, M. (2017). Down-regulation of the Wnt/β-catenin signaling pathway by Cacnb4. *Molecular Biology of the Cell, 28*, 3699–3708.

Rodriguez, C., Jacobs, E. J., Deka, A., Patel, A. V., Bain, E. B., Thun, M. J., & Calle, E. E. (2009). Use of blood-pressure-lowering medication and risk of prostate cancer in the Cancer Prevention Study II Nutrition Cohort. *Cancer Causes & Control, 20*, 671–679.

Ruan, J., Liu, X., Xiong, X., Zhang, C., Li, J., Zheng, H., Huang, C., Shi, Q., & Weng, Y. (2015). miR-107 promotes the erythroid differentiation of leukemia cells via the downregulation of Cacna2d1. *Molecular Medicine Reports, 11*, 1334–1339.

Saltzman BS, Weiss NS, Sieh W, Fitzpatrick AL, McTiernan A, Daling JR, Li CI (2013) Use of antihypertensive medications and breast cancer risk. Cancer Causes & Control 24:365–371 https://doi.org/10.1007/s10552-012-0122-8.

Scholl UI, Nelson-Williams C, Yue P, Grekin R, Wyatt RJ, Dillon MJ, Couch R, Hammer LK, Harley FL, Farhi A, Wang W-H, Lifton RP (2012) Hypertension with or without adrenal hyperplasia due to different inherited mutations in the potassium channel KCNJ5. Proceedings of the National Academy of Sciences 109:2533–2538 https://doi.org/10.1073/pnas.1121407109.

Scholl, U. I., et al. (2013). Somatic and germline CACNA1D calcium channel mutations in aldosterone-producing adenomas and primary aldosteronism. *Nature Genetics, 45*, 1050–1054.

Setlur SR et al. (2008) Estrogen-dependent signaling in a molecularly distinct subclass of aggressive prostate cancer. JNCI: Journal of the National Cancer Institute 100:815–825 https://doi.org/10.1093/jnci/djn150.

Sharma, S., et al. (2019). Ca and CACNA1H mediate targeted suppression of breast cancer brain metastasis by AM RF EMF. *eBioMedicine, 44*, 194–208.

Shi, Z.-Z., Shang, L., Jiang, Y.-Y., Shi, F., Xu, X., Wang, M.-R., & Hao, J.-J. (2015). Identification of genomic biomarkers associated with the clinicopathological parameters and prognosis of esophageal squamous cell carcinoma. *Cancer Biomarkers, 15*, 755–761.

Shiozaki, A., Katsurahara, K., Kudou, M., Shimizu, H., Kosuga, T., Ito, H., Arita, T., Konishi, H., Komatsu, S., Kubota, T., Fujiwara, H., Okamoto, K., & Otsuji, E. (2021). Amlodipine and verapamil, voltage-gated Ca channel inhibitors, suppressed the growth of gastric cancer stem cells. *Annals of Surgical Oncology, 28*, 5400–5411.

Slater, A. A., Alokail, M., Gentle, D., Yao, M., Kovacs, G., Maher, E. R., & Latif, F. (2013). DNA methylation profiling distinguishes histological subtypes of renal cell carcinoma. *Epigenetics, 8*, 252–267.

Somervaille, T. C. P., Matheny, C. J., Spencer, G. J., Iwasaki, M., Rinn, J. L., Witten, D. M., Chang, H. Y., Shurtleff, S. A., Downing, J. R., & Cleary, M. L. (2009). Hierarchical maintenance of MLL myeloid leukemia stem cells employs a transcriptional program shared with embryonic rather than adult stem cells. *Cell Stem Cell, 4*, 129–140.

Sousa, S. R., Vetter, I., Ragnarsson, L., & Lewis, R. J. (2013). Expression and pharmacology of endogenous Cav channels in SH-SY5Y human neuroblastoma cells. *PLoS One, 8*, e59293.

Squecco, R., Tani, A., Zecchi-Orlandini, S., Formigli, L., & Francini, F. (2015). Melatonin affects voltage-dependent calcium and potassium currents in MCF-7 cell line cultured either in growth or differentiation medium. *European Journal of Pharmacology, 758*, 40–52.

Sui, X., Geng, J.-H., Li, Y.-H., Zhu, G.-Y., & Wang, W.-H. (2018). Calcium channel α2δ1 subunit (CACNA2D1) enhances radioresistance in cancer stem-like cells in non-small cell lung cancer cell lines. *Cancer Management and Research, 10*, 5009–5018.

Sun Y-H (2006) Androgens induce increases in intracellular calcium via a G protein-coupled receptor in LNCaP prostate cancer cells. Journal of Andrology 27:671–678 https://doi.org/10.2164/jandrol.106.000554.

Suo A, Childers A, D'Silva A, Petersen LF, Otsuka S, Dean M, Li H, Enwere EK, Pohorelic B, Klimowicz A, Souza IA, Hamid J, Zamponi GW, Bebb D (2018) Cav3.1 overexpression is associated with negative characteristics and prognosis in non-small cell lung cancer. Oncotarget 9:8573–8583 https://doi.org/10.18632/oncotarget.24194.

Taniguchi-Ponciano, K., Gomez-Apo, E., Chavez-Macias, L., Vargas, G., Espinosa-Cardenas, E., Ramirez-Renteria, C., Ferreira-Hermosillo, A., Sosa, E., Silva-Román, G., Peña-Martínez, E., Andonegui-Elguera, S., Vargas-Chavez, S., Santiago-Andres, Y., Peralta, R., Marrero-Rodríguez, D., & Mercado, M. (2020). Molecular alterations in non-functioning pituitary adenomas. *Cancer Biomarkers, 28*, 193–199.

Taylor, J. M., & Simpson, R. U. (1992). Inhibition of cancer cell growth by calcium channel antagonists in the athymic mouse. *Cancer Research, 52*, 2413–2418.

Taylor, J. T., Huang, L., Pottle, J. E., Liu, K., Yang, Y., Zeng, X., Keyser, B. M., Agrawal, K. C., Hansen, J. B., & Li, M. (2008). Selective blockade of T-type Ca^{2+} channels suppresses human breast cancer cell proliferation. *Cancer Letters, 267*, 116–124.

Thorell, K., Bergman, A., Carén, H., Nilsson, S., Kogner, P., Martinsson, T., & Abel, F. (2009). Verification of genes differentially expressed in neuroblastoma tumours: A study of potential tumour suppressor genes. *BMC Medical Genomics, 2*, 53.

Tomoshige, K., Matsumoto, K., Tsuchiya, T., Oikawa, M., Miyazaki, T., Yamasaki, N., Mishima, H., Kinoshita, A., Kubo, T., Fukushima, K., Yoshiura, K.-I., & Nagayasu, T. (2015). Germline mutations causing familial lung cancer. *Journal of Human Genetics, 60*, 597–603.

Toyota, M., Ho, C., Ohe-Toyota, M., Baylin, S. B., & Issa, J. P. (1999). Inactivation of CACNA1G, a T-type calcium channel gene, by aberrant methylation of its 5 CpG island in human tumors. *Cancer Research, 59*, 4535–4541.

Valerie, N. C. K., Dziegielewska, B., Hosing, A. S., Augustin, E., Gray, L. S., Brautigan, D. L., Larner, J. M., & Dziegielewski, J. (2013). Inhibition of T-type calcium channels disrupts Akt signaling and promotes apoptosis in glioblastoma cells. *Biochemical Pharmacology, 85*, 888–897.

Vig M, Kinet J-P (2009) Calcium signaling in immune cells. Nature Immunology 10:21–27 https://doi.org/10.1038/ni.f.220.

Visa, A., Sallán, M. C., Maiques, O., Alza, L., Talavera, E., López-Ortega, R., Santacana, M., Herreros, J., & Cantí, C. (2019). T-Type Ca3.1 channels mediate progression and chemotherapeutic resistance in glioblastoma. *Cancer Research, 79*, 1857–1868.

Wanajo A, Sasaki A, Nagasaki H, Shimada S, Otsubo T, Owaki S, Shimizu Y, Eishi Y, Kojima K, Nakajima Y, Kawano T, Yuasa Y, Akiyama Y (2008) Methylation of the calcium channel-related gene, CACNA2D3, is frequent and a poor prognostic factor in gastric cancer Gastroenterology 135:580–590.e3 https://doi.org/10.1053/j.gastro.2008.05.041.

Wang, X. T., Nagaba, Y., Cross, H. S., Wrba, F., Zhang, L., & Guggino, S. E. (2000). The mRNA of L-type calcium channel elevated in colon cancer: Protein distribution in normal and cancerous colon. *The American Journal of Pathology, 157*, 1549–1562.

Wang, C.-Y., Lai, M.-D., Phan, N. N., Sun, Z., & Lin, Y.-C. (2015). Meta-analysis of public microarray datasets reveals voltage-gated calcium gene signatures in clinical cancer patients. *PLoS One, 10*, e0125766.

Warnier, M., Roudbaraki, M., Derouiche, S., Delcourt, P., Bokhobza, A., Prevarskaya, N., & Mariot, P. (2015). CACNA2D2 promotes tumorigenesis by stimulating cell proliferation and angiogenesis. *Oncogene, 34*, 5383–5394.

Weaver EM, Zamora FJ, Hearne JL, Martin-Caraballo M (2015a) Posttranscriptional regulation of T-type Ca 2 channel expression by interleukin-6 in prostate cancer cells. Cytokine 76:309–320 https://doi.org/10.1016/j.cyto.2015.07.004.

Weaver, E. M., Zamora, F. J., Puplampu-Dove, Y. A., Kiessu, E., Hearne, J. L., & Martin-Caraballo, M. (2015b). Regulation of T-type calcium channel expression by sodium butyrate in prostate cancer cells. *European Journal of Pharmacology, 749*, 20–31.

Wong, D. J., Liu, H., Ridky, T. W., Cassarino, D., Segal, E., & Chang, H. Y. (2008). Module map of stem cell genes guides creation of epithelial cancer stem cells. *Cell Stem Cell, 2*, 333–344.

Wong, A. M. G., Kong, K. L., Chen, L., Liu, M., Wong, A. M. G., Zhu, C., Tsang, J. W.-H., & Guan, X.-Y. (2013). Characterization of CACNA2D3 as a putative tumor suppressor gene in the development and progression of nasopharyngeal carcinoma. *International Journal of Cancer, 133*, 2284–2295.

Xiang, Z., Huang, X., Wang, J., Zhang, J., Ji, J., Yan, R., Zhu, Z., Cai, W., & Yu, Y. (2018). Cross-database analysis reveals sensitive biomarkers for combined therapy for ERBB2+ gastric cancer. *Frontiers in Pharmacology, 9*, 861.

Xiong, Y., Xie, C.-R., Zhang, S., Chen, J., & Yin, Z.-Y. (2019). Detection of a novel panel of somatic mutations in plasma cell-free DNA and its diagnostic value in hepatocellular carcinoma. *Cancer Management and Research, 11*, 5745–5756.

Yu, X.-J., et al. (2015). Characterization of somatic mutations in air pollution-related lung cancer. *eBioMedicine, 2*, 583–590.

Yu, D., Holm, R., Goscinski, M. A., Trope, C. G., Nesland, J. M., & Suo, Z. (2016). Prognostic and clinicopathological significance of Cacna2d1 expression in epithelial ovarian cancers: A retrospective study. *American Journal of Cancer Research, 6*, 2088–2097.

Yu J, Wang S, Zhao W, Duan J, Wang Z, Chen H, Tian Y, Wang D, Zhao J, An T, Bai H, Wu M, Wang J (2018) Mechanistic exploration of cancer stem cell marker voltage-dependent calcium channel α2δ1 subunit-mediated chemotherapy resistance in small-cell lung cancer. Clinical Cancer Research 24:2148–2158 https://doi.org/10.1158/1078-0432.ccr-17-1932.

Zhang, Y., Zhang, J., Jiang, D., Zhang, D., Qian, Z., Liu, C., & Tao, J. (2012). Inhibition of T-type Ca²⁺ channels by endostatin attenuates human glioblastoma cell proliferation and migration. *British Journal of Pharmacology, 166*, 1247–1260.

Zhang, X., Zhang, M., Hou, Y., Xu, L., Li, W., Zou, Z., Liu, C., Xu, A., & Wu, S. (2016). Single-cell analyses of transcriptional heterogeneity in squamous cell carcinoma of urinary bladder. *Oncotarget, 7*, 66069–66076.

Zhang, Y., Cruickshanks, N., Yuan, F., Wang, B., Pahuski, M., Wulfkuhle, J., Gallagher, I., Koeppel, A. F., Hatef, S., Papanicolas, C., Lee, J., Bar, E. E., Schiff, D., Turner, S. D., Petricoin, E. F., Gray, L. S., & Abounader, R. (2017). Targetable T-type calcium channels drive glioblastoma. *Cancer Research, 77*, 3479–3490.

Zhang, Y., Li, L., Liang, P., Zhai, X., Li, Y., & Zhou, Y. (2018). Differential expression of microRNAs in medulloblastoma and the potential functional consequences. *Turkish Neurosurgery, 28*, 179–185.

Zhang, Y., Zhao, W., Li, S., Lv, M., Yang, X., Li, M., & Zhang, Z. (2019). CXCL11 promotes self-renewal and tumorigenicity of α2δ1 liver tumor-initiating cells through CXCR3/ERK1/2 signaling. *Cancer Letters, 449*, 163–171.

Zhao, W., Wang, L., Han, H., Jin, K., Lin, N., Guo, T., Chen, Y., Cheng, H., Lu, F., Fang, W., Wang, Y., Xing, B., & Zhang, Z. (2013). 1B50-1, a mAb raised against recurrent tumor cells, targets liver tumor-initiating cells by binding to the calcium channel α2δ1 subunit. *Cancer Cell, 23*, 541–556.

Zhu, G., Liu, Z., Epstein, J. I., Davis, C., Christudass, C. S., Carter, H. B., Landis, P., Zhang, H., Chung, J.-Y., Hewitt, S. M., Miller, M. C., & Veltri, R. W. (2015). A novel quantitative multiplex tissue immunoblotting for biomarkers predicts a prostate cancer aggressive phenotype. *Cancer Epidemiology, Biomarkers & Prevention, 24*, 1864–1872.

Part IV

Pharmacology of VGCCs

Pharmacology of Voltage-Gated Calcium Channels at Atomic Resolution

William A. Catterall

Abstract

Voltage-gated calcium channels are evolutionarily related transmembrane signaling proteins that mediate calcium entry during action potentials and other forms of cellular depolarization in order to trigger neurotransmission, secretion, contraction, gene transcription, and other physiological processes. Calcium channels are the molecular targets for several major classes of drugs used in the treatment of cardiovascular disorders, pain, and epilepsy. Recent advances in the structural biology of these proteins using X-ray crystallography and cryogenic electron microscopy have given new insights into the molecular basis of their function and pharmacology. Here I review the major classes of drugs and neurotoxins that act on voltage-gated calcium channels and elucidate their complex pharmacology at the atomic level. New understanding of the diseases and therapeutics of these channel proteins will arise from the emerging mechanistic principles derived from these recent structural analyses.

Keywords

Dihydropyridines · Diltiazem · Verapamil · Conotoxin · Agatoxin · Gabapentin · Z944

Calcium Channel Pharmacology

Neurotoxins and calcium channel antagonist drugs inhibit members of each of the three major calcium channel families, often with high specificity and potency (Triggle, 2003; Zamponi et al., 2015). The L-type calcium currents conducted by calcium channels of the Ca_V1 family are blocked by three chemically distinct classes of drugs: phenylalkylamines, benzothiazepines, and dihydropyridines (Fleckenstein, 1983; Hockerman et al., 1997b; Zamponi et al., 2015; Godfraind, 2017). These drugs are all used in the therapy of cardiovascular disorders. The phenylalkylamine verapamil and the benzothiazepine diltiazem are used primarily in the treatment of atrial arrhythmias, whereas dihydropyridines such as nifedipine and amlodipine are used primarily in the treatment of hypertension and angina pectoris (Fleckenstein, 1983; Hockerman et al., 1997b; Triggle, 2003; Sampson & Kass, 2011; Zamponi et al., 2015; Godfraind, 2017).

N-type calcium currents conducted by $Ca_V2.2$ channels, P/Q-type calcium currents conducted by $Ca_V2.1$ channels, and R-type calcium currents conducted by $Ca_V2.3$ channels trigger exocytosis

W. A. Catterall (✉)
Department of Pharmacology, University of Washington, Seattle, WA, USA
e-mail: wcatt@uw.edu

© The Author(s), under exclusive license to Springer Nature Switzerland AG 2022
G. W. Zamponi, N. Weiss (eds.), *Voltage-Gated Calcium Channels*,
https://doi.org/10.1007/978-3-031-08881-0_25

of neurotransmitters from presynaptic nerve terminals (Olivera et al., 1994; Zamponi et al., 2015). These calcium channels are blocked by peptide neurotoxins, which paralyze the prey of cone snails, spiders, and other venomous animals (Olivera et al., 1994; Zamponi et al., 2015). They are specifically blocked by drugs that are either in current use or in active development for the treatment of epilepsy and chronic pain (Zamponi et al., 2015).

T-type calcium currents conducted by Ca_V3 channels are important in repetitively firing cells such as the specialized cardiomyocytes of the sino-atrial (SA) node in the heart and the reticular neurons and relay neurons in the thalamus (Perez-Reyes, 2003). Drugs that inhibit these calcium channels are in development for the treatment of cardiac arrhythmia, epilepsy, and chronic pain (Tringham et al., 2012; Bourinet et al., 2014; Zamponi et al., 2015).

Major advances in understanding the complex effects of these drugs on voltage-gated calcium channels have come from detailed electrophysiological studies, and the structural basis for their specific actions has been elucidated through X-ray crystallography and cryogenic electron microscopy (cryo-EM). In the sections below, I review the functional mechanisms for the actions of these drugs and neurotoxins, and then present recent work on the structures of their drug-receptor complexes on calcium channels.

State-Dependent Drug Block of Ca_V1 Channels

Among the classical voltage-gated calcium channel blockers, the phenylalkylamines and benzothiazepines are used primarily for cardiac arrhythmia (Triggle, 1999; Sampson & Kass, 2011; Godfraind, 2017). They have strongly frequency-dependent block, which enhances their action on calcium channels in rapidly firing injured cardiac myocytes that are responsible for arrhythmia relative to uninjured myocytes contracting at a normal rate (Hondeghem & Katzung, 1984). They are also effective in reducing the rate of rise of action potentials in the atrioventricular

(AV) node. Therefore, they are effective in preventing re-entry of action potentials into the atria from the AV node and in slowing the conduction of action potentials from the atria to the ventricles via the AV node (Triggle, 1999; Sampson & Kass, 2011; Godfraind, 2017). In contrast, dihydropyridines are primarily used for hypertension and angina pectoris (Triggle, 1999; Godfraind, 2017). They have strongly voltage-dependent block, which is driven by high-affinity binding to voltage-dependent calcium channels in the inactivated state (Bean, 1984). They also bind with higher affinity to the isoform of $Ca_V1.2$ expressed in vascular smooth muscle (Welling et al., 1997). Therefore, through these two mechanisms, they preferentially inhibit calcium channels in continuously depolarized vascular smooth muscle cells that sustain contraction of blood vessels in hypertension and angina pectoris (Godfraind, 2017). State-dependent binding and action is essential for the clinical use of these calcium antagonist drugs.

The state-dependent block of sodium channels by local anesthetic and antiarrhythmic drugs is well described by the classical Modulated Receptor Hypothesis (Hille, 1977; Hondeghem & Katzung, 1977; Bean et al., 1983; Hille, 2001). In this model, drug block is frequency dependent because the receptor site is located in the pore and is more rapidly accessible for drug binding when the pore is open; therefore, generation of action potentials at high frequency increases drug block (Hille, 1977). Drug block is voltage dependent because these drugs bind to the inactivated state of sodium channels with high affinity; therefore, sodium channels in damaged, depolarized cells are preferentially blocked (Hille, 1977; Hondeghem & Katzung, 1984). Together, these mechanisms allow local anesthetic and antiarrhythmic drugs to prevent pain and cardiac arrhythmias without blocking normal sensory and cardiac function (Kanaya et al., 1983; Bean, 1984; Hondeghem & Katzung, 1984). Frequency-dependent block by phenylalkylamines and benzothiazepines results from binding in the pore, which is opened during each action potential and provides rapid drug access to their receptor site(s) (Kanaya et al., 1983). Voltage-dependent block by dihydropyridines results from preferential binding to the inactivated conformation of calcium channels (Bean, 1984). These two characteristic forms of block of Ca_V1 channels are illustrated in Fig. 1a, b for the model calcium channel Ca_VAb, a calcium-selective derivative of the ancestral bacterial sodium channel Na_VAb (Tang et al., 2014). Ca_VAb has surprisingly high affinity for calcium-antagonist drugs, which leads to frequency- and voltage-dependent block at concentrations <10-fold higher than in mammalian cardiac calcium channels (Fig. 1; Ren et al., 2001; Tang et al., 2016). The phenylalkylamine verapamil enters the pore when it is open and progressively blocks the ionic current during single depolarizations (Fig. 1a, left). Repetitive stimulations at higher concentrations generate cumulative open channel block (Fig. 1a, right). In contrast, the dihydropyridine nimodipine does not cause rapid pore block during single depolarizing pulses that open the pore (Fig. 1b, left). However, repetitive depolarizations at increasing concentrations cause cumulative voltage-dependent block (Fig. 1b, right). Together, the frequency dependence and voltage dependence of drug action determine the clinical uses of these calcium-antagonist drugs for cardiac arrhythmia vs. hypertension and angina pectoris (Hondeghem & Katzung, 1984; Sampson & Kass, 2011; Godfraind, 2017).

Drug Receptor Sites for State-Dependent Block of Calcium Channels

The initial molecular-mapping studies of the receptor site for calcium-antagonist drugs were based on photoaffinity labeling with photoreactive derivatives of phenylalkylamines and dihydropyridines (Striessnig et al., 1990, 1991; Nakayama et al., 1991). A photoreactive derivative of verapamil labeled the intracellular end of the S6 segment in domain IV (Striessnig et al., 1990). In contrast, photoreactive derivatives of dihydropyridines specifically labeled the outer half of both the IIIS6 and IVS6 segments (Nakayama et al., 1991; Striessnig et al., 1991).

These results led to a domain-interface model for dihydropyridine binding in which dihydropyridines bind in a voltage-dependent manner to a site on the lipid-facing surface of the pore domain that is formed by the IIIS6 and IVS6 segments (Fig. 1c; Catterall & Striessnig, 1992). In contrast, pore-blocking phenylalkylamines were proposed to bind within the pore to a site on the IVS6 segment that is accessible from the intracellular end of the pore in the open state (Fig. 1c). This model of drug binding was confirmed and extended by extensive mapping of the receptor sites for phenylalkylamines (Hockerman et al., 1995; Hockerman et al., 1997a), benzothiazepines (Brauns et al., 1995; Kraus et al., 1996), and dihydropyridines (Mitterdorfer et al., 1996; Peterson et al., 1996) by site-directed mutagenesis, and the receptor site for dihydropyridines was constructed *de novo* by inserting mutations of nine or ten amino acid residues into the drug-binding IIIS6 and IVS6 transmembrane segments of drug-insensitive Ca_V2 channels (Hockerman et al., 1997c; Ito et al., 1997; Sinnegger et al., 1997). Altogether, these photoaffinity labeling and site-directed mutagenesis studies gave a detailed two-dimensional view of these separate, but closely spaced, drug receptor sites on calcium channels (Hockerman et al., 1997b; Hofmann et al., 1999; Striessnig, 1999).

Structures of the Drug Receptor Sites on Ca_VAb Channels

X-ray crystallography revealed two distinct receptor sites for phenylalkylamines and dihydropyridines on Ca_VAb in three dimensions (Fig. 2a; Tang et al., 2016). As expected, verapamil binds in the pore just at the intracellular exit from the ion selectivity filter into the central cavity (Fig. 2a, bottom; Tang et al., 2016). Its charged amino group projects upward into the pore, forming a complex with the backbone carbonyls of Thr175 at the intracellular end of the ion selectivity filter. Its two flanking aromatic moieties make hydrophobic interactions on either side of the ion-conducting pathway through the ion selectivity filter, staunching ion flow through

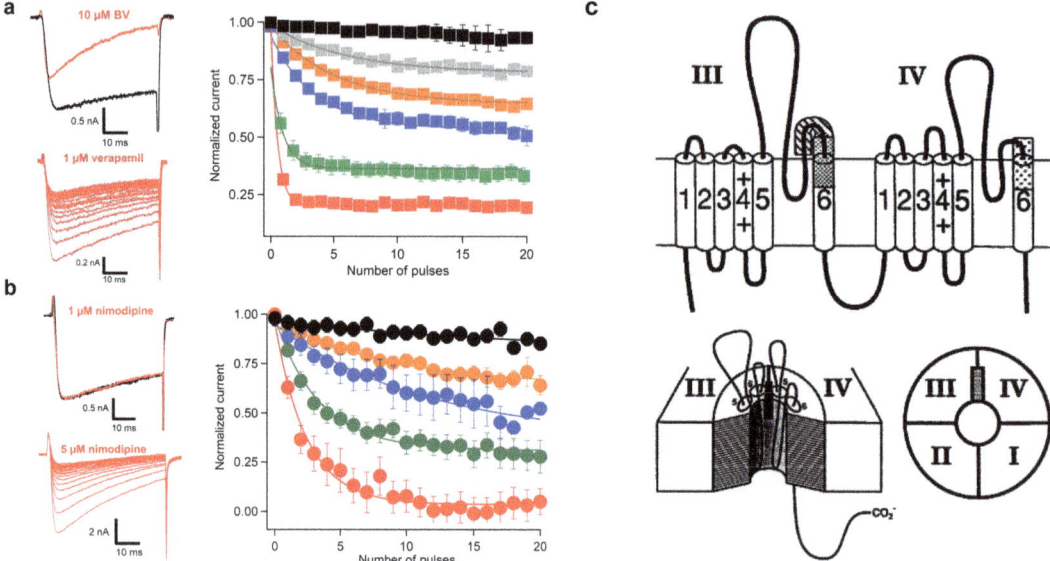

Fig. 1 State-dependent block and molecular mapping of the receptor sites for calcium antagonist drugs. (**a**) Left, top. Barium currents through Ca$_V$Ab channels during depolarization from −120 mV to 0 mV for 50 ms in control (black) and in the presence of 10 μM Br-verapamil. Left, bottom. Use-Dependent Block of Ca$_V$Ab by 1 μM verapamil. Right. Amplitudes of peak barium currents during repetitive stimulation at 1 Hz at the following concentrations of Br-verapamil: 0 μM (black), 0.1 μM (gray), 0.5 μM (orange), 1 μM (blue), 5 μM (green), and 10 μM (red). (Adapted from Tang et al. 2016). (**b**) Left, top.

Barium currents through Ca$_V$Ab channels during depolarization from −120 mV to 0 mV for 50 ms in control (black) and 1 μM nimodipine (red). Left, bottom. Use-dependent block of Ca$_V$Ab by 5 μM nimodipine. Right. Amplitudes of peak barium currents during repetitive stimulation at the following concentrations of nimodipine: 5 nM (black), 25 nM (orange), 100 nM (blue), 1 μM (green), and 5 μM (red). (Adapted from Tang et al. 2016). (**c**) Domain Interface Model for dihydropyridine binding. Hatched segments show the sequences that were photoaffinity labeled with photoreactive drugs. (Adapted from Hockerman et al 1997b)

the selectivity filter like a bandage (Fig. 2a, bottom, inset; Tang et al., 2016). X-ray crystallography studies indicated that the benzothiazepine diltiazem also binds in the pore of Ca$_V$Ab in a position that partially overlaps the phenylalkylamine binding site (Tang et al., 2019). Moreover, the binding poses for verapamil and diltiazem partially overlap that of the sodium channel-blocking local anesthetic and antiarrhythmic drugs (Tang et al., 2016; Lenaeus et al., 2017; Gamal El-Din et al., 2018), suggesting that frequency-dependent block by both sodium and calcium channel drugs results in a similar drug-receptor complex.

In contrast to the phenylalkylamine and benzothiazepine receptor sites, dihydropyridines such as amlodipine and nimodipine bind to a site on the external lipid-facing surface of the pore module, between two voltage sensors (Fig. 2a,

top; Tang et al., 2016). Surprisingly, only a single dihydropyridine molecule bound to the Ca$_V$Ab homotetramer and induced a quaternary conformational change that prevented binding to the other three analogous positions in the tetrameric structure (Tang et al., 2016). This quaternary conformational change disrupts the fourfold symmetry of Ca$_V$Ab and causes one Ca^{2+} ion to bind directly to a carboxyl side chain in one of the coordination sites in the outer selectivity filter, effectively blocking the pore by tightly binding this Ca^{2+} ion. These results show that binding of dihydropyridines to a site on the lipid-facing surface of the pore module can effectively block Ca^{2+} conductance by allosterically inducing high-affinity binding of Ca^{2+} in the pore. This indirect allosteric mode of block was previously inferred from ligand-binding studies showing that binding of Ca^{2+} in the ion-selectivity filter is

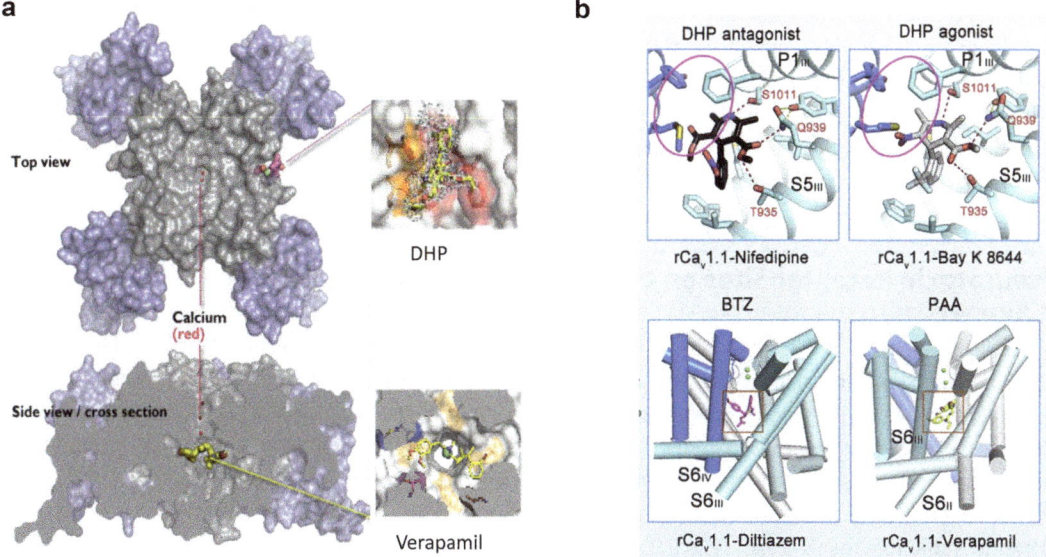

Fig. 2 Receptor sites for calcium antagonist drugs. (a) Calcium antagonist drugs at work. *Top view*. CaᵥAb with the DHP amlodipine bound. Pore module, gray; voltage sensors, blue. *Inset*. The dihydropyridine amlodipine bound in its receptor site. Electron density, blue mesh; amlodipine, yellow sticks. *Side view*. Cross section through CaᵥAb with the phenylalkylamine (PAA) verapamil (yellow sticks) bound. *Inset*. Ca^{2+}, green; verapamil. (Adapted from Tang et al. 2016). (b) High resolution views of the indicated calcium agonist and calcium antagonist drugs bound to Caᵥ1.1. (Adapted from Zhao et al., 2019a). Nearby amino acid residues and transmembrane segments are noted

required for high-affinity binding of dihydropyridines to their separate receptor sites (Mitterdorfer et al., 1995; Peterson & Catterall, 1995), which implied that drug binding would also strengthen and stabilize the binding of Ca^{2+} in the pore through reciprocal energetic interactions.

Structures of the Drug Receptor Sites on Caᵥ1.1 Channels

As described in chapter "Subunit Architecture and Atomic Structure of Voltage Gated Ca2+ Channels", the skeletal muscle calcium channel Caᵥ1.1 has been an essential molecular model for biochemical, molecular biological, and structural studies of Caᵥ channels in general. Its high abundance allowed purification of large amounts in homogeneous form as required for structural studies (Wu et al., 2016). Moreover, its similar pharmacology to Caᵥ1.2 channels has provided a model mammalian Caᵥ channel for structural studies of the channel protein and its

receptor sites for calcium-antagonist drugs (Fig. 2b; Wu et al., 2016; Zhao et al., 2019a). As for CaᵥAb (Tang et al., 2016, 2019), the receptor sites for verapamil and diltiazem are located in the pore in overlapping positions at the exit from the ion selectivity filter into the central cavity (Fig. 2b; Zhao et al., 2019a). Both drugs would physically block the influx of cations in these overlapping binding positions. Dihydroyridine agonists and antagonists bind in almost identical poses in a site formed at the intersection of the S5 and S6 segments in Domains III and IV (Fig. 2b; Zhao et al., 2019a), consistent with the Domain Interface Model for dihydropyridine binding developed from early photoaffinity labeling and site-directed mutagenesis studies (Hockerman et al., 1997b, c; Ito et al., 1997; Sinnegger et al., 1997). However, the binding position for dihydropyridines differs by approximately one helical turn between CaᵥAb and Caᵥ1.1, suggesting significant evolution in the structure of this drug-receptor site in the calcium channel between bacteria and humans (Zhao et al., 2019a). The detailed structural information now available for calcium-antagonist drugs binding to mammalian Ca_V1 channels may provide a molecular template for the design of new generations of these highly effective drugs.

Neurotoxin Receptor Sites on Ca_V2 Channels

Voltage-gated calcium channels in the Ca_V2 family mediate N-type and P/Q-type calcium currents that trigger the release of neurotransmitters at presynaptic nerve terminals (Olivera et al., 1994; Chapter "Pharmacology and Structure-Function of Venom Peptide Inhibitors of N-Type (Cav2.2) Calcium Channels" by Md. Mahadhi Hasan et al.). Their function is regulated by a wide range of neurotransmitters and hormones working through intracellular second-messenger signaling pathways (Hille, 1994; Chapter "Pharmacology and Structure-Function of Venom Peptide Inhibitors of N-Type (Cav2.2) Calcium Channels" by Md. Mahadhi Hasan et al.). Because of their crucial roles in neuromuscular transmission, they are the molecular targets for many classes of polypeptide neurotoxins from cone snails, spiders, and other venomous species (McCleskey et al., 1987; Olivera et al., 1994).

Conotoxins The first toxin discovered to work specifically on Ca_V2 channels was ω-conotoxin GVI-A from the Philippine cone snail *Conus geographus* (Reynolds et al., 1986). Subsequent work has characterized additional conotoxins that are specific for $Ca_V2.2$ channels (Olivera et al., 1987; Hirning et al., 1988; Chapter "Pharmacology and Structure-Function of Venom Peptide Inhibitors of N-Type (Cav2.2) Calcium Channels" by Md. Mahadhi Hasan et al.), as well as a large family of conotoxins with many other molecular targets (Puillandre et al., 2012). Conotoxins are encoded in large preproteins, which undergo disulfide crosslinking, proteolytic processing, and other posttranslational modifications to reach their mature structures (Bulaj & Olivera, 2008). The mature toxins are highly specific for their wide range of molecular targets and

bind to their diverse receptor sites with high affinity (Puillandre et al., 2012; Chapter "Pharmacology and Structure-Function of Venom Peptide Inhibitors of N-Type (Cav2.2) Calcium Channels" by Md. Mahadhi Hasan et al.).

Agatoxins $Ca_V2.1$ calcium channels mediate P/Q-type calcium currents (Llinás et al., 1989). ω-Agatoxin-IIIA from the funnel web spider *Agelenopsis aperta* was the first widely studied toxin that specifically blocks P/Q-type calcium currents (Uchitel et al., 1992). It was shown to block neurotransmission at mammalian neuromuscular junctions and at many synapses in the spinal cord and brain (Mintz et al., 1992a, b; Uchitel et al., 1992). These channels are also the molecular targets of some cone snail toxins, such as ω-conotoxin MVII-C (Hillyard et al., 1992; Chapter _26).

Structure of the $Ca_V2.2$ Complex with Ziconotide

$Ca_V2.2$ channels are localized at the nerve terminals of the primary afferent nociceptive neurons, which bring pain information into the spinal cord and activate ascending neurotransmission at synapses in the dorsal horn (Westenbroek et al., 1992, 1998). For this reason, ω-conotoxin GVI-A and related conotoxins prevent pain sensation in many nociceptive circuits and substantially reduce pain in animal models, including rodents and primates (Miljanich & Ramachandran, 1995; Bowersox et al., 1996). Drug discovery efforts led to the formulation of ziconotide, a synthetic derivative of ω-conotoxin GVI-A that is a potent analgesic when administered by intrathecal infusion into the spinal cord (Miljanich, 2004; Snutch, 2005). It is used for the treatment of extreme pain in advanced-stage cancer and other medical conditions. Structure-function studies using site-directed mutagenesis revealed that ω-conotoxin GVI-A binds to the outer mouth of the ion selectivity filter of the $Ca_V2.2$ channel and physically blocks it (Ellinor et al., 1994). Based on this informa-

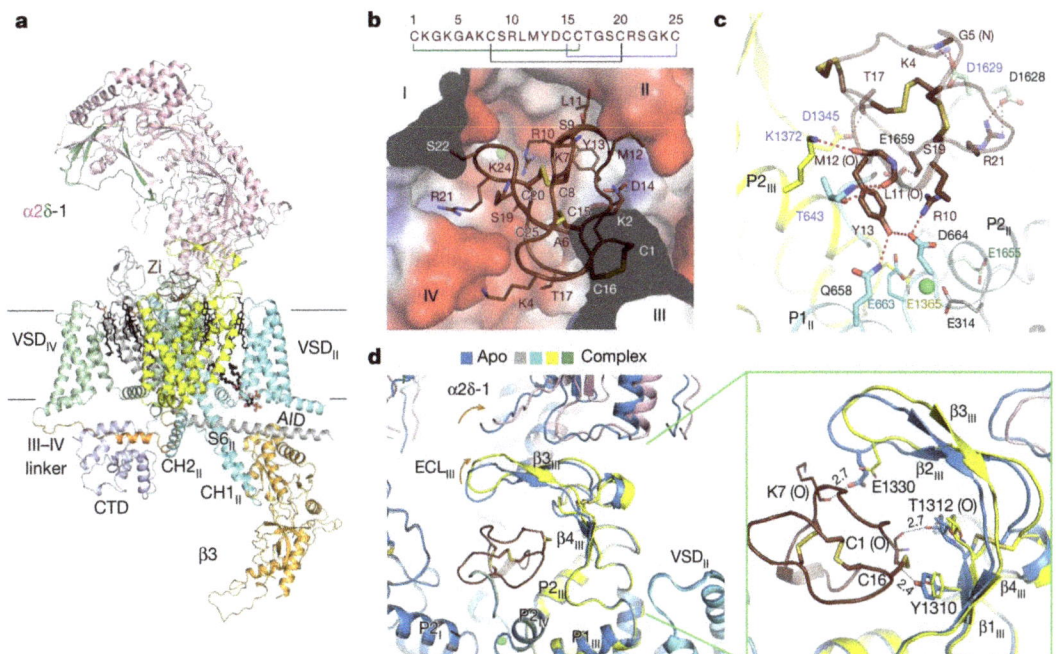

Fig. 3 Specific outer pore blockade of Ca$_v$2.2 by ziconotide. (**a**) Overall structure of the Ca$_v$2.2–ziconotide complex at an average resolution of 3.0 Å. CTD, C-terminal domain; Zi, ziconotide. The resolved lipid, cholesterol, and cholesterol hemisuccinate molecules are shown as black sticks. (**b**) Ziconotide is caged by the extracellular linkers with interacting residues Glu314, Glu663, Glu1365 and Glu1655. The sequence of ziconotide is shown with the three disulfide bonds indicated. The surface electrostatic potential, shown in semitransparent presentation, was calculated in PyMol. (**c**) Specific coordination of ziconotide by the α1 subunit of Ca$_v$2.2. The residues that are not conserved in Ca$_v$ channels are labelled blue. (**d**) Comparison of the conformation of the ziconotide receptor site in the absence and presence of bound drug. (Adapted from Gao et al. 2021)

tion, Gao et al. expressed and purified the human Ca$_v$2.2 channel with an α2δ-1 subunit, a β3 subunit, and ziconotide bound and determined its structure at high resolution by cryo-EM (Gao et al., 2021). The resulting ziconotide/channel complex gave a detailed atomic view of the bound drug and the molecular interactions it makes with its receptor site (Fig. 3). The overall structures of apo-Ca$_v$2.2 and apo-Ca$_v$1.1 can be superimposed with a root mean square deviation (RMSD) of 1.10 Å over 1,728 Cα atoms, and the binding interaction with the α2δ-1 subunit remains essentially identical. On the cytosolic side, the intracellular AID helix following the S6 transmembrane segment in Domain I folds parallel to the membrane plane and interacts with the β3 subunit. Ziconotide, which blocks Ca$_v$2.2 with a half-maximal inhibitory concentration of ~1 nM, binds tightly in the electronegative cavity that

surrounds the entrance to the selectivity filter (Fig. 3a). Specific recognition is mediated by charged and polar residues on the P1 and P2 helices, and the extracellular linkers in Domains II, III, and IV (Fig. 3b, c; Gao et al., 2021). Ziconotide directly seals the outer entrance to the vestibule of the selectivity filter by neutralizing its electronegativity and physically blocking the ion permeation pathway. A bound Ca^{2+} ion is shown as a green sphere, and the EEEE motif in the high field-strength site that determines Ca^{2+} selectivity is shown as thin sticks in Fig. 3c. Sequence comparison showed that four of the eight ziconotide-interacting residues in Ca$_v$2.2 (Thr643, Asp1345, Lys1372, and Asp1629) are not conserved in other Ca$_v$ channels (Fig. 3c, d), providing a structural map of the determinants of drug-binding specificity. The mutations Y13A and R10A substantially reduce pore blocking by ziconotide, and Y13R abolishes its activity completely, consistent with the direct interactions of these amino acid residues with drugs bound in the drug receptor site in Ca$_V$2.2. Evidently, the high-affinity interacting surface of ziconotide is large and engages several different amino acid side chains. Nevertheless, development of small molecule inhibitors that engage a portion of this binding surface with high-affinity interactions may yield Ca$_V$2.2-specific drugs compatible with oral dosing and broader clinical application than ziconotide.

Drugs Acting on the Auxiliary α$_2$ Subunits of Ca$_V$2 Channels

Calcium channels have multiple auxiliary subunits that are required to fine-tune their functional properties and to support maturation and cell-surface expression of the channel complex (Chapter "Subunit Architecture and Atomic Structure of Voltage Gated Ca^{2+} Channels" by William A. Catterall; Figs. 1a, b and 2a, b; Isom et al., 1994; Gurnett & Campbell, 1996; Davies et al., 2007). However, only a single class of drugs, the gabapentinoid calcium-channel antagonists, gabapentin, and pregabalin, act on the auxiliary subunits (Davies et al., 2007; Hendrich et al., 2008). These drugs are used in the treatment of epilepsy and chronic pain. They bind adjacent to the Von Willebrand Factor Type-A homology domain on the extracellular surface of the α$_2$ subunit and modulate the cell surface expression of Ca$_V$2.2 channels, which conduct N-type Ca^{2+} currents that are required for the release of neurotransmitters in the brain and in nociceptive pathways in the spinal cord (Hendrich et al., 2008). Drug binding disrupts normal recycling of these Ca$_V$2.2 channels to the cell surface and thereby reduces nociceptive signaling from the periphery to the central nervous system (Bauer et al., 2009). Although the structure of the Von Willebrand Factor Type-A homology domain has been modeled based on its sequence homology, there are no direct structural studies of the

binding and action of gabapentinoid drugs to date.

Drug Receptor Sites on Ca$_V$3 Channels

Ca$_V$3 channels are important targets for the treatment of epilepsy, chronic pain, and potentially cardiac arrhythmia (Weiss & Zamponi, 2020). Absence epilepsy in GAERS rats arises from inappropriate repetitive firing in the thalamus and cerebral cortex driven by the trisynaptic circuit of inhibitory thalamic reticular neurons, excitatory thalamic relay neurons, and excitatory cortical pyramidal cells, whose normal circadian activity drives sleep spindles and REM sleep (Hosford, 1995; Proft et al., 2017). Specific mutations in Ca$_V$3.2 channels cause idiopathic generalized epilepsy (Khosravani et al., 2005). Type-type calcium currents conducted by Ca$_V$3 channels are also implicated in chronic pain (Bourinet et al., 2016) and cardiac arrhythmia (Torrente et al., 2020).

Extensive drug discovery and development efforts have been devoted to Ca$_V$3.1 and Ca$_V$3.2 in academic laboratories and at Zalicus Inc. and its corporate precursor Neuromed Inc. (Powell et al., 2014). Z944 is the most advanced drug to emerge from these studies, and it is effective in animal models of chronic pain (Lee, 2014; LeBlanc et al., 2016) and absence epilepsy (Tringham et al., 2012). Recent structural studies have revealed the molecular basis for the inhibition of Ca$_V$3.1 channels by this drug (Zhao et al., 2019b). As shown in Fig. 4a, b, Z944 binds in the central cavity of Ca$_V$3.1 channels, covering the outlet from the ion selectivity filter in a similar way as verapamil and diltiazem binding to Ca$_V$Ab and Ca$_V$1 channels. In this position, Z944 would block ion permeation physically and completely inhibit channel function. A detailed analysis of the specific interactions reveals some overlap with the positions of amino acid residues in the Ca$_V$Ab and Ca$_V$1.1 channels that bind the pore blockers verapamil and diltiazem (Fig. 4c–f).

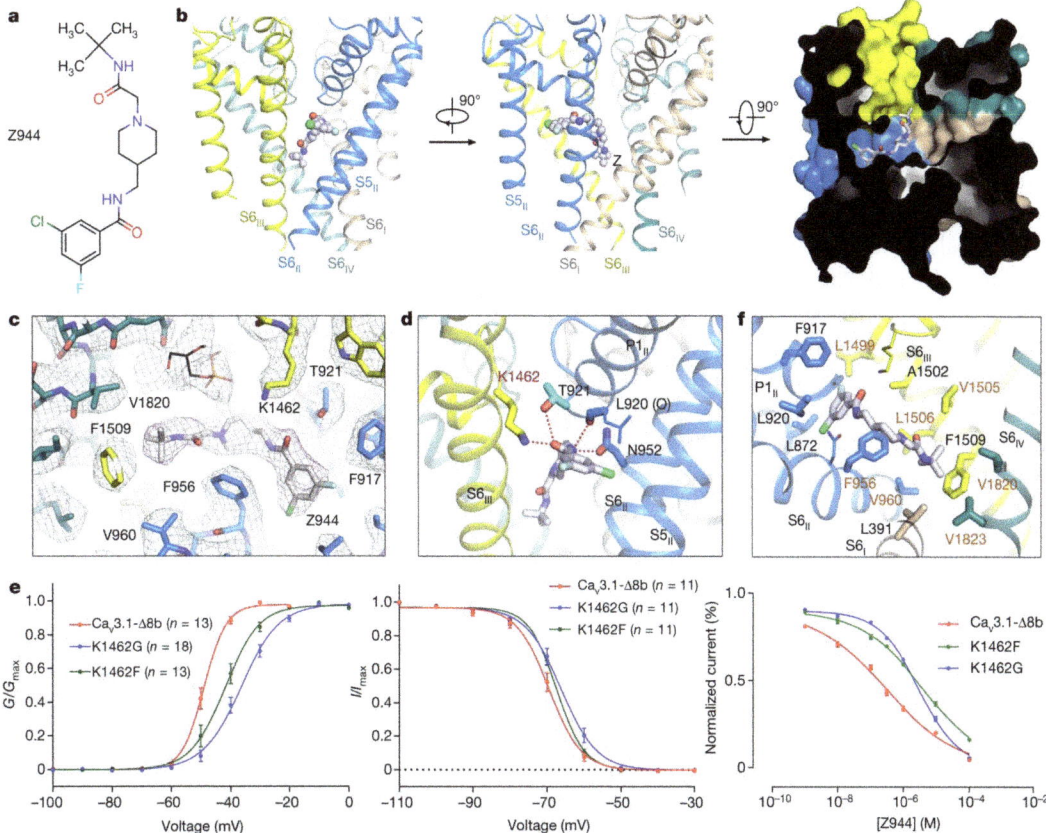

Fig. 4 Specific inner pore blockade of Ca$_V$3 channels. (a) Chemical structure of Z944. (b) Structural basis for pore blockade by Z944. Left. Cartoon views of Z944 bound in the pore with transmembrane segments S5 and S6 in the indicated domains shown in ribbons. Right. A cut-open surface presentation viewed from the extracellular side. Z944, colored silver, is shown as spheres. (c) EM map for Z944 and surrounding residues contoured at 7σ. A nearby lipid is shown as black sticks. (d) Specific coordination of Z944 by polar residues. The potential electrostatic interactions are indicated by red dashed lines. (e) Functional validation of the coordination of Z944 by the T-type specific Lys. The locus corresponding to Lys1462 is replaced by Phe in Ca$_V$1 and Gly in Ca$_V$2 channels. Voltage-dependent activation (left) and inactivation (center). The right panel shows that single point mutations K1462F (green) or K1462G (blue) resulted in a change of the IC$_{50}$ from 311 ± 25.6 nM to 4.2 ± 0.3 μM or 3.1 ± 0.2 μM, respectively. G/G_{max} and I/I_{max} represent normalized conductance and ionic current, respectively. (f) Z944 is surrounded by hydrophobic residues on the S6 tetrahelical bundle. Residues that are not conserved in Ca$_V$1 and Ca$_V$2 channels are labelled in orange. (Adapted from Gao et al. 2021)

Mutation of some of these key residues reduces the affinity for block of ion current, confirming the essential role of these amino acid side chains in drug binding (Fig. 4e). These overlapping regions of drug binding in Ca$_V$Ab, Ca$_V$1, and Ca$_V$3 channels are also similar to the binding sites for local anesthetic and antiarrhythmic drugs that have been characterized in Na$_V$Ab and the mammalian cardiac sodium channel Na$_V$1.5 (Gamal El-Din et al., 2018; Jiang et al., 2020, 2021; Li et al., 2021), revealing a common mechanism for the molecular pharmacology of these diverse sodium- and calcium-channel blockers.

Conclusion

From the initial discoveries of the calcium antagonist drugs in the 1970s through drug screening studies with vascular smooth muscle (Fleckenstein, 1983; Godfraind, 2017), the pharmacology of calcium channels has grown as a field to include clinical uses in cardiovascular diseases such as hypertension, angina pectoris, and cardiac arrhythmia, and in neurological diseases such as epilepsy and chronic pain. These drugs were instrumental in the initial identification, purification, and reconstitution of skeletal muscle calcium channels (see chapter 2 "Subunit Architecture and Atomic Structure of Voltage Gated Ca2+ Channels"), which eventually resulted in the determination of high-resolution structures of all three families of calcium channels with drugs bound in their receptor sites as reviewed here. I look forward to exciting new developments in calcium-channel drug discovery and pharmacology based on these basic research findings.

Acknowledgements Research in my laboratory on sodium and calcium channels is supported by research grants from the National Institutes of Health: R35 NS111573 and R01 HL112808. Special thanks to Dr. Ning Zheng (Professor of Pharmacology, University of Washington, and Investigator of the Howard Hughes Medical Institute), whose deeps insights into structural biology have been crucial in all of our studies of sodium- and calcium-channel structure.

References

Bauer, C. S., Nieto-Rostro, M., Rahman, W., Tran-Van-Minh, A., Ferron, L., Douglas, L., Kadurin, I., Sri Ranjan, Y., Fernandez-Alacid, L., Millar, N. S., Dickenson, A. H., Lujan, R., & Dolphin, A. C. (2009). The increased trafficking of the calcium channel subunit alpha2delta-1 to presynaptic terminals in neuropathic pain is inhibited by the alpha2delta ligand pregabalin. *The Journal of Neuroscience, 29*(13), 4076–4088. https://doi.org/10.1523/JNEUROSCI.0356-09.2009

Bean, B. P. (1984). Nitrendipine block of cardiac calcium channels: High-affinity binding to the inactivated state. *Proceedings of the National Academy of Sciences of the United States of America, 81*, 6388–6392.

Bean, B. P., Cohen, C. J., & Tsien, R. W. (1983). Lidocaine block of cardiac sodium channels. *The Journal of General Physiology, 81*, 613–642.

Bourinet, E., Altier, C., Hildebrand, M. E., Trang, T., Salter, M. W., & Zamponi, G. W. (2014). Calcium-permeable ion channels in pain signaling. *Physiological Reviews, 94*(1), 81–140. https://doi.org/10.1152/physrev.00023.2013

Bourinet, E., Francois, A., & Laffray, S. (2016). T-type calcium channels in neuropathic pain. *Pain, 157*(Suppl 1), S15–S22. https://doi.org/10.1097/j.pain.0000000000000469

Bowersox, S. S., Gadbois, T., Singh, T., Pettus, M., Wang, Y. X., & Luther, R. R. (1996). Selective N-type neuronal voltage-sensitive calcium channel blocker, SNX-111, produces spinal antinociception in rat models of acute, persistent and neuropathic pain. *The Journal of Pharmacology and Experimental Therapeutics, 279*, 1243–1249.

Brauns, T., Cai, Z. W., Kimball, S. D., Kang, H. C., Haugland, R. P., Berger, W., Berjukov, S., Hering, S., Glossmann, H., & Striessnig, J. (1995). Benzothiazepinone binding domain of purified L-type calcium channels: Direct labeling using a novel fluorescent diltiazem analogue. *Biochemistry, 34*, 3461–3469.

Bulaj, G., & Olivera, B. M. (2008). Folding of conotoxins: Formation of the native disulfide bridges during chemical synthesis and biosynthesis of Conus peptides. *Antioxidants & Redox Signaling, 10*(1), 141–155. https://doi.org/10.1089/ars.2007.1856

Catterall, W. A., & Striessnig, J. (1992). Receptor sites for Ca^{2+} channel antagonists. *Trends in Pharmacological Sciences, 13*, 256–262.

Davies, A., Hendrich, J., Van Minh, A. T., Wratten, J., Douglas, L., & Dolphin, A. C. (2007). Functional biology of the alpha-2-delta subunits of voltage-gated calcium channels. *Trends in Pharmacological Sciences, 28*(5), 220–228. https://doi.org/10.1016/j.tips.2007.03.005

Ellinor, P. T., Zhang, J. F., Horne, W. A., & Tsien, R. W. (1994). Structural determinants of the blockade of N-type calcium channels by a peptide neurotoxin. *Nature, 372*(6503), 272–275. https://doi.org/10.1038/372272a0

Fleckenstein, A. (1983). History of calcium antagonists. *Circulation Research, 52*(2 Pt 2), I3–I16.

Gamal El-Din, T. M., Lenaeus, M. J., Zheng, N., & Catterall, W. A. (2018). Fenestrations control resting-state block of a voltage-gated sodium channel. *Proceedings of the National Academy of Sciences of the United States of America.* https://doi.org/10.1073/pnas.1814928115

Gao, S., Yao, X., & Yan, N. (2021). Structure of human Cav2.2 channel blocked by the painkiller ziconotide. *Nature, 596*(7870), 143–147. https://doi.org/10.1038/s41586-021-03699-6

Godfraind, T. (2017). Discovery and development of calcium channel blockers. *Frontiers in Pharmacology, 8*, 286. https://doi.org/10.3389/fphar.2017.00286

Gurnett, C. A., & Campbell, K. P. (1996). Transmembrane auxiliary subunits of voltage-dependent ion channels. *The Journal of Biological Chemistry, 271*, 27975–27978.

Hendrich, J., Van Minh, A. T., Heblich, F., Nieto-Rostro, M., Watschinger, K., Striessnig, J., Wratten, J., Davies, A., & Dolphin, A. C. (2008). Pharmacological disruption of calcium channel trafficking by the alpha2delta ligand gabapentin. *Proceedings of the National Academy of Sciences of the United States of America, 105*(9), 3628–3633. https://doi.org/10.1073/pnas.0708930105

Hille, B. (1977). Local anesthetics: Hydrophilic and hydrophobic pathways for the drug-receptor reaction. *The Journal of General Physiology, 69*, 497–515.

Hille, B. (1994). Modulation of ion-channel function by G-protein-coupled receptors. *Trends in Neurosciences, 17*, 531–536.

Hille, B. (2001). *Ionic channels of excitable membranes* (3rd ed.). Sinauer Associates Inc.

Hillyard, D. R., Monje, V. D., Mintz, I. M., Bean, B. P., Nadasdi, L., Ramachandran, J., Miljanich, G., Azimi-Zoonooz, A., McIntosh, J. M., Cruz, L. J., & et, a.l. (1992). A new Conus peptide ligand for mammalian presynaptic Ca²⁺ channels. *Neuron, 9*(1), 69–77.

Hirning, L. D., Fox, A. P., McCleskey, E. W., Olivera, B. M., Thayer, S. A., Miller, R. J., & Tsien, R. W. (1988). Dominant role of N-type Ca²⁺ channels in evoked release of norepinephrine from sympathetic neurons. *Science, 239*, 57–61.

Hockerman, G. H., Johnson, B. D., Scheuer, T., & Catterall, W. A. (1995). Molecular determinants of high affinity phenylaklyamine block of L-type calcium channels. *The Journal of Biological Chemistry, 270*, 22119–22122.

Hockerman, G. H., Johnson, B. D., Abbott, M. R., Scheuer, T., & Catterall, W. A. (1997a). Molecular determinants of high affinity phenylalkylamine block of L-type calcium channels in transmembrane segment IIIS6 and the pore region of the α1 subunit. *The Journal of Biological Chemistry, 272*, 18759–18765.

Hockerman, G. H., Peterson, B. Z., Johnson, B. D., & Catterall, W. A. (1997b). Molecular determinants of drug binding and action on L-type calcium channels. *Annual Review of Pharmacology and Toxicology, 37*, 361–396.

Hockerman, G. H., Peterson, B. Z., Sharp, E., Tanada, T. N., Scheuer, T., & Catterall, W. A. (1997c). Construction of a high-affinity receptor site for dihydropyridine agonists and antagonists by single amino acid substitutions in a non-L-type Ca²⁺ channel. *Proceedings of the National Academy of Sciences of the United States of America, 94*, 14906–14911.

Hofmann, F., Lacinová, L., & Klugbauer, N. (1999). Voltage-dependent calcium channels: From structure to function. *Reviews of Physiology, Biochemistry and Pharmacology, 139*, 33–87.

Hondeghem, L. M., & Katzung, B. G. (1977). Timed- and voltage-dependent interactions of antiarrhythmic drugs with cardiac sodium channels. *Biochimica et Biophysica Acta, 472*, 373–398.

Hondeghem, L. M., & Katzung, B. G. (1984). Antiarrhythmic agents: The modulated receptor mechanism of action of sodium and calcium channel blocking drugs. *Annual Review of Pharmacology and Toxicology, 24*, 387–423.

Hosford, D. A. (1995). Models of primary generalized epilepsy. *Current Opinion in Neurology, 8*(2), 121–125. https://doi.org/10.1097/00019052-199504000-00006

Isom, L. L., De Jongh, K. S., & Catterall, W. A. (1994). Auxiliary subunits of voltage-gated ion channels. *Neuron, 12*(6), 1183–1194.

Ito, H., Klugbauer, N., & Hofmann, F. (1997). Transfer of the high affinity dihydropyridine sensitivity from L-type to non-L-type calcium channel. *Molecular Pharmacology, 52*, 735–740.

Jiang, D., Shi, H., Tonggu, L., Gamal El-Din, T. M., Lenaeus, M. J., Zhao, Y., Yoshioka, C., Zheng, N., & Catterall, W. A. (2020). Structure of the cardiac sodium channel. *Cell, 180*(1), 122–134.e10. https://doi.org/10.1016/j.cell.2019.11.041

Jiang, D., Banh, R., Gamal El-Din, T. M., Tonggu, L., Lenaeus, M. J., Pomès, R., Zheng, N., & Catterall, W. A. (2021). Open-state structure and pore gating mechanism of the cardiac sodium channel. *Cell, 184*(20), 5151–5162.e11. https://doi.org/10.1016/j.cell.2021.08.021

Kanaya, S., Arlock, P., Katzung, B. G., & Hondeghem, L. M. (1983). Diltiazem and verapamil preferentially block inactivated cardiac calcium channels. *Journal of Molecular and Cellular Cardiology, 15*, 145–148.

Khosravani, H., Bladen, C., Parker, D. B., Snutch, T. P., McRory, J. E., & Zamponi, G. W. (2005). Effects of Cav3.2 channel mutations linked to idiopathic generalized epilepsy. *Annals of Neurology, 57*(5), 745–749. https://doi.org/10.1002/ana.20458

Kraus, R., Reichl, B., Kimball, S. D., Grabner, M., Murphy, B. J., Catterall, W. A., & Striessnig, J. (1996). Identification of benz(othi)azepine-binding regions within L-type calcium channel a1 subunits. *The Journal of Biological Chemistry, 271*, 20113–20118.

LeBlanc, B. W., Lii, T. R., Huang, J. J., Chao, Y. C., Bowary, P. M., Cross, B. S., Lee, M. S., Vera-Portocarrero, L. P., & Saab, C. Y. (2016). T-type calcium channel blocker Z944 restores cortical synchrony and thalamocortical connectivity in a rat model of neuropathic pain. *Pain, 157*(1), 255–263. https://doi.org/10.1097/j.pain.0000000000000362

Lee, M. (2014). Z944: A first in class T-type calcium channel modulator for the treatment of pain. *Journal of the Peripheral Nervous System, 19*(Suppl 2), S11–S12. https://doi.org/10.1111/jns.12080_2

Lenaeus, M. J., Gamal El-Din, T. M., Ing, C., Ramanadane, K., Pomes, R., Zheng, N., & Catterall, W. A. (2017). Structures of closed and open states of a voltage-gated sodium channel. *Proceedings of the National Academy of Sciences of the United States of America, 114*(15), E3051–E3060. https://doi.org/10.1073/pnas.1700761114

Li, Z., Jin, X., Wu, T., Huang, G., Wu, K., Lei, J., Pan, X., & Yan, N. (2021). Structural basis for pore blockade of the human cardiac sodium channel Na. *Angewandte Chemie (International Ed. in English), 60*(20), 11474–11480. https://doi.org/10.1002/anie.202102196

Llinás, R., Sugimori, M., Lin, J. W., & Cherksey, B. (1989). Blocking and isolation of a calcium channel from neurons in mammals and cephalopods utilizing a toxin fraction (FTX) from funnel-web spider poison. *Proceedings of the National Academy of Sciences of the United States of America, 86*, 1689–1693.

McCleskey, E. W., Fox, A. P., Feldman, D. H., Cruz, L. J., Olivera, B. M., Tsien, R. W., & Yoshikami, D. (1987). ω-Conotoxin: Direct and persistent blockade of specific types of calcium channels in neurons but not muscle. *Proceedings of the National Academy of Sciences of the United States of America, 84*, 4327–4331.

Miljanich, G. P. (2004). Ziconotide: Neuronal calcium channel blocker for treating severe chronic pain. *Current Medicinal Chemistry, 11*(23), 3029–3040. https://doi.org/10.2174/0929867043363884

Miljanich, G. P., & Ramachandran, J. (1995). Antagonists of neuronal calcium channels: Structure, function, and therapeutic implications. *Annual Review of Pharmacology and Toxicology, 35*, 707–734.

Mintz, I. M., Adams, M. E., & Bean, B. P. (1992a). P-type calcium channels in rat central and peripheral neurons. *Neuron, 9*, 85–95.

Mintz, I. M., Venema, V. J., Swiderek, K. M., Lee, T. D., Bean, B. P., & Adams, M. E. (1992b). P-type calcium channels blocked by the spider toxin omega-Aga-IVA. *Nature, 355*(6363), 827–829.

Mitterdorfer, J., Sinnegger, M. J., Grabner, M., Striessnig, J., & Glossmann, H. (1995). Coordination of Ca²⁺ by the pore region glutamates is essential for high-affinity dihydropyridine binding to the cardiac Ca²⁺ channel alpha-1 subunit. *Biochemistry, 34*, 9350–9355.

Mitterdorfer, J., Wang, Z. Y., Sinnegger, M. J., Hering, S., Striessnig, J., Grabner, M., & Glossmann, H. (1996). Two amino acid residues in the IIIS5 segment of L-type calcium channels differentially contribute to 1,4-dihydropyridine sensitivity. *The Journal of Biological Chemistry, 271*, 30330–30335.

Nakayama, H., Taki, M., Striessnig, J., Catterall, W. A., & Kanaoka, Y. (1991). Identification of 1,4-dihydropyridine binding regions within the α1 subunit of skeletal muscle Ca²⁺ channels by photoaffinity labeling with diazipine. *Proceedings of the National Academy of Sciences of the United States of America, 88*, 9203–9207.

Olivera, B. M., Cruz, L. J., de Santos, V., LeCheminant, G. W., Griffin, D., Zeikus, R., McIntosh, J. M., Galyean, R., Varga, J., Gray, W. R., & Rivier, J. (1987). Neuronal calcium channel antagonists. Discrimination between calcium channel subtypes using ω-conotoxin from *Conus magus* venom. *Biochemistry, 26*, 2086–2090.

Olivera, B. M., Miljanich, G. P., Ramachandran, J., & Adams, M. E. (1994). Calcium channel diversity and neurotransmitter release: The omega-conotoxins and omega-agatoxins. *Annual Review of Biochemistry, 63*, 823–867.

Perez-Reyes, E. (2003). Molecular physiology of low-voltage-activated t-type calcium channels. *Physiological Reviews, 83*(1), 117–161.

Peterson, B. Z., & Catterall, W. A. (1995). Calcium binding in the pore of L-type calcium channels modulates high affinity dihydropyridine binding. *The Journal of Biological Chemistry, 270*, 18201–18204.

Peterson, B. Z., Tanada, T. N., & Catterall, W. A. (1996). Molecular determinants of high affinity dihydropyridine binding in L-type calcium channels. *The Journal of Biological Chemistry, 271*, 5293–5296.

Powell, K. L., Cain, S. M., Snutch, T. P., & O'Brien, T. J. (2014). Low threshold T-type calcium channels as targets for novel epilepsy treatments. *British Journal of Clinical Pharmacology, 77*(5), 729–739. https://doi.org/10.1111/bcp.12205

Proft, J., Rzhepetskyy, Y., Lazniewska, J., Zhang, F. X., Cain, S. M., Snutch, T. P., Zamponi, G. W., & Weiss, N. (2017). The Cacna1h mutation in the GAERS model of absence epilepsy enhances T-type Ca. *Scientific Reports, 7*(1), 11513. https://doi.org/10.1038/s41598-017-11591-5

Puillandre, N., Koua, D., Favreau, P., Olivera, B. M., & Stöcklin, R. (2012). Molecular phylogeny, classification and evolution of conopeptides. *Journal of Molecular Evolution, 74*(5-6), 297–309. https://doi.org/10.1007/s00239-012-9507-2

Ren, D., Navarro, B., Xu, H., Yue, L., Shi, Q., & Clapham, D. E. (2001). A prokaryotic voltage-gated sodium channel. *Science, 294*(5550), 2372–2375.

Reynolds, I. J., Wagner, J. A., Snyder, S. H., Thayer, S. A., Olivera, B. M., & Miller, R. J. (1986). Brain voltage-sensitive calcium channel subtypes differentiated by omega-conotoxin fraction GVIA. *Proceedings of the National Academy of Sciences of the United States of America, 83*(22), 8804–8807. https://doi.org/10.1073/pnas.83.22.8804

Sampson, K. J., & Kass, R. K. (2011). Antiarrhythmic drugs. In Goodman & Gilman's (Eds.), *Pharmacological basis of therapeutics* (pp. 815–848).

Sinnegger, M. J., Wang, Z. Y., Grabner, M., Hering, S., Striessnig, J., Glossmann, H., & Mitterdorfer, J. (1997). Nine L-type amino acid residues confer full 1,4-dihydropyridine sensitivity to the neuronal calcium channel α1A subunit – Role of L-type MET. *The Journal of Biological Chemistry, 272*, 27686–27693.

Snutch, T. P. (2005). Targeting chronic and neuropathic pain: The N-type calcium channel comes of age. *NeuroRx, 2*(4), 662–670. https://doi.org/10.1602/neurorx.2.4.662

Striessnig, J. (1999). Pharmacology, structure and function of cardiac L-type calcium channels. *Cellular Physiology and Biochemistry, 9*(4-5), 242–269.

Striessnig, J., Glossmann, H., & Catterall, W. A. (1990). Identification of a phenylalkylamine binding region within the α1 subunit of skeletal muscle Ca²⁺ channels. *Proceedings of the National Academy of Sciences of the United States of America, 87*, 9108–9112.

Striessnig, J., Murphy, B. J., & Catterall, W. A. (1991). The dihydropyridine receptor of L-type Ca²⁺ channels: Identification of binding domains for (+)-[³H]

PN200-110 and [³H]azidopine within the alpha-1 subunit. *Proceedings of the National Academy of Sciences of the United States of America, 88*, 10769–10773.

Tang, L., Gamal El-Din, T. M., Payandeh, J., Martinez, G. Q., Heard, T. M., Scheuer, T., Zheng, N., & Catterall, W. A. (2014). Structural basis for Ca²⁺ selectivity of a voltage-gated calcium channel. *Nature, 505*(7481), 56–61. https://doi.org/10.1038/nature12775

Tang, L., Gamal El-Din, T. M., Swanson, T. M., Pryde, D. C., Scheuer, T., Zheng, N., & Catterall, W. A. (2016). Structural basis for inhibition of a voltage-gated Ca²⁺ channel by Ca²⁺ antagonist drugs. *Nature, 537*, 117–121. https://doi.org/10.1038/nature19102

Tang, L., Gamal El-Din, T. M., Lenaeus, M. J., Zheng, N., & Catterall, W. A. (2019). Structural basis for diltiazem block of a voltage-gated Ca²⁺ channel. *Molecular Pharmacology, 96*(4), 485–492. https://doi.org/10.1124/mol.119.117531

Torrente, A. G., Mesirca, P., Bidaud, I., & Mangoni, M. E. (2020). Channelopathies of voltage-gated L-type Cav1.3/α. *Pflügers Archiv, 472*(7), 817–830. https://doi.org/10.1007/s00424-020-02421-1

Triggle, D. J. (1999). The pharmacology of ion channels: With particular reference to voltage-gated Ca²⁺ channels. *European Journal of Pharmacology, 375*, 311–325.

Triggle, D. J. (2003). Drug targets in the voltage-gated calcium channel family: Why some are and some are not. *Assay and Drug Development Technologies, 1*(5), 719–733.

Tringham, E., Powell, K. L., Cain, S. M., Kuplast, K., Mezeyova, J., Weerapura, M., Eduljee, C., Jiang, X., Smith, P., Morrison, J. L., Jones, N. C., Braine, E., Rind, G., Fee-Maki, M., Parker, D., Pajouhesh, H., Parmar, M., O'Brien, T. J., & Snutch, T. P. (2012). T-type calcium channel blockers that attenuate thalamic burst firing and suppress absence seizures. *Science Translational Medicine, 4*(121), 121ra19. https://doi.org/10.1126/scitranslmed.3003120

Uchitel, O. D., Protti, D. A., Sanchez, V., Cherksey, B. D., Sugimori, M., & Llinás, R. (1992). P-type voltage-dependent calcium channel mediates presynaptic calcium influx and transmitter release in mamma-

lian synapses. *Proceedings of the National Academy of Sciences of the United States of America, 89*, 3330–3333.

Weiss, N., & Zamponi, G. W. (2020). Genetic T-type calcium channelopathies. *Journal of Medical Genetics, 57*(1), 1–10. https://doi.org/10.1136/jmedgenet-2019-106163

Welling, A., Ludwig, A., Zimmer, S., Klugbauer, N., Flockerzi, V., & Hofmann, F. (1997). Alternatively spliced IS6 segments of the alpha 1C gene determine the tissue-specific dihydropyridine sensitivity of cardiac and vascular smooth muscle L-type Ca²⁺ channels. *Circulation Research, 81*(4), 526–532. https://doi.org/10.1161/01.res.81.4.526

Westenbroek, R. E., Hell, J. W., Warner, C., Dubel, S. J., Snutch, T. P., & Catterall, W. A. (1992). Biochemical properties and subcellular distribution of an N-type calcium channel alpha-1 subunit. *Neuron, 9*(6), 1099–1115.

Westenbroek, R. E., Hoskins, L., & Catterall, W. A. (1998). Localization of Ca²⁺ channel subtypes on rat spinal motor neurons, interneurons, and nerve terminals. *The Journal of Neuroscience, 18*, 6319–6330.

Wu, J., Yan, Z., Li, Z., Qian, X., Lu, S., Dong, M., Zhou, Q., & Yan, N. (2016). Structure of the voltage-gated calcium channel Cav1.1 at 3.6 A resolution. *Nature, 537*(7619), 191–196. https://doi.org/10.1038/nature19321

Zamponi, G. W., Striessnig, J., Koschak, A., & Dolphin, A. C. (2015). The physiology, pathology, and pharmacology of voltage-gated calcium channels and their future therapeutic potential. *Pharmacological Reviews, 67*(4), 821–870. https://doi.org/10.1124/pr.114.009654

Zhao, Y., Huang, G., Wu, J., Wu, Q., Gao, S., Yan, Z., Lei, J., & Yan, N. (2019a). Molecular basis for ligand modulation of a mammalian voltage-gated calcium channel. *Cell, 177*(6), 1495–1506.e12. https://doi.org/10.1016/j.cell.2019.04.043

Zhao, Y., Huang, G., Wu, Q., Wu, K., Li, R., Lei, J., Pan, X., & Yan, N. (2019b). Cryo-EM structures of apo and antagonist-bound human Cav3.1. *Nature, 576*(7787), 492–497. https://doi.org/10.1038/s41586-019-1801-3

Pharmacology and Structure-Function of Venom Peptide Inhibitors of N-Type (Ca$_v$2.2) Calcium Channels

Md. Mahadhi Hasan, David J. Adams, and Richard J. Lewis

Abstract

Venoms contain a complex mixture of mostly small peptides that are released from animals to facilitate prey capture and defensive strategies. An important class of neurotoxic venom peptides selectively inhibit specific types of voltage-gated ion channels, including N-type calcium channels (Ca$_V$2.2), that play a key role in the neurotransmission associated with pain pathways. In recognition of their importance as research tools and validated clinical potential, in this chapter, we highlight venom peptides that inhibit Ca$_V$2.2 (N-type calcium channels) and discuss their structure-activity relationships (SAR). Recent advances in our understanding of the mechanism of binding of Ca$_V$2.2-selective ω-conotoxins revealed from cryo-EM structures of the MVIIA-Ca$_V$2.2 complex allow the rational development of modified ω-conotoxins better able to treat severe pain conditions.

Keywords

Conotoxins · Spider toxins · Venom peptides · Pain

Introduction

Venoms contain a complex mixture of organic and inorganic molecules, salts, proteins, and peptides that are released from animals to facilitate prey capture and defensive strategies. The composition of these venoms can range from relatively simple to complex mixtures of hundreds to thousands of individual components with diverse pharmacology. Broadly, venoms can be categorized as cytotoxic or neurotoxic, often inducing pain upon envenomation, although many venoms are paralytic and are often non-toxic to mammals. Many neurotoxic venom peptides modulate voltage-gated ion channels, including (i) pore blockers that can directly inhibit the flow of cations through the ion conduction pore, or (ii) gating modifiers that interact with regions of the channel that control channel opening and closing. Many of these calcium channel modulators are

M. M. Hasan
Institute for Molecular Bioscience, The University of Queensland, Brisbane, QLD, Australia

Pharmacy Discipline, Khulna University, Khulna, Bangladesh

D. J. Adams
Illawarra Health & Medical Research Institute (IHMRI), University of Wollongong, Wollongong, NSW, Australia

R. J. Lewis (✉)
Institute for Molecular Bioscience, The University of Queensland, Brisbane, QLD, Australia
e-mail: r.lewis@imb.uq.edu.au

© The Author(s), under exclusive license to Springer Nature Switzerland AG 2022
G. W. Zamponi, N. Weiss (eds.), *Voltage-Gated Calcium Channels*,
https://doi.org/10.1007/978-3-031-08881-0_26

small peptides (10–40 amino acids) that are highly selective for different classes of Ca_Vs.

Voltage-gated calcium (Ca_V) channels transduce membrane potential changes into Ca^{2+} influx to initiate many a diverse range of physiological responses in different types of cells, including contraction in cardiac, smooth, and skeletal muscle cells, hormone secretion in endocrine cells, synaptic transmission across neurons, enzyme activity, and gene transcription in hepatic cells (Catterall, 2011). Early studies showed that high voltage-activated calcium channels are a multi-subunit complex of α_1, $\alpha_2\delta$, β, and γ subunits (Takahashi et al., 1987; Tanabe et al., 1987; Curtis & Catterall, 1986; Flockerzi et al., 1986). The α_1 subunit forms the main voltage-sensitive, calcium-selective pore, whereas the $\alpha_2\delta$, β, and γ operate as auxiliary subunits (Catterall, 2000). After analyzing the biochemical, glycosylation, and hydrophobic properties of these subunits, a model of the calcium channel was established, comprising the pore-forming α_1 subunit along with a disulfide-linked $\alpha_2\delta$ subunit, an intracellular β subunit, and a transmembrane γ subunit (Takahashi et al., 1987). The first purified calcium channel was dihydropyridine-sensitive $Ca_V1.1$ (Tanabe et al., 1987), which has been the prototype for structural, functional, and mechanistic investigations of Ca_V channels (Bannister & Beam, 2013). Recently, the rabbit $Ca_V1.1$ and human $Ca_V2.2$ structures were solved and refined to 3.6 Å and 3.1 Å resolution, respectively, using cryo-electron microscopy (Wu et al., 2016; Gao et al., 2021), providing a breakthrough in our understanding of the structure-function of these large integral membrane proteins. For additional information on calcium channel structure, please see chapter "Subunit Architecture and Atomic Structure of Voltage Gated Ca2+ Channels" by Dr. Catterall.

In recognition of their importance as research tools and clinical potential, in this chapter, we highlight venom peptides that inhibit $Ca_V2.2$ (N-type calcium channels) and discuss their structure-activity relationships (SAR).

Selective $Ca_V2.2$ Inhibitors

A diverse array of venom peptides that potently inhibit $Ca_V2.2$ are listed in Table 1. While mostly non-selective $Ca_V2.2$ channel inhibitors have been identified from spider and snake venoms, ω-conotoxins from the cone snail venoms remain the best-studied source of subtype selective inhibitors (Lewis et al., 2012; Bourinet & Zamponi, 2016). ω-Conotoxins are polar and highly water-soluble small basic peptides typically ranging in size between twenty and thirty amino acids with a net charge between +5 and +7. Their rigid backbone structure is achieved through the formation of three disulfide bonds formed between conserved cysteine residues arranged C^1-C^2-C^3C^1-C^2-C^3 to generate a globular cysteine knot structure (Lewis et al., 2012). The first $Ca_V2.2$ channel inhibiting ω-conotoxins was identified from fish-hunting cone snails, including *Conus geographus* (Olivera et al., 1994; McCleskey et al., 1987), *Conus magus* (Olivera et al., 1985), *Conus catus* (Lewis et al., 2000), and *Conus consors* (Favreau et al., 2001). Recently, ω-conotoxins were identified for the first time from a worm-hunting cone snail (*Conus moncuri*). Interestingly, these were the most potent ω-conotoxins at fish Ca_Vs, suggesting they may have arisen from the defensive venom of ancestral vermivorous cone snails (Sousa et al., 2018).

Most piscivorous cone snails utilize a mix of ω-conotoxins. For example, multiple ω-conotoxins have been identified from the venom of *Conus geographus,* including GVIA, GVIB, GVIC, GVIIA, and GVIIB (Olivera et al., 1994). Among these, GVIA is highly selective $Ca_V2.2$ inhibition, whereas the others are not completely characterized (Olivera et al., 1985; Ramírez et al., 2017). The well-studied ω-GVIA is a 27-amino-acid peptide that irreversibly inhibits the $Ca_V2.2$ channels unless strong (non-physiological) membrane hyperpolarization is applied to enhance its dissociation from the channel (Olivera et al., 1985). Since its discovery, GVIA has been widely used as a pharmacological tool for the electrophysiological characterization of the $Ca_V2.2$ channel.

Table 1 Venom-derived peptides inhibiting Ca$_v$2.2 (N-type calcium channels) expressed in mammalian neurons

Peptide	Source	Activity	IC$_{50}$	References
Kurtoxin	*Parabuthus transvaalicus*	Gating modifier	460 nM	Sidach and Mintz (2002)
ω-Filistatoxin-Kh1a	*Filista hibernalis*	Pore blocker	2.3 nM	Pringos et al. (2011)
ω-Conotoxin GVIA	*Conus geographus*	Pore blocker	2 pM	Favreau et al. (2001) and Ramírez et al. (2017)
ω-Conotoxin GVIIA	*Conus geographus*	Pore blocker	22.9 nM	Miljanich et al. (1993a, b) and Ramírez et al. (2017)
ω-Conotoxin MVIIA	*Conus catus*	Pore blocker	8 nM	Lee et al. (2010) and Ramírez et al. (2017)
ω-Conotoxin MVIIB	*Conus catus*	Pore blocker	101 pM	Ramírez et al. (2017)
ω-Conotoxin MVIIC	*Conus magus*	Pore blocker	7 nM	Lewis et al. (2000) and Ramírez et al. (2017)
ω-Conotoxin SVIA	*Conus striatus*	Pore blocker	1.46 μM	Miljanich et al. (1993a, b) and Ramírez et al. (2017)
ω-Conotoxin SVIB	*Conus striatus*	Pore blocker	1.09 nM	Miljanich et al. (1993a, b) and Ramírez et al. (2017)
ω-Conotoxin MVIIA/SO-3	*Conus striatus*	Pore blocker	160 nM	Wang et al. (2016)
ω-Conotoxin CVIA	*Conus catus*	Pore blocker	0.6 nM	Lewis et al. (2000) and Ramírez et al. (2017)
ω-Conotoxin CVIB	*Conus catus*	Pore blocker	8 nM	Lewis et al. (2000), Berecki et al. (2010) and Ramírez et al. (2017)
ω-Conotoxin CVIC	*Conus catus*	Pore blocker	7.6 nM	Lewis et al. (2000) and Ramírez et al. (2017)
ω-Conotoxin CVID	*Conus catus*	Pore blocker	0.07 nM	Lewis et al. (2000) and Ramírez et al. (2017)
ω-Conotoxin CVIE	*Conus catus*	Pore blocker	2.6 nM	Berecki et al. (2010) and Ramírez et al. (2017)
ω-Conotoxin CVIF	*Conus catus*	Pore blocker	19.9 nM	Berecki et al. (2010) and Ramírez et al. (2017)
ω-Conotoxin FVIA	*Conus fulman*	Pore blocker	11.5 nM	Lee et al. (2010) and Ramírez et al. (2017)
ω-Conotoxin RVIA	*Conus radiatus*	Pore blocker	229 nM	Miljanich et al. (1993a, b) and Ramírez et al. (2017)
ω-Conotoxin TVIA	*Conus tulipa*	Pore blocker	228 pM	Miljanich et al. (1993a, b) and Ramírez et al. (2017)
ω-Conotoxin CnVIIA	*Conus consors*	Pore blocker	2.3–3.7 pM	Favreau et al. (2001) and Ramírez et al. (2017)
ω-Agatoxin IIA	*Agelenopsis aperta*	Pore blocker	10 nM	Adams et al. (1990) and Bindokas and Adams (1989)
ω-Agatoxin IIIA	*Agelenopsis aperta*	Pore blocker	1.4 nM	Ertel et al. (1994) and Olivera et al. (1994)
ω-Agatoxin IIIB	*Agelenopsis aperta*	Pore blocker	140 nM	Ertel et al. (1994) and Yan and Adams (2000)
ω-Agatoxin IIID	*Agelenopsis aperta*	Pore blocker	35 nM	Ertel et al. (1994)
α-Conotoxin Eu1.6	*Conus eburneus*	Not known	1 nM	Liu et al. (2018)
SNX-325	*Segestria florentina*	Pore blocker	3–30 nM	Newcomb et al. (1995)

(continued)

Table 1 (continued)

Peptide	Source	Activity	IC_{50}	References
Huwentoxin-1	*Selenocosmia huwena*	Pore blocker	~100 nM	Peng et al. (2001)
ω-Ctenitoxin-Pr1a(PRTx3-7)	*Phoneutria reidyi*	Gating modifier	436 nM	Peng et al. (2001)
PnTx3–6	*Phoneutria nigriventer*	Pore blocker	122 nM	Vieira et al. (2005)
ω-Phonetoxin IIA	*Phoneutria nigriventer*	Not known	0.16 nM	Ertel et al. (1994) and Yan and Adams (2000)

Conus magus is another important source of ω-conotoxins, including MVIIA, MVIIB, MVIIC, and MVIID (Olivera et al., 1985, 1987; Hillyard et al., 1992; Monje et al., 1993). Among them, MVIIA and MVIIB selectively inhibit the $Ca_V2.2$ channel, whereas despite high sequence homology, MVIIC is more selective to the $Ca_V2.1$ channel, showing that a small change in the sequence is enough to switch Ca_V channel subtype selectivity (Woppmann et al., 1994; Grantham et al., 1994). ω-Conotoxin MVIIA (ziconotide or Prialt) is a clinically approved intrathecal drug for pain based on its potent antinociceptive effect in a wide range of neuropathic pain models including the sciatic nerve ligation (SNL) and chronic constriction injury (CCI) models (Bowersox et al., 1996; Chaplan et al., 1994; Chang et al., 2015), as well as inflammatory (Malmberg & Yaksh, 1994; Bowersox et al., 1996) and postoperative pain (Wang et al., 2000a), where it produced dose-dependent analgesia when administered spinally. However, the analgesic effect of ziconotide was less profound for acute pain models (Bowersox et al., 1996; Wang et al., 2000b), suggesting that ziconotide may interfere with the pain processing associated with central (spinal) sensitization. At least three intrathecal clinical trials of ziconotide have shown a significant lowering of pain scores in patients with a variety of refractory neuropathic pain syndromes (Staats et al., 2004; Wallace et al., 2006; Rauck et al., 2006). In 2004, ziconotide was the first Ca_V channel peptide inhibitor to gain approval from FDA for intrathecal administration for treating intractable cancer pain in humans. Although ziconotide does not develop tolerance or dependence like opioid analgesics, it has a narrow therapeutic window

arising from dose-limiting CNS side effects (Schmidtko et al., 2010).

Multiple ω-conotoxins have also been isolated from *Conus catus* (Lewis et al., 2000) and *Conus striatus* venoms (Ramilo et al., 1992), which have shown potent inhibition of $Ca_V2.2$ channels. At least six ω-conotoxins, CVIA–F, have now been identified from the venoms of *Conus catus* (Lewis et al., 2000; Berecki et al., 2010). Although most inhibited both $Ca_V2.1$ and $Ca_V2.2$ channels, ω-conotoxins CVID and CVIE showed high selectivity to $Ca_V2.2$ channels (Lewis et al., 2000; Berecki et al., 2010). Indeed, CVID showed a better efficacy-side effect profile in preclinical studies and marginal therapeutic advantage in limited clinical studies (Malmberg & Yaksh, 1995; Smith et al., 2002; Scott et al., 2002; Lewis et al., 2000). ω-Conotoxins SVIA and SVIB from *Conus striatus* venom were potent in fish but in mice, their activity was significantly lower than observed for MIIA (Ramilo et al., 1992). A relatively low dose (300 pmol/g) of SVIB was lethal to mice, whereas GVIA, MVIIA, and SVIA were non-lethal even in high doses (Ramilo et al., 1992), suggesting differences in subtype selectivity and/or tissue distribution, which was supported by radioligand binding studies (Woppmann et al., 1994; Nielsen et al., 1996).

ω-Conotoxins have also been identified from fish-hunting *Conus fulmen* (FVIA) (Lee et al., 2010), *Conus consors* (CnVIIA-C), *Conus radiatus* (RVIA) (Ramírez et al., 2017), and *Conus tulipa* (TVIA) (Miljanich & Ramachandran, 1995). Both FVIA and CnVIIA selectively inhibited $Ca_V2.2$ channels despite CnVIIA containing an unusual sequence (SSSKGR) in loop 4 of its structure (Lee et al., 2010; Ramírez et al., 2017).

Although dose-dependent analgesic effect and reversible binding suggests FVIA to be a suitable analgesic lead, no follow-up study has been reported. Recently, a novel ω-conotoxin, Bu8, found in the fish-hunting cone snail *Conus bullatus*, has been reported to selectively block Ca$_V$2.2 and display potent analgesic activity (Chen et al., 2021). In animal pain models, it exhibits strong analgesic activity with fewer side effects and lower toxicity than MVIIA, which has been attributed to its faster on- and off-rates for binding (Chen et al., 2021).

Two novel ω-conotoxins, MoVIA and MoVIB from *Conus moncuri*, are the first selective Ca$_V$2.2 channel inhibitors identified from vermivorous cone snails (Sousa et al., 2018). Both potently inhibit the Ca$_V$2.2 channel and displace radiolabeled ω-conotoxin GVIA (^{125}I-GVIA) (Sousa et al., 2018), confirming that worm-hunting cone snails can also be a source of analgesic ω-conotoxins.

Non-selective Ca$_V$2.2 Inhibitors

A diverse array of Ca$_V$2.2 channel modifiers have also been identified from spiders, scorpions, centipedes, and assassin bugs that inhibit or activate Ca$_V$2.2 channels mostly non-selectively. ω-Agatoxin from the American funnel-web spider *Agenelopsis aperta* was the first reported Ca$_V$ channel inhibitor identified from spiders (Adams et al., 1993; Olivera et al., 1994). They have been classified into four different families (ω-agatoxin I-IV) depending on their structure and pharmacology (Adams, 2004). Type II ω-agatoxins include Aa2a, which preferentially blocks Ca$_V$2.2, whereas type III ω-agatoxins including ω-agatoxin-Aa3a (Bindokas & Adams, 1989) inhibit high-threshold Ca$_V$ channels Ca$_V$2.1, Ca$_V$2.2, and Ca$_V$2.3 (Adams et al., 1990). In contrast, type I and type IV ω-agatoxins are selective for Ca$_V$1 and Ca$_V$2.1, respectively, although ω-agatoxin IVA blocks Ca$_V$2.2 at higher concentrations (Sidach & Mintz, 2000; Williams et al., 1994).

Other non-selective spider toxins that bind to the Ca$_V$2.2 channel include ω-grammotoxin SIA from the Chilean tarantula *Grammostola spatulate* (McDonough et al., 1997a, b), huwentoxin-I from the Chinese spider *Selenocosmia huwena* (Peng et al., 2001), Phα1β, PhTx3-3, and PhTx3-4 from the Brazilian spider *Phoneutria nigriventer* (Cordeiro Mdo et al., 1993), and SNX-325 from the European spider *Segestria Florentina* (Newcomb et al., 1995).

The mechanism of action of Ca$_V$2.2 channel inhibition by spider toxins remains incompletely characterized, although they are likely gating modifiers. Radioligand binding studies showed that ω-conotoxin GVIA cannot displace ω-grammotoxin SIA from its binding site (Lampe et al., 1993), indicating that SIA does not block the pore, supporting a gating modifier mode of action. Interestingly, ω-grammotoxin SIA inhibition was additive with ω-agatoxin IVA at Ca$_V$2.2 (McDonough et al., 1997a, b), suggesting that these two toxins might target different sites or domains of the Ca$_V$2.2 channel. However, SNX-325 displaced ω-conotoxin MVIIA from Ca$_V$2.2 (Newcomb et al., 1995), suggesting that some spider toxins might act as pore blockers.

Ca$_V$ channel modulators are also found in the venom of venomous species other than cone snails and spiders, including kurtoxin from the venom of the scorpion *Parabuthus transvaalicus* (Sidach & Mintz, 2002), calcicludine from the venom of the green mamba *Dendroaspis angusticeps* (Stotz et al., 2000), ω-SLPTX-Ssm1a and ω-SLPTX-Ssm2a from the centipede *Scolopendra subspinipes mutilans* (Yang et al., 2012), glycerotoxin from the marine blood worm *Glycera convulata* (Meunier et al., 2002), and Adol, Ptul, and lobl from assassin bug venom (Corzo et al., 2001). However, these peptides are non-selective for the Ca$_V$2.2 channel, or in the case of glycerotoxin, are Ca$_V$2.2 channel activators.

Structure-Activity Studies of ω-Conotoxins at Ca$_V$2.2

Our initial understanding of the binding site for ω-conotoxins on Ca$_V$2.2 was provided by Ellinor et al. (1994), who showed that deletion of the domain III S5-S6 region of the α$_{1B}$ subunit reduces the on-rate for ω-conotoxin GVIA block. A single cluster of residues at the domain III S5-S6 region had the greatest effect on the on-rate but effects of this region on the off-rate were not considered. Feng et al. (2001a, b) later showed that all high voltage-activated calcium channels contained a putative EF-hand motif in the domain III S5-S6 region. This EF-hand motif comprised a central glycine residue flanked by negatively charged residues. The replacement of these highly conserved negative residues with positively charged residues significantly changed the permeation characteristics of the channel, with the central glycine also influencing the reversibility of ω-conotoxin binding (Feng et al., 2001a, b).

Recently, the cryo-EM structure of $Ca_V2.2$ has been solved in the presence and absence of ω-conotoxin MVIIA (Gao et al., 2021) (Fig. 1a). This study definitively revealed that MVIIA is positioned in an electronegative space comprising E314, E663, E1365, and E1655 from the P1 and P2 helices of domains I, II, III, and IV (Fig. 1b). In this pose, MVIIA blocks the entrance to the selectivity filter from the extracellular side by interacting with a cluster of polar and positively charged residues on the P1 and P2 helices and the extracellular loops of domains II, III, and IV (Gao et al., 2021) to occlude Ca^{2+} ion conduction. Surprisingly, none of the residues identified from domain I are directly involved in MVIIA binding (Gao et al., 2021), suggesting that these affected binding indirectly.

The cryo-EM structure of $Ca_V2.2$ with MVIIA revealed that MVIIA uses Arg10 and Tyr13 to connect to the Asp664 positioned at the start of the P2 of domain II (Fig. 1b) (Gao et al., 2021). This is consistent with previous findings showing that the Y13A, R10A, and Y13R mutants of MVIIA lost pore-blocking activity (Kim et al., 1995; Sousa et al., 2018; Nadasdi et al., 1995). The orientation of Tyr13 is also important since D-Tyr13 analogues of MVIIA had 3-4 orders of magnitude lower affinity than the native conotoxin (Nielsen et al., 1999a, b). Another residue of MVIIA that interacts with a pore loop residue is Ser19. It interacts with Glu1659, which is placed on the beginning of the P2 of domain IV

(Fig. 1b) (Gao et al., 2021). Residues from extracellular loops are also involved in MVIIA binding, with Thr17, Arg21, and Lys4 interacting with Asp1345, Asp1628, and Asp1629, respectively (Fig. 1b). Interestingly, Asp 1345 is located on the ECL of domain III, and Asp1628 and Asp1629 are located on the ECL of domain IV, revealing the broad extent of the interacting surface. Of the eight residues in $Ca_V2.2$ that interact with MVIIA, four (Thr643, Asp1345, Lys1372, and Asp1629) are unique to $Ca_V2.2$ and potentially contribute to the subtype selectivity of MVIIA.

The $α_2δ$ subunit of the $Ca_V2.2$ channel has also been found to be important in ω-conotoxin binding (Mould et al., 2004; Motin et al., 2007; Lewis et al., 2000). The absence of the $α_2δ_1$ subunit reduced the affinity of MVIIA and CVID three to fourfold for $Ca_V2.2$ in *Xenopus* oocyte studies (Mould et al., 2004). However, the cryo-EM structure of the $Ca_V2.2$ channel with MVIIA did not show any direct interaction with the $α_2δ$ subunit of the $Ca_V2.2$ channel (Gao et al., 2021). The only significant change after MVIIA binding is a small upward movement of $α_2δ_1$ subunit and ECL of domain III of $α_{1B}$ subunit (Fig. 2), which appears necessary to incorporate MVIIA into the binding site (Gao et al., 2021). Therefore, $α_2δ_1$ subunit may play a role in MVIIA binding by energetically favoring the upward movement of the ECL of domain III to accommodate MVIIA binding.

The key residues for the binding of ω-conotoxin to $Ca_V2.2$ are the highly conserved Tyr13 and Lys2 (Flinn et al., 1999; Lew et al., 1997; Kim et al., 1994). The replacement of Tyr13 with Phe significantly reduced the binding affinity in MVIIA (Nadasdi et al., 1995), GVIA (Kim et al., 1994), and CVID (Adams et al., 2003), suggesting that these structurally distinct ω-conotoxins share a similar overlapping binding site above the selectivity filter. In contrast, ω-conotoxins MoVIA and MoVIB have an Arg instead of Tyr at 13 position but still potently block $Ca_V2.2$, with MoVIB-[R13Y] showing reduced activity compared to MoVIB (Sousa et al., 2018), indicating that 13 Arg may have evolved before Tyr13 in ancestral ω-conotoxins from vermivorous species. Lys2Ala changes also

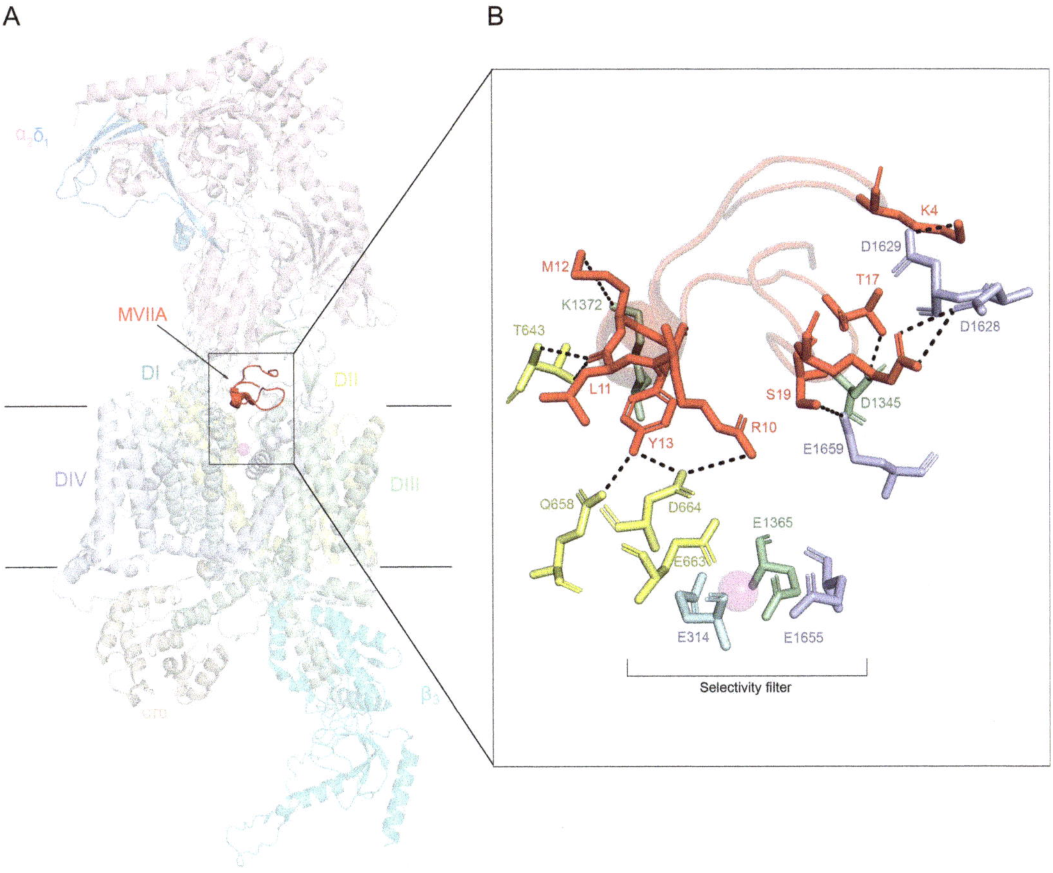

A

B

Fig. 1 Binding mode of ω-conotoxin MVIIA. (Gao et al., 2021). (**a**) MVIIA occludes the pore on top of the selectivity filter of the Ca$_v$2.2 channel. Different subunits and domains I-IV are distinguished by color. (**b**) Pairwise interactions between MVIIA and Ca$_v$2.2, with the selectivity filter (E314, E663, E1365, and E1655) of the Ca$_v$2.2 channel highlighted below MVIIA

Fig. 2 ω-Conotoxin MVIIA binding stabilizes the α$_2$δ$_1$ subunit and the extracellular loop of domain III outward (extracellularly) compared to the unbound (apo) state. (Gao et al., 2021)

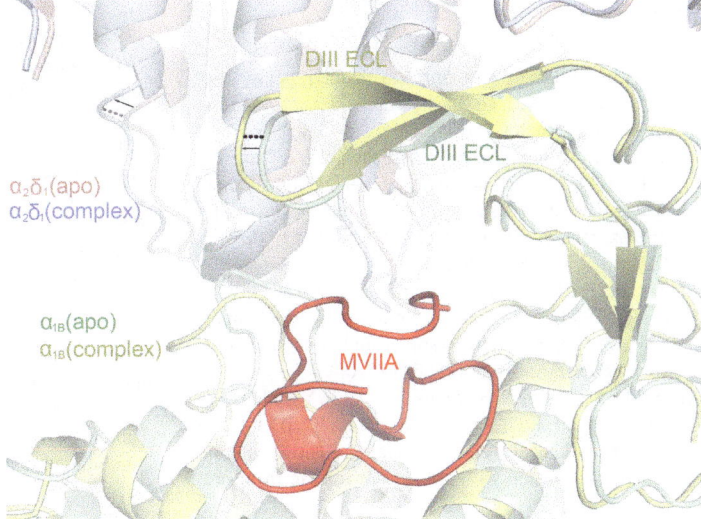

significantly reduced the affinity of GVIA (Kim et al., 1995; Sato et al., 1993; Lew et al., 1997) and MVIIA (Kim et al., 1995; Nadasdi et al., 1995) for $Ca_V2.2$ along with arginine, ornithine, leucine, and norleucine replacements, suggesting that the length of the side chain and charge at position 2 influence binding (Flinn et al., 1999). Surprisingly, Lys2 did not show any direct interaction with $Ca_V2.2$ in the MVIIA-$Ca_V2.2$ cryo-EM structure (Gao et al., 2021), suggesting that Lys2 might provide structural integrity by interacting with Asp 14 in loop 2 of MVIIA.

Conclusions

Venom peptides provided a rich source of calcium channel modulators, with inhibitors of $Ca_V2.2$ showing most clinical promise. The most selective $Ca_V2.2$ inhibitors identified to date have been isolated from the venoms of cone snails, with several entering clinical trials and MVIIA approved as an intrathecal treatment for severe (neurological) pain. Recent advances in our understanding of the mechanism of binding of ω-conotoxins to $Ca_V2.2$ revealed from cryo-EM structures pave the way for the rational development of ω-conotoxin analogues with improved potency and selectivity for therapeutically relevant calcium channels.

References

Adams, M. E. (2004). Agatoxins: Ion channel specific toxins from the American funnel web spider, *Agelenopsis aperta. Toxicon, 43*, 509–525.

Adams, M. E., Bindokas, V. P., Hasegawa, L., & Venema, V. J. (1990). Omega-agatoxins: Novel calcium channel antagonists of two subtypes from funnel web spider *Agelenopsis aperta* venom. *The Journal of Biological Chemistry, 265*, 861–867.

Adams, M. E., Myers, R. A., Imperial, J. S., & Olivera, B. M. (1993). Toxityping rat brain calcium channels with .omega.-toxins from spider and cone snail venoms. *Biochemistry, 32*, 12566–12570.

Adams, D. J., Smith, A. B., Schroeder, C. I., Yasuda, T., & Lewis, R. J. (2003). Omega-conotoxin CVID inhibits a pharmacologically distinct voltage-sensitive calcium channel associated with transmitter release from pre-ganglionic nerve terminals. *The Journal of Biological Chemistry, 278*, 4057–4062.

Bannister, R. A., & Beam, K. G. (2013). $Ca_V1.1$: The atypical prototypical voltage-gated Ca^{2+} channel. *Biochimica et Biophysica Acta, s1828*, 1587–1597.

Berecki, G., Motin, L., Haythornthwaite, A., Vink, S., Bansal, P., Drinkwater, R., Wang, C. I., Moretta, M., Lewis, R. J., Alewood, P. F., Christie, M. J., & Adams, D. J. (2010). Analgesic ω-conotoxins CVIE and CVIF selectively and voltage-dependently block recombinant and native N-type calcium channels. *Molecular Pharmacology, 77*, 139–148.

Bindokas, V. P., & Adams, M. E. (1989). omega-Aga-I: A presynaptic calcium channel antagonist from venom of the funnel web spider, *Agelenopsis aperta. Journal of Neurobiology, 20*, 171–188.

Bourinet, E., & Gerald Zamponi, W. (2016). Block of voltage-gated calcium channels by peptide toxins. *Neuropharmacology, 127*, 109–115.

Bowersox, S. S., Gadbois, T., Singh, T., Pettus, M., Wang, Y. X., & Luther, R. R. (1996). Selective N-type neuronal voltage-sensitive calcium channel blocker, SNX-111, produces spinal antinociception in rat models of acute, persistent and neuropathic pain. *The Journal of Pharmacology and Experimental Therapeutics, 279*, 1243–1249.

Catterall, W. A. (2000). Structure and regulation of voltage-gated Ca^{2+} channels. *Annual Review of Cell and Developmental Biology, 16*, 521–555.

Catterall, W. A. (2011). Voltage-gated calcium channels. *Cold Spring Harbor Perspectives in Biology, 3*, a003947.

Chang, E., Chen, X., Kim, M., Gong, N., Bhatia, S., & Luo, Z. D. (2015). Differential effects of voltage-gated calcium channel blockers on calcium channel alpha-2-delta-1 subunit protein-mediated nociception. *European Journal of Pain (London, England), 19*, 639–648.

Chaplan, S. R., Pogrel, J. W., & Yaksh, T. L. (1994). Role of voltage-dependent calcium channel subtypes in experimental tactile allodynia. *The Journal of Pharmacology and Experimental Therapeutics, 269*, 1117–1123.

Chen, J., Liu, X., Yu, S., Liu, J., Chen, R., Zhang, Y., Jiang, L., & Dai, Q. (2021). A novel ω-conotoxin Bu8 inhibiting N-type voltage-gated calcium channels displays potent analgesic activity. *Acta Pharmaceutica Sinica B, 11*, 2685–2693.

Cordeiro Mdo, N., de Figueiredo, S. G., Valentim Ado, C., Diniz, C. R., von Eickstedt, V. R., Gilroy, J., & Richardson, M. (1993). Purification and amino acid sequences of six Tx3 type neurotoxins from the venom of the Brazilian 'armed' spider *Phoneutria nigriventer. Toxicon, 31*, 35–42.

Corzo, G., Adachi-Akahane, S., Nagao, T., Kusui, Y., & Nakajima, T. (2001). Novel peptides from assassin bugs *Hemiptera reduviidae*: Isolation, chemical and biological characterization. *FEBS Letters, 499*, 256–261.

Curtis, B. M., & Catterall, W. A. (1986). Reconstitution of the voltage-sensitive calcium channel purified from skeletal muscle transverse tubules. *Biochemistry, 25,* 3077–3083.

Ellinor, P. T., Zhang, J. F., Horne, W. A., & Tsien, R. W. (1994). Structural determinants of the blockade of N-type calcium channels by a peptide neurotoxin. *Nature, 372,* 272–275.

Ertel, E. A., Warren, V. A., Adams, M. E., Griffin, P. R., Cohen, C. J., & Smith, M. M. (1994). Type III omega-agatoxins: A family of probes for similar binding sites on L- and N-type calcium channels. *Biochemistry, 33,* 5098–5108.

Favreau, P., Gilles, N., Lamthanh, H., Bournaud, R., Shimahara, T., Bouet, F., Laboute, P., Letourneux, Y., Ménez, A., Molgó, J., & Le Gall, F. (2001). A new ω-Conotoxin that targets N-type voltage-sensitive calcium channels with unusual specificity. *Biochemistry, 40,* 14567–14575.

Feng, Z. P., Hamid, J., Doering, C., Bosey, G. M., Snutch, T. P., & Zamponi, G. W. (2001a). Residue Gly1326 of the N-type calcium channel α$_{1B}$ subunit controls reversibility of ω-conotoxin GVIA and MVIIA block. *The Journal of Biological Chemistry, 276,* 15728–15735.

Feng, Z.-P., Hamid, J., Doering, C., Jarvis, S. E., Bosey, G. M., Bourinet, E., Snutch, T. P., & Zamponi, G. W. (2001b). Amino acid residues outside of the pore region contribute to N-type calcium channel permeation. *Journal of Biological Chemistry, 276,* 5726–5730.

Flinn, J. P., Pallaghy, P. K., Lew, M. J., Murphy, R., Angus, J. A., & Norton, R. S. (1999). Roles of key functional groups in ω-conotoxin GVIA. *European Journal of Biochemistry, 262,* 447–455.

Flockerzi, V., Oeken, H. J., Hofmann, F., Pelzer, D., Cavalie, A., & Trautwein, W. (1986). Purified dihydropyridine-binding site from skeletal muscle t-tubules is a functional calcium channel. *Nature, 323,* 66–68.

Gao, S., Yao, X., & Yan, N. (2021). Structure of human Cav2.2 channel blocked by the painkiller ziconotide. *Nature, 596,* 143–147.

Grantham, C. J., Bowman, D., Bath, C. P., Bell, D. C., & Bleakman, D. (1994). ω-Conotoxin MVIIC reversibly inhibits a human N-type calcium channel and calcium influx into chick synaptosomes. *Neuropharmacology, 33,* 255–258.

Hillyard, D. R., Monje, V. D., Mintz, I. M., Bean, B. P., Nadasdi, L., Ramachandran, J., Miljanich, G., Azimi-Zoonooz, A., McIntosh, J. M., Cruz, L. J., et al. (1992). A new Conus peptide ligand for mammalian presynaptic Ca^{2+} channels. *Neuron, 9,* 69–77.

Kim, J. I., Takahashi, M., Ogura, A., Kohno, T., Kudo, Y., & Sato, K. (1994). Hydroxyl group of Tyr13 is essential for the activity of ω-conotoxin GVIA, a peptide toxin for N-type calcium channel. *The Journal of Biological Chemistry, 269,* 23876–23878.

Kim, J. I., Takahashi, M., Ohtake, A., Wakamiya, A., & Sato, K. (1995). Tyr13 is essential for the activity of omega-conotoxin MVIIA and GVIA, specific

N-type calcium channel blockers. *Biochemical and Biophysical Research Communications, 206,* 449–454.

Lampe, R. A., Defeo, P. A., Davison, M. D., Young, J., Herman, J. L., Spreen, R. C., Horn, M. B., Mangano, T. J., & Keith, R. A. (1993). Isolation and pharmacological characterization of ω-grammotoxin SIA, a novel peptide inhibitor of neuronal voltage-sensitive calcium channel responses. *Molecular Pharmacology, 44,* 451–460.

Lee, S., Kim, Y., Back, S. K., Choi, H. W., Lee, J. Y., Jung, H. H., Ryu, J. H., Suh, H. W., Na, H. S., Kim, H. J., Rhim, H., & Kim, J. I. (2010). Analgesic effect of highly reversible ω-conotoxin FVIA on N type Ca^{2+} channels. *Molecular Pain, 6,* 97.

Lew, M. J., Flinn, J. P., Pallaghy, P. K., Murphy, R., Whorlow, S. L., Wright, C. E., Norton, R. S., & Angus, J. A. (1997). Structure-function relationships of ω-Conotoxin GVIA: Synthesis, structure, calcium channel binding, and functional assay of alanine-substituted analogues. *Journal of Biological Chemistry, 272,* 12014–12023.

Lewis, R. J., Nielsen, K. J., Craik, D. J., Loughnan, M. L., Adams, D. A., Sharpe, I. A., Luchian, T., Adams, D. J., Bond, T., Thomas, L., Jones, A., Matheson, J. L., Drinkwater, R., Andrews, P. R., & Alewood, P. F. (2000). Novel omega-conotoxins from Conus catus discriminate among neuronal calcium channel subtypes. *The Journal of Biological Chemistry, 275,* 35335–35344.

Lewis, R. J., Dutertre, S., Vetter, I., & Christie, M. J. (2012). Conus venom peptide pharmacology. *Pharmacological Reviews, 64,* 259.

Liu, Z., Bartels, P., Sadeghi, M., Tianpeng, D., Dai, Q., Zhu, C., Yu, S., Wang, S., Dong, M., Sun, T., Guo, J., Peng, S., Jiang, L., Adams, D. J., & Dai, Q. (2018). A novel α-conopeptide Eu1.6 inhibits N-type (Ca$_v$2.2) calcium channels and exhibits potent analgesic activity. *Scientific Reports, 8,* 1004.

Malmberg, A. B., & Yaksh, T. L. (1994). Voltage-sensitive calcium channels in spinal nociceptive processing: Blockade of N- and P-type channels inhibits formalin-induced nociception. *The Journal of Neuroscience, 14,* 4882–4890.

Malmberg, A. B., & Yaksh, T. L. (1995). Effect of continuous intrathecal infusion of omega-conopeptides, N-type calcium-channel blockers, on behavior and antinociception in the formalin and hot-plate tests in rats. *Pain, 60,* 83–90.

McCleskey, E. W., Fox, A. P., Feldman, D. H., Cruz, L. J., Olivera, B. M., Tsien, R. W., & Yoshikami, D. (1987). Omega-conotoxin: Direct and persistent blockade of specific types of calcium channels in neurons but not muscle. *Proceedings of the National Academy of Sciences, 84,* 4327.

McDonough, S. I., Mintz, I. M., & Bean, B. P. (1997a). Alteration of P-type calcium channel gating by the spider toxin omega-Aga-IVA. *Biophysical Journal, 72,* 2117–2128.

McDonough, S. I., Lampe, R. A., Keith, R. A., & Bean, B. P. (1997b). Voltage-dependent inhibition of N-

and P-type calcium channels by the peptide toxin ω-Grammotoxin-SIA. *Molecular Pharmacology, 52,* 1095–1104.

Meunier, F. A., Feng, Z. P., Molgó, J., Zamponi, G. W., & Schiavo, G. (2002). Glycerotoxin from Glycera convoluta stimulates neurosecretion by up-regulating N-type Ca^{2+} channel activity. *The EMBO Journal, 21,* 6733–6743.

Miljanich G. P., Bitner, R. S., Fox, J. A., Valentino, K. L., Yamashiro, D. H., & Tsubokawa, M. (1993a). Method of reducing neuronal damage using omega conotoxin peptides. In *U.S. Patent 5,424,218 A.*

Miljanich, G. P., Bowersox S. S., Fox J. A., Valentino K. L., Bitner R. S., & Yamashiro D. H. (1993b). Compositions for delayed treatment of ischemia-related neuronal damage. In *WO: Patent 1993010145 A1: NEUREX CORP.*

Miljanich, G. P., & Ramachandran, J. (1995). Antagonists of neuronal calcium channels: Structure, function, and therapeutic implications. *Annual Review of Pharmacology and Toxicology, 35,* 707–734.

Monje, V. D., Haack, J. A., Naisbitt, S. R., Miljanich, G., Ramachandran, J., Nasdasdi, L., Olivera, B. M., Hillyard, D. R., & Gray, W. R. (1993). A new Conus peptide ligand for Ca channel subtypes. *Neuropharmacology, 32,* 1141–1149.

Motin, L., Yasuda, T., Schroeder, C. I., Lewis, R. J., & Adams, D. J. (2007). ω-Conotoxin CVIB differentially inhibits native and recombinant N- and P/Q-type calcium channels. *The European Journal of Neuroscience, 25,* 435–444.

Mould, J., Yasuda, T., Schroeder, C. I., Beedle, A. M., Doering, C. J., Zamponi, G. W., Adams, D. J., & Lewis, R. J. (2004). The $\alpha_2\delta$ auxiliary subunit reduces affinity of ω-conotoxins for recombinant N-type (Cav2.2) calcium channels. *The Journal of Biological Chemistry, 279,* 34705–34714.

Nadasdi, L., Yamashiro, D., Chung, D., Tarczy-Hornoch, K., Adriaenssens, P., & Ramachandran, J. (1995). Structure-activity analysis of a Conus peptide blocker of N-type neuronal calcium channels. *Biochemistry, 34,* 8076–8081.

Newcomb, R., Palma, A., Fox, J., Gaur, S., Lau, K., Chung, D., Cong, R., Bell, J. R., Horne, B., Nadasdi, L., et al. (1995). SNX-325, a novel calcium antagonist from the spider *Segestria florentina. Biochemistry, 34,* 8341–8347.

Nielsen, K. J., Thomas, L., Lewis, R. J., Alewood, P. F., & Craik, D. J. (1996). A consensus structure for ω-Conotoxins with different selectivities for voltage-sensitive calcium channel subtypes: Comparison of MVIIA, SVIB and SNX-202. *Journal of Molecular Biology, 263,* 297–310.

Nielsen, K. J., Adams, D. A., Alewood, P. F., Lewis, R. J., Thomas, L., Schroeder, T., & Craik, D. J. (1999a). Effects of chirality at Tyr13 on the structure-activity relationships of omega-conotoxins from Conus magus. *Biochemistry, 38,* 6741–6751.

Nielsen, K. J., Adams, D., Thomas, L., Bond, T., Alewood, P. F., Craik, D. J., & Lewis, R. J. (1999b). Structure-

activity relationships of ω-conotoxins MVIIA, MVIIC and 14 loop splice hybrids at N and P/Q-type calcium channels. *Journal of Molecular Biology, 289,* 1405–1421.

Olivera, B. M., Gray, W. R., Zeikus, R., McIntosh, J. M., Varga, J., Rivier, J., de Santos, V., & Cruz, L. J. (1985). Peptide neurotoxins from fish-hunting cone snails. *Science, 230,* 1338–1343.

Olivera, B. M., Cruz, L. J., de Santos, V., LeCheminant, G. W., Griffin, D., Zeikus, R., McIntosh, J. M., Galyean, R., Varga, J., Gray, W. R., et al. (1987). Neuronal calcium channel antagonists. Discrimination between calcium channel subtypes using omega-conotoxin from *Conus magus* venom. *Biochemistry, 26,* 2086–2090.

Olivera, B. M., Miljanich, G. P., Ramachandran, J., & Adams, M. E. (1994). Calcium channel diversity and neurotransmitter release: The ω-conotoxins and ω-agatoxins. *Annual Review of Biochemistry, 63,* 823–867.

Peng, K., Chen, X.-D., & Liang, S.-P. (2001). The effect of Huwentoxin-I on Ca2+ channels in differentiated NG108-15 cells, a patch-clamp study. *Toxicon, 39,* 491–498.

Pringos, E., Vignes, M., Martinez, J., & Rolland, V. (2011). Peptide neurotoxins that affect voltage-gated calcium channels: A close-up on ω-Agatoxins. *Toxins (Basel), 3,* 17–42.

Ramilo, C. A., Zafaralla, G. C., Nadasdi, L., Hammerland, L. G., Yoshikami, D., Gray, W. R., Kristipati, R., Ramachandran, J., & Miljanich, G. (1992). Novel .alpha.- and .omega.-conotoxins and *Conus striatus* venom. *Biochemistry, 31,* 9919–9926.

Ramírez, D., Gonzalez, W., Fissore, R. A., & Carvacho, I. (2017). Conotoxins as tools to understand the physiological function of voltage-gated calcium (Ca(V)) channels. *Marine Drugs, 15,* 313.

Rauck, R. L., Wallace, M. S., Leong, M. S., Minehart, M., Webster, L. R., Charapata, S. G., Abraham, J. E., Buffington, D. E., Ellis, D., & Kartzinel, R. (2006). A randomized, double-blind, placebo-controlled study of intrathecal ziconotide in adults with severe chronic pain. *Journal of Pain and Symptom Management, 31,* 393–406.

Sato, K., Park, N. G., Kohno, T., Maeda, T., Kim, J. I., Kato, R., & Takahashi, M. (1993). Role of basic residues for the binding of omega-conotoxin GVIA to N-type calcium channels. *Biochemical and Biophysical Research Communications, 194,* 1292–1296.

Schmidtko, A., Lötsch, J., Freynhagen, R., & Geisslinger, G. (2010). Ziconotide for treatment of severe chronic pain. *Lancet, 375,* 1569–1577.

Scott, D. A., Wright, C. E., & Angus, J. A. (2002). Actions of intrathecal omega-conotoxins CVID, GVIA, MVIIA, and morphine in acute and neuropathic pain in the rat. *European Journal of Pharmacology, 451,* 279–286.

Sidach, S. S., & Mintz, I. M. (2000). Low-affinity blockade of neuronal N-type Ca channels by the spider toxin ω-Agatoxin-IVA. *The Journal of Neuroscience, 20*, 7174–7182.

Sidach, S. S., & Mintz, I. M. (2002). Kurtoxin, A gating modifier of neuronal high- and low-threshold Ca channels. *The Journal of Neuroscience, 22*, 2023–2034.

Smith, M. T., Cabot, P. J., Ross, F. B., Robertson, A. D., & Lewis, R. J. (2002). The novel N-type calcium channel blocker, AM336, produces potent dose-dependent antinociception after intrathecal dosing in rats and inhibits substance P release in rat spinal cord slices. *Pain, 96*, 119–127.

Sousa, S. R., McArthur, J. R., Brust, A., Bhola, R. F., Rosengren, K. J., Ragnarsson, L., Dutertre, S., Alewood, P. F., Christie, M. J., Adams, D. J., Vetter, I., & Lewis, R. J. (2018). Novel analgesic ω-conotoxins from the vermivorous cone snail Conus moncuri provide new insights into the evolution of conopeptides. *Scientific Reports, 8*, 13397.

Staats, P. S., Yearwood, T., Charapata, S. G., Presley, R. W., Wallace, M. S., Byas-Smith, M., Fisher, R., Bryce, D. A., Mangieri, E. A., Luther, R. R., Mayo, M., McGuire, D., & Ellis, D. (2004). Intrathecal ziconotide in the treatment of refractory pain in patients with cancer or AIDS: A randomized controlled trial. *JAMA, 291*, 63–70.

Stotz, S. C., Spaetgens, R. L., & Zamponi, G. W. (2000). Block of voltage-dependent calcium channel by the green mamba toxin calcicludine. *The Journal of Membrane Biology, 174*, 157–165.

Takahashi, M., Seagar, M. J., Jones, J. F., Reber, B. F., & Catterall, W. A. (1987). Subunit structure of dihydropyridine-sensitive calcium channels from skeletal muscle. *Proceedings of the National Academy of Sciences of the United States of America, 84*, 5478–5482.

Tanabe, T., Takeshima, H., Mikami, A., Flockerzi, V., Takahashi, H., Kangawa, K., Kojima, M., Matsuo, H., Hirose, T., & Numa, S. (1987). Primary structure of the receptor for calcium channel blockers from skeletal muscle. *Nature, 328*, 313–318.

Vieira, L. B., Kushmerick, C., Hildebrand, M. E., Garcia, E., Stea, A., Cordeiro, M. N., Richardson, M., Gomez, M. V., & Snutch, T. P. (2005). Inhibition of high voltage-activated calcium channels by spider toxin PnTx3-6. *Journal of Pharmacology and Experimental Therapeutics, 314*, 1370.

Wallace, M. S., Charapata, S. G., Fisher, R., Byas-Smith, M., Staats, P. S., Mayo, M., McGuire, D., & Ellis, D. (2006). Intrathecal ziconotide in the treatment of chronic nonmalignant pain: A randomized, double-blind, placebo-controlled clinical trial. *Neuromodulation, 9*, 75–86.

Wang, Y. X., Gao, D., Pettus, M., Phillips, C., & Bowersox, S. S. (2000a). Interactions of intrathecally administered ziconotide, a selective blocker of neuronal N-type voltage-sensitive calcium channels, with morphine on nociception in rats. *Pain, 84*, 271–281.

Wang, Y.-X., Pettus, M., Gao, D., Phillips, C., & Scott Bowersox, S. (2000b). Effects of intrathecal administration of ziconotide, a selective neuronal N-type calcium channel blocker, on mechanical allodynia and heat hyperalgesia in a rat model of postoperative pain. *Pain, 84*, 151–158.

Wang, F., Yan, Z., Liu, Z., Wang, S., Qiaoling, W., Shuo, Y., Ding, J., & Dai, Q. (2016). Molecular basis of toxicity of N-type calcium channel inhibitor MVIIA. *Neuropharmacology, 101*, 137–145.

Williams, M. E., Marubio, L. M., Deal, C. R., Hans, M., Brust, P. F., Philipson, L. H., Miller, R. J., Johnson, E. C., Harpold, M. M., & Ellis, S. B. (1994). Structure and functional characterization of neuronal alpha 1E calcium channel subtypes. *Journal of Biological Chemistry, 269*, 22347–22357.

Woppmann, A., Ramachandran, J., & Miljanich, G. P. (1994). Calcium channel subtypes in rat brain: Biochemical characterization of the high-affinity receptors for ω-conopeptides SNX-230 (synthetic MVIIC), SNX-183 (SVIB), and SNX-111 (MVIIA). *Molecular and Cellular Neuroscience, 5*, 350–357.

Wu, J., Yan, Z., Li, Z., Qian, X., Lu, S., Dong, M., Zhou, Q., & Yan, N. (2016). Structure of the voltage-gated calcium channel Cav1.1 at 3.6 A resolution. *Nature, 537*, 191–196.

Yan, L., & Adams, M. E. (2000). The spider toxin ω-Aga IIIA defines a high affinity site on neuronal high voltage-activated calcium channels. *Journal of Biological Chemistry, 275*, 21309–21316.

Yang, S., Liu, Z., Xiao, Y., Li, Y., Rong, M., Liang, S., Zhang, Z., Yu, H., King, G. F., & Lai, R. (2012). Chemical punch packed in venoms makes centipedes excellent predators. *Molecular & Cellular Proteomics, 11*, 640–650.

From Calcium Channels to New Therapeutics

Stefan I. McDonough

Abstract

An ultimate goal of biomedical research into calcium channels is to enable better treatment of disease, most prominently through the development of experimental therapeutics for unaddressed medical needs. As described throughout this book, voltage-gated calcium channels unquestionably are key control points for a diversity of physiological and pathophysiological mechanisms, particularly in excitable cells. Leveraging such biological information into successful therapeutics requires many scientific, clinical, and strategic steps, starting from robust validation for the role of an individual target in disease. Calcium channels are the targets for several drugs with a trajectory that now can be seen from conception to maturity, showing some key determinants of success. This chapter reviews the development of nifedipine and dihydropyridines, pregabalin and gabapentin, and ziconotide, which are clinically effective via inhibition of Cav1 L-type channels, Cavα2δ1/2, and Cav2.2 N-type channels, respectively. Emerging tools and resources for drug discovery are highlighted, particularly the genetic and genomic resources increasingly used to find molecular intervention points from an untargeted perspective. Finally, this chapter surveys the ongoing development of experimental therapeutics targeting Cav3 T-type, Cav1, and other calcium channels, including experimental treatments for Parkinson's, essential tremor, psychiatric diseases, and epilepsy, in light of the drug discovery flow path.

Keywords

Dihydropyridine · Pregabalin · Ziconotide · T-type calcium channel · Epilepsy

Nifedipine

The dihydropyridine class of drugs is an unqualified success, scientifically and in clinical use. Nifedipine, the prototype dihydropyridine inhibitor of L-type calcium channels, particularly Cav1.2, was approved for therapeutic use in the USA in 1981 and held steady at about five million prescriptions per year throughout the 2010s (ClinCalc resource, using data derived from US Department of Health and Human Services Medical Expenditure Panel Survey, clincalc.com/DrugStats). Nifedipine and amlodipine (and nimodipine, isradipine, and other dihydropyridine drugs, as well as L-type blockers of different chemical classes) are used to treat angina,

S. I. McDonough (✉)
Atalanta Therapeutics, Boston, MA, USA
e-mail: smcdonough@atalantatx.com

© The Author(s), under exclusive license to Springer Nature Switzerland AG 2022
G. W. Zamponi, N. Weiss (eds.), *Voltage-Gated Calcium Channels*,
https://doi.org/10.1007/978-3-031-08881-0_27

hypertension, subarachnoid hemorrhage, and many other cardiac indications. Nifedipine and amlodipine both are on the World Health Organization's list of essential medicines (2019). A full accounting of the properties of dihydropyridines, L-type blockers, and their clinical use is beyond the scope of this chapter, but Kazda (1991) has written a beautifully informative review on the discovery of nifedipine. Research into the dependence of cardiac contractility on extracellular calcium ions, the pharmacological modulation of the calcium-dependence of cardiac contractility, and the search for coronary vasodilators converged to produce nifedipine and other compounds, analoging from the starting point of compounds that were vasodilators on isolated heart (Bossert & Vater, 1989). The effects of dihydropyridines were originally discovered in the 1960s in direct testing on canines, in a search for orally delivered compounds that increased oxygen content in the coronary sinus. Key to eventual success was the translatability to patients of this screening model. The first dihydropyridine vasodilators increased oxygen saturation at low doses, and also produced potent and powerful lowering of blood pressure and negative inotropy, effects possibly unsuitable for patients with cardiovascular diseases. Further pharmacology led to nifedipine, which lacked a depressant effect on cardiac nodes and had a much larger effect on vasodilation than on inotropy. Nifedipine was also found to decrease arterial spasms, an effect recognized eventually as beneficial to clinical angina, and to have minimal effects on the blood pressure of normotensive patients. This work led eventually to a clinical indication of hypertension (Kazda, 1991). Molecular understanding of a seeming paradox of how dihydropyridines exerted selective action on hypertensive smooth muscle came in 1984 with Bean's description that nitrendipine potency on calcium channels of voltage-clamped dog ventricular myocytes increased almost three log units with holding voltages that produced channel inactivation, interpreted via the modulated receptor hypothesis as state-dependent binding (Bean, 1984; Sun & Triggle, 1995). Experiments on hypertensive rat strains suggested nifedipine also

may exert effects via the kidney, via enhancement of sodium excretion via sodium-calcium exchange and reduction of volume load (Kazda et al., 1982; Luckhaus et al., 1985). Nifedipine has also been explored for beneficial effects on the establishment of atherosclerosis and of coronary artery disease (Fleckenstein et al., 1990; Lichtlen et al., 1990). Meta-analysis of nifedipine's real-world outcomes introduced some caution in its use, particularly dosing levels in patients with coronary heart disease (Furberg et al., 1995).

Dihydropyridines are an excellent example of how successive iterations of a drug can leverage mechanistic understanding and appropriate driver models to improve efficacy, tolerability, and use in combination with other drugs (Wang et al., 2017). Succeeding generations of dihydropyridines have achieved better control of the therapeutic effect, fewer adverse effects, better tolerance in patients with heart disease, and broader therapeutic applications. The second-generation amlodipine, approved in 1990, is the most prescribed dihydropyridine monotherapy by far, with 75 million USA prescriptions in 2019, the sixth-most prescribed of all drugs of all classes (ClinCalc DrugStats resource). Keys are its intrinsically longer half-life, increased oral bioavailability, and slower binding kinetics and onset of action. The latter effect likely leads to less activation of the sympathetic nervous system upon vasodilation and so to less cardiac load (Burges & Moisey, 1994) and is dependent on nifedipine formulation (Toal et al., 2012). The next-generation cilnidipine (Chandra & Ramesh, 2013), approved in some Asian and European countries, is an inhibitor of N-type and L-type calcium channels (Uneyama et al., 1997; Fujii et al., 1997), and so may minimize feedback cardiac effects of vasodilation through direct inhibition of sympathetic activation (Mori et al., 2002).

Over 50 years of effort on dihydropyridines has produced an extraordinarily well-characterized toolkit of clinical compounds and knowledge of what patients to treat and how to treat them. One future avenue often mentioned is to generate compounds selective for Cav1.3 over Cav1.2, but this likely would require an intensive in vitro screening and medicinal chemistry campaign, using functional assays that are sufficiently difficult that experienced labs using similar techniques can arrive at different conclusions (McDonough, 2013; Ortner & Striessnig, 2016). It seems likely that future development of dihydropyridines will come from refining usage of existing compounds and drugs or applying them to new diseases, including CNS disorders. New indications may appear as large databases with genetic information, disease histories and longitudinal trajectories, and drug prescription information become available.

Gabapentin and Pregabalin

Gabapentin (Neurontin) has been available in the USA since 1993 and is labeled for epilepsy with partial onset seizures in adults and in pediatric patients 3 years and older, and for post-herpetic neuralgia pain. Initial approval was for epilepsy, and approval for postherpetic neuralgia followed in 2002 following demonstrated analgesic efficacy in several controlled clinical trials (Backonja & Glanzman, 2003). It has been estimated that of the over 40 million gabapentin prescriptions per year (ClinCalc resource), the majority are for off-label neurological indications (Peckham et al., 2018). Gabapentin originally was synthesized in hopes of mimicking the overall inhibitory actions of the neurotransmitter GABA itself while adding hydrophobicity to produce CNS distribution suitable for dosing (Maneuf et al., 2006). Gabapentin showed efficacy in multiple animal models of seizure, with a pattern distinct from existing anti-epileptics (McLean, 1995). Following initial hypotheses that gabapentin indeed acted via GABA receptors (Ng et al., 2001; Bertrand et al., 2001), gabapentin has been shown to affect neither GABAA ion channels nor GABAB GPCRs directly (Lanneau et al., 2001; Jensen et al., 2002; Martin et al., 2002; Hara & Sata, 2007), although downstream modulation of the GABAergic system may contribute to its clinical effects (Maneuf et al., 2006; Yu et al., 2019). Rather, the high-affinity binding site for gabapentin in brain (Suman-Chauhan et al., 1993;

Thurlow et al., 1993) was determined to be a calcium-channel $\alpha2\delta$ subunit (Gee et al., 1996), with residues within $\alpha2$ and δ each required for high-affinity binding (see also chapter "Regulation of Calcium Channels and Synaptic Function by Auxiliary $\alpha2\delta$ Subunits" by Dolphin and Obermair). Conflicting data on whether gabapentin inhibits the actual functional calcium channel were resolved by the demonstration that chronic but not acute gabapentin exposure reduces functional calcium-channel currents at the plasma membrane, via an intracellular mechanism that requires gabapentin uptake into the cell via system L transporters (Hendrich et al., 2008). The most plausible explanation for gabapentin efficacy is that it reduces neurotransmitter release via calcium channels, accounting for its effects on pain as well as on epilepsy (Hendrich et al., 2008). Gabapentin may have a selective effect on the insular cortex and brainstem in chronic pain syndromes (Iannetti et al., 2005).

Pregabalin (Lyrica) is approved for epilepsy and for chronic pain, backed by extensive series of large and well-controlled clinical trials (reviewed in Derry et al., 2019). First FDA approval was in 2004, for neuropathic pain. Pregabalin initially was aimed at increasing GABAergic signaling, not by itself being a GABA analog, but via inhibition of the GABA aminotransferase enzyme that breaks down GABA while sparing the L-glutamic acid decarboxylase (GAD) enzyme that produces GABA from L-glutamate (Silverman, 2008). Thus inhibiting GABA breakdown while sparing its production was hypothesized to lead to an overall increase in GABA levels. When tested, however, molecules in the lead chemical series actually were activators of GAD. As this also was consistent with the goal of raising GABA levels, compounds were tested in rodents. Dosed i.v., the series was indeed found to contain highly effective anticonvulsants, as measured by reduction in tonic extensor seizures of the hindlimbs produced by corneal electroshock stimulus, without side effects of ataxia (Silverman et al., 1991; Taylor et al., 1992). Efficacy in the chemical series, however, did not track with GAD inhibition, and pregabalin subsequently also was found to bind to α2δ subunits (Bryans & Wustrow, 1999). Its efficacy was found to depend on crossing the blood-brain barrier via the system L transporter (Verrey, 2003). Field and colleagues demonstrated in 2006 that the analgesic efficacy of pregabalin in the mouse formalin pain model and of pregabalin and gabapentin in the mouse nerve injury model were abolished by a point mutation in α2δ1. The mutation itself did not affect response to pain, nor the analgesic effects of morphine or amitriptyline. This strongly suggested α2δ1 as the molecular target mediating analgesia by pregabalin and gabapentin (Field et al., 2006). Key to the success of Lyrica were its excellent drug properties. Gabapentin delivered orally has saturable absorption and a nonlinear relation between dosing and plasma concentration at higher doses, due to a saturable transport process in the intestine (Stewart et al., 1993). By contrast, pregabalin has over 90% bioavailability regardless of dose (Elwes & Binnie, 1996), which allows for better control of dose titration and therapeutic index and may lead to greater efficacy (Bockbrader et al., 2010; Schulze-Bonhage, 2013). Efficacy is seen within a day or two (Silverman, 2008), and pregabalin neither inhibits nor induces liver enzymes, particularly cytochrome P450s, suggesting no interference with the metabolism of other drugs (Perruca, 2006). Over 100 clinical trials were executed with pregabalin in the late 1990s and early 2000s (Tassone et al., 2007) for diverse clinical indications. Lyrica now carries a label for neuropathic pain associated with diabetic peripheral neuropathy, post-herpetic neuralgia, fibromyalgia, and neuropathic pain associated with spinal cord injury. Lyrica also is labeled for adult partial onset seizures and carries a clear dosing recommendation for each indication to guide physicians.

Can pregabalin or gabapentin be improved? Although quite similar from an in vitro perspective and in many drug-like parameters, the linear dose-exposure relationship of pregabalin and understanding of its clinical use in pain were enough of an advantage to sustain blockbuster sales for Lyrica even following loss of patent expiry for gabapentin in 2004. It is conceivable

that pregabalin could be improved, since it binds to α2δ2 in addition to α2δ1 (Marais et al., 2001; Hendrich et al., 2008), and α2δ2 binding could govern CNS side effects (Barclay et al., 2001; Brodbeck et al., 2002; Ivanov et al., 2004). It is possible that a pregabalin analog with selectivity for α2δ1 over α2δ2 might retain its efficacy while having an increased therapeutic window. However, as noted for ziconotide (below), preclinical models for CNS side effects are difficult, meaning that the hypothesis of an increased therapeutic window likely could only be determined with a large head-to-head clinical trial. Data are not yet public, but Novoassay SA lists NVA1309 on its website as a successor to pregabalin in late preclinical studies, with selective binding to α2δ1 over α2δ2 and exposure in the spinal cord but not the brain. Patent expiration of pregabalin in 2019 means patient uptake and reimbursement for a next generation drug may turn on a strong advantage seen in head-to-head trials with pregabalin.

Ziconotide

Ziconotide arose from pioneering biology on the venom of cone snails, in the Olivera laboratory at the University of Utah (Olivera, 2021). The resulting drug, Prialt (for primary alternative to morphine), was approved in 2004 and carries a label for the treatment of severe chronic pain refractory to other treatments delivered via intrathecal infusion with a microinfusion pump. Note that this is a broad label and is not limited to individual etiologies of pain. The key observation leading to the development of ziconotide (originally termed ω-CTx-MVIIA, and SNX-111 in its early development by Neurex) was high-affinity binding of the radiolabeled peptide to the outer dorsal horn of the rat spinal cord (Gohil et al., 1994). This led Neurex to advance it as a therapeutic for pain, delivered intrathecally, effective via inhibition of presynaptic Cav2.2 N-type channels to block nociceptive transmission into or within the spinal cord. Ziconotide originally also was reported to have widespread specific binding within the brain (Gohil et al., 1994), suggesting that it might interfere with synaptic transmission or neuronal excitability in brain as well as in spinal cord. Ziconotide showed strong efficacy in multiple animal models of pain, notably complete reversal of the second phase of formalin-induced flinching, including with a seven-day chronic intrathecal infusion, a condition under which morphine was ineffective (Malmberg & Yaksh, 1994). In clinical trials, ziconotide showed efficacy as monotherapy in three placebo-controlled clinical trials (Staats et al., 2004; Rauck et al., 2006; Wallace et al., 2006), in each case in patients with chronic pain uncontrolled by existing therapeutics. In these trials, ziconotide showed efficacy in pain arising from cancer or HIV neuropathy, nonmalignant severe chronic pain from diverse systems, and mixed malignant and nonmalignant pain (mostly neuropathic pain from failed back surgery syndrome). Efficacy subsequently was confirmed with meta-analysis of these three trials (Brookes et al., 2017). Treatment-related adverse events were seen in almost all patients in all trials, with dizziness, confusion, ataxia, abnormal gait, memory impairment, and asthenia the most frequently reported in the last of these three trials (Rauck et al., 2006), a trial designed to minimize side effects through lower and slower dosing. Efficacy has been demonstrated in many additional pain etiologies in open-label studies, with 38.5% of patients experiencing over 30% reduction in pain scores at 18 months but almost all experiencing CNS side effects, as reported in a long-term prospective study of ziconotide in clinical practice (McDowell et al., 2020). The efficacy and side effects seen in these open-label studies are generally consistent with the original placebo-controlled trials.

Despite the estimate of almost 20 million people in the United States with high-impact chronic pain (Dahlhamer et al., 2018), the uptake of ziconotide has been fairly modest, with sales of approximately $20 million yearly reported in company financial statements. Its original developer, Neurex, was sold to Elan for an announced $700 million in stock equivalent in 1998, with ziconotide as the main asset. This far exceeds the subsequent pricing of tens of millions of dollars for Prialt to Jazz Pharmaceuticals in 2010, and

finally $80 million again to sell Prialt to TerSera in 2018 (company press releases). By comparison, about ten million Lyrica prescriptions per year are written in the United States alone (ClinCalc resource). This comparatively small uptake does not detract from the scientific and medical achievement of ziconotide, nor from its efficacy in those patients who do use it. But despite its sometimes superb efficacy and non-addictive profile (Staats et al., 2001; Vitale et al., 2008), ziconotide is not the game-changer hoped for, particularly in the modern days of a crisis of opioid addiction that makes the need for non-addictive analgesics all the more apparent. The National Health Service of England, United Kingdom, as of 2016 does not recommend ziconotide for the most common chronic pain patient population of cancer pain (NHS England Clinical Commissioning Policy Reference 16011/P).

Factors determining the use of a drug in the real world are difficult to determine and to quantitate, but the now over 15 years of clinical history with ziconotide suggests challenges in its clinical use. The side-effect profile is extensive and may be a deterrent for use of ziconotide outside specialty care structures with strong patient support (Brookes et al., 2017). Ziconotide carries a boxed warning on the label for neuropsychiatric side effects, and one case study has been reported of psychosis in a psychiatrically normal patient immediately following up-titration of ziconotide dosing (Thompson et al., 2006; Brookes et al., 2017; Burdge et al., 2018). Accordingly, the risk of exacerbation or initiation of neuropsychiatric symptoms may require specialist profiling of ziconotide patients. Ziconotide also presents the hurdle of pump implantation (although some have seen promise with bolus dosing, Mohammed et al. 2013), and the slow onset of efficacy and narrow therapeutic window are challenges to dose titration, with "start low, go slow" the clinician's mantra (Brookes et al., 2017). Among intrathecal agents, ziconotide has been proposed as a first-line therapy (Deer et al., 2019), but patients may already be using intrathecal analgesics including baclofen, clonidine, bupivacaine, and morphine (Kress et al., 2009; Wallace et al., 2016), additional clinical variables which the pumps approved for ziconotide may not be rated to deliver (Brookes et al., 2017).

It is possible that further clinical development of ziconotide might resolve some of these questions, although the investment case for a therapeutic late in its commercial trajectory is less common. As to future iterations of Cav2.2 blockers, a reasonable inference from ziconotide is that to mitigate CNS side effects, any future therapeutics will need to minimize inhibition of Cav2.2 in the brain. This potentially could be achieved via route of administration, direct chemical selectivity, or some form of functional selectivity. For functional selectivity, state-dependent inhibition by small molecules is an appealing prospect to target selectively channels associated with the pathological hyperexcitable state, analogous to the selectivity of nifedipine for inactivated channels. The precise pattern of state dependence needed, however, is unknown, and so hard to engineer into a small molecule. Moreover, many of the clinical CNS side effects seen with ziconotide are difficult to pick up with preclinical models. The profile, then, for orally-available small molecules that would leverage the efficacy of ziconotide while having easier dosing and fewer side effects is inhibition of spinal cord Cav2.2 in the hyperexcitable pain state, plus sufficient selectivity within the calcium-channel family to avoid CNS side effects (and peripheral side effects, if dosed p.o. or i.v.). In the absence of improvements in technology, clinical trials might be the earliest test readout. By inference from the peak of patent applications in the early 2000s, despite the prize of a nonaddictive analgesic, the challenges for small-molecule inhibitors of CNS Cav2.2 proved steep (McDonough, 2013). A lead in this class, Z160, failed small efficacy trials for lumbosacral radiculopathy and for post-herpetic neuralgia (NCT01655849 and NCT01757873) (Zamponi et al., 2009, 2015), but does nevertheless provide a benchmark for future small-molecule Cav2.2 inhibitors. An intriguing possibility is that local administration of Cav2.2 inhibitors at the skin might produce analgesia for specific clinical syndromes, perhaps with a different clinical niche than topical lidocaine or capsaicin (DuBreuil et al., 2021). Another hope for the future is that the high-resolution cryo-electron microscopy structure of Cav2.2 complexed with ziconotide (Gao et al., 2021) might provide in silico insights for targeting of Cav2.2.

Learnings and Emerging Resources for Drug Development

One lesson from these three classes of drugs is that neuronal calcium channels present on-target liabilities for the development of systemically administered drugs in general. It seems clear that inhibition of Cav2.2 in the brain leads to significant CNS side effects; inhibition of Cav2.2 in sympathetic neurons can lead to impaired baroreceptor reflexes; and inhibition of Cav1.2 channels can be a cardiovascular liability for some patient populations. Cav1.2 is included by scientists at Roche as a priority assay for early de-risking of all small molecules (Bendels et al., 2019). Pregabalin and gabapentin also have dose-limiting CNS side effects, and perhaps these are on-target effects on Cavα2δ mediated by too much reduction in calcium currents and neurotransmission. What percentage calcium current modulation, of what subtypes, and in what tissues produces prohibitive clinical side effects is difficult to know, and CNS side effects have so many potential causes that routine early screening of small molecules on Cavα2δ or calcium channels probably is not productive. Splice-form specificity offers biological opportunities for more selective compounds (Zamponi et al., 2015), but given assay and engineering challenges may be better suited for nucleic acid approaches rather than small molecules.

Together, the dihydropyridine antihypertensives, gabapentin and pregabalin, and ziconotide offer a rather interesting lesson for the discovery and development of first-in-class therapeutics in the modern era. In each case, the key data, and in the case of dihydropyridines the original screening, were with animal pharmacology. None of the targets of these drugs are apparent from subsequent human genomic or genetic studies as the highest-impact and safest targets for their corresponding indications, and even if so, each would be difficult from an assay standpoint to screen and prosecute de novo. One might even avoid Cav1.2 entirely as a drug target for hypertension, since variants in *CACNA1C* have been associated with rare cardiac and neurodevelopmental defects, as well as with diastolic blood pressure (Splawski et al., 2004; Evangelou et al., 2018; Gakenheimer-Smith et al., 2021), and so Cav1.2 might be regarded as having an intractable on-target liability. For nifedipine, gabapentin and pregabalin, the molecular target was not even known at the time of approval. Ziconotide was targeted at pain from knowledge of its binding to calcium channels within the spinal cord and the inference that it governed nociceptive transmission, and not by deliberately screening Cav2.2 as a target. (Interestingly, after efforts to optimize potency, the original ω-CTx-MVIIA sequence remained the most potent; Miljanich, 2004.) What made these models of hypertension, epilepsy, and pain translate successfully into human clinical indications? Notably, the preclinical effects of each of these now therapeutics were large and unmistakable, and these large effects were on the eventual clinical endpoint, whether a behavioral readout of disease (seizure, pain) or a quantitative endophenotype (blood pressure). Systems assays may offer intrinsic advantage for small molecule calcium-channel pharmacology since therapeutic properties of state dependence are so difficult to define and to translate into a molecular assay.

From one point of view, success in these preclinical models might justify sustained efforts into new preclinical models that apply to current unmet medical needs, such as Parkinson's, schizophrenia, and tardive dyskinesia (see below). Animal models of disease, however, can be difficult to validate without an existing therapeutic, which to some extent obviates their application to first-in-class drug discovery, and appropriate quantitative readouts for these mechanistically complex diseases are difficult. Particularly in indications such as cardiometabolic diseases, which have effective therapies, the bar for regulatory approval and for payer reimbursement has moved higher, and clinical trials aim to demonstrate not just modulation of an endophenotype such as LDL levels or blood pressure but reduction in cardiac events and increases in life years (Sabatine et al., 2018; Azari et al., 2020). Predicting these ultimate human endpoints with preclinical models or even early human data may not suffice. To pick a dismaying example, the small-molecule myosin stabilizer omecamtiv mecarbil (Planelles-Herrero et al., 2017) increased ventricular ejection fraction in preclinical species and even in humans (Cleland et al., 2011; Teerlink et al., 2016), yet showed little or no reduction of cardiac events or overall mortality in chronic heart failure patients with low ejection fraction (Teerlink et al., 2021). Translational models are probably most straightforward for diseases driven by high-penetrance or Mendelian variants, where the molecular defect in some cases can be replicated in animals or in reductive systems, as is the case for the potentiators of CFTR channel opening or CFTR expression now used to treat cystic fibrosis (Van Goor et al., 2009, 2011). Even models based on a high-penetrance disease-associated human variant, however, can be challenging to extend to humans. For example, compounds that improve behavior and even survival in rodents overexpressing human disease-causing variants in *SOD1* have not translated into clinical success in SOD1-ALS (Lambrechts et al., 2003; Azzouz et al., 2004; Storkebaum et al., 2005; Scott et al., 2008; Gouel et al., 2019). Among cellular models, high-content imaging data, combined with machine learning, is an active area to search for compounds that correct a cellular defect that reflects disease (Mullard, 2019). Cellular models of disease without knowledge of the target can make for a more difficult medicinal chemistry campaign, but such models can be high throughput and do enable polypharmacology. Machine learning likewise is being applied to refine behavioral readouts of disease in animals, particularly of neurological disease (Hausmann et al., 2021). Whatever the model used preclinically, for first-in-class drug discovery, the translation from a reductive model to the precise human population—including patient age, drug treatment history, co-morbidities, and clinical endpoint—will remain a matter of data, skill, intuition, and luck particular to each disease.

Beyond initial clinical success, extensive and thorough clinical trials can be key to overall success, as they guide physicians on the exact indications, patient population, dosing levels and routes, synergy with existing medications, and possible side effects of a new therapeutic. Clinical experience also can orient a compound to the most appropriate disease, as was the case for pregabalin (discovery for epilepsy, widest application to pain) and nifedipine (original cardiac concept refined to hypertension). Not only do thorough clinical trials determine the best patient experience, but careful clinical data can enable iterative improvements of a drug, as in the case of dihydropyridines. Given its efficacy and the unmet need for chronic pain, it seems possible that ziconotide might have reached more patients following larger and wider clinical studies. To some extent, however, the demonstration of ziconotide efficacy in mechanistically mixed pain populations necessarily came at the expense of detailed risk-benefit profile and treatment information for individual clinical pain syndromes. In this light, 45 controlled clinical studies of pregabalin in neuropathic pain with almost 12,000 participants as of 2019 (Derry et al., 2019) and the decades of work to refine the pharmacology of dihydropyridine compounds do not seem unwarranted.

Target discovery and drug development are now moving to take advantage of extraordinary emerging resources in human biology: genetics, proteomics, and medical and health records. These offer the promise of identifying drug targets, quantitative biomarkers, and possibly the best treatment populations in an unbiased, genome-wide manner, with data based on actual human pathophysiology. Drugging a known target, as opposed to working with cellular or animal systems, enables powerful screening, in silico, and structural approaches. A recent retrospective study shows greater success for drugs aimed at targets that are genetic drivers of the corresponding indication, with a fourfold increase in overall success in drugging targets

that are high-penetrance or Mendelian genetic drivers of disease (King et al., 2019). Increasingly, genetic resources are available, for whole populations, individual diseases, ultra-rare genetics, and populations with high rates of homozygosity (inter alia, Trujillano et al., 2017; Karczewski et al., 2020; Singh et al., 2020; Szustakowski et al., 2021; Iwaki et al., 2021). Paired with phenotyping and in appropriate partnership with the involved communities (Garrison et al., 2019), such genes may be targets for future high-impact therapeutics, as seen in the genetic association of *SOST* with bone density and the corresponding therapeutic romosozumab to treat osteoporosis (Balemans et al., 2001; Lewiecki, 2020). Mendelian randomization analyses of pQTLs and other parameters from omics datasets are used to discover disease-causal biomarkers that can be translational or diagnostic (e.g., Blauw et al., 2018, 2019). Challenges in integrating these human resources to drug discovery include applying the lifetime risk measure of most genetic studies to making therapeutics that treat patients who are already sick, obtaining accurate phenotyping from diagnostic codes and unstructured information, adequate statistical powering and replication datasets, and, particularly, translation of genetic datasets to a drug target that governs large effect size not just on its own but within the landscape of existing therapeutics. Logistically such resources present hurdles in collaboration among multiple institutions and stakeholders, and in informed consent standards. Operationalizing such resources to enable new therapeutics requires effective collaboration and communication among applied mathematicians and statisticians, clinicians, data specialists, biologists, legal professionals, and industrial scientists (as well as dedicated funding). The promise, however, is discovery of targets for novel human therapeutics based on unbiased interrogation of the actual human disease, a scenario in which animal data are used more to verify in vivo target engagement than to model disease efficacy. It is a work in progress in general for how best to convert human genomic, proteomic, and medical information into new therapies and improved health, but calcium channels are part of this equation, in no small part because of the rich pharmacology and biological information available for the class.

Experimental Therapeutics

Studies of human biology with unbiased approaches have indeed pointed to several calcium channels as genetic drivers of neurological disease, including autism spectrum disorder, hereditary ataxias and migraine, schizophrenia, major depressive disorder, bipolar disorder, hearing loss, and multiple forms of epilepsy (Terwindt et al., 1998; Andrade et al., 2019), and medical record analyses may show effects of widely used calcium-channel blockers in new indications. Several new classes of experimental therapeutics that target calcium channels are in development, based on a variety of types of data for target validation. As illustrated above, there is no single method of drug discovery that is best or that unequivocally produces success. Ideally one would like to see evidence linking modulation of the target directly to a large effect on the clinical endpoints of interest, a direct mechanistic link between target and disease, confidence in the safety of a target, a strategic niche in light of existing and emerging treatments, a clear medical niche supported by reimbursers, and of course the feasibility of achieving the target product profile. The remainder of this chapter surveys the key rationale and hurdles for some of the recent experimental therapeutics targeted at calcium channels.

Dihydropyridines in Neurology Cav1.3 was proposed as a target for Parkinson's disease in 2007 based on pioneering work showing a switch of the spiking of neurons of the substantia nigra pars compacta (SNc) from calcium based (Fujimura & Matsuda, 1989; Grace & Onn, 1989) to sodium based upon application of the dihydropyridine isradipine (Chan et al., 2007). Subsequent work showed possible calcium-mediated downstream degradation of the mitochondrial membrane potential, suggesting that decreasing calcium influx with isradipine might maintain spiking while improving neuronal survival (Guzman et al., 2010). This led to the hypothesis that isradipine might be an effective treatment for Parkinson's via inhibition of Cav1.3, Cav1.2, or both. Tested in mice, isradipine delivered subcutaneously for 28 days via an implanted minipump at a dose of 3 mg/kg per day (in DMSO/PEG300 vehicle) protected SNc tyrosine hydroxylase-positive fibers and cells from destruction by subsequent 6-OHDA toxin injection (Ilijic et al., 2011). Notably, this dosing produced plasma concentrations of isradipine consistent with its clinical use as an antihypertensive, suggesting that dosing close to that already used for hypertension might be beneficial in Parkinson's. Isradipine has been characterized as a high-affinity binder of Cav1.3 (Koschak et al., 2001) that exerts partial inhibition of Cav1.3 currents (Xu & Lipscombe, 2001), and it was approved by the FDA in 1989 for clinical use as an antihypertensive. A notable boost to the theory that L-type channels of some sort are involved in Parkinson's came from retrospective analyses of medical records, which show on balance that among antihypertensive medications, use of dihydropyridines correlates with a decreased risk of Parkinson's disease (Tseng et al., 2021).

Regrettably, a modestly sized ($n = 336$, randomized 1:1 to $n = 170$ in the treatment arm, with 137 completing initial and final visits at the target dose) clinical trial of isradipine in Parkinson's did not meet the primary endpoint on the Parkinson's disease rating scale. Nor did isradipine have effect on secondary endpoints, nor show a strong trend toward efficacy, including as a function of plasma isradipine levels (Biglan et al., 2017; The Parkinson Study Group STEADY-III Investigators, 2020; Vinuto et al., 2021). The study compared 5 mg of immediate-release isradipine twice daily to placebo over a 36-month period and was powered to detect approximately a 25% change in the rate of disease worsening. This dose was proportionately much lower than the daily 3 mg/kg used for proof-of-concept in mice. Trial participants were early-diagnosis Parkinson's patients and were not on dopaminergic medications for at least the first 3 months of the trial.

It is a guessing game to assign a cause to failure of a clinical trial, particularly a first-in-class treatment of as complex a disease as Parkinson's, without knowing the unbound concentration of the therapeutic at the target tissue, whether the therapeutic bound the target, and whether the therapeutic modulated the function of the target. One possibility for this negative outcome is that isradipine was not dosed at a sufficiently high level to inhibit the target in the brain (whether the target was Cav1.2, Cav1.3, or both). Analysis of the Parkinson's clinical trial suggested that at some points, plasma concentrations of isradipine were somewhat below the preclinical IC50 for neuroprotection in the mouse challenge model (Venuto et al., 2021). Without a measure of isradipine in cerebrospinal fluid or a mechanistic biomarker for function; however, these analyses can only infer effects on Cav1.3 (and Cav1.2) in the brain. Another possibility is that the patients were no longer amenable to treatment by this mechanism. Although no PET measurements of dopamine were made in this trial, a majority of nigral dopaminergic neurons may already be gone at diagnosis of Parkinson's (Fearnley & Lees, 1991). Finally, one must look at the validation for the target and mechanism. Success for this hypothesis would depend on how well isradipine alteration of spiking, prevention of acute damage in the 6-OHDA challenge model of neuronal death in mice, and lowered Parkinson's risk in hypertensive patients treated with dihydropyridines each predict effects of this particular dose of isradipine on already-established clinical Parkinson's. Note the mouse challenge model necessarily does not reflect the endpoints of the clinical trial, and medical record datasets of the incidence of Parkinson's in hypertensive patients prescribed dihydropyridines may reflect mostly effects of nifedipine, amlodipine, and cilnidipine rather than isradipine itself. Amlodipine in particular is an outlier among dihydropyridines in multiple pharmacokinetic parameters including its behavior in the CNS (Uchida et al.,

1997), and cilnidipine has significant activity against Cav2.2 (Uneyama et al., 1999), adding a caution to grouping dihydropyridines as a single class. Moreover, there is a complex relation between hypertension and Parkinson's (Hou et al., 2018), meaning the patient population in medical records might not well represent a Parkinson's population. Further efforts to drug L-type channels for Parkinson's by pharmaceutical companies probably would have to come in the context of an overall Parkinson's strategy and require stronger evidence of the exact target product profile, particularly evidence that inhibition of Cav1.3 would have a large interventional effect on established disease. The Parkinson's meta-GWAS of all European populations available as of 2019 does not show a locus containing *CACNA1D* (Nalls et al., 2019), although this addresses lifetime risk rather than disease progression. As of this writing, improved symptomatic treatment via selective D1/D5 agonism (tavapadon, in multiple efficacy trials, see press releases from Cerevel, Inc.), and potential disease-modifying targets including LRRK2, GBA, and SNCA seem the most promising paths for the treatment of Parkinson's.

Dihydropyridines also have been nominated as candidates for the treatment of neuropsychiatric disorders (Ortner & Striessnig, 2016), and here there are robust data from multiple genetic studies showing involvement of *CACNA1C* in lifetime risk of human neuropsychiatric disease (Andrade et al., 2019). To repurpose a dihydropyridine to neuropsychiatry, the challenges would be similar to those with applying dihydropyridines to Parkinson's: how to verify that an inhibitor will have high effect size in already-established disease, including in light of existing medications; biomarkers for target engagement and target coverage; and selecting the correct intervention time, patient population, and drug dosage in the treatment population. Preclinical models and genetic studies of neuropsychiatric disease generally have not given straightforward opportunities to hope for a large effect size in a broad population (Legge et al., 2021), and so the drug discovery field currently is moving toward defining mechanistic subsets of neuropsychiatric disease.

T-type Calcium Channels T-type calcium channels are widespread in multiple tissue types and yet seem to be surprisingly safe to inhibit. Global knockout mice for *Cacna1g*, *Cacna1h*, and *Cacna1l* are generally healthy; CRISPR and siRNA knockdown of T-type expression does not affect the growth of a wide range of human cell lines (depmap.broadinstitute.org); and many (nonselective) clinical therapeutics inhibit T-type calcium channels, notably the cerebral vasodilator nimodipine and anti-epileptics zonisamide, ethosuximide, and valproate (Zamponi, 2016). T-type calcium channels control burst firing in many neuronal systems (Huguenard, 1996), and individual isoforms have evidence for involvement in epilepsy, sleep, chronic pain, neuropsychiatric disease, and cognitive disorders (Cheong & Shin, 2013). Inhibition of one or more T-type calcium channels is the basis of several emerging strategies to control neuronal excitability more selectively than with broad-spectrum sodium-channel inhibitors. Chemical matter for T-type channels seems to be abundant (Nam, 2018), and several therapies based on mechanistic hypotheses are in clinical trials.

One illustrative small-molecule effort sparked by genetic data is the effort to make an activator of Cav3.3 based on the replicated involvement of common variants at *CACNA1I* in lifetime genetic risk for schizophrenia (Purcell et al., 2014; Pardinas et al., 2018; Lam et al., 2019). Looking at rare variants, studies of 105 schizophrenia probands within trios found two probands carrying rare coding mutations in *CACNA1I* (Gulsuner et al., 2013), and one of the two has since been determined to be a functional hypomorph (Andrade et al., 2016). If this allele indeed is the main factor that drives schizophrenia in the affected proband, this suggests that modulation of Cav3.3 may produce a high effect size in schizophrenia. Involvement of *CACNA1I* in idiopathic schizophrenia is suggested by GWASs of common variants. Accordingly, the loss-of-function rare variant found in one schizophrenic proband is interpreted as evidence that increasing *CACNA1I* function with a pharmacological Cav3.3 potentiator might confer therapeutic benefit to a wider population (Stanley Center for Therapeutics website, Broad Institute). Key here is whether the single rare *CACNA1I* allele does indeed drive schizophrenia, which is difficult to ascertain from an individual patient or individual family due to the many candidate rare variants within families (Li et al., 2021). Note that *CACNA1I* protein-truncating and missense variants are, if anything, under-represented in schizophrenics compared to the general population (schema.broadinstitute.org). Given the challenge in finding translational systems that reflect the clinical endpoints of a schizophrenia study, likely key decision-turning data for this hypothesis will need to come from direct clinical studies.

The hypothesis that inhibition of T-type calcium channels can produce analgesia is a longstanding hypothesis with extensive support in the literature based on effects in animal pain models of nonselective pharmacological inhibitors, genetic knockdown of T-type channels, and knockout of T-type channel genes, particularly *Cacna1h* knockout mice (Bourinet et al., 2016). In direct clinical trials, however, a six-week trial of the Cav3.2 blocker ABT-639 following washout of all other pain medications did not show efficacy in a cohort of patients with painful diabetic neuropathic pain, despite efficacy in multiple preclinical models of pain (Ziegler et al., 2015). The primary endpoint was a weekly mean of 24-h pain score. The positive-control pregabalin achieved efficacy over placebo in the first 2 weeks but not in later weeks of this trial, consistent with other clinical experience with pregabalin. The 100 mg dose of ABT-639 demonstrated an acceptable safety profile. ABT-639 also showed no effect in clinical studies in experimental pain induced by capsaicin, although again the therapeutic was well-tolerated (Wallace et al., 2016). These studies, although not achieving clinical efficacy, produced valuable information about the use of T-type inhibitors in the clinic and may have directed investment into T-type inhibitors toward additional indications. As with isradipine, it is difficult to infer a cause for a clinical failure, particularly for a first-in-class mechanism of disease, absent information about free drug concentration at the site of action, receptor occupancy, and receptor modulation. Mechanistic drivers of clinical pain syndromes may differ among indications, and, unlike Parkinson's, genetic studies of pain lack a clear case/control structure. Well-powered genetic studies of pain, however, have been enabled with emerging large-scale omics resources and with participant questionnaires (McIntosh et al., 2016; Jones et al., 2016). T-type channels do not figure prominently in these, although exome sequencing of trigeminal neuralgia patients finds an excess burden of *CACNA1H* variants (Dong et al., 2020).

Multiple small-molecule inhibitors of T-type channels are in clinical trials for tardive dyskinesia and essential tremor disorders, based on mechanistic hypotheses. Essential tremor is thought to be a common, mechanistically heterogeneous disorder (Deuschl & Elble, 2009), with possible involvement from T-type channels in pacemaking neurons in the inferior olivary nucleus (Handforth et al., 2010) and in the synchrony of firing of cerebellar Purkinje neurons (Lin et al., 2014). CX-8998 is a moderately potent inhibitor of all three subtypes of T-type channel as measured with FLIPR and with electrophysiology (Shipe et al., 2008). CX-8998 showed efficacy in an inbred rat model of absence epilepsy (Lee et al., 2019). A clinical trial of CX-8998 for absence epilepsy is listed as having concluded in March 2018 (NCT03406702). In a subsequent small placebo-controlled trial for essential tremor, CX-8998 showed efficacy after 15 and 28 days of treatment by investigator rating, but not by independent rating of patients recorded by video, with dizziness as a main treatment-related adverse event (Papapetropoulos et al., 2021). PRAX-944 (formerly Z944) is also a T-type inhibitor (Cain et al., 2015) in clinical studies for the treatment of essential tremor, with a dosing, pharmacokinetics, and safety trial underway in patients (NCT05021978) and an efficacy trial registered (NCT05021991). The T-type inhibitor NBI-827104 also is in clinical studies, with small dose-finding, safety, tolerability, and efficacy trials listed for essential tremor

(NCT04880616) and for pediatric epileptic encephalopathy with continuous spike and wave during sleep (NCT04625101). Keys to success of these therapies compared to other mechanisms in development for treating essential tremor (Ondo, 2020) will be how well the mechanistic hypothesis of T-type inhibition translates into clinical disease endpoints, and the therapeutic index, given that CNS side effects will be difficult to de-risk preclinically. Note that the T-type blocker ethosuximide given twice daily at 500 mg did not show efficacy in a trial of seven patients with essential tremor (Gironell & Marin-Lahoz, 2016). Publication of detailed profiles of these experimental compounds and of ethosuximide side-by-side in the same hands, on neuronal and on cloned channels, with electrophysiology assays, will be helpful in understanding their clinical effects and any differentiation among them. Should these trials not turn out as hoped, the combination of sequencing and medical records in large populations, such as increasingly available from the UK Biobank, may enable the direction of existing molecules to clinical indications with genetic evidence for the involvement of T-type channels. Indications driven by abnormal excitability give the opportunity for fast clinical trials, compared to clinical trials in neurodegenerative or metabolic disease.

Variants in T-type channels also cause rare genetic forms of disease including epilepsy, autism, schizophrenia, and possibly pain (Weiss & Zamponi, 2020). Characterization of the variants involved is essential to ascribing disease causality to the rare variant with confidence and to shaping possible treatments. Diseases driven by gain-of-function variants may possibly be treated with existing T-type inhibitors, whereas loss-of-function variants that cause disease are more problematic to treat (see *CACNA1I* above). Alternatively, antisense or short interfering RNA therapies might provide knockdown of pathogenic (and wildtype) alleles at the transcript level, with high specificity for a particular calcium-channel type or even splice variant. Challenges for oligonucleotide therapies in CNS disorders include an intrathecal dosing route (making these therapies most suitable for severe and intractable diseases), a therapeutic window for the knockdown of wildtype as well as variant alleles, and distribution within the brain.

Conclusions

Calcium channels are chemically tractable key control points in the biology of multiple systems, offering multiple angles for therapeutic intervention (Zamponi, 2016), and this is reflected in the continued medical prominence of dihydropyridines and of pregabalin and gabapentin. Clinical experience from these drugs and from ziconotide shows that a large effect in the right preclinical model can indeed predict clinical efficacy. Each of these compounds has been an indispensable tool for understanding the receptors they target and the diseases they are used to treat, and much more research is likely to come. Promising angles for new therapeutics that target Cav1.3, Cav2.2, and Cav3 channels so far have not borne out, but clinical trials have shown valuable information on human dosing and safety, and clinical tests of Cav3 inhibitors are underway for the treatment of several diseases. A stable of clinic-ready compounds of known function will be tremendously valuable as science opens up new therapeutic areas for calcium channels, such as control by the peripheral nervous system of tumors and of inflammation (Monje et al., 2020). 2020 saw a record $20 billion in venture investment, plus almost $13 billion in biopharma initial public offerings (Evaluate Vantage Pharma, Biotech, and Medtech 2020 in review). Testing a clinical asset in a plausible indication is a straightforward investment case, and one might hope that good clinical molecules for Cav3 and other calcium channels might find a use even if ongoing trials do not meet primary endpoint. Powerful resources for the mechanistic understanding of human diseases are rapidly becoming available, driven by advances in genetics, proteomics, and digital health, and through understanding of the logistics of the required partnerships among basic and industrial scientists, funders, and patient organizations. These resources may find unanticipated uses for existing calcium-channel drugs by correlating drug usage with phenotype, or new diseases that are driven by calcium channels may be found via genetics. As seen with the established and experimental therapeutics targeting calcium channels, to bring basic scientific advances to best use in patients, 20 years is not an unreasonable amount of time.

References

Andrade, A., Brennecke, A., Mallat, S., Brown, J., Gomez-Rivadeneira, J., Czepiel, N., & Londrigan, L. (2019). Genetic associations between voltage-gated calcium channels and psychiatric disorders. *International Journal of Molecular Sciences, 20*(14), 3537. https://doi.org/10.3390/ijms20143537

Andrade, A., Hope, J., Allen, A., Yorgan, V., Lipscombe, D., & Pan, J. Q. (2016). A rare schizophrenia risk variant of CACNA1I disrupts Cav3.3 channel activity. *Scientific Reports, 6*, 34233. https://doi.org/10.1038/srep34233

Azari, S., Rezapour, A., Omidi, N., Alipour, V., Behzadifar, M., Safari, H., Tajdini, M., & Bragazzi, N. L. (2020). Cost-effectiveness analysis of PCSK9 inhibitors in cardiovascular diseases: A systematic review. *Heart Failure Reviews, 25*, 1077–1088.

Azzouz, M., Ralph, G. S., Storkebaum, E., Walmsley, L. E., Mitrophanous, K. A., Kingsman, S. M., Carmeliet, P., & Mazarakis, N. D. (2004). VEGF delivery with retrogradely transported lentivector prolongs survival in a mouse ALS model. *Nature, 429*, 413–417.

Backonja, M., & Glanzman, R. L. (2003). Gabapentin dosing for neuropathic pain: Evidence from randomized, placebo-controlled clinical trials. *Clinical Therapeutics, 25*, 81–104.

Balemans, W., Ebeling, M., Patel, N., Van Hul, E., Olson, P., Dioszegi, M., Lacza, C., Wuyts, W., Van Den Ende, J., Willems, P., Paes-Alves, A. F., Hill, S., Bueno, M., Ramos, F. J., Tacconi, P., Dikkers, F. G., Stratakis, C., Lindpaintner, K., Vickery, B., … Van Hul, W. (2001). Increased bone density in sclerosteosis is due to the deficiency of a novel secreted protein (SOST). *Human Molecular Genetics, 10*, 537–543.

Barclay, J., Balaguero, N., Mione, M., Ackerman, S. L., Letts, V. A., Brodbeck, J., Canti, C., Meir, A., Page, K. M., Kusumi, K., Perez-Reyes, E., Lander, E. S., Frankel, W. N., Gardiner, R. M., Dolphin, A. C., & Rees, M. (2001). Ducky mouse phenotype of epilepsy and ataxia is associated with mutations in the Cacna2d2 gene and decreased calcium channel current in cerebellar Purkinje cells. *The Journal of Neuroscience, 21*, 6095–6104.

Bean, B. P. (1984). Nitrendipine block of cardiac calcium channels: high-affinity binding to the inactivated state. *Proceedings. National Academy of Sciences. United*

States of America, 81(20), 6388–6392. https://doi.org/10.1073/pnas.81.20.6388

Bertrand, S., Ng, G. Y., Purisai, M. G., Wolfe, S. E., Severidt, M. W., Nouel, D., Robitaille, R., Low, M. J., O'Neill, G. P., Metters, K., Lacaille, J. C., Chronwall, B. M., & Morris, S. J. (2001). The anticonvulsant, antihyperalgesic agent gabapentin is an agonist at brain gamma-aminobutyric acid type B receptors negatively coupled to voltage-dependent calcium channels. *The Journal of Pharmacology and Experimental Therapeutics, 298*, 15–24.

Bendels, S., Bissantz, C., Fasching, B. et al. (2019). Safety screening in early drug discovery: An optimized assay panel. *Journal of Pharmacological and Toxicological Methods, 99*. https://doi.org/10.1016/j.vascn.2019.106609

Biglan, K. M., Oakes, D., Lang, A. E., Hauser, R. A., Hodgeman, K., Greco, B., Lowell, J., Rockhill, R., Shoulson, I., Venuto, C., Young, D., Simuni, T., & Parkinson Study Group S-PDIIII. (2017). A novel design of a Phase III trial of isradipine in early Parkinson disease (STEADY-PD III). *Annals of Clinical Translational Neurology, 4*, 360–368.

Blauw, L. L., Li-Gao, R., Noordam, R., de Mutsert, R., Trompet, S., Berbee, J. F. P., Wang, Y., van Klinken, J. B., Christen, T., van Heemst, D., Mook-Kanamori, D. O., Rosendaal, F. R., Jukema, J. W., Rensen, P. C. N., & Willems van Dijk, K. (2018). CETP (Cholesteryl Ester Transfer Protein) concentration: A genome-wide association study followed by Mendelian randomization on coronary artery disease. *Circulation: Genomic and Precision Medicine, 11*, e002034.

Blauw, L. L., Noordam, R., Soidinsalo, S., Blauw, C. A., Li-Gao, R., de Mutsert, R., Berbee, J. F. P., Wang, Y., van Heemst, D., Rosendaal, F. R., Jukema, J. W., Mook-Kanamori, D. O., Wurtz, P., Willems van Dijk, K., & Rensen, P. C. N. (2019). Mendelian randomization reveals unexpected effects of CETP on the lipoprotein profile. *European Journal of Human Genetics, 27*, 422–431.

Bockbrader, H. N., Wesche, D., Miller, R., et al. (2010). A comparison of the pharmacokinetics and pharmacodynamics of pregabalin and gabapentin. *Clinical Pharmacokinetics, 49*, 661–669.

Bossert, F., & Vater, W. (1989). 1,4-Dihydropyridines—A basis for developing new drugs. *Medicinal Research Reviews, 9*, 291–324.

Bourinet, E., Francois, A., & Laffray, S. (2016). T-type calcium channels in neuropathic pain. *Pain, 157*(Suppl 1), S15–S22.

Brodbeck, J., Davies, A., Courtney, J. M., Meir, A., Balaguero, N., Canti, C., Moss, F. J., Page, K. M., Pratt, W. S., Hunt, S. P., Barclay, J., Rees, M., & Dolphin, A. C. (2002). The ducky mutation in Cacna2d2 results in altered Purkinje cell morphology and is associated with the expression of a truncated alpha 2 delta-2 protein with abnormal function. *The Journal of Biological Chemistry, 277*, 7684–7693.

Brookes, M. E., Eldabe, S., & Batterham, A. (2017). Ziconotide monotherapy: A systematic review of randomised controlled trials. *Current Neuropharmacology, 15*, 217–231.

Bryans, J. S., & Wustrow, D. J. (1999). 3-substituted GABA analogs with central nervous system activity: A review. *Medicinal Research Reviews, 19*, 149–177.

Burges, R., & Moisey, D. (1994). Unique pharmacologic properties of amlodipine. *The American Journal of Cardiology, 73*(3), A2–A9. https://doi.org/10.1016/0002-9149(94)90268-2

Burdge, G., Leach, H., & Walsh, K. (2018). Ziconotide-induced psychosis: A case report and literature review. *The Mental Health Clinician, 8*, 242–246.

Cain, S. M., Tyson, J. R., Jones, K. L., & Snutch, T. P. (2015). Thalamocortical neurons display suppressed burst-firing due to an enhanced Ih current in a genetic model of absence epilepsy. *Pflügers Archiv, 467*, 1367–1382.

Chan, C. S., Guzman, J. N., Ilijic, E., Mercer, J. N., Rick, C., Tkatch, T., Meredith, G. E., & Surmeier, D. J. (2007). 'Rejuvenation' protects neurons in mouse models of Parkinson's disease. *Nature, 447*, 1081–1086.

Chandra, K. S., & Ramesh, G. (2013). The fourth-generation calcium channel blocker: Cilnidipine. *Indian Heart Journal, 65*, 691–695.

Cheong, E., & Shin, H. S. (2013). T-type Ca2+ channels in normal and abnormal brain functions. *Physiological Reviews, 93*, 961–992.

Cleland, J. G., Teerlink, J. R., Senior, R., Nifontov, E. M., Mc Murray, J. J., Lang, C. C., Tsyrlin, V. A., Greenberg, B. H., Mayet, J., Francis, D. P., Shaburishvili, T., Monaghan, M., Saltzberg, M., Neyses, L., Wasserman, S. M., Lee, J. H., Saikali, K. G., Clarke, C. P., Goldman, J. H., … Malik, F. I. (2011). The effects of the cardiac myosin activator, omecamtiv mecarbil, on cardiac function in systolic heart failure: A double-blind, placebo-controlled, crossover, dose-ranging phase 2 trial. *Lancet, 378*, 676–683.

Dahlhamer, J., Lucas, J., Zelaya, C., Nahin, R., Mackey, S., DeBar, L., Kerns, R., Von Korff, M., Porter, L., & Helmick, C. (2018). Prevalence of chronic pain and high-impact chronic pain among adults – United States, 2016. *MMWR. Morbidity and Mortality Weekly Report, 67*(36), 1001–1006. https://doi.org/10.15585/mmwr.mm6736a2

Deer, T. R., Pope, J. E., Hanes, M. C., & McDowell, G. C. (2019). Intrathecal therapy for chronic pain: A review of morphine and ziconotide as firstline options. *Pain Medicine, 20*, 784–798.

Derry, S., Bell, R. F., Straube, S., Wiffen, P. J., Aldington, D., & Moore, R. A. (2019). Pregabalin for neuropathic pain in adults. *Cochrane Database of Systemic Reviews 2019, 2019*(1), Art. No.:CD007076.

Deuschl, G., & Elble, R. (2009). Essential tremor—Neurodegenerative or nondegenerative disease towards a working definition of ET. *Movement Disorders, 24*, 2033–2041.

Dong, W., Jin, S. C., Allocco, A., Zeng, X., Sheth, A. H., Panchagnula, S., et al. (2020). Exome sequencing implicates impaired GABA signaling and neuronal ion transport in trigeminal neuralgia. *iScience, 23*(10), 101552.

DuBreuil, D. M., Lopez Soto, E. J., Daste, S., Meir, R., Li, D., Wainger, B., Fleischmann, A., & Lipscombe, D. (2021). Heat but not mechanical hypersensitivity depends on voltage-gated CaV2.2 calcium channel activity in peripheral axon terminals innervating skin. *The Journal of Neuroscience, 41*, 7546–7560.

Elwes, R. D., & Binnie, C. D. (1996). Clinical pharmacokinetics of newer antiepileptic drugs. Lamotrigine, vigabatrin, gabapentin and oxcarbazepine. *Clinical Pharmacokinetics, 30*, 403–415.

Evangelou, E., Warren, H. R., Mosen-Ansorena, D., Mifsud, B., Pazoki, R., Gao, H., Ntritsos, G., Dimou, N., Cabrera, C. P., Karaman, I., Ng, F. L., Evangelou, M., Witkowska, K., Tzanis, E., Hellwege, J. N., Giri, A., Velez Edwards, D. R., Sun, Y. V., Cho, K., et al. (2018). Genetic analysis of over 1 million people identifies 535 new loci associated with blood pressure traits. *Nature Genetics, 50*, 1412–1425.

Fearnley, J. M., & Lees, A. J. (1991). Ageing and Parkinson's disease: Substantia nigra regional selectivity. *Brain, 114*(Pt 5), 2283–2301.

Field, M. J., Cox, P. J., Stott, E., Melrose, H., Offord, J., Su, T. Z., Bramwell, S., Corradini, L., England, S., Winks, J., Kinloch, R. A., Hendrich, J., Dolphin, A. C., Webb, T., & Williams, D. (2006). Identification of the alpha2-delta-1 subunit of voltage-dependent calcium channels as a molecular target for pain mediating the analgesic actions of pregabalin. *Proceedings of the National Academy of Sciences of the United States of America, 103*, 17537–17542.

Fleckenstein, A., Fleckenstein-Grun, G., Frey, M., & Thimm, F. (1990). Experimental antiarteriosclerotic effects of calcium antagonists. *Journal of Clinical Pharmacology, 30*, 151–154.

Fujii, S., Kameyama, K., Hosono, M., Hayashi, Y., & Kitamura, K. (1997). Effect of cilnidipine, a novel dihydropyridine Ca++-channel antagonist, on N-type Ca++ channel in rat dorsal root ganglion neurons. *The Journal of Pharmacology and Experimental Therapeutics, 280*, 1184–1191.

Fujimura, K., & Matsuda, Y. (1989). Autogenous oscillatory potentials in neurons of the guinea pig substantia nigra pars compacta in vitro. *Neuroscience Letters, 104*, 53–57.

Furberg, C. D., Psaty, B. M., & Meyer, J. V. (1995). Nifedipine. Dose-related increase in mortality in patients with coronary heart disease. *Circulation, 92*, 1326–1331.

Gakenheimer-Smith, L., Meyers, L., Lundahl, D., Menon, S. C., Bunch, T. J., Sawyer, B. L., Tristani-Firouzi, M., & Etheridge, S. P. (2021). Expanding the phenotype of CACNA1C mutation disorders. *Molecular Genetics & Genomic Medicine, 9*, e1673.

Gao, S., Yao, X., & Yan, N. (2021). Structure of human Cav2.2 channel blocked by the painkiller ziconotide. *Nature, 596*, 143–147.

Garrison, N. A., Hudson, M., Ballantyne, L. L., Garba, I., Martinez, A., Taualii, M., Arbour, L., Caron, N. R.,

& Rainie, S. C. (2019). Genomic research through an indigenous lens: Understanding the expectations. *Annual Review of Genomics and Human Genetics, 20*, 495–517.

Gee, N. S., Brown, J. P., Dissanayake, V. U., Offord, J., Thurlow, R., & Woodruff, G. N. (1996). The novel anticonvulsant drug, gabapentin (Neurontin), binds to the alpha2delta subunit of a calcium channel. *The Journal of Biological Chemistry, 271*, 5768–5776.

Gironell, A., & Marin-Lahoz, J. (2016). Ethosuximide for essential tremor: An open-label trial. *Tremor and Other Hyperkinetic Movements (N Y), 6*, 378.

Gohil, K., Bell, J. R., Ramachandran, J., & Miljanich, G. P. (1994). Neuroanatomical distribution of receptors for a novel voltage-sensitive calcium-channel antagonist, SNX-230 (omega-conopeptide MVIIC). *Brain Research, 653*, 258–266.

Gouel, F., Rolland, A. S., Devedjian, J. C., Burnouf, T., & Devos, D. (2019). Past and future of neurotrophic growth factors therapies in ALS: From single neurotrophic growth factor to stem cells and human platelet lysates. *Frontiers in Neurology, 10*, 835.

Grace, A. A., & Onn, S. P. (1989). Morphology and electrophysiological properties of immunocytochemically identified rat dopamine neurons recorded in vitro. *The Journal of Neuroscience, 9*, 3463–3481.

Gulsuner, S., Walsh, T., Watts, A. C., Lee, M. K., Thornton, A. M., Casadei, S., Rippey, C., Shahin, H., Consortium on the Genetics of S, Group PS, Nimgaonkar, V. L., Go, R. C., Savage, R. M., Swerdlow, N. R., Gur, R. E., Braff, D. L., King, M. C., & McClellan, J. M. (2013). Spatial and temporal mapping of de novo mutations in schizophrenia to a fetal prefrontal cortical network. *Cell, 154*, 518–529.

Guzman, J. N., Sanchez-Padilla, J., Wokosin, D., Kondapalli, J., Ilijic, E., Schumacker, P. T., & Surmeier, D. J. (2010). Oxidant stress evoked by pacemaking in dopaminergic neurons is attenuated by DJ-1. *Nature, 468*, 696–700.

Handforth, A., Homanics, G. E., Covey, D. F., Krishnan, K., Lee, J. Y., Sakimura, K., Martin, F. C., & Quesada, A. (2010). T-type calcium channel antagonists suppress tremor in two mouse models of essential tremor. *Neuropharmacology, 59*, 380–387.

Hara, K., & Sata, T. (2007). Inhibitory effect of gabapentin on N-methyl-D-aspartate receptors expressed in Xenopus oocytes. *Acta Anaesthesiologica Scandinavica, 51*, 122–128.

Hausmann, S. B., Vargas, A. M., Mathis, A., & Mathis, M. W. (2021). Measuring and modeling the motor system with machine learning. *Current Opinion in Neurobiology, 70*, 11–23.

Hendrich, J., Van Minh, A. T., Heblich, F., Nieto-Rostro, M., Watschinger, K., Striessnig, J., Wratten, J., Davies, A., & Dolphin, A. C. (2008). Pharmacological disruption of calcium channel trafficking by the alpha2delta ligand gabapentin. *Proceedings of the National Academy of Sciences of the United States of America, 105*, 3628–3633.

Hou, L., Li, Q., Jiang, L., Qiu, H., Geng, C., Hong, J. S., Li, H., & Wang, Q. (2018). Hypertension and diagnosis of Parkinson's disease: A meta-analysis of Cohort studies. *Frontiers in Neurology, 9*, 162.

Huguenard, J. R. (1996). Low-threshold calcium currents in central nervous system neurons. *Annual Review of Physiology, 58*, 329–348.

Iannetti, G. D., Zambreanu, L., Wise, R. G., Buchanan, T. J., Huggins, J. P., Smart, T. S., Vennart, W., & Tracey, I. (2005). Pharmacological modulation of pain-related brain activity during normal and central sensitization states in humans. *Proceedings of the National Academy of Sciences of the United States of America, 102*, 18195–18200.

Ilijic, E., Guzman, J. N., & Surmeier, D. J. (2011). The L-type channel antagonist isradipine is neuroprotective in a mouse model of Parkinson's disease. *Neurobiology of Disease, 43*, 364–371.

Ivanov, S. V., Ward, J. M., Tessarollo, L., McAreavey, D., Sachdev, V., Fananapazir, L., Banks, M. K., Morris, N., Djurickovic, D., Devor-Henneman, D. E., Wei, M. H., Alvord, G. W., Gao, B., Richardson, J. A., Minna, J. D., Rogawski, M. A., & Lerman, M. I. (2004). Cerebellar ataxia, seizures, premature death, and cardiac abnormalities in mice with targeted disruption of the Cacna2d2 gene. *The American Journal of Pathology, 165*, 1007–1018.

Iwaki, H., Leonard, H. L., Makarious, M. B., Bookman, M., Landin, B., Vismer, D., Casey, B., Gibbs, J. R., Hernandez, D. G., Blauwendraat, C., Vitale, D., Song, Y., Kumar, D., Dalgard, C. L., Sadeghi, M., Dong, X., Misquitta, L., Scholz, S. W., Scherzer, C. R., … Uniformed Services University of the Health Sciences A, Group APWGSW and consortium AP. (2021). Accelerating medicines partnership: Parkinson's disease. Genetic resource. *Movement Disorders, 36*, 1795–1804.

Jensen, A. A., Mosbacher, J., Elg, S., Lingenhoehl, K., Lohmann, T., Johansen, T. N., Abrahamsen, B., Mattsson, J. P., Lehmann, A., Bettler, B., & Brauner-Osborne, H. (2002). The anticonvulsant gabapentin (neurontin) does not act through gamma-aminobutyric acid-B receptors. *Molecular Pharmacology, 61*, 1377–1384.

Jones, A. V., Hockley, J. R. F., Hyde, C., Gorman, D., Sredic-Rhodes, A., Bilsland, J., McMurray, G., Furlotte, N. A., Hu, Y., Hinds, D. A., Cox, P. J., & Scollen, S. (2016). Genome-wide association analysis of pain severity in dysmenorrhea identifies association at chromosome 1p13.2, near the nerve growth factor locus. *Pain, 157*, 2571–2581.

Karczewski, K. J., Francioli, L. C., Tiao, G., Cummings, B. B., Alfoldi, J., Wang, Q., Collins, R. L., Laricchia, K. M., Ganna, A., Birnbaum, D. P., Gauthier, L. D., Brand, H., Solomonson, M., Watts, N. A., Rhodes, D., Singer-Berk, M., England, E. M., Seaby, E. G., Kosmicki, J. A., … MacArthur, D. G. (2020). The mutational constraint spectrum quantified from variation in 141,456 humans. *Nature, 581*, 434–443.

Kazda, S. (1991). The story of nifedipine. In P. R. Lichtlen & A. Reale (Eds.), *Adalat*. Springer. https://doi.org/10.1007/978-3-642-85498-9_3

Kazda, S., Garthoff, B., Dycka, J., & Iwai, J. (1982). Prevention of malignant hypertension in salt loaded "S" Dahl rats with the calcium antagonist nifedipine. *Clinical and Experimental Hypertension. Part A, 4*, 1231–1241.

Kress, H. G., Simpson, K. H., Marchettini, P., Ver Donck, A., & Varrassi, G. (2009). Intrathecal therapy: what has changed with the introduction of ziconotide. *Pain Practice, 9*, 338–347. https://doi.org/10.1111/j.1533-2500.2009.00308.x

King, E. A., Davis, J. W., & Degner, J. F. (2019). Are drug targets with genetic support twice as likely to be approved? Revised estimates of the impact of genetic support for drug mechanisms on the probability of drug approval. *PLoS Genetics, 15*, e1008489.

Koschak, A., Reimer, D., Huber, I., Grabner, M., Glossmann, H., Engel, J., & Striessnig, J. (2001). alpha 1D (Cav1.3) subunits can form l-type Ca2+ channels activating at negative voltages. *The Journal of Biological Chemistry, 276*, 22100–22106.

Lam, M., Chen, C. Y., Li, Z., Martin, A. R., Bryois, J., Ma, X., Gaspar, H., Ikeda, M., Benyamin, B., Brown, B. C., Liu, R., Zhou, W., Guan, L., Kamatani, Y., Kim, S. W., Kubo, M., Kusumawardhani, A., Liu, C. M., Ma, H., … Huang, H. (2019). Comparative genetic architectures of schizophrenia in East Asian and European populations. *Nature Genetics, 51*, 1670–1678.

Lambrechts, D., Storkebaum, E., Morimoto, M., Del-Favero, J., Desmet, F., Marklund, S. L., Wyns, S., Thijs, V., Andersson, J., van Marion, I., Al-Chalabi, A., Bornes, S., Musson, R., Hansen, V., Beckman, L., Adolfsson, R., Pall, H. S., Prats, H., Vermeire, S., … Carmeliet, P. (2003). VEGF is a modifier of amyotrophic lateral sclerosis in mice and humans and protects motoneurons against ischemic death. *Nature Genetics, 34*, 383–394.

Lanneau, C., Green, A., Hirst, W. D., Wise, A., Brown, J. T., Donnier, E., Charles, K. J., Wood, M., Davies, C. H., & Pangalos, M. N. (2001). Gabapentin is not a GABAB receptor agonist. *Neuropharmacology, 41*, 965–975.

Lee, M. S., Newbold, E. J., & Papapetropoulos, S. (2019). Selective T-type calcium channel modulator CX-8998 fully suppresses seizures in the GAERS genetic model of epilepsy at human achievable concentrations (P1.5-026). *Neurology, 92*(15 Suppl), P1.5-026.

Legge, S. E., Santoro, M. L., Periyasamy, S., Okewole, A., Arsalan, A., & Kowalec, K. (2021). Genetic architecture of schizophrenia: A review of major advancements. *Psychological Medicine, 51*, 2168–2177.

Lewiecki, E. M. (2020). Romosozumab, clinical trials, and real-world care of patients with osteoporosis. *The Annals of Translational Medicine, 8*, 974.

Li, S., DeLisi, L. E., & McDonough, S. I. (2021). Rare germline variants in individuals diagnosed with schizophrenia within multiplex families. *Psychiatry Research, 303*, 114038.

Lichtlen, P. R., Hugenholtz, P. G., Rafflenbeul, W., Hecker, H., Jost, S., Nikutta, P., & Deckers, J. W. (1990). Retardation of coronary artery disease in humans by the calcium-channel blocker nifedipine: Results of the INTACT study (International Nifedipine Trial on Antiatherosclerotic Therapy). *Cardiovascular Drugs and Therapy, 4*(Suppl 5), 1047–1068.

Lin, C. Y., Louis, E. D., Faust, P. L., Koeppen, A. H., Vonsattel, J. P., & Kuo, S. H. (2014). Abnormal climbing fibre-Purkinje cell synaptic connections in the essential tremor cerebellum. *Brain, 137*, 3149–3159.

Luckhaus, G., Nash, G., Garthoff, B., Kazda, S., & Feller, W. (1985). Healing of malignant hypertensive arteriopathy in Dahl rats by nifedipine. *Arzneimittel-Forschung, 35*, 115–121.

Malmberg, A. B., & Yaksh, T. L. (1994). Voltage-sensitive calcium channels in spinal nociceptive processing: Blockade of N- and P-type channels inhibits formalin-induced nociception. *The Journal of Neuroscience, 14*, 4882–4890.

Maneuf, Y. P., Luo, Z. D., & Lee, K. (2006). alpha2delta and the mechanism of action of gabapentin in the treatment of pain. *Seminars in Cell & Developmental Biology, 17*, 565–570.

Marais, E., Klugbauer, N., & Hofmann, F. (2001). Calcium channel alpha(2)delta subunits-structure and gabapentin binding. *Molecular Pharmacology, 59*, 1243–1248.

Martin, D. J., McClelland, D., Herd, M. B., Sutton, K. G., Hall, M. D., Lee, K., Pinnock, R. D., & Scott, R. H. (2002). Gabapentin-mediated inhibition of voltage-activated Ca2+ channel currents in cultured sensory neurones is dependent on culture conditions and channel subunit expression. *Neuropharmacology, 42*, 353–366.

McDonough, S. I. (2013). Calcium ion channels: Challenges and successes in drug discovery. *Wiley Interdisciplinary Reviews: Membrane Transport and Signaling, 2*(2), 85–104. https://doi.org/10.1002/wmts.71

McDowell, G. C., Saulino, M. F., Wallace, M., Grigsby, E. J., Rauck, R. L., Kim, P., Vanhove, G. F., Ryan, R., Huang, I. Z., & Deer, T. (2020). Effectiveness and safety of intrathecal ziconotide: Final results of the patient registry of intrathecal ziconotide management (PRIZM). *Pain Medicine, 21*, 2925–2938.

McIntosh, A. M., Hall, L. S., Zeng, Y., Adams, M. J., Gibson, J., Wigmore, E., Hagenaars, S. P., Davies, G., Fernandez-Pujals, A. M., Campbell, A. I., Clarke, T. K., Hayward, C., Haley, C. S., Porteous, D. J., Deary, I. J., Smith, D. J., Nicholl, B. I., Hinds, D. A., Jones, A. V., … Hocking, L. J. (2016). Genetic and environmental risk for chronic pain and the contribution of risk variants for major depressive disorder: A family-based mixed-model analysis. *PLoS Medicine, 13*, e1002090.

McLean, M. J. (1995). "Gabapentin" Epilepsia 36(s2) S73–S86. https://doi.org/10.1111/j.1528-1157.1995.tb06001.x

Miljanich, G. P. (2004). Ziconotide: Neuronal calcium channel blocker for treating severe chronic pain. *Current Medicinal Chemistry, 11*, 3029–3040.

Mohammed, S. I., Eldabe, S., Simpson, K. H., Brookes, M., Madzinga, G., Gulve, A., Baranidharan, G., Radford, H., Crowther, T., Buchser, E., Perruchoud, C., & Batterham, A. M. (2013). Bolus intrathecal injection of ziconotide (Prialt(R)) to evaluate the option of continuous administration via an implanted intrathecal drug delivery (ITDD) system: A Pilot study. *Neuromodulation, 16*, 576–581; discussion 582.

Monje, M., Borniger, J. C., D'Silva, N. J., Deneen, B., Dirks, P. B., Fattahi, F., Frenette, P. S., Garzia, L., Gutmann, D. H., Hanahan, D., Hervey-Jumper, S. L., Hondermarck, H., Hurov, J. B., Kepecs, A., Knox, S. M., Lloyd, A. C., Magnon, C., Saloman, J. L., Segal, R. A., … Winkler, F. (2020). Roadmap for the emerging field of cancer neuroscience. *Cell, 181*, 219–222.

Mori, Y., Nishida, M., Shimizu, S., Ishii, M., Yoshinaga, T., Ino, M., Sawada, K., & Niidome, T. (2002). Ca(2+) channel alpha(1B) subunit (Ca(V) 2.2) knockout mouse reveals a predominant role of N-type channels in the sympathetic regulation of the circulatory system. *Trends in Cardiovascular Medicine, 12*, 270–275.

Mullard, A. (2019). Machine learning brings cell imaging promises into focus. *Nature Reviews. Drug Discovery, 18*, 653–655.

Nalls, M. A., Blauwendraat, C., Vallerga, C. L., Heilbron, K., Bandres-Ciga, S., Chang, D., Tan, M., Kia, D. A., Noyce, A. J., Xue, A., Bras, J., Young, E., von Coelln, R., Simon-Sanchez, J., Schulte, C., Sharma, M., Krohn, L., Pihlstrom, L., Siitonen, A., … Me Research T, System Genomics of Parkinson's Disease C and International Parkinson's Disease Genomics C. (2019). Identification of novel risk loci, causal insights, and heritable risk for Parkinson's disease: A meta-analysis of genome-wide association studies. *Lancet Neurology, 18*, 1091–1102.

Nam, G. (2018). T-type calcium channel blockers: A patent review (2012-2018). *Expert Opinion on Therapeutic Patents, 28*, 883–901.

Ng, G. Y., Bertrand, S., Sullivan, R., Ethier, N., Wang, J., Yergey, J., Belley, M., Trimble, L., Bateman, K., Alder, L., Smith, A., McKernan, R., Metters, K., O'Neill, G. P., Lacaille, J. C., & Hebert, T. E. (2001). Gamma-aminobutyric acid type B receptors with specific heterodimer composition and postsynaptic actions in hippocampal neurons are targets of anticonvulsant gabapentin action. *Molecular Pharmacology, 59*, 144–152.

Olivera, B. M. (2021). A serendipitous path to pharmacology. *Annual Review of Pharmacology and Toxicology, 61*, 9–23.

Ondo, W. G. (2020). Current and emerging treatments of essential tremor. *Neurologic Clinics, 38*, 309–323.

Ortner, N. J., & Striessnig, J. (2016). L-type calcium channels as drug targets in CNS disorders. *Channels (Austin, Tex.), 10*, 7–13.

Papapetropoulos, S., Lee, M. S., Versavel, S., Newbold, E., Jinnah, H. A., Pahwa, R., Lyons, K. E., Elble, R.,

Ondo, W., Zesiewicz, T., Hedera, P., Handforth, A., Elder, J., & Versavel, M. (2021). A phase 2 proof-of-concept, randomized, placebo-controlled trial of CX-8998 in essential tremor. *Movement Disorders, 36*, 1944–1949.

Pardinas, A. F., Holmans, P., Pocklington, A. J., Escott-Price, V., Ripke, S., Carrera, N., Legge, S. E., Bishop, S., Cameron, D., Hamshere, M. L., Han, J., Hubbard, L., Lynham, A., Mantripragada, K., Rees, E., MacCabe, J. H., McCarroll, S. A., Baune, B. T., Breen, G., ... Walters, J. T. R. (2018). Common schizophrenia alleles are enriched in mutation-intolerant genes and in regions under strong background selection. *Nature Genetics, 50*, 381–389.

Parkinson Study Group S-PDIIII. (2020). Isradipine versus placebo in early Parkinson disease: A randomized trial. *Annals of Internal Medicine, 172*, 591–598.

Peckham, A. M., Evoy, K. E., Ochs, L., & Covvey, J. R. (2018). Gabapentin for off-label use: Evidence-based or cause for concern? *Substance Abuse, 12*, 1178221818801311.

Perucca, E. (2006). Clinically relevant drug interactions with antiepileptic drugs. *The British Journal of Clinical Pharmacology, 61*(3), 246–255.

Planelles-Herrero, V. J., Hartman, J. J., Robert-Paganin, J., Malik, F. I., & Houdusse, A. (2017). Mechanistic and structural basis for activation of cardiac myosin force production by omecamtiv mecarbil. *Nature Communications, 8*, 190.

Purcell, S. M., Moran, J. L., Fromer, M., Ruderfer, D., Solovieff, N., Roussos, P., O'Dushlaine, C., Chambert, K., Bergen, S. E., Kahler, A., Duncan, L., Stahl, E., Genovese, G., Fernandez, E., Collins, M. O., Komiyama, N. H., Choudhary, J. S., Magnusson, P. K., Banks, E., ... Sklar, P. (2014). A polygenic burden of rare disruptive mutations in schizophrenia. *Nature, 506*, 185–190.

Rauck, R. L., Wallace, M. S., Leong, M. S., Minehart, M., Webster, L. R., Charapata, S. G., Abraham, J. E., Buffington, D. E., Ellis, D., Kartzinel, R., & Ziconotide 301 Study G. (2006). A randomized, double-blind, placebo-controlled study of intrathecal ziconotide in adults with severe chronic pain. *Journal of Pain and Symptom Management, 31*, 393–406.

Sabatine, M. S., Wiviott, S. D., Im, K., Murphy, S. A., & Giugliano, R. P. (2018). Efficacy and safety of further lowering of low-density lipoprotein cholesterol in patients starting with very low levels: A meta-analysis. *JAMA Cardiology, 3*, 823–828.

Schulze-Bonhage, A. (2013). Pharmacokinetic and pharmacodynamic profile of pregabalin and its role in the treatment of epilepsy. *Expert Opinion on Drug Metabolism & Toxicology, 9*, 105–115.

Scott, S., Kranz, J. E., Cole, J., Lincecum, J. M., Thompson, K., Kelly, N., Bostrom, A., Theodoss, J., Al-Nakhala, B. M., Vieira, F. G., Ramasubbu, J., & Heywood, J. A. (2008). Design, power, and interpretation of studies in the standard murine model of ALS. *Amyotrophic Lateral Sclerosis, 9*, 4–15.

Shipe, W. D., Barrow, J. C., Yang, Z. Q., Lindsley, C. W., Yang, F. V., Schlegel, K. A., Shu, Y., Rittle, K. E., Bock, M. G., Hartman, G. D., Tang, C., Ballard, J. E., Kuo, Y., Adarayan, E. D., Prueksaritanont, T., Zrada, M. M., Uebele, V. N., Nuss, C. E., Connolly, T. M., ... Renger, J. J. (2008). Design, synthesis, and evaluation of a novel 4-aminomethyl-4-fluoropiperidine as a T-type Ca2+ channel antagonist. *Journal of Medicinal Chemistry, 51*, 3692–3695.

Silverman, R. B. (2008). From basic science to blockbuster drug: The discovery of Lyrica. *Angewandte Chemie, 47*(19), 3500–3504.

Silverman, R. B., Andruszkiewicz, R., Nanavati, S. M., Taylor, C. P., & Vartanian, M. G. (1991). 3-Alkyl-4-aminobutyric acids: The first class of anticonvulsant agents that activates L-glutamic acid decarboxylase. *Journal of Medicinal Chemistry, 34*, 2295–2298.

Singh, K., Bijarnia-Mahay, S., Ramprasad, V. L., Puri, R. D., Nair, S., Sharda, S., Saxena, R., Kohli, S., Kulshreshtha, S., Ganguli, I., Gujral, K., & Verma, I. C. (2020). NGS-based expanded carrier screening for genetic disorders in North Indian population reveals unexpected results—A Pilot study. *BMC Medical Genetics, 21*, 216.

Splawski, I., Timothy, K. W., Sharpe, L. M., Decher, N., Kumar, P., Bloise, R., Napolitano, C., Schwartz, P. J., Joseph, R. M., Condouris, K., Tager-Flusberg, H., Priori, S. G., Sanguinetti, M. C., & Keating, M. T. (2004). Cav1.2 calcium channel dysfunction causes a multisystem disorder including arrhythmia and autism. *Cell, 119*(1), 19–31.

Staats, P. S., Luthardt, F., Shipley, J., Jackson, C., & Fischer, K. (2001). Long-term intrathecal ziconotide therapy: A case study and discussion. *Neuromodulation, 4*, 121–126.

Staats, P. S., Yearwood, T., Charapata, S. G., Presley, R. W., Wallace, M. S., Byas-Smith, M., Fisher, R., Bryce, D. A., Mangieri, E. A., Luther, R. R., Mayo, M., McGuire, D., & Ellis, D. (2004). Intrathecal ziconotide in the treatment of refractory pain in patients with cancer or AIDS: A randomized controlled trial. *JAMA, 291*, 63–70.

Stewart, B. H., Kugler, A. R., Thompson, P. R., & Bockbrader, H. N. (1993). A saturable transport mechanism in the intestinal absorption of gabapentin is the underlying cause of the lack of proportionality between increasing dose and drug levels in plasma. *Pharmaceutical Research, 10*, 276–281.

Storkebaum, E., Lambrechts, D., Dewerchin, M., Moreno-Murciano, M. P., Appelmans, S., Oh, H., Van Damme, P., Rutten, B., Man, W. Y., De Mol, M., Wyns, S., Manka, D., Vermeulen, K., Van Den Bosch, L., Mertens, N., Schmitz, C., Robberecht, W., Conway, E. M., Collen, D., ... Carmeliet, P. (2005). Treatment of motoneuron degeneration by intracerebroventricular delivery of VEGF in a rat model of ALS. *Nature Neuroscience, 8*, 85–92.

Suman-Chauhan, N., Webdale, L., Hill, D. R., & Woodruff, G. N. (1993). Characterisation of [3H]gabapentin binding to a novel site in rat brain: Homogenate binding studies. *European Journal of Pharmacology, 244*, 293–301.

Sun, J., & Triggle, D. J. (1995). Calcium channel antagonists: Cardiovascular selectivity of action. *The Journal of Pharmacology and Experimental Therapeutics, 274*(1), 419–426.

Szustakowski, J. D., Balasubramanian, S., Kvikstad, E., Khalid, S., Bronson, P. G., Sasson, A., Wong, E., Liu, D., Wade Davis, J., Haefliger, C., Katrina Loomis, A., Mikkilineni, R., Noh, H. J., Wadhawan, S., Bai, X., Hawes, A., Krasheninina, O., Ulloa, R., Lopez, A. E., … Team U-ER. (2021). Advancing human genetics research and drug discovery through exome sequencing of the UK Biobank. *Nature Genetics, 53*, 942–948.

Tassone, D. M., Boyce, E., Guyer, J., & Nuzum, D. (2007). Pregabalin: A novel gamma-aminobutyric acid analogue in the treatment of neuropathic pain, partial-onset seizures, and anxiety disorders. *Clinical Therapeutics, 29*, 26–48.

Taylor, C. P., Vartanian, M. G., Andruszkiewicz, R., & Silverman, R. B. (1992). 3-alkyl GABA and 3-alkylglutamic acid analogues: Two new classes of anticonvulsant agents. *Epilepsy Research, 11*, 103–110.

Teerlink, J. R., Felker, G. M., McMurray, J. J., Solomon, S. D., Adams, K. F., Jr., Cleland, J. G., Ezekowitz, J. A., Goudev, A., Macdonald, P., Metra, M., Mitrovic, V., Ponikowski, P., Serpytis, P., Spinar, J., Tomcsanyi, J., Vandekerckhove, H. J., Voors, A. A., Monsalvo, M. L., Johnston, J., … Investigators, C.-H. (2016). Chronic Oral Study of Myosin Activation to Increase Contractility in Heart Failure (COSMIC-HF): A phase 2, pharmacokinetic, randomised, placebo-controlled trial. *Lancet, 388*, 2895–2903.

Teerlink, J. R., Diaz, R., Felker, G. M., McMurray, J. J. V., Metra, M., Solomon, S. D., Adams, K. F., Anand, I., Arias-Mendoza, A., Biering-Sorensen, T., Bohm, M., Bonderman, D., Cleland, J. G. F., Corbalan, R., Crespo-Leiro, M. G., Dahlstrom, U., Echeverria, L. E., Fang, J. C., Filippatos, G., … Investigators, G.-H. (2021). Cardiac myosin activation with omecamtiv mecarbil in systolic heart failure. *The New England Journal of Medicine, 384*, 105–116.

Terwindt, G. M., Ophoff, R. A., Haan, J., Sandkuijl, L. A., Frants, R. R., & Ferrari, M. D. (1998). Migraine, ataxia and epilepsy: A challenging spectrum of genetically determined calcium channelopathies. Dutch Migraine Genetics Research Group. *European Journal of Human Genetics, 6*, 297–307.

Thompson, J. C., Dunbar, E., & Laye, R. R. (2006). Treatment challenges and complications with ziconotide monotherapy in established pump patients. *Pain Physician, 9*, 147–152.

Thurlow, R. J., Brown, J. P., Gee, N. S., Hill, D. R., & Woodruff, G. N. (1993). [3H]gabapentin may label a system-L-like neutral amino acid carrier in brain. *European Journal of Pharmacology, 247*, 341–345.

Toal, C. B., Meredith, P. A., & Elliott, H. L. (2012). Long-acting dihydropyridine calcium-channel blockers and sympathetic nervous system activity in hypertension:

A literature review comparing amlodipine and nifedipine GITS. *Blood Pressure, 21*(Suppl 1), 3–10.

Trujillano, D., Oprea, G. E., Schmitz, Y., Bertoli-Avella, A. M., Abou Jamra, R., & Rolfs, A. (2017). A comprehensive global genotype-phenotype database for rare diseases. *Molecular Genetics & Genomic Medicine, 5*, 66–75.

Tseng, Y. F., Lin, H. C., Chao, J. C., Hsu, C. Y., & Lin, H. L. (2021). Calcium Channel blockers are associated with reduced risk of Parkinson's disease in patients with hypertension: A population-based retrospective Cohort study. *Journal of the Neurological Sciences, 424*, 117412.

Uchida, S., Yamada, S., Nagai, K., Deguchi, Y., & Kimura, R. (1997). Brain pharmacokinetics and in vivo receptor binding of 1,4-dihydropyridine calcium channel antagonists. *Life Sciences, 61*, 2083–2090.

Uneyama, H., Takahara, A., Dohmoto, H., Yoshimoto, R., Inoue, K., & Akaike, N. (1997). Blockade of N-type Ca2+ current by cilnidipine (FRC-8653) in acutely dissociated rat sympathetic neurones. *British Journal of Pharmacology, 122*, 37–42.

Uneyama, H., Uchida, H., Konda, T., Yoshimoto, R., & Akaike, N. (1999). Selectivity of dihydropyridines for cardiac L-type and sympathetic N-type Ca2+ channels. *European Journal of Pharmacology, 373*, 93–100.

Van Goor, F., Hadida, S., Grootenhuis, P. D., Burton, B., Cao, D., Neuberger, T., Turnbull, A., Singh, A., Joubran, J., Hazlewood, A., Zhou, J., McCartney, J., Arumugam, V., Decker, C., Yang, J., Young, C., Olson, E. R., Wine, J. J., Frizzell, R. A., … Negulescu, P. (2009). Rescue of CF airway epithelial cell function in vitro by a CFTR potentiator, VX-770. *Proceedings of the National Academy of Sciences of the United States of America, 106*, 18825–18830.

Van Goor, F., Hadida, S., Grootenhuis, P. D. J., & Negulescu, P. A. (2011). Correction of the F508del-CFTR protein processing defect in vitro by the investigational drug VX-809. *108*, 18843–18848. https://doi.org/10.1073/pnas.1105787108

Venuto, C. S., Yang, L., Javidnia, M., Oakes, D., Surmeier, D. J., & Simuni, T. (2021). Isradipine plasma pharmacokinetics and exposure-response in early Parkinson's disease. *Annals of Clinical Translational Neurology, 8*(3), 603–612.

Verrey, F. (2003). System L: Heteromeric exchangers of large, neutral amino acids involved in directional transport. *Pflügers Archiv, 445*, 529–533.

Vinuto, C. S., Yang, L. Javidnia, M., et al. (2021). Isradipine plasma pharmacokinetics and exposure-response in early Parkinson's disease. *Annals of Clinical and Translational Neurology, 8*(3), 603–612. https://doi.org/10.1002/acn3.51300

Vitale, V., Battelli, D., Gasperoni, E., & Monachese, N. (2008). Intrathecal therapy with ziconotide: Clinical experience and considerations on its use. *Minerva Anestesiologica, 74*, 727–733.

Wallace, M. S., Charapata, S. G., Fisher, R., Byas-Smith, M., Staats, P. S., Mayo, M., McGuire, D., Ellis, D.,

& Ziconotide Nonmalignant Pain Study G. (2006). Intrathecal ziconotide in the treatment of chronic non-malignant pain: A randomized, double-blind, placebo-controlled clinical trial. *Neuromodulation, 9*, 75–86.

Wallace, M., Duan, R., Liu, W., Locke, C., & Nothaft, W. (2016). A randomized, double-blind, placebo-controlled, crossover study of the T-type calcium channel blocker ABT-639 in an intradermal capsaicin experimental pain model. *Pain Medicine, 17*(3), 551–560. https://doi.org/10.1093/pm/pnv068

Wang, A. L., Iadecola, C., & Wang, G. (2017). New generations of dihydropyridines for treatment of hypertension. *Journal of Geriatric Cardiology, 14*, 67–72.

Weiss, N., & Zamponi, G. W. (2020). Genetic T-type calcium channelopathies. *Journal of Medical Genetics, 57*, 1–10.

Xu, W., & Lipscombe, D. (2001). Neuronal Ca(V)1.3alpha(1) L-type channels activate at relatively hyperpolarized membrane potentials and are incompletely inhibited by dihydropyridines. *The Journal of Neuroscience, 21*, 5944–5951.

Yu, J., Wang, D.-S., Bonin, R. P., Penna, A., Alavian-Ghavanini, A., Zurek, A. A., Rauw, G., Baker, G. B.,

Orser, B. A. (2019). "Gabapentin increases expression of δ subunit-containing GABAA receptors." *eBioMedicine, 42*, 203–213.

Zamponi, G. W. (2016). Targeting voltage-gated calcium channels in neurological and psychiatric diseases. *Nature Reviews Drug Discovery, 15*(1), 19–34.

Zamponi, G. W., Feng, Z. P., Zhang, L., Pajouhesh, H., Ding, Y., Belardetti, F., Pajouhesh, H., Dolphin, D., Mitscher, L. A., & Snutch, T. P. (2009). Scaffold-based design and synthesis of potent N-type calcium channel blockers. *Bioorganic & Medicinal Chemistry Letters, 19*, 6467–6472.

Zamponi, G. W., Striessnig, J., Koschak, A., & Dolphin, A. C. (2015). The physiology, pathology, and pharmacology of voltage-gated calcium channels and their future therapeutic potential. *Pharmacological Reviews, 67*, 821–870.

Ziegler, D., Duan, W. R., An, G., Thomas, J. W., & Nothaft, W. (2015). A randomized double-blind, placebo-, and active-controlled study of T-type calcium channel blocker ABT-639 in patients with diabetic peripheral neuropathic pain. *Pain, 156*, 2013–2020.

Index

A

Absence epilepsy, 95, 203, 298, 309, 314, 326, 327, 335, 336, 534, 537, 542–546, 668, 698

Acrosome reaction (AR), 126, 163, 315, 337, 387, 601, 609–610, 615, 622–623

Active zone, 84, 135, 136, 145, 256, 283, 285, 297, 334, 360–363, 366–368, 370, 374, 432

Active zone proteins, 297, 334, 362–366, 369

Agatoxin, 143, 666

Aging, 260, 310, 319, 583

$\alpha 2\delta$, 12, 34, 75, 94, 116, 162, 196, 257, 372, 390, 459, 476, 495, 637, 676, 689

α_1-subunit, 12, 32, 56, 74, 94, 116, 162, 256, 363, 462, 497, 636, 667, 676

Alternative splicing, 14, 48, 51–53, 55–57, 59, 61–66, 131, 132, 134, 142, 171, 203, 204, 225, 226, 256, 260, 266, 267, 272, 275, 278, 283, 450, 451, 459, 462, 464, 477, 478, 483, 493, 497, 564, 580

Ancillary subunit, 116, 118, 133, 134, 145, 196, 205, 206

Arachidonic acid (AA), 166, 167, 169, 176, 178, 298, 606

A-type, 146, 203, 237–240, 242, 243, 247–250, 402, 431, 542, 555, 579, 583, 584

B

Behaviour, 643, 648, 649

β cell, 426–443

β subunit, 13, 19, 32, 34, 37–41, 52, 53, 73–86, 99, 101, 121–123, 125, 129, 130, 133, 138, 140–142, 165, 196, 273, 334, 362, 363, 365, 366, 369, 372, 390, 454, 461, 476, 579, 642, 646, 676

Bioenergetics, 576, 578, 581, 582

Brain disorders, 327, 516

Burst-firing, 314, 315, 326, 328, 333, 335, 336, 534–544, 546, 697

C

Ca^{2+} channel, 4, 32, 73, 162, 196, 218, 255, 372, 441, 475, 539, 576, 600, 636, 665, 676

CACHD0, 124–125, 140, 141, 145, 146

CACHD1, 123–125, 128, 141, 144–146

Ca^{2+} currents, 2–11, 14, 16, 18, 19, 48, 51, 54, 64, 76, 116, 125–126, 128, 130–135, 138–140, 142, 145, 165, 197, 202, 204, 206, 217–220, 265, 266, 270, 273–276, 279, 281, 284, 292, 294, 298, 300, 308, 316, 322, 326, 329–331, 334, 335, 337, 369, 372, 389, 407, 408, 429, 435, 437, 440, 465, 477, 480–483, 540, 542, 602–608, 644, 668

Calcium, 1, 32, 33, 36, 37, 42, 55, 75, 76, 94, 99, 103, 106, 107, 116–125, 128–134, 144, 163, 164, 167, 171, 174–176, 185, 196, 201, 238–242, 244–250, 360–362, 364, 367, 370–373, 387, 402–416, 426, 450–459, 461, 462, 464–467, 482–484, 493, 495, 518, 523, 534–536, 554–560, 580, 583, 600, 601, 603–607, 610, 614, 620, 636, 662–665, 668–670, 676, 688, 689, 692, 693, 695, 698, 699

Calcium channel, 1–16, 31–42, 47–66, 73, 74, 94–107, 116–135, 138, 139, 144–146, 196–208, 237–250, 360, 362, 369, 371, 372, 375, 386–394, 403, 405, 449–467, 492, 493, 495, 497, 498, 502, 518, 536, 544, 554–557, 559, 560, 562, 575–581, 606, 661–670, 675–682, 687–699

Calcium channel auxiliary subunit, 73–86, 121, 196, 476

Calcium channelopathy, vi, 20, 477, 482–484

Calcium channels structure and function, 20

Calcium currents, 12, 31, 32, 38, 54, 78, 82, 83, 96, 99, 105, 116, 118, 126, 129, 169, 229, 243, 244, 371, 387, 392, 394, 404, 450, 452, 453, 457–459, 461, 462, 464–466, 493–496, 519, 534, 541, 545, 546, 557, 661, 662, 666, 668, 692

Calcium signaling, 462, 554, 555

Calcium transporters, 403, 406, 407

Calmodulation, 199, 218, 220, 223–227, 229, 479

Calmodulin, 51, 118–120, 127–131, 138, 143, 144, 169, 171, 176, 177, 181, 199, 200, 217–229, 239, 242, 249, 256, 259, 274, 278, 283, 298, 365, 371, 395, 434, 456, 460, 477–479, 606–607

Cancer, 11, 496, 636–650, 666, 678, 691

Ca^{2+} regulation, 218, 220, 227–229

Ca-spikes, 2–4

CatSper channels, 599–623

CatSper regulation, 609, 613, 616, 618–620

Ca_v1, 14, 38, 61, 74, 96, 116, 162, 206, 221, 256, 404, 428, 459, 475, 560, 576, 603, 637, 662, 679

Milton Keynes UK
Ingram Content Group UK Ltd.
UKHW050908131123
432466UK00003B/25

9 783031 088834